a LANGE medical book

1989

Review of Medical Physiology

Fourteenth Edition

William F. Ganong, MD
Jack and DeLoris Lange Professor of Physiology
University of California
San Francisco

Prentice-Hall International Inc.

Notice: Our knowledge in clinical sciences is constantly changing. As new information becomes available, changes in treatment and in the use of drugs become necessary. The authors and the publisher of this volume have taken care to make certain that the doses of drugs and schedules of treatment are correct and compatible with the standards generally accepted at the time of publication. The reader is advised to consult carefully the instruction and information material included in the package insert of each drug or therapeutic agent before administration. This advice is especially important when using new or infrequently used drugs.

89 90 91 92 / 10 9 8 7 6 5 4 3 2 1

Prentice-Hall International (UK) Limited, *London*
Prentice-Hall of Australia, Pty. Limited, *Sydney*
Prentice-Hall Canada, Inc., *Toronto*
Prentice-Hall Hispanoamericana, S.A., *Mexico*
Prentice-Hall of India Private Limited, *New Delhi*
Prentice-Hall of Japan, Inc., *Tokyo*
Simon & Schuster Asia Pte. Ltd., *Singapore*
Editora Prentice-Hall do Brasil Ltd., *Rio de Janeiro*
Prentice-Hall, *Englewood Cliffs, New Jersey*

ISBN: 0-8385-8429-2
ISSN: 0892-1253

PRINTED IN THE UNITED STATES OF AMERICA

Table of Contents

SECTION III. FUNCTIONS OF THE NERVOUS SYSTEM

SECTION IV. ENDOCRINOLOGY, METABOLISM, & REPRODUCTIVE FUNCTION

SECTION VII. RESPIRATION

SECTION V. GASTROINTESTINAL FUNCTION

SECTION VI. CIRCULATION

SECTION VIII. FORMATION & EXCRETION OF URINE

Preface

This book is designed to provide a concise summary of mammalian and, particularly, of human physiology which medical students and others can use as a text by itself or supplement with readings in current texts, monographs, and reviews. Pertinent aspects of general and comparative physiology are also included. Summaries of relevant anatomic considerations will be found in each section, but this book is written primarily for those who have some knowledge of anatomy, chemistry, and biochemistry. It also includes an appendix that contains, among other things, a discussion of statistical methods, a glossary of abbreviations and symbols commonly used in physiology, and several useful tables. The index is comprehensive and specifically designed for ease in locating important terms, topics, and concepts.

Examples from clinical medicine are given where pertinent to illustrate physiologic points. Physicians desiring to use this book as a review will find short discussions of important symptoms produced by disordered function in several sections.

It has not been possible to be complete and concise without also being dogmatic. I believe, however, that the conclusions presented without a detailed discussion of the experimental data on which they are based are those supported by the bulk of the currently available evidence. Much of this evidence can be found in the papers cited in the credit lines of the illustrations. Further discussions of particular subjects and information on subjects not considered in detail in this book can be found in the references listed at the end of each section. Information about serial review publications that provide up-to-date discussions of various physiologic subjects is included in a note on general references in the appendix.

In the interest of brevity and clarity, I have in most instances omitted the names of the many investigators whose work made possible the view of physiology presented here. This is in no way intended to slight their contributions, but including their names and specific references to original papers would greatly increase the length of this book.

In this fourteenth edition, as in previous editions, the entire book has been thoroughly revised, eliminating errors, incorporating suggestions of readers, updating concepts, and discarding material that is no longer relevant. In this way, the book has been kept as up-to-date and accurate as possible without materially increasing its length. Since the last edition, there have been spectacular advances in knowledge about receptors and transmembrane signalling. Consequently, there is expanded coverage of receptor families, G proteins, and regulation of ion channels in Chapters 1, 2, and 4. In addition, advances in this area that are relevant to particular endocrine glands and other organs have been incorporated into the chapters on those organs. Another field that has begun to move rapidly is learning and memory, with important contributions by both molecular neurobiologists and cognitive psychologists. These contributions have led to revision of Chapter 16 and updated coverage in Chapter 4. Chapter 12 has also been rewritten to reflect current concepts on the control of posture and movement. The fourteenth edition also includes new information on many other topics, including diabetes mellitus, osteoporosis, regulation of growth, control of the proliferation of red and white blood cells and platelets, immune mechanisms, endothelial effects on vascular tone, and several aspects of pulmonary function.

Students, instructors, and others who are reviewing or studying for special examinations may be interested in the study guide that I have written to go with this book:

Physiology: A Study Guide, 3rd ed.
Appleton & Lange, 1989

The guide provides objectives, essay questions, and multiple-choice questions of various types for each of the chapters in *Review of Medical Physiology*. The third edition of the study guide is being published at the same time as this fourteenth edition of *Review of Medical Physiology*.

I am greatly indebted to many individuals who helped with the preparation of this book. Those to whom I wish to extend special thanks for their help with the fourteenth edition include Drs. Jim Hudspeth, Larry Swanson, Ira Goldfine, Sanford Gips, and Donald Fawcett. I am also indebted to my wife, who, as always, labored long hours typing corrections. Many associates and friends provided unpublished illustrative materials, and numerous authors and publishers generously granted permission to reproduce illustrations from other books and journals. I also wish to thank all the students and others who took the time to write to me offering helpful criticisms and suggestions. Such comments are always welcome, and I solicit additional corrections and criticisms, which may be addressed to me at

Department of Physiology
University of California
San Francisco, CA 94143-0444, USA

Since the book was first published in 1963, the following translations have been published and are currently available: Portuguese (fourth edition), German (fourth edition), Italian (sixth edition), Spanish (eleventh edition), Japanese (tenth edition), Indonesian (second edition), Greek (second edition), and French. Translations have also been published in Chinese, Czech, Polish, Serbo-Croatian, and Turkish but are no longer in print. A new translation into Serbo-Croatian is under way. The book has also appeared in various foreign English-language editions and has been recorded in English on tape for use by the blind. The tape recording is available from Recording for the Blind, Inc., 20 Roszel Road, Princeton, NJ 08540, USA. The German edition is also being recorded on tape.

William F. Ganong

San Francisco
June 1989

Section I. Introduction

The General & Cellular Basis of Medical Physiology

1

INTRODUCTION

In unicellular organisms, all vital processes occur in a single cell. As the evolution of multicellular organisms has progressed, various cell groups have taken over particular functions. In humans and other vertebrate animals, the specialized cell groups include a gastrointestinal system to digest and absorb food; a respiratory system to take up O_2 and eliminate CO_2; a urinary system to remove wastes; a cardiovascular system to distribute food, O_2, and the products of metabolism; a reproductive system to perpetuate the species; and nervous and endocrine systems to coordinate and integrate the functions of the other systems. This book is concerned with the way these systems function and the way each contributes to the functions of the body as a whole.

This chapter presents general concepts and principles that are basic to the function of all the systems. It also reviews fundamental aspects of cell physiology. Additional aspects of cellular and molecular biology are considered in the relevant chapters on the various organs.

BODY FLUID COMPARTMENTS

Organization of the Body

The cells that make up the bodies of all but the simplest multicellular animals, both aquatic and terrestrial, exist in an "internal sea" of **extracellular fluid (ECF)** enclosed within the integument of the animal. From this fluid, the cells take up O_2 and nutrients; into it, they discharge metabolic waste products. The ECF is more dilute than present-day seawater, but its composition closely resembles that of the primordial oceans in which, presumably, all life originated.

In animals with a closed vascular system, the ECF is divided into 2 components: the **interstitial fluid** and the circulating **blood plasma.** The plasma and the cellular elements of the blood, principally red blood cells, fill the vascular system, and together they constitute the **total blood volume.** The interstitial fluid is that part of the ECF that is outside the vascular system, bathing the cells. The special fluids lumped together as transcellular fluids are discussed below. About a third of the **total body water (TBW)** is extracellular; the remaining two-thirds are intracellular **(intracellular fluid).**

Size of the Fluid Compartments

In the average young adult male, 18% of the body weight is protein and related substances, 7% is mineral, and 15% is fat. The remaining 60% is water. The distribution of this water is shown in Fig. 1–1.

The intracellular component of the body water accounts for about 40% of body weight and the extracellular component for about 20%. Approximately 25% of the extracellular component is in the vascular system (plasma = 5% of body weight) and 75% outside the blood vessels (interstitial fluid = 15% of body weight). The total blood volume is about 8% of body weight.

Measurement of Body Fluid Volumes

It is theoretically possible to measure the size of each of the body fluid compartments by injecting substances that will stay in only one compartment and then calculating the volume of fluid in which the test substance is distributed (the **volume of distribution** of the injected material). The volume of distribution is equal to the amount injected (minus any that has been removed from the body by metabolism or excretion during the time allowed for mixing) divided by the concentration of the substance in the sample. *Example:* 150 mg of sucrose is injected into a 70-kg man. The plasma sucrose level after mixing is 0.01 mg/mL, and 10 mg has been excreted or metabolized during the mixing period. The volume of distribution of the sucrose is

$$\frac{150\text{ mg} - 10\text{ mg}}{0.01\text{ mg/mL}} = 14{,}000\text{ mL}$$

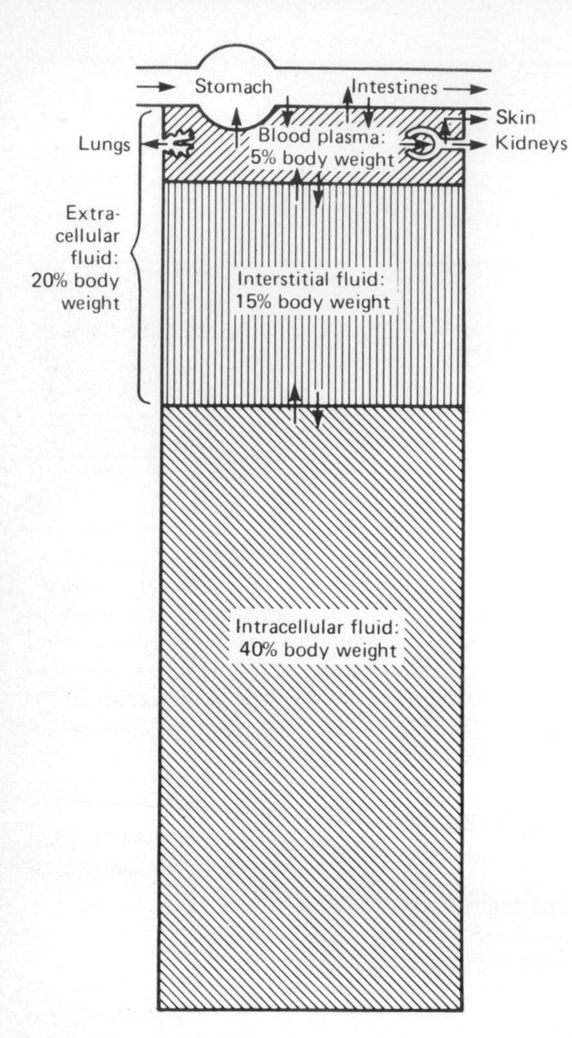

Figure 1–1. Body fluid compartments. Arrows represent fluid movement. Transcellular fluids, which constitute a very small percentage of total body fluids, are not shown. (Modified and reproduced, with permission, from Gamble JL: *Chemical Anatomy, Physiology, and Pathology of Extracellular Fluid,* 6th ed. Harvard Univ Press, 1954.)

Since 14,000 mL is the space in which the sucrose was distributed, it is also called the **sucrose space.**

Volumes of distribution can be calculated for any substance that can be injected into the body provided the concentration in the body fluids and the amount removed by excretion and metabolism can be accurately measured.

Although the principle involved in such measurements is simple, there are a number of complicating factors that must be considered. The material injected must be nontoxic, must mix evenly throughout the compartment being measured, and must have no effect of its own on the distribution of water or other substances in the body. In addition, either it must be unchanged by the body during the mixing period, or the amount changed must be known. The material also should be relatively easy to measure.

Plasma Volume, Total Blood Volume & Red Cell Volume

Plasma volume has been measured by using dyes that become bound to plasma protein—particularly Evans blue (T-1824). Plasma volume can also be measured by injecting serum albumin labeled with radioactive iodine. Suitable aliquots of the injected solution and plasma samples obtained after injection are counted in a scintillation counter. An average value is 3500 mL (5% of the body weight of a 70-kg man, assuming unit density).

If one knows the plasma volume and the hematocrit (ie, the percentage of the blood volume that is made up of cells), the **total blood volume** can be calculated by multiplying the plasma volume by

$$\frac{100}{100 - \text{hematocrit}}$$

Example: The hematocrit is 38 and the plasma volume 3500 mL. The total blood volume is

$$3500 \times \frac{100}{100 - 38} = 5645 \text{ mL}$$

The **red cell volume** (volume occupied by all the circulating red cells in the body) can be determined by subtracting the plasma volume from the total blood volume. It may also be measured independently by injecting tagged red blood cells and, after mixing has occurred, measuring the fraction of the red cells that is tagged. A commonly used tag is ^{51}Cr, a radioactive isotope of chromium that is attached to the cells by incubating them in a suitable chromium solution. Isotopes of iron and phosphorus (^{59}Fe and ^{32}P) and antigenic tagging have also been employed.

Extracellular Fluid Volume

The ECF volume is difficult to measure because the limits of this space are ill defined and because few substances mix rapidly in all parts of the space while remaining exclusively extracellular. The lymph cannot be separated from the ECF and is measured with it. Many substances enter the cerebrospinal fluid (CSF) slowly because of the blood-brain barrier (see Chapter 32). Equilibration is slow with joint fluid and aqueous humor and with the ECF in relatively avascular tissues such as dense connective tissue, cartilage, and some parts of bone. Substances that distribute in ECF appear in glandular secretions and in the contents of the gastrointestinal tract. Because they are separated from the rest of the ECF, these fluids—as well as CSF, the fluids in the eye, and a few other special fluids—are called **transcellular fluids.** Their volume is relatively small.

Perhaps the most accurate measurement of ECF volume is that obtained by using inulin, a polysaccharide with a molecular weight of 5200. Radioactive inulin has been prepared by substituting ^{14}C for one of the carbon atoms of the molecule; and when this material is used, inulin levels are easily determined by counting the samples with suitable radiation detectors. Mannitol and sucrose have also been used to measure ECF volume. Because Cl^-, for example, is largely extracellular in location, radioactive isotopes ($^{36}Cl^-$ and $^{38}Cl^-$) have been used for determination of ECF volume. However, some Cl^- is known to be intracellular, and consequently, volumes determined using Cl^- are greater than the actual ECF volumes. The same objection applies to $^{82}Br^-$, which interchanges with Cl^- in the body. A generally accepted value for ECF volume is 20% of the body weight, or about 14 L in a 70-kg man (3.5 L = plasma; 10.5 L = interstitial fluid).

Interstitial Fluid Volume

The interstitial fluid space cannot be measured directly, since it is difficult to sample interstitial fluid and since substances that equilibrate in interstitial fluid also equilibrate in plasma. The volume of the interstitial fluid can be calculated by subtracting the plasma volume from the ECF volume. The ECF volume/intracellular fluid volume ratio is larger in infants and children than it is in adults, but the absolute volume of ECF in children is, of course, smaller than it is in adults. Therefore, dehydration develops more rapidly and is frequently more severe in children than in adults.

Intracellular Fluid Volume

The intracellular fluid volume cannot be measured directly, but it can be calculated by subtracting the ECF volume from the total body water (TBW). TBW can be measured by the same dilution principle used to measure the other body spaces. Deuterium oxide (D_2O, heavy water) is most frequently used. D_2O has properties that are slightly different from H_2O but in equilibration experiments for measuring body water it gives accurate results. Tritium oxide and aminopyrine have also been used for this purpose.

The water content of lean body tissue is constant at 71–72 mL/100 g of tissue, but since fat is relatively free of water, the ratio of TBW to body weight varies with the amount of fat present. In young men, water constitutes about 60% of body weight. The values for women are somewhat lower. In both sexes, the values tend to decrease with age (Table 1–1).

UNITS FOR MEASURING CONCENTRATION OF SOLUTES

In considering the effects of various physiologically important substances and the interactions between them, the number of molecules, electrical charges, or particles of a substance per unit volume of a particular body fluid are often more meaningful than simply the weight of the substance per unit volume. For this reason, concentrations are frequently expressed in moles, equivalents, or osmoles.

Table 1–1. TBW (as percentage of body weight) in relation to age and sex.

Age	Male	Female
10–18	59%	57%
18–40	61%	51%
40–60	55%	47%
Over 60	52%	46%

Moles

A mole is the gram-molecular weight of a substance, ie, the molecular weight of the substance in grams. Each mole (mol) consists of approximately 6×10^{23} molecules. The millimole (mmol) is 1/1000 of a mole, and the micromole (μmol) is 1/1,000,000 of a mole. Thus, 1 mol of NaCl = 23 + 35.5 g = 58.5 g, and 1 mmol = 58.5 mg. The mole is the standard unit for expressing the amount of substances in the SI unit system (see Appendix).

It is worth noting that the molecular weight of a substance is the ratio of the mass of one molecule of the substance to the mass of 1⁄12 the mass of an atom of carbon-12. Since molecular weight is a ratio, it is dimensionless. The dalton (Da) is a unit of mass equal to 1⁄12 the mass of an atom of carbon-12, and 1000 Da = 1 kilodalton (kDa). The kilodalton, which is often expressed simply as K, is a useful unit for expressing the molecular mass of proteins. Thus, for example, one can speak of a 64K protein or state that the molecular mass of the protein is 64,000 Da. However, since molecular weight is a dimensionless ratio, it is incorrect to say that the molecular weight of the protein is 64K.

Equivalents

The concept of electrical equivalence is important in physiology because many of the important solutes in the body are in the form of charged particles. One equivalent (eq) is 1 mol of an ionized substance divided by its valence. One mole of NaCl dissociates into 1 eq of Na^+ and 1 eq of Cl^-. One equivalent of Na^+ = 23 g/1 = 23 g; but 1 eq of Ca^{2+} = 40 g/2 = 20 g. The milliequivalent (meq) is 1/1000 of 1 eq.

Electrical equivalence is not necessarily the same as chemical equivalence. A gram equivalent is that weight of a substance which is chemically equivalent to 8.000 g of oxygen. The normality (N) of a solution is the number of gram equivalents in 1 liter. A 1 N solution of hydrochloric acid contains 1 + 35.5 g/L = 36.5 g/L.

Osmoles

When one is dealing with concentrations of osmotically active particles, the amounts of these particles are usually expressed in osmoles. Osmosis is discussed in detail in a later section of this chapter. One osmole (osm) equals the molecular weight of the substance in grams divided by the number of freely moving particles each molecule liberates in solution. The milliosmole (mosm) is 1/1000 of 1 osm.

The **osmolal concentration** of a substance in a fluid is measured by the degree to which it depresses the freezing point, 1 mol/L of ideal solute depressing the freezing point 1.86 Celsius degrees. The number of milliosmoles per liter in a solution equals the freezing point depression divided by 0.00186. The **osmolarity** is the number of osmoles per liter of solution—eg, plasma—whereas the **osmolality** is the number of osmoles per kilogram of solvent. Therefore, osmolarity is affected by the volume of the various solutes in the solution and the temperature, while the osmolality is not. Osmotically active substances in the body are dissolved in water, and the density of water is 1, so osmolal concentrations can be expressed as osmoles per liter (osm/L) of water. In this book, osmolal (rather than osmolar) concentrations are considered, and osmolality is expressed in mosm/L (of water).

pH

The maintenance of a stable hydrogen ion concentration in the body fluids is essential to life. The pH of a solution is the logarithm to the base 10 of the reciprocal of the H^+ concentration ($[H^+]$), ie, the negative logarithm of the $[H^+]$. The pH of water at 25 °C, in which H^+ and OH^- ions are present in equal numbers, is 7.0 (Fig 1–2). For each pH unit less than 7.0, the $[H^+]$ is increased tenfold; for each pH unit above 7.0, it is decreased tenfold.

Buffers

Intracellular and extracellular pH are generally maintained at very constant levels. For example, the pH of the ECF is 7.40, and in health, this value usually varies less than ±0.05 pH unit. Body pH is stabilized by the **buffering capacity** of the body fluids. A buffer is a substance that has the ability to bind or release H^+ in solution, thus keeping the pH of the solution relatively constant despite the addition of considerable quantities of acid or base. One buffer in the body is carbonic acid. This acid is only partly dissociated into H^+ and bicarbonate: $H_2CO_3 \rightleftharpoons H^+ + HCO_3^-$. If H^+ is added to a solution of carbonic acid, the equilibrium shifts to the left and most of the added H^+ is removed from solution. If OH^- is added, H^+ and OH^- combine, taking H^+ out of solution. However, the decrease is countered by more dissociation of H_2CO_3, and the decline in H^+ concentration is minimized. Other buffers include the blood proteins and the proteins in cells. The quantitative aspects of buffering and the respiratory and renal adjustments that operate with buffers to maintain a stable ECF pH of 7.40 are discussed in Chapters 35 and 39.

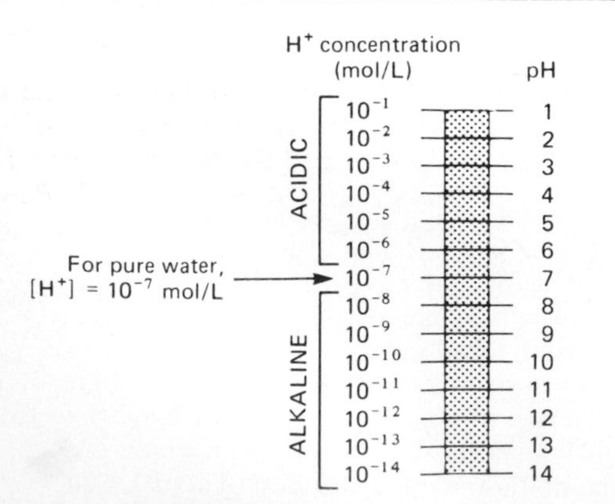

Figure 1–2. pH. (Reproduced, with permission, from Alberts B et al: *Molecular Biology of the Cell.* Garland, 1983.)

COMPOSITION OF BODY FLUIDS

The distribution of electrolytes in the various compartments of the body fluid is shown in Fig 1–3. The figures for the intracellular phase ("cell fluid") are approximations. The composition of intracellular fluid varies somewhat depending upon the nature and function of the cell.

It is apparent from Fig 1–3 that electrolyte concentrations differ markedly in the various compartments. The most striking differences are the relatively low content of protein anions in interstitial fluid compared to intracellular fluid and plasma, and the fact that Na^+ and Cl^- are largely extracellular, whereas most of the K^+ is intracellular.

FORCES PRODUCING MOVEMENT OF SUBSTANCES BETWEEN COMPARTMENTS

The differences in composition of the various body fluid compartments are due in large part to the nature of the barriers separating them. The membranes of the cells separate the interstitial fluid and intracellular fluid, and the capillary wall separates the interstitial fluid from the plasma. The primary forces producing movement of water and other molecules across these barriers are diffusion, facilitated diffusion, solvent drag, filtration, osmosis, active transport, and the processes of exocytosis and endocytosis.

It should be noted that next to biologic membranes in living animals, there is a layer of relatively unstirred fluid. Solutes cross this **unstirred layer** by diffusion, and diffusion across it must be taken into account in considering transport across membranes. Its role is currently a matter of some debate, but in some instances it represents a significant portion of the resistance to movement of a given solute across a membrane.

Diffusion

Diffusion is the process by which a gas or a substance in solution expands, because of the motion of its particles, to fill all of the available volume. The particles (molecules or ions) of a substance dissolved in a solvent are in continuous random movement. In regions where they are abundant, they frequently collide. They therefore tend to spread from regions of high concentration to regions of low concentration

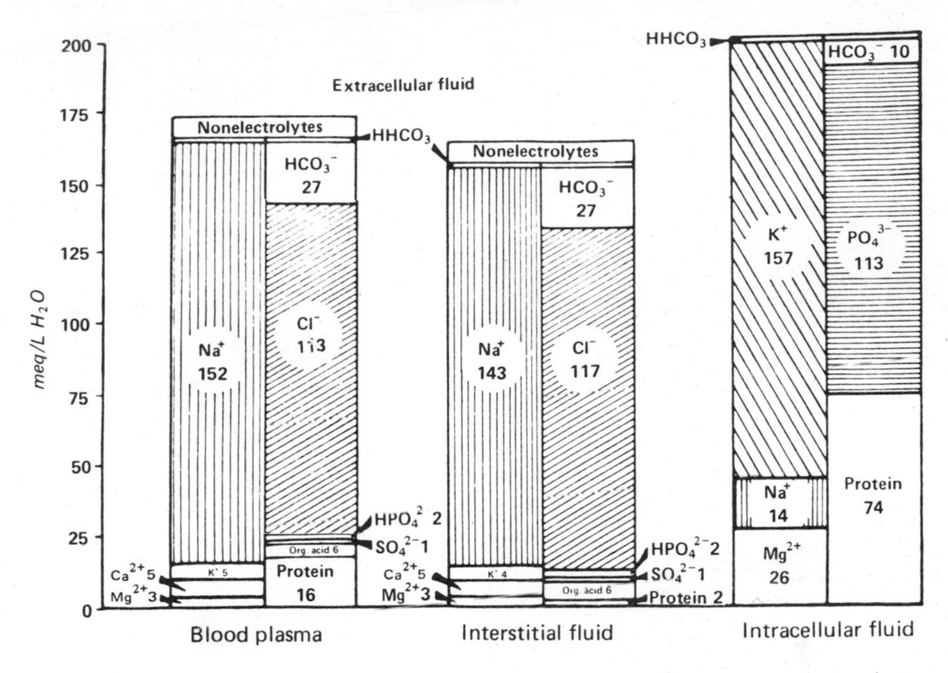

Figure 1–3. Electrolyte composition of human body fluids. Note that the values are in meq/L of water, not of body fluid. (Reproduced, with permission, from Leaf A, Newburgh LH: *Significance of the Body Fluids in Clinical Medicine,* 2nd ed. Thomas, 1955.)

until the concentration is uniform throughout the solution. The time required for equilibrium by diffusion is proportionate to the square of the diffusion distance. Solute particles are, of course, moving both into and out of the region of high concentration; however, since more molecules move out than in, there is a **net flux** of the molecular species to the region of low concentration. The magnitude of the diffusing tendency from one region to another is directly proportionate to the difference in concentration of the substance in the 2 regions (the **concentration gradient,** or **chemical gradient**) and the cross-sectional area across which diffusion is taking place, and inversely proportionate to the thickness of the boundary **(Fick's law of diffusion).** Diffusion of ions is also affected by their electrical charge. Whenever there is a difference in electrical potential between 2 regions, positively charged ions move along this **electrical gradient** to the more negatively charged region; negatively charged ions move in the opposite direction.

In the body, diffusion occurs not only within fluid compartments but also from one compartment to another provided the barrier between the compartments is permeable to the diffusing substances. The diffusion rate (J) of most solutes across the barriers is determined by the permeability coefficient (P), the area across which diffusion takes place (A), and the concentration difference ($C_1 - C_2$):

$$J = PA(C_1 - C_2)$$

In general, the P values are much smaller than the P value for water, but diffusion is still a major force affecting the distribution of water and solutes.

Donnan Effect

When there is an ion on one side of a membrane that cannot diffuse through the membrane, the distribution of other ions to which the membrane is permeable is affected in a predictable way. For example, the negative charge of a nondiffusible anion hinders diffusion of the diffusible cations and favors diffusion of the diffusible anions. Consider the following situation,

X	m	Y
K^+	|	K^+
Cl^-	|	Cl^-
$Prot^-$	|	

in which the membrane (m) between compartments X and Y is impermeable to $Prot^-$ but freely permeable to K^+ and Cl^-. Assume that the concentrations of the anions and of the cations on the 2 sides are initially equal. Cl^- tends to diffuse down its concentration gradient from Y to X, and some K^+ moves with the negatively charged Cl^-, helping to maintain electroneutrality on side Y. Therefore, at equilibrium,

$$[K^+_X] > [K^+_Y]$$

Furthermore,

$$[K^+_X] + [Cl^-_X] + [Prot^-_X] > [K^+_Y] + [Cl^-_Y]$$

ie, there are more osmotically active particles on side X than on side Y.

Donnan and Gibbs showed that in the presence of a nondiffusible ion, the diffusible ions distribute themselves so that at equilibrium, their concentration ratios are equal:

$$\frac{[K^+X]}{[K^+Y]} = \frac{[Cl^-Y]}{[Cl^-X]}$$

In the case of single cations and anions of the same valence, the product of the concentration of the diffusible ions on one side equals that on the other side. Thus,

$$[K^+X][Cl^-X] = [K^+Y][Cl^-Y]$$

Electroneutrality requires that the sum of the anions on each side of the membrane equal the sum of the cations on that side. However, there is a slight excess of cations on side X and a slight excess of diffusible anions on side Y at equilibrium, and therefore a difference in electrical potential exists between X and Y. It should be emphasized that the difference between the number of anions and the number of cations on either side of the membrane is extremely small relative to the total numbers of anions and cations present.

The **Donnan effect** on the distribution of diffusible ions is important in the body because of the presence in cells and in plasma, but not in interstitial fluid, of large quantities of nondiffusible protein anions (see Chapters 35 and 38).

Solvent Drag

When there is net solvent flow in one direction **(bulk flow),** the solvent tends to drag along some molecules of solute. This force is called **solvent drag.** In most situations in the body, its effects are very small.

Filtration

Filtration is the process by which fluid is forced through a membrane or other barrier because of a difference in pressure on the 2 sides. The amount of fluid filtered in a given interval is proportionate to the difference in pressure, the surface area of the membrane, and the permeability of the membrane. Molecules that are smaller in diameter than the pores of the membrane pass through with the fluid, and larger molecules are retained. Filtration of small molecules across the capillary walls occurs when the hydrostatic pressure in the vessels is greater than that in the extravascular tissues, and this pressure gradient is greater than any inwardly directed osmotic pressure gradient (see below).

Osmosis

When a substance is dissolved in water, the concentration of water molecules in the solution is less than that in pure water, since the addition of solute to water results in a solution that occupies a greater volume than does the water alone. If the solution is placed on one side of a membrane that is permeable to water but not to the solute, and an equal volume of water is placed on the other, water molecules diffuse down their concentration gradient into the solution (Fig 1–4). This process—the diffusion of **solvent** molecules into a region in which there is a higher concentration of a **solute** to which the membrane is impermeable—is called **osmosis.** It is an immensely important factor in physiologic processes. The tendency for movement of solvent molecules to a region of greater solute concentration can be prevented by applying pressure to the more concentrated solution. The pressure necessary to prevent solvent migration is the **effective osmotic pressure** of the solution.

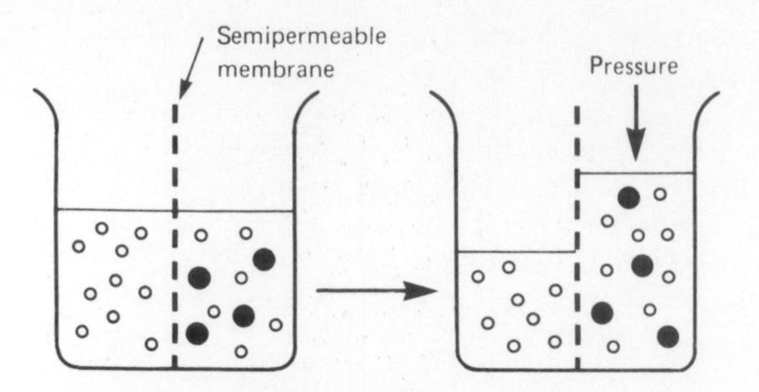

Figure 1–4. Diagrammatic representation of osmosis. Water molecules are represented by small open circles, solute molecules by large solid circles. In the diagram on the left, water is placed on one side of a membrane permeable to water but not to solute, and an equal volume of a solution of the solute is placed on the other. Water molecules move down their concentration gradient into the solution, and as shown in the diagram on the right, the volume of the solution increases. As indicated by the arrow on the right, the osmotic pressure is the pressure that would have to be applied to prevent the movement of the water molecules.

Osmotic pressure, like vapor pressure lowering, freezing point depression, and boiling point elevation, depends upon the number rather than the type of particles in a solution; ie, it is a fundamental colligative property of solutions. In an **ideal solution,** osmotic pressure (P) is related to temperature and volume in the same way as the pressure of a gas:

$$P = \frac{nRT}{V}$$

when n is the number of particles, R the gas constant, T the absolute temperature, and V the volume. If T is held constant, it is clear that the osmotic pressure is proportionate to the number of particles in solution per unit volume of solution. If the solute is a nonionizing compound such as glucose, the osmotic pressure is a function of the number of glucose molecules present. If the solute ionizes and forms an ideal solution, each ion is an osmotically active particle. For example, NaCl would dissociate into Na^+ and Cl^- ions, so that each mole in solution would supply 2 osm. One mole of Na_2SO_4 would dissociate into Na^+, Na^+, and SO_4^{2-}, supplying 3 osm. However, the body fluids are not ideal solutions, and although the dissociation of strong electrolytes is complete, the number of particles free to exert an osmotic effect is reduced owing to interactions between the ions. Thus, it is actually the effective concentration (**activity** in the body fluids) rather than the number of equivalents of an electrolyte in solution that determines its osmotic effect. This is why, for example, 1 mmol of NaCl per liter in the body fluids contributes somewhat less than 2 mosm of osmotically active particles per liter. The more concentrated the solution, the greater the deviation from an ideal solution.

It should be noted that although a homogeneous solution contains osmotically active particles and can be said to have an osmotic pressure, it can only exert an osmotic pressure when it is in contact with another solution across a membrane permeable to the solvent but not to the solute.

Osmolal Concentration of Plasma: Tonicity

The freezing point of normal human plasma averages —0.54 °C, which corresponds to an osmolal concentration in plasma of 290 mosm/L. This is equivalent to an osmotic pressure against pure water of 7.3 atmospheres. The osmolality might be expected to be higher than this, because the sum of all the cation and anion equivalents in plasma is over 300. It is not this high because plasma is not an ideal solution, and ionic interactions reduce the number of particles free to exert an osmotic effect. Except when there has been insufficient time after a sudden change in composition for equilibrium to occur, all fluid compartments of the body are apparently in or nearly in osmotic equilibrium. The term **tonicity** is used to describe the osmolality of a solution relative to plasma. Solutions that have the same osmolality as plasma are said to be **isotonic;** those with greater osmolality are **hypertonic;** and those with greater osmolality are **hypotonic.** All solutions that are initially isosmotic with plasma—ie, have the same actual osmotic pressure or freezing point depression as plasma—would remain isotonic if it were not for the fact that some solutes diffuse into cells and others are metabolized. Thus, a 0.9% saline solution remains isotonic because there is no net movement of the osmotically active particles in the solution into cells and the particles are not metabolized. However, urea diffuses rapidly into cells, so that the osmolality drops when cells are suspended in an aqueous solution that initially contains 290 mosm of urea per liter. Similarly, a 5% glucose solution is isotonic when initially infused intravenously, but glucose is metabolized, so the net effect is that of infusing a hypotonic solution.

It is important to note the relative contributions of the various plasma components to the total osmolal concentration of plasma. All but about 20 of the 290 mosm in each liter of normal plasma are contributed by Na^+ and its accompanying anions, principally Cl^- and HCO_3^-. Other cations and anions make a relatively small contribution. Although the concentration of the plasma proteins is large when expressed in grams per liter, they normally contribute less than 2 mosm/L because of their very large molecular weights. The major nonelectrolytes of plasma are glucose and urea, which in the steady state are in equilibrium with cells. Their contributions to osmolality are normally about 5 mosm/L each but can become quite large in hyperglycemia or uremia. The total plasma osmolality is important in assessing dehydration, overhydration, and other fluid and electrolyte abnormalities. Hyperosmolality can cause coma (hyperosmolar coma; see Chapter 19). Because of the predominant role of the major solutes and the deviation of plasma from an ideal solution, one can ordinarily approximate the plasma osmolality within a few milliosmoles per liter by using the following formula, in which the constants convert the clinical units to millimoles of solute per liter:

$$\underset{(\text{mosm/L})}{\text{Osmolality}} = 2\underset{(\text{meq/L})}{[\text{Na}^+]} + 0.055\underset{(\text{mg/dL})}{[\text{Glucose}]} + 0.36\underset{(\text{mg/dL})}{[\text{BUN}]}$$

BUN is the blood urea nitrogen. The formula is also useful in calling attention to abnormally high concentrations of other solutes. An observed plasma osmolality (measured by freezing point depression) that greatly exceeds the value predicted by this formula probably indicates the presence of a foreign substance such as ethanol, mannitol (sometimes injected to shrink swollen cells osmotically), or poisons such as ethylene glycol or methanol (components of antifreeze).

Nonionic Diffusion

Some weak acids and bases are quite soluble in cell membranes in the undissociated form, whereas in the ionic form they cross membranes with difficulty. Consequently, if molecules of the undissociated substance diffuse from one side of the membrane to the other and then dissociate, there is appreciable net movement of the undissociated substance from one side of the membrane to another. This phenomenon, which occurs in the gastrointestinal tract (see Chapter 25) and kidneys (see Chapter 38), is called **nonionic diffusion.**

Carrier-Mediated Transport

In addition to moving across cell membranes by diffusion and osmosis, ions and large nonionized molecules are transported by carrier molecules in the membranes. When such **carrier-mediated transport** is from an area of greater concentration of the transported molecules to an area of lesser concentration, energy is not required, and the process is called **facilitated diffusion.** When, on the other hand, the transport is from an area of lesser to an area of greater concentration and cannot be explained by movement down an electrical gradient, the process requires energy and is referred to as **active transport.** Active transport is carried out by protein "pumps" in cell membranes, and the energy is supplied by the metabolism of the cells through adenosine triphosphate (ATP; see below and Chapter 17). Active transport is of major importance throughout the body.

Transport of Proteins & Other Large Molecules

In certain situations proteins enter cells, and a number of hormones and other substances secreted by cells are proteins or large polypeptides. Proteins and other large molecules enter cells by the process of endocytosis and are secreted by exocytosis (see below). These processes provide an explanation of how large molecules can enter and leave cells without disrupting cell membranes.

Transport Across Epithelia

In various parts of the body, electrolytes and nonelectrolytes are transported across epithelial surfaces. Examples of such surfaces are the mucosa of the stomach and small intestine, the mucosa of the gallbladder, and the cells that make up the renal tubules. In these situations, the cells must have different transport mechanisms in their luminal and basolateral membranes. For example, Na^+ moves from the tubular lumen into renal tubular cells in exchange for H^+, and is pumped from the other side of the cells into the interstitial fluid in exchange for K^+ (see Chapter 38). Another example is the Na^+ - dependent transport of glucose from the intestine (see Chapter 25). Some epithelia (eg, renal proximal tubule, intestine, gallbladder) have leaky junctions between cells which permit Na^+, K^+, and Cl^- to flow back into the tubular lumen. Others (eg, stomach, urinary bladder) have tight junctions where such backflow does not occur. In either case, it is necessary to maintain the total number of ions in cells more or less constant so that osmotic changes do not cause shrinkage or swelling. It now appears that there is "cross talk" between luminal and basolateral membranes of cells in epithelia so that permeabilities are adjusted and ionic balance is maintained.

THE CAPILLARY WALL

The structure of the capillary wall, the barrier between the plasma and the interstitial fluid, varies from one vascular bed to another (see Chapter 30). However, in skeletal muscle and many other organs, water and relatively small solutes are the only substances that cross the wall with ease. The apertures in the wall (the junctions between the endothelial cells) are too small to permit plasma proteins and other colloids to pass through in significant quantities. The colloids have a high molecular weight but are present in large amounts. Small amounts cross the capillary wall by vesicular transport (see below), but their effect is slight. Therefore, the capillary wall behaves like a membrane impermeable to colloids, which exert an osmotic pressure of about 25 mm Hg. The colloid osmotic pressure due to the plasma colloids is called the **oncotic pressure.** Filtration across the capillary membrane due to the hydrostatic pressure head in the vascular system is opposed by the oncotic pressure. The way the balance between the hydrostatic and oncotic pressures controls exchanges across the capillary wall is considered in detail in Chapter 30.

There are vesicles in the cytoplasm of endothelial cells, and tagged protein molecules injected into the bloodstream have been found in the vesicles and in the interstitium. This indicates that small amounts of protein are transported out of capillaries across endothelial cells by endocytosis followed by exocytosis on the interstitial side of the cells. The transport mechanism makes use of coated vesicles and is called **vesicular transport** or **cytopempsis**.

SODIUM & POTASSIUM DISTRIBUTION & TOTAL BODY OSMOLALITY

The foregoing discussion of the various compartments of the body fluids and the barriers between them facilitates consideration of the total body stores of the principal cations, Na^+ and K^+.

Total Body Sodium

The total amount of exchangeable Na^+ in the body (Na_E)—as opposed to its concentration in any particular body fluid—can be determined by the same dilution principle used to measure the body fluid compartments. A radioactive isotope of sodium (usually ^{24}Na) is injected, and after equilibration, the fraction of the sodium in the body that is radioactive is determined. The fraction of a substance that is radioactive—ie, the concentration of radioactive molecules divided by the concentration of radioactive plus nonradioactive molecules—is the **specific activity** (SA) of the substance. The SA of sodium in plasma after the injection of ^{24}Na, for example, is

$$\frac{^{24}\text{Na (counts/min/L)}}{^{24}\text{Na + nonradioactive Na (meq/L)}}$$

The total exchangeable body sodium (Na_E) =

$$\frac{^{24}\text{Na injected} - {^{24}\text{Na excreted}}}{\text{SA of plasma}}$$

The average normal value for Na_E in healthy adults is 41 meq/kg, whereas the total amount of Na^+ in the body is about 58 meq/kg. Therefore, approximately 17 meq/kg is not available for exchange. The vast majority of this nonexchangeable Na^+ is the hydroxyapatite crystal lattice of bone. The distribution of body Na^+, which is mostly extracellular, is summarized in Table 1–2.

Table 1–2. Distribution of sodium in the body in meq/kg body weight and percentage of total body sodium.*

	meq/kg	% of Total
Total	58	100
Exchangeable body sodium (Na_E)	41	70.7
Total intracellular	5.2	9.0
Total extracellular	52.8	91.0
Plasma	6.5	11.2
Interstitial fluid	16.8	29.0
Dense connective tissue and cartilage	6.8	11.7
Bone sodium		
Exchangeable	6.4	11.0
Nonexchangeable	14.8	25.5
Transcellular	1.5	2.6

* Data courtesy of IS Edelman.

Total Body Potassium

The total exchangeable body K^+ can be determined in the same way as total exchangeable body Na^+, using radioactive potassium (^{42}K). The average value in young adult men is about 45 meq/kg body weight. It is somewhat less in women, and it declines slightly with advancing age. The distribution of body K^+, which is mostly intracellular, is shown in Table 1–3. About 10% of the total body K^+ is bound, and the remaining 90% is exchangeable.

Table 1–3. Distribution of body potassium.

Location	% of Total
Plasma	0.4
Interstitial fluid	1.0
Dense connective tissue and cartilage	0.4
Bone	7.6
Transcellular	1.0
Intracellular	89.6

Interrelationships of Sodium & Potassium

Since the salts of sodium and potassium dissociate in the body to such a great extent, and since they are so plentiful, they determine in large part the osmolality of the body fluids. A change in the amount of electrolyte in one compartment is followed by predictable changes in the volume and electrolyte concentration in the others because the various compartments are in osmotic equilibrium. Fig 1–5 shows, for example, the changes in osmolal concentration and volume of the intracellular and extracellular fluid spaces that follow removal of 500 mosm of extracellular electrolyte. Loss of electrolyte in excess of water from the ECF leads to hypotonicity of the ECF relative to the intracellular fluid. Consequently,

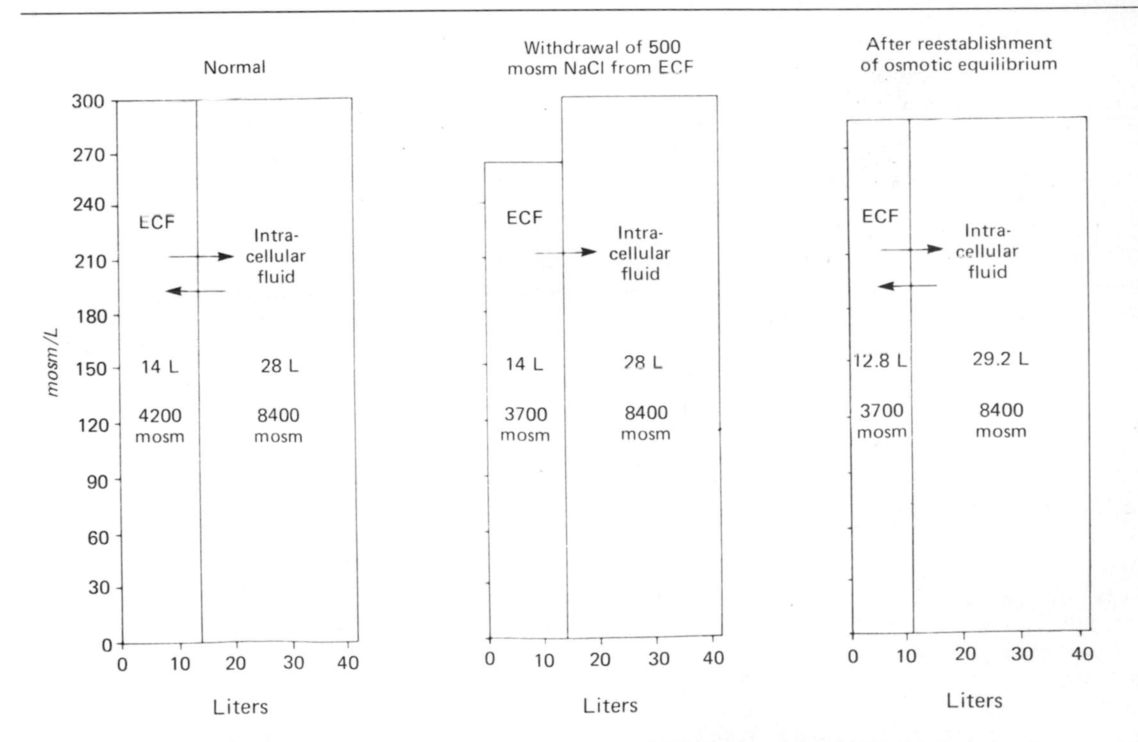

Figure1–5. Response of body fluid to removal of 500 mosm from ECF without change in total body water in a 70-kg man. The hypotonicity of the ECF causes water to enter cells until osmotic equilibrium is reestablished. Total mosm in body after removal of 500 mosm = (14 × 300) + (28 × 300) − 500 = 12,100 mosm. New osmotic equilibrium = 12,100/(28 + 14) = 288 mosm/L. This means that the new volume of the ECF will be (4200 − 500)/288 = 12.8 L and the new volume of the intracellular fluid will be 8400/288 = 29.2 L. (After Darrow DC and Yannet H. Modified and reproduced, with permission, from Gamble JL: *Chemical Anatomy, Physiology, and Pathology of Extracellular Fluid,* 4th ed. Harvard Univ Press, 1942.)

water moves into the cells by osmosis until osmotic equilibrium between the ECF and the intracellular fluid is again attained. The net result is a decrease in ECF volume and an increase in intracellular fluid volume.

Total Body Osmolality

Since Na^+ is the principal cation of the plasma, the plasma osmotic pressure correlates well with the plasma Na^+ level. However, plasma Na^+ does not necessarily correlate well with Na_E. Total body osmolality reflects the electrolyte concentration in the body, ie, the total exchangeable body sodium plus the total exchangeable body potassium divided by the total body water:

$$\frac{Na_E + K_E}{TBW}$$

Because this latter figure has been found to correlate well with plasma Na^+, plasma Na^+ is a relatively accurate measure of total body osmolality. It is worth noting that plasma Na^+ can be affected by changes in any of 3 body components: Na_E itself, K_E, or TBW. A detailed discussion of the effects of disease on electrolyte and water balance is beyond the scope of this book, but the facts discussed above are basic to any consideration of pathologic water and electrolyte changes.

Plasma K^+ levels are not a good indicator of the total body K^+, since most of the K^+ is in the cells. There is a correlation between the K^+ and H^+ content of plasma, the 2 rising and falling together.

FUNCTIONAL MORPHOLOGY OF THE CELL

Revolutionary advances in the understanding of cell structure and function have been made through use of the techniques of modern cellular and molecular biology. Cell biology per se is beyond the scope of this book. However, a basic knowledge of cell biology is essential to an understanding of the organ systems in the body and the way they function.

The specialization of the cells in the various organs is very great, and no cell can be called "typical" of all cells in the body. However, a number of structures (**organelles**) are common to most cells. These structures are shown in Fig 1–6.

Cell Membrane

The membrane that surrounds the cell is a remarkable structure. It is not only semipermeable, allowing

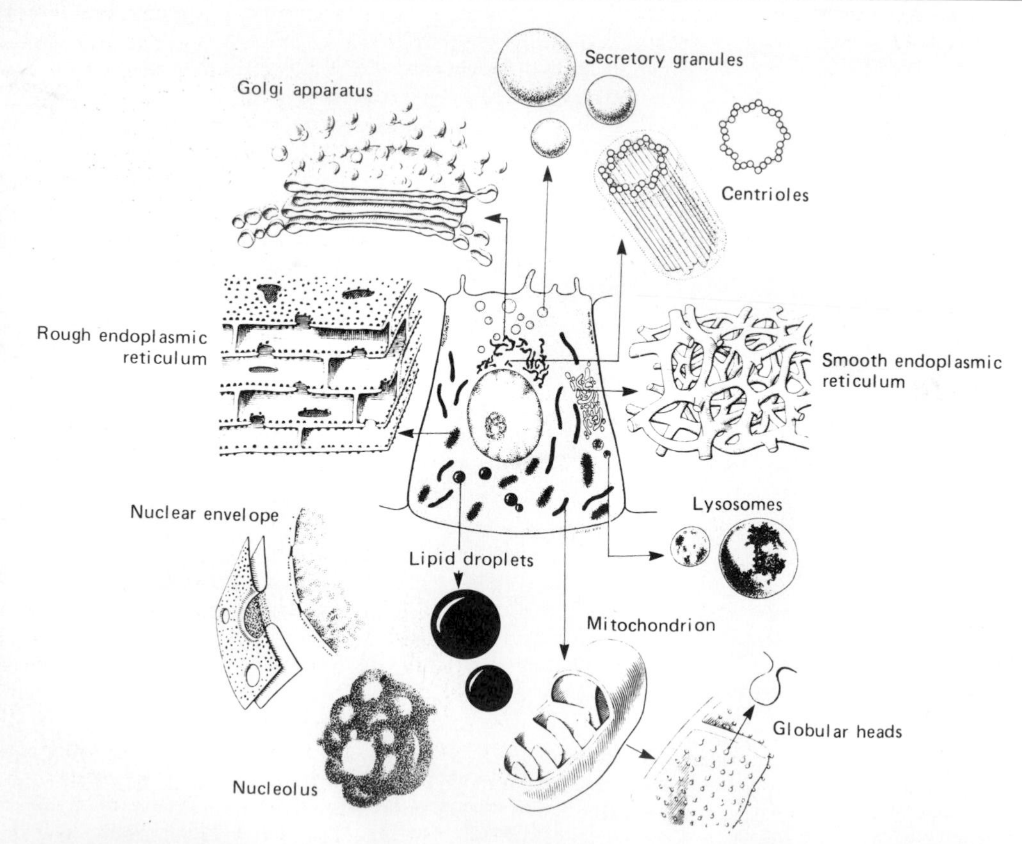

Figure 1–6. Diagram showing a hypothetical cell in the center as seen with the light microscope. It is surrounded by various organelles. (After Bloom and Fawcett. Reproduced, with permission, from Junqueira LC, Carneiro J, Kelley, RO: *Basic Histology,* 6th ed. Appleton & Lange, 1989.)

some substances to pass through it and excluding others, but its permeability can be varied. It is generally referred to as the **plasma membrane.** The nucleus is also surrounded by a membrane, and the organelles are surrounded by or made up of a membrane.

Although the chemical structure of membranes and their properties vary considerably from one location to another, they have certain common features. They are generally about 7.5 nm (75 Angstrom units) thick. They are made up primarily of protein and lipids. The chemistry of proteins and lipids is discussed in Chapter 17. The major lipids are phospholipids such as phosphatidylcholine and phosphatidylethanolamine. The shape of the phospholipid molecule is roughly that of a clothespin (Fig 1–7). The head end of the molecule contains the phosphate portion and is relatively soluble in water (polar, **hydrophilic**). The tails are relatively insoluble (nonpolar, **hydrophobic**). In the membrane, the hydrophilic ends of the molecules are exposed to the aqueous environment that bathes the exterior of the cells and the aqueous cytoplasm; the hydrophobic ends meet in the water-poor interior of the membrane. In **prokaryotes** (cells like bacteria in which there is no nucleus), phospholipids are generally the only membrane lipids, but in **eukaryotes** (cells containing nuclei), cell membranes also contain cholesterol (in animals) or other steroids (in plants). The cholesterol/phospholipid ratio in the membrane is maintained at a nearly constant level by a variety of regulatory mechanisms.

There are many different proteins embedded in the membrane. They exist as separate globular units and stud the inside and outside of the membrane in a random array (Fig 1–7). Some are located in the inner surface of the membrane; some are located on the outer surface; and some extend through the membrane (**transmembrane proteins**). In general, the uncharged, hydrophobic portions of the protein molecules are located in the interior of the membrane and the charged, hydrophilic portions are located on the surfaces. Some of the proteins contain lipids (lipoproteins) and some contain carbohydrates (glycoproteins). There are 5 types of proteins in the membrane. In addition to **structural proteins,** there are proteins that function as **pumps,** actively transporting ions across the membrane. Other proteins function as passive **channels** for ions that can be opened or closed by changes in the conformation of the protein. A fourth group of proteins functions as **receptors** that bind neurotransmitters and hormones, initiating physiologic changes inside the cell. A fifth group functions as **enzymes,** catalyzing reactions at the surfaces of the membrane.

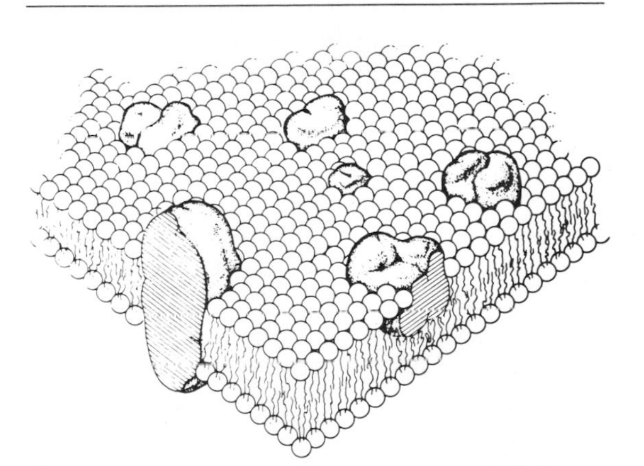

Figure 1–7. Biologic membrane. The phospholipid molecules each have 2 fatty acid chains (wavy lines) attached to a phosphate head (open circle). Proteins are shown as irregular shaded globules. Some are on the outside of the membrane, others are on the inside (not shown), and still others span the membrane. (Reproduced, with permission, from Singer SJ, Nicolson GL: The fluid mosaic model of the structure of cell membranes. *Science* 1972;**175**:720. Copyright 1972 by the American Association for the Advancement of Science.)

Other cell surface proteins are attached to the cell membrane by glycosylated forms of phosphatidylinositol. Proteins held by these **glycosylphosphatidylinositol anchors** include enzymes such as alkaline phosphatase, various antigens, and a number of cell adhesion molecules.

The protein structure—and particularly the enzyme content—of biologic membranes varies not only from cell to cell but also within the same cell. For example, there are different enzymes embedded in cell membranes than in mitochondrial membranes; in epithelial cells, the enzymes in the cell membrane on the mucosal surface differ from those in the cell membrane on the lateral margins of the cells; ie, the cells are **polarized.** The membranes are dynamic structures, and their constituents are being constantly renewed at different rates. Some proteins move laterally in the membrane. For example, receptors move in the membrane and aggregate at sites of endocytosis (see below).

Underlying most cells is a thin, fuzzy layer plus some fibrils that collectively make up the **basement membrane** or, more properly, the **basal lamina.** The material that makes up the basal lamina has been shown to be made up of a collagen derivative plus 2 glycoproteins.

Intercellular Connections

Two types of junctions form between the cells that make up tissues: junctions that fasten the cells to one another and to surrounding tissues, and junctions that permit transfer of ions and other molecules from one cell to another. The junctions that tie cells together and endow tissues with strength and stability include **tight junctions,** which are also known as the **zonulae occludens. Desmosomes, hemidesmosomes,** and probably **zonulae adherens** (Fig 1–8) also hold tissues together. The junctions by which molecules are transferred are **gap junctions.**

Tight junctions characteristically surround the apical margins of the cells in epithelia such as the

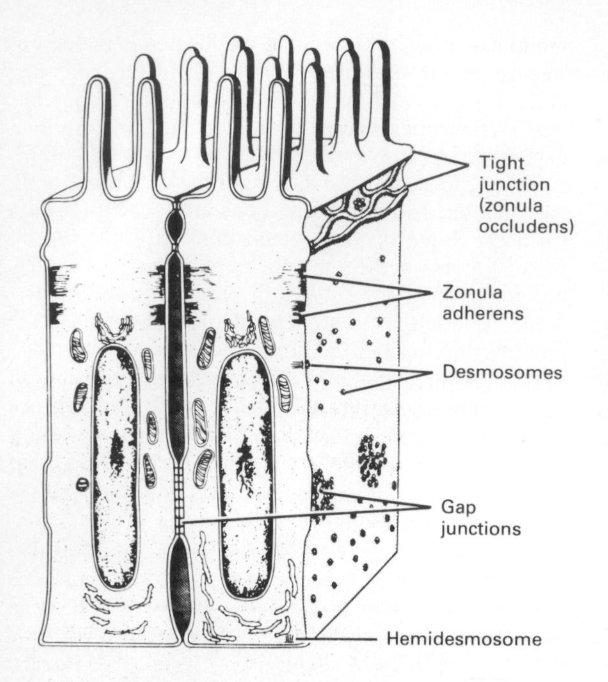

Figure 1–8. Intercellular junctions in the mucosa of the small intestine. The various types of desmosomes are not shown in detail. (Modified from Weinstein RS, McNutt SN: Cell junctions. *N Engl J Med* 1972;**286**:521.)

intestinal mucosa, the walls of the renal tubules, and the choroid plexus. They are made up of ridges—half from one cell and half from the other—that adhere so strongly at cell junctions that they almost obliterate the space between the cells. The ridges are made up of proteins, and 2 of these, **cingulin** and **ZO-1,** have been isolated and studied. In addition to tying the cells together, tight junctions form a barrier to the movement of ions and other solutes from one side of the epithelium to the other. The magnitude of this barrier varies, and some epithelia are more "leaky" to solutes than others. Another function of tight junctions is maintenance of cell polarity. The ridges prevent the lateral movement of proteins in the cell membrane, and consequently they act as a fence, keeping proteins inserted into the apical membrane in the apical region.

In epithelial cells, the zonula adherens is usually a continuous structure on the basal side of the zonula occludens, and it is a major site of attachment for intracellular **actin fibers** (see below and Chapter 3). There is little that is visible under the microscope on the external surfaces of adjacent zonulae adherens, but there is evidence that the cell adhesion molecule **L-CAM,** an adhesive protein also known as **uromorulin,** is found in this location.

Desmosomes are spotlike patches characterized by apposed thickenings of the membranes of 2 adjacent cells. Attached to the thickened area in each cell are cytoplasmic fibrils made up of actin, some running parallel to the membrane and others radiating away from it. Between the 2 membrane thickenings, there is filamentous material in the intercellular space. Hemidesmosomes are half desmosomes that attach cells to an underlying basal lamina without apparent specialization of the basal lamina.

At gap junctions, the intercellular space narrows from 25 nm to 3 nm, and hexagonal arrays of protein units called **connexons** in the membrane of each cell are lined up with one another (Fig 1–9). Each connexon is made up of 6 subunits surrounding a channel that, when lined up with the channel in the corresponding connexon in the adjacent cell, permits substances to pass between the cells without entering the extracellular fluid. The diameter of the channel is normally about 2 nm, which permits the passage of ions, sugars, amino acids, and other solutes with molecular weights up to about 1000. Gap junctions thus permit the rapid propagation of electrical activity from cell to cell (see Chapter 4) and the exchange of various chemical messengers. The diameter of each channel is regulated by intracellular Ca^{2+}, and an increase in Ca^{2+} causes the subunits to slide together, reducing the diameter of the channel. The diameter may also be regulated by pH and voltage.

Mitochondria

Although their morphology varies somewhat from cell to cell, each mitochondrion (Fig 1–6) is in essence a sausage-shaped structure. It is made up of an outer membrane and an inner membrane that is folded to form shelves **(cristae).** The space between the 2 membranes is called the intracristal space, and the space inside the inner membrane is called the matrix space. The mitochondria are the power-generating units of the cell and are most plentiful and best developed in parts of cells where energy-requiring processes take place. The chemical reactions occur-

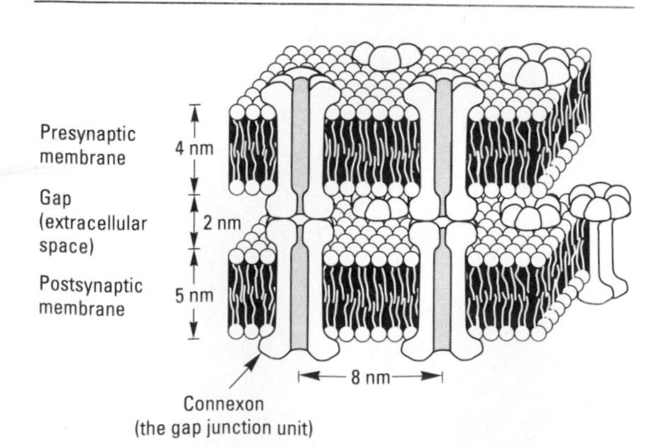

Figure 1–9. Gap junction. Note that each connexon is made up of 6 subunits, and that each connexon in the membrane of one cell lines up with a connexon in the membrane of the neighboring cell, forming a channel through which substances can pass from one cell to another without entering the ECF. (Reproduced, with permission, from Kandel ER, Schwartz JH, (editors): *Principles of Neural Science,* 2nd ed. Elsevier, 1985.)

ring in them are discussed in detail in Chapter 17. The outer membrane of each mitochondrion is studded with the enzymes concerned with biologic oxidations, providing raw materials for the reactions occurring inside the mitochondrion. In the interior, the matrix space contains the enzymes that convert the products of carbohydrate, protein, and fat metabolism to CO_2 and water via the citric acid cycle. In this process, electrons are transferred along the respiratory-enzyme chain and there is synthesis of the high-energy phosphate compound **adenosine triphosphate (ATP)** by the process of **oxidative phosphorylation.** ATP is the principal energy source for energy-requiring actions in plants and animals. The enzymes responsible for electron transfer and oxidative phosphorylation are embedded in the inner membrane, and they form characteristic repeating units visible under the electron microscope. Each unit is made up of a basepiece, a stalk, and a spherical headpiece (Fig 1–6). The basepieces contain the enzymes of the electron transfer chain, and the stalks and headpieces contain adenosine triphosphatase and other enzymes concerned with the synthesis and metabolism of ATP.

Mitochondrial DNA is discussed below.

Lysosomes

In the cytoplasm of the cell, there are large, somewhat irregular structures surrounded by membrane. The interior of these structures, which are called **lysosomes,** is more acidic than the rest of the cytoplasm, and they may contain fragments of other cell structures. Some of the granules of the granulocytic white blood cells are lysosomes. Each lysosome contains a variety of enzymes (Table 1–4) that would cause the destruction of most cellular components if the enzymes were not separated from the rest of the cell by the membrane of the lysosome.

Table 1–4. Some of the enzymes found in lysosomes and the cell components that are their substrates.

Enzyme	Substrate
Ribonuclease	RNA
Deoxyribonuclease	DNA
Phosphatase	Phosphate esters
Glycosidases	Complex carbohydrates: glycosides and polysaccharides
Arylsulfatases	Sulfate esters
Collagenase	Proteins
Cathepsins	Proteins

The lysosomes function as a form of digestive system for the cell. Exogenous substances such as bacteria that become engulfed by the cell end up in membrane-lined vacuoles. A vacuole of this type **(phagocytic vacuole)** may merge with a lysosome, permitting the contents of the vacuole and the lysosome to mix within a common membrane. Some of the products of the "digestion" of the engulfed material are absorbed through the walls of the vacuole, and the remnants are dumped from the cell (exocytosis; see below). The lysosomes also engulf worn-out components of the cell in which they are located, forming **autophagic vacuoles.** When a cell dies, lysosomal enzymes cause autolysis of the rem-

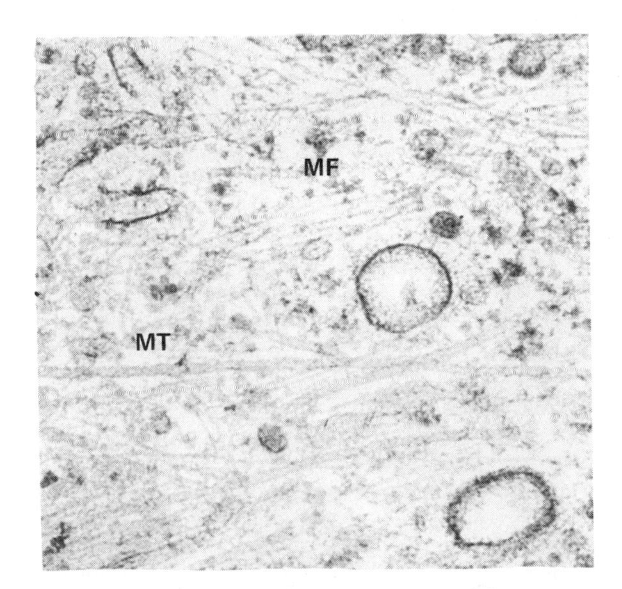

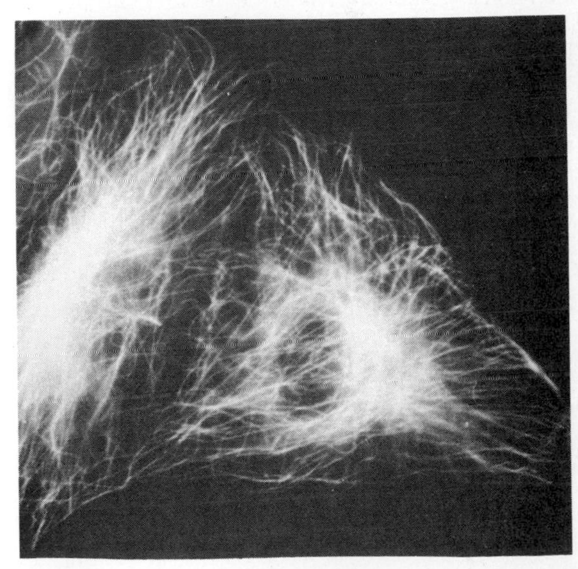

Figure 1–10. *Left:* Electron micrograph of the cytoplasm of a fibroblast, showing microfilaments (MF) and microtubules (MT). (Reproduced, with permission, from Junqueira LC, Carneiro J, Kelley RO: *Basic Histology,* 6th ed. Appleton & Lange, 1989.) *Right:* Distribution of microtubules in fibroblasts. The cells are treated with a fluorescently labeled antibody to tubulin, making microtubules visible as the light-colored structures. (Reproduced, with permission, from Connolly J et al: Immunofluorescent staining of cytoplasmic and spindle microtubules in mouse fibroblasts with antibody to τ protein. *Proc Natl Acad Sci USA* 1977;**74**:2437.)

nants. In vitamin A intoxication and certain other conditions, lysosomal enzymes are released to the exterior of the cell, with resultant breakdown of intercellular material. There is evidence that in gout, phagocytes ingest uric acid crystals, and that such ingestion triggers the extracellular release of lysosomal enzymes that contribute to the inflammatory response in the joints. When one of the lysosomal enzymes is congenitally absent, the lysosomes become engorged with one of the materials they normally degrade. This eventually disrupts the cells that contain the defective lysosomes and leads to one of the **lysosomal storage diseases.** More than 25 such diseases have been described. They are generally rare, but they include such widely known disorders as Tay-Sachs disease.

Microfilaments, Intermediate Filaments & Microtubules

Most, if not all, eukaryotic cells contain **microfilaments,** long solid fibers 4–6 nm in diameter, and **microtubules,** long hollow structures with 5-nm walls surrounding a cavity 15 nm in diameter (Fig 1–10).

Microfilaments are made up of **actin,** the protein that by its interaction with myosin brings about contraction of muscle (see Chapter 3). Actin and its mRNA (see below) are present in all types of cells. Myosin is difficult to observe in cells other than muscle and is not arranged with actin in orderly arrays, but it is also present. Actin microfilaments reach to the tips of the microvilli on the epithelial cells of the intestinal mucosa and bring about their contraction, probably via interaction with myosin filaments located at their bases. Microfilaments are also found in association with desmosomes and zonulae adherens (see above), in bundles under the plasma membrane, and scattered in a seemingly random fashion in the cytoplasm. Actin molecules polymerize and depolymerize in vivo, with polymerization often occurring at one end of a microfilament and depolymerization at the other. Thus, the microfilaments are dynamic structures.

Microtubules are made up of 2 globular protein subunits, α and β tubulin. The subunits unite to form dimers (Fig 1–11), and the dimers aggregate to form long tubes made up of stacked rings, with each ring usually containing 13 subunits. The tubules also contain other proteins that facilitate their formation. The assembly of microtubules is facilitated by warmth and various other factors, and disassembly is facilitated by cold and other factors. Both processes occcur simultaneously in vitro. Assembly is prevented by colchicine and vinblastine.

Microtubules have been called the skeleton of the cell, but because of their constant assembly and disassembly, they are a very dynamic skeleton. They provide the tracks for transport of vesicles, organelles such as secretory granules, and mitochondria from one part of the cell to another. Cross bridges apparently form between the microtubules and the transported organelles. Hydrolysis of ATP provides the energy for transport, possibly by causing the cross bridges to bend. Microtubules can transport in both directions, and, indeed, the same microtubule has been seen transporting 2 particles in opposite directions.

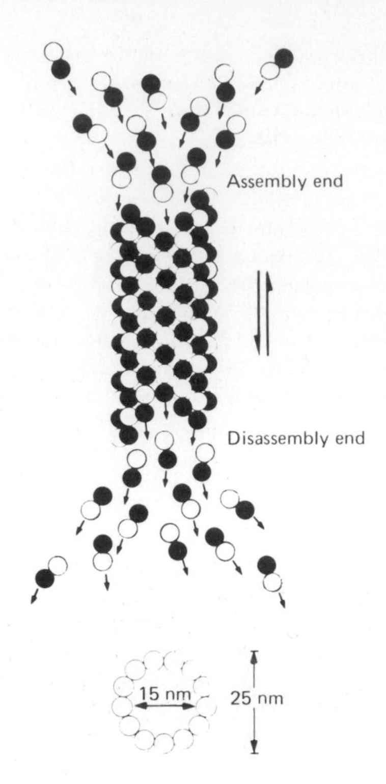

Figure 1–11. Assembly and disassembly of a microtubule by aggregation and disaggregation of dimers made up of α and β tubulin. (From Sloboda RD: The role of microtubules in cell structure and cell division. *Am Sci* 1980; 68:290. Reprinted by permission of *American Scientist,* journal of Sigma Xi, the Scientific Research Society.)

Cells also contain **intermediate filaments** that are 8–10 nm in diameter and are made up of rodlike subunits. Some of these filaments connect the nuclear membrane to the cell membrane. They appear to be part of the cytoskeleton, but their exact function is not known.

Centrioles

In the cytoplasm of most cells there are 2 short cylinders called **centrioles.** The centrioles are located near the nucleus, and they are arranged so that they are at right angles to each other. Microtubules in groups of 3 run longitudinally in the walls of the centriole (Fig 1–6). There are 9 of these triplets spaced at regular intervals around the circumference.

The centrioles are concerned with the movement of the chromosomes during cell division. They duplicate themselves at the start of mitosis, and the pairs move apart to form the poles of the mitotic spindle. In multinucleate cells, there is a pair of centrioles near each nucleus. The mitotic spindle is made up of microtubules and microfilaments. There is also some evidence that centrioles are concerned with the movement of whole cells.

Cilia

There are various types of projections from cells. True **cilia,** motile processes that unicellular organisms use to propel themselves through the water and multicellular organisms use to propel mucus and other substances over the surface of various epithelia, resemble centrioles in having an array of 9 tubular structures in their walls, but they have in addition a pair of microtubules in the center, and there are 2 rather than 3 microtubules in each of the 9 circumferential structures. The **basal granule,** on the other hand, the structure to which a cilium is anchored, has 9 circumferential triplets, like a centriole.

Peroxisomes

If cells are homogenized and the resulting suspension is centrifuged, various cellular components can be isolated. The nuclei sediment first, followed by the mitochondria. High-speed centrifugation that generates forces of 100,000 times gravity or more causes a fraction made up of granules called the **microsomes** to sediment. This fraction includes the ribosomes, but it also contains other granular material that can be isolated by further ultracentrifugation and other techniques. The **peroxisomes** are found in the granular material. These organelles are about 0.5 μm in diameter and are surrounded by a membrane. They contain various oxidases which catalyze reactions that generate hydrogen peroxide, and **catalase,** an enzyme that catalyzes the conversion of hydrogen peroxide to water. Various peroxisomes have various substrates. In mammals, peroxisomes are most common in the liver and kidneys and may be involved in gluconeogenesis.

Nucleus & Related Structures

A nucleus is present in all eukaryotic cells that divide. If a cell is cut in half, the anucleate portion eventually dies without dividing. The nucleus is made up in large part of the **chromosomes,** the structures in the nucleus that carry a complete blueprint for all the heritable species and individual characteristics of the animal. Each chromosome is made up of a giant molecule of **deoxyribonucleic acid (DNA)** that in multicellular organisms is virtually covered with proteins. The molecule has been likened to a string of beads; the beads are balls of histone proteins with DNA wrapped around them, and the string is more DNA with other proteins attached to it. This complex of DNA and proteins is called **chromatin.** During cell division, the pairs of chromosomes become visible, but between cell divisions, only clumps of chromatin can be discerned in the nucleus. The ultimate units of heredity are the **genes** on the chromosomes (see below), and each gene is a portion of the DNA molecule.

During normal cell division by **mitosis,** the chromosomes duplicate themselves and then divide in such a way that each daughter cell receives a full complement **(diploid number)** of chromosomes. During their final maturation, germ cells undergo a division in which half the chromosomes go to each daughter cell (see Chapter 23). This reduction division **(meiosis)** is actually a 2-stage process; but the important consideration is that, as a result of it, mature sperms and ova contain half the normal number (the **haploid number**) of chromosomes. When a sperm and ovum unite, the resultant cell **(zygote)** has a full (diploid) complement of chromosomes, one-half from the female parent and one-half from the male.

The nucleus of most cells contains a **nucleolus** (Fig 1–6), a patchwork of granules rich in **ribonucleic acid (RNA).** In some cells, the nucleus contains several of these structures. Nucleoli are most prominent and numerous in growing cells. They are the site of synthesis of ribosomes, the structures in the cytoplasm in which proteins are synthesized (see below).

The nucleus is surrounded by a **nuclear membrane (envelope)** (Fig 1–6). This membrane is double, and spaces between the 2 folds are called **perinuclear cisterns.** The nuclear membrane is apparently quite permeable, since it permits passage of molecules as large as RNA from the nucleus to the cytoplasm. There are areas of discontinuity in the nuclear membrane, but these "pores" are closed by a thin, homogeneous membrane.

Endoplasmic Reticulum

The endoplasmic reticulum is a complex series of tubules in the cytoplasm of the cell (Fig 1–6). The tubule walls are made up of membrane. In **rough** or **granular endoplasmic reticulum,** granules called **ribosomes** are attached to the cytoplasmic side of the membrane, whereas in **smooth** or **agranular endoplasmic reticulum,** the granules are absent. Free ribosomes are also found in the cytoplasm. The granular endoplasmic reticulum is concerned with protein synthesis and the initial folding of polypeptide chains with the formation of disulfide bonds. The agranular endoplasmic reticulum is the site of steroid synthesis in steroid-secreting cells and the site of detoxification processes in other cells. As the sarcoplasmic reticulum (see Chapter 3), it plays an important role in skeletal and cardiac muscle.

Ribosomes

The ribosomes are about 15 nm in diameter. Each is made up of a large and a small subunit called, on the basis of their rates of sedimentation in the ultracentrifuge, the 50S and the 30S subunits. The ribosomes contain about 65% RNA and 35% protein. They are the sites of protein synthesis. The ribosomes that become attached to the endoplasmic reticulum synthesize proteins such as hormones that are secreted by the cell, proteins that are segregated in lysosomes, and proteins that are inserted in cell membranes. The polypeptide chains that form these proteins are extruded into the endoplasmic reticulum. The free ribosomes synthesize cytoplasmic proteins such as hemoglobin (see Chapter 27) and the proteins found in peroxisomes and mitochondria.

Golgi Apparatus

The Golgi apparatus is a collection of membrane-enclosed sacs (cisterns) that are stacked like dinner plates (Fig 1–6). There are usually about 6 sacs in each apparatus, but there may be more. One or more Golgi apparatuses are present in all eukaryotic cells, usually near the nucleus. The Golgi apparatus is a polarized structure, with *cis* and *trans* sides. Membranous vesicles containing newly synthesized proteins bud off from the granular endoplasmic reticulum and fuse with the cistern on the *cis* side of the apparatus. The proteins are then passed, probably via other vesicles, to the medial cisterns and finally to the cistern on the *trans* side, from which vesicles branch off into the cytoplasm (Fig 1–12). The initial glycosylation of proteins occurs with the attachment of preformed oligosaccharides in the endoplasmic reticulum, but these oligosaccharides are altered to a variety of different carbohydrate moieties in the Golgi apparatus. The alterations provide the code for the ultimate destination of the protein. Proteins destined for lysosomes have phosphate groups added and then bind to one of the 2 types of **mannose-6-phosphate receptors (MPRs)** in the Golgi. They shuttle from the Golgi to a vesicular acidified compartment (prelysosome), where the proteins are released and transferred to lysosomes. The MPRs return to the Golgi to be used again. Proteins destined to be secreted by the cell have their sugar residues modified in a more extensive fashion, and glycoproteins destined for the cell membrane have their sugar residues modifed in a different, equally extensive way.

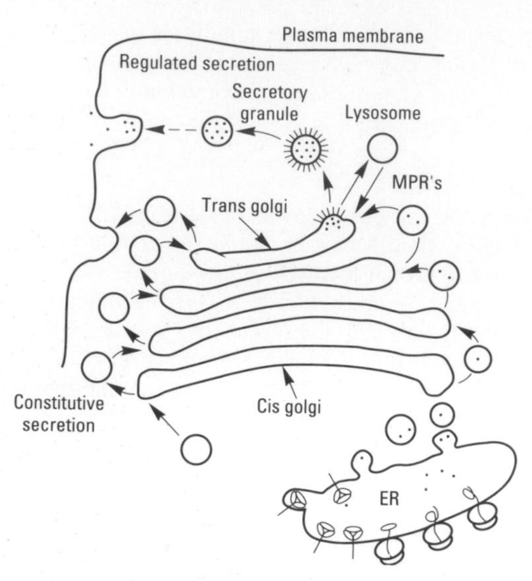

Figure 1–12. Processing of proteins from the endoplasmic reticulum (ER) through the Golgi apparatus to secretory granules and vesicles in the cytoplasm. Note that the cell has 2 secretory pathways, one involving storage of the secreted protein with further processing in secretory granules (nonconstitutive secretion) and the other involving transit via vesicles directly to the cell membrane without appreciable storage (constitutive secretion). MPRs, mannose-6-phosphate receptors.

Structure & Function of DNA & RNA

DNA is found in bacteria, in the nuclei of eukaryotic cells, and in mitochondria. It is made up of 2 extremely long nucleotide chains containing adenine (A), guanine (G), thymine (T), and cytosine (C) (Fig 1–13). The chemistry of these purine and pyrimidine bases and of nucleotides is discussed in Chapter 17. The chains are bound together by hydrogen bonding between the bases, with adenine bonding to thymine and guanine to cytosine. The resultant double-helical structure of the molecule is shown in Fig 1–14. An indication of the complexity of the molecule is the fact that the DNA in the human genome (the total genetic message) is made up of 2.5×10^9 base pairs.

DNA is the component of the chromosomes that carries the "genetic message," the blueprint for all the heritable characteristics of the cell and its descendants. Each chromosome is in effect a segment of the DNA double helix. The genetic message is coded by the sequence of purine and pyrimidine bases in the nucleotide chains. The text of the message is the order in which the amino acids are lined up in the proteins manufactured by the cell. The message is transferred to the sites of protein synthesis in the cytoplasm by RNA. The proteins formed include all the enzymes, and these in turn control the metabolism of the cell. A gene has been defined as the amount of information necessary to specify a single peptide molecule, although this is an oversimplification, since it is now known that single genes code for more than one protein. Genes also contain promoters and enhancers, DNA sequences that facilitate the formation of RNA. Mutations occur when the base sequence in the DNA is altered by x-rays, cosmic rays, or other mutagenic agents. Of course, the pattern of DNA is unique for any individual. Indeed, analysis of this pattern, which may now be done using a small sample of blood, semen, or even hair, has an accuracy test at least as great as that of fingerprinting for identifying an individual, and **DNA fingerprinting** is rapidly becoming an important forensic tool in police work.

At the time of each somatic cell division **(mitosis),** the 2 DNA chains separate, each serving as a template for the synthesis of a new complementary chain. **DNA polymerase** catalyzes this reaction. One of the double helices thus formed goes to one daughter cell and one to the other, so the amount of DNA in each daughter cell is the same as that in the parent cell. In

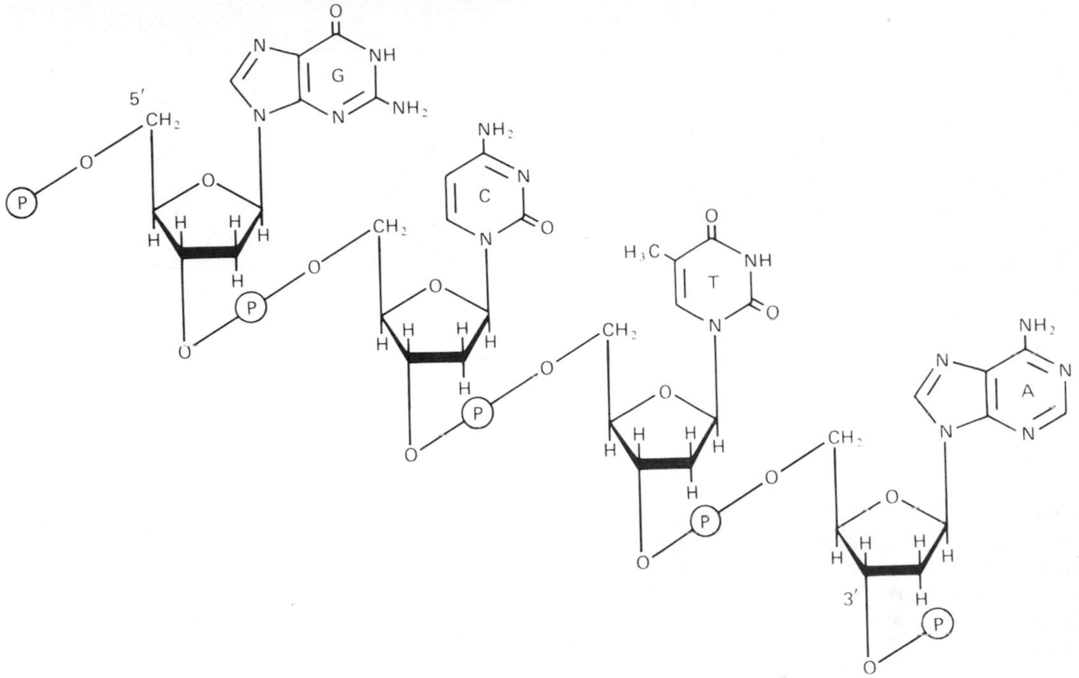

Figure 1–13. Segment of the structure of the DNA molecule in which the purine and pyrimidine bases adenine (A), thymine (T), cytosine (C), and guanine (G) are held together by a phosphodiester backbone between 2′-deoxyribosyl moieties attached to the nucleobases by an N-glycosidic bond. Note that the backbone has a polarity (ie, a 5′ and a 3′ direction). (Reproduced, with permission, from Murray RK et al: *Harper's Biochemistry,* 21st ed. Appleton & Lange, 1987.)

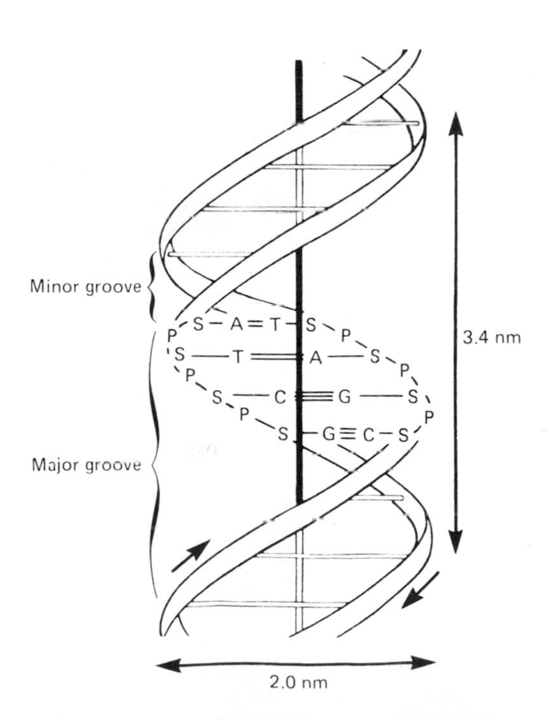

Figure 1–14. Double-helical structure of DNA, with adenine (A) bonding to thymine (T) and cytosine (C) to guanine (G). (Reproduced, with permission, from Murray RK et al: *Harper's Biochemistry,* 21st ed. Appleton & Lange, 1987.)

germ cells, reduction division **(meiosis)** takes place during maturation. The net result is that one DNA chain goes to each mature germ cell; consequently, each mature germ cell contains half the amount of chromosomal material found in somatic cells. Therefore, when a sperm unites with an ovum, the resulting zygote has the full complement of DNA, half of which came from the father and half from the mother. The chromosomal events that occur at the time of fertilization are discussed in detail in Chapter 23.

The strands of the DNA double helix not only replicate themselves, but they serve as templates by lining up complementary bases for the formation in the nucleus of **messenger RNA (mRNA)** and **soluble (transfer) RNA (tRNA).** This process is called **transcription** (Fig 1–15) and is catalyzed by **RNA polymerase.** In prokaryotes, the RNAs move without modification out of the nucleus to the ribosomes. However, in eukaryotes, the mRNA that is transcribed is really a pre-mRNA, and it generally undergoes considerable **posttranscriptional modification** before moving out of the nucleus and becoming associated with the ribosomes. In the ribosomes, mRNA dictates the formation of the polypeptide chain of a protein **(translation).** tRNA attaches the amino acids to mRNA. The mRNA molecules are smaller than the DNA molecules, and each represents a transcription of a small segment of the DNA chain. There are additional small RNAs in the cell. The

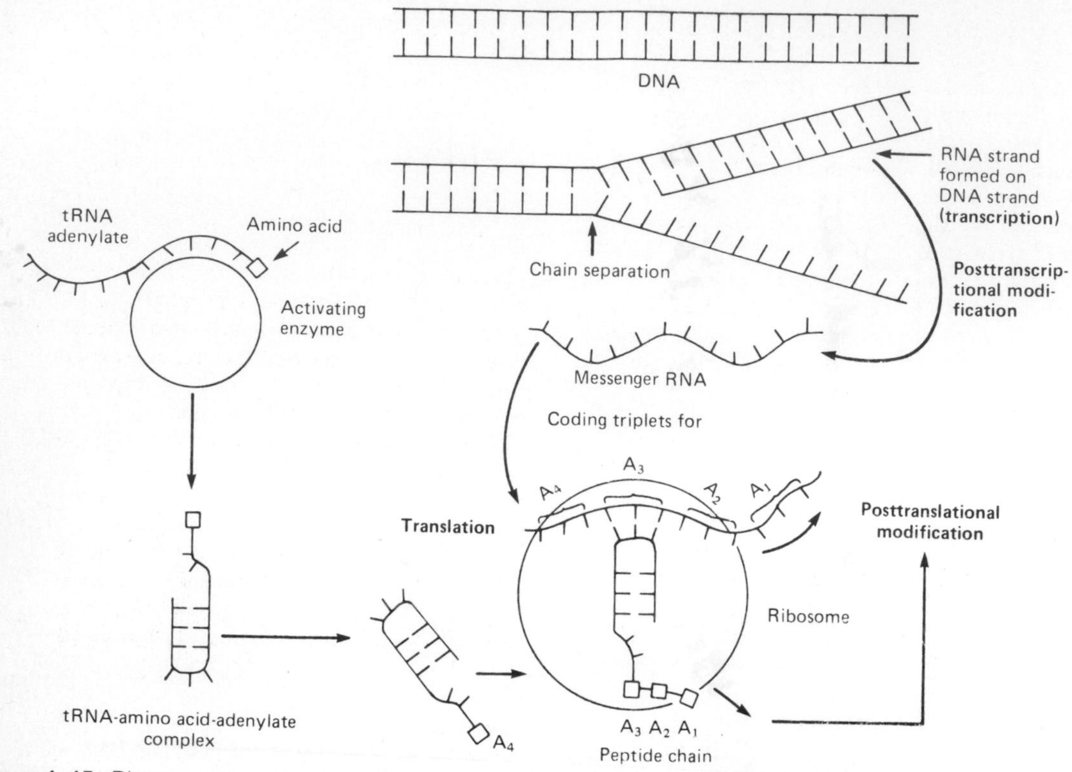

Figure 1–15. Diagrammatic outline of protein synthesis. The nucleic acids are represented as lines with multiple short projections representing the individual bases.

molecules of tRNA contain only 70–80 nitrogenous bases, compared with hundreds in mRNA and as many as 500 million in DNA. tRNA and mRNA are single-stranded, and they contain the base uracil (U) in place of thymine. A third distinct type of RNA is the RNA in the ribosomes **(ribosomal RNA).**

It is worth noting that DNA is responsible for the maintenance of the species; it passes from one generation to the next in germ cells. RNA, on the other hand, is responsible for the production of the individual; it transcribes the information coded in the DNA and forms a mortal individual, a process that has been called "budding off from the germ line."

Mitochondrial DNA

The mitochondria contain DNA and can synthesize protein. Many investigators believe that mitochondria were once autonomous microorganisms that developed a symbiotic relation with and became incorporated into ancestral eukaryotic cells. However, the mitochondrial DNA alone does not contain enough genetic information to code for all mitochondrial components; for example, there are 16,569 nucleotides in human mitochondrial DNA, a minute fraction of the number of nucleotides in the nucleus. Many mitochondrial proteins are synthesized in the cytoplasm and migrate into the mitochondria.

Genes

Information is accumulating at an accelerating rate about the structure of genes and their regulation. A typical gene is shown in Fig 1–16. It is made up of a strand of DNA that includes a transcription unit and flanking regions at either end of this unit. In

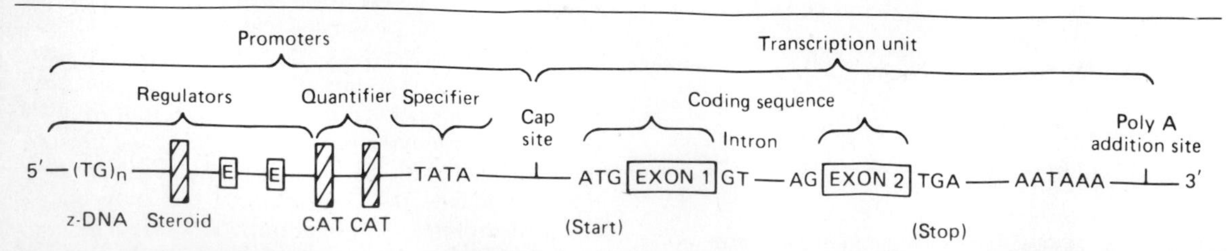

Figure 1–16. Diagram showing organization of a gene. The segment of DNA, from the 5′ end to the 3′ end, is divided into a transcription unit and a promoter region. Key base sequences are indicated along the segment. CAT, sequences related to amplitude of transcription; E, enhancer sequences. (Reproduced, with permission, from Habener JF: Regulation of polypeptide-hormone biosynthesis at the level of the genome. *Am J Physiol* 1985;**249**:C191.)

eukaryotes, unlike prokaryotes, the portions of the genes that dictate the formation of proteins are usually broken into several segments **(exons)** separated by segments that are not translated **(introns).** A pre-mRNA is formed from the DNA, and then the introns and sometimes some of the exons are eliminated in the nucleus by posttranscriptional processing, so that the final mRNA which enters the cytoplasm is made up of exons (Fig 1–17). More than one mRNA can be formed from the same gene, the difference in mRNAs being the inclusion of different exons. The physiologic function of the introns is still unsettled, although they may foster changes in the genetic message and thus aid evolution.

The **transcription unit** of the gene starts with the **cap site** and extends about 20 bases beyond an AATAAA sequence to a poly(A) addition site. The **promoter region** on the 5′ side of the cap site contains various DNA segments that regulate gene expression. These include the sequence TATAA called the **TATA box,** which ensures that transcription starts at the proper point. There are sequences that regulate the amplitude of transcription (quantifiers) and enhancer sequences, as well as regulator sequences that respond to specific regulatory signals such as steroid hormones. Finally, there is a region, called z-DNA, that may determine tissue-specific expression.

There is currently a great deal of interest in **oncogenes,** genes which are carried in the genomes of cancer cells and seem to be responsible for producing many of their malignant properties. These genes are derived by somatic mutation from closely related proto-oncogenes, which are normal genes that control growth. Over 40 different oncogenes have been described. Some may exert their effects by altering receptors on the cell surface so that the cells are continuously stimulated by various ligands. Others may alter second messenger systems within the cell (see below) so that the response to normal stimulation is excessive. Still others may stimulate cells to produce growth factors that act on the cells themselves (autocrine stimulation). However, none of these mechanisms are as yet established with certainty.

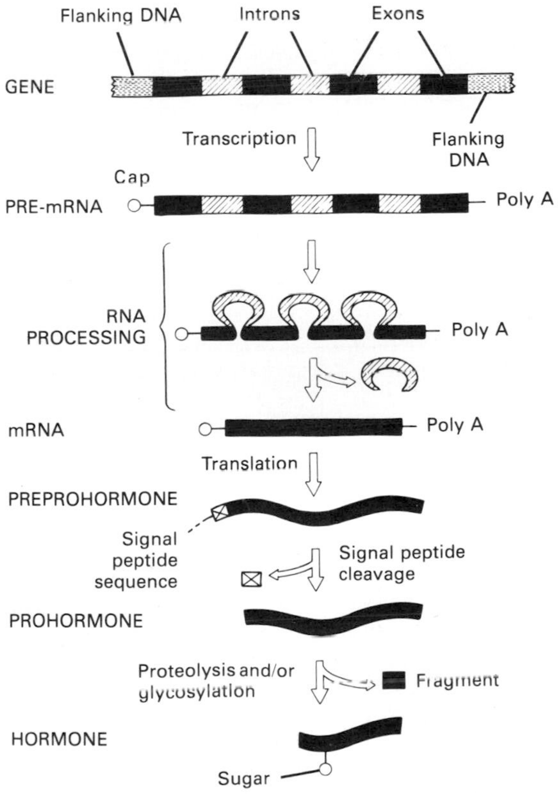

Figure 1–17. Transcription, posttranscriptional modification of mRNA, translation in the ribosomes, and posttranslational processing in the formation of hormones and other proteins. (Modified and reproduced, with permission, from Baxter JD: Principles of endocrinology. In: *Cecil Textbook of Medicine,* 16th ed. Wyngaarden JB, Smith LH Jr [editors]. Saunders, 1982.)

Regulation of Gene Expression

Each nucleated somatic cell in the body contains the full genetic message, yet there is great differentiation and specialization in the functions of the various types of adult cells, and only small parts of the message are normally transcribed. Thus, the genetic message is normally maintained in a repressed state. What turns on genes in one cell and not in other cells? What turns on genes in a cell at one stage of development and not at other, inappropriate stages? What maintains orderly growth in cells and prevents the uncontrolled growth that we call cancer? Obviously, the various parts of the promoter sequences in genes play an important role. It now appears that the promoter regions are turned on not only by steroid hormones and other signals but also by proteins manufactured by other genes in the cell (*trans* regulation). In addition, genes are expressed to a lesser degree when they are condensed into chromatin ("closed") in any given cell. Another factor is the degree to which the DNA is methylated. Methylation of DNA can be inhibited by 5-azacytidine, and this drug has been shown to bring about transcription of previously inactive genes.

Protein Synthesis

The process of protein synthesis is a complex but fascinating one that, as noted above, involves 4 steps: transcription, posttranscriptional modification, translation, and posttranslational modification. The various steps are summarized in simplified form in Figs 1–15 and 1–17.

When suitably activated, transcription of the gene starts at the cap site and ends about 20 bases beyond the signal sequence AATAAA. The RNA transcript is capped in the nucleus by addition of 7-methylguanosine triphosphate to the 5′ end; this cap is necessary for proper binding to the ribosome (see below). A **poly(A) tail** of about 100 bases is added to the untranslated segment at the 3′ end. The function of the poly(A) tail is unsettled, but it may help maintain the stability of the mRNA. The pre-mRNA formed by capping and addition of the poly(A) tail is then processed by elimination of the introns (Fig 1–17), and once this posttranscriptional modification is complete, the mature mRNA moves to the cytoplasm. Posttranscriptional modification of the pre-mRNA is a regulated process, and as noted above, differential splicing can occur, with the formation of more than one mRNA from a single pre-mRNA.

When a definitive mRNA reaches a ribosome in the cytoplasm, it dictates the formation of a polypeptide chain. Amino acids in the cytoplasm are activated by combination with an enzyme and adenosine monophosphate (adenylate), and each **activated amino acid** then combines with a specific molecule of tRNA. There is at least one tRNA per 20 unmodified amino acids found in large quantities in the body proteins of animals (see Chapter 17), but there is more than one tRNA for some amino acids. The tRNA-amino acid-adenylate complex is next attached to the mRNA template, a process that occurs in the ribosomes. This process is shown diagrammatically in Fig 1–15. The tRNA "recognizes" the proper spot to attach on the mRNA template because it has on its active end a set of 3 bases that are complementary to a set of 3 bases in a particular spot on the mRNA chain. The genetic code is made up of such **triplets,** sequences of 3 purine or pyramidine bases or both; each triplet stands for a particular amino acid.

Translation starts in the ribosomes with an AUG (transcribed from ATG in the gene) which codes for methionine. The N-terminal amino acids is then added, and the chain is lengthened one amino acid at a time. The mRNA attaches to the 30S subunit of the ribosome during protein synthesis; the polypeptide chain being formed attaches to the 50S subunit; and the tRNA attaches to both. As the amino acids are added in the order dictated by the triplet code, the ribosome moves along the mRNA molecule like a bead on a string. Translation stops at one of 3 stop, or nonsense, codons—UGA, UAA, or UAG—and the polypeptide chain is released. The tRNA molecules are used again. The mRNA molecules are also reused approximately 10 times before being replaced.

Typically, there is more than one ribosome on a given mRNA chain at one time. The mRNA chain plus its collection of ribosomes is visible under the electron microscope as an aggregation of ribosomes called a **polyribosome (polysome).**

Posttranslational Modification

After the polypeptide chain is formed, it is modified to the final protein by one or more of a combination of reactions that include hydroxylation, carboxylation, glycosylation, or phosphorylation of amino acid residues; cleavage of peptide bonds that converts a larger polypeptide to a smaller form; and the folding and packaging of the protein into its ultimate, often complex configuration.

It has been claimed that a typical eukaryotic cell synthesizes about 100,000 different proteins during its lifetime. How do these proteins get to the right locations in the cell? Synthesis starts in the free ribosomes. Peptides enter the endoplasmic reticulum

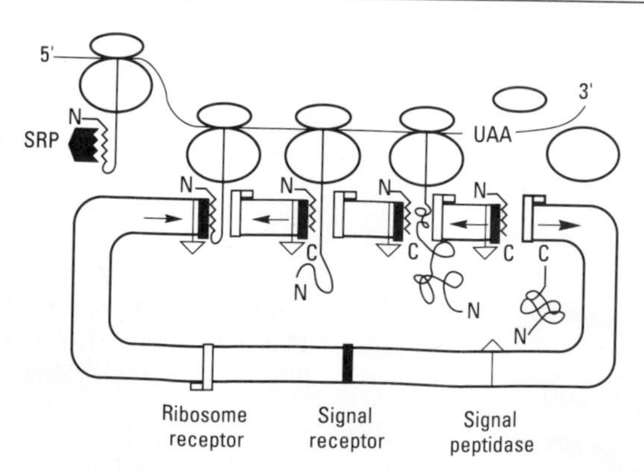

Figure 1–18. Translation of protein into endoplasmic reticulum according to the signal hypothesis. The ribosomes synthesizing a protein move along the mRNA from the 5′ to the 3′ end. When the signal peptide of a protein destined for secretion, the cell membrane, or lysosomes emerges from the large unit of the ribosome, it binds to a signal recognition particle (SRP), and this arrests further translation until it binds to the signal receptor on the endoplasmic reticulum. This frees the SRP, which is recycled in the cytoplasm. Binding also occurs to the ribosome receptor, and a tunnel opens, permitting the growing protein chain to enter the endoplasmic reticulum. The signal peptide is removed by signal peptidase. At the termination of protein synthesis, the 2 subunits of the ribosome dissociate, the carboxy terminus enters the endoplasmic reticulum, and the tunnel units disassemble. (Reproduced, with permission, from Perara E, Linappa VR: Transport of proteins into and across the endoplasmic reticulum membrane. In *Protein Transfer and Organelle Biogenesis.* Das RC, Robbins PW [editors]. Academic, 1988.)

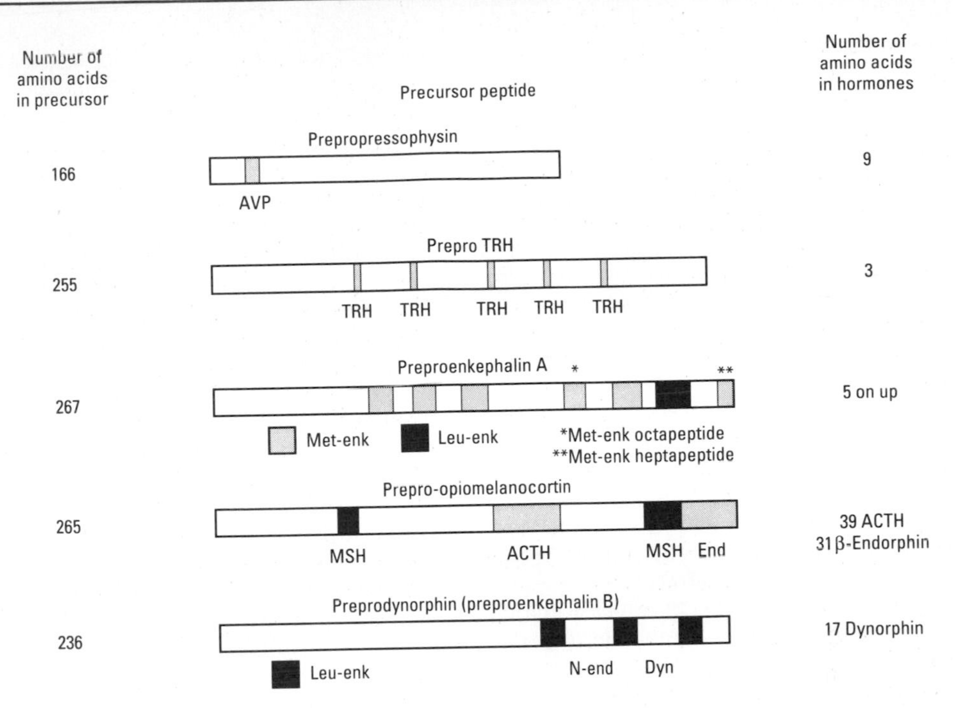

Figure 1–19. Examples of large precursors (preprohormones) for small peptide hormones. TRH, thyrotropin releasing hormone. AVP, arginine vasopressin; Met-enk, Met-enkephalin; Leu-enk, leu-enkephalin; MSH, melanocyte-stimulating hormone; ACTH, adrenocorticotropic hormone; End, β-endorphin; Dyn, dynorphin; N-end neoendorphin.

if the developing peptide chain has at its N terminal a **leader sequence (signal peptide)** made up of 15–30 predominantly hydrophobic amino acid residues. The signal peptide, once synthesized, binds to a **signal recognition particle (SRP),** a complex molecule made up of 6 polypeptides and 7S RNA, one of the small RNAs. The SRP stops translation until it binds to an SRP receptor in the endoplasmic reticulum. The ribosome also binds, and the signal peptide leads the growing peptide chain into the cavity of the endoplasmic reticulum (Fig 1–18). The signal peptide is next cleaved from the rest of the peptide by a signal peptidase while the rest of the peptide chain is still being synthesized.

The signals that direct nascent proteins to some of the other parts of the cell are fashioned in the Golgi apparatus and, as discussed above, involve specific modifications of the carbohydrate residues on glycoproteins.

It should be noted that many and perhaps all proteins that are secreted by cells are synthesized as larger proteins, and polypeptide sequences are cleaved off from them during maturation. In the case of the hormones, these larger forms are called **preprohormones** and **prohormones** (Fig 1–17 and 1–19). Parathyroid hormone is an example. It is synthesized as a molecule containing 115 amino acid residues (preproparathyroid hormone). The leader sequence, 25 amino acid residues at the N terminal, is rapidly removed to form proparathyroid hormone. Before secretion, an additional 6 amino acids are removed from the N terminal to form the secreted molecule. The function of the 6-amino-acid fragment is unknown.

Effects of Antibiotiotics on Protein Synthesis

Antibiotics act in a variety of ways, but many act by inhibiting protein synthesis at one or another of the

Table 1–5. Mechanisms by which antibiotics and related drugs inhibit protein synthesis.

Agent	Effect
Chloramphenicol	Prevents normal association of mRNA with ribosomes.
Streptomycin, neomycin, kanamycin	Cause misreading of genetic code.
Cycloheximide, tetracycline	Inhibit transfer of tRNA-amino acid complex to polypeptide.
Puromycin	Puromycin-amino acid complex substitutes for tRNA-amino acid complex and prevents addition of further amino acids to polypeptide.
Mithramycin, mitomycin C, dactinomycin (actinomycin D)	Bind to DNA, preventing polymerization of RNA on DNA.
Chloroquine, colchicine, novobiocin	Inhibit DNA polymerase.
Nitrogen mustards, eg, mechlorethamine (Mustargen)	Bind to guanine in base pairs.
Diphtheria toxin	Prevents ribosome from moving on mRNA.

steps described above (Table 1–5). Some of them have this effect primarily in bacteria, but others inhibit protein synthesis in the cells of other animals, including mammals. This fact makes antibiotics of great value for research as well as for treatment of infections.

Exocytosis

Proteins that are secreted by cells move from the endoplasmic reticulum to the Golgi apparatus, and from the *trans* Golgi, they are extruded into secretory granules or vesicles (Fig 1–12). The granules and vesicles move to the cell membrane. Their membrane then fuses to the cell membrane (Fig 1–20), and the area of fusion breaks down. This leaves the contents of the granules or vesicles outside the cell and the cell membrane intact. The extrusion process is called **exocytosis.** It requires Ca^{2+} and energy, but the details of the mechanism responsible for the breakdown of the membrane are unknown.

Note that there are 2 pathways by which secretion from the cell occurs (Fig 1–12). In the **nonconstitutive pathway,** proteins initially enter secretory granules coated with clathrin (see below). Processing of prohormones to the mature hormones occurs in these granules. The clathrin coat is then lost, and the secretory granules move to the cell membrane. The other pathway, the **constitutive pathway,** involves the prompt transport of proteins to the cell membrane, with little or no processing or storage. The nonconstitutive pathway is sometimes called the **regulated pathway,** but this term is misleading because constitutive secretion is also regulated.

Endocytosis

Endocytosis is the reverse of exocytosis. One form of endocytosis, called **phagocytosis** (''cell eating''), is the process by which bacteria, dead tissue, or other bits of material visible under the microscope are engulfed by cells such as the polymorphonuclear leukocytes of the blood. The material makes contact with the cell membrane, which then invaginates. The invagination is pinched off, leaving the engulfed material in the membrane-enclosed vacuole and the cell membrane intact. **Pinocytosis** (''cell drinking'') is essentially the same process, the only difference being that the substances ingested are in solution and hence not visible under the microscope. In the cell, the membrane around a **pinocytic** or **phagocytic vacuole** generally fuses with that of a lysosome, mixing the ''digestive'' enzymes in the lysosome with the contents of the vacuole. It has also been assumed that the membrane around vacuoles can be digested away, but the ultimate fate of the vacuoles is uncertain.

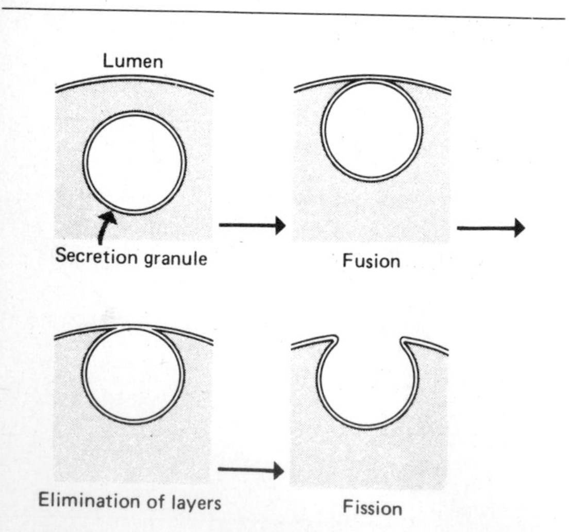

Figure 1–20. Sequence of events in exocytosis.

There are 2 kinds of endocytosis: **constitutive endocytosis,** a continuous process that is not induced; and **receptor-mediated endocytosis,** which is produced for the most part via specialized **coated pits** on the cell membrane and triggered by various ligands binding to their receptors on the cell surface. Some receptors are localized to the coated pits. Others move laterally in the cell membrane once a ligand has bound to them and aggregate in the coated pits. The pit then pinches off, forming a coated vesicle. The vesicle then becomes an **endosome,** and the pH inside it is lowered by protein pumps. This frees the receptors, which recycle to the membrane. The endosome then fuses with a lysosome. The coated appearance of the pits and vesicles is due to the presence in their walls of a protein lattice made up in large part of the polypeptide **clathrin,** which has a molecular weight of 18,000. However, it now appears on the basis of experiments with mutant cells that clathrin is not necessary for survival and that considerable endocytosis goes on in its absence. Receptor-mediated endocytosis can occur at a more rapid rate than constitutive endocytosis, and the former is more specific in the sense that the particular molecules which triggered it are concentrated on their receptors in the coated pits and hence the coated vesicles. Receptor-mediated endocytosis is responsible for the internalization of low-density lipoproteins—an important part of cellular metabolism of cholesterol (see Chapter 17)—and for the uptake of many substances, including insulin, epidermal growth factor, nerve growth factor, diphtheria toxin, and a number of different viruses.

Although most endocytotic vesicles fuse with lysosomes, some bypass the lysosomes and move back to the cell membrane, fuse with it, and discharge their contents by exocytosis (see above and Chapter 30).

It is apparent that exocytosis adds to the total amount of membrane surrounding the cell, and if membrane were not removed elsewhere at an equivalent rate, the cell would enlarge. However, removal of cell membrane occurs by endocytosis, and such exocytosis-endocytosis coupling (Fig 1–21) maintains the surface area of the cell at its normal size.

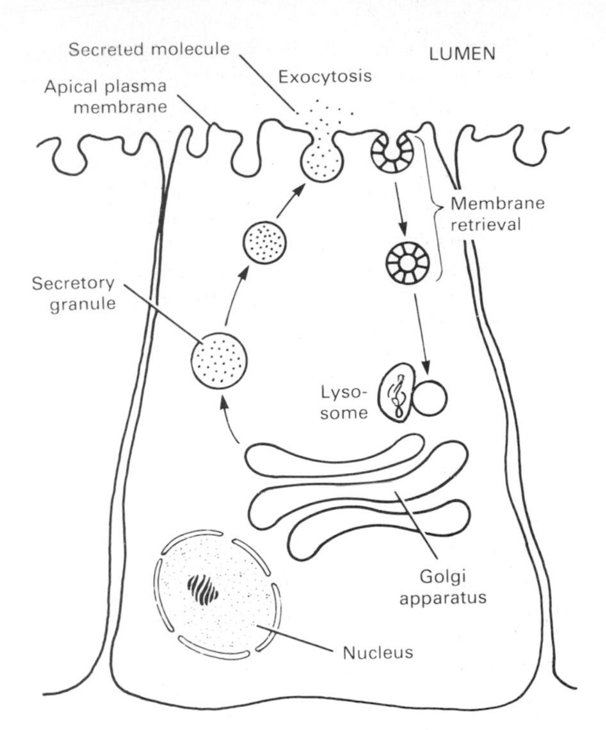

Figure 1–21. Membrane recycling, as it is postulated to occur in secretory epithelial cells. Membrane from the Golgi apparatus surrounds secretory vesicles, and this membrane becomes part of the cell membrane. At least in some instances, membrane is retrieved via coated pits *(right)*. (Modified and reproduced, with permission, from Alberts B et al: *Molecular Biology of the Cell*. Garland, 1983.)

TRANSPORT ACROSS CELL MEMBRANES & MEMBRANE POTENTIALS

Ion Distribution Across Cell Membranes

The unique properties of the cell membranes are responsible for the differences in the composition of intracellular and interstitial fluid. Average values for the composition of intracellular fluid in humans are shown in Fig 1–3, and specific values for mammalian neurons are shown in Table 1–6.

Table 1–6. Concentration of some ions inside and outside mammalian spinal motor neurons.*

Ion	Concentration ($mmol/L\ H_2O$) Inside Cell	Concentration ($mmol/L\ H_2O$) Outside Cell	Equilibrium Potential (mV)
Na^+	15.0	150.0	+60
K^+	150.0	5.5	−90
Cl^-	9.0	125.0	−70

Resting membrane potential = −70 mV

*Data from Mommaerts WFHM, in: *Essentials of Human Physiology*. Ross G (editor). Year Book, 1978.

Membrane Potentials

There is a potential difference across the membranes of most if not all cells, with the inside of the cells negative to the exterior. By convention, this **resting membrane potential (steady potential)** is written with a minus sign, signifying that the inside is negative relative to the exterior. Its magnitude varies considerably from tissue to tissue, ranging from −9 to −100 mV.

Membrane Permeability & Membrane Transport Proteins

Cell membranes are practically impermeable to intracellular protein and other organic anions, which make up most of the intracellular anions and are usually represented by the symbol A^-. When considering membrane permeability to smaller molecules, it is important to distinguish between the lipid bilayer itself and the numerous and various transport proteins embedded in it. The lipid bilayer is highly permeable to water. Its permeability to other substances depends on their size (Table 1–7), lipid solubility, and charge. Molecules such as O_2 and N_2 that are nonpolar, ie, hydrophobic, dissolve in the bilayer and cross with ease. Small uncharged polar (hydrophilic) molecules such as CO_2 and urea also diffuse rapidly across lipid bilayers, whereas large uncharged polar molecules such as glucose diffuse much more slowly and the diffusion of charged particles, ie, ions, is extremely slow. Thus, for example, the permeability coefficient—an index of the rate of movement of a substance across a membrane—is about 10^{-2} cm/s for water in a lipid bilayer. The corresponding value for urea is about 10^{-6}; for glucose, about 10^{-7} for Cl^-, about 10^{-10}; and for K^+ and Na^+, about 10^{-12}.

However, ions and many other small molecules also utilize transport proteins to cross cell membranes. These transmembrane proteins are of various types (Fig 1–22).

Some transport proteins are simply a **channel** with an aqueous center that permits ions to diffuse into or

Table 1–7. Size of hydrated ions and other substances of biologic interest.*

Substance	Atomic or Molecular Weight	Radius (nm)
Cl^-	35	0.12
K^+	39	0.12
H_2O	18	0.12
Ca^{2+}	40	0.15
Na^+	23	0.18
Urea	60	0.23
Li^+	7	0.24
Glucose	180	0.38
Sucrose	342	0.48
Inulin	5000	0.75
Albumin	69,000	7.50

*Data from Moore EW: *Physiology of Intestinal Water and Electrolyte Absorption*. American Gastroenterological Association, 1976.

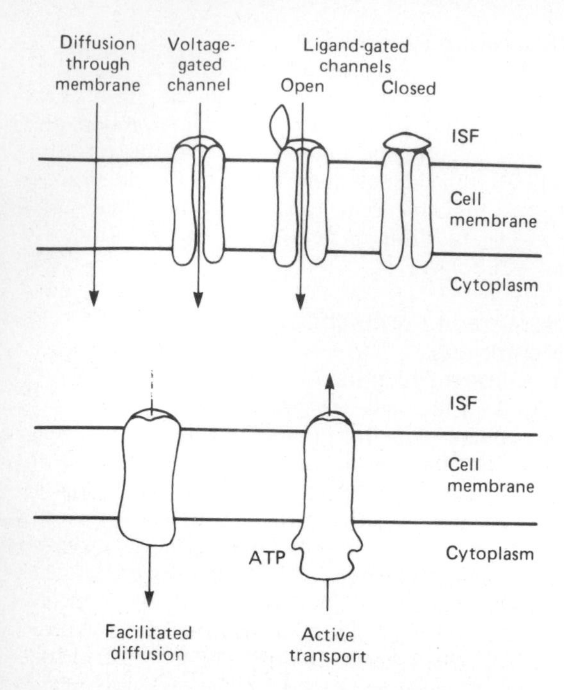

Figure 1–22. Transport of ions and other small molecules across cell membranes. *Top:* Diffusion pathways, showing slow diffusion through membrane or rapid diffusion through voltage-gated or ligand-gated channel proteins that span the membrane. *Bottom:* Transport proteins that carry substances down their electrochemical gradients (facilitated diffusion) or against these gradients (active transport).

out of a cell. An important technique that has permitted major advances in knowledge about channels is **patch clamping.** A micropipette is placed on the membrane of a cell and forms a tight seal to the membrane, which can then be pulled loose. The patch of membrane left at the pipette tip usually contains only a few channels, and they can be studied in detail. Alternatively, the patch can be sucked out, with the micropipette still attached to the rest of the cell membrane, providing direct access to the interior of the cell.

Some channels are continuously open, whereas others are **gated;** ie, they have gates that open or close. Some are gated by alterations in membrane potential **(voltage-gated),** whereas others are opened or closed when they bind a ligand **(ligand-gated).** The ligand is often external, eg, a neurotransmitter or a hormone. However, it can also be internal; there are now well-documented instances in which intracellular Ca^{2+}, cyclic AMP, or one of the G proteins produced in cells (see below) binds directly to channels and activates them. It is possible that some channels are also opened by mechanical stretch. A typical voltage-gated channel is the sodium channel (see below), and a typical ligand-gated channel is the acetylcholine receptor (see Chapter 4). Other transport proteins are **carriers** that bind ions and other molecules and then change their configuration, moving the bound molecule from one side of the cell membrane to the other. When carrier proteins move substances in the direction of their chemical or electrical gradients, no energy input is required and the process is called **facilitated diffusion.** A typical example is glucose transport by the glucose transporter, which moves glucose down its concentration gradient from the ECF to the cytoplasm of the cell (see Chapter 19). Other carriers transport substances against their electrical and chemical gradients. This form of transport requires energy and is called **active transport.** In animal cells, the energy is provided almost exclusively by hydrolysis of ATP (see above and Chapter 17). Not surprisingly, therefore, the carrier molecules are ATPases, enzymes that catalyze the hydrolysis of ATP. The most important of these ATPases is **sodium-potassium-activated adenosine triphosphatase** (Na^+-K^+ ATPase), which is also known as the Na^+-K^+ pump. There are in addition H^+-K^+ ATPases in the gastric mucosa (see Chapter 26) and the renal tubules (see Chapter 38). In mitochondria, this pump operates in reverse to synthesize ATP (see Chapter 17). Some cell membranes contain ATPases that transport Ca^{2+}.

Some of the carrier proteins are called **uniports** because they transport only one substance. Others are called **symports** because transport requires the binding of more than one substance to the transport protein (Fig 1–23) and the substances are transported across the membrane together. An example is the symport in the intestinal mucosa that is responsible for the cotransport by facilitated diffusion of Na^+ and glucose from the intestinal lumen into mucosal cells (see Chapter 25). Other transporters are called **antiports** because they exchange one substance for another. The Na^+-K^+ ATPase mentioned above is a typical antiport; it moves 3 Na^+ out of the cell in exchange for each 2 K^+ that it moves into the cell.

Sodium & Other Ion Channels

The chemistry of the voltage-gated Na^+ channel has now been studied in considerable detail, and the

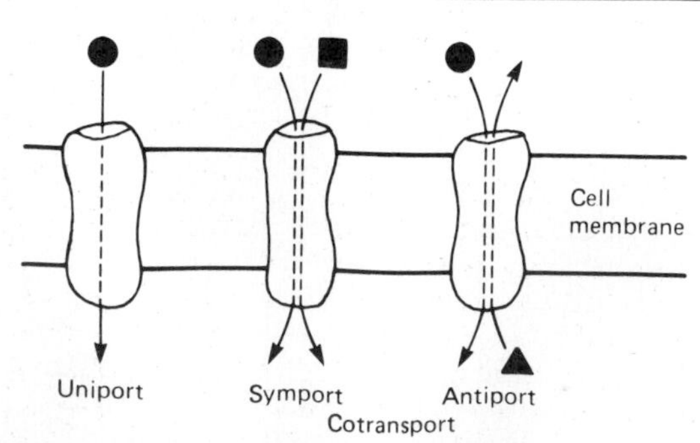

Figure 1–23. Transport proteins. The substance transported (●) may cross the membrane by itself, or its transport may be coupled to that of another substance transported in the same direction (■; symport) or the opposite direction (▲; antiport).

cDNA for it has been cloned. In the electric eel, the Na^+ channel is a single protein made up of 1820 amino acid residues, but in rat and rabbit skeletal muscle and in rat brain, this protein, which has a molecular weight of 200,000, is associated with one or 2 smaller peptides that have molecular weights of 33,000–45,000. The large protein has 4 subunits, each made up of 6 membrane-spanning segments. The subunits surround an aqueous pore about 0.5 nm in diameter. The toxins tetrodotoxin (TTX) and saxitoxin (STX) bind to Na^+ channels and block them. The number and distribution of Na^+ channels in tissue can be determined by tagging TTX or STX with a suitable label and analyzing the distribution of the label.

There are many different K^+ channels in the membranes of various types of cells (Table 2–1). Some of these can be blocked by tetraethyl ammonium (TEA) or 4-aminopyridine (4AP). There are also Ca^{2+}, Cl^-, H^+, and other types of channels, and efforts are currently being made to isolate and characterize them.

Na^+-K^+ ATPase

As noted above, Na^+-K^+ATPase catalyzes the hydrolysis of ATP to ADP and uses the energy to extrude 3 Na^+ from the cell and take 2 K^+ into the cell for each mole of ATP hydrolyzed. Consequently, the pump is said to have a **coupling ratio** of 3/2. Its activity is inhibited by ouabain and related digitalis glycosides used in the treatment of heart failure. It is made up of two α subunits, each with a molecular weight of about 95,000, and two β subunits, each with a molecular weight of about 40,000. Separation of the subunits leads to loss of ATPase activity. The α subunits contain binding sites for ATP and ouabain, whereas the β subunits are glycoproteins (Fig 1–24). Application of ATP by micropipette to the inside of the membrane increases transport, whereas application of ATP to the outside of the membrane has no effect. Conversely, ouabain inhibits transport when applied to the outside but not to the inside of the membrane. Consequently, the α subunits must extend through the cell membrane. A proposed structure for Na^+-K^+ ATPase is shown in Fig 1–24. The protein probably exists in 2 conformational states. In one, 3 Na^+ bind to sites accessible only from the inside of the membrane. This triggers hydrolysis of ATP, and the protein changes its conformation so that the 3 Na^+ are extruded into the ECF. In the second conformation, 2 K^+ bind to sites accessible only from the outside of the membrane. This triggers a return to the original conformation while extruding 2 K^+ into the interior of the cell. It appears that Na^+ binding is associated with phosphorylation of the protein and K^+ binding with dephosphorylation.

Effects of Na^+ & K^+ Transport

It should be noted that since Na^+-K^+ ATPase transports 3 Na^+ out of the cell for each 2 K^+ it transports in, it is an **electrogenic pump;** ie, it produces net movement of positive charge out of the cell. The amount of Na^+ provided to the pump can be the rate-limiting factor in its operation; consequently, the amount of Na^+ extruded by the cell is partly regulated in a feedback fashion by the amount of Na^+ in the cell.

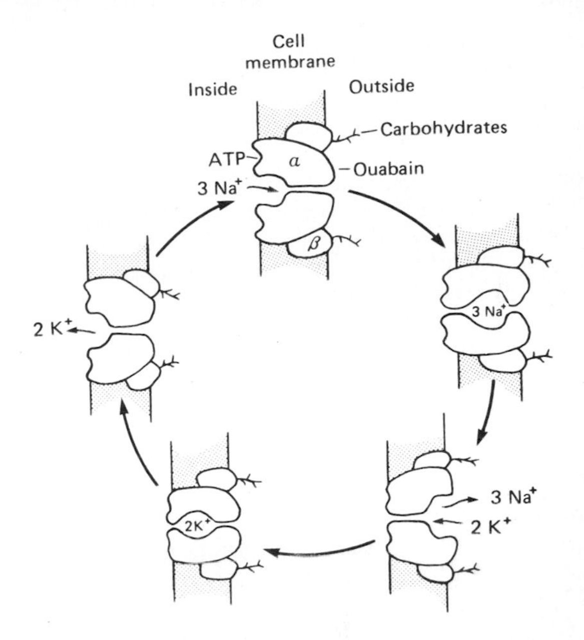

Figure 1–24. Proposed structure of Na^+-K^+ ATPase. The molecule is made up of 2 α and 2 β subunits. Each α unit has an intracellular catalytic site for ATP and an extracellular site to which cardiac glycosides such as ouabain bind. There are also 3 binding sites for Na^+ and 2 binding sites for K^+. The mechanism by which transport of the ions occurs is unknown but probably involves conformation changes with resulting shifts in the relations of the subunits to each other like those shown here. (Modified from a drawing by S Hootman.)

Na^+-K^+ ATPase is found in all parts of the body. In some tissues, the active transport of Na^+ is coupled to the transport of other substances **(secondary active transport).** For example, the luminal membranes of mucosal cells in the small intestine contain a symport that transports glucose into the cell only if Na^+ binds to the protein and is transported down its electrochemical gradient at the same time. The Na^+-glucose symport has been identified by cloning and sequencing of its cDNA. It is a 662-amino-acid protein that appears to span the cell membrane 11 times. From the cells, the glucose diffuses into the blood. The electrochemical gradient for Na^+ is maintained by the active transport of Na^+ out of the mucosal cell into ECF (see Chapter 25). Other nutrients are reabsorbed in a similar fashion. In the brain, transport by Na^+-K^+ ATPase is coupled to the reuptake of neurotransmitters. In the heart, Na^+-K^+ ATPase indirectly affects Ca^{2+} transport because an antiport in the membranes of cardiac muscle cells exchanges intracellular Ca^{2+} for extracellular Na^+ on an electrically neutral 1 for 2 basis. The rate of this exchange is proportionate to the concentration gradient for Na^+ across the cell membrane. If the operation of the Na^+-K^+ ATPase is inhibited (eg, by ouabain), less intracellular Ca^{2+} is extruded and intracellular Ca^{2+} is increased. This facilitates the contraction of cardiac muscle and is the probable explanation of the positively inotropic effect of digitalis glycosides (see Chapter 3).

Active transport of Na^+ and K^+ is one of the major energy-using processes in the body. On the average, it accounts for 33% of the energy utilized by cells, and in neurons, it accounts for 70%. Thus, it accounts for a large part of the basal metabolism. Furthermore, there is a direct link between Na^+ and K^+ transport and metabolism; the greater the rate of pumping, the more ADP is formed, and the available supply of ADP determines the rate at which ATP is formed by oxidative phosphorylation (see Chapter 17).

In animals, the maintenance of normal cell volume and pressure depends on Na^+ and K^+ pumping. In the absence of such pumping, Cl^- and Na^+ would enter the cells down their concentration gradients, and water would follow along the osmotic gradient thus created, causing the cells to swell until the pressure inside them balanced the influx. This does not occur, and the osmolality of the cells remains the same as that of the interstitial fluid because Na^+ and K^+ are actively transported.

Forces Acting on Ions

With the background provided in the preceding paragraphs, the forces acting across the cell membrane on each ion can be analyzed. Chloride ions are present in higher concentration in the ECF than in the cell interior, and they tend to diffuse along this **concentration gradient** into the cell. The interior of the cell is negative to the exterior, and chloride ions are pushed out of the cell along this **electrical gradient.** An equilibrium is reached at which Cl^- influx and Cl^- efflux are equal. The membrane potential at which this equilibrium exists is the **equilibrium potential.** Its magnitude can be calculated from the **Nernst equation,** as follows:

$$E_{Cl} = \frac{RT}{FZ_{Cl}} \ln \frac{[Cl_o^-]}{[Cl_i^-]}$$

where

E_{Cl} = equilibrium potential for Cl^-
R = gas constant
T = absolute temperature
F = the faraday (number of coulombs per mole of charge)
Z_{Cl} = valence of $Cl^-(-1)$
$[Cl_o^-]$ = Cl^- concentration outside the cell
$[Cl_i^-]$ = Cl^- concentration inside the cell

Converting from the natural log to the base 10 log and replacing some of the constants with numerical values, the equation becomes

$$E_{Cl} = 61.5 \log \frac{[Cl_i^-]}{[Cl_o^-]} \text{ at } 37° C$$

Note that in converting to the simplified expression, the concentration ratio is reversed because the −1 valence of Cl^- has been removed from the expression.

E_{Cl}, calculated from the values in Table 1–6, is −70 mV, a value identical to the measured resting membrane potential of −70 mV. Therefore, no forces other than those represented by the chemical and electrical gradients need be invoked to explain the distribution of Cl^- across the membrane.

A similar equilibrium potential can be calculated for K^+.

$$E_K = \frac{RT}{FZ_K} \ln \frac{[K_o^+]}{[K_i^+]} = 61.5 \log \frac{[K_o^+]}{[K_i^+]} \text{ at } 37° C$$

where

E_K = equilibrium potential for K^+
Z_K = valence of $K^+(+1)$
$[K_o^+]$ = K^+ concentration outside the cell
$[K_i^+]$ = K^+ concentration inside the cell
R, T, and F as above

In this case, the concentration gradient is outward and the electrical gradient inward. In mammalian spinal motor neurons, E_K is −90 mV (Table 1–6). Since the resting membrane potential is −70 mV, there is somewhat more K^+ in the neurons than can be accounted for by the electrical and chemical gradients.

The situation for Na^+ is quite different from that for K^+ and Cl^-. The direction of the chemical gradient for Na^+ is inward, to the area where it is in lesser concentration, and the electrical gradient is in the same direction. E_{Na} is +60 mV (Table 1–6). Since neither E_K nor E_{Na} are at the membrane potential, one would expect the cell to gradually gain Na^+ and lose K^+ (Fig 1–25) if only passive electrical and chemical forces were acting across the membrane. However, the intracellular concentration of Na^+ and K^+ remains constant because there is active transport of Na^+ out of the cell against its electrical and concentration gradients, and this transport is coupled to active transport of K^+ into the cell. The passive and active fluxes across the membrane of a motor neuron at rest are summarized in Fig 1–25.

The magnitude of the membrane potential at any given time depends, of course, upon the distribution of Na^+, K^+, and Cl^- and the permeability of the membrane to each of these ions. An equation that describes this relationship with considerable accuracy is the **Goldman constant-field equation:**

$$V = \frac{RT}{F} \ln \left(\frac{P_K \cdot [K_o^+] + P_{Na} \cdot [Na_o^+] + P_{Cl} \cdot [Cl_i^-]}{P_K \cdot [K_i^+] + P_{Na} \cdot [Na_i^+] + P_{Cl} \cdot [Cl_o^-]} \right)$$

where V is the membrane potential, R the gas constant, T the absolute temperature, F the faraday, and P_{K+}, P_{Na+}, and P_{Cl-} the permeability of the membrane to K^+, Na^+, and Cl^-, respectively. The

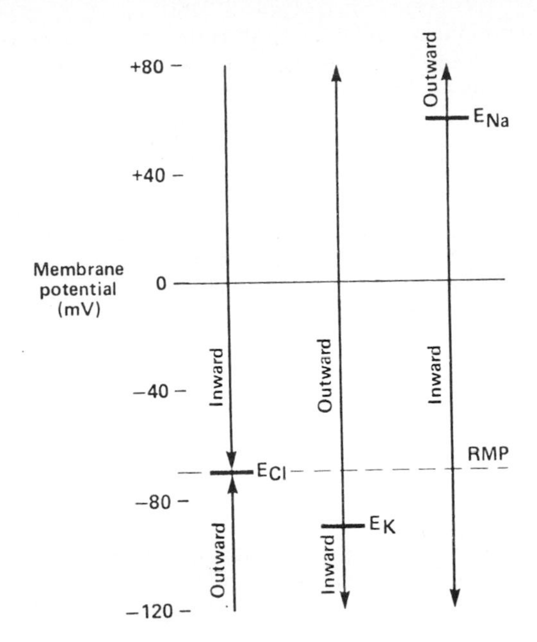

Figure 1–25. Summary of the net passive fluxes of ions across the membrane of mammalian motor neurons at various membrane potentials. RMP, resting membrane potential. (Courtesy of AM Thompson.)

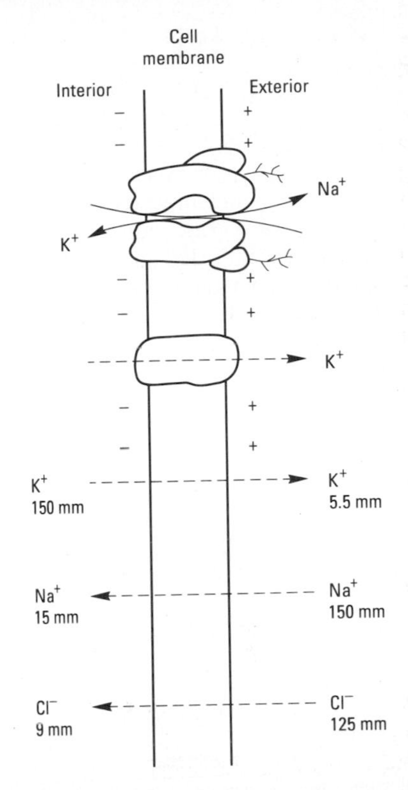

Figure 1–26. Origin of the membrane potential. Na^+ is actively transported out of the cell and K^+ into it by Na^+-K^+ ATPase (top). K^+ leaves the cell in part by a K^+ leak channel (next to top), which makes the membrane more permeable to K^+ than to Na^+. Diffusion of K^+ out of the cell down its concentration gradient makes the interior of the cell negative relative to the exterior. The ion concentrations shown are those found in skeletal muscle.

brackets signify concentration, and i and o refer to the inside and outside of the cell. Since $P_{Na}{}^+$ is low relative to $P_K{}^+$ in the resting cells, Na^+ contributes little to the value of V. As would be predicted from the Goldman equation, changes in external Na^+ produce little change in the resting membrane potential whereas increases in external K^+ decrease it.

The situation in muscle and various other types of cells is similar. In muscle, for example, the resting membrane potential is about −90mV. E_{Cl} is −86 mV, E_K is −100 mV, and E_{Na} is +55 mV. In red blood cells, on the other hand, the membrane potential is low (−9 mV) and near the Cl^- equilibrium potential, whereas both Na^+ and K^+ are actively transported across the cell membrane.

Ca^{2+} should also be mentioned in the present context. In mammals, the ionized calcium concentration in ECF is about 1.2 mmol/L, whereas the intracellular free Ca^{2+} concentration is normally much lower, about 100 nmol/L. Thus, both the electrical and chemical gradients are directed inward.

Genesis of the Membrane Potential

The distribution of ions across the cell membrane and the nature of this membrane provide the explanation for the membrane potential. Two transport proteins are responsible for the resting membrane potential: a K^+ leak channel that permits K^+ to diffuse out of the cell, and Na^+-K^+ATPase (Fig 1–26). The concentration gradient for K^+ facilitates its diffusion out of the cell, but its electrical gradient is in the opposite (inward) direction. Consequently, a balance is reached in which the tendency of K^+ to move out of the cell is balanced by its tendency to move into the cell, and at that equilibrium there is a slight excess of cations on the outside and anions on the inside. This condition is maintained by Na^+-K^+ATPase, which pumps K^+ back into the cell and keeps the intracellular concentration of Na^+ low. The Na^+-K^+ pump is also electrogenic, because it pumps 3 Na^+ out of the cell for every 2 K^+ it pumps in; thus, it also contributes a small amount to the membrane potential by itself. It should be emphasized that the number of ions responsible for the membrane potential is a minute fraction of the total number present and that the total concentrations of positive and negative ions are equal everywhere except along the membrane. Na^+ influx does not compensate for the K^+ efflux because, owing to the K^+ leak channel, the membrane is more permeable to K^+ than to Na^+.

Variations in Membrane Potential

If the resting membrane potential is decreased by the passage of a current through the membrane, the electrical gradient that keeps K^+ inside the cell is decreased, and there is an increase in K^+ diffusion out of the cell. This K^+ efflux and the simultaneous movement of Cl^- into the cell result in a net movement of positive charge out of the cell, with consequent restoration of the resting membrane potential. When the membrane potential increases, these ions move in the opposite direction. These processes occur in all polarized cells and tend to keep the resting membrane potential of the cells constant within narrow limits. However, in nerve and muscle cells, reduction of the membrane potential triggers a voltage-dependent increase in Na^+ permeability. This unique feature permits these cells to generate self-propagating impulses that are transmitted along their membranes for great distances. These impulses are considered in detail in Chapter 2.

INTERCELLULAR COMMUNICATION

Cells communicate with each other via chemical messengers. Within a given tissue, some chemicals move from cell to cell via gap injunctions (see above) without entering the ECF. In addition, cells are affected by chemical messengers secreted into the ECF. There are 3 general types of intercellular communication mediated in this fashion: (1) **neural communication,** in which neurotransmitters are released at synaptic junctions from nerve cells and act across a narrow synaptic cleft on a postsynaptic cell (see Chapter 4); (2) **endocrine communication,** in which hormones reach cells via the circulating blood (see Chapters 18–24); and (3) **paracrine communication**, in which the products of cells diffuse in the ECF to affect neighboring cells that may be some distance away (Fig 1–27). In addition, there is evidence that in some situations, cells secrete chemical messengers that bind to receptors on the same cell, ie, the cell that secreted the messenger **(autocrine communication).** It is worth noting that in various parts of the body, the same chemical messenger can function as a neurotransmitter, a paracrine mediator, a hormone secreted by neurons into the blood (neural hormone), and a hormone secreted by gland cells into the blood.

Chemical messengers bind to protein receptors on the surface of the cell or in some instances in the cytoplasm or the nucleus, triggering sequences of intracellular changes that produce their physiologic effects.

Radioimmunoassay

Many of the chemical messengers are polypeptides or proteins, whereas others are amino acids, amines, or steroids. Antibodies to the polypeptides and proteins can be produced; using special techniques, it is possible to make antibodies to the other messengers as well. The antibodies can then be used to measure the messengers in body fluids and in tissue extracts by **radioimmunoassay.** This technique depends on the fact that the naturally occurring, unlabeled ligand and added radioactive ligand compete to bind to an antibody to the ligand. The greater the amount of unlabeled ligand in the specimen being analyzed, the more it competes and the smaller the amount of radioactive ligand that binds to the antibody. Radioimmunoassays are extensively used in research and in clinical medicine.

Receptors for Hormones, Neurotransmitters, & Other Ligands

Many of the receptors for chemical messengers have now been isolated and characterized. These proteins are not static components of the cell, but their numbers increase and decrease in response to various stimuli, and their properties change with changes in physiologic conditions. When a hormone or neurotransmitter is present in excess, the number

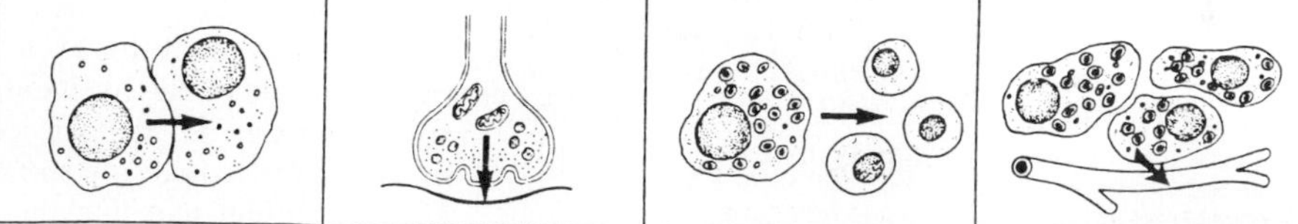

	GAP JUNCTIONS	SYNAPTIC	PARACRINE	ENDOCRINE
Message transmission	Directly from cell to cell	Across synaptic cleft	By diffusion in interstitial fluid	By circulating body fluids
Local or general	Local	Local	Locally diffuse	General
Specificity depends on	Anatomic location	Anatomic location and receptors	Receptors	Receptors

Figure 1–27. Intercellular comunication by chemical mediators. (Reproduced, with permission, from Greenspan FS, Forsham PH [editors]: *Basic & Clinical Endocrinology,* 3rd ed. Appleton & Lange, 1989.)

of active receptors generally decreases **(down regulation)**, whereas in the presence of a deficiency of the chemical messenger, there is an increase in the number of active receptors **(up regulation).** Angiotensin II in its actions on the adrenal cortex is an exception; it increases rather than decreases the number of its receptors in the adrenal. There are also cross-stimulation effects, in which one substance affects the number or the affinity of receptors for other substances, or both. These effects on receptors are important in explaining such phenomena as denervation hypersensitivity (see Chapter 4), tolerance to morphine (see Chapter 7), and decreased sensitivity to insulin in diabetes (see Chapter 19). Receptor-mediated endocytosis is responsible for down regulation in some instances; ligands bind to their receptors in the membrane, and the ligand-receptor complexes move laterally in the membrane to coated pits, where they are taken into the cell by endocytosis **(internalization).** This decreases the number of receptors in the membrane. Some receptors are recycled after internalization, whereas others are replaced only by de novo synthesis in the cell. Internalization is too slow a process to explain the rapid actions of hormones, but it has been argued that some of the slower actions of hormones may be due to the action of internalized hormone within the cell.

Mechanisms by Which Chemical Messengers Act

Although the number of newly described mechanisms by which chemical messengers exert their intracellular effects has been increasing, there is still a relatively small number of mechanisms by which physiologic effects are brought about. The principal mechanisms are summarized in Table 1–8. Proteins, polypeptides, and most other ligands in the ECF are called "first messengers," because they bind to surface receptors and trigger the release of intracellular mediators called "second messengers." Thyroid and steroid hormones are exceptions; they enter cells and, after binding to cytoplasmic or nuclear receptors, increase transcription of selected mRNAs.

Stimulation of Transcription

Steroid and thyroid hormones act by inducing the synthesis of enzymes in their target cells. The hormones enter cells and bind to specific receptor proteins. The binding changes the conformation of the receptor, exposing each receptor's DNA-binding domain (Fig 1–28). The receptor-hormone complex moves to DNA, where it binds to enhance elements in the untranslated 5′ flanking portions of certain genes (Fig 1–28). The estrogen and the triiodothyronine (T_3) receptors are in the nucleus before they bind hormones. The T_3 receptor also binds thyroxine (T_4), but with less affinity. The glucocorticoid receptor is located mainly in the cytoplasm but migrates promptly to the nucleus as soon as it binds its ligand. The initial location of the other receptors that act in this fashion is not yet known. In any case, binding of the receptor-hormone complex to DNA increases transcription of mRNAs encoded by the gene to which it binds. The mRNAs are translated in the ribosomes, with the production of increased quantities of proteins that alter cell function. (Fig 1–28).

Structure of Receptors

The structures of the human glucocorticoid, mineralocorticoid, and estrogen receptors are shown in Fig 1–29. One of the 2 known T_3 receptors is also shown in the figure; for unknown reason, there are 2 similar T_3 receptors, encoded by genes on separate chromosomes. All these receptors are part of a large family of receptors that have in common a highly conserved cysteine-rich DNA binding domain; a ligand-binding domain at or near the C terminal of the receptor; and a relatively variable, poorly conserved N-terminal region. Other receptors in the family include the other T_3 receptor (hT_3R_2) and the receptors for progesterone, androgen, and 1,25'dihydroxy-cholecalciferol. It has recently been demonstrated that retinoic acid is a morphogen that dictates the formation of the limbs during embryonic life, and the retinoic acid receptor is also a member of this family. Finally, the protein encoded by the *erb*A oncogene is closely related to the T_3 receptors. Thus, this family of receptors occupies a prominent place in regulatory physiology. The cysteines in the DNA-binding

Table 1–8. Principal mechanisms by which chemical messengers in the ECF bring about changes in cell function.

Mechanism	Examples
Open or close ion channels in cell membrane	Acetylcholine on nicotinic cholinergic receptor
Increase tyrosine kinase activity of cytoplasmic portions of transmembrane receptors	Insulin, EGF, PDGF
Activate phospholipase C with intracellular production of DAG, IP_3, and other inositol phosphates	Angiotensin II, norepinephrine via α_1, adrenergic receptor, vasopressin via V_1 receptor
Activate or inhibit adenylate cyclase, causing increased or decreased intracellular production of cyclic AMP	Norepinephrine via β_1 adrenergic receptor, (increased cyclic AMP); Norepinephrine via α_2 adrenergic receptor (decreased cyclic AMP)
Increase cyclic GMP in cell	ANP
Act via cytoplasmic or nuclear receptors to increase transcription of selected mRNAs	Thyroid hormones, retinoic acid, steroid hormones

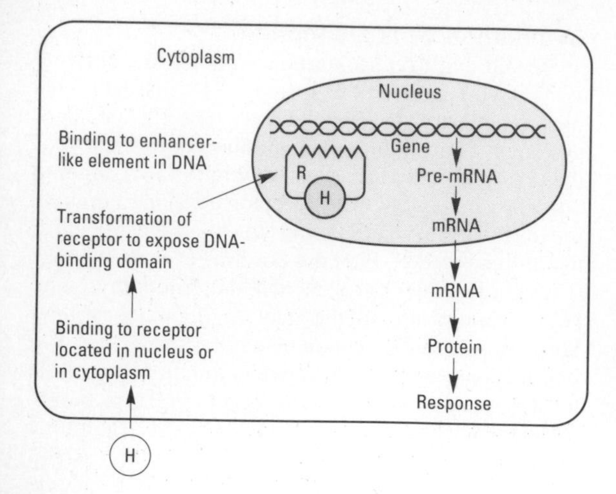

Figure 1–28. Mechanism of action of steroid and thyroid hormones. H, hormone; R, receptor.

domains of the receptors form characteristically shaped complexes with zinc ions that have been called "zinc fingers," but the role of these fingers in the binding to DNA and the action of the receptor-steroid complex is unsettled.

Intracellular Ca^{2+}

The free Ca^{2+} concentration in the cytoplasm is maintained at about 100 nmol/L. The Ca^{2+} concentration in the interstitial fluid is about 1200 nmol/L, so there is a marked inwardly directed concentration gradient as well as an inwardly directed electrical gradient. Much of the intracellular Ca^{2+} is bound by mitochondria and the endoplasmic reticulum (Fig 1–30), and these organelles provide a store from which Ca^{2+} can be mobilized to increase the concentration of free Ca^{2+} in the cytoplasm. Cytoplasmic Ca^{2+} is also taken up by calcium-binding proteins, and when these bind Ca^{2+}, they become activated and produce many different physiologic effects (see below).

Ca^{2+} enters cells through 2 kinds of channels: voltage-gated Ca^{2+} channels activated by depolarization, and ligand-gated Ca^{2+} channels which in various types of cells are activated by many different neurotransmitters and hormones. Ca^{2+} is pumped out of cells in exchange for 2 H^+ by a Ca^{2+}-H^+ ATPase. It is also transported out of cells by an antiport driven by the Na^+ gradient that exchanges 2 Na^+ for each Ca^{2+}.

Many hormones and other agents act by increasing cytoplasmic Ca^{2+}. The increase is produced by releasing Ca^{2+} from intracellular stores—primarily the endoplasmic reticulum—or by increasing the entry of Ca^{2+} into cells, or by both mechanisms. The increase in cytoplasmic Ca^{2+} then produces its effect by way of calcium-binding proteins.

Calcium-Binding Proteins

The principal Ca^{2+}-binding proteins in cells are **calmodulin** and **troponin C,** although other proteins also bind Ca^{2+}. Calmodulin is an acidic polypeptide with 4 Ca^{2+}-binding domains (Fig 1–31). It contains 148 amino acid residues, has a molecular weight of 16,700, and is unique in that residue 115 is trimethylated lysine. It binds Ca^{2+}, and the calmodulin-Ca^{2+} complex then binds to various enzymes, activating them. It is found in a wide variety of different cells in mammals and in the cells of simple invertebrates and plants. In addition, its structure has been remarkably

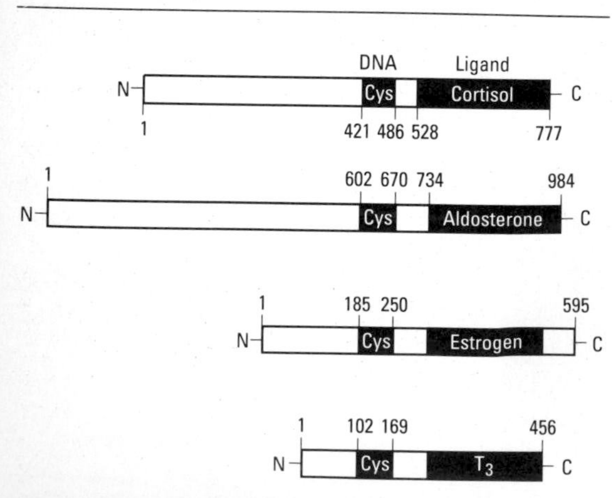

Figure 1–29. Structure of human glucocorticoid, mineralocorticoid, and estrogen receptors, and one of 2 T_3 receptors, hT_3R_β. Note that each receptor has a cysteine-rich DNA-binding domain and a ligand-binding domain at or near the C terminus, with considerable variability in the N-terminal part of the protein.

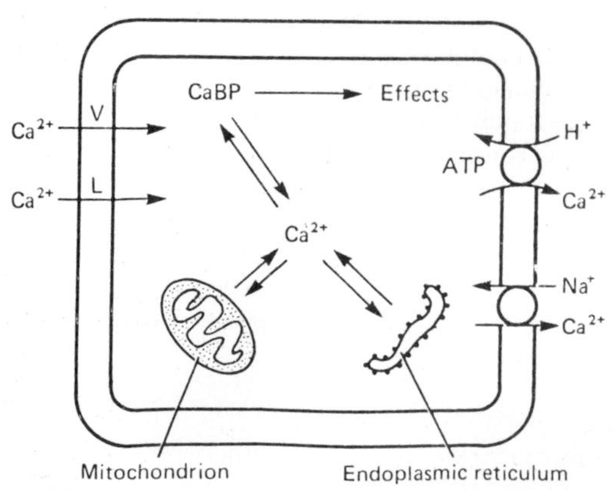

Figure 1–30. Ca^{2+} metabolism in mammalian cells. Cytoplasmic Ca^{2+} is in equilibrium with Ca^{2+} bound to mitochondria and the endoplasmic reticulum. Calcium-binding proteins (CaBP) bind cytoplasmic Ca^{2+} and, when activated in this fashion, bring about a variety of physiologic effects. Ca^{2+} enters the cell via voltage-gated (V) and ligand-gated (L) Ca^{2+} channels. It is transported out of the cell by a Ca^{2+}-H^+ ATPase and a Na^+-Ca^{2+} antiport.

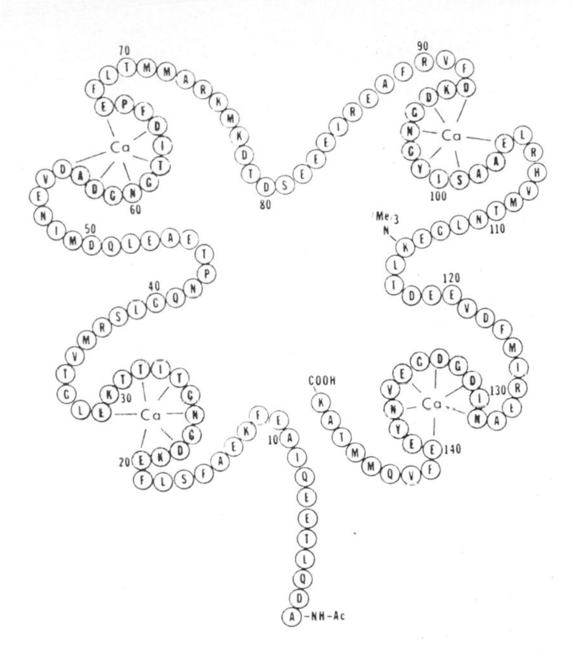

Figure 1–31. Structure of calmodulin from bovine brain. The abbreviations for the amino acid residues are those listed in the section on abbreviations and symbols in the appendix. Note the 4 calcium domains (dark residues) flanked on either side by stretches of α helix. (Reproduced, with permission, from Cheung WY: Calmodulin: An overview. *Fed Proc* 1982;**41**:2253.)

Table 1–9. Some of the many enzymes and cellular processes regulated by calmodulin.*

Myosin light chain kinase
Phosphodiesterase
Gap junction permeability
Phospholipase A_2
Ca^{2+} ATPase
Microtubule disassembly
Cellular secretion
Membrane phosphorylation
Neurotransmitter release
NAD kinase
Ca^{2+}-dependent protein kinase
Guanylate cyclase
Phosphorylase kinase

*Modified from Cheung WY: Calmodulin plays a pivotal role in cellular regulation. *Science* 1980; **207**:19.

preserved through the course of evolution; for instance, antibodies against mammalian calmodulin cross-react with calmodulin in coelenterates and cotton seeds. This indicates that it must perform vital intracellular functions. Some of the many cell processes in which it appears to be involved are summarized in Table 1–9. It is present in spindles during mitosis, and since it catalyzes disassembly of microtubules, it may move the chromosomes apart during cell division by shortening the microtubules that attach the chromosomes to the spindle poles. It activates **myosin light chain kinase,** and this enzyme catalyzes the phosphorylation of myosin, which in turn brings about contraction of smooth muscle. In skeletal and cardiac muscle, contraction is initiated by Ca^{2+} binding to troponin C rather than calmodulin (see Chapter 3), but troponin C has remarkable structural similarities to calmodulin.

Inositol Triphosphate & Diacylglycerol as Second Messengers

The link between membrane binding of a ligand that acts via Ca^{2+} and the prompt increase in cytoplasmic Ca^{2+} is generally **inositol triphosphate** (IP_3). When one of these ligands binds to its receptor, activation of the receptor produces activation of phospholipase C on the inner surface of the membrane via a nucleotide regulatory protein called G_p. It is becoming clear that a variety of different receptors on the cell surface are coupled to their catalytic units via nucleotide regulatory proteins (G proteins; see below).

Phospholipase C catalyzes the hydrolysis of phosphatidylinositol 4,5-diphosphate (PIP_2) to form IP_3 and diacylglycerol (Fig 1–32). The IP_3 diffuses to the endoplasmic reticulum, where it triggers the release of Ca^{2+} into the cytoplasm (Fig 1–33). Inositol tetrakisphosphate appears to be formed as well, and this compound may increase the permeability of the cell membrane to Ca^{2+}. The diacylglycerol also plays a role as a second messenger; it stays in the cell membrane, where it activates **protein kinase C.** Protein kinases are intracellular enzymes that phosphorylate proteins, bringing about changes in the configuration of the proteins which alter their function. Protein kinase C has been implicated in the production of many different cellular responses (Table 1–10). In some instances, protein kinase C and Ca^{2+}-binding protein responses interact with or potentiate each other. Protein kinase C is now known to be a family of related calcium phospholipid-dependent protein kinases, rather than a single entity.

The precursor of PIP_2 is phosphatidylinositol (Fig 1–32). This phospholipid is found in relatively small amounts in the inner lamella of the cell membrane. It is first converted to phosphatidyl 4-phosphate (PIP) and then to PIP_2, the derivative that is hydrolyzed to form IP_3 and diacylglycerol. The IP_3 is metabolized by stepwise dephosphorylation to inositol. Diacylglycerol is converted to phosphatidic acid and then to cytosine diphosphate (CDP) diacylglycerol, which combines with inositol to form phosphatidylinositol, completing the cycle.

G Proteins

As noted above, many different receptors on the cell surface are coupled to their catalytic units or other effectors via nucleotide regulatory proteins (G proteins) that bind guanosine triphosphate (GTP). GTP is the guanosine analogue of ATP (see Chapter 17). All these proteins are made up of α, β, and subunits. The α subunit is normally bound to guanosine diphosphate (GDP). When a ligand binds

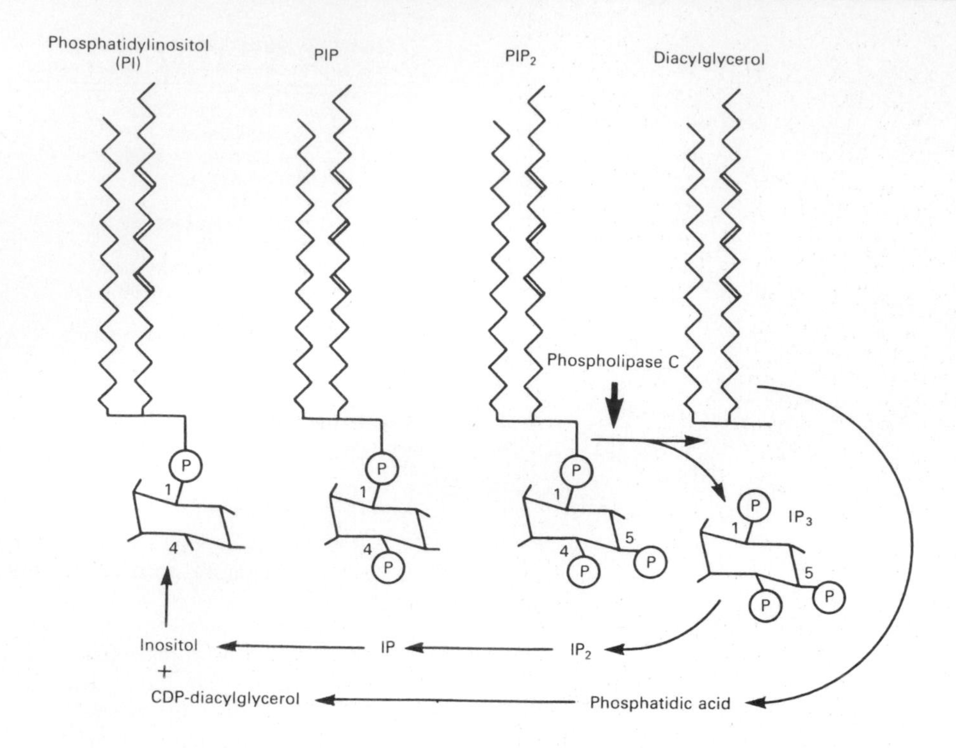

Figure 1–32. Metabolism of phosphatidylinositol in cell membranes. Phosphatidylinositol is successively phosphorylated to form phosphatidylinositol 4-phosphate (PIP), then phosphatidylinositol 4,5-diphosphate (PIP_2). Phospholipase C catalyzes the breakdown of PIP_2 to inositol 1,4,5-triphosphate (IP_3) and diacylglycerol. IP_3 can be converted to inositol tetrakisphosphate (inositol 1,3,43,5, P_4) and this can be converted in turn to 1,3,4, P_3 (not shown). IP_3 is also dephosphorylated to inositol, and diacylglycerol is metabolized to cytosine diphosphate (CDP) diacylglycerol, CDP-diacylglycerol and inositol then combine to form phosphatidylcholine, completing the cycle. (Modified from Berridge MJ: Inositol triphosphate and diacylglycerol as second messengers. *Biochem J* 1984;**220**:345.)

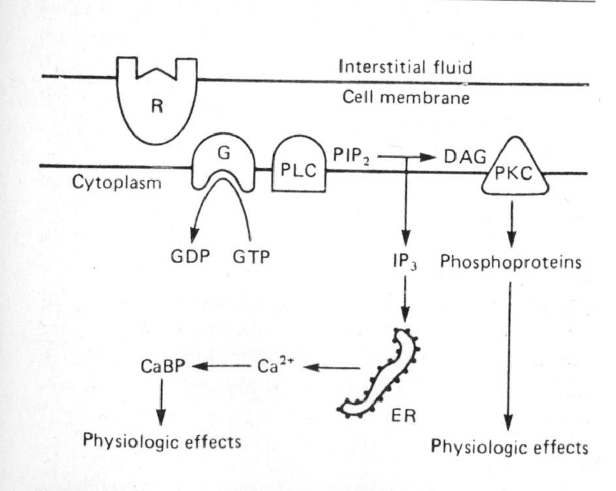

Figure 1–33. Diagrammatic representation of release of inositol triphosphate (IP_3) and diacylglycerol (DAG) as second messengers. Binding of ligand to receptor (R) activates phospholipase C (PLC) via a G protein, G_p. The resulting hydrolysis of PIP_2 produces IP_3, which releases Ca^{2+} from the endoplasmic reticulum (ER), and DAG, which activates protein kinase C (PKC). CaBP, Ca^{2+}-binding proteins.

Table 1–10. Some of the many physiologic responses that appear to be mediated by protein kinase C.*

Cell	Response
Endocrine systems	
Adrenal medulla	Catecholamine secretion
Adrenal cortex	Aldosterone secretion
B cells of pancreas	Insulin secretion
Pituitary	Secretion of growth hormone, LH, prolactin, TSH
Parathyroid	Secretion of parathyroid hormone
Leydig cells	Steroidogenesis
Exocrine systems	
Pancreas	Amylase
Parotid	Amylase and mucin secretion
Alveolar cells	Surfactant secretion
Nervous system	
Neuromuscular junction	Acetylcholine secretion
Caudate nucleus	Acetylcholine secretion
Inflammation and immune system	
Platelets	Thromboxane synthesis
Neutrophils	Superoxide generation
Mast cells	Release of histamine
Lymphocytes	Activation of T and B cells

*Modified from Nishizuka Y: Studies and perspectives on protein kinase C. *Science* 1986;**233**:305.

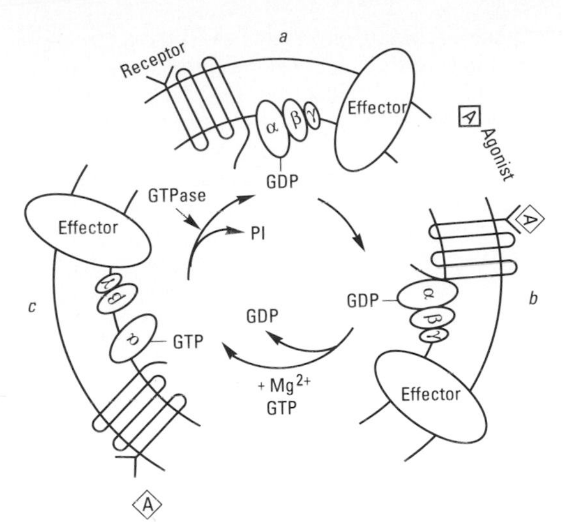

Figure 1–34. G protein signal transduction. *a*, unliganded state; *b*, ligand A binds to receptor, and G-protein–receptor interaction permits GTP to replace GDP on α subunits; *c*, α-GTP separates from the βγ subunit, and it and /or βγ interacts with effectors such as adenylate cyclase or a K^+ channel. The intrinsic GTPase activity of the subunit then hydrolyzes the GTP to GDP, and α-GDP and βγ reassociate, completing the cycle. (Reproduced, with permission, from Near FJ, Clapham DE: Roles of G protein subunits in transmembrane signaling. *Nature* 1988;333:129.)

to a G-coupled receptor, GTP replaces GDP (Fig 1–34) and the α subunit separates from the combined β and γ subunits. Apparently, the β and γ subunits never separate in the native state, although both α and βγ may interact with effectors. The intrinsic GTPase activity of the α subunit then leads to conversion of GTP back to GDP, and α again combines with βγ, restoring the resting state. The α subunits are markedly heterogeneous in structure, whereas the β and γ subunits are more homogeneous. They may anchor the α subunits to the cell membrane, although all the subunits normally stay within the membrane.

The list of G proteins that couple ligands to intracellular events is becoming large. It includes not only G_p, which couples receptors to phospholipase C, but also G_s and G_i, which couples receptors exerting stimulatory and inhibitory effects, respectively, on adenylate cyclase (see below), and G_t, which couples rhodopsin to phosphodiesterase in rods (see Chapter 8) In addition, it is now known that a G protein called G_k is activated by binding of acetylcholine to muscarinic cholinergic receptor in the heart, and that apparently its α and its βγ subunits can both open K^+ channels, probably by a direct action on the channels.

Cyclic AMP

Another important second messenger is cyclic AMP (Fig 1–35). Some of the many effects mediated by this compound are listed in Table 1–11. Cyclic AMP, which is also known as **cAMP,** is cyclic adenosine 3′,5′-monophosphate. It is formed from

Adenosine triphosphate (ATP)

Adenylate cyclase

Cyclic AMP

Phosphodiesterase

5′-Adenosine monophosphate (5′-AMP)

Figure 1–35. Formation and metabolism of cyclic AMP.

ATP by the action of the enzyme **adenylate cyclase** and converted to physiologically inactive 5′-AMP by the action of the enzyme phosphodiesterase. Cyclic AMP activates a cyclic AMP-dependent protein kinase **(protein kinase A)** that, like protein kinase C, catalyzes the phosphorylation of proteins. Phosphorylation of serine and, to a lesser extent, threonine residues of these proteins generally changes their conformation and, in enzymes, increases or decreases

Table 1–11. Some of the hormones and related substances that act by altering intracellular cyclic AMP.

Increase cyclic AMP	
Adrenocorticotropic hormone	Lipotropin
Calcitonin	Melanocyte-stimulating hormone
Catecholamines (β_1, β_2 receptors)	Nerve growth factor
Chorionic gonadotropin	Parathyroid hormone
Follicle-stimulating hormone	Prostaglandin E_1
Glucagon	Thyrotropin-releasing hormone
Luteinizing hormone	Thyrotropin (TSH)
Luteinizing hormone-releasing hormone	Vasopressin (V_2 receptor)
Decrease cyclic AMP	
Catecholamines (α_2 receptors)	Dopamine (D_2 receptors)
Somatostatin	

their activity, thus producing changes in metabolism. A typical example is the activation of phosphorylase in the liver by epinephrine via cyclic AMP and protein kinase A (Fig 17–13). It is now clear that phosphorylation and dephosphorylation of proteins are major mechanisms for regulating cellular processes. The phosphatases responsible for dephosphorylation are widespread and relatively few in number. As a general rule, enzymes in biodegradative pathways are activated by phosphorylation, whereas enzymes in biosynthetic pathways are inactivated by phosphorylation. Phosphodiesterase is inhibited by methylxanthines such as caffeine and theophylline; consequently, these compounds augment hormonal and transmitter effects mediated via cyclic AMP.

Activation of Adenylate Cyclase

Five components are involved in the mechanism by which ligands bring about changes in the intracellular concentration of cyclic AMP: a catalytic unit, **adenylate cyclase,** which catalyzes the conversion of ATP to cyclic AMP; a stimulatory and an inhibitory receptor; and a stimulatory and an inhibitory G protein that links the receptor to the catalytic unit (Fig 1–36). When the appropriate ligand binds to the stimulatory receptor, the α subunit of G_s activates adenylate cyclase. Conversely, when the appropriate ligand binds to the inhibitory receptor, the α subunit of G_i: inhibits adenylate cyclase. G_s mediates the excitory effects of many different ligands on adenylate cyclase, yet the effects of chemical messengers are specific. For example, ACTH stimulates adrenocorticosteroid secretion, whereas TSH does not; and TSH stimulates thyroid hormone secretion, whereas ACTH does not. The specificity of responses depends on the receptors associated with the adenylate cyclase; in each cell, the receptor has a high degree of specificity for the substance or substances that normally stimulate it. Some cyclic AMP escapes from cells upon stimulation by certain hormones, but the amounts are small compared with the intracellular concentration, and only small amounts of extracellular cyclic AMP enter cells. Intracellularly, the specificity of the response to cyclic AMP depends on the nature and the amounts of the phosphorylatable proteins in the particular cell. In a similar fashion, specificity is imparted to the Ca^{2+}-mediated responses and the protein kinase C-mediated responses by the specificity of the receptors and the calcium-binding and phosphorylatable proteins in the target cells.

Evidence is accumulating that Ca^{2+} can interact with cyclic AMP and that the diacylglycerol–protein kinase C systems interact with the cyclic AMP–protein kinase A systems. In some instances, the cyclic AMP system inhibits the diacylglycerol system, whereas in others, the diacylglycerol system facilitates the cyclic AMP system.

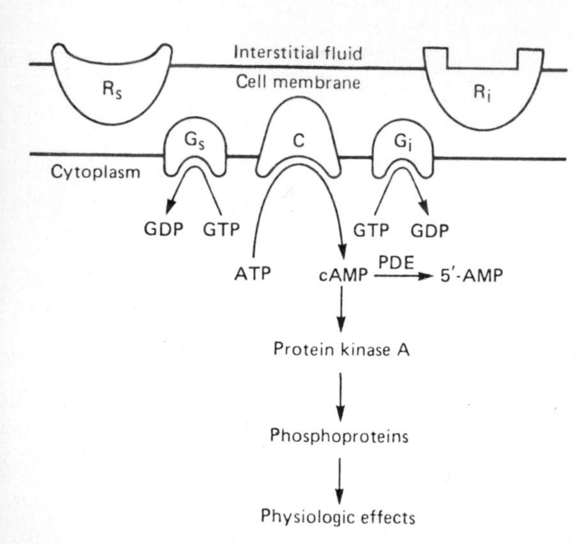

Figure 1–36. The cyclic AMP system. Activation of the catalytic unit (C), adenylate cyclase, catalyzes the conversion of ATP to cyclic AMP. Cyclic AMP activates protein kinase A, which phosphorylates protein, producing physiologic effects. Stimulatory ligands bind to the stimulatory receptor (R_s) and activate C via G_s, the stimulatory nucleotide regulatory protein. Inhibitory ligands inhibit C via the inhibitory receptor (R_i) and the inhibitory G protein (G_i). (Modified and reproduced, with permission, from Spiegel AM et al: Clinical implications of guanine nucleotide-binding proteins as receptor-effector couplers. Reprinted by permission of the *New England Journal of Medicine* 1985;312:26.)

Receptor Diseases

Disordered function of receptors and the membrane proteins with which they interact causes a variety of diseases. For example, in pseudohypoparathyroidism and nephrogenic diabetes insipidus, parathyroid hormone and vasopressin fail to produce the increase in cyclic AMP in their target organs that they produce in normal individuals. There are several forms of pseudohypoparathyroidism, and one of them appears to be due to a deficiency of the G protein. In addition, certain diseases are due to the production of antibodies against receptors. Thus, antibodies against TSH receptors cause Graves' disease (see Chapter 18), antibodies against nicotinic acetylcholine receptors cause myasthenia gravis, and antibodies against insulin receptors cause the diabetes that occurs in acanthosis nigricans (see Chapter 19).

HOMEOSTASIS

The actual environment of the cells of the body is the interstitial component of the ECF. Since normal cell function depends upon the constancy of this fluid, it is not surprising that in multicellular animals, an immense number of regulatory mechanisms have evolved to maintain it. To describe "the various physiologic arrangements which serve to restore the normal state, once it has been disturbed," W.B.

Cannon coined the term **homeostasis.** The buffering properties of the body fluids and the renal and respiratory adjustments to the presence of excess acid or alkali are examples of homeostatic mechanisms. There are countless other examples, and a large part of physiology is concerned with regulatory mechanisms that act to maintain the constancy of the internal environment. Many of these regulatory mechanisms operate on the principle of negative feedback; deviations from a given normal set point are detected by a sensor, and signals from the sensor trigger compensatory changes that continue until the set point is again reached.

AGING

Aging is a general physiologic process that is as yet poorly understood. It affects cells and the systems made up of them, as well as tissue components such as collagen, and numerous theories have been advanced to explain it. One theory holds that tissues age as a result of random mutations in the DNA of somatic cells, with consequent introduction of cumulative abnormalities. Others hold that cumulative abnormalities are produced by increased cross-linkage of collagen, other proteins, and DNA, possibly as the end result of the nonenzymatic combination of glucose with amino groups on these molecules. A third theory envisions aging as the cumulative result of damage to tissues by free radicals formed in them. Some investigators have speculated that in mammals, there is a biologic clock, possibly located in the hypothalamus, that is responsible for aging via hormonal or other pathways. However, it seems fair to say that at present, despite intensive research, the mechanisms of aging are still unknown.

REFERENCES
Section I: Introduction

Alberts B et al: *Molecular Biology of the Cell.* Garland, 1983.

Andreoli T, Hoffman JF, Fanestil DD (editors): *Membrane Physiology.* Plenum, 1980.

Bennett MVL, Spray DC (editors): *Gap Junctions.* Cold Spring Harbor, 1985.

Berridge MJ, Irvine RF: Inositol triphosphate, a novel second messenger in cellular signal transduction. *Nature* 1984; **312:**315.

Bretscher M: Endocytosis: Relation to capping and cell locomotion. *Science* 1984;**224:**681.

Cannon WB: *The Wisdom of the Body.* Norton, 1932.

Cerami A, Vlassara H, Brownlee M: Glucose and aging. *Sci Am* (May) 1987;**256:**90.

Cohn JP: The molecular biology of aging. *Bioscience* 1987;**37:**99.

Darnell JE Jr: The processing of mRNA. *Sci Am* (Oct) 1983; **249:**72.

Darnell J, Lodish H, Baltimore D: *Molecular Cell Biology* Sci Am Books, 1986.

Dautry-Varsat A, Lodish HF: How receptors bring proteins and particles into cells. *Sci Am* (May) 1984;**250:**52.

de Duve C: *A Guided Tour of the Living Cell. 2 vols. Freeman, 1985.*

Diamond JM: Transcellular cross-talk between epithelial cell membranes. *Nature* 1982;**300:**683.

Drlica K: *Understanding DNA and Gene Cloning: A Guide for the Curious.* Wiley, 1984.

Evans RM: The steroid and thyroid hormone receptor superfamily. *Science* 1988; **240**:889.

Fawcett DW: *Bloom and Fawcett—A Textbook of Histology,* 11th ed. Saunders, 1986.

Gale EF et al: *The Molecular Basis of Antibiotic Action,* 2nd ed. Wiley, 1981.

Gennari FJ: Serum osmolality. *N Engl J Med* 1984;**310:**102.

Glauman H, Bullard FJ (editors): *Lysosomes.* Academic, 1987.

Griffiths G, Simons K: The trans Golgi network: Sorting at the exit site of the Golgi complex. *Science* 1986;**234:**430.

Gumbiner B: Structure, biochemistry, and assembly of epithelial tight junctions. *Am J Physiol* 1987;**253:**c749.

Ingebritsen TS, Cohen P: Protein phosphates: Properties and role in cellular regulation. *Science* 1983;**221:**331.

Kelly R: Pathways of protein secretion in eukaryotes. *Science* 1985;**230:**25.

Lazarides E: Intermediate filaments as mechanical integrators of cellular space. *Nature* 1980;**283:**249.

Low MG, Saltiel AR: Structure and functional roles of glycosylphosphatidylinositol in membranes. *Science* 1988;**239:**268.

Maniatis T, Reed R: The role of small nuclear ribonucleoprotein particles in pre-mRNA splicing. *Nature* 1987;**325:**673.

Margolis RL, Wilson L: Microtubule treadmills: Possible molecular machinery. *Nature* 1981;**293:**705.

Means AR, Tash JS, Chafouleas JG: Physiological implications of the presence, distribution and regulation of calmodulin in eukaryotic cells. *Physiol Rev* 1982;**62:**1.

Nishizuka Y: The role of protein kinase C in cell surface signal transduction and tumour promotion. *Nature* 1984;**308:**693.

Nomura M: The control of ribosome synthesis. *Sci Am* (Jan) 1984;**250:**102.

Palmiter RD, Brinster RL: Germ-line transformation of mice. *Annu Rev Genetics* 1986;**20:**465.

Petersen OH, Maruyama Y: Calcium-activated potassium channels and their role in secretion. *Nature* 1984; **307:**693.

Pollard TD, Cooper JA: Actin and actin-binding proteins: A critical evaluation of mechanisms and function. *Annu Rev Biochem* 1986;**55:**987.

Rothman JE: The compartmental organization of the Golgi apparatus. *Sci Am* (Sept) 1986;**253:**74.

Rozengurt E: Early signals in the mitogenic response *Science* 1986;**234:**161.

Samuels HH et al: Regulation of gene expression by thyroid hormone. *Annu Rev Physiol* 1989;**51.** [in press].

Schliwa M: *The Cytoskeleton.* Springer-Verlag, 1986.

Schuurmans Stekhoven F, Bonting SL: Transport adenosine triphosphatases: Properties and functions. *Physiol Rev* 1981;**61:**1.

Stein WD: *Transport and Diffusion Across Cell Membranes*. Academic Press, 1986.
Stevens CF: Biophysical analysis of the function of receptors. *Annu Rev Physiol* 1980;**42;**643.
Sweadner KJ, Goldin SM: Active transport of sodium and potassium ions: Mechanisms, function, and regulation. *N Engl J Med* 1980;**302:**77.
Tzogoloff A: *Mitochondria*. Plenum, 1982.
Unwin N, Henderson R: The structure of proteins in biological membranes. *Sci Am* (Feb) 1984;**250:**78.
Valentine GH: *The Chromosomes and their Disorders. An Introduction for Clinicians*. Heinemann, 1986.
Vogel F, Motulsky AG: *Human Genetics: Problems and Approaches*, 2nd ed. Springer-Verlag, 1987.
Walter P, Lingappa VR: Mechanisms of protein translocation across the endoplasmic reticulum. *Annu Rev Cell Biol* 1986;**2:**499.
Watson JD, Tooze J, Kuntz DT: *Recombinant DNA: A Short Course*. Freeman, 1984.
Watson JD, et al: *Molecular Biology of the Gene*, 4th ed. 2 vols. Benjamin/Cummings, 1987.
Weinberg RA: The action of oncogenes in the cytoplasm and nucleus. *Science* 1985;**230:**770.
Wheatley DN: *The Centriole: A Central Enigma of Cell Biology*. Elsevier, 1982.

Section II. Physiology of Nerve & Muscle Cells

Excitable Tissue: Nerve 2

INTRODUCTION

The human nervous system contains about 10^{12} (1 trillion) neurons. It also contains 10–50 times this number of glial cells. The neurons, the basic building blocks of the nervous system, have evolved from primitive neuroeffector cells that respond to various stimuli by contracting. In more complex animals, contraction has become the specialized function of muscle cells, whereas integration and transmission of nerve impulses have become the specialized functions of neurons. This chapter is concerned with the ways these neurons are excited and the way they integrate and transport impulses.

NERVE CELLS

Morphology

Neurons in the cerebral nervous system come in many different shapes and sizes (Fig 2–1). However, most have the same parts as the typical spinal motor neuron illustrated in Fig 2–2. This cell has 5–7 processes called **dendrites** that extend out from the

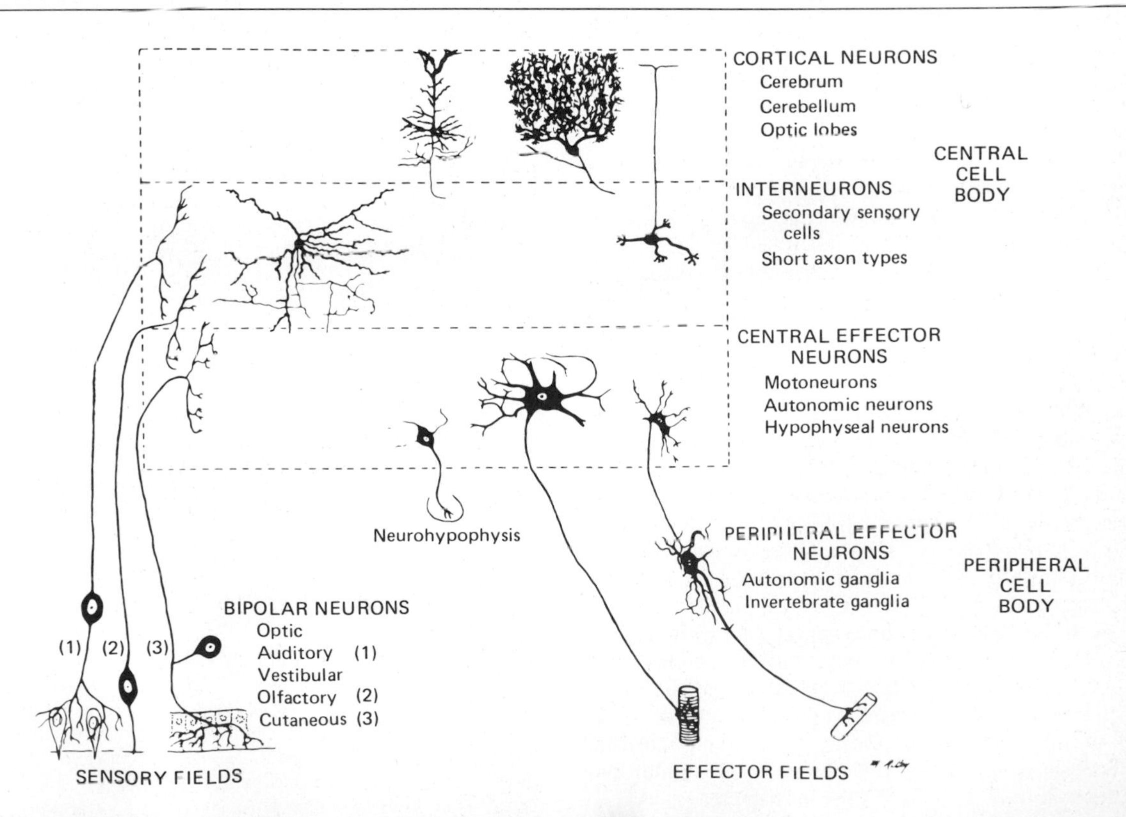

Figure 2–1. Types of neurons in the mammalian nervous system. (Reproduced, with permission, from Bodian D: Introductory survey of neurons. *Cold Spring Harbor Symp Quant Biol* 1952; 17:1.)

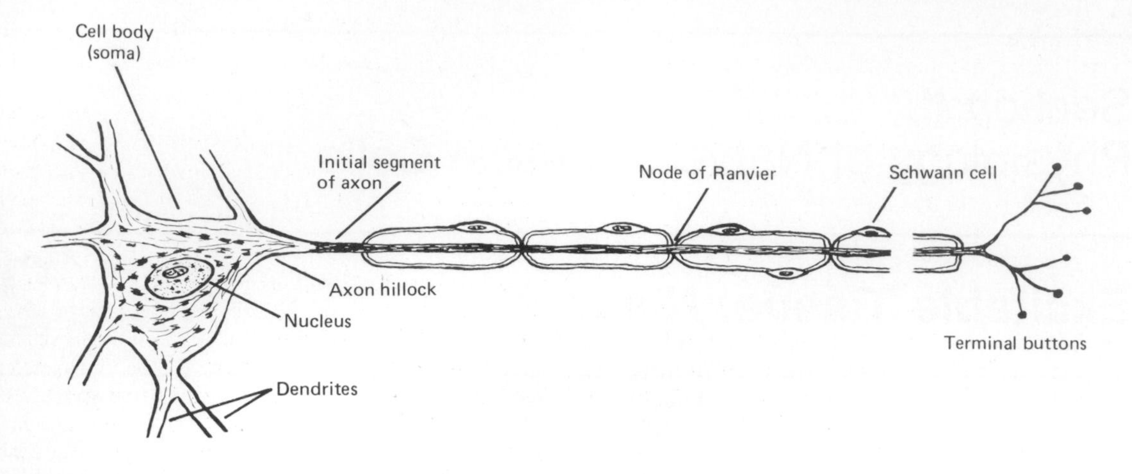

Figure 2–2. Motor neuron with myelinated axon.

cell body and arborize extensively. It also has a long fibrous **axon** that originates from a somewhat thickened area of the cell body, the **axon hillock.** The first portion of the axon is called the **initial segment.** The axon divides into terminal branches, each ending in a number of **synaptic knobs.** The knobs are also called **terminal buttons** or **axon telodendria.** They contain granules or vesicles in which the synaptic transmitters secreted by the nerves are stored (see Chapter 4). The neuron is **myelinated;** ie, a short distance from its origin, the axon acquires a sheath of **myelin,** a protein-lipid complex made up of many layers of the cell membrane of Schwann cells (Fig 2–3). Schwann cells are glialike cells found along peripheral nerves. The myelin sheath envelopes the axon except at its ending and at the **nodes of Ranvier,** periodic 1-μm constrictions that are about 1 mm apart. The insulating function of myelin is discussed below. Not all mammalian neurons are myelinated; some are **unmyelinated,** ie, are simply surrounded by Schwann cells without the wrapping of the Schwann cell membrane around the axon that produces myelin. Most neurons in invertebrates are unmyelinated. In the central nervous system of mammals, most neurons are myelinated, but the cells that form the myelin are oligodendrogliocytes rather than Schwann cells (Fig 2–3). Furthermore, unlike the Schwann cell, which forms the myelin between 2 nodes of Ranvier on a single neuron, oligodendrogliocytes send off multiple processes that form myelin on many neighboring axons.

The dimensions of some neurons are truly remarkable. For spinal neurons supplying the muscles of the foot, for example, it has been calculated that if the cell body were the size of a tennis ball, the dendrites of the cell would fill an average-sized living room and the axon would be up to 1.6 km (almost a mile) long although only 13 mm (½ in) in diameter.

In multiple sclerosis, a crippling disease of unknown cause, there is patchy destruction of myelin in the central nervous system. The loss of myelin is associated with delayed or blocked conduction in the demyelinated axons.

The conventional terminology used above for the parts of a neuron works well enough for spinal motor neurons and interneurons, but there are problems in terms of "dendrites" and "axons" when it is applied to other types of neurons found in the nervous system. From a functional point of view (see below and Chapters 4 and 5), neurons generally have 4

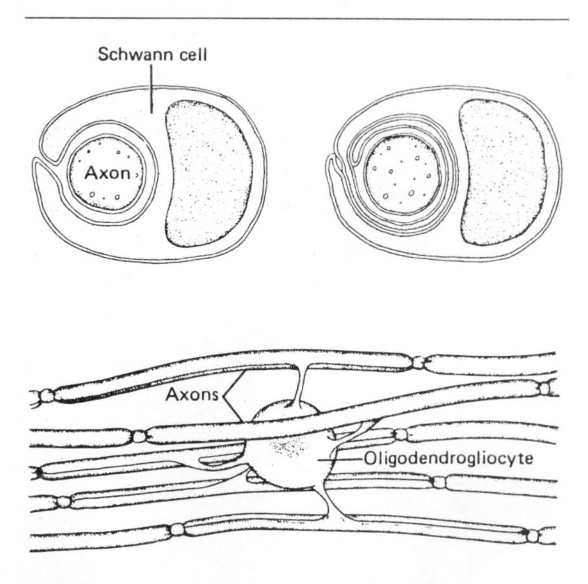

Figure 2–3. *Top:* Relation of Schwann cells to axons in peripheral nerves. *Left:* Unmyelinated axon. *Right:* Myelinated axon. Note that the cell membrane of the Schwann cell has wrapped itself around and around the axon. *Bottom:* Myelination of axons in the central nervous system by oligodendrogliocytes. One oligodendrogliocyte sends processes to up to 40 axons.

important zones. There is a receptor, or dendritic zone, where multiple local potential changes generated by synaptic connections are integrated (Fig 2–4); a site where propagated action potentials are generated (the initial segment in spinal motor neurons, the initial node of Ranvier in cutaneous sensory neurons); an axonal process that transmits propagated impulses to the nerve endings; and the nerve endings, where action potentials cause the release of synaptic transmitters. The cell body is often located at the dendritic zone end of the axon, but it can be within the axon (eg, auditory neurons) or attached to the side of the axon (eg, cutaneous neurons). Its location makes no difference as far as the receptor function of the dendritic zone and the transmission function of the axon are concerned. It should be noted that integration of activity is not the only function of dendrites. Some neurons in the central nervous system have no axons, and local potentials pass from one dendrite to another (see below).

Protein Synthesis & Axoplasmic Transport

Nerve cells are secretory cells, but they differ from other secretory cells in that the secretory zone is often at the end of the axon, far removed from the cell body. There are no ribosomes in axons and nerve terminals, and all necessary proteins are synthesized in the endoplasmic reticulum and Golgi apparatus of the cell body and then transported along the axon to the synaptic knobs by the process of **axoplasmic flow.** Thus, the cell body maintains the functional and anatomic integrity of the axon; if the axon is cut, the part distal to the cut degenerates **(Wallerian degeneration).** Transport occurs along microtubules and can be divided into 3 components: fast transport, at about 400 mm/day, which moves organelles such as synaptic vesicles, is dependent on ATP, and is independent of action potentials; a slower transport, at 6–10 mm/day, which moves some cytoskeletal elements, clathrin, and soluble proteins; and an even slower transport, at 0.5–3 mm/day, which moves the neurofibrillar proteins that make up the cytoskeleton and tubulin. There is also **retrograde transport** from the endings to the cell body at about 200 mm/day. Synaptic vesicles recycle in the membrane, but some used vesicles are carried back to the cell body and deposited in lysosomes. Some of the material taken up at the ending by endocytosis, including nerve growth factor (see below) and various viruses, is also transported back to the cell body.

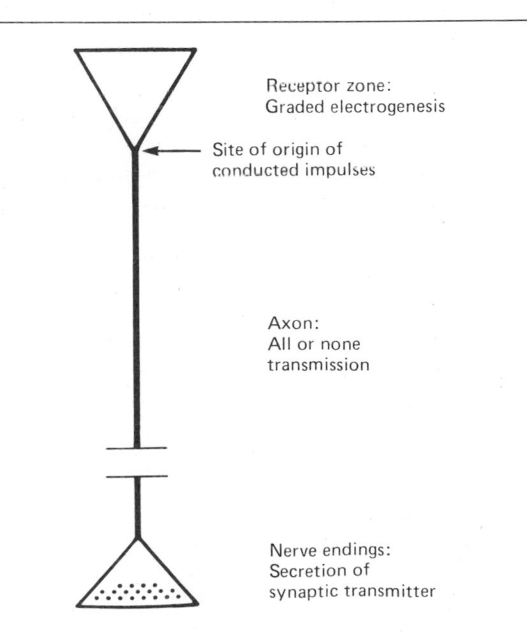

Figure 2–4. Functional organization of neurons. Nonconducted local potentials are integrated in the receptor zone, and action potentials are initiated at a site close to the receptor zone (arrow). The action potentials are conducted along the axon to the nerve endings where they cause release of synaptic transmitters.

Excitation

Nerve cells have a low threshold for excitation. The stimulus may be electrical, chemical, or mechanical. Two types of physiochemical disturbances are produced: local, nonpropagated potentials called, depending on their location, **synaptic, generator,** or **electrotonic potentials;** and propagated disturbances, the **action potentials,** or **nerve impulses.** These are the only responses of neurons and other excitable tissues, and they are the universal language of the nervous system. They are due to changes in the conduction of ions across the cell membrane that are produced by alterations in ion channels.

The impulse is normally transmitted **(conducted)** along the axon to its termination. Nerves are not "telephone wires" that transmit impulses passively; conduction of nerve impulses, although rapid, is much slower than that of electricity. Nerve tissue is in fact a relatively poor passive conductor, and it would take a potential of many volts to produce a signal of a fraction of 1 volt at the other end of a 1-meter axon in the absence of active processes in the nerve. Conduction is an active, self-propagating process, and the impulse moves along the nerve at a constant amplitude and velocity. The process is often compared to what happens when a match is applied to one end of a train of gunpowder; by igniting the powder particles immediately in front of it, the flame moves steadily down the train to its end.

ELECTRICAL PHENOMENA IN NERVE CELLS

For over 100 years, it has been known that there are electrical potential changes in a nerve when it conducts impulses, but it was not until suitable equipment was developed that these electrical events could be measured and studied in detail. Special instruments are necessary because the events are rapid, being measured in **milliseconds (ms);** and the potential changes are small, being measured in **millivolts (mV).** In addition to development of microelectrodes with a tip diameter of less than 1 μm, the principal advances that made detailed study of the electrical activity in nerves possible were the development of electronic amplifiers and the cathode-ray oscilloscope. Modern amplifiers magnify potential changes 1000 times or more, and the cathode-ray oscilloscope provides an almost inertialess and almost instantaneously responding "lever" for recording electrical events.

The Cathode-Ray Oscilloscope

The cathode-ray oscilloscope (CRO) is used to measure the electrical events in living tissue. A cathode emits electrons when a high voltage is applied across it and a suitable anode in a vacuum. In the CRO, the electrons are directed into a focused beam that strikes the face of the glass tube in which the cathode is located (Fig 2–5). The face is coated with one of a number of substances (phosphors) that emit light when struck by electrons. A vertical metal plate is placed on either side of the electron beam. When a voltage is applied across these plates, the negatively charged electrons are drawn toward the positively charged plate and repelled by the negatively charged plate. If the voltage applied to the vertical plates (X plates) is increased slowly and then reduced suddenly and increased again, the beam moves steadily toward the positive plate, snaps back to its former position, and moves toward the positive plate again. Application of a "saw-tooth voltage" of this type thus causes the beam to sweep across the face of the tube, and the speed of the sweep is proportionate to the rate of rise of the applied voltage.

Another set of plates (Y plates) is arranged horizontally, with one plate above and one below the beam. Voltages applied to these plates deflect the beam up and down as it sweeps across the face of the tube, and the magnitude of the vertical deflection is proportionate to the potential difference between the horizontal plates. When these plates are connected to electrodes on a nerve, any changes in potential occurring in the nerve are recorded as vertical deflections of the beam as it moves across the tube.

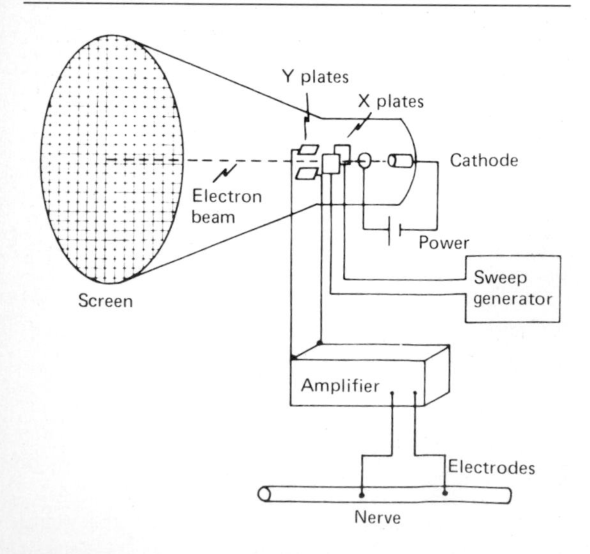

Figure 2–5. Cathode-ray oscilloscope. Simplified diagram of principal connections when arranged to record potential changes in nerve.

Recording From Single Axons

With the CRO, the electrical events occurring in a piece of peripheral nerve dissected from a laboratory animal can be demonstrated. However, such preparations contain many axons, and it is important to be able to study the properties of a single axon. Mammalian axons are relatively small (20 μm or less in diameter) and are difficult to separate from other axons, but giant unmyelinated nerve cells exist in a number of invertebrate species. Such giant cells are found, for example, in crabs (*Carcinus*) and cuttlefish (*Sepia*), but the largest known axons are found in the squid (*Loligo*). The neck region of the muscular mantle of the squid contains single axons up to 1 mm in diameter. The fundamental properties of these long axons are similar to those of mammalian axons.

Resting Membrane Potential

When 2 electrodes are connected through a suitable amplifier to a CRO and placed on the surface of a single axon, no potential difference is observed. However, if one electrode is inserted into the interior of the cell, a constant potential difference is observed, with the inside negative relative to the outside of the cell at rest. This **resting membrane potential** is found in almost all cells. Its genesis is discussed in Chapter 1. In neurons, it is usually about −70 mV.

Latent Period

If the axon is stimulated and a conducted impulse occurs, a characteristic series of potential changes known as the **action potential** is observed as the impulse passes the exterior electrode.

When the stimulus is applied, there is a brief irregular deflection of the baseline, the **stimulus artifact.** This artifact is due to current leakage from the stimulating electrodes to the recording electrodes. It usually occurs despite careful shielding, but it is of value because it marks on the cathode-ray screen the point at which the stimulus was applied.

The stimulus artifact is followed by an isopotential interval **(latent period)** that ends with the start of the action potential and corresponds to the time it takes the impulse to travel along the axon from the site of stimulation to the recording electrodes. Its duration is proportionate to the distance between the stimulating and recording electrodes and the speed of conduction of the axon. If the duration of the latent period and the distance between the electrodes are known, the speed of conduction in the axon can be calculated. For example, assume that the distance between the cathodal stimulating electrode and the exterior electrode in Fig 2–6 is 4 cm. The cathode is normally the stimulating electrode (see below). If the latent period is 2 ms long, the speed of conduction is 4 cm/2 ms, or 20 m/s.

Action Potential

The first manifestation of the approaching action potential is a beginning depolarization of the membrane. After an initial 15 mV of depolarization, the rate of depolarization increases. The point at which this change in rate occurs is called the **firing level.** Thereafter, the tracing on the oscilloscope rapidly reaches and **overshoots** the isopotential (zero potential) line to approximately +35 mV. It then reverses and falls rapidly toward the resting level. When repolarization is about 70% completed, the rate of repolarization decreases and the tracing approaches the resting level more slowly. The sharp rise and rapid fall are the **spike potential** of the axon, and the slower fall at the end of the process is the **after-depolarization.** After reaching the previous resting level, the tracing overshoots slightly in the hyperpolarizing direction to form the small but prolonged **after-hyperpolarization.** When recorded with one electrode in the cell, the action potential is called **monophasic,** because it is primarily in one direction.

The proportions of the tracing in Fig 2–6 are intentionally distorted to illustrate the various components of the action potential. A tracing with the components plotted on exact temporal and magnitude scales for a mammalian neuron is shown in Fig 2–7. Note that the rise of the action is so rapid that it fails to show clearly the change in depolarization rate at the firing level, and also that the after-hyperpolarization is only about 1–2 mV in amplitude although it lasts about 40 ms. The duration of the after–depolarization is about 4 ms in this instance. It is shorter and less prominent in many other neurons. Changes may occur in the after–polarizations without changes in the rest of the action potential. For example, if the nerve has been conducting repetitively for a long time, the after–hyperpolarization is usually quite large.

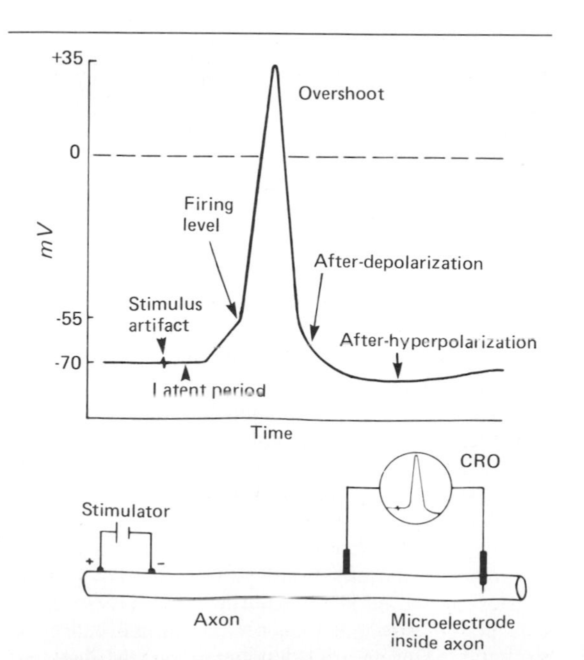

Figure 2–6. Action potential in a neuron recorded with one electrode inside the cell.

"All or None" Law

If an axon is arranged for recording as shown in Fig 2–6, with the recording electrodes at an appreciable distance from the stimulating electrodes, it is possible to determine the minimal intensity of stimulating current **(threshold intensity)** that will just produce an impulse. This threshold varies with the experimental conditions and the type of axon, but once it is reached, a full-fledged action potential is produced. Further increases in the intensity of a stimulus produce no increment or other change in the action potential as long as the other experimental conditions remain constant. The action potential fails to occur if the stimulus is subthreshold in magnitude, and it occurs with a constant amplitude and form regardless of the strength of the stimulus if the stimulus is at or above threshold intensity. The action potential is therefore "all or none" in character and is said to obey the **all or none law.**

Strength-Duration Curve

In a preparation such as that in Fig 2–6, stimuli of extremely short duration will not excite the axon no matter how intense they may be. With stimuli of

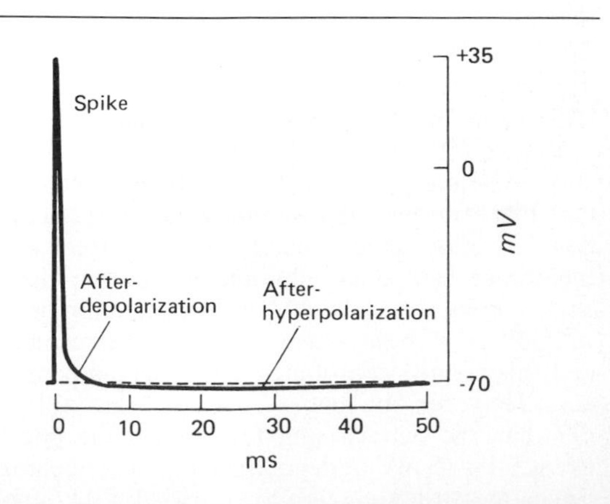

Figure 2–7. Diagram of complete action potential of large mammalian myelinated fiber, drawn without time or voltage distortion to show proportions of components.

longer duration, threshold intensity is related to the duration of stimulus. With weak stimuli, a point is reached where no response occurs no matter how long the stimulus is applied. The relationship applies only to currents that rise to peak intensity rapidly. Slowly rising currents fail to fire the nerve because the nerve adapts to the applied stimulus, a process called **accommodation.**

Electrotonic Potentials, Local Response, & Firing Level

Although subthreshold stimuli do not produce an action potential, they do have an effect on the membrane potential. This can be demonstrated by placing recording electrodes within a few millimeters of a stimulating electrode and applying subthreshold stimuli of fixed duration. Application of such currents with a cathode leads to a localized depolarizing potential change that rises sharply and decays exponentially with time. The magnitude of this response drops off rapidly as the distance between the stimulating and recording electrodes is increased. Conversely, an anodal current produces a hyperpolarizing potential change of similar duration. These potential changes are called **electrotonic potentials,** those produced at a cathode being **catelectrotonic** and those at an anode **anelectrotonic.** They are passive changes in membrane polarization caused by addition or subtraction of charge by the particular electrode. At low current intensities producing up to about 7mV of depolarization or hyperpolarization, their size is proportionate to the magnitude of the stimulus. With stronger stimuli, this relationship remains constant for anelectrotonic responses but not for responses at the cathode. The cathodal responses are greater than would be expected from the magnitude of the applied current. Finally, when the cathodal stimulation is great enough to produce about 15 mV of depolarization, ie, at a membrane potential of −55 mV, the membrane potential suddenly begins to fall rapidly, and propagated action potential occurs. The disproportionately greater response at the cathode to stimuli of sufficient strength to produce 7–15 mV of depolarization indicates active participation by the membrane in the process and is called the **local response** (Fig 2–8). The point at which a runaway spike potential is initiated is the **firing level.** Thus, cathodal currents that produce up to 7 mV of depolarization have a purely passive effect on the membrane caused by addition of negative charges. Those producing 7–15 mV of depolarization produce in addition a slight active change in the membrane, and this change contributes to the depolarizing process. However, the repolarizing forces are still stronger than the depolarizing forces, and the potential decays. At 15 mV of depolarization, the depolarizing forces are strong enough to overwhelm the repolarizing processes, and an action potential results.

Stimulation normally occurs at the cathode, because cathodal stimuli are depolarizing. Anodal currents, by taking the membrane potential farther away from the firing level, inhibit impulse formation. However, cessation of an anodal current may lead to an overshoot of the membrane potential in the depolarizing direction. This rebound is sometimes large enough to cause the nerve to fire at the end of an anodal stimulus.

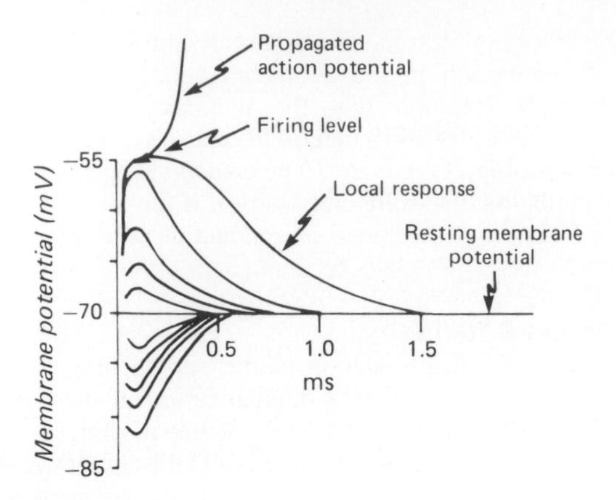

Figure 2–8. Electrotonic potentials and local response. The changes in the membrane potential of a neuron following application of stimuli of 0.2, 0.4, 0.6, 0.8, and 1.0 times threshold intensity are shown superimposed on the same time scale. The responses below the horizontal line are those recorded near the anode, and the responses above the line are those recorded near the cathode. The stimulus of threshold intensity was repeated twice. Once it caused a propagated action potential (top line), and once it did not.

Changes in Excitability During Electrotonic Potentials & the Action Potential

During the action potential as well as during catelectrotonic and anelectrotonic potentials and the local response, there are changes in the threshold of the neuron to stimulation. Hyperpolarizing anelectrotonic responses elevate the threshold and catelectrotonic potentials lower it as they move the membrane potential closer to the firing level. During the local response, the threshold is also lowered, but during the rising and much of the falling phases of the spike potential, the neuron is refractory to stimulation. This **refractory period** is divided into an **absolute refractory period,** corresponding to the period from the time the firing level is reached until repolarization is about one-third complete, and a **relative refractory period,** lasting from this point to the start of after-depolarization. During the absolute refractory period, no stimulus, no matter how strong, will excite the nerve, but during the relative refractory period, stronger than normal stimuli can cause excitation. During after-depolarization, the threshold is again decreased, and during after-hyperpolarization, it is increased. These changes in threshold are correlated with the phases of the action potential in Fig 2–9.

Electrogenesis of the Action Potential

The nerve cell membrane is polarized at rest, with positive charges lined up along the outside of the membrane and negative charges along the inside. During the action potential, this polarity is abolished and for a brief period is actually reversed (Fig 2–10). Positive charges from the membrane ahead of and behind the action potential flow into the area of negativity represented by the action potential ("current sink"). By drawing off positive charges, this flow decreases the polarity of the membrane ahead of the action potential. Such electrotonic depolarization initiates a local response, and when the firing level is reached, a propagated response occurs that in turn electrotonically depolarizes the membrane in front of it. This sequence of events moves regularly along an unmyelinated axon to its end. Thus, the self-propagating nature of the nerve impulse is due to circular current flow and successive electrotonic depolarizations to the firing level of the membrane ahead of the action potential. Once initiated, a moving impulse does not depolarize the area behind it to the firing level, because this area is refractory.

The action potentials produced at synaptic junctions and sensory receptors also depend upon electrotonic depolarization of the nerve cell membrane to the firing level. The details of the process by which the receptor zone of a neuron serves as a large current sink that draws off positive charges from the axon until the firing level is reached are described in Chapter 4.

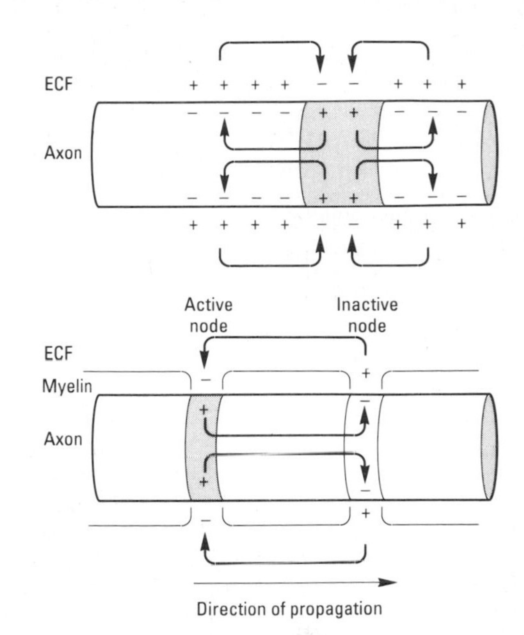

Figure 2–10. Local current flow (movement of positive charges) around an impulse in axon. *Top:* Unmyelinated axon. *Bottom:* Myelinated axon.

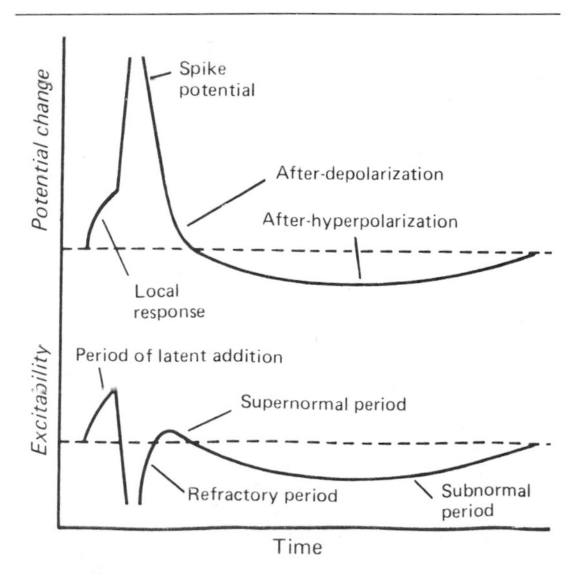

Figure 2–9. Relative changes in excitability of a nerve cell membrane during the passage of an impulse. Note that excitability is the reciprocal of threshold. (Modified and reproduced, with permission, from Morgan CT: *Physiological Psychology.* McGraw-Hill, 1943.)

Saltatory Conduction

Conduction in myelinated axons depends upon a similar pattern of circular current flow. However, myelin is an effective insulator, and current flow through it is negligible. Instead, depolarization in myelinated axons jumps from one node of Ranvier to the next, with the current sink at the active node serving to electrotonically depolarize to the firing level the node ahead of the action potential (Fig 2–10). This jumping of depolarization from node to node is called **saltatory conduction.** It is a rapid process, and myelinated axons conduct up to 50 times faster than the fastest unmyelinated fibers.

Orthodromic & Antidromic Conduction

An axon can conduct in either direction. When an action potential is initiated in the middle of it, 2 impulses traveling in opposite directions are set up by electrotonic depolarization on either side of the initial current sink.

In a living animal, impulses normally pass in one direction only, ie, from synaptic junctions or receptors along axons to their termination. Such conduction is called **orthodromic.** Conduction in the opposite direction is called **antidromic.** Since synapses, unlike axons, permit conduction in one direction only, any antidromic impulses that are set up fail to pass the first synapse they encounter (see Chapter 4) and die out at that point.

Biphasic Action Potentials

The descriptions of the resting membrane potential and action potential outlined above are based on recording with 2 electrodes, one on the surface of the axon and the other inside the axon. If both recording electrodes are placed on the surface of the axon, there is no potential difference between them at rest. When the nerve is stimulated and an impulse is conducted past the 2 electrodes, a characteristic sequence of potential changes results. As the wave of depolarization reaches the electrode nearest the stimulator, this electrode becomes negative relative to the other electrode (Fig 2–11). When the impulse passes to the portion of the nerve between the 2 electrodes, the potential returns to zero and then, as it passes the second electrode, the first electrode becomes positive relative to the second. It is conventional to connect the leads in such a way that when the first electrode becomes negative relative to the second, an upward deflection is recorded. Therefore, the record shows an upward deflection followed by an isoelectric interval and then a downward deflection. This sequence is called a **biphasic action potential** (Fig 2–11). The duration of the isoelectric interval is proportionate to the speed of conduction of the nerve and the distance between the 2 recording electrodes.

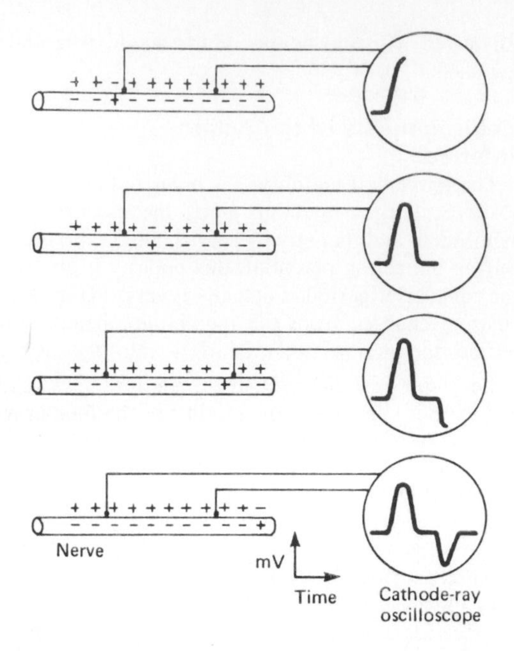

Figure 2–11. Biphasic action potential. Both recording electrodes are on the outside of the nerve membrane.

Conduction in a Volume Conductor

Because the body fluids contain large quantities of electrolytes, the nerves in the body function in a conducting medium that is often called a **volume conductor.** The monophasic and biphasic action potentials described above are those seen when an axon is stimulated in a nonconducting medium outside the body. The potential changes observed during extracellular recording in a volume conductor are basically similar to these action potentials, but they are complicated by the effects of current flow in the volume conductor. These effects are complex and are influenced by such factors as the orientation of the electrodes relative to the direction the action potential is moving and the distance between the recording electrode over active tissue and the indifferent electrode. In general, when an action potential is recorded in a volume conductor, there are positive deflections before and after the negative spike.

IONIC BASIS OF EXCITATION & CONDUCTION

The cell membranes of nerves, like those of other cells, contain many different types of ion channels (Table 2–1). Some of these are passive, ie, continu-

Table 2–1. Some of the major types of ion channels found in neurons.*

Type	Characteristics†
Na^+	Depolarization opens, then inactivates; blocked by TTX, STX.
Ca^{2+}	Depolarization opens slowly.
K^+ (delayed rectifier)	Depolarization opens slowly, then inactivates; blocked by TEA, 4AP.
K^+ (inward rectifier)	Hyperpolarization opens; blocked by Ca^{2+}, Ba^{2+}.
K^+ ("A" current)	Depolarization opens, inactivates rapidly; blocked by 4AP.
K^+ ("M")	Depolarization opens slowly; blocked by muscarinic agents.
K^+ (Ca^{2+}-activated)	Depolarization, cytoplasmic Ca^{2+} opens.
K^+ (s)	Closed when phosphorylated via serotonin and cyclic AMP in *Aplysia*.
Cation (Ca^{2+}-activated)	Cytoplasmic Ca^{2+} opens.
Cl^- (Ca^{2+}-activated)	Cytoplasmic Ca^{2+} opens.
Cl^-	Hyperpolarization opens.
H^+	Depolarization opens; pH-sensitive.
Transmitter-activated	Numerous types.

*Courtesy of MD Cahalan.
†TTX, tetrodotoxin; STX, saxitoxin; TEA, tetraethylammonium; 4AP, 4-aminopyridine.

ally open, whereas others are voltage-gated and still others are ligand-gated. It is the behavior of these channels, and particularly the Na^+ and K^+ channels, that explains the electrical events in nerves.

Ionic Basis of Resting Membrane Potential

The ionic basis of the resting membrane potential is discussed in Chapter 1. In nerves, as in other tissues, Na^+ is actively transported out of the cell and K^+ is actively transported into the cell (Fig 1–26). K^+ diffuses out of the cell down its concentration gradient through K^+ channels, and Na^+ diffuses back in, but since the permeability of the membrane to K^+ is much greater than it is to Na^+ at rest, the passive K^+ efflux is much greater than the passive Na^+ influx. Since the membrane is impermeable to most of the anions in the cell, the K^+ efflux is not accompanied by an equal flux of anions and the membrane is maintained in a polarized state, with the outside positive relative to the inside.

Ionic Fluxes During the Action Potential

The changes in membrane conductance of Na^+ and K^+ that occur during the action potentials are shown in Fig 2–12. The conductance of an ion is the reciprocal of its electrical resistance in the membrane and is a measure of the membrane permeability to that ion.

A slight decrease in resting membrane potential leads to increased K^+ efflux and Cl^- influx, restoring the resting membrane potential. However, when depolarization exceeds 7 mV, the voltage-gated Na^+ channels start to open at an increased rate **(Na^+ channel activation)**, and when the firing level is reached, the influx of Na^+ along its inwardly directed concentration and electrical gradients is so great that it temporarily swamps the repolarizing forces.

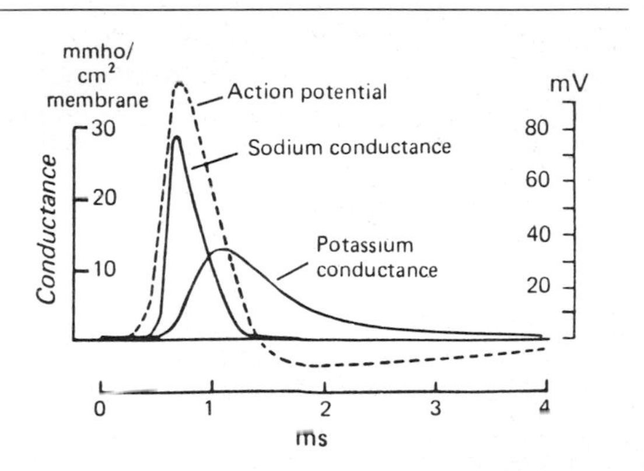

Figure 2–12. Changes in Na^+ and K^+ conductance during the action potential in giant squid axon. The dotted line represents the action potential superimposed on the same time coordinate. Note that the initial electrotonic depolarization initiates the change in Na^+ conductance, which in turn adds to the depolarization. (Redrawn and reproduced, with permission, from Hodgkin AL: Ionic movements and electrical activity in giant nerve fibers. *Proc R Soc Lond [Biol]* 1958; 148:1.)

The equilibrium potential for Na^+ in mammalian neurons, calculated by using the Nernst equation, is about +60 mV. The membrane potential moves toward this value but does not reach it during the action potential, primarily because the increase in Na^+ conductance is short-lived. The Na^+ channels rapidly enter a closed state called the **inactivated state** and remain in this state for a few milliseconds before returning to the resting state. In addition, the direction of the electrical gradient for Na^+ is reversed during the overshoot because the membrane potential is reversed, and this limits Na^+ influx. A third factor producing repolarization is the opening of voltage-gated K^+ channels. This opening is slower and more prolonged than the opening of the Na^+ channels, and consequently, much of the increase in K^+ conductance comes after the increase in Na^+ conductance. The net movement of positive charge out of the cell due to K^+ efflux at this time helps complete the process of repolarization. The slow return of the K^+ channels to the closed state also explains the after-hyperpolarization.

Decreasing the external Na^+ concentration decreases the size of the action potential but has little effect on the resting membrane potential. The lack of much effect on the resting membrane potential would be predicted from the Goldman equation (see Chapter 1), since the permeability of the membrane to Na^+ at rest is relatively low. Conversely, increasing the external K^+ concentration decreases the resting membrane potential.

Although Na^+ enters the nerve cell and K^+ leaves it during the action potential, the number of ions involved is minute relative to the total numbers present. The fact that the nerve gains Na^+ and loses K^+ during activity has been demonstrated experimentally, but significant differences in ion concentrations can be measured only after prolonged, repeated stimulation.

The slower opening and delayed closing of the voltage-gated K^+ channels also explains accommodation. If depolarization occurs rapidly, the opening of the Na^+ channels overwhelms the repolarizing forces, but if the induced depolarization is produced slowly, the opening of K^+ channels balances the gradual opening of Na^+ channels, and an action potential does not occur.

A decrease in extracellular Ca^{2+} increases the excitability of nerve and muscle cells by decreasing the amount of depolarization necessary to initiate the changes in the Na^+ and K^+ conductance that produce the action potential. Conversely, an increase in extracellular Ca^{2+} "stabilizes the membrane" by decreasing excitability. The concentration and electrical gradients for Ca^{2+} are directed inward (see Chapter 1), and Ca^{2+} enters neurons during the action potential. The early phase of Ca^{2+} entry is blocked by the Na^+ channel blocker TTX, and even though Na^+ permeability is much greater than Ca^{2+} permeability, Ca^{2+} may be entering via the Na^+ channels. An additional late phase of Ca^{2+} entry is unaffected by TTX and the K^+ channel blocker TEA and apparently occurs via a different voltage-sensitive Ca^{2+} pathway. Ca^{2+} entering during the delayed phase plays an important role in the secretion of synaptic transmitters, a Ca^{2+}-dependent process (see Chapter 4). In addition, Ca^{2+} entry contributes to depolarization, and in some instances in invertebrates it is primarily responsible for the action potential.

Distribution of Ion Channels in Myelinated Neurons

As noted in Chapter 1, various substances that bind to Na^+ and K^+ channels can be labeled and used to identify the locations of the channels in the cell membrane. Na^+ channels are highly concentrated in the nodes of Ranvier and the initial segment in myelinated neurons. The number of Na^+ channels per square micrometer of membrane in myelinated mammalian neurons has been estimated to be 50–75 in the cell body, 350–500 in the initial segment, less than 25 on the surface of the myelin, 2000–12,000 at the nodes of Ranvier, and 20–75 at the axon terminals. Along the axons of unmyelinated neurons, the number is about 110. In many myelinated neurons, the Na^+ channels are flanked by K^+ channels. However, in mammalian myelinated (as opposed to unmyelinated) neurons, most of the repolarization at the nodes results from the opening of voltage-independent rather than voltage-gated K^+ channels.

Energy Sources & Metabolism of Nerve

The major part of the energy requirement of nerve—about 70%—is the portion used to maintain polarization of the membrane by the action of Na^+-K^+ ATPase. During maximal activity, the metabolic rate of the nerve doubles; by comparison, that of skeletal muscle increases as much as 100-fold. Inhibition of lactic acid production does not influence nerve function.

Like muscle, nerve has a resting heat while inactive, an initial heat during the action potential, and a recovery heat that follows activity. However, in nerve, the recovery heat after a single impulse is about 30 times the initial heat. There is some evidence that the initial heat is produced during the after-depolarization rather than the spike. The metabolism of muscle is discussed in detail in Chapter 3.

PROPERTIES OF MIXED NERVES

Peripheral nerves in mammals are made up of many axons bound together in a fibrous envelope called the **epineurium.** Potential changes recorded from such nerves therefore represent an algebraic summation of the all or none action potentials of many axons. The thresholds of the individual axons in the nerve and their distance from the stimulating electrodes vary. With subthreshold stimuli, none of the axons are stimulated and no response occurs. When the stimuli are of threshold intensity, axons with low thresholds fire and a small potential change is observed. As the intensity of the stimulating current is increased, the axons with higher thresholds are also discharged. The electrical response increases proportionately until the stimulus is strong enough to excite all of the axons in the nerve. The stimulus that produces excitation of all the axons is the **maximal stimulus,** and further application of greater, **supramaximal** stimuli produces no further increase in the size of the observed potential.

Compound Action Potentials

Another property of mixed nerves, as opposed to single axons, is the appearance of multiple peaks in the action potential. The multipeaked action potential is called a **compound action potential.** Its shape is due to the fact that a mixed nerve is made up of families of fibers with varying speeds of conduction. Therefore, when all the fibers are stimulated, the activity in fast-conducting fibers arrives at the recording electrodes sooner than the activity in slower fibers; and the farther away from the stimulating electrodes the action potential is recorded, the greater is the separation between the fast and slow fiber peaks. The number and size of the peaks vary with the types of fibers in the particular nerve being studied. If less than maximal stimuli are used, the shape of the compound action potential also depends upon the number and type of fibers stimulated.

Erlanger and Gasser divided mammalian nerve fibers into A, B, and C groups, further subdividing the A group into α, β, γ and δ fibers. The relative latencies of the electrical activity due to each of these components are shown in Fig 2–13. It should be emphasized that the drawing is not the compound action potential of any particular peripheral nerve; none of the peripheral nerves show all the components illustrated in this composite diagram because none contain all of the fiber types.

NERVE FIBER TYPES & FUNCTION

By comparing the neurologic deficits produced by careful dorsal root section and other nerve-cutting experiments with the histologic changes in the nerves, the functions and histologic characteristics of each of the families of axons responsible for the various peaks of the compound action potential have been established. In general, the greater the diameter of a given nerve fiber, the greater its speed of conduction. The larger axons are primarily concerned with proprioceptive sensation and somatic motor function, while the smaller axons subserve pain and tempera-

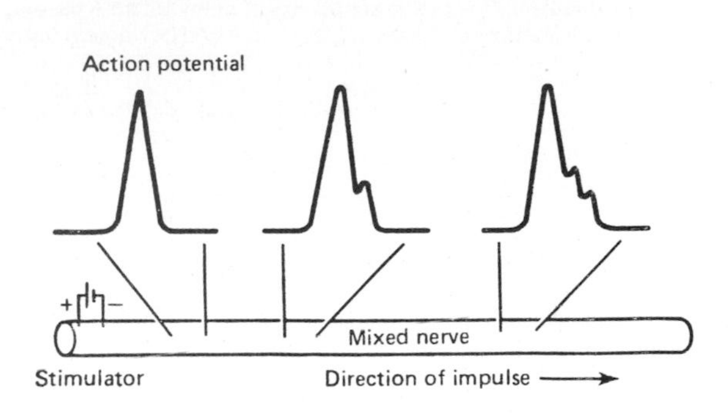

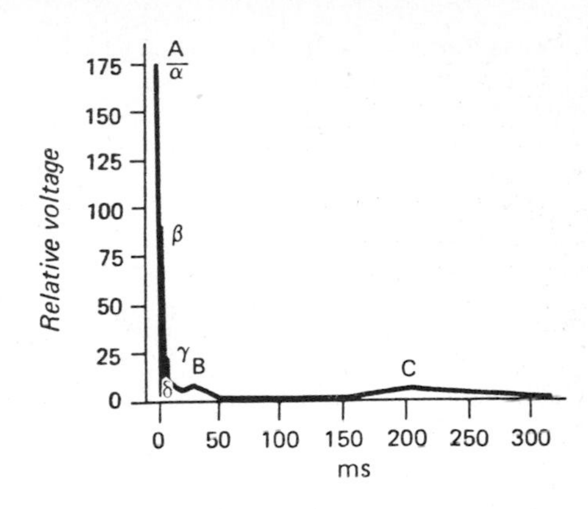

Figure 2–13. Compound action potential. ***Left:*** Record obtained with recording electrodes at varous distances from the stimulating electrodes along a mixed nerve. ***Right:*** Reconstruction of a compound action potential to show relative sizes and time relationships of the components. (Redrawn and reproduced, with permission, from Erlanger J, Gasser HS: *Electrical Signs of Nervous Activity.* Univ of Pennsylvania Press, 1937.)

ture sensations and autonomic function. In Table 2–2, the various fiber types are listed with their diameters, electrical characteristics, and functions. There is evidence that the dorsal root C fibers conduct impulses generated by touch and other cutaneous receptors in addition to pain and temperature receptors, but only pain and temperature are relayed to consciousness. The other fibers presumably are concerned with reflex responses integrated in the spinal cord and brain stem.

Further research has shown that not all the classically described lettered components are homogeneous, and a numerical system (Ia, Ib, II, III, IV) has been used by some physiologists to classify sensory fibers. Unfortunately, this has led to some confusion. A comparison of the number system and the letter system is shown in Table 2–3.

In addition to variations in speed of conduction and fiber diameter, the various classes of fibers in peripheral nerves differ in their sensitivity to hypoxia and anesthetics (Table 2–4). This fact has clinical as well as physiologic significance. Local anesthetics depress transmission in the group C fibers before they affect the touch fibers in the A group. Conversely, pressure on a nerve can cause loss of conduction in motor, touch, and pressure fibers while pain sensation remains relatively intact. Patterns of this type are sometimes seen in individuals who sleep with their arms under their heads for long periods, causing compression of the nerves in the arms. Because of the association of deep sleep with alcoholic intoxication, the syndrome is commonest on weekends and has acquired the interesting name Saturday night or Sunday morning paralysis.

Human peripheral nerve fibers are also classified on a physio-anatomic basis. This classification (Table 2–5) divides nerves into afferent and efferent categories and further subdivides them according to whether they have somatic or visceral and general or special functions. The term special is applied to nerves that supply the organs of the special senses and the musculature of branchiomeric origin, ie, musculature arising from the branchial arches during embryonic development.

Table 2–2. Nerve fiber types in mammalian nerve.

Fiber Type	Function	Fiber Diameter (μm)	Conduction Velocity (m/s)	Spike Duration (ms)	Absolute Refractory Period (ms)
A α	Proprioception; somatic motor	12–20	70–120	0.4–0.5	0.4–1
β	Touch, pressure	5–12	30–70		
γ	Motor to muscle spindles	3–6	15–30		
δ	Pain, cold, touch	2–5	12–30		
B	Preganglionic autonomic	<3	3–15	1.2	1.2
C dorsal root	Pain, temperature, some mechanoreception, reflex responses	0.4–1.2	0.5–2	2	2
sympathetic	Postganglionic sympathetics	0.3–1.3	0.7–2.3	2	2

Table 2–3. Numerical classification sometimes used for sensory neurons.

Number	Origin	Fiber Type
I a	Muscle spindle, annulospiral ending	A α
b	Golgi tendon organ	A α
II	Muscle spindle, flower-spray ending; touch, pressure	A β
III	Pain and cold receptors; some touch receptors	A δ
IV	Pain, temperature, and other receptors	Dorsal root C

Table 2–4. Relative susceptibility of mammalian A, B, and C nerve fibers to conduction block produced by various agents.

	Most Susceptible	Inter-mediate	Least Susceptible
Sensitivity to hypoxia	B	A	C
Sensitivity to pressure	A	B	C
Sensitivity to cocaine and local anes-thetics	C	B	A

NERVE GROWTH FACTOR

Nerve growth factor (NGF) is a protein that is necessary for the growth and maintenance of sympathetic neurons and some sensory neurons. It is present in a broad spectrum of animal species, including humans, and is found in many different tissues. In male mice, there is a particularly high concentration in the submaxillary salivary glands, and the level is reduced by castration to that seen in females. The factor is a dimer made up of 2 α, 2 β, and 2 γ subunits. The β subunits, each of which has a molecular weight of 13,200, have all the nerve growth-promoting activity, the α subunits have trypsinlike activity, and the γ subunits are serine proteases. The function of the proteases is unknown. Antiserum against NGF has been prepared, and injection of this antiserum in newborn animals leads to near total destruction of the sympathetic ganglia; it thus produces an **immunosympathectomy.** The structure of a β unit of NGF somewhat resembles that of insulin, and it appears to be one of a number of different hormonelike protein factors that stimulate the growth of various tissues in the body (see Chapter 22). It is picked up by neurons in the extracerebral organs they innervate and is transported in retrograde fashion from the endings of the neurons to their cell bodies. It now appears that it is also present in the brain and is responsible for the growth and maintenance of cholinergic neurons in the basal forebrain and the striatum.

Several other substances that maintain certain kinds

Table 2–5. Types of fibers in peripheral and cranial nerves.

Afferent (A)	Somatic (S)	General (G)	Sensory from cutaneous and muscle receptors
		Special (S)	Vision, hearing
	Visceral (V)	General (G)	Sensation from viscera
		Special (S)	Smell, taste
Efferent (E)	Somatic (S)	General (G)	Motor to skeletal muscle
		Special (S)	Efferent to eye, ear
	Visceral (V)	General (G)	Motor to viscera (autonomic nerves)
		Special (S)	Motor to muscles of facial field (branchiomeric musculature)

Thus, GVA = general visceral afferent; SVE = special visceral efferent; etc.

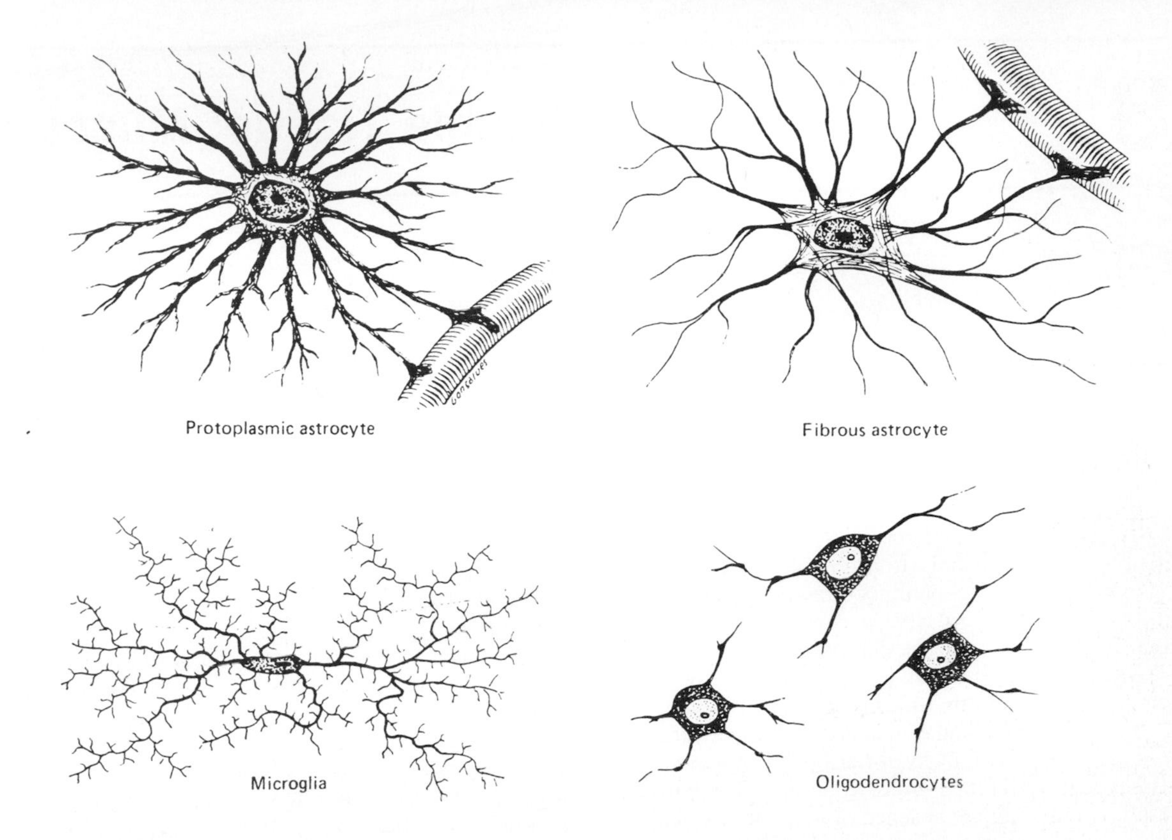

Figure 2–14. Glial cells. (Reproduced, with permission, from Junqueira LC, Carneiro J, Kelley RO: *Basic Histology,* 6th ed. Appleton & Lange, 1989.)

of nerve cells have also been isolated, including neuroleukin, a 56K protein produced by the brain and by T lymphocytes.

GLIA

In addition to neurons, the nervous system contains glial cells (neuroglia) (Fig 2–14). Glial cells are very numerous; as noted above, there are 10–50 times as many glial cells as neurons. The Schwann cells that invest axons in peripheral nerves are classified as glia. In the central nervous system, there are 3 types of glia. **Microglia** are scavenger cells that enter the nervous system from the blood vessels. **Oligodendrogliocytes** are involved in myelin formation (Fig 2–3). **Astrocytes are found** throughout the brain, and many send end-feet to blood vessels (see Chapter 32). They have a membrane potential that varies with the external K^+ concentration but do not generate propagated potentials. They appear to produce substances that are trophic to neurons, and they probaby help maintain the appropriate concentration of substances in the interstitial fluid by taking up K^+ and the neurotransmitters glutamate and gamma-aminobutyrate (GABA; see Chapter 4). They also induce capillaries to form tight junctions and are responsible for the production of the blood-brain barrier in mammals.

3 Excitable Tissue: Muscle

INTRODUCTION

Muscle cells, like neurons, can be excited chemically, electrically, and mechanically to produce an action potential that is transmitted along their cell membrane. They contain contractile proteins and, unlike neurons, they have a contractile mechanism that is activated by the action potential.

Muscle is generally divided into 3 types, **skeletal, cardiac,** and **smooth,** although smooth muscle is not a homogeneous single category. Skeletal muscle comprises the great mass of the somatic musculature. It has well-developed cross-striations, does not normally contract in the absence of nervous stimulation, lacks anatomic and functional connections between individual muscle fibers, and is generally under voluntary control. Cardiac muscle also has cross-striations, but it is functionally syncytial in character and contracts rhythmically in the absence of external innervation owing to the presence in the myocardium of pacemaker cells that discharge spontaneously. Smooth muscle lacks cross-striations. The type found in most hollow viscera is functionally syncytial in character and contains pacemakers that discharge irregularly. The type found in the eye and in some other locations is not spontaneously active and resembles skeletal muscle. There are contractile proteins similar to those in muscle in many other cells, and it appears that these proteins are responsible for cell motility, mitosis, and the movement of various components within cells (see Chapter 1).

SKELETAL MUSCLE

MORPHOLOGY

Organization

Skeletal muscle is made up of individual muscle fibers that are the "building blocks" of the muscular system in the same sense that the neurons are the building blocks of the nervous system. Most skeletal muscles begin and end in tendons, and the muscle fibers are arranged in parallel between the tendinous ends, so that the force of contraction of the units is additive. Each muscle fiber is a single cell, multinucleated, long, and cylindric. There are no syncytial bridges between cells.

The muscle fibers are made up of fibrils, as shown in Fig 3–1, and the fibrils are divisible into individual filaments. The filaments are made up of the contractile proteins.

The contractile mechanism in skeletal muscle depends on the proteins **myosin** (molecular weight 460,000), **actin** (molecular weight 43,000), **tropomyosin** (molecular weight 70,000), and **troponin.** Troponin is made up of 3 subunits, **troponin I, troponin T,** and **troponin C.** The 3 subunits have molecular weights ranging from 18,000 to 35,000. Other proteins play important roles in linking excitation to contraction.

Striations

The cross-striations characteristic of skeletal muscle are due to differences in the refractive indexes of the various parts of the muscle fiber. The parts of the cross-striations are identified by letters (Fig 3–2). The light I band is divided by the dark Z line, and the dark A band has the lighter H band in its center. A transverse M line is seen in the middle of the H band, and this line plus the narrow light areas on either side of it are sometimes called the pseudo-H zone. The area between 2 adjacent Z lines is called a **sarcomere.** The arrangement of thick and thin filaments that is responsible for the striations is diagrammed in Fig 3–3. The thick filaments, which are about twice the diameter of the thin filaments, are made up of myosin; the thin filaments are made up of actin, tropomyosin, and troponin. The thick filaments are lined up to form the A bands, whereas the array of thin filaments forms the less dense I bands. The lighter H bands in the center of the A bands are the regions where, when the muscle is relaxed, the thin filaments do not overlap the thick filaments. The Z lines transect the fibrils and connect to the thin filaments. If a transverse section through the A band is examined under the electron microscope, each thick filament is found to be surrounded by 6 thin filaments in a regular hexagonal pattern.

Myosin is a complex protein made up of 6 different polypeptides (Fig 3–3), 2 heavy chains and 4 light

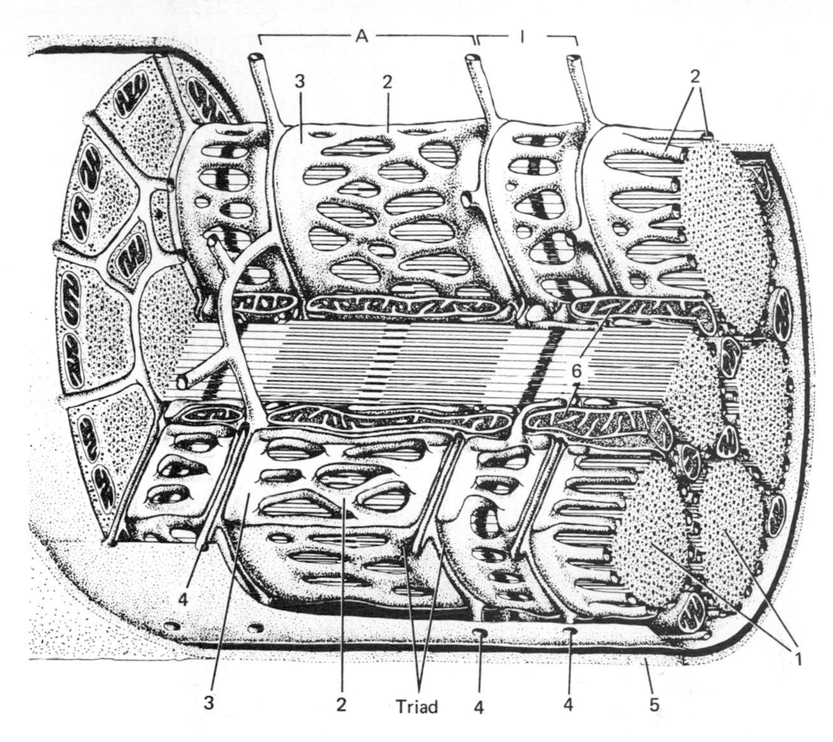

Figure 3–1. Mammalian skeletal muscle. A single muscle fiber surrounded by its sarcolemma has been cut away to show individual myofibrils. (1). The cut surface of the myofibrils shows the arrays of thick and thin filaments. The sarcoplasmic reticulum (2) with its terminal cisterns (3) surrounds each myofibril. The T system of tubules (4), which invaginates from the sarcolemma, contacts the myofibrils between the A and I bands twice in every sarcomere. The T system and the adjacent cisterns of the sarcoplasmic reticulum constitute a triad. A basal lamina (5) surrounds the sarcolemma. (6) Mitochondria. (Modified and reproduced, with permission, from Krstić RV: *Ultrastructure of the Mammalian Cell.* Springer-Verlag, 1979.)

chains. The myosin molecules are asymmetric, with the C-terminal portions forming enlarged globular heads. The heads contain an actin-binding site and a catalytic site that hydrolyzes ATP (see below). The molecules are arranged as shown in Fig 3–3, and the heads of the myosin molecules form cross-bridges to the actin molecules. The myosin molecules are arranged symmetrically on either side of the center of the sarcomere, and it is this arrangement that creates the light areas in the pseudo-H zone. The M line is due to a central bulge in each of the thick filaments. At these points, there are slender cross-connections that hold the thick filaments in proper array. There are several hundred myosin molecules in each thick segment.

The thin filaments are made up of 2 chains of globular units that form a long double helix. Tropomyosin molecules are long filaments located in the groove between the 2 chains in the actin (Fig 3–3). Each thin filament contains 300–400 actin molecules and 40–60 tropomyosin molecules. Troponin molecules are small globular units located at intervals along the tropomyosin molecules. Troponin T binds the other troponin components to tropomyosin, troponin I inhibits the interaction of myosin with actin (see below), and troponin C contains the binding sites for the Ca^{2+} that initiates contraction.

Sarcotubular System

The muscle fibrils are surrounded by structures made up of membrane that appear in electron photomicrographs as vesicles and tubules. These structures form the **sarcotubular system,** which is made up of a **T system** and a **sarcoplasmic reticulum.** The T system of transverse tubules, which is continuous with the membrane of the muscle fiber, forms a grid perforated by the individual muscle fibrils (Fig 3–1). The space between the 2 layers of the T system is an extension of the extracellular space. The sarcoplasmic reticulum, which forms an irregular curtain around each of the fibrils, has enlarged **terminal cisterns** in close contact with the T system at the junctions between the A and I bands. At these points of contact, the arrangement of the central T system with a cistern of the sarcoplasmic reticulum on either side has led to the use of the term **triads** to describe the system. The function of the T system is the rapid transmission of the action potential from the cell membrane to all the fibrils in the muscle. The sarcoplasmic reticulum is concerned with Ca^{2+} movement and muscle metabolism (see below).

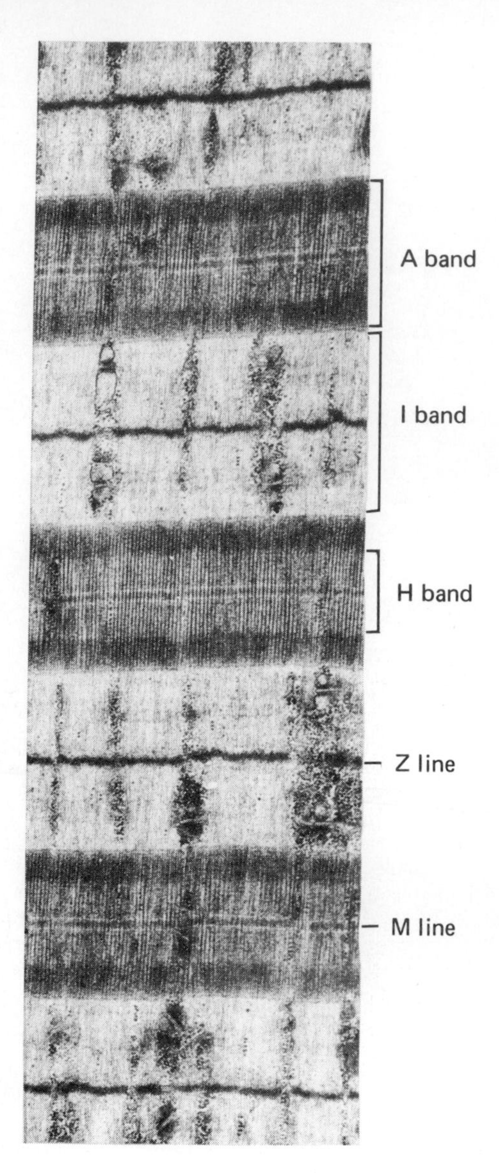

Figure 3–2. Electron micrograph of human gastrocnemius muscle. The various bands and lines are identified on the right. (x13,500.) (Courtesy of SM Walker and GR Schrodt.)

ELECTRICAL PHENOMENA & IONIC FLUXES

Electrical Characteristics of Skeletal Muscle

The electrical events in skeletal muscle and the ionic fluxes underlying them are similar to those in nerve, although there are quantitative differences in timing and magnitude. The resting membrane potential of skeletal muscle is about –90mV. The action potential lasts 2–4 ms and is conducted along the muscle fiber at about 5 m/s. The absolute refractory period is 1–3 ms long, and the after-polarizations, with their related changes in threshold to electrical stimulation, are relatively prolonged. The initiation of impulses at the myoneural junction is discussed in Chapter 4.

Although the electrical properties of the individual fibers in a muscle do not differ sufficiently to produce anything resembling a compound action potential, there are slight differences in the thresholds of the various fibers. Furthermore, in any stimulation experiment, some fibers are farther from the stimulating electrodes than others. Therefore, the size of the action potential recorded from a whole-muscle preparation is proportionate to the intensity of the stimulating current between threshold and maximal current intensities.

Ion Distribution & Fluxes

The distribution of ions across the muscle fiber membrane is similar to that across the nerve cell membrane. The values for the various ions and their equilibrium potentials are shown in Table 3–1. As in nerves, depolarization is a manifestation of Na^+ influx, and repolarization is a manifestation of K^+ efflux (as described in Chapter 2 for nerves).

CONTRACTILE RESPONSES

It is important to distinguish between the electrical and mechanical events in muscle. Although one response does not normally occur without the other, their physiologic basis and characteristics are different. Muscle fiber membrane depolarization normally starts at the motor end-plate, the specialized structure under the motor nerve ending (see Chapter 4); the action potential is transmitted along the muscle fiber and initiates the contractile response.

The Muscle Twitch

A single action potential causes a brief contraction followed by relaxation. This response is called a

Table 3–1. Steady-state distribution of ions in the intracellular and extracellular compartments of mammalian skeletal muscle, and the equilibrium potentials for these ions.* A^- represents organic anions. The value for intracellular Cl^- is calculated from the membrane potential, using the Nernst equation.

Ion	Concentration, mmol/L		Equilibrium Potential (mV)
	Intracellular Fluid	Extracellular Fluid	
Na^+	12	145	+65
K^+	155	4	−95
H^+	13×10^{-5}	3.8×10^{-5}	−32
Cl^-	3.8	120	−90
HCO_3^-	8	27	−32
A^-	155	0	. . .

Membrane potential = −90 mV

* Data from Ruch TC, Patton HD (editors): *Physiology and Biophysics*, 19th ed. Saunders, 1965.

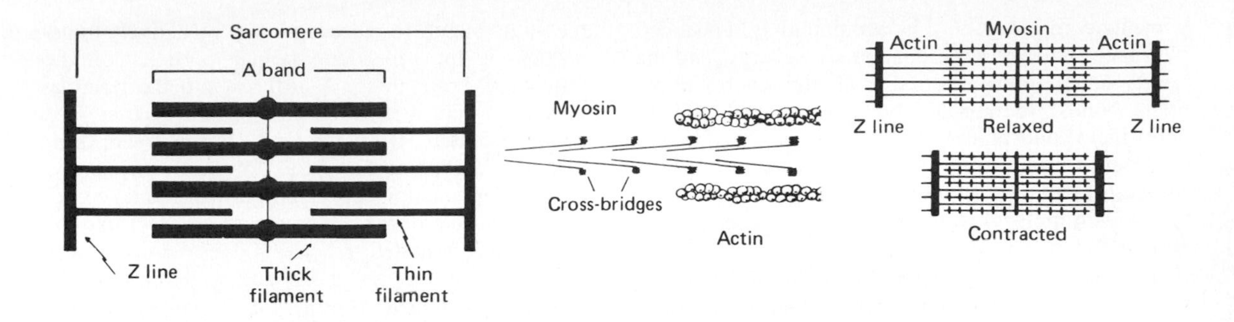

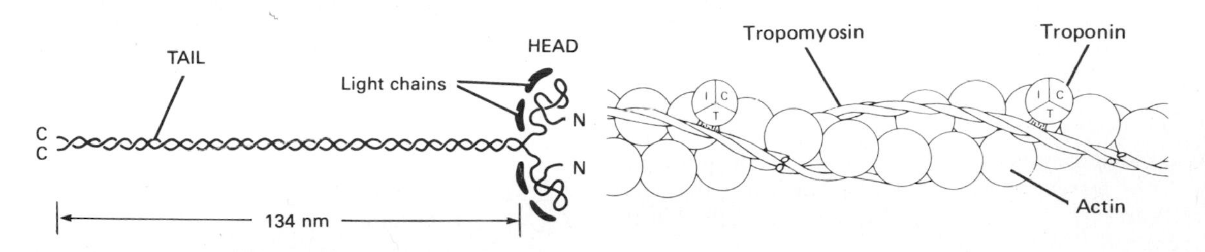

Figure 3–3. *Top left:* Arrangement of thin (actin) and thick (myosin) filaments in skeletal muscle. *Top center:* Detail of structure of myosin and actin. *Top right:* Sliding of actin on myosin during contraction so that Z lines move closer together. *Bottom left:* Myosin molecule, made up of 2 intertwined heavy chains and 4 light chains forming part of the heads of the molecule (Modified from Sheetz MP: Muscle-bound bacteria and weak worms. *Nature* 1988;**331:**212.) *Bottom right:* Diagrammatic representation of the arrangement of actin, tropomyosin, and the 3 subunits of troponin (I, C, and T: see text).

muscle twitch. In Fig 3–4, the action potential and the twitch are plotted on the same time scale. The twitch starts about 2 ms after the start of depolarization of the membrane, before repolarization is complete. The duration of the twitch varies with the type of muscle being tested (see below). ''Fast'' muscle fibers, primarily those concerned with fine, rapid, precise movement, have twitch durations as short as 7.5 ms. ''Slow'' muscle fibers, principally those involved in strong, gross, sustained movements, have twitch durations up to 100 ms.

Figure 3–4. The electrical and mechanical responses of a mammalian skeletal muscle fiber to a single maximal stimulus. The electrical response (mV potential change) and the mechanical response (T, tension in arbitrary units) are plotted on the same abscissa (time).

Molecular Basis of Contraction

The process by which the shortening of the contractile elements in muscle is brought about is a sliding of the thin filaments over the thick filaments. The width of the A bands is constant, whereas the Z lines move closer together when the muscle contracts and farther apart when it is stretched (Fig 3–3). As the muscle shortens, the thin filaments from the opposite ends of the sarcomere approach each other; when the shortening is marked, these filaments overlap.

The sliding during muscle contraction is produced by breaking and re-forming of the cross-linkages between actin and myosin. The heads of the myosin molecules link to actin at a 90-degree angle, produce movement of myosin on actin by swiveling, and then disconnect and reconnect at the next linking site, repeating the process in serial fashion (Fig 3–5). Each single cycle of attaching, swiveling, and detaching shortens the muscle 1%. Each thick filament has about 500 myosin heads, and each of these cycles about 5 times per second during a rapid contraction.

The immediate source of energy for muscle contraction is ATP. Hydrolysis of the bonds between the phosphate residues of this compound is associated with the release of a large amount of energy, and the bonds are therefore referred to as high-energy phosphate bonds. In muscle, the hydrolysis of ATP to adnosine diphosphate (ADP) is catalyzed by the contractile protein myosin; this **adenosine triphosphatase (ATPase)** activity is found in the heads of the myosin molecules, where they are in contact with actin.

The process by which depolarization of the muscle fiber initiates contraction is called **excitation-contraction coupling.** The action potential is transmitted to all the fibrils in the fiber via the T system. It triggers the release of Ca^{2+} from the terminal cisterns, the lateral sacs of the sarcoplasmic reticulum next to the T system. The Ca^{2+} initiates contraction. The link between the T system and the cisterns involves a number of proteins, including dihydropyridine receptors in the T system. **Dystrophin,** the large protein that is absent in Duchenne muscular dystrophy, is also associated with the triads, but its exact function is still unknown.

Ca^{2+} initiates contraction by binding to troponin C. In resting muscle, troponin I is tightly bound to actin, and tropomyosin covers the sites where myosin heads bind to actin. Thus, the troponin-tropomyosin complex constitutes a "relaxing protein" that inhibits the interaction between actin and myosin. When the Ca^{2+} released by the action potential binds to troponin C, the binding of troponin I to actin is presumably weakened, and this permits the tropomyosin to move laterally (Fig 3–5). This movement uncovers binding sites for the myosin heads, so that ATP is split and contraction occurs. Seven myosin binding sites are uncovered for each molecule of troponin that binds a calcium ion.

Shortly after releasing Ca^{2+}, the sarcoplasmic reticulum begins to reaccumulate it by actively transporting it into the longitudinal portions of the reticulum. From there, it diffuses into the terminal cisterns, where it is stored until released by the next action potential. Once the Ca^{2+} concentration outside the reticulum has been lowered sufficiently, chemical interaction between myosin and actin ceases and the muscle relaxes. If transport of Ca^{2+} into the reticulum is inhibited, relaxation does not occur even though there are no more action potentials; the resulting sustained contraction is called a **contracture.** ATP provides the energy for the active transport of Ca^{2+} into the sarcoplasmic reticulum. Thus, both contraction and relaxation of muscle require ATP.

The events involved in muscle contraction and relaxation are summarized in Table 3–2.

Table 3–2. Sequence of events in contraction and relaxation of skeletal muscle. (Steps 1–6 in contraction are discussed in Chapter 4.)

Steps in contraction
(1) Discharge of motor neuron.
(2) Release of transmitter (acetylcholine) at motor end-plate.
(3) Binding of acetylcholine to nicotinic acetylcholine receptors.
(4) Increased Na^+ and K^+ conductance in end-plate membrane.
(5) Generation of end-plate potential.
(6) Generation of action potential in muscle fibers.
(7) Inward spread of depolarization along T tubules.
(8) Release of Ca^{2+} from terminal cisterns of sarcoplasmic reticulum and diffusion to thick and thin filaments.
(9) Binding of Ca^{2+} to troponin C, uncovering myosin binding sites on actin.
(10) Formation of cross-linkages between actin and myosin and sliding of thin on thick filaments, producing shortening.

Steps in relaxation
(1) Ca^{2+} pumped back into sarcoplasmic reticulum.
(2) Release of Ca^{2+} from troponin.
(3) Cessation of interaction between actin and myosin.

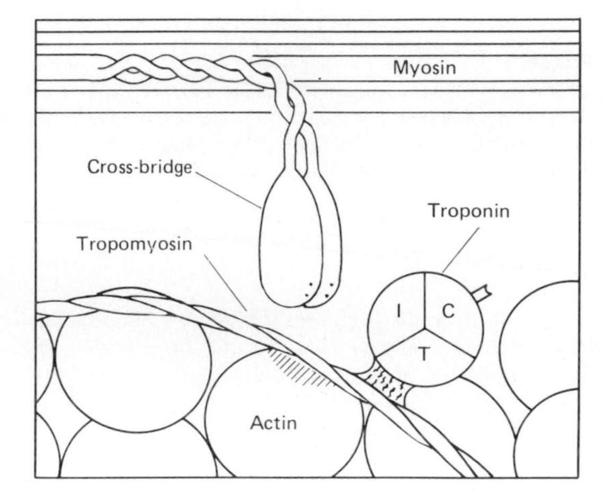

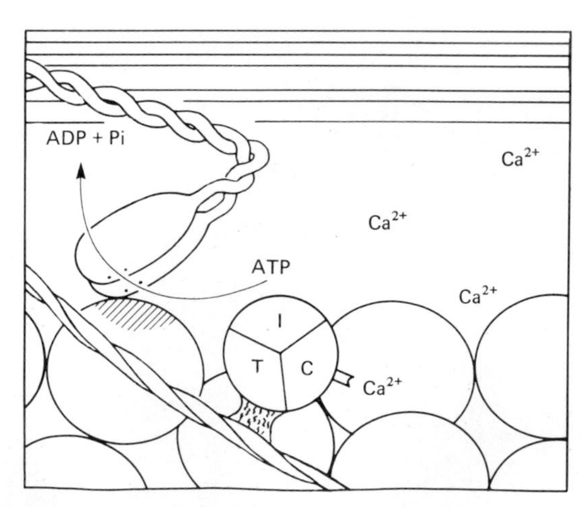

Figure 3–5. Initiation of muscle contraction by Ca^{2+}. The cross-bridges (heads of myosin molecules) attach to binding sites on actin (striped areas) and swivel when tropomyosin is displaced laterally by binding of Ca^{2+} to troponin C. There is now some debate about whether the whole head or only the neck swivels.

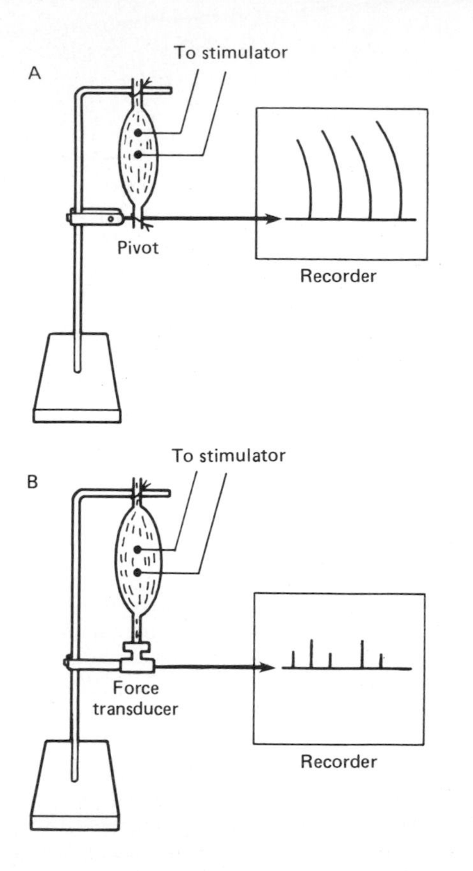

Figure 3–6. *A:* Muscle preparation arranged for recording isotonic contractions. *B:* Preparation arranged for recording isometric contractions. In A, the muscle is fastened to a writing lever that swings on a pivot. In B, it is attached to an electronic transducer that measures the force generated without permitting the muscle to shorten.

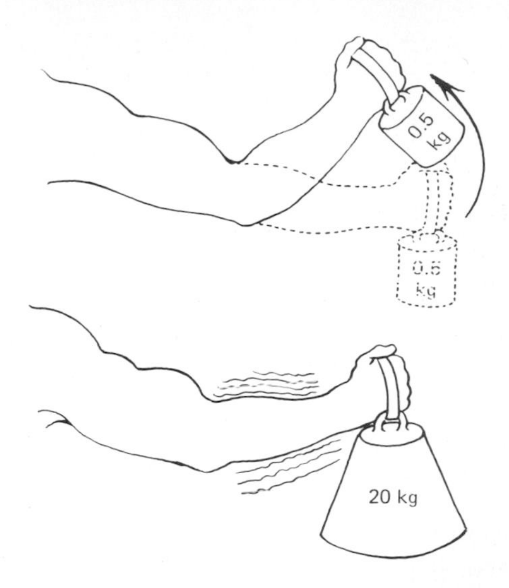

Figure 3–7. *Top:* Isotonic (free) contraction. Biceps shortens freely, weight is lifted. *Bottom:* Isometric contraction. Biceps generates force but cannot shorten and raise weight.

Types of Contraction

Muscular contraction involves shortening of the contractile elements, but because muscles have elastic and viscous elements in series with the contractile mechanism, it is possible for contraction to occur without an appreciable decrease in the length of the whole muscle. Such a contraction is called **isometric** (''same measure'' or length). Contraction against a constant load, with approximation of the ends of the muscle, is **isotonic** (''same tension'').

A whole-muscle preparation arranged for recording isotonic contractions is shown in Fig 3–6A. The muscle lifts the lever, and the distance the lever moves indicates the degree of shortening. In this situation, the muscle does external work, since the lever is being moved a certain distance. The muscle preparation in Fig 3–6B is arranged for recording isometric contractions. It is attached to an electronic force transducer, and the force that is generated is recorded. Since the product of force times distance in this situation is essentially zero, no external work is done by the muscle. In other situations, it is possible for muscles to do negative work while contracting (Fig 3–7). This happens, for example, when a heavy weight is lowered onto a table. In this case, the biceps muscle actively resists the descent of the object, but the net effect of the effort is to lengthen the biceps muscle while it is contracting.

Summation of Contractions

The electrical response of a muscle fiber to repeated stimulation is like that of nerve. The fiber is electrically refractory only during the rising and part of the falling phase of the spike potential. At this time, the contraction initiated by the first stimulus is just beginning. However, because the contractile mechanism does not have a refractory period, repeated stimulation before relaxation has occurred produces additional activation of the contractile elements and a response that is added to the contraction already present. This phenomenon is known as **summation of contractions.** The tension developed during summation is considerably greater than that during the single muscle twitch. With rapidly repeated stimulation, activation of the contractile mechanism occurs repeatedly before any relaxation has occurred, and the individual responses fuse into one continuous contraction. Such a response is called a **tetanus (tetanic contraction).** It is a **complete tetanus** when there is no relaxation between stimuli and an **incomplete tetanus** when there are periods of incomplete relaxation between the summated stimuli. During a complete tetanus, the tension developed is about 4 times that developed by the individual twitch contractions. The development of an incomplete and a complete tetanus in response to stimuli of increasing frequency is shown in Fig 3–8.

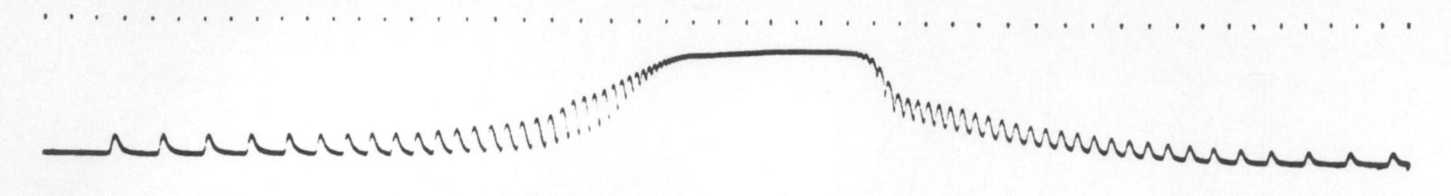

Figure 3–8. Tetanus. Isometric tension of a single muscle fiber during continuously increasing and decreasing stimulation frequency. Dots at top are at intervals of 0.2 seconds. (Reproduced, with permission, from Buchthal. F: *Dan Biol Med* 1942;17:1.)

The stimulation frequency at which summation of contractions occurs is determined by the twitch duration of the particular muscle being studied. For example, if the twitch duration is 10 ms, frequencies less than 1/10 ms (100/s) cause discrete responses interrupted by complete relaxation, and frequencies greater than 100/s cause summation.

Treppe

When a series of maximal stimuli is delivered to skeletal muscle at a frequency just below the tetanizing frequency, there is an increase in the tension developed during each twitch until, after several contractions, a uniform tension per contraction is reached. This phenomenon is known as **treppe,** or the "staircase" phenomenon (German *Treppe,* staircase). It also occurs in cardiac muscle. Treppe is believed to be due to increased availability of Ca^{2+} for binding to troponin C. It should not be confused with summation of contractions and tetanus.

Relation Between Muscle Length, Tension, & Velocity of Contraction

Both the tension that a muscle develops when stimulated to contract isometrically (the **total tension**) and the **passive tension** exerted by the unstimulated muscle vary with the length of the muscle fiber. This relationship can be studied in a whole skeletal muscle preparation such as that shown in Fig 3–6B. The length of the muscle can be varied by changing the distance between its 2 attachments. At each length, the passive tension is measured, the muscle is then stimulated electrically, and the total tension is measured. The difference between the 2 values at any length is the amount of tension actually generated by the contractile process, the **active tension.** The records obtained by plotting passive tension and total tension against muscle length are shown in Fig 3–9. Similar curves are obtained when single muscle fibers are studied. Passive tension rises slowly at first and then rapidly as the muscle is stretched. Rupture of the muscle occurs when it is stretched to about 3 times its **equilibrium length,** ie, the length of the relaxed muscle cut free from its bony attachments.

The total tension curve rises to a maximum and then declines until it reaches the passive tension curve, ie, until no additional tension is developed upon further stimulation. The length of the muscle at which the active tension is maximal is usually called its **resting length.** The term comes originally from experiments demonstrating that the length of many of the muscles in the body at rest is the length at which they develop maximal tension. Other definitions of resting length are sometimes used, but that given here is the most widely accepted and has the greatest physiologic validity.

The observed length-tension relation in skeletal muscle is explained by the sliding filament mechanism of muscle contraction. When the muscle fiber contracts isometrically, the tension developed is proportionate to the number of cross-linkages between the actin and the myosin molecules. When muscle is stretched, the overlap between actin and myosin is reduced and the number of cross-linkages is reduced. Conversely, when the muscle is appreciably shorter than resting length, the thin filaments overlap, and this also reduces the cross-linkages.

The velocity of muscle contraction varies inversely with the load on the muscle. At a given load, the velocity is maximal at the resting length and declines if the muscle is shorter or longer than this length.

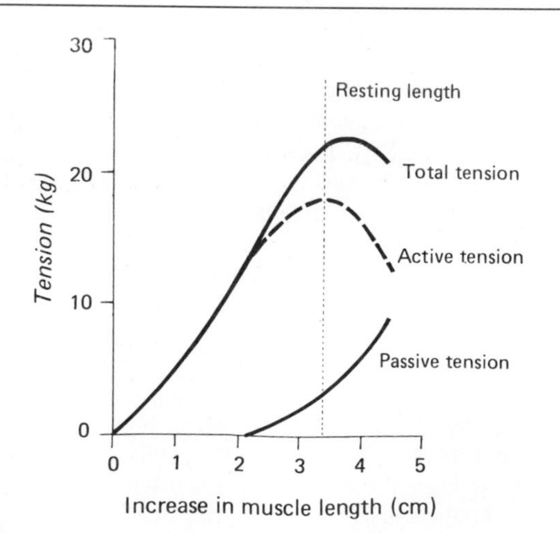

Figure 3–9. Length-tension for human triceps muscle. The passive tension curve measures the tension exerted by this skeletal muscle at each length when it is not stimulated. The total tension curve represents the tension developed when the muscle contracts isometrically in response to a maximal stimulus. The active tension is the difference between the two.

ENERGY SOURCES & METABOLISM

Muscle contraction requires energy, and muscle has been called "a machine for converting chemical into mechanical energy." The immediate source of this energy is the energy-rich organic phosphate derivatives in muscle; the ultimate source is the intermediary metabolism of carbohydrate and lipids. The hydrolysis of ATP to provide the energy for contraction has been discussed above.

Phosphorylcreatine

ATP is resynthesized from ADP by the addition of a phosphate group. Under normal conditions, the energy for this endothermic reaction is supplied by the breakdown of glucose to Co_2 and H_2O, but there also exists in muscle another energy-rich phosphate compound that can supply this energy. This compound is **phosphorylcreatine** (Fig 17–21), which is hydrolyzed to creatine and phosphate groups with the release of considerable energy. At rest, some ATP in the mitochondria transfers its phosphate to creatine, so that a phosphorylcreatine store is built up. During exercise, the phosphorylcreatine is hydrolyzed at the junction between the myosin heads and the actin, forming ATP from ADP and thus permitting contraction to continue. The synthesis of phosphorylcreatine is catalyzed by the isoenzyme of creatine phosphokinase in the mitochondria, and its hydrolysis by the isoenzyme of creatine phosphokinase in the myosin heads.

Carbohydrate Breakdown

Much of the energy for phosphorylcreatine and ATP resynthesis comes from the breakdown of glucose to CO_2 and H_2O. An outline of the major metabolic pathways involved is presented in Chapter 17. For the purposes of the present discussion, it is sufficient to point out that glucose in the bloodstream enters cells, where it is degraded through a series of chemical reactions to pyruvate. Another source of intracellular glucose, and consequently of pyruvate, is glycogen, the carbohydrate polymer that is especially abundant in liver and skeletal muscle. When adequate O_2 is present, pyruvate enters the citric acid cycle and is metabolized—through this cycle and the so-called respiratory enzyme pathway—to CO_2 and H_2O. This process is called **aerobic glycolysis.** The metabolism of glucose or glycogen to CO_2 and H_2O liberates sufficient energy to form large quantities of ATP from ADP. If O_2 supplies are insufficient, the pyruvate formed from glucose does not enter the tricarboxylic acid cycle but is reduced to lactate. This process of **anaerobic glycolysis** is associated with the net production of much smaller quantities of energy-rich phosphate bonds, but it does not require the presence of O_2. Skeletal muscle also takes up free fatty acids (FFA) from the blood and oxidizes them to CO_2 and H_2O. Indeed, FFA are probably the major substrates for muscle at rest and during recovery after contraction. The various reactions involved in supplying energy to skeletal muscle are summarized in Fig 3–10.

$$ATP + H_2O \rightarrow ADP + H_3PO_4 + 7.3\text{ kcal}$$

$$\text{Phosphorylcreatine} + ADP \rightleftharpoons \text{Creatine} + ATP$$

$$\text{Glucose} + 2\,ATP\ (\text{or glycogen} + 1\,ATP) \xrightarrow{\text{Anaerobic}} 2\ \text{Lactic acid} + 4\,ATP$$

$$\text{Glucose} + 2\,ATP\ (\text{or glycogen} + 1\,ATP) \xrightarrow{\text{Oxygen}} 6\,CO_2 + 6\,H_2O + 40\,ATP$$

$$FFA \xrightarrow{\text{Oxygen}} CO_2 + H_2O + ATP$$

Figure 3–10. Energy released by hydrolysis of 1 mol of ATP, and reactions responsible for resynthesis of ATP. The amount of ATP formed per mole of free fatty acid (FFA) oxidized is large but varies with the size of the FFA. For example, complete oxidation of 1 mol of palmitic acid generates 140 mol of ATP.

The Oxygen Debt Mechanism

During muscular exercise, the muscle blood vessels dilate and blood flow is increased so that the available O_2 supply is increased. Up to a point, the increase in O_2 consumption is proportionate to the energy expended, and all the energy needs are met by aerobic processes. However, when muscular exertion is very great, aerobic resynthesis of energy stores cannot keep pace with their utilization. Under these conditions, phosphorylcreatine is still used to resynthesize ATP. Some ATP synthesis is accomplished by using the energy released by the anaerobic breakdown of glucose to lactate. However, use of the anaerobic pathway is self-limiting because in spite of rapid diffusion of lactate into the bloodstream, enough accumulates in the muscles to eventually exceed the capacity of the tissue buffers and produce an enzyme-inhibiting decline in pH. However, for short periods, the presence of an anaerobic pathway for glucose breakdown permits muscular exertion of a far greater magnitude than would be possible without it. For example, in a 100-meter dash that takes 10 seconds, 85% of the energy consumed is derived anaerobically; in a 2-mile race that takes 10 minutes, 20% of the energy is derived anaerobically; and in a long-distance race that takes 60 minutes, only 5% of the energy comes from anaerobic metabolism.

After a period of exertion is over, extra O_2 is consumed to remove the excess lactate, replenish the ATP and phosphorylcreatine stores, and replace the small amounts of O_2 that have come from myoglobin. The amount of extra O_2 consumed is proportionate to the extent to which the energy demands during exertion exceeded the capacity for the aerobic synthesis of energy stores, ie, the extent to which an **oxygen debt** was incurred. The O_2 debt is measured experimentally by determining O_2 consumption after exercise until a constant, basal consumption is reached and subtracting the basal consumption from the total. The amount of this debt may be 6 times the basal O_2 consumption, which indicates that the subject is capable of 6 times the exertion that would have been possible without it. Obviously, the maximal debt can be incurred rapidly or slowly; violent exertion is possible for only short periods, whereas less strenuous exercise can be carried on for longer periods.

Trained athletes are able to increase the O_2 consumption of their muscles to a greater degree than untrained individuals. Consequently, they are capable of greater exertion without increasing their lactate production, and they contract smaller oxygen debts for a given amount of exertion.

Rigor

When muscle fibers are completely depleted of ATP and phosphorylcreatine, they develop a state of extreme rigidity called **rigor.** When this occurs after death, the condition is called rigor mortis. In rigor, almost all of the myosin heads attach to actin but in an abnormal, fixed, and resistant way.

Heat Production in Muscle

Thermodynamically, the energy supplied to a muscle must equal its energy output. The energy output appears in work done by the muscle, in energy-rich phosphate bonds formed for later use, and in heat. The overall mechanical efficiency of skeletal muscle (work done/total energy expenditure) ranges up to 50% while lifting a weight during isotonic contraction and is essentially 0% during isometric contraction. Energy storage in phosphate bonds is a small factor. Consequently, heat production is considerable. The heat produced in muscle can be measured accurately with suitable thermocouples.

Resting heat, the heat given off at rest, is the external manifestation of basal metabolic processes. The heat produced in excess of resting heat during contraction is called the **initial heat.** This is made up of **activation heat,** the heat that muscle produces whenever it is contracting, and **shortening heat,** which is proportionate in amount to the distance the muscle shortens. Shortening heat is apparently due to some change in the structure of the muscle during shortening.

Following contraction, heat production in excess of resting heat continues for as long as 30 minutes. This **recovery heat** is the heat liberated by the metabolic processes that restore the muscle to its precontraction state. The recovery heat of muscle is approximately equal to the initial heat, ie, the heat produced during recovery is equal to the heat produced during contraction.

If a muscle that has contracted isotonically is restored to its previous length, extra heat in addition to recovery heat is produced **(relaxation heat).** External work must be done on the muscle to return it to its previous length, and relaxation heat is mainly a manifestation of this work.

Table 3–3. Classification of fiber types in skeletal muscles.*

	Type I	Type IIB	Type IIA
Other names	Slow oxidative; red.	Fast glycolytic; white.	Fast oxidative; red.
Myosin isoenzyme ATPase activity	Slow.	Fast.	Fast.
Ca^{2+} pumping capacity of sarcoplasmic reticulum	Moderate.	High.	High.
Diameter	Moderate.	Large.	Small.
Glycolytic capacity	Moderate.	High.	High.
Oxidative capacity (correlates with content of mitochondria, capillary density, myoglobin content)	High.	Low.	High.

*Modified from Murphy RA: Muscle. In: *Psychology,* 2nd ed. Berne RM, Levy MN (editors). Mosby, 1988.

Fiber Types in Muscle

There are variations in the myosin ATPase activity, metabolism, and contractile properties of the different fibers that make up muscles. The properties of the 3 main types of fibers found in skeletal muscle in mammals are summarized in Table 3–3. In humans, type IIA fibers are infrequent, and in general, slow muscles are type I and fast muscles are type IIB.

Muscles containing many type I fibers are called **red muscles** because they are darker than other muscles. The red muscles, which respond slowly and have a long latency, are adapted for long, slow, posture-maintaining contractions. The long muscles of the back are red muscles. **White muscles,** which contain mostly type IIB fibers, have short twitch durations and are specialized for fine, skilled movement. The extraocular muscles and some of the hand muscles are fast muscles.

Protein Isoforms in Muscle & Their Genetic Control

The differences in the fibers that make up muscles stem from differences in the proteins in them. Myosin, actin, and tropomyosin show relatively marked heterogeneity from one skeletal muscle to another, and there is also heterogeneity in cardiac and smooth muscle. Actin and tropomyosin each have α and β isoforms, ie, forms with similar biologic activity but different structures, and there are at least 7 different isoforms of the myosin heavy chains (MHC). The MHC isoforms, which differ somewhat in myosin ATPase activity, are encoded by a family of genes that are probably clustered on the same chromosome. In rats, there is an embryonic MHC, a fetal MHC, a fast IIA MHC, a fast IIB MHC, and an MHC that is unique to extraocular muscles. In addition, an α MHC and a β MHC are found in the heart. The expression of the MHCs is precisely regulated during development and, in adults, from one muscle to another. In addition, changes in muscle function can be produced by alterations in activity, innervation, and hormonal milieu, and these changes are generally produced by alterations in the transcription of MHC genes. Examples include conversion of fast to slow skeletal muscle when muscle innervation is changed (see below) and marked alterations in the isoforms in skeletal and cardiac muscle in hypothyroidism.

PROPERTIES OF MUSCLES IN THE INTACT ORGANISM

Effects of Denervation

In the intact animal or human, healthy skeletal muscle does not contract except in response to stimulation of its motor nerve supply. Destruction of this nerve supply causes muscle atrophy. It also leads to abnormal excitability of the muscle and increases its sensitivity to circulating acetylcholine (denervation hypersensitivity; see Chapter 4). Fine, irregular contractions of individual fibers (**fibrillations**) appear. If the motor nerve regenerates, these disappear. Such contractions usually are not visible grossly and should not be confused with **fasciculations,** which are jerky, visible contractions of groups of muscle fibers that occur as a result of pathologic discharge of spinal motor neurons.

The Motor Unit

Since the axons of the spinal motor neurons supplying skeletal muscle each branch to innervate several muscle fibers, the smallest possible amount of muscle that can contract in response to the excitation of a single motor neuron is not one muscle fiber but all the fibers supplied by the neuron. Each single motor neuron and the muscle fibers it innervates constitute a **motor unit.** The number of muscle fibers in a motor unit varies. In muscles such as those of the hand and those concerned with motion of the eye—ie, muscles concerned with fine, graded, precise movement—there are 3–6 muscle fibers per motor unit. On the other hand, values of 120–165 fibers per unit have been reported in cat leg muscles, and some of the large muscles of the back in humans probably contain even more.

Each spinal motor neuron innervates only one kind of muscle fiber, so that all of the muscle fibers in a motor unit are of the same type. On the basis of the type of muscle fiber they innervate (Table 3–3), and thus on the basis of the duration of their twitch contraction, motor units are divided into fast and slow units. In general, slow muscle units are innervated by small, slowly conducting motor neurons and fast units by large, rapidly conducting motor neurons (**size principle).** In large limb muscles, the small, slow units are first recruited in most movements, are resistant to fatigue, and are the most frequently used units. The fast units, which are more easily fatigued, are generally recruited with more forceful movements.

The differences between types of muscle units are

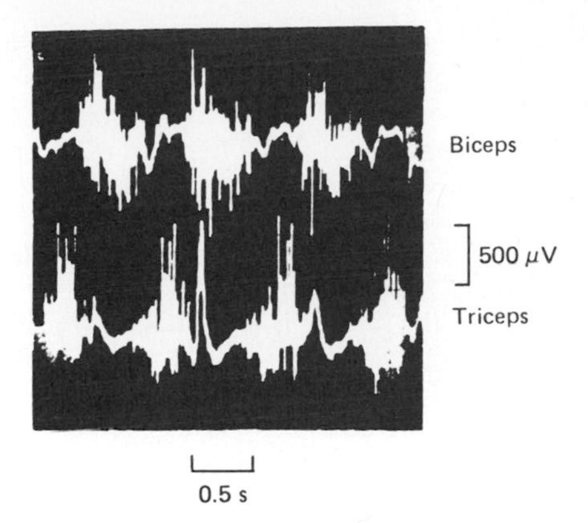

Figure 3–11. Electromyographic tracings from human biceps and triceps muscles during alternate flexion and extension of the elbow. (Courtesy of BC Garoutte.)

not inherent but are determined by, among other things, their activity (see below). When the nerve to a slow muscle is cut and replaced with the nerve to a fast muscle, the nerve regenerates and innervates the slow muscle. However, the muscle becomes fast and there are corresponding changes in its MHC isoforms and myosin ATPase activity. There has been speculation that this change is due to trophic substances flowing from the nerve to the muscle, but it is apparently due instead to changes in the pattern of activity of the muscle. In stimulation experiments, it has been demonstrated that changes in the expression of MHC genes and consequently of MHC isoforms can be produced by changes in the pattern of electrical activity used to stimulate the muscle.

Denervation of skeletal muscle leads to atrophy and flaccid paralysis, with the appearance of fibrillations. These effects are the classic consequences of a **lower motor neuron lesion.** The muscle also becomes hypersensitive to acetylcholine (denervation hypersensitivity; see Chapter 4).

Electromyography

Activation of motor units can be studied by **electromyography,** the process of recording the electrical activity of muscle on a cathode-ray oscilloscope. This may be done in unanesthetized humans by using small metal disks on the skin overlying the muscle as the pick up electrodes or by using hypodermic needle electrodes. The record obtained with such electrodes is the **electromyogram (EMG).** With needle electrodes, it is usually possible to pick up the activity of single muscle fibers. A typical EMG is shown in Fig 3–11.

Factors Responsible for Grading of Muscular Activity

It has been shown by electromyography that there is little if any spontaneous activity in the skeletal muscles of normal individuals at rest. With minimal voluntary activity a few motor units discharge, and with increasing voluntary effort more and more are brought into play. This process is sometimes called **recruitment of motor units.** Gradation of muscle response is therefore in part a function of the number of motor units activated. In addition, the frequency of discharge in the individual nerve fibers plays a role, the tension developed during a tetanic contraction being greater than that during individual twitches. The length of the muscle is also a factor. Finally, the motor units fire asynchronously, ie, out of phase with each other. This asynchronous firing causes the individual muscle fiber responses to merge into a smooth contraction of the whole muscle.

The Strength of Skeletal Muscles

Human skeletal muscle can exert 3–4 kg of tension per square centimeter of cross-sectional area. This figure is about the same as that obtained in a variety of experimental animals and seems to be constant for all mammalian species. Since many of the muscles in humans have a relatively large cross-sectional area, the tension they can develop is quite large. The gastrocnemius, for example, not only supports the weight of the whole body during climbing but resists a force several times this great when the foot hits the ground during running or jumping. An even more striking example is the gluteus maximus, which can exert a tension of 1200 kg. The total tension that could be developed by all muscles in the body of an adult man is approximately 22,000 kg (nearly 25 tons).

Body Mechanics

Body movements are generally organized in such a way that they take maximal advantage of the physiologic principles outlined above. For example, the attachments of the muscles in the body are such that many of them are normally at or near their resting length when they start to contract. In muscles that extend over more than one joint, movement at one joint may compensate for movement at another in such a way that relatively little shortening of the muscle occurs during contraction. Nearly isometric contractions of this type permit development of maximal tension per contraction. The hamstring muscles extend from the pelvis over the hip joint and the knee joint to the tibia and fibula. Hamstring contraction produces flexion of the leg on the thigh. If the thigh is flexed on the pelvis at the same time, the lengthening of the hamstrings across the hip joint tends to compensate for the shortening across the knee joint. In the course of walking and other activities, the body moves in a way that takes advantage of this.

Such factors as momentum and balance are integrated into body movement in ways that make possible maximal motion with minimal muscular exertion. In walking, each limb passes rhythmically through a support or stance phase when the foot is on the ground and a swing phase when the foot is off the ground. The support phases of the 2 legs overlap, so that there are 2 periods of double support during each cycle. There is a brief burst of activity in the leg flexors at the start of each step, and then the leg is swung forward with little more active muscular contraction. Therefore, the muscles are active for only a fraction of each step, and walking for long periods causes relatively little fatigue.

A young adult walking at a comfortable pace moves at a velocity of about 80m/min and generates a power output of 150–175 watts per step. A group of young adults asked to walk at their most comfortable rate selected a velocity close to 80m/min, and it was found that they had selected the velocity at which their energy output was minimal. Walking more rapidly or more slowly took more energy.

Even though walking is a complex activity, it is common knowledge that it is carried out more or less automatically. Experiments in animals indicate that it is organized in preprogrammed nerve pathways within the spinal cord and that it is activated by some sort of command signal in a fashion that is probably analogous to the initiation of patterns of activity by the discharge of command neurons in invertebrates.

CARDIAC MUSCLE

MORPHOLOGY

The striations in cardiac muscle are similar to those in skeletal muscle, and Z lines are present. The muscle fibers branch and interdigitate, but each is a complete unit surrounded by a cell membrane. Where the end of one muscle fiber abuts on another, the membranes of both fibers parallel each other through an extensive series of folds. These areas, which always occur at Z lines, are called **intercalated disks** (Fig 3–12). They provide a strong union between fibers, maintaining cell-to-cell cohesion, so that the pull of one contractile unit can be transmitted along its axis to the next. Along the sides of the muscle fibers next to the disks, the cell membranes of adjacent fibers fuse for considerable distances. These gap junctions provide low-resistance bridges for the spread of excitation from one fiber to another (see Chapter 1). They permit cardiac muscle to function as if it were a syncytium, even though there are no protoplasmic bridges between cells. The T system in cardiac muscle is located at the Z lines rather than at the A-I junction (as it is in mammalian skeletal muscle).

Like skeletal muscle, cardiac muscle contains myosin, actin, tropomyosin, and troponin. It also shows heterogeneity and contains large numbers of elongated mitochondria in close contact with the fibrils.

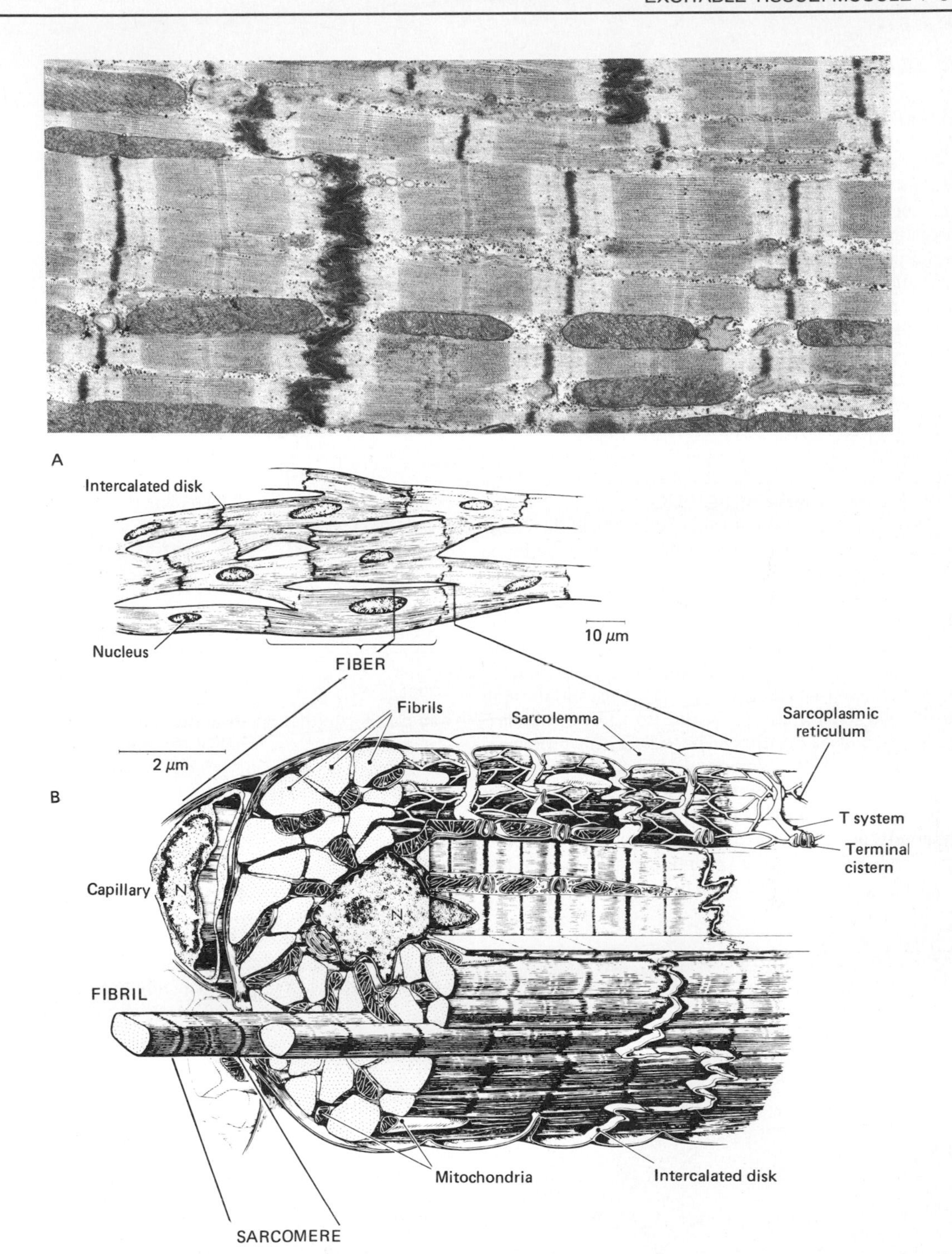

Figure 3–12. *Top:* Electron photomicrograph of cardiac muscle. The fuzzy thick lines are intercalated disks (x12,000). (Reproduced, with permission, from Bloom W. Fawcett DW: *A Textbook of Histology,* 10th ed. Saunders, 1975). ***Bottom:*** Diagram of cardiac muscle as seen under the light microscope ***(A)*** and the electron microscope ***(B)***. N, nucleus. (Reproduced, with permission, from Braunwald E. Ross J. Sonnenblick EH: Mechanisms of contraction of the normal and failing heart. *N Engl J Med* 1967;277:794. Courtesy of Little, Brown, Inc.)

ELECTRICAL PROPERTIES

Resting Membrane & Action Potentials

The resting membrane potential of individual mammalian cardiac muscle cells is about −90 mV (interior negative to exterior). Stimulation produces a propagated action potential that is responsible for initiating contraction. Depolarization proceeds rapidly and an overshoot is present, as in skeletal muscle and nerve, but this is followed by a plateau before the membrane potential returns to the baseline (Fig 3–13). In mammalian hearts, depolarization lasts about 2 ms, but the plateau phase and repolarization last 200 ms or more. Repolarization is therefore not complete until the contraction is half over. With extracellular recording, the electrical events include a spike and a later wave that bear a resemblance to the QRS complex and T wave of the ECG.

As in other excitable tissues, changes in the external K^+ concentration affect the resting membrane potential of cardiac muscle, whereas changes in the external Na^+ concentration affect the magnitude of the action potential. The initial rapid depolarization and the overshoot (phase 0) are due to a rapid increase in Na^+ conductance similar to that occurring in nerve and skeletal muscle (Fig 3–14). The initial rapid repolarization (phase 1) is due to closure of Na^+ channels and Cl^- influx. The subsequent prolonged plateau (phase 2) is due to a slower but prolonged opening of voltage-gated Ca^{2+} channels. Final repolarization (phase 3) is due to closure of the Ca^{2+} channels and prolonged opening of K^+ channels. This restores the resting potential (phase 4). The fast Na^+ channel in cardiac muscle probably has 2 gates, an outer gate that opens at the start of depolarization, at a membrane potential of −70 to −80 mV, and a second inner gate that then closes and precludes further influx until the action potential is over (Na^+ channel inactivation). The slow Ca^{2+} channel is activated at a membrane potential of −30 to −40 mV.

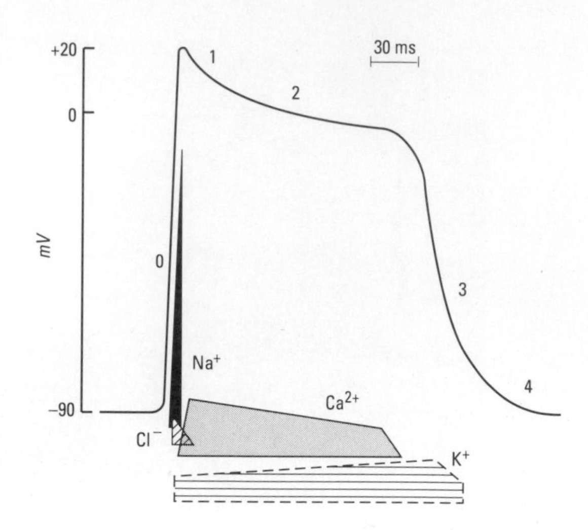

Figure 3–14. Phases of the action potential of a cardiac muscle fiber and the corresponding changes in ionic conductances across the muscle membrane. 0, depolarization; 1, initial rapid repolarization; 2, plateau phase; 3, late rapid repolarization; 4, baseline (Modified from Shepherd JT, Vanhoutte PM: *The Human Cardiovascular System: Facts and Concepts.* Raven, 1979.)

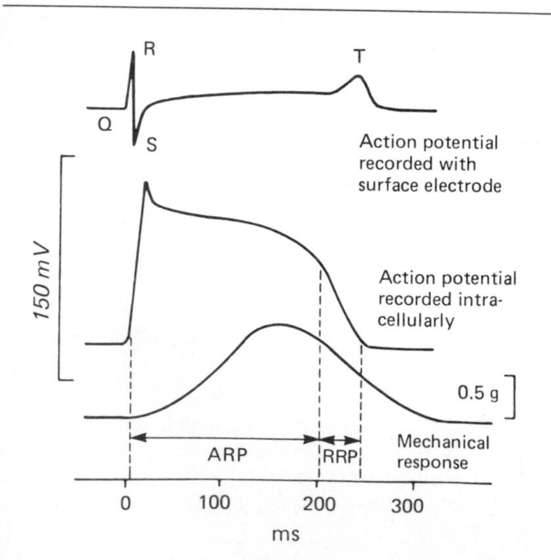

Figure 3–13. Action potentials and contractile response of mammalian cardiac muscle fiber plotted on the same time axis. ARP, absolute refractory period; RRP, relative refractory period.

In cardiac muscle, the repolarization time decreases as the cardiac rate increases. At a cardiac rate of 75 beats per minute, the duration of the action potential (0.25 second) is almost 70% longer than it is at a cardiac rate of 200 beats per minute (0.15 second).

MECHANICAL PROPERTIES

Contractile Response

The contractile response of cardiac muscle begins just after the start of depolarization and lasts about 1½ times as long as the action potential (Fig 3–13). The role of Ca^{2+} in excitation-contraction coupling is similar to its role in skeletal muscle (see above). However, the sequence of events is different. In cardiac muscle, depolarization due to opening of Na^+ channels activates the Ca^{2+} channels, and it is the resulting influx of Ca^{2+} from the ECF that triggers release of Ca^{2+} from the sarcoplasmic reticulum.

During phases 0–2 and about half of phase 3 (until the membrane potential reaches approximately −50 mV during repolarization), cardiac muscle cannot be excited again; ie, it is in its **absolute refractory period** (Fig 3–13). It remains relatively refractory until phase 4. Therefore, tetanus of the type seen in skeletal muscle cannot occur. Of course, tetanization of cardiac muscle for any length of time would have lethal consequences, and in this sense, the fact that cardiac muscle cannot be tetanized is a safety feature. Ventricular muscle is said to be in the "vulnerable period" just at the end of the action potential, because stimulation at this time will sometimes initiate ventricular fibrillation.

Isoforms

Cardiac muscle is generally slow and has relatively low ATPase activity. Its fibers are dependent on oxidative metabolism and hence on a continuous supply of O_2. In rats, there are α and β MHCs in the cardiac muscle fibers. In the ventricle, α MHC predominates in adulthood, but in hypothyroid animals, it is largely replaced by β MHC, which has a lower myosin ATPase activity. Thyroid deficiency also alters the MHC patterns in skeletal muscle, but the pattern of change varies from muscle to muscle.

Correlation Between Muscle Fiber Length & Tension

The relation between initial fiber length and total tension in cardiac muscle is similar to that in skeletal muscle; there is a resting length at which the tension developed upon stimulation is maximal. In the body, the initial length of the fibers is determined by the degree of diastolic filling of the heart, and the pressure developed in the ventricle is proportionate to the total tension developed **(Starling's law of the heart;** see Chapter 29). As the diastolic filling increases, the force of contraction of the ventricles is increased (Fig 3–15). The homeostatic value of this response is discussed in Chapter 29.

The force of contraction of cardiac muscle is also increased by catecholamines (see Chapters 13 and 20), and this increase occurs without a change in muscle length. The increase, which is called the positively inotropic effect of catecholamines, is mediated via β-adrenergic receptors and cyclic AMP (see Chapter 1). Cyclic AMP activates protein kinase A, and this leads to phosphorylation of the voltage-dependent Ca^{2+} channels, causing them to spend more time in the open state. Cyclic AMP also increases the active transport of Ca^{2+} to the sarcoplasmic reticulum, thus accelerating relaxation and consequently shortening systole. This is important when the cardiac rate is increased, because it permits adequate diastolic filling (see Chapter 29). Digitalis glycosides increase cardiac contractions by inhibiting the Na^+-K^+ ATPase in cell membranes of the muscle fibers. The resultant increase in intracellular Na^+ increases the efflux of Na^+ in exchange for Ca^{2+} via the Ca^{2+}-Na^+antiport in the cell membrane (see Chapter 1).

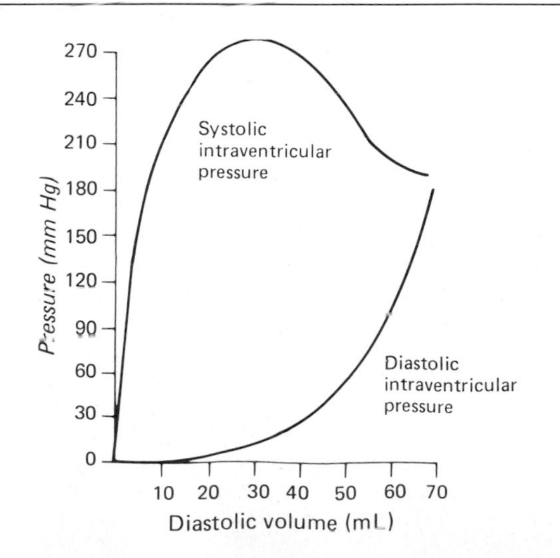

Figure 3–15. "Length-tension" relationship for cardiac muscle. The values are for dog heart.

METABOLISM

Mammalian hearts have an abundant blood supply, numerous mitochondria, and a high content of myoglobin, a muscle pigment that may function as an O_2 storage mechanism (see Chapter 35). Normally, less than 1% of the total energy liberated is provided by anaerobic metabolism. During hypoxia, this figure may increase to nearly 10%; but under totally anaerobic conditions, the energy liberated is inadequate to sustain ventricular contractions. Under basal conditions, 35% of the caloric needs of the human heart are provided by carbohydrate, 5% by ketones and amino acids, and 60% by fat. However, the proportions of substrates utilized vary greatly with the nutritional state. After ingestion of large amounts of glucose, more lactate and pyruvate are used; during prolonged starvation, more fat is used. Circulating free fatty acids normally account for almost 50% of the lipid utilized. In untreated diabetics, the carbohydrate utilization of cardiac muscle is reduced and that of fat increased. The factors affecting the O_2 consumption of the human heart are discussed in Chapter 29.

PACEMAKER TISSUE

The heart continues to beat after all nerves to it are sectioned; indeed, if the heart is cut into pieces, the pieces continue to beat. This is because of the presence in the heart of specialized pacemaker tissue that can initiate repetitive action potentials. The pacemaker tissue makes up the conduction system that normally spreads impulses throughout the heart (see Chapter 28).

Pacemaker tissue is characterized by an unstable membrane potential. Instead of having a steady value between impulses, the membrane potential declines steadily after each action potential until the firing level is reached and another action potential is trig-

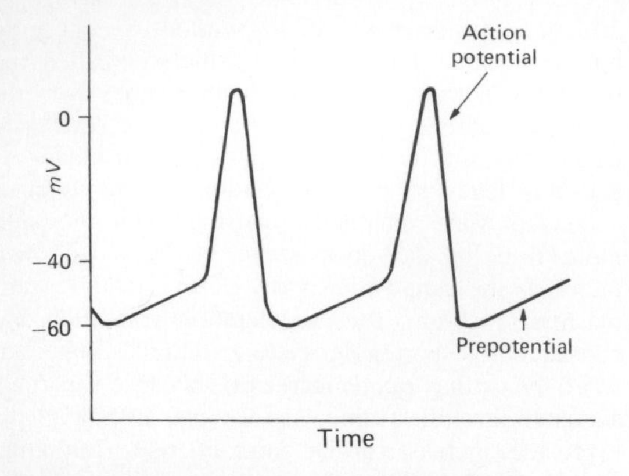

Figure 3–16. Diagram of the membrane potential of pacemaker tissue. Note that the resting membrane potential of pacemaker tissue is somewhat less than the membrane potential of atrial and ventricular muscle.

gered. This slow depolarization between action potentials is called a **pacemaker potential (prepotential)** (Figs 3–16 and 28–1). The steeper its slope, the higher the rate at which the pacemaker fires. Some agents that modify the firing rate of pacemakers do so by changing the slope of the prepotential, although others act by altering the membrane potential and thus changing the amount of time required to reach the firing level. The prepotential is due primarily to a slow decrease in K^+ permeability and consequently reduced K^+ efflux, with a resultant reduction in membrane potential. Prepotentials are not seen in atrial and ventricular muscle cells, and in these cells, K^+ permeability is constant during diastole.

SMOOTH MUSCLE

MORPHOLOGY

Smooth muscle is distinguished anatomically from skeletal and cardiac muscle because it lacks visible cross-striations. There is a sarcoplasmic reticulum, but it is poorly developed. The muscle contains actin, myosin, and tropomyosin but apparently does not contain troponin. In intestinal smooth muscle, the contractile units are made up of small bundles of interdigitating thick and thin filaments that are irregularly shaped and randomly arranged. When the muscle contracts, the thick and thin filaments slide on each other. There is heterogeneity in the structures of the muscle proteins, as there is in skeletal and cardiac muscle. In general, smooth muscles contain few mitochondria and depend to a large degree on glycolysis for their metabolic needs.

Types

There are various types of smooth muscle in the body. In general, smooth muscle can be divided into **visceral smooth muscle** and **multi-unit smooth muscle.** Visceral smooth muscle occurs in large sheets, has low-resistance bridges between individual muscle cells, and functions in a syncytial fashion. The bridges, like those in cardiac muscle, are junctions where the membranes of the 2 adjacent cells fuse to form a single membrane. Visceral smooth muscle is found primarily in the walls of hollow viscera. The musculature of the intestine, the uterus, and the ureters are examples. Multi-unit smooth muscle is made up of individual units without interconnecting bridges. It is found in structures such as the iris of the eye, in which fine, graded contractions occur. It is usually not under voluntary control, but it has many functional similarities to skeletal muscle.

VISCERAL SMOOTH MUSCLE

Electrical & Mechanical Activity

Visceral smooth muscle is characterized by the instability of its membrane potential and by the fact that it shows continuous, irregular contractions that are independent of its nerve supply. This maintained state of partial contraction is called **tonus** or **tone.** The membrane potential has no true "resting" value, being relatively low when the tissue is active and higher when it is inhibited, but in periods of relative quiescence it averages about −50 mV. Superimposed on the membrane potential are waves of various types (Fig 3–17). There are slow sine wave-like fluctua-

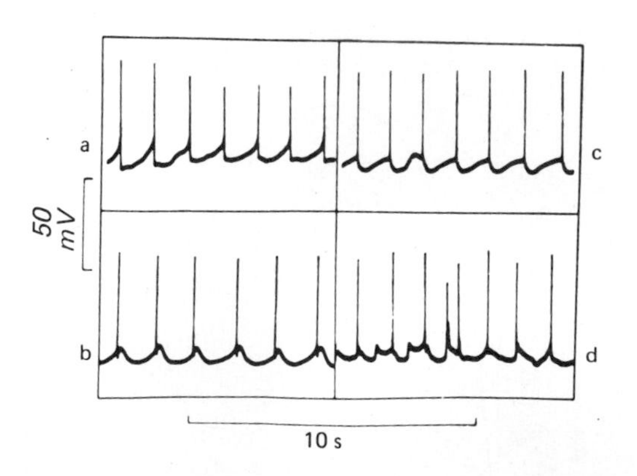

Figure 3–17. Spontaneous electrical activity in individual smooth muscle cells of teniae coli of guinea pig colon. *a:* Pacemaker type; *b:* sinusoidal waves with action potentials on the rising phases; *c:* sinusoidal waves with action potentials on the falling phases; *d:* mixture of pacemaker, sinusoidal, and action potentials. (Reproduced, with permission, from Bulbring E: Physiology and pharmacology of intestinal smooth muscle. *Lectures on the Scientific Basis of Medicine* 1957;7:374.)

tions a few millivolts in magnitude and spikes that sometimes overshoot the zero potential line and sometimes do not. In many tissues, the spikes have a duration of about 50 ms. However, in some tissues the action potentials have a prolonged plateau during repolarization, like the action potentials in cardiac muscle. The spikes may occur on the rising or falling phases of the sine wave oscillations. There are, in addition, pacemaker potentials similar to those found in the cardiac pacemakers. However, in visceral smooth muscle, these potentials are generated in multiple foci that shift from place to place. Spikes generated in the pacemaker foci are conducted for some distance in the muscle. Because of the continuous activity, it is difficult to study the relation between the electrical and mechanical events in visceral smooth muscle, but in some relatively inactive preparations, a single spike can be generated. The muscle starts to contract about 200 ms after the start of the spike and 150 ms after the spike is over. The peak contraction is reached as long as 500 ms after the spike. Thus, the excitation-contraction coupling in visceral smooth muscle is a very slow process compared with that in skeletal and cardiac muscle, in which the time from initial depolarization to initiation of contraction is less than 10 ms.

Ca^{2+} is involved in the initiation of contraction of smooth muscle, as it is in skeletal muscle. Ca^{2+} enters via voltage-gated Ca^{2+} channels. However, the myosin in smooth muscle must be phosphorylated for activation of the myosin ATPase. Phosphorylation and dephosphorylation of myosin also occur in skeletal muscle, but phosphorylation is not necessary for activation of the ATPase. In smooth muscle, Ca^{2+} binds to calmodulin (see Chapter 1), and the resulting complex activates **myosin light chain kinase,** the enzyme that catalyzes the phosphorylation of myosin. Actin then slides on myosin, producing contraction. This is in contrast to skeletal and cardiac muscle, where contraction is triggered by the binding of Ca^{2+} to troponin C.

Visceral smooth muscle is unique in that, unlike other types of muscle, it contracts when stretched in the absence of any extrinsic innervation. Stretch is followed by a decline in membrane potential, an increase in the frequency of spikes, and a general increase in tone.

If epinephrine or norepinephrine is added to a preparation of intestinal smooth muscle arranged for recording of intracellular potentials in vitro, the membrane potential usually becomes larger, the spikes decrease in frequency, and the muscle relaxes (Fig 3–18). Norepinephrine is the chemical mediator released at noradrenergic nerve endings (see Chapter 4), and stimulation of the noradrenergic nerves to the preparation produces inhibitory potentials (see Chapter 4). Stimulation of the noradrenergic nerves to the intestine inhibits contractions in vivo. Norepinephrine exerts both α and β actions (see Chapter 4) on the muscle. The β action, reduced muscle tension in response to excitation, is mediated via cyclic AMP

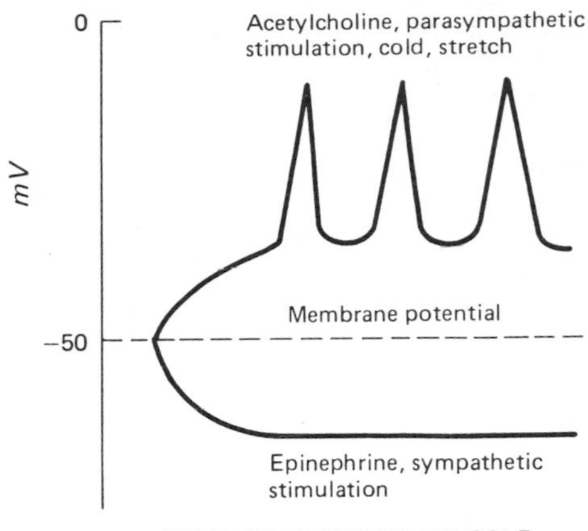

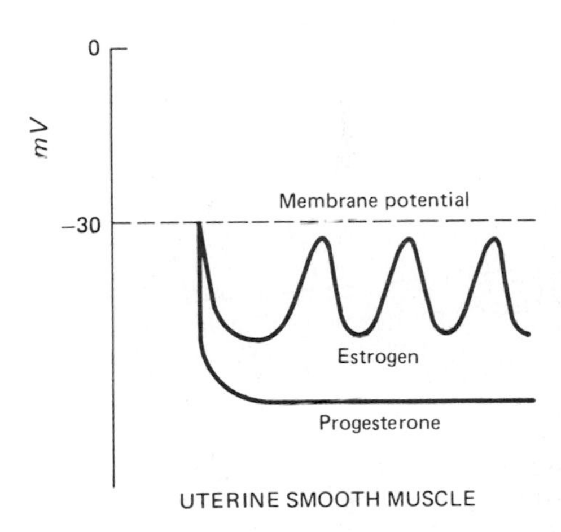

Figure 3–18. *Top:* Effects of various agents on the membrane potential of intestinal smooth muscle. ***Bottom:*** Effects of estrogens and of progesterone on the membrane potential of the uterus of the ovariectomized rabbit or rat.

(see Chapter 1) and is probably due to increased intracellular binding of Ca^{2+}. The α action, which is also inhibition of contraction, is associated with increased Ca^{2+} efflux from the muscle cells.

Acetylcholine has an effect opposite to that of norepinephrine on the membrane potential and contractile activity of intestinal smooth muscle. If acetylcholine is added to the fluid bathing a smooth-muscle preparation in vitro, the membrane potential decreases and the spikes become more frequent (Fig 3–18). The muscle becomes more active, with an increase in tonic tension and the number of rhythmic contractions. The depolarization is apparently due to increased Ca^{2+} entry into the cell. In the intact animal, stimulation of cholinergic nerves causes release of acetylcholine, excitatory potentials (see Chapter 4), and increased intestinal contractions. In vitro, similar effects are produced by cold and stretch.

Function of the Nerve Supply to Smooth Muscle

The effects of acetylcholine and norepinephrine on visceral smooth muscle serve to emphasize 2 of its important properties: (1) its spontaneous activity in the absence of nervous stimulation, and (2) its sensitivity to chemical agents released from nerves locally or brought to it in the circulation. In mammals, visceral muscle usually has a dual nerve supply from the 2 divisions of the autonomic nervous system. The structure and function of the contacts between these nerves and smooth muscle are discussed in Chapter 4. The function of the nerve supply is not to initiate activity in the muscle but rather to modify it. Stimulation of one division of the autonomic nervous system usually increases smooth muscle activity, whereas stimulation of the other decreases it. However, in some organs, noradrenergic stimulation increases and cholinergic stimulation decreases smooth muscle activity; in others, the reverse is true.

Other chemical agents also affect smooth muscle. An interesting example is the uterus. Uterine smooth muscle is relatively inexcitable during diestrus and in the ovariectomized animal. During estrus or in the estrogen-treated ovariectomized animal, excitability is enhanced, and tonus and spontaneous contractions occur. However, estrogen increases rather than decreases the membrane potential (Fig 3–18). Progesterone increases the membrane potential even further and inhibits the electrical and contractile activity of uterine muscle (see Chapter 23).

Relation of Length to Tension; Plasticity

Another special characteristic of smooth muscle is the variability of the tension it exerts at any given length. If a piece of visceral smooth muscle is stretched, it first exerts increased tension (see above). However, if the muscle is held at the greater length after stretching, the tension gradually decreases. Sometimes the tension falls to or below the level exerted before the muscle was stretched. It is consequently impossible to correlate length and developed tension accurately, and no resting length can be assigned. In some ways, therefore, smooth muscle behaves more like a viscous mass than a rigidly structured tissue, and it is this property that is referred to as the **plasticity** of smooth muscle.

The consequences of plasticity can be demonstrated in intact humans. For example, the tension exerted by the smooth muscle walls of the bladder can be measured at varying degrees of distention as fluid is infused into the bladder via a catheter, as shown in Fig 38–25. Initially there is relatively little increase in tension as volume is increased, because of the plasticity of the bladder wall. However, a point is eventually reached at which the bladder contracts in a forceful fashion.

MULTI-UNIT SMOOTH MUSCLE

Unlike visceral smooth muscle, multi-unit smooth muscle is nonsyncytial and contractions do not spread widely through it. Because of this, the contractions of multi-unit smooth muscle are more discrete, fine, and localized than those of visceral smooth muscle. Like visceral smooth muscle, multi-unit smooth muscle is very sensitive to circulating chemical substances and is normally activated by chemical mediators (acetylcholine and norepinephrine) released at the endings of its motor nerves. Norepinephrine in particular tends to persist and to cause repeated firing of the muscle after a single stimulus rather than a single action potential. Therefore, the contractile response produced is usually an irregular tetanus rather than a single twitch. When a single twitch response is obtained, it resembles the twitch contraction of skeletal muscle except that its duration is 10 times as long.

Synaptic & Junctional Transmission

4

INTRODUCTION

The all or none type of conduction seen in axons and skeletal muscle has been discussed in Chapters 2 and 3. Impulses are transmitted from one nerve cell to another cell at **synapses** (Fig 4–1). These are the junctions where the axon or some other portion of one cell (the **presynaptic cell**) terminates on the dendrites, soma, or axon of another neuron (Fig 4–2) or in some cases a muscle or gland cell **(the postsynaptic cell).** It is worth noting that dendrites as well as axons can be presynaptic or postsynaptic. Transmission at most synaptic junctions is chemical; the impulse in the presynaptic axon causes secretion of **neurotransmitter** such as acetylcholine or serotonin. This chemical mediator binds to receptors on the surface of the postsynaptic cell, and this triggers intracellular events that open or close channels in the membrane of the postsynaptic cell. At some of the junctions, however, transmission is electrical, and at a few conjoint synapses, it is both electrical and chemical. In any case, transmission is not a simple jumping of one action potential from the presynaptic to the postsynaptic cell. The effects of discharge at individual synaptic endings can be excitatory or inhibitory, and when the postsynaptic cell is a neuron, the summation of all the excitatory and inhibitory effects determines whether an action potential is generated. Thus, synaptic transmission is a complex process that permits the grading and adjustment of neural activity necessary for normal function.

In electrical synapses, the membranes of the presynaptic and postsynaptic neurons come close together, and gap junctions form between the cells (see Chapter 1). Like the intercellular junctions in other tissues, these junctions form low-resistance bridges through which ions pass with relative ease. Electrical and conjoint synapses occur in mammals, and there is electrical coupling, for example, between some of the neurons in the lateral vestibular nucleus. However, most synaptic transmission is chemical. Unless otherwise specified, consideration in this chapter is limited to chemical transmission.

Transmission from nerve to muscle resembles chemical synaptic transmission from one neuron to another. The **myoneural junction,** the specialized area where a motor nerve terminates on a skeletal muscle fiber, is the site of a stereotyped transmission process. The contacts between autonomic neurons and smooth and cardiac muscle are less specialized, and transmission in these locations is a more diffuse process.

SYNAPTIC TRANSMISSION

FUNCTIONAL ANATOMY

There is considerable variation in the anatomic structure of synapses in various parts of the mammalian nervous system. The ends of the presynaptic fibers are generally enlarged to form **terminal buttons (synaptic knobs)** (Fig 4–1). Endings are commonly located on dendrites (Fig 4–2) and sometimes on **dendritic spines,** which are small knobs projecting from dendrites. In some instances, the terminal branches of the axon of the presynaptic neuron form a basket or net around the soma of the postsynaptic cell (''basket cells'' of the cerebellum and autonomic ganglia). In other locations, they intertwine with the dendrites of the postsynaptic cell (climbing fibers of the cerebellum) or end on the dendrites directly (apical dendrites of cortical pyramids) or on the axons (axo-axonal endings). On average, each neuron divides to form 1000 synaptic endings, and since there are 10^{12} neurons in the human brain, it follows that there are about 10^{15} synapses. In the spinal cord, the presynaptic endings are closely applied to the soma and the proximal portions of the dendrites of the postsynaptic neuron. The number of synaptic knobs applied to a single spinal motor neuron has been calculated to be about 10,000, with 2000 on the cell body and 8000 on the dendrites. Indeed, there are so many knobs that the neuron appears to be encrusted with them. The portion of the membrane covered by any single synaptic knob is small, but the synaptic knobs are so numerous that, in aggregate, the area covered by them all is often 40% of the soma membrane area (Fig 4–1) and 75% of the dendritic membrane area. It has been calculated that in the cerebral cortex, 98% of the synapses are on dendrites and only 2% are on cell bodies. The ratio of synapses to neurons in the human forebrain has been calculated to be 40,000:1.

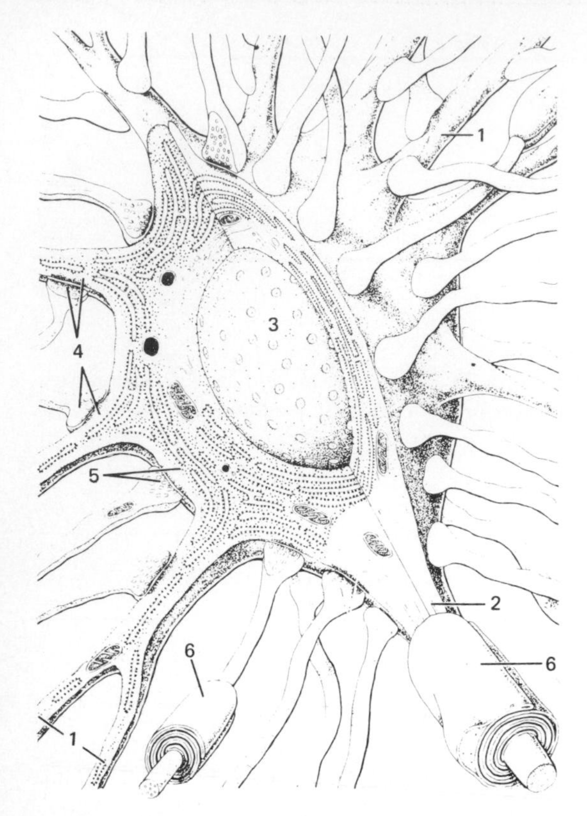

Figure 4–1. Synapses on a typical motor neuron. The neuron has dendrites (1), an axon (2), and a prominent nucleus (3). Note that rough endoplasmic reticulum extends into the dendrites but not into the axon. Many different axons converge on the neuron, and their terminal buttons form axodendritic (4) and axosomatic (5) synapses. (6) Myelin sheath. (Reproduced, with permission, from Krstić RV: *Ultrastructure of the Mammalian Cell.* Springer-Verlag, 1979.)

Synaptic Knobs

Under the electron microscope, the synaptic knobs at synapses where transmission is chemical are found to be separated from the soma of the postsynaptic cell by a definite **synaptic cleft** 30–50 nm wide. The synaptic knob and the soma each have an intact membrane. Inside the knob, there are many mitochondria and small vesicles or granules (Figs 4–2, 4–3), the latter being especially numerous in the part of the knob closest to the synaptic cleft. The vesicles or granules contain small "packets" of the chemical transmitter responsible for synaptic transmission (see below). They vary in morphology depending on the particular transmitter they contain.

The transmitter is released from the synaptic knobs when action potentials pass along the axon to the endings. The membranes of the vesicles or granules fuse to the nerve cell membrane, and the area of fusion breaks down, releasing the contents by the process of exocytosis (see Chapter 1). Ca^{2+} triggers this process, and the action potential opens voltage-gated Ca^{2+} channels in the synaptic knob. Ca^{2+} enters the ending, and exocytosis is increased. The action of Ca^{2+} is terminated by its rapid sequestration within the ending. The amount of transmitter released is proportionate to the Ca^{2+} influx.

At least at the myoneural junction (see below) and at some synaptic junctions in the brain, transmitter is released along dense bars **(active zones)** that extend considerable distances along the membrane of the presynaptic terminal. Membrane particles are lined up, usually along both sides of these bars, and these particles may be Ca^{2+} channels. At most synapses, there is also some degree of postsynaptic specialization in the form of a postsynaptic density or web under the membrane of the postsynaptic cell.

Convergence & Divergence

Only a few of the synaptic knobs on a postsynaptic neuron are endings of any single presynaptic neuron. The inputs to the cell are multiple. In spinal motor neurons, for example, some inputs come directly from the dorsal root, some from the long descending spinal tracts, and many from **interneurons,** the short interconnecting neurons of the spinal cord. Thus, many presynaptic neurons **converge** on any single postsynaptic neuron. Conversely, the axons of most presynaptic neurons divide into many branches that **diverge** to end on many postsynaptic neurons. Convergence and divergence are the anatomic substrates for facilitation, occlusion, and reverberation (see below). Since there are 10^{12} neurons in the human brain and, on the average, each neuron has 1000 inputs converging on it and diverges to form 1000

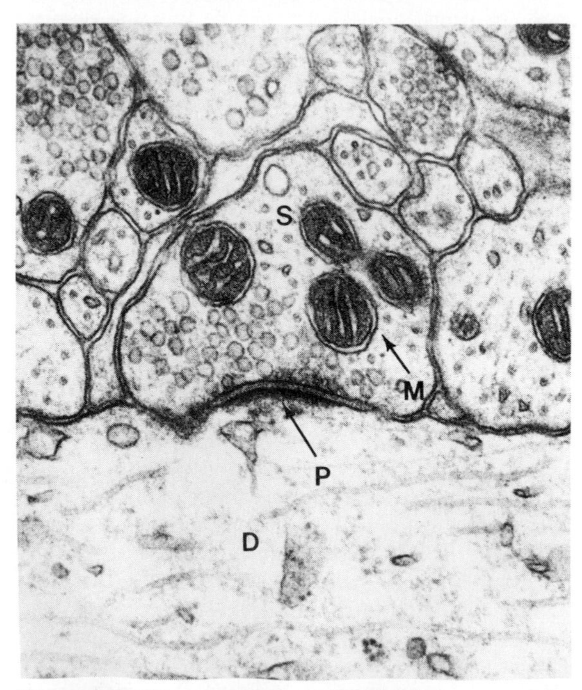

Figure 4–2. Electron photomicrograph of synaptic knob (S) ending on a dendrite (D) in the central nervous system. P, postsynaptic thickening; M, mitochondrion × 56,000. (Courtesy of DM McDonald.)

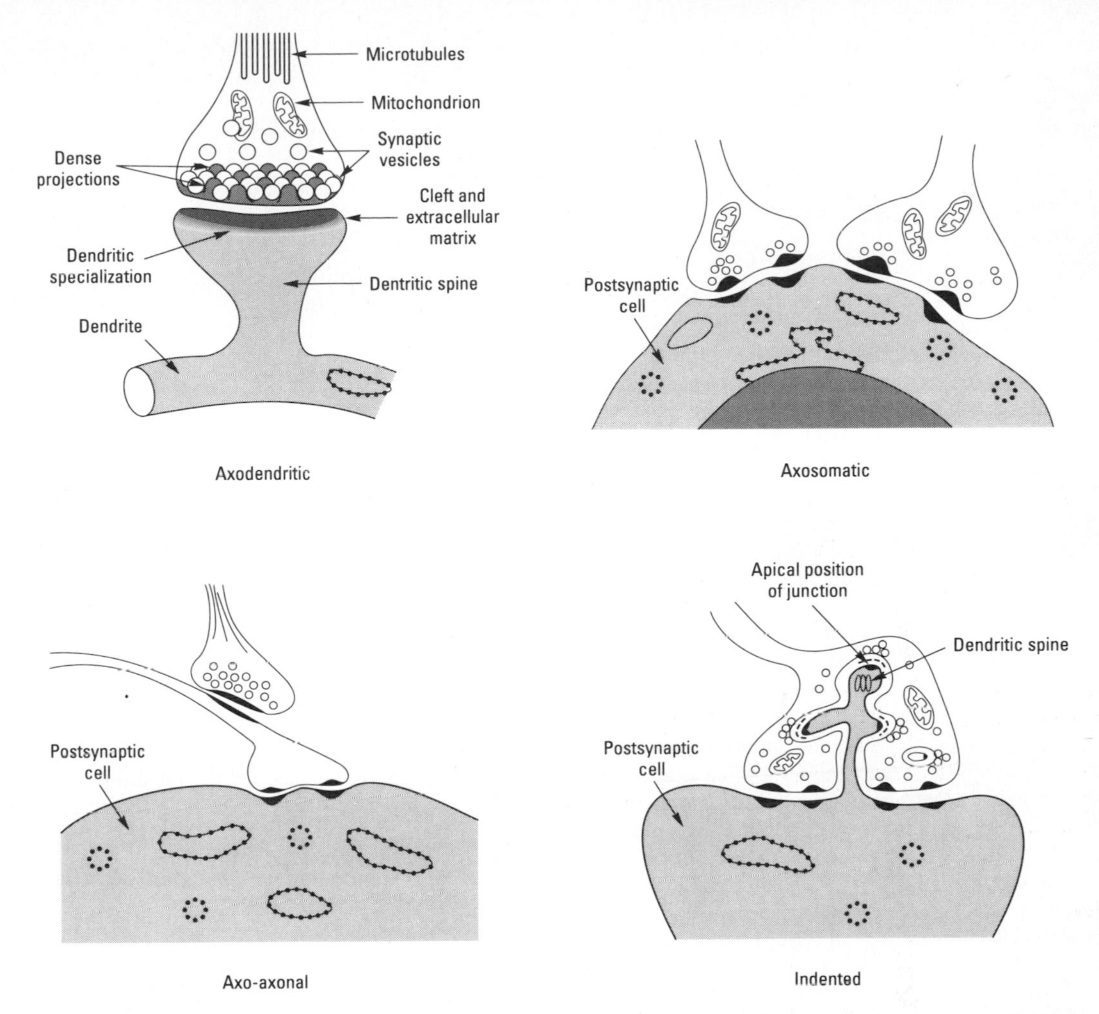

Figure 4–3. Various types of synapses. Gray type I synapses have the synaptic vesicles and active zones spread across the presynaptic terminal and a prominent postsynaptic density, whereas Gray type II synapses have a narrower synaptic cleft and symmetrical collections of synaptic vesicles on either side of the presynaptic knob. Forms that are intermediate between these 2 are also seen.

synaptic junctions on other neurons, the number of possible paths an impulse can take is astronomically large.

One-Way Conduction

Synapses generally permit conduction of impulses in one direction only, from the presynaptic to the postsynaptic neurons. An impulse conducted antidromically up the axons of the ventral root dies out after depolarizing the cell bodies of the spinal motor neurons. Since axons will conduct in either direction with equal facility, the one-way gate at the synapses is necessary for orderly neural function. Chemical mediation at synaptic junctions explains one-way conduction. The mediator is located in the synaptic knobs of the presynaptic fibers, and very little if any is present in the postsynaptic membrane. Therefore, an impulse arriving at the postsynaptic membrane cannot release synaptic mediator. Progression of impulse traffic occurs only when the action potential arrives in the presynaptic terminals and causes secretion of stored chemical transmitter.

Synaptic Development

A fascinating question that has attracted a great deal of attention is how, during development, neurons find the "right" targets and make the "right" synaptic connections. It now seems clear that a number of different factors are involved. It is well established that chemical signals permit neurons to recognize their targets and ignore other neurons in the same region, and there are trophic interactions between pre- and postsynaptic cells once the connections have been made. However, it has been demonstrated that in many situations, many neurons make synaptic connections with other neurons or skeletal muscles and then the "inappropriate" connections disappear, leaving only the connections found in the adult. Many neurons die during development, and it may be the most active neurons and synaptic junctions that persist. In addition, there is competition between neurons for synaptic sites, and nerve endings from adjacent neurons grow into a brain area that has been denervated.

ELECTRICAL EVENTS AT SYNAPSES

Synaptic activity in the spinal cord has been studied in detail by inserting a microelectrode into the soma of a motor neuron and recording the electrical events that follow stimulation of the excitatory and inhibitory inputs to these cells. Activity at other synapses has also been studied, and many of the events occurring at these synapses are similar to those occurring at spinal synapses.

Penetration of an anterior horn cell is achieved by advancing a microelectrode through the ventral portion of the spinal cord. Puncture of a cell membrane is signaled by the appearance of a steady 70-mV potential difference between the microelectrode and an electrode outside the cell. The cell can be identified as a spinal motor neuron by stimulating the appropriate ventral root and observing the electrical activity of the cell. Such stimulation initiates an antidromic impulse (see Chapter 2) that is conducted to the soma and stops at this point. Therefore, the presence of an action potential in the cell after antidromic stimulation indicates that the cell that has been penetrated is a motor neuron rather than an interneuron. Activity in some of the presynaptic terminals impinging on the impaled spinal motor neuron (Fig 4–4) can be initiated by stimulating the dorsal roots.

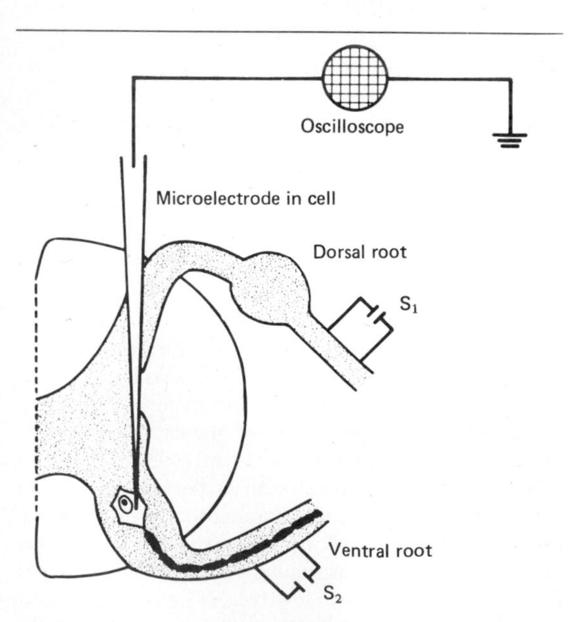

Figure 4–4. Arrangement of recording electrodes and stimulators for studying synaptic activity in spinal motor neurons in mammals. One stimulator (S_2) is used to produce antidromic impulses for identifying the cell; the other (S_1) is used to produce orthodromic stimulation via reflex pathways.

Excitatory Postsynaptic Potentials

Single stimuli applied to the sensory nerves in the experimental situation described above characteristically do not lead to the formation of a propagated action potential in the postsynaptic neuron. Instead, the stimulation produces either a transient, partial depolarization or a transient hyperpolarization.

The initial depolarizing response produced by a single stimulus to the proper input begins about 0.5 ms after the afferent impulse enters the spinal cord. It reaches its peak 1–1.5 ms later and then declines exponentially, with a **time constant** (time required for the response to decay to 1/e, or 1/2.718 of its maximum) that varies depending on the transmitter and the properties of the postsynaptic membrane. During this potential, the excitability of the neuron to other stimuli is increased, and consequently the potential is called an **excitatory postsynaptic potential (EPSP).**

The EPSP is due to depolarization of the postsynaptic cell membrane immediately under the active synaptic knob. The area of inward current flow thus created is so small that it will not drain off enough positive charges to depolarize the whole membrane. Instead, an EPSP is inscribed. The EPSP due to activity in one synaptic knob is small, but the depolarizations produced by each of the active knobs summate.

Summation may be **spatial** or **temporal.** When activity is present in more than one synaptic knob at the same time, spatial summation occurs and activity in one synaptic knob is said to **facilitate** activity in another to approach the firing level. Temporal summation occurs if repeated afferent stimuli cause new EPSPs before previous EPSPs have decayed. Obviously, the longer the time constant for the EPSP, the greater the opportunity for summation. Spatial and temporal facilitation are illustrated in Fig 4–5. The EPSP is therefore not an all or none response but is proportionate in size to the strength of the afferent stimulus. If the EPSP is large enough to reach the firing level of the cell, a full-fledged action potential is produced.

Synaptic Delay

When an impulse reaches the presynaptic terminals, there is an interval of at least 0.5 ms, the **synaptic delay,** before a response is obtained in the postsynaptic neuron. The delay following maximal stimulation of the presynaptic neuron corresponds to the latency of the EPSP and is due to the time it takes for the synaptic mediator to be released and to act on the membrane of the postsynaptic cell. Because of it, conduction along a chain of neurons is slower if there are many synapses in the chain than if there are only a few. Since the minimum time for transmission across one synapse is 0.5 ms, it is also possible to

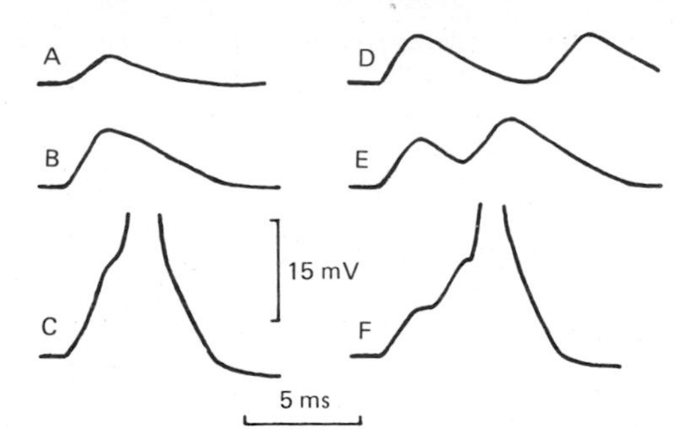

Figure 4–5. Spatial summation (A–C) and temporal summation (D–F) of EPSPs. Records are potential changes recorded with one electrode inside the postsynaptic cell. In A–C, afferent volleys of increasing strength were delivered. In C, the firing level was reached and an action potential generated. In D–F, 2 different volleys of the same strength were delivered, but the time interval between them was shortened. In F, the firing level was reached and an action potential generated.

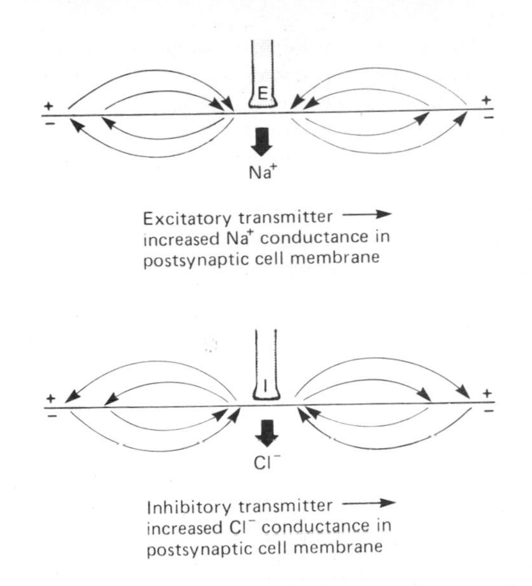

Figure 4–6. Summary of events occurring at synapses in mammals.

determine whether a given reflex pathway is monosynaptic or polysynaptic (contains more than one synapse) by measuring the delay in transmission from the dorsal to the ventral root across the spinal cord.

Ionic Basis of EPSPs

When transmitters that exert an excitatory effect cross the synaptic cleft and bind to appropriate postsynaptic receptors, they usually trigger the opening of chemically gated ion channels clustered in large numbers in the membrane of the postsynaptic cell under the synaptic knob. It is important to remember that whereas axons typically have only voltage-gated Na^+ channels and K^+ channels, the cell bodies, dendrites, and axonal endings have many different kinds of chemically gated channels. The type of postsynaptic response produced by a transmitter depends on the type of channel that is associated with and activated by that particular transmitter.

The production of EPSPs by acetylcholine at nicotinic synapses where acetylcholine is an excitatory transmitter provides a good example of the operation of these mechanisms. Acetylcholine binds to nicotinic receptors, and this triggers the opening of channels that permit Na^+ and other small cations to pass with relative ease. Na^+ moves along its concentration and electrical gradients into the cell (see Chapter 2), and an EPSP is produced (Fig 4–6). However, the area in which this influx occurs is so small that the repolarizing forces are able to overcome its influence, and runaway depolarization of the whole membrane does not result. If more excitatory synaptic knobs are active, more Na^+ enters and the depolarizing potential is greater. If Na^+ influx is great enough, the firing level is reached and a propagated action potential results.

EPSPs can also be produced by agents that close K^+ channels.

Inhibitory Postsynaptic Potentials

An EPSP is usually produced by afferent stimulation, but stimulation of certain presynaptic fibers regularly initiates a hyperpolarizing response in spinal motor neurons. This response often begins 1–1.25 ms after the afferent stimulus enters the cord, reaches its peak in 1.5–2 ms, and declines exponentially with a time constant of about 3 ms (Fig 4–6). Longer hyperpolarizing potentials are also observed (see below). During this potential, the excitability of the neuron to other stimuli is decreased; consequently, it is called an **inhibitory postsynaptic potential (IPSP).** Spatial summation of IPSPs occurs, as shown by the increasing size of the response as the strength of an inhibitory afferent volley is increased. Temporal summation also occurs. This type of inhibition is called **postsynaptic** or **direct inhibition.**

Ionic Basis of IPSPs

The IPSP is often due to a localized increase in membrane permeability to Cl^- but not to Na^+. When an inhibitory synaptic knob becomes active, the released transmitter triggers the opening of Cl^- channels in the area of the postsynaptic cell membrane under the knob. Cl^- moves down its concentration gradient. The net effect is the transfer of negative charge into the cell, so that the membrane potential increases. However, the permeability change is short-lived, and resting conditions are rapidly restored. IPSPs can also be produced by localized opening of K^+ channels, with movement of K^+ out of the postsynaptic cell.

The decreased excitability of the nerve cell during the IPSP is due in part to moving the membrane potential away from the firing level. Consequently, more excitatory (depolarizing) activity is necessary to reach the firing level. The fact that an IPSP is mediated by Cl^- can be demonstrated by repeating the stimulus while varying the resting membrane potential of the postsynaptic cell and holding it with a voltage clamp. When the membrane potential is set at E_{Cl}, the potential disappears (Fig 4–7), and at more negative membrane potentials, it becomes positive. However, it is still inhibitory, because the total Cl^- conductance is increased, and it moves only toward E_{cl}, which is still greater than the resting membrane potential.

IPSPs can also be produced by closure of Na^+ or Ca^{2+} channels.

Neurons Responsible for Postsynaptic Inhibition

Stimulation of certain sensory nerve fibers known to pass directly to motor neurons in the spinal cord produces EPSPs in these neurons and IPSPs in other neurons. In the inhibitory pathways, a single interneuron is inserted between the afferent dorsal root fiber and the motor neuron. This special interneuron, called a Golgi bottle neuron, is short and plump and has a thick axon. Its synaptic transmitter is glycine, and when this amino acid is secreted from its synaptic knobs to the proximal dendrites or cell body of the postsynaptic neuron, an IPSP is produced owing to an increase in the conductance of the postsynaptic cell membrane to Cl^-. In this way, excitatory input is "converted" into inhibitory input.

Slow Postsynaptic Potentials

In addition to the EPSPs and IPSPs described above, slow EPSPs and IPSPs have been described in autonomic ganglia, cardiac and smooth muscle, and cortical neurons. These postsynaptic potentials have a latency of 100–500 ms and last several seconds. The EPSPs are generally due to decreases in K^+ conductance, and the IPSPs are due to increases in K^+ conductance. In sympathetic ganglia, there is also a late slow EPSP that has a latency of 1–5 seconds and lasts 10–30 minutes. This potential is also due, at least in part, to decreased K^+ conductance, and the transmitter responsible for the potential is a peptide very closely related to LHRH, the hormone secreted by neurons in the hypothalamus that stimulates LH secretion (see Chapter 14).

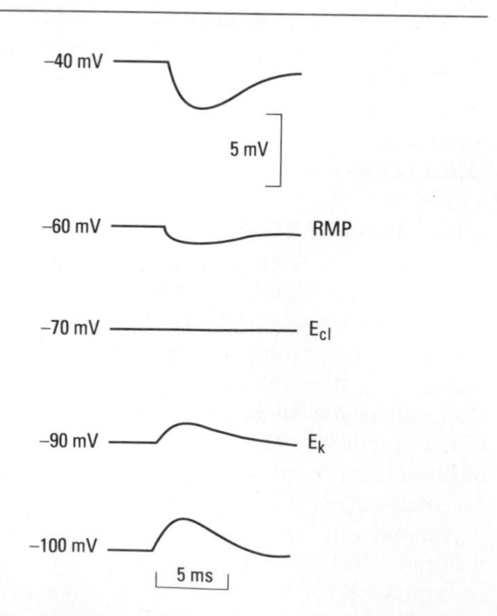

Figure 4–7. IPSPs produced by stimulating inhibitory input to a neuron when its membrane potential is set at various values with a voltage clamp. RMP, testing membrane potential of this neuron. Note that when the voltage is set at E_{Cl}, the IPSP disappears, and at greater membrane voltages, it becomes positive.

Generation of the Action Potential in the Postsynaptic Neuron

The constant interplay of excitatory and inhibitory activity on the postsynaptic neuron produces a fluctuating membrane potential that is the algebraic sum of the hyperpolarizing and depolarizing activity. The soma of the neuron thus acts as a sort of integrator. When the 10–15 mV of depolarization sufficient to reach the firing level is attained, a propagated spike results. However, the discharge of the neuron is slightly more complicated than this. In motor neurons, the portion of the cell with the lowest threshold for the production of a full-fledged action potential is the **initial segment,** the portion of the axon at and just beyond the axon hillock. This unmyelinated segment is depolarized or hyperpolarized electrotonically by the current sinks and sources under the excitatory and inhibitory synaptic knobs. It is the first part of the neuron to fire, and its discharge is propagated in 2 directions; down the axon and back into the soma. Retrograde firing of the soma in this fashion probably has value in "wiping the slate clean" for subsequent renewal of the interplay of excitatory and inhibitory activity on the cell.

Function of the Dendrites

Dendrites usually do not conduct like the axons. Action potentials are generated in some of the dendrites, but they are usually part of the "receptor membrane" of the neuron (see Chapter 2), the site of current sources or sinks that electrotonically change the membrane potential of the initial segment or other locus from which the action potentials are generated. In the central nervous system and the retina, there are in addition neurons with dendrites but no axons. These cells transmit action potentials or spread EPSPs and IPSPs from one neuron to another without a propagated action potential. In the olfactory bulb, for example, action potentials in mitral cells activate granule cell dendrites via dendrodendritic synapses. The dendrite then discharges an inhibitory transmitter on the same mitral cell, and as the electrical activity spreads in the granule cell, transmitter is released on other nearby mitral cells. Such microcircuits appear to be common in various parts of the brain.

When the dendritic tree of a neuron is extensive and has multiple presynaptic knobs ending on it, there is room for a great interplay of inhibitory and excitatory activity. Current flow to and from the dendrites waxes and wanes. The role of the dendrites in the genesis of the EEG is discussed in Chapter 11.

Electrical Transmission

At synaptic junctions where transmission is electrical, the impulse reaching the presynaptic terminal generates an EPSP in the postsynaptic cell that, because of the low-resistance bridge between the two, has a much shorter latency than the EPSP at a synapse where transmission is chemical. In conjoint synapses, there are both a short-latency response and a longer-latency, chemically mediated, postsynaptic response.

INHIBITION & FACILITATION AT SYNAPSES

Direct & Indirect Inhibition

Postsynaptic inhibition during the course of an IPSP is also called **direct inhibition** because it is not a consequence of previous discharges of the postsynaptic neuron. Various forms of **indirect inhibition,** inhibition due to the effects of previous postsynaptic neuron discharge, also occur. For example, the postsynaptic cell can be refractory to excitation because it has just fired and is in its refractory period. During after-hyperpolarization it is also less excitable; and in spinal neurons, especially after repeated firing, this after-hyperpolarization may be large and prolonged.

Postsynaptic Inhibition in the Spinal Cord

The various pathways in the nervous system that are known to mediate postsynaptic inhibition are discussed in Chapter 6, but one illustrative example is presented here. Afferent fibers form the muscle spindles (stretch receptors) in skeletal muscle are known to pass directly to the spinal motor neurons of the motor units supplying the same muscle. Impulses in this afferent supply cause EPSPs and, with summation, propagated responses in the postsynaptic motor neurons. At the same time, IPSPs are produced in motor neurons supplying the antagonistic muscles. This latter response is mediated by branches of the afferent fibers that end on Golgi bottle neurons. These interneurons, in turn, secrete an inhibitory transmitter and end on the proximal dendrites or cell bodies of the motor neurons that supply the antagonist (Fig 4–8). Therefore, activity in the afferent fibers from the muscle spindles excites the motor neurons supplying the muscle from which the impulses come and inhibits those supplying its antagonists **(reciprocal innervation).**

Presynaptic Inhibition & Facilitation

Another type of inhibition occurring in the central nervous system is **presynaptic inhibition,** a process that reduces the amount of neurotransmitter secreted when action potentials arrive at excitatory synaptic knobs. The neurons responsible for postsynaptic and presynaptic inhibition are compared in Fig 4–9. As noted above, an action potential that reaches an ending opens voltage-gated Ca^{2+} channels in the ending, and the resulting Ca^{2+} influx triggers exocytosis of the neurotransmitter-containing vesicles or

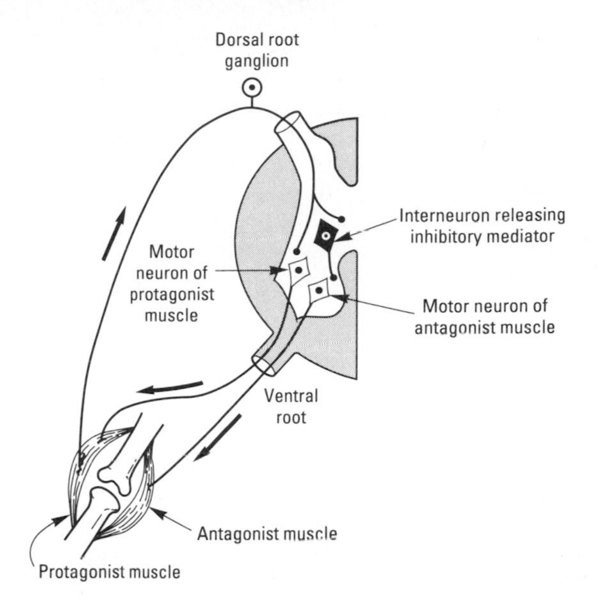

Figure 4–8. Diagram illustrating the probable anatomic connections responsible for inhibiting the antagonists to a muscle contracting in response to stretch. Activity is initiated in the spindle in the protagonist muscle. Impulses pass directly to the motor neurons supplying the same muscle and, via branches, to inhibitory interneurons that end on the motor neurons of the antagonist muscle.

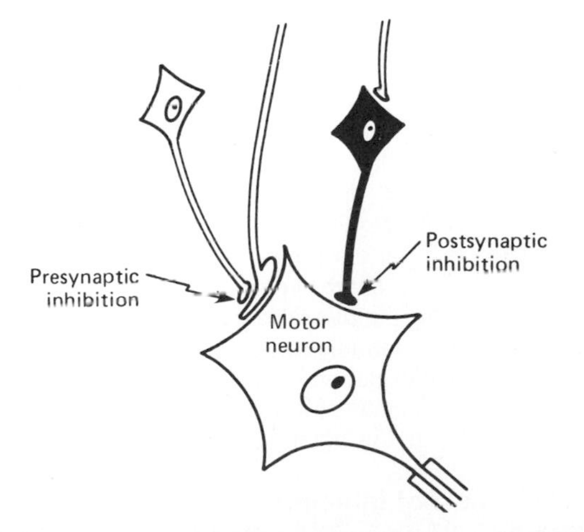

Figure 4–9. Arrangement of neurons producing presynaptic and postsynaptic inhibition. The neuron producing presynaptic inhibition is shown ending on an excitatory synaptic knob. Many of these neurons actually end higher up along the axon of the excitatory cell.

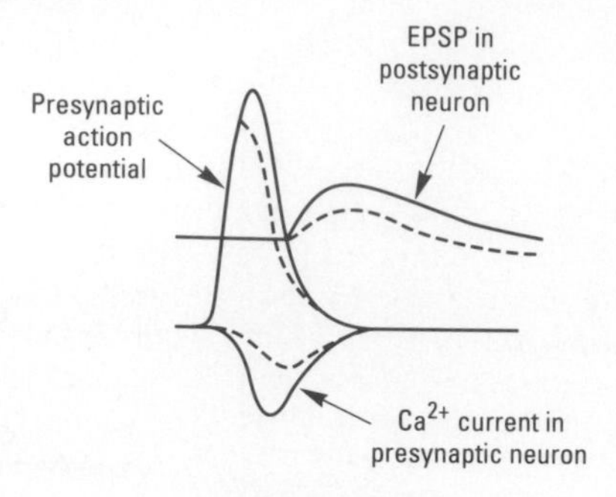

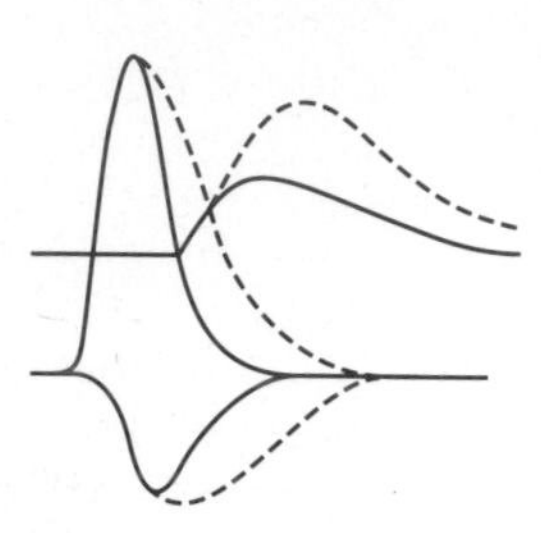

Figure 4–10. Effects of presynaptic inhibition and facilitation on presynaptic action potential, Ca^{2+} current in ending, and EPSP in postganglionic cell. In each case, the solid lines are the controls and the dashed lines the charges seen during inhibition and facilitation.

granules. If the size of the action potential is reduced (Fig 4–10), the Ca^{2+} channels are open for a shorter time and less transmitter is released. GABA, the transmitter responsible for presynaptic inhibition, has this effect; it increases Cl^- conductance, but in this instance, the membrane potential is decreased, reducing the size of the action potential because E_{Cl} is less than the resting membrane potential. Alternatively, in some situations, the transmitter released at the axo-axonic ending may depress Ca^{2+} channels. Conversely, **presynaptic facilitation** is produced when the action potential is prolonged (Fig 4–10) and the Ca^{2+} channels are open for a longer period. The molecular events responsible for the production of presynaptic facilitation mediated by serotonin in the sea snail *Aplysia* have been worked out in detail (Table 4–1). Serotonin released at an axo-axonic ending increases intraneuronal cyclic AMP, and the resulting phosphorylation of one group of K^+ channels closes the channels, slowing depolarization and prolonging the action potential.

Organization of Inhibitory Systems

Presynaptic and postsynaptic inhibition are usually produced by stimulation of certain systems converging on a given postsynaptic neuron ("afferent inhibition"). Neurons may also inhibit themselves in a negative feedback fashion ("negative feedback inhibition"). For instance, spinal motor neurons regularly give off a recurrent collateral, which synapses with an inhibitory interneuron that terminates on the cell body of the spinal neuron and other spinal motor neurons (Fig 4–11). This particular inhibitory neuron is sometimes called the Renshaw cell after its discoverer. Impulses generated in the motor neuron activate the inhibitory interneuron to secrete inhibitory mediator, and this slows or stops the discharge of the motor neuron. Similar inhibition via recurrent collaterals is seen in the cerebral cortex and limbic system. Presynaptic inhibition due to descending pathways that terminate on afferent pathways in the dorsal horn may be involved in the "gating" of pain transmission (see Chapter 7).

Another type of inhibition is seen in the cerebellum. In this part of the brain, stimulation of basket cells produces IPSPs in the Purkinje cells. However, the basket cells and the Purkinje cells are excited by the same excitatory input. This arrangement, which has been called "feed-forward inhibition," presumably limits the duration of the excitation produced by any given afferent volley.

Summation & Occlusion

The interplay between excitatory and inhibitory influences at synaptic junctions in a nerve net illus-

Table 4–1. Sequence of event producing presynaptic facilitation in *Aplysia*.

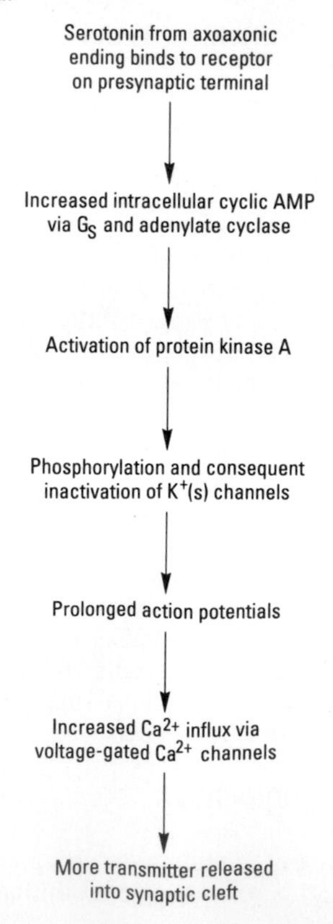

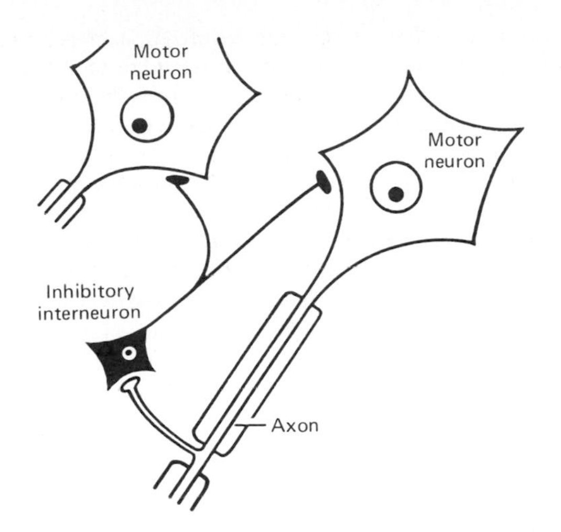

Figure 4–11. Negative feedback inhibition of a spinal motor neuron via an inhibitory interneuron (Renshaw cell).

trates the integrating and modulating activity of the nervous system.

In the hypothetical nerve net shown in Fig 4–12, neurons A and B converge on X, and neuron B diverges on X and Y. A stimulus applied to A or to B will set up an EPSP in X. If A and B are stimulated at the same time, 2 areas of depolarization will be produced in X and their actions will sum. The resultant EPSP in X will be twice as large as that produced by stimulation of A or B alone, and the membrane potential may well reach the firing level of X. The effect of the depolarization caused by the impulse in A is facilitated by that due to activity in B, and vice versa; spatial facilitation has taken place. In this case, Y has not fired, but its excitability has been increased, and it is easier for activity in neuron C to fire Y during the EPSP. Y is therefore said to be in the **subliminal fringe** of X. More generally stated, neurons are in the subliminal fringe if they are not discharged by an afferent volley (not in the **discharge zone**) but do have their excitability increased. The neurons that have few active knobs ending on them are in the subliminal fringe, and those with many are in the discharge zone. However, this does not mean that with increasing strength of afferent stimulus the discharge zone becomes the same size as the subliminal fringe. During stimulation of a dorsal root, the number of neurons discharged increases to a maximum with increasing strength of stimulation of the root, but so does the size of the subliminal fringe. Inhibitory impulses show similar temporal and spatial facilitation and subliminal fringe effects.

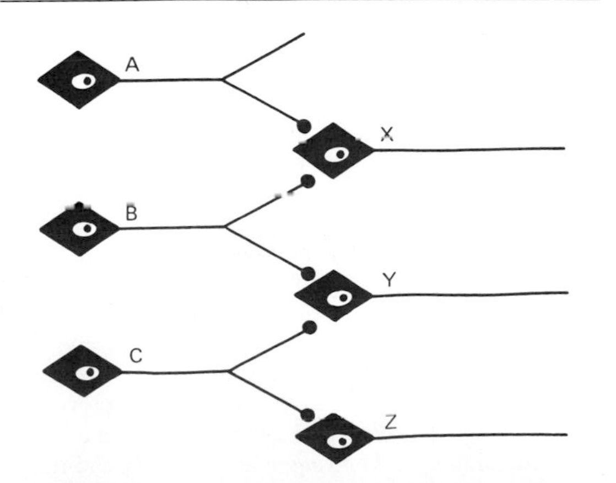

Figure 4–12. Simple nerve net. Neurons A, B, and C have excitatory endings on neurons X, Y, and Z.

If neuron B in Fig 4–12 is stimulated repetitively, X and Y will discharge as a result of temporal summation of the EPSPs produced. If C is stimulated repetitively, Y and Z will discharge. If B and C are fired repetitively at the same time, X, Y, and Z will discharge. Thus, the response to stimulation of B and C together is not as great as the sum of responses to stimulation of B and C separately, because B and C both end on neuron Y. This decrease in expected response, due to presynaptic fibers sharing postsynaptic neurons, is called **occlusion.**

Excitatory and inhibitory subliminal effects and occlusive phenomena can have pronounced effects on transmission in any given pathway. Because of these effects, temporal patterns in peripheral nerves are usually altered as they pass through synapses on the way to the brain. These effects may also explain such important phenomena as referred pain (see Chapter 7).

Synaptic Plasticity & Learning

Long-term changes in synaptic function can occur as a result of the history of discharge at a synapse. Four classes of synaptic chance can occur: 2 are homosynaptic, ie, they result from changes in the presynaptic neuron, whereas 2 are heterosynaptic, ie, they are mediated by other neurons that converge on the synapse. All are propably mediated by changes in the concentration of Ca^{2+} within the presynaptic terminal.

One form of homosynaptic plastic change is **posttetanic potentiation,** an enhanced postsynaptic potential in response to stimulation. This enhancement lasts for minutes to hours and occurs after a brief (tetanizing) train of stimuli in the presynaptic neuron. The tetanizing stimulation causes Ca^{2+} to accumulate in the presynaptic neuron to such a degree that the intracellular binding sites that keep cytoplasmic Ca^{2+} low are overwhelmed. Conversely, homosynaptic **low-frequency depression** occurs following a period of prolonged low-frequency stimulation of the presynaptic neuron and is due to decreased intracellular Ca^{2+}. This is the mechanism by which the response to a given stimulus decreases and eventually disappears when the stimulus is benign and is repeated over and over **(habituation).** The 2 heterosynaptic forms of plastic change are presynaptic inhibition and presynaptic facilitation, which is also called **sensitization.** Sensitization is the prolonged occurrence of augmented postsynaptic responses after a stimulus to which an animal has become habituated is paired once or twice with a noxious stimulus. At least in the sea snail *Aplysia,* the noxious stimulus causes discharge of serotonergic neurons that end on the presynaptic endings of other neurons. This initiates the sequence of events summarized in Table 4–1.

All 4 of the forms of plastic change described above represent changes produced by experience and consequently are primitive forms of learning. Sensitization is also related to classical conditioning (see Chapter 16).

CHEMICAL TRANSMISSION OF SYNAPTIC ACTIVITY

Implications

The fact that transmission at most if not all synapses is chemical is of great physiologic and pharmacologic importance. Nerve endings have been called biologic transducers that convert electrical energy into chemical energy. In broad terms, this conversion process involves the synthesis of the transmitter agents, their storage in the synaptic knobs, and their release by the nerve impulses into the synaptic cleft. The secreted transmitters then act on appropriate receptors on the membrane of the postsynaptic cell and are rapidly removed from the synaptic cleft by diffusion, metabolism, and, at least for amine transmitters, reuptake into the presynaptic neuron. All these processes, plus the postreceptor events in the postsynaptic neuron, are regulated by many physiologic factors and at least in theory can be altered by drugs. Therefore pharmacologists should be able to develop drugs that regulate not only somatic and visceral motor activity but also emotions, behavior, and all the other complex functions of the brain.

Chemistry of Transmitters

The nature of the chemical mediators at many synapses is not known. However, the number of identified synaptic mediators is constantly increasing, and so is the number of locations in the nervous system in which given transmitters are strongly suspected or known. Evidence that a substance is a transmitter includes its uneven distribution in the nervous system and a parallel distribution of the enzymes responsible for the substance's synthesis and catabolism. Additional evidence is provided by differential centrifugation of brain tissue, which has demonstrated the presence of many suspected mediators in fractions known to contain nerve endings. It has also been shown that certain of the mediators are released by the brain in vitro and that glutamate and some other central nervous system mediators excite single neurons when applied to their membranes by means of a micropipette (microiontophoresis). Many transmitters have now been localized in nerve endings by special staining techniques. One of the most useful and important techniques has been **immunocytochemistry,** in which antibodies to a given substance are labeled and applied to brain and other tissues. The antibodies bind to the substance, and the location of the substance is then determined by locating the label with the light microscope or electron microscope. Histochemical and immunocytochemical techniques have been used to localize serotonin, dopamine, norepinephrine, epinephrine, the enzymes that synthesize and metabolize these substances, and many different peptides.

Identified neurotransmitters can be divided into broad categories or families based on their chemical structure; some are amines, others are amino acids, and many are polypeptides. It is worth noting that most of these substances are not only released into synaptic clefts, where they produce highly localized effects. In other situations, they diffuse in the ECF and exert effects at some distance from their site of release (paracrine communication; see Chapter 1). In some cases, they are also released by neurons into the bloodstream as hormones. A somewhat arbitrary compilation of most of the substances currently known or suspected to be synaptic mediators is presented in Table 4–2.

It is now known that many and possibly all neurons contain more than one transmitter; ie, they contain **cotransmitters.** Often, but not always, an amine is colocalized with one or more polypeptides. This phenomenon is discussed in more detail below.

Receptors

Cloning and related molecular biology techniques have permitted spectacular recent advances in knowledge about the structure and function of receptors for neurotransmitters and other chemical messengers. The individual receptors, along with their ligands, are discussed in the following parts of this chapter. However, 3 themes have emerged that should be mentioned in this introductory paragraph.

First, in every instance studied in detail to date, it has become clear that for each ligand, there are many subtypes of receptors. Thus, for example, norepinephrine acts on α_1, α_2, β_1, and β_2 receptors, and there is now some evidence that there are 2 kinds of α_1 receptors. Obviously, this multiplies and makes more selective in any given cell the possible effects of a given ligand.

Second, there are receptors on the pre-as well as the postsynaptic elements at many synaptic junctions. These **presynaptic receptors,** or **autoreceptors,** often inhibit further secretion of the ligand, providing feedback control. For example, norepinephrine acts on α_2 presynaptic receptors to inhibit norepinephrine secretion (Fig 4–13). However, autoreceptors can also facilitate the release of neurotransmitters.

Third, although there are many ligands and many subtypes of receptors for each ligand, the receptors tend to group in large families as far as structure and

Table 4–2. Synaptic transmitter agents and "neural hormones" in mammals.

Substance	Locations
Acetylcholine	Myoneural junction; preganglionic autonomic endings, postganglionic parasympathetic endings, postganglionic sympathetic sweat gland and muscle vasodilator endings; many parts of brain; endings of some amacrine cells in retina.
Amines	
Norepinephrine	Most postganglionic sympathetic endings; cerebral cortex, hypothalamus, brain stem, cerebellum, spinal cord.
Dopamine	SIF cells in sympathetic ganglia; striatum, median eminence, and other parts of hypothalamus; limbic system; parts of neocortex; endings of some interneurons in retina.
Epinephrine	Hypothalamus, thalamus, periaqueductal gray, spinal cord.
Serotonin	Hypothalamus, limbic system, cerebellum, spinal cord; retina, gastrointestinal tract.
Histamine	Hypothalamus.
Amino acids	
Glycine	Neurons mediating direct inhibition; retina.
Gamma-aminobutyrate (GABA)	Cerebellum; cerebral cortex; neurons mediating presynaptic inhibition; retina.
Glutamate	Cerebral cortex, brain stem.
Aspartate	Spinal cord.
Polypeptides	
Substance P	Endings of primary afferent neurons mediating nociception; many parts of brain; retina; gastrointestinal tract.
Vasopressin	Posterior pituitary; medulla; spinal cord.
Oxytocin	Posterior pituitary; medulla; spinal cord.
Hypothalamic hypophyseotropic hormones (see Chapter 14)	
CRH	Median eminence of hypothalamus; other part of brain.
TRH	Median eminence of hypothalamus; other parts of brain; retina; gastrointestinal tract.
GRH	Median eminence of hypothalamus.
Somatostatin	Median eminence of hypothalamus; other parts of brain; substantia gelatinosa; retina; gastrointestinal tract.
LHRH	Median eminence of hypothalamus; circumventricular organs; preganglionic autonomic endings; retina.
Enkephalins	Substantia gelatinosa, many other parts of CNS; retina; gastrointestinal tract.
β-Endorphin, other derivatives of pro-opiomelanocortin	Hypothalamus, thalamus, brain stem; retina.
Cholecystokinin (CCK) octapeptide	Cerebral cortex; hypothalamus; retina.
Vasoactive intestinal polypeptide (VIP)	Postganglionic cholinergic neurons; some sensory neurons; hypothalamus; cerebral cortex; gastrointestinal tract; retina.
Neurotensin	Hypothalamus; gastrointestinal tract; retina.
Gastrin-releasing peptide (GRP)	Hypothalamus; gastrointestinal tract.
Gastrin	Hypothalamus; medulla oblongata.
Glucagon	Hypothalamus; retina.
Motilin	Neurohypophysis; cerebral cortex, cerebellum.
Secretin	Hypothalamus, thalamus, olfactory bulb, brain stem, cerebral cortex, septum, hippocampus, striatum.
Calcitonin-gene-related peptide (CGRP) α	Endings of primary afferent neurons; taste pathways, sensory nerves.
CGRPβ	Gastrointestinal tract
Neuropeptide Y	Noradrenergic, adrenergic, and other neurons in medulla, periaqueductal gray, hypothalamus.
Inhibin	Brain stem.
Angiotensin II	Hypothalamus, amygdala, brain stem, spinal cord.
FMRF amide	Hypothalamus, brain stem
Galanin	Hypothalamus.
Atrial natriuretic peptide (ANP)	Hypothalamus, brain stem.
Brain natriuretic peptide (BNP)	Hypothalamus, brain stem.

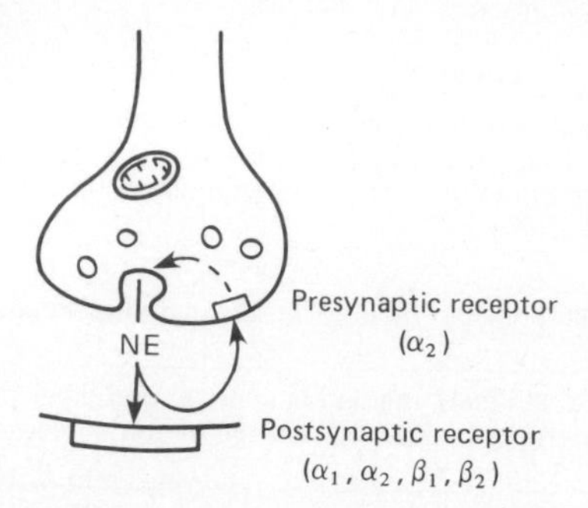

Figure 4–13. Presynaptic and postsynaptic receptors at the ending of a noradrenergic neuron. The presynaptic receptor that is shown is α_2; the postsynaptic receptors can be α_1, α_2, β_1, or β_2.

function are concerned. Many act via G proteins to bring about changes in intracellular metabolism, and it is striking that these receptors appear to have a common structure; they are large proteins that span the cell membrane 7 times. Thus, for example, the β_2 adrenergic receptor and rhodopsin have a similar structure (Fig 4–14), and a similar, pattern with 7 membrane-spanning domains occurs in the β_1 and α_1 adrenergic receptors, the other 3 human visual pigments (see Chapter 8), all 4 subtypes of muscarinic receptors, at least 2 kinds of serotonin receptors $5HT_{1A}$ and $5HT_{1C}$, and the receptor for one of the polypeptides, the tachykinin substance K. Another group of receptors are ion channels that are opened or closed by their ligands. Included in this group are the nicotinic cholinergic receptor, the GABA receptor, and the 3 kinds of glutamate receptors.

It is obvious that synaptic physiology is a rapidly expanding, complex field that cannot be covered in detail in this book. However, it is appropriate to summarize information about some of the principal neurotransmitters and their receptors.

Acetylcholine

The relatively simple structure of acetylcholine, which is the acetyl ester of choline, is shown in Fig 4–15. It exists, largely enclosed in small, clear synaptic vesicles, in high concentration in the terminal buttons of cholinergic neurons. Neurons that release acetylcholine are known as **cholinergic** neurons. The arrival of an impulse at a synaptic knob increases the Ca^{2+} permeability of the membrane, and the resultant Ca^{2+} influx causes secretion of acetylcholine into the synaptic cleft by the process of exocytosis.

Acetylcholine Receptors

Acetylcholine receptors can be divided into 2 main types on the basis of their pharmacologic properties. Muscarine, the alkaloid responsible for the toxicity of toadstools, has little effect on the receptors in autonomic ganglia but mimics the stimulatory action of acetylcholine on smooth muscle and glands. These actions of acetylcholine are therefore called **muscarinic actions,** and the receptors involved are **muscarinic receptors.** They are blocked by the drug atropine. In sympathetic ganglia, small amounts of acetylcholine stimulate postganglionic neurons and large amounts block transmission of impulses from pre- to postganglionic neurons. These actions are unaffected by atropine but mimicked by nicotine. Consequently, these actions of acetylcholine are **nicotinic actions** and the receptors are **nicotinic receptors.** The receptors in the motor end-plates of skeletal muscle (see below) are also nicotinic but are not identical to those in sympathetic ganglia, since they respond differently to certain other drugs.

Nicotinic acetylcholine receptors have been studied in detail in mammalian skeletal muscle and in the

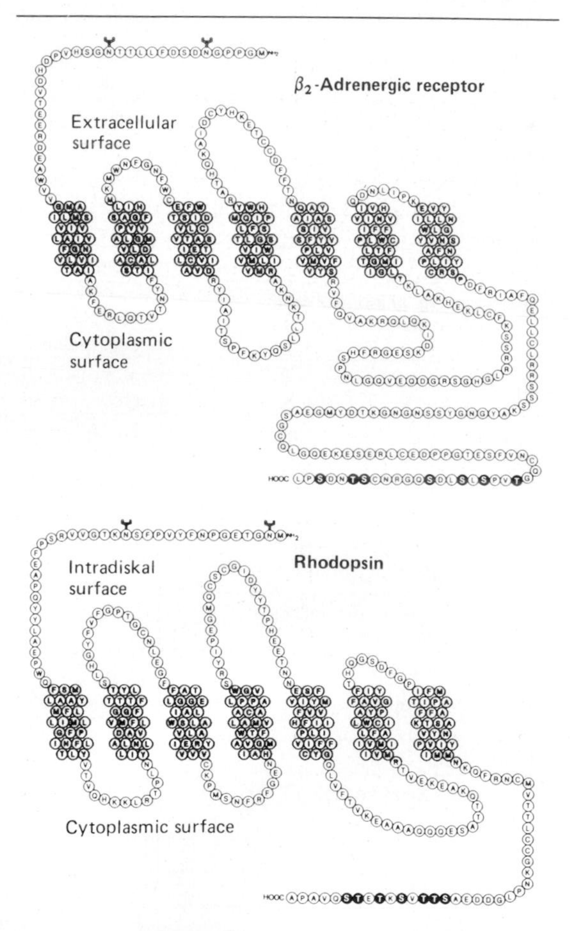

Figure 4–14. Structure of the β_2-adrenergic receptor and rhodopsin. The individual amino acid residues are identified by their single-letter codes (see Table 17–2), and the black residues are sites of phosphorylation. The y-shaped symbols on N residues identify glycosylation sites. Note the extracellular N terminal, the intracellular C terminal, and the 7 membrane-spanning portions of each protein. β_1 and α_2 receptors have a similar structure. (Reproduced, with permission, from Benovic JL et al: Light-dependent phosphorylation of rhodopsin by β-adrenergic receptor kinase. Reprinted by permission from *Nature* 1986;*321:*869. Copyright © 1986, Macmillan Journals Limited.)

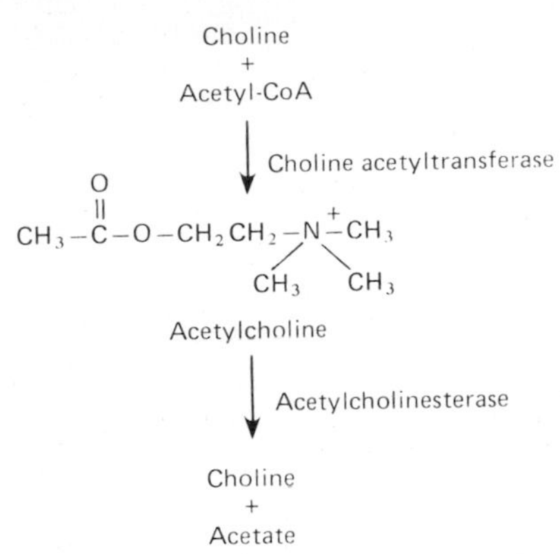

Figure 4–15. Biosynthesis and catabolism of acetylcholine.

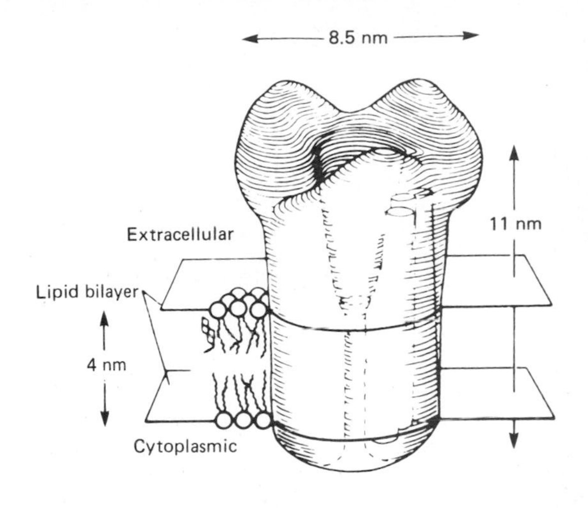

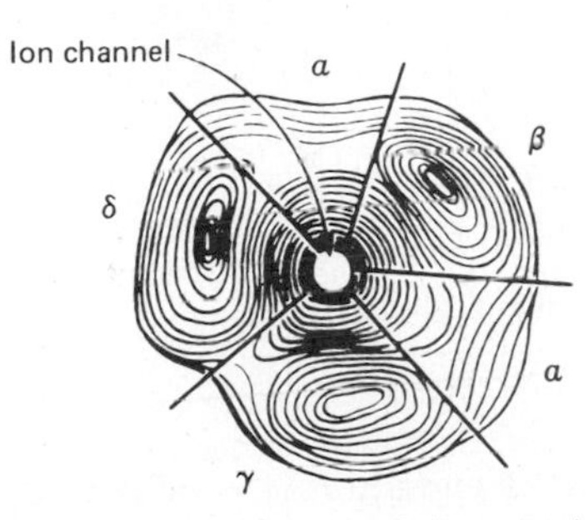

Figure 4–16. Diagram of fetal nicotinic acetylcholine receptor as viewed from side *(above)* and from top *(below)*. α, β, γ, δ: receptor subunits. (Reproduced, with permission, from McCarthy MP et al: Molecular biology of the acetylcholine receptor. *Annu Rev Neurosci* 1986;9:383. Reproduced, with permission, from the *Annual Review of Neuroscience,* vol 9. © 1986 by Annual Reviews Inc.)

electric organ of the electric eel, an organ made up of a large collection of motor end-plates. In the electric eel and in fetal muscle in mammals, the nicotinic acetylcholine receptor is a protein with a molecular weight of slightly more than 250,000. It is made up of 5 subunits: 2 identical α subunits and one β, one γ, and one δ subunit. These 5 subunits extend through the cell membrane (Fig 4–16) and are arranged in a nearly symmetric fashion around a channel that is wide outside the cell and narrows as it goes through the cell membrane. There is a binding site for acetylcholine on each α subunit, and when 2 acetylcholine molecules bind in this fashion, they induce a configurational change in the protein so that the channel opens. This increases the conductance of Na^+ and other cations, and the resulting influx of Na^+ produces a depolarizing potential.

The subunits of the nicotinic receptor described above are coded by 4 different genes. In adult mammals, the γ subunit is replaced by the ϵ subunit, which is apparently coded by yet another gene. This change does not alter its pentagonal structure but decreases its open time and increases its conductance. The nature of the signal that brings about the switch from the fetal γ subunit to the adult ϵ subunit is unsettled. The half-life of nicotinic receptors in the cell membrane is about 1 week.

Muscarinic acetylcholine receptors are quite different from nicotinic acetylcholine receptors. Two types of muscarinic receptors can be distinguished: M_1 receptors, which bind the drug pirenzepine, are found in the central nervous system and elsewhere; and M_2 receptors, which do not bind pirenzepine to as great a degree, are found in the heart and other locations. M_1 receptors probably act via phospholipase C to mediate decreases in K^+ conductance and therefore exert excitatory effects, whereas M_2 receptors act via inhibition of adenylate cyclase to increase K^+ conductance and therefore produce inhibitory effects. The muscarinic receptors have now been cloned, and it has been found that 4 instead of 2 different receptors are encoded by 4 different genes. All have the characteristic 7 membrane-spanning domains (see above) and act via G proteins, but the physiologic and pharmacologic characteristics of the 2 additional receptors are not yet known.

Cholinesterases

Acetylcholine must be rapidly removed from the synapse if repolarization is to occur. The removal occurs by way of hydrolysis of acetylcholine to choline and acetate, a reaction catalyzed by the enzyme **acetylcholinesterase.** This enzyme is also called **true** or **specific cholinesterase.** Its greatest affinity is for acetylcholine, but it also hydrolyzes other choline esters. There are a variety of esterases in the body. One found in plasma is capable of hydrolyzing acetylcholine but has different properties from acetylcholinesterase. It is therefore called **pseudocholinesterase** or **nonspecific cholinesterase.** The plasma moiety is partly under endocrine control and is affected by variations in liver function. On the other hand, the specific cholinesterase at nerve endings is highly localized. Hydrolysis of acetylcholine by this enzyme is rapid enough to explain the

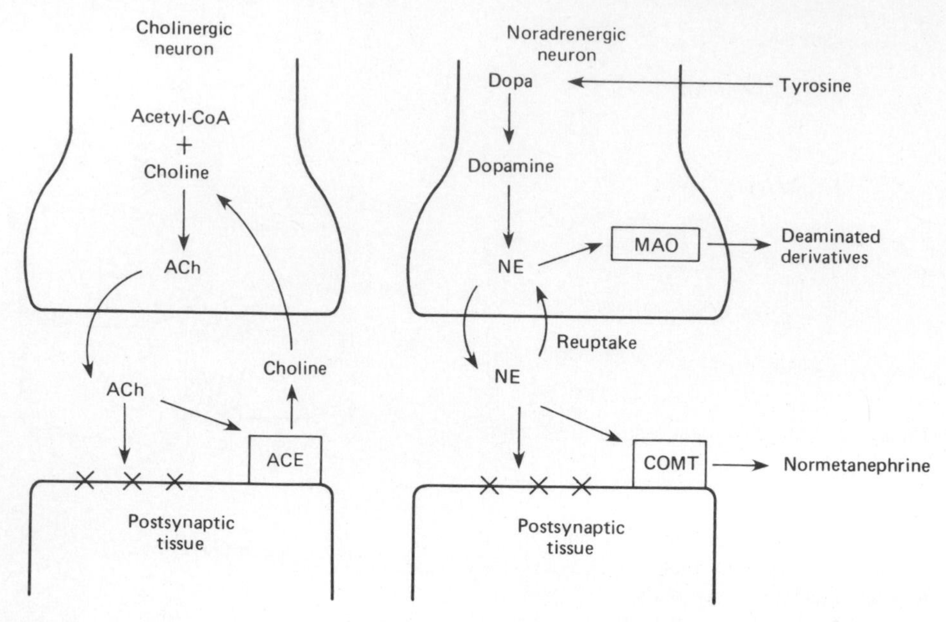

Figure 4–17. Comparison of the biochemical events at cholinergic endings with those at noradrenergic endings. ACh, acetylcholine; ACE, acetylcholinesterase; NE, norepinephrine; X, receptor. Note that monoamine oxidase (MAO) is intracellular, so that some norepinephrine is being constantly deaminated in noradrenergic endings. Catechol-O-methyltransferase (COMT) acts on norepinephrine after it is secreted.

observed changes in Na^+ conductance and electrical activity during synaptic transmission.

Acetylcholine Synthesis

Synthesis of acetylcholine involves the reaction of choline with acetate. There is an active uptake of choline into cholinergic neurons (Fig 4–17). Choline is also synthesized in neurons. The acetate is activated by the combination of acetate groups with reduced coenzyme A. The reaction between active acetate (acetyl-coenzyme A, acetyl-CoA) and choline is catalyzed by the enzyme **choline acetyltransferase.** This enzyme is found in high concentration in the cytoplasm of cholinergic nerve endings; indeed, its localization is so specific that the presence of a high concentration in any given neural area has been taken as evidence that the synapses in that area are cholinergic.

Norepinephrine & Epinephrine

The chemical transmitter at most sympathetic postganglionic endings is norepinephrine (levarterenol). It is stored in the synaptic knobs of the neurons that secrete it in characteristic vesicles that have a dense core (granulated vesicles). There are 2 populations of granulated vesicles: small vesicles about 40 nm in diameter and large vesicles about 75 nm in diameter. Norepinephrine and its methyl derivative, epinephrine, are secreted by the adrenal medulla (see Chapter 20), but epinephrine is not a mediator at postganglionic sympathetic endings. The endings of sympathetic postganglionic neurons in smooth muscle are discussed below; each neuron has multiple varicosities along its course, and each of these varicosities appears to be a site at which norepinephrine is liberated. There are also norepinephrine-secreting, dopamine-secreting, and epinephrine-secreting neurons in the brain (see Chapter 15). Norepinephrine-secreting neurons are properly called **noradrenergic neurons,** although the term adrenergic neurons is also applied. However, it seems appropriate to reserve the term **adrenergic neurons** for epinephrine-secreting neurons. Dopamine-secreting neurons are called **dopaminergic neurons.**

Biosynthesis & Release of Catecholamines

The principal **catecholamines** found in the body—norepinephrine, epinephrine, and dopamine—are formed by hydroxylation and decarboxylation of the amino acids phenylalanine and tyrosine (Fig 4–18). **Phenylalanine hydroxylase** is found primarily in the liver. Tyrosine is transported into catecholamine-secreting neurons and adrenal medullary cells by a concentrating mechanism. It is converted to dopa and then to dopamine in the cytoplasm of the cells by **tyrosine hydroxylase** and **dopa decarboxylase.** The decarboxylase, which is also called aromatic L-amino acid decarboxylase, is very similar but probably not identical to 5-hydroxytryptophan decarboxylase. The dopamine then enters the granulated vesicles, within which it is converted to norepinephrine by **dopamine β-hydroxylase.** L-Dopa is the isomer involved, but the norepinephrine that is formed is in the D configuration. This is true even though it is levorotatory (l, or [−]). Dextrorotatory (+) norepinephrine is much

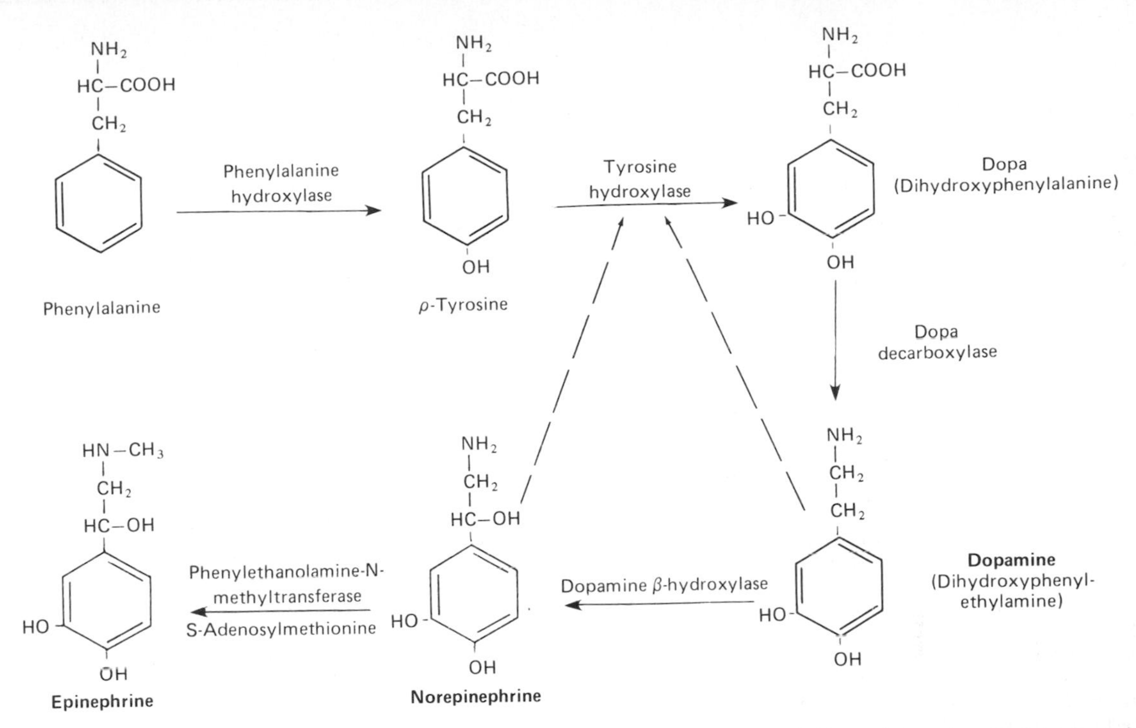

Figure 4–18. Biosynthesis of catecholamines. The dashed lines indicate inhibition of tyrosine hydroxylase by norepinephrine and dopamine. Tetrahydrobiopterin is a cofactor for the action of phenylalanine hydroxylase and tyrosine hydroxylase.

less active. The rate-limiting step in synthesis is the conversion of tyrosine to dopa. Tyrosine hydroxylase, which catalyzes this step, is subject to feedback inhibition by dopamine and norepinephrine, thus providing internal control of the synthetic process. The cofactor for tyrosine hydroxylase is **tetrahydrobiopterin,** which is converted to dihydrobiopterin when tyrosine is converted to dopa.

Some neurons and adrenal medullary cells also contain the cytoplasmic enzyme **phenylethanolamine-N-methyltransferase (PNMT),** which catalyzes the conversion of norepinephrine to epinephrine. In these cells, norepinephrine apparently leaves the vesicles, is converted to epinephrine, and then enters other storage vesicles.

In granulated vesicles, norepinephrine and epinephrine are bound to ATP and associated with proteins called **chromogranins,** the function of which is unknown. In some but not all noradrenergic neurons, the large granulated vesicles also contain neuropeptide Y (see below). The most abundant chromogranin is an acidic protein called **chromogranin A.** This protein or a protein closely related to it is also found in various cells that secrete polypeptides, and it may play some sort of general role in hormone storage or secretion. Plasma levels of chromogramin A are increased in patients with a variety of different peptide hormone-secreting tumors, and measurement of plasma chromogramin A may be of value in the diagnosis of these tumors.

The catecholamines are held in the granulated vesicles by an active transport system, and the action of this transport system is inhibited by the drug reserpine.

Catecholamines are released from autonomic neurons and adrenal medullary cells by exocytosis (see Chapter 1). Since they are present in the granulated vesicles, ATP, chromogranins, and the dopamine β-hydroxylase that is not membrane-bound are released with norepinephrine and epinephrine. The half-life of circulating dopamine β-hydroxylase is much longer than that of the catecholamines, and circulating levels of this substance are affected by genetic and other factors in addition to the rate of sympathetic activity. Circulating levels of chromogranin A appear to be a better index of sympathetic activity.

Some of the norepinephrine in nerve endings is manufactured there, but some is also norepinephrine that has been secreted and then taken up again into the noradrenergic neurons. An active **reuptake mechanism** is characteristic of noradrenergic neurons. Circulating norepinephrine and epinephrine are also picked up in small amounts by noradrenergic neurons in the autonomic nervous system. In this regard, noradrenergic neurons differ from cholinergic neurons. Acetylcholine is not taken up to any appreciable degree, but the choline formed by the action of acetylcholinesterase is actively taken up and recycled (Fig 4–17).

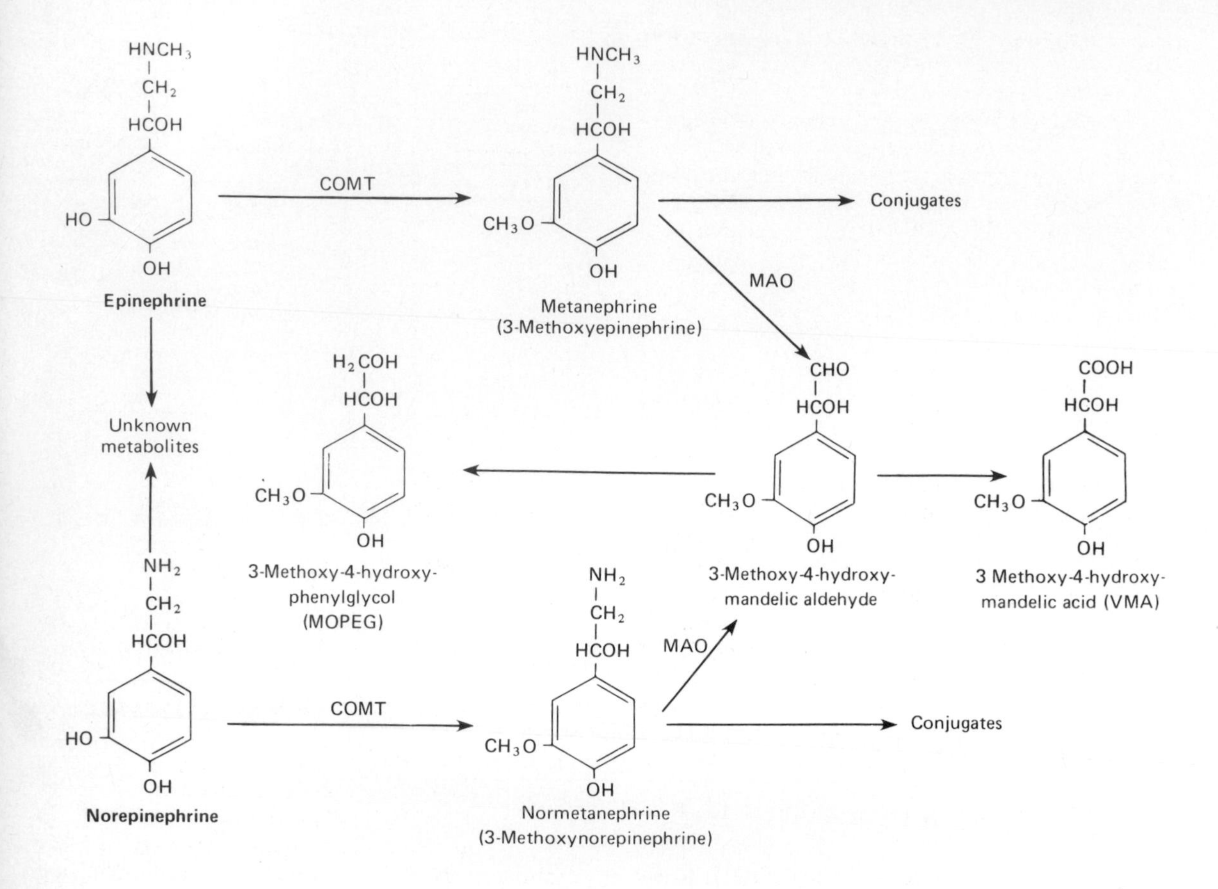

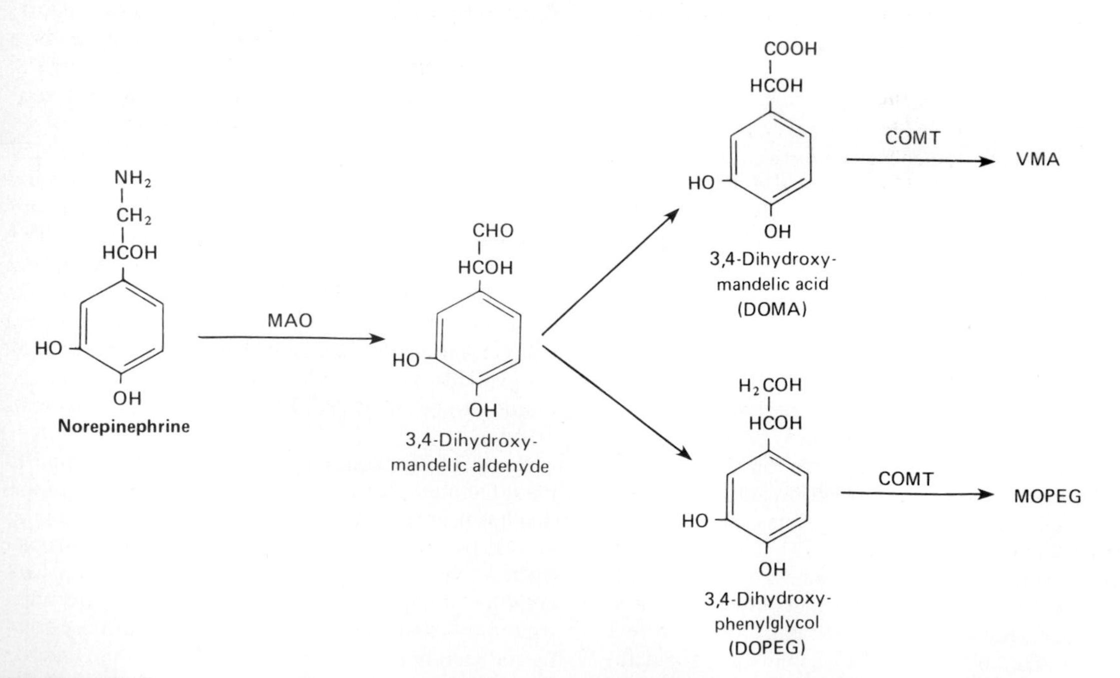

Figure 4–19. *Top:* Catabolism of circulating epinephrine and norepinephrine. The main site of catabolism is the liver. The conjugates are mostly glucuronides and sulfates. MOPEG is also conjugated. ***Bottom:*** Catabolism of norepinephrine in noradrenergic nerve endings. The acid and the glycol enter the circulation and may be subsequently O-methylated to VMA and MOPEG. Epinephrine in nerve endings is presumably catabolized in the same way. MAO, monoamine oxidase; COMT, catechol-O-methyltransferase.

The pathways involved in the metabolism of phenylalanine are of considerable clinical importance because they are the site of several **inborn errors of metabolism,** diseases caused by congenital absence of various specific enzymes. **Phenylpyruvic oligophrenia,** a disorder characterized by severe mental deficiency and the accumulation in the blood and tissues of large amounts of phenylalanine and its keto acid derivatives, is usually due to congenital deficiency or absence of phenylalanine hydroxylase (Fig 4–18). However, it can also be produced by deficiency or absence or pteridine reductase, the enzyme that regenerates the biopterin cofactor needed for the hydroxylation, or by deficient biopterin synthesis. If the disease is due to phenylalaine hydroxylase deficiency, mental retardation is largely due to accumulation of phenylalanine and its derivatives in the blood, and it can be treated with considerable success by markedly reducing the amount of phenylalanine in the diet.

Catabolism of Catecholamines

Epinephrine and norepinephrine are metabolized to biologically inactive products by oxidation and methylation. The former reaction is catalyzed by **monoamine oxidase (MAO)** and the latter by **catechol-O-methyltransferase (COMT)** (Figs 4–17 and 4–19). MAO is located on the outer surface of the mitochondria. It is widely distributed, being particularly plentiful in the nerve endings at which catecholamines are secreted. COMT is also widely distributed, with high concentrations in the liver and kidneys, but it is not found in nerve endings. Consequently, there are 2 different patterns of catecholamine metabolism.

Circulating epinephrine and norepinephrine are for the most part O-methylated, and measurement of the O-methylated derivatives normetanephrine and metanephrine in the urine is a good index of the rate of secretion of norepinephrine and epinephrine. The O-methylated derivatives that are not excreted are largely oxidized, and VMA (Fig 4–19) is the most plentiful catecholamine metabolite in the urine. Small amounts of the O-methylated derivatives are also conjugated to sulfates and glucuronides.

In the noradrenergic nerve endings, on the other hand, some of the norepinephrine is being constantly converted by MAO to the physiologically inactive deaminated derivatives, 3,4-dihydroxymandelic acid (DOMA) and its corresponding glycol (DOPEG). These compounds enter the circulation and may subsequently be converted to their corresponding O-methyl derivatives (Fig 4–19).

The reuptake of released norepinephrine mentioned above is a major mechanism by which norepinephrine is removed from the vicinity of autonomic endings. The hypersensitivity of the sympathetically denervated structures (see below) is probably explained in part on this basis. After the noradrenergic neurons are cut, their endings degenerate; consequently, there is no reuptake, and more of a given dose of norepinephrine is available to stimulate the receptors of the autonomic effectors.

Table 4–3. Adrenergic and dopaminergic receptors.

Type	Response to Catecholamines*	Mechanism of Action†
α_1	NE > E > ISO	Ca^{2+} ↑
α_2	NE > E > ISO	AC ↓
β_1	ISO > E = NE	AC ↑
β_2	ISO > E > NE	AC ↑
D_1	DA > E = NE	AC ↑
D_2	DA > E = NE	AC ↓

*NE, norepinephrine; E, epinephrine; ISO, isoproterenol, a synthetic catecholamine; DA, dopamine.
†AC ↑, adenylate cyclase stimulated; AC ↓, adenylate cyclase inhibited.

Alpha & Beta Receptors

The effects produced by epinephrine and norepinephrine can be separated into 4 categories on the basis of their different sensitivities to certain drugs (Table 4–3). These differences are in turn due to the existence of 4 types of catecholamine receptors (Fig 4–13). The β_2 receptors are larger than the β_1 receptors. The effects of both β_1 and β_2 receptor stimulation are brought about by activation of adenylate cyclase via G_s, with a consequent increase in intracellular cyclic AMP (see Chapter 1). Alpha$_1$ receptors produce their effects by activating phospholipase C, whereas α_2 receptors produce their effects by inhibiting adenylate cyclase via G_i and thus decreasing intracellular cyclic AMP. The β_1, β_2 and α_2 receptors all have the characteristic pattern of 7 membrane-spanning domains (Fig 4–14).

Dopamine

In the small, intensely fluorescent (SIF) cells in autonomic ganglia (see Chapter 13) and in certain parts of the brain (see Chapter 15), catecholamine synthesis stops at dopamine (Fig 4–18), and this catecholamine is secreted as a synaptic transmitter. Released dopamine acts on dopamine receptors, of which there are at least 2 types. One type, the D_1 receptor, activates dopamine-sensitive adenylate cyclase via G_s; a second type, the D_2 receptor, inhibits adenylate cyclase via G_i. There is some pharmacologic evidence for a D_3 receptor. Released dopamine is recaptured via an active reuptake mechanism and inactivated by monoamine oxidase and catechol-O-methyltransferase (Fig 4–20) in a manner analogous to the inactivation of norepinephrine. DOPAC and HVA are also conjugated, primarily to sulfates.

Serotonin

Serotonin (5-hydroxytryptamine, 5-HT) is present in highest concentration in blood platelets and in the gastrointestinal tract, where it is found in the entero-

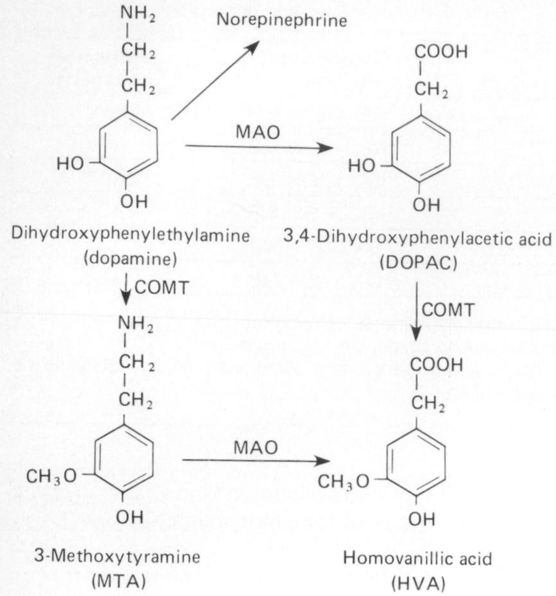

Figure 4–20. Catabolism of dopamine, MAO, monoamine oxidase; COMT, catechol-O-methyltransferase. As in other oxidative deaminations catalyzed by monoamine oxidase, aldehydes are formed first and then oxidized in the presence of aldehyde dehydrogenase to the corresponding acids (DOPAC and HVA). The aldehydes are also reduced to 3,4-dihydroxyphenylethanol (DOPET) and 3-methoxy-4-hydroxyphenylethanol. Some of the DOPAC and HVA form sulfate conjugates.

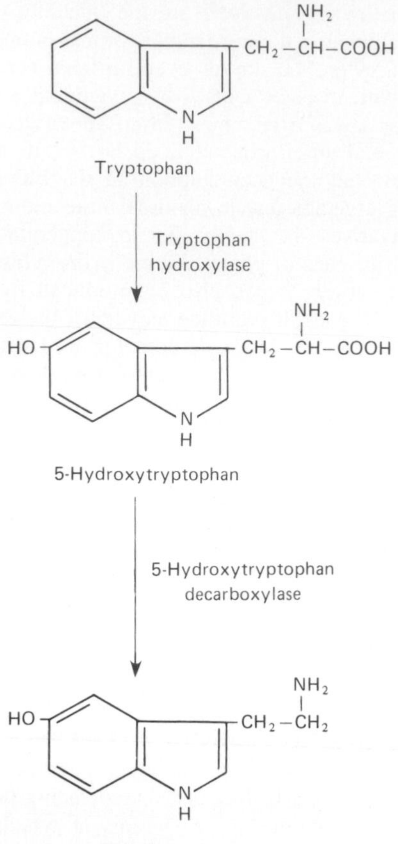

Figure 4–21. Biosynthesis of serotonin (5-hydroxytryptamine). The enzyme that catalyzes the decarboxylation of 5-hydroxytryptophan is very similar but probably not identical to the enzyme that catalyzes the decarboxylation of dopa. Tetrahydrobiopterin is a cofactor for the action of tryptophan hydroxylase.

chromaffin cells and the myenteric plexus (see Chapter 26). Lesser amounts are found in the brain and in the retina.

Serotonin is formed in the body to hydroxylation and decarboxylation of the essential amino acid tryptophan (Fig 4–21). Normally, the hydroxylase is not saturated; consequently, increased intake of tryptophan in the diet can increase brain serotonin content. After release from serotonergic neurons, much of the released serotonin is recaptured by an active reuptake mechanism and inactivated by monoamine oxidase (Figs 4–22 and 4–23) to form 5-hydroxyindoleacetic acid (5-HIAA). This substance is the principal urinary metabolite of serotonin, and urinary output of 5-HIAA is used as an index of the rate of serotonin metabolism in the body. In the pineal gland, serotonin is converted to melatonin (see Chapter 24).

Six types of serotonin receptors have now been described: $5HT_{1A}$, $5HT_{1B}$, $5HT_{1C}$, and $5HT_{1D}$ receptors, at least some of which are presynaptic; $5HT_2$ receptors, which mediate platelet aggregation, smooth muscle contraction, and various effects in the brain; and $5HT_3$ receptors, which are also present in peripheral tissues and the brain and presumably mediate other brain functions. At least 2 of these receptors, the $5HT_{1A}$ and $5HT_{1C}$ receptors, have the typical 7 membrane-spanning domains that are characteristic of receptors coupled to G proteins.

Histamine

There are large amounts of histamine in the anterior and posterior lobes of the pituitary and the adjacent hypothalamus. The heparin-containing tissue cells called **mast cells** have a high histamine content, and most of the histamine in the posterior pituitary is in mast cells, although the histamine in the anterior pituitary and the hypothalamus is not. In the hypothalamus, the limbic system, and other parts of the cortex, histamine appears to be a neurotransmitter.

Histamine is formed by decarboxylation of the amino acid histidine (Fig 4–24). The enzyme that catalyzes this step differs from the L-aromatic amino acid decarboxylases that decarboxylate 5-hydroxytryptophan and L-dopa. Histamine is converted to methylhistamine or, alternatively, to imidazoleacetic

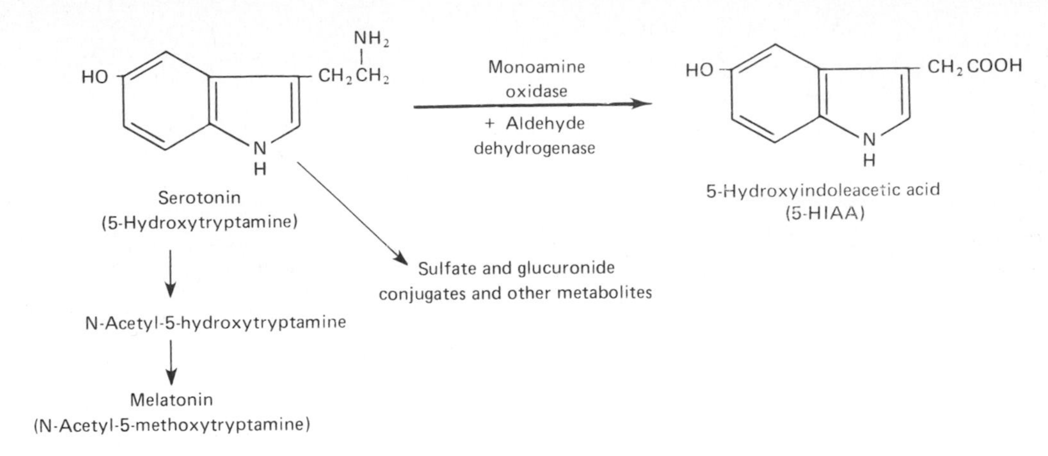

Figure 4–22. Catabolism of serotonin. The major metabolite is 5-HIAA. In oxidative deaminations catalyzed by monoamine oxidase, an aldehyde is formed first and then in the presence of aldehyde dehydrogenase oxidized to the corresponding acid. Some of the aldehyde is also reduced to the corresponding alcohol (5-hydroxytryptophol). The details of the formation of melatonin are shown in Fig 24–9.

acid. The latter reaction is quantitatively less important in humans. It requires the enzyme **diamine oxidase (histaminase)** rather than monoamine oxidase, even though monoamine oxidase catalyzes the oxidation of methylhistamine to methylimidazoleacetic acid.

There are 3 known types of histamine receptors, H_1, H_2, and H_3, and all 3 are found in peripheral tissues and the brain. Most if not all the H_3 receptors are presynaptic, and they mediate inhibition of histamine secretion. H_1 receptors activate phospholipase C, and H_2 receptors increase intracellular cyclic AMP. The function of the histaminergic systems in the brain is uncertain, but histamine has been related to arousal and regulation of the secretion of some anterior pituitary hormones.

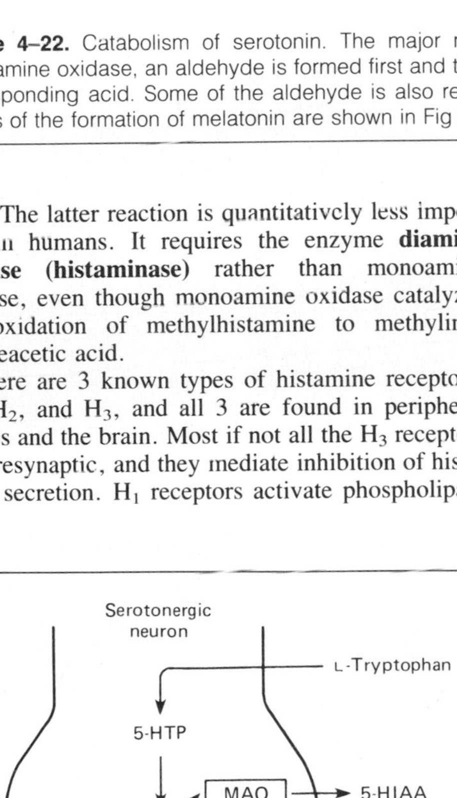

Gamma-Aminobutyric Acid

Gamma-aminobutyric acid (GABA) is an inhibitory mediator in the brain and the retina and is the mediator responsible for presynaptic inhibition (see above). It acts by increasing Cl^- conductance, and its action is antagonized by picrotoxin.

GABA, which exists as γ-aminobutyrate in the body fluids, is formed by decarboxylation of glutamate (Fig 4–25). The enzyme that catalyzes this reaction is **glutamate decarboxylase** (GAD), which has been demonstrated by immunocytochemical techniques to be present in nerve endings in many parts of the brain. GABA is metabolized primarily by transamination to succinic semialdehyde and thence to succinate in the citric acid cycle (see Chapter 17). **GABA transaminase (GABA-T)** is the enzyme that catalyzes the transamination. Pyridoxal phosphate, a derivative of the B complex vitamin pyridoxine, is a cofactor for GAD and GABA-T. However, the decarboxylation, unlike the transamination, is essentially irreversible. Consequently, the GABA content of the brain is reduced in pyridoxine deficiency. Pyridoxine deficiency is associated with signs of neural hyperexcitability and convulsions, although pyridoxine treatment is unfortunately of no value in most clinical cases of idiopathic epilepsy.

There are 2 very different GABA receptors. The $GABA_B$ receptor acts via a G protein to increase conductance in a K^+ channel. On the other hand, the $GABA_A$ receptor is a Cl^- channel. It is made up of 2 α and 2 β subunits, with each subunit spanning the cell membrane 4 times (Fig 4-26). GABA binds to the β subunits, increasing Cl^- conductance. However,

Serotonergic neuron — L-Tryptophan — 5-HTP — MAO — 5-HIAA — 5-HT — Reuptake — 5-HT — Postsynaptic tissue

Figure 4–23. Biochemical events at serotonergic synapses. Compare with Fig 4–14. 5-HTP, 5-hydroxytryptophan; 5-HT, 5-hydroxytryptamine (serotonin); 5-HIAA, 5-hydroxyindoleacetic acid; MAO, monoamine oxidase; X, serotonin receptor.

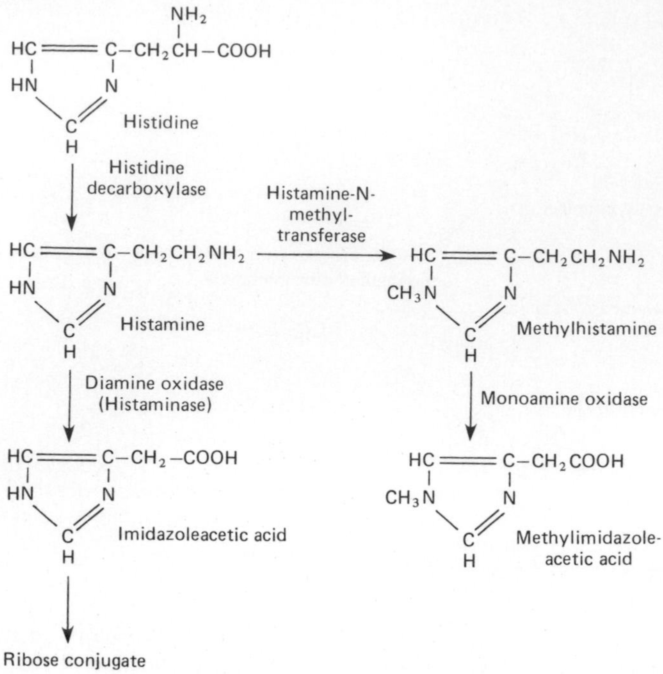

NH_2

HC═C–CH_2CH–COOH

HN N

C

H

Histidine

Histidine decarboxylase

HC═C–$CH_2CH_2NH_2$

HN N

C

H

Histamine

Histamine-N-methyl-transferase

HC═C–$CH_2CH_2NH_2$

CH_3N N

C

H

Methylhistamine

Diamine oxidase (Histaminase)

Monoamine oxidase

HC═C–CH_2–COOH

HN N

C

H

Imidazoleacetic acid

HC═C–CH_2COOH

CH_3N N

C

H

Methylimidazole-acetic acid

Ribose conjugate

Figure 4–24. Synthesis and catabolism of histamine.

the effects of GABA on Cl^- conductance are facilitated by the benzodiazepines, drugs that have marked antianxiety activity and are also effective muscle relaxants, anticonvulsants, and sedatives. The most widely used benzodiazepines in the USA are chlordiazepoxide (Librium) and flurazepam (Dalmane), but in these and other forms, they are used throughout the world. The benzodiazepines exert their facilitatory effect by binding to the α subunits. Barbiturates and alcohol also act by facilitating Cl^- conductance through the Cl^- channel. The search in brain tissue for an endogenous ligand for the benzodiazepine receptor has led to isolation of a polypeptide, but this polypeptide increases rather than decreases anxiety.

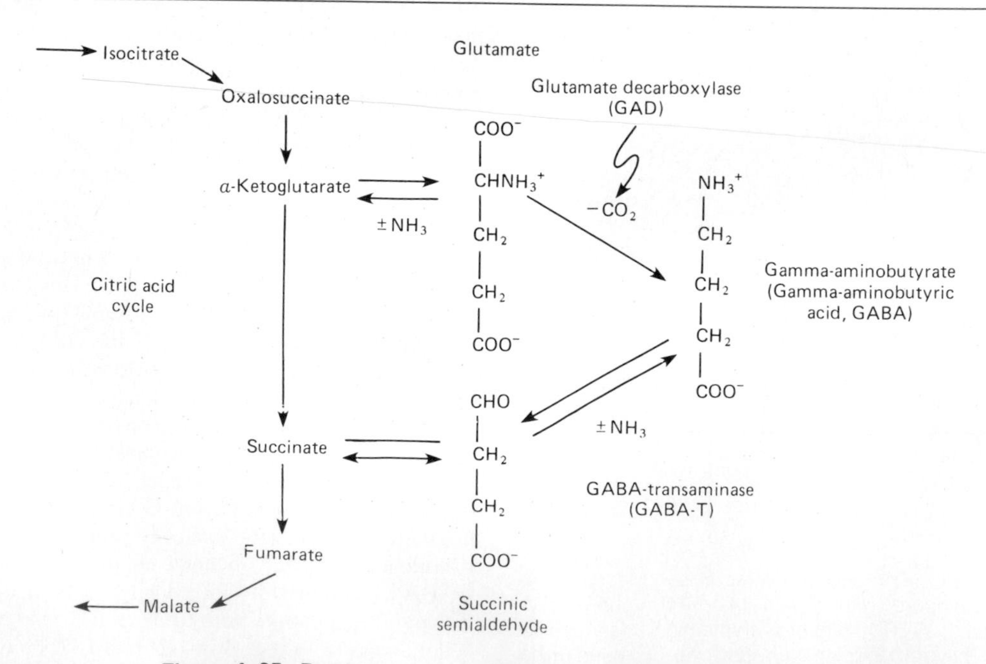

Figure 4–25. Formation and metabolism of gamma-aminobutyrate.

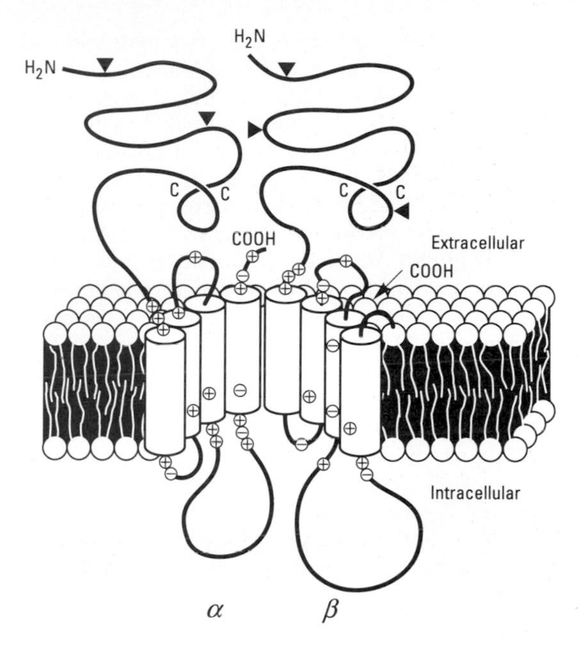

Figure 4–26. Probable structure of the δ and β subunits of the $GABA_A$ receptor in a cell membrane. The triangles identify potential sites for N-glycosylation. The receptor, which is made up of 2 δ and 2 β subunits, forms a Cl^- channel. (Reproduced, with permission, from Schofield PR et al: Sequence and functional expression of the $GABA_A$ receptor shows a ligand-gated receptor superfamily. *Nature* 1987;**328**:221.)

Its significance and physiologic role remain to be determined.

Glutamate & Aspartate

Glutamate and **aspartate** depolarize many different mammalian neurons when delivered directly on their cell membranes by iontophoresis. Glutamate is very widely used as an excitatory transmitter in the brain, and it has been calculated that it is the transmitter responsible for 75% of the excitatory transmission in the brain. Three types of glutamate receptors have been identified and named for the drugs that preferentially excite them: **N-methyl-D-aspartate (NMDA) receptors, kainate receptors,** and **quisqualate receptors.** Aspartate also binds to NMDA and quisqualate receptors. The kainate and quisqualate receptors are channels for cations, and when activated, they permit Na^+ influx and K^+ efflux. The NMDA receptor is also a cation channel, but it is unique in several ways. First, glycine binds to it to facilitate its function, and recent research indicates that glycine is essential for its normal response to glutamate. Second, at normal membrane potential, the channel is blocked by an Mg^{2+} ion and becomes unblocked only when the membrane becomes partially depolarized. Third, phencylidine and ketamine, which produce amnesia and a feeling of dissociation from the environment, bind to another site inside the channel and are also released only upon depolarization. There is a high concentration of NMDA receptors in the hippocampus, and blockade of these receptors prevents long-term potentiation, an enhanced discharge of neural pathways after a brief period of rapid stimulation (see above). Thus, these receptors seem to be involved in memory and learning. Their unique voltage dependence also raises the possibility that they are involved in epilepsy.

Glutamate and some of its synthetic cogeners are unique in that they can produce such marked stimulation that neurons die. It is for this reason that microinjections of these "excitotoxins" are used to produce discrete lesions that destroy neuronal cells bodies without affecting neighboring axons. An interesting hypothesis holds that when a part of the brain is ischemic, as it is, for example, after thrombosis of a cerebral artery, there is marked release of glutamate. It is argued that much of the brain damage is due to the action of this transmitter as an excititoxin, rather than to a lack of O_2 per se. Evidence in favor of this hypothesis is the sparing of brain tissue that occurs when vascular obstruction is produced after administration of agents that block NMDA receptors.

It has recently been suggested that the 3 types of glutamate receptors are actually substates of a complex single receptor.

Glycine

By its action on the NMDA receptors, glycine has an excitatory effect in the brain. However, glycine is the mediator responsible for direct inhibition in the spinal cord, and, like GABA, it acts by increasing Cl^- conductance. Its action is antagonized by strychnine and tetanus toxin. The clinical picture of convulsions and muscular hyperactivity produced by tetanus toxin and strychnine emphasizes the importance of postsynaptic inhibition in normal neural function. The glycine receptor responsible for inhibition is a Cl^- channel formed by a large glycoprotein that contains 3 polypeptide subunits. One of these binds strychnine and has an appreciable amino acid sequence homology to proteins in the nicotinic cholinergic receptor.

One conceptual objection to glycine and other amino acids as transmitters is the fact that these amino acids probably occur not only in neurons but in most if not all living cells. However, the necessary specificity for chemical transmission is provided not by unique chemicals but by specialized neuronal mechanisms for the storage, release, and postsynaptic action of a particular substance. Thus, almost any small diffusible substance could be a transmitter.

Substance P

Substance P is a polypeptide that is found in appreciable quantities in the intestine, where it may be a chemical mediator in the myenteric reflex (Chapter 26). Its structure is shown in Table 26–1. In the nervous system, it is found in nerve endings in many different locations. Little is known about its synthesis and catabolism. However, it is produced by primary afferent neurons in tissue culture, and it is almost certainly the transmitter released in the substantia gelatinosa by the neurons mediating nociception. Upon subcutaneous injection, substance P produces vasodilation, and its release from the peripheral ends of the primary afferent neurons is probably responsible for the axon reflex (see Chapter 32).

The preprotachykinin gene that codes for substance P can be processed to form 2 different mRNAs. By elimination of one set of introns during posttranscriptional processing, the mRNA for substance P is produced. However, in some parts of the nervous system and other tissues, different splicing occurs, with the production of an mRNA that instead produces the related polypeptide **substance K.** It is also possible that the same neuron can switch production from one mRNA to another during development or in response to environmental stimuli. Another example of alternative posttranscriptional processing producing 2 different mRNAs from a single gene coding for a neural polypeptide is the calcitonin gene, which can produce calcitonin or CGRP (see below). Many other examples have now been documented.

The substance K receptor is the first neural polypeptide receptor to be cloned. As noted above, it has the 7 membrane-spanning domains characteristic of rhodopsin; other visual pigments; muscarinic cholinergic receptors; and β_1, β_2, and α_2 adrenergic receptors. The one difference is that its N terminal appears to be blocked by binding to a fatty acid. The function of substance K remains to be determined.

Opioid Peptides

The brain and the gastrointestinal tract contain receptors that bind morphine. The search for endogenous ligands for these receptors led to the discovery of 2 closely related pentapeptides called **enkephalins** (Table 4–4) that bind to these opiate receptors. These and other peptides that bind to opiate receptors are called **opioid peptides.** The enkephalins are found in nerve endings in the gastrointestinal tract and many different parts of the brain, and they appear to function as synaptic transmitters. They are found in the substantia gelatinosa, and they have analgesic activity when injected into the brain stem. They also decrease intestinal motility (see Chapter 26).

Like other small peptides, the opioid peptides are synthesized as part of larger precursor molecules (see Chapter 1). At least 18 active opioid peptides have been identified. Unlike other peptides, however, the opioid peptides have a number of different precursors. Each has a prepro form and a pro form from which the signal peptide has been cleaved. The 3 precursors that have been characterized and the opioid peptides they produce are shown in Table 4–4. **Proenkephalin** was first identified in the adrenal medulla (see Chapter 20), but it is also the precursor for met-enkephalin and leu-enkephalin in the brain. Each proenkephalin molecule contains 4 met-enkephalins, one leu-eukephalin, one octapeptide, and one heptapeptide. **Pro-opiomelanocortin,** a large precursor molecule found in the anterior and intermediate lobes of the pituitary gland and the brain, contains β-endorphin, a polypeptide of 31 amino acid residues that has met-enkephalin at its N terminal (see Chapter 22). Other shorter endorphins may also be produced, and the precursor molecule also produces ACTH and MSHs. There are separate enkephalin-secreting and β-endorphin-secreting systems of neurons in the brain (see Chapter 15). Beta-endorphin is also secreted into the bloodstream by the pituitary gland. A third precursor molecule is **prodynorphin,** a protein that contains 3 leu-enkephalin residues associated with dynorphin and neoendorphin. Dynorphin 1–17 is found in the duodenum and dynorphin 1–8 in the posterior pituitary and hypothalamus. Alpha- and β-neoendorphin are also found in the hypothalamus. The reasons for the existence of multiple opioid peptide precursors and for the presence of the peptides in the circulation as well as in the brain and the gastrointestinal tract are presently unknown.

Opiate receptors have been studied in detail, and 5 different subtypes, δ, κ, ξ, ϵ, and μ receptors, have been characterized. They differ in pharmacologic properties, distribution in the brain and elsewhere, and affinity for various opioid peptides. The μ receptor seems to be most concerned with the central pain mechanisms and is the receptor for which the morphine antagonist naloxone has greatest affinity. However, the natural ligand for μ receptor is probably β-endorphin, whereas the natural ligands for δ receptors are enkephalins and the natural ligands for κ receptors are dynorphins. ϵ receptors also bind β endorphin.

Table 4–4. Opioid Peptides and their precursors.*

Precursor	Opioid Peptides	Structures
Proenkephalin (see Chapter 20)	Met-enkephalin	Tyr-Gly-Gly-Phe-Met_5
	Leu-enkephalin	Tyr-Gly-Gly-Phe-Leu_5
	Octapeptide	Tyr-Gly-Gly-Phe-Met-Arg-Gly-Leu_8
	Heptapeptide	Tyr-Gly-Gly-Phe-Met-Arg-Phe_7
Pro-opiomelanocortin (see Chapter 22)	β-Endorphin	See Chapter 22.
	Other endorphins	See Chapter 22.
Prodynorphin	Dynorphin 1–8	Tyr-Gly-Gly-Phe-Leu-Arg-Arg-Ile_8
	Dynorphin 1–17	Tyr-Gly-Gly-Phe-Leu-Arg-Arg-Ile-Arg-Pro-Lys-Leu-Lys-Trp-Asp-Asn-Gln_{17}
	α-Neoendorphin	Tyr-Gly-Gly-Phe-Leu-Arg-Lys-Tyr-Pro-Lys_{10}
	β-Neoendorphin	Tyr-Gly-Gly-Phe-Leu-Arg-Lys-Tyr-Pro_9

*Modified from Rossier J: Opioid peptides have found their roots. *Nature* 1981;**298:**221.

It is interesting that in addition to opioid peptides, small amounts of endogenously produced morphine and codeine are found in the mammalian brain; however, the significance of this finding is unknown.

Enkephalins are metabolized primarily by 2 peptidases: enkephalinase A, which splits the Gly-Phe bond, and enkephalinase B, which splits the Gly-Gly bond. Aminopepidase, which splits the Tyr-Gly bond, also contributes to their metabolism.

Other Polypeptides

The hypophyseotropic hormones (see Chapter 14) are found in many different parts of the nervous system, and it seems clear that many and perhaps all of them function as neurotransmitters as well as hormones. Vasopressin and oxytocin are not only secreted as hormones but are present in neurons that project to the brain stem and spinal cord. Insulinlike molecules are found in the brain, but the structures of the mRNAs that produce them indicate that they do not have the same structure as insulin. They may be growth factors. The gastrointestinal hormones VIP and the C-terminal octapeptide fragment of CCK are also found in the brain, the former in highest concentration in the hypothalamus and the latter in highest concentration in the cerebral cortex. Gastrin, neurotensin, and gastrin-releasing peptide are also found in the gastrointestinal tract and brain. The hypothalamus contains both gastrin 17 and gastrin 34 (see Chapter 26). VIP produces vasodilation and is found in vasomotor nerve fibers. The functions of these peptides in the nervous system are unknown.

Calcitonin-gene-related peptide (CGRP) is a polypeptide that in rats and humans exists in 2 forms, CGRPα and CGRPβ. These 2 forms differ by only one amino acid residue, yet they are encoded by different genes. In rats, and presumably in humans, CGRPβ is present in the gastrointestinal tract, whereas CGRPα is found in primary afferent neurons, neurons by which taste impulses project to the thalamus, and neurons in the medial forebrain bundle. It is also present along with substance P in the branches of primary afferent neurons that end near blood vessels. CGRPα and the calcium-lowering hormone calcitonin (see Chapter 21) are both products of the calcitonin gene. However, in the thyroid gland, splicing produces the mRNA that codes for calcitonin, whereas in the brain, alternative splicing produces the mRNA that codes for CGRPα.

Neuropeptide Y is a polypeptide containing 36 amino acid residues that is closely related to pancreatic polypeptide (see Chapter 19). It is present in many parts of the brain and the autonomic nervous system. In the autonomic nervous system, although not in the brain, much of it is located in noradrenergic neurons, and it probably functions as a cotransmitter with norepinephrine (see below).

Cotransmitters

Numerous examples have now been described in which neurons contain and secrete 2 and even 3 transmitters. The **cotransmitters** in these situations are often a catecholamine or serotonin plus a polypeptide, but situations in which neurons contain 2 different polypeptides have also been described. Many cholinergic neurons contain VIP, and many noradrenergic and adrenergic neurons contain ATP and neuropeptide Y. The physiologic significance of cotransmitters is still obscure. However, the VIP secreted with acetylcholine potentiates the postsynaptic actions of acetylcholine, and neuropeptide Y potentiates some of the actions of norepinephrine.

Adenosine

There are adenosine receptors in the brain, and neurons that may secrete this purine project from the hypothalamus to the forebrain.

Prostaglandins

Prostaglandins—fatty acid derivatives found in many cells (see Chapter 17)—are also found in the nervous system. They are present in nerve ending fractions of brain homogenates and are released from neural tissue in vitro. However, they appear to exert their effects by modulating reactions mediated by cyclic AMP rather than by functioning as synaptic transmitters.

NEUROMUSCULAR TRANSMISSION

THE MYONEURAL JUNCTION

Anatomy

As the axon supplying a skeletal muscle fiber approaches its termination, it loses its myelin sheath and divides into a number of terminal buttons or end-feet (Fig 4–27). The end-feet contain many small, clear vesicles that contain acetylcholine, the transmitter at these junctions. The endings fit into depressions in the **motor end-plate,** the thickened portion of the muscle membrane of the junction. Underneath the nerve ending, the muscle membrane of the end-plate is thrown into folds, the **palisades.** The space between the nerve and the thickened muscle membrane is comparable to the synaptic cleft at synapses. The whole structure is known as the **myoneural junction.** Only one nerve fiber ends on each end-plate, with no convergence of multiple inputs.

Sequence of Events During Transmission

The events occurring during transmission of impulses from the motor nerve to the muscle (Table 3–2) are somewhat similar to those occurring at other synapses. The impulse arriving in the end of the

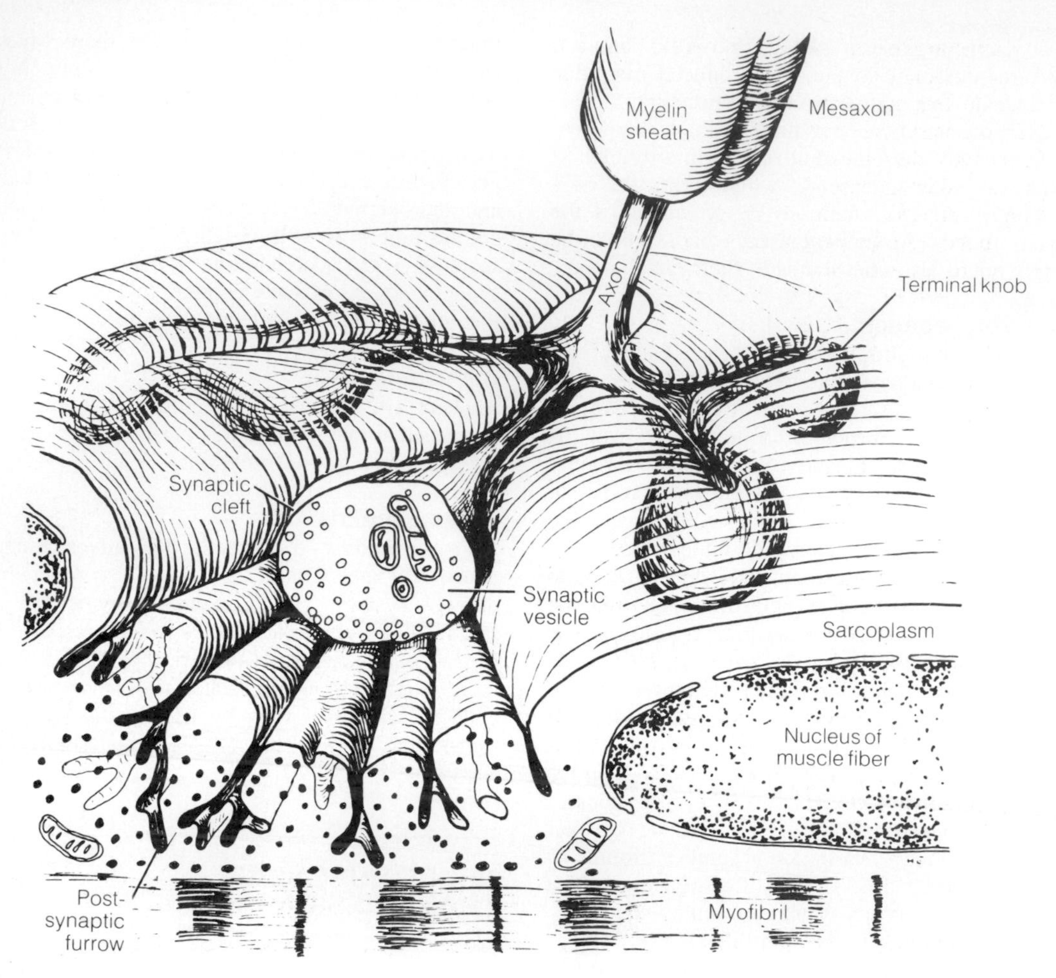

Figure 4–27. Myoneural junction. The drawing shows the terminal ends of a motor neuron axon buried in the end-plate cytoplasm and separated by the synaptic cleft from the much folded end-plate membrane. (Reproduced, with permission, from Elias H, Pauly JE, Burns ER: *Histology and Human Microanatomy*, 4th ed. Wiley, 1978.)

motor neuron increases the permeability of the endings to Ca^{2+}. Ca^{2+} enters the endings and triggers a marked increase in exocytosis of the acetylcholine-containing vesicles. The acetylcholine diffuses to the nicotinic acetylcholine receptors (Fig 4–16) on the folds of the membrane of the motor end-plate. Binding of acetylcholine to these receptors increases the Na^+ and K^+ conductance of the membrane, and the resultant influx of Na^+ produces a depolarizing potential, the **end-plate potential.** The current sink created by this local potential depolarizes the adjacent muscle membrane to its firing level. Action potentials are generated on either side of the end-plate and are conducted away from the end-plate in both directions along the muscle fiber. The muscle action potential, in turn, initiates muscle contraction, as described in Chapter 3.

End-Plate Potential

An average end-plate contains about 50 million acetylcholine receptors. Each nerve impulse releases about 60 acetylcholine vesicles, and each vesicle contains about 10,000 molecules of the neurotransmitter. This amount is enough to activate about 10 times the number of acetylcholine receptors needed to produce a full end-plate potential. Therefore, a propagated response in the muscle is regularly produced, and this large response obscures the end-plate potential. However, the end-plate potential can be seen if the 10-fold safety factor is overcome and the potential is reduced to a size that is insufficient to fire the adjacent muscle membrane. This can be accomplished by administration of small doses of curare, a drug that competes with acetylcholine for binding to the acetylcholine receptors. The response is then recorded only at the end-plate region and decreases exponentially away from it. Under these conditions, end-plate potentials can be shown to undergo temporal summation.

Quantal Release of Transmitter

Small quanta (''packets'') of acetylcholine are released randomly from the nerve cell membrane at rest, each producing a minute depolarizing spike called a **miniature end-plate potential,** which is about 0.5 mV in amplitude. The number of quanta of acetylcholine released in this way varies directly with the Ca^{2+} concentration and inversely with the Mg^{2+} concentration at the end-plate. When a nerve impulse reaches the ending, the number of quanta released increases by several orders of magnitude, and the result is the large end-plate potential that exceeds the firing level of the muscle fiber. The nerve impulse increases the permeability of the ending to Ca^{2+}, and the Ca^{2+} is responsible for the increased quantal release.

Quantal release of acetylcholine similar to that seen at the myoneural junction has been observed at other cholinergic synapses, and similar processes are presumably operating at noradrenergic and other synaptic junctions.

Myasthenia Gravis

Myasthenia gravis is a serious and sometimes fatal disease in which skeletal muscles are weak and tire easily. It is now known that the disease is caused by the formation of circulating antibodies to the acetylcholine receptors. These antibodies destroy some of the receptors and bind others to neighboring receptors, triggering their removal by endocytosis (see Chapter 1). The reason for the development of autoimmunity to acetylcholine receptors in this disease is still unknown.

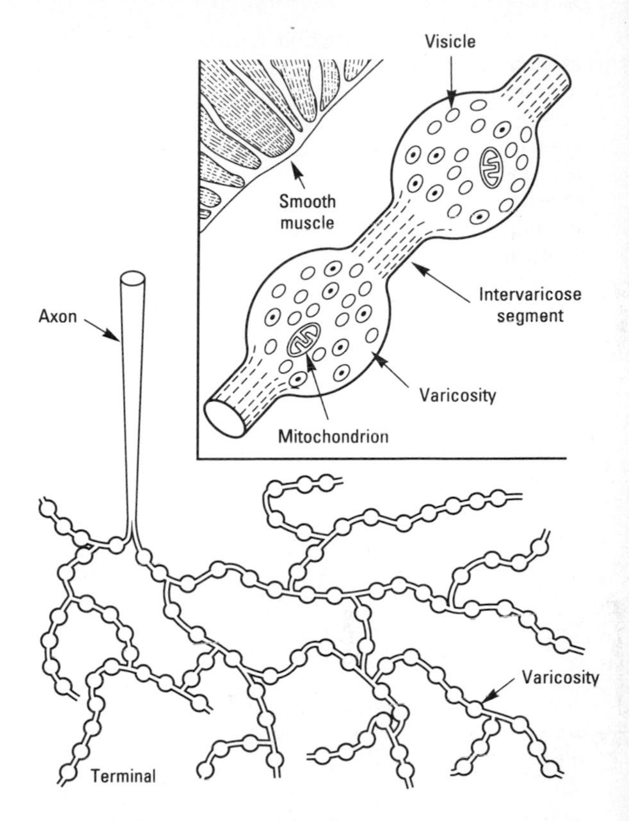

Figure 4–28. Endings of postganglionic autonomic neurons on smooth muscle. (Reproduced, with permission, from Kandel ER, Schwartz JH [editors]: *Principles of Neural Science,* 2nd ed. Elsevier, 1985.)

NERVE ENDINGS IN SMOOTH & CARDIAC MUSCLE

Anatomy

The postganglionic neurons in the various smooth muscles that have been studied in detail branch extensively and come in close contact with the muscle cells (Fig 4–28). Some of these nerve fibers contain clear vesicles and are cholinergic, whereas others contain the characteristic dense-core granules within vesicles of 2 sizes that are known to contain norepinephrine. There are no recognizable end-plates or other postsynaptic specializations. The nerve fibers run along the membranes of the muscle cells and sometimes groove their surfaces. The multiple branches of the noradrenergic and, presumably, the cholinergic neurons are beaded with enlargements **(varicosities)** that are not covered by Schwann cells and contain synaptic vesicles (Fig 4–28). In noradrenergic neurons, the varicosities are 1–2 μm in channels and about 5μm apart, with up to 20,000 varicosities per neuron. Transmitter is apparently liberated at each varicosity, ie, at many locations along each axon. This arrangement permits one neuron to innervate many effector cells. The type of contact in which a neuron forms a synapse on the surface of another neuron or a smooth muscle cell and then passes on to make similar contacts with other cells is called a **synapse en passant.**

In the heart, cholinergic and noradrenergic nerve fibers end on the sinoatrial node, the atrioventricular node, and the bundle of His. Noradrenergic fibers also innervate the ventricular muscle. The exact nature of the endings on nodal tissue is not known. In the ventricle, the contacts between the noradrenergic fibers and the cardiac muscle fibers resemble those found in smooth muscle.

Electrical Responses

In smooth muscles in which noradrenergic discharge is excitatory, stimulation of the noradrenergic nerves produces discrete partial depolarizations that look like small end plate potentials and are called **excitatory junction potentials (EJPs).** These potentials summate with repeated stimuli. Similar EJPs are seen in tissues excited by cholinergic discharges. In tissues inhibited by noradrenergic stimuli, hyperpolarizing **inhibitory junction potentials (IJPs)** have been produced by stimulation of the noradrenergic nerves.

These electrical responses are observed in many smooth muscle cells when a single nerve is stimulated, but their latency varies. This finding is consistent with the synapse en passant arrangement described above, but it could also be explained by transmission of the junction responses from cell to cell across low-resistance junctions or by diffusion of transmitter from its site of release to many smooth muscle cells. Miniature excitatory junction potentials similar to the miniature end-plate potentials in the skeletal muscle have been observed in some smooth muscle preparations, but they show considerable variation in size and duration. They may represent responses to single packets of transmitter, with the variation due to diffusion of the transmitter for variable distances.

DENERVATION HYPERSENSITIVITY

When the motor nerve to skeletal muscle is cut and allowed to degenerate, the muscle gradually becomes extremely sensitive to acetylcholine. This **denervation hypersensitivity** or **supersensitivity** is also seen in smooth muscle. Smooth muscle, unlike skeletal muscle, does not atrophy when denervated, but it becomes hyperresponsive to the chemical mediator that normally activates it. Denervated exocrine glands, except for sweat glands, also become hypersensitive. A good example of denervation hypersensitivity is the response of the denervated iris. If the postganglionic sympathetic nerves to one pupil are cut in an experimental animal and, after several weeks, norepinephrine is injected intravenously, the denervated pupil dilates widely. A much smaller, less prolonged response is observed on the intact side.

The reactions triggered by section of an axon are summarized in Fig 4–29. Hypersensitivity of the post-synaptic receptors to the transmitter previously secreted by the axon endings is a general phenomenon, largely due to the synthesis or activation of more receptors. There is in addition orthograde degeneration (**wallerian degeneration;** see Chapter 2) and retrograde degeneration of the axon stump to the nearest collateral **(sustaining collateral).** A series of changes occur in the cell body that include a decrease in Nissl substance (chromatolysis). The nerve then starts to regrow, with multiple small branches projecting along the path the axon previously followed (regenerative sprouting).

When higher centers in the nervous system are destroyed, the activity of the lower centers they control is generally increased ("release phenomenon"). The increased activity may be due in part to denervation hypersensitivity of the lower centers. Indeed, one theory holds that many of the signs and symptoms of neurologic disease are due to denervation hypersensitivity of various groups of neurons in the brain.

Hypersensitivity is limited to the structures immediately innervated by the destroyed neurons and fails to develop in neurons and muscle farther "downstream." Suprasegmental spinal cord lesions do not lead to hypersensitivity of the paralyzed skeletal

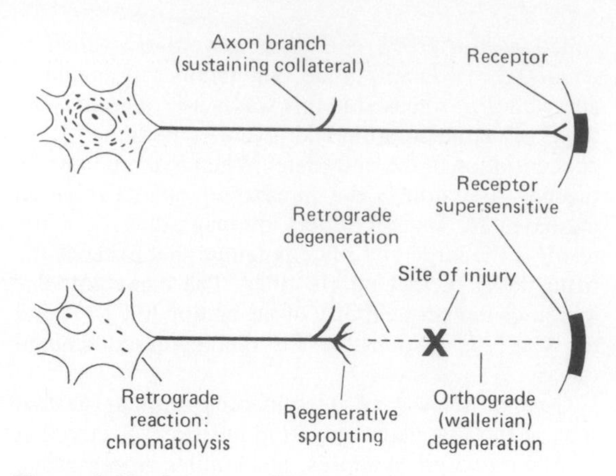

Figure 4–29. Summary of changes occurring in a neuron and the structure it innervates when its axon is crushed or cut at the point marked X. (Modified from D Ries.)

muscles to acetylcholine, and destruction of the preganglionic autonomic nerves to visceral structures does not cause hypersensitivity of the denervated viscera. This fact has practical implications in the treatment of diseases due to spasm of the blood vessels in the extremities. For example, if the upper extremity is sympathectomized by removing the upper part of the ganglion chain and the stellate ganglion, the hypersensitive smooth muscle in the vessel walls is stimulated by circulating norepinephrine, and episodic vasospasm continues to occur. However, if preganglionic sympathectomy of the arm is performed by cutting the ganglion chain below the third ganglion (to interrupt ascending preganglionic fibers) and the white rami of the first 3 thoracic nerves, no hypersensitivity results.

The causes of denervation hypersensitivity are probably multiple. As noted in Chapter 1, a deficiency of a given chemical messenger generally produces an increase in the number of its receptors. In denervated skeletal muscle, there is an increase in the area of the muscle membrane sensitive to acetylcholine. Normally, only the end-plate region contains acetylcholine receptors, and they are of the adult type (see above). After denervation, the sensitivity of the end-plate is no greater, but acetylcholine receptors of the fetal type appear over large portions of the muscle membrane. These disappear, and the sensitivity returns to normal if the nerve regrows. A similar spread of acetylcholine sensitivity has been demonstrated in denervated postganglionic cholinergic neurons. At endings where norepinephrine is normally secreted, another factor is lack of re-uptake of liberated catecholamines; the nerve endings in normal tissue take up large amounts of norepinephrine, and after they have degenerated, norepinephrine reaching the receptors from other sites has a greater effect than it otherwise would.

Initiation of Impulses in Sense Organs

5

INTRODUCTION

Information about the internal and external environment reaches the central nervous system via a variety of **sensory receptors.** These receptors are transducers that convert various forms of energy in the environment into action potentials in neurons. The characteristics of these receptors, the way they generate impulses in afferent neurons, and the general principles or "laws" that apply to sensation are considered in this chapter.

SENSE ORGANS & RECEPTORS

It is worth noting that the term receptor is used in physiology to refer not only to sensory receptors but, in a very different sense, to proteins that bind neuro-transmitters, hormones, and other substances with great affinity and specificity as a first step in initiating specific physiologic responses (see Chapter 1). The sensory receptor may be part of a neuron or a specialized cell that generates action potentials in neurons. The receptor is often associated with non-neural cells that surround it, forming a **sense organ.** The forms of energy converted by the receptors include, for example, mechanical (touch-pressure), thermal (degrees of warmth), electromagnetic (light), and chemical energy (odor, taste, and O_2 content of blood). The receptors in each of the sense organs are adapted to respond to one particular form of energy at a much lower threshold than other receptors respond to this form of energy. The particular form of energy to which a receptor is most sensitive is called its **adequate stimulus.** The adequate stimulus for the rods and cones in the eye, for example, is light. Receptors do respond to forms of energy other than their adequate stimulus, but the threshold for these nonspecific responses is much higher. Pressure on the eyeball will stimulate the rods and cones, for example, but the threshold of these receptors to pressure is much higher than the threshold of the pressure receptors in the skin.

THE SENSES

Sensory Modalities

Because the sensory receptors are specialized to respond to one particular form of energy and because many variables in the environment are perceived, it follows that there must be many different types of receptors. We learn in elementary school that there are "5 senses," but the inadequacy of this dictum is apparent if we list the major sensory modalities and their receptors in humans. The list (Table 5–1) includes at least 11 conscious senses. There are, in addition, a large number of sensory receptors which relay information that does not reach consciousness. For example, the muscle spindles provide information about muscle length, and other receptors provide information about such variables as the arterial blood pressure, the temperature of the blood in the head, and the pH of the cerebrospinal fluid. The existence of other receptors of this type is suspected, and future research will undoubtedly add to the list of "unconscious senses." Furthermore, any listing of the senses is bound to be arbitrary. The rods and cones, for example, respond maximally to light of different wavelengths, and there are different cones for each of the 3 primary colors. There are 4 different modalities of taste—sweet, salt, sour, and bitter—and each is subserved by a more or less distinct type of taste bud. Sounds of different pitches are heard primarily because different groups of hair cells in the organ of Corti are activated maximally by sound waves of different frequencies. Whether these various responses to light, taste, and sound should be considered separate senses is a semantic question that in the present context is largely academic.

Classifications of Sense Organs

Numerous attempts have been made to classify the senses into groups, but none has been entirely successful. Traditionally, the special senses are smell, vision, hearing, rotational and linear acceleration, and taste; the cutaneous senses are those with receptors in the skin; and the visceral senses are those concerned with perception of the internal environment. Pain from visceral structures is usually classified as a visceral sensation. Another classification of the various receptors divides them into (1) teleceptors ("distance receivers"), the receptors concerned with events at a distance; (2) exteroceptors, those concerned with the external environment near at hand; (3) interoceptors, those concerned with the internal environment; and (4) proprioceptors, those which provide information about the position of the body in space at any given instant. However, the conscious

Table 5–1. Principal sensory modalities. (The first 11 are conscious sensations.)

Sensory Modality	Receptor	Sense Organ
Vision	Rods and cones	Eye
Hearing	Hair cells	Ear (organ of Corti)
Smell	Olfactory neurons	Olfactory mucous membrane
Taste	Taste receptor cells	Taste bud
Rotational acceleration	Hair cells	Ear (semicircular canals)
Linear acceleration	Hair cells	Ear (utricle and saccule)
Touch-pressure	Nerve endings	Various*
Warmth	Nerve endings	Various*
Cold	Nerve endings	Various*
Pain	Naked nerve endings	. . .
Joint position and movement	Nerve endings	Various*
Muscle length	Nerve endings	Muscle spindle
Muscle tension	Nerve endings	Golgi tendon organ
Arterial blood pressure	Nerve endings	Stretch receptors in carotid sinus and aortic arch
Central venous pressure	Nerve endings	Stretch receptors in walls of great veins, atria
Inflation of lung	Nerve endings	Stretch receptors in lung parenchyma
Temperature of blood in head	Neurons in hypothalamus	. . .
Arterial P_{O_2}	Nerve endings?	Carotid and aortic bodies
pH of CSF	Receptors on ventral surface of medulla oblongata	. . .
Osmotic pressure of plasma	Cells in OVLT and possibly other circumventricular organs in anterior hypothalamus	. . .
Arteriovenous blood glucose difference	Cells in hypothalamus (glucostats)	. . .

* See text.

component of proprioception ("body image") is actually synthesized from information coming not only from receptors in and around joints but from cutaneous touch and pressure receptors as well. Certain other special terms are sometimes used. Because pain fibers have connections that mediate strong and prepotent withdrawal reflexes (see Chapter 6) and because pain is initiated by potentially noxious or damaging stimuli, pain receptors are sometimes called **nociceptors.** The term **chemoreceptor** is used to refer to those receptors which are stimulated by a change in the chemical composition of the environment in which they are located. These include receptors for taste and smell as well as visceral receptors such as those sensitive to changes in the plasma level of O_2, pH, and osmolality.

Cutaneous Sense Organs

There are 4 cutaneous senses: touch-pressure (pressure is sustained touch), cold, warmth, and pain. The skin contains various types of sensory endings. These include naked nerve endings, expanded tips on sensory nerve terminals, and encapsulated endings. The expanded endings include Merkel's disks and Ruffini endings (Fig 5–1), whereas the encapsulated endings include pacinian corpuscles, Meissner's corpuscles, and Krause's end-bulbs. Ruffini endings and pacinian corpuscles are also found in deep fibrous tissues. In addition, sensory nerves end around hair follicles. However, none of the expanded or encapsulated endings appear to be necessary for cutaneous sensation. Their distribution varies in different regions of the body, and it has been repeatedly demonstrated that all 4 sensory modalities can be elicited from areas that on histologic examination contain only naked nerve endings. Where they are present, the expanded or encapsulated endings appear to function as mechanoreceptors that respond to tactile stimuli; Meissner's and pacinian corpuscles are rapidly adapting touch receptors, and Merkel's disks and Ruffini endings are slowly adapting touch receptors. The nerve endings around hair follicles mediate touch, and movements of hairs initiate tactile sensations. It should be emphasized that although cutaneous sensory receptors lack histologic specificity, they are physiologically specific. Thus, any given ending signals one and only one kind of cutaneous sensation.

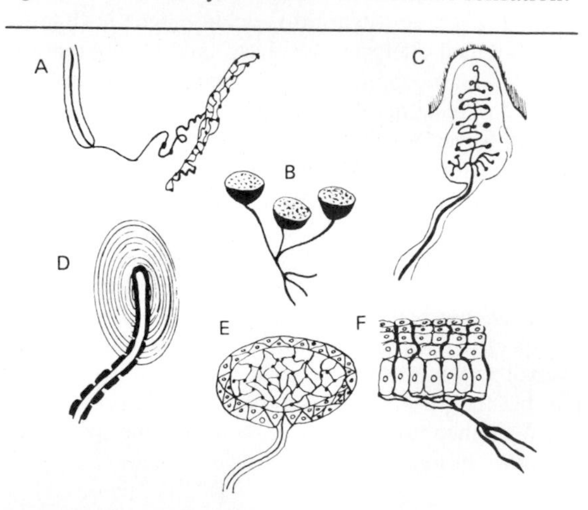

Figure 5–1. Sensory receptors in the skin. Ruffini endings (A) and Merkel's disks (B) are expanded ends of sensory nerve fibers. Meissner's corpuscles (C), pacinian corpuscles (D), and Krause's end-bulbs (E) are encapsulated endings. (F) naked curve endings between cells in tissue.

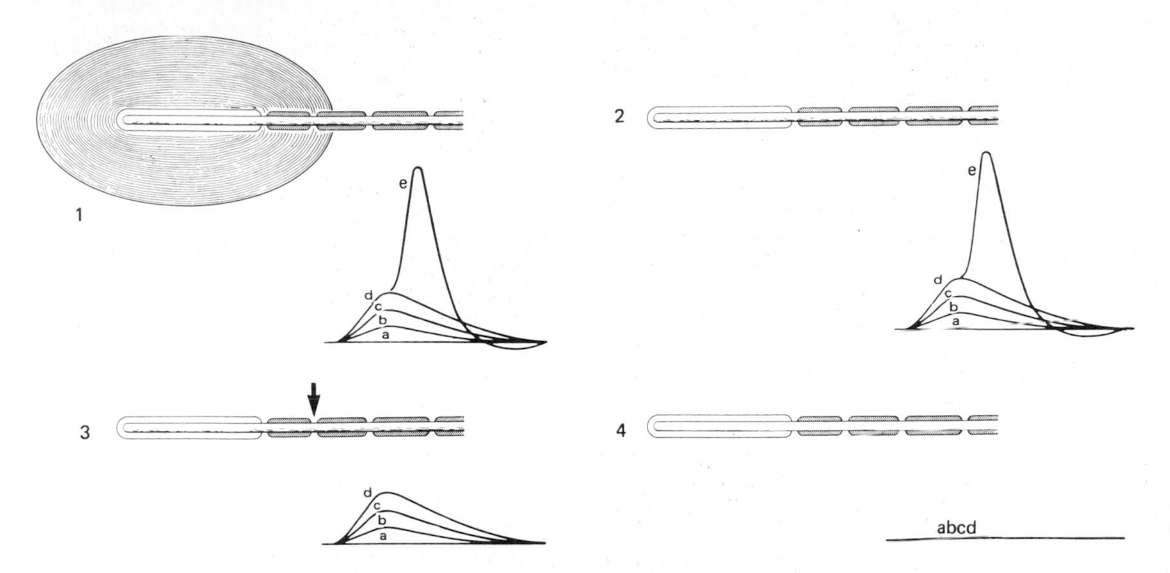

Figure 5–2. Demonstration that the generator potential in a pacinian corpuscle originates in the nonmyelinated nerve terminal. In 1, the electrical responses to a pressure of x (record a), 2x (b), 3x (c), and 4x (d) were recorded. The strongest stimulus produced an action potential in the sensory nerve (e). In 2, the same responses persisted after removal of the connective tissue capsule. In 3, the generator responses persisted but the action potential was absent when the first node of Ranvier was blocked with pressure or narcotics (arrow). In 4, all responses disappeared when the sensory nerve was cut and allowed to degenerate before the experiment. (Reproduced, with permission, from Loewenstein WR: Biological transducers. *Sci Am* [Aug] 1960;**203**:98. Copyright © 1960 by Scientific American, Inc. All rights reserved.)

ELECTRICAL & IONIC EVENTS IN RECEPTORS

Anatomic Relationships

The problem of how receptors convert energy into action potentials in the sensory nerves has been the subject of intensive study. In the complex sense organs such as those concerned with vision, hearing, equilibrium, and taste, there are separate receptor cells and synaptic junctions between receptors and afferent nerves. However, in most of the cutaneous sense organs, the receptors are specialized, histologically modified ends of sensory nerve fibers.

Pacinian corpuscles, which are touch receptors, have been studied in detail. Because of their relatively large size and accessibility in the mesentery of experimental animals, they can be isolated, studied with microelectrodes, and subjected to microdissection. Each capsule consists of the straight, unmyelinated ending of a sensory nerve fiber, 2μm in diameter, surrounded by concentric lamellas of connective tissue that give the organ the appearance of a minute cocktail onion. The myelin sheath of the sensory nerve begins inside the corpuscle. The first node of Ranvier is also located inside, whereas the second is usually near the point at which the nerve fiber leaves the corpuscle (Fig 5–2).

Generator Potentials

Recording electrodes can be placed on the sensory nerve as it leaves a pacinian corpuscle and graded pressure applied to the corpuscle. When a small amount of pressure is applied, a nonpropagated depolarizing potential resembling an EPSP is recorded. This is called the **generator potential** or **receptor potential.** As the pressure is increased, the magnitude of the receptor potential increases. When the magnitude of the generator potential is about 10 mV, an action potential is generated in the sensory nerve. As the pressure is further increased, the generator potential becomes even larger and the sensory nerve fires repetitively. Eventually, in the pacinian corpuscle, the size of the generator potential reaches a maximum, but its rate of rise continues to increase as the magnitude of the applied pressure is increased.

Source of the Generator Potential

By microdissection techniques, it has been shown that removal of the connective tissue lamellas from the unmyelinated nerve ending in a pacinian corpuscle does not abolish the generator potential. When the first node of Ranvier is blocked by pressure or narcotics, the generator potential is unaffected but conducted impulses are abolished (Fig 5–2). When the sensory nerve is sectioned and the nonmyelinated terminal is allowed to degenerate, no generator potential is formed. These and other experiments have established that the generator potential is produced in the unmyelinated nerve terminal. The nerve terminals also appear to be the site of origin of the generator potential in Merkel's disks. The generator potential electronically depolarizes the first node of Ranvier. The receptor therefore converts mechanical energy into an electrical response, the magnitude of which is proportionate to the intensity of the stimulus. The generator potential in turn depolarizes the sensory nerve at the first node of Ranvier. Once the firing level is reached, an action potential is produced and the membrane repolarizes. If the generator potential is great enough, the neuron fires again as soon as it repolarizes, and it continues to fire as long as the generator potential is large enough to bring the membrane potential of the node to the firing level. Thus, the node converts the graded response of the receptor into action potentials, the frequency of which is proportionate to the magnitude of the applied stimuli.

Similar generator potentials in the muscle spindle have been studied. The relationship between the stimulus intensity and the size of the generator potential and between the stimulus intensity and the frequency of the action potentials in the afferent nerve fiber from a spindle is shown in Fig 5–3. The frequency of the action potentials is generally related to the intensity of the stimulus by a power function (see below). Generator potentials have also been observed in the organ of Corti, the olfactory and taste organs, and other sense organs. Presumably, they are in most cases the mechanism by which sensory nerve fibers are activated.

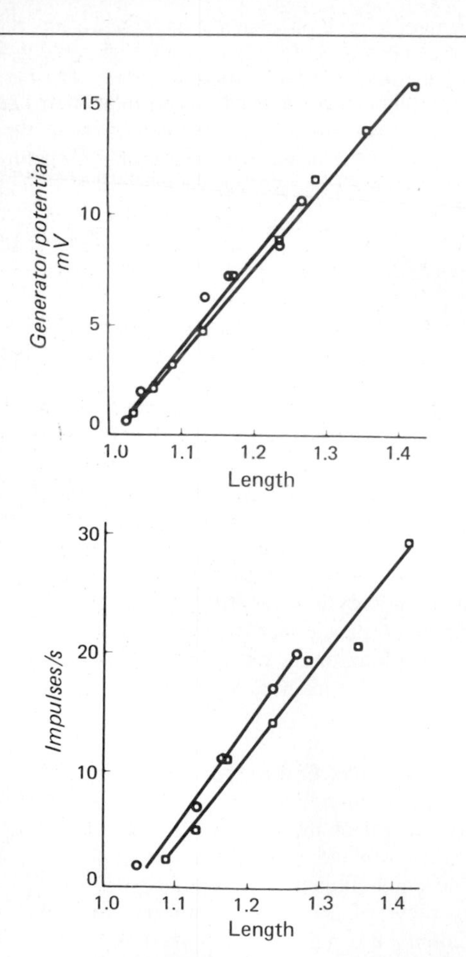

Figure 5–3. Relation between muscle length and size of generator potential *(top)* and impulse frequency *(bottom)* in crayfish stretch receptor. Squares and circles indicate values in 2 different preparations. (Reproduced, with permission, from Terzuolo CA, Washizu Y: Relation between stimulus strength, generator potential, and impulse frequency in stretch receptor of crustacea. *J Neurophysiol* 1962:25:56.)

Ionic Basis of Excitation

The biophysical events underlying the generator potential have been clarified in a few receptors. In some receptors, mechanical distortion opens channels in the receptor membrane. The resultant influx of Na^+ produces the generator potential, and presumably the number of channels opened is proportionate to the intensity of the stimulus. In other receptors, such as the rods and cones, different mechanisms are responsible, and in many instances, the mechanisms is still unknown.

Adaptation

When a maintained stimulus of constant strength is applied to a receptor, the frequency of the action potentials in its sensory nerve declines over time. This phenomenon is known as **adaptation.** The degree to which adaptation occurs varies with the type of sense organ (Fig 5–4). Touch adapts rapidly, and its receptors are called **phasic receptors.** Application of a maintained pressure to a pacinian corpuscle produces a generator potential that decays rapidly. On the other hand, the carotid sinus, the muscle spindles, and the organs for cold, pain, and lung

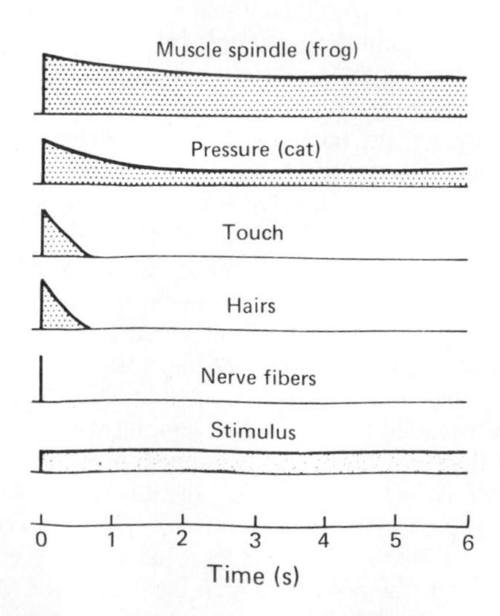

Figure 5–4. Adaptation. The height of the curve in each case indicates the frequency of the discharge in afferent nerve fibers at various times after beginning sustained stimulation. (Reproduced, with permission, from Adrian ED: *Basis of Sensation.* Christophers, 1928.)

inflation adapt very slowly and incompletely; the receptors involved are termed **tonic receptors.** This correlates with the fact that the generator potential of the muscle spindle is prolonged and decays very slowly when a steady stimulus is applied to it. Maintained pressure applied to the outside of a pacinian corpuscle causes steady displacement in the outer lamellas, but the lamellas near the nerve fiber slip back to their original position, ending the distortion of the nerve ending and causing the generator potential to decline. However, this is not the only factor producing adaptation; there is still a slow decline in the number of action potentials generated over time by a steady stimulus after removal of the outer lamellas from the pacinian corpuscle. This decline is due to accommodation of the sensory nerve fiber to the generator potential.

The slow, incomplete adaptation of the carotid sinus and the organs for muscle stretch, pain, and cold is of some value to the animal. Muscle stretch plays a role in prolonged postural adjustments. The sensations of pain and cold are initiated by potentially noxious stimuli, and they would lose some of their warning value if their receptors showed marked adaptation. Carotid and aortic receptors operate continuously in the regulation of blood pressure, and adaptation of these receptors would limit the precision with which the regulatory system operates.

"CODING" OF SENSORY INFORMATION

There are variations in the speed of conduction and other characteristics of sensory nerve fibers (see Chapter 2), but action potentials are similar in all nerves. The action potentials in the nerve from a touch receptor, for example, are essentially identical to those in the nerve from a warmth receptor. This raises the question of why stimulation of a touch receptor causes a sensation of touch and not of warmth. It also raises the question of how it is possible to tell whether the touch is light or heavy.

Doctrine of Specific Nerve Energies

The sensation evoked by impulses generated in a receptor depends upon the specific part of the brain they ultimately activate. The specific sensory pathways are discrete from sense organ to cortex. Therefore, when the nerve pathways from a particular sense organ are stimulated, the sensation evoked is that for which the receptor is specialized no matter how or where along the pathway the activity is initiated. This principle, first enunciated by Müller, has been given the rather cumbersome name of the **doctrine of specific nerve energies.** For example, if the sensory nerve from a pacinian corpuscle in the hand is stimulated by pressure at the elbow or by irritation from a tumor in the brachial plexus, the sensation evoked is one of touch. Similarly, if a fine enough electrode could be inserted into the appropriate fibers of the dorsal columns of the spinal cord, the thalamus, or the postcentral gyrus of the cerebral cortex, the sensation produced by stimulation would be touch. This doctrine has been questioned from time to time, especially by those who claim that pain is produced by overstimulation of a variety of receptors. However, the overstimulation hypothesis has been largely discredited, and the principle of specific nerve energies remains one of the cornerstones of sensory physiology.

Projection

No matter where a particular sensory pathway is stimulated along its course to the cortex, the conscious sensation produced is referred to the location of the receptor. This principle is called the **law of projection.** Cortical stimulation experiments during neurosurgical procedures on conscious patients illustrate this phenomenon. For example, when the cortical receiving area for impulses from the left hand is stimulated, the patient reports sensation in the left hand, not in the head. Another dramatic example is seen in amputees. These patients may complain, often bitterly, of pain and proprioceptive sensations in the absent limb ("phantom limb"). These sensations are due in part to pressure on the stump of the amputated limb. This pressure initiates impulses in nerve fibers that previously came from sense organs in the amputated limb, and the sensations evoked are projected to where the receptors used to be.

Intensity Discrimination

There are 2 ways in which information about intensity of stimuli is transmitted to the brain: by variation in the frequency of the action potentials generated by the activity in a given receptor, and by variation in the number of receptors activated. It has long been taught that the magnitude of the sensation felt is proportionate to the log of the intensity of the stimulus (**Weber-Fechner law**). It now appears, however, that a power function more accurately describes this relation. In other words, $R = KS^A$, where R is the sensation felt, S is the intensity of the stimulus, and, for any specific sensory modality, K and A are constants. The frequency of the action potentials a stimulus generates in a sensory nerve fiber is also related to the intensity of the initiating stimulus by a power function. An example of this relation is shown in Fig 5–3, in which the exponent is approximately 1.0. Another example is shown in Fig 5–5, in which the calculated exponent is 0.52. Current evidence indicates that in the central nervous system the relation between stimulus and sensation is linear; consequently, it appears that for any given sensory modality, the relation between sensation and stimulus intensity is determined primarily by the properties of the peripheral receptors themselves.

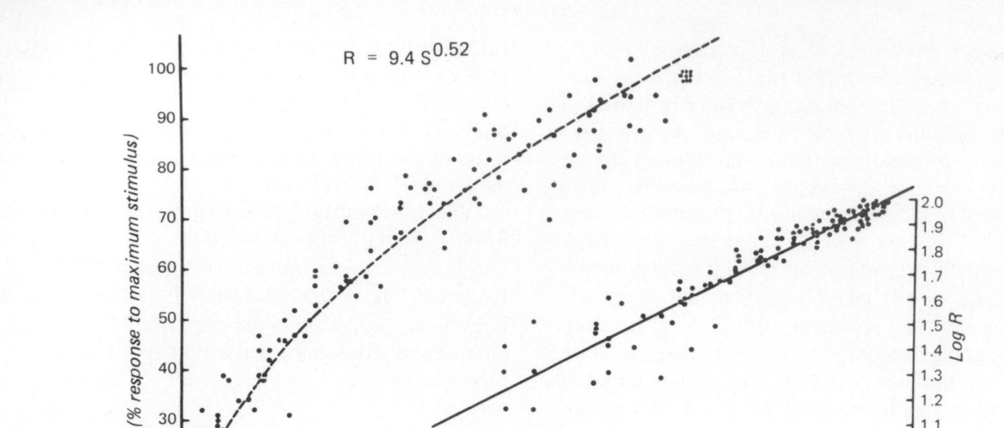

Figure 5–5. Relation between magnitude of touch stimulus (S) and frequency of action potentials in sensory nerve fibers (R). Dots are individual values from cats plotted on linear coordinates ***(left)*** and log-log coordinates ***(right)***. The equation shows the calculated power function relationship between R and S. (Reproduced, with permission, from Werner G. Mountcastle VB: Neural activity in mechanoreceptive cutaneous afferents. Stimulus-response relations, Weber functions, and information transmission. *J. Neurophysiol* 1965:**28**:359.)

Sensory Units

The term sensory unit is applied to a single sensory axon and all its peripheral branches. The number of these branches varies, but they may be numerous, especially in the cutaneous senses. The **receptive field** of a sensory unit is the area from which a stimulus produces a response in that unit. In the cornea and adjacent sclera of the eye, the surface area supplied by a single sensory unit is 50–200 mm^2. Generally, the areas supplied by one unit overlap and interdigitate with the areas supplied by others.

Recruitment of Sensory Units

As the strength of a stimulus is increased, it tends to spread over a large area and generally not only activates the sense organs immediately in contact with it but "recruits" those in the surrounding area as well. Furthermore, weak stimuli activate the receptors with the lowest thresholds, whereas stronger stimuli also activate those with higher thresholds. Some of the receptors activated are part of the same sensory unit, and impulse frequency in the unit therefore increases. Because of overlap and interdigitation of one unit with another, however, receptors of other units are also stimulated, and consequently more units fire. In this way, more afferent pathways are activated, and this is interpreted in the brain as an increase in intensity of the sensation.

REFERENCES
Section II: Physiology of Nerve & Muscle Cells

Akil H et al: Endogenous opioids: Biology and function. *Annu Rev Neurosci* 1984;7:223.

Basmajian JV: Electromyography comes of age. *Science* 1972;**176:**603.

Bessman SP, Geiger PJ: Transport of energy in muscle: The phosphorylcreatine shuttle. *Science* 1981;**211:**448.

Braunwald E, Ross J, Sonnenblick EH: *Mechanisms of Contraction of the Normal and Failing Heart,* 2nd ed. Little, Brown, 1976.

Bülbring E, Tomita T: Catecholamine action on smooth muscle. *Pharm Rev* 1987;**39:**49.

Bülbring E et al (editors): *Smooth Muscle: An Assessment of Current Knowledge*. Arnold, 1982.

Catterall WA: The molecular basis of neural excitability. *Science* 1984;**223:**653.

Chesselet M-F: Presynaptic regulation of neurotransmitter release in the brain: Facts and hypothesis. *Neuroscience* 1984;**12:**347.

Cooper JR, Bloom FE, Roth RH: *The Biochemical Basis of Neuropharmacology*, 5th ed. Oxford Univ Press, 1986.

Cowan WM, Cuénod M (editors): *Use of Axonal Transport for Studies of Neuronal Connectivity*. Elsevier, 1975.

Easter SS Jr et al: The changing view of neural specificity. *Science* 1985;**230:**507.

Fleming WW, McPhillips JJ, Westfall DP: Postjunctional supersensitivity and subsensitivity of excitable tissues to drugs. *Ergeb Physiol* 1973;**68:**55.

Freed WJ, de Medinaceli L, Wyatt RJ: Promoting functional plasticity in the damaged nervous system. *Science* 1985;**227:**1544.

Goodman CS, Bastiani MJ: How embryonic nerve cells recognize one another. *Sci Am* (Dec) 1984;**251:**58.

Greene LA, Shooter EM: The nerve growth factor: Biochemistry, synthesis, and mechanism of action. *Annu Rev Neurosci* 1980;**3:**353.

Hartzell HC: Mechanisms of slow postsynaptic potentials. *Nature* 1981;**291:**539.

Hille B: *Ionic Channels of Excitable Membranes*. Sinauer Associates, 1984.

Iggo A, Andres KH: Morphology of cutaneous receptors. *Annu Rev Neurosci* 1982;**5:**1.

Jessell TM: Adhesion molecules and the hierarchy of neural development. *Neuron* 1988;**1:**3.

Jolesz F, Strefert A: Development, innervation, and activity pattern-induced changes in skeletal muscle. *Annu Rev Physiol* 1981;**43:**531.

Kandel ER (editor): *Handbook of Physiology*. Section 1, Part 1. *The Nervous System*. American Physiological Society, 1977.

Katz B: Quantal mechanism of neural transmitter release. *Science* 1971;**173:**123.

Kuhar MJ, De Souza EB, Unnerstall JR: Neurotransmitter receptor mapping by radioautography and other methods. *Annu Rev Neurosci* 1986;**9:**27.

Langer SZ: Presynaptic regulation of the release of catecholamines, *Pharmacol Rev* 1980;**32:**337.

Levi-Montalcini R: The nerve growth factor 35 years later. *Science* 1987;**237:**1154.

Loewenstein WR (editor): Principles of receptor physiology. In: *Handbook of Sensory Physiology*. Vol. 1. Springer-Verlag, 1971.

Matthews PB: Where does Sherrington's "muscular sense" originate? Muscles, joints, corollary discharges? *Annu Rev Neurosci* 1982;**5:**189.

McCarthy MP et al: Molecular biology of the acetylcholine receptor. *Annu Rev Neurosci* 1986;**9:**383.

McMahon TA: *Muscles, Reflexes, and Locomotion*. Princeton Univ Press, 1984.

Miller RJ: Multiple calcium channels and neuronal function. *Science* 1987;**235:**46.

Morell P, Norton WT: Myelin. *Sci Am* (May) 1980;**242:**88.

Motulsky HJ, Insel PA: Adrenergic receptors in man. *N Engl J Med* 1982;**307:**18.

Nakanishi S: Substance P precursor and kininogen: their structure, gene organizations, and regulation. *Physiol Rev* 1987;**67:**1117.

Nestler EJ, Greengard P: Protein phosphorylation in the brain. *Nature* 1983;**305:**583.

Noble D, Powell T: *Electrophysiology of Single Cardiac Cells*. Academic, 1987.

Palka J (editor): Developmental neurobiology, *Bioscience* 1984;**34:**294.

Peachey LD (editor): *Handbook of Physiology*, Section 10. *Skeletal Muscle*. American Physiological Society, 1983.

Pernow B: Substance P. *Pharmacol Rev* 1983;**35:**85.

Purves D, Lichtman JW: Elimination of synapses in the developing nervous system. *Science* 1980;**210:**153.

Reuter H: Calcium channel modulation by neurotransmitters, enzymes and drugs. *Nature* 1983;**301:**569.

Robinson MB, Coylce JT: Glutamate and related excitatory neurotransmitters: from basic science to clinical application. *FASEB J* 1987;**1:**446.

Salpeter MM (editor): *The Vertebrate Neuromuscular Junction*. Plenum, 1987.

Scriver CH, Clow CL: Phenylketonuria: Epitome of human biochemical genetics. *N Engl J Med* 1980;**303:**1336.

Seeman P: Brain dopamine receptors. *Pharmacol Rev* 1980;**32:**229.

Shepherd GM: Microcircuits in the nervous system. *Sci Am* (Feb) 1978;**238:**93.

Smith TW: Digitalis mechanism of action and clinical use. *N Eng J Med* 1988;**318:**358.

Snyder S: Adenosine as a neuromodulator. *Annu Rev Neurosci* 1985;**8:**103.

Stevens CF: The neuron. *Sci Am* (Sept) 1979;**241:**54.

Stiles GL. Caron MG, Lefkowitz RJ: β-Adrenergic receptors: Biochemical mechanisms of physiological regulation. *Physiol Rev* 1984;**64:**661.

Stracher A (editor): *Muscle and Nonmuscle Motility*. Academic Press, 1983.

Symposium: Coordination of metabolism and contractility by phosphorylation in cardiac, skeletal, and smooth muscle. *Fed Proc* 1983;**42:**7.

Wamsley JK: Opioid receptors: Autoradiography. *Pharmacol Rev* 1983;**35:**69.

Watson WE: Physiology of neuroglia. *Physiol Rev* 1974; **54:**245.

Section III. Functions of the Nervous System

6 Reflexes

INTRODUCTION

The basic unit of integrated neural activity is the **reflex arc.** This arc consists of a sense organ, an afferent neuron, one or more synapses in a central integrating station or sympathetic ganglion, an efferent neuron, and an effector. In mammals, the connection between afferent and efferent somatic neurons is generally in the brain or spinal cord. The afferent neurons enter via the dorsal roots or cranial nerves and have their cell bodies in the dorsal root ganglia or in the homologous ganglia on the cranial nerves. The efferent fibers leave via the ventral roots or corresponding motor cranial nerves. The principle that in the spinal cord the dorsal roots are sensory and the ventral roots are motor is known as the **Bell-Magendie law.**

In previous chapters, the function of each of the components of the reflex arc has been considered in detail. As noted in Chapters 2 and 3, impulses generated in the axons of the afferent and efferent neurons and in muscle are "all or none" in character. On the other hand, there are 3 junctions or junction-like areas in the reflex arc where responses are graded (Fig 6–1). These are the receptor-afferent neuron region, the synapse or synapses between the afferent and efferent neurons, and the myoneural junction. At each of these points, a nonpropagated potential proportionate in size to the magnitude of the incoming stimulus is generated. The graded potentials serve to electrotonically depolarize the adjacent nerve or muscle membrane and set up all or none responses. The number of action potentials in the afferent nerve is proportionate to the magnitude of the applied stimulus at the sense organ. There is also a rough correlation between the magnitude of the stimulus and the frequency of action potentials in the efferent nerve; however, the connection between the afferent and efferent neurons is usually in the central nervous system, and activity in the reflex arc is modified by the multiple inputs converging on the efferent neurons.

The simplest reflex arc is one with a single synapse between the afferent and efferent neurons. Such arcs are **monosynaptic,** and reflexes occurring in them are **monosynaptic reflexes.** Reflex arcs in which one or more interneurons are interposed between the afferent and efferent neurons are **polysynaptic,** the number of synapses in the arcs varying from 2 to many hundreds. In both types, but especially in

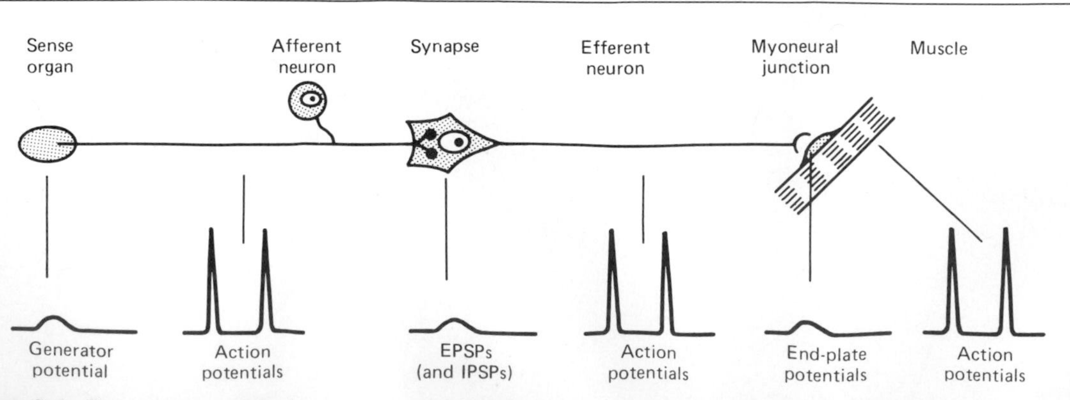

Figure 6–1. The reflex arc. Note that at the receptor and at each of the junctions in the arc there is a nonpropagated graded response that is proportionate to the magnitude of the stimulus, whereas in the portions of the arc specialized for transmission (axons, muscle membrane) the responses are all or none action potentials.

polysynaptic reflex arcs, activity is modified by spatial and temporal facilitation, occlusion, subliminal fringe effects, and other effects.

MONOSYNAPTIC REFLEXES: THE STRETCH REFLEX

When a skeletal muscle with an intact nerve supply is stretched, it contracts. This response is called the **stretch reflex.** The stimulus that initiates the reflex is stretch of the muscle, and the response is contraction of the muscle being stretched. The sense organ is the muscle spindle. The impulses originating in the spindle are conducted in the central nervous system by fast sensory fibers that pass directly to the motor neurons which supply the same muscle. The neurotransmitter at the central synapse is glutamate. Stretch reflexes are the best known and studied monosynaptic reflexes in the body.

Clinical Examples

Tapping the patellar tendon elicits the **knee jerk,** a stretch reflex of the quadriceps femoris muscle, because the tap on the tendon stretches the muscle. A similar contraction is observed if the quadriceps is stretched manually. Stretch reflexes can also be elicited from most of the large muscles of the body. Tapping on the tendon of the triceps brachii, for example, causes an extensor response at the elbow due to reflex contraction of the triceps; tapping on the Achilles tendon causes an ankle jerk due to reflex contraction of the gastrocnemius; and tapping on the side of the face causes a stretch reflex in the masseter. Other examples of stretch reflexes are listed in neurology textbooks.

Structure of Muscle Spindles

Each muscle spindle consists of 2–10 muscle fibers enclosed in a connective tissue capsule. These fibers are more embryonal in character and have less distinct striations than the rest of the fibers in the muscle. They are called **intrafusal fibers** to distinguish them from the **extrafusal fibers,** the regular contractile units of the muscle. The intrafusal fibers are in parallel with the rest of the muscle fibers because the ends of the capsule of the spindle are attached to the tendons at either end of the muscle or to the sides of the extrafusal fibers.

There are 2 types of intrafusal fibers in mammalian muscle spindles. The first type contains many nuclei in a dilated central area and is therefore called a **nuclear bag fiber** (Fig 6–2). The second type, the **nuclear chain fiber,** is thinner and shorter and lacks a definite bag. The ends of the nuclear chain fibers connect to the sides of the nuclear bag fibers. The ends of the intrafusal fibers are contractile, whereas the central portions probably are not.

There are 2 kinds of sensory endings in each spindle. The **primary,** or **annulospiral, endings** are the terminations of rapidly conducting group Ia afferent fibers that wrap around the center of the nuclear bag and nuclear chain fibers. The **secondary,** or **flower-spray, endings** are terminations of group II sensory fibers and are located nearer the ends of the intrafusal fibers, but only on nuclear chain fibers.

The spindles have a motor nerve supply of their own. These nerves are 3–6 μm in diameter, constitute about 30% of the fibers in the ventral roots, and belong in Erlanger and Gasser's A γ group. Because of their characteristic size, they are called the γ **efferents of Leksell** or the **small motor nerve system.** There also appears to be a sparse innervation of spindles by branches of the fibers that innervate extrafusal fibers in the same muscle. The function of this β **innervation** is unknown.

The endings of the γ efferent fibers are of 2 histologic types. There are motor end-plates **(plate endings)** on the nuclear bag fibers, and there are endings that form extensive networks **(trail endings)** primarily on the nuclear chain fibers. It is known that the spindles receive 2 functional types of innerva-

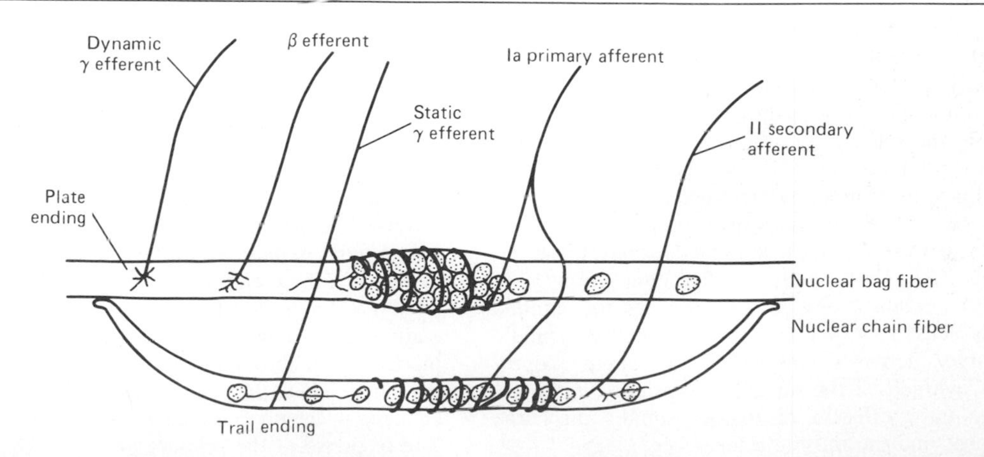

Figure 6–2. Diagram of muscle spindle. (Modified and reproduced, with permission, from Stein RB: Peripheral control of movement. *Physiol Rev* 1974;**54**:215.)

tion—dynamic γ efferents and static γ efferents (see below)—and it is reasonable to hypothesize that the dynamic γ efferents terminate at the plate endings whereas the static γ efferents terminate at the trail endings.

Central Connections of Afferent Fibers

It can be proved experimentally that the Ia fibers from the primary endings end directly on motor neurons supplying the extrafusal fibers of the same muscle. The time between the application of the stimulus and the response is the **reaction time.** In humans, the reaction time for a stretch reflex such as the knee jerk is 19–24 ms. Weak stimulation of the sensory nerve from the muscle, known to stimulate only Ia fibers, causes a contractile response with a similar latency. Since the conduction velocities of the afferent and efferent fiber types are known and the distance from the muscle to the spinal cord can be measured, it is possible to calculate how much of the reaction time was taken up by conduction to and from the spinal cord. When this value is subtracted from the reaction time, the remainder, called the **central delay,** is at the time taken for the reflex activity to traverse the spinal cord. In humans, the central delay for the knee jerk is 0.6–0.9 ms, and figures of similar magnitude have been found in experimental animals. Since the minimal synaptic delay is 0.5 ms (see Chapter 4), only one synapse could have been traversed.

Muscle spindles also make connections that cause muscle contraction via polysynaptic pathways, and the afferents involved are probably those from the secondary endings. However, group II fibers also make monosynaptic connections to the motor neurons and make a small contribution to the stretch reflex.

Function of Muscle Spindles

When the muscle spindle is stretched, the primary endings are distorted and receptor potentials are generated. These in turn set up action potentials in the sensory fibers at a frequency that is proportionate to the degree of stretching. The spindle is in parallel with the extrafusal fibers, and when the muscle is passively stretched, the spindles are also stretched. This initiates reflex contraction of the extrafusal fibers in the muscle. On the other hand, the spindle afferents characteristically stop firing when the muscle is made to contract by electrical stimulation of the nerve fibers to the extrafusal fibers because the muscle shortens while the spindle does not (Fig 6–3).

Thus, the spindle and its reflex connections constitute a feedback device that operates to maintain muscle length; if the muscle is stretched, spindle discharge increases and reflex shortening is produced, whereas if the muscle is shortened without a change in γ efferent discharge, spindle discharge decreases and the muscle relaxes.

Primary endings on the nuclear bag fibers and nuclear chain fibers are both stimulated when the spindle is stretched, but the pattern of response differs. The nerves from the endings in the nuclear bag region discharge most rapidly while the muscle is being stretched and less rapidly during sustained stretch (Fig 6–4). The nerves from the primary endings on the nuclear chain fibers discharge at an increased rate throughout the period when a muscle is stretched. Thus, the primary endings respond to both changes in length and changes in the rate of stretch. The response of the primary ending to the phasic as well as the tonic events in the muscle is important because the prompt, marked phasic response helps to

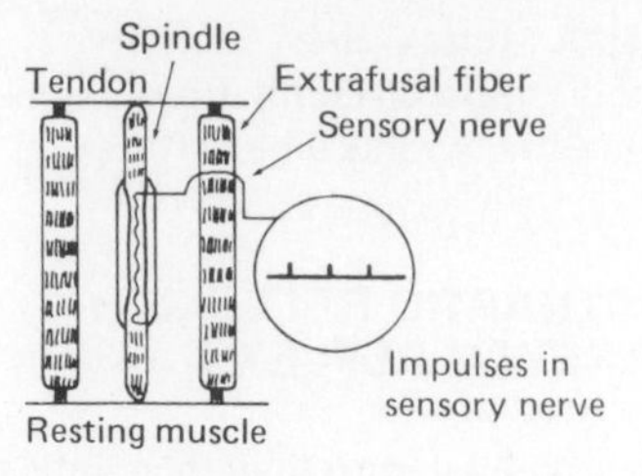

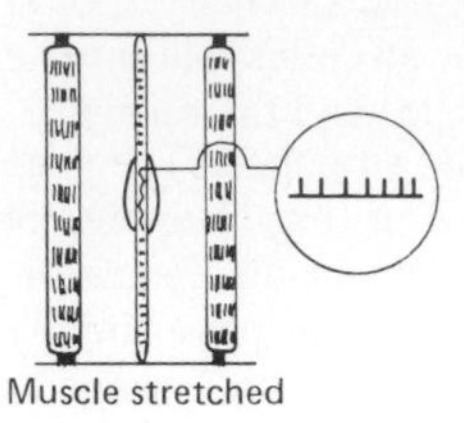

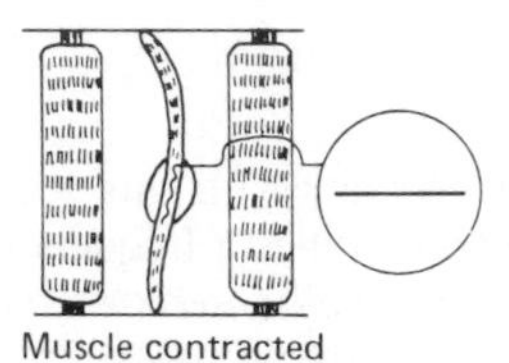

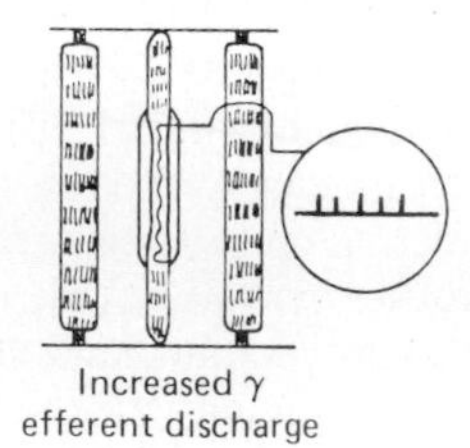

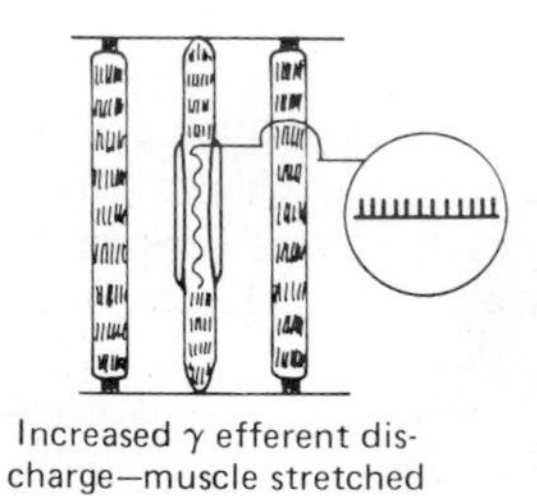

Figure 6–3. Effect of various conditions on muscle spindle discharge.

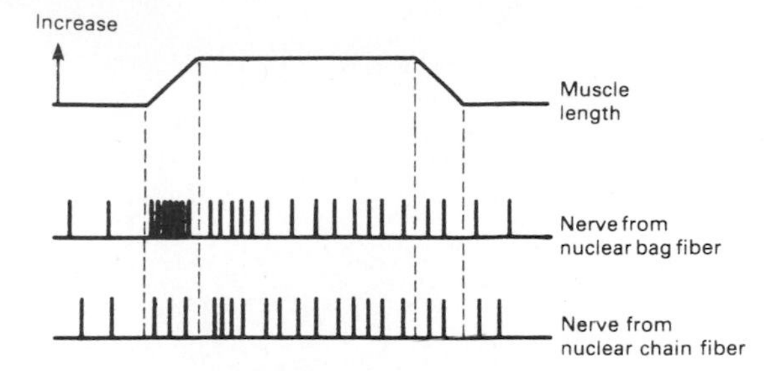

Figure 6–4. Response of spindle afferents to muscle stretch. The bottom 2 lines represent the number of discharges in afferent nerves from the primary endings on nuclear bag and nuclear chain fibers as the muscle is stretched and then permitted to return to its original length.

dampen oscillation due to conduction delays in the feedback loop regulating muscle length. There is normally a small oscillation in this feedback loop. This **physiologic tremor** has a frequency of approximately 10 Hz. However, the tremor would be worse if it were not for the sensitivity of the spindle to velocity of stretch.

Effects of Gamma Efferent Discharge

Stimulation of the γ efferent system produces a very different picture from that produced by stimulation of the extrafusal fibers. Such stimulation does not lead directly to detectable contraction of the muscles because the intrafusal fibers are not strong enough or plentiful enough to cause shortening. However, stimulation does cause the contractile ends of the intrafusal fibers to shorten and therefore stretches the nuclear bag portion of the spindles, deforming the annulospiral endings and initiating impulses in the Ia fibers. This in turn can lead to reflex contraction of the muscle. Thus, muscle can be made to contract via stimulation of the α motor neurons that innervate the extrafusal fibers or the γ efferent neurons that initiate contraction indirectly via the stretch reflex.

When the rate of γ efferent discharge is increased, the intrafusal fibers are shorter than the extrafusal ones. If the whole muscle is stretched during stimulation of the γ efferents, additional action potentials are generated owing to the additional stretch of the nuclear bag region, and the rate of discharge in the Ia fibers is further increased (Fig 6–3). Increased γ efferent discharge thus increases spindle sensitivity, and the sensitivity of the spindles to stretch varies with the rate of γ efferent discharge.

There is considerable evidence of increased γ efferent discharge along with the increased discharge of the α motor neurons that initiates movements. Because of this "α-γ linkage," the spindle shortens with the muscle, and spindle discharge may continue throughout the contraction. In this way, the spindle remains capable of responding to stretch and reflexly adjusting motor neuron discharge throughout the contraction.

The existence of dynamic and static γ efferents is mentioned above. Stimulation of the former, which may end via plate endings on nuclear bag fibers, increases spindle sensitivity to rate of change of stretch. Stimulation of the latter, probably via trail endings on nuclear chain fibers, increases spindle sensitivity to steady, maintained stretch. It is thus possible to adjust separately the spindle responses to phasic and tonic events.

Control of Gamma Efferent Discharge

The motor neurons of the γ efferent system are regulated to a large degree by descending tracts from a number of areas in the brain. Via these pathways, the sensitivity of the muscle spindles and hence the threshold of the stretch reflexes in various parts of the body can be adjusted and shifted to meet the needs of postural control (see Chapter 12).

Other factors also influence γ efferent discharge. Anxiety causes an increased discharge, a fact that probably explains the hyperactive tendon reflexes sometimes seen in anxious patients. Stimulation of the skin, especially by noxious agents, increases γ efferent discharge to ipsilateral flexor muscle spindles while decreasing that to extensors and produces the opposite pattern in the opposite limb. It is well known that trying to pull the hands apart when the flexed fingers are hooked together facilitates the knee jerk reflex (Jendrassik's maneuver), and this may also be due to increased γ efferent discharge initiated by afferent impulses from the hands.

Reciprocal Innervation

When a stretch reflex occurs, the muscles that antagonize the action of the muscle involved (antagonists) relax. This phenomenon is said to be due to **reciprocal innervation.** Impulses in the Ia fibers from the muscle spindles of the protagonist muscle cause postsynaptic inhibition of the motor neurons to the antagonists. The pathway mediating this effect appears to be bisynaptic. A collateral from each Ia fiber passes in the spinal cord to an inhibitory interneuron (Golgi bottle neuron) that synapses directly on one of the motor neurons supplying the antagonist muscles. This example of postsynaptic inhibition is discussed in Chapter 4, and the pathway is illustrated in Fig. 4–8.

Inverse Stretch Reflex

Up to a point, the harder a muscle is stretched, the stronger is the reflex contraction. However, when the tension becomes great enough, contraction suddenly ceases and the muscle relaxes. This relaxation in response to strong stretch is called the **inverse stretch reflex** or **autogenic inhibition.**

The receptor for the inverse stretch reflex is in the **Golgi tendon organ** (Fig 6–5). This organ consists of

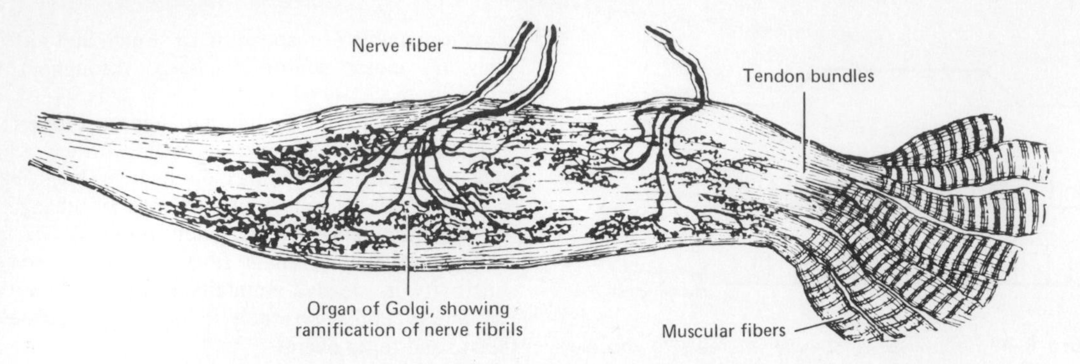

Figure 6–5. Golgi tendon organ. (Reproduced, with permission, from Goss CM [editor]: *Gray's Anatomy of the Human Body,* 29th ed. Lea & Febiger, 1973.)

a netlike collection of knobby nerve endings among the fascicles of a tendon. There are 3–25 muscle fibers per tendon organ. The fibers from the Golgi tendon organs make up the Ib group of myelinated, rapidly conducting sensory nerve fibers. Stimulation of these Ib fibers leads to the production of IPSPs on the motor neurons that supply the muscle from which the fibers arise. The Ib fibers end in the spinal cord on inhibitory interneurons that, in turn, terminate directly on the motor neurons (Fig 6–6). They also make excitatory connections with motor neurons supplying antagonists to the muscle.

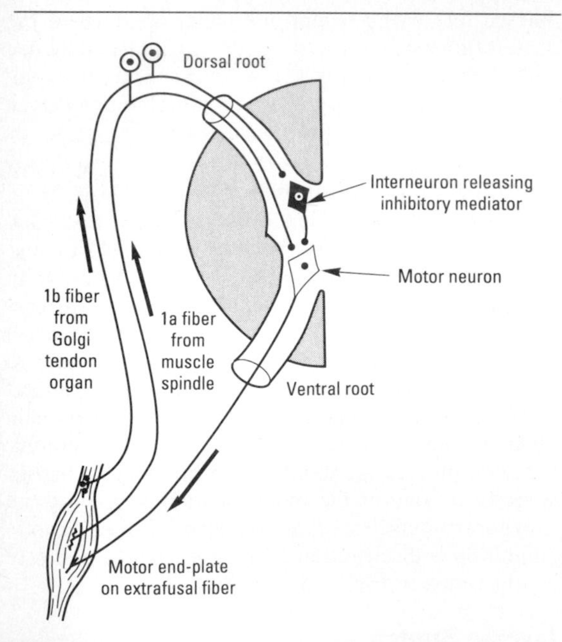

Figure 6–6. Diagram illustrating the pathways responsible for the stretch reflex and the inverse stretch reflex. Stretch stimulates the spindle, and impulses pass up the Ia fiber to excite the motor neuron. It also stimulates the Golgi tendon organ, and impulses passing up the Ib fiber activate the interneuron to release the inhibitory mediator glycine. With strong stretch, the resulting hyperpolarization of the motor neuron is so great that it stops discharging.

Since the Golgi tendon organs, unlike the spindles, are in series with the muscle fibers, they are stimulated by both passive stretch and active contraction of the muscle. The threshold of the Golgi tendon organs is low. The degree of stimulation by passive stretch is not great, because the more elastic muscle fibers take up much of the stretch, and this is why it takes a strong stretch to produce relaxation. However, discharge is regularly produced by contraction of the muscle, and the Golgi tendon organ thus functions as a transducer in a feedback circuit that regulates muscle force in a fashion analogous to the spindle feedback circuit that regulates muscle length.

The importance of the primary endings in the spindles and the Golgi tendon organs in regulating the velocity of the muscle contraction, muscle length, and muscle force is illustrated by the fact that section of the afferent nerves to a limb causes the limb to hang loosely at the side in a semiparalyzed state. The organization of the system is shown in Fig 6–7, and the interaction of spindle discharge, tendon organ discharge, and reciprocal innervation in determining the rate of discharge of a motor neuron is shown in Fig 6–8.

Muscle Tone

The resistance of a muscle to stretch is often referred to as its **tone** or **tonus.** If the motor nerve to a muscle is cut, the muscle offers very little resistance and is said to be **flaccid.** A **hypertonic (spastic)** muscle is one in which the resistance to stretch is high because of hyperactive stretch reflexes. Somewhere between the states of flaccidity and spasticity is the ill-defined area of normal tone. The muscles are generally **hypotonic** when the rate of γ efferent discharge is low and hypertonic when it is high.

Lengthening Reaction

When the muscles are hypertonic, the sequence of moderate stretch→muscle contraction, strong stretch→muscle relaxation is clearly seen. Passive flexion of the elbow, for example, meets immediate resistance due to the stretch reflex in the triceps muscle. Further stretch activates the inverse stretch

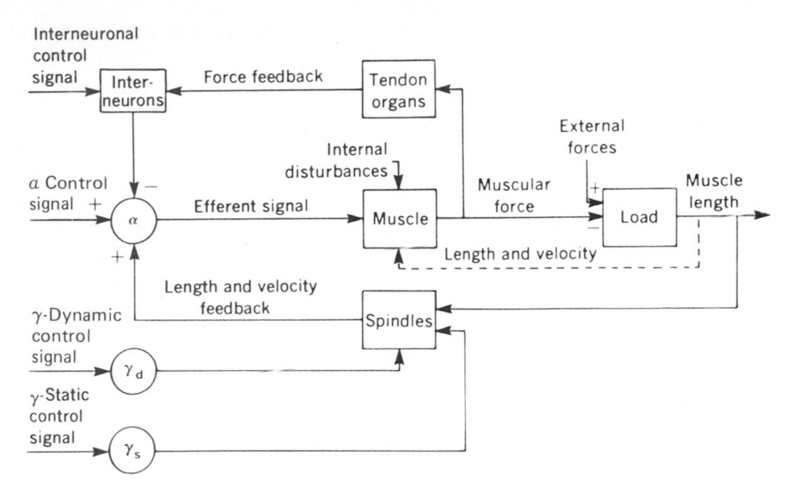

Figure 6–7. Block diagram of peripheral motor control system. The dashed line indicates the nonneural feedback from muscle that limits length and velocity via the inherent mechanical properties of muscle. γ_d, dynamic γ motor neurons; γ_s static γ motor neurons. (Reproduced, with permission, from Houk J in: *Medical Physiology,* 13th ed. Mountcastle VB [editor]. Mosby, 1974.)

reflex. The resistance to flexion suddenly collapses, and the arm flexes. Continued passive flexion stretches the muscle again, and the sequence may be repeated. This sequence of resistance followed by give when a limb is moved passively is known clinically as the **clasp-knife effect** because of its resemblance to the closing of a pocket knife. The physiologic name for it is the **lengthening reaction** because it is the response of a spastic muscle (in the example cited, the triceps) to lengthening.

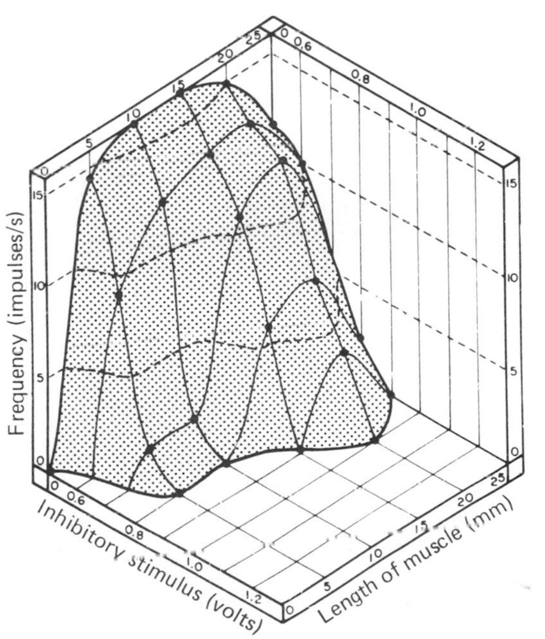

Figure 6–8. Discharge frequency of a motor neuron supplying a leg muscle in a cat. Discharge frequency is plotted against muscle length at various magnitudes of stimulation of the nerve from the antagonist to the muscle. (Reproduced, with permission, from Henneman E et al: Excitability and inhibitability of motor neurons of different sizes. *J Neurophysiol* 1965;**28**:599.)

Clonus

Another finding characteristic of states in which increased γ efferent discharge is present is **clonus.** This neurologic sign is the occurrence of regular, rhythmic contractions of a muscle subjected to sudden, maintained stretch. Ankle clonus is a typical example. This is initiated by brisk, maintained dorsiflexion of the foot, and the response is rhythmic plantar flexion at the ankle. The stretch reflex—inverse stretch reflex sequence described above may contribute to this response. However, it can occur on the basis of synchronized motor neuron discharge without Golgi tendon organ discharge. The spindles of the tested muscle are hyperactive, and the burst of impulses from them discharges all the motor neurons supplying the muscle at once. The consequent muscle contraction stops spindle discharge. However, the stretch has been maintained, and as soon as the muscle relaxes it is again stretched and the spindles stimulated.

POLYSYNAPTIC REFLEXES: THE WITHDRAWAL REFLEX

Polysynaptic reflex paths branch in a complex fashion (Fig 6–9). The number of synapses in each of their branches is variable. Because of the synaptic delay incurred at each synapse, activity in the branches with fewer synapses reaches the motor neurons first, followed by activity in the longer pathways. This causes prolonged bombardment of the motor neurons from a single stimulus and consequently prolonged

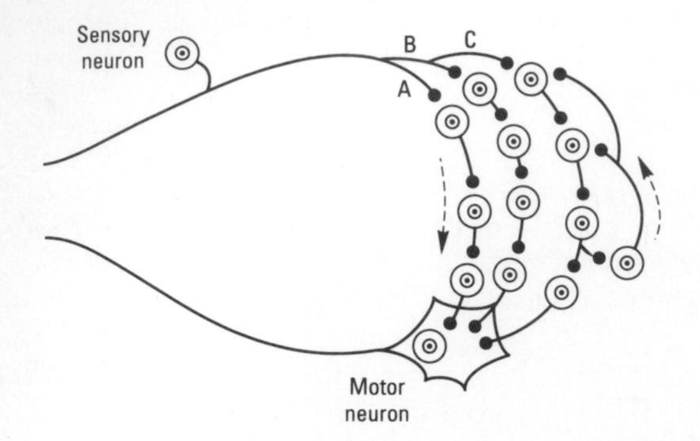

Figure 6–9. Diagram of polysynaptic connections between afferent and efferent neurons in the spinal cord. The dorsal root fiber activates pathway A with 3 interneurons, pathway B with 4 interneurons, and pathway C with 4 interneurons. Note that one of the interneurons in pathway C connects to a neuron that doubles back to other interneurons, forming reverberating circuits.

responses. Furthermore, as shown in Fig 6–9, at least some of the branch pathways turn back on themselves, permitting activity to reverberate until it becomes unable to cause a propagated transsynaptic response and dies out. Such **reverberating circuits** are common in the brain and spinal cord.

Withdrawal Reflex

The withdrawal reflex is a typical polysynaptic reflex that occurs in response to a noxious and usually painful stimulation of the skin or subcutaneous tissues and muscle. The response is flexor muscle contraction and inhibition of extensor muscles, so that the part stimulated is flexed and withdrawn from the stimulus. When a strong stimulus is applied to a limb, the response includes not only flexion and withdrawal of that limb but also extension of the opposite limb. This **crossed extensor response** is properly part of the withdrawal reflex. It can also be shown in experimental animals that strong stimuli generate activity in the interneuron pool which spreads to all 4 extremities. This is difficult to demonstrate in normal animals but is easily demonstrated in an animal in which the modulating effects of impulses from the brain have been abolished by prior section of the spinal cord **(spinal animal).** For example, when the hind limb of a spinal cat is pinched, the stimulated limb is withdrawn, the opposite hind limb extended, the ipsilateral forelimb extended, and the contralateral forelimb flexed. This spread of excitatory impulses up and down the spinal cord to more and more motor neurons is called **irradiation of the stimulus,** and the increase in the number of active motor units is called **recruitment of motor units.**

Importance of the Withdrawal Reflex

Flexor responses can be produced by innocuous stimulation of the skin or by stretch of the muscle, but strong flexor responses with withdrawal are initiated only by stimuli that are noxious or at least potentially harmful to the animal. These stimuli are therefore called **nociceptive stimuli.** Sherrington pointed out the survival value of the withdrawal response. Flexion of the stimulated limb gets it away from the source of irritation, and extension of the other limb supports the body. The pattern assumed by all 4 extremities puts the animal in position to run away from the offending stimulus. Withdrawal reflexes are **prepotent,** ie, they preempt the spinal pathways from any other reflex activity taking place at the moment.

Many of the characteristics of polysynaptic reflexes can be demonstrated by studying the withdrawal reflex in the laboratory. A weak noxious stimulus to one foot evokes a minimal flexion response; stronger stimuli produce greater and greater flexion as the stimulus irradiates to more and more of the motor neuron pool supplying the muscles of the limb. Stronger stimuli also cause a more prolonged response. A weak stimulus causes one quick flexion movement; a strong stimulus causes prolonged flexion and sometimes a series of flexion movements. This prolonged response is due to prolonged, repeated firing of the motor neurons. The repeated firing is called **after-discharge** and is due to continued bombardment of motor neurons by impulses arriving by complicated and circuitous polysynaptic paths.

As the strength of a noxious stimulus is increased, the reaction time is shortened. Spatial and temporal facilitation occurs at synapses in the polysynaptic pathway. Stronger stimuli produce more action potentials per second in the active branches and cause more branches to become active; summation of the EPSPs to the firing level therefore occurs more rapidly.

Local Sign

The exact flexor pattern of the withdrawal reflex in a limb varies with the part of the limb that is stimulated. If the medial surface of the limb is stimulated, for example, the response will include some abduction, whereas stimulation of the lateral surface will produce some adduction with flexion. The reflex response in each case generally serves to effectively remove the limb from the irritating stimulus. This dependence of the exact response on the location of the stimulus is called **local sign.** The degree to which local sign determines the particular pattern is illustrated in Fig 6–10.

Fractionation & Occlusion

Another characteristic of the withdrawal response is the fact that supramaximal stimulation of any of the sensory nerves from a limb never produces as strong a contraction of the flexor muscles as that elicited by direct electrical stimulation of the muscles themselves. This indicates that the afferent inputs **fractionate** the motor neuron pool, ie, each input goes to only part of the motor neuron pool for the flexors of that particular extremity. On the other hand, if all the sensory inputs are dissected out and stimulated one

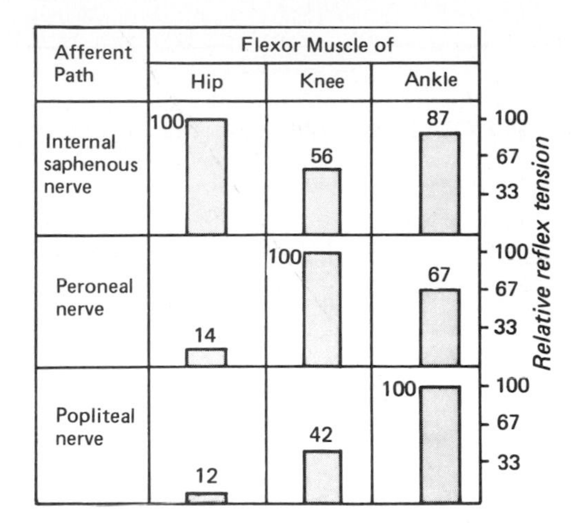

Figure 6–10. The importance of local sign in determining the character of the withdrawal response in a leg. When afferent fibers in each of the 3 nerves on the left were stimulated, hip, knee, and ankle flexors contracted but the relative tension developed in each case (shaded bars) varied.

after the other, the sum of the tension developed by stimulation of each is greater than that produced by direct electrical stimulation of the muscle or stimulation of all inputs at once. This indicates that the various afferent inputs share some of the motor neurons and that occlusion (see Chapter 4) occurs when all inputs are stimulated at once.

Other Polysynaptic Reflexes

There are many polysynaptic reflexes in addition to the withdrawal reflex, all with similar properties. Such reflexes as the abdominal and cremasteric reflexes are forms of the withdrawal reflex. Others include visceral components. Numerous polysynaptic reflexes that relate to specific regulatory functions are described in other sections of this book, and comprehensive lists can be found in neurology textbooks.

GENERAL PROPERTIES OF REFLEXES

It is apparent from the preceding description of the properties of monosynaptic and polysynaptic reflexes that reflex activity is stereotyped and specific in terms of both the stimulus and the response; a particular stimulus elicits a particular response.

Adequate Stimulus

The stimulus that triggers a reflex is generally very precise. This stimulus is called the **adequate stimulus** for the particular reflex. A dramatic example is the scratch reflex in the dog. This spinal reflex is adequately stimulated by multiple linear touch stimuli such as those produced by an insect crawling across the skin. The response is vigorous scratching of the area stimulated. (Incidentally, the precision with which the scratching foot goes to the site of the irritant is a good example of local sign.) If the multiple touch stimuli are widely separated or not in a line, the adequate stimulus is not produced and no scratching occurs. Fleas crawl, but they also jump from place to place. This jumping separates the touch stimuli so that an adequate stimulus for the scratch reflex is not produced. It is doubtful if the flea population would long survive without the ability to jump.

Final Common Path

The motor neurons that supply the extrafusal fibers in skeletal muscles are the efferent side of the reflex arc. All neural influences affecting muscular contraction ultimately funnel through them to the muscles, and they are therefore called the **final common paths.** Numerous inputs converge on them. Indeed, the surface of the average motor neuron accommodates about 10,000 synaptic knobs. There are at least 5 inputs from the same spinal segment to a typical spinal motor neuron. In addition to these, there are excitatory and inhibitory inputs, generally relayed via interneurons, from other levels of the spinal cord and multiple long descending tracts from the brain. All of these pathways converge on and determine the activity in the final common paths.

Central Excitatory & Inhibitory States

The spread up and down the spinal cord of subliminal fringe effects from excitatory stimulation has already been mentioned. Direct and presynaptic inhibitory effects can also be widespread. These effects are generally transient. However, the spinal cord also shows prolonged changes in excitability, possibly because of activity in reverberating circuits or prolonged effects of synaptic mediators. The terms **central excitatory state** and **central inhibitory state** have been used to describe prolonged states in which excitatory influences overbalance inhibitory influences and vice versa. When the central excitatory state is marked, excitatory impulses irradiate not only to many somatic areas of the spinal cord but also to autonomic areas. In chronically paraplegic humans, for example, a mild noxious stimulus may cause, in addition to prolonged withdrawal-extension patterns in all 4 limbs, urination, defecation, sweating, and blood pressure fluctuations **(mass reflex).**

Habituation & Sensitization of Reflex Responses

The fact that reflex responses are stereotyped does not exclude the possibility of their being modified by experience. Examples include habituation and sensitization, which are discussed in terms of synaptic function in Chapter 4 and in terms of their relation to learning and memory in Chapter 16.

7 Cutaneous, Deep, & Visceral Sensation

INTRODUCTION

The sense organs for mechanical stimulation (touch and pressure), warmth, cold, and pain have been discussed in Chapter 5, and the types of neurons that carry impulses generated in them to the central nervous system are listed in Chapter 2. These primary afferent neurons have their cell bodies in the dorsal root ganglia or equivalent ganglia in cranial nerves. They enter the spinal cord or brain stem and make polysynaptic reflex connections to motor neurons at many levels as well as connections that relay impulses to the cerebral cortex. Each of the sensations they mediate is considered in this chapter.

PATHWAYS

The dorsal horns are divided on the basis of histologic characteristics into laminas I–VI, with I being the most superficial and VI the deepest. Lamina II and part of lamina III make up the **substantia gelatinosa,** a lightly stained area near the top of each dorsal horn. There are 3 types of primary afferent fibers: large myelinated A β fibers that transmit impulses generated by mechanical stimuli; small myelinated A δ fibers, some of which transmit impulses from cold receptors and nociceptors that mediate fast pain (see below) and some of which transmit impulses from mechanoreceptors; and small unmyelinated C fibers that are concerned primarily with pain and temperature. However, there are also a few C fibers that transmit impulses from mechanoreceptors. The orderly distribution of these fibers in the dorsal columns and the various layers of the dorsal horn is shown in Fig 7–1.

The principal direct pathways to the cerebral cortex for the cutaneous senses are shown in Fig 7–2. Fibers mediating fine touch and proprioception ascend in the dorsal columns to the medulla, where they synapse in the gracile and cuneate nuclei. The second-order neurons from the gracile and cuneate nuclei cross the midline and ascend in the medial lemniscus to end in the ventral posterior nucleus and related specific sensory relay nuclei of the thalamus (see Chapter 11). This ascending system is frequently called the **dorsal column** or **lemniscal system.**

Other touch fibers, along with those mediating temperature and pain, synapse on neurons in the dorsal horn. The axons from these neurons cross the midline and ascend in the anterolateral quadrant of the spinal cord, where they form the **anterolateral system** of ascending fibers. In general, touch is associated with the ventral spinothalamic tract whereas pain and temperature are associated with the lateral spinothalamic tract, but there is no rigid localization of function. Some of the fibers of the anterolateral system end in the specific relay nuclei of the thalamus; others project to the midline and intralaminar nonspecific projection nuclei. There is a major input from the anterolateral systems into the mesencephalic reticular formation. Thus, sensory input activates the reticular activating system, which in turn maintains the cortex in the alert state (see Chapter 11).

Collaterals from the fibers that enter the dorsal columns pass to the dorsal horn. These collaterals may modify the input into other cutaneous sensory systems, including the pain system. The dorsal horn represents a "gate" in which impulses in the sensory nerve fibers are translated into impulses in ascending tracts, and it now appears that passage through this gate is dependent on the nature and pattern of impulses reaching the substantia gelatinosa and its environs. This gate is also affected by impulses in

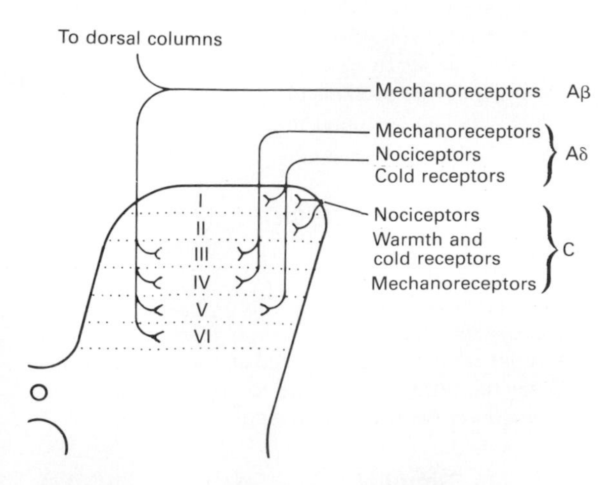

Figure 7–1. Schematic representation of the terminations of the 3 types of primary afferent neurons in the various layers of the dorsal horn of the spinal cord. (Modified from Salt TH, Hill RE: Neurotransmitter candidates of somatosensory primary afferent fibers. *Neuroscience* 1983; **10:** 1033.)

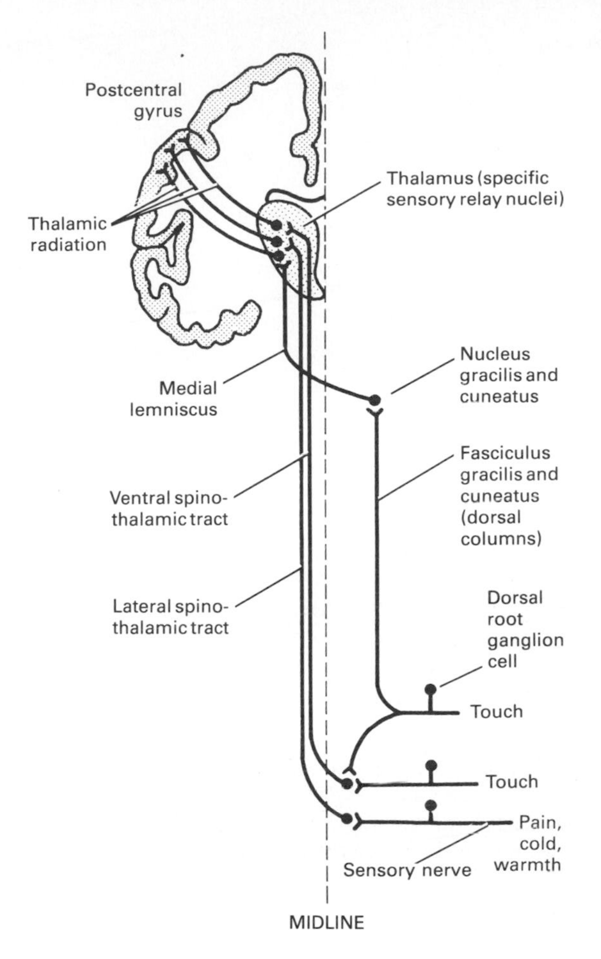

Figure 7–2. Touch, pain, and temperature pathways from the trunk and limbs. The anterolateral system (ventral and lateral spinothalamic and related ascending tracts) also projects to the mesencephalic reticular formation and the nonspecific thalamic nuclei.

descending tracts from the brain. The relation of the gate to pain is discussed below.

Axons of the spinothalamic tracts from sacral and lumbar segments of the body are pushed laterally by axons crossing the midline at successively higher levels. On the other hand, sacral and lumbar dorsal column fibers are pushed medially by fibers from higher segments (Fig 7–3). Consequently, both of these ascending systems are laminated, with cervical, thoracic, lumbar, and sacral segments represented from medial to lateral in the anterolateral pathways and sacral to cervical segments from medial to lateral in the dorsal columns. Because of this lamination, tumors arising outside the spinal cord first compress the spinothalamic fibers from sacral and lumbar areas, causing the early symptom of loss of pain and temperature sensation in the sacral region. Intraspinal tumors cause anesthesia first in higher segments.

The fibers within the lemniscal and anterolateral systems are joined in the brain stem by fibers mediating sensation from the head. Pain and temperature impulses are relayed via the spinal nucleus of the trigeminal nerve, and touch and proprioception mostly via the main sensory and mesencephalic nuclei of this nerve.

Cortical Representation

From the specific sensory nuclei of the thalamus, neurons project in a highly specific way to the 2 somatic sensory areas of the cortex: somatic sensory area I(SI) in the postcentral gyrus and somatic sensory area II(SII) in the wall of the sylvian fissure (lateral cerebral sulcus). In addition, SI projects to SII. SI corresponds to areas 1, 2, and 3 of Brodmann, the histologist who painstakingly divided the cerebral cortex into numbered areas based on their histologic characteristics.

The arrangement of the thalamic fibers to SI is such that the parts of the body are represented in order along the postcentral gyrus, with the legs on top and the head at the foot of the gyrus (Fig 7–4). Not only is there detailed localization of the fibers from the various parts of the body in the postcentral gyrus, but the size of the cortical receiving area for impulses from a particular part of the body is proportionate to the number of receptors in the part. The relative sizes of the cortical receiving areas are shown dramatically in Fig 7–5, in which the proportions of the homunculus have been distorted to correspond to the size of the cortical receiving areas for each. Note that the cortical areas for sensation from the trunk and back are small, whereas very large areas are concerned with impulses from the hand and the parts of the mouth concerned with speech.

Studies of the sensory receiving area emphasize the very discrete nature of the point-for-point localization of peripheral areas in the cortex and provide further evidence for the validity of the doctrine of specific nerve energies (see Chapter 5). Stimulation of the various parts of the postcentral gyrus gives rise to sensations projected to appropriate parts of the body. The sensations produced are usually those of numbness, tingling, or a sense of movement, but with fine enough electrodes it has been possible to produce relatively pure sensations of touch, warmth, cold, and pain. The cells in the postcentral gyrus appear to be organized in vertical columns, like cells in the visual cortex (see Chapter 8). The cells in a given column are all activated by afferents from a given part of the body, and all respond to the same sensory modality.

It is now clear that projections of afferents on the postcentral gyrus and other cortical areas are not innate and immutable but can be changed by experience and reflect use of the represented area. For example, if a digit is amputated in a monkey, the cortical representation of the neighboring digits spreads into the cortical area that was formerly occupied by the representation of the amputated digit. Conversely, if the cortical area representing a digit is removed, the somatosensory map of the digit moves

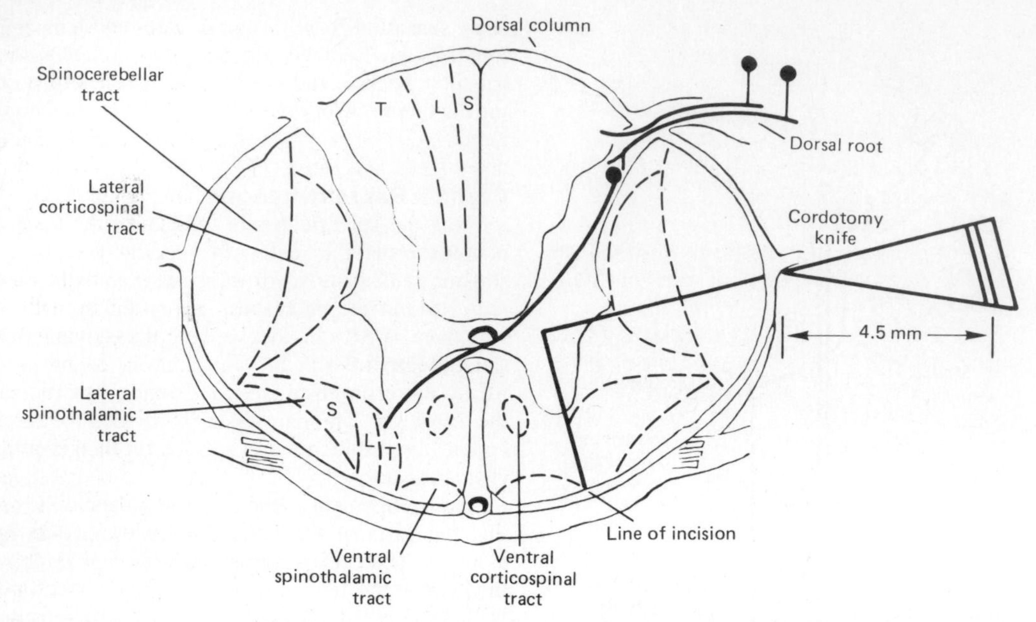

Figure 7–3. Major spinal pathways. The solid line on the right represents the line of incision in performing an anterolateral cordotomy. Note the lamination of the tracts. S, sacral; L, lumbar; T, thoracic.

to the surrounding cortex. These findings illustrate the plasticity of the brain and its ability to adapt.

Somatic sensory area II is located in the superior wall of the sylvian fissure. The head is represented at the inferior end of the postcentral gyrus, and the feet at the bottom of the sylvian fissure. The representation of the body parts is not as complete or detailed as it is in the postcentral gyrus.

Effects of Cortical Lesions

Ablation of SI in animals causes deficits in position sense and ability to discriminate size and shape. Ablation of SII causes deficits in learning based on tactile discrimination. Ablation of SI causes deficits in sensory processing in SII, whereas ablation of SII has no gross effect on processing in SI. Thus, it

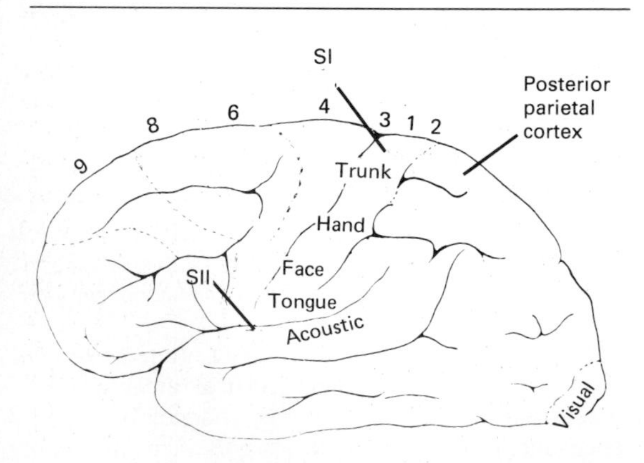

Figure 7–4. Brain areas concerned with somatic sensation, and some of the cortical receiving areas for other sensory modalities in the human brain. The numbers are those of Brodmann's cortical areas. The auditory (acoustic) area is actually located in the sylvian fissure on the top of the superior temporal gyrus and is not normally visible.

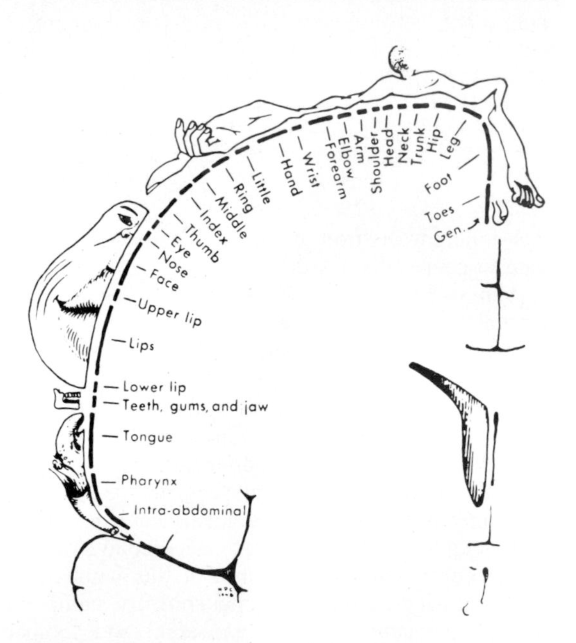

Figure 7–5. Sensory homunculus, drawn overlying a coronal section through the postcentral gyrus. Gen., genitalia. (Reproduced, with permission, from Penfield W, Rasmussen G: *The Cerebral Cortex of Man,* Macmillan, 1950.)

seems clear that SI and SII process sensory information in series rather than in parallel and that SII is concerned with further elaboration of sensory data. SI also projects to the posterior parietal cortex (Fig 7–4), and lesions of this association area produce complex abnormalities of spatial orientation on the contralateral side of the body (see Chapter 16).

It is worth emphasizing that in experimental animals and humans, cortical lesions do not abolish somatic sensation. Proprioception and fine touch are most affected by cortical lesions. Temperature sensibility is less affected, and pain sensibility is only slightly affected. Thus, perception is possible in the absence of the cortex. Upon recovery, pain sensibility returns first, followed by temperature sense and, finally, proprioception and fine touch.

Principles of Sensory Physiology

The important general principles that relate to the physiology of sensory systems have been discussed in detail in Chapter 5. Each sense organ is specialized to convert one particular form of energy into action potentials in the sensory nerves. Each modality has a discrete pathway to the brain, and the sensation perceived as well as the part of the body to which it is localized is determined by the particular part of the brain activated. Differences in intensity of a given sensation are signaled in 2 ways: by changes in the frequency of action potentials in the sensory nerves, and by changes in the number of receptors activated. An increase in the intensity of stimulation of a sense organ has very little (if any) effect on the quality of the sensation produced.

Another principle that applies to cutaneous sensation is that of punctate representation. If the skin is carefully mapped, millimeter by millimeter, with a fine hair, a sensation of touch is evoked from spots overlying touch receptors. None is evoked from the intervening areas. Similarly, pain and temperature sensations are produced by stimulation of the skin only over the spots where the sense organs for these modalities are located.

TOUCH

As noted in Chapter 5, touch is present in areas that have no specialized receptors, but in areas where they exist, Meissner's and pacinian corpuscles are rapidly adapting and Merkel's disks and Ruffini endings are slowly adapting touch receptors. Touch receptors are most numerous in the skin of the fingers and lips and relatively scarce in the skin of the trunk. There are many receptors around hair follicles in addition to those in the subcutaneous tissues of hairless areas. When a hair is moved, it acts as a lever with its fulcrum at the edge of the follicle, so that slight movements of the hairs are magnified into relatively potent stimuli to the nerve endings around the follicles. The stiff vibrissae on the snouts of some animals are highly developed examples of hairs that act as levers to magnify tactile stimuli. The pacinian corpuscles are found in the subcutaneous tissues, muscles, and joints as well as the skin. They also abound in the mesentery, where their function is not known.

The group II sensory fibers that transmit impulses from touch receptors to the central nervous system are 5–12 μm in diameter and have conduction velocities of 30–70 m/s. Some touch impulses are also conducted via C fibers.

Touch information is transmitted in both the lemniscal and anterolateral pathways, so that only very extensive lesions completely interrupt touch sensation. However, there are differences in the type of touch information transmitted in the 2 systems. When the dorsal columns are destroyed, vibratory sensation and proprioception are lost, the touch threshold is elevated, and the number of touch-sensitive areas in the skin is decreased. In addition, localization of touch sensation is impaired. An increase in touch threshold and a decrease in the number of touch spots in the skin are also observed after interrupting the spinothalamic tracts, but the touch deficit is slight and touch localization remains normal. The information carried in the lemniscal system is concerned with the detailed localization, spatial form, and temporal pattern of tactile stimuli. The information carried in the spinothalamic tracts, on the other hand, is concerned with poorly localized, gross tactile sensations.

PROPRIOCEPTION

Proprioceptive information is transmitted up the spinal cord in the dorsal columns. A good deal of the proprioceptive input goes to the cerebellum, but some passes via the medial lemnisci and thalamic radiations to the cortex. Diseases of the dorsal columns produce ataxia because of the interruption of proprioceptive input to the cerebellum.

There is some evidence that proprioceptive information passes to consciousness in the anterolateral columns of the spinal cord. Conscious awareness of the positions of the various parts of the body in space depends in part upon impulses from sense organs in and around the joints. The organs involved are slowly adapting "spray" endings, structures that resemble Golgi tendon organs, and probably pacinian corpuscles in the synovia and ligaments. Impulses from these organs, touch receptors in the skin and other tissues, and muscle spindles are synthesized in the cortex into a conscious picture of the position of the body in space. Microelectrode studies indicate that many of the neurons in the sensory cortex respond to particular movements, not just to touch or static position. In this regard, the sensory cortex is organized like the visual cortex (see Chapter 8).

TEMPERATURE

There are 2 types of temperature sense organs: those responding maximally to temperatures slightly above body temperature, and those responding maximally to temperatures slightly below body temperature. The former are the sense organs for what we call warmth, and the latter for what we call cold. However, the adequate stimuli are actually 2 different degrees of warmth, since cold is not a form of energy.

Mapping experiments show that there are discrete cold-sensitive and warmth-sensitive spots in the skin. There are 4–10 times as many cold spots as warm. The temperature sense organs are naked nerve endings that respond to absolute temperature, not the temperature gradient across the skin. Cold receptors respond from 10–40°C and warm receptors from 30–45°C. The afferents for cold are Aδ and C fibers, whereas the afferents for warmth are C fibers (Fig 7–1). These afferents relay information to the postcentral gyrus via the lateral spinothalamic tract and the thalamic radiation.

Because the sense organs are located subepithelially, it is the temperature of the subcutaneous tissues that determines the responses. Cool metal objects feel colder than wooden objects of the same temperature because the metal conducts heat away from the skin more rapidly, cooling the subcutaneous tissues to a greater degree. Below a skin temperature of 20°C and above 40°C, there is no adaptation, but between 20 and 40°C there is adaptation, so that the sensation produced by a temperature change gradually fades to one of thermal neutrality. Above 45°C, tissue damage begins to occur, and the sensation becomes one of pain.

PAIN

The sense organs for pain are the naked nerve endings found in almost every tissue of the body. Pain impulses are transmitted to the central nervous system by 2 fiber systems. One system is made up of small myelinated A δ fibers 2–5 μm in diameter, which conduct at rates of 12–30 m/s. The other consists of unmyelinated C fibers 0.4–1.2 μm in diameter. These latter fibers are found in the lateral division of the dorsal roots and are often called dorsal root C fibers. They conduct at the low rate of 0.5–2 m/s. Both fiber groups end in the dorsal horn; the former terminate primarily on neurons in laminas I and V, whereas the dorsal root C fibers terminate on neurons in laminas I and II. Some of the axons of the dorsal horn neurons end in the spinal cord and brain stem. Others enter the anterolateral system, including the lateral spinothalamic tract. Some of the anterolateral system neurons project to the specific sensory relay nuclei of the thalamus and from there to the postcentral gyrus. However, many end in the reticular system, which projects to the midline and intralaminar nonspecific projection nuclei of the thalamus and from there to many different parts of the cortex. Other anterolateral neurons end in the periaqueductal gray, an area known to be concerned with pain (see below). There is abundant evidence that the synaptic transmitter secreted by the primary afferent fibers subserving pain sensation is substance P.

Fast & Slow Pain

The presence of 2 pain pathways, one slow and one fast, explains the physiologic observation that there are 2 kinds of pain. A painful stimulus causes a "bright," sharp, localized sensation followed by a dull, intense, diffuse, and unpleasant feeling. These 2 sensations are variously called fast and slow pain or first and second pain. The farther from the brain the stimulus is applied, the greater the temporal separation of the 2 components. This and other evidence make it clear that fast pain is due to activity in the A δ pain fibers, while slow pain is due to activity in the C pain fibers.

Subcortical Perception & Affect

There is considerable evidence that sensory stimuli are perceived in the absence of the cerebral cortex, and this is especially true of pain. The cortical receiving areas are apparently concerned with the discriminative, exact, and meaningful interpretation of pain, but perception alone does not require the cortex.

Pain was called by Sherrington the "physical adjunct of an imperative protective reflex." Stimuli that are painful generally initiate potent withdrawal and avoidance responses. Furthermore, pain is unique among the senses in that it is associated with a strong emotional component. Information transmitted via the special senses may secondarily evoke pleasant or unpleasant emotions, depending largely upon past experience, but pain alone has a "built-in" unpleasant affect. Present evidence indicates that this affective response depends upon connections of the pain pathways in the thalamus. Damage to the thalamus may be associated with a peculiar overreaction to painful stimuli known as the **thalamic syndrome.** In this condition, usually due to blockage of the thalamogeniculate branch of the posterior cerebral artery with consequent damage to the posterior thalamic nuclei, minor stimuli lead to prolonged, severe, and very unpleasant pain. Such bouts of pain may occur spontaneously or at least without evident external stimuli. Another interesting fact is that at least in some cases pain can be dissociated from its unpleasant subjective affect by cutting the deep connections between the frontal lobes and the rest of the brain (**prefrontal lobotomy**). After this operation patients report that they feel pain but that it "doesn't bother" them. The operation is sometimes useful in the treatment of intractable pain caused by terminal cancer.

Lobotomy is, of course, only one of the many neurosurgical procedures employed to relieve intractable pain in terminal cancer patients, and it produces extensive personality changes (see Chapter 16). Other operations are summarized in Fig 7–6. One of the most extensively used is **anterolateral cordotomy.** In this procedure, a knife is inserted into the lateral

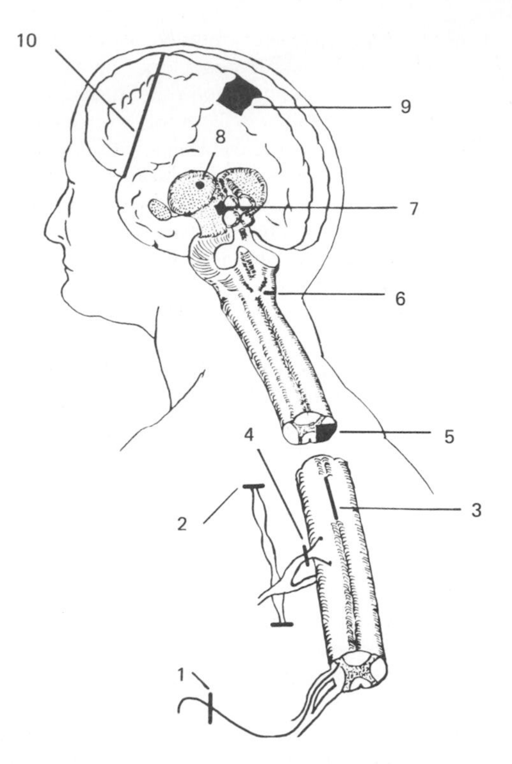

Figure 7–6. Diagram of various surgical procedures designed to alleviate pain. 1, nerve section; 2, sympathectomy (for visceral pain); 3, myelotomy to section spinothalamic fibers in anterior white commissure; 4, posterior rhizotomy; 5, anterolateral cordotomy; 6, medullary tractotomy; 7, mesencephalic tractotomy; 8, thalamotomy; 9, gyrectomy; 10, prefrontal lobotomy.

aspect of the spinal cord and swept anteriorly and laterally as shown in Fig 7–3. When properly performed, this procedure cuts the lateral spinothalamic and other anterolateral system pain fibers while leaving most of the ventral spinothalamic touch fibers intact. Even if there is damage to the touch fibers in the anterolateral quadrants, touch is impaired very little, because the dorsal column touch system is intact. The use of dorsal column and periaqueductal gray stimulation to control pain is discussed below.

Deep Pain

The main difference between superficial and deep sensibility is the different nature of the pain evoked by noxious stimuli. Unlike superficial pain, deep pain is poorly localized, nauseating, and frequently associated with sweating and changes in blood pressure. Pain can be elicited experimentally from the periosteum and ligaments by injecting hypertonic saline into them. The pain produced in this fashion initiates reflex contraction of nearby skeletal muscles. This reflex contraction is similar to the muscle spasm associated with injuries to bones, tendons, and joints. The steadily contracting muscles become ischemic, and ischemia stimulates the pain receptors in the muscles (see below). The pain in turn initiates more spasm, setting up a vicious circle.

Adequate Stimulus

Pain receptors are specific, and pain is not produced by overstimulation of other receptors. On the other hand, the adequate stimulus for pain receptors is not as specific as that for others, because they can be stimulated by a variety of strong stimuli. For example, pain receptors respond to warmth, but it has been calculated that their threshold for thermal energy is over 100 times that of the warmth receptors. Pain receptors also respond to electrical, mechanical, and, especially, chemical energy.

It has been suggested that pain is chemically mediated and that stimuli which provoke it have in common the ability to liberate a chemical agent that stimulates the nerve endings. The chemical agent might be a kinin (see Chapter 31) or histamine, both of which cause pain on local injection.

Muscle Pain

If a muscle contracts rhythmically in the presence of an adequate blood supply, pain does not usually result. However, if the blood supply to a muscle is occluded, contraction soon causes pain. The pain persists after the contraction until blood flow is reestablished. If a muscle with a normal blood supply is made to contract continuously without periods of relaxation, it also begins to ache because the maintained contraction compresses the blood vessels supplying the muscle.

These observations are difficult to interpret except in terms of the release during contraction of a chemical agent (Lewis's **"P factor"**) that causes pain when its local concentration is high enough. When the blood supply is restored, the material is washed out or metabolized. The identity of the P factor is not settled, but it could be K^+.

Clinically, the substernal pain that develops when the myocardium becomes ischemic during exertion (angina pectoris) is a classic example of the accumulation of P factor in a muscle. Angina is relieved by rest because this decreases the myocardial O_2 requirement and permits the blood supply to remove the factor. Intermittent claudication, the pain produced in the leg muscles of persons with occlusive vascular disease, is another example. It characteristically comes on while the patient is walking and disappears upon resting.

Hyperalgesia

In pathologic conditions, the sensitivity of the pain receptors is altered. There are 2 important types of alteration, primary and secondary hyperalgesia. In the area surrounding an inflamed or injured area, the threshold for pain is lowered so that trivial stimuli cause pain. This phenomenon, **primary hyperalgesia,** is seen in the area of the **flare,** the region of vasodilation around the injury In the area of actual

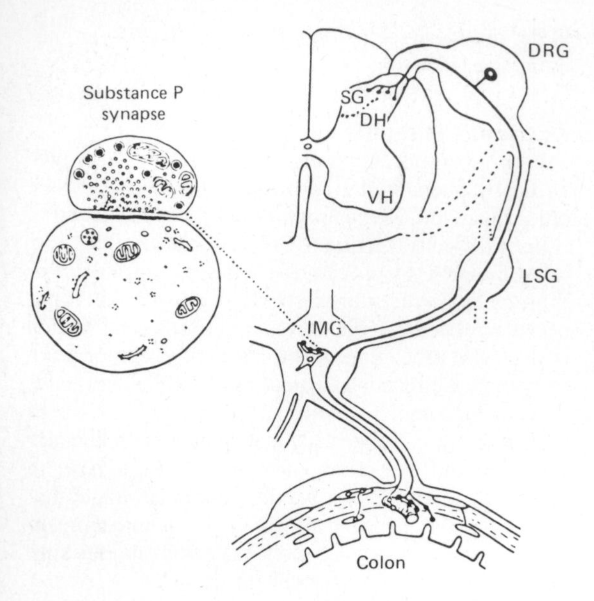

Figure 7–7. Schematic representation of sensory substance P-containing fiber from colon. Note the collateral that synapses with a postganglionic neuron in the inferior mesenteric ganglion (IMG) and the cell body in the dorsal root ganglion (DRG). SG, substantia gelatinosa; DH, dorsal horn; VH, ventral horn; LSG, lumbar paravertebral sympathetic ganglion. (Reproduced, with permission, from Matthews MR, Cuello AC: Substance P-immunoreactive peripheral branches of sensory neurons innervate guinea pig sympathetic neurons. *Proc Natl Acad Sci USA* 1982;**79**:1668.)

tissue damage, the vasodilation and, presumably, the pain are due to substances liberated from injured cells, but the flare in surrounding undamaged tissue is due to substance P liberated by antidromic impulses in primary afferent fibers (see Chapter 32).

Another aberration of sensation following injury is **secondary hyperalgesia.** In the area affected, the threshold for pain is actually elevated, but the pain produced is unpleasant, prolonged, and severe. The area from which this response is obtained extends well beyond the site of injury, and the condition does not last as long as primary hyperalgesia. It is probably due to some sort of central facilitation by impulses from the injured area of the pathways responsible for the unpleasant affect component of pain. Such facilitation or alteration of pathways may be a spinal subliminal fringe effect, or it may occur at the thalamic or even the cortical level.

DIFFERENCES BETWEEN SOMATIC & VISCERAL SENSORY MECHANISMS

The autonomic nervous system, like the somatic, has afferent components, central integrating stations, and effector pathways. The visceral afferent mechanisms play a major role in homeostatic adjustments. In the viscera, there are a number of special receptors—osmoreceptors, baroreceptors, chemoreceptors, etc—that respond to changes in the internal environment. The afferent nerves from these receptors make reflex connections that are intimately concerned with regulating the function of the various systems with which they are associated, and their physiology is discussed in the chapters on these systems.

The receptors for pain and the other sensory modalities present in the viscera are similar to those in skin, but there are marked differences in their distribution. There are no proprioceptors in the viscera, and few temperature and touch sense organs. If the abdominal wall is infiltrated with a local anesthetic, the abdomen can be opened and the intestines can be handled, cut, and even burned without eliciting any discomfort. Pain receptors are present in the viscera, however, and although they are more sparsely distributed than in somatic structures, certain types of stimuli cause severe pain.

Afferent fibers from visceral structures reach the central nervous system via sympathetic and parasympathetic pathways (Fig 7–7). Their cell bodies are located in the dorsal roots and the homologous cranial nerve ganglia. Specifically, there are visceral afferents in the facial, glossopharyngeal, and vagus nerves; in the thoracic and upper lumbar dorsal roots; and in the sacral roots (Fig 7–8). There may also be visceral afferent fibers from the eye in the trigeminal nerve. It is worth noting that at least some substance P-containing afferents make connections via collaterals to postganglionic sympathetic neurons in collateral sympathetic ganglia such as the inferior mesenteric ganglion, as shown in Fig 7–7. These connections may play a part in reflex control of the viscera independent of the central nervous system.

In the central nervous system, visceral sensation travels along the same pathways as somatic sensation in the spinothalamic tracts and thalamic radiations, and the cortical receiving areas for visceral sensation are intermixed with the somatic receiving areas in the postcentral gyri.

VISCERAL PAIN

Pain from visceral structures is poorly localized, unpleasant, and associated with nausea and autonomic symptoms. It often radiates or is referred to other areas.

Stimulation of Pain Fibers

Because there are relatively few pain receptors in the viscera, visceral pain is poorly localized. However, as almost everyone knows from personal experience, visceral pain can be very severe. The receptors in the walls of the hollow viscera are especially sensitive to distention of these organs. Such distention can be produced experimentally in the gastrointestinal tract by inflation of a swallowed balloon attached to a tube. This produces pain that waxes and

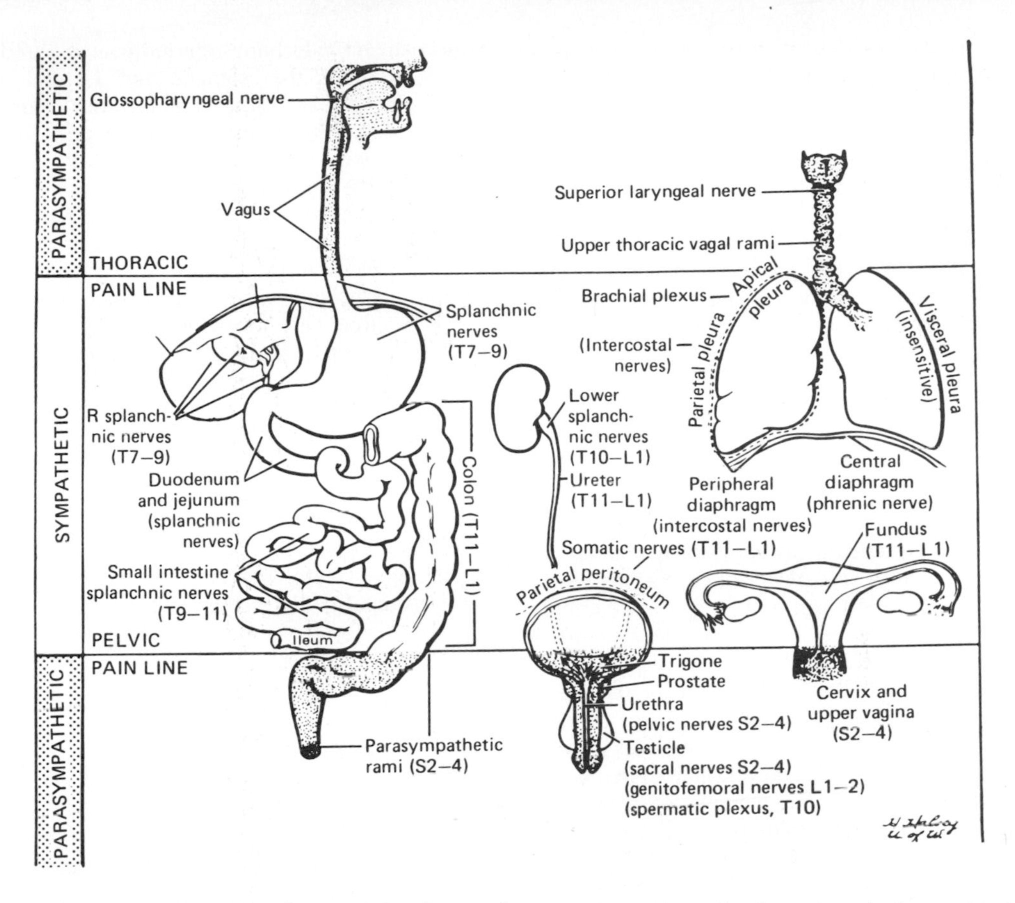

Figure 7–8. Pain innervation of the viscera. Pain afferents from structures above the thoracic pain line and below the pelvic pain line traverse parasympathetic pathways. (After White JC. Reproduced, with permission, from Ruch TC in: *Physiology and Biophysics,* 19th ed. Ruch TC, Patton HD [editors]. Saunders, 1965.)

wanes (intestinal colic) as the intestine contracts and relaxes on the balloon. Similar colic is produced in intestinal obstruction by the contractions of the dilated intestine above the obstruction. When a viscus is inflamed or hyperemic, relatively minor stimuli cause severe pain. This is probably a form of primary hyperalgesia similar to that which occurs in somatic structures. Traction on the mesentery is also claimed to be painful, but the significance of this observation in the production of visceral pain is not clear. Visceral pain is particularly unpleasant not only because of the affective component it has in common with all pain but also because so many visceral afferents excited by the same process that causes the pain have reflex connections that initiate nausea, vomiting, and other autonomic effects.

Pathways

Pain impulses from most of the thoracic and abdominal viscera are conducted through the sympathetic nervous system and the dorsal roots of the first thoracic to the second lumbar spinal nerves. This is why sympathectomy effectively abolishes pain from the heart, stomach, and intestines. However, pain impulses from the esophagus, trachea, and pharynx are mediated via vagal and glossopharyngeal afferents, and pain impulses from the structures deep in the pelvis are transmitted in the sacral parasympathetic nerves (Fig 7–8).

Muscle Spasm & Rigidity

Visceral pain, like deep somatic pain, initiates reflex contraction of nearby skeletal muscle. This reflex spasm is usually in the abdominal wall and makes the abdominal wall rigid. It is most marked when visceral inflammatory processes involve the peritoneum. However, it can occur without such involvement. The anatomic details of the reflex pathways by which impulses from diseased viscera initiate skeletal muscle spasm are still obscure. The spasm protects the underlying inflamed structures from inadvertent trauma. Indeed, this reflex spasm is sometimes called "guarding."

The classic signs of inflammation in an abdominal viscus are pain, tenderness, autonomic changes such as hypotension and sweating, and spasm of the abdominal wall. From the preceding discussion, the genesis of each of these signs is apparent. The tenderness is due to the heightened sensitivity of the pain receptors in the viscus, the autonomic changes due to activation of visceral reflexes, and the spasm due to reflex contraction of skeletal muscle in the abdominal wall.

REFERRAL & INHIBITION OF PAIN

Referred Pain

Irritation of a viscus frequently produces pain which is felt not in the viscus but in some somatic structure that may be a considerable distance away. Such pain is said to be **referred** to the somatic structure. Deep somatic pain may also be referred, but superficial pain is not. When visceral pain is both local and referred, it sometimes seems to spread (**radiate**) from the local to the distant site.

Obviously, a knowledge of referred pain and the common sites of pain referral from each of the viscera is of great importance to the physician. Perhaps the best-known example is referral of cardiac pain to the inner aspect of the left arm. Other dramatic examples include pain in the tip of the shoulder owing to irritation of the central portion of the diaphragm and pain in the testicle due to distention of the ureter. Additional instances abound in the practice of medicine, surgery, and dentistry. However, sites of reference are not stereotyped, and unusual reference sites occur with considerable frequency. Heart pain, for instance, may be purely abdominal, may be referred to the right arm, and may even be referred to the neck. Referred pain can be produced experimentally by stimulation of the cut end of a splanchnic nerve.

Dermatomal Rule

When pain is referred, it is usually to a structure that developed from the same embryonic segment or dermatome as the structure in which the pain originates. This principle is called the **dermatomal rule.** For example, during embryonic development, the diaphragm migrates from the neck region to its adult location in the abdomen and takes its nerve supply, the phrenic nerve, with it. One-third of the fibers in the phrenic nerve are afferent, and they enter the spinal cord at the level of the second to fourth cervical segments, the same location at which afferents from the tip of the shoulder enter. Similarly, the heart and the arm have the same segmental origin, and the testicle has migrated with its nerve supply from the primitive urogenital ridge from which the kidney and ureter also developed.

Role of Convergence in Referred Pain

Not only do the nerves from the visceral structures and the somatic structures to which pain is referred enter the nervous system at the same level, but there are many more sensory fibers in the peripheral nerves than there are axons in the lateral spinothalamic tracts. Therefore, a considerable degree of convergence of peripheral sensory fibers on the spinothalamic neurons must occur. One theory of the mechanism underlying referred pain is based on this fact. It holds that somatic and visceral afferents converge on the same spinothalamic neurons (Fig 7–9). Since somatic pain is much more common than visceral pain, the brain has "learned" that activity arriving in a given pathway is caused by a pain stimulus in a particular somatic area. When the same pathway is stimulated by activity in visceral afferents, the signal reaching the brain is no different and the pain is projected to the somatic area.

Past experience does play an important role in referred pain. Although pain originating in an inflamed abdominal viscus is usually referred to the midline, in patients who have had previous abdominal surgery the pain of abdominal disease is frequently referred to their surgical scars. Pain originating in the maxillary sinus is usually referred to nearby teeth, but in patients with a history of traumatic dental work such pain is regularly referred to the previously traumatized teeth. This is true even when the teeth are a considerable distance away from the sinus.

Facilitation Effects

Another theory of the origin of referred pain holds that, owing to subliminal fringe effects, incoming impulses from visceral structures lower the threshold of spinothalamic neurons receiving afferents from somatic areas, so that minor activity in the pain pathways from the somatic areas—activity which would normally die out in the spinal cord—passes on to the brain.

If convergence alone were the explanation for referred pain, anesthetizing the somatic area of reference with procaine should have no effect on the

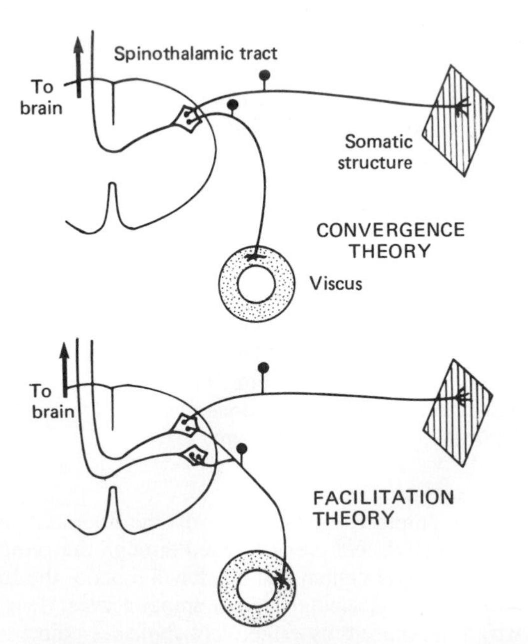

Figure 7–9. Diagram of convergence and facilitation theories of referred pain.

pain, whereas if subliminal fringe effects were responsible, the pain should disappear. The effects of local anesthesia in the area of reference vary. When the pain is severe, it is usually unaffected, but when the pain is mild, it may be completely abolished. Therefore, it appears that both convergence and facilitation play a role in the pathogenesis of referred pain.

Central Inhibition & Counterirritants

It is well known that soldiers wounded in the heat of battle may feel no pain until the battle is over (**stress analgesia**). Many people have learned from practical experience that touching or shaking an injured area decreases the pain of the injury. Acupuncture has been used for 4000 years to prevent or relieve pain, and, using this technique, it is possible in some instances to perform major surgery without any other type of anesthesia. These and other observations make it clear that pain transmission and perception are subject to inhibition or modification.

Inhibition in central sensory pathways may explain the efficacy of counterirritants. Stimulation of the skin over an area of visceral inflammation produces some relief of the pain due to the visceral disease. The old-fashioned mustard plaster works on this principle.

Gating in the Dorsal Horn

One site of inhibition of pain transmission is the dorsal horn, the "gate" through which pain impulses reach the lateral spinothalamic system. Stimulation of large fiber afferents from an area from which pain is being initiated reduces the pain. Collateral fibers from the dorsal column touch fibers enter the substantia gelatinosa, and it has been postulated that impulses in these collaterals or interneurons on which they end inhibit transmission from the dorsal root pain fibers to the spinothalamic neurons. The mechanism involved appears to be presynaptic inhibition (see Chapter 4) at the endings of the primary afferents that transmit pain impulses. Chronic stimulation of the dorsal columns with implanted stimulators has been used clinically as a method of relieving intractable pain. The pain relief is presumably due to antidromic rather than orthodromic conduction, with impulses passing via the collaterals to the "gate," since section of the dorsal columns enhances rather than reduces responses to noxious stimuli.

Action of Morphine & Enkephalins

Morphine relieves pain and produces euphoria. The receptors that bind morphine and the "body's own morphines," the opioid peptides, are discussed in Chapter 4. There are opioid receptors in the substantia gelatinosa that may be on the substance P-containing terminals of nociceptive afferents and inhibit substance P release. It has been argued that enkephalin-containing neurons terminate presynaptically at the site of the receptors on these afferents, but no presynaptic terminals have been identified. Therefore, the mechanism by which opioids act remains unsettled. In addition, injections of morphine into the periaqueductal gray of the midbrain relieve pain by activating descending pathways that produce inhibition of primary afferent transmission in the dorsal horn. There is evidence that this activation occurs via projections from the periaqueductal gray to the nearby raphe magnus nucleus and related serotonergic nuclei (see Chapter 15) and that descending serotonergic fibers from the raphe magnus nucleus mediate the inhibition. However, the mechanism by which serotonin inhibits transmission in the dorsal horn is still unknown. Self-stimulation of the periaqueductal gray with implanted electrodes relieves intractable pain in humans and increases the amount of β-endorphin in their cerebrospinal fluid. Prolonged exposure of receptors to the neurotransmitter that normally binds to them decreases the number of receptors (see Chapter 1), and the development of tolerance to morphine may be due to a decrease in the number of enkephalin receptors.

There is some evidence that acupuncture exerts its analgesic effect by causing release of enkephalins, and the analgesia is said to be blocked by the morphine antagonist naloxone. In addition, there appears to be a component of stress analgesia that is mediated by endogenous opioids, because in experimental animals, some forms of stress analgesia are prevented by naloxone. However, other forms are unaffected, so other components are also involved.

OTHER SENSATIONS

Itch & Tickle

Relatively mild stimulation, especially if produced by something that moves across the skin, produces itch and tickle. Itch spots can be identified on the skin by careful mapping; like the pain spots, they are in regions in which there are many naked endings of unmyelinated fibers. Itch persists along with burning pain in nerve block experiments when only C fibers are conducting, and itch, like pain, is abolished by section of the spinothalamic tracts. However, the distributions of itch and pain are different; itching occurs only in the skin, eyes, and certain mucous membranes and not in deep tissues or viscera. Furthermore, low-frequency stimulation of pain fibers produces pain, not itch, and high-frequency stimulation of itch spots on the skin may merely increase the intensity of the itching without producing pain. These observations indicate that the C fiber system responsible for itching is not the same as that responsible for pain. It is interesting that a tickling sensation is usually regarded as pleasurable, whereas itching is annoying, and pain is unpleasant.

Itching can be produced not only by repeated local mechanical stimulation of the skin but by a variety of chemical agents. Histamine produces intense itching, and injuries cause its liberation in the skin. However, in most instances of itching, histamine does not appear to be the responsible agent; doses of histamine that are too small to produce itching still produce redness and swelling on injection into the skin, and severe itching frequently occurs without any visible change in the skin. The kinins cause severe itching. It is interesting in this regard that itch powder, which is made up of the spicules from the pods of the tropical plant cowhage, contains a proteolytic enzyme, and the powder presumably acts by liberating itch-producing peptides.

"Synthetic Senses"

The cutaneous senses for which separate receptors exist are touch, warmth, cold, pain, and possibly itching. Combinations of these sensations, patterns of stimulation, and, in some cases, cortical components are synthesized into the sensations of vibratory sensation, 2-point discrimination, and stereognosis.

Vibratory Sensibility

When a vibrating tuning fork is applied to the skin, a buzzing or thrill is felt. The sensation is most marked over bones, but it can be felt when the tuning fork is placed in other locations. The receptors involved are the receptors for touch, especially pacinian corpuscles, but a time factor is also necessary. A pattern of rhythmic pressure stimuli is interpreted as vibration. The impulses responsible for the vibrating sensation are carried in the dorsal columns. Degeneration of this part of the spinal cord occurs in poorly controlled diabetes, pernicious anemia, some vitamin deficiencies, and occasionally other conditions; depression of the threshold for vibratory stimuli is an early symptom of this degeneration. Vibratory sensation and proprioception are closely related; when one is depressed, so is the other.

Two-Point Discrimination

The minimal distance by which 2 touch stimuli must be separated to be perceived as separate is called the **2-point threshold.** It depends upon touch plus the cortical component of identifying one or 2 stimuli. Its magnitude varies from place to place on the body and is smallest where the touch receptors are most abundant. Points on the back, for instance, must be separated by 65 mm or more before they can be distinguished as separate points, whereas on the fingers 2 stimuli can be resolved if they are separated by as little as 3 mm. On the hands, the magnitude of the 2-point threshold is approximately the diameter of the area of skin supplied by a single sensory unit. However, the peripheral neural basis of discriminating 2 points is not completely understood, and in view of the extensive interdigitation and overlapping of the sensory units, it is probably complex.

Stereognosis

The ability to identify objects by handling them without looking at them is called **stereognosis.** Normal persons can readily identify objects such as keys and coins of various denominations. This ability obviously depends upon relatively intact touch and pressure sensation, but it also has a large cortical component. Impaired stereognosis is an early sign of damage to the cerebral cortex and sometimes occurs in the absence of any defect in touch and pressure sensation when there is a lesion in the parietal lobe posterior to the postcentral gyrus.

Vision 8

INTRODUCTION

The eyes are complex sense organs that have evolved from primitive light-sensitive spots on the surface of invertebrates. Within its protective casing, each eye has a layer of receptors, a lens system for focusing light on these receptors, and a system of nerves for conducting impulses from the receptors to the brain. The way these components operate to set up conscious visual images is the subject of this chapter.

ANATOMIC CONSIDERATIONS

The principal structures of the eye are shown in Fig 8-1. The outer protective layer of the eyeball, the **sclera,** is modified anteriorly to form the transparent **cornea,** through which light rays enter the eye. Inside the sclera is the **choroid,** a pigmented layer that contains many of the blood vessels which nourish the structures in the eyeball. Lining the posterior two-thirds of the choroid is the **retina,** the neural tissue containing the receptor cells.

The **crystalline lens** is a transparent structure held in place by a circular **lens ligament (zonule).** The zonule is attached to the thickened anterior part of the choroid, the **ciliary body.** The ciliary body contains circular muscle fibers and longitudinal fibers that attach near the corneoscleral junction. In front of the lens is the pigmented and opaque **iris,** the colored portion of the eye. The iris contains circular muscle fibers that constrict and radial fibers that dilate the **pupil.** Variations in the diameter of the pupil can produce up to 5-fold changes in the amount of light reaching the retina.

The space between the lens and the retina is filled primarily with a clear gelatinous material called the **vitreous (vitreous humor). Aqueous humor,** a clear liquid, is produced in the ciliary body by diffusion and active transport and flows through the pupil to fill the anterior chamber of the eye. It is normally reabsorbed through a network of trabeculae into the **canal of Schlemm,** a venous channel at the junction

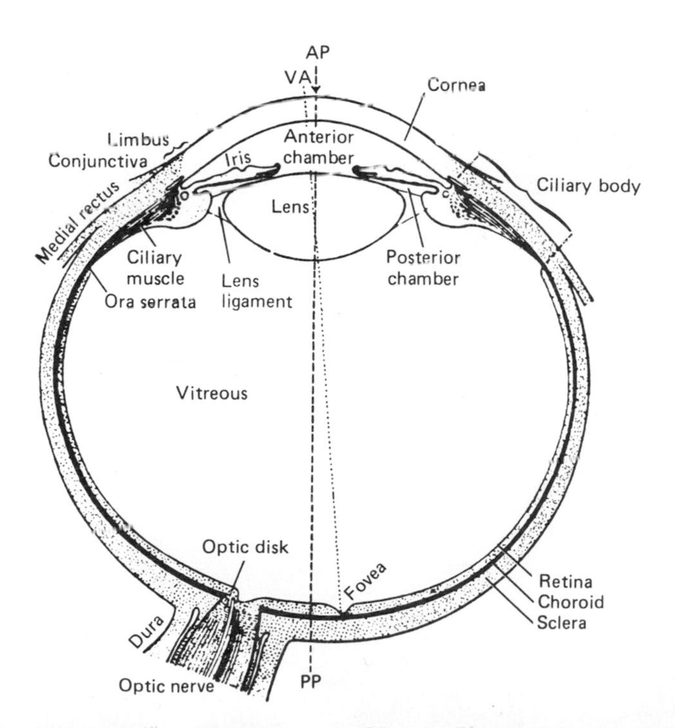

Figure 8–1. Horizontal section of the right eye. AP, anterior pole; PP, posterior pole; VA, visual axis. (Reproduced, with permission, from Warwick R: *Eugene Wolff's Anatomy of the Eye and Orbit,* 7th ed. Saunders, 1977.)

between the iris and the cornea (anterior chamber angle). Obstruction of this outlet leads to increased intraocular pressure and the serious eye disease **glaucoma.** One cause is decreased permeability through the trabeculae **(open-angle glaucoma),** and another is forward movement of the iris, obliterating the angle **(angle-closure glaucoma).**

Retina

The retina extends anteriorly almost to the ciliary body. It is organized in 10 layers and contains the **rods** and **cones,** which are the visual receptors, plus 4 types of neurons: **bipolar cells, ganglion cells, horizontal cells,** and **amacrine cells** (Fig 8–2). The rods and cones, which are next to the choroid, synapse with bipolar cells, and the bipolar cells synapse with ganglion cells. The axons of the ganglion cells converge and leave the eye as the optic nerve. Horizontal cells connect receptor cells to the other receptor cells in the outer plexiform layer. Amacrine cells connect ganglion cells to one another in the inner plexiform layer and in some instances may be inserted between bipolar cells and ganglion cells. They have no axons, and their processes make both pre- and postsynaptic connections with neighboring neural elements. There is considerable overall convergence of receptors on bipolar cells and of bipolar cells on ganglion cells (see below).

Since the receptor layer of the retina is apposed to the choroid, light rays must pass through the ganglion cell and bipolar cell layers to reach the rods and cones. The pigmented layer of choroid next to the retina absorbs light rays, preventing the reflection of rays back through the retina. Such reflection would produce blurring of the visual images.

The neural elements of the retina are bound together by glial cells called Müller cells. The processes of these cells form an internal limiting membrane on the inner surface of the retina and an external limiting membrane in the receptor layer.

The optic nerve leaves the eye and the retinal blood vessels enter it at a point 3 mm medial to and slightly above the posterior pole of the globe. This region is visible through the ophthalmoscope as the **optic disk** (Fig 8–3). There are no visual receptors overlying the disk, and consequently this spot is blind (the **blind spot**). At the posterior pole of the eye, there is a yellowish pigmented spot, the **macula lutea.** This

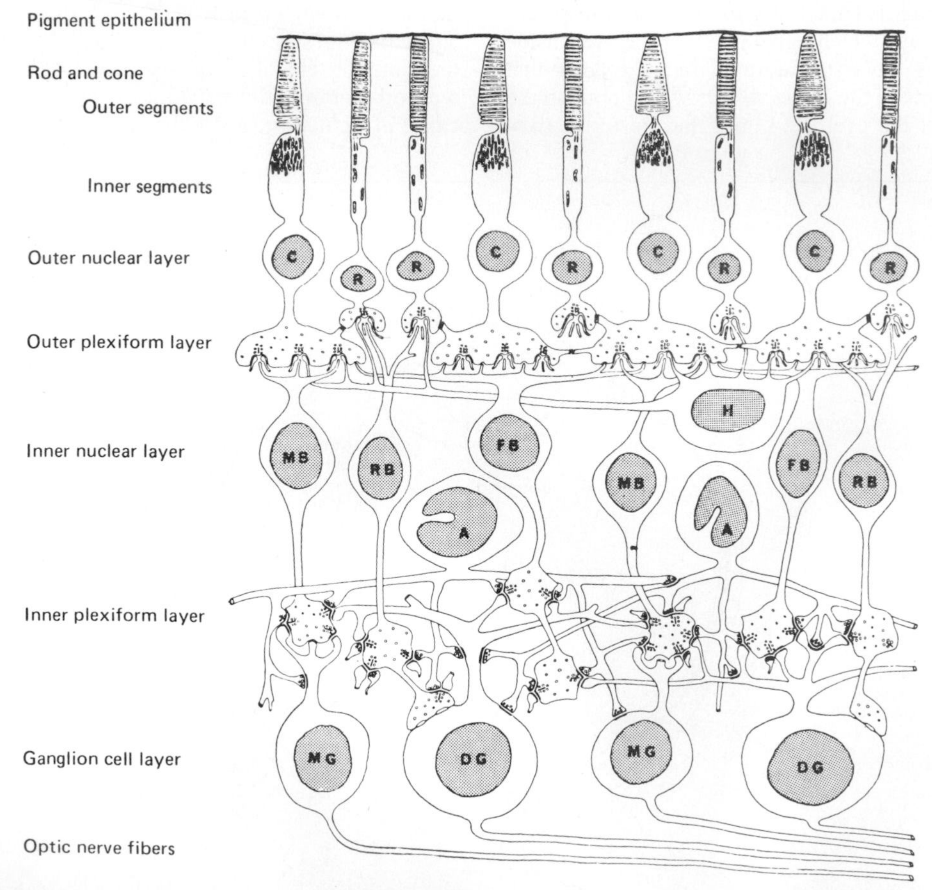

Figure 8–2. Neural components of the retina. C, cone; R, rod; MB, RB, and FB, midget, rod, and flat bipolar cells; DG and MG, diffuse and midget ganglion cells; H, horizontal cells; A, amacrine cells. (Reproduced, with permission, from Dowling JE, Boycott, BB: Organization of the primate retina: Electron microscopy. *Proc R Soc Lond* [*Biol*] 1966;**166**:80.)

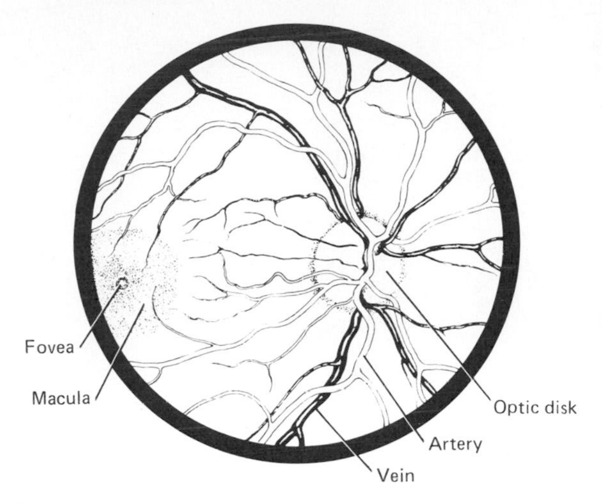

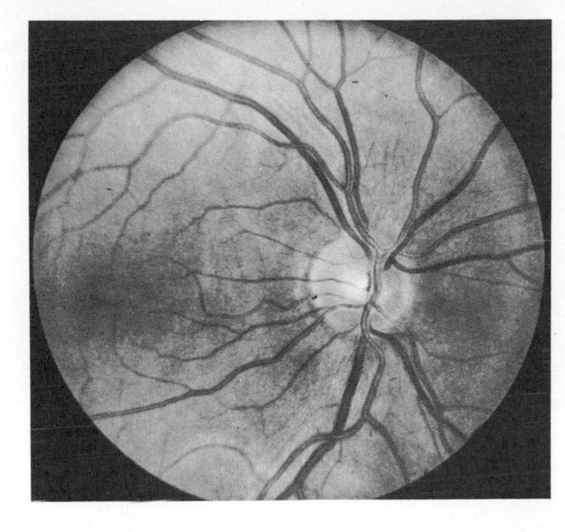

Figure 8–3. Retina seen through the ophthalmoscope in a normal human. Diagram at left identifies the landmarks in the photograph on the right. (Reproduced, with permission, from Vaughan D, Asbury T: *General Ophthalmology,* 11th ed. Appleton & Lange, 1986.)

marks the location of the **fovea centralis,** a thinned-out, rod-free portion of the retina where the cones are densely packed and there are very few cells and no blood vessels overlying the receptors. The fovea is highly developed in humans. It is the point where visual acuity is greatest. When attention is attracted to or fixed on an object, the eyes are normally moved so that light rays coming from the object fall on the fovea.

The arteries, arterioles, and veins in the superficial layers of the retina near its vitreous surface can be seen through the ophthalmoscope. Since this is the one place in the body where arterioles are readily visible, ophthalmoscopic examination is of great value in the diagnosis and evaluation of diabetes mellitus, hypertension, and other diseases that affect blood vessels. The retinal vessels supply the bipolar and ganglion cells, but the receptors are nourished for the most part by the capillary plexus in the choroid. This is why retinal detachment is so damaging to the receptor cells.

Neural Pathways

The axons of the ganglion cells pass caudally in the **optic nerve** and **optic tract** to end in the **lateral geniculate body,** a part of the thalamus (Fig 8–4). The fibers from each nasal hemiretina decussate in the **optic chiasm.** In the geniculate body, the fibers from the nasal half of one retina and the temporal half of the other synapse on the cells whose axons form the **geniculocalcarine tract.** This tract passes to the occipital lobe of the cerebral cortex.

The primary visual receiving area **(visual cortex,** Brodmann's area 17), is located principally on the sides of the calcarine fissure (Fig 8–5). The organization of the primary visual cortex is discussed below.

Branches of the ganglion cell axons pass from the optic tract to the pretectal region of the midbrain and the superior colliculus, where they form connections that mediate visual reflexes. Other axons pass directly from the optic chiasm to the suprachiasmatic nuclei in the hypothalamus, where they form connections that mediate light entrainment of a variety of endocrine and other circadian rhythms (see Chapter 14).

The brain areas activated by visual stimuli have been investigated in monkeys by means of radioactive 2-deoxyglucose (see Chapter 32). Activation occurs not only in the occipital lobe but also in parts of the inferior temporal cortex, the posteroinferior parietal cortex, and portions of the frontal lobe. The subcortical structures activated in addition to the lateral geniculate body include the superior colliculus, pulvinar, caudate nucleus, putamen, claustrum, and amygdala.

Receptors

Each rod and cone is divided into an outer segment, an inner segment that includes a nuclear region, and a synaptic zone (Fig 8–6). The outer segments are modified cilia and are made up of regular stacks of flattened saccules or disks composed of membrane. These saccules and disks contain the photosensitive pigment. The inner segments are rich in mitochondria. The rods are named for the thin, rodlike appearance of their outer segments. Cones generally have thick inner segments and conical outer segments, although their morphology varies from place to place in the retina. In cones, the saccules are formed in the outer segments by infoldings of the cell membrane, but in rods, the disks are separated from the cell membrane.

Rod outer segments are being constantly renewed by formation of new disks at the inner edge of the

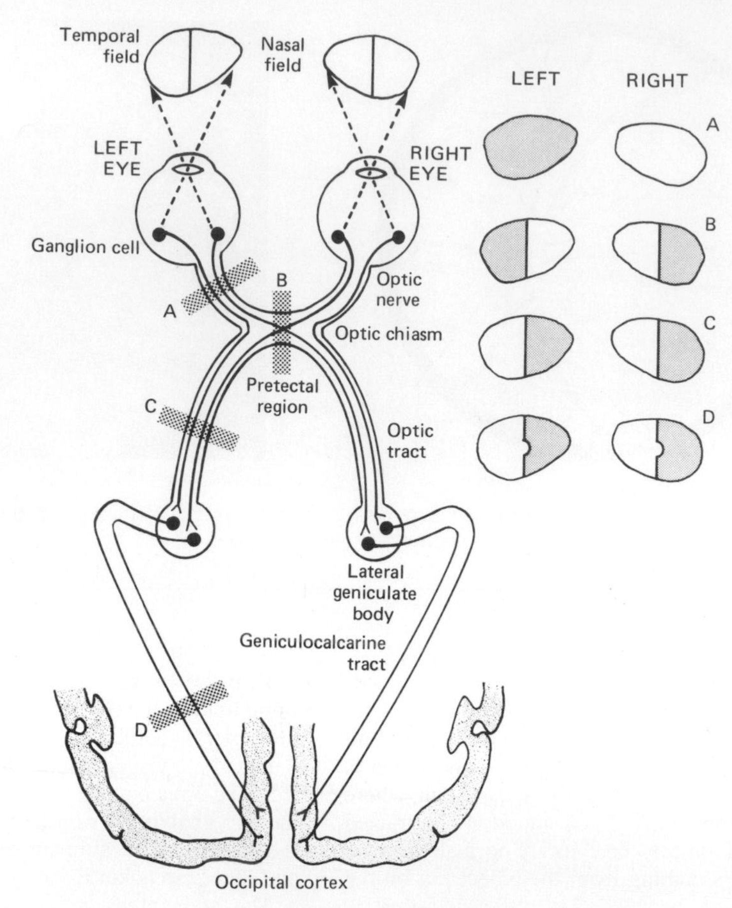

Figure 8–4. Visual pathways. Transection of the pathways at the locations indicated by the letters causes the visual field defects shown in the diagrams on the right (see text). Occipital lesions may spare the fibers from the macula (as in D) because of the separation in the brain of these fibers from the others subserving vision (Fig 8–5).

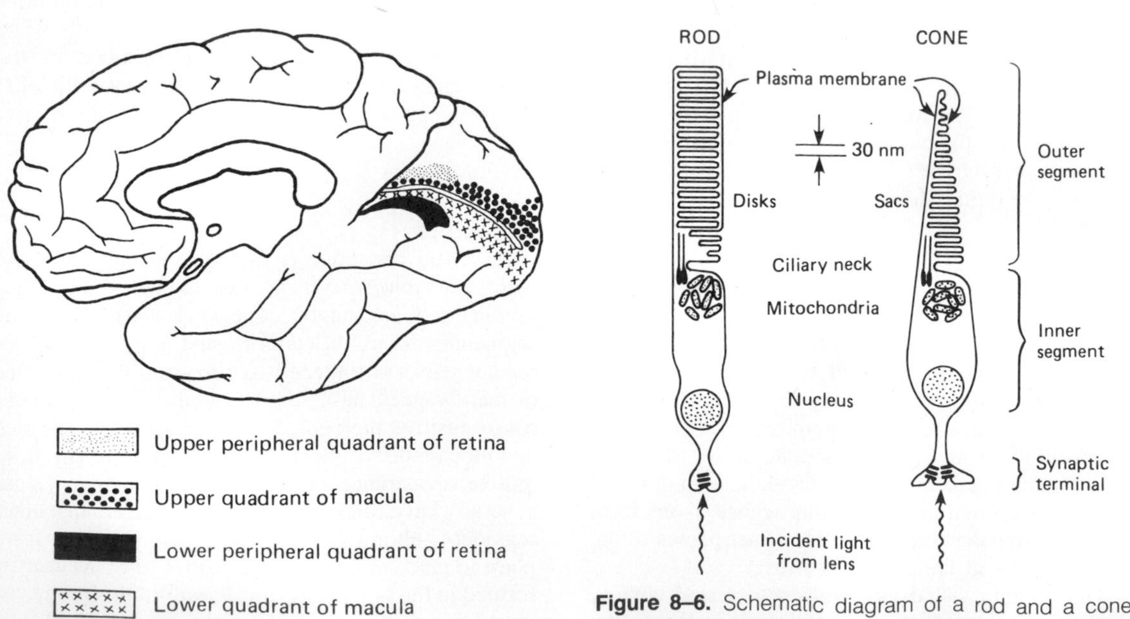

Figure 8–5. Medial view of human cerebral hemisphere showing projection of the retina on the calcarine fissure.

Figure 8–6. Schematic diagram of a rod and a cone. (Reproduced, with permission, from Lamb TD: Electrical responses of photoreceptors. In: *Recent Advances in Physiology,* No. 10. Baker PF [editor]. Churchill Livingstone, 1984.)

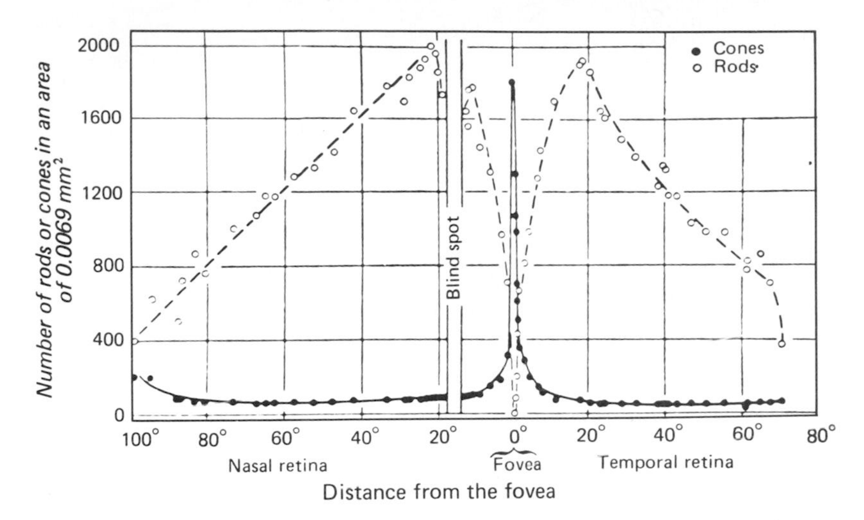

Figure 8–7. Rod and cone density along the horizontal meridian through the human retina. A plot of the relative acuity of vision in the various parts of the light-adapted eye would parallel the cone density curve; a similar plot of relative acuity of the dark-adapted eye would parallel the rod density curve. (Modified and reproduced, with permission, from Østerberg G: Topography of the layer of rods and cones in the human retina. *Acta Ophthalmol [Kbh]* 1935;13:[Suppl 6].

segment and phagocytosis of old disks from the outer tip by cells of the pigment epithelium. In the disease retinitis pigmentosa, the phagocytic process is defective, and a layer of debris accumulates between the receptors and the pigment epithelium. Cone renewal is a more diffuse process and appears to occur at multiple sites in the outer segments.

The fovea contains no rods, and each foveal cone has a single midget bipolar cell connecting it to a single ganglion cell, so that each foveal cone is connected to a single fiber in the optic nerve. In other portions of the retina, rods predominate (Fig 8–7), and there is a good deal of convergence. Flat bipolar cells (Fig 8–2) make synaptic contact with several cones, and rod bipolar cells make synaptic contact with several rods. Since there are approximately 6 million cones and 120 million rods in each human eye but only 1.2 million nerve fibers in each optic nerve, the overall convergence of receptors through bipolar cells on ganglion cells is about 105:1.

The rods are extremely sensitive to light and are the receptors for night vision **(scotopic vision).** The scotopic visual apparatus is not capable of resolving the details and boundaries of objects or determining their color. The cones have a much higher threshold, but the cone system has a much greater acuity and is the system responsible for vision in bright light **(photopic vision)** and for color vision. There are thus 2 kinds of inputs to the central nervous system from the eye: input from the rods and input from the cones. The existence of these 2 kinds of input, each working maximally under different conditions of illumination, is called the **duplicity theory.**

Eye Muscles

The eye is moved within the orbit by 6 ocular muscles (Fig 8–8). These are innervated by the oculomotor, trochlear, and abducens nerves. The muscles and the directions in which they move the eyeball are discussed at the end of this chapter.

Protection

The eye is well protected from injury by the body walls of the orbit. The cornea is moistened and kept clear by tears that course from the **lacrimal gland** in the upper portion of each orbit across the surface of the eye to empty via the **lacrimal duct** into the nose. Blinking helps keep the cornea moist.

THE IMAGE-FORMING MECHANISM

The eyes convert energy in the visible spectrum into action potentials in the optic nerve. The wavelengths of visible light are approximately 397–723 nm. The images of objects in the environment are focused on the retina. The light rays striking the retina generate potentials in the rods and cones. Impulses initiated in the retina are conducted to the cerebral cortex, where they produce the sensation of vision.

Principles of Optics

Light rays are bent (refracted) when they pass from one medium into a medium of a different density, except when they strike perpendicular to the interface. Parallel light rays striking a biconvex lens (Fig 8–9) are refracted to a point **(principal focus)** behind the lens. The principal focus is on a line passing through the centers of curvature of the lens, the **principal axis.** The distance between the lens and the principal focus is the **principal focal distance.** For practical purposes, light rays from an object that

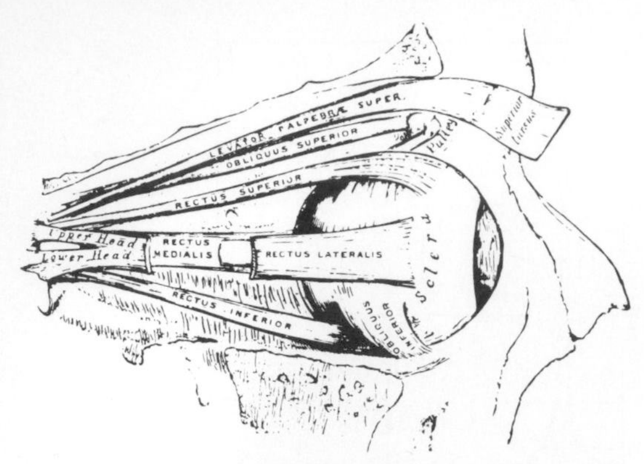

Figure 8–8. Muscles of the right orbit. The 6 muscles that move the eyeball and the levator palpebrae superioris, which raises the upper lid, are shown. (Reproduced, with permission, from Goss CM [editor]: *Gray's Anatomy of the Human Body,* 29th ed. Lea & Febiger, 1973.)

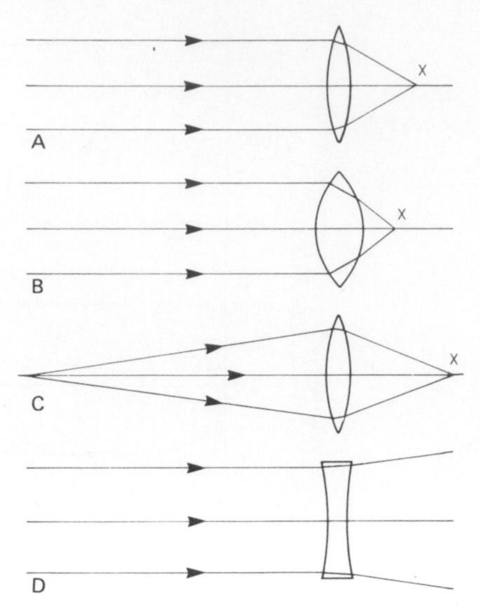

Figure 8–9. Refraction of light rays by lenses: *A:* Biconvex lens. *B:* Biconvex lens of greater strength than *A. C:* Same lens as *A,* showing effect on light rays from a near point. *D:* Biconcave lens. The center line in each case is the principal axis. X is the principal focus.

strike a lens more than 20 ft (6 m) away are considered to be parallel. The rays from an object closer than 20 ft are diverging and are therefore brought to a focus farther back on the principal axis than the principal focus (Fig 8–9). Biconcave lenses cause light rays to diverge.

The greater the curvature of a lens, the greater its refractive power. The refractive power of a lens is conveniently measured in **diopters,** the number of diopters being the reciprocal of the principal focal distance in meters. For example, a lens with a principal focal distance of 0.25 m has a refractive power of 1/0.25, or 4 diopters. The human eye has a refractive power of approximately 66.7 diopters at rest.

Accommodation

When the ciliary muscle is relaxed, parallel light rays striking the optically normal **(emmetropic)** eye are brought to a focus on the retina. As long as this relaxation is maintained, rays from objects closer than 6 m from the observer are brought to a focus behind the retina, and consequently the objects appear blurred. The problem of bringing diverging rays from close objects to a focus on the retina can be solved by increasing the distance between the lens and the retina or by increasing the curvature or refractive power of the lens. In bony fish, the problem is solved by increasing the length of the eyeball, a solution analogous to the manner in which the images of objects closer than 6 m are focused on the film of a camera by moving the lens away from the film. In mammals, the problem is solved by increasing the curvature of the lens.

The process by which the curvature of the lens is increased is called **accommodation.** At rest, the lens is held under tension by the lens ligaments. Because the lens substance is malleable and the lens capsule has considerable elasticity, the lens is pulled into a flattened shape. When the gaze is directed at a near object, the ciliary muscle contracts. This decreases the distance between the edges of the ciliary body and relaxes the lens ligaments, so that the lens springs into a more convex shape. In young individuals, the change in shape may add as many as 12 diopters to the refractive power of the eye. The relaxation of the lens ligaments produced by contraction of the ciliary muscle is due partly to the sphincterlike action of the circular muscle fibers in the ciliary body and partly to the contraction of longitudinal muscle fibers that attach anteriorly, near the corneoscleral junction. When these fibers contract, they pull the whole ciliary body forward and inward. This motion brings the edges of the ciliary body closer together.

The change in lens curvature during accommodation affects principally the anterior surface of the lens (Fig 8–10). This can be demonstrated by a simple experiment first described many years ago. If an observer holds an object in front of the eyes of an individual who is looking into the distance, 3 reflections of the object are visible in the subject's eye. A clear, small upright image is reflected from the cornea; a larger, fainter upright image is reflected from the anterior surface of the lens; and a small inverted image is reflected from the posterior surface of the lens. If the subject then focuses on an object nearby, the large, faint upright image becomes smaller and moves toward the other upright image, whereas the other 2 images change very little. The change in size of the image is due to the increase in curvature of the reflecting surface, the anterior surface of the lens (Fig 8–10). The fact that the small upright image does not change and the inverted image changes very little

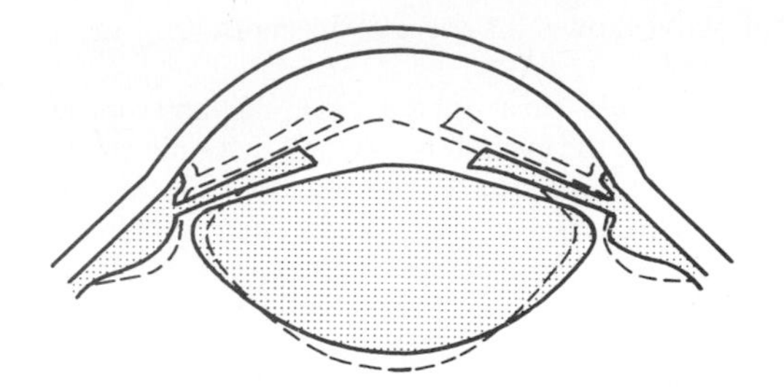

Figure 8–10. Accommodation. The solid lines represent the shape of the lens, iris, and ciliary body at rest, and the dotted lines represent the shape during accommodation.

shows that the corneal curvature is unchanged and that the curvature of the posterior lens surface is changed very little by accommodation.

Near Point

Accommodation is an active process, requiring muscular effort, and can therefore be tiring. Indeed, the ciliary muscle is one of the most used muscles in the body. The degree to which the lens curvature can be increased is, of course, limited, and light rays from an object very near the individual cannot be brought to a focus on the retina even with the greatest of effort. The nearest point to the eye at which an object can be brought into clear focus by accommodation is called the **near point of vision.** The near point recedes throughout life, slowly at first and then rapidly with advancing age, from approximately 9 cm at age 10 to approximately 83 cm at age 60. This recession is due principally to increasing hardness of the lens, with a resulting loss of accommodation (Fig 8–11) due to the steady decrease in the degree to which the curvature of the lens can be increased. By the time a normal individual reaches age 40–45, the loss of accommodation is usually sufficient to make reading and close work difficult. This condition, which is known as **presbyopia,** can be corrected by wearing glasses with convex lenses.

The Near Response

In addition to accommodation, the visual axes converge and the pupil constricts when an individual looks at a near object. This 3-part response—accommodation, convergence, and pupillary constriction—is called the **near response.**

Other Pupillary Reflexes

When light is directed into one eye, the pupil constricts **(pupillary light reflex).** The pupil of the other eye also constricts **(consensual light reflex).** The optic nerve fibers that carry the impulses initiating pupillary responses end in the pretectal region and the superior colliculi. The pathway for the light reflex, which presumably passes from the pretectal region to the oculomotor nuclei (Edinger-Westphal nuclei) bilaterally, is different from that for accommodation. In some pathologic conditions, the pupillary response to light may be absent while the response to accommodation remains intact. This phenomenon, the so-called **Argyll Robertson pupil,** is said to be due to a destructive lesion in the tectal region.

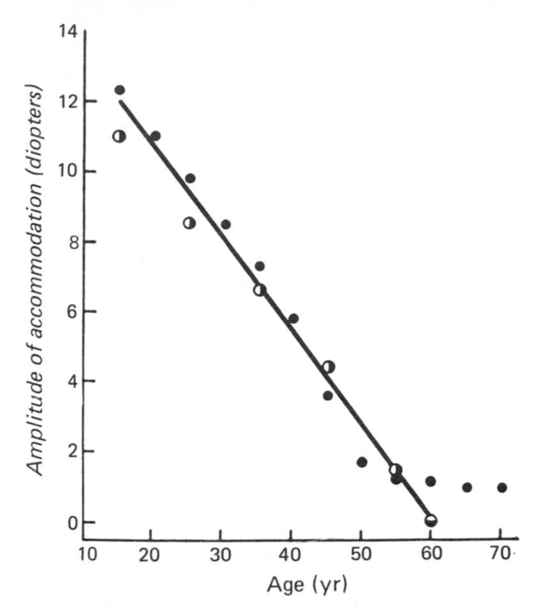

Figure 8–11. Decline in the amplitude of accommodation in the human with advancing age. The different symbols identify data from different studies. (Reproduced, with permission, from Fisher RF: Presbyopia and the changes with age in the human crystalline lens. *J Physiol* 1973; 228:765.)

Retinal Image

In the eye, light is actually refracted at the anterior surface of the cornea and at the anterior and posterior surfaces of the lens. The process of refraction can be represented diagrammatically, however, without introducing any appreciable error, by drawing the rays of light as if all refraction occurs at the anterior surface of the cornea. Fig 8–12 is a diagram of such a "reduced" or "schematic" eye. In this diagram, the **nodal point** (optical center of the eye) coincides with the junction of the middle and posterior third of the lens, 15 mm from the retina. This is the point through which the light rays from an object pass without refraction. All other rays entering the pupil from each point on the object are refracted and brought to a focus on the retina. If the height of the object (AB) and its distance from the observer (Bn) are known, the size of its retinal image can be calculated, because AnB and anb in Fig 8–12 are similar triangles. The angle AnB is the **visual angle** subtended by object AB. It should be noted that the retinal image is inverted. The connections of the

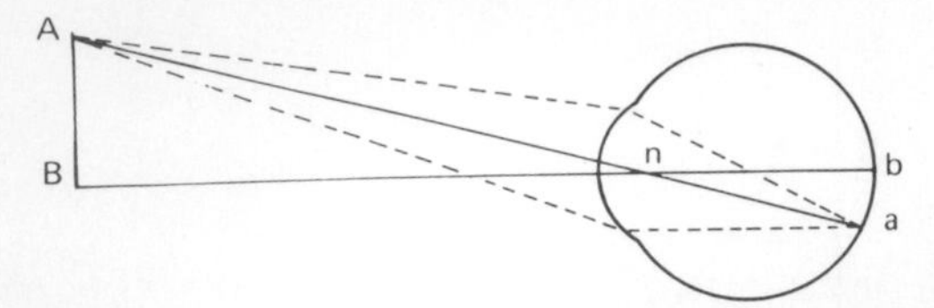

Figure 8–12. Reduced eye, n, nodal point. AnB and anb are similar triangles. In this reduced eye, the nodal point is 15 mm from the retina. All refraction is assumed to take place at the surface of the cornea, 5 mm from the nodal point, between a medium of density 1.000 (air) and a medium of density 1.333 (water). The dotted lines represent rays of light diverging from A and refracted at the cornea so that they are focused on the retina at a.

retinal receptors are such that from birth any inverted image on the retina is viewed right side up and projected to the visual field on the side opposite to the retinal area stimulated. This perception is present in infants and is innate. If retinal images are turned right side up by means of special lenses, the objects viewed look as if they were upside down.

Common Defects of the Image-Forming Mechanism

In some individuals, the eyeball is shorter than normal and parallel rays of light are brought to a focus behind the retina. This abnormality is called **hyperopia** or farsightedness (Fig 8–13). Sustained accommodation, even when viewing distant objects, can partially compensate for the defect, but the prolonged muscular effort is tiring and may cause headaches and blurring of vision. The prolonged convergence of the visual axes associated with the accommodation may lead eventually to squint **(strabismus)** (see below). The defect can be corrected by using glasses with convex lenses, which aid the refractive power of the eye in shortening the focal distance.

In **myopia** (nearsightedness), the anteroposterior diameter of the eyeball is too long. Myopia appears to be genetic in origin, but abnormal visual experiences such as excessive close work can accelerate its development. This defect can be corrected by glasses with biconcave lenses, which make parallel light rays diverge slightly before they strike the eye.

Astigmatism is a common condition in which the curvature of the cornea is not uniform. When the curvature in one meridian is different from that in others, light rays in that meridian are refracted to a different focus, so that part of the retinal image is blurred. A similar defect may be produced if the lens is pushed out of alignment or the curvature of the lens is not uniform, but these conditions are rare. Astigmatism can usually be corrected with cylindric lenses placed in such a way that they equalize the refraction in all meridians. **Presbyopia** has been mentioned above.

THE PHOTORECEPTOR MECHANISM: GENESIS OF ACTION POTENTIALS

The potential changes that initiate action potentials in the retina are generated by the action of light on photosensitive compounds in the rods and cones. When light is absorbed by these substances, their structure changes, and this change triggers a sequence of events that initiates neural activity.

Electrical Responses of Retinal Cells

The eye is unique in that the receptor potentials of the photoreceptors and the electrical responses of

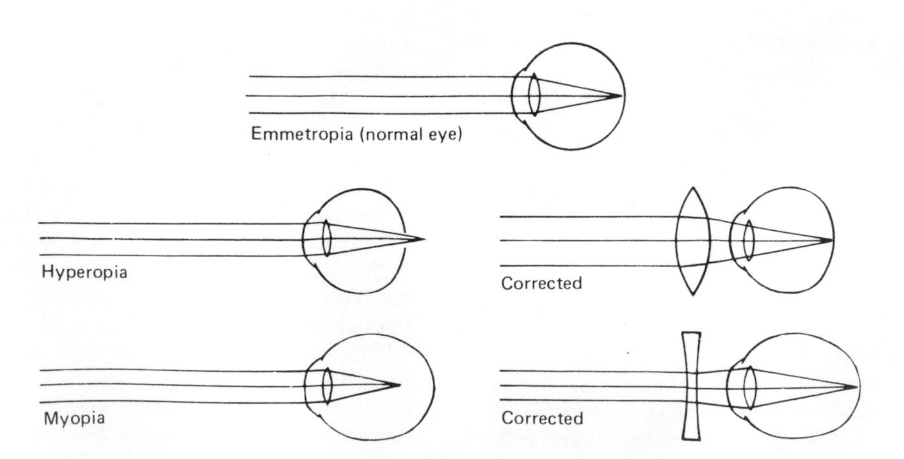

Figure 8–13. Common defects of the optical system of the eye. In hyperopia, the eyeball is too short, and light rays come to a focus behind the retina. A biconvex lens corrects this by adding to the refractive power of the lens of the eye. In myopia, the eyeball is too long, and light rays focus in front of the retina. Placing a biconcave lens in front of the eye causes the light rays to diverge slightly before striking the eye, so that they are brought to a focus on the retina.

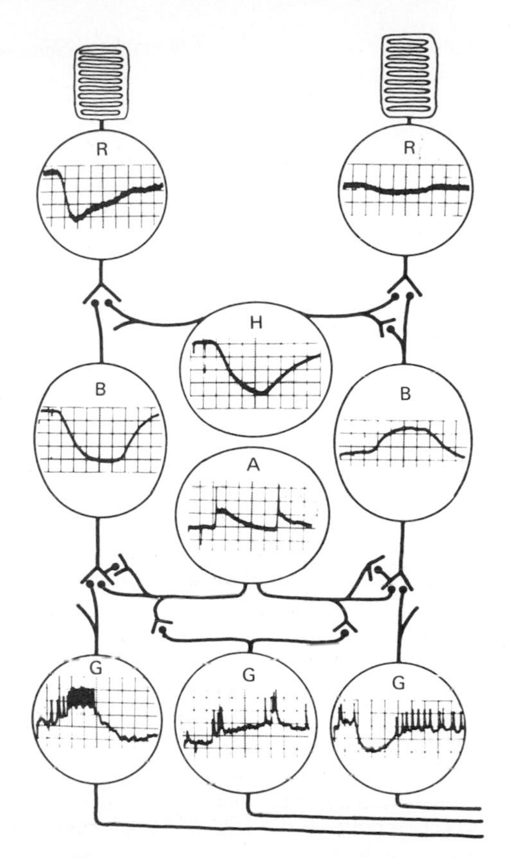

Figure 8–14. Intracellularly recorded responses of cells in retina to light. The synaptic connections of the cells are also indicated. The rod (R) on the left is receiving a light flash, whereas the rod on the right is receiving steady, low-intensity illumination. H, horizontal cell; B, bipolar cell; A, amacrine cell; G, ganglion cell. (Reproduced, with permission, from Dowling JE: Organization of vertebrate retinas. *Invest Ophthalmol* 1970;**9**:655.)

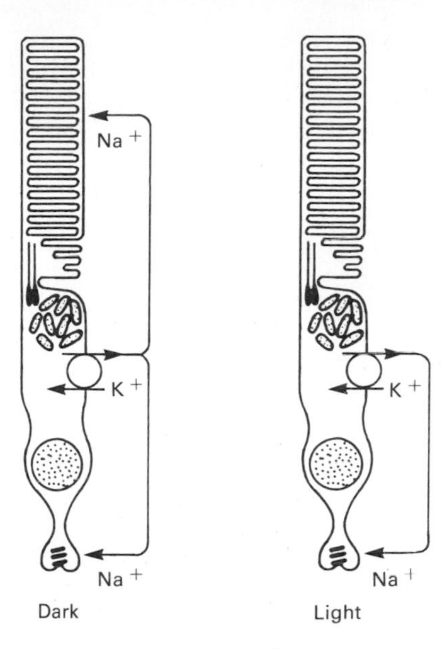

Figure 8–15. Effect of light on current flow in visual receptors. In the dark, Na^+ channels in the outer segment are held open by cGMP. Light leads to increased conversion of cGMP to 5′-GMP, and some of the channels close. This produces hyperpolarization of the synaptic terminal of the photoreceptor. (Modified from O'Brien DF: The chemistry of vision. *Science* 1982;**218**:961.)

most of the other natural elements in the retina are local, graded potentials, and it is only in the ganglion cells that all or none action potentials transmitted over appreciable distances are generated. The responses of the rods, cones, and horizontal cells are hyperpolarizing (Fig 8–14), and the responses of the bipolar cells are either hyperpolarizing or depolarizing, whereas amacrine cells produce depolarizing potentials and spikes that may act as generator potentials for the propagated spikes produced in the ganglion cells.

The cone receptor potential has a sharp onset and offset, whereas the rod receptor potential has a sharp onset and slow offset. The curves relating the amplitude of receptor potentials to stimulus intensity have similar shapes in rods and cones, but the rods are much more sensitive. Therefore, rod responses are proportionate to stimulus intensity at levels of illumination that are below the threshold for cones. On the other hand, cone responses are proportionate to stimulus intensity at high levels of illumination when the rod responses are maximal and cannot change. This is why cones generate good responses to changes in light intensity above background but do not represent absolute illumination well, whereas rods detect absolute illumination.

Ionic Basis of Photoreceptor Potentials

Na^+ channels in the outer segments of the rods and cones are open in the dark, so current flows from the inner to the outer segment (Fig 8–15). Current also flows to the synaptic ending of the photoreceptor. The Na^+-K^+ exchange pump in the inner segment maintains ionic equilibrium. When light strikes the outer segment, some of the Na^+ channels are closed, and the result is a hyperpolarizing receptor potential. The hyperpolarization reduces the release of synaptic transmitter, and this generates a signal which ultimately leads to action potentials in ganglion cells. The action potentials are transmitted to the brain. It is worth emphasizing that there is a steady release of transmitter in the dark, and the hyperpolarization induced by light decreases transmitter release.

Photosensitive Compounds

The photosensitive compounds in the eyes of humans and most other mammals are made up of a protein called an **opsin,** and **retinene$_1$,** the aldehyde of vitamin A_1 (Fig 8–16). The term retinene$_1$ is used to distinguish this compound from retinene$_2$, which is found in the eyes of some animal species. Since the retinenes are aldehydes, they are also called **retinals.** The A vitamins themselves are alcohols and are therefore called **retinols.**

Rhodopsin

The photosensitive pigment in the rods is called **rhodopsin** or **visual purple.** Its opsin is called **scotopsin.** Rhodopsin has a peak sensitivity to light at a wavelength of 505 nm.

Human rhodopsin has a molecular weight of 41,000. It is found in the membranes of the rod disks and makes up 90% of the total protein in these membranes. Like the β_2-adrenergic receptor and the muscarinic acetylcholine receptor, which it resembles, it passes through the membrane 7 times (Fig 4–14). Retinene$_1$ is attached to 3 of these loops in the membrane in a position parallel to the surface of the membrane.

In the dark, the retinene$_1$ in rhodopsin is in the 11-*cis* configuration. The only action of light is to change the shape of the retinene, converting it to the all-*trans* isomer (Fig 8–16). This activation of rhodopsin triggers formation of a series of intermediates, one of which, **metarhodopsin II,** appears to be the key compound in initiating the closure of the Na^+ channels. The final step is the separation of retinene$_1$ from the opsin (bleaching). Some of the rhodopsin is regenerated directly, while some of the retinene$_1$ is reduced by the enzyme alcohol dehydrogenase in the presence of NADH to vitamin A_1, and this in turn reacts with scotopsin to form rhodopsin (Fig 8–16). All of these reactions except the formation of the all-trans isomer of retinene$_1$ are independent of light, proceeding equally well in light or darkness. The amount of rhodopsin in the receptors therefore varies inversely with the incident light.

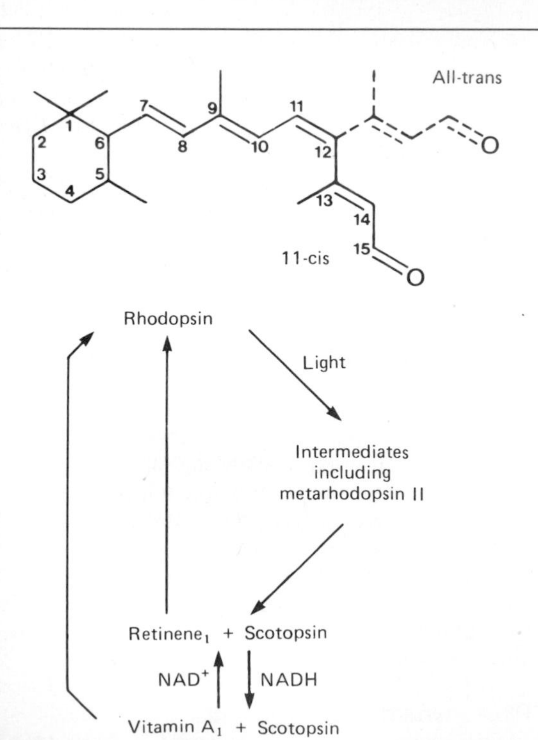

Figure 8–16. *Top:* Structure of retinene$_1$, showing the 11-*cis* configuration (unbroken lines) and the all-*trans* configuration produced by light (dashed lines). ***Bottom:*** Effects of light on rhodopsin.

Linkage Between Rhodopsin & Na^+ Channels

Activation of rhodopsin activates **transducin,** a G protein also known as G_t (see Chapters 1 and 4). Transducin binds GTP (Fig 8–17), and this in turn activates a phosphodiesterase that catalyzes the conversion of cyclic GMP (cGMP) to 5′-GMP. cGMP in the cytoplasm of photoreceptors acts directly on Na^+ channels to maintain them in the open position, and the reduction in cGMP leads to channel closure and hyperpolarization. This cascade of reactions amplifies the light signal and helps explain the remarkable sensitivity of rod photoreceptors: these receptors are capable of producing a detectable response to as little as one photon of light.

The details of the interaction between rhodopsin, transducin, and phosphodiesterase have now been elucidated. Like other G proteins, transducin has 3 subunits. Activation of rhodopsin triggers the replacement of GDP with GTP on the α subunit of transducin. The GTP-containing α subunit is released and activates phosphodiesterase by removing an inhibitory constraint on the molecule. The GTP on the α subunit is then cleaved to GDP, and the unit reassociates the β and γ subunits of transducin, terminating the reaction. Amplification occurs at the first and third steps of the cascade; each activated rhodopsin activates 500 molecules of transducin, and each activated phosphodiesterase hydrolyzes several thousand molecules of cGMP.

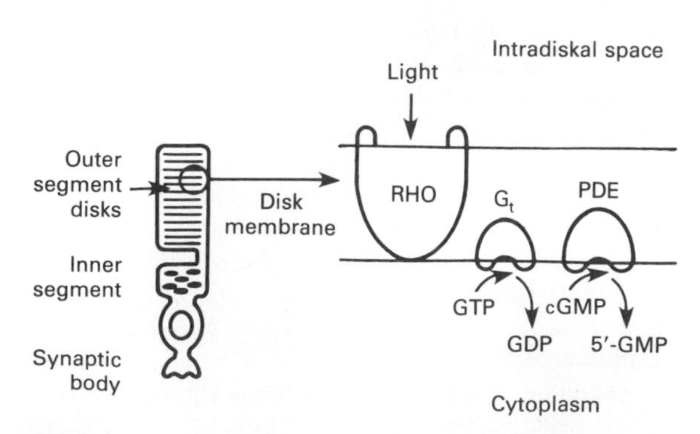

Figure 8–17. Initial steps in phototransduction in rods. Light activates rhodopsin (RHO), which activates transducin (G_t) to bind GTP. This activates phosphodiesterase (PDE), which catalyzes conversion of cGMP to 5′-GMP. (Reproduced, with permission, from Spiegel AM et al: Clinical implications of guanine nucleotide-binding proteins as receptor-effector couplers. Reprinted by permission of the *New England Journal of Medicine* 1985;**312**:26.)

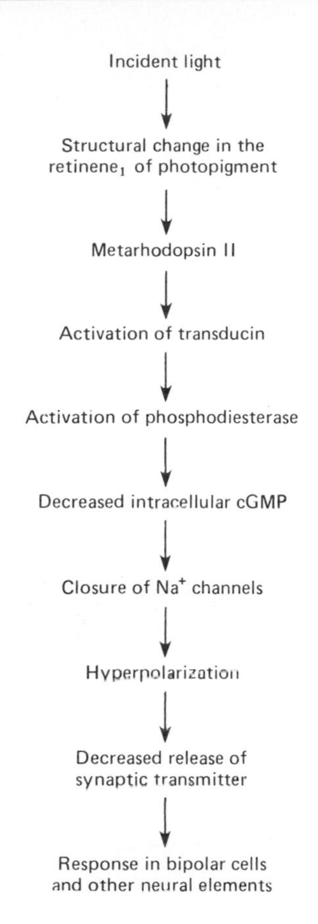

Figure 8–18. Probable sequence of events involved in phototransduction in rods and cones.

Cone Pigments

There are 3 different kinds of cones in primates. These receptors subserve color vision and respond maximally to light at wavelengths of 440, 535, and 565 nm (see below). Each contains retinene$_1$ and an opsin. The opsin resembles rhodopsin and spans the cone membrane 7 times but has a characteristic structure in each type of cone. As noted above, the cell membrane of cones is invaginated to form the saccules, but there are no separate intracellular disks like those in rods. The details of the responses of cones to light have not been completely worked out, but they are probably similar to those in rods. Light activates retinene$_1$, and this activates a transducin, the exact structure of which differs somewhat from rod transducin and varies with the type of cone. Transducin in turn activates phosphodiesterase, catalyzing the conversion of cGMP to 5′-GMP. This results in closure of Na^+ channels between the extracellular fluid and the cone cytoplasm, a rise in intracellular Na^+, and hyperpolarization of the cone outer segments.

The sequence of events in photoreceptors by which incident light leads to production of a signal in the next succeeding neural unit in the retina is summarized in Fig 8–18.

Synaptic Mediators in the Retina

A great variety of different synaptic transmitters are found in the retina. These include acetylcholine, dopamine, serotonin, GABA, glycine, substance P, somatostatin, TRH, LHRH, enkephalins, β-endorphin, CCK, VIP, neurotensin, and glucagon (see Chapter 4). Amacrine cells are the only cells that secrete acetylcholine in the retina. Peptides are found in different populations of amacrine cells, and there are in addition cholinergic, dopaminergic, and serotonergic amacrine cells, each with a different shape and, presumably, function. One of the drugs that inhibits monoamine oxidase, the enzyme which catalyzes the oxidation of 5-hydroxytryptamine and dopamine, is said to impair red-green color discrimination.

Image Formation

In a sense, the processing of visual information in the retina involves the formation of 3 images. The first image, formed by the action of light on the photoreceptors, is changed to a second image in the bipolar cells, and this in turn is converted to a third image in the ganglion cells. In the formation of the second image, the signal is altered by the horizontal cells, and in the formation of the third, it is altered by the amacrine cells. There is little change in the impulse pattern in the lateral geniculate bodies, so the third image reaches the occipital cortex.

A characteristic of the bipolar and ganglion cells (as well as the lateral geniculate cells and the cells in layer IV of the visual cortex) is that they respond best to a small, circular stimulus and that, within their receptive field, an annulus of light around the center (surround illumination) inhibits the response to the central spot (Fig 8–19). The center can be excitory with an inhibitory surround (an "on center" cell) or inhibitory with an excitory surround (an "off center" cell). The inhibition of the center response by the surround is probably due to inhibitory feedback from one photoreceptor to another mediated via horizontal cells. Thus, activation of nearby photoreceptors by addition of the annulus triggers horizontal cell hyperpolarization, which in turn inhibits the response of the centrally activated photoreceptors. The inhibition of the response to central illumination by an increase in surrounding illumination is an example of **lateral** or **afferent inhibition**—that form of inhibition in which activation of a particular neural unit is associated with inhibition of the activity of nearby units. It is a general phenomenon in mammalian sensory systems and helps to sharpen the edges of a stimulus and improve discrimination.

The electrical responses of bipolar cells are graded, slow potential changes. One type of bipolar cell depolarizes on steady illumination of the retina with a spot of light, whereas another type hyperpolarizes. The ganglion cells, which produce propagated spikes, are also of 2 types in terms of their center responses: "on center" cells that increase their discharge and

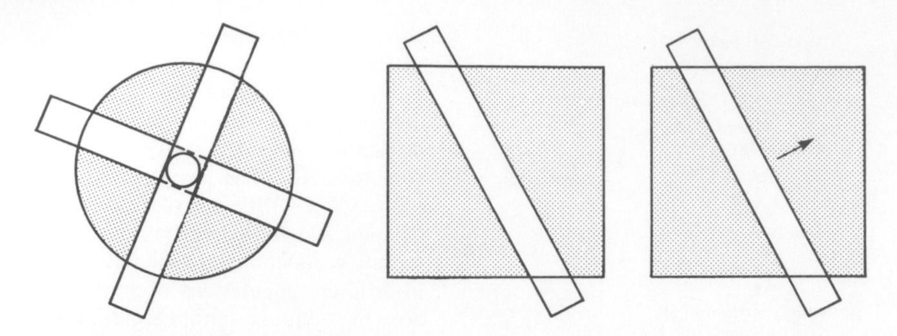

Figure 8–19. Receptive fields of cells in visual pathways. *Left:* Ganglion cells, lateral geniculate cells, and cells in layer IV of cortical area 17 have circular fields with an excitatory center and an inhibitory surround, or an inhibitory center and an excitatory surround. There is no preferred orientation of a linear stimulus. *Center:* Simple cells respond best to a linear stimulus with a particular orientation in a particular part of the cell's receptive field. *Right:* Complex cells respond to linear stimuli with a particular orientation, but they are less selective in terms of location in the receptive field and often respond maximally when the stimulus is moved laterally, as indicated by the arrow. (Modified from Hubel DH: The visual cortex of normal and deprived monkeys. *Am Sci* 1979;**67**:532.)

"off center" cells that decrease their discharge upon illumination of the center of their receptive fields. The amacrine cells depolarize when a stimulus is turned on and when it is turned off.

Electroretinogram

The electrical activity of the eye has been studied by recording fluctuations in the potential difference between an electrode in the eye and another on the back of the eye. At rest, there is a 6-mV potential difference between the front and the back of the eye, with the front positive. When light strikes the eye, a characteristic sequence of potential changes follows. The record of this sequence is the **electroretinogram (ERG).** Turning on the light stimulus elicits the **a and b waves,** as shown in Fig 8–20, and also the **c wave,** which is so slow that with short stimuli its peak occurs after the end of the stimulus. When the stimulus is turned off, there is also a small negative off-deflection. The c wave is generated in the pigment epithelium. The neural components summate to give the other deflections of the ERG.

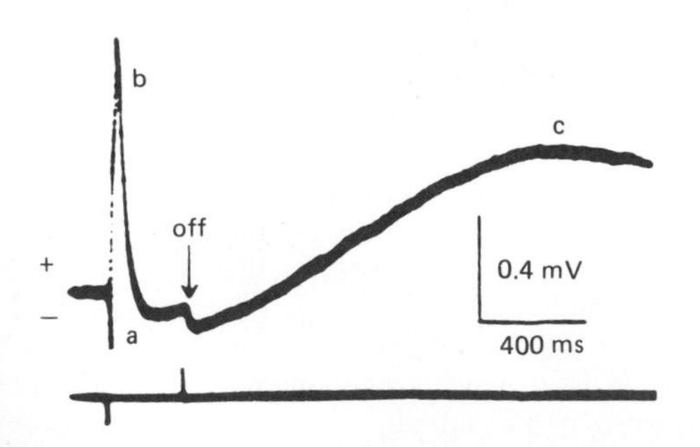

Figure 8–20. Electroretinogram recorded with an electrode in the vitreous humor of a cat. Note that, by convention in electroretinography, positive deflections are upward. The onset and termination of the stimulus are indicated by the vertical lines in the lower record. (Reproduced, with permission, from Brown KT: The electroretinogram: Its components and their origins. *Vision Res* 1968;**8**:633.)

Electroretinograms can be recorded in humans with one electrode on the cornea and the other on the skin of the head. Electroretinographic records are helpful in the diagnosis of certain ophthalmologic disorders, but most of these diseases are more readily diagnosed by simpler methods.

RESPONSES IN THE VISUAL PATHWAYS & CORTEX

Pathways to the Cortex

The ganglion cell axons project a detailed spatial representation of the retina on the lateral geniculate body. Each geniculate body contains 6 well-defined layers. Layers 3–6 have small cells and are called parvocellular, whereas layers 1 and 2 have large cells and are called magnocellular. On each side, layers 1, 4, and 6 receive input from the contralateral eye, whereas layers 2, 3, and 5 receive input from the ipsilateral eye. In each layer, there is a precise point-for-point representation of the retina, and all 6 layers are in register so that along a line perpendicular to the layers, the receptive fields of the cells in each layer are almost identical.

Two kinds of ganglion cells can be distinguished in the retina: large ganglion cells (Y cells), which add responses from different kinds of cones and are concerned with movement and stereopsis; and small ganglion cells (X cells), which subtract input from one type of cone from input from another and are concerned with color vision. The large ganglion cells project to the magnocellular portion of the lateral geniculate, whereas the small ganglion cells project to the parvocellular portion.

Visual Cortex

Just as the ganglion cell axons project a detailed spatial representation of the retina on the lateral geniculate body, the body projects a similar point-for-point representation on the visual cortex (Fig 8–5). In the visual cortex, there are many nerve cells associated with each fiber. Like the rest of the neocortex, the visual cortex has 6 layers (see Chapter 11). The axons from the lateral geniculate neurons end on pyramidal cells in layer IV and specifically in its deepest part, layer IV C. The axons from the magnocellular portion of the geniculate body end more superficially in layer IV C than the axons from the parvocellular portion.

The cells in layer IV C in turn project primarily to more superficial layers of the cortex, particularly layer IV B and layers II and III. Layers II and III contain clusters of cells about 0.2 mm in diameter that, unlike the neighboring cells, contain a high concentration of the mitochondrial enzyme cytochrome oxidase. The clusters have been named **blobs.** They are arranged in a mosaic in the visual cortex and are concerned with color vision (see below). The areas between the blobs are called interblob regions.

The visual association areas next to the primary visual cortex (primarily Brodmann's area 18) can be divided into thick stripes, thin stripes, and pale stripes on the basis of characteristic staining reactions. Layer IV B of the visual cortex projects to thick stripes, the blobs to thin stripes, and the interblob regions to pale stripes.

Like the ganglion cells, the lateral geniculate neurons and the neurons in layer IV of the visual cortex respond to stimuli in their receptive fields with on centers and inhibitory surrounds or off centers and excitatory surrounds. A bar of light covering the center is an effective stimulus for them because it stimulates all the center and relatively little of the surround (Fig 8–19). However, the bar has no preferred orientation and, as a stimulus, is equally effective at any angle.

The responses of the neurons in other layers of the visual cortex are strikingly different. So-called **simple cells** in these locations respond to bars of light, lines, or edges, but only when they have a particular orientation. When, for example, a bar of light is rotated as little as 10 degrees from the preferred orientation, the firing rate of the simple cell is usually decreased, and if the stimulus is rotated much more, the response disappears. There are also **complex cells**, which resemble simple cells in requiring a preferred orientation of a linear stimulus but are less dependent upon the location of a stimulus in the visual field than the simple cells and the cells in layer IV. They often respond maximally when a linear stimulus is moved laterally without a change in its orientation (Fig 8–19).

If a microelectrode is inserted perpendicularly into the visual cortex and passed through the various layers, the orientation preference of the neurons is the same. However, cells a few millimeters lateral to the electrode have different orientations. Thus, the visual cortex, like the somatosensory cortex (see Chapter 7), is arranged in vertical columns that are concerned with orientation **(orientation columns).** Each is about 1 mm in diameter. However, the orientation preferences of neighboring columns differ in a systematic way; as one moves from column to column across the cortex, there are sequential changes in orientation preference of 5–10 degrees (Fig 8–21). Thus, it is possible to speculate that for each ganglion cell receptive field in the visual field, there is a collection of columns in a small area of visual cortex representing the possible preferred orientations at small intervals throughout the full 360 degrees. The simple and complex cells have been called **feature detectors**

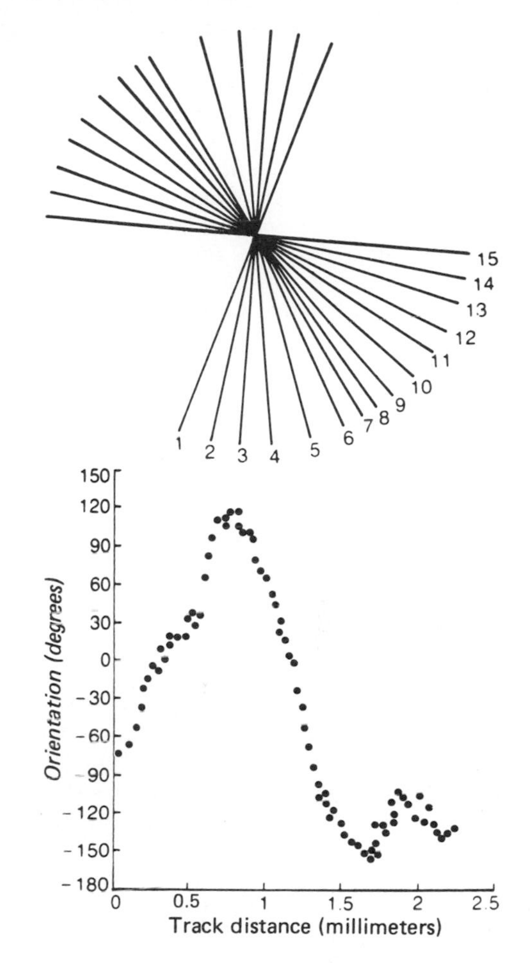

Figure 8–21. *Top:* Orientation preferences of 15 neurons encountered as a microelectrode penetrated the visual cortex obliquely. The preferred orientation changed steadily in a counterclockwise direction. ***Bottom:*** Results of a similar experiment plotted against distance the electrode traveled. In this case, there were a number of reversals in the direction of rotation. (Modified and reproduced, with permission, from Hubel DH, Wiesel TN: Sequence regularity of orientation columns in the monkey striate cortex. *J Comp Neurol* 1974;**158**:267.)

because they respond to and analyze certain features of the stimulus. Feature detectors are also found in the cortical areas for other sensory modalities.

The orientation columns can be mapped with the aid of radioactive 2-deoxyglucose. The uptake of this glucose derivative is proportional to the activity of neurons (see Chapter 32). When this technique is employed in animals exposed to uniformly oriented visual stimuli such as vertical lines, the brain shows a remarkable array of intricately curved but evenly spaced orientation columns over a large area of the visual cortex.

Another feature of the visual cortex is the presence of **ocular dominance columns.** The geniculate cells, the cells of layer IV, and the simple cells receive input from only one eye. However, about half the complex cells receive an input from both eyes. The inputs are identical or nearly so in terms of the portion of the visual field involved and the preferred orientation. However, they differ in strength, so that between the cells to which the input is totally from the ipsilateral or the contralateral eye, there is a spectrum of cells influenced to varying degrees by both eyes. The cells influenced by one eye are in vertical ocular dominance columns that alternate with columns of cells influenced by the other eye. The ocular dominance columns can be mapped by injecting a large amount of a radioactive amino acid into one eye. The amino acid is incorporated into protein and transported by axoplasmic flow to the ganglion cell terminals, across the geniculate synapses, and along the geniculocalcarine fibers to the visual cortex. Layer IV becomes evenly labeled, but above and below this layer in the cortex, labeled columns 0.8 mm in diameter alternate with unlabeled columns receiving input from the uninjected eye. The result is a vivid pattern of stripes that covers much of the visual cortex (Fig 8–22) and is separate from and independent of the grid of orientation columns. Ocular dominance columns can also be demonstrated by radioautography following injection of 2-deoxyglucose with one eye of the subject closed. The reason for the existence of the rigorously ordered, complex binocular input to some complex cells is unknown but may have something to do with binocular stereoscopic vision.

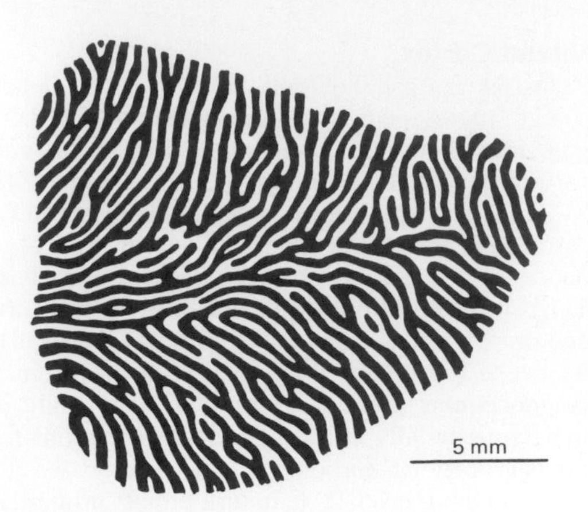

Figure 8–22. Reconstruction of ocular dominance columns in a subdivision of layer IV of a portion of right visual cortex of a rhesus monkey. Dark stripes represent one eye, light stripes the other. (Reproduced, with permission, from LeVay S, Hubel DH, Wiesel TN: The pattern of ocular dominance columns in macaque visual cortex revealed by a reduced silver stain. *J Comp Neurol* 1975;**159**:559.)

COLOR VISION

Characteristics of Color

Colors have 3 attributes: **hue, intensity,** and **saturation** (degree of freedom from dilution with white). For any color there is a **complementary color** that, when properly mixed with it, produces a sensation of white. Black is the sensation produced by the absence of light, but it is probably a positive sensation, because the blind eye does not "see black"; it "sees nothing." Such phenomena as successive and simultaneous contrasts, optical tricks that produce a sensation of color in the absence of color, negative and positive after-images, and various psychologic aspects of color vision are also pertinent. Detailed discussion of these phenomena can be found in textbooks of physiologic optics.

Another observation of basic importance is the demonstration that the sensation of white, any spectral color, and even the extraspectral color, purple, can be produced by mixing various proportions of red light (wavelength 723–647 nm), green light (575–492 nm), and blue light (492–450 nm). Red, green, and blue are therefore called the **primary colors.**

A third important point is that, as shown by Land, the color perceived depends in part on the color of other objects in the visual field. Thus, for example, a red object is seen as red if the field is illuminated with green or blue light but as pale pink or white if the field is illuminated with red light.

Retinal Mechanisms

The **Young-Helmholtz theory** of color vision in humans postulates the existence of 3 kinds of cones, each containing a different photopigment and maximally sensitive to one of the 3 primary colors, with the sensation of any given color being determined by the relative frequency of the impulses from each of these cone systems. The correctness of this theory has now been demonstrated by the identification and chemical characterization of each of the 3 pigments by recombinant DNA techniques. One pigment (the blue-sensitive or short-wave pigment) absorbs light maximally in the blue-violet portion of the spectrum (Fig 8–23). Another (the green-sensitive or middle-wave pigment) absorbs maximally in the green portion. The third (the red-sensitive or long-wave pigment) absorbs maximally in the yellow portion. Blue, green, and red are the primary colors, but the cones with their maximal sensitivity in the yellow portion of the spectrum are sensitive enough in the red portion to

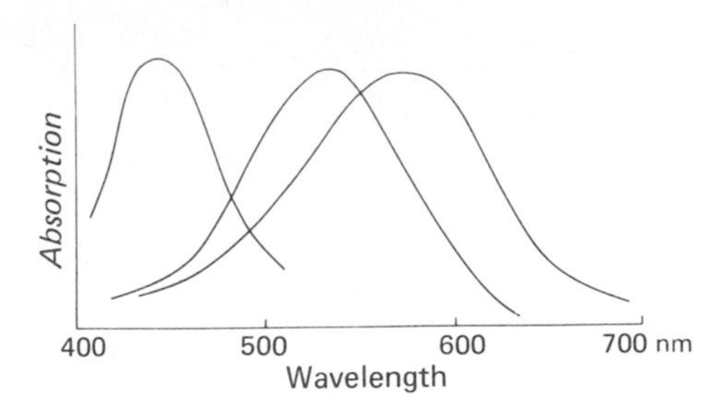

Figure 8–23. Absorption spectrums of the 3 cone pigments in the human retina. The pigment that peaks at 440 nm senses blue, and the pigment that peaks at 535 nm senses green. The remaining pigment peaks in the yellow portion of the spectrum, at 565 nm, but its spectrum extends far enough into the long wavelengths to sense red. (Reproduced, with permission, from Michael CR: Color vision. *N Engl J Med* 1973;**288**:724.)

respond to red light at a lower threshold than green. This is all the Young-Helmholtz theory requires.

The gene for human rhodopsin is on chromosome 3, and the gene for the blue-sensitive cone pigment is on chromosome 7. The other 2 cone pigments are coded by genes arranged in tandem on the q arm of the X chromosome. The green-sensitive and red-sensitive pigments are very similar in structure; their opsins show 96% homology of amino acid sequences, whereas each of these pigments has only about 43% homology with the opsin of blue-sensitive pigment, and all 3 have about 41% homology with rhodopsin. Many mammals are **dichromats;** ie, they have only 2 cone pigments, a short-wave and a long-wave pigment. Old World monkeys, apes, and humans are **trichromats,** with separate middle- and long-wave pigments—in all probability because there was duplication of the ancestral long-wave gene followed by divergence. There is more than one copy of the gene for the green-sensitive pigment on the q arm of the X chromosome, and there may be more than one copy of the gene for the red-sensitive pigment as well.

Neural Mechanisms

As noted above, color sensation is mediated by the small ganglion cells, which subtract input from one type of cone from input from another type of cone. Information carried by the small cells is relayed in the parvocellular portion of the internal geniculate body to the deep portion of layer IV C of the visual cortex. From there, the information passes to the blobs in layers II and III. The neurons in the blobs lack orientation specificity but respond to colors. Like the ganglion and geniculate cells, they are center-surround cells. Many are **double opponent cells,** which, for example, are stimulated by a green center and inhibited by a green surround and are inhibited by a red center and stimulated by a red surround. From the blobs, color information is relayed to thin stripes in the visual association area and from there to a separate area concerned with color at the anterior edge of the visual cortex (visual area 4). This area contains color-specific columns (Fig 8–24). The cells respond weakly when the visual field is illuminated by light of only one color but respond vigorously to a center stimulus when the surround is illuminated by light of a different wavelength.

Color Blindness

There are numerous tests for detecting color blindness. The most commonly used routine tests are the yarn matching test and the Ishihara charts. In the former test, the subject is presented with a skein of yarn and asked to pick out the ones that match it from a pile of variously colored skeins. The Ishihara charts and similar polychromatic plates are plates on which are printed figures made up of colored spots on a background of similarly shaped colored spots. The figures are intentionally made up of colors that are liable to look the same as the background to an individual who is color-blind.

Some color-blind individuals are unable to distinguish certain colors, whereas others have only a color weakness. The suffix -anomaly denotes color weakness and the suffix -anopia color blindness. The prefixes prot-, deuter-, and tri- refer to defects of the

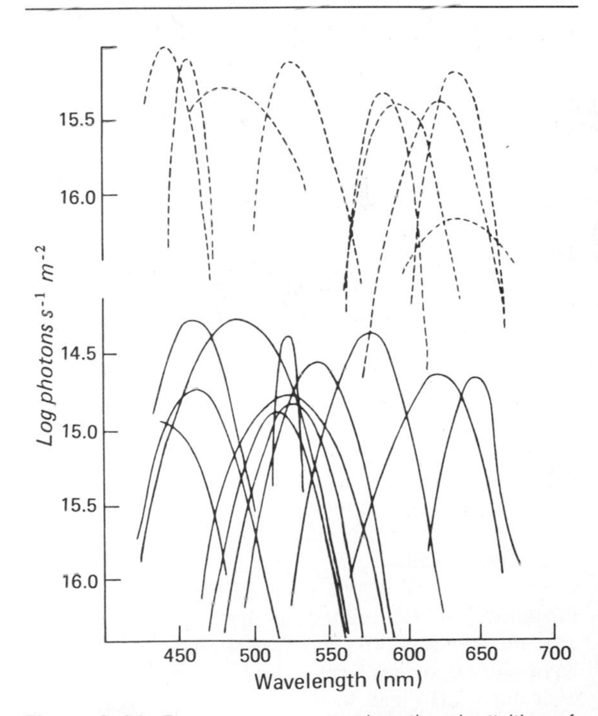

Figure 8–24. Representative wavelength selectivities of cells in the anterior portion of the fourth visual cortex area of rhesus monkeys. The solid lines represent the action spectrums of cells excited by light of the indicated wavelengths, whereas the dashed lines represent cells inhibited by light of the indicated wavelengths. (Reproduced, with permission, from Zeki S: The representation of colours in the cerebral cortex. *Nature* 1980;**284**:412. Copyright © 1980, Macmillan Journals Limited.)

red, green, and blue cone systems, respectively. Individuals with normal color vision and those with protanomaly, deuteranomaly, and tritanomaly are called **trichromats;** they have all 3 cone systems, but one may be weak. **Dichromats** are individuals with only 2 cone systems; they may have protanopia, deuteranopia, or tritanopia. **Monochromats** have only one cone system. Dichromats can match their color spectrum by mixing only 2 primary colors, and monochromats match theirs by varying the intensity of only one. Apparently monochromats see only black and white and shades of gray.

Inheritance of Color Blindness

Abnormal color vision is present as an inherited abnormality in Caucasian populations in about 8% of the males and 0.4% of the females. Tritanomaly and tritanopia are rare and show no sexual selectivity. However, about 2% of the color-blind males are dichromats who have protanopia or deuteranopia, and about 6% are anomalous trichromats in whom the red-sensitive or the green-sensitive pigment is shifted in its spectral sensitivity. These abnormalities are inherited as recessive and X-linked characteristics; ie, they are due to an abnormal gene on the X chromosome. Since all of the male's cells except germ cells contain one X and one Y chromosome in addition to the 44 somatic chromosomes (see Chapter 23), color blindness is present in males if the X chromosome has the abnormal gene. On the other hand, the normal female's cells have 2 X chromosomes, one from each parent, and since these abnormalities are recessive, females show a defect only when both X chromosomes contain the abnormal gene. However, female children of a man with X-linked color blindness are carriers of the color blindness and pass the defect on to half of their sons. Therefore, X-linked color blindness skips generations and appears in males of every second generation. Hemophilia, Duchenne muscular dystrophy, and a variety of other inherited disorders are caused by mutant genes on the X chromosome.

The common occurrence of deuteranomaly and protanomaly is probably due to the fact that the genes for the green-sensitive and red-sensitive cone pigments are located near each other in tandem on the q arm of the X chromosome and are prone to recombination (unequal crossing over) during development of the germ cells. Different combinations of introns may also occur, with both processes producing opsins that have shifted spectral sensitivities. Unequal crossing over could also lead to loss of functional genes.

Visual Perception

Although much is still unknown about visual perception, information from a number of different disciplines permits consolidation of the data outlined in the preceding 2 sections into a hypothesis that a 3-part system is responsible for visual sensations. One system is concerned with the perception of shape; a second is concerned with the perception of color; and a third is concerned with the perception of movement, location, and spatial organization. The pathways involved are summarized in Fig 8–25. It is interesting that from the visual association cortex, the pathways for each of the systems appear to project to different parts of the brain: the movement system to the middle portion of the temporal lobe (MT area), the color system to visual area 4, and the shape system to an as yet unknown but presumably separate area. Somewhere else in the brain, the information from these 3 areas is combined into a single integrated visual perception.

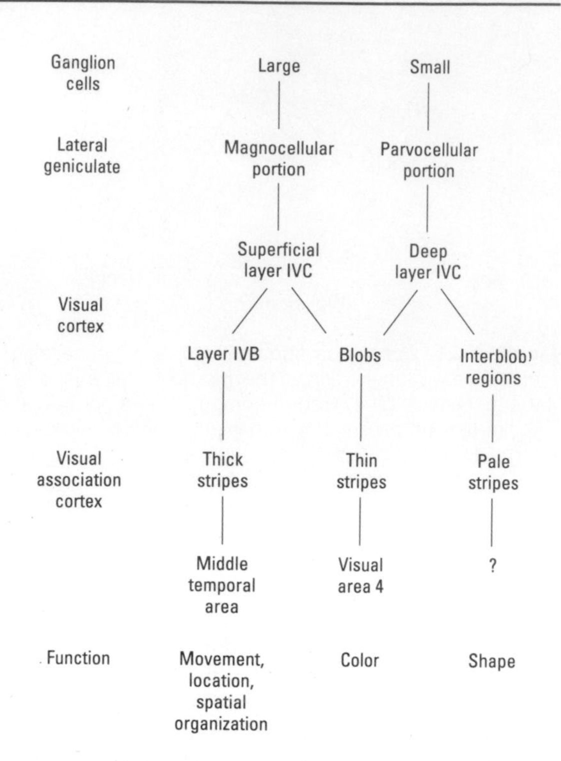

Figure 8–25. Organization of visual pathways into 3 systems that process information in parallel. Information from the system that detects movement, the system that detects color, and the system that detects shape is then integrated into a single visual perception.

OTHER ASPECTS OF VISUAL FUNCTION

Dark Adaptation

The truly remarkable range of luminance to which the human eye responds is summarized in Fig 8–26. If a person spends a considerable length of time in brightly lighted surroundings and then moves to a dimly lighted environment, the retinas slowly become more sensitive to light as the individual becomes "accustomed to the dark." This decline in visual threshold is known as **dark adaptation.** It is nearly maximal in about 20 minutes, although there is some further decline over longer periods. On the other

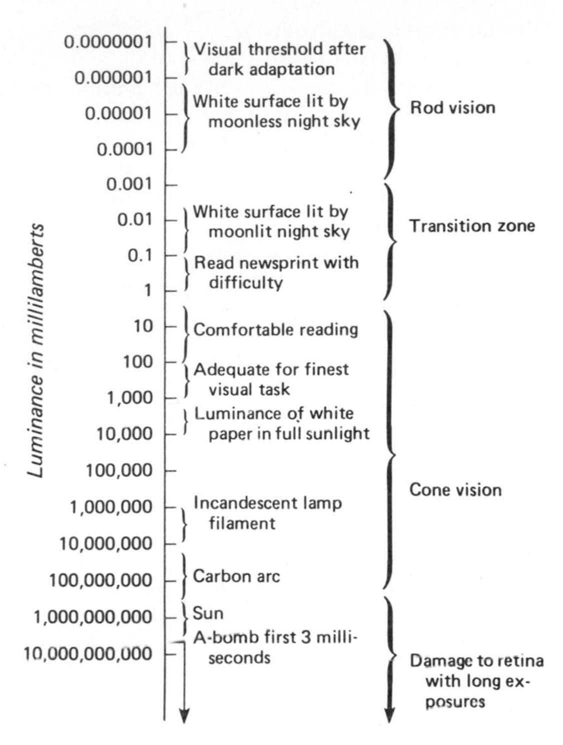

Figure 8–26. Range of luminance to which the human eye responds, with the receptive mechanisms involved. (Reproduced, with permission, by courtesy of Campbell FW, from Bell GH, Emslie-Smith D, Paterson CR: *Textbook of Physiology and Biochemistry,* 9th ed. Churchill Livingstone, 1976.)

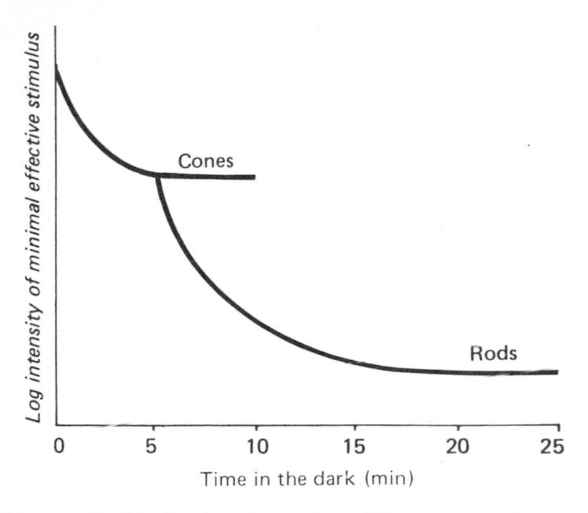

Figure 8–27. Dark adaptation. The curve shows the change in the intensity of a stimulus necessary to just excite the retina in dim light as a function of the time the observer has been in the dark.

hand, when one passes suddenly from a dim to a brightly lighted environment, the light seems intensely and even uncomfortably bright until the eyes adapt to the increased illumination and the visual threshold rises. This adaptation occurs over a period of about 5 minutes and is called **light adaptation,** although, strictly speaking, it is merely the disappearance of dark adaptation.

There are actually 2 components to the dark adaptation response (Fig 8–27). The first drop in visual threshold, rapid but small in magnitude, is known to be due to dark adaptation of the cones because when only the foveal, rod-free portion of the retina is tested, the decline proceeds no further. In the peripheral portions of the retina, a further drop occurs owing to adaptation of the rods. The total change in threshold between the light-adapted and the fully dark-adapted eye is very great.

Radiologists, aircraft pilots, and others who need maximal visual sensitivity in dim light can avoid having to wait 20 minutes in the dark to become dark-adapted if they wear red goggles when in bright light. Light wavelengths in the red end of the spectrum stimulate the rods to only a slight degree while permitting the cones to function reasonably well. Therefore, a person wearing red glasses can see in bright light during the time it takes for the rods to become dark-adapted.

The time required for dark adaptation is determined in part by the time required to build up the rhodopsin stores. In bright light, much of the pigment is continuously being broken down, and some time is required in dim light for accumulation of the amounts necessary for optimal rod function. However, dark adaptation also occurs in the cones, and additional factors are undoubtedly involved.

Effect of Vitamin Deficiencies on the Eye

In view of the importance of vitamin A in the synthesis of retinene$_1$, it is not surprising that avitaminosis A produces visual abnormalities. Among these, one of the earliest to appear is night blindness **(nyctalopia).** This fact first called attention to the role of vitamin A in rod function, but is now clear that concomitant cone degeneration occurs as vitamin A deficiency develops. Prolonged deficiency is associated with anatomic changes in the rods and cones followed by degeneration of the neural layers of the retina. Treatment with vitamin A can restore retinal function if given before the receptors are destroyed.

Other vitamins, especially those of the B complex, are necessary for the normal functioning of the retina and other neural tissues. Nicotinamide is part of the nicotinamide adenine dinucleotide (NAD^+) molecule, and this coenzyme plays a role in the interconversion of retinene and vitamin A in the rhodopsin cycle.

Physiologic Nystagmus

Even when a subject stares fixedly at a stationary object, the eyeballs are not still; there are continuous jerky motions and other movements. This **physiologic nystagmus** appears to have an important function. Although individual visual receptors do not adapt rapidly to constant illumination, their neural connections do. Indeed, it has been shown that if, by means of an optical lever system, the image of an object is fixed so that it falls steadily on the same spot in the retina, the object disappears from view. Continuous visualization of objects apparently requires that the retinal images be continuously and rapidly shifted from one receptor to another.

Visual Acuity

Physiologic nystagmus is one of the many factors that determine **visual acuity.** This parameter of vision should not be confused with **visual threshold.** Visual threshold is the minimal amount of light that elicits a sensation of light; visual acuity is the degree to which the details and contours of objects are perceived. Although there is evidence that other measures are more accurate, visual acuity is usually defined in terms of the **minimum separable**—ie, the shortest distance by which 2 lines can be separated and still be perceived as 2 lines. Clinically, visual acuity is often determined by use of the familiar Snellen letter charts viewed at a distance of 20 ft (6 m). The individual being tested reads aloud the smallest line distinguishable. The results are expressed as a fraction. The numerator of the fraction is 20, the distance at which the subject reads the chart. The denominator is the greatest distance from the chart at which a normal individual can read the smallest line the subject can read. Normal visual acuity is 20/20; a subject with 20/15 visual acuity has better than normal vision (not farsightedness); and one with 20/100 visual acuity has subnormal vision. The Snellen charts are designed so that the height of the letters in the smallest line a normal individual can read at 20 feet subtends a visual angle of 5 minutes. Each line in the letters subtends 1 minute of arc, and the lines in the letters are separated by 1 minute of arc. Thus, the minimum separable in a normal individual corresponds to a visual angle of about 1 minute.

Visual acuity is a complex phenomenon and is influenced by a large variety of factors. These include optical factors such as the state of the image-forming mechanisms of the eye, retinal factors such as the state of the cones, and stimulus factors including the illumination, brightness of the stimulus, contrast between the stimulus and the background, and the length of time the subject is exposed to the stimulus.

Critical Fusion Frequency

The time-resolving ability of the eye is determined by measuring the **critical fusion frequency (CFF),** the rate at which stimuli can be presented and still be perceived as separate stimuli. Stimuli presented at a higher rate than the CFF are perceived as a continuous stimulus. Motion pictures move because the frames are presented at a rate above the CFF. Consequently, movies begin to flicker when the projector slows down.

Visual Fields & Binocular Vision

The visual field of each eye is the portion of the external world visible out of that eye. Theoretically, it should be circular, but actually it is cut off medially by the nose and superiorly by the roof of the orbit (Fig 8–28). Mapping the visual fields is important in neurologic diagnosis. The peripheral portions of the visual fields are mapped with an instrument called a **perimeter,** and the process is referred to as **perimetry.** One eye is covered while the other is fixed on a central point. A small target is moved toward this central point along selected meridians, and, along each, the location where the target first becomes visible is plotted in degrees of arc away from the central point (Fig 8–28). The central visual fields are mapped with a **tangent screen,** a black felt screen across which a white target is moved. By noting the locations where the target disappears and reappears, the blind spot and any **objective scotomas** (blind spots due to disease) can be outlined.

The central parts of the visual fields of the 2 eyes coincide; therefore, anything in this portion of the field is viewed with **binocular vision.** The impulses set up in the 2 retinas by light rays from an object are fused at the cortical level into a single image **(fusion).** The points on the retina on which the image of an object must fall if it is to be seen binocularly as a single object are called **corresponding points.** If one eye is gently pushed out of line while staring fixedly at an object in the center of the visual field, double vision **(diplopia)** results; the image on the retina of the eye that is displaced no longer falls on the corresponding point.

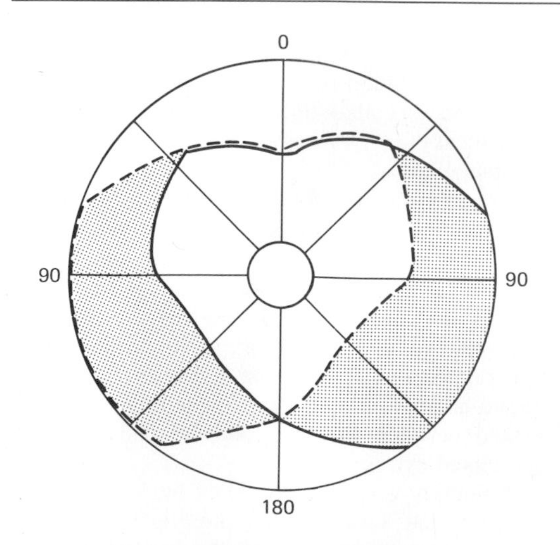

Figure 8–28. Monocular and binocular visual fields. The dashed line encloses the visual field of the left eye; the solid line, that of the right eye. The common area (heart-shaped clear zone in the center) is viewed with binocular vision. The shaded areas are viewed with monocular vision.

Binocular vision is often assigned an important role in the perception of depth. Actually, depth perception is to a large degree monocular, depending upon the relative sizes of objects, their shadows, and, for moving objects, their movement relative to one another (movement parallax). However, binocular vision does add some appreciation of depth and proportion.

Effect of Lesions in the Optic Pathways

The anatomy of the pathways from the eyes to the brain is shown in Fig 8–4. Lesions along these pathways can be localized with a high degree of accuracy by the effects they produce in the visual fields.

The fibers from the nasal half of each retina decussate in the optic chiasm, so that the fibers in the optic tracts are those from the temporal half of one retina and the nasal half of the other. In other words, each optic tract subserves half of the field of vision. Therefore, a lesion that interrupts one optic nerve causes blindness in that eye, but a lesion in one optic tract causes blindness in half of the visual field (Fig 8–4). This defect is classified as a **homonymous** (same side of both visual fields) **hemianopia** (half-blindness). Lesions affecting the optic chiasm, such as pituitary tumors expanding out of the sella turcica, cause destruction of the fibers from both nasal hemiretinas and produce a **heteronymous** (opposite sides of the visual fields) **hemianopia.** Since the fibers from the maculas are located posteriorly in the optic chiasm, hemianopic scotomas develop before there is complete loss of vision in the 2 hemiretinas. Selective visual field defects are further classified as bitemporal, binasal, and right or left.

The optic nerve fibers from the upper retinal quadrants subserving vision in the lower half of the visual field terminate in the medial half of the lateral geniculate body, while the fibers from the lower retinal quadrants terminate in the lateral half. The geniculocalcarine fibers from the medial half of the lateral geniculate terminate on the superior lip of the calcarine fissure, while those from the lateral half terminate on the inferior lip. Furthermore, the fibers from the lateral geniculate body that subserve macular vision separate from those that subserve peripheral vision and end more posteriorly on the lips of the calcarine fissure (Fig 8–5). Because of this anatomic arrangement, occipital lobe lesions may produce discrete quadrantic visual field defects (upper and lower quadrants of each half visual field). **Macular sparing,** ie, loss of peripheral vision with intact macular vision, is also common with occipital lesions (Fig 8–4), because the macular representation is separate from that of the peripheral fields and very large relative to that of the peripheral fields. Therefore, occipital lesions must extend considerable distances to destroy macular as well as peripheral vision. Bilateral destruction of the occipital cortex in humans causes essentially complete blindness, although in mammals other than primates, considerable vision (especially rod vision) remains.

The fibers to the pretectal region that subserve the reflex pupillary constriction produced by shining a light into the eye leave the optic tracts near the geniculate bodies. Therefore, blindness with preservation of the pupillary light reflex is usually due to a lesion behind the optic tracts.

EYE MOVEMENTS

The direction in which each of the eye muscles moves the eye and the definitions of the terms used in describing eye movements are summarized in Table 8–1. Since the oblique muscles pull medially (Fig 8–8), their actions vary with the position of the eye. When the eye is turned nasally, the obliques elevate and depress it, whereas the superior and inferior recti rotate it; when the eye is turned temporally, the superior and inferior recti elevate and depress it and the obliques rotate it.

Since much of the visual field is binocular, it is clear that a very high order of coordination of the movements of the 2 eyes is necessary if visual images are to fall at all times on corresponding points in the 2 retinas and diplopia is to be avoided.

There are 4 types of eye movements, each controlled by a different neural system but sharing the same final common path, the motor neurons that supply the external ocular muscles (Fig 8–29). **Saccades,** sudden jerky movements, occur as the gaze shifts from one object to another. **Smooth pursuit movements** are tracking movements of the eyes as they follow moving objects. **Vestibular movements,** adjustments that occur in response to stimuli initiated in the semicircular canals, maintain visual fixation as the head moves. **Convergence movements** bring the visual axes toward each other as attention is focused on objects near the observer. The similarity to a man-made tracking system on an unstable platform such as a ship is apparent: saccadic movements seek out visual targets; pursuit movements follow them as

Table 8–1. Actions of the external ocular muscles.*

Muscle	Primary Action	Secondary Action
Lateral rectus	Abduction	None
Medial rectus	Adduction	None
Superior rectus	Elevation	Adduction, intorsion
Inferior rectus	Depression	Adduction, extorsion
Superior oblique	Depression	Intorsion, abduction
Inferior oblique	Elevation	Extorsion, abduction

*Abduction and adduction refer to rotation of the eyeball around the vertical axis, with the pupil moving away from or toward the midline, respectively; elevation and depression refer to rotation around the transverse horizontal axis, with the pupil moving up or down; and torsion refers to rotation around the anteroposterior horizontal axis, with the top of the pupil moving toward the nose (intorsion) or away from the nose (extorsion). (Reproduced, with permission, from Vaughan D, Asbury T: *General Ophthalmology,* 11th ed. Appleton & Lange, 1986.)

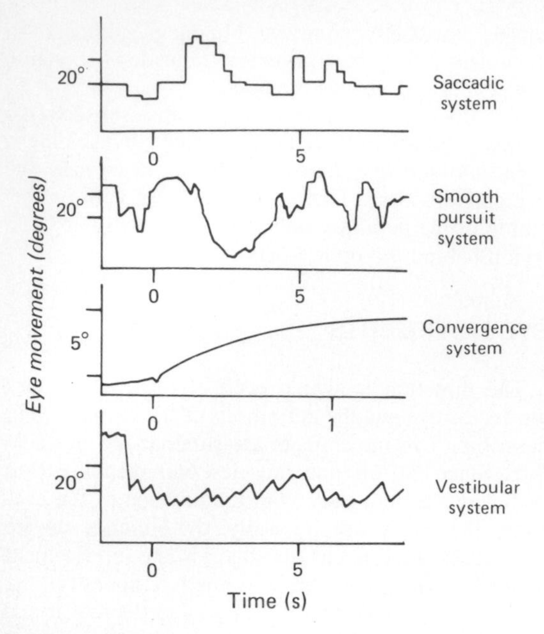

Figure 8–29. Types of eye movements. (Modified and reproduced, with permission, from Robinson DA: Eye movement control in primates. *Science* 1968;**161**:1219. Copyright 1968 by the American Association for the Advancement of Science.)

they move about; and vestibular movements stabilize the tracking device as the platform on which the device is mounted (ie, the head) moves about. In primates, these eye movements depend on an intact visual cortex.

Strabismus

Abnormalities of the coordinating mechanisms can be due to a variety of causes. When the visual axes no longer are maintained in a position that keeps the visual images on corresponding retinal points, **strabismus** (squint) is said to be present. Successful treatment of some types of strabismus is possible by careful surgical shortening of some of the eye muscles, by eye muscle training exercises, and by the use of glasses with prisms that bend the light rays sufficiently to compensate for the abnormal position of the eyeball. However, subtle defects in depth perception persist, and it has been suggested that congenital abnormalities of the visual tracking mechanisms may cause both the strabismus and the defective depth perception.

When visual images chronically fall on noncorresponding points in the 2 retinas in children under age 6, one is eventually suppressed **(suppression scotoma)** and diplopia disappears. This suppression is a cortical phenomenon, and it usually does not develop in adults. It is important to institute treatment before age 6 in children with one visual image suppressed, because if the suppression persists, there is permanent loss of visual acuity in the eye generating the suppressed image. A similar suppression with subsequent permanent loss of visual acuity can occur in children in whom vision in one eye is blurred or distorted owing to a refractive error. The loss of vision in these cases is called **amblyopia ex anopsia,** a term that refers to uncorrectable loss of visual acuity which is not directly due to organic disease of the eye. In infant monkeys, covering one eye with a patch for 3 months causes a loss of ocular dominance columns; input from the remaining eye spreads to take over all the cortical cells, and the patched eye becomes functionally blind. Comparable changes presumably occur in children with strabismus.

Hearing & Equilibrium

9

INTRODUCTION

Receptors for 2 sensory modalities, hearing and equilibrium, are housed in the ear. The external ear, the middle ear, and the cochlea of the inner ear are concerned with hearing. The semicircular canals, the utricle, and the saccule of the inner ear are concerned with equilibrium. In each case, the sensory receptors involved are hair cells, and there are 6 groups of hair cells in each inner ear: one in each of the 3 semicircular canals, one in the utricle, one in the saccule, and one in the cochlea.

ANATOMIC CONSIDERATIONS

External & Middle Ear

The external ear funnels sound waves to the **external auditory meatus.** In some animals, the ears can be moved like radar antennas to seek out sound. From the meatus, the **external auditory canal** passes inward to the **tympanic membrane** (eardrum) (Fig 9–1).

The middle ear is an air-filled cavity in the temporal bone that opens via the auditory (eustachian) tube into the nasopharynx and through the nasopharynx to the exterior. The tube is usually closed, but during swallowing, chewing, and yawning it opens, keeping the air pressure on the 2 sides of the eardrum equalized. The 3 **auditory ossicles,** the **malleus, incus,** and **stapes,** are located in the middle ear. The **manubrium** (handle of the malleus) is attached to the back of the tympanic membrane. Its head is attached to the wall of the middle ear and its short process is attached to the incus, which in turn articulates with the head of the stapes. The stapes is named for its resemblance to a stirrup. Its **foot plate** is attached by an annular ligament to the walls of the **oval window** (Fig 9–2). Two small skeletal muscles, the **tensor tympani** and the **stapedius,** are also located in the middle ear. Contraction of the former pulls the manubrium of the malleus medially and decreases the vibrations of the tympanic membrane; contraction of the latter pulls the foot plate of the stapes out of the oval window.

Inner Ear

The inner ear **(labyrinth)** is made up of 2 parts, one within the other. The **bony labyrinth** is a series of channels in the petrous portion of the temporal bone. Inside these channels, surrounded by a fluid called **perilymph,** is the **membranous labyrinth.** The membranous labyrinth more or less duplicates the shape of the bony channels (Fig 9–3). It is filled

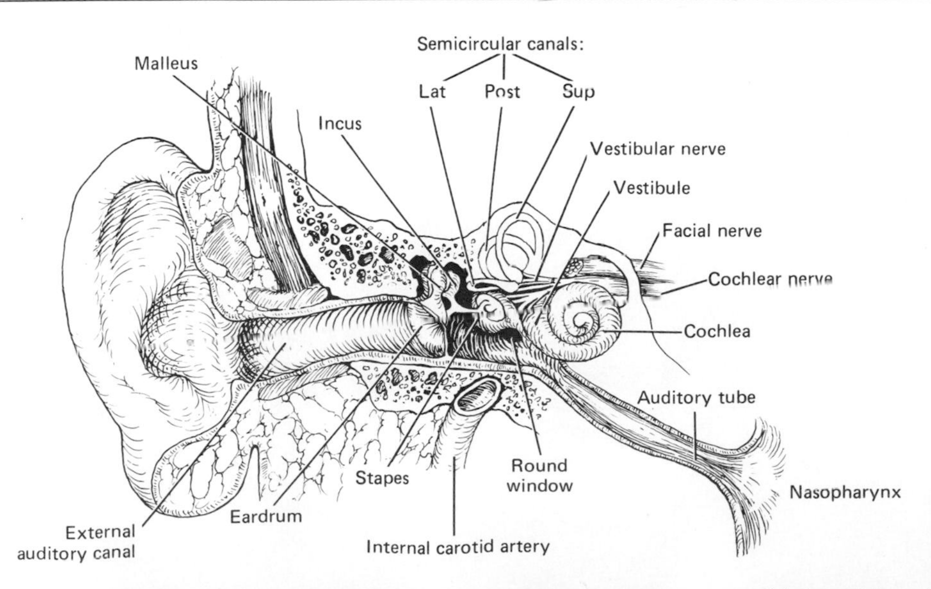

Figure 9–1. The human ear. To make the relationships clear, the cochlea has been turned slightly and the middle ear muscles have been omitted. Sup, superior; Post, posterior; Lat, lateral.

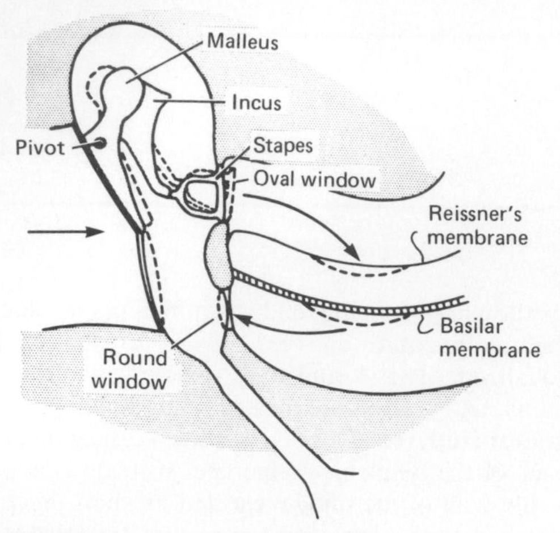

Figure 9–2. Schematic representation of the auditory ossicles and the way their movement translates movements of the tympanic membrane into a wave in the fluid of the inner ear. The wave is dissipated at the round window. The movements of the ossicles, the membranous labyrinth, and the round window are indicated by dashed lines. (Redrawn and reproduced, with permission, from an original by Netter FH in *Ciba Clinical Symposia.* Copyright © Ciba Pharmaceutical Co., 1962.)

with a fluid called **endolymph,** and there is no communication between the spaces filled with endolymph and those filled with perilymph.

Cochlea

The cochlear portion of the labyrinth is a coiled tube which in humans is 35 mm long and makes 2¾ turns. Throughout its length, the basilar membrane and Reissner's membrane divide it into 3 chambers **(scalae)** (Fig 9–4). The upper **scala vestibuli** and the lower **scala tympani** contain perilymph and communicate with each other at the apex of the cochlea through a small opening called the **helicotrema.** At the base of the cochlea, the scala vestibuli ends at the oval window, which is closed by the foot plate of the stapes. The scala tympani ends at the **round window,** a foramen on the medial wall of the middle ear that is closed by the flexible **secondary tympanic membrane.** The **scala media,** the middle cochlear chamber, is continuous with the membranous labyrinth and does not communicate with the other 2 scalae. It contains endolymph (Figs 9–3 and 9–4).

Organ of Corti

Located on the basilar membrane is the organ of Corti, the structure that contains the hair cells which are the auditory receptors. This organ extends from the apex to the base of the cochlea and consequently has a spiral shape. The processes of the hair cells pierce the tough, membranelike **reticular lamina** that is supported by the **rods of Corti** (Fig 9–4). The hair cells are arranged in 4 rows: 3 rows of **outer hair cells** lateral to the tunnel formed by the rods of Corti, and one row of **inner hair cells** medial to the tunnel. There are 20,000 outer hair cells and 3500 inner hair cells in each human cochlea. Covering the rows of hair cells is a thin, viscous but elastic **tectorial membrane** in which the tips of the hairs of the outer but not the inner hair cells are embedded. The cell bodies of the afferent neurons that arborize around the bases of the hair cells are located in the **spiral ganglion** within the **modiolus,** the bony core around which the cochlea is wound. Ninety to 95% of these

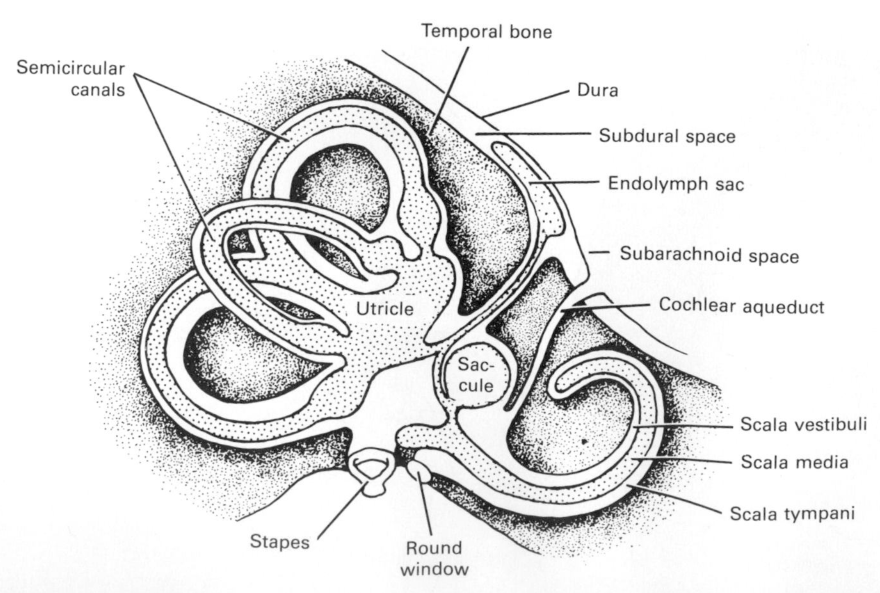

Figure 9–3. Relationship between the membranous and osseous labyrinths (diagrammatic). The cochlear aqueduct is not patent in primates.

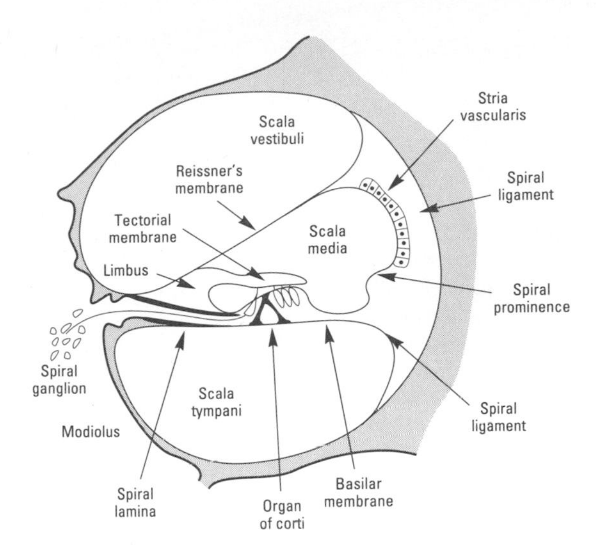

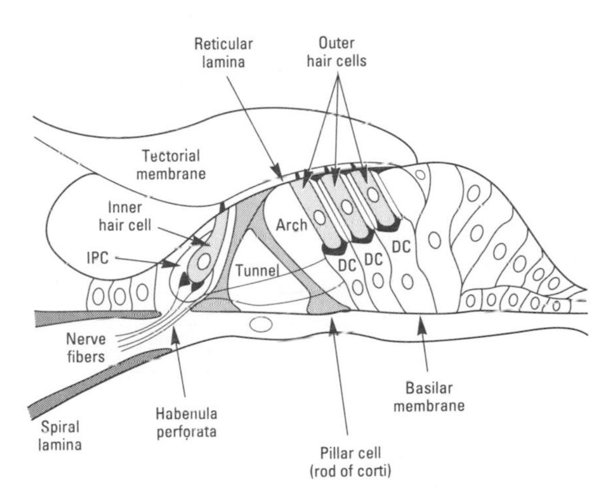

Figure 9–4. *Top:* Cross section of the cochlea, showing the organ of Corti and the 3 scala of the cochlea. *Bottom:* Structure of the organ of Corti, as it appears in the basal turn of the cochlea. DC, outer phalangeal cells (Deiters' cells) supporting outer hair cells; IPC, inner phalangeal cell supporting inner hair cell. (Reproduced, with permission, from Pickels JO: *An Introduction to the Physiology of Hearing,* 2nd ed. Academic, 1988.)

afferent neurons innervate the inner hair cells; only 5–10% innervate the more numerous outer hair cells, and each neuron innervates several of these outer cells. In addition, most of the efferent fibers in the auditory nerve (see below) terminate on the outer hair cells rather than on the inner hair cells. The axons of the neurons that innervate the hair cells form the auditory (cochlear) division of the vestibulocochlear acoustic nerve and terminate in the **dorsal** and **ventral cochlear nuclei** of the medulla oblongata. The total number of afferent and efferent fibers in each auditory nerve is approximately 28,000.

In the cochlea, there are tight junctions between the hair cells and the adjacent phalangeal cells; these prevent endolymph from reaching the bases of the cells. However, the basilar membrane is relatively permeable to perilymph in the scala tympani, and consequently, the tunnel of the organ of Corti and the bases of the hair cells are bathed in perilymph. Because of similar tight junctions, the arrangement is similar for the hair cells in other parts of the middle ear, ie, the processes of the hair cells are bathed in endolymph, whereas their bases are bathed in perilymph.

Central Auditory Pathways

From the cochlear nuclei, axons carrying auditory impulses pass via a variety of pathways to the **inferior colliculi,** the centers for auditory reflexes, and via the **medial geniculate body** in the thalamus to the **auditory cortex.** Others enter the reticular formation (Fig 9–5). Information from both ears converges on each superior olive, and at all higher levels most of the neurons respond to inputs from both sides. The primary auditory cortex, Brodmann's area 41, is in the superior portion of the temporal lobe. In humans, it is located in the floor of the lateral cerebral fissure (Fig 7–4) and is not normally visible on the surface of the brain. There are several additional auditory receiving areas, just as there are several receiving areas for cutaneous sensation (see Chapter 7). The auditory association areas adjacent to the primary auditory receiving area are widespread, extending onto the insula. The **olivocochlear bundle** is a prominent bundle of efferent fibers in each auditory nerve that arises from both the ipsilateral and the contralateral superior olivary complex and ends primarily around the bases of the outer hair cells of the organ of Corti.

Semicircular Canals

On each side of the head, the semicircular canals are perpendicular to each other, so that they are oriented in the 3 planes of space. Inside the bony canals, the membranous canals are suspended in perilymph. A receptor structure, the **crista ampullaris,** is located in the expanded end **(ampulla)** of each of the membranous canals. Each crista consists of hair cells and sustentacular cells surmounted by a gelatinous partition **(cupula)** that closes off the ampulla (Fig 9–6). The processes of the hair cells are embedded in the cupula, and the bases of the hair cells are in close contact with the afferent fibers of the vestibular division of the vestibulocochlear nerve.

Utricle & Saccule

Within each membranous labyrinth, on the floor of the utricle, there is an **otolithic organ (macula).** Another macula is located on the wall of the saccule in a semivertical position. The maculas contain sustentacular cells and hair cells, surmounted by an otolithic membrane in which are embedded crystals of calcium carbonate, the **otoliths** (Fig 9–7). The otoliths, which are also called **otoconia** or **ear dust,** range from 3 to 19 μm in length in humans and are

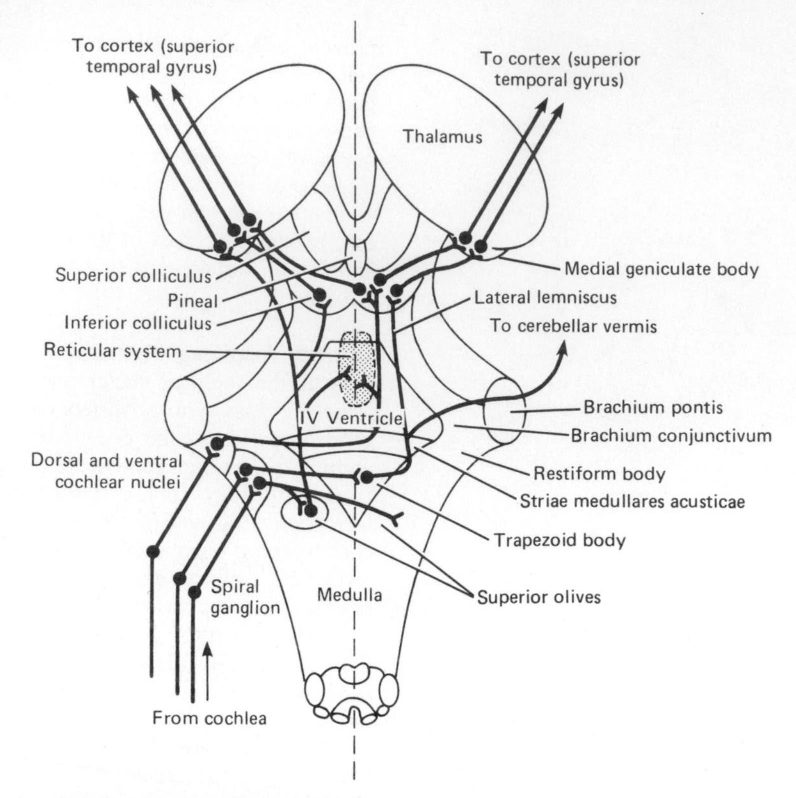

Figure 9–5. Simplified diagram of main auditory pathways superimposed on a dorsal view of the brain stem. Cerebellum and cerebral cortex removed.

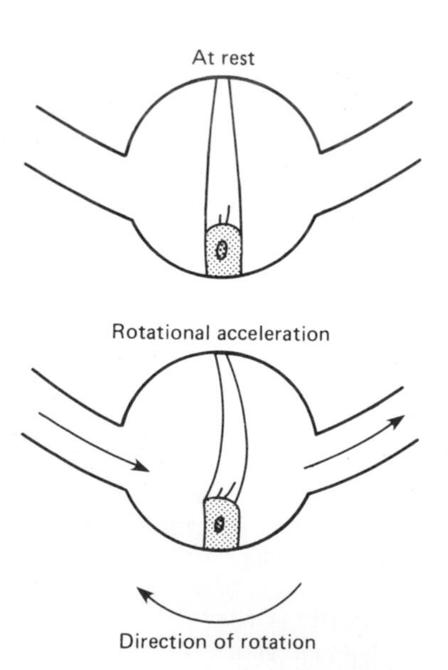

Figure 9–6. Diagrammatic representation of ampullar crista. The cupula on the top of the crista closes off the ampulla and is flexible. Because of its inertia, the endolymph is displaced in a direction opposite to the direction of rotation during rotational acceleration (see below). This bends the hair cell processes, altering their permeability and changing the membrane potential of the hair cells.

more dense than the endolymph. The processes of the hair cells are embedded in the membrane. The nerve fibers from the hair cells join those from the cristae in the vestibulocochlear nerve.

Neural Pathways

The cell bodies of the 19,000 neurons supplying the cristae and maculas on each side are located in the vestibular ganglion. Each vestibular nerve terminates in the ipsilateral 4-part vestibular nucleus and in the flocculonodular lobe of the cerebellum. Second-order neurons pass down the spinal cord from the vestibular nuclei in the vestibulospinal tracts and ascend through the **medial longitudinal fasciculi** to the motor nuclei of the cranial nerves concerned with the control of eye movement. There are also anatomically poorly defined pathways by which impulses from the vestibular receptors are relayed via the thalamus to the cerebral cortex (Fig 9–8).

HAIR CELLS

Structure

The hair cells in the inner ear have a common structure (Fig 9–9). Each is embedded in an epithelium made up of sustentacular cells. The basal end is in close contact with afferent neurons. Projecting from the apical end are 30–150 rod-shaped processes,

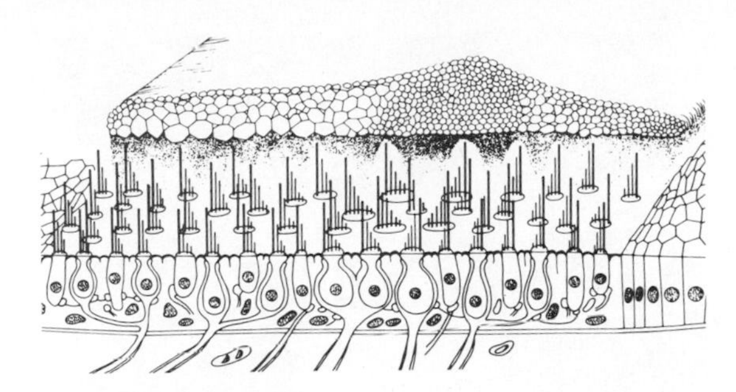

Figure 9–7. Saccular macula of a guinea pig. The otoliths are embedded in a gelatinous membrane that sits on the hair cells. Each hair cell has a large kinocilium and smaller stereocilia. Some of the afferent vestibular nerve fibers (bottom) surround the bases of hair cells, forming calices (type I cells). Others (type II cells) merely contact the hair cells. The structure of the human saccular macula is similar. (Reproduced, with permission, from Lindeman HH: Morphology of vestibular sensory receptors. *Ergeb Anat Entwicklungs* 1969;42:1.)

or hairs. Except in the cochlea, one of these, the **kinocilium,** is a true but nonmotile cilium with 9 pairs of microtubules around its circumference and a central pair of microtubules (see Chapter 1). It is one of the largest processes and has a clubbed end. The kinocilium is lost in the hair cells of the cochlea in adult mammals. However, the other processes, which are called **stereocilia,** are present in all hair cells. They have cores composed of parallel filaments of actin. Within the circular clump of processes on each cell, there is an orderly structure: Along an axis toward the kinocilium, the stereocilia increase progressively in height; along the perpendicular axis, all the stereocilia are the same height.

Electrical Responses

The membrane potential of the hair cells is about −60 mV. When the stereocilia are pushed toward the kinocilium, the membrane potential is decreased to about −50 mV. When the bundle of processes is pushed in the opposite direction, the cell is hyperpolarized. Displacing the processes in a direction perpendicular to this axis provides no change in membrane potential, and displacing the processes in directions that are intermediate between these 2 directions produces depolarization or hyperpolarization that is proportionate to the degree to which the direction is toward or away from the kinocilium. Thus, the hair processes provide a mechanism for generating changes in membrane potential proportional to the direction of displacement.

Genesis of Action Potentials in Afferent Nerve Fibers

As noted above, the processes of the hair cells project into the endolymph whereas the bases are bathed in perilymph. This arrangement is necessary for the normal production of generator potentials,

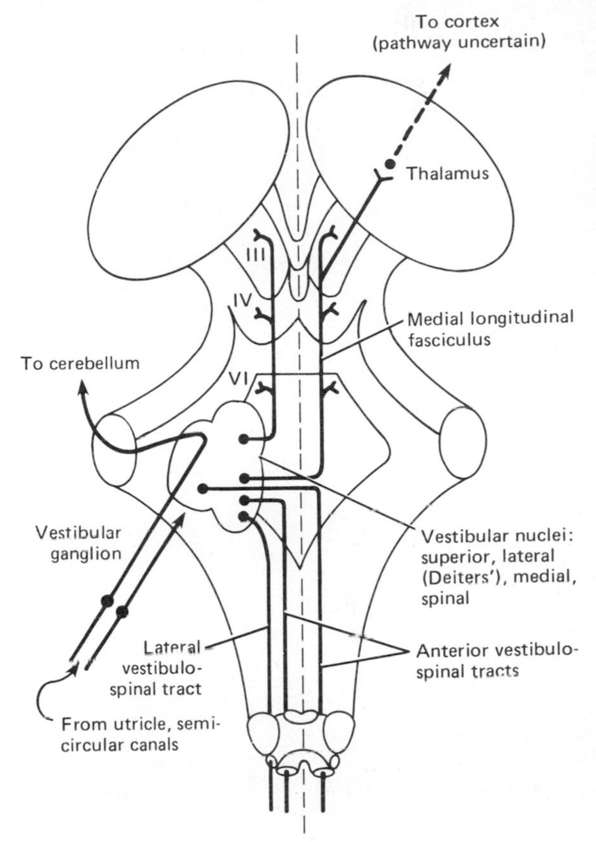

Figure 9–8. Principal vestibular pathways superimposed on a dorsal view of the brain stem. Cerebellum and cerebral cortex removed.

although it is not known exactly why. The perilymph is formed mainly from plasma. Entry of mannitol and sucrose from plasma into perilymph in the scala tympani is slower than entry into the perilymph in the scala vestibuli, and there are small differences in composition between these 2 fluids, but both resemble extracellular fluid (Fig 9–10). On the other hand, endolymph is formed by the stria vascularis and has a high concentration of K^+ and a low concentration of Na^+. Cells in the stria vascularis have a high concentration of Na^+-K^+ ATPase. In addition, it appears that there is a unique electrogenic K^+ pump in the stria vascularis, which accounts for the fact that the scala media is electrically positive relative to the scala vestibuli and scala tympani.

Current evidence indicates that displacement of the stereocilia toward the kinocilium increases their permeability to ions at their apices. K^+ enters the hair cell, producing a generator potential. This in turn increases the influx of Ca^{2+} through voltage-gated Ca^{2+} channels, and the increase in intracellular Ca^{2+} triggers release of a synaptic transmitter from the hair cells that depolarizes the adjacent afferent neuron. The identity of the transmitter has not been established, but there is solid evidence for its existence. Conversely, displacement of the stereocilia away

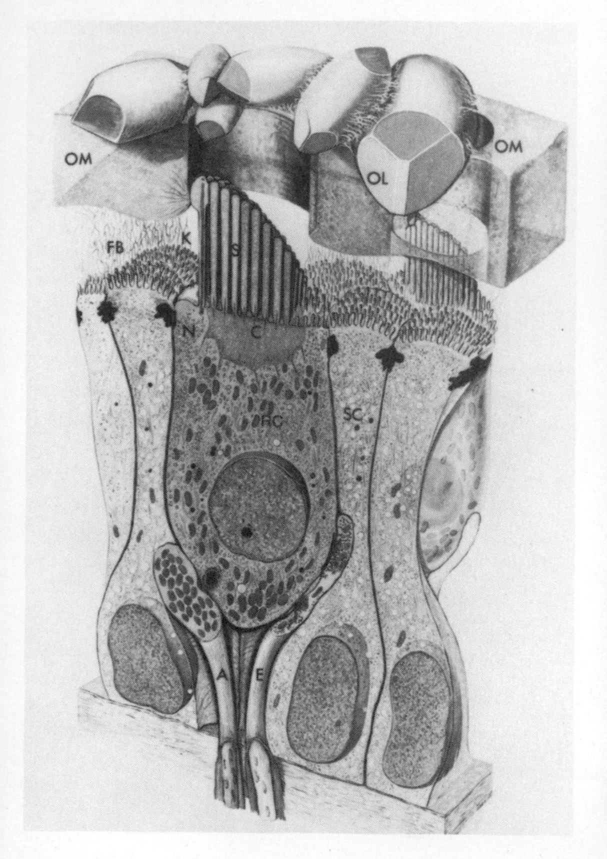

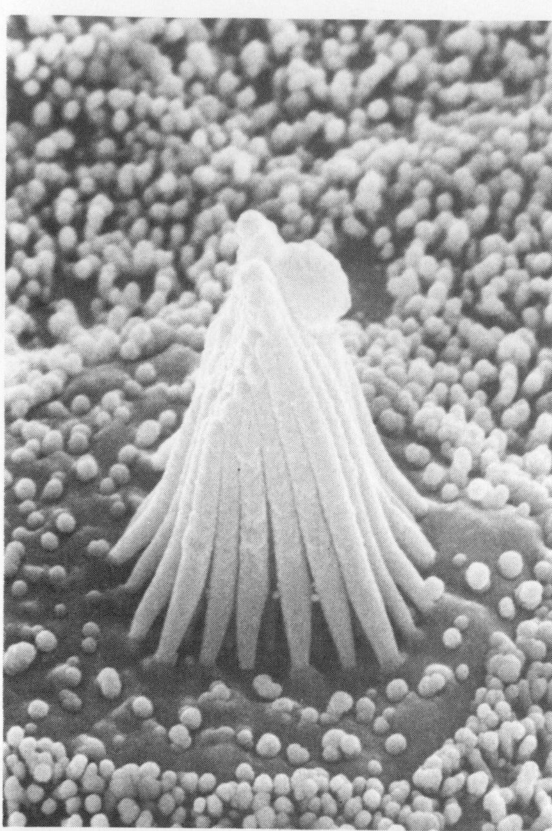

Figure 9–9. *Left:* Structure of a hair cell in the saccule of a frog, showing its relation to the otolithic membrane (OM). K, kinocilium; S, stereocilia; RC, hair cell with afferent (A) and efferent (E) nerve fibers; OL, otolith; SC, supporting cell. (Reproduced, with permission, from Hillman DE: Morphology of peripheral and central vestibular systems. In: Llinás R, Precht W [editors]: *Frog Neurobiology*. Springer-Verlag, 1976.) ***Right:*** Scanning electron photomicrograph of processes on a hair cell in the saccule of a frog. The otolithic membrane has been removed. The small projections around the hair cell are microvilli on supporting cells. (Courtesy of AJ Hudspeth.)

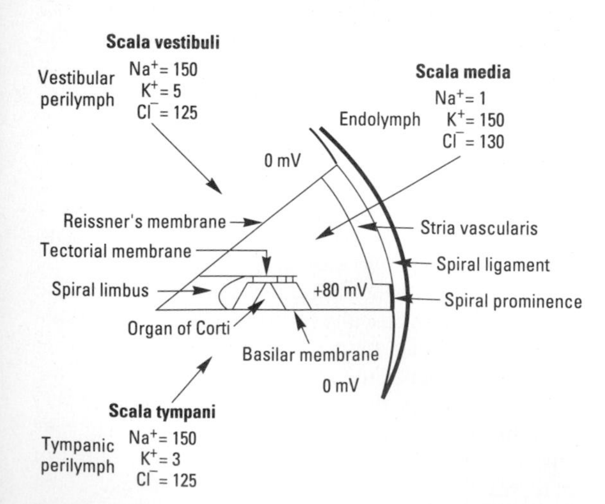

Figure 9–10. Composition of perilymph in the scala vestibuli, perilymph in the scala tympani, and endolymph. Values for Na^+, K^+, and Cl^- are in mmol/L. (Modified and reproduced, with permission, from Sterkers O, Ferrary E, Amiel C: How are inner ear fluids formed? *News Physiol Sci* 1987;2:176.)

from the kinocilium decreases their resting permeability. K^+ influx is decreased, hyperpolarizing the cell. The molecular basis of the direction-related change in permeability of the stereocilia is unknown.

HEARING

Sound Waves

Sound is the sensation produced when longitudinal vibrations of the molecules in the external environment, ie, alternate phases of condensation and rarefaction of the molecules, strike the tympanic membrane. A plot of these movements as changes in pressure on the tympanic membrane per unit of time is a series of waves (Fig 9–11), and such movements in the environment are generally called sound waves. The waves travel through air at a speed of approximately 344 m/s (770 mi/h) at 20 °C at sea level. The speed of sound increases with temperature and with altitude. Other media in which humans occasionally find themselves also conduct sound waves, but at

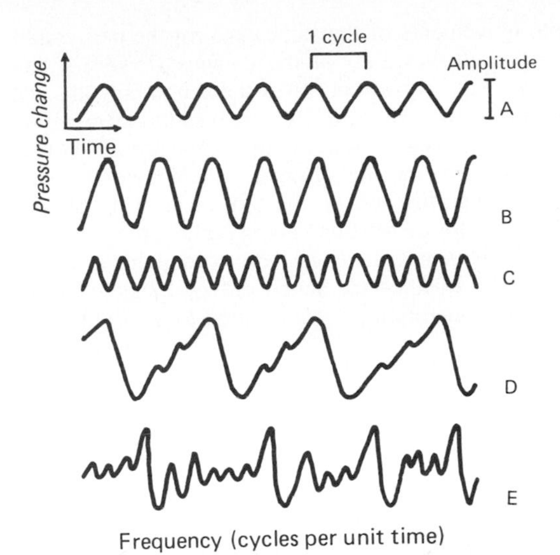

Figure 9–11. Characteristics of sound waves. *A* is the record of a pure tone. *B* has a greater amplitude and is louder than A. *C* has the same amplitude as A but a greater frequency, and its pitch is higher. *D* is a complex wave form that is regularly repeated. Such patterns are perceived as musical sounds, whereas waves like that shown in *E*, which have no regular pattern, are perceived as noise.

different speeds. For example, the speed of sound in fresh water is 1450 m/s at 20 °C and even greater in salt water.

Generally speaking, the **loudness** of a sound is correlated with the **amplitude** of a sound wave and its **pitch** with the **frequency** (number of waves per unit of time). The greater the amplitude, the louder the sound; and the greater the frequency, the higher the pitch. However, pitch is determined by other poorly understood factors in addition to frequency, and frequency affects loudness, since the auditory threshold is lower at some frequencies than others (see below). Sound waves that have repeating patterns, even though the individual waves are complex, are perceived as musical sounds; aperiodic nonrepeating vibrations cause a sensation of noise. Most musical sounds are made up of a wave with a primary frequency that determines the pitch of the sound plus a number of harmonic vibrations **(overtones)** that give the sound its characteristic **timbre** (quality). Variations in timbre permit us to identify the sounds of the various musical instruments even though they are playing notes of the same pitch.

The amplitude of a sound wave can be expressed in terms of the maximum pressure change or the root mean square pressure at the eardrum, but a relative scale is more convenient. The **decibel scale** is such a scale. The intensity of a sound in **bels** is the logarithm of the ratio of the intensities of that sound and a standard sound:

$$\text{bel} = \log \frac{\text{intensity of sound}}{\text{intensity of standard sound}}$$

The intensity is proportionate to the square of the sound pressure. Therefore,

$$\text{bel} = 2 \log \frac{\text{pressure of sound}}{\text{pressure of standard sound}}$$

A decibel is 0.1 bel. The standard sound reference level adopted by the Acoustical Society of America corresponds to 0 decibels at a pressure level of 0.000204 dyne/cm^2, a value that is just at the auditory threshold for the average human. In Fig 9–12, the decibel levels of various common sounds are compared. It is important to remember that the decibel scale is a log scale. Therefore, a value of 0 decibels does not mean the absence of sound but a sound level of an intensity equal to that of the standard. Furthermore, the 0–140 decibel range from threshold intensity to an intensity that is potentially damaging to the organ of Corti actually represents a 10^{14} (100 trillion)-fold variation in sound intensity.

The sound frequencies audible to humans range from about 20 to a maximum of 20,000 cycles per second (cps, Hz). In other animals, notably bats and dogs, much higher frequencies are audible. The threshold of the human ear varies with the pitch of the sound (Fig 9–13), the greatest sensitivity being in the 1000–4000 Hz range. The pitch of the average male voice in conversation is about 120 Hz and that of the average female voice about 250 Hz. The number of pitches that can be distinguished by an average individual is about 2000, but trained musicians can improve on this figure considerably. Pitch discrimination is best in the 1000–3000 Hz range and is poor at high and low pitches.

Masking

It is common knowledge that the presence of one sound decreases an individual's ability to hear other sounds. This phenomenon is known as **masking.** It is

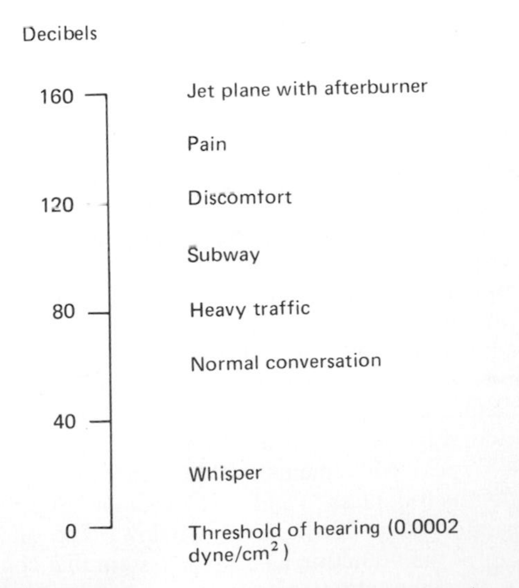

Figure 9–12. Decibel scale for common sounds.

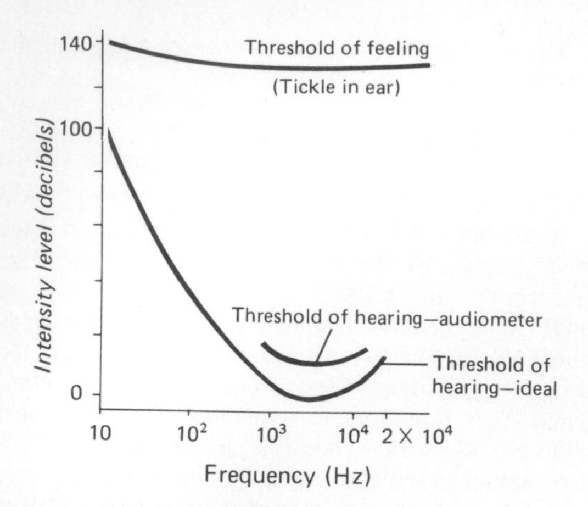

Figure 9–13. Human audibility curve. The middle curve is that obtained by audiometry under the usual conditions. The lower curve is that obtained under ideal conditions. At about 140 decibels (top curve), sounds are felt as well as heard.

believed to be due to the relative or absolute refractoriness of previously stimulated auditory receptors and nerve fibers to other stimuli. The degree to which a given tone masks other tones is related to its pitch. The masking effect of the background noise in all but the most carefully soundproofed environments raises the auditory threshold a definite and measurable amount.

Sound Transmission

The ear converts sound waves in the external environment into action potentials in the auditory nerves. The waves are transformed by the eardrum and auditory ossicles into movements of the foot plate of the stapes. These movements set up waves in the fluid of the inner ear. The action of the waves on the organ of Corti generates action potentials in the nerve fibers.

Functions of the Tympanic Membrane & Ossicles

In response to the pressure changes produced by sound waves on its external surface, the tympanic membrane moves in and out. The membrane therefore functions as a **resonator** that reproduces the vibrations of the sound source. It stops vibrating almost immediately when the sound wave stops, ie, it is very nearly **critically damped.** The motions of the tympanic membrane are imparted to the manubrium of the malleus. The malleus rocks on an axis through the junction of its long and short processes, so that the short process transmits the vibrations of the manubrium to the incus. The incus moves in such a way that the vibrations are transmitted to the head of the stapes. Movements of the head of the stapes swing its foot plate to and fro like a door hinged at the posterior edge of the oval window. The auditory ossicles thus function as a lever system that converts the resonant vibrations of the tympanic membrane into movements of the stapes against the perilymph-filled scala vestibuli of the cochlea (Figs 9–2 and 9–14). This system increases the sound pressure that arrives at the oval window, because the lever action of the malleus and incus multiplies the force 1.3 times and the area of the tympanic membrane is much greater than the area of the foot plate of the stapes. There are losses of sound energy owing to resistance, but it has been calculated that, at frequencies below 3000 Hz, 60% of the sound energy incident on the tympanic membrane is transmitted to the fluid in the cochlea.

Tympanic Reflex

When the middle ear muscles—the tensor tympani and the stapedius—contract, they pull the manubrium of the malleus inward and the foot plate of the stapes outward. This decreases sound transmission. Loud sounds initiate a reflex contraction of these muscles generally called the **tympanic reflex.** Its function is protective, preventing strong sound waves from causing excessive stimulation of the auditory receptors. However, the reaction time for the reflex is 40–160 ms, so it does not protect against brief intense stimulation such as that produced by gunshots.

Bone & Air Conduction

Conduction of sound waves to the fluid of the inner ear via the tympanic membrane and the auditory ossicles, the main pathway for normal hearing, is called **ossicular conduction.** Sound waves also initiate vibrations of the secondary tympanic membrane that closes the round window. This process, unimportant in normal hearing, is **air conduction.** A third type of conduction, **bone conduction,** is the transmission of vibrations of the bones of the skull to the fluid of the inner ear. Considerable bone conduction occurs when tuning forks or other vibrating bodies are applied directly to the skull. This route also plays a role in transmission of extremely loud sounds.

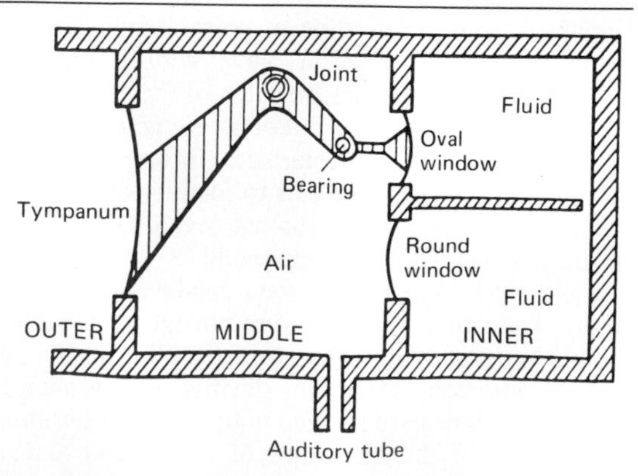

Figure 9–14. Diagrammatic representation of the transmission of vibrations from the outer to the inner ear. (Reproduced, with permission, from Lippold OCJ, Winton FR: *Human Physiology,* 6th ed. Churchill, 1972.)

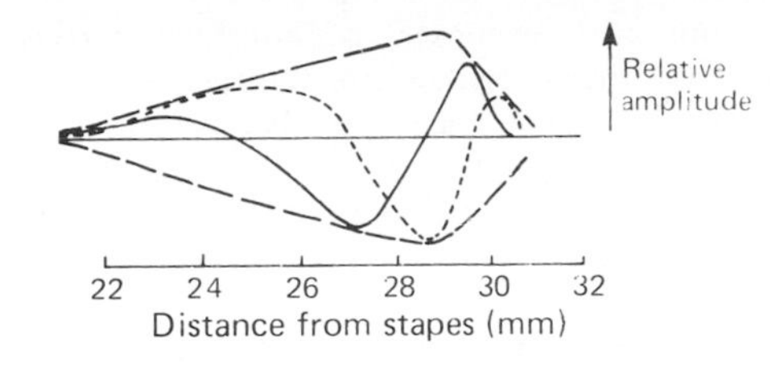

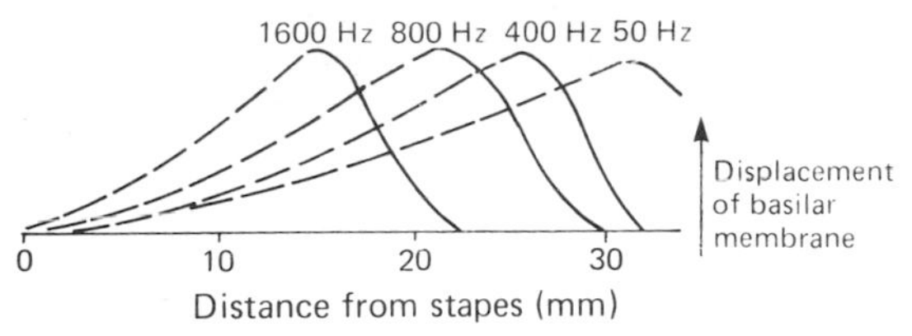

Figure 9–15. Traveling waves. *Top:* The solid and the short-dashed lines represent the wave at 2 instants of time. The long-dashed line shows the "envelope" of the wave formed by connecting the wave peaks at successive instants. *Bottom:* Displacement of the basilar membrane by the waves generated by stapes vibration of the frequencies shown at the top of each curve.

Traveling Waves

The movements of the foot plate of the stapes set up a series of traveling waves in the perilymph of the scala vestibuli. A diagram of such a wave is shown in Fig 9–15. As the wave moves up the cochlea, its height increases to a maximum and then drops off rapidly. The distance from the stapes to this point of maximum height varies with the frequency of the vibrations initiating the wave. High-pitched sounds generate waves that reach maximum height near the base of the cochlea; low-pitched sounds generate waves that peak near the apex. The bony walls of the scala vestibuli are rigid, but Reissner's membrane is flexible. The basilar membrane is not under tension, and it also is readily depressed into the scala tympani by the peaks of waves in the scala vestibuli. Displacements of the fluid in the scala tympani are dissipated into air at the round window. Therefore, sound produces distortion of the basilar membrane, and the site at which this distortion is maximal is determined by the frequency of the sound wave. The tops of the hair cells in the organ of Corti are held rigid by the reticular lamina, and the hairs of the outer hair cells are embedded in the tectorial membrane (Fig 9–4). When the stapes moves, both membranes move in the same direction, but they are hinged on different axes, so there is a shearing motion that bends the hairs. The hairs of the inner hair cells are probably not attached to the tectorial membrane, but they are apparently bent by fluid moving between the tectorial membrane and the underlying hair cells.

Functions of the Inner & Outer Hair Cells

The inner hair cells appear to be the primary sensory cells that generate action potentials in the auditory nerves, and presumably they are stimulated by the fluid movements noted above.

The outer hair cells, on the other hand, are innervated by cholinergic efferent fibers from the superior olivary complexes, and there is evidence that the outer hair cells are motile. It appears that these hair cells improve hearing by influencing the vibration patterns of the basilar membrane, but the exact way they alter the patterns is unknown.

Action Potentials in Auditory Nerve Fibers

The frequency of the action potentials in single auditory nerve fibers is proportionate to the loudness of the sound stimuli. At low sound intensities, each axon discharges to sounds of only one frequency, and this frequency varies from axon to axon depending upon the part of the cochlea from which the fiber originates. At higher sound intensities, the individual axons discharge to a wider spectrum of sound frequencies (Fig 9–16)—particularly to frequencies lower than that at which threshold simulation occurs. The "response area" of each unit, the area above the line that defines its threshold at various frequencies, resembles the shape of the traveling wave in the cochlea.

The major determinant of the pitch perceived when a sound wave strikes the ear is the place in the organ of Corti that is maximally stimulated. The traveling wave set up by a tone produces peak depression of the

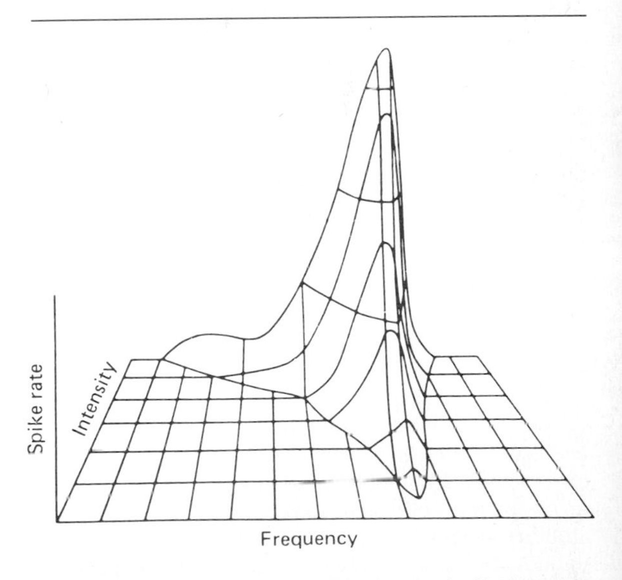

Figure 9–16. Relation of discharge rate (spike rate) in auditory nerve fiber to frequency and intensity of sound. Because the diagram represents the type of response seen in many different individual fibers, no numbers are given on the scales. (Modified and reproduced, with permission, from Kiang NYS: Peripheral neural processing of auditory information. In: *The Nervous System,* Vol 3. Part 2. *Handbook of Physiology.* Brookhart JM, Mountcastle VB [editors]. American Physiological Society, 1984.)

basilar membrane, and consequently maximal receptor stimulation, at one point. The distance between this point and the stapes is inversely related to the pitch of the sound, low tones producing maximal stimulation at the apex of the cochlea and high tones producing maximal stimulation at the base. The pathways from the various parts of the cochlea to the brain are distinct. An additional factor involved in pitch perception at sound frequencies of less than 2000 Hz may be the pattern of the action potentials in the auditory nerve. When the frequency is low enough, the nerve fibers begin to respond with an impulse to each cycle of a sound wave. The importance of this **volley effect,** however, is limited; the frequency of the action potentials in a given auditory nerve fiber determines principally the loudness, rather than the pitch, of a sound.

Although the pitch of a sound depends primarily on the frequency of the sound wave, loudness also plays a part; low tones (below 500 Hz) seem lower and high tones (above 4000 Hz) seem higher as their loudness increases. Duration also affects pitch to a minor degree. The pitch of a tone cannot be perceived unless it lasts for more than 0.01 second, and with durations between 0.01 and 0.1 second, pitch rises as duration increases.

Auditory Responses of Neurons in the Medulla Oblongata

The responses of individual second-order neurons in the cochlear nuclei to sound stimuli are like those of the individual auditory nerve fibers. The frequency at which sounds of the lowest intensity evoke a response varies from unit to unit; with increased sound intensities, the band of frequencies to which a response occurs becomes wider. The major difference between the responses of the first- and second-order neurons is the presence of a sharper "cutoff" on the low-frequency side in the medullary neurons. This greater specificity of the second-order neurons is probably due to some sort of inhibitory process in the brain stem, but how it is achieved is not known.

Auditory Cortex

The pathways from the cochlea to the auditory cortex are described in the first section of this chapter. Impulses ascend from the dorsal and ventral cochlear nuclei through complex paths that are both crossed and uncrossed. In animals, there is an organized pattern of tonal localization in the primary auditory cortex, as if the cochlea had been unrolled upon it. In humans, low tones are represented anterolaterally and high tones posteromedially in the auditory cortex. Individual neurons in the auditory cortex respond to such parameters as the onset, duration, and repetition rate of an auditory stimulus and particularly to the direction from which it came. In this way, they are analogous to some of the neurons in the visual cortex (see Chapter 8). In laboratory mammals, destruction of the auditory cortex not only fails to cause deafness but also fails to obliterate established conditioned responses to sound of a given frequency. On the other hand, responses to a given sequence of 3 or more tones are absent. Thus, the auditory cortex is concerned with recognition of tonal patterns, with analysis of properties of sounds, and with sound localization.

Sound Localization

Determination of the direction from which a sound emanated depends upon detecting the difference in time between the arrival of the stimulus in the 2 ears and the consequent difference in phase of the sound waves on the 2 sides; it also depends upon the fact that the sound is louder on the side closest to the source. The time difference is said to be the most important factor at frequencies below 3000 Hz and the loudness difference the most important at frequencies above 3000 Hz. Many neurons in the auditory cortex receive input from both ears, and they respond maximally or minimally when the time of arrival of a stimulus at one ear is delayed by a fixed period relative to the time of arrival at the other ear. This fixed period varies from neuron to neuron. Sounds coming from directly in front of the individual presumably differ in quality from those coming from behind, because the external ears are turned slightly forward. Sound localization is markedly disrupted by lesions of the auditory cortex in laboratory mammals and humans.

Deafness

Clinical deafness may be due to impaired sound transmission in the external or middle ear **(conduction deafness)** or to damage to the hair cells or neural pathways **(nerve deafness).** Among the causes of conduction deafness are plugging of the external auditory canals with wax or foreign bodies, destruction of the auditory ossicles, thickening of the eardrum following repeated middle ear infections, and abnormal rigidity of the attachments of the stapes to the oval window. One cause of nerve deafness is toxic degeneration of the hair cells produced by streptomycin and gentamicin, which are concentrated in the endolymph. Damage to the outer hair cells by antibiotics or prolonged exposure to noise is associated with hearing loss. Other causes include tumors of the vestibulocochlear nerve and cerebellopontine angle, and vascular damage in the medulla.

Conduction and nerve deafness can be distinguished by a number of simple tests with a tuning fork. Three of these tests, named for the men who developed them, are outlined in Table 9–1. The Weber and Schwabach tests demonstrate the important masking effect of environmental noise on the auditory threshold.

Audiometry

Auditory acuity is commonly measured with an **audiometer.** This device presents the subject with pure tones of various frequencies through earphones. At each frequency, the threshold intensity is deter-

Table 9–1. Common tests with a tuning fork to distinguish between nerve and conduction deafness.

	Weber	Rinne	Schwabach
Method	Base of vibrating tuning fork placed on vertex of skull.	Base of vibrating tuning fork placed on mastoid process until subject no longer hears it, then held in air next to ear.	Bone conduction of patient compared with that of normal subject.
Normal	Hears equally on both sides.	Hears vibration in air after bone conduction is over.	
Conduction deafness (one ear)	Sound louder in diseased ear because masking effect of environmental noise is absent on diseased side.	Vibrations in air not heard after bone conduction is over.	Bone conduction better than normal (conduction defect excludes masking noise).
Nerve deafness (one ear)	Sound louder in normal ear.	Vibration heard in air after bone conduction is over.	Bone conduction less than normal.

mined and plotted on a graph as a percentage of normal hearing. This provides an objective measurement of the degree of deafness and a picture of the tonal range most affected.

Fenestration Procedures

A common form of conduction deafness is that due to **otosclerosis,** a disease in which the attachments of the foot plate of the stapes to the oval window become abnormally rigid. In patients with this disease, air conduction (see above) can be utilized to bring about some restoration of hearing. A membrane-covered outlet from the bony labyrinth is created so that waves set up by vibrations of the secondary tympanic membrane can be dissipated. In the "fenestration operation," an outlet is created by drilling a hole in the horizontal semicircular canal and covering it with skin.

VESTIBULAR FUNCTION

Responses to Rotational Acceleration

Rotational acceleration in the plane of a given semicircular canal stimulates its crista. The endolymph, because of its inertia, is displaced in a direction opposite to the direction of rotation. The fluid pushes on the cupula, deforming it. This bends the processes of the hair cells (Fig 9–6). When a constant speed of rotation is reached, the fluid spins at the same rate as the body and the cupula swings back into the upright position. When rotation is stopped, deceleration produces displacement of the endolymph in the direction of the rotation, and the cupula is deformed in a direction opposite to that during acceleration. It returns to midposition in 25–30 seconds. Movement of the cupula in one direction commonly causes increased impulse traffic in single nerve fibers from its crista, whereas movement in the opposite direction commonly inhibits neural activity (Fig 9–17).

Rotation causes maximal stimulation of the semicircular canals most nearly in the plane of rotation. Since the canals on one side of the head are a mirror image of those on the other side, the endolymph is displaced toward the ampulla on one side and away from it on the other. The pattern of stimulation reaching the brain therefore varies with the direction as well as the plane of rotation. Linear acceleration probably fails to displace the cupula and therefore does not stimulate the cristae. However, there is considerable evidence that when one part of the labyrinth is destroyed other parts take over its functions. Experimental localization of labyrinthine functions is therefore difficult.

The tracts that descend from the vestibular nuclei into the spinal cord are concerned primarily with postural adjustments (see Chapter 12); the ascending connections to cranial nerve nuclei are largely concerned with eye movements.

Nystagmus

The characteristic jerky movement of the eye observed at the start and end of a period of rotation is called **nystagmus.** It is actually a reflex that main-

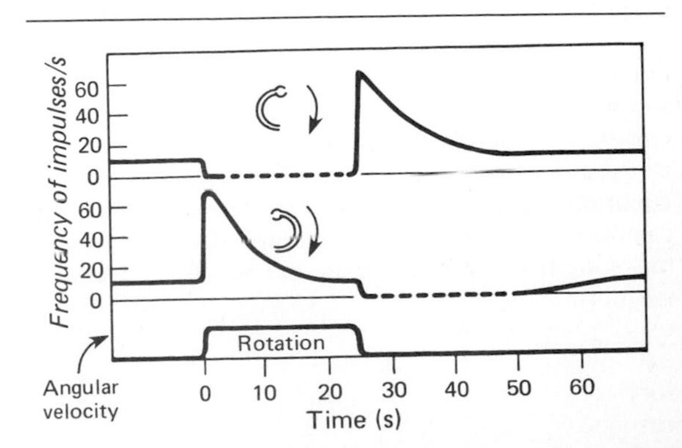

Figure 9–17. Ampullary responses to rotation. Average time course of impulse discharge from ampulla of 2 semicircular canals during rotational acceleration, steady rotation, and deceleration. (Reproduced, with permission, from Adrian ED: Discharges from vestibular receptors in the cat. *J Physiol [Lond]* 1943;**101**:389.)

tains visual fixation on stationary points while the body rotates, although it is not initiated by visual impulses and is present in blind individuals. When rotation starts, the eyes move slowly in a direction opposite to the direction of rotation, maintaining visual fixation **(vestibulo-ocular reflex, VOR).** When the limit of this movement is reached, the eyes quickly snap back to a new fixation point and then again move slowly in the other direction. The slow component is initiated by impulses from the labyrinths; the quick component is triggered by a center in the brain stem. Nystagmus is frequently horizontal, ie, the eyes move in the horizontal plane; but it can also be vertical, when the head is tipped sidewise during rotation, or rotatory, when the head is tipped forward. By convention, the direction of eye movement in nystagmus is identified by the direction of the quick component. The direction of the quick component during rotation is the same as that of the rotation, but the **postrotatory nystagmus** that occurs owing to displacement of the cupula when rotation is stopped is in the opposite direction.

Responses to Linear Acceleration

In mammals, the utricular and saccular maculas respond to linear acceleration. In general, the utricle responds to horizontal acceleration and the saccule to vertical acceleration. The otoliths are more dense than the endolymph, and acceleration in any direction causes them to be displaced in the opposite direction, distorting the hair cell processes and generating activity in the nerve fibers. The maculas also discharge tonically in the absence of head movement, because of the pull of gravity on the otoliths. The impulses generated from these receptors are partly responsible for reflex righting of the head and other important postural adjustments discussed in Chapter 12.

Although most of the responses to stimulation of the maculas are reflex in nature, vestibular impulses also reach the cerebral cortex. These impulses are presumably responsible for conscious perception of motion and supply part of the information necessary for orientation in space. The nausea, blood pressure changes, sweating, pallor, and vomiting that are the well-known accompaniments of excessive vestibular stimulation are probably due to reflexes mediated via vestibular connections in the brain stem. **Vertigo** is the sensation of rotation in the absence of actual rotation.

Caloric Stimulation

The semicircular canals can be stimulated by instilling water that is hotter or colder than body temperature into the external auditory meatus. The temperature difference sets up convection currents in the endolymph, with consequent motion of the cupula. This technique of **caloric stimulation,** which is sometimes used diagnostically, causes some nystagmus, vertigo, and nausea. To avoid these symptoms when irrigating the ear canals in the treatment of ear infections, it is important to be sure that the fluid used is at body temperature.

Orientation in Space

Orientation in space depends in large part upon input from the vestibular receptors, but visual cues are also important. Pertinent information is also supplied by impulses from proprioceptors in joint capsules, which supply data about the relative position of the various parts of the body, and impulses from cutaneous exteroceptors, especially touch and pressure receptors. These 4 inputs are synthesized at a cortical level into a continuous picture of the individual's orientation in space.

Effects of Labyrinthectomy

The effects of unilateral labyrinthectomy are presumably due to unbalanced discharge from the remaining normal side. They vary from species to species and with the speed at which the labyrinth is destroyed. Rats develop abnormal body postures and roll over and over as they continuously attempt to right themselves. Postural changes are less marked in humans, but the symptoms are distressing. Motion aggravates the symptoms, but efforts to minimize stimulation by lying absolutely still are constantly thwarted by waves of nausea, vomiting, and even diarrhea. Fortunately, compensation for the loss occurs, and after 1–2 months the symptoms disappear completely.

Symptoms of this type are absent after bilateral destruction of the labyrinths, but defects in orientation are present. For this reason, diving is hazardous for persons who lack vestibular function. The only means these individuals have of finding the surface is vision, because in the water the pressure sensed by the cutaneous exteroceptors is equal all over the body. If for any reason vision is obscured, a person whose vestibular apparatus is impaired has no mechanism for self-orientation. It is therefore possible to swim away from rather than toward the surface and, consequently, to drown.

Smell & Taste

10

INTRODUCTION

Smell and taste are generally classified as visceral senses because of their close association with gastrointestinal function. Physiologically, they are related to each other. The flavors of various foods are in large part a combination of their taste and smell. Consequently, foods may taste "different" if one has a cold that depresses the sense of smell. Both taste and smell receptors are chemoreceptors that are stimulated by molecules in solution in mucus in the case of the nose and saliva in the case of the mouth. However, these 2 senses are anatomically quite different. The smell receptors are distance receptors (teleceptors); the smell pathways have no relay in the thalamus; and there is no neocortical projection area for olfaction. The taste pathways pass up the brain stem to the thalamus and project to the postcentral gyrus along with those for touch and pressure sensibility from the mouth.

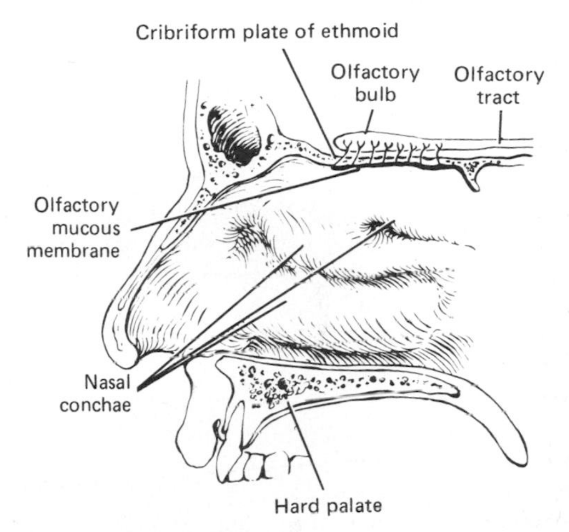

Figure 10–1. Olfactory mucous membrane. (Reproduced, with permission, from deGroot J, Chusid JG: *Correlative Neuroanatomy,* 20th ed. Appleton & Lange, 1988.)

SMELL

RECEPTORS & PATHWAYS

Olfactory Mucous Membrane

The olfactory receptors are located in a specialized portion of the nasal mucosa, the yellowish-pigmented **olfactory mucous membrane.** In dogs and other animals in which the sense of smell is highly developed (macrosmatic animals), the area covered by this membrane is large; in microsmatic animals such as humans, it is small, covering an area of 5 cm^2 in the roof of the nasal cavity near the septum (Fig 10–1). It contains supporting cells and neuroblastlike progenitor cells that form the olfactory receptor neurons. Interspersed between these cells are 10–20 million receptor cells. Each olfactory receptor is a neuron, and the olfactory mucous membrane is said to be the place in the body where the nervous system is closest to the external world. The neurons have short, thick dendrites with expanded ends called olfactory rods (Fig 10–2). From these rods, cilia project to the surface of the mucus. The cilia are unmyelinated processes about 2 μm long and 0.1 μm in diameter. There are 10–20 cilia per receptor neuron. The axons of the olfactory receptor neurons pierce the cribriform plate of the ethmoid bone and enter the olfactory bulbs.

The olfactory mucous membrane is constantly covered by mucus. This mucus is produced by Bowman's glands, which are just under the basal lamina of the membrane, and there is some evidence that the mucus contains proteins that aid the transport of odoriferous molecules to the olfactory bulbs.

Olfactory Bulbs

In the olfactory bulbs, the axons of the receptors terminate among the dendrites of the **mitral cells** (Fig 10–3) to form the complex globular synapses called **olfactory glomeruli.** An average of 26,000 receptor cell axons converge on each glomerulus. The tufted cells and the periglomerular short axon cells participate in the formation of glomeruli. In the next layer,

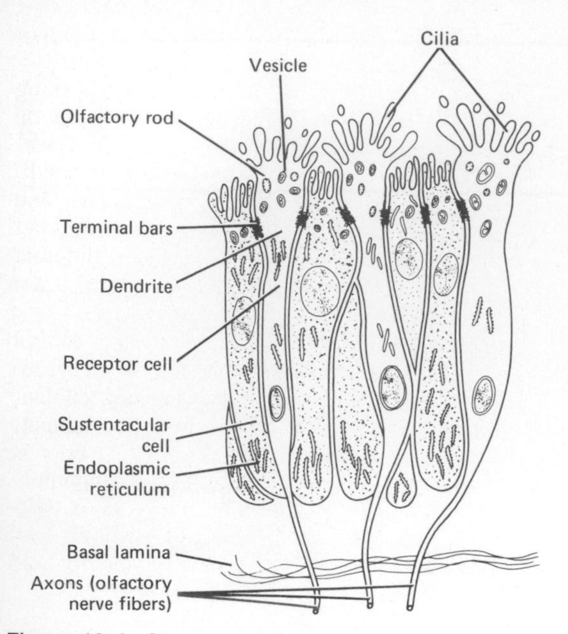

Figure 10–2. Structure of the olfactory mucous membrane.

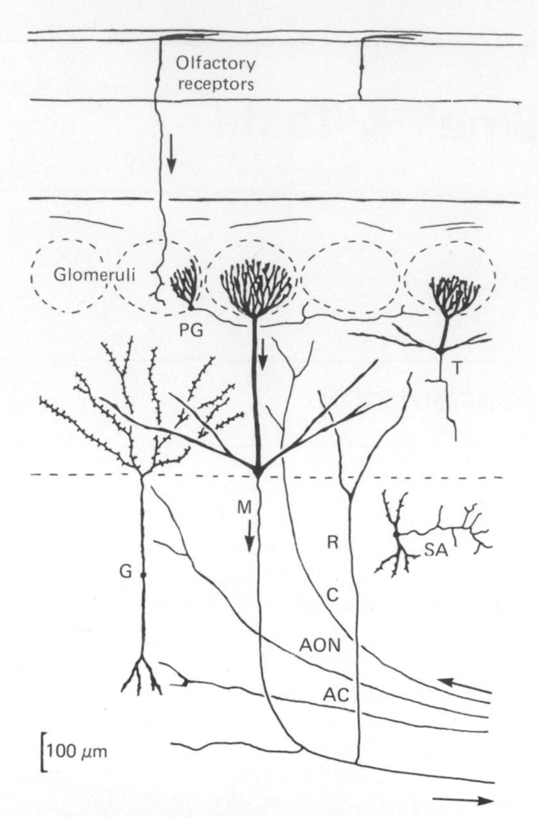

Figure 10–3. Neural elements in the olfactory bulb. In addition to olfactory nerve fibers, which terminate in the olfactory glomeruli, inputs include centrifugal fibers (C) from the nucleus of the diagonal band; ipsilateral fibers from the anterior olfactory nucleus (AON); and contralateral fibers from the anterior commissure (AC). Neurons include mitral cells (M) with recurrent collaterals (R), tufted cells (T), granule cells (G), periglomerular short-axon cells (PG), and deep short-axon cells (SA). (Reproduced, with permission, from Shepherd GM: *The Synaptic Organization of the Brain.* Oxford Univ Press, 1974.)

the dendrites of the mitral and granule cells form extensive reciprocal synapses. The axons of the mitral cells pass posteriorly through the **intermediate olfactory stria** and **lateral olfactory stria** to the **olfactory cortex.** The olfactory cortex includes the anterior olfactory nucleus, prepyriform cortex, olfactory tubercle, corticomedial amygdala, and transitional entorhinal cortex. These structures are part of the limbic system (see Chapter 15). The mitral cell axons terminate on the apical dendrites of pyramidal cells in the olfactory cortex. Evoked potentials have been recorded in the amygdaloid nuclei in response to olfactory stimuli in humans.

There are 3 inputs from other parts of the brain into the olfactory bulb in addition to the extrinsic input from the olfactory mucous membrane via the olfactory nerves. The terminations of these efferent fibers are shown in Fig 10–3. One of the central inputs arises from the nucleus of the horizontal limb of the diagonal band (centrifugal fibers). Another arises from the ipsilateral anterior olfactory nucleus just posterior to the bulb. Stimulation of efferent fibers in the olfactory striae decreases the electrical activity of the olfactory bulbs. Presumably, therefore, inhibitory mechanisms analogous to those in the eye, ear, and other sensory systems exist in the olfactory system. The third efferent input arises from the contralateral anterior olfactory nucleus, reaching the olfactory bulb via the anterior commissure. The functional significance of this peculiar fiber arrangement is not known. However, there is evidence that the olfactory commissural pathways are involved in the transfer of olfactory memories from one side to the other and that memories formed on one side before the pathways develop can be transferred to the other side after they mature.

PHYSIOLOGY OF OLFACTION

Stimulation of Receptors

Olfactory receptors respond only to substances that are in contact with the olfactory epithelium and are dissolved in the thin layer of mucus that covers it. The olfactory thresholds for the respresentative substances shown in Table 10–1 illustrate the remarkable sensitivity of the olfactory receptors to some substances. For example, methyl mercaptan, the substance that gives garlic its characteristic odor, can be smelled at a concentration of less than 500 pg/L of air. On the other hand, discrimination of differences in the intensity of any given odor is poor. The

Table 10–1. Some olfactory thresholds.*

Substance	mg/L of Air
Ethyl ether	5.83
Chloroform	3.30
Pyridine	0.03
Oil of peppermint	0.02
Iodoform	0.02
Butyric acid	0.009
Propyl mercaptan	0.006
Artificial musk	0.00004
Methyl mercaptan	0.0000004

* Data from Allison VC, Katz SH: *J Ind Chem* 1919;**11**:336.

concentration of an odor-producing substance must be changed by about 30% before a difference can be detected. The comparable visual discrimination threshold is a 1% change in light intensity.

Odoriferous molecules bind to receptors on the cilia of olfactory receptor neurons. The activated receptors then activate adenylate cyclase via G_s, producing an increase in intracellular cyclic AMP (Fig 10–4). The cyclic AMP binds to and opens Na^+ channels in a fashion that is analogous to the opening of channels by cyclic GMP in the rods (see Chapter 8). The resultant influx of Na^+ produces a receptor potential. The receptor potential depolarizes the initial segment of the axon to the firing level, opening voltage-gated channels in this area and initiating transmitted impulses.

Odor-producing molecules are generally small, containing from 3–4 to 18–20 carbon atoms, and molecules with the same number of carbon atoms but different structural configurations have different odors. Relatively high water and lipid solubility are characteristic of substances with strong odors.

There is a pronounced degree of inhibitory control within the olfactory pathways. The reciprocal synaptic connections between mitral and granule cell dendrites mediate inhibitory control of mitral cell output. In the olfactory cortex, the response to an odor is excitation of pyramidal cells followed by inhibition. The pyramidal cells are then subject to self-reexcitation via long axon collaterals, and this may explain the propensity for rhythmic activity and seizures in the olfactory cortex.

Discrimination of Different Odors

Humans can distinguish between 2000 and 4000 different odors. The physiologic basis of this olfac-

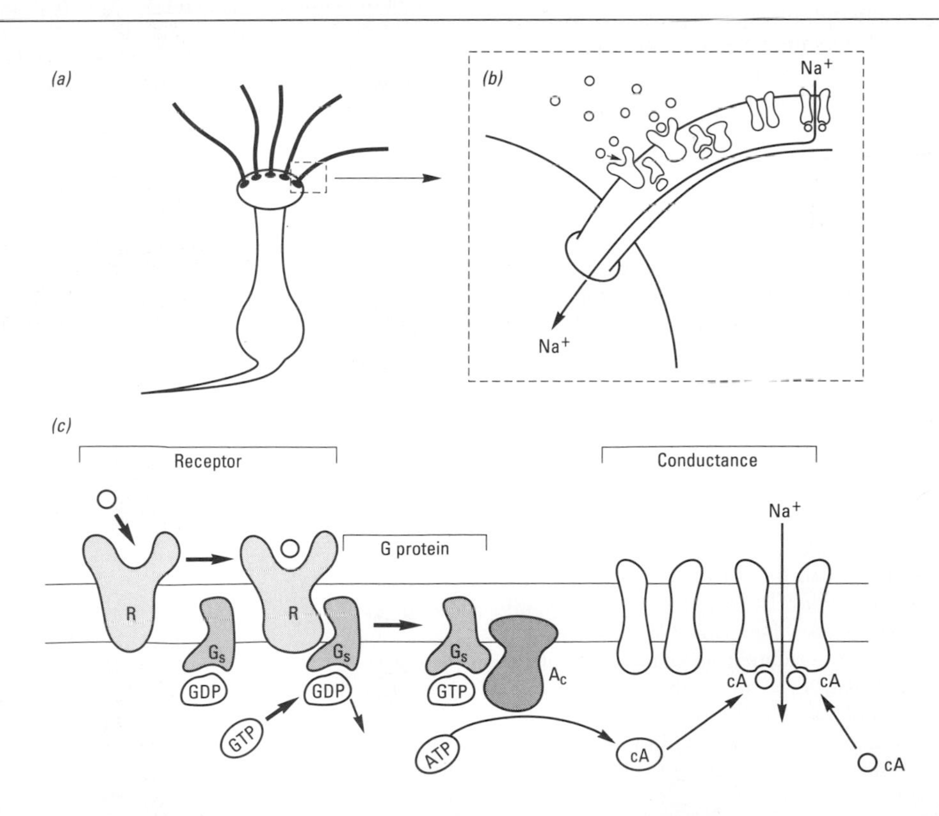

Figure 10–4. Signal transduction in olfactory mucosa. Odoriferous molecules bind to receptors on cilia of olfactory receptors *(a),* producing an influx of Na^+ *(b).* The molecular events that trigger the influx are shown in *(c).* An odor-producing molecule binds to the receptor (R). It in turn activates G_s, and this activates adenylate cyclase (A_c). The cyclic AMP (cA) that is formed binds to and opens the Na^+ channel. (Reproduced, with permission, from Anonymous: Another cyclic-nucleotide-gated conductance. *Nature* 1987;**325**:389.)

tory discrimination is not known. However, different odors produce different spatial patterns of increased metabolic activity in the olfactory bulb, as determined by the 2-deoxyglucose method (see Chapter 32), and it may be that each particular odor depends on the spatial pattern of stimulation of receptors in the olfactory mucous membrane. Different odors also produce different patterns of increased metabolic activity in the olfactory cortex. The direction from which a smell comes appears to be indicated by the slight difference in the time of arrival of odoriferous molecules in the 2 nostrils.

There is a close relationship between smell and sexual function in many species of animals, and the perfume ads are ample evidence that a similar relationship exists in humans. The sense of smell is said to be more acute in women than in men, and in women it is most acute at the time of ovulation.

Sniffing

The portion of the nasal cavity containing the olfactory receptors is poorly ventilated. Most of the air normally moves quietly through the lower part of the nose with each respiratory cycle, although eddy currents pass some air over the olfactory mucous membrane. These eddy currents are probably set up by convection as cool air strikes the warm mucosal surfaces. The amount of air reaching this region is greatly increased by sniffing, an action that includes contraction of the lower part of the nares on the septum to help deflect the airstream upward. Sniffing is a semireflex response that usually occurs when a new odor attracts attention.

Role of Pain Fibers in the Nose

Naked endings of many trigeminal pain fibers are found in the olfactory mucous membrane. They are stimulated by irritating substances, and an irritative, trigeminally mediated component is part of the characteristic ''odor'' of such substances as peppermint, menthol, and chlorine. These endings are also responsible for initiating sneezing, lacrimation, respiratory inhibition, and other reflex responses to nasal irritants.

Adaptation

It is common knowledge that when one is continuously exposed to even the most disagreeable odor, perception of the odor decreases and eventually ceases. This sometimes beneficent phenomenon is due to the fairly rapid adaptation that occurs in the olfactory system. It is specific for the particular odor being smelled, and the threshold for other odors is unchanged. Olfactory adaptation is in part a central phenomenon, but it is also due to a change in the receptors.

Abnormalities

Abnormalities of olfaction include **anosmia** (absense of the sense of smell), **hyposmia** (diminished olfactory sensitivity), and **dysosmia** (distorted sense of smell). Olfactory thresholds increase with advancing age, and more than 75% of humans over the age of 80 have an impaired ability to identify smells.

TASTE

RECEPTOR ORGANS & PATHWAYS

Taste Buds

The taste buds, the sense organs for taste, are ovoid bodies measuring 50–70 μm. Each taste bud is made up of 4 types of cells (Fig 10–5): basal cells; type I cells, which are granulated and are probably sustentacular cells; type II cells, of unknown function; and type III cells, which are the gustatory receptor cells that make synaptic connections to sensory nerve fibers. The type I, II, and III cells have microvilli, which project into the taste pore, an opening in the lingual epithelium. The necks of all these cells are connected to each other and to the surrounding epithelial cells by tight junctions, so that the only part of the gustatory receptor exposed to the fluids in the oral cavity is its apical crown of microvilli. Each taste bud is innervated by about 50 nerve fibers, and conversely, each nerve fiber receives input from an

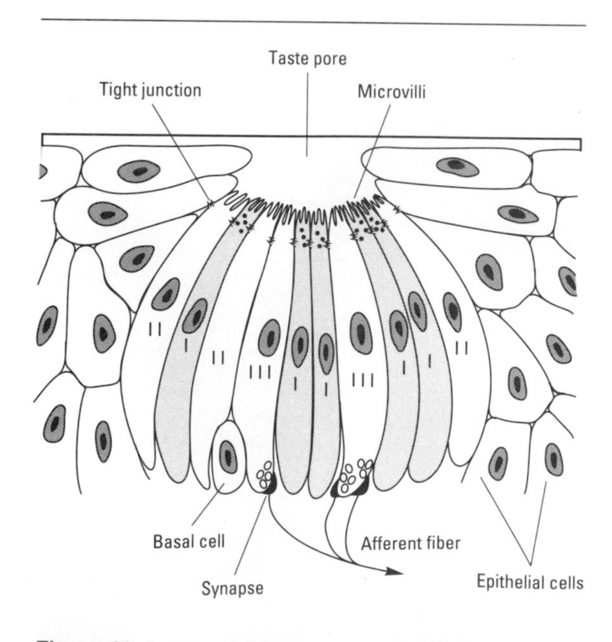

Figure 10–5. Taste bud, showing type I, II, and III cells, taste pore, and synaptic connections of type III cells to the sensory nerves that mediate taste. (Modified from Junqueira LC, Carneiro J, Kelley RO: *Basic Histology*, 6th ed., Appleton & Lange, 1989.)

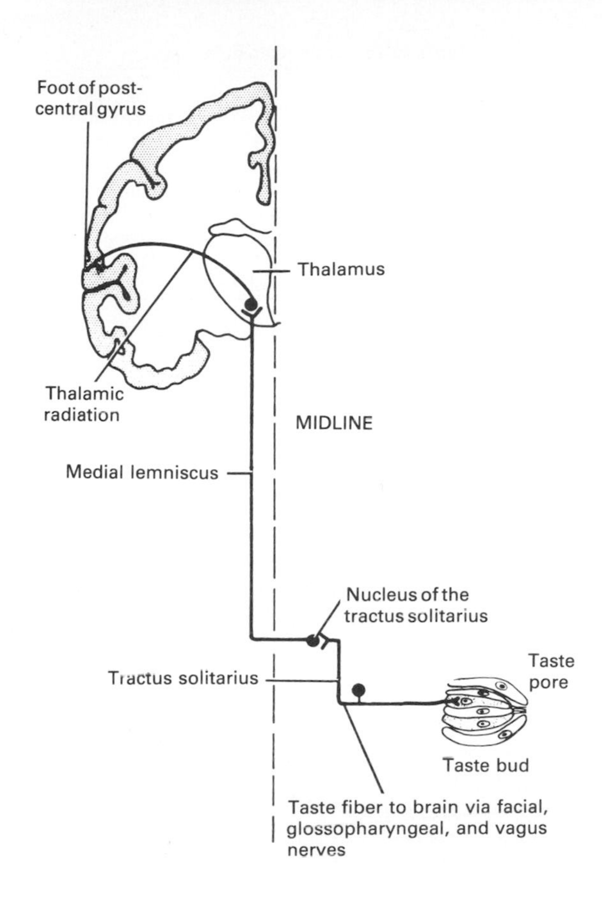

Figure 10–6. Diagram of taste pathways.

average of 5 taste buds. If the sensory nerve is cut, the taste buds it innervates degenerate and eventually disappear. However, if the nerve regenerates, the cells in the neighborhood become organized into new taste buds, presumably as a result of some sort of chemical inductive effect from the regenerating fiber.

In humans, the taste buds are located in the mucosa of the epiglottis, palate, and pharynx and in the walls of the **fungiform** and **vallate papillae** of the tongue. The fungiform papillae are rounded structures most numerous near the tip of the tongue; the vallate papillae are prominent structures arranged in a V on the back of the tongue. There are up to 5 taste buds per fungiform papilla, and they are usually located at the top of the papilla. The larger vallate papillae each contain up to 100 taste buds, usually located along the sides of the papillae. The small conical **filiform papillae** that cover the dorsum of the tongue do not usually contain taste buds. There are a total of about 10,000 taste buds.

Taste Pathways

The sensory nerve fibers from the taste buds on the anterior two-thirds of the tongue travel in the chorda tympani branch of the facial nerve, and those from the posterior third of the tongue reach the brain stem via the glossopharyngeal nerve. The fibers from areas other than the tongue reach the brain stem via the vagus nerve. On each side, the myelinated but relatively slow-conducting taste fibers in these 3 nerves unite in the medulla oblongata to enter the **nucleus of the tractus solitarius** (Fig 10–6). There they synapse on second-order neurons, the axons of which cross the midline and join the medial lemniscus, ending with the fibers for touch, pain, and temperature sensibility in the specific sensory relay nuclei of the thalamus. Impulses are relayed from there to the taste projection area in the cerebral cortex at the foot of the postcentral gyrus. Taste does not have a separate cortical projection area but is represented in the portion of the postcentral gyrus that subserves cutaneous sensation from the face.

PHYSIOLOGY OF TASTE

Basic Taste Modalities

In humans there are 4 basic tastes: sweet, sour, bitter, and salt. Bitter substances are tasted on the back of the tongue, sour along the edges, sweet at the tip, and salt on the dorsum anteriorly (Fig 10–7). Sour and bitter substances are also tasted on the palate along with some sensitivity to sweet and salt. All 4 modalities can be sensed on the pharynx and epiglottis. The taste buds are not histologically different in the different areas, but the existence of physiologic differences has been demonstrated by recording the electrical activity of nerve fibers from single taste buds in animals. These studies show that some taste buds respond only to bitter stimuli whereas others respond only to salt, sweet, or sour stimuli. Some respond to more than one modality, and some respond to all 4.

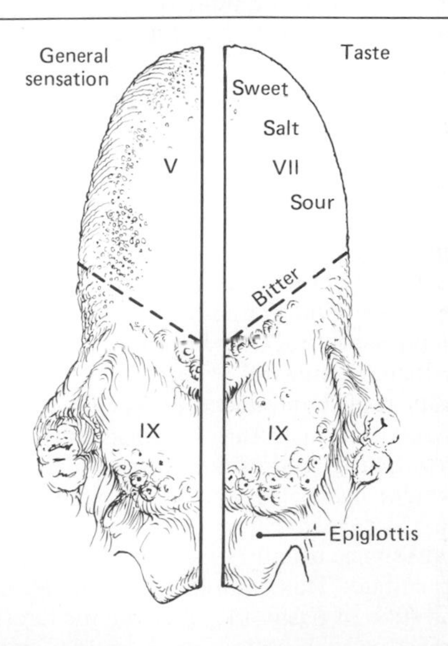

Figure 10–7. Sensory innervation of the tongue. The numbers refer to cranial nerves.

Receptor Stimulation

The gustatory receptor cells are chemoreceptors that respond to substances dissolved in the oral fluids bathing them. These substances act on the exposed microvilli in the taste pore to evoke generator potentials in the receptor cells, which generate action potentials in the sensory neurons. The way the molecules in solution produce generator potentials is unsettled and appears to vary from one gustatory modality to another. Salt stimuli probably depolarize salt receptor cells by influx of Na^+ through passive, ungated channels, because application of the sodium channel-blocking diuretic amiloride (see Chapter 38) directly to the tongue in humans abolishes the ability to taste salt. Acids, which taste sour, probably depolarize receptor cells by H^+ blocking of K^+ channels. Substances that taste bitter probably bind to membrane receptors and, via G_p (see Chapter 1), activate phospholipase C, with a resulting increase in intracellular IP_3 and release of Ca^{2+} from the endoplasmic reticulum. Substances that taste sweet appear to bind to membrane receptors and, via G_s, activate adenylate cyclase, with a resulting increase in intracellular cyclic AMP. The cyclic AMP acts via protein kinase A to reduce K^+ conductance by phosphorylating a K^+ channel. Cyclic nucleotides also affect conductances in visual and olfactory receptors, but in these cases, they bind directly to channels instead of phosphorylating them.

Taste Thresholds & Intensity Discriminations

The ability of humans to discriminate differences in the intensity of tastes, like intensity discrimination in olfaction, is relatively crude. A 30% change in the concentration of the substance being tasted is necessary before an intensity difference can be detected. The threshold concentrations of substances to which the taste buds respond vary with the particular substance (Table 10–2).

Table 10–2. Some taste thresholds.

Substance	Taste	Threshold Concentration (μmol/L)
Hydrochloric acid	Sour	100
Sodium chloride	Salt	2000
Strychnine hydrochloride	Bitter	1.6
Glucose	Sweet	80,000
Sucrose	Sweet	10,000
Saccharin	Sweet	23

Substances Evoking Primary Taste Sensations

Acids taste **sour.** The H^+, rather than the associated anion, stimulates the receptors. For any given acid, sourness is generally proportionate to the H^+ concentration, but organic acids are often more sour for a given H^+ concentration than are mineral acids. This is probably because they penetrate cells more rapidly than the mineral acids.

A **salty** taste is produced by Na^+. Some organic compounds also taste salty; for example, the dipeptides lysyltaurine and ornithyltaurine taste salty, and on a weight basis, lysyltaurine is more potent than NaCl.

The substance usually used to test the **bitter** taste is quinine sulfate. This compound can be detected in a concentration of 8 μmol/L, although the threshold for strychnine hydrochloride is even lower (Table 10–2). Other organic compounds, especially morphine, nicotine, caffeine, and urea, taste bitter. Inorganic salts of magnesium, ammonium, and calcium also taste bitter. The taste is due to the cation. Thus, there is no apparent common feature of the molecular structure of substances that taste bitter.

Most **sweet** substances are organic. Sucrose, maltose, lactose, and glucose are the most familiar examples, but polysaccharides, glycerol, some of the alcohols and ketones, and a number of compounds with no apparent relation to any of these, such as chloroform, beryllium salts, and various amides of aspartic acid, also taste sweet. Two proteins isolated from African berries, thaumatin and monellin, are 100,000 times as sweet as sucrose. Artificial sweeteners such as saccharin and aspartame are in demand as sweetening agents in reducing diets because they produce satisfactory sweetening in amounts that are a minute fraction of the amount of calorie-rich sucrose required for the same purpose. Lead salts also taste sweet.

Flavor

The almost infinite variety of tastes so dear to the gourmet are mostly synthesized from the 4 basic taste components. In some cases, a desirable taste includes an element of pain stimulation (eg, "hot" sauces). In addition, smell plays an important role in the overall sensation produced by food, and the consistency (or texture) and temperature of foods also contribute to their "flavor."

Variation & After-Effects

There is considerable variation in the distribution of the 4 basic taste buds in various species and, within a given species, from individual to individual. In humans, there is an interesting variation in ability to taste **phenylthiocarbamide (PTC).** In dilute solution, PTC tastes sour to about 70% of the Caucasian population but is tasteless to the other 30%. Inability to taste PTC is inherited as an autosomal recessive trait. Testing for this trait is of considerable value in studies of human genetics.

Taste exhibits after-reactions and contrast phenomena that are similar in some ways to visual after-images and contrasts. Some of these are chemical "tricks," but others may be true central phenomena. A taste modifier protein, **miraculin,** has been discovered in a plant. When applied to the tongue, this protein makes acids taste sweet.

Animals, including humans, form particularly strong aversions to novel foods if eating the food is followed by illness. The survival value of such aversions is apparent in terms of avoiding poisons. The facility with which this type of learning occurs is so great that it has led some investigators to argue that its occurrence is reason to challenge classic learning theory.

Abnormalities

Abnormalities of taste include **ageusia** (absence of the sense of taste), **hypogeusia** (diminished taste sensitivity), and **dysgeusia** (disturbed sense of taste). Many different diseases can produce hypogeusia. In addition, drugs such as captopril and penicillamine, which contain sulfhydryl groups, cause temporary loss of taste sensation. The reason for this effect of sulfhydryl compounds is not known.

11 Arousal Mechanisms, Sleep, & the Electrical Activity of the Brain

INTRODUCTION

Most of the various sensory pathways described in Chapters 7–10 relay impulses from sense organs via 3- and 4-neuron chains to particular loci in the cerebral cortex. The impulses are responsible for perception and localization of individual sensations. Impulses in these systems also relay via collaterals to the **reticular activating system (RAS)** in the brain stem reticular formation. Activity in this system produces the conscious, alert state that makes perception possible. Other systems are responsible for drowsiness and sleep. This chapter is concerned with the way these systems operate and the correlations between behavioral states and the electroencephalogram (EEG).

THE RETICULAR FORMATION & THE RETICULAR ACTIVATING SYSTEM

The **reticular formation,** the phylogenetically old reticular core of the brain, occupies the midventral portion of the medulla and midbrain. It is made up of myriads of small neurons arranged in complex, intertwining nets. Located within it are centers that regulate respiration, blood pressure, heart rate, and other vegetative functions. In addition, it contains ascending and descending components that play important roles in the adjustment of endocrine secretion, the regulation of sensory input, and consciousness. The components concerned with consciousness and the modulation of sensory input are discussed in this chapter. The vegetative functions of the reticular core are considered in the chapters on the endocrine, gastrointestinal, cardiovascular, and respiratory systems.

The reticular activating system is a complex polysynaptic pathway. Collaterals funnel into it not only from the long ascending sensory tracts but also from the trigeminal, auditory, and visual systems and the olfactory system. The complexity of the neuron net and the degree of convergence in it abolish modality specificity, and most reticular neurons are activated with equal facility by different sensory stimuli. The system is therefore **nonspecific,** whereas the classic sensory pathways are **specific** in that the fibers in them are activated only by one type of sensory stimulation. Part of the RAS bypasses the thalamus to project diffusely to the cortex. Another part of the RAS ends in the intralaminar and related thalamic nuclei, and from them is projected diffusely and nonspecifically to the whole neocortex (Fig 11–1). The RAS is intimately concerned with the electrical activity of the cortex.

THE THALAMUS & THE CEREBRAL CORTEX

Thalamic Nuclei

On developmental and topographic grounds, the thalamus can be divided into 3 parts: the epithalamus, the dorsal thalamus, and the ventral thalamus. The **epithalamus** has connections to the olfactory system, and the projections and functions of the **ventral thalamus** are undetermined. The **dorsal thalamus** can be divided into nuclei that project diffusely to the whole neocortex and the nuclei that project to specific discrete portions of the neocortex and limbic system.

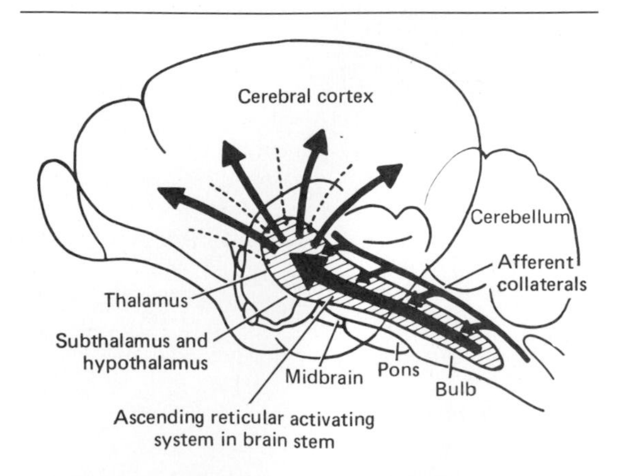

Figure 11–1. Diagram of ascending reticular system projected on a sagittal section of the cat brain. (Reproduced, with permission, from Starzl TE, Taylor CW, Magoun HW: Collateral afferent excitation of reticular formation of brain stem. *J Neurophysiol* 1951;**14**:479.)

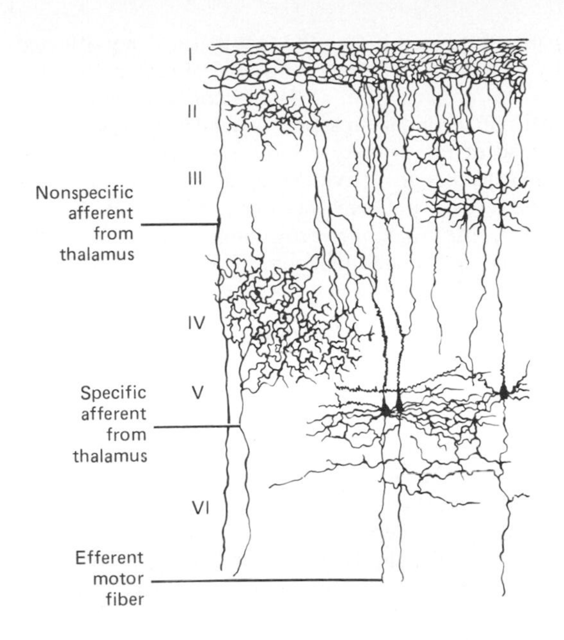

Figure 11–2. Neuronal connections in the neocortex. On the left are afferent fibers from the thalamus. The numbers identify the cortical layers. Note the extensive dendritic processes of the cells, especially those in the deep layers.

The nuclei that project to all parts of the neocortex are the midline and intralaminar nuclei. They are called collectively the **nonspecific projection nuclei.** They receive input from the reticular activating system. Impulses responsible for the diffuse secondary response and the alerting effect of reticular activation (see below) are relayed through them. The nuclei of the dorsal thalamus that project to specific areas can be divided into 3 groups: the specific sensory relay nuclei, the nuclei concerned with efferent control mechanisms, and the nuclei concerned with complex integrative functions. The **specific sensory relay nuclei** include the medial and lateral geniculate bodies, which relay auditory and visual impulses to the auditory and visual cortices, and the ventrobasal group of nuclei, which relay somatesthetic information to the postcentral gyrus. The **nuclei concerned with efferent control mechanisms** include several nuclei that are concerned with motor function. They receive input from the basal ganglia and the cerebellum and project to the motor cortex. Also included in this group are the anterior nuclei, which receive afferents from the mamillary bodies and project to the limbic cortex. This is part of the limbic circuit, which appears to be concerned with recent memory and emotion (see Chapters 15 and 16). The **nuclei concerned with complex integrative functions** are the dorsolateral nuclei that project to the cortical association areas and are concerned primarily with functions such as language.

Cortical Organization

The neocortex is generally arranged in 6 layers (Fig 11–2). The neurons are mostly pyramidal cells with extensive vertical dendritic trees (Figs 11–2, 11–3) that may reach to the cortical surface. The axons of these cells usually give off recurrent collaterals that turn back and synapse on the superficial portions of the dendritic trees. Afferents from the specific nuclei of the thalamus terminate primarily in cortical layer IV, whereas the nonspecific afferents are distributed to layers I–IV.

EVOKED CORTICAL POTENTIALS

The electrical events that occur in the cortex after stimulation of a sense organ can be monitored with an exploring electrode connected to another electrode at an indifferent point some distance away. A characteristic response is seen in animals under barbiturate anesthesia. If the exploring electrode is over the primary receiving area for the particular sense, a surface-positive wave appears with a latency of 5–12 ms. This is followed by a small negative wave and then by a larger, more prolonged positive deflection with a latency of 20–80 ms. This sequence of

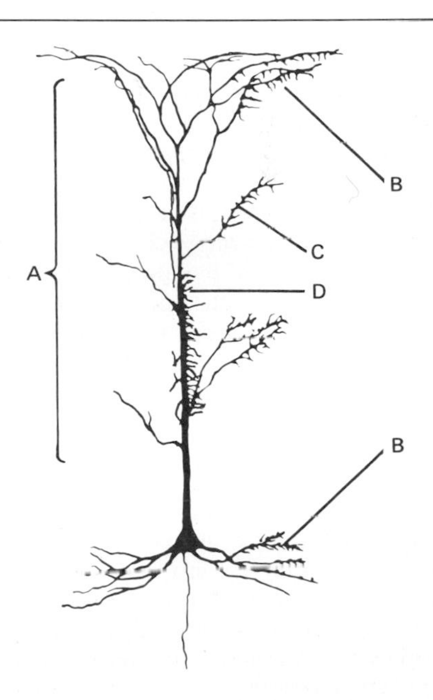

Figure 11–3. Neocortical pyramidal cell, showing the distribution of neurons that terminate on it. *A,* nonspecific afferents from the reticular formation and the thalamus; *B,* recurrent collaterals of pyramidal cell axons; *C,* commissural fibers from mirror image sites in contralateral hemisphere; *D,* specific afferents from thalamic sensory relay nuclei. (Reproduced, with permission, from Chow KL, Leiman AL: The structural and functional organization of the neocortex. *Neurosci Res Program Bull* 1970;8:157.)

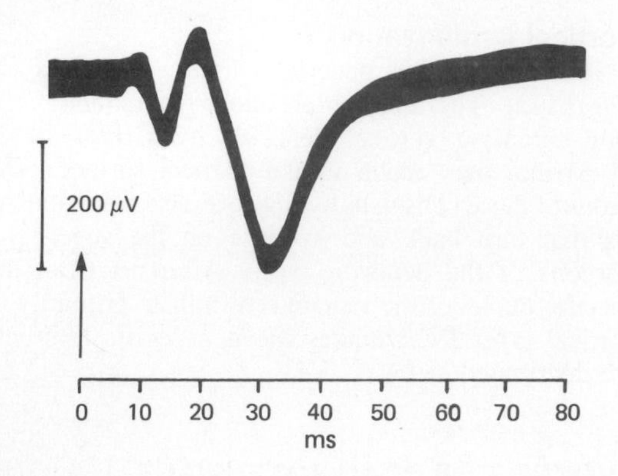

Figure 11–4. Response evoked in the contralateral sensory cortex by stimulation (at the arrow) of the sciatic nerve in a cat under barbiturate anesthesia. Upward deflection is surface-negative.

potential changes is illustrated in Fig 11–4. The first positive-negative wave sequence is the **primary evoked potential;** the second is the **diffuse secondary response.**

Primary Evoked Potential

The primary evoked potential is highly specific in its location and can be observed only where the pathways from a particular sense organ end. Indeed, it is so discrete that it has been used to map the specific cortical sensory areas. It was thought at one time that the primary response was due to activity ascending to the cortex in the thalamic radiations. However, an electrode on the pial surface of the cortex samples activity to a depth of only 0.3–0.6 mm. The primary response is negative rather than positive when it is recorded with a microelectrode inserted in layers II–VI of the underlying cortex, and the negative wave within the cortex is followed by a positive wave. This indicates depolarization on the dendrites and somas of the cells in the cortex, followed by hyperpolarization. The positive-negative wave sequence recorded from the surface of the cortex is due to the fact that the superficial cortical layers are positive relative to the initial negativity, then negative relative to the deep hyperpolarization. In unanesthetized animals, the primary evoked potential is largely obscured by the spontaneous activity of the brain, but it can be demonstrated with special techniques. It is somewhat more diffuse in unanesthetized animals but still well localized compared with the diffuse secondary response.

Diffuse Secondary Response

The surface-positive diffuse secondary response is sometimes followed by a negative wave or series of waves. Unlike the primary response, the secondary response is not highly localized. It appears at the same time over most of the cortex and in many other parts of the brain. The uniform latency in different parts of the cortex and the fact that it is not affected by a circular cut through the cortical gray matter that isolates an area from all lateral connections indicate that the secondary response cannot be due to lateral spread of the primary evoked response. It therefore must be due to activity ascending from below the cortex. The pathway involved appears to be the nonspecific thalamic projection system from the midline and related thalamic nuclei.

THE ELECTROENCEPHALOGRAM

The background electrical activity of the brain in unanesthetized animals was described in the 19th century, but it was first analyzed in a systematic fashion by the German psychiatrist Hans Berger, who introduced the term **electroencephalogram (EEG)** to denote the record of the variations in potential recorded from the brain. The EEG can be recorded with scalp electrodes through the unopened skull or with electrodes on or in the brain. The term **electrocorticogram (ECoG)** is sometimes used to refer to the record obtained with electrodes on the pial surface of the cortex.

EEG records may be **bipolar** or **unipolar.** Bipolar records show fluctuations in potential between 2 cortical electrodes; unipolar records show potential differences between a cortical electrode and a theoretically indifferent electrode on some part of the body distant from the cortex.

Alpha Rhythm

In an adult human at rest with mind wandering and eyes closed, the most prominent component of the

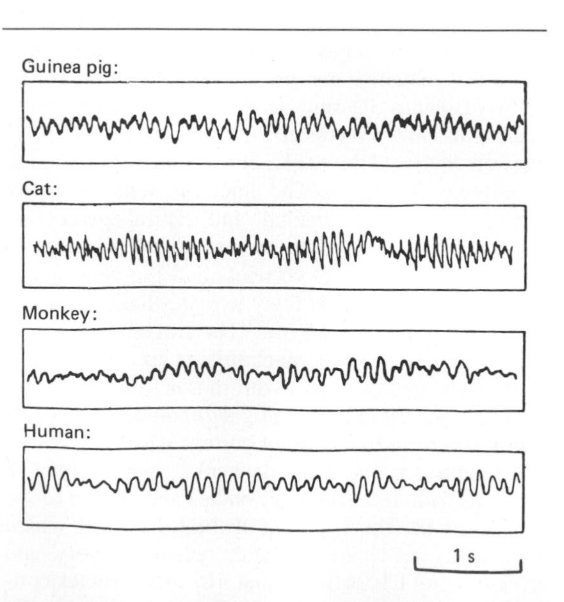

Figure 11–5. EEG records showing alpha rhythm from 4 different species. (Reproduced, with permission, from Brazier MAB: *Electrical Activity of the Nervous System,* 2nd ed. Pitman, 1960.)

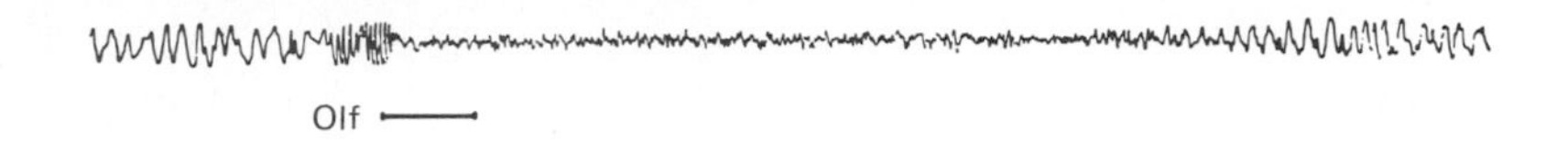

Figure 11–6. Desynchronization of the cortical EEG of a rabbit by an olfactory stimulus (indicated by the line after Olf). (Reproduced, with permission, from Green JD, Arduini A: Hippocampal electrical activity during arousal. *J Neurophysiol* 1954;**17:**533.)

EEG is a fairly regular pattern of waves at a frequency of 8–12/s and an amplitude of about 50 μV when recorded from the scalp. This pattern is the **alpha rhythm.** It is most marked in the parieto-occipital area, although it is sometimes observed in other locations. A similar rhythm has been observed in a wide variety of mammalian species (Fig 11–5). In the cat, it is slightly more rapid than in the human, and there are other minor variations from species to species, but in all mammals the pattern is remarkably similar.

Other Rhythms

In addition to the dominant rhythm, 18–30/s patterns of lower voltage are sometimes seen over the frontal regions. This rhythm, the **beta rhythm,** may be a harmonic of the alpha. A pattern of large, regular 4–7/s waves called the **theta rhythm** occurs in normal children and is generated in the hippocampus in experimental animals (see below). Large, slow waves with a frequency of less than 4/s are sometimes called **delta waves.**

Variations in the EEG

In humans, the frequency of the dominant EEG rhythm at rest varies with age. In infants, there is fast, betalike activity, but the occipital rhythm is a slow 0.5–2/s pattern. During childhood this latter rhythm speeds up, and the adult alpha pattern gradually appears during adolescence. The frequency of the alpha rhythm is decreased by a low blood glucose level, a low body temperature, a low level of adrenal glucocorticoid hormones, and a high arterial partial pressure of CO_2 (P_{CO_2}). It is increased by the reverse conditions. Forced overbreathing to lower the arterial P_{CO_2} is sometimes used clinically to bring out latent EEG abnormalities.

Alpha Block

When the eyes are opened, the alpha rhythm is replaced by fast, irregular low-voltage activity with no dominant frequency. This phenomenon is called **alpha block.** A breakup of the alpha pattern is also produced by any form of sensory stimulation (Fig 11–6) or mental concentration such as solving arithmetic problems. A common term for this replacement of the regular alpha rhythm with irregular low-voltage activity is **desynchronization,** because it represents a breaking up of the synchronized activity of neural elements responsible for the wave pattern. Because desynchronization is produced by sensory stimulation and is correlated with the aroused, alert state, it is also called the **arousal** or **alerting response.**

Sleep Patterns

There are 2 different kinds of sleep: **rapid eye movement (REM) sleep** and **non-REM (NREM)** or **slow-wave sleep.** NREM sleep is divided into 4 stages. A person falling asleep first enters stage 1, which is characterized by low-amplitude, fast-frequency EEG activity (Fig 11–7). Stage 2 is marked by the appearance of sleep spindles. These are bursts of alphalike, 10–14/s, 50-μV waves. In stage 3, the pattern is one of slower frequency and increased amplitude of the EEG waves. Maximum slowing with large waves is seen in stage 4. Thus, the characteristic of deep sleep is a pattern of rhythmic slow waves, indicating **synchronization.**

REM Sleep

The high-amplitude slow waves seen in the EEG during sleep are sometimes replaced by rapid, low-voltage, irregular EEG activity, which resembles that seen in alert animals and humans. However, sleep is not interrupted; indeed, the threshold for arousal by sensory stimuli and by stimulation of the reticular formation is elevated. The condition has been called **paradoxical sleep.** There are rapid, roving movements of the eyes during paradoxical sleep, and for this reason it is also called REM sleep. There are no such movements in slow-wave sleep, and consequently it is often called NREM sleep. Another characteristic of REM sleep is the occurrence of large phasic potentials, occurring in groups of 3–5, that originate in the pons and pass rapidly to the lateral geniculate body and thence to the occipital cortex. For this reason, they are called **ponto-geniculo-occipital (PGO) spikes.** There is a marked reduction in skeletal muscle tone during REM sleep (Fig 11–7) despite the rapid eye movements and PGO spikes. The hypotonia is due to increased activity of the reticular inhibiting area in the medulla (see Chapter 12), which brings about decreases in stretch and polysynaptic reflexes by way of both pre- and post-synaptic inhibition.

PHYSIOLOGIC BASIS OF THE EEG, CONSCIOUSNESS, & SLEEP

The presence of rhythmic waves in the alpha and slow-wave sleep patterns indicates a priori that neural

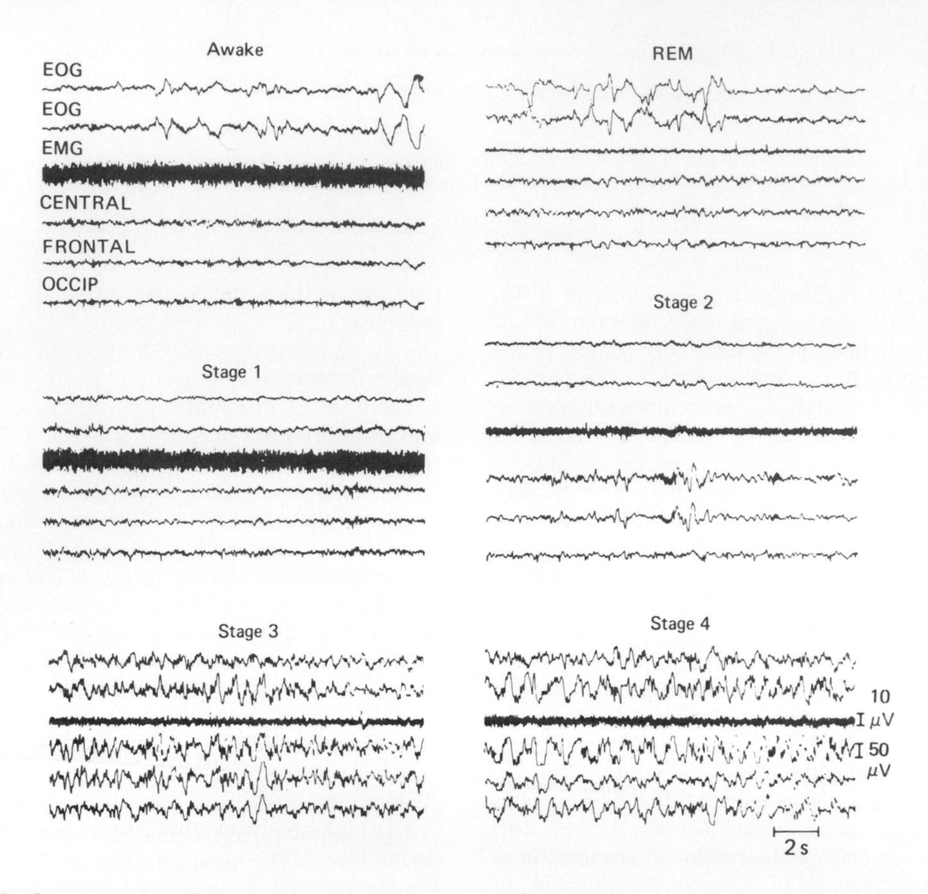

Figure 11–7. Sleep stages, EOG, electro-oculogram registering eye movements; EMG, electromyogram registering skeletal muscle activity; central, frontal, occip, 3 EEG leads. Note the low muscle tone with extensive eye movements in REM. (Reproduced, with permission, from Kales A et al: Sleep and dreams: Recent research on clinical aspects. *Ann Intern Med* 1968;**68**:1078.)

components of some sort are discharging rhythmically, since random discharges of individual units would cancel out and no waves would be produced. The main questions to be answered are which units are discharging synchronously and what mechanisms are responsible for desynchronization and synchronization.

Source of the EEG

The activity recorded in the EEG is mostly that of the most superficial layers of the cortical gray substance, and there are relatively few cell bodies in these layers. The potential changes in the cortical EEG are due to current flow in the fluctuating dipoles formed on the dendrites of the cortical cells and the cell bodies. The dendrites of the cortical cells are a forest of similarly oriented, densely packed units in the superficial layers of the cerebral cortex (Fig 11–2). As pointed out in Chapter 4, dendrites may be conducting processes, but they generally do not produce propagated all or none spikes; rather, they are the site of nonpropagated hypopolarizing and hyperpolarizing local potential changes. As excitatory and inhibitory endings on the dendrites of each cell become active, current flows into and out of these current sinks and sources from the rest of the dendritic processes and the cell body. The cell-dendrite relationship is therefore that of a constantly shifting dipole. Current flow in this dipole would be expected to produce wavelike potential fluctuations in a volume conductor (Fig 11–8). When the sum of the dendritic activity is negative relative to the cell, the cell is hypopolarized and hyperexcitable; when it is positive, the cell is hyperpolarized and less excitable. Information is also transmitted electrotonically from one dendrite to another (see Chapter 4). The cerebellar cortex and the hippocampus are 2 other parts of the central nervous system where many complex, parallel dendritic processes are located subpially over a layer of cells. In both areas, there is a characteristic rhythmic fluctuation in surface potential similar to that observed in the cortical EEG.

Desynchronizing Mechanisms

Desynchronization, the replacement of a rhythmic EEG pattern with irregular low-voltage activity, is produced by stimulation of the specific sensory systems up to the level of the midbrain, but stimulation

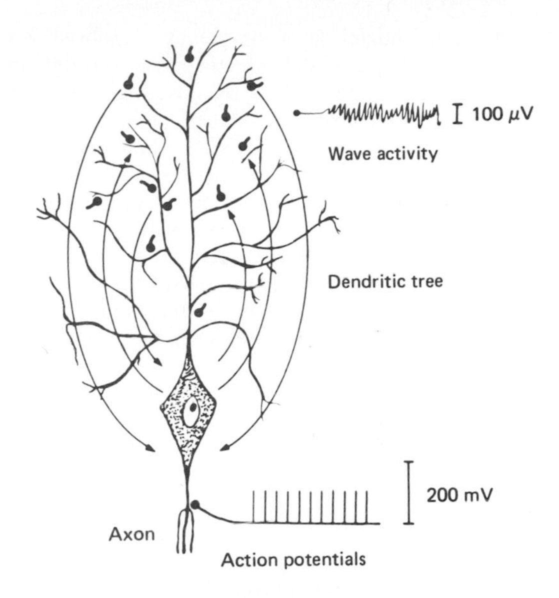

Figure 11–8. Diagrammatic comparison of the electrical responses of the axon and the dendrites of a large cortical neuron. Current flow to and from active synaptic knobs on the dendrites produces wave activity, while all or none action potentials are transmitted down the axon.

of these systems above the midbrain, stimulation of the specific sensory relay nuclei of the thalamus, or stimulation of the cortical receiving areas themselves does not produce desynchronization. On the other hand, high-frequency stimulation of the reticular formation in the midbrain tegmentum and of the nonspecific projection nuclei of the thalamus desynchronizes the EEG and arouses a sleeping animal. Large lesions of the lateral and superior portions of the midbrain that interrupt the medial lemnisci and other ascending specific sensory systems fail to prevent the desynchronization produced by sensory stimulation, but lesions in the midbrain tegmentum that disrupt the RAS without damaging the specific systems are associated with a synchronized pattern that is unaffected by sensory stimulation (Fig 11–9). Animals with the former type of lesion are awake; those with the latter type are comatose for long periods. It therefore appears that the ascending activity responsible for desynchronization following sensory stimulation passes up the specific sensory systems to the midbrain, enters the RAS via collaterals, and continues through the thalamus and the nonspecific thalamic projection system to the cortex.

It should be pointed out that although arousal and EEG desynchronization generally occur together, they do not always coexist. The presence of desynchronization in paradoxical sleep is discussed in detail below. In addition, strong nociceptive peripheral stimulation can produce arousal without desynchronization in animals with lesions of the midbrain tegmentum and desynchronization without arousal in animals with lesions in the posterior hypothalamus.

Effect of RAS Stimulation on Cortical Neurons

It is not known how impulses ascending in the RAS break up synchronized cortical activity or produce arousal. Stimulation of the RAS inhibits burst activity, the recruiting response (see below), and the cortical waves produced by the application of strychnine to certain parts of the brain. Since RAS stimulation arouses and alerts the animal, one might expect it to produce a heightened rather than a depressed excitability of cortical neurons. However, some of the neurons in the cortical sensory projection areas fire randomly during sleep, whereas they discharge only in response to specific stimuli when the animal is awake. It is possible that RAS activity abolishes this random firing by producing some degree of cortical inhibition, leaving these neurons free to respond only to specific sensory signals. The net effect of RAS activity, therefore, might be an increase in the "signal/noise ratio" at the expense of a slight decrease in absolute excitability.

Arousal Following Cortical Stimulation

Electrical stimulation of certain portions of the cerebral cortex causes increased reticular activity and EEG arousal. The most effective cortical loci in the monkey are the superior temporal gyrus and the orbital surface of the frontal lobe. Stimulation of these regions wakes a sleeping animal but causes no movements and has few visible effects in the conscious animal. Stimulation of other cortical areas, even with strong currents, does not produce electrical or behavioral arousal. These observations indicate that a system of **corticofugal fibers** passes from the cortex to the reticular formation, providing a pathway by which intracortical events can initiate arousal. The

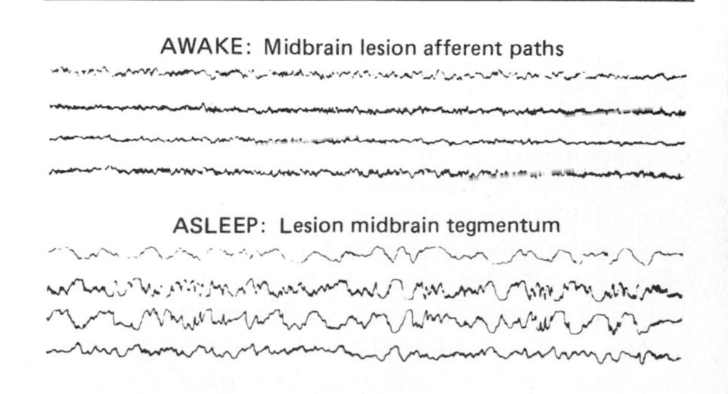

Figure 11–9. Typical EEG records of cat with lesion of lemniscal pathways *(top)* and cat with lesion of midbrain tegmentum *(bottom)*. (Reproduced, with permission, from Lindsley DB et al: Behavioral and EEG changes following chronic brain stem lesions in the cat. *Electroencephalogr Clin Neurophysiol* 1950;2:483.)

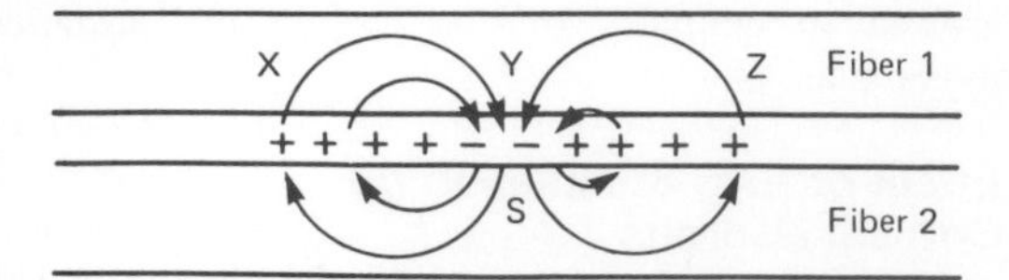

Figure 11–10. Effect of activity in one nerve fiber on the activity of a neighboring fiber in a volume conductor. Current flows into the area of depolarization (S) on fiber 2 from the surrounding membrane. At X and Z on the membrane of fiber 1, positive charges pile up, hyperpolarizing the membrane; at Y, positive charges are pulled off, partly depolarizing it.

system may be responsible for the alerting responses to emotions and related psychic phenomena that occur in the absence of any apparent external stimulus.

Alerting Effects of Epinephrine & Related Compounds

Epinephrine and norepinephrine produce EEG arousal and behavioral alerting by lowering the threshold of reticular neurons in the brain stem. Therefore, one of the effects of the mass sympathetic discharge with increased adrenal medullary secretion in emergency situations (see Chapter 13) is reinforcement of the alert, attentive state necessary for effective action. The effect may be direct, or it may be secondary to increased blood pressure.

Synchronizing Mechanisms

The wave pattern of the alpha rhythm indicates that the activity of many of the dendritic units is synchronized. Two factors contribute to this synchronization: the synchronizing effect on each unit of activity in neighboring parallel fibers, and rhythmic discharge of impulses from the thalamus.

If 2 nerve fibers are placed side by side in a volume conductor, depolarization of one of them rarely if ever causes a propagated discharge in the other but does influence its excitability. This is because external current flow into the area of depolarization in the active fiber hyperpolarizes the membrane of the neighboring fiber at 2 points and partially depolarizes it at another. Current flow in the opposite direction creates areas of positivity in fibers adjacent to sites where inhibitory endings are active (Fig 11–10). In a large population of similarly oriented neural processes, this formation of current sinks and sources in neighboring fibers tends to synchronize current flow.

The dendritic potentials of the cerebral cortex are also influenced by projections from the thalamus. A circular cut around a piece of cortex does not alter its synchronized activity if the blood supply is intact, but the rhythmicity is markedly decreased if the deep connections of the cortical "button" are severed. Large lesions of the thalamus disrupt the EEG synchrony on the side of the lesion. Stimulation of the thalamic nuclei at a frequency of about 8/s produces a characteristic 8/s response throughout most of the ipsilateral cortex. Because the amplitude of this response waxes and wanes, it is called the **recruiting response** (Fig 11–11). The recruiting response has many similarities to the alpha rhythm and to **burst activity,** the occurrence of short trains of alphalike waves superimposed on slower rhythms. The records of burst activity such as sleep spindles are essentially identical when recorded simultaneously from the cortex and the thalamus, which suggests that they are due to activity projected from the thalamus or reverberating activity between the thalamus and the cortex. One theory holds that synchrony is produced by a thalamic circuit which includes a pathway for **recurrent collateral inhibition.** Each time a given central neuron discharges, it activates via a collateral branch, an inhibitory interneuron that produces IPSPs in the discharging neuron and its neighbors. The thalamic neurons are also postulated to be hyperexcitable after a period of inhibition and to discharge spontaneously during the rebound **(post-inhibitory rebound excitation).** Therefore, they discharge rhythmically.

EEG synchrony and slow-wave sleep (see below) can be produced by stimulation of at least 3 subcortical regions. The region discussed in the preceding paragraph is the **diencephalic sleep zone** in the posterior hypothalamus and the nearby intralaminar and anterior thalamic nuclei. The stimulus frequency must be about 8/s; faster stimuli produce arousal. This finding need not be confusing; the important point is that low-frequency stimulation produces one response, whereas high-frequency stimulation produces another. The second zone is the **medullary synchronizing zone** in the reticular formation of the medulla oblongata at the level of the nucleus of the tractus solitarius. Stimulation of this zone, like stimulation of the diencephalic sleep zone, produces synchrony and sleep if the frequency is low but arousal if the frequency is high. The mechanism by which synchrony is produced is not known, but it presumably involves pathways that ascend to the thalamus. The third synchronizing region is the **basal forebrain sleep zone.** This zone includes the preoptic area and the diagonal band of Broca. It differs from the other 2 zones in that stimulation of the basal forebrain zone produces synchrony and sleep whether the stimulating frequency is high or low. There is some evidence that its effects are mediated by descending pathways which inhibit neurons in the ascending reticular formation.

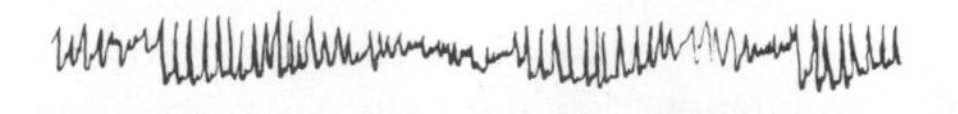

Figure 11–11. Recruiting response in cortex, produced by 8/s stimulation of intralaminar thalamic region.

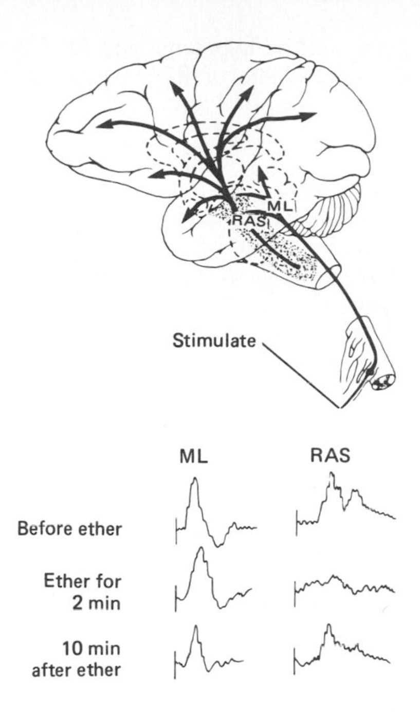

Figure 11–12. Effect of ether anesthesia on lemniscal (ML) and reticular (RAS) response to stimulation of the sciatic nerve.

The RAS & General Anesthesia

The mechanism by which drugs produce general anesthesia remains an intriguing mystery. One puzzling aspect is why so many drugs with so many totally different structures can produce this state. However, one feature that correlates very well with ability to produce anesthesia is lipid solubility. Another fact is that many of the neurons in the brain are hyperpolarized during anesthesia. Conduction at synaptic junctions is depressed. This raises the possibility that general anesthetics can produce unconsciousness by depressing conduction in the RAS. The effect of ether on the response of the reticular formation and the medial lemnisci to a sensory stimulus is shown in Fig 11–12. Similar blocking effects are produced by anesthetic doses of barbiturates. There are hundreds of synapses in the reticular pathways, whereas the specific systems have only 2–4. At each synapse, the EPSPs must build up until the firing level of the postsynaptic neurons is reached (see Chapter 4). Any agent that has a moderate depressant effect on this process will block conduction in multisynaptic paths before blocking conduction in those with only a few synapses.

Concomitants of REM Sleep

Humans aroused at a time when they show the EEG characteristics of REM sleep generally report that they were dreaming, whereas individuals wakened from slow-wave sleep do not. This observation and other evidence indicate that REM sleep and dreaming are closely associated. The teeth-grinding **(bruxism)** that occurs in some individuals is also associated with dreaming. REM sleep is found in all species of mammals and birds that have been studied, but it probably does not occur in other classes of animals.

If humans are awakened every time they show REM sleep, they become somewhat anxious and irritable. If they are then permitted to sleep without interruption, they show a great deal more than the normal amount of paradoxical sleep for a few nights. The same "rebound" effect is seen in animals. These observations led some investigators to conclude that dreaming is necessary to maintain mental health. However, it is now established that prolonged REM deprivation has no adverse psychological effects.

Distribution of Sleep Stages

In a typical night of sleep, a young adult first enters NREM sleep, passes through stages 1 and 2, and spends 70–100 minutes in stages 3 and 4. Sleep then lightens, and an REM period follows. This cycle is repeated at intervals of about 90 minutes throughout the night (Fig 11–13). The cycles are similar, although there is less stage 3 and 4 sleep and more REM sleep toward morning. Thus, there are 4–6 REM periods per night. At all ages, REM sleep constitutes about

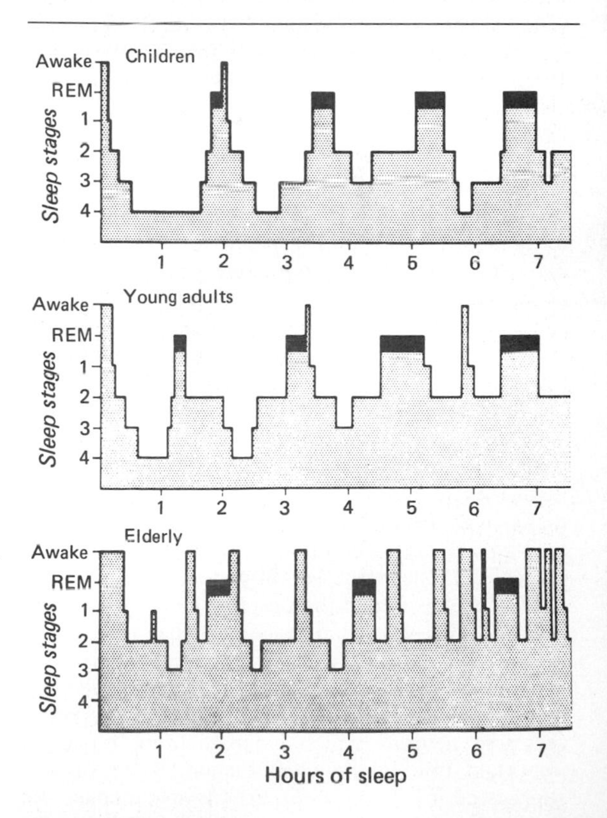

Figure 11–13. Normal sleep cycles at various ages. REM sleep is indicated by the dark areas. (Reproduced, with permission, from Kales AM, Kales JD: Sleep disorders. *N Engl J Med* 1974;**290**:487.)

25% of total sleep time. Children have more stage 3 and 4 sleep than young adults, and old people have much less.

Genesis of Slow-Wave Sleep

Slow-wave sleep is produced in part by the absence of desynchronizing activity transmitted via the ascending reticular system. However, it is actively produced as well by the synchronizing zones. The synchronizing activity of the thalamus is apparently influenced by ascending activity from the pons and hindbrain, since stimulation of the medullary sleep zone produces synchrony and sleep (see above). The role of the basal forebrain zone in the production of normal sleep is unknown. Stimulation of afferents from mechanoreceptors in the skin at rates of 10/s or less also produces sleep in animals, apparently via the brain stem, and it is of course common knowledge that regularly repeated monotonous stimuli put humans to sleep.

There has been considerable debate about the relation of serotonergic neurons in the brain (see Chapter 15) to sleep, but it now appears that serotonin agonists suppress sleep and that the serotonin antagonist ritanserin increases slow-wave sleep in humans. There is some evidence that adenosine causes sleep and that caffeine and other methylxanthines cause wakefulness by blocking adenosine receptors. Several investigators have argued that a peptide produced in the brain is responsible for sleep. However, there is disagreement about the chemistry of this putative **sleep peptide,** and its physiologic role, if any, is uncertain.

Genesis of REM Sleep

The mechanism that triggers REM sleep is located in the pontine reticular formation. Lesions of the oral and caudal pontine reticular nuclei abolish the desynchronized EEG activity, usually without affecting slow-wave sleep or arousal. Other evidence also indicates that the desynchronization is brought about by pathways which are different from those mediating desynchronization during the aroused waking state. The rapid eye movements are mediated via the vestibular nuclei. The structures that activate the medullary reticular inhibitory area to produce hypotonia are located in the mediolateral pontine tegmentum and include the locus ceruleus. PGO spikes originate in the lateral pontine tegmentum.

PGO spikes herald the onset of REM sleep, but it is not known what initiates the discharges from the pons. The available evidence indicates that no single synaptic transmitter is responsible for REM sleep but that the locus ceruleus and the norepinephrine-secreting neurons which emanate from it play an important role in the phenomenon. REM sleep is suppressed by brain stem lesions that deplete the forebrain of norepinephrine. However, drugs that inhibit monoamine oxidase increase brain norepinephrine and decrease REM sleep. Reserpine, which depletes serotonin and catecholamines, blocks slow-wave sleep and the hypotonia and EEG desynchronization characteristic of REM sleep but increases PGO spike activity. Barbiturates decrease the amount of REM sleep.

Sleep Disorders

Sleepwalking **(somnambulism)** and bed-wetting **(nocturnal enuresis)** have been shown to occur during slow-wave sleep or, more specifically, during arousal from slow-wave sleep. They are not associated with REM sleep. Episodes of sleepwalking are more common in children than adults and occur predominantly in males. They may last several minutes. Somnambulists walk with their eyes open and avoid obstacles, but when awakened they cannot recall the episodes.

Narcolepsy is a disease of unknown cause in which there is an eventually irresistible urge to sleep during daytime activities. In some cases, it has been shown to start with the sudden onset of REM sleep. REM sleep almost never occurs without previous slow-wave sleep in normal individuals.

Clinical Uses of the EEG

Patients with tumors or other lesions that interrupt the RAS are generally comatose. This system is especially vulnerable at the top of the midbrain and in the posterior hypothalamus, where it is pushed medially by other fiber systems. Relatively small lesions in this location may cause prolonged coma while producing very few other symptoms.

The EEG is sometimes of value in localizing pathologic processes. When a collection of fluid overlies a portion of the cortex, activity over this area may be damped. This fact may aid in diagnosing and localizing conditions such as subdural hematomas. Lesions in the cortex cause local formation of irregular or slow waves that can be picked up in the EEG leads. Epileptogenic foci sometimes generate high-voltage waves that can be localized.

The waves recorded from corresponding anatomic loci on the 2 hemispheres are normally remarkably similar in timing and shape. This is presumably due to a central subcortical "pacemaker" that keeps the

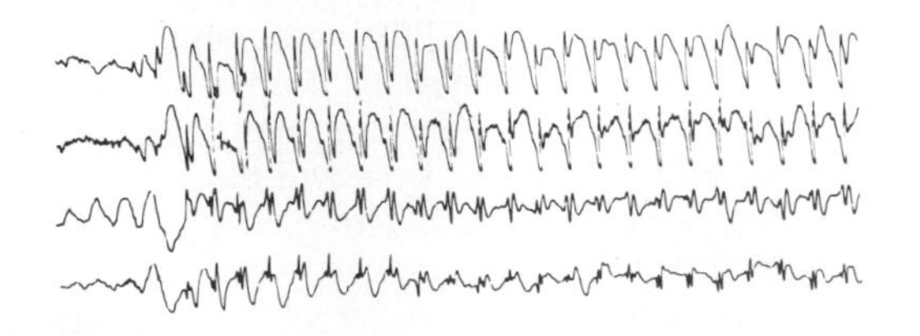

Figure 11–14. Petit mal epilepsy. Record of 4 cortical EEG leads from a 6-year-old boy who during the recording had one of his "blank spells" in which he was transiently unaware of his surroundings and blinked his eyelids. (Reproduced, with permission, from DeGroot J, Chusid JG: *Correlative Neuroanatomy,* 20th ed. Appleton & Lange, 1988.)

activity of both hemispheres in phase. A lesion in one hemisphere distorts the pattern; consequently, if records from homologous points on the 2 hemispheres are out of phase with each other, focal cortical damage is indicated.

In epileptics, there are characteristic EEG patterns during seizures; between attacks, however, abnormalities are often difficult to demonstrate. **Grand mal** seizures, epileptic attacks in which an aura precedes a generalized convulsion with tonic muscular contraction and clonic jerks, have a characteristic EEG pattern. There is fast activity during the tonic phase. Slow waves, each preceded by a spike, occur at the time of each clonic jerk. For a while after the attack, slow waves are present. Similar changes are seen in experimental animals during convulsions produced by electric shocks. **Petit mal** epileptic attacks, seizures characterized by a momentary loss of responsiveness, are associated with 3/s doublets, each consisting of a typical spike and rounded wave (Fig 11–14). **Psychomotor seizures,** attacks in which emotional lability and stereotyped behavior are seen owing to discharge from a temporal lobe focus, and various types of diencephalic seizures that originate in the brain stem are not associated with any typical EEG changes.

12 Control of Posture & Movement

INTRODUCTION

Somatic motor activity depends ultimately upon the pattern and rate of discharge of the spinal motor neurons and homologous neurons in the motor nuclei of the cranial nerves. These neurons, the final common paths to skeletal muscle, are bombarded by impulses from an immense array of pathways. There are many inputs to each spinal motor neuron from the same spinal segment (see Chapter 6). Numerous suprasegmental inputs also converge on these cells from other spinal segments, the brain stem, and the cerebral cortex. Some of these inputs end directly on the motor neurons, but many exert their effects via interneurons or via the γ efferent system to the muscle spindles and back through the Ia afferent fibers to the spinal cord. It is the integrated activity of these multiple inputs from spinal, medullary, midbrain, and cortical levels that regulates the posture of the body and makes coordinated movement possible.

The inputs converging on the motor neurons subserve 3 semidistinct functions: they bring about voluntary activity; they adjust body posture to provide a stable background for movement; and they coordinate the action of the various muscles to make movements smooth and precise. The patterns of voluntary activity are planned within the brain, and the commands are sent to the muscles primarily via the **corticospinal and corticobulbar system.** Posture is continually adjusted not only before but also during movement by **posture-regulating systems.** Movement is smoothed and coordinated by the medial and intermediate portions of the cerebellum (the **spinocerebellum**) and its connections. The **basal ganglia** and the **lateral portions of the cerebellum (neocerebellum)** are part of a feedback circuit to the premotor and motor cortex that is concerned with planning and organizing voluntary movement.

GENERAL PRINCIPLES

Organization

Although there is much about the control of voluntary movement that is still unknown, there is considerable evidence in monkeys and humans for the general control scheme shown in Fig 12–1. Commands for voluntary movement originate in cortical association areas. The movements are planned in the cortex as well as in the basal ganglia and the lateral portions of the cerebellar hemispheres, both of which funnel information to the premotor and motor cortex by way of the thalamus. Motor commands from the motor cortex are relayed in large part via the corticospinal tracts and the corresponding corticobulbar tracts to motor neurons in the brain stem. However, collaterals from these pathways and a few

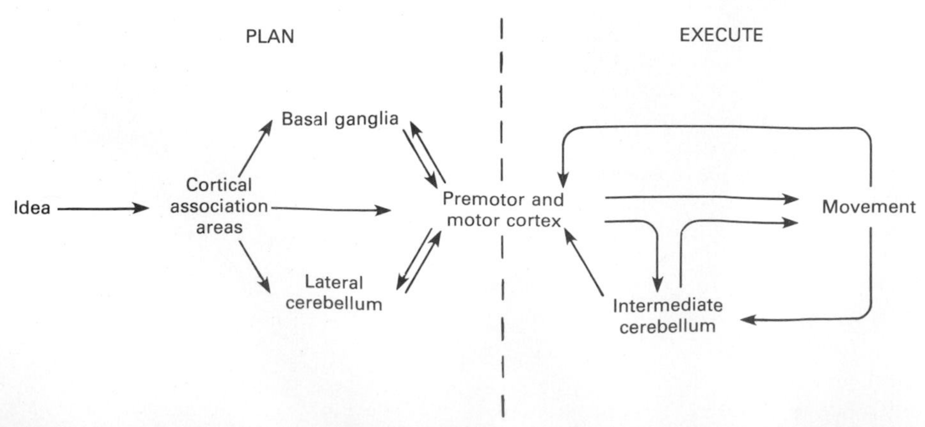

Figure 12–1. Control of voluntary movement. (Modified from McGeer PL, McGeer EG: The control of movement by the brain. *Trends in Neuroscience* [Nov] 1980; pages 3–4.)

direct connections from the motor cortex end on brain stem nuclei, which also project to motor neurons in the brain stem and spinal cord, and these pathways can also mediate voluntary movement. Movement sets up alterations in sensory input from muscles, tendons, joints, and the skin. This feedback information, which adjusts and smooths movement, is relayed directly to the motor cortex and to the spinocerebellum. The spinocerebellum projects in turn to the brain stem. The main brain stem pathways that are concerned with posture and coordination are the rubrospinal, reticulospinal, tectospinal, and vestibulospinal tracts and corresponding projections to motor neurons in the brain stem.

Encephalization

In species in which the cerebral cortex is highly developed, the process of **encephalization,** ie, the relatively greater role of the cerebral cortex in various functions, is an important phenomenon. For this reason, the effects of cortical ablation in primates are generally more severe and sometimes quite different from those in cats, dogs, and other laboratory animals. Because this chapter is concerned primarily with motor function in humans, pathologic and clinical observations in humans and experiments in laboratory primates are emphasized.

Control of Axial and Distal Muscles

Another theme that is important in motor control is that in the brain stem and spinal cord, medial or ventral pathways and neurons are concerned with the control of muscles of the trunk and proximal portions of the limbs, whereas lateral pathways are concerned with the control of muscles in the distal portions of the limbs. The axial muscles are concerned with postural adjustments and gross movements, whereas the distal limb muscles are those that mediate fine, skilled movements. Thus, for example, the neurons in the medial portion of the ventral horn innervate the proximal limb muscles, particularly the flexors, whereas the lateral ventral horn neurons innervate the distal limb muscles. Similarly, the ventral corticospinal tract and the medial descending paths from the brain stem (the testospinal, reticulospinal, and vestibulospinal tracts) are concerned with adjustments of proximal muscles and posture, whereas the lateral corticospinal tract and the rubrospinal tract are concerned with distal limb muscles and, particularly in the case of the lateral corticospinal tract, with skilled voluntary movements. Phylogenetically, the medial pathways are old, whereas the lateral pathways are new.

Other Terms

Because the fibers of the lateral corticospinal tract form the pyramids in the medulla, the corticospinal pathways have often been referred to as the **pyramidal system.** The rest of the descending brain stem and spinal pathways that do not pass through the pyramids and are concerned with postural control have been called the **extrapyramidal system.** However, the ventral corticospinal pathway does not go through the pyramids, many pyramidal fibers are concerned with other functions, and the system that used to be called extrapyramidal is made up of many different pathways with multiple functions. Consequently, the terms pyramidal and extrapyramidal are misleading, and it seems wise to drop them.

In addition, the motor system has often been divided into **upper** and **lower motor neurons.** Lesions of the lower motor neurons—the spinal and cranial motor neurons that directly innervate the muscles—are associated with flaccid paralysis, muscular atrophy, and absence of reflex responses. The syndrome of spastic paralysis and hyperactive stretch reflexes in the absence of muscle atrophy is said to be due to destruction of the ''upper motor neurons,'' the neurons in the brain and spinal cord that activate the motor neurons. However, there are 3 types of ''upper motor neurons'' to consider. Lesions in many of the posture-regulating pathways cause spastic paralysis, but lesions limited to the pyramidal tracts produce weakness (**paresis**) rather than paralysis, and the affected musculature is generally hypotonic. Cerebellar lesions produce incoordination. The unmodified term upper motor neuron is therefore confusing and should not be used.

CORTICOSPINAL AND CORTICOBULBAR SYSTEM

ANATOMY

Tracts

The nerve fibers that cross the midline in the medullary pyramids and form the **lateral corticospinal tract** make up about 80% of the fibers in the corticospinal pathway. The remaining 20% make up the **anterior** or **ventral corticospinal tract** (Fig 12–2), which does not cross the midline until the level at which it synapses with motor neurons. In addition, this tract contains corticospinal neurons that end on the same side of the body. The ventral pathway, which is the oldest phylogenetically, ends primarily on interneurons. These interneurons synapse on neurons in the medial portion of the ventral horn that control axial and proximal limb muscles. Conversely, the lateral corticospinal pathway innervates lateral neurons in the ventral horn that are concerned with distal limb muscles and hence with skilled movements. In humans, the neurons of this phylogenetically new system end directly on the lateral motor neurons.

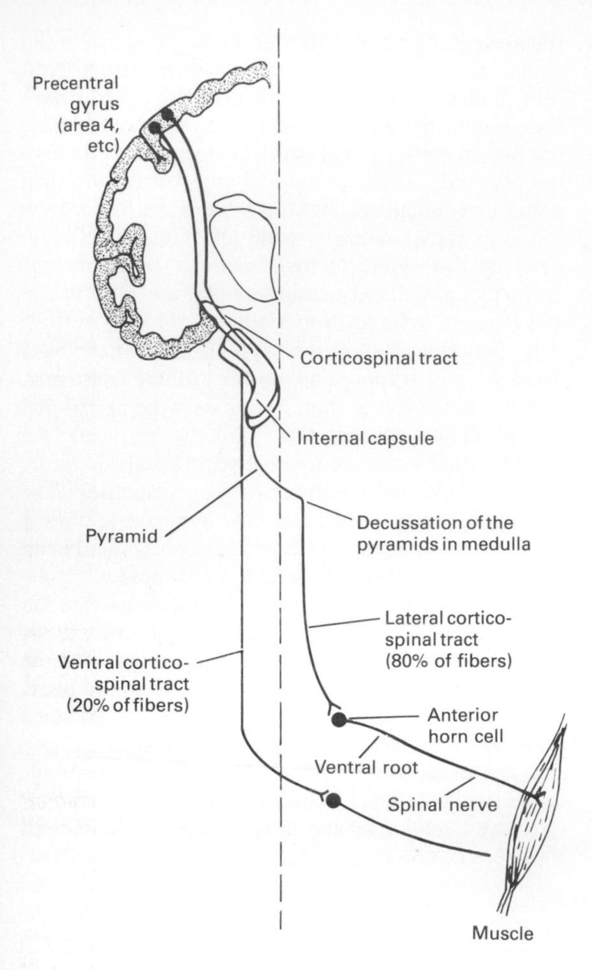

Figure 12–2. The corticospinal tracts.

Cortical Motor Areas

The cortical areas from which the corticospinal and corticobulbar system originates are generally held to be those where stimulation produces prompt discrete movement. There are 4 areas in the cortex that produce such movements. The best known is the **motor cortex** in the precentral gyrus (Fig 12–3). However, there is a **supplementary motor area** on and above the superior bank of the cingulate sulcus on the medial side of the hemisphere and the premotor cortex below it on the lateral surface of the brain (Fig 12–3). Motor responses are also produced by stimulation of somatic sensory area I in the postcentral gyrus and by stimulation of somatic sensory area II in the wall of the sylvian fissure (see Chapter 7). These observations fit with the fact that 30% of the fibers making up the corticospinal and corticobulbar tracts come from the motor cortex but 30% come from the premotor cortex and 40% from the parietal lobe, especially the somatic sensory area.

By means of stimulation experiments in patients undergoing craniotomy under local anesthesia, it has been possible to outline most of the motor projection from the motor cortex. The various parts of the body are represented in the precentral gyrus, with the feet at the top of the gyrus and the face at the bottom (Fig 12–4). The facial area is represented bilaterally, but the rest of the representation is unilateral, the cortical motor area controlling the musculature on the opposite side of the body. The cortical representation of each body part is proportionate in size to the skill with

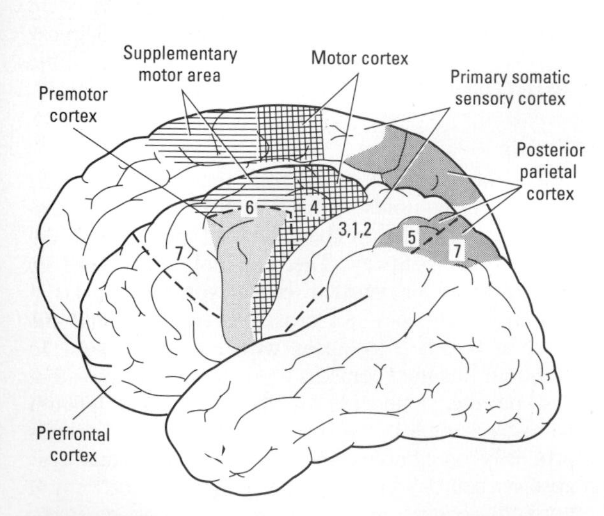

Figure 12–3. Medial (*above*) and lateral (*below*) view of human cerebral cortex showing motor cortex (Brodmann's area 4) and other areas concerned with control of voluntary movement, along with the numbers assigned to the regions by Brodmann. (Reproduced, with permission, from Kandel ER, Schwartz JH [editors]: *Principles of Neural Science,* 2nd ed. Elsevier, 1985.)

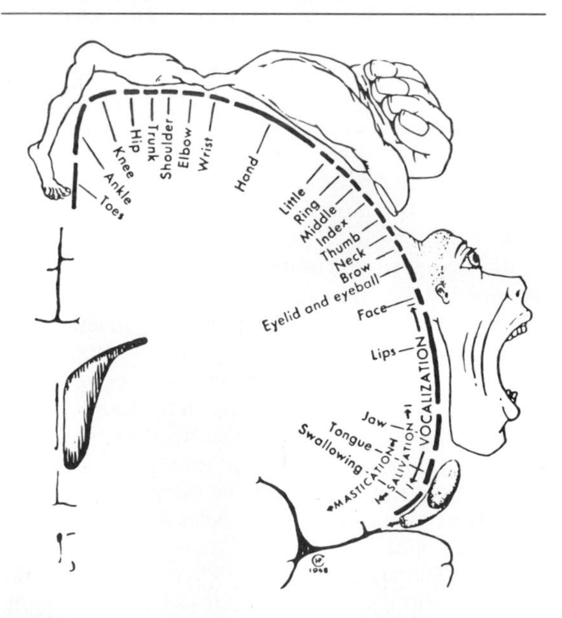

Figure 12—4. Motor homunculus. The figure represents, on a coronal section of the percentral gyrus, the location of the cortical representation of the various parts. The size of the various parts in proportionate to the amount of cortical area devoted to them. (Reproduced, with permission, from Penfield W. Rasmussen G: *The Cerebral Cortex of Man.* Macmillan, 1950.)

which the part is used in fine, voluntary movement. The areas involved in speech and hand movements are especially large in the cortex; use of the pharynx, lips, and tongue to form words and of the fingers and apposable thumbs to manipulate the environment are activities in which humans are especially skilled.

Studies of local blood flow in the brains of unanesthetized humans during motor activity, using the xenon washout or 2-deoxyglucose technique (see Chapter 32), provide additional information about motor function. Increased blood flow correlates with increased neuronal activity. When one foot is moved, for example, the foot region of the contralateral motor cortex shows increased blood flow, and when the fingers are moved, the hand area shows increased flow. When subjects count to themselves without speaking, the motor cortex is quiescent, but when they speak the numbers aloud as they count, blood flow increases in the motor cortex and the supplementary motor area. Thus, the supplementary motor area as well as the motor cortex is involved in voluntary movement when the movements being performed are complex and involve planning.

The conditions under which the human stimulation studies were performed precluded stimulation of the banks of the sulci and other inaccessible areas. In monkeys, meticulous study has shown that the hand area is not interposed between the trunk and face areas. Instead, there is a regular representation of the body, with the axial musculature and the proximal portions of the limbs represented along the anterior edge of the precentral gyrus and the distal part of the limbs along the posterior edge. The cells in the cortical motor areas are arranged in columns. The cells in each column receive fairly extensive sensory input from the peripheral area in which they produce movement, providing the basis for feedback control of movement. Some of this input may be direct, and some is relayed from somatic sensory area I in the postcentral gyrus.

Supplementary Motor Area

For the most part, the supplementary motor area projects to the motor cortex. It appears to be involved primarily in programming motor sequences. Lesions of this area in monkeys produce awkwardness in performing complex activities and difficulty with bimanual coordination. As noted above, blood flow increases in the human supplementary motor area when movements performed are complex and involve organizing motor discharge.

Premotor Cortex

The premotor cortex projects to the brain stem areas concerned with postural control and to the motor cortex as well as providing part of the corticospinal and corticobulbar output. Its function is still incompletely understood, but it may be concerned with setting posture at the start of a planned movement and with getting the individual ready to perform.

Posterior Parietal Cortex

In addition to providing fibers that run in the corticospinal and corticobulbar tracts, the somatic sensory area and related portions of the posterior parietal lobe project to the premotor area. Lesions of the somatic sensory area cause defects in motor performance that are characterized by inability to execute learned sequences of movements such as eating with a knife and fork. Some of the neurons in area 5 (Fig 12–3) are concerned with aiming the hands toward an object and manipulating it, whereas some of the neurons in area 7 are concerned with hand-eye coordination.

FUNCTION

Role in Movement

The corticospinal and corticobulbar system is the primary pathway for the initiation of skilled voluntary movement. This does not mean that movement—even skilled movement—is impossible without it. Nonmammalian vertebrates have essentially no corticospinal and corticobulbar system, but they move with great agility. Cats and dogs stand, walk, run, and even eat if food is presented to them after complete destruction of this system. It is only in primates that relatively marked deficits are produced.

Careful section of the pyramids producing highly selective destruction of the lateral corticospinal tract in laboratory primates produces prompt and sustained loss of the ability to grasp small objects between 2 fingers and to make isolated movements of the wrists. However, the animal can still use the hand in a gross fashion and can stand and walk. These deficits are consistent with loss of control of the distal musculature of the limbs, which is concerned with fine skilled movements. However, some control of distal musculature persists through the collaterals to the rubrospinal tract (see above). On the other hand, lesions of the ventral corticospinal tract produce axial muscle deficits that cause difficulty with balance, walking, and climbing. The corticospinal fibers from the parietal lobe, which are phylogenetically the oldest, end in the dorsal rather than the ventral horn and are presumably concerned with direct sensory-motor coordination.

Effects on Stretch Reflexes

Section of the pyramids in monkeys produces prolonged hypotonia and flaccidity rather than spasticity. The anatomic arrangements in humans are such that disease processes rarely, if ever, damage the corticospinal and corticobulbar tracts without also destroying posture-regulating pathways. When spasticity is present, it is probably due to damage to these latter pathways rather than to the corticospinal and corticobulbar tracts.

Damage to the lateral corticospinal tract in humans produces the **Babinski sign:** dorsiflexion of the great toe and fanning of the other toes when the lateral aspect of the sole of the foot is scratched. Except in infancy, the normal response to this stimulation is plantar flexion of all the toes. The Babinski sign is believed to be a flexor withdrawal reflex that is normally held in check by the lateral corticospinal system. It is of value in the localization of disease processes, but its physiologic significance is unknown.

Transcortical Load-Compensating Reflex

When a human is performing a motor task such as pushing a lever and the resistance to pushing is suddenly increased, there is a prompt increase in the force of the movement. This increase is due in part to monosynaptic stretch reflexes, but most of it has a latency of more than 50 ms, considerably longer than the latency of stretch reflexes. On the other hand, its latency is shorter than the time required for a voluntary response, which is about 100 ms. In conscious monkeys, the increase is associated with increased firing of motor neurons in the motor cortex, and it appears to be a transcortical load-compensating reflex initiated by feedback information to the cortex from the affected limb. This long-loop reflex is probably important in adjusting motor responses and making them appropriate.

POSTURE-REGULATING SYSTEMS

The posture-regulating mechanisms are multiple. They involve a whole series of nuclei and many structures, including the spinal cord, the brain stem, and the cerebral cortex. They are concerned not only with static posture but also, in concert with the corticospinal and corticobulbar systems, with the initiation and control of movement.

Integration

At the spinal cord level, afferent impulses produce simple reflex responses. At higher levels in the nervous system, neural connections of increasing complexity mediate increasingly complicated motor responses. This principle of levels of motor integration is illustrated in Table 12–1. In the intact animal, the individual motor responses are fitted into or "submerged" in the total pattern of motor activity. When the neural axis is transected, the activities integrated below the section are cut off or **released** from the "control of higher brain centers" and often appear to be accentuated. Release of this type, long a cardinal principle in neurology, may in some situations be due to removal of an inhibitory control by higher neural centers. A more important cause of the apparent hyperactivity is loss of differentiation of the reaction, so that it no longer fits into the broader pattern of motor activity. An additional factor may be denervation hypersensitivity of the centers below the transection, but the role of this component remains to be determined.

Postural Control

It is impossible to separate postural adjustments from voluntary movement in any rigid way, but it is possible to differentiate a series of postural reflexes (Table 12–2) that not only maintain the body in an upright, balanced position but also provide the constant adjustments necessary to maintain a stable postural background for voluntary activity. These adjustments include maintained **static** reflexes and dynamic, short-term **phasic** reflexes. The former involve sustained contraction of the musculature, whereas the latter involve transient movements. Both are integrated at various levels in the central nervous system from the spinal cord to the cerebral cortex and are effected largely through various motor pathways. A major factor in postural control is variation in the threshold of the spinal stretch reflexes caused in turn by changes in the excitability of motor neurons and,

Table 12–1. Summary of levels involved in various neural functions.

	Preparation						
Functions	Normal	Decorticate	Midbrain	Hindbrain (Decerebrate)	Spinal	Decerebellate	Level of Integration
Initiative, memory, etc	+	0	0	0	0	+	Cerebral cortex required
Conditioned reflexes	+	+*	0	0	0	+	Cerebral cortex facilitates
Emotional responses	+	++	0	0	0	+	Hypothalamus, limbic system
Locomotor reflexes	+	++	+	0	0	Incoordinate	Midbrain, thalamus
Righting reflexes	+	+	++	0	0	Incoordinate	Midbrain
Antigravity reflexes	+	+	+	++	0	Incoordinate	Medulla
Respiration	+	+	+	+	0	+	Lower medulla
Spinal reflexes†	+	+	+	+	++	+	Spinal cord

Legend: 0 = absent; + = present; ++ = accentuated.
*Conditioned reflexes are more difficult to establish in decorticate than in normal animals.
†Other than stretch reflexes.

Table 12–2. Principal postural reflexes.

Reflex	Stimulus	Response	Receptor	Integrated In
Stretch reflexes	Stretch	Contraction of muscle	Muscle spindles	Spinal cord, medulla
Positive supporting (magnet) reaction	Contact with sole or palm	Foot extended to support body	Proprioceptors in distal flexors	Spinal cord
Negative supporting reaction	Stretch	Release of positive supporting reaction	Proprioceptors in extensors	Spinal cord
Tonic labyrinthine reflexes	Gravity	Contraction of limb extensor muscles	Otolithic organs	Medulla
Tonic neck reflexes	Head turned: (1) To side (2) Up (3) Down	Change in pattern of extensor contraction (1) Extension of limbs on side to which head is turned (2) Hind legs flex (3) Forelegs flex	Neck proprioceptors	Medulla
Labyrinthine righting reflexes	Gravity	Head kept level	Otolithic organs	Midbrain
Neck righting reflexes	Stretch of neck muscles	Righting of thorax and shoulders, then pelvis	Muscle spindles	Midbrain
Body on head righting reflexes	Pressure on side of body	Righting of head	Exteroceptors	Midbrain
Body on body righting reflexes	Pressure on side of body	Righting of body even when head held sideways	Exteroceptors	Midbrain
Optical righting reflexes	Visual cues	Righting of head	Eyes	Cerebral cortex
Placing reactions	Various visual, exteroceptive, and proprioceptive cues	Foot placed on supporting surface in position to support body	Various	Cerebral cortex
Hopping reactions	Lateral displacement while standing	Hops, maintaining limbs in position to support body	Muscle spindles	Cerebral cortex

indirectly, by changes in the rate of discharge in the γ efferent neurons to muscle spindles.

SPINAL INTEGRATION

The responses of animals and humans after spinal cord transection of the cervical region illustrate the integration of reflexes at the spinal level. The individual spinal reflexes are discussed in detail in Chapter 6.

Spinal Shock

In all vertebrates, transection of the spinal cord is followed by a period of **spinal shock** during which all spinal reflex responses are profoundly depressed. During it, the resting membrane potential of the spinal motor neurons is 2–6 mV greater than normal. Subsequently, reflex responses return and become relatively hyperactive. The duration of spinal shock is proportionate to the degree of encephalization of motor function in the various species. In frogs and rats it lasts for minutes; in dogs and cats it lasts for 1–2 hours; in monkeys it lasts for days; and in humans it usually lasts for a minimum of 2 weeks.

The cause of spinal shock is uncertain. Cessation of tonic bombardment of spinal neurons by excitatory impulses in descending pathways undoubtedly plays a role, but the subsequent return of reflexes and their eventual hyperactivity also need to be explained. The recovery of reflex excitability may be due to the development of denervation hypersensitivity to the mediators released by the remaining spinal excitatory endings. Another possibility for which there is some evidence is the sprouting of collaterals from existing neurons, with the formation of additional excitatory endings on interneurons and motor neurons.

The first reflex response to appear as spinal shock wears off in humans is frequently a slight contraction of the leg flexors and adductors in response to a noxious stimulus. In some patients, the knee jerks come back first. The interval between cord transection and the beginning return of reflex activity is about 2 weeks in the absence of any complications, but if complications are present it is much longer. It is not known why infection, malnutrition, and other complications of cord transection inhibit spinal reflex activity.

Complications of Cord Transection

The problems in the management of paraplegic and quadriplegic humans are complex. Like all immobilized patients, they develop a negative nitrogen balance and catabolize large amounts of body protein. The weight of the body compresses the circulation to the skin over bony prominences, so that unless the patient is moved frequently the skin breaks down at these points and **decubitus ulcers** form. The ulcers heal poorly and are prone to infection because of body protein depletion. The tissues broken down include the protein matrix of bone, and calcium is released in large amounts. The hypercalcemia leads to hypercalciuria, and calcium stones often form in the urinary tract. The stones and the paralysis of bladder function both cause urinary stasis, which predisposes to urinary infection. Therefore, the prognosis in patients with transected spinal cords used to be very poor, and death from septicemia, uremia, or inanition was the rule. Since World War II, however, the use of antibiotics and meticulous attention to nutrition, fluid balance, skin care, bladder function, and general nursing care have made it possible for many of these patients to survive and lead meaningful lives.

Responses in Chronic Spinal Animals & Humans

Once the spinal reflexes begin to reappear after spinal shock, their threshold steadily drops. In chronically quadriplegic humans, the threshold of the withdrawal reflex is especially low. Even minor noxious stimuli may cause not only prolonged withdrawal of one extremity but marked flexion-extension patterns in the other 3 limbs. Repeated flexion movements may occur for prolonged periods, and contractures of the flexor muscles develop. Stretch reflexes are also hyperactive, as are more complex reactions based on this reflex as well. For example, if a finger is placed on the sole of the foot of an animal after the spinal cord has been transected **(spinal animal),** the limb usually extends, following the examining finger. This **magnet reaction (positive supporting reaction)** involves proprioceptive as well as tactile afferents and transforms the limb into a rigid pillar to resist gravity and support the animal. Its disappearance is also in part an active phenomenon **(negative supporting reaction)** initiated by stretch of the extensor muscles. On the basis of the positive supporting reaction, spinal cats and dogs can be made to stand, albeit awkwardly, for as long as 2–3 minutes.

Locomotion Generator

Not only can spinal cats and dogs be made to stand, but a circuit intrinsic to the spinal cord produces walking movements when stimulated in a suitable fashion. Thus, the **pattern generator** for walking is located in the spinal cord. However, this does not mean that spinal animals can walk without stimulation; the pattern generator has to be turned on by tonic discharge of a discrete area in the midbrain, the mesencephalic locomotor region. In addition, of course, walking in the intact animal is altered, adjusted, and made appropriate by other motor pathways that descend from the brain.

Autonomic Reflexes

Reflex contractions of the full bladder and rectum occur in spinal animals and humans, although the bladder is rarely emptied completely. Hyperactive bladder reflexes can keep the bladder in a shrunken state long enough for hypertrophy and fibrosis of its wall to occur. Blood pressure is generally normal at rest, but the precise feedback regulation normally supplied by the baroreceptor reflexes is absent and wide swings in pressure are common. Bouts of sweating and blanching of the skin also occur.

Sexual Reflexes

Other reflex responses are present in the spinal animal, but in general they are only fragments of patterns that are integrated in the normal animal into purposeful sequences. The sexual reflexes are an example. Coordinated sexual activity depends upon a series of reflexes integrated at many neural levels and is absent after cord transection. However, genital manipulation in male spinal animals and humans produces erection and even ejaculation. In female spinal dogs, vaginal stimulation causes tail deviation and movement of the pelvis into the copulatory position.

Mass Reflex

In chronic spinal animals, afferent stimuli irradiate from one reflex center to another. When even a relatively minor noxious stimulus is applied to the skin, it may irradiate to autonomic centers and produce evacuation of the bladder and rectum, sweating, pallor, and blood pressure swings in addition to the withdrawal response. This distressing **mass reflex** can sometimes be used to give paraplegic patients a degree of bladder and bowel control. They can be trained to initiate urination and defecation by stroking or pinching their thighs, thus producing an intentional mass reflex.

MEDULLARY COMPONENTS

In experimental animals in which the hindbrain and spinal cord are isolated from the rest of the brain by transection of the brain stem at the superior border of the pons, the most prominent finding is marked spasticity of the body musculature. The operative procedure is called **decerebration,** and the resulting pattern of spasticity is called **decerebrate rigidity.** Decerebration produces no phenomenon akin to spinal shock, and the rigidity develops as soon as the brain stem is transected.

Mechanism of Decerebrate Rigidity

On analysis, decerebrate rigidity is found to be spasticity due to diffuse facilitation of stretch reflexes (see Chapter 6). The facilitation is due to 2 factors: increased general excitability of the motor neuron pool and increase in the rate of discharge in the γ efferent neurons.

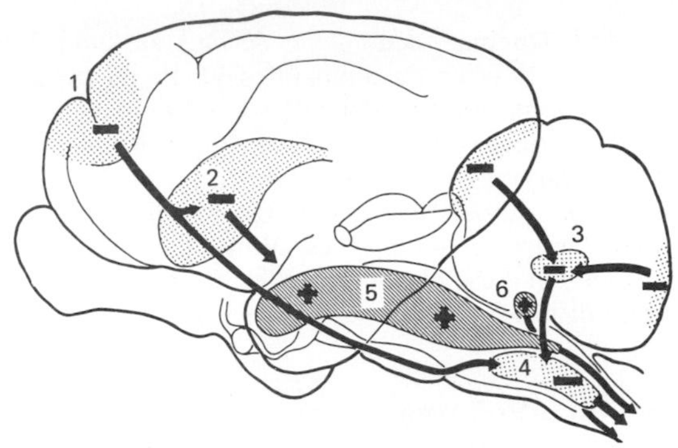

Figure 12–5. Areas in the cat brain where stimulation produces facilitation (plus signs) or inhibition (minus signs) of stretch reflexes. *1,* motor cortex; *2,* basal ganglia; *3,* cerebellum; *4,* reticular inhibitory area; *5,* reticular facilitatory area; *6,* vestibular nuclei. (Reproduced, with permission, from Lindsley DB, Schreiner LH, Magoun HW: An electromyographic study of spasticity. *J Neurophysiol* 1949;**12**:197.)

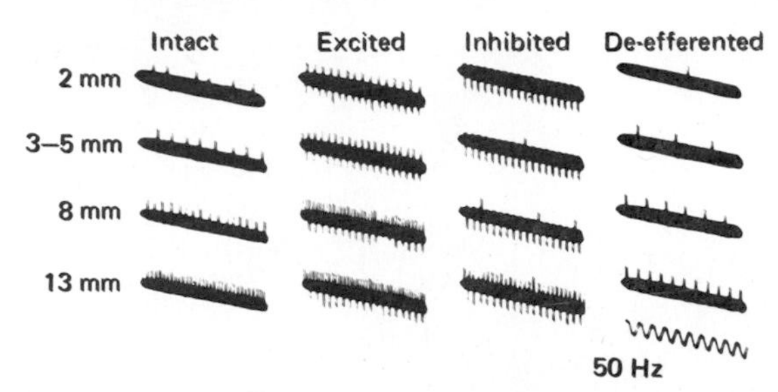

Figure 12–6. Response of single afferent fiber from muscle spindle to various degrees of muscle stretch. Numbers at left indicate amount of stretch. The upward deflections are action potentials; the downward deflections are stimulus artifacts. Records were obtained before brain stimulation (first column), during stimulation of brain areas facilitating (second column) and inhibiting (third column) stretch reflexes, and after section of the motor nerve (fourth column). (Reproduced, with permission, from Eldred E, Granit R, Merton PA: Supraspinal control of muscle spindles. *J Physiol [Lond]* 1953;**122**:498.)

Supraspinal Regulation of Stretch Reflexes

The brain areas that facilitate and inhibit stretch reflexes are shown in Fig 12–5. These areas generally act by increasing or decreasing spindle sensitivity (Fig 12–6). The large facilitatory area in the brain stem reticular formation discharges spontaneously, possibly in response to afferent input like the reticular activating system. However, the smaller brain stem area that inhibits γ efferent discharge does not discharge spontaneously but is driven instead by fibers from the cerebral cortex and the cerebellum. The inhibitory area in the basal ganglia may act through descending connections, as shown in Fig 12–5, or by stimulating the cortical inhibitory center. From the reticular inhibitory and facilitatory areas, impulses descend in the lateral funiculus of the spinal cord. When the brain stem is transected at the level of the top of the pons, 2 of the 3 inhibitory areas that drive the reticular inhibitory center are removed. Discharge of the facilitatory area continues, but that of the inhibitory area is decreased. Consequently, the balance of facilitatory and inhibitory impulses converging on the γ efferent neurons shifts toward facilitation. Gamma efferent discharge is increased, and stretch reflexes become hyperactive. The cerebellar inhibitory area is still present, and in decerebrate animals, removal of the cerebellum increases the rigidity. The influence of the cerebellum is complex, however, and destruction of the cerebellum in humans produces hypotonia rather than spasticity.

The vestibulospinal and some related descending pathways are also facilitatory to stretch reflexes and promote rigidity. Unlike the reticular pathways, they pass primarily in the anterior funiculus of the spinal cord, and the rigidity due to increased discharge in them is not abolished by deafferentation of the muscles. This indicates that this rigidity is due to a direct action on the α motor neurons to increase their excitability, rather than an effect mediated through the small motor nerve system, which would, of course, be blocked by deafferentation.

Significance of Decerebrate Rigidity

In dogs and cats, the spasticity produced by decerebration is most marked in the extensor muscles. Sherrington pointed out that these are the muscles with which the cat and dog resist gravity; the decerebrate posture in these animals is, as he put it, "a caricature of the normal standing position." What has been uncovered by decerebration, then, are the tonic, static postural reflex mechanisms that support the animal against gravity. Additional evidence that this is the correct interpretation of the phenomenon comes from the observation that decerebration in the sloth, an arboreal animal that hangs upside down from branches most of the time, causes rigidity in flexion. In humans, the pattern in true decerebrate rigidity is extensor in all 4 limbs, like that in cats and dogs. Apparently, human beings are not far enough removed from their quadruped ancestors to have changed the pattern in their upper extremities even though the main antigravity muscles of the arms in the upright position are flexors. However, decerebrate rigidity is rare in disease states, and the defects that produce it are usually incompatible with life. The more common pattern of extensor rigidity in the legs and moderate flexion in the arms is actually **decorticate rigidity** due to lesions of the cerebral cortex, with most of the brain stem intact (Fig 12–7).

Tonic Labyrinthine Reflexes

In the decerebrate animal, the pattern of rigidity in the limbs varies with the position. No righting responses are present, and the animal stays in the position in which it is put. If the animal is placed on

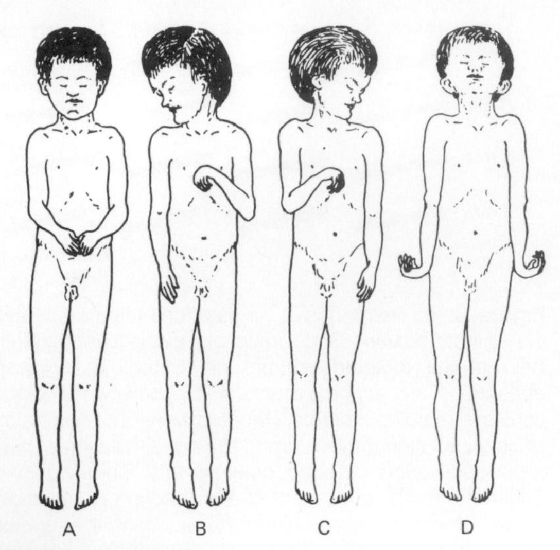

Figure 12–7. Human decorticate rigidity *(A–C)* and true decerebrate rigidity *(D)*. In *A* the patient is lying supine with head unturned. In *B* and *C,* the tonic neck reflex patterns produced by turning of the head to the right or left are shown. (Reproduced, with permission, from Fulton JF [editor]: *Textbook of Physiology,* 17th ed. Saunders, 1955.)

its back, the extension of all 4 limbs is maximal. As the animal is turned to either side, the rigidity decreases, and when it is prone, the rigidity is minimal though still present. These changes in rigidity, the **tonic labyrinthine reflexes,** are initiated by the action of gravity on the otolithic organs and are effected via the vestibulospinal tracts. They are rather surprising in view of the role of rigidity in standing, and their exact physiologic significance remains obscure.

Tonic Neck Reflexes

If the head of a decerebrate animal is moved relative to the body, changes in the pattern of rigidity occur. If the head is turned to one side, the limbs on that side (''jaw limbs'') become more rigidly extended while the contralateral limbs become less so. This is the position often assumed by a normal animal looking to one side. Flexion of the head causes flexion of the forelimbs and continued extension of the hindlimbs, the posture of an animal looking into a hole in the ground, Extension of the head causes flexion of the hindlimbs and extension of the forelimbs, the posture of an animal looking over an obstacle. These responses are the **tonic neck reflexes.** They are initiated by stretch of the proprioceptors in the upper part of the neck, and they can be sustained for long periods.

MIDBRAIN COMPONENTS

After section of the neural axis at the superior border of the midbrain **(midbrain animal),** extensor rigidity like that seen in the decerebrate animal is present only when the animal lies quietly on its back. In the decerebrate animal, the rigidity, which is a static postural reflex, is prominent because there are no modifying phasic postural reflexes. Chronic midbrain animals can rise to the standing position, walk, and right themselves. While the animals are engaged in these phasic activities, the static phenomenon of rigidity is not seen.

Righting Reflexes

Righting reflexes operate to maintain the normal standing position and keep an animal's head upright. These reflexes are a series of responses integrated for the most part in the nuclei of the midbrain.

When the midbrain animal is held by its body and tipped from side to side, the head stays level in response to the **labyrinthine righting reflexes.** The stimulus is tilting of the head, which stimulates the otolithic organs; the response is compensatory contraction of the neck muscles to keep the head level. If the animal is laid on its side, the pressure on that side of the body initiates reflex righting of the head even if the labyrinths have been destroyed. This is the **body on head righting reflex.** If the head is righted by either of these mechanisms and the body remains tilted, the neck muscles are stretched. Their contraction rights the thorax and initiates a wave of similar stretch reflexes that pass down the body, righting the abdomen and the hindquarters **(neck righting reflexes).** Pressure on the side of the body may cause body righting even if the head is prevented from righting **(body on body righting reflex).**

In cats, dogs, and primates, visual cues can initiate **optical righting reflexes** that right the animal in the absence of labyrinthine or body stimulation. Unlike the other righting reflexes, these responses depend upon an intact cerebral cortex.

Grasp Reflex

When a primate in which the brain tissue above the thalamus has been removed lies on its side, the limbs next to the supporting surface are extended. The upper limbs are flexed, and the hand on the upper side grasps firmly any object brought in contact with it **(grasp reflex).** This whole response is probably a supporting reaction that steadies the animal and aids in pulling upright.

Other Midbrain Responses

Animals with intact midbrains show pupillary light reflexes if the optic nerves are also intact. Nystagmus, the reflex response to rotational acceleration described in Chapter 9, is also present. If a blindfolded animal is lowered rapidly, its forelegs extend and its toes spread. This response to linear acceleration is a **vestibular placing reaction** that prepares the animal to land on the floor.

CORTICAL COMPONENTS

Effects of Decortication

Removal of the cerebral cortex **(decortication)** produces little motor deficit in many species of mammals. In primates, the deficit is more severe, but movement is still possible. Decorticate animals have all the reflex patterns of midbrain animals. In addition, decorticate animals are easier to maintain than midbrain animals because temperature regulation and other visceral homeostatic mechanisms integrated in the hypothalamus (see Chapter 14) are present. The most striking defect is inability to react in terms of past experience. With certain special types of training, conditioned reflexes can be established in the absence of the cerebral cortex, but under normal laboratory conditions, there is no evidence that learning or conditioning occurs.

Decorticate Rigidity

Moderate rigidity is present in the decorticate animal as a result of the loss of the cortical area that inhibits γ efferent discharge via the reticular formation. Like the rigidity present after transection of the neural axis anywhere above the top of the midbrain, this **decorticate rigidity** is obscured by phasic postural reflexes and is seen only when the animal is at rest. Decorticate rigidity is seen on the hemiplegic side in humans after hemorrhages or thromboses in the internal capsule. Probably because of their anatomy, the small arteries in the internal capsule are especially prone to rupture or thrombotic obstruction, so this type of decorticate rigidity is common. Sixty percent of intracerebral hemorrhages occur in the internal capsule, as opposed to 10% in the cerebral cortex, 10% in the pons, 10% in the thalamus, and 10% in the cerebellum.

The exact site of origin in the cerebral cortex of the fibers that inhibit stretch reflexes is a subject of debate. Under certain experimental conditions, stimulation of the anterior edge of the precentral gyrus is said to cause inhibition of stretch reflexes and cortically evoked movements. This region, which also projects to the basal ganglia, has been named area 4s or the **suppressor strip.** Four other suppressor regions (Brodmann's areas 2, 8, 19, and 24) have also been described. However, the stimulation experiments on which the hypothesis of discrete suppressor areas is based have not been uniformly confirmed.

Placing & Hopping Reactions

Two types of postural reactions are seriously disrupted by decortication, the **hopping** and **placing reactions.** The former are the hopping movements that keep the limbs in position to support the body when a standing animal is pushed laterally. The latter are the reactions that place the foot firmly on a supporting surface. They can be initiated in a blindfolded animal held suspended in the air by touching the supporting surface with any part of the foot. Similarly, when the snout or vibrissae of a suspended animal touch a table, the animal immediately places both forepaws on the table; and if one limb of a standing animal is pulled out from under it, the limb is promptly replaced on the supporting surface. The vestibular placing reaction has already been mentioned. In cats, dogs, and primates, the limbs are extended to support the body when the animal is lowered toward a surface it can see.

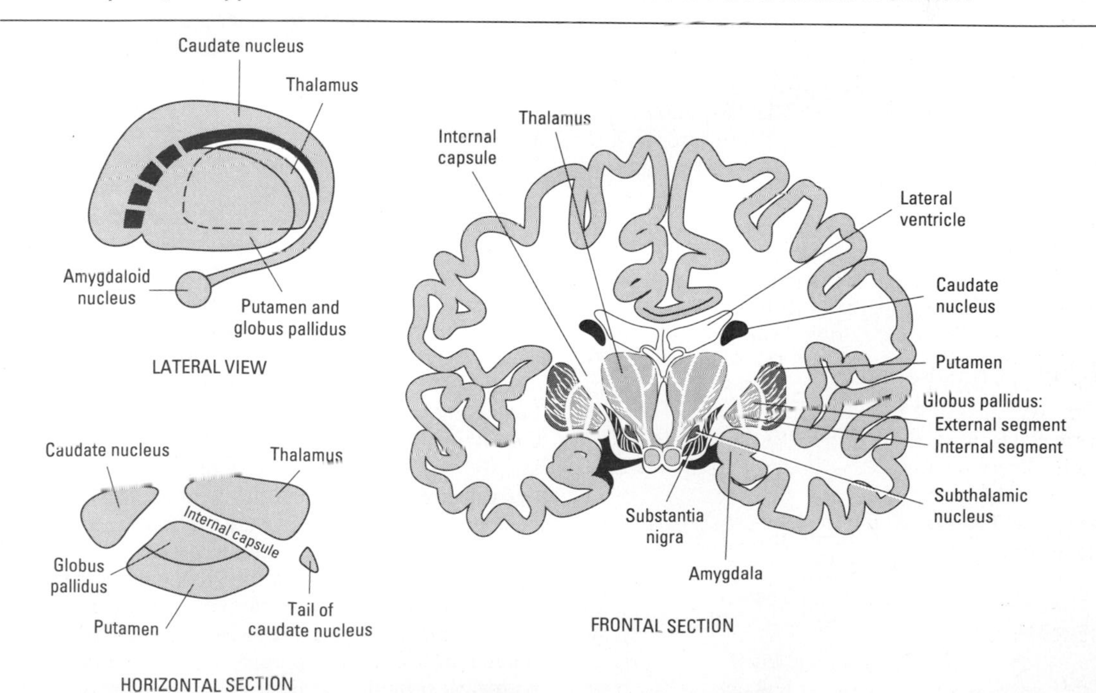

Figure 12–8. The basal ganglia.

BASAL GANGLIA

Anatomic Considerations

The term **basal ganglia** is generally applied to the **caudate nucleus, putamen,** and **globus pallidus**—the 3 large nuclear masses underlying the cortical mantle (Fig 12–8)—and the functionally related **subthalamic nucleus** (body of Luys) and **substantia nigra** on each side. The globus pallidus is divided into an external and an internal segment. Parts of the thalamus are intimately related to the basal ganglia. The caudate nucleus and the putamen are sometimes called the **striatum;** the putamen and the globus pallidus are sometimes called the **lenticular nucleus** (Table 12–3).

The main afferent connections to the basal ganglia terminate in the striatum (Fig 12–9). They include the **corticostriate projection** from all parts of the cerebral cortex and a projection from the centromedian nucleus of the thalamus, which in turn has its own input from the cortex.

The connections between the parts of the basal ganglia include a dopaminergic nigrostriatal projection from the substantia nigra to the striatum and a corresponding GABA-ergic projection from the striatum to the substantia nigra. The caudate nucleus and the putamen project to both segments of the globus pallidus. The external segment of the globus pallidus projects to the subthalamic nucleus, which in turn projects to both segments of the globus pallidus and the substantia nigra.

The principal output from the basal ganglia is from the internal segment of the globus pallidus via the **thalamic fasciculus** to the ventral lateral, ventral anterior, and centromedian nuclei of the thalamus. From the thalamic nuclei, fibers project to the prefrontal and premotor cortex. The substantia nigra also projects to the thalamus. There are a few additional projections to the habenula, superior colliculus and brain stem. However, the main feature of the connections of basal ganglia is that the cerebral cortex projects to the striatum, the striatum to the internal segment of the globus pallidus, the internal segment of the globus pallidus to the thalamus, and the thalamus back to the cortex, completing a loop.

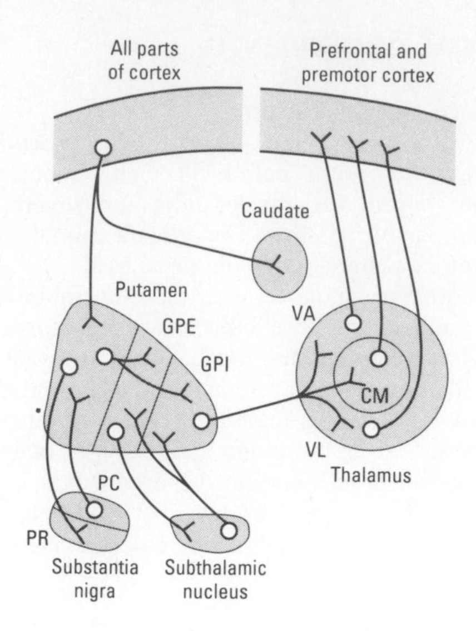

Figure 12–9. Principal connections of the basal ganglia. All parts of the cortex project to the striatium. There is a dopaminergic pathway from the pars compacta (PC) of the substantia nigra to the striatum and a return GABA-ergic pathway to the pars reticularis (PR) of the substantia nigra. In addition, the external segment of the globus pallidus (GPE) projects to the subthalamic nucleus, and this nucleus projects to GPE and the internal segment of the globus pallidus (GPI). GPI projects via the thalamic fasciculus to the ventral anterior (VA), ventral lateral (VL), and centromedian (CM) nuclei of the thalamus. These nuclei in turn project to the prefrontal and premotor cortex, completing the corticocortical loop.

Metabolic Considerations

The metabolism of the basal ganglia is unique in a number of ways. These structures have a high O_2 consumption. The copper content of the substantia nigra and the nearby locus ceruleus is particularly high. In Wilson's disease, a genetic autosomal recessive disorder of copper metabolism in which the plasma copper-binding protein, **ceruloplasmin,** is usually low, there is chronic copper intoxication and severe degeneration of the lenticular nucleus.

Table 12–3. The basal ganglia.

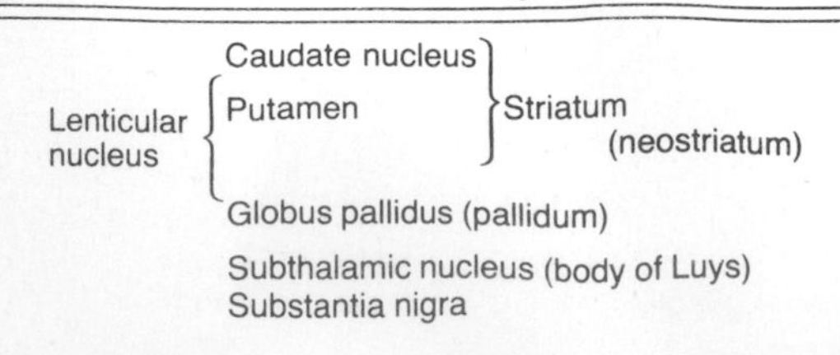

Caudate nucleus } Striatum (neostriatum)
Lenticular nucleus { Putamen } Striatum (neostriatum)
Lenticular nucleus { Globus pallidus (pallidum)
Subthalamic nucleus (body of Luys)
Substantia nigra

Function

Our knowledge of the precise functions of the basal ganglia is still rudimentary. Lesions in the basal ganglia of animals have relatively little effect. However, recording studies have made it clear that neurons in the basal ganglia, like those in the lateral portions of the cerebellar hemispheres, discharge before movements begin. These observations plus careful analysis of the effects of diseases of the basal ganglion in humans and the effects of drugs that destroy dopaminergic neurons in animals (see below) have led to the concept that the basal ganglia are involved in the planning and programming of movement or, more broadly, in the processes by which an abstract thought is converted into voluntary action (Fig 12–1). They discharge via the thalamus to areas related to the motor cortex, and the corticospinal pathways provide the final common pathway to the motor neurons.

Diseases of the Basal Ganglia in Humans

It is interesting that even though lesions in the basal ganglia in experimental animals have little apparent effect, disease processes affecting these ganglia in humans produce marked and characteristic abnormalities of motor function. Disorders of movement associated with diseases of the basal ganglia in humans are of 2 general types: **hyperkinetic** and **hypokinetic.** The hyperkinetic conditions, those in which there is excessive and abnormal movement, include chorea, athetosis, and ballism. In Parkinson's disease, there are both hyperkinetic and hypokinetic features.

Hyperkinetic Disorders

Chorea is associated with degeneration of the caudate nucleus. It is characterized by rapid, involuntary "dancing" movements. Choreiform and athetotic movements have been likened to the start of voluntary movements occurring in an involuntary, disorganized way. **Athetosis** is due to lesions in the lenticular nucleus and is characterized by continuous slow, writhing movements. In **ballism,** the involuntary movements are flailing, intense, and violent. They appear when the subthalamic nuclei are damaged, and a sudden onset of the movements on one side of the body **(hemiballism)** due to hemorrhage in the contralateral subthalamic nucleus is one of the most dramatic syndromes in clinical medicine.

Huntington's Disease

Huntington's disease is associated with loss of cholinergic and GABA-secreting neurons in the striatum as a result of an abnormal gene on the short arm of chromosome 4. It is characterized by dominant inheritance, onset usually between the ages of 30 and 50 of chorea and dementia, and steady progression to death in 15–20 years. At present, there is no effective treatment for the disease and the biochemical abnormality responsible for the loss of cells in the striatum is unknown.

Parkinson's Disease (Paralysis Agitans)

In the syndrome originally described by James Parkinson and named for him, the nigrostriatal system of dopaminergic neurons degenerates. Parkinsonism was a common late complication of the type of influenza that was epidemic during World War I, and it occurs today in idiopathic form in elderly individuals. The dopamine content of the caudate nucleus and putamen is about 50% of normal. The norepinephrine level in the hypothalamus is also reduced, but not to so great a degree. There is a steady loss of dopamine and dopamine receptors with age in the basal ganglia in normal individuals, and it is apparently acceleration of these losses that precipitates parkinsonism. Parkinsonism is also seen as a complication of treatment with the phenothiazine group of tranquilizer drugs, and other drugs that block D_2 dopamine receptors. It can be produced in rapid and dramatic form by injection of 1-methyl-4-phenyl-1,2,5,6-tetrahydropyridine (MPTP). This effect was discovered by chance when a drug dealer in California supplied some of his clients with a homemade preparation of "synthetic heroin" that contained MPTP. MPTP is metabolized by the enzyme monoamine oxidase B in various brain cells to produce a toxic metabolite, MPP^+. In rodents, this metabolite is rapidly removed from the brain, but in primates, it is removed at a lower rate and is taken up by dopaminergic neurons in the substantia nigra, which it destroys without affecting other dopaminergic neurons to any appreciable degree. Consequently, MPTP can be used to produce parkinsonism in monkeys, and its availability has accelerated research on the function of the basal ganglia.

The hallmarks of parkinsonism are **akinesia** or **poverty of movement** (a hypokinetic feature) and the hyperkinetic features **rigidity** and **tremor.** The absence of motor activity and the difficulty in initiating voluntary movements are striking. There is a decrease in **associated movements,** the normal, unconscious movements such as swinging of the arms during walking, the panorama of facial expressions related to the emotional content of thought and speech, and the multiple "fidgety" actions and gestures that occur in all of us. The rigidity is different from spasticity because there is increased motor neurons discharge to both the agonist and antagonist muscles. Passive motion of an extremity meets with a plastic, dead-feeling resistance that has been likened to bending a lead pipe and is therefore called **lead-pipe rigidity.** Sometimes there is a series of "catches" during passive motion **(cog-wheel rigidity),** but the sudden loss of resistance seen in a spastic extremity is absent. The tremor, which is present at rest and disappears with activity, is due to regular, alternating, 8/s contractions of antagonistic muscles.

An important consideration in Parkinson's disease is the balance between the output of cholinergic neurons and the output of dopaminergic neurons in the striatum. Some improvement is produced by decreasing the cholinergic influence with anticholinergic drugs. More dramatic improvement is produced by administration of L-dopa. Unlike dopamine, this dopamine precursor crosses the blood-brain barrier (see Chapter 15) and helps repair the dopamine deficiency Bromocriptine and several other synthetic dopamine agonists that penetrate brain tissue have also proved to be of value. Some patients have shown improvement when the medulla of one of their adrenal glands is transplanted to the ventricular surface of the caudate nucleus. The transplanted gland cells may provide an ongoing source of dopamine or they may provide a growth factor that induces neural repair. Implantation of tissue from the basal ganglia of fetuses is also being explored. However, all these transplantation procedures are still in the experimental stage, and their therapeutic value is not yet established.

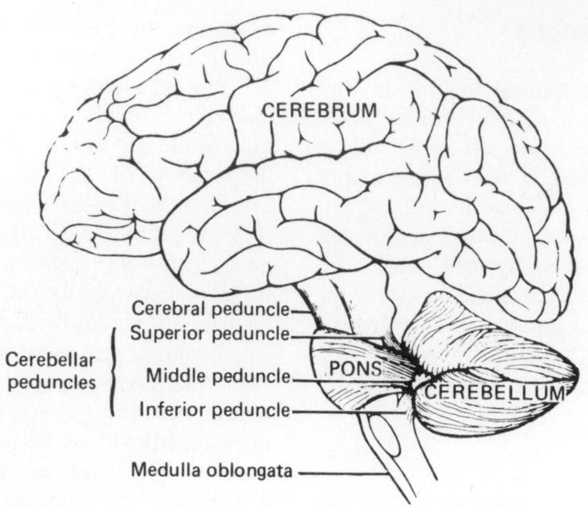

Figure 12–10. Diagrammatic representation of the principal parts of the brain. The parts are distorted to show the cerebellar peduncles and the way the cerebellum, pons, and middle peduncle form a "napkin ring" around the brain stem. (Reproduced, with permission, from Goss CM [editor]: *Gray's Anatomy of the Human Body,* 27th ed. Lea & Febiger, 1959.)

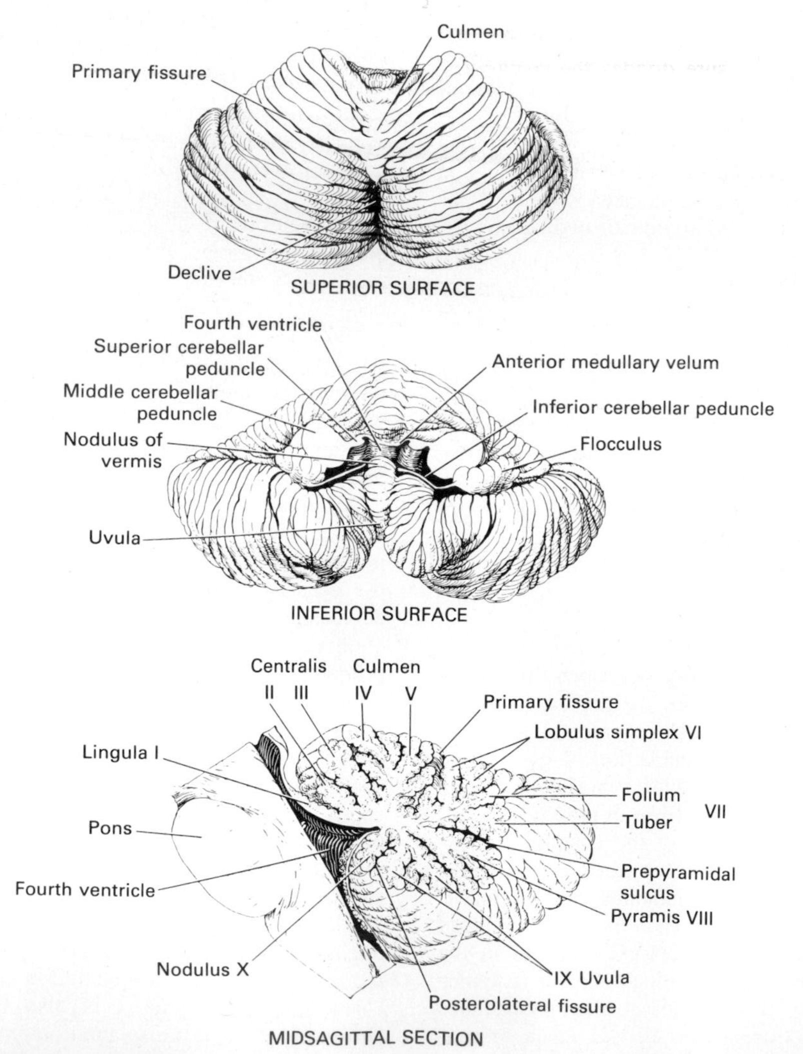

Figure 12–11. Superior and inferior views and sagittal section of the human cerebellum. The 10 principal lobules are identified by name and by number (I–X).

CEREBELLUM

ANATOMIC & FUNCTIONAL ORGANIZATION

Anatomic Divisions

The cerebellum sits astride the main sensory and motor systems in the brain stem (Fig 12–10). It is connected to the brain stem on each side by a **superior peduncle** (brachium conjunctivum), **middle peduncle** (brachium pontis), and **inferior peduncle** (restiform body). The medial **vermis** and lateral **cerebellar hemispheres** are more extensively folded and fissured than the cerebral cortex; the cerebellum weighs only 10% as much as the cerebral cortex, but its surface area is about 75% of that of the cerebral cortex. Anatomically, the cerebellum is divided into 3 parts by 2 transverse tissues. The posterolateral fissure separates the medial nodulus and the lateral flocculus on either side from the rest of the cerebellum, and the primary fissure divides the remainder into an anterior and a posterior lobe. Lesser fissures divide the vermis into smaller sections, so that it contains 10 primary lobules numbered I–X from superior to inferior. These lobules are identified by name and number in Fig 12–11.

Functional Divisions

From a functional point of view, the cerebellum is also divided into 3 parts, but in a different way. The nodulus in the vermis and the flanking flocculus in the hemisphere on each side form the **flocculonodular lobe** (Fig 12–12). This lobe, which is phylogenetically the oldest part of the cerebellum, has vestibular connections and is concerned with equilibrium. The rest of the vermis and the adjacent medial portions of the hemispheres form the **spinocerebellum,** the region that receives proprioceptive input from the body as well as a copy of the "motor plan" from the motor cortex. By comparing plan with performance, it smooths and coordinates movements that are ongoing (Fig 12–13). The vermis projects to the brain stem area concerned with control of axial and proximal limb muscles, whereas the hemispheres project the brain stem areas concerned with control of distal limb muscles. The lateral portions of the cerebellar hemispheres are called the **neocerebellum.** They are the newest from a phylogenetic point of view, reaching their greatest development in humans. They interact with the motor cortex in planning and programming movements.

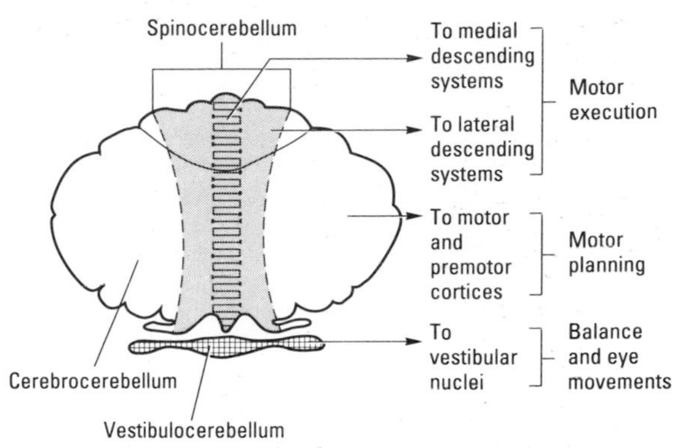

Figure 12–12. Functional division of the cerebellum. (Reproduced, with permission, from Kandel ER, Schwartz JH [editors]: *Principles of Neural Science,* 2nd ed. Elsevier, 1985.)

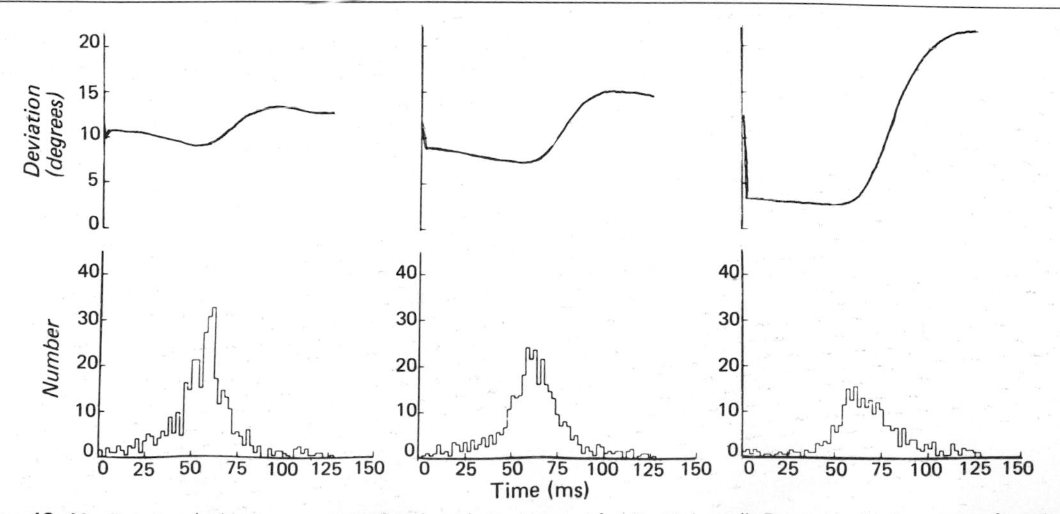

Figure 12–13. Relation between eye movement and discharge of a Purkinje cell. Deviation of the eyes of a monkey recorded in degrees on the top and the number of spikes per 2 ms from the Purkinje cell at the bottom. Note that the larger saccadic eye movement, the smaller the number of spikes in the Purkinje cell before the movement started. (Reproduced, with permission, from Llinás RR: Motor aspects of cerebellar control. *Physiologist* 1974;17:19.)

Table 12–4. Function of principal afferent systems to the cerebellum.*

Afferent Tracts	Transmits
Vestibulocerebellar	Vestibular impulses from labyrinths direct and via vestibular nuclei
Dorsal spinocerebellar	Proprioceptive and exteroceptive impulses from body
Ventral spinocerebellar	Proprioceptive and exteroceptive impulses from body
Cuneocerebellar	Proprioceptive impulses, especially from head and neck
Tectocerebellar	Auditory and visual impulses via inferior and superior colliculi
Olivocerebellar	Proprioceptive input from whole body via relay in inferior olive
Pontocerebellar	Impulses from motor and other parts of cerebral cortex via pontine nuclei

*Several other pathways transmit impulses from nuclei in the brain stem to the cerebellar cortex and to the deep nuclei, including a serotonergic input from the raphe nuclei and a noradrenergic input from the locus ceruleus.

Connections

The cerebellum has an external **cerebellar cortex** separated by white matter from the **deep cerebellar nuclei.** Its afferent input goes to the cortex and via collaterals, to the deep nuclei. There are 4 deep nuclei: the **dentate,** the **globose,** the **emboliform** and the **fastigial** nuclei. The globose and the emboliform are sometimes lumped together as the **interpositus nucleus.** The cerebellar cortex projects to these nuclei, and they provide the only output from the spinocerebellum and the neocerebellum. The medial portion of the spinocerebellum projects to the fastigial nuclei and from there to the brain stem. The adjacent hemispheric portions of the spinocerebellum project to the emboliform and globose nuclei and from them to the brain stem. The neocerebellum projects to the dentate nucleus and from there either directly or indirectly to the ventrolateral nucleus of the thalamus.

The afferent pathways to the cerebellum are summarized in Table 12–4. They transmit proprioceptive and sensory information from all parts of the body. One portion of the proprioceptive input is relayed via the inferior olive, and the olivocerebellar fibers form the excitatory climbing fiber input (see below). In addition, information is relayed to the cerebellum from all the motor areas in the cerebral cortex via the pontine nuclei. There is considerable localization of the sensory input, with at least 2 point-for-point representations in the spinocerebellum.

Neural Circuits

The cerebellar cortex contains only 5 types of neurons: Purkinje, granule, basket, stellate, and Golgi cells. It has 3 layers (Fig 12–14): an external molecular layer, a Purkinje cell layer that is only one cell thick, and an internal granular layer. The Purkinje cells are among the biggest neurons in the body. They

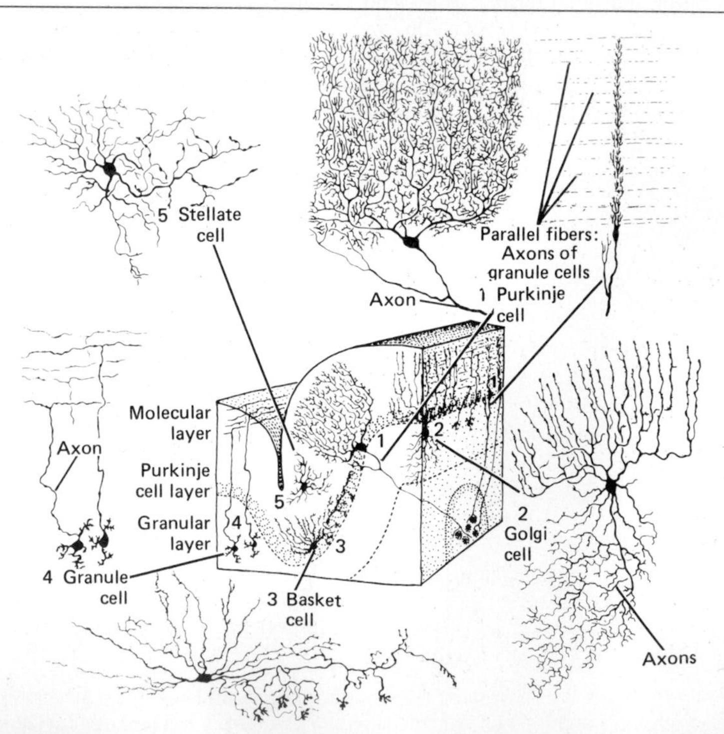

Figure 12–14. Location and structure of neurons in the cerebellar cortex. (Reproduced, with permission, from Kuffler SW, Nicholls JG, Martin AR: *From Neuron to Brain,* 2nd ed. Sinauer Associates, 1984.)

have very extensive dendritic arbors that extend throughout the molecular layer. Their axons, which are the only output from the cerebellar cortex, pass to the deep nuclei. The cerebellar cortex also contains **granule cells,** which receive input from the mossy fibers (see below) and innervate the Purkinje cells. The granule cells have their cell bodies in the granular layer. Each sends an axon to the molecular layer, where the axon bifurcates to form a T. The branches of the T are straight and run long distances. Consequently, they are called **parallel fibers.** The dendritic trees of the Purkinje cells are markedly flattened (Fig 12–14) and oriented at right angles to the parallel fibers. The parallel fibers thus make synaptic contact with the dendrites of many Purkinje cells, and the parallel fibers and Purkinje dendritic trees form a grid of remarkably regular proportions.

The other 3 types of neurons in the cerebellar cortex are in effect inhibitory interneurons. The **basket cells** (Fig 12–14) are located in the molecular layer. They receive input from the parallel fibers, and each projects to many Purkinje cells. Their axons form a basket around the cell body and axon hillock of each Purkinje cell they innervate. The **stellate cells** are similar to the basket cells but more superficial in location. The **Golgi cells** are located in the granular layer. Their dendrites, which project into the molecular layer, receive input from the parallel fibers. Their cell bodies receive input via collaterals from the incoming climbing fibers (see below) and the Purkinje cells. Their axons project to the dendrites of the granule cells.

There are 2 main sources of input to the cerebellar cortex: **climbing fibers** and **mossy fibers.** The climbing fibers come from the inferior olivary nuclei, and each projects to the primary dendrites of a Purkinje cell, around which it entwines like a climbing plant. The mossy fibers constitute most of the other incoming fibers. They end on the dendrites of granule cells in complex synaptic junctions called **glomeruli.** The glomeruli also contain the inhibitory endings of the Golgi cells mentioned above.

The fundamental circuits of the cerebellar cortex are thus relatively simple (Fig 12–15). Climbing fiber inputs exert a strong excitatory effect on single Purkinje cells, whereas mossy fiber inputs exert a weak excitatory effect on many Purkinje cells via the granule cells. The basket and stellate cells are also excited by granule cells via the parallel fibers, and their output inhibits Purkinje cell discharge. Golgi cells are excited by the mossy fiber collaterals, Purkinje cell collaterals, and parallel fibers, and they inhibit transmission from mossy fibers to granule cells. The transmitter secreted by the stellate, basket, Golgi, and Purkinje cells appears to be GABA, whereas the granule cells probably secrete glutamate.

The output of the Purkinje cells is in turn inhibitory to the deep cerebellar nuclei. These nuclei also receive excitatory inputs via collaterals from the mossy and climbing fibers, and they may also receive other excitatory inputs. It is interesting, in view of their inhibitory Purkinje cell input, that the output of the deep cerebellar nuclei to the brain stem and thalamus is always excitatory. Thus, the entire cerebellar circuitry seems to be concerned solely with modulating or timing the excitatory output of the deep cerebellar nuclei to the brain stem and thalamus.

Flocculonodular Lobe

The phylogenetically oldest part of the cerebellum, the **flocculonodular lobe,** has vestibular connections. Animals in which it has been destroyed walk in a staggering fashion on a broad base. They tend to fall and are reluctant to move without support. Similar defects are seen in children as the earliest signs of a midline cerebellar tumor arises from cell rests in the nodule and early in its course produces damage that is generally localized to the flocculonodular lobe.

Motion Sickness

Selective ablation of the flocculondular lobe in dogs abolishes the syndrome of **motion sickness,** whereas extensive lesions in other parts of the cerebellum and the rest of the brain fail to affect it. This "disease," a ubiquitous if minor complication of modern travel, is a consequence of excessive and repetitive labyrinthine stimulation due to the motion of the vehicle. It is also a problem of considerable magnitude in space flight. In humans, it can generally be controlled by a number of antiemetic drugs.

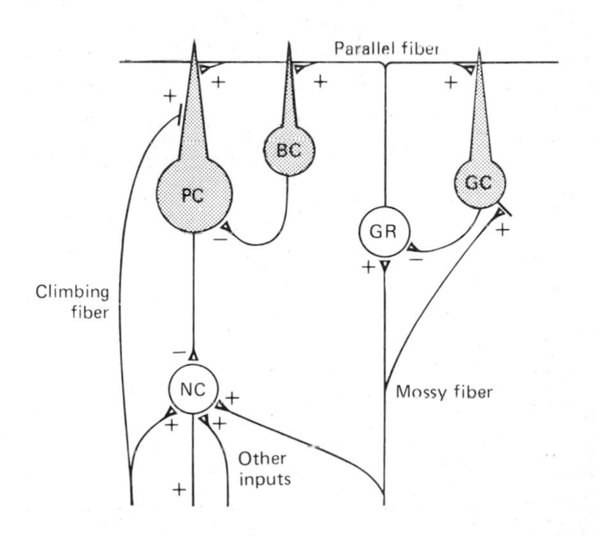

Figure 12–15. Diagram of neural connections in the cerebellum. Shaded neurons are inhibitory, and + and − signs indicate whether endings are excitatory or inhibitory. BC, basket cell; GC, Golgi cell; GR, granule cell; NC, nuclear cell; PC, Purkinje cell. The connections of the stellate cells are similar to those of the basket cells, except that they end for the most part on Purkinje cell dendrites. (Modified from Eccles JC, Itoh M, Szentágothai J: *The Cerebellum as a Neuronal Machine.* Springer, 1967.)

Effects on Stretch Reflexes

Stimulation of the cerebellar areas that receive proprioceptive input sometimes inhibits and sometimes facilitates movements evoked by stimulation of the cerebral cortex. Lesions in folia I–VI and the paramedian areas in experimental animals cause spasticity localized to the part of the body that is represented in the part of the cerebellum destroyed. However, hypotonia is characteristic of cerebellar destruction in humans.

Effects on Movement

Except for the changes in stretch reflexes, experimental animals and humans with lesions of the cerebellar hemispheres show no abnormalities as long as they are at rest. However, pronounced abnormalities are apparent when they move. There is no paralysis and no sensory deficit, but all movements are characterized by a marked **ataxia,** a defect defined as incoordination due to errors in the rate, range, force, and direction of movement. With circumscribed lesions, the ataxia may be localized to one part of the body. If only the cortex of the cerebellum is involved, the movement abnormalities gradually disappear as **compensation** occurs. Lesions of the cerebellar nuclei produce more generalized defects, and the abnormalities are permanent. For this reason, care should be taken to avoid damaging the nuclei when surgical removal of the parts of the cerebellum is necessary.

Other signs of cerebellar deficit in humans provide additional illustrations of the importance of the cerebellum in the control of movement. The common denominator of most cerebellar signs is inappropriate rate, range, force, and direction of movement. Ataxia is manifest not only in the wide-based, unsteady, "drunken" gait of patients but also in defects of the skilled movements involved in the production of speech, so that slurred or **scanning speech** results. Other voluntary movements are also highly abnormal. For example, attempting to touch an object with a finger results in overshooting to one side or the other. This **dysmetria,** which is also called **past-pointing,** promptly initiates a gross corrective action, but the correction overshoots to the other side. Consequently, the finger oscillates back and forth. This oscillation is the **intention tremor** of cerebellar disease. Unlike the resting tremor of parkinsonism, it is absent at rest; however, it appears whenever the patient attempts to perform some voluntary action. Another characteristic of cerebellar disease is inability to "put on the brakes," to stop movement promptly. Normally, for example, flexion of the forearm against resistance is quickly checked when the resistance force is suddenly broken off. The patient with cerebellar disease cannot brake the movement of the limb, and the forearm flies backward in a wide arc. This abnormal response is known as the **rebound phenomenon,** and similar impairment is detectable in other motor activities. This is one of the important reasons these patients show **adiadochokinesia,** the inability to perform rapidly alternating opposite movements such as repeated pronation and supination of the hands. Finally, patients with cerebellar disease have difficulty performing actions that involve simultaneous motion at more than one joint. They dissect such movements and carry them out one joint at a time, a phenomenon known as **decomposition of movement.**

The Cerebellum and Learning

The cerebellum is concerned with learned adjustments that make coordination easier when a given task is performed over and over. The basis of this learning is probably the input via the olivary nuclei. It is worth noting in this regard that each Purkinje cell receives inputs from 250,000 to 1 million mossy fibers, but each has only a single climbing fiber from the inferior olive, and this fiber makes 2000–3000 synapses on the Purkinje cell. Climbing fiber activation produces a large, complex spike in the Purkinje cell; and this spike in some way produces long-term modification of the pattern of mossy fiber input to that particular Purkinje cell. Climbing fiber activity is increased when a new movement is being learned, and selective lesions of the olivary complex abolish the ability to produce long-term adjustments in certain motor responses. The role of the cerebellum in the development of conditioned reflexes is discussed in Chapter 16.

Mechanisms

Although the functions of the flocculonodular lobe, spinocerebellum, and neocerebellum are relatively clear and the cerebellar circuits are simple, the exact ways their different parts carry out their functions are still unknown. The relation of the electrical events in the cerebellum to its function in motor control is another interesting problem. The cerebellar cortex has a basic, 150–300/s, 200-μV electrical rhythm and, superimposed on this, a 1000–2000/s component of smaller amplitude. The frequency of the basic rhythm is thus more than 10 times as great as that of the similarly recorded cerebral cortical EEG, and the amplitude is considerably less. Incoming stimuli generally alter the amplitude of the cerebellar rhythm, like a broadcast signal modulating a carrier frequency in radio transmission. However, the significance of these electrical phenomena in terms of cerebellar function is not known.

The Autonomic Nervous System 13

INTRODUCTION

The autonomic nervous system, like the somatic nervous system, is organized on the basis of the reflex arc. Impulses initiated in visceral receptors are relayed via afferent autonomic pathways to the central nervous system, integrated within it at various levels, and transmitted via efferent pathways to visceral effectors. This organization deserves emphasis because the functionally important afferent components have often been ignored. The visceral receptors and afferent pathways have been considered in Chapter 5 and 7 and the major autonomic effector, smooth muscle, in Chapter 3. The efferent pathways to the viscera are the subject of this chapter. Autonomic integration in the central nervous system is considered in Chapter 14.

ANATOMIC ORGANIZATION OF AUTONOMIC OUTFLOW

The peripheral motor portions of the autonomic nervous system are made up of **preganglionic** and **postganglionic neurons** (Figs 13–1 and 13–2). The cell bodies of the preganglionic neurons are located in the intermediolateral gray (visceral efferent) column of the spinal cord or the homologous motor nuclei of the cranial nerves. Their axons are mostly myelinated, relatively slow-conducting B fibers. The axons synapse on the cell bodies of postganglionic neurons that are located in all cases outside the central nervous system. Each preganglionic axon diverges to an average of 8–9 postganglionic neurons. In this way, autonomic output is diffused. The axons of the postganglionic neurons, mostly unmyelinated C fibers, end on the visceral effectors.

Anatomically, the autonomic outflow is divided into 2 components: the **sympathetic** and **parasympathetic** divisions of the autonomic nervous system.

Sympathetic Division

The axons of the sympathetic preganglionic neurons leave the spinal cord with the ventral roots of the first thoracic to the third or fourth lumbar spinal nerves. They pass via the **white rami communicantes** to the **paravertebral sympathetic ganglion chain,** where most of them end on the cell bodies of the postganglionic neurons. The axons of some of the

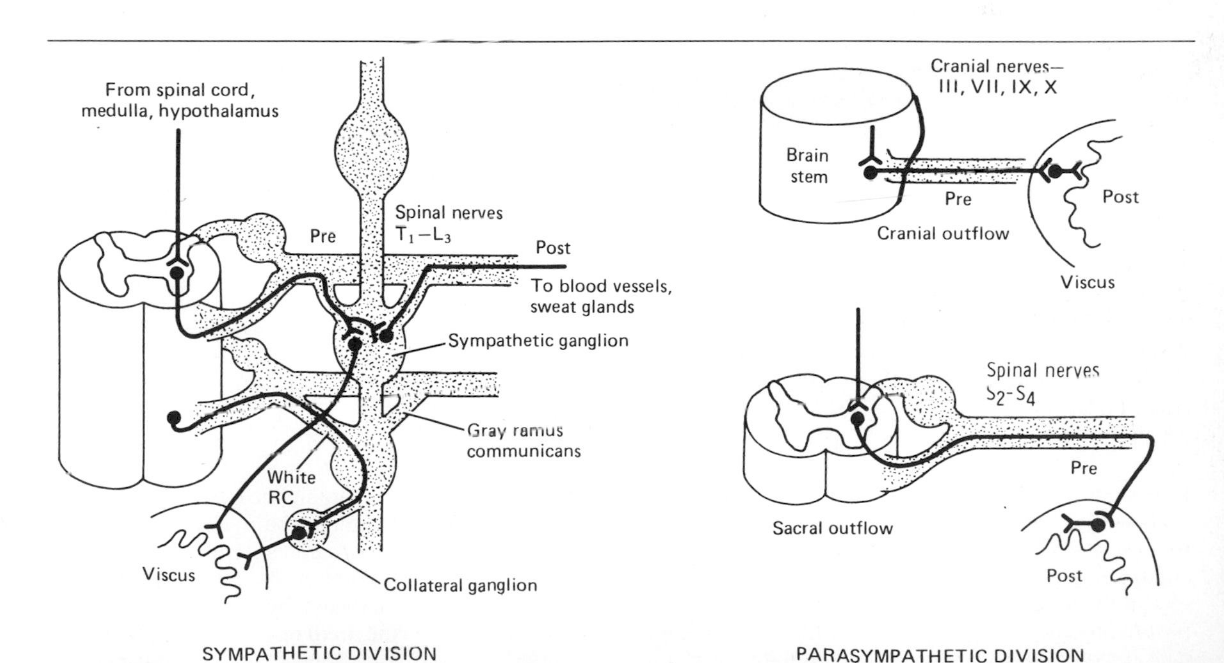

Figure 13–1. Autonomic nervous system. Pre, preganglionic neuron; Post, postganglionic neuron; RC, ramus communicans.

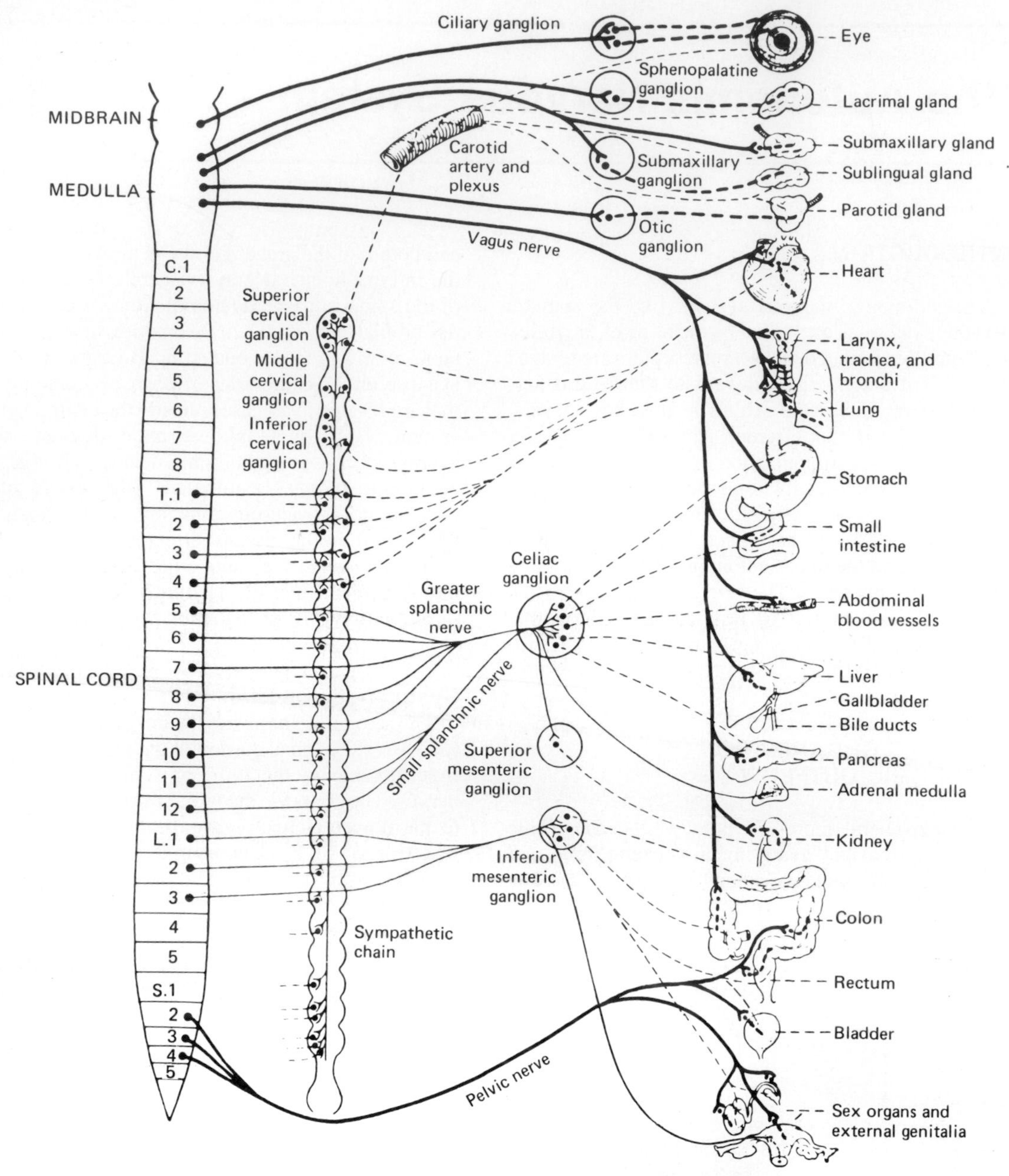

Figure 13–2. Diagram of the efferent autonomic pathways. Preganglionic neurons are shown as solid lines, postganglionic neurons as dashed lines. The heavy lines are parasympathetic fibers; the light lines are sympathetic. (Modified and reproduced, with permission, from Youmans W: *Fundamentals of Human Physiology,* 2nd ed. Year Book, 1962.)

postganglionic neurons pass to the viscera in the various sympathetic nerves. Others reenter the spinal nerves via the **gray rami communicantes** from the chain ganglia and are distributed to autonomic effectors in the areas supplied by these spinal nerves. The postganglionic sympathetic nerves to the head originate in the **superior, middle,** and **stellate** ganglia in the cranial extension of the sympathetic ganglion chain and travel to the effectors with the blood vessels. Some preganglionic neurons pass through the paravertebral ganglion chain and end on postganglionic neurons located in **collateral ganglia** close to the viscera. Parts of the uterus and the male genital tract are innervated by a special system of **short noradrenergic neurons** with cell bodies in ganglia in or near these organs, and the preganglionic fibers to these postganglionic neurons presumably go all the way to the organs (Fig 13–2).

Parasympathetic Division

The **cranial outflow** of the parasympathetic division supplies the visceral structures in the head via the oculomotor, facial, and glossopharyngeal nerves, and those in the thorax and upper abdomen via the vagus nerves. The **sacral outflow** supplies the pelvic viscera via the pelvic branches of the second to fourth sacral spinal nerves. The preganglionic fibers in both outflows end on short postganglionic neurons located on or near the visceral structures. (Fig 13–2).

CHEMICAL TRANSMISSION AT AUTONOMIC JUNCTIONS

Transmission at the synaptic junctions between pre- and postganglionic neurons and between the postganglionic neurons and the autonomic effectors is chemically mediated. The principal transmitter agents involved are **acetylcholine** and **norepinephrine,** although **dopamine** is also secreted by interneurons in the sympathetic ganglia and **LHRH** is secreted by some of the preganglionic neurons. LHRH mediates a slow excitatory response (see below). In addition, there are cotransmitters in autonomic neurons, and VIP is released with acetylcholine and ATP and neuropeptide Y with norepinephrine. The chemistry of all these transmitters and the receptors on which they act are discussed in Chapters 4 and 14.

Chemical Divisions of the Autonomic Nervous System

On the basis of the chemical mediator released, the autonomic nervous system can be divided into cholinergic and noradrenergic divisions. The neurons that are cholinergic are (1) all preganglionic neurons, (2) the anatomically parasympathetic postganglionic neurons, (3) the anatomically sympathetic postganglionic neurons which innervate sweat glands, and (4) the anatomically sympathetic neurons which end on blood vessels in skeletal muscles and produce vasodilation when stimulated (sympathetic vasodilator nerves; see Chapter 31). The remaining postganglionic sympathetic neurons are noradrenergic. The adrenal medulla is essentially a sympathetic ganglion in which the postganglionic cells have lost their axons and become specialized for secretion directly into the bloodstream. The cholinergic preganglionic neurons to these cells have consequently become the secretomotor nerve supply of this gland.

Transmission in Sympathetic Ganglia

At least in experimental animals, the responses produced in postganglionic neurons by stimulation of their preganglionic innervation include not only a rapid depolarization **(fast EPSP)** that generates action potentials (Table 13–1) but also a prolonged inhibitory postsynaptic potential **(slow IPSP),** a prolonged excitatory postsynaptic potential **(slow EPSP),** and a late slow EPSP (see Chapter 4). The late slow EPSP is very prolonged, lasting minutes rather than milliseconds. These 3 responses apparently modulate and regulate transmission through the sympathetic ganglia. The initial depolarization is produced by acetylcholine via a nicotinic receptor. The slow IPSP is probably produced by dopamine, which is secreted by an interneuron within the ganglion. The interneuron is excited by activation of an M_1 muscarinic receptor. The interneurons that secrete dopamine are the small, intensely fluorescent cells **(SIF cells)** in the ganglia. The production of the slow IPSP does not appear to be mediated via cyclic AMP, suggesting that a D_2 receptor is involved (see Chapter 4). The slow EPSP is produced by acetylcholine acting on a muscarinic receptor on the membrane of the postganglionic neuron. The late slow EPSP is produced by LHRH or a peptide closely resembling it.

Table 13–1. Fast and slow responses of postganglionic neurons in sympathetic ganglia.

	Duration	Mediator	Receptor
Fast EPSP	30 ms	Acetylcholine	Nicotinic cholinergic
Slow IPSP	2 s	Dopamine	D_2
Slow EPSP	30 s	Acetylcholine	M_1 cholinergic
Late slow EPSP	4 min	LHRH	LHRH

RESPONSES OF EFFECTOR ORGANS TO AUTONOMIC NERVE IMPULSES

General Principles

The effects of stimulation of the noradrenergic and cholinergic postganglionic nerve fibers to the viscera are listed in Table 13–2. The smooth muscle in the walls of the hollow viscera is generally innervated by both noradrenergic and cholinergic fibers, and activity in one of these systems increases the intrinsic activity of the smooth muscle whereas activity in the other decreases it. However, there is no uniform rule about which system stimulates and which inhibits. In the case of sphincter muscles, both noradrenergic and cholinergic innervations are excitatory, but one supplies the constrictor component of the sphincter and the other the dilator.

There is usually no acetylcholine in the circulating blood, and the effects of localized cholinergic discharge are generally discrete and of short duration because of the high concentration of acetylcholinesterase at cholinergic nerve endings. Norepinephrine spreads farther and has a more prolonged action than acetylcholine. Norepinephrine, epinephrine, and dopamine are all found in plasma (see Chapter 20). The epinephrine and some of the dopamine come from the adrenal medulla, but most of the norepinephrine diffuses into the bloodstream from noradrenergic nerve endings.

Table 13–2. Responses of effector organs to autonomic nerve impulses and circulating catecholamines.*

Effector Organs	Cholinergic Impulses Response	Noradrenergic Impulses	
		Receptor Type	Response
Eye			
Radial muscle of iris	...	α	Contraction (mydriasis)
Sphincter muscle of iris	Contraction (miosis)		...
Ciliary muscle	Contraction for near vision	β	Relaxation for far vision
Heart			
S-A node	Decrease in heart rate; vagal arrest	β_1	Increase in heart rate
Atria	Decrease in contractility and (usually) increase in conduction velocity	β_1	Increase in contractility and conduction velocity
A-V node and conduction system	Decrease in conduction velocity; A-V block	β_1	Increase in conduction velocity
Ventricles	...	β_2	Increase in contractility and conduction velocity
Arterioles			
Coronary, skeletal muscle, pulmonary, abdominal viscera, renal	Dilation	α	Constriction
		β_2	Dilation
Skin and mucosa, cerebral, salivary glands	...	α	Constriction
Systemic veins	...	α	Constriction
		β_2	Dilation
Lung			
Bronchial muscle	Contraction	β_2	Relaxation
Bronchial glands	Stimulation	?	Inhibition(?)
Stomach			
Motility and tone	Increase	α, β_2	Decrease (usually)
Sphincters	Relaxation (usually)	α	Contraction (usually)
Secretion	Stimulation		Inhibition (?)
Intestine			
Motility and tone	Increase	α, β_2	Decrease
Sphincters	Relaxation (usually)	α	Contraction (usually)
Secretion	Stimulation		Inhibition (?)
Gallbladder and ducts	Contraction		Relaxation
Urinary bladder			
Detrusor	Contraction	β	Relaxation (usually)
Trigone and sphincter	Relaxation	α	Contraction
Ureter			
Motility and tone	Increase (?)	α	Increase (usually)
Uterus	Variable†	α, β_2	Variable†
Male sex organs	Erection	α	Ejaculation
Skin			
Pilomotor muscles	...	α	Contraction
Sweat glands	Generalized secretion	α	Slight, localized secretion‡
Spleen capsule	...	α	Contraction
		β_2	Relaxation
Adrenal medulla	Secretion of epinephrine and norepinephrine		...
Liver	...	α, β_2	Glycogenolysis

* Modified from Gilman AG et al (editors): *Goodman and Gilman's The Pharmacological Basis of Therapeutics,* 7th ed. Macmillan, 1985.

† Depends on stage of menstrual cycle, amount of circulating estrogen and progesterone, pregnancy, and other factors.

‡ On palms of hands and in some other locations ("adrenergic sweating").

Table 13–2 (cont'd). Responses of effector organs to autonomic nerve impulses and circulating catecholamines.*

Effector Organs	Cholinergic Impulses Response	Noradrenergic Impulses	
		Receptor Type	Response
Pancreas			
Acini	Increased secretion	α	Decreased secretion
Islets	Increased insulin and glucagon secretion	α	Decreased insulin and glucagon secretion
		β_2	Increased insulin and glucagon secretion
Salivary glands	Profuse, watery secretion	α	Thick, viscous secretion
		β_2	Amylase secretion
Lacrimal glands	Secretion		...
Nasopharyngeal glands	Secretion		...
Adipose tissue	...	β_1	Lipolysis
Juxtaglomerular cells	...	β_1	Increased renin secretion
Pineal gland	...	β	Increased melatonin synthesis and secretion

Cholinergic Discharge

In a general way, the functions promoted by activity in the cholinergic division of the autonomic nervous system are those concerned with the vegetative aspects of day-to-day living. For example, cholinergic action favors digestion and absorption of food by increasing the activity of the intestinal musculature, increasing gastric secretion, and relaxing the pyloric sphincter. For this reason, and to contrast it with the "catabolic" noradrenergic division, the cholinergic division is sometimes called the **anabolic nervous system.**

The function of the VIP released from postganglionic cholinergic neurons is unsettled, but there is evidence that it facilitates the postsynaptic actions of acetylcholine. Since VIP is a vasodilator, it may also increase blood flow in target organs.

Noradrenergic Discharge

The noradrenergic division discharges as a unit in emergency situations. The effects of this discharge are of considerable value in preparing the individual to cope with the emergency, although it is important to avoid the teleologic fallacy involved in the statement that the system discharges in order to do this. For example, noradrenergic discharge relaxes accommodation and dilates the pupils (letting more light into the eyes), accelerates the heartbeat and raises the blood pressure (providing better perfusion of the vital organs and muscles), and constricts the blood vessels of the skin (which limits bleeding from wounds). Noradrenergic discharge also leads to lower thresholds in the reticular formation (reinforcing the alert, aroused state) and elevated blood glucose and free fatty acid levels (supplying more energy). On the basis of effects like these, Cannon called the emergency-induced discharge of the noradrenergic nervous system the "preparation for flight or fight."

The emphasis on mass discharge in stressful situations should not obscure the fact that the noradrenergic autonomic fibers also subserve other functions. For example, tonic noradrenergic discharge to the arterioles maintains arterial pressure, and variations in this tonic discharge are the mechanism by which carotid sinus feedback regulation of blood pressure is effected. In addition, sympathetic discharge is decreased in fasting animals and increased when fasted animals are refed. These changes may explain the decrease in blood pressure and metabolic rate produced by fasting and the opposite changes produced by feeding.

The small granulated vesicles in postganglionic noradrenergic neurons contain ATP and norepinephrine, and the large granulated vesicles contain neuropeptide Y. There is evidence that low-frequency stimulation promotes release of ATP whereas high-frequency stimulation causes release of neuropeptide Y. However, the functions of the released ATP and neuropeptide Y are unsettled.

Autonomic Pharmacology

The junctions in the peripheral autonomic motor pathways are a logical site for pharmacologic manipulation of visceral function because transmission across them is chemical. The transmitter agents are synthesized, stored in the nerve endings, and released near the neurons, muscle cells, or gland cells on which they act. They bind to receptors on these cells, thus initiating their characteristic actions, and they are then removed from the area by reuptake or metabolism. Each of these steps can be stimulated or inhibited, with predictable consequences. In noradrenergic endings, certain drugs also cause the formation of compounds that replace norepinephrine in the granules, and these weak or inactive "false

Table 13–3. Some drugs and toxins that affect sympathetic activity. Only the principal actions are listed. Note that guanethidine is believed to have 2 principal actions.

Site of Action	Compounds that Augment Sympathetic Activity	Compounds that Depress Sympathetic Activity
Sympathetic ganglia	**Stimulate postganglionic neurons** Nicotine Dimethyphenylpiperazinium **Inhibit acetylcholinesterase** DFP (diisopropyl fluorophosphate) Physostigmine (eserine) Neostigmine (Prostigmin) Parathion	**Block conduction** Hexamethonium (C-6) Mecamylamine (Inversine) Pentolinium Trimethaphan (Arfonad) High concentrations of acetylcholine, anticholinesterase drugs
Endings of postganglionic neurons	**Release norepinephrine** Tyramine Ephedrine Amphetamine	**Block norepinephrine synthesis** Metyrosine (Demser) **Interfere with norepinephrine storage** Reserpine Guanethidine (Ismelin) **Prevent norepinephrine release** Bretylium (Bretylol) Guanethidine (Ismelin) **Form false transmitters** Methyldopa (Aldomet)
α Receptors	**Stimulate α_1 receptors** Methoxamine (Vasoxyl) Phenylephrine (Neo-Synephrine) **Stimulates α_2 receptors** Clonidine* (Catapres)	**Block α receptors** Phenoxybenzamine (Dibenzyline) Phentolamine (Regitine) Prazosin (Minipress) (blocks α_1) Yohimbine (blocks α_2)
β Receptors	**Stimulates β receptors** Isoproterenol (Isuprel)	**Block β receptors** Propranolol (Inderal) and others (block β_1 and β_2) Atenolol (Tenormin) and others (block β_1) Butoxamine (blocks β_2)

*Clonidine stimulates α_2 receptors in the periphery, but along with other α_2 agonists that cross the blood-brain barrier, it also stimulates α_2 receptors in the brain that decrease sympathetic output. Therefore, the overall effect is decreased sympathetic discharge.

transmitters'' are released instead of norepinephrine by the action potentials reaching the endings.

Some of the drugs and toxins that affect the activity of the sympathetic nervous system and the mechanisms by which they produce their effects are listed in Table 13–3. Compounds with muscarinic actions include congeners of acetylcholine and drugs that inhibit acetylcholinesterase. Among the latter are the insecticide parathion and diisopropyl fluorophosphate (DFP), a component of the so-called nerve gases, which kill by producing massive inhibition of acetylcholinesterase. Atropine and scopolamine block muscarinic receptors. The drugs that block the effects of norepinephrine on visceral effectors are referred to as adrenergic blocking agents, peripheral sympathetic blocking agents, adrenolytic agents, or sympatholytic agents. Circulating catecholamines are inactivated by COMT and MAO and by reuptake into noradrenergic endings. Monoamine oxidase inhibitors increase the catecholamine content of the brain but do not affect circulating levels of these amines. When COMT is inhibited, there is a slight prolongation of the physiologic effects of catecholamines; but even when both MAO and COMT are inhibited, the metabolism of epinephrine and norepinephrine is still rapid. However, inhibition of reuptake prolongs the half-life of circulating catecholamines.

Central Regulation of Visceral Function

14

INTRODUCTION

The levels of autonomic integration within the central nervous system are arranged, like their somatic counterparts, in a hierarchy. Simple reflexes such as contraction of the full bladder are integrated in the spinal cord (see Chapter 12). More complex reflexes are the subject of this chapter. Those that regulate respiration and blood pressure are integrated in the medulla oblongata. Those that control pupillary responses to light and accommodation are integrated in the midbrain. The complex autonomic mechanisms that maintain the chemical constancy and temperature of the internal environment are integrated in the hypothalamus. The hypothalamus also functions with the limbic system as a unit that regulates emotional and instinctual behavior, and these aspects of hypothalamic function are discussed in the next chapter.

MEDULLA OBLONGATA

Control of Respiration, Heart Rate, & Blood Pressure

The medullary centers for the autonomic reflex control of the circulation, heart, and lungs are called the **vital centers** because damage to them is usually fatal. The afferent fibers to these centers originate in a number of instances in highly specialized visceral receptors. The specialized receptors include not only those of the carotid and aortic sinuses and bodies but also receptor cells that are located in the medulla itself. The motor responses are graded and delicately adjusted and include somatic as well as visceral components. The details of the reflexes themselves are discussed in the chapters on the regulation of the circulation and respiration.

Other Medullary Autonomic Reflexes

Swallowing, coughing, sneezing, gagging, and vomiting are also reflex responses integrated in the medulla oblongata. The swallowing reflex is initiated by the voluntary act of propelling the oral contents toward the back of the pharynx (see Chapter 26). Coughing is initiated by irritation of the lining of the trachea and extrapulmonary bronchi. The glottis closes and strong contraction of the respiratory muscles builds up intrapulmonary pressure, whereupon the glottis suddenly opens, causing an explosive discharge of air (see Chapter 36). Sneezing is a somewhat similar response to irritation of the nasal epithelium. It is initiated by stimulation of pain fibers in the trigeminal nerves.

Vomiting

Vomiting is another example of the way visceral reflexes integrated in the medulla include coordinated and carefully timed somatic as well as visceral components. Vomiting starts with salivation and the sensation of nausea. Reverse peristalsis empties material from the upper part of the small intestine into the stomach. The glottis closes, preventing aspiration of vomitus into the trachea. The breath is held in midinspiration. The muscles of the abdominal wall contract, and because the chest is held in a fixed position, the contraction increases intro-abdominal pressure. The esophagus and gastric cardiac sphincter relax, and the gastric contents are ejected.

The "vomiting center" in the reticular formation of the medulla at the level of the olivary nuclei controls these activities (Fig 14–1).

Afferents

Irritation of the mucosa of the upper gastrointestinal tract causes vomiting. Impulses are relayed from the mucosa to the vomiting center over visceral afferent pathways in the sympathetic nerves and vagi. Other afferents presumably reach the vomiting center from the diencephalon and limbic system, because emetic responses to emotionally charged stimuli also occur. Thus, we speak of "nauseating smells" and "sickening sights."

Chemoreceptor cells in the medulla initiate vomiting when they are stimulated by certain circulating chemical agents. The **chemoreceptor trigger zone** in which these cells are located (Fig 14–1) is in or near the **area postrema,** a V-shaped band of tissue on the lateral walls of the fourth ventricle near the obex. This structure is one of the circumventricular organs (see Chapter 32) and is more permeable to many substances than the underlying medulla. Lesions of the area postrema have little effect on the vomiting response to gastrointestinal irritation but abolish the vomiting that follows injection of apomorphine and a

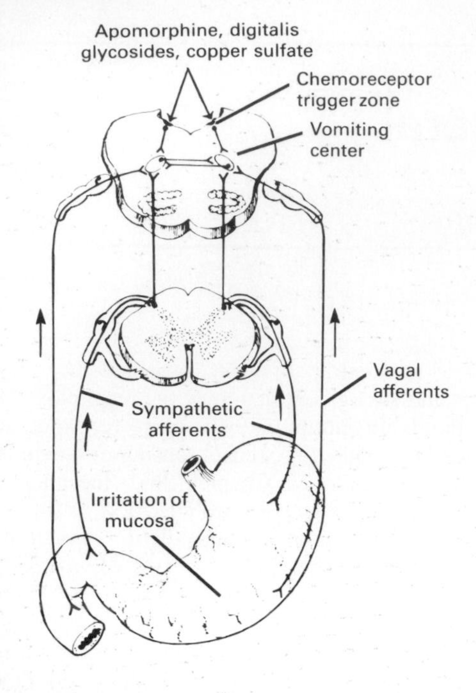

Figure 14–1. Afferent pathways for the vomiting reflex, showing the chemoreceptor trigger zone in the medulla. (Redrawn and reproduced, with permission, from: *Research in the Service of Medicine.* Vol 44. Searle & Co., 1956.)

number of other emetic drugs. Such lesions also decrease vomiting in uremia and radiation sickness, both of which may be associated with endogenous production of circulating emetic substances. A chemoreceptor mechanism stimulated by circulating toxins would explain vomiting in many clinical disorders.

HYPOTHALAMUS

ANATOMIC CONSIDERATIONS

The hypothalamus is that portion of the anterior end of the diencephalon which lies below the hypothalamic sulcus and in front of the interpeduncular nuclei. It is divided into a variety of nuclei and nuclear areas (Fig 14–2).

Relation to the Pituitary Gland

There are neural connections between the hypothalamus and the posterior lobe of the pituitary gland and vascular connections between the hypothalamus and the anterior lobe. Embryologically, the posterior pituitary arises as an evagination of the floor of the third ventricle. It is made up in large part of the

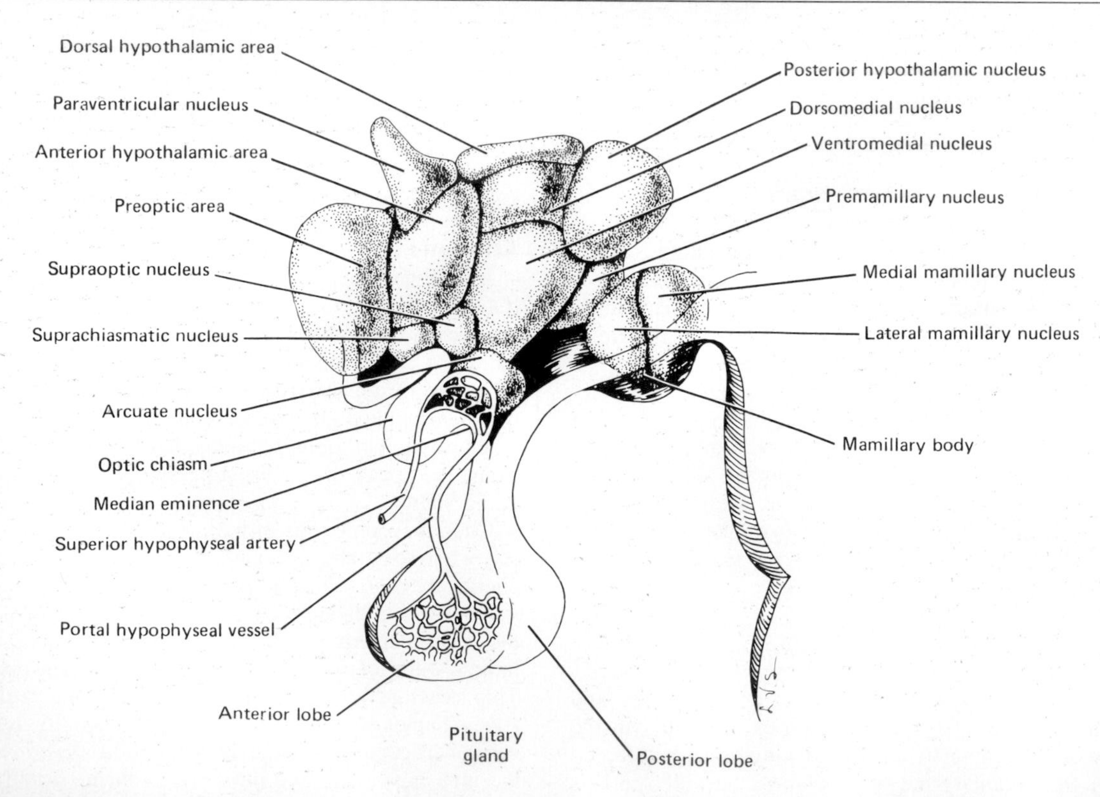

Figure 14–2. The human hypothalamus, with a superimposed diagrammatic representation of the portal-hypophyseal vessels.

endings of axons that arise from cell bodies in the supraoptic and paraventricular nuclei and pass to the posterior pituitary via the **hypothalamohypophyseal tract.** Most of the supraoptic fibers end in the posterior lobe itself, whereas some of the paraventricular fibers end in the median eminence. The anterior and intermediate lobes of the pituitary arise in the embryo from Rathke's pouch, an evagination from the roof of the pharynx (Fig 22–1). Sympathetic nerve fibers reach the anterior lobe from its capsule, and parasympathetic fibers reach it from the petrosal nerves, but very few nerve fibers pass to it from the hypothalamus. However, the **portal hypophyseal vessels** form a direct vascular link between the hypothalamus and the anterior pituitary. Arterial twigs from the carotid arteries and circle of Willis form a network of fenestrated capillaries called the **primary plexus** on the ventral surface of the hypothalamus (Fig 14–3). Capillary loops also penetrate the median eminence. The capillaries drain into the sinusoidal portal hypophyseal vessels that carry blood down the pituitary stalk to the capillaries of the anterior pituitary. This system begins and ends in capillaries without going through the heart and is therefore a true portal system. In birds and some mammals, including humans, there is no other anterior hypophyseal arterial supply except capsular vessels and anastomotic connections from the capillaries of the posterior pituitary. In other mammals, some blood reaches the anterior lobe through a separate set of anterior hypophyseal arteries; but in all vertebrates, a large fraction of the anterior lobe blood supply is carried by the portal vessels. The **median eminence** is generally defined as the portion of the ventral hypothalamus from which the portal vessels arise. This region is "outside the blood-brain barrier" (see Chapter 32).

Afferent & Efferent Connections of the Hypothalamus

The principal afferent and efferent neural pathways to and from the hypothalamus are listed in Table 14–1. Most of the fibers are unmyelinated. Many connect the hypothalamus to the limbic system. There are also important connections between the hypothalamus and nuclei in the midbrain tegmentum, pons, and hindbrain.

Norepinephrine-secreting neurons with their cell bodies in the hindbrain end in many different parts of the hypothalamus (Fig 15–6). Paraventricular neurons that probably secrete oxytocin and vasopressin project in turn to the hindbrain and the spinal cord. Neurons that secrete epinephrine have their cell bodies in the hindbrain and end in the ventral hypothalamus. There is an intrahypothalamic system of dopamine-secreting neurons which have their cell bodies in the arcuate nucleus and end on or near the capillaries that form the portal vessels in the median eminence. Serotonin-secreting neurons project to the hypothalamus from the raphe nuclei.

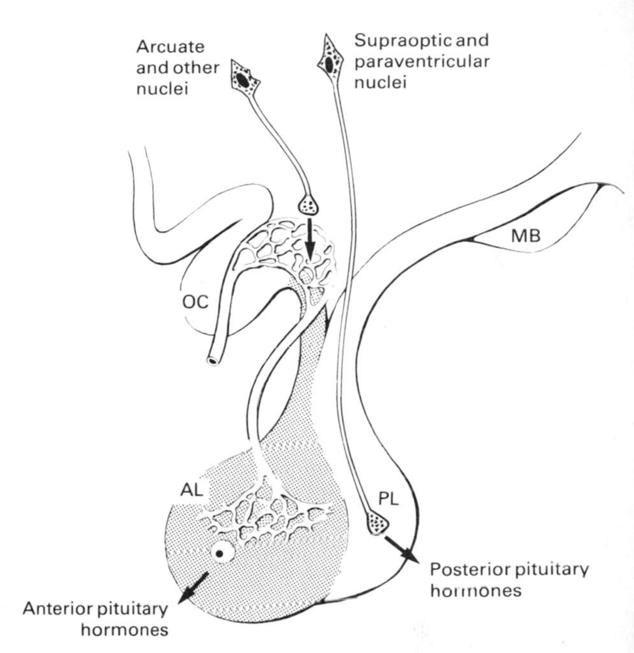

Figure 14–3. Secretion of hypothalamic hormones. The hormones of the posterior lobe (PL) are released into the general circulation from the endings of supraoptic and paraventricular neurons, whereas hypophyseotropic hormones are secreted into the portal hypophyseal circulation from the endings of arcuate and other hypothalamic neurons. AL, anterior lobe; MB, mamillary bodies; OC, optic chiasm.

HYPOTHALAMIC FUNCTION

The major functions of the hypothalamus are summarized in Table 14–2. Some are fairly clear-cut visceral reflexes, and others include complex behavioral and emotional reactions; but all involve a particular response to a particular stimulus. It is important to keep this pattern of stimulus-integration-response in mind in considering hypothalamic function.

RELATION OF HYPOTHALAMUS TO AUTONOMIC FUNCTION

Many years ago, Sherrington called the hypothalamus "the head ganglion of the autonomic system." Stimulation of the hypothalamus produces autonomic responses, but there is little evidence that the hypothalamus is concerned with the regulation of visceral function per se. Rather, the autonomic responses triggered in the hypothalamus are part of more complex phenomena such as rage and other emotions.

"Parasympathetic Center"

Stimulation of the superior anterior hypothalamus occasionally causes contraction of the urinary bladder, a parasympathetic response. Largely on this

Table 14–1. Principal pathways to and from the hypothalamus.

Tract		Description
Medial forebrain bundle	A, E	Connects limbic lobe and midbrain via lateral hypothalamus, where fibers enter and leave it; includes direct amygdalohypothalamic fibers, which are sometimes referred to as a separate pathway.
Fornix	A, E	Connects hippocampus to hypothalamus, mostly mamillary bodies.
Stria terminalis	A	Connects amygdala to hypothalamus, especially ventromedial region.
Mamillary peduncle	A	Diverges from sensory pathways in midbrain to enter hypothalamus; may be the pathway by which sensory stimuli enter.
Ventral noradrenergic bundle	A	Axons of noradrenergic neurons projecting from nucleus of tractus solitarius and other hindbrain nuclei to paraventricular nuclei and other parts of hypothalamus.
Dorsal noradrenergic bundle	A	Axons of noradrenergic neurons projecting from locus ceruleus to dorsal hypothalamus.
Serotonergic neurons	A	Axons of serotonin-secreting neurons projecting from raphe nuclei to hypothalamus.
Adrenergic neurons	A	Axons of epinephrine-secreting neurons from medulla to ventral hypothalamus.
Retinohypothalamic fibers	A	Optic nerve fibers to suprachiasmatic nuclei from optic chiasm.
Thalamohypothalamic and pallidohypothalamic fibers	A	Connections from thalamus and lenticular nucleus.
Periventricular system (including dorsal longitudinal fasciculus of Schütz)	A, E	Interconnects hypothalamus and midbrain; efferent projections to spinal cord, afferent from sensory pathways.
Mamillothalamic tract of Vicq d'Azyr	E	Connects mamillary nuclei to anterior thalamic nuclei.
Mamillotegmental tract	E	Connects hypothalamus to reticular portions of midbrain.
Hypothalamohypophyseal tract (supraopticohypophyseal and paraventriculohypophyseal tracts)	E	Axons of neurons in supraoptic and paraventricular nuclei that end in pituitary stalk and posterior pituitary.
Neurons containing vasopressin, oxytocin	E	Run from paraventricular nucleus to nucleus of tractus solitarius, other brain stem nuclei, intermediolateral column of spinal cord; also from paraventricular nucleus to central nucleus of amygdala.
Neurons containing CRH	E	Run from paraventricular nucleus to brain stem and spinal cord.

A = principally afferent; E = principally efferent.

basis, the statement is often made that there is a "parasympathetic center" in the anterior hypothalamus. However, bladder contraction can also be elicited by stimulation of other parts of the hypothalamus, and hypothalamic stimulation causes very few other parasympathetic responses. Thus, there is very little evidence that a localized "parasympathetic center" exists. Stimulation of the hypothalamus can cause cardiac arrhythmias, and there is reason to believe that these are due to simultaneous activation of vagal and sympathetic nerves to the heart.

Sympathetic Responses

Stimulation of various parts of the hypothalamus, especially the lateral areas, produces a rise in blood pressure, pupillary dilation, piloerection, and other signs of diffuse noradrenergic discharge. The stimuli that trigger this pattern of responses in the intact animal are not regulatory impulses from the viscera but emotional stiumuli, especially rage and fear. Noradrenergic responses are also triggered as part of the reactions that conserve heat (see below).

Low-voltage electrical stimulation of the middorsal portion of the hypothalamus causes vasodilation in muscle. Associated vasoconstriction in the skin and elsewhere maintains blood pressure at a fairly constant level. This observation and other evidence support the conclusion that the hypothalamus is a way station on the so-called cholinergic sympathetic vasodilator system which originates in the cerebral cortex. It may be this system that is responsible for the dilation of muscle blood vessels at the start of exercise (see Chapter 31).

Stimulation of the dorsomedial nuclei and posterior hypothalamic areas produces increased secretion of epinephrine and norepinephrine from the adrenal medulla. Increased adrenal medullary secretion is one of the physical changes associated with rage and fear and may occur when the cholinergic sympathetic vasodilator system is activated. It has been claimed that there are separate hypothalamic centers for the control of epinephrine and norepinephrine secretion. Differential secretion of one or the other of these adrenal medullary catecholamines does occur in certain situations (see Chapter 20), but the selective increases are small.

Table 14–2. Summary of principal hypothalamic regulatory mechanisms.

Function	Afferents From	Integrating Areas
Temperature regulation	Cutaneous cold receptors; temperature-sensitive cells in hypothalamus	Anterior hypothalamus, response to heat; posterior hypothalamus, response to cold
Neuroendocrine control of: Catecholamines	Emotional stimuli, probably via limbic system	Dorsomedial and posterior hypothalamus
Vasopressin	Osmoreceptors, "volume receptors," others	Supraoptic and paraventricular nuclei
Oxytocin	Touch receptors in breast, uterus, genitalia	Supraoptic and paraventricular nuclei
Thyroid-stimulating hormone (thyrotropin, TSH) via TRH	Temperature receptors in infants, perhaps others.	Paraventricular nuclei and neighboring areas.
Adrenocorticotropic hormone (ACTH) and β-lipotropin (β-LPH) via CRH	Limbic system (emotional stimuli); reticular formation ("systemic" stimuli); hypothalamic or anterior pituitary cells sensitive to circulating blood cortisol level; suprachiasmatic nuclei (diurnal rhythm)	Paraventricular nuclei
Follicle-stimulating hormone (FSH) and luteinizing hormone (LH) via LHRH	Hypothalamic cells sensitive to estrogens; eyes, touch receptors in skin and genitalia of reflex ovulating species	Preoptic area, other areas
Prolactin via PIH and PRH	Touch receptors in breasts, other unknown receptors	Arcuate nucleus, other areas (hypothalamus inhibits secretion)
Growth hormone via somatostatin and GRH	Unknown receptors	Periventricular nucleus, arcuate nucleus
"Appetitive" behavior Thirst	Osmoreceptors, subfornical organ	Lateral superior hypothalamus
Hunger	"Glucostat" cells sensitive to rate of glucose utilization	Ventromedial satiety center, lateral hunger center, also limbic components
Sexual behavior	Cells sensitive to circulating estrogen and androgen, others	Anterior ventral hypothalamus, plus, in the male, piriform cortex
Defensive reactions Fear, rage	Sense organs and neocortex, paths unknown	Diffuse, in limbic system and hypothalamus
Control of various endocrine and activity rhythms	Retina via retinohypothalamic fibers	Suprachiasmatic nuclei

RELATION TO SLEEP

Lesions of the posterior hypothalamus cause prolonged sleep, and stimulation of the dorsal hypothalamus in conscious animals causes them to go to sleep. These observations have led to considerable speculation about the existence of "sleep centers" and "wakefulness centers" in the hypothalamus, but study of the functions of the RAS and nonspecific projection nuclei of the thalamus (see Chapter 11) has provided alternative explanations. The posterior hypothalamic lesions that cause coma involve fibers of the RAS as they pass to the thalamus and cortex. Therefore, the sleep-producing effects of posterior hypothalamic lesions can be explained by damage to the RAS rather than destruction of a hypothetical hypothalamic "wakefulness center." The stimulating frequency that produces sleep is approximately 8/s. As noted in Chapter 11, stimulation of the thalamus, the orbital surface of the frontal lobe, and some parts of the brain stem at a frequency of 8/s produces sleep. Thus, the reported effects of hypothalamic stimulation are probably due to stimulation of a diffuse system, the low-frequency stimulation of which produces sleep. On the basis of these considerations, it seems unlikely that the hypothalamus plays any direct or unique role in the regulation of sleep.

RELATION TO CYCLIC PHENOMENA

Lesions of the suprachiasmatic nuclei disrupt the circadian rhythm in the secretion of ACTH (see Chapter 20) and melatonin (see Chapter 24). In addition, these lesions interrupt estrous cycles and activity patterns in laboratory animals. The suprachiasmatic nuclei receive an important input from the eyes via the retinohypothalamic fibers, and it appears that these nuclei normally function to entrain various body rhythms to the 24-hour light-dark cycle. There is a prominent serotonergic input from the raphe nuclei to the suprachiasmatic nuclei, but the exact relation of this input to their function is not known.

HUNGER

Feeding & Satiety

Food intake is generally regulated with great precision. If animals are made obese by force-feeding and then permitted to eat as they wish, their spontaneous food intake decreases until their weight falls to the control level (Fig 14–4). Conversely, if animals are starved and then permitted to eat freely, their spontaneous food intake increases until they regain the lost weight.

Hypothalamic regulation of the appetite for food depends primarily upon the interaction of 2 areas: a lateral **"feeding center"** in the bed nucleus of the medial forebrain bundle at its junction with the pallidohypothalamic fibers, and a medial **"satiety center"** in the ventromedial nucleus. Stimulation of the feeding center evokes eating behavior in conscious animals, and its destruction causes severe, fatal anorexia in otherwise healthy animals. Stimulation of the ventromedial nucleus causes cessation of eating, whereas lesions in this region cause hyperphagia and, if the food supply is abundant, the syndrome of **hypothalamic obesity** (Fig 14–5). Destruction of the feeding center in rats with lesions of the satiety center causes anorexia, which indicates that the satiety center functions by inhibiting the feeding center (Fig 14–6). It appears that the feeding center is chronically active and that its activity is transiently inhibited by activity in the satiety center after the ingestion of food. However, it is not certain that the feeding center and the satiety center simply control the desire for food. For example, rats with ventromedial lesions gain weight for a while, but their food intake then levels off. After their intake reaches a plateau, their appetite mechanism operates to maintain their new, higher weight. One theory that has been advanced to explain these observations is that it is the set point for body weight rather than food intake per se which is regulated by the hypothalamic centers.

Afferent Mechanisms

There is considerable debate about the signals that are sensed by the satiety and feeding center to regulate food intake. The activity of the satiety center is probably governed in part by the level of glucose utilization of cells within the center. These cells have therefore been called **glucostats.** It has been postulated that when their glucose utilization is low—and consequently when the arteriovenous blood glucose difference across them is low—their activity is decreased. Under these conditions, the activity of the feeding center is unchecked, and the individual is hungry. When utilization is high, the activity of the glucostats is increased, the feeding center is inhibited, and the individual feels sated. This **glucostatic hypothesis** of appetite regulation is supported by an appreciable body of experimental data. Other factors undoubtedly affect appetite, but the glucostatic hypothesis has the merit of explaining the increased appetite in diabetes, in which the blood sugar is high but the glucose utilization of the cells is low because of the insulin deficiency. The objection has been raised that most neural tissue does not require insulin to metabolize glucose. However, the region of the ventromedial nucleus has been shown to be different from the rest of the brain in that its rate of glucose utilization does vary with the amount of insulin in the circulation.

Relatively large amounts of radioactive glucose are taken up by cells of the ventromedial nuclei. This

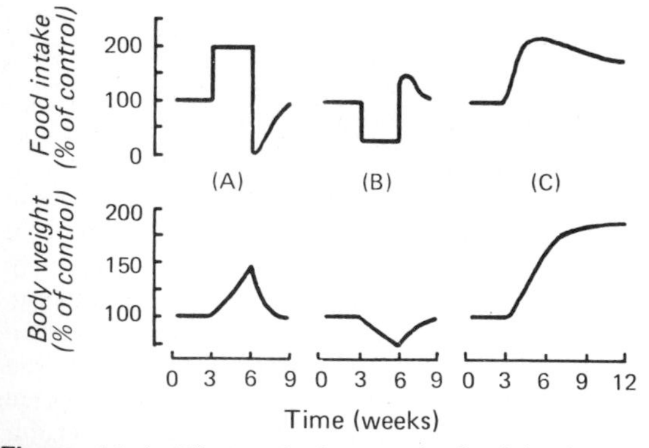

Figure 14–4. Effects of changes in food intake and ventromedial hypothalamic lesions on spontaneous food intake and body weight. In A, rats were force-fed for weeks 3–6, then permitted free access to food. In B, rats were partially starved for weeks 3–6, then permitted free access to food. In C, bilateral ventromedial hypothalamic lesions were produced at 3 weeks and the rats allowed free access to food throughout. (Reproduced, with permission, from Stricker EM: Hyperphagia, *N Engl J Med* 1978; 298:1010.)

Figure 14–5. Hypothalamic obesity. The animal on the right, in which bilateral lesions were placed in the ventromedial nuclei 4 months previously, weighs 1080 g. The control animal on the left weighs 520 g. (Reproduced, with permission, from Stevenson JAF, in: *The Hypothalamus.* Haymaker W, Anderson E, Nauta WJH [editors]. Thomas, 1969.)

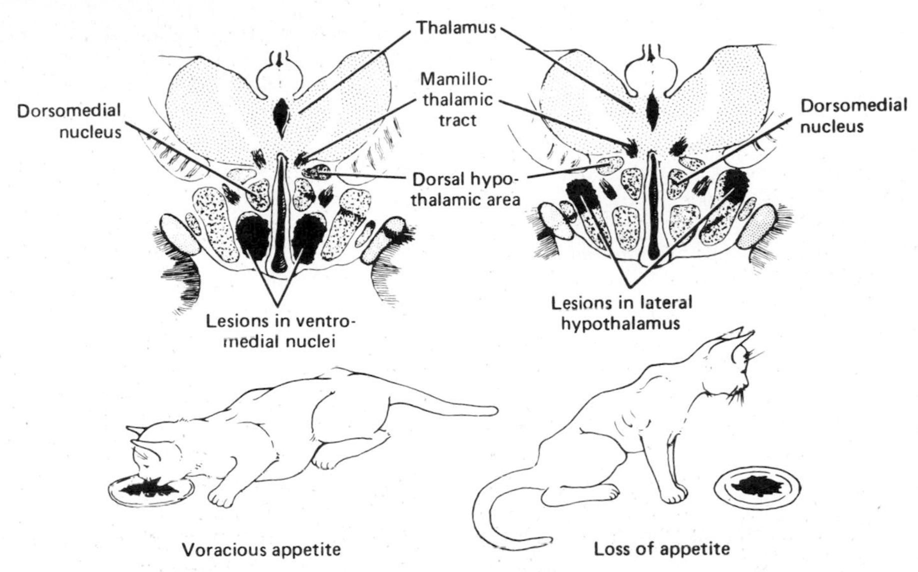

Figure 14–6. Diagrammatic summary of the effects of hypothalamic lesions on feeding. (Redrawn from an original by Netter FH in: Ciba Clinical Symposia. Copyright© 1956, Ciba Pharmaceutical Co. Reproduced with permission.)

affinity for glucose probably explains why injections of gold thioglucose produce obesity in mice. The ventromedial cells presumably take up the toxic gold-substituted glucose molecules, which destroy the cells and thus prevent inhibition of the lateral feeding center. Animals that have received gold thioglucose injections and have become obese have demonstrable lesions in the ventromedial nuclei.

Other Factors Regulating Food Intake

The limbic system is also involved in the neural regulation of appetite. Lesions of the amygdaloid nuclei produce moderate hyperphagia. However, unlike animals with ventromedial hypothalamic lesions, animals with amygdaloid lesions will eat adulterated or tainted food. They are omniphagic and attempt to eat all sorts of objects (see Chapter 15). This suggests that fundamentally different mechanisms underlie the hyperphagia.

Amphetamine inhibits appetite, and there is extensive binding of this sympathomimetic amine in the hypothalamus. However, it does not appear to act by releasing catecholamines, and its mechanism of action is presently unknown.

There is evidence that the size of body fat depots is sensed by either neural or hormonal signals that are relayed to the brain, and that appetite is controlled in this fashion **(lipostatic hypothesis).** In addition, there is some evidence that protein intake is regulated independently of body weight. A cold environment stimulates and a hot environment depresses appetite. Food in the gastrointestinal tract may cause the secretion of gastrointestinal or other hormones that inhibit further food intake, and cholecystokinin (CCK; see Chapter 26) and calcitonin (see Chapter 21) have been shown to decrease appetite. However, a physiologic role for these or other gastrointestinal hormones in the regulation of food intake has not been established. Some of the hormones appear to act in the abdomen, because their effects are abolished by subdiaphragmatic vagotomy. Distention of the gastrointestinal tract inhibits and contractions of the empty stomach **(hunger contractions)** stimulate appetite, but denervation of the stomach and intestines does not affect the amount of food eaten. In animals with esophageal fistulas, chewing and swallowing food causes some satiety even though the food never reaches the stomach. Especially in humans, cultural factors, environment, and past experiences relative to the sight, smell, and taste of food also affect food intake.

It now appears that defective neural control of brown fat may contribute to the obesity of animals with ventromedial hypothalamic lesions produced electrolytically or by injection of gold thioglucose. Brown fat, a special form of body fat that has an extensive sympathetic innervation and produces more heat and less ATP than other tissues (see Chapter 17), is normally stimulated by cold and food intake. In animals with ventromedial lesions, the response to cold is normal but food fails to produce the normal increase in sympathetic discharge to brown fat, and this abnormality has the effect of maintaining a high efficiency of food utilization and promoting obesity. In addition, the increased heat generated by food intake in normal animals may contribute to satiety, and this signal is reduced or absent in animals with ventromedial lesions.

The net effect of all the appetite-regulating mechanisms in normal adult animals and humans is an adjustment of food intake to the point where caloric intake balances energy expenditures, with the result that body weight is maintained. In hyperthyroidism and diabetes mellitus, hunger increases as energy output increases. However, the link between energy needs and food intake is indirect, and in some situations, intake is not correlated with immediate energy expenditure. During growth and after exercise, food intake is increased. During recovery from debilitating illness, food intake is increased in a catch-up fashion until lost weight is regained. How these long-term adjustments are brought about is not known.

THIRST

Another appetitive mechanism under hypothalamic control is thirst. Appropriately placed hypothalamic lesions diminish or abolish fluid intake, in some instances without any change in food intake, and electrical stimulation of the hypothalamus causes drinking.

Drinking is regulated by plasma osmolality and ECF volume in much the same fashion as vasopressin secretion (see below). Water intake is increased by increased effective osmotic pressure of the plasma (Fig 14–7), by decreases in ECF volume, and by psychologic and other factors. Osmolality acts via **osmoreceptors,** receptors that sense the osmolality of the body fluids. These osmoreceptors are located in the anterior hypothalamus.

Decreases in ECF volume also stimulate thirst by a pathway independent of that mediating thirst in response to increased plasma osmolality (Fig 14–8). Thus, hemorrhage causes increased drinking even though there is no change in the osmolality of the plasma. The effect of ECF volume depletion on thirst is mediated in part via the renin-angiotensin system (see Chapter 24). Renin secretion is increased by hypovolemia, and there is a resultant increase in circulating angiotensin II. The angiotensin II acts on the **subfornical organ,** a specialized receptor area in the diencephalon (Fig 32–5), to stimulate the neural areas concerned with thirst. There is some evidence that it acts on the **organum vasculosum of the lamina terminalis (OVLT)** as well. These areas are highly permeable and are 2 of the circumventricular organs located "outside the blood-brain barrier" (see Chapter 32). However, drugs which block the action of angiotensin II do not completely block the thirst response to hypovolemia, and it appears that the baroreceptors in the heart and blood vessels are also involved.

Whenever the sensation of thirst is obtunded, either by direct damage to the diencephalon or by depressed or altered states of consciousness, patients stop drinking adequate amounts of fluid. Dehydration results if appropriate measures are not instituted to maintain water balance. If the protein intake is high, the products of protein metabolism cause an osmotic diuresis (see Chapter 38), and the amounts of water required to maintain hydration are large. Most cases of hypernatremia are actually due to simple dehydration in patients with psychoses or cerebral disease who do not or cannot increase their water intake when their thirst mechanism is stimulated.

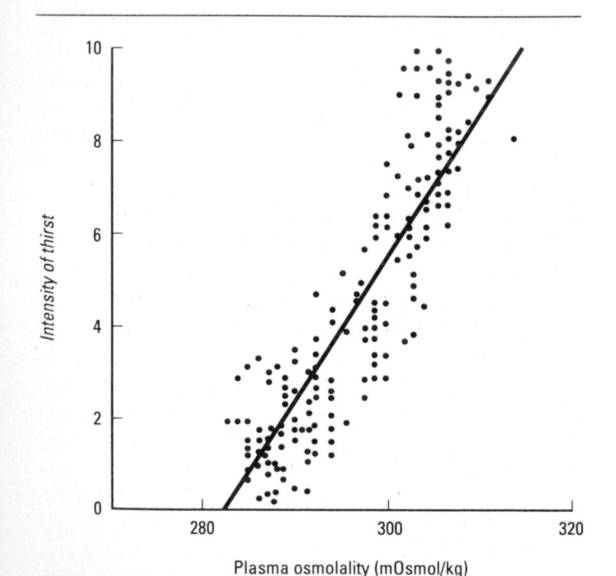

Figure 14–7. Relation of plasma osmolality to thirst in healthy adult humans during infusion of hypertonic saline. The intensity of thirst is measured on a special analog scale. (Reproduced, with permission, from Thompson CJ, Bland J, Burd J, Baylis PH: The osmotic thresholds for thirst and vasopressin release are similar in healthy man. *Clin Sci Lond* 1986; 71:651.)

Other Factors Regulating Water Intake

A number of other well-established factors contribute to the regulation of water intake. Psychologic and social factors are important. Dryness of the pharyngeal mucous membrane causes a sensation of thirst.

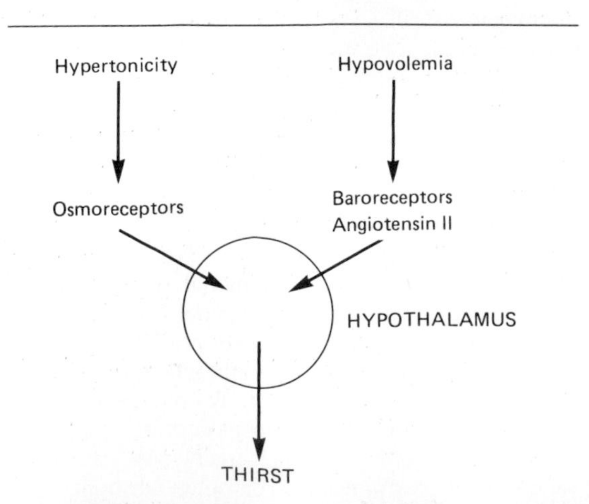

Figure 14–8. Diagrammatic representation of the way changes in plasma osmolality and changes in ECF volume affect thirst by separate pathways.

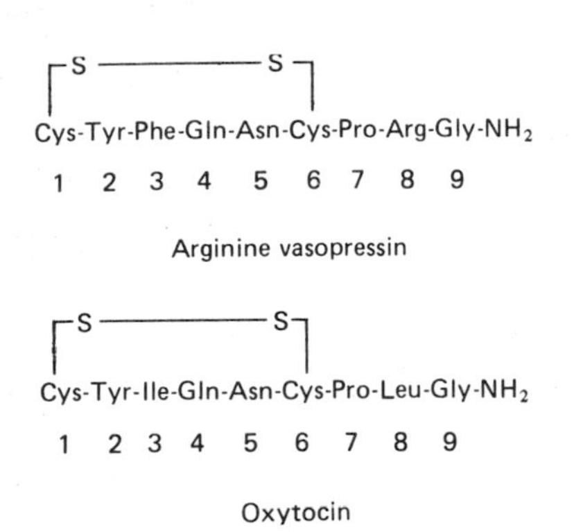

Figure 14–9. Arginine vasopressin and oxytocin.

Patients in whom fluid intake must be restricted sometimes get appreciable relief of thirst by sucking ice chips or a wet cloth.

Dehydrated dogs, cats, camels, and some other animals rapidly drink just enough water to make up their water deficit. They stop drinking before the water is absorbed (while their plasma is still hypertonic), so some kind of pharyngeal or gastrointestinal "metering" must be involved. There is some evidence that humans have a similar metering ability.

CONTROL OF POSTERIOR PITUITARY SECRETION

Vasopressin & Oxytocin

In most mammals, the hormones secreted by the posterior pituitary gland are **oxytocin** and **arginine vasopressin** (Fig 14–9). In hippopotamuses and most pigs, arginine in the vasopressin molecule is replaced by lysine to form **lysine vasopressin.** The posterior pituitaries of some species of pigs and marsupials contain a mixture of arginine and lysine vasopressin. The posterior lobe hormones are monapeptides (considering each half cystine as a single amino acid) with a disulfide ring at one end.

Biosynthesis, Intraneuronal Transport, & Secretion

The hormones of the posterior pituitary gland are synthesized in the cell bodies of the magnocellular neurons in the supraoptic and paraventricular nuclei and transported down the axons of these neurons to their endings in the posterior lobe, where they are secreted in response to electrical activity in the endings. Some of the neurons make oxytocin and others make vasopressin, and oxytocin-containing and vasopressin-containing cells are found in both nuclei.

Oxytocin and vasopressin are typical **neural hormones,** ie, hormones secreted into the circulation by nerve cells. This type of neural regulation is compared with other types in Fig 14–10. The term **neurosecretion** was originally coined to describe the secretion of hormones by neurons, but the term is somewhat misleading, because it is probable that all neurons secrete chemical messengers (see Chapter 1).

Like other peptide hormones, the posterior lobe hormones are synthesized as part of larger precursor molecules. Vasopressin and oxytocin each have a characteristic **neurophysin** associated with them in the granules in the neurons that secrete them, neurophysin I in the case of oxytocin and neurophysin II in the case of vasopressin. The neurophysins were originally thought to be binding polypeptides, but it now appears that they are simply parts of the precursor molecules. The precursor for arginine vasopressin, **prepropressophysin,** contains a leader sequence of 19 amino acid residues followed by arginine vasopressin, vasopressin neurophysin, and a glycopeptide (Fig 14–11). **Prepro-oxyphysin,** the precursor for oxytocin, is a similar but smaller molecule that lacks the glycopeptide.

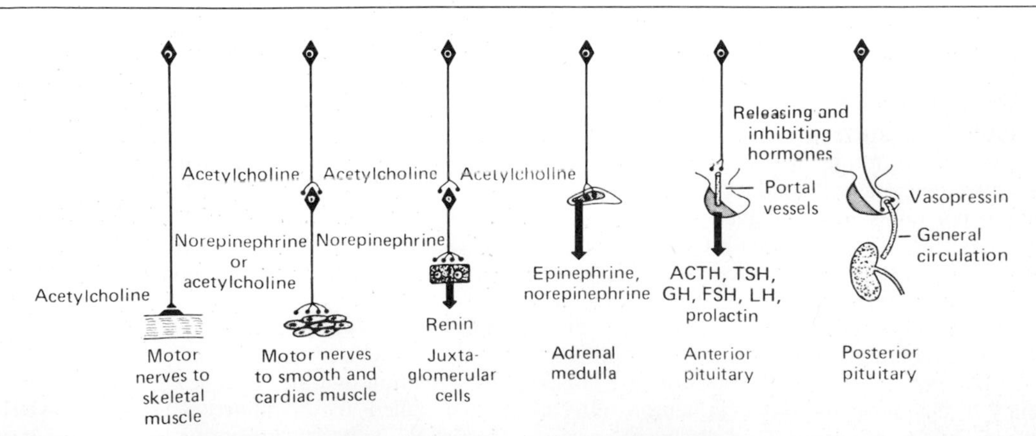

Figure 14–10. Neural control mechanisms. In the 2 situations on the left, neurotransmitters act at nerve endings on muscle; in the 2 in the middle, neurotransmitters regulate the secretion of endocrine glands; and in the 2 on the right, neurons secrete hormones into the hypophyseal portal or general circulation.

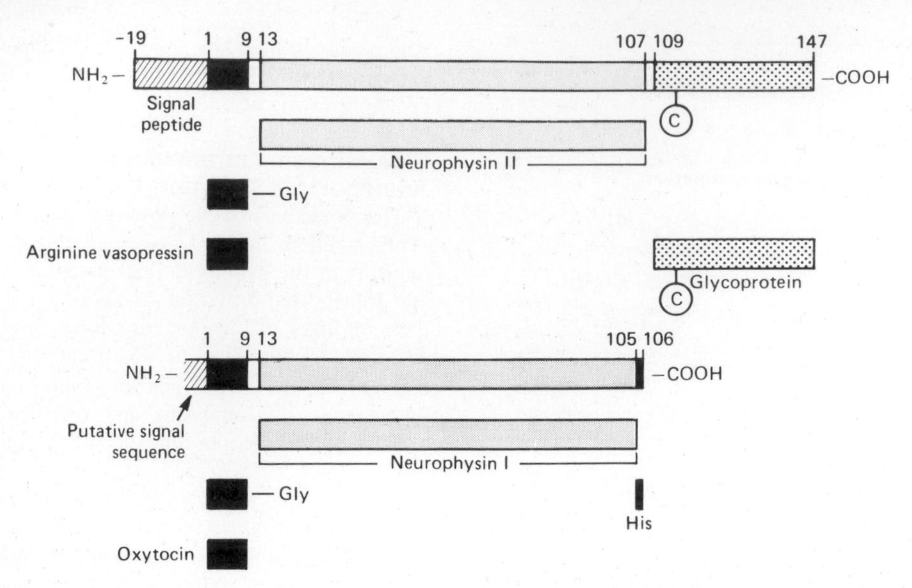

Figure 14–11. Structure of prepropressophysin *(top)* and prepro-oxyphysin *(bottom)*. Gly in the 10 position of both peptides is necessary for amidation of the Gly residue in position 9. Numbers of amino acid residues at the top start from the N terminal of the secreted peptide. C, carbohydrate. (Reproduced, with permission, from Ivell R, Schmale H, Richter D: Vasopressin and oxytocin precursors as models of preprohormones. *Neuroendocrinology* 1983;**37**:235. S. Karger A.G., Basel.)

The precursor molecules are synthesized in the ribosomes of the cell bodies of the neurons. They have their leader sequences removed in the endoplasmic reticulum, are packaged into secretory granules in the Golgi apparatus, and are transported down the axons by axoplasmic flow to the endings in the posterior pituitary . The secretory granules, called **Herring bodies,** are easy to stain in tissue sections, and they have been extensively studied. Cleavage of the precursor molecules occurs as they are being transported, and the storage granules in the endings contain free vasopressin or oxytocin and the corresponding neurophysin. In the case of vasopressin, the glycopeptide is also present. All these products are secreted, but the functions of the components other than the established posterior pituitary hormones are unknown.

Electrical Activity of Magnocellular Neurons

The oxytocin-secreting and vasopressin-secreting neurons also generate and conduct action potentials, and action potentials reaching their endings trigger release of hormone from them by Ca^{2+}-dependent exocytosis. At least in anesthetized rats, these neurons are silent at rest or discharge at low, irregular rates (0.1–3 spikes per second). However, their response to stimulation varies (Fig 14–12). Stimulation of the nipples causes a synchronous, high-frequency discharge of the oxytocin neurons after a latency of 10–15 minutes. This discharge causes release of a pulse of oxytocin and consequent milk ejection (see below). On the other hand, stimulation of the vasopressin-secreting neurons by a stimulus such as hemorrhage causes an initial steady increase in firing rate followed by a prolonged pattern of phasic discharge in which periods of a high-frequency discharge alternate with periods of electrical quiescence **(phasic bursting).** These phasic bursts are generally not synchronous in different vasopressin-secreting neurons. They are well suited to maintain a prolonged increase in the output of vasopressin, as opposed to the synchronous, relatively short, high-frequency discharge of oxytocin-secreting neurons in response to stimulation of the nipples.

Vasopressin & Oxytocin in Other Locations

Vasopressin-secreting neurons are found in the suprachiasmatic nuclei, and vasopressin and oxytocin are also found in the endings of neurons that project from the paraventricular nuclei to the brain stem and spinal cord. These neurons may be involved in cardiovascular control. In addition, it now appears that vasopressin and oxytocin are synthesized in the gonads (see Chapter 23) and the adrenal cortex and there is oxytocin in the thymus. The functions of the peptides in these organs are unknown.

Effects of Vasopressin

Because one of its principal physiologic effects is the retention of water by the kidney, vasopressin is often called the antidiuretic hormone (ADH). It increases the permeability of the collecting ducts of the kidney, so that water enters the hypertonic interstitium of the renal pyramids (see Chapter 38).

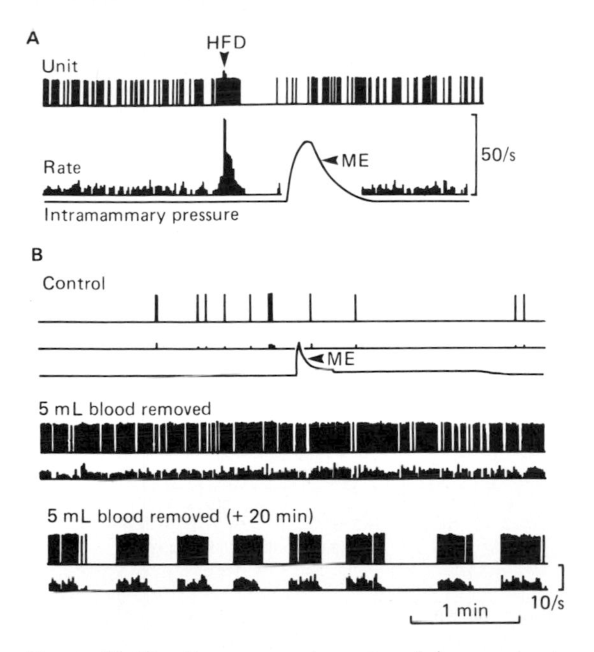

Figure 14–12. Responses of magnocellular neurons to stimulation. *A.* Response of an oxytocin-secreting neuron. The tracings show individual extracellularly-recorded action potentials, discharge rates, and intramammary duct pressure. HFD, high-frequency discharge; ME, milk ejection. Stimulation of nipples started before onset of recording *B.* Responses of a vasopressin-secreting neuron, showing no change in slow firing rate in response to stimulation of nipples and a prompt increase in firing rate when 5 mL of blood was drawn, followed by typical phasic discharge. The tracings show individual action potentials, discharge rates, and, in the case of nipple stimulation, intramammary pressure. (Tracings from Wakerly JB: Hypothalamic neurosecretory function: Insights from electrophysiological studies of the magnocellular nuclei. *IBRO News* 1985;4:15.)

The urine becomes concentrated, and its volume decreases. The overall effect is therefore retention of water in excess of solute; consequently, the effective osmotic pressure of the body fluids is decreased. In the absence of vasopressin, the urine is hypotonic to plasma, urine volume is increased, and there is a net water loss. Consequently, the osmolality of the body fluids rises.

In large doses, vasopressin elevates arterial blood pressure by an action on the smooth muscle of the arterioles, but the amount of endogenous vasopressin in the circulation of normal individuals does not normally affect blood pressure. However, vasopressin is a potent vasoconstrictor in vitro, and the reason that it does not increase blood pressure when small doses are injected in vivo is that it acts on the brain to cause a concurrent decrease in cardiac output. Hemorrhage is a potent stimulus to vasopressin secretion. There is controversy about whether the blood pressure fall after hemorrhage is significantly greater or more prolonged when the posterior pituitary is absent than when it is present, but the fall is more marked in animals that have been treated with synthetic peptides which block the pressor action of vasopressin. Consequently, it appears that vasopressin does play a role in blood pressure homeostasis. The vasculature may be hyper-responsive to vasopressin in some forms of hypertension.

Vasopressin Receptors & Mechanism of Action

There are at least 2 kinds of vasopressin receptors. The receptors that mediate the vasoconstrictor action of vasopressin are called **V_1 receptors.** They are found in blood vessels, including renal vessels, and they are also found in glomerular mesangial cells and in the brain. When activated, they increase PIP_2 metabolism (see Chapter 1), hence increasing cytoplasmic Ca^{2+} without affecting adenylate cyclase.

The vasopressin receptors that mediate the antidiuretic effect of the peptide are called **V_2 receptors,** and they are found in the nephrons on the blood side of the tubular cells in the thick ascending limb of Henle and the collecting duct (see Chapter 38). They activate adenylate cyclase, producing an increase in intracellular cyclic AMP. In the collecting ducts, the increase in cyclic AMP produces a striking increase in the permeability of the membrane on the luminal side of the cells to water, urea, and some other solutes. The effect may be mediated via an increased formation of microtubules and microfilaments. It is known that vasopressin causes aggregation of particles in the cell membrane, but the relation of this aggregation to the changes in permeability is unsettled.

Receptors for the other effects of vasopressin have not been fully characterized, and there may be additional types in the body.

Synthetic Agonists & Antagonists

Synthetic peptides that have selective actions and are more active than naturally occurring vasopressin and oxytocin have now been produced by altering the amino acid residues. For example, 1-deamino-8-D-arginine vasopressin (DDAVP, desmopressin) has very high antidiuretic activity with little pressor activity, making it valuable in the treatment of vasopressin deficiency (see below). Antagonists that selectively block the pressor or antidiuretic activity of vasopressin have also been synthesized.

Metabolism

Circulating vasopressin is rapidly inactivated, principally in the liver and kidneys. It has a **biologic half-life** (time required for inactivation of half a given amount) of approximately 18 minutes in humans. Its effects on the kidney develop rapidly but are of short duration.

Table 14–3. Summary of stimuli affecting vasopressin secretion.

Vasopressin Secretion Increased	Vasopressin Secretion Decreased
Increased effective osmotic pressure of plasma	Decreased effective osmotic pressure of plasma
Decreased extracellular fluid volume	Increased extracellular fluid volume
Pain, emotion, "stress," exercise	Alcohol
Nausea and vomiting standing	Butorphanol, oxilorphan
Morphine, nicotine, barbiturates	
Chlorpropamide, clofibrate, carbamazepine	
Angiotensin II	

Control of Vasopressin Secretion: Osmotic Stimuli

Vasopressin is stored in the posterior pituitary and released into the bloodstream by impulses in the nerve fibers that contain the hormone. The factors affecting its secretion are summarized in Table 14–3. When the effective osmotic pressure of the plasma is increased above the normal 285 mosm/kg, the rate of discharge of these neurons increases and vasopressin secretion is increased (Fig 14–13). At 285 mosm/kg, plasma vasopressin is at or near the limits of detection by available assays, but there is probably a further decrease when plasma osmolality is below this level. Vasopressin secretion is regulated by osmoreceptors located in the anterior hypothalamus. They are outside the blood-brain barrier and appear to be located in a circumventricular organ, possibly the organum vasculosum of the lamina terminalis (see Chapter 32). The osmotic threshold for thirst (Fig 14–7) is the same as the threshold for increased vasopressin secretion (Fig 14–12), but it is still uncertain whether the same osmoreceptors mediate both effects.

Vasopressin secretion is thus controlled by a delicate feedback mechanism that operates continuously to defend the osmolality of the plasma. Significant changes in secretion occur when osmolality is changed as little as 1%. In this way, the osmolality of the plasma in normal individuals is maintained very close to 290 mosm/L.

Volume Effects

ECF volume also affects vasopressin secretion. Vasopressin secretion is increased when ECF volume is low and decreased when ECF volume is high (Table 14–3). There is an inverse relationship between the rate of vasopressin secretion and the rate of discharge in afferents from stretch receptors in the low- and high-pressure portions of the vascular system. The low-pressure receptors are those in the great veins, right and left atria, and pulmonary vessels; the high-pressure receptors are those in the carotid sinuses and aortic arch (see Chapter 31). The exponential increases in vasopressin secretion produced by decreases in blood pressure are documented in Fig

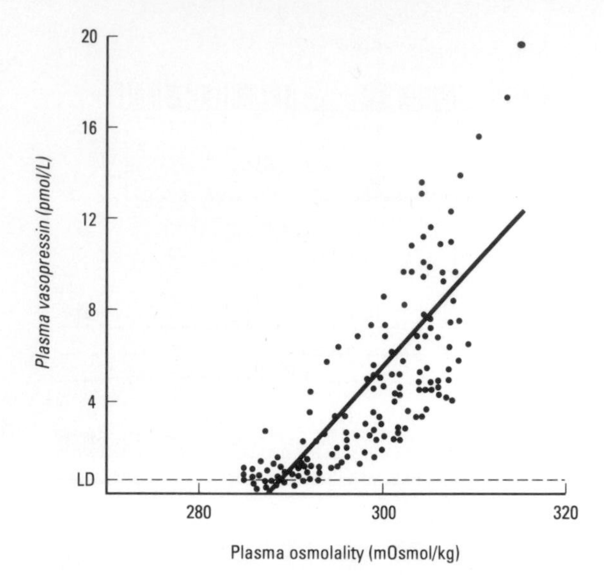

Figure 14–13. Relation between plasma osmolality and plasma vasopressin in healthy adult humans during infusion of hypertonic saline. LD, limit of detection. (Reproduced, with permission, from Thompson CJ, Bland J, Burd J, Baylis PH: The osmotic thresholds for thirst and vasopressin are similar in healthy man. *Clin Sci Lond* 1986; **71**:651.)

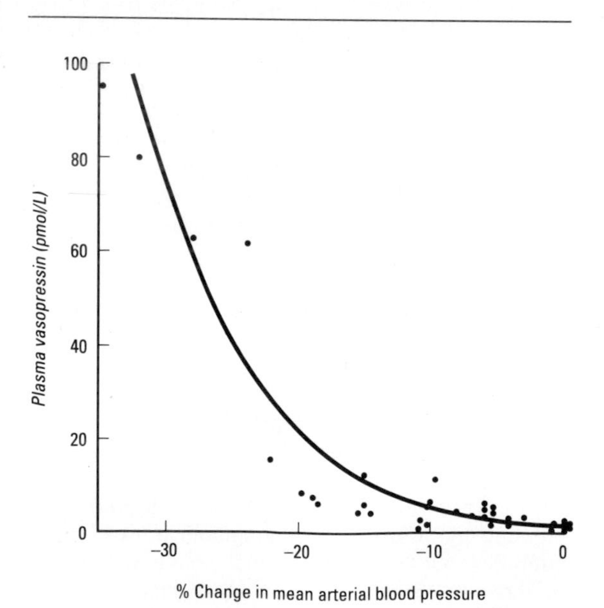

Figure 14–14. Relation of mean arterial blood pressure to plasma vasopressin in healthy adult humans in whom a progressive decline in blood pressure was induced by infusion of graded doses of the ganglionic blocking drug trimethaphan. Notice that the relation is exponential rather than linear. (Drawn from data in Baylis PH: Osmoregulation and control of vasopressin secretion in healthy humans. *Am J Physiol* 1987;**253**:R671.)

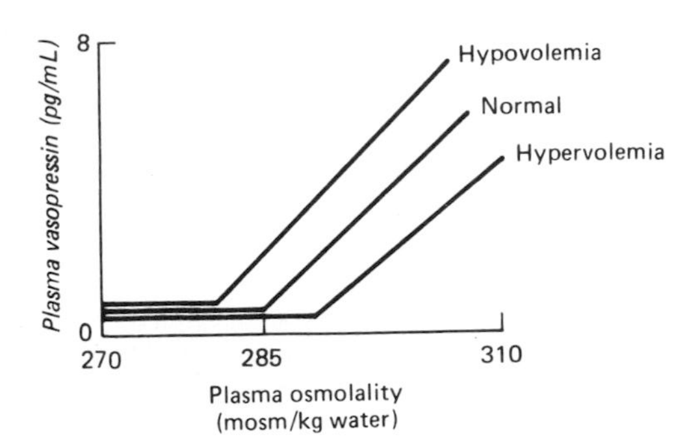

Figure 14–15. Effect of hypovolemia and hypervolemia on the relation between plasma vasopressin and osmolality in dogs. (Modified from Quillen EW Jr, Cowley AW Jr: Influence of volume changes on osmolality-vasopressin relationships in conscious dogs. *Am J Physiol* 1983; 244:H73.)

14–14. However, the low-pressure receptors monitor the fullness of the vascular system, and moderate decreases in blood volume that decrease central venous pressure without lowering arterial pressure can also increase plasma vasopressin.

Thus, the low-pressure receptors are the primary mediators of volume effects on vasopressin secretion. Angiotensin II may reinforce the response to hypovolemia and hypotension by acting on the brain to increase vasopressin secretion.

It is worth noting that hypovolemia and hypotension produced by conditions such as hemorrhage release greater amounts of vasopressin than increases in plasma osmolality, and in the presence of hypovolemia, the osmotic response curve is shifted to the left (Fig 14–15). The result is water retention and reduced plasma osmolality. The hyponatremia that is sometimes seen after surgery is partly explained on this basis.

Other Stimuli Affecting Vasopressin Secretion

A variety of stimuli in addition to osmotic pressure changes and ECF volume aberrations increase vasopressin secretion. These include pain, nausea, surgical stress, some emotions, and a number of drugs (Table 14–3), including morphine, nicotine, and large doses of barbiturates. Alcohol and the opiate antagonists butorphanol and oxilorphan decrease vasopressin secretion.

Clinical Implications

In various clinical conditions, volume and other nonosmotic stimuli bias the osmotic control of vasopressin secretion. In patients with hypersecretion of vasopressin as a result of hypovolemia or pain, a high fluid intake can cause water intoxication. Water retention occurs in addition to salt retention in patients with edema due to congestive heart failure, cirrhosis of the liver, and nephrosis. The cause of excess vasopressin secretion in diseases of the heart, liver, and kidneys is not known.

In the **syndrome of "inappropriate" hypersecretion of antidiuretic hormone (SIADH),** vasopressin is responsible for the hyponatremia that occurs with high levels of salt excretion in some patients with cerebral or pulmonary disease. In these cases of "cerebral salt wasting" and "pulmonary salt wasting," water retention is sufficient to expand ECF volume. This inhibits the secretion of aldosterone by the adrenal cortex (see Chapter 20), and salt is lost in the urine. Hypersecretion of vasopressin in patients with pulmonary diseases such as lung cancer may be due in part of the interruption of inhibitory impulses in vagal afferents from the stretch receptors in the atria and great veins. However, a significant number of lung tumors and some other cancers secrete vasopressin. Patients with inappropriate hypersecretion of vasopressin have been successfully treated with demeclocycline, an antibiotic that reduces the renal response to vasopressin. If the increased amounts of vasopressin do not come from a tumor, secretion can be reduced by treatment with butorphanol and oxilorphan.

Diabetes insipidus is the syndrome that results when vasopressin deficiency develops because of disease processes in the supraoptic and paraventricular nuclei, the hypothalamohypophyseal tract, or the posterior pituitary gland. It has been estimated that 30% of the clinical cases are due to neoplastic lesions of the hypothalamus, either primary or metastatic, 30% are posttraumatic, 30% are idiopathic, and the remaining 10% are due to vascular lesions, infections, or systemic diseases such as sarcoidosis which affect the hypothalamus. The disease that develops after surgical removal of the posterior lobe of the pituitary may be temporary if only the distal ends of the supraoptic and paraventricular fibers are damaged, because the fibers recover and begin again to secrete vasopressin. The symptoms of diabetes insipidus are passage of large amounts of dilute urine **(polyuria)** and the drinking of large amounts of fluid **(polydipsia),** provided the thirst mechanism is intact. It is the polydipsia that keeps these patients healthy. If their sense of thirst is depressed for any reason and their intake of dilute fluid decreases, they develop dehydration that can be fatal. In patients who have an incomplete rather than total vasopressin deficiency, treatment with drugs that increase vasopressin secretion such as clofibrate has proved to be of value. Chlorpropamide is also of value because it increases the renal response to vasopressin.

Another cause of diabetes insipidus is inability of the kidneys to respond to vasopressin. In this condition, which is known as **nephrogenic diabetes insipidus,** vasopressin fails to increase renal cyclic AMP.

The amelioration of diabetes insipidus produced by the development of concomitant anterior pituitary insufficiency is discussed in Chapter 22.

Effects of Oxytocin

Oxytocin acts primarily on the breasts and uterus, although it may also be involved in luteolysis (see Chapter 23).

In mammals, oxytocin causes contraction of the **myoepithelial cells,** smooth-muscle-like cells that line the ducts of the breast. This squeezes the milk out of the alveoli of the lactating breast into the large ducts (sinuses) and thence out the nipple **(milk ejection).** Many hormones acting in concert are responsible for breast growth and the secretion of milk into the ducts (see Chapter 23), but milk ejection in most species requires oxytocin.

The Milk Ejection Reflex

Milk ejection is normally initiated by a neuroendocrine reflex. The receptors involved are the touch receptors, which are plentiful in the breast—especially around the nipple. Impulses generated in these receptors are relayed from the somatic touch pathways to the supraoptic and paraventricular nuclei. Discharge of the oxytocin-containing neurons causes secretion of oxytocin from the posterior pituitary (Fig 14–12). The infant suckling at the breast stimulates the touch receptors, the nuclei are stimulated, oxytocin is released, and the milk is expressed into the sinuses, ready to flow into the mouth of the waiting infant. In lactating women, genital stimulation and emotional stimuli also produce oxytocin secretion, sometimes causing milk to spurt from the breasts.

Other Actions of Oxytocin

Oxytocin causes contraction of the smooth muscle of the uterus. The sensitivity of the uterine musculature to oxytocin is enhanced by estrogen and inhibited by progestrone. In late pregnancy, the uterus becomes very sensitive to oxytocin coincident with a marked increase in the number of oxytocin receptors (see Chapter 23). Oxytocin secretion is increased during labor. After dilation of the cervix, descent of the fetus down the birth canal initiates impulses in the afferent nerves that are relayed to the supraoptic and paraventricular nuclei, causing secretion of sufficient oxytocin to enhance labor (Fig 23–39). Oxytocin may also play some role in initiating labor. However, the mechanisms controlling the onset of labor are immensely complex, and neuroendocrine reflexes are certainly not the only mechanisms responsible for delivery.

Oxytocin may also act on the nonpregnant uterus to facilitate sperm transport. The passage of sperm up the female genital tract to the uterine tubes, where fertilization normally takes place, depends not only on the motile powers of the sperm but, at least in some species, on uterine contractions as well. The genital stimulation involved in coitus releases oxytocin, but it has not been proved that it is oxytocin which initiates the rather specialized uterine contractions that transport the sperm.

The secretion of oxytocin is increased by stressful stimuli and, like that of vasopressin, is inhibited by alcohol.

CONTROL OF ANTERIOR PITUITARY SECRETION

Anterior Pituitary Hormones

The anterior pituitary secretes 6 hormones: **adrenocorticotropic hormone** (corticotropin, ACTH), **thyroid-stimulating hormone** (thyrotropin, TSH), **growth hormone, follicle-stimulating hormone** (FSH), **luteinizing hormone** (LH), and **prolactin** (luteotropic hormone, LTH). An additional polypeptide, **β-lipotropin** (β-LPH), is secreted with ACTH, but its physiologic role is unknown. The

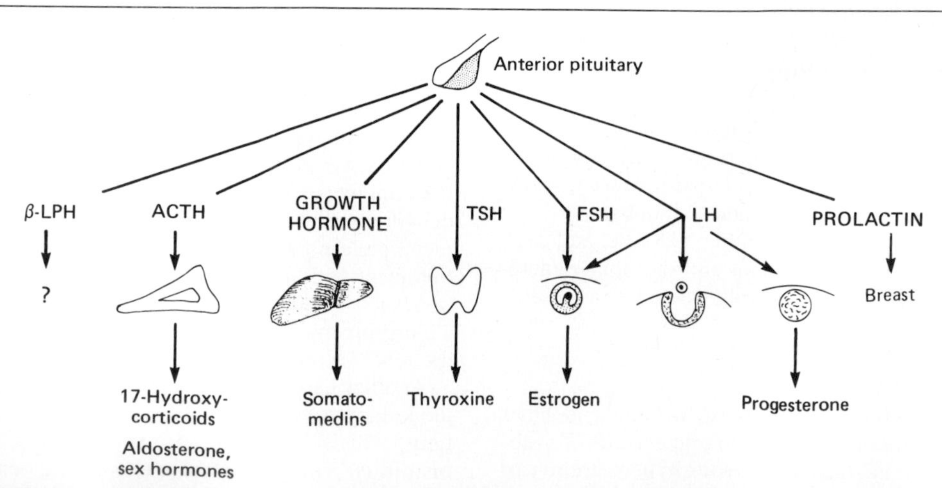

Figure 14–16. Anterior pituitary hormones. In women, FSH and LH act in sequence on the ovary to produce growth of the ovarian follicle, which secretes estrogen; ovulation; and formation and maintenance of the corpus luteum, which secretes estrogen and progesterone. In men, FSH and LH control the functions of the testes (see Chapter 23). Prolactin stimulates lactation.

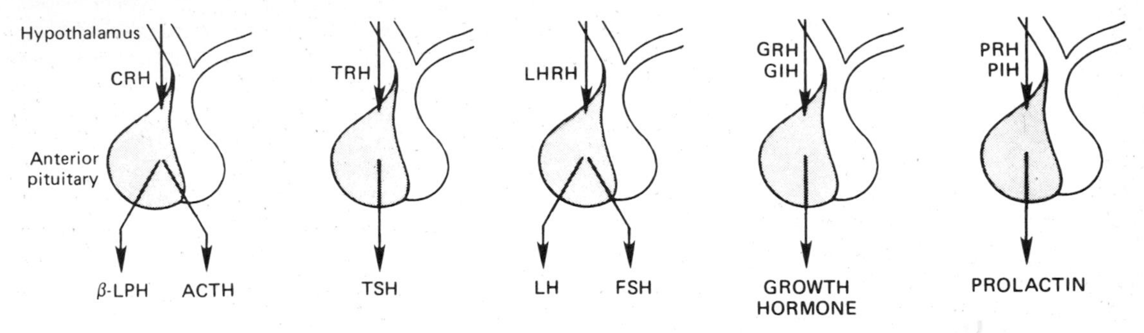

Figure 14–17. Effects of hypophyseotropic hormones on the secretion of anterior pituitary hormones.

actions of these hormones are summarized in Fig 14–16, and synonyms and abbreviations for them are listed in Table 22–1. The hormones are discussed in detail in the chapters on the endocrine system. The hypothalamus plays an important stimulatory role in regulating the secretion of ACTH, β-LPH, TSH, growth hormone, FSH, and LH. It also regulates prolactin secretion, but its effect is predominantly inhibitory rather than stimulatory.

Nature of Hypothalamic Control

Anterior pituitary secretion is controlled by chemical agents carried in the portal hypophyseal vessels from the hypothalamus to the pituitary. These substances have generally been referred to as releasing and inhibiting factors, but they are now commonly called hypophyseotropic hormones. The latter term seems appropriate, since they are secreted into the bloodstream and act at a distance from their site of origin. Consequently, the term hormone is used in this book.

Hypophyseotropic Hormones

There are 6 established hypothalamic releasing and inhibiting hormones: **corticotropin-releasing hormone (CRH); thyrotropin-releasing hormone (TRH); growth hormone-releasing hormone (GRH); growth hormone-inhibiting hormone (GIH;** also called **somatostatin); luteinizing hormone-releasing hormone (LHRH); and prolactin-inhibiting hormone** (PIH). (Fig 14–17). In addition, it seems likely that there is **a prolactin-releasing hormone (PHR),** although this has not been isolated. LHRH stimulates the secretion of FSH as well as LH, and it is uncertain whether there is a separate **follicle-stimulating hormone-releasing hormone (FRH).** The proponents of a single gonadotropin-regulating hormone refer to LHRH as GnRH **(gonadotropin-releasing hormone).** The structures of the 6 hypophyseotropic hormones are shown in Fig 14–18. The structures of the preprohormones for TRH, LHRH, somatostatin, CRH, and GRH are known. PreproTRH contains 5 copies of TRH (Fig 1–19). Several others may contain other hormonally active peptides in addition to the hypophyseotropic hormones.

The area from which the hypothalamic releasing and inhibiting hormones are secreted is the median eminence of the hypothalamus. This region contains few nerve cell bodies, but there are many nerve endings in close proximity to the capillary loops from which the portal vessels originate.

The locations of the cell bodies of the neurons that project to the external layer of the median eminence

TRH	(pyro)Glu-His-Pro-NH_2
LHRH	(pyro)Glu-His-Trp-Ser-Tyr-Gly-Leu-Arg-Pro-Gly-NH_2
Somatostatin	Ala-Gly-Cys-Lys-Asn-Phe-Phe-Trp-Lys-Thr-Phe-Thr-Ser-Cys (S–S bridge between the two Cys)
CRH	Ser-Glu-Glu-Pro-Pro-Ile-Ser-Leu-Asp-Leu-Thr-Phe-His-Leu-Leu-Arg-Glu-Val-Leu-Glu-Met-Ala-Arg-Ala-Glu-Gln-Leu-Ala-Gln-Gln-Ala-His-Ser-Asn-Arg-Lys-Leu-Met-Glu-Ile-Ile-NH_2
GRH	Tyr-Ala-Asp-Ala-Ile-Phe-Thr-Asn-Ser-Tyr-Arg-Lys-Val-Leu-Gly-Gln-Leu-Ser-Ala-Arg-Lys-Leu-Leu-Gln-Asp-Ile-Met-Ser-Arg-Gln-Gln-Gly-Glu-Ser-Asn-Gln-Glu-Arg-Gly-Ala-Arg-Ala-Arg-Leu-NH_2
PIH	Dopamine (Fig 4–15)

Figure 14–18. Structure of hypophyseotriopic hormones in humans. The structure of somatostatin that is shown is the tetradecapeptide (somatostatin 14, SS14). In addition, preprosomatostatin is the source of an N-terminal extended polypeptide containing 28 amino acid residues (SS28) and a polypeptide containing 12 amino acid residues [SS28(1–12)] that are found in tissues.

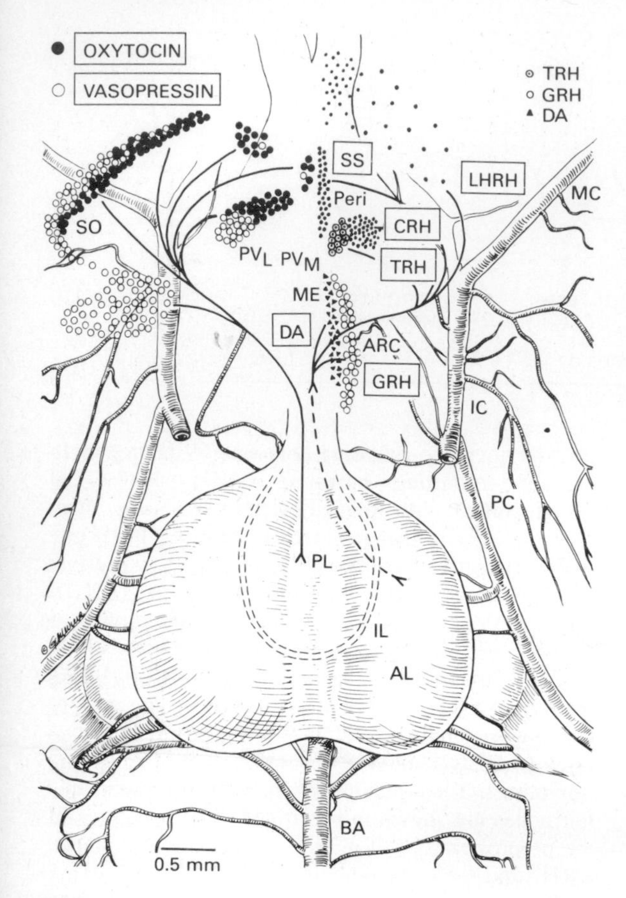

Figure 14–19. Location of all bodies of hypophyseotropic hormone-secreting neurons projected on a ventral view of the hypothalamus and pituitary of the rat. AL, anterior lobe; ARC, arcuate nucleus; BA, basilar artery; IC, internal carotid; IL, intermediate lobe; MC, middle cerebral; ME, median eminence; PC, posterior cerebral; peri, periventricular nucleus; PL, posterior lobe; PV_L and PV_M, lateral and medial portions of paraventricular nucleus; SO, supraoptic nucleus. The names of the hormones are enclosed in the boxes. (Courtesy of LW Swanson and ET Cunningham Jr.)

and secrete the hypophyseotropic hormones are shown in Fig 14–19, which also shows the location of the neurons secreting oxytocin and vasopressin. The LHRH-secreting neurons are primarily in the medial preoptic area,the somatostatin-secreting neurons are in the periventricular nuclei, the TRH-secreting and CRH-secreting neurons are in the medial parts of the paraventricular nuclei, and the GRH-secreting and dopamine-secreting neurons are in the arcuate nuclei.

Most, if not all, of the hypophyseotropic hormones affect the secretion of more than one anterior pituitary hormone (Fig 14–17). The FSH-stimulating activity of LHRH has been mentioned above. TRH stimulates the secretion of prolactin as well as TSH. Somatostatin inhibits the secretion of TSH as well as growth hormone. It does not normally inhibit the secretion of the other anterior pituitary hormones, but it inhibits the abnormally elevated secretion of ACTH in patients with Nelson's syndrome. CRH stimulates the secretion of ACTH and β-LPH (see Chapter 22).

Hypophyseotropic hormones function as neurotransmitters in other parts of the brain, the retina, and the autonomic nervous system (see Chapter 4). In addition, somatostatin is found in the pancreas (see Chapter 19), GRH is secreted by pancreatic tumors, and somatostatin and TRH are found in the gastrointestinal tract (see Chapter 26).

Significance & Clinical Implications

Research delineating the multiple neuroendocrine regulatory functions of the hypothalamus is important because it explains how endocrine secretion is made appropriate to the demands of a changing environment. The nervous system receives information about changes in the internal and external environment from the sense organs. It brings about adjustments to these changes through effector mechanisms that include not only somatic movement but also changes in the rate at which hormones are secreted.There are many examples of the operation of the neuroendocrine effector pathways. In birds and many mammals, the increasing number of daylight hours in the spring stimulates gonadotropin secretion, activating the quiescent gonads and starting the breeding season. In some avian species, the sight of a member of the opposite sex in a mating dance apparently produces the gonadotopin secretion necessary to cause ovulation. It has been claimed that Eskimo women cease to ovulate during the long winter night and that regular menstrual cycles and sexual vigor return in the spring. In more temperate climates, menstruation in women is a year-round phenomenon, but its regularity can be markedly affected by somatic and emotional stimuli. The disappearance of regular periods when young girls move away from home (sometimes called ''boarding school amenorrhea'') and the inhibition of menstruation that can be produced by fear of pregnancy are examples of the many ways in which psychic phenomena affect endocrine secretion.

The manifestations of hypothalamic disease are neurologic defects, endocrine changes,and metabolic abnormalities such as hyperphagia and hyperthermia. The relative frequencies of the signs and symptoms of hypothalamic disease in one large series of cases are shown in Table 14–4. The possibility of hypothalamic pathology should be kept in mind in evaluating all patients with pituitary dysfunction, especially those with isolated deficiencies of single pituitary tropic hormones. Many of these deficiencies are proving to be due to hypothalamic rather than primary pituitary disease.

TEMPERATURE REGULATION

In the body, heat is produced by muscular exercise, assimilation of food, and all the vital processes

Table 14–4. Symptoms and signs in 60 autopsied cases of hypothalamic disease.*

Symptoms and Signs	Percentage of Cases
Endocrine and metabolic findings	
Precocious puberty	40
Hypogonadism	32
Diabetes insipidus	35
Obesity	25
Abnormalities of temperature regulation	22
Emaciation	18
Bulimia	8
Anorexia	7
Neurologic findings	
Eye signs	78
Pyramidal and sensory deficits	75
Headache	65
Extrapyramidal signs	62
Vomiting	40
Psychic disturbances, rage attacks, etc	35
Somnolence	30
Convulsions	15

*Data from Bauer HG: Endocrine and other clinical manifestations of hypothalamic disease. *J Clin Endocrinol* 1954;**14**:13. See also Kahana L et al: Endocrine manifestations of intracranial extrasellar lesions. *J Clin Endocrinol* 1962;**22**:304.

that contribute to the basal metabolic rate (see Chapter 17). It is lost from the body by radiation, conduction, and vaporization of water in the respiratory passages and on the skin. Small amounts of heat are also removed in the urine and feces. The balance between heat production and heat loss determines the body temperature. Because the speed of chemical reactions varies with the temperature and because the enzyme systems of the body have narrow temperature ranges in which their function is optimal, normal body function depends upon a relatively constant body temperature.

Invertebrates generally cannot adjust their body temperatures and so are at the mercy of the environment. In vertebrates, mechanisms for maintaining body temperature by adjusting heat production and heat loss have evolved. In reptiles, amphibia, and fish, the adjusting mechanisms are relatively rudimentary, and these species are called "cold-blooded" **(poikilothermic)** because their body temperature fluctuates over a considerable range. In birds and mammals, the "warm-blooded" **(homeothermic)** animals, a group of reflex responses that are primarily integrated in the hypothalamus operate to maintain body temperature within a narrow range in spite of wide fluctuations in environmental temperature. The hibernating mammals are a partial exception. While awake, they are homeothermic, but during hibernation, their body temperature falls.

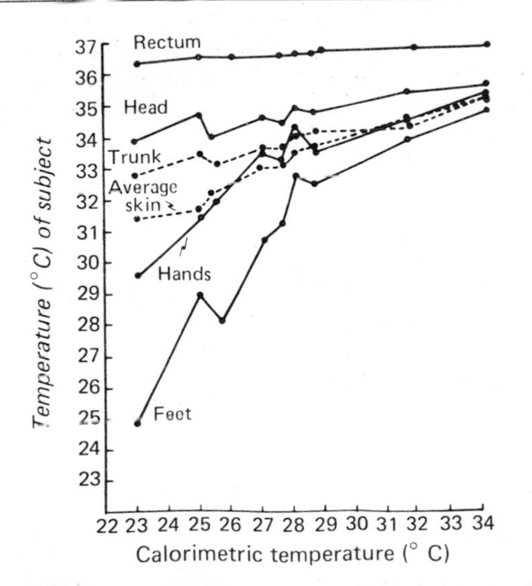

Figure 14–20. Temperature of various parts of the body of a naked subject at various ambient temperatures in a calorimeter. (Redrawn and reproduced, with permission, from Hardy JD, DuBois EF: Basal metabolism, radiation, convection and vaporization at temperatures of 22–35 °C. *J Nutr* 1938;**15**:477.)

Normal Body Temperature

In homeothermic animals, the actual temperature at which the body is maintained varies from species to species and, to a lesser degree, from individual to individual. In humans, the traditional normal value for the oral temperature is 37°C (98.6 °F), but in one large series of normal young adults, the morning oral temperature averaged 36.7 °C, with a standard deviation of 0.2 °C. Therefore, 95% of all young adults would be expected to have a morning oral temperature of 36.3-37.1 °C (97.3-98.8 °F) (mean ±1.96 standard deviations; see Appendix). Various parts of the body are at different temperatures and the magnitude of the temperature difference between the parts varies with the environmental temperature (Fig 14–20). The extremities are generally cooler than the rest of the body. The temperature of the scrotum is carefully regulated at 32 °C. The rectal temperature is representative of the temperature at the core of the body and varies least with changes in environmental temperature. The oral temperature is normally 0.5 °C lower than the rectal temperature, but it is affected by many factors, including ingestion of hot or cold fluids, gum-chewing, smoking, and mouth breathing.

The normal human core temperature undergoes a regular circadian fluctuation of 0.5-0.7 °C. In individuals who sleep at night and are awake during the day (even when hospitalized at bed rest), it is lowest at about 6:00 AM and highest in the evenings (Fig 14–21). It is lowest during sleep, slightly higher in the awake but relaxed state, and rises with activity. In women, there is an additional monthly cycle of temperature variation characterized by a rise in basal temperature at the time of ovulation (see Chapter 23 and Fig 23–28). Temperature regulation is less precise in young children, and they may normally have a temperature that is 0.5 °C or so above the established norm for adults.

During exercise, the heat produced by muscular

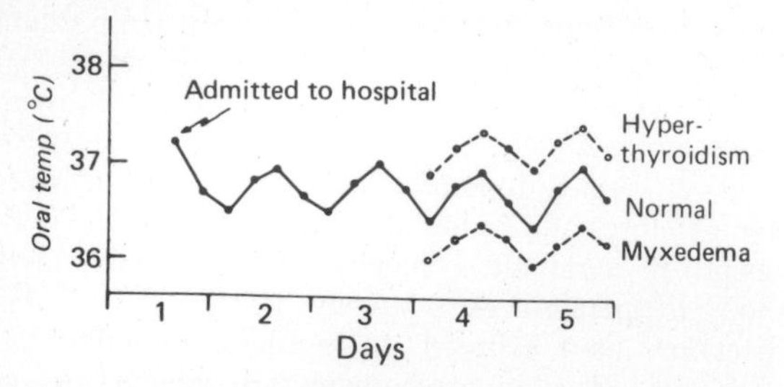

Figure 14–21. Typical temperature chart of a hospitalized patient who does not have a febrile disease. Note the slight rise in temperature, due to excitement and apprehension, at the time of admission to the hospital, and the regular circadian temperature cycle.

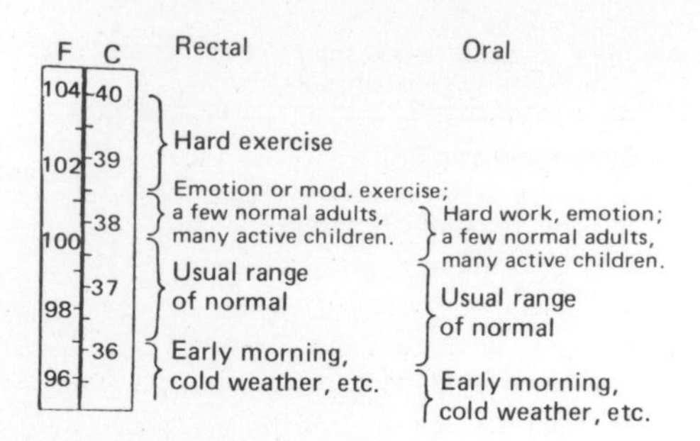

Figure 14–22. Ranges in rectal and oral temperatures seen in normal humans. (Reproduced, with permission, form DuBois EF: *Fever and the Regulation of Body Temperature.* Thomas, 1948.)

contraction accumulates in the body, and the rectal temperature normally rises as high as 40 °C (104 °F). This rise is due in part to the inability of the heat-dissipating mechanisms to handle the greatly increased amount of heat produced, but there is evidence that in addition there is an elevation of the body temperature at which the heat-dissipating mechanisms are activated during exercise. Body temperature also rises slightly during emotional excitement, probably owing to unconscious tensing of the muscles (Fig 14–22). It is chronically elevated by as much as 0.5 °C when the metabolic rate is high, as in hyperthyroidism, and lowered when the metabolic rate is low, as in myxedema (Fig 14–21). Some apparently normal adults chronically have a temperature above the normal range (constitutional hyperthermia).

Heat Production

Heat production and energy balance are discussed in Chapter 17. A variety of basic chemical reactions contribute to body heat production at all times. Ingestion of food increases heat production because of the specific dynamic action of the food (see Chapter 17), but the major source of heat is the contraction of skeletal muscle (Table 14–5). Heat production can be varied by endocrine mechanisms in the absence of food intake or muscular exertion. Epinephrine and norepinephrine produce a rapid but short-lived increase in heat production; thyroid hormones produce a slowly developing but prolonged increase. Furthermore, sympathetic discharge is decreased during fasting and increased by feeding (see Chapter 13).

A source of considerable heat, particularly in infants, is a special type of fat, **brown fat,** that is located between and around the scapulas and in other parts of the body. This fat has a high rate of metabolism, and its function has been likened to that of an electric blanket (see Chapter 17).

Table 14–5. Body heat production and heat loss.

Body heat is produced by:	
Basic metabolic processes	
Food intake (specific dynamic action)	
Muscular activity	
Body heat is lost by:	**Percentage of Heat Lost at 21 °C**
Radiation and conduction	70
Vaporization of sweat	27
Respiration	2
Urination and defecation	1

Heat Loss

The processes by which heat is lost from the body when the environmental temperature is below body temperature are listed in Table 14-5. **Conduction** is heat exchange between objects or substances at different temperatures that are in contact with one another. A basic characteristic of matter is that above absolute zero, its molecules are in motion, with the amount of motion proportionate to the temperature. These molecules collide with the molecules in cooler objects, transferring thermal energy to them. The amount of heat transferred is proportionate to the temperature difference between the objects in contact **(thermal gradient).** Conduction is aided by **convection,** the movement of molecules away from the area of contact. Thus, for example, an object in contact with air at a different temperature changes the specific gravity of the air, and since warm air rises and cool air falls, a new supply of air is brought into contact with the object. Of course, convection is greatly aided if the object moves about in the medium or the medium moves past the object, eg, if a subject swims through water or a fan blows air through a room. **Radiation** is transfer of heat by infrared electromagnetic radiation from one object to another at a different temperature with which it is not in contact. When an individual is in a cold environment, heat is lost by conduction to the surrounding air and by radiation to cool objects in the vicinity. Conversely, of course, heat is transferred to an individual and the heat load is increased by these processes when the environmental temperature is above body temperature. Note that because of radiation, an individual can feel chilly in a room with cold walls even though the room is relatively warm. On a cold but sunny day, the heat of the sun reflected off bright objects exerts an appreciable warming effect. It is the heat reflected from the snow, for example, that makes it possible to ski in fairly light clothes even though the air temperature is below freezing.

Since conduction occurs from the surface of one object to the surface of another, the temperature of the skin determines to a large extent the degree to which body heat is lost or gained. The amount of heat reaching the skin from the deep tissues can be varied by changing the blood flow to the skin. When the cutaneous vessels are dilated, warm blood wells into the skin, whereas in the maximally vasoconstricted state, heat is held centrally in the body. The rate at which heat is transferred from the deep tissues to the skin is called the **tissue conductance.** Birds have a layer of feathers next to the skin, and most mammals have a significant layer of hair or fur. Heat is conducted from the skin to the air trapped in this layer and from the trapped air to the exterior. When the thickness of the trapped layer is increased by fluffing the feathers or erection of the hairs **(horripilation),** heat transfer across the layer is reduced and heat losses (or, in a hot environment, heat gains) are decreased. ''Goose pimples'' are the result of horripilation in humans; they are the visible manifestation of cold-induced contraction of the piloerector muscles attached to the rather meager hair supply. Humans usually supplement this layer of hair with a layer of clothes. Heat is conducted from the skin to the layer of air trapped by the clothes, from the inside of the clothes to the outside, and from the outside of the clothes to the exterior. The magnitude of the heat transfer across the clothing, a function of its texture and thickness, is the most important determinant of how warm or cool the clothes feel, but other factors, especially the size of the trapped layer of warm air, are important also. Dark clothes absorb radiated heat, and light-colored clothes reflect it back to the exterior.

The other major process transferring heat from the body in humans and those animals that sweat is vaporization of water on the skin and mucous membranes of the mouth and respiratory passages. Vaporization of 1g of water removes about 0.6 kcal of heat. A certain amount of water is vaporized at all times. This **insensible water loss** amounts to 50 mL/h in humans. When sweat secretion is increased, the degree to which the sweat vaporizes depends upon the humidity of the environment. It is common knowledge that one feels hotter on a humid day. This is due in part to the decreased vaporization of sweat, but even under conditions in which vaporization of sweat is complete, an individual in a humid environment feels warmer than an individual in a dry environment. The reason for this difference is not known, but it seems related to the fact that in the humid environment sweat spreads over a greater area of skin before it evaporates. During muscular exertion in a hot environment, sweat secretion reaches values as high as 1600 mL/h, and in a dry atmosphere, most of this sweat is vaporized. Heat loss by vaporization of water therefore varies from 30 to over 900 kcal/h.

Some mammals lose heat by **panting.** This rapid, shallow breathing greatly increases the amount of water vaporization in the mouth and respiratory passages and therefore the amount of heat lost. Because the breathing is shallow, it produces relatively little change in the composition of alveolar air (see Chapter 34).

The relative contribution of each of the processes that transfer heat away from the body (Table 14–5) varies with the environmental temperature. At 21 °C, vaporization is a minor component in humans at rest. As the environmental temperature approaches body temperature, radiation losses decline and vaporization losses increase.

Temperature-Regulating Mechanisms

The reflex and semireflex thermoregulatory responses are listed in Table 14–6. They include autonomic, somatic, endocrine, and behavioral changes. One group of responses increases heat loss and decreases heat production; the other decreases heat loss and increases heat production. In general, exposure to heat stimulates the former group of responses and inhibits the latter, whereas exposure to cold does the opposite.

Curling up ''in a ball'' is a common reaction to cold in animals and has a counterpart in the position some people assume on climbing into a cold bed. Curling up decreases the body surface exposed to the environment. Shivering is an involuntary response of the skeletal muscles, but cold also causes a semiconscious general increase in motor activity. Examples include foot stamping and dancing up and down on a cold day. Increased catecholamine secretion is an important endocrine response to cold; adrenal medullectomized rats die faster than normal controls when exposed to cold. TSH secretion is increased by cold

Table 14–6. Temperature-regulating mechanisms.

Mechanisms activated by cold:	
Shivering Hunger Increased voluntary activity Increased secretion of norepinephrine and epinephrine	Increase heat production
Cutaneous vasoconstriction Curling up Horripilation	Decrease heat loss
Mechanisms activated by heat:	
Cutaneous vasodilation Sweating Increased respiration	Increase heat loss
Anorexia Apathy and inertia	Decrease heat production

and decreased by heat in laboratory animals, but the change in TSH secretion produced by cold in adult humans is small and of questionable significance. It is common knowledge that activity is decreased in hot weather—the "it's too hot to move" reaction.

Thermoregulatory adjustments involve local responses as well as more general reflex responses. When cutaneous blood vessels are cooled, they become more sensitive to catecholamines and the arterioles and venules constrict. This local effect of cold directs blood away from the skin. Another heat-conserving mechanism that is important in animals living in cold water is heat transfer from arterial to venous blood in the limbs. The deep veins **(venae comitantes)** run alongside the arteries supplying the limbs, and heat is transferred from the warm arterial blood going to the limbs to the cold venous blood coming from the extremities **(counter-current exchange;** see Chapter 38). This keeps the tips of the extremities cold but conserves body heat.

The reflex responses activated by cold are controlled from the posterior hypothalamus. Those activated by warmth are controlled primarily from the anterior hypothalamus, although some thermoregulation against heat still occurs after decerebration at the level of the rostral midbrain. Stimulation of the anterior hypothalamus causes cutaneous vasodilation and sweating, and lesions in this region cause hyperthermia, with rectal temperatures sometimes reaching 43 °C (109.4 °F). Posterior hypothalamic stimulation causes shivering, and the body temperature of animals with posterior hypothalamic lesions falls toward that of the environment.

There is some evidence that in primates and humans serotonin is a synaptic mediator in the centers controlling the mechanisms activated by cold, and norepinephrine plays a similar role in those activated by heat. However, there are marked species variations in the temperature responses to these amines. Peptides may also be involved, but the details of the central synaptic connections concerned with thermoregulation are still unknown.

Afferents

The signals that activate the hypothalamic temperature-regulating centers come from 2 sources: temperature-sensitive cells in the anterior hypothalamus, and cutaneous temperature receptors, especially cold receptors. Present evidence indicates that the stimuli which activate the defense against high temperatures in humans come mainly from the temperature-sensitive cells in the hypothalamus. This conclusion grew out of research in which the temperature of the back of the nasal cavity and the interior of the ear, near the hypothalamus, was correlated with thermoregulatory responses to changing environmental temperatures (Fig 14–23). The response of body heat production to cooling is modified by interactions between cutaneous and central stimuli. Heat production is increased when head temperature falls below a given threshold value, but the threshold for the

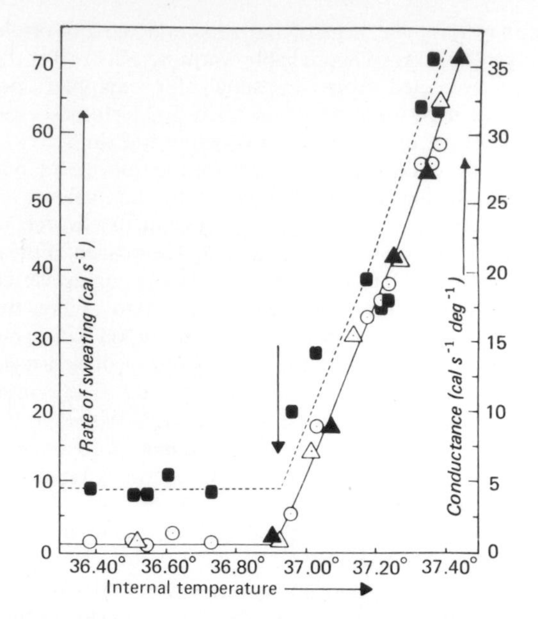

Figure 14–23. Quantitative relations in humans between the temperature of the interior of the head (internal temperature) and cutaneous blood flow (squares, scale at the right), and sweating (circles and triangles, scale at the left). The arrow points to the sharp threshold at which these parameters start to rise. In this subject, the threshold was at 36.9 °C. (Reproduced with permission, from Benzinger TH: Receptor organs and quantitative mechanisms of human temperature control in a warm environment. *Fed Proc* 1960;**19**:32.)

response is lower and its magnitude decreased when the skin temperature is increased.

Fever

Fever is perhaps the oldest and most universally known hallmark of disease. It occurs not only in mammals but also in birds, reptiles, amphibia, and fish. When it occurs in homeothermic animals, the thermoregulatory mechanisms behave as if they were adjusted to maintain body temperature at a higher than normal level, ie, "as if the thermostat had been reset" to a new point above 37 °C. The temperature receptors then signal that the actual temperature is below the new set point, and temperature-raising mechanisms are activated. This usually produces chilly sensations due to cutaneous vasoconstriction and occasionally enough shivering to produce a shaking chill. However, the nature of the response depends on the ambient temperature. The temperature rise in experimental animals injected with a pyrogen is due mostly to increased heat production if they are in a cold environment and mostly to decreased heat loss if they are in a warm environment.

The pathogenesis of fever is summarized in Fig 14–24. Toxins from bacteria such as endotoxin act on monocytes, macrophages, and Kupffer cells to produce **interleukin-1** (IL-1; see Chapter 27), a polypeptide also known as **endogenous pyrogen (EP).**

IL-1 has widespread effects in the body. It enters the brain and produces fever by a direct action on the preoptic area of the hypothalamus. It also acts on lymphocytes to activate the immune system, stimulates release of neutrophils for the bone marrow, and causes proteolysis in muscle. Diverse other agents, including the steroid etiocholanolone, a metabolite of testosterone (see Chapter 23), also cause the production of IL-1. Its production in peripheral blood requires energy and is blocked by inhibitors of protein synthesis (see Chapter 1).

The fever produced by IL-1 may be due to local release of prostaglandins. Injection of prostaglandins into the hypothalamus produces fever. In addition, the antipyretic effect of aspirin is exerted directly on the hypothalamus, and aspirin inhibits prostaglandin synthesis. However, there is some evidence that the prostaglandins are not involved, and the question of their role in the production of fever is a matter of much debate.

The benefit of fever to the organism is uncertain, although it is presumably beneficial, because it has evolved and persisted as a response to infections and other diseases. Many microorganisms grow best within a relatively narrow temperature range, and a rise in temperature inhibits their growth. In addition, antibody production is increased when body temperature is elevated. Before the advent of antibiotics, fevers were artificially induced for the treatment of neurosyphilis and proved to be beneficial. Hyperthermia benefits individuals infected with anthrax, pneumococcal pneumonia, leprosy, and various fungal, rickettsial, and viral diseases. Hyperthermia also slows the growth of some tumors. However, very high temperatures are harmful. When the rectal temperature is over 41 °C (106 °F) for prolonged periods, some permanent brain damage results. When it is over 43 °C, heat stroke develops, and death is common.

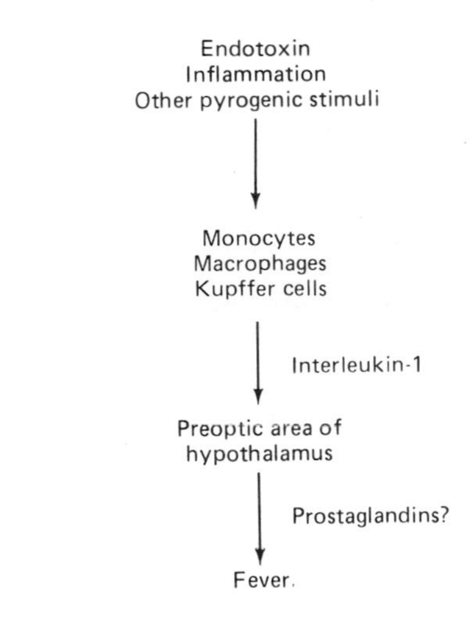

Figure 14–24. Pathogenesis of fever.

Hypothermia

In hibernating mammals, body temperature drops to low levels without causing any ill effects that are demonstrable upon subsequent arousal. This observation led to experiments on induced hypothermia. When the skin or the blood is cooled enough to lower the body temperature in nonhibernating animals and in humans, metabolic and physiologic processes slow down. Respiration and heart rate are very slow, blood pressure is low, and consciousness is lost. At rectal temperatures of about 28 °C, ability to spontaneously return the temperature to normal is lost, but the individual continues to survive and, if rewarmed with external heat, returns to survive and, if rewarmed with external heat, returns to a normal state. If care is taken to prevent the formation of ice crystals in the tissues, the body temperature of experimental animals can be lowered to subfreezing levels without producing any damage that is detectable after subsequent rewarming.

Humans tolerate body temperatures of 21–24 °C (70–75 °F) without permanent ill effects, and induced hypothermia has been used extensively in surgery. In hypothermic patients, the circulation can be stopped for relatively long periods because the 0_2 needs of the tissues are greatly reduced. Blood pressure is low, and bleeding is minimal. It is possible under hypothermia to stop and open the heart and to perform other procedures, especially brain operations, that would have been impossible without cooling.

15 Neural Basis of Instinctual Behavior & Emotions

INTRODUCTION

Emotions have both mental and physical components. They involve **cognition,** an awareness of the sensation and usually its cause; **affect,** the feeling itself; **conation,** the urge to take action; and **physical changes** such as hypertension, tachycardia, and sweating. The hypothalamus and limbic systems are intimately concerned with emotional expression and with the genesis of emotions.

This chapter reviews the physiologic basis of emotion, sexual behavior, fear, rage, and motivation. It also considers the relation of major neurotransmitter systems in the brain to these processes.

ANATOMIC CONSIDERATIONS

The term **limbic lobe** or **limbic system** is applied to the part of the brain that consists of a rim of cortical tissue around the hilus of the cerebral hemisphere and a group of associated deep structures—the amygdala, the hippocampus, and the septal nuclei (Figs 15–1 and 15–2). The region was formerly called the rhinencephalon because of its relation to olfaction, but only a small part of it is actually concerned with smell.

Histology

The limbic cortex is phylogenetically the oldest part of the cerebral cortex. Histologically, it is made up of a primitive type of cortical tissue called **allocortex,** surrounding the hilus of the hemisphere, and a second ring of a transitional type of cortex called **juxtallocortex** between the allocortex and the rest of the cerebral hemisphere. The cortical tissue of the remaining nonlimbic portions of the hemisphere is called **neocortex** and is the most highly developed type. The actual extent of the allocortical and juxtallocortical areas has changed little as mammals have evolved, but these regions have been overshadowed by the immense growth of the neocortex, which reaches its greatest development in humans (Fig 15–1).

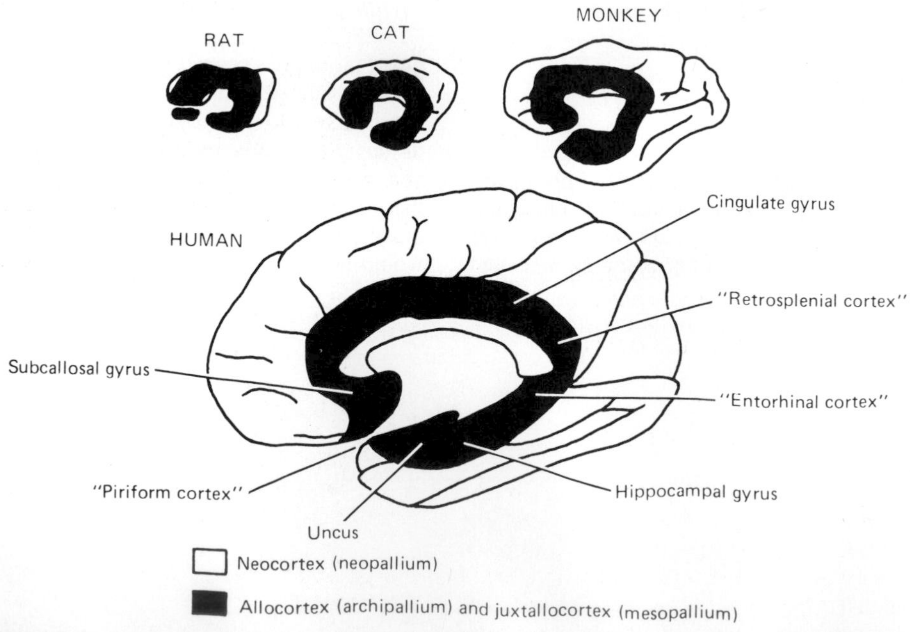

Figure 15–1. Relation of the limbic cortex to the neocortex in rats, cats, monkeys, and humans.

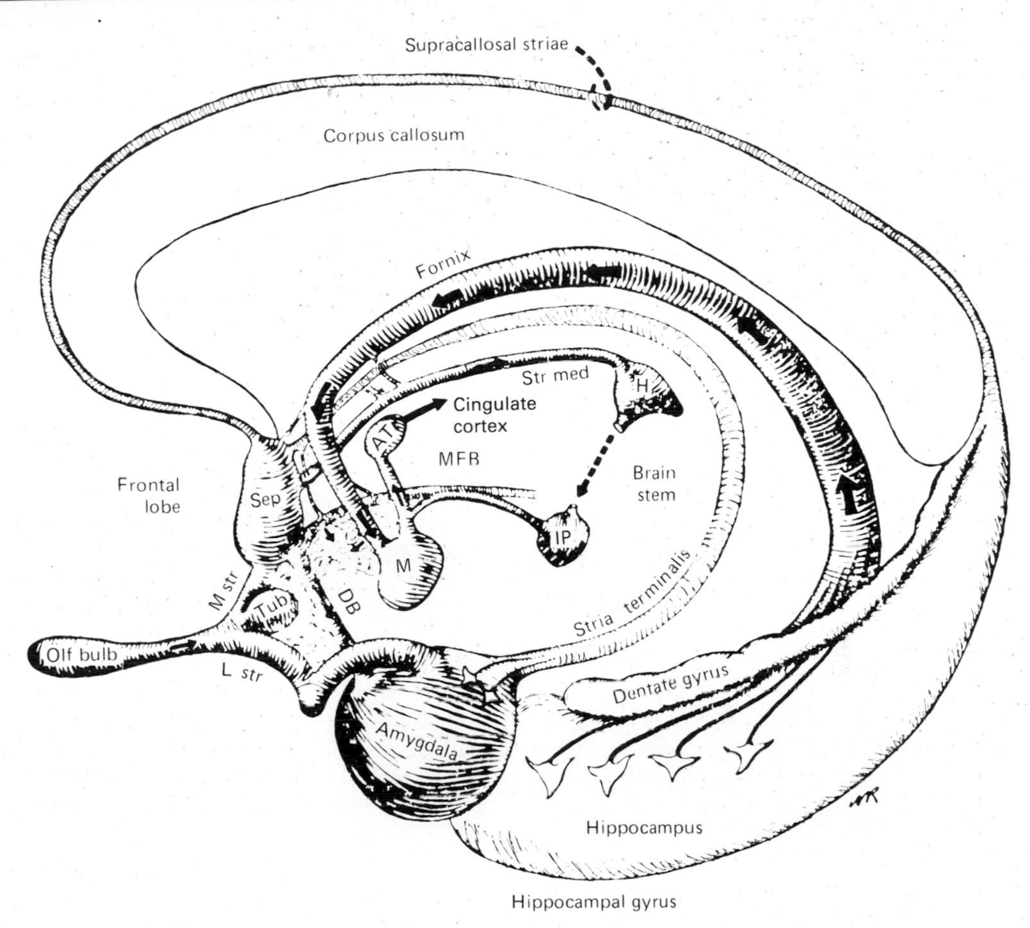

Figure 15–2. Diagram of the principal connections of the limbic system. M str, L str, medial and lateral olfactory striae; Str med, stria medullaris; Tub, olfactory tubercle; DB, diagonal band of Broca; Sep, septum; AT, anterior nuclei of the thalamus; M, mamillary body; H, habenula; IP, interpeduncular nucleus; MFB, medial forebrain bundle. The arrows identify the so called Papez circuit. (Modified from MacLean PD: Psychosomatic disease and the visceral brain. *Psychosom Med* 1949;11:338.)

Afferent & Efferent Connections

The major connections of the limbic system are shown in Fig 15–2. The fornix connects the hippocampus to the mamillary bodies, which are in turn connected to the anterior nuclei of the thalamus by the mamillothalamic tract. The anterior nuclei of the thalamus project to the cingulate cortex, and from the cingulate cortex there are connections to the hippocampus, completing a complex closed circuit. This circuit was originally described by Papex and has been called the Papez circuit.

Correlations Between Structure & Function

One characteristic of the limbic system is the paucity of the connections between it and the neocortex. Nauta has aptly stated that "the neocortex sits astride the limbic system like a rider on a horse without reins." Actually, there are a few connections, and from a functional point of view, neocortical activity does modify emotional behavior and vice versa. However, one of the characteristics of emotion is that it cannot be turned on and off at will.

Another characteristic of limbic circuits is their prolonged after-discharge following stimulation. This may explain in part the fact that emotional responses are generally prolonged rather than evanescent and outlast the stimuli that initiate them.

LIMBIC FUNCTIONS

Stimulation and ablation experiments indicate that in addition to its role in olfaction (see Chapter 10), the limbic system is concerned with feeding behavior. Along with the hypothalamus, it is also concerned with sexual behavior, the emotions of rage and fear, and motivation.

Autonomic Responses & Feeding Behavior

Limbic stimulation produces autonomic effects, particularly changes in blood pressure and respiration. These responses are elicited from many limbic structures, and there is little evidence of localization of autonomic responses. This suggests that the autonomic effects are part of more complex phenomena, particularly emotional and behavioral responses. Stimulation of the amygdaloid nuclei causes movements such as chewing and licking and other activities related to feeding. Lesions in the amygdala cause moderate hyperphagia, with indiscriminate ingestion of all kinds of food. The relation of this type of omniphagia to the hypothalamic mechanisms regulating appetite is discussed in Chapter 14.

SEXUAL BEHAVIOR

Mating is a basic but complex phenomenon in which many parts of the nervous system are involved. Copulation itself is made up of a series of reflexes integrated in spinal and lower brain stem centers, but the behavioral components that accompany it, the urge to copulate, and the coordinated sequence of events in the male and female that lead to pregnancy are regulated to a large degree in the limbic system and hypothalamus. Learning plays a part in the development of mating behavior, particularly in humans and other primates, but in nonprimate mammals, courtship and successful mating can occur with no previous sexual experience. The basic responses are therefore innate and are undoubtedly present in all mammals. However, in humans, the sexual functions have become extensively encephalized and conditioned by social and psychic factors. The basic physiologic mechanisms of sexual behavior in animals are therefore considered first and then compared to the responses in humans.

Relation to Endocrine Function

In nonprimate mammals, removal of the gonads leads eventually to decreased or absent sexual activity in both the male and the female—although the loss is slow to develop in the males of some species. Injections of gonadal hormones in castrate animals revive sexual activity. Testosterone in the male and estrogen in the female have the most marked effect. Large doses of progesterone are also effective in the female, while in the presence of smaller doses of progesterone, the dose of estrogen necessary to produce sexual activity is lowered. In primates, on the other hand, progesterone has no facilitatory effect and is even somewhat inhibitory. Large doses of testosterone and other androgens in castrate females initiate female behavior, and large doses of estrogens in castrate males trigger male mating responses. It is unsettled why responses appropriate to the sex of the animal occur when the hormones of the opposite sex are injected.

Effects of Hormones in Humans

In adult women, ovariectomy does not necessarily reduce libido (defined in this context as sexual interest and drive) or sexual ability. Postmenopausal women continue to have sexual relations, often without much change in frequency from their premenopausal pattern. This persistence is probably due to secretion of steroids from the adrenal cortex that are converted to circulating estrogens but may also be due to the greater degree of encephalization of sexual functions in humans and their relative emancipation from instinctual and hormonal control. Treatment with sex hormones increases sexual interest and drive in humans. Testosterone, for example, increases libido in males, and so does estrogen used to treat diseases such as carcinoma of the prostate. The behavioral pattern present before treatment is stimulated but not redirected. Thus, administration of testosterone to homosexuals intensifies their homosexual drive but does not convert it to a heterosexual drive.

Neural Control in the Male

In male animals, removal of the neocortex generally inhibits sexual behavior. Partial cortical ablations also produce some inhibition, the degree of the inhibition being independent of the coexisting motor deficit and most marked when the lesions are in the frontal lobes. On the other hand, cats and monkeys with bilateral limbic lesions localized to the piriform cortex overlying the amygdala (Fig 15–3) develop a marked intensification of sexual activity. These ani-

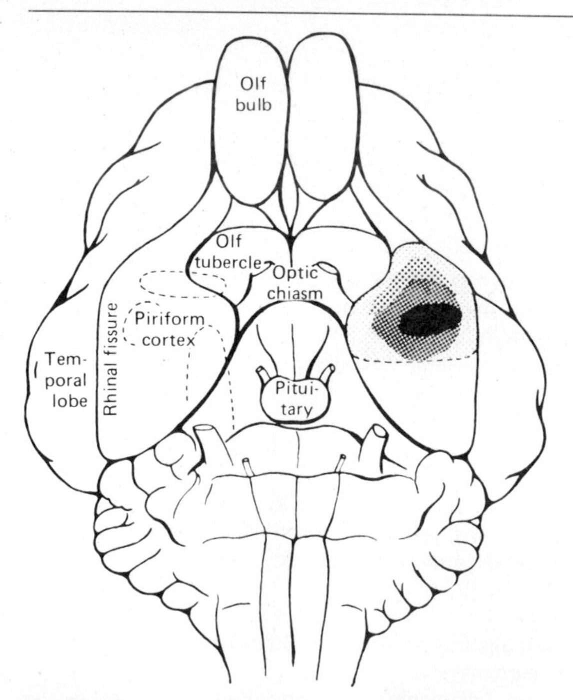

Figure 15–3. Site of lesions producing hypersexuality in male cats. The incidence of hypersexuality was proportionate to the intensity of the stippling, and when the black area was destroyed, hypersexuality was always present. Olf, olfactory. (Reproduced, with permission, from Green J et al: Rhinencephalic lesions and behavior in cats. *J Comp Neurol* 1957;**108**:505.)

mals not only mount adult females; they also mount immature females and other males and attempt to copulate with animals of other species and with inanimate objects. Despite some claims to the contrary, such behavior is clearly abnormal in the species studied. The behavior is dependent upon the presence of testosterone but is not due to any increase in its secretion.

The hypothalamus is also involved in the control of sexual activity in males. Stimulation along the medial forebrain bundle and in neighboring hypothalamic areas causes penile erection with considerable emotional display in monkeys. In castrated rats, intrahypothalamic implants of testosterone restore the complete pattern of sexual behavior, and in intact rats, appropriately placed anterior hypothalamic lesions abolish interest in sex.

The extent to which the findings in male animals with periamygdaloid lesions are applicable to men is, of course, difficult to determine, but there are reports of hypersexuality in men with bilateral lesions in or near the amygdaloid nuclei.

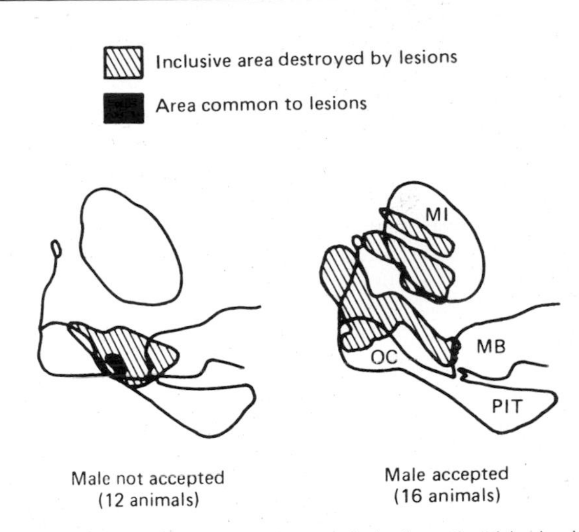

Figure 15–4. Sites of hypothalamic lesions that blocked behavioral heat without affecting ovarian cycles in ewes. MI, massa intermedia; MB, mamillary body; OC, optic chiasm; PIT, pituitary. (From data of MT Clegg and WF Ganong.)

Sexual Behavior in the Female

In mammals, the sexual activity of the male is more or less continuous, but in many species, the sexual activity of the female is cyclic. Most of the time, the female avoids the male and repulses his sexual advances. Periodically, however, there is an abrupt change in behavior and the female seeks out the male, attempting to mate. These short episodes of **heat** or **estrus** are so characteristic that the sexual cycle in mammalian species that do not menstruate is named the **estrous cycle.**

This change in female sexual behavior is brought on by a rise in the circulating blood estrogen level. Some animals, notably the rabbit and the ferret, come into heat and remains estrous until pregnancy or pseudopregnancy results. In these species, ovulation is due to a neuroendocrine reflex. Stimulation of the genitalia and other sensory stimuli at the time of copulation provoke release from the pituitary of the LH that makes the ovarian follicles rupture. In many other species, spontaneous ovulation occurs at regular intervals, and the periods of heat coincide with its occurrence. This is true in monkeys and apes. In captivity, these species mate at any time; but in the wild state, the females accept the male more frequently at the time of ovulation. In women, sexual activity occurs throughout the menstrual cycle, and according to one report, there is slightly more activity near the time of menstruation. However, other careful studies indicate that in women, as in other primates, there is more spontaneous female-initiated sexual activity at about the time of ovulation.

In monkeys, the sex drive of the male is greater when he is exposed to a female at the time of ovulation than when he is exposed to a female at another time of her cycle. The "message" sent by the female to the male in this situation is olfactory, and the substances responsible are certain fatty acids in the vaginal secretions. These fatty acids are also found in increased amounts in human vaginal secretions at about midcycle. Substances produced by an animal that act at a distance to produce behavioral or other physiologic changes in another animal of the same species have been called **pheromones.** The sex attractants of certain insects are particularly well known examples, but it appears from the evidence cited above that pheromones also operate in the regulation of sexual behavior in primates.

Neural Control in the Female

In female animals, removal of the neocortex and the limbic cortex abolishes active seeking out of the male ("enticement reactions") during estrus, but other aspects of heat are unaffected. Amygdaloid and periamygdaloid lesions do not produce hypersexuality as they do in the male. However, discrete anterior hypothalamic lesions abolish behavioral heat (Fig 15–4) without affecting the regular pituitary-ovarian cycle (see Chapter 23).

Implantation of minute amounts of estrogen in the anterior hypothalamus causes heat in ovariectomized rats (Fig 23–31). Implantation into other parts of the brain and outside the brain does not have this effect. Apparently, therefore, some element in the hypothalamus is sensitive to circulating estrogen and is stimulated by the hormone to initiate estrous behavior.

Effects of Sex Hormones in Infancy on Adult Behavior

In female experimental animals, exposure to sex steroids in utero or during early postnatal development causes marked abnormalities of sexual behavior when the animals reach adulthood. Female rats treated with a single relatively small dose of androgen before the fifth day of life do not have normal heat periods when they mature; they generally will not mate, even though they have cystic ovaries that secrete enough estrogen to cause the animals to have a persistently estrous type of vaginal smear (see Chapter 23). These rats do not show the cyclic release of pituitary gonadotropins characteristic of the adult female, but rather the tonic, steady secretion characteristic of the adult male; their brains have been "masculinized" by the single brief exposure to androgens. Conversely, male rats castrated at birth develop the female pattern of cyclic gonadotropin secretion and show considerable female sexual behavior when given doses of ovarian hormones that do not have this effect in intact males. Thus, the development of a "female hypothalamus" depends simply on the absence of androgens in early life rather than on exposure to female hormones.

Rats are particularly immature at birth, and animals of other species in which the young are more fully developed at birth do not show these changes when exposed to androgens during the postnatal period. However, these animals develop genital abnormalities when exposed to androgens in utero (see Chapter 23). Female monkeys exposed to androgens in utero do not lose the female pattern of gonadotropin secretion but do develop abnormalities of sexual behavior in adulthood.

Exposure to androgens in human females in utero does not change the cyclic pattern of gonadotropin secretion in adulthood (see Chapter 23). However, there is evidence that masculinizing effects on behavior do occur.

Maternal Behavior

Maternal behavior is depressed by lesions of the cingulate and retrosplenial portions of the limbic cortex in animals. Hormones do not appear to be necessary for its occurrence, but prolactin, which is secreted in large amounts during pregnancy and lactation, facilitates it.

FEAR & RAGE

Fear and rage are in some ways closely related emotions. The external manifestations of the **fear, fleeing,** or **avoidance reaction** in animals are autonomic responses such as sweating and pupillary dilation, cowering, and turning the head from side to side to seek escape. The **rage, fighting,** or **attack reaction** is associated in the cat with hissing, spitting, growling, piloerection, pupillary dilation, and well-directed biting and clawing. Both reactions—and sometimes mixtures of the two—can be produced by hypothalamic stimulation. When an animal is threatened, it usually attempts to flee. If cornered, an animal fights. Thus, fear and rage reactions are probably related instinctual protective responses to threats in the environment.

Fear

The fear reaction can be produced in conscious animals by stimulation of the hypothalamus and the amygdaloid nuclei. Conversely, the fear reaction and its autonomic and endocrine manifestations are absent in situations in which they would normally be evoked when the amygdalae are destroyed. A dramatic example is the reaction of monkeys to snakes. Monkeys are normally terrified by snakes. After bilateral temporal lobectomy, monkeys approach snakes without fear, pick them up, and even eat them.

Rage & Placidity

Most animals, including humans, maintain a balance between rage and its opposite, the emotional state that for lack of a better name is referred to here as placidity. Major irritations make normal individuals "lose their temper," but minor stimuli are ignored. In animals with certain brain lesions, this balance is altered. Some lesions produce a state in which the most minor stimuli evoke violent episodes of rage; others produce a state in which the most traumatic and anger-provoking stimuli fail to ruffle the animal's abnormal calm.

Rage responses to minor stimuli are observed after removal of the neocortex and after destruction of the ventromedial hypothalamic nuclei and septal nuclei in animals with intact cerebral cortices. On the other hand, bilateral destruction of the amygdaloid nuclei in monkeys causes a state of abnormal placidity. Stimulation of some parts of the amygdala in cats produces rage. The placidity produced by amygdaloid lesions in animals is converted into rage by subsequent destruction of the ventromedial nuclei of the hypothalamus.

Rage can also be produced by stimulation of an area extending back through the lateral hypothalamus to the central gray area of the midbrain, and the rage response usually produced by amygdaloid stimulation is abolished by ipsilateral lesions in the lateral hypothalamus or rostral midbrain.

Gonadal hormones appear to affect aggressive behavior. In male animals, aggression is decreased by castration and increased by androgens. It is also conditioned by social factors; it is more prominent in males that live with females and increases when a stranger is introduced into an animal's territory.

"Sham Rage"

It was originally thought that rage attacks in animals with diencephalic and forebrain lesions represented only the physical, motor manifestations of anger, and the reaction was therefore called "sham rage." This now appears to be incorrect. Although rage attacks in animals with diencephalic lesions are induced by minor stimuli, they are usually directed with great accuracy at the source of the irritation. Furthermore, hypothalamic stimulation that produces the fear-rage reaction is apparently unpleasant to animals, because they become conditioned against the place where the experiments are conducted and try to avoid the experimental sessions. They can easily be taught to press a lever or perform some other act to avoid a hypothalamic stimulus that produces the manifestations of fear or rage. It is difficult if not impossible to form conditioned reflex responses (see Chapter 16) by stimulation of purely motor systems, and it is also difficult if the unconditioned stimulus does not evoke either a pleasant or unpleasant feeling. The fact that hypothalamic stimulation is a potent unconditioned stimulus for the formation of conditioned avoidance responses and the fact that the avoidance responses are extremely persistent indicate that the stimulus is unpleasant. There is therefore little doubt that rage attacks include the mental as well as the physical manifestations of rage, and the term "sham rage" should be dropped.

Significance & Clinical Correlates

It is tempting on the basis of the evidence cited above to speculate that there are 2 intimately related mechanisms in the hypothalamus and limbic system, one promoting placidity and the other rage. If this is true, the emotional state is probably determined by afferent impulses that adjust the balance between them. An arrangement of this sort would be analogous to the systems controlling feeding and body temperature.

Although emotional responses are much more complex and subtle in humans than in animals, the neural substrates are probably the same. It is doubtful if placidity would be recognized as a clinical syndrome in our culture, but rage attacks in response to trivial stimuli have been observed many times in patients with brain damage. They are a complication of pituitary surgery when there is inadvertent damage to the base of the brain. They also follow a number of diseases of the nervous system, especially epidemic influenza and encephalitis, which destroy neurons in the limbic system and hypothalamus. Stimulation of the amygdaloid nuclei and parts of the hypothalamus in conscious humans produces sensations of anger and fear. In Japan, bilateral amygdaloid lesions have been produced in agitated, aggressive mental patients. The patients are said to have become placid and manageable, and it is of some interest that they were reported not to have developed hypersexuality or memory loss (see Chapter 16).

MOTIVATION

If an animal is placed in a box with a pedal or bar that can be pressed, the animal sooner or later accidentally presses it. Olds and his associates showed that if the bar is connected in such a way that each press delivers a stimulus to an electrode implanted in certain parts of the brain, the animal returns to the bar and presses it again and again. Pressing the bar soon comes to occupy most of the animal's time. Some animals go without food and water and others endure strong adversive stimuli to press the bar for brain stimulation, and some will continue until they fall over exhausted. Rats press the bar 5000–12,000 times per hour, and monkeys have been clocked at 17,000 bar presses per hour. On the other hand, when the electrode is in certain other areas, the animals avoid pressing the bar, and stimulation of these areas is a potent unconditioned stimulus for the development of conditioned avoidance responses.

The points where stimulation leads to repeated bar pressing are located in a medial band of tissue passing from the frontal cortex through the hypothalamus to the midbrain tegmentum (Fig 15–5). The highest rates are obtained from points in the medial forebrain bundle, with high rates also obtained from points in the tegmentum and septal nuclei. The points where stimulation is avoided are in the lateral portion of the posterior hypothalamus, the dorsal midbrain, and the entorhinal cortex. The latter points are sometimes close to points where bar pressing is repeated, but they are part of a separate system. The areas where bar pressing is repeated are much more extensive than those where it is avoided. It has been calculated that in rats repeated pressing is obtained from 35% of the brain, avoidance from 5%, and indifferent responses (neither repetition nor avoidance) from 60%.

It is obvious that some effect of the stimulation causes the animals to stimulate themselves again and again, but what the animals feel is, of course, unknown. There are a number of reports of bar-pressing experiments in humans with chronically implanted electrodes. Most of the subjects were schizophrenics or epileptics, but a few were patients with visceral neoplasms and intractable pain. Like animals, humans press the bar repeatedly. They generally report that the sensations evoked are pleasurable, using phrases like "relief of tension" and "a quiet, relaxed feeling" to describe the experience. However, they rarely report "joy" or "ecstasy," and some persons with the highest self-stimulation rates cannot tell why they keep pushing the bar. When the

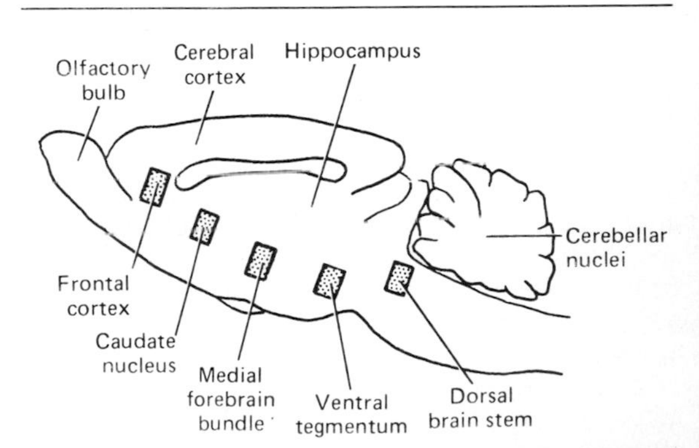

Figure 15–5. Areas where stimulation leads to repeated bar pressing projected on a parasagital view of the rat brain. The regions where high rates of self-stimulation are produced are indicated by the rectangles. (Modified from Routtenberg A: The reward system of the brain. *Sci Am* [Nov] 1978;**239**:154.)

electrodes are in the areas where stimulation is avoided, patients report sensations ranging from vague fear to terror. It is probably wise, therefore, to avoid vivid terms and call the brain systems involved the **reward** or **approach system** and the **punishment** or **avoidance system.**

Stimulation of the approach system provides a potent motivation for learning mazes or performing other tasks. Drugs that block dopamine receptors reduce the rate of self-stimulation whereas dopamine agonists increase it; norepinephrine does not appear to be involved.

Studies of the kind described above provide physiologic evidence that behavior is motivated not only by reduction or prevention of an unpleasant effect but also by primary rewards such as those produced by stimulation of the approach system of the brain. The implications of this fact are great in terms of the classical drive-reduction theory of motivation, in terms of the disruption and facilitation of ongoing behavior, and in terms of normal and abnormal emotional responses.

BRAIN CHEMISTRY, BEHAVIOR, & SYNAPTIC TRANSMISSION IN THE CENTRAL NERVOUS SYSTEM

Drugs that modify human behavior include **hallucinogenic agents,** drugs that produce hallucinations and other manifestations of the psychoses; **tranquilizers,** drugs that allay anxiety and various psychiatric symptoms; and **psychic energizers,** antidepressant drugs that elevate mood and increase interest and drive. These and many other drugs act by modifying transmission at synaptic junctions in the brain, and their discovery has stimulated great interest in the nature and properties of the transmitter agents involved. The chemistry of the known and suspected synaptic transmitters and their distribution in the peripheral as well as the central nervous system are discussed in Chapter 4.

The monoamine transmitters in the brain are related to emotional responses, and their distribution has been mapped using histochemical techniques.

Serotonin

Serotonin is found in relatively high concentrations in a number of areas in the brain. The serotonin-containing neurons have their cell bodies in the midline raphe nuclei of the brain stem and project to portions of the hypothalamus, the limbic system, the neocortex, and the spinal cord (Fig 15–6).

The hallucinogenic agent lysergic acid diethylamide (LSD) is a serotonin agonist that produces its effects by activating 5-HT_2 receptors (see Chapter 4) in the brain. The transient hallucinations and other mental aberrations produced by this drug were discovered when the chemist who synthesized it inhaled some by accident. Its discovery called attention to the correlation between behavior and variations in brain

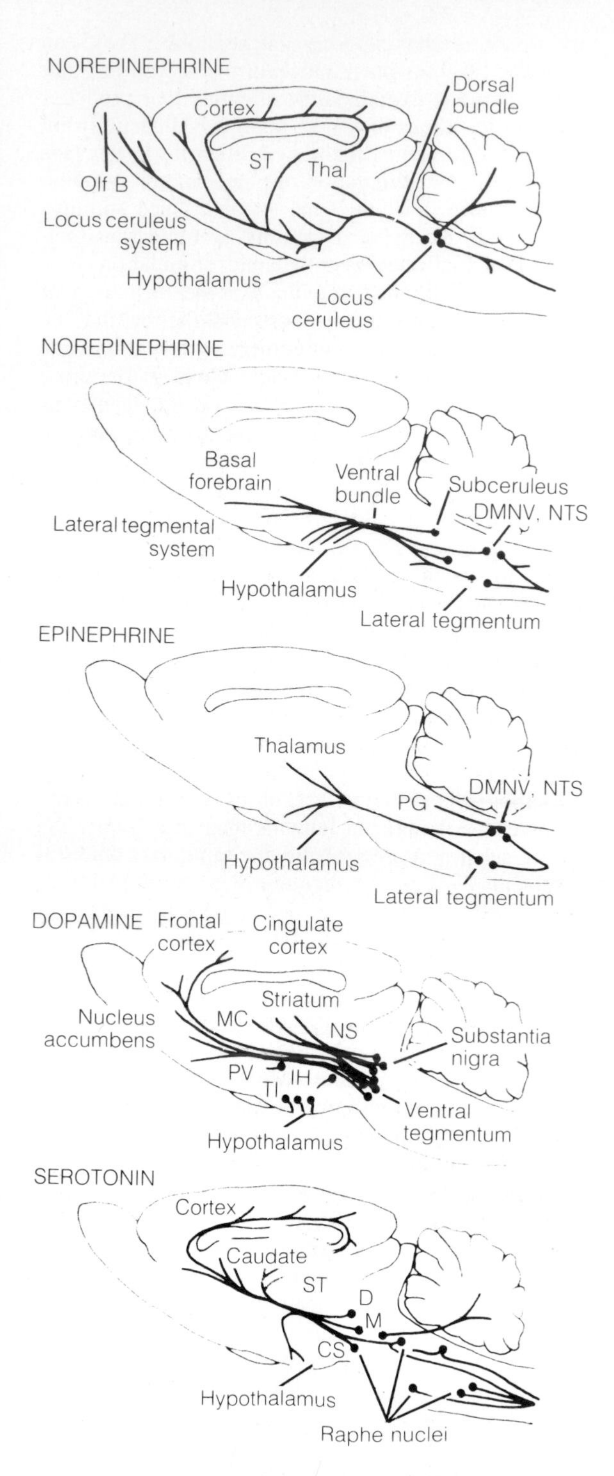

Figure 15–6. Aminergic pathways in rat brain. The pathways in humans are similar. The 2 principal noradrenergic systems (locus ceruleus and lateral tegmental) are shown separately. Olf B, olfactory bulb; Thal, thalamus; ST, stria terminalis; DMNV, dorsal motor nucleus of vagus; NTS, nucleus of tractus solitarius; PG, periaqueductal gray; NS, nigrostriatal system; MC, mesocortical system; PV, periventricular system; IH, incertohypothalamic system; TI, tuberoinfundibular system; D, M, and CS, dorsal, medial, and central superior raphe nuclei.

serotonin content. Psilocin, a substance found in certain mushrooms, and N,N-dimethyltryptamine (DMT) are also hallucinogenic, and, like serotonin, they are derivatives of tryptamine. 2,5-Dimethoxy-4-methylamphetamine (DOM) and mescaline and its congeners, the other true hallucinogens, are phenylethylamines rather than indolamines. However, all these hallucinogens appear to exert their effects by binding to $5HT_2$ receptors.

Selective depletion of brain serotonin can be produced by administering *p*-chlorophenylalanine, a compound that blocks conversion of tryptophan to 5-hydroxytryptophan (Fig 4–21). This is the rate-limiting step in serotonin biosynthesis. In animals, *p*-chlorophenylalanine produces prolonged wakefulness, suggesting that serotonin plays a role in sleep. However, wakefulness is not produced in humans, and no clear-cut psychic changes are produced even by large doses. Serotonergic neurons discharge rapidly in the awake state, slowly during drowsiness, more slowly with bursts during sleep, and not at all during REM sleep (see Chapter 11).

Other functions for brain serotonin have been proposed. Serotonin may play an excitatory role in the regulation of prolactin secretion (see Chapter 23). There is some evidence that descending serotonergic fiber systems inhibit transmission in pain pathways in the dorsal horns. In addition, there is a prominent serotonergic innervation of the suprachiasmatic nuclei of the hypothalamus, and serotonin may be involved in the regulation of circadian rhythms (see Chapter 14).

Norepinephrine

The distribution of norepinephrine in the brain is compared with that of serotonin in Table 15–1. The cell bodies of the norepinephrine-containing neurons are located in the locus ceruleus and other nuclei in the pons and medulla. From the locus ceruleus, the axons of the noradrenergic neurons form a **locus ceruleus system.** They descend into the spinal cord, enter the cerebellum, and ascend to innervate the paraventricular, supraoptic, and periventricular nuclei of the hypothalamus, the thalamus, the basal telencephalon, and the entire neocortex (Fig 15–6). From cell bodies in the dorsal motor nucleus of the vagus, the nucleus of the tractus solitarius, and areas in the dorsal and lateral tegmentum, the axons of the noradrenergic neurons form a **lateral tegmental system** that projects the spinal chord, the brain stem, all of the hypothalamus, and the basal telencephalon.

Drugs that increase extracellular norepinephrine in the brain elevate mood, and drugs that decrease extracellular norepinephrine cause depression. Manic-depressive illness runs in families, and at least in some of these families, there is a genetic abnormality close to or in the gene on chromosome 11 that codes for tyrosine hydroxylase, the rate-limiting enzyme in catecholamine biosynthesis (see Chapter 4). On the other hand, it has not been possible to prove that extracellular norepinephrine is deficient in depressed patients, and there is some evidence that depression is due instead to down-regulation of β-adrenergic receptors or even $5HT_2$ receptors in the brain.

In the cerebellum, noradrenergic neurons inhibit Purkinje cells, and there is evidence that these inhibitory effects are mediated via β receptors and cyclic AMP.

The norepinephrine-containing neurons in the hypothalamus are involved in regulation of the secretion of vasopressin and oxytocin, and they adjust the secretion of the hypophyseotropic hormones that regulate the secretion of anterior pituitary hormones (see Chapter 14). There is some evidence that norepinephrine is involved in the control of food intake. Along with serotonin, it may be involved in the regulation of body temperature.

Epinephrine

There is a system of phenylethanolamine-N-methyltransferase (PNMT)-containing neurons with cell bodies in the medulla that project to the hypothalamus. Those neurons secrete epinephrine, but their function is uncertain. Epinephrine-secreting neurons also project to the thalamus, periaqueductal gray, and spinal cord. There are appreciable quantities of tyramine in the central nervous system, but no function has been assigned to this agent.

Table 15–1. Amine and substance P content of selected portions of the human brain. Data compiled from various authors.

	Norepinephrine	Dopamine	Serotonin	Histamine	Substance P
	(μg/g of Fresh Tissue)				(units/g)
Amygdala	0.21	0.6	0.26	*	*
Caudate nucleus	0.09	3.5	0.33	0.5	85
Putamen	0.12	3.7	0.32	0.7	*
Globus pallidus	0.15	0.5	0.23	0.6	*
Thalamus	0.13	0.3	0.26	0.4	12
Hypothalamus	1.25	0.8	0.29	2.5	102
Substantia nigra	0.21	0.9	0.55	*	699

* No data.

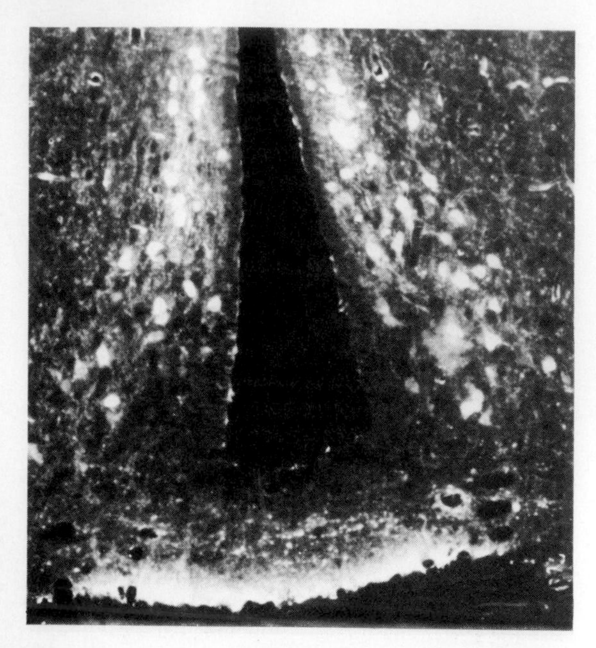

Figure 15–7. Dopaminergic neurons in the hypothalamus of the rat. α-Methylnorepinephrine injected before sacrifice to intensify fluorescence. Transverse section. Cell bodies can be seen above in the arcuate nucleus on either side of the third ventricle, and the dense nerve terminals can be seen below in the external layer of the median eminence. (Reproduced, with permission, from Hökfelt T, Fuxe K: On the morphology and the neuroendocrine role of the hypothalamic catecholamine neurons. In: *Brain-Endocrine Interaction.* Knigge K, Scott D, Weindl A [editors]. Karger, 1972.)

Dopamine

Many dopaminergic neurons have their cell bodies in the midbrain (Fig 15–6). They project from the substantia nigra to the striatum (**nigrostriatal system**) and from other portions of the midbrain to the olfactory tubercle, nucleus accumbens, related limbic areas, and the frontal, cingulate, entorhinal, and perirhinal cortex (**mesocortical system**). A separate intra-hypothalamic system of dopaminergic neurons (**tuberoinfundibular system**) projects from cell bodies in the arcuate nucleus to the external layer of the median eminence of the hypothalamus (Fig 15–7). Some dopaminergic neurons also project from the arcuate nucleus to the intermediate and posterior lobe of the pituitary (**tuberohypophyseal system**). There is in addition an **incertohypothalamic system,** with cell bodies in the zona incerta and endings in the septum and dorsal hypothalamus, and a **periventricular system** made up of short dopaminergic neurons around the third and fourth ventricle. There are dopaminergic neurons in the **olfactory bulb** and the **retina.** Recent studies with positron emission tomography (PET) scanning (see Chapter 32) in normal humans show that there is a steady loss of dopamine receptors in the basal ganglia with age. The loss is greater in men than in women.

There is increasing evidence that dopamine is involved in the pathogenesis of schizophrenia. PET scanning of schizophrenics has shown that their brains contain elevated levels of D_2 receptors even if they have never received drugs. Amphetamine, which stimulates the secretion of norepinephrine and dopamine, produces a psychosis that resembles schizophrenia when administered in large doses. On the other hand, the phenothiazine tranquilizers are effective in the relief of the symptoms of schizophrenia, and their antipsychotic activity parallels their ability to block D_2 dopamine receptors.

The nigrostriatal system is related to motor function. Parkinson's disease, the motor disorder caused by degeneration of this system, is discussed in Chapter 12.

Dopamine is the prolactin-inhibiting hormone secreted in the hypothalamus. Its role in the regulation of prolactin secretion is discussed in Chapter 23.

Acetylcholine

Acetylcholine is distributed throughout the central nervous system, with high concentrations in the cerebral cortex, thalamus, and various nuclei in the basal forebrain. The distribution of choline acetyltransferase and acetylcholinesterase parallels that of acetylcholine. Most of the acetylcholinesterase is in neurons, but some is found in glia. Pseudocholinesterase is found in many parts of the central nervous system.

The development of antibodies specific for choline acetyltransferase has permitted mapping the cholinergic pathways in the brain by immunocytochemical techniques. In addition to motor neurons and preganglionic autonomic neurons, which were previously known to be cholinergic, there are a modest number of cholinergic interneurons in the cerebral cortex. However, most of the cholinergic neurons are subcortical. There are cholinergic neurons that project from nuclei in the basal forebrain to the cerebral cortex, hippocampus, and related areas; and there are cholinergic neurons in the brain stem that project to the thalamus. In monkeys, cholinergic neurons in the medial septal nucleus and the nucleus of the vertical limb of the diagonal band on each side project to the hippocampus and are involved in memory. Cholinergic neurons in the nucleus of the horizontal limb of the diagonal band project to the olfactory bulb. There is also a large projection from the nucleus basalis of Meynert and adjacent nuclei to the amygdala and the entire neocortex, and these projections are probably involved in motivation, perception, and cognition. There is extensive cell loss in this projection in Alzheimer's disease (see Chapter 16). Cholinergic neurons in the pontopeduncular and lateral tegmental nuclei project through the pontomesencephalic reticular formation to the thalamus and are presumably involved in the attention and arousal functions of the ascending reticular system (see Chapter 11).

Acetylcholine has been linked directly or indirectly to a variety of brain functions. Injection of acetylcholine derivatives into the hypothalamus and parts of the limbic system causes drinking. In blinded rats, acetylcholinesterase activity is decreased in the superior colliculi and elevated in the occipital cortex. Cortical levels of acetylcholinesterase are greater in rats raised in a complex environment than in rats raised in isolation, but the significance of this type of correlation is uncertain. Acetylcholine is an excitatory transmitter in the basal ganglia, whereas dopamine is an inhibitory transmitter in these structures. In parkinsonism, the loss of dopamine alters the cholinergic-dopaminergic balance, and anticholinergic drugs are of benefit along with L-dopa in the treatment of the disease.

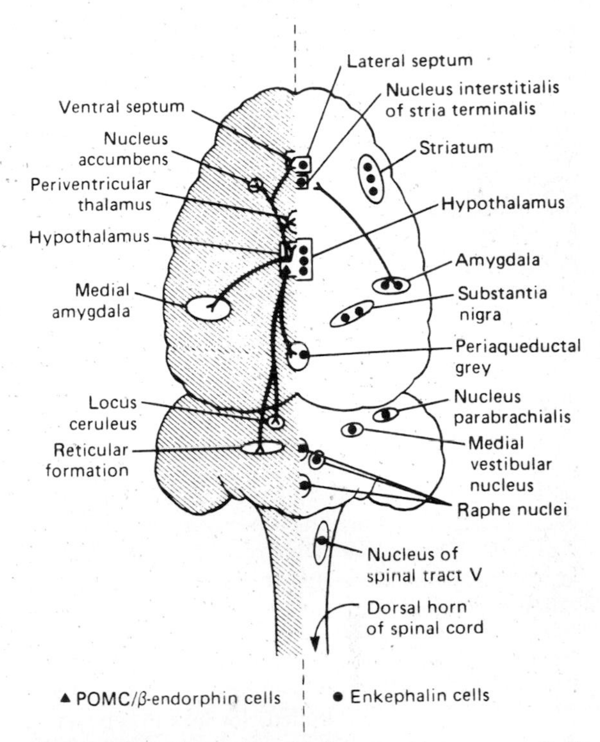

Figure 15–8. Distribution of β-endorphin neurons (left) and enkephalin neurons (right) in the brain. (Reproduced, with permission, from Barchas JD et al: Behavioral neurochemistry: Neuroregulatory and behavioral states. *Science* 1978;**200**:964. Copyright 1978 by the American Association for the Advancement of Science.)

Opioid Peptides

There are 3 types of opioid-peptide-secreting neurons in the brain, each producing one of the 3 opioid peptide precurser molecules (see Chapter 4). Proenkephalin-producing neurons are scattered throughout the brain, whereas pro-opiomelanocortin-producing neurons have their cell bodies in the arcuate nuclei and project to the thalamus and parts of the brain stem (Fig 15–8). Prodynorphin-producing neurons are located primarily in the hypothalamus, limbic system, and brain stem. The peptides they secrete are involved in various functions including, presumably, the phenomenon of tolerance and addiction produced by morphine, but the details are uncertain.

Other Transmitters

The relation of GABA to anxiety and the anxiolytic actions of benzodiazepines are discussed in Chapter 4. Peptides other than enkephalins and β-endorphin are probably involved in emotional and other behavioral responses, but the details remain to be determined.

16 "Higher Functions of the Nervous System": Conditioned Reflexes, Learning, & Related Phenomena

INTRODUCTION

In previous chapters, somatic and visceral input to the brain and output from it have been described. The organization of the nervous system is summarized in Fig 16–1. Topics that have been discussed include specific inputs and outputs; the functions of the reticular core in maintaining an alert, awake state; the functions of the limbic-midbrain circuit in the maintenance of homeostatic equilibriums; and the regulation of instinctual and emotional behavior. There remain the phenomena called, for lack of a better or more precise term, the "higher functions of the nervous system"; learning, memory, judgment, language, and the other functions of the mind. These phenomena are the subject of this chapter.

METHODS

Many of the phenomena of the mind are difficult to study because they are characteristic of humans and are much less developed, if at all, in other species. In addition, it is worth remembering that the brain of the rhesus monkey is only one-fourth the size of the brain of the chimpanzee, our nearest primate relative, and the chimpanzee brain is in turn one-fourth the size of

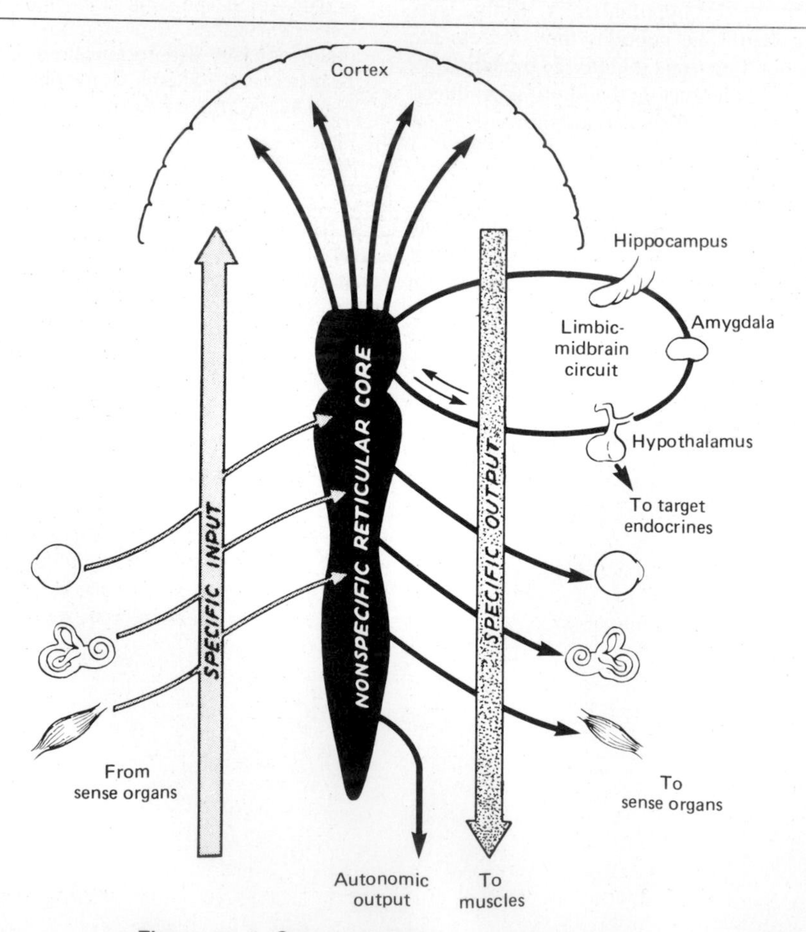

Figure 16–1. Organization of the nervous system.

the human brain. However, there are similarities between them, and much has been learned from the study of laboratory primates, particularly monkeys. In addition, valuable data can be and have been obtained in humans. One method is to correlate clinical observations with the site and extent of brain pathology determined by imaging and other diagnostic techniques or at autopsy. Another is stimulation of the exposed cerebral cortex in conscious humans undergoing neurosurgical procedures under local anesthesia, or in a few instances, stimulation of implanted electrodes. A third method is brain imaging by PET and other scanning procedures in normal humans (see Chapters 8 and 32). These methods have been refined by techniques that allow subtraction of nonspecific background changes when a subject is performing a particular task, greatly enhancing resolution and localization.

It should be emphasized that some forms of learning and memory are not unique to primates or even vertebrates and probably occur in all parts of the nervous system and in all animals with nervous systems. Studies of the molecular and cellular events underlying habituation, sensitization, and classical conditioning (see Chapter 4 and below) in species such as the sea snail *Aplysia* have been very fruitful, and indeed it is now possible to begin to talk about a "molecular biology of learning and memory."

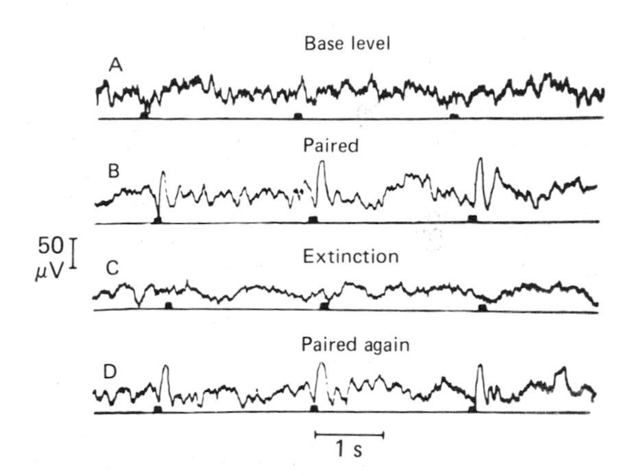

Figure 16–2. Records of the electrical activity of the hippocampus in a monkey exposed repeatedly to a tone stimulus (black signal marks). *A:* Control. *B:* After the tone was paired with a reward of food. *C:* After extinction of this response. *D:* After again pairing the tone with food. (Reproduced, with permission, from Hearst E et al: Some electrophysiological correlates of conditioning in the monkey. *Electroencephalogr Clin Neurophysiol* 1960;**12**:137.)

LEARNING AND MEMORY

Learning may be defined as the ability to alter behavior on the basis of past experience, and memory is the ability to recall past experience at the conscious or unconscious level. The 2 are obviously closely related and need to be considered together.

Forms of Learning

Habituation is a simple form of learning in which a neutral stimulus is repeated many times. The first time it is applied, it is novel and evokes a reaction (the orienting reflex or "what is it?" response). However, it evokes less and less electrical response as it is repeated. Eventually, the subject becomes habituated to the stimulus and ignores it. **Sensitization** is in a sense the opposite reaction. A repeated stimulus produces a greater response if it is coupled one or more times with an unpleasant or a pleasant stimulus. An example of sensitization in a monkey is shown in Fig 16-2. It is common knowledge that similar intensification of the **arousal value** of stimuli occurs in humans. The mother who sleeps through many kinds of noise but wakes promptly when her baby cries is one example. The hospital resident who is unaware of the calls on the loudspeaker unless his or her own name is called is another.

Habituation and sensitization are examples of nonassociative learning. In nonassociative learning, the organism learns about a single stimulus. In associative learning, the organism learns about the relation of one stimulus to another. A classic example of associative learning is a conditioned reflex.

Conditioned Reflexes

A conditioned reflex is a reflex response to a stimulus that previously elicited little or no response, acquired by repeatedly pairing the stimulus with another stimulus that normally does produce the response. In Pavlov's classic experiments, the salivation normally induced by placing meat in the mouth of a dog was studied. A bell was rung just before the meat was placed in the dog's mouth, and this was repeated a number of times until the animal would salivate when the bell was rung even though no meat was placed in its mouth. In this experiment, the meat placed in the mouth was the **unconditioned stimulus** (US), the stimulus that normally produces a particular innate response. The **conditioned stimulus** (CS) was the bell-ringing. After the CS and US had been paired a sufficient number of times, the CS produced the response originally evoked only by the US. This is so-called classical conditioning. An immense number of somatic, visceral, and other neural changes can be made to occur as conditioned reflex responses. Conditioning of visceral responses is often called **biofeedback.** The changes that can be produced include alterations in heart rate and blood pressure, and conditioned decreases in blood pressure have been advocated for the treatment of hypertension. However, the depressor responses that are produced in this fashion are small.

If the CS is presented repeatedly without the US, the conditioned reflex eventually dies out. This process is called **extinction** or **internal inhibition.** If the animal is disturbed by an external stimulus immediately after the CS is applied, the conditioned response may not occur (**external inhibition**). However, if the conditioned reflex is **reinforced** from time to time by again pairing the CS and US, the conditioned reflex persists indefinitely.

When a conditioned reflex is first established, it can be evoked not only by the CS but also by similar stimuli. However, if only the CS is reinforced and the similar stimuli are not, the animal can be taught to discriminate between different signals with great accuracy. The elimination of the response to other stimuli is an example of internal inhibition. By means of such **discriminative conditioning,** dogs can be taught, for example, to distinguish between a tone of 800 Hz and one of 812 Hz. Most of the data on pitch discrimination, color vision, and other sensory discriminations in animals have been obtained in this way.

For conditioning to occur, the CS must precede the US. If the CS follows the US, no conditioned response develops. The conditioned response follows the CS by the time interval that separated the CS and US during training. The delay between stimulus and response may be as long as 90 seconds.

As noted in Chapter 15, conditioned reflexes are difficult to form unless the US is associated with a pleasant or unpleasant affect. Stimulation of the brain reward system is a powerful US (pleasant or **positive reinforcement**), and so is stimulation of the avoidance system or a painful shock to the skin (unpleasant or **negative reinforcement**).

Operant conditioning is a form of conditioning in which the animal is taught to perform some task (''operate on the environment'') in order to obtain a reward or avoid punishment. The US is the pleasant or unpleasant event, and the CS is a light or some other signal that alerts the animal to perform the task. Conditioned motor responses that permit an animal to avoid an unpleasant event are called **conditioned avoidance reflexes.** For example, an animal is taught that by pressing a bar it can prevent an electric shock to the feet. Another example is **food aversion conditioning.** An animal exposed to the taste of a food develops a strong aversion to the food if the tasting is coupled with injection of a drug that produces nausea or illness. Similar aversion responses occur in humans. These conditioned responses are very strong, can sometimes be learned with a single pairing of the CS and the US, and, unlike other conditioned responses, will develop when the CS and US are separated by an hour or more. The survival value of food aversion conditioning is obvious in terms of avoiding poisons, and it is not surprising that the brain is probably genetically ''programmed'' to facilitate the development of food aversion responses.

Intercortical Transfer of Learning

If a cat or monkey is conditioned to respond to a visual stimulus, with one eye covered and then tested with the blindfold transferred to the other eye, it performs the conditioned response. This is true even if the optic chiasm has been cut, making the visual input from each eye go only to the ipsilateral cortex. If, in addition to the optic chiasm, the anterior and posterior commissures and the corpus callosum are sectioned (''split brain animal''), no transfer of learning occurs. Similar results have been obtained in humans in whom the corpus callosum is congenitally absent or in whom it has been sectioned surgically in an effort to control epileptic seizures. This demonstrates that the neural coding necessary for ''remembering with one eye what has been learned with the other'' has been transferred somehow to the opposite cortex via the commissures. There is evidence for similar transfer of information acquired through other sensory pathways. ''Split brain animals'' can even be trained to respond to different and conflicting stimuli, one with one eye and another with the other—literally an example of not letting the right side know what the left side is doing. Attempts at such training in normal animals and humans lead to confusion, but they do not faze the animals with split brains.

Memory

It is obvious that there are various types of memory, and that they range in complexity from the primitive types that underlie habituation and sensitization to the complex memories of humans. It is useful in this regard to divide memory into **reflexive** or **habit memory** and **declarative** or **recognition memory.** The former is the largely and often totally unconscious memory that underlies nonassociative learning and such forms of associative learning as classical conditioned reflexes. The latter is conscious recall, which involves evaluation and comparison and is often established by a single experience or association. These 2 types probably depend on processing in different parts of the brain (see below) and obviously vary in complexity. However, they are related in that declarative memory can be converted into reflexive memory by constant repetition. Thus, for example, athletes develop moves and responses that with training become ''instinctive,'' and many aspects of behavior as complex as driving a car or playing a piano become habit responses.

In discussing declarative memory in humans, it is important to distinguish between recent and remote memories. Three mechanisms are actually involved: (1) one mediates immediate recall of the events of the moment; (2) one mediates memories of recent events that occurred seconds to hours or days before; and (3) one mediates memories of the remote past. The second mechanism is responsible for ''consolidation of the memory trace,'' a process that eventually encodes the memory in a remarkably resistant form. Recent memory is impaired by various neurologic diseases and injuries, but remote memories persist in the presence of severe brain damage.

Once long-term memories have been established, they can be recalled or accessed by a large number of different associations. For example, the memory of a vivid scene can be evoked not only by a similar scene but also by a sound or smell associated with the scene and by words such as scene, vivid, and view. Thus, there must be multiple routes or keys to each stored memory. In addition, many memories have an emotional component or "color"; ie, in simplest terms, memories can be pleasant or unpleasant.

Biologic Bases of Learning & Memory

It has been argued that learning and memory involve formation of new synaptic contacts in the nervous system. This is difficult to disprove, but it now seems likely that most if not all instances involve instead biochemical changes in existing pathways, which lead to facilitated or, in the case of habituation, inhibited postsynaptic responses. This does not mean that there are no morphologic changes associated with learning. For example, in *Aplysia,* 40% of the relevant sensory terminals normally contain active zones, whereas in habituated animals, 10% have active zones, and in sensitized animals, 65% have active zones. Changes occur in mammals. For example, rats exposed to visually complex environments and trained to perform various tasks have thicker, heavier cerebral cortices than those of control rats exposed to monotonously uniform environments. Mice raised in darkness and then exposed to light develop additional spines on the dendrites of their pyramidal cells. However, these changes are manifestations of the initial molecular changes.

The biochemical events involved in habituation and sensitization in *Aplysia* and other invertebrates have been worked out in considerable detail. As described in Chapter 4, habituation is due to a decrease in Ca^{2+} in the sensory endings that mediate the response to a particular stimulus, and sensitization is due to prolongation of the action potential in these endings with a resultant increase in intracellular Ca^{2+} that facilitates release of neurotransmitter by exocytosis. Posttetanic potentiation, a learninglike facilitation of transmission, is also due to accumulation of intracellular Ca^{2+} (see Chapter 4).

Classical conditioning also occurs in *Aplysia* and in mammals in the isolated spinal cord. In *Aplysia,* the US leaves free Ca^{2+} in the cell; this binds to calmodulin, leading to a long-term change in adenylate cyclase, so that when this enzyme activated by the CS, more cyclic AMP is produced. This in turn closes K^+ channels and prolongs action potentials by the mechanism described in Chapter 4.

The events underlying learning and memory in primates, including humans, are undoubtedly chemical as well, but they also involve complex circuits in the brain, and considerable progress has been made in understanding these circuits.

It is clear that declarative memory involves the cerebral cortex. For example, Penfield has reported that stimulation of portions of the temporal lobes in humans evokes detailed memories of events that occurred in the remote past, often beyond the power of voluntary recall. Stimulation of other parts of the temporal lobes sometimes causes a change in interpretation of one's surroundings. For example, when the stimulus is applied, the subject may feel strange in a familiar place or may feel that what is happening now has happened before. The occurrence of a sense of familiarity or a sense of strangeness in appropriate situations probably helps the normal individual adjust to the environment. In strange surroundings, one is alert and on guard, whereas in familiar surroundings, vigilance is relaxed. An inappropriate feeling of familiarity with new events or in new surroundings is known clinically as the ***déjà vu* phenomenon,** from the French words meaning "already seen." The phenomenon occurs from time to time in normal individuals, but it also may occur as an aura (a sensation immediately preceding a seizure) in patients with temporal lobe epilepsy.

Other data pertinent to the physiology of memory are the clinical and experimental observations showing that there is frequently a loss of memory for the events immediately preceding brain concussion or electroshock therapy (**retrograde amnesia**). In humans, this amnesia encompasses longer periods than it does in experimental animals—sometimes days, weeks, and even years—but remote memory is not affected. In animals, acquisition of learned responses—or at least their retrieval—is prevented if, within 5 minutes after each training session, the animals are anesthetized, given electroshock treatment, subjected to hypothermia, or treated with antibiotics that inhibit protein synthesis. Such treatment 4 hours after the training sessions has no effect on acquisition. Thus, as in humans, there is a period of "encoding" or "consolidation" of memory during which the memory trace is vulnerable. Following this period, there is a stable and remarkably resistant memory engram, which presumably is due to a long-term change in the synapses in a given neural circuit.

Site of the Encoding Process

There is considerable evidence that the encoding process involves the hippocampus and its connections (Fig 16–3). In humans, bilateral destruction of the ventral hippocampus or disease processes that destroy the CA1 neurons in it cause striking defects in recent memory. So do bilateral lesions of the same area in monkeys. Humans with such destruction have intact remote memory and immediate recall. In addition, some forms of learning are unaffected. These humans perform adequately in terms of conscious memory as long as they concentrate on what they are doing. However, if they are distracted for even a very short period, all memory of what they were doing and proposed to do is lost. They are thus capable of new learning and retain old prelesion memories, but they cannot form new long-term memories.

The connections of the hippocampus to the diencephalon are also involved in memory. Some alcoholics with brain damage develop impairment of recent memory, and the memory loss correlates well with the presence of pathologic changes in the mamil-

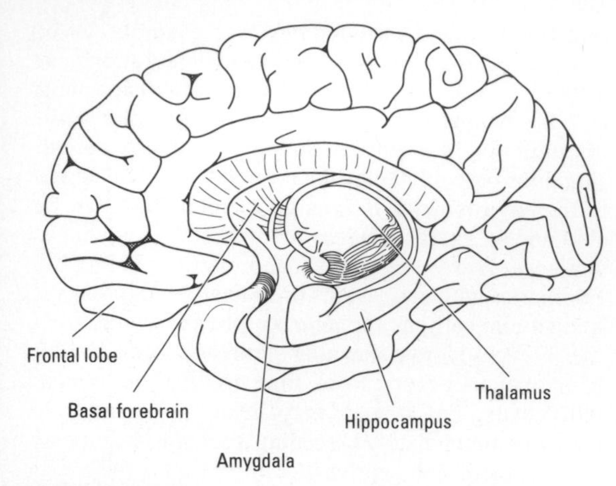

Figure 16–3. Brain areas concerned with encoding long-term memories.

lary bodies, which have extensive efferent connections to the hippocampus via the formix. The mamillary bodies project to the anterior thalamus via the mamillothalamic tract, and in monkeys, lesions of the thalamus cause loss of recent memory. From the thalamus, the fibers concerned with memory project to the prefrontal cortex and from there to the basal forebrain. From the basal forebrain, there is a diffuse cholinergic projection to all the neocortex, the amygdala, and the hippocampus from the **nucleus basalis of Meynert.** There is a severe loss of these cholinergic fibers in Alzheimer's disease (see below).

The amygdala is also involved in declarative memory. Lesions of the amygdala in monkeys slow but do not abolish learning. However, when lesions of the amygdala are combined with lesions of the hippocampus, the amnesia is more severe than with hippocampal lesions alone. The amygdala may mediate the association of memories formed through different senses. In addition, it has extensive connections with the hypothalamus, and since the hypothalamus is concerned with emotion, it may be that amygdalohypothalamic pathways give memories their emotional flavor.

Encoding Reflexive Memory

There is considerable evidence that although pathways for declarative and reflexive memories are connected, the reflexive memory pathways are separate. A number of investigators have argued that they involve the striatum, and some reflexive tasks are disrupted by lesions of the basal ganglia. The amygdala may mediate conditioned cardiovascular responses. However, there is other evidence that the cerebellum is involved. For example, the vestibulo-occular reflex (VOR), which maintains visual fixation while the head is moving (see Chapter 9), can be adjusted to new eye positions, and this plasticity is abolished by lesions of the flocculus. In addition, conditioning of an eye blink reflex by using a puff of air on the corneas as the US and a tone as the CS is prevented by lesions of the interpositus nucleus. In this case, it appears that impulses set up by the US act via the inferior olive and climbing fibers to the cerebellar cortex to alter the Purkinje cell response to the tone arriving via the pontine nuclei and mossy fibers. Climbing-fiber-mediated modification of mossy-fiber-driven Purkinje cell discharge is also responsible for plastic changes in the VOR and learned muscle movements.

Alzheimer's Disease & Senile Dementia

Alzheimer's disease is characterized by progressive loss of memory and cognitive function in middle age. Similar deterioration in elderly individuals is called senile dementia of the Alzheimer type and accounts for 50–60% of cases of senile dementia. Since 10–15% of the population over age 65 have some degree of dementia, the condition is a common and serious problem. In addition to the loss of cholinergic nerve terminals in the cerebral cortex and hippocampus in Alzheimer's disease and senile dementia of the Alzheimer type, there is a severe loss of cholinergic neurons in the nucleus basalis of Meynert and related nuclei that contain the cell bodies of cholinergic neurons which project to the hippocampus, the amygdala, and all of the neocortex (see Chapter 15). The cause of this apparently selective degeneration is unknown. Modest temporary improvement can sometimes be obtained in Alzheimer's disease with physostigmine, a drug that inhibits acetylcholinesterase and hence decreases the breakdown of acetylcholine, but this treatment has no effect on the underlying degenerative process.

It is interesting that selective degeneration with aging can occur in 3 different types of cells in the central nervous system, causing 3 different progressive, crippling, and eventually fatal diseases. As noted above, degeneration of the cholinergic neurons in the basal forebrain is associated with Alzheimer's disease; degeneration of the dopaminergic neurons in the substantia nigra is associated with Parkinson's disease (see Chapter 12); and degeneration of the cholinergic motor neurons in the brain stem and spinal cord is associated with one form of **amyotrophic lateral sclerosis.** This last disease is often called Lou Gehrig's disease because Gehrig, a famous American baseball player, died of it.

Drugs That Facilitate Memory

A variety of central nervous system stimulants have been shown to improve learning in animals when administered immediately before or after the learning sessions. These include caffeine, physostigmine, amphetamine, nicotine, and the convulsants picrotoxin, strychnine, and pentylenetetrazol (Metrazol). They seem to act by facilitating consolidation of the memory trace. In senile humans, small doses of pentylenetetrazol appear to improve memory and general awareness.

FUNCTIONS OF THE NEOCORTEX

Memory and learning are functions of large parts of the brain, but the centers controlling some of the other "higher functions of the nervous system," particularly the mechanisms related to language, are more or less localized to the neocortex. It is interesting that speech and other intellectual functions are especially well developed in humans—the animal species in which the neocortical mantle is most highly evolved.

Anatomic Considerations

There are 3 living species with brains larger than a human's (the porpoise, the elephant, and the whale), but in humans, the ratio between brain weight and body weight far exceeds that of any of their animal relatives. From the comparative point of view, the most prominent gross feature of the human brain is the immense growth of the 3 major **association areas:** the **frontal,** in front of the premotor area; the **parieto-occipital-temporal,** between the somasthetic and visual cortices, extending into the posterior portion of the temporal lobe; and the **temporal,** extending from the lower portion of the temporal lobe to the limbic system (Fig 16–4). The association areas are part of the 6-layered neocortical mantle of gray matter that spreads over the lateral surfaces of the cerebral hemispheres from the concentric allocortical and juxtallocortical rings around the hilus (see Chapter 15).

The neuronal connections within the neocortex form a complicated network (Fig 11–2). The descending axons of the larger cells in the pyramidal cell layer give off collaterals that feed back via association neurons to the dendrites of the cells from which they originate, laying the foundation for complex feedback control. The recurrent collaterals also connect to neighboring cells. The large, complex dendrites of the deep cells receive ascending fibers, nonspecific thalamic and reticular afferents, and association fibers ending in all layers. Specific thalamic afferents end in layer IV of the cortex.

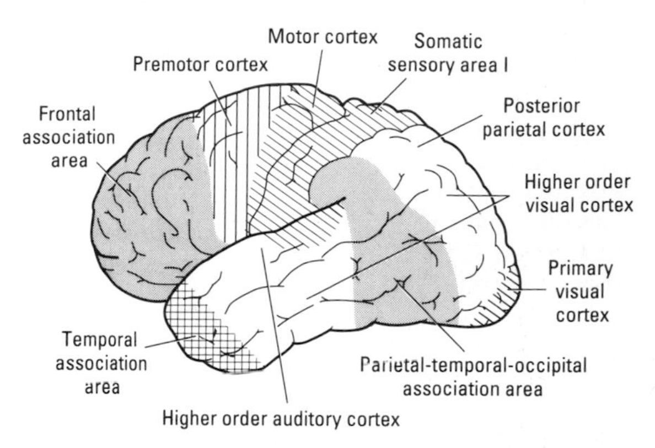

Figure 16–4. Lateral view of the human cerebral cortex, showing the primary sensory and motor areas and the association areas.

Complementary Specialization of the Hemispheres Versus "Cerebral Dominance"

One group of functions more or less localized to the neocortex in humans consists of those related to language, ie, to understanding the spoken and printed word and to expressing ideas in speech and writing. It is a well-established fact that human language functions depend more on one cerebral hemisphere than on the other. This hemisphere is concerned with categorization and symbolization and has often been called the **dominant hemisphere.** However, it is clear that the other hemisphere is not simply less developed or "nondominant"; instead, it is specialized in the area of spatiotemporal relations. It is this hemisphere that is concerned, for example, with the recognition of faces, the identification of objects by their form, and the recognition of musical themes. Consequently, the concept of "cerebral dominance" and a dominant and nondominant hemisphere has been replaced by a concept of complementary specialization of the hemispheres, one for language functions and sequential-analytic processes (the **categorical hemisphere)** and one for visuospatial relations (the **representational hemisphere**).

Lesions in the categorical hemisphere produce language disorders, whereas extensive lesions of the representational hemisphere do not. Instead, lesions in the representational hemisphere produce **astereognosis**—inability to identify objects by feeling them—and other agnosias. **Agnosia** is the general term used for the inability to recognize objects by a particular sensory modality even though the sensory modality itself is intact. Lesions producing these defects are generally in the parietal lobe. Especially when they are in the representational hemisphere, lesions of the inferior parietal lobule, a region in the posterior part of the parietal lobe that is close to the occipital lobe, cause **unilateral inattention and neglect.** Individuals with such lesions do not have any apparent primary visual, auditory, or somesthetic defects, but they ignore stimuli from the contralateral portion of their bodies or the space around these portions. This leads to failure to care for half their bodies and, in extreme cases, to situations in which individuals shave half their faces, dress half their bodies, or read half of each page.

Hemispheric specialization extends to other parts of the cortex as well. Patients with lesions in the categorical hemisphere are disturbed about their disability and often depressed, whereas patients with lesions in the representational hemisphere are sometimes unconcerned and even euphoric. In addition, patients with representational hemisphere lesions have trouble recognizing emotions in other individuals. Patients with lesions in the temporal lobe of the categorical hemisphere have loss of recent verbal memory, whereas patients with lesions of the temporal lobe of the representational hemisphere have loss of recent memory for visual and spatial material.

Hemispheric specialization is related to handedness. Handedness appears to be genetically determined. In about 4% of right-handed individuals, who constitute 91% of the human population, the left hemisphere is the dominant or categorical hemisphere; in approximately 15% of left-handed individuals, the right hemisphere is the categorical hemisphere and in 15%, there is no clear lateralization. However, in the remaining 70% of left-handers, the left hemisphere is the categorical hemisphere. In adults, the characteristic defects produced by lesions in either the categorical or the representational hemisphere are long-lasting. However, in young children subjected to hemispherectomy for brain tumors or other diseases, the functions of the missing hemisphere may be largely taken over by the remaining hemisphere no matter which hemisphere is removed. It is interesting that learning disabilities such as **dyslexia,** an impaired ability to learn to read, are 12 times as common in left-handers as they are in right-handers, possibly because some fundamental abnormality in the left hemisphere led to a switch in handedness early in development. However, the spatial talents of left-handers may be well above average; a disproportionately large number of artists, musicians, and mathematicians are left-handed.

There are anatomic differences between the 2 hemispheres that may correlate with the functional differences. For example, it can be shown not only anatomically but also by computerized tomography (CT scanning) that the right frontal lobe is normally thicker than the left and that the left occipital lobe is wider and protrudes across the midline. Portions of the upper surface of the left temporal lobe are regularly larger in right-handed individuals. This is particularly true of the region known as the **planum temporale** (Fig 16–5). In addition, there are chemical differences between the 2 sides of the brain. For example, there is a higher concentration of dopamine in the nigrostriatal pathway on the left side in right-handed humans and a higher concentration on the right in left-handers. The physiologic significance of these differences is not known.

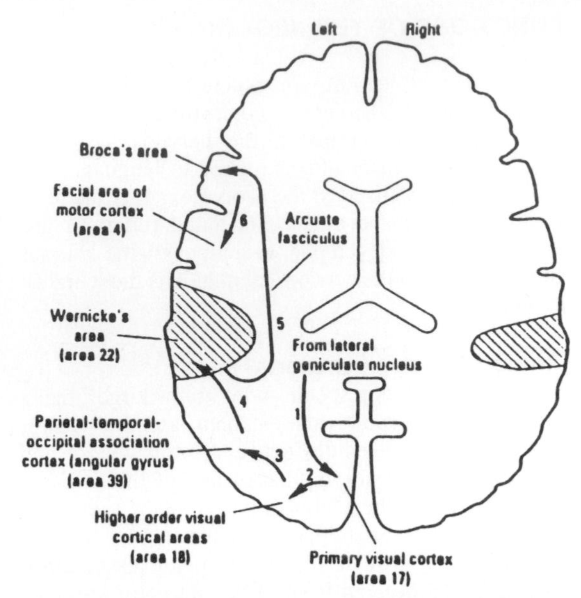

Figure 16–5. Path taken by impulses when a subject names a visual object. The location of the planum temporale, which is larger on the left side, is indicated by the shaded areas on this horizontal section of the human brain. (Modified from Patton HD et al: *Introduction to Basic Neurology.* Saunders, 1976.)

Prosopagnosia

Bilateral lesions of the undersurface of the occipital and temporal lobes (Fig 16–6) cause a unique agnosia: inability to recognize people by their faces. This condition is called **prosopagnosia.** Patients with this abnormality can recognize forms and reproduce them. They can recognize people by their voices, and they show autonomic responses when they see familiar as opposed to unfamiliar faces. However, they cannot identify the familiar people they see.

Physiology & Pathology of Language Disorders

Although knowledge about the physiology of language functions is still incomplete, there is evidence that a region at the posterior end of the superior temporal gyrus called **Wernicke's area** (Fig 16–7) in the categorical hemisphere is concerned with comprehension of auditory and visual information. It projects via the **arcuate fasciculus** to **Broca's area** (area 44) in the frontal lobe immediately in front of the inferior end of the motor cortex. Broca's area in the categorical hemisphere processes the information received from Wernicke's area into a detailed and coordinated pattern for vocalization and then projects the pattern to the motor cortex, which initiates the

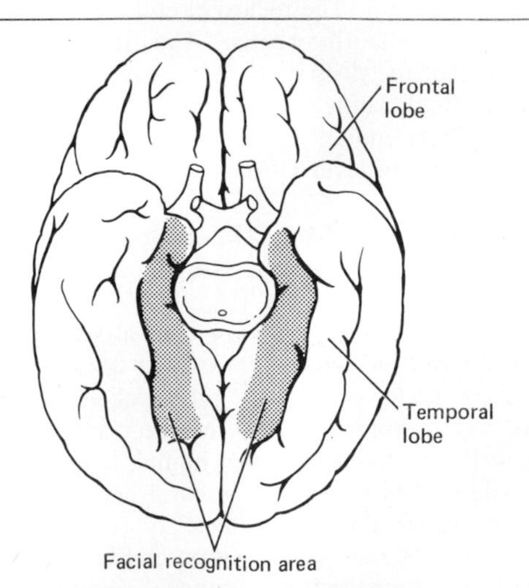

Figure 16–6. Basal view of the human brain, showing the facial recognition areas where bilateral lesions cause prosopagnosia. (Modified and reproduced, with permission, from Geschwind N: Specializations of the human brain. *Sci Am* [Sept] 1979;**241**:180. Copyright © 1979 by Scientific American. All rights reserved.)

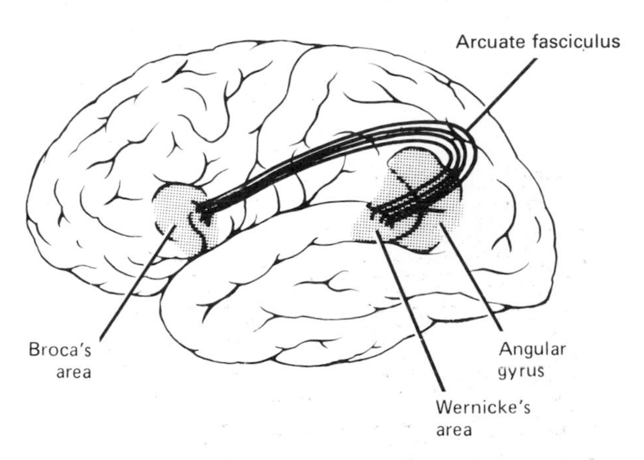

Figure 16–7. Location of areas that in the categorical hemisphere are concerned with language functions.

appropriate movements of the lips, tongue, and larynx to produce speech. The probable sequence of events that occurs when a subject names a visual object is shown in Fig 16–5. The angular gyrus behind Wernicke's area appears to process information from words that are read in such a way that they can be converted into the auditory forms of the words in Wernicke's area.

Abnormalities of language functions that are not due to defects of vision or hearing or to motor paralysis are called **aphasias.** They are due to lesions in the categorical hemisphere; comparable lesions in the representational hemisphere have little if any effect. The most common cause is embolism or thrombosis of a cerebral blood vessel. Many different classifications of the aphasias have been published, and their nomenclature is chaotic. However, they can be divided into **fluent** and **nonfluent aphasias.** In nonfluent aphasia, the lesion is in Broca's area (Table 16–1). Speech is slow, and words are hard to come by. Patients with severe damage to this area are limited to 2 or 3 words with which to express the whole range of meaning and emotion. Sometimes the words retained are those which were being spoken at the time of the injury or vascular accident that caused the aphasia.

Table 16–1. Aphasias. Characteristic responses of patients with lesions in various areas when shown a picture of a chair.*

Type of Aphasia and Site of Lesion	Characteristic Naming Errors
Nonfluent (Broca's area)	"Tssair"
Fluent (Wernicke's area)	"Stool" or "choss" (neologism)
Conduction (arcuate fasciculus)	"Flair . . . no, swair . . . tair."
Anomic (angular gyrus)	"I know what it is . . . I have a lot of them."

*Modified from Goodglass H: Disorders of naming following brain injury. *Am Sci* 1980;**68:**647.

In fluent aphasia, the lesion is in Wernicke's area or the arcuate fasciculus. In this condition, speech itself is normal and sometimes the patients talk excessively. However, what they say is full of jargon and neologisms that make little sense. When the lesion is in Wernicke's area, the patient also fails to comprehend the meaning of spoken or written words, so other aspects of the use of language are also compromised. When, more infrequently, the lesion is confined to the arcuate fasciculus, comprehension is normal but a form of fluent aphasia called conduction aphasia is present, because Wernicke's area is disconnected from Broca's area.

When there is a lesion damaging the angular gyrus in the categorical hemisphere without affecting Wernicke's or Broca's areas, there is no difficulty with speech or the understanding of auditory information, but there is trouble understanding written language or pictures, because visual information is not processed and transmitted to Wernicke's area. The result is a condition called **anomic aphasia.**

The isolated lesions that cause the selective defects described above occur in some patients, but brain destruction is often more general. Consequently, more than one form of aphasia is often present. Frequently, the aphasia is general (**global**), involving both receptive and expressive functions. Writing is abnormal in all aphasias in which speech is abnormal, but the neural circuits involved are not known. In addition, deaf subjects lose their ability to communicate in sign language if they develop a lesion in the categorical hemisphere.

The Frontal Lobes

Some insight into the other functions of the various parts of the cerebral cortex is gained by ablation studies. Bilateral removal of the neocortical portions of the frontal lobes in primates produces, after a period of apathy, hyperactivity and constant pacing back and forth. General intelligence is little affected, and results of tests involving immediate responses to environmental stimuli are normal. However, responses requiring the use of previously acquired information are abnormal. In humans, frontal lobectomy leads to deficiencies in the temporal ordering of events. For example, lobectomized humans have difficulty remembering how long ago they saw a particular stimulus card. Interestingly, left frontal lobectomy causes the biggest deficit in tests involving word stimuli, while right frontal lobectomy causes the biggest deficit in tests involving picture stimuli. Frontal lobectomy also abolishes the experimental neurosis.

Experimental Neurosis

As noted above, animals can be conditioned to respond to one stimulus and not to another even when the 2 stimuli are very much alike. However, when the stimuli are so nearly identical that they cannot be distinguished, the animal becomes upset, whines, fails to cooperate, and tries to escape. Pavlov called these symptoms the **experimental neurosis.** One may quarrel about whether this reaction is a true neurosis in the psychiatric sense, but the term is convenient.

Frontal lobectomized animals are still capable of discriminating between like stimuli up to a point; but when they can no longer discriminate, their failure does not upset them. As a result of these experiments in animals, **prefrontal lobotomy** and various other procedures aimed at cutting the connections between the frontal lobes and the deeper portions of the brain were used to treat various mental diseases in humans.

In some mental patients, tensions resulting from real or imagined failures of performance and the tensions caused by delusions, compulsions, and phobias are so great as to be incapacitating. Successful lobotomy reduces the tension. The delusions and other symptoms are still there, but they no longer bother the patient. A similar lack of concern over severe pain led to the use of lobotomy in treating patients with intractable pain (see Chapter 7). Unfortunately, this lack of concern often extends to other aspects of the environment, including relations with associates, social amenities, and even toilet habits. Furthermore, the relief from the suffering associated with pain usually lasts less than a year.

The effects that lobotomy can have on the personality were well described more than 120 years ago by the physician who cared for a man named Phineas P. Gage. Gage was a construction foreman who was packing blasting powder into a hole with a tamping iron. The powder exploded, driving the tamping iron through his face and out the top of his skull, transecting his frontal lobes. After the accident, he became, in the words of his physician, ". . . fitful, irreverent, indulging at times in the grossest profanity (which was not previously his custom), manifesting but little deference to his fellows, impatient of restraint or advice when it conflicts with his desires, at times pertinaciously obstinate yet capricious and vacillating, devising many plans for future operation which are no sooner arranged than they are abandoned in turn for others appearing more feasible. . . . His mind was radically changed, so that his friends and acquaintances said he was no longer Gage."*

*From Harlow JM: Recovery from the passage of an iron bar through the head. *Mass Med Soc Publications* 1868;**2:**329.

This description is classic, but it cannot be said to be typical. The effect of lobotomy in humans are highly variable from patient to patient, and complications are frequent. For these reasons and because the desirable effects of lobotomy can generally be achieved with tranquilizers and other drugs, lobotomies are now performed rarely, if ever, for the treatment of mental disease.

Temporal Lobes

The effects of bilateral temporal lobectomy were first described by Klüver and Bucy. Because of removal of limbic structures, temporal lobectomized monkeys ("Klüver-Bucy animals") are docile and hyperphagic and the males are hypersexual. The animals also demonstrate visual agnosia and a remarkable increase in oral activity. The monkeys repeatedly pick up all moveable objects in their environment. They manipulate each object in a compulsive way; mouth, lick, and bite it; and then, unless it is edible, discard it. However, discarded objects are picked up again in a few minutes as if the animal had never seen them before and subjected to the same manipulation and oral exploration. It has been suggested that the cause of this pattern of behavior may be an inability to identify objects. It could well be a manifestation of a memory loss due to hippocampal ablation. In addition, the animals are easily distracted. They heed every stimulus, whether it is novel or not, and usually approach, explore, manipulate, and, if possible, bite its source. This failure to ignore peripheral stimuli is called **hypermetamorphosis.**

Clinical Implications & Significance

Various parts of the syndrome described by Klüver and Bucy in monkeys are seen in humans with temporal lobe disease. Impaired recent memory follows bilateral damage to the hippocampus. Hypersexuality may develop in some individuals with bilateral damage in the amygdaloid nuclei and piriform cortex. It is obvious, however, that in the present state of our knowledge the abnormalities seen with temporal lobe lesions and, more generally, those produced by other neocortical lesions cannot be fitted into any general hypothesis of intellectual function. Future research may provide such a synthesis and a better understanding of the neural basis of mental phenomena.

REFERENCES
Section III: Functions of the Nervous System

Arnold AP, Gorski R: Gonadal steroid induction of structural sex differences in the central nervous system. *Annu Rev Neurosci* 1984;**7:**413.

Barrie JG, Huang S-C, Phelps ME: In vivo assessment of neurotransmitter biochemistry in humans. *Annu Rev Pharmacol Toxicol* 1988;**28:**213.

Bashbaum AI, Fields HL: Endogenous pain control systems: Brain stem spinal pathways and endorphin circuitry. *Annu Rev Neurosci* 1984;**7:**309.

Besson J-M, Chaouch A: Peripheral and spinal mechanisms of nociception. *Physiol Rev* 1987;**67:**67.

Bjorklund A, Stenevi U: Regeneration of monoaminergic

and cholinergic neurons in the mammalian central nervous system. *Physiol Rev* 1979;**59:**62.

Bligh J: The central neurology of mammalian thermoregulation. *Neuroscience* 1979;**4:**1213.

Bradbury M: *The Concept of a Blood-Brain Barrier*. Wiley, 1979.

Bronisch FW: *The Clinically Important Reflexes*. Grune & Stratton, 1952.

Brooks VB: *The Neural Basis of Motor Control*. Oxford Univ Press, 1986.

Byrne JH: Cellular analysis of associative learning. *Physiol Rev* 1987;**67:**329.

Coyle JT, Price DL, DeLong M: Alzheimer's disease: A disorder of cortical cholinergic innervation. *Science* 1983;**219:**1184.

Damasio AR, Geschwind N: The neural basis of language. *Annu Rev Neurosci* 1984;**7:**127.

deCaro G, Epstein AN, Massi M: *The Physiology of Thirst and Sodium Appetite*. Plenum, 1986.

Dowling JE: *The Retina*. Harvard Univ Press, 1987.

Edelman GM, Gall WE, Cowan WM (editors): *Molecular Bases of Neural Development*. Wiley, 1985.

Empson J: *Human Brainwaves: The Psychological Significance of the Electroencephalogram*. Macmillan, 1986.

Evarts EV: Brain mechanisms of movement. *Sci Am* (Sept) 1979;**241:**164.

Finger TE, Silver WL (editors): *Neurobiology of Taste and Smell*, Wiley, 1987.

Frégnac Y, Imbert M: Development of neuronal selectivity in primary visual cortex of cat. *Physiol Rev* 1984;**64:**325.

Ganong WF: The brain as an endocrine organ. *Acta Physiol Lat Am* 1983;**32:**31.

Geschwind N: The apraxias: Neural mechanisms of disorders of learned movement. *Am Sci* 1975;**63:**188.

Getchell TV: Functional properties of vertebrate olfactory receptor neurons. *Physiol Rev* 1986;**66:**772.

Goelet P et al: The long and short of long-term memory: A molecular frame. *Nature* 1986;**322:**419.

Goldman-Rakic PS: Topography of cognition: parallel distributed networks in primate association cortex. *Annu Rev Neurosci* 1988;**11:**137.

Goodman DS: Vitamin A and retinoids in health and disease. *N Engl J Med* 1984;**310:**1023.

Herzog DB, Copeland PM: Eating disorders. *N Engl J Med* 1985;**313:**295.

Hobson JA: *The Dreaming Brain*. Basic Books, 1988.

Hubel DH: Exploration of the primary visual cortex, 1955–78. *Nature* 1982;**299:**515.

Hudspeth AJ: The cellular basis of hearing: The biophysics of hair cells. *Science* 1985;**230:**745.

Hudspeth AJ: The hair cells of the inner ear. *Sci Am* (Jan) 1983;**248:**54.

Hughes J (editor): Opioid peptides. *Br Med Bull* 1983;**39:**1.

Hyvarinen J: The posterior parietal lobe of the primate brain. *Physiol Rev* 1982;**62:**1060.

Ingvar DH, Lassen NA (editors): *Brain Work*. Munksgaard, 1976.

Iverson LL, Iverson SD, Snyder SH (editors): *Handbook of Psychopharmacology*. Vols 1–14. Plenum, 1975–1978.

Jacobs BL: How hallucinogenic drugs work. *Am Sci* 1987;**75:**386.

Kandel ER, Schwartz JH (editors): *Principles of Neural Science*, 2nd ed. Elsevier, 1985.

Katzman R: Alzheimer's disease. *N Engl J Med* 1986; **314:**964.

Kopin IT: Catecholamine metabolism: Basic metabolic aspects and clinical significance. *Pharmacol Rev* 1985; **37:**333.

Krieger DT: Brain peptides: What, where, and why? *Science* 1983;**222:**975.

Kuffler SW, Nicholls JG, Martin AR: *From Neuron to Brain*, 2nd ed. Sinauer Associates, 1984.

Landsberg L, Young JB: Fasting, feeding and regulation of the sympathetic nervous system. *N Engl J Med* 1978; **298:**1295.

Le Douarin N: *The Neural Crest*. Cambridge Univ Press, 1983.

Le Magnen J: Body energy balance and food intake: A Neuroendocrine regulatory mechanism. *Physiol Rev* 1983;**63:**314.

Leporé F, Ptito M, Jasper HH (editors): *Two Hemispheres—One Brain*. A.R. Liss, 1986.

Livingstone MS: Art, illusion, and the visual system. *Sci Am* (Jan) 1988;**258:**78.

MacKay WA, Murphy JT: Cerebellar modulation of reflex gain. *Prog Neurobiol* 1979:**13:**362.

Martin JB: Molecular genetics: applications to the clinical neurosciences. *Science* 1987;**238:**765.

Masland RH: The functional architecture of the retina. *Sci Am* (Dec) 1986;**255:**102.

Mendelson WB: *Human Sleep: Research and Clinical Care*. Plenum, 1987.

Moore RY, Bloom F: Central catecholamine neuronal systems: Anatomy and physiology of dopaminergic systems. *Annu Rev Neurosci* 1978:**1:**129.

Moore-Ede MC, Czeisler CA, Richardson GS: Circadian time-keeping in health and disease. *N Engl J Med* 1983:**309:**469.

Nathans J: The genes for color vision. *Sci Am* (Feb) 1989; **260:**42.

Nichol SE, Gottesman II: Clues to the genetics and neurobiology of schizophrenia. *Am Sci* 1983;**71:**398.

Nicholls DG, Locke RM: Thermogenic mechanisms in brown fat. *Physiol Rev* 1984;**64:**1.

Olton DS, Gamzu E, Corking S (editors): Memory dysfunctions: An integration of animal and human research from preclinical and clinical perspectives. *Ann NY Acad Sci* 1985;**944:**1.

Parker DE: The vestibular apparatus. *Sci Am* (Nov) 1980;**243:**118.

Peroutka SJ: 5-Hydroxytryptamine receptor subtypes. *Annu Rev Neurosci* 1988;**11:**45.

Pavlov IP: *Conditioned Reflexes*. Oxford Univ Press, 1928.

Pickels JO: *An Introduction to the Physiology of Hearing*, 2nd ed. Academic Press, 1988.

Rakic P: *Local Circuit Neurons*. MIT Press, 1976.

Reis DJ, Doba N: The central nervous system and neurogenic hypertension. *Prog Cardiovasc Dis* 1974;**17:**51.

Roper SD: The cell biology of taste receptors. *Annu Rev Neurosci* 1989;**12:** [In press.]

Rodbard D: The role of regional body temperature in the pathogenesis of disease. *N Eng J Med* 1981;**305:**808.

Schiffman SS: Taste and smell in disease. *N Engl J Med* 1983;**308:**1275.

Sherrington CS: *The Integrative Action of the Nervous System*. Cambridge Univ Press, 1947.

Snyder SH: *Drugs and the Brain*. Freeman, 1986.

Sokoloff L: *Metabolic Probes of Central Nervous System Activity in Experimental Animals and Man*. Sinauer Associates, 1984.

Squire LR: *Memory and Brain*. Oxford Univ Press, 1987.

Stein JF: Role of the cerebellum in the visual guidance of movement, *Nature* 1986;**323:**217.

Stryer L: The molecules of visual excitation. *Sci Am* (July) 1987;**257:**42.

Swanson LW, Sawchenko PE: Hypothalamic integration: Organization of the paraventricular and supraoptic nuclei. *Annu Rev Neurosci* 1983:**6:**269.

Tallman JF et al: Receptors for the age of anxiety: Pharmacology of the benzodiazepines. *Science* 1980;**207:**274.

Terman GW et al: Intrinsic mechanisms of pain inhibition: Activation by stress. *Science* 1984;**226:**1270.

Thoenen H, Edgar D: Neurotrophic factors. *Science* 1985;**229:**238.

Thompson RF: *The Brain.* Freeman, 1985.

Thompson RF: The neurobiology of learning and memory. *Science* 1986;**233:**941.

Trayburn P, Nicholls DG (editors): *Brown Adipose Tissue,* Arnold, 1986.

Tsukahara N: Synaptic plasticity in the mammalian central nervous system. *Annu Rev Neurosci* 1981;**4:**351.

Weiner RW, Ganong WF: Role of brain monoamines and histamine in regulation of anterior pituitary secretion. *Physiol Rev* 1978;**58:**905.

Weisel TN, Gilbert CD: The Sharpey-Schafer lecture: Morphological basis of visual function. *Q J Exp Physiol* 1983;**68:**525.

Willis WD Jr: *The Pain System.* Karger, 1985.

Zerbe R, Stropes L, Robertson G: Vasopressin function in the syndrome of inappropriate antidiuresis. *Annu Rev Med* 1980;**31:**315.

Zwislocki JJ: Sound analysis in the ear: A history of discoveries. *Am Sci* 1981;**69:**184.

Section IV. Endocrinology, Metabolism, & Reproductive Function

Energy Balance, Metabolism, & Nutrition 17

INTRODUCTION

The endocrine system, like the nervous system, adjusts and correlates the activities of the various body systems, making them appropriate to the changing demands of the external and internal environment. Endocrine integration is brought about by **hormones,** chemical messengers produced by ductless glands that are transported in the circulation to target cells, where they regulate the metabolic processes. The term **metabolism,** meaning literally "change," is used to refer to all the chemical and energy transformations that occur in the body.

The animal organism oxidizes carbohydrates, proteins, and fats, producing principally CO_2, H_2O, and the energy necessary for life processes. CO_2, H_2O, and energy are also produced when food is burned outside the body. However, in the body, oxidation is not a one-step, semiexplosive reaction but a complex, slow, stepwise process called **catabolism,** which liberates energy in small, usable amounts. Energy can be stored in the body in the form of special energy-rich phosphate compounds and in the form of proteins, fats, and complex carbohydrates synthesized from simpler molecules. Formation of these substances by processes that take up rather than liberate energy is called **anabolism.** This chapter sets the stage for consideration of endocrine function by providing a brief summary of the production and utilization of energy and the metabolism of carbohydrates, proteins, and fats.

ENERGY METABOLISM

Metabolic Rate

The amount of energy liberated by the catabolism of food in the body is the same as the amount liberated when food is burned outside the body. The energy liberated by catabolic processes in the body appears as external work, heat, and energy storage:

$$\text{Energy output} = \text{External work} + \text{Energy storage} + \text{Heat}$$

The amount of energy liberated per unit of time is the **metabolic rate.** Isotonic muscle contractions perform work at a peak efficiency approximating 50%:

$$\text{Efficiency} = \frac{\text{Work done}}{\text{Total energy expended}}$$

Essentially all of the energy of isometric contractions appears as heat, because little or no external work (the force multiplied by the distance the force moves a mass) is done (see Chapter 3). Energy is stored by forming energy-rich compounds. The amount of energy storage varies, but in fasting individuals it is zero or negative. Therefore, in an individual who is not moving and who has not eaten recently, essentially all of the energy output appears as heat. When food is burned outside the body, all of the energy liberated also appears as heat.

Calories

The standard unit of heat energy is the **calorie** (cal), defined as the amount of heat energy necessary to raise the temperature of 1 g of water 1 degree, from 15 to 16 °C. This unit is also called the gram calorie, small calorie, or standard calorie. The unit commonly used in physiology and medicine is the **Calorie (kilo-calorie, kcal),** which equals 1000 cal.

Calorimetry

The energy released by combustion of foodstuffs outside the body can be measured directly (**direct calorimetry**) by oxidizing the compounds in an apparatus such as a **bomb calorimeter,** a metal vessel surrounded by water inside an insulated container. The food is ignited by an electric spark. The change in the temperature of the water is a measure of the calories produced. Similar measurements of the energy released by combustion of compounds in living animals and humans are much more complex, but calorimeters have been constructed that can physically accommodate human beings. The heat produced by their bodies is measured by the change in temperature of the water circulating through the calorimeter.

The caloric values of the common foodstuffs, as measured in a bomb calorimeter, are found to be 4.1 kcal/g of carbohydrate, 9.3 kcal/g of fat, and 5.3 kcal/g of protein. In the body, similar values are obtained for carbohydrate and fat, but the oxidation of protein is incomplete, the end products of protein catabolism being urea and related nitrogenous compounds in addition to CO_2 and H_2O (see below). Therefore, the caloric value of protein in the body is only 4.1 kcal/g.

Indirect Calorimetry

Energy production can also be calculated, at least in theory, by measuring the products of the energy-producing biologic oxidations—ie, the CO_2, H_2O, and the end products of protein catabolism produced—or by measuring the O_2 consumed. This is **indirect calorimetry.** It is difficult to measure all of the end products, but measurement of O_2 consumption is relatively easy. Since O_2 is not stored and since its consumption, except when an O_2 debt is being incurred or repaid, always keeps pace with immediate needs, the amount of O_2 consumed per unit of time is proportionate to the energy liberated.

One problem with using O_2 consumption as a measure of energy output is that the amount of energy released per mole of O_2 consumed varies slightly with the type of compound being oxidized. The approximate energy liberation per liter of O_2 consumed is 4.82 kcal, and for many purposes this value is accurate enough. More accurate measurements require data on the foods being oxidized. Such data can be obtained from an analysis of the respiratory quotient and the nitrogen excretion.

Respiratory Quotient (RQ)

The **respiratory quotient** (RQ), which is also called the **respiratory exchange ratio** (R), is the ratio of the volume of CO_2 produced to the volume of O_2 consumed per unit of time. It can be calculated for reactions outside the body, for individual organs and tissues, and for the whole body. The RQ of carbohydrate is 1.00 and that of fat is about 0.70. This is because H and O are present in carbohydrate in the same proportions as in water, whereas in the various fats, extra O_2 is necessary for the formation of H_2O.

Carbohydrate:

$$C_6H_{12}O_6 + 6O_2 \longrightarrow 6CO_2 + 6H_2O$$
(glucose)

$$RQ = 6/6 = 1.00$$

Fat:

$$2C_{51}H_{98}O_6 + 145O_2 \longrightarrow 102CO_2 + 98H_2O$$
(tripalmitin)

$$RQ = 102/145 = 0.703$$

Determining the RQ of protein in the body is a complex process, but an average value of 0.82 has been calculated.

The approximate amounts of carbohydrate, protein, and fat being oxidized in the body at any given time can be calculated from the CO_2 expired, the O_2 inspired, and the urinary nitrogen excretion. However, the values calculated in this fashion are only approximations. Furthermore, the volume of CO_2 expired and the volume of O_2 inspired may vary with factors other than metabolism. For example, if the individual hyperventilates, the RQ rises because CO_2 is being blown off. During exercise, the RQ may reach 2.00 because of hyperventilation and blowing off of CO_2 while contracting an O_2 debt. After exercise, it falls for a while to 0.50 or less. In metabolic acidosis, the RQ rises because respiratory compensation for the acidosis causes the amount of CO_2 expired to rise (see Chapter 39). In severe acidosis, the RQ may be greater than 1.00. In metabolic alkalosis, the RQ falls.

The O_2 consumption and CO_2 production of an organ can be calculated by multiplying its blood flow per unit of time by the arteriovenous differences for O_2 and Co_2 across the organ, and the RQ can then be calculated. Data on the RQ of individual organs are of considerable interest in drawing inferences about the metabolic processes occurring in them. For example, the RQ of the brain is regularly 0.97 to 0.99, indicating that its principal but not its only fuel is carbohydrate. During secretion of gastric juice, the stomach has a negative RQ because it takes up more CO_2 from the arterial blood than it puts into the venous blood (see Chapter 26).

Measuring the Metabolic Rate

In determining the metabolic rate, O_2 consumption is usually measured with some form of oxygen-filled spirometer and a CO_2 absorbing system. Such a device is illustrated in Fig 17–1. The spirometer bell is connected to a pen that writes on a rotating drum as the bell moves up and down. The slope of a line joining the ends of each of the spirometer excursions is proportionate to the O_2 consumption. The amount of O_2 (in milliliters) consumed per unit of time is corrected to standard temperature and pressure (see Chapter 34) and then converted to energy production by multiplying by 4.82 kcal/L of O_2 consumed.

Factors Affecting the Metabolic Rate

The metabolic rate is affected by many factors (Table 17–1). The most important is muscular exertion. O_2 consumption is elevated not only during

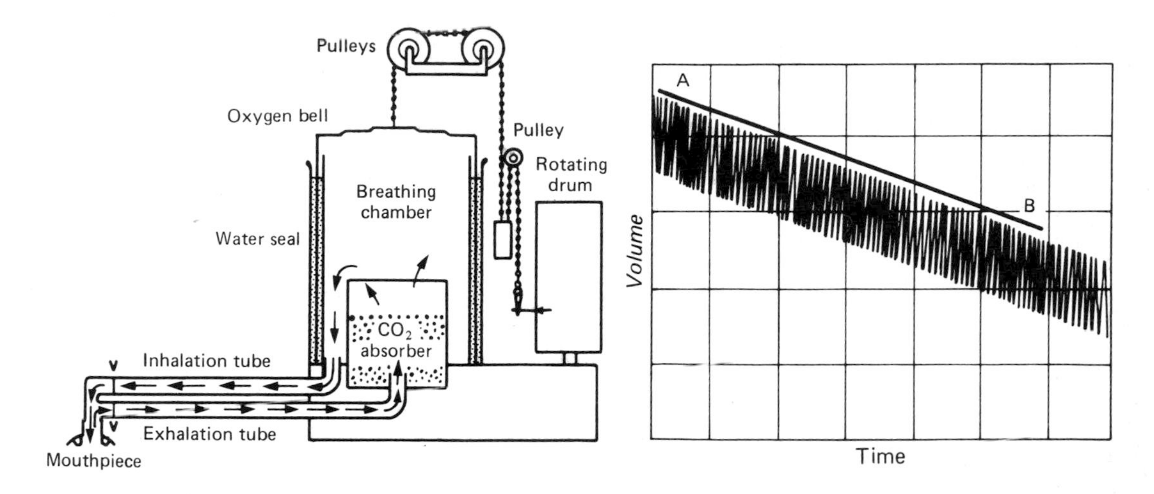

Figure 17–1. Diagram of modified Benedict apparatus, a recording spirometer used for measuring human O_2 consumption, and record obtained with it. The slope of the line AB is proportionate to the O_2 consumption. V, V: one-way check valves.

exertion but also for as long afterward as is necessary to repay the O_2 debt (see Chapter 3). Recently ingested foods also increase the metabolic rate because of their **specific dynamic action (SDA).** The SDA of a food is the obligatory energy expenditure that occurs during its assimilation into the body. An amount of protein sufficient to provide 100 kcal increases the metabolic rate a total of 30 kcal; a similar amount of carbohydrate increases it 6 kcal; and a similar amount of fat, 4 kcal. This means, of course, that the amount of calories available from the 3 foods is in effect reduced by this amount; the energy used in their assimilation must come from the food itself or the body energy stores. The cause of the SDA is uncertain. It may be due in part to increased sympathetic discharge after feeding, with increased release of epinephrine and norepinephrine and a consequent increase in metabolic rate (see below). The SDA of proteins may also be related to the process of deamination of the constituent amino acids in the liver after their absorption. The SDA of fats may be due to a direct stimulation of metabolism by liberated fatty acids. That of carbohydrate may be a manifestation of the extra energy required to form glycogen. The stimulating effects of food on metabolism may last 6 hours or more.

Table 17–1. Factors affecting the metabolic rate.

Muscular exertion during or just before measurement
Recent ingestion of food
High or low environmental temperature
Height, weight, and surface area
Sex
Age
Emotional state
Body temperature
Pregnancy or menstruation
Circulating levels of thyroid hormones
Circulating epinephrine and norepinephrine levels

Another factor that stimulates metabolism is the environmental temperature. The curve relating the metabolic rate to the environmental temperature is U-shaped. When the environmental temperature is lower than body temperature, heat-conserving mechanisms such as shivering are activated, and the metabolic rate rises. When the temperature is high enough to raise the body temperature, there is a general acceleration of metabolic processes, and the metabolic rate also rises.

The metabolic rate determined at rest in a room at a comfortable temperature 12–14 hours after the last meal is called the **basal metabolic rate (BMR).** Actually, the rate is not truly "basal"; the metabolic rate during sleep is lower than the "basal" rate. What the term basal actually denotes is a set of widely known and accepted standard conditions.

The BMR of a man of average size is about 2000 kcal/d. Large animals have higher absolute BMRs, but the ratio of BMR to body weight in small animals is much greater. One variable that correlates well with the metabolic rate in different species is the body surface area. This would be expected, since heat exchange occurs at the body surface. However, the line relating heat production to body weight is actually steeper than this; body surface area correlates to body weight raised to the 0.67 power, whereas heat production correlates to body weight raised to the 0.75 power (Fig 17–2). It has been argued that this difference represents the greater heat production of larger animals supporting their masses against gravity. Nevertheless, BMRs in humans are regularly related to body surface area.

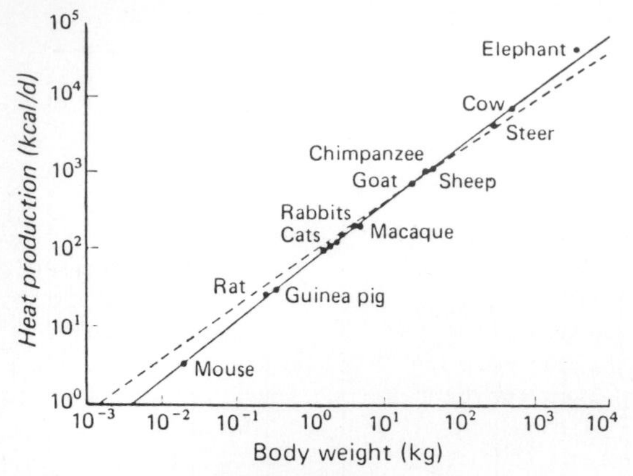

Figure 17–2. Correlation between metabolic rate and body weight, plotted on logarithmic scales. The slope of the solid line is 0.75. The dashed line represents the way surface area increases with weight for geometrically similar shapes and has a slope of 0.67. (Modified from Kleiber M and reproduced, with permission, from McMahon TA: Size and shape in biology. *Science* 1973;**179**:1201. Copyright 1973 by the American Association for the Advancement of Science.)

The relationship between weight and height and body surface area in humans can be expressed by the following formula:

$$S = 0.007184 \times W^{0.425} \times H^{0.725}$$

where

S = Surface area in m²
W = Body weight in kg
H = Height in cm

Nomograms constructed from this formula are available for easy estimation of body surface area. A normal BMR in an adult male is about 40 kcal/m²/h. For convenience, the BMR is usually expressed as a percentage of increase or decrease above or below a set of generally used standard normal values. Thus, a value of +65 means that the individual's BRM is 65% above the standard for that age and sex.

In females, the BMR at all ages is slightly lower than in males. The rate is high in children and declines with age. Anxiety and tension elevate the BMR because they cause increased epinephrine secretion and increased tensing of the muscles, even when the individual is quiet. On the other hand, apathetic, depressed patients may have low BMRs. An increase in body temperature speeds up chemical reactions, and the BMR rises approximately 14% for each Celsius degree of fever (7% per Fahrenheit degree). The stimulatory effects of catecholamines and thyroid hormones on the BMR are discussed in the chapters on those hormones. During prolonged starvation, the BMR falls. Sympathetic function is also depressed, and the decrease in circulating catecholamines may contribute to the decrease in BMR. In addition, the circulating levels of biologically active thyroid hormones fall (see Chapter 18). Conversely, sympathetic function and BMR increase after feeding. The decrease in metabolic rate during starvation is the explanation of why, when an individual reduces, weight loss is initially rapid and then slows down.

Energy Balance

The first law of thermodynamics, the principle that states that energy is neither created nor destroyed when it is converted from one form to another, applies to living organisms as well as inanimate systems. One may therefore speak of an **energy balance** between caloric intake and energy output. If the caloric content of the food ingested is less than the energy output—ie, if the balance is negative—endogenous stores are utilized. Glycogen, body protein, and fat are catabolized, and the individual loses weight. If the caloric value of the food intake exceeds energy loss due to heat and work and the food is properly digested and absorbed—ie, if the balance is positive—energy is stored, and the individual gains weight.

Except in humans and some hibernating and domesticated animals, the appetite mechanism regulates food intake with such precision that obesity is rare. Appetite is regulated as if it were determined by caloric needs, although the link between caloric consumption and appetite is probably indirect. The mechanisms regulating food intake are discussed in Chapter 14.

To balance basal output so that the energy-consuming tasks essential for life can be performed, the average adult must take in about 2000 kcal/d. Caloric requirements above the basal level depend upon the individual's activity. The average sedentary student (or professor) needs another 500 kcal, whereas a lumberjack needs up to 3000 additional kilocalories per day. Children require a lower absolute intake because of their smaller size, but a larger relative intake is essential to promote growth.

INTERMEDIARY METABOLISM

General Considerations

The end products of the digestive processes discussed in Chapters 25 and 26 are for the most part amino acids, fat derivatives, and the hexoses fructose, galactose, and glucose. These compounds are absorbed and metabolized in the body by various routes. The details of their metabolism are the concern of biochemistry and are not considered here. However, an outline of carbohydrate, protein, and fat metabolism is included for completeness and because some knowledge of the pathways involved is essential to an understanding of the action of thyroid, pancreatic, and adrenal hormones.

General Plan of Metabolism

The short-chain fragments produced by hexose, amino acid, and fat catabolism are very similar. From this **common metabolic pool** of intermediates, carbohydrates, proteins, and fats can be synthesized, or the fragments can enter the citric acid cycle, a sort of final common pathway of catabolism, in which they are broken down to hydrogen atoms and CO_2. The hydrogen atoms are oxidized to form water by a chain of flavoprotein and cytochrome enzymes.

Energy Transfer

The energy liberated by catabolism is not used directly by cells but is applied instead to the formation of ester bonds between phosphoric acid residues and certain organic compounds. Because the energy of bond formation in these phosphates is particularly high, relatively large amounts of energy (10–12 kcal/mol) are released when the bond is hydrolyzed. Compounds containing such bonds are called **high-energy phosphate compounds.** Not all organic phosphates are of the high-energy type. Many, like glucose 6-phosphate, are low-energy phosphates that on hydrolysis liberate 2–3 kcal/mol. Some of the intermediates formed in carbohydrate metabolism are high-energy phosphates, but the most important high-energy phosphate compound is **adenosine triphosphate (ATP).** This ubiquitous molecule (Fig 17–3) is the energy storehouse of the body. Upon hydrolysis to adenosine diphosphate (ADP), it liberates energy directly to such processes as muscle contraction, active transport, and the synthesis of many chemical compounds. Loss of another phosphate to form adenosine monophosphate (AMP) releases more energy. Another energy-rich phosphate compound found in muscle is creatine phosphate (phosphorylcreatine, CrP; Fig 17–21). Other important phosphorylated compounds, at least some of which can serve as energy donors, include the triphosphate derivatives of pyrimidine or purine bases other than adenine (Fig 17–22). These include the guanine derivative, guanosine triphosphate (GTP); the cytosine derivative, cytidine triphosphate (CTP); the uracil derivative, uridine triphosphate (UTP); and the hypoxanthine derivative, inosine triphosphate (ITP). Many catabolic reactions are associated with the formation of energy-rich phosphates.

Another group of high-energy compounds are the thioesters, the acyl derivatives of mercaptans. **Coenzyme A (CoA)** is a widely distributed mercaptan containing adenine, ribose, pantothenic acid, and thioethanolamine (Fig 17–4). Reduced CoA (usually abbreviated HS-CoA) reacts with acyl groups (R–CO–) to form R–CO–S–CoA derivatives. A prime example is the reaction of HS-CoA with acetic acid to form acetylcoenzyme A (acetyl-CoA), a compound of pivotal importance in intermediary metabolism. Because acetyl-CoA has a much higher energy content than acetic acid, it combines readily with substances in reactions that would otherwise require outside energy. Acetyl-CoA is therefore often called "active acetate." From the point of view of energetics, formation of 1 mol of any acyl-CoA compound is equivalent to the formation of 1 mol of ATP.

Biologic Oxidations

Oxidation is the combination of a substance with O_2, or loss of hydrogen, or loss of electrons. The corresponding reverse processes are called reduction. Biologic oxidations are catalyzed by enzymes, a particular protein enzyme being responsible in most cases for one particular reaction. Cofactors (simple ions) or coenzymes (organic, nonprotein substances) are accessory substances that usually act as carriers

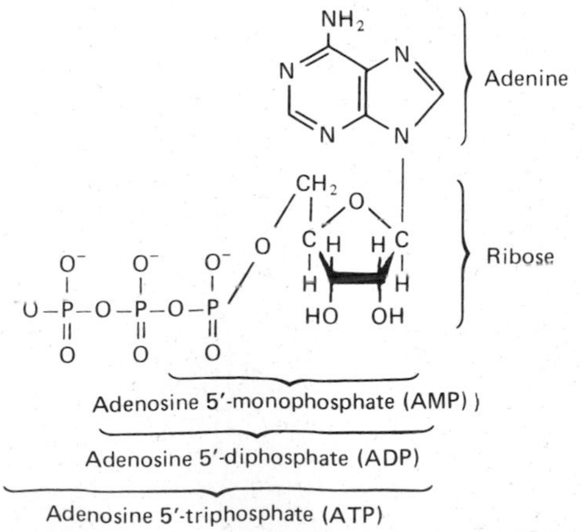

Adenine
Ribose
Adenosine 5′-monophosphate (AMP)
Adenosine 5′-diphosphate (ADP)
Adenosine 5′-triphosphate (ATP)

Figure 17–3. Energy-rich adenosine derivatives. (Reproduced, with permission, from Murray RK et al: *Harper's Biochemistry,* 21st ed. Appleton & Lange, 1988.)

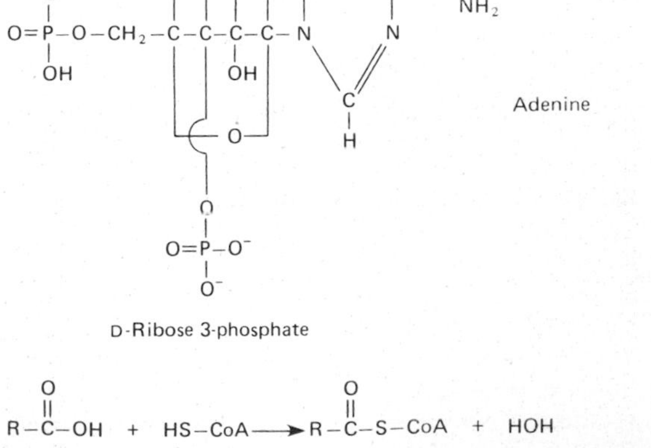

Pantothenic acid
Thioethanolamine
Diphosphate
Adenine
D-Ribose 3-phosphate

$$R{-}\overset{O}{\overset{\|}{C}}{-}OH + HS{-}CoA \longrightarrow R{-}\overset{O}{\overset{\|}{C}}{-}S{-}CoA + HOH$$

Figure 17–4. *Top:* Formula of reduced CoA (HS-CoA). *Bottom:* Formula for reaction of CoA with biologically important compounds to form thioesters. R, rest of molecule.

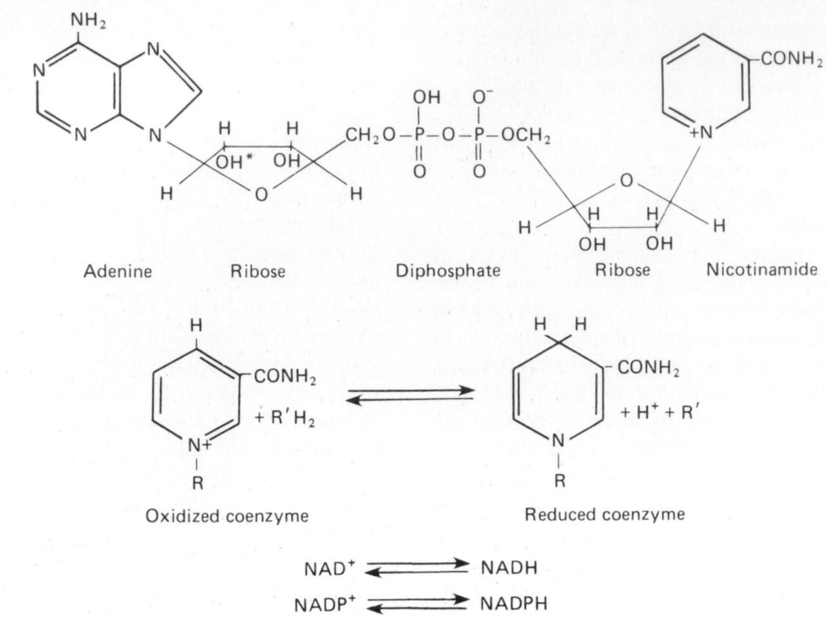

Adenine Ribose Diphosphate Ribose Nicotinamide

Oxidized coenzyme Reduced coenzyme

$NAD^+ \rightleftharpoons NADH$

$NADP^+ \rightleftharpoons NADPH$

Figure 17–5. *Top:* Formula of oxidized form of nicotinamide adenine dinucleotide (NAD^+). Nicotinamide adenine dinucleotide phosphate ($NADP^+$) has an additional phosphate group at the location marked by the asterisk. *Bottom:* Reaction by which NAD^+ and $NADP^+$ become reduced to form NADH and NADPH. R, remainder of molecule; R, hydrogen donor.

for products of the reaction. Unlike the enzymes, the coenzymes may catalyze a variety of reactions.

A number of coenzymes serve as hydrogen acceptors. One common form of biologic oxidation is removal of hydrogen from an R–OH group, forming R=O. In such dehydrogenation reactions, nicotinamide adenine dinucleotide (**NAD^+**) and nicotinamide adenine dinucleotide phosphate (**$NADP^+$**) pick up hydrogen, forming dihydronicotinamide adenine dinucleotide (**NADH**) and dihydronicotinamide adenine dinucleotide phosphate (**NADPH**) (Fig 17–5). The hydrogen is then transferred to the flavoprotein-cytochrome system, reoxidizing the NAD^+ and $NADP^+$.

The flavoprotein-cytochrome system is a chain of enzymes that transfers hydrogen to oxygen, forming water. This process occurs in the mitochondria. Each enzyme in the chain is reduced and then reoxidized as the hydrogen is passed down the line (Fig 17–6). Each of the enzymes is a protein with an attached nonprotein prosthetic group. The flavoprotein prosthetic group is a derivative of the B complex vitamin riboflavin. The prosthetic groups of the cytochromes contain iron in a porphyrin configuration that resembles hemoglobin (see Chapter 27).

Oxidative Phosphorylation

The transfer of hydrogen from NADH to flavoprotein is associated with the formation of ATP from ADP, and the further transfer along the flavoprotein-cytochrome system generates 2 more molecules of ATP per pair of protons transferred. Production of ATP associated with oxidation in this situation is called **oxidative phosphorylation.** The

H_2 → NAD^+ → NADH + H^+; FAD H_2 / FAD; Co Q / Co Q H_2; Cyt b reduced / Cyt b oxidized; Cyt c_1 oxidized / Cyt c_1 reduced; Cyt c reduced / Cyt c oxidized; Cyt a oxidized / Cyt a reduced; Cyt a_3 reduced / Cyt a_3 oxidized; 1/2 O_2 → H_2O; R; Succinate, fumarate; ADP + P_i → ATP (×3)

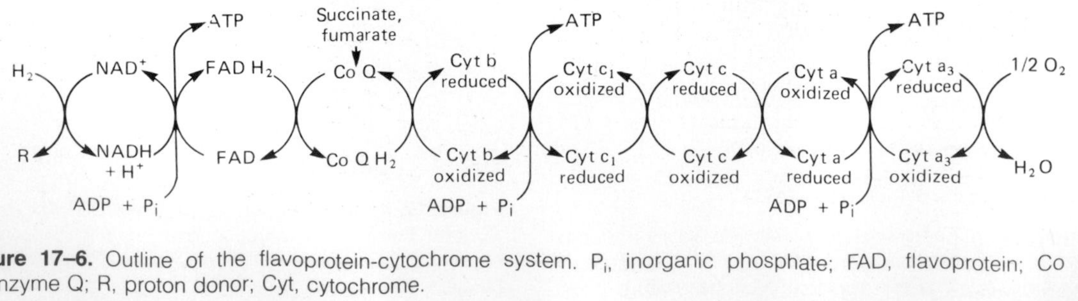

Figure 17–6. Outline of the flavoprotein-cytochrome system. P_i, inorganic phosphate; FAD, flavoprotein; Co Q, coenzyme Q; R, proton donor; Cyt, cytochrome.

mechanism is chemiosmotic and involves the transfer of protons across an insulating membrane (the inner membrane that forms the cristae of the mitochondria), the transfer being driven by oxidation in the respiratory chain. This creates an electrochemical potential difference across the membrane, and the transport of protons from the intracristal space back into the matrix space (see Chapter 1) drives a reversible ATPase in the membrane in the direction that converts ADP and inorganic phosphate (Pi) to ATP (Fig 17–7). The process depends on an adequate supply of ADP and consequently is under a form of feedback control; the more rapid the utilization of ATP in the tissues, the greater the accumulation of ADP and, consequently, the more rapid the rate of oxidative phosphorylation.

In addition to its function in energy transfer, ATP is the precursor of cyclic adenosine 3′,5′-monophosphate. The function of this compound is discussed in Chapter 1.

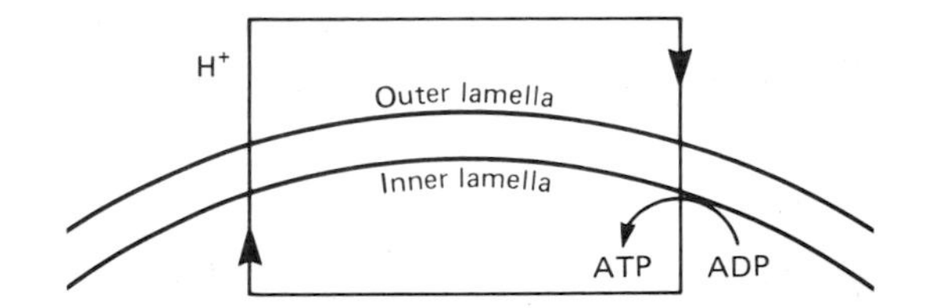

Figure 17–7. Simplified diagram of transport of protons across the inner and outer lamellas of the inner mitochondrial membrane by the electron transport system (flavoprotein-cytochrome system), with return movement of protons down the protein gradient generating ATP.

CARBOHYDRATE METABOLISM

Dietary carbohydrates are for the most part polymers of hexoses, of which the most important are galactose, fructose, and glucose (Fig 17–8). Most of the monosaccharides occurring in the body are the D isomers. The principal product of carbohydrate digestion and the principal circulating sugar is glucose. The normal fasting level of glucose in peripheral venous blood is 70–110 mg/dL (3.9–5.6 mmol/L). In arterial blood, the glucose level is 15–30 mg/dL higher than in venous blood.

Once it enters the cells, glucose is normally phosphorylated to form glucose 6-phosphate. The enzyme that catalyzes this reaction is **hexokinase.** In the liver, there is in addition an enzyme called **glucokinase,** which has greater specificity for glucose and which, unlike hexokinase, is increased by insulin and decreased in starvation and diabetes. The glucose 6-phosphate is either polymerized into glycogen or catabolized. The steps involved are outlined in Fig 17–9. The process of glycogen formation is called **glycogenesis,** and glycogen breakdown is called **glycogenolysis.** Glycogen, the storage form of glucose, is present in most body tissues, but the major supplies are in the liver and skeletal muscle. The breakdown of glucose to pyruvate or lactate (or both) is called **glycolysis.** Glucose catabolism proceeds via cleavage to trioses, or via oxidation and decarboxylation to pentoses. The pathway to pyruvate through the trioses is the **Embden-Meyerhof pathway,** and that through 6-phosphogluconate and the pentoses is the **direct oxidative pathway (hexosemonophosphate shunt)** (Fig 17–9). Pyruvate is converted to acetyl-CoA. Interconversions between carbohydrate, protein, and fat (see below) include conversion of the glycerol from fats to dihydroxyacetone phosphate and conversion of a number of amino acids with carbon skeletons resembling intermediates in the Embden-Meyerhof pathway and citric acid cycle to these intermediates by deamination. In this way, and by conversion of lactate to glucose, nonglucose molecules can be converted to glucose **(gluconeogenesis).** Glucose can be converted to fats through acetyl-CoA, but since the conversion of pyruvate to acetyl-CoA, unlike most reactions in glycolysis, is irreversible (Fig 17–10), fats are not converted to glucose via this pathway. There is therefore very little net conversion of fats to carbohydrate in the body because, except for the quantitatively unimportant production from glycerol, there is no pathway for conversion.

Citric Acid Cycle

The **citric acid cycle** (Krebs cycle, tricarboxylic acid cycle) is a sequence of reactions in which acetyl-CoA is metabolized to CO_2 and H atoms. Acetyl-CoA is first condensed with the anion of a 4-carbon acid, oxaloacetate, to form citrate and HS-CoA. In a series of 7 subsequent reactions, 2 Co_2 molecules are split off, regenerating oxaloacetate (Fig 17–10). Four pairs of H atoms are transferred to the flavoprotein-cytochrome chain, producing 12 ATP and 4 H_2O, of which $2H_2O$ are used in the cycle. The citric acid cycle is the common pathway for oxidation to CO_2 and H_2O of carbohydrate, fat, and some amino acids. The major entry into it is through acetyl-CoA, but pyruvate also enters by taking up CO_2 to form oxaloacetate, and a number of amino acids can be converted to citric acid cycle intermedi-

D-Glucose	D-Galactose	D-Fructose
H–C=O	H–C=O	CH_2OH
H–C–OH	H–C–OH	C=O
HO–C–H	HO–C–H	HO–C–H
H–C–OH	HO–C–H	H–C–OH
H–C–OH	H–C–OH	H–C–OH
CH_2OH	CH_2OH	CH_2OH

Figure 17–8. Structure of principal dietary hexoses. The naturally occurring D isomers are shown.

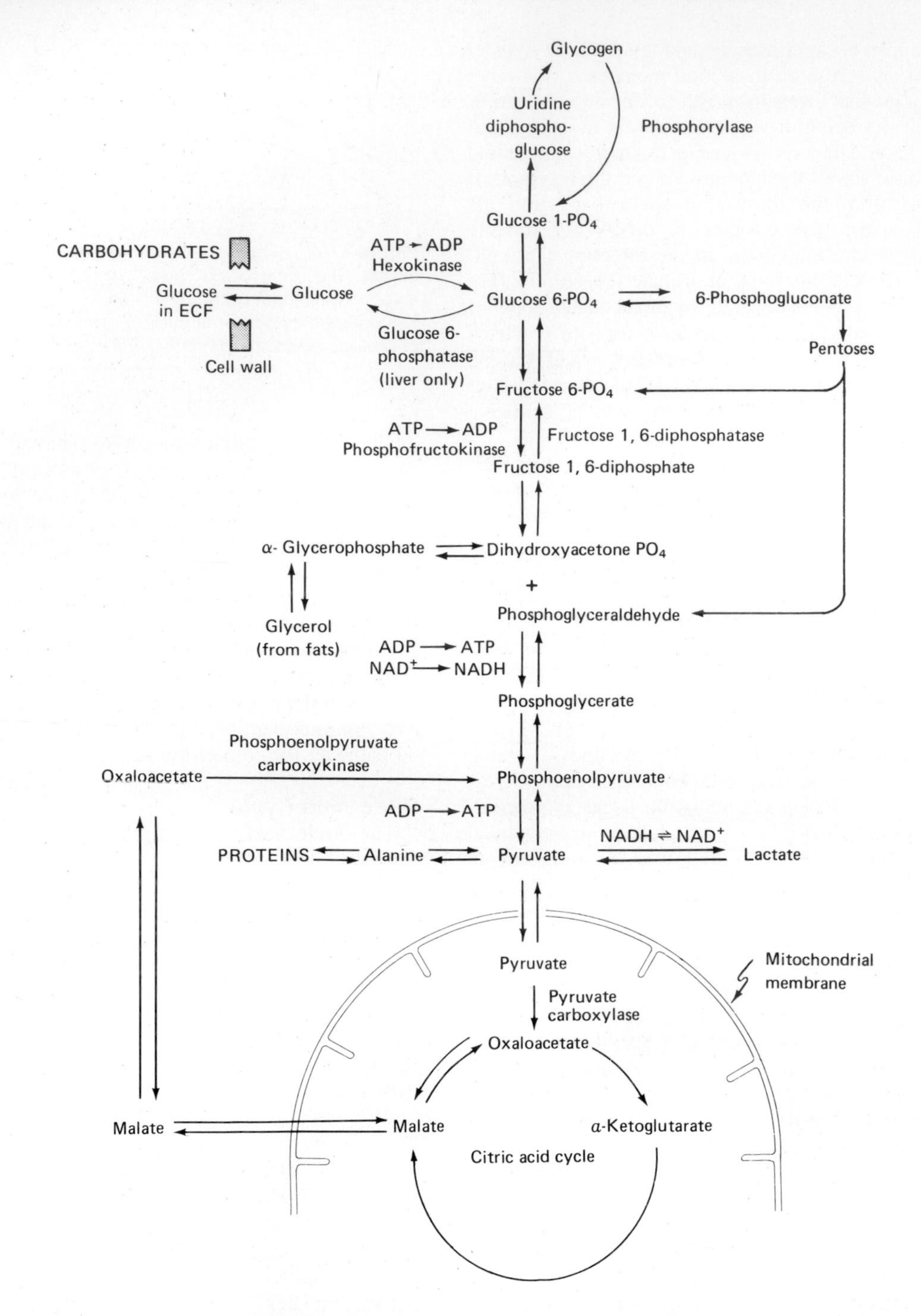

Figure 17–9. Outline of the metabolism of carbohydrate in cells, showing some of the principal enzymes involved.

ates by deamination. The combination of pyruvate with CO_2 to form oxaloacetate is but one of a considerable number of metabolic reactions in which CO_2 is a building block rather than a waste substance. The citric acid cycle requires O_2 and does not function under anaerobic conditions.

Glycolysis to pyruvate occurs outside the mitochondria. Pyruvate then enters the mitochondria and is metabolized. Oxidative phosphorylation occurs only in the mitochondria. Within the mitochondria, the enzymes involved in this process are arranged in orderly sequences along the shelves and inner walls.

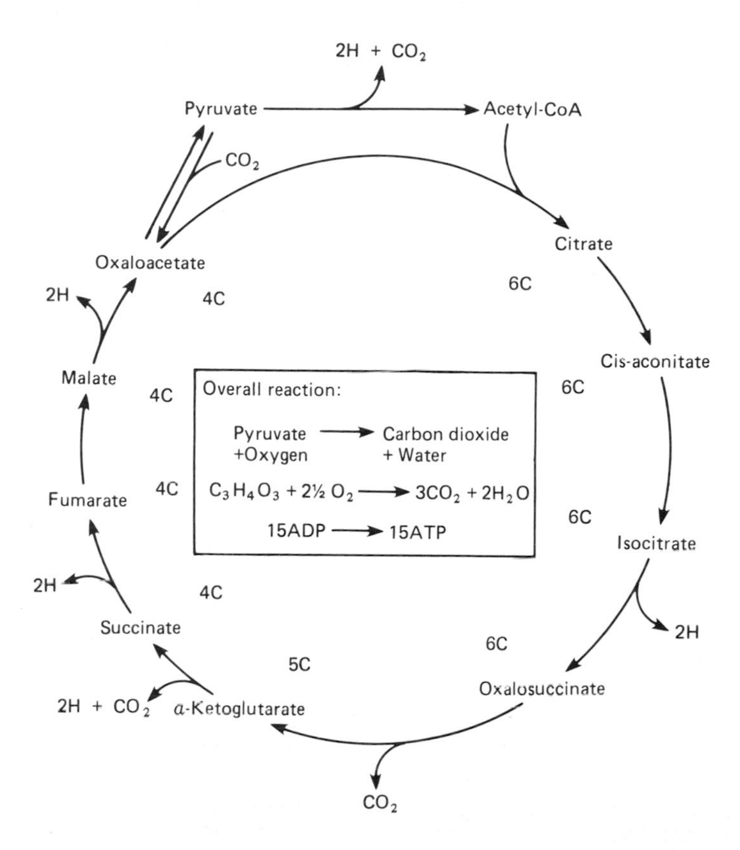

Figure 17–10. Citric acid cycle. The numbers in the circle (6C, 5C, 4C) indicate the number of carbon atoms in each of the acid intermediates. Note that 2 H atoms are transferred from the pyruvate → acetyl-CoA reaction and 8 H atoms are transferred for each turn of the cycle proper, generating by oxidative phosphorylation a total of 15 ATP.

Energy Production

The net production of energy-rich phosphate compounds during the metabolism of glucose and glycogen to pyruvate depends on whether metabolism occurs via the Embden-Meyerhof pathway or the hexosemonophosphate shunt. The conversion of 1 mol of phosphoglyceraldehyde to phosphoglycerate generates 1 mol of ATP, and the conversion of 1 mol of phosphoenolpyruvate to pyruvate another. Since each mole of glucose 6-phosphate produces, via the Embden-Meyerhof pathway, 2 mol of phosphoglyceraldehyde, 4 mol of ATP are generated per mole of glucose metabolized to pyruvate. All these reactions occur in the absence of O_2 and consequently represent anaerobic production of energy. However, 1 mol of ATP is used in forming fructose 1,6-diphosphate from fructose 6-phosphate and 1 mol in phosphorylating glucose when it enters the cell. Consequently, when pyruvate is formed anaerobically from glycogen, there is a *net* production of 3 mol of ATP per mole of glucose 6-phosphate; however, when pyruvate is formed from 1 mol of blood glucose, the net gain is only 2 mol of ATP.

A supply of NAD^+ is necessary for the conversion of phosphoglyceraldehyde to phosphoglycerate. Under aerobic conditions, NADH is oxidized via the flavoprotein-cytochrome chain, regenerating NAD^+ and forming an additional 6 mol of ATP for each 2 mol of phosphoglycerate. Under anaerobic conditions (anaerobic glycolysis), a block of glycolysis at the phosphoglyceraldehyde conversion step might be expected to develop as soon as the available NAD^+ is converted to NADH. However, pyruvate can accept hydrogen from NADH, forming NAD^+ and lactate.

$$\text{Pyruvate} + \text{NADH} \rightleftharpoons \text{Lactate} + \text{NAD}^+$$

In this way, glucose metabolism and energy production can continue for a while without O_2. The lactate that accumulates is converted back to pyruvate when the O_2 supply is restored, NADH transferring its hydrogen to the flavoprotein-cytochrome chain.

During aerobic glycolysis, the net production of ATP is 19 times as great as the 2 ATPs formed under anaerobic conditions. Not only are an additional 6 moles of ATP formed by NADH oxidation via the flavoprotein-cytochrome chain, but similar oxidation of the 2 mol of NADH formed by the conversion of 2 mol of pyruvate to acetyl-CoA produces 6 mol of ATP, and each turn of the strictly aerobic citric acid cycle generates 12 mol of ATP. Therefore, the net production per mole of blood glucose metabolized

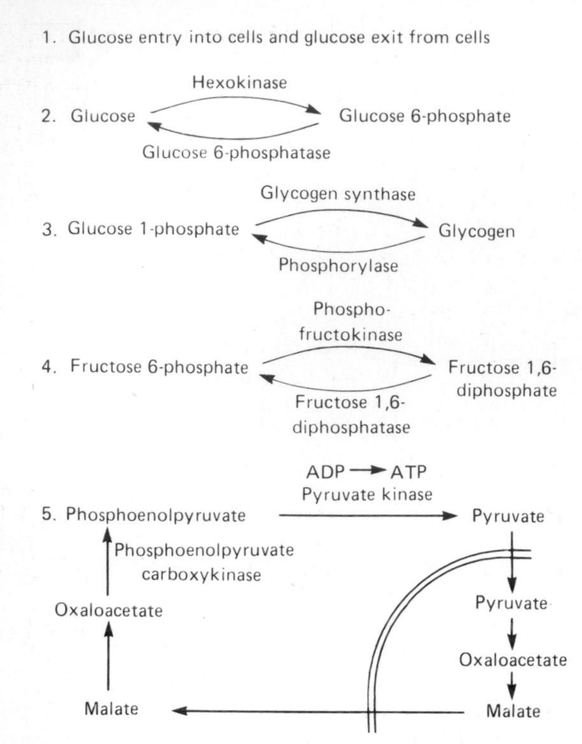

Figure 17–11. Five examples of "directional flow valves" in carbohydrate metabolism: reactions that proceed in one direction by one mechanism and in the other direction by a different mechanism. The double line in 5 represents the mitochondrial membrane. Pyruvate is converted to malate in mitochondria, and the malate diffuses out of the mitochondria to the cytosol, where it is converted to phosphoenolpyruvate.

aerobically via the Embden-Meyerhof pathway and citric acid cycle is 38 mol (2 + [2 × 3] + [2 × 3] + [2 × 12]) of ATP.

Glucose oxidation via the hexosemonophosphate shunt generates large amounts of NADPH. A supply of this reduced coenzyme is essential for many metabolic processes. The pentoses formed in the process are building blocks for nucleotides (see below). The amount of ATP generated depends upon the amount of NADPH converted to NADH and then oxidized.

"Directional Flow Valves"

Metabolism is regulated by a variety of hormones and other factors. To bring about any net change in a particular metabolic process, regulatory factors obviously must drive a chemical reaction in one direction. Most of the reactions in intermediary metabolism are freely reversible, but there are a number of "directional flow valves," ie, reactions that proceed in one direction under the influence of one enzyme or transport mechanism and in the opposite direction under the influence of another. Five examples in the intermediary metabolism of carbohydrate are shown in Fig 17–11. The different pathways for fatty acid synthesis and catabolism (see below) are another example. Regulatory factors exert their influence on metabolism by acting directly or indirectly at these "directional flow valves."

Phosphorylase

Glycogen breakdown is regulated by several hormones. Glycogen is synthesized from glucose 1-phosphate via uridine diphosphoglucose (UDPG; Fig 17–12), with the enzyme **glycogen synthase** catalyzing the final step. Glycogen is a branched glucose polymer with 2 types of glucoside linkage. Cleavage of the 1:4α linkage in the polymer chain is catalyzed by **phosphorylase,** while cleavage of the 1:6α linkages at branching points is catalyzed by another enzyme.

Phosphorylase is activated in part by the action of epinephrine on β-adrenergic receptors in the liver. This in turn initiates a sequence of reactions that provides a classic example of hormonal action via cyclic AMP (Fig 17–13). Protein kinase A is activated by cyclic AMP (Fig 1–36) and catalyzes the transfer of a phosphate group to phosphorylase kinase, converting it to its active form. The phosphorylase kinase in turn catalyzes the phosphorylation and consequent activation of phosphorylase. Inactive phosphorylase is known as phosphorylase b (dephosphophosphorylase), and activated phosphorylase as phosphorylase a (phosphophosphorylase).

Activation of protein kinase A by cyclic AMP not only increases glycogen breakdown but also inhibits glycogen synthesis. Glycogen synthase (Fig 17–12) is active in its dephosphorylated form and inactive when phosphorylated, and it is phosphorylated along with phosphorylase kinase when protein kinase is activated.

Glycogen is also broken down by the action of catecholamines on α-adrenergic receptors in the liver. This breakdown is mediated by intracellular Ca^{2+} and involves an activation of phosphorylase kinase that is independent of cyclic AMP. Large doses of vasopressin and angiotensin II can also cause glycogenolysis by this mechanism, but it is doubtful if these hormones have any normal physiologic role in glucose homeostasis.

Because the liver contains the enzyme **glucose 6-phosphatase,** much of the glucose 6-phosphate that is formed in this organ can be converted to glucose and enter the bloodstream, raising the blood glucose level. The kidneys can also contribute to the elevation. Other tissues do not contain this enzyme, so in them a large proportion of the glucose 6-phosphate is catabolized via the Embden-Meyerhof pathway and hexosemonophosphate shunt pathway. Increased glucose catabolism in skeletal muscle causes a rise in the blood lactate level (see Chapter 3).

By stimulating adenylate cyclase, epinephrine causes activation of the phosphorylase in liver and skeletal muscle. The consequences of this activation are a rise in the blood glucose and lactate levels. Glucagon has a similar action, but it exerts its effect

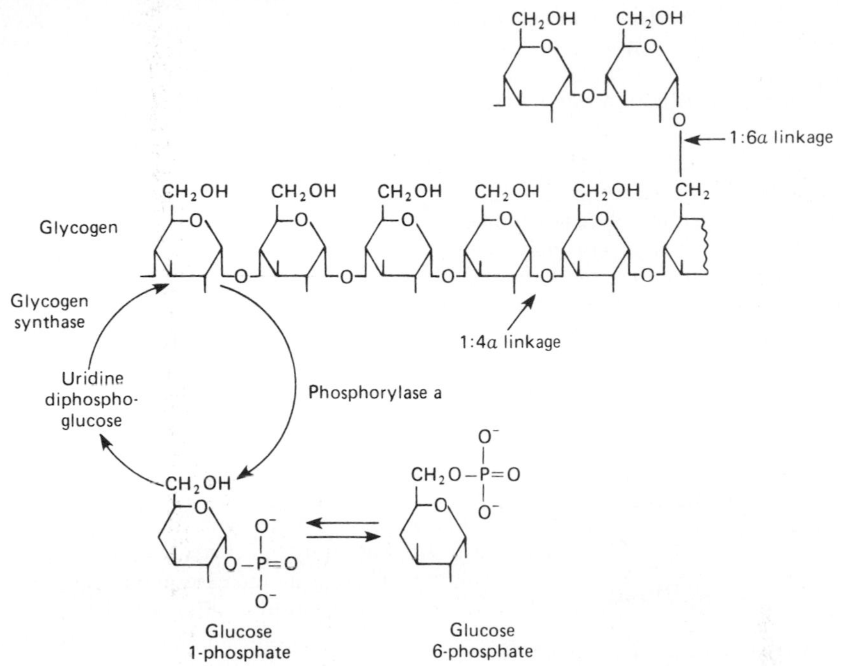

Glycogen

Glycogen synthase

Uridine diphospho-glucose

Phosphorylase a

1:4α linkage

1:6α linkage

Glucose 1-phosphate

Glucose 6-phosphate

Figure 17–12. Glycogen formation and breakdown. The activation of phosphorylase a is summarized in Fig 17–13.

only on the phosphorylase in the liver. Consequently, glucagon causes a rise in blood glucose without any change in blood lactate.

McArdle's Disease

In the clinical condition known as **McArdle's disease** or **myophosphorylase deficiency glycogenosis,** glycogen accumulates in skeletal muscles because of a deficiency of muscle phosphorylase. Patients with this disease develop muscle pain and stiffness on exertion, and they have a greatly reduced exercise tolerance; they cannot break down their muscle glycogen to provide the energy for muscle contraction (see Chapter 3), and the glucose reaching their muscles from the bloodstream is sufficient only for the demands of very mild exercise. They respond with a normal rise in blood glucose when given glucagon or epinephrine, which indicates that their hepatic phosphorylase is normal.

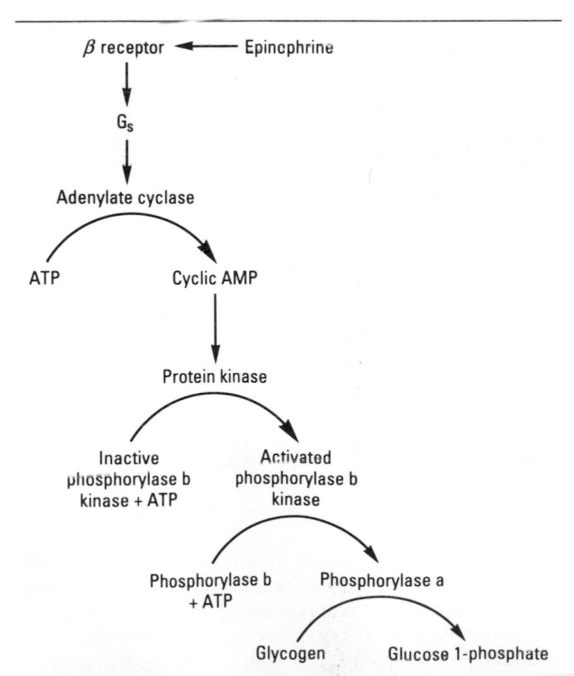

Figure 17–13. Cascade of reactions by which epinephrine activates phosphorylase. Glucagon has a similar action in liver but not in skeletal muscle.

The "Hepatic Glucostat"

There is a net uptake of glucose by the liver when the blood glucose is high and a net discharge when it is low. The liver thus functions as a sort of "glucostat," maintaining a constant circulating glucose level. This function is not automatic; glucose uptake and glucose discharge are affected by the actions of numerous hormones. Endocrine regulation of the blood glucose level and of carbohydrate metabolism in general is discussed in Chapter 19.

Renal Handling of Glucose

In the kidneys, glucose is freely filtered; but at normal blood glucose levels, all but a very small amount is reabsorbed in the proximal tubules (see Chapter 38). When the amount filtered increases, reabsorption increases, but there is a limit to the amount of glucose the proximal tubules can reabsorb. When the tubular maximum for glucose (Tm_G) is exceeded, appreciable amounts of glucose appear in the urine **(glycosuria).** The **renal threshold** for glucose, the arterial blood level at which glycosuria appears, is reached when the venous blood glucose concentration is usually about 180 mg/dL, but it may be higher if the glomerular filtration rate is low.

Glycosuria

Glycosuria occurs when the blood glucose is elevated because of relative insulin deficiency (diabetes mellitus) or because of excessive glycogenolysis after physical or emotional trauma. In some individuals, the glucose transport mechanism in the renal tubules is congenitally defective, so that glycosuria is present at normal blood glucose levels. This condition is called **renal glycosuria. Alimentary glycosuria,** glycosuria after ingestion of a high-carbohydrate meal, has been claimed to occur in normal individuals, but many of these people actually have mild diabetes mellitus. The maximal rate of glucose absorption from the intestine is about 120 g/h.

Factors Determining the Blood Glucose Level

The blood glucose level at any given time is determined by the balance between the amount of glucose entering the bloodstream and the amount leaving it. The principal determinants are therefore the dietary intake; the rate of entry into the cells of muscle, adipose tissue, and other organs; and the glucostatic activity of the liver (Fig 17–14). Five percent of ingested glucose is promptly converted into glycogen in the liver, and 30–40% is converted into fat. The remainder is metabolized in muscle and other tissues. During fasting, liver glycogen is broken down and the liver adds glucose to the bloodstream. With more prolonged fasting, glycogen is depleted and there is increased gluconeogenesis from amino acids and glycerol in the liver. There is a modest decline in blood glucose to about 65 mg/dL in men and about 40 mg/dL in premenopausal women, but gluconeogenesis prevents the occurrence of more severe hypoglycemia, even during prolonged starvation. The cause of the lower fasting blood glucose concentrations in women is unknown, but it is interesting that values similar to those seen in women are seen in fasting prepuberal boys.

Carbohydrate Homeostasis in Exercise

In a 70-kg man, carbohydrate reserves total about 2500 kcal, stored in 400 g of muscle glycogen, 100 g of liver glycogen, and 20 g of glucose in extracellular fluid. In contrast, 112,000 kcal (about 80% of body fuel supplies) are stored in fat and the remainder in protein. Resting muscle utilizes fatty acids for its metabolism. In the fasting human at rest, the brain accounts for 70–80% of the glucose utilized, and red blood cells account for most of the rest.

During exercise, the caloric needs of muscle are initially met by glycogenolysis in muscle and increased uptake of muscle glucose. Blood glucose

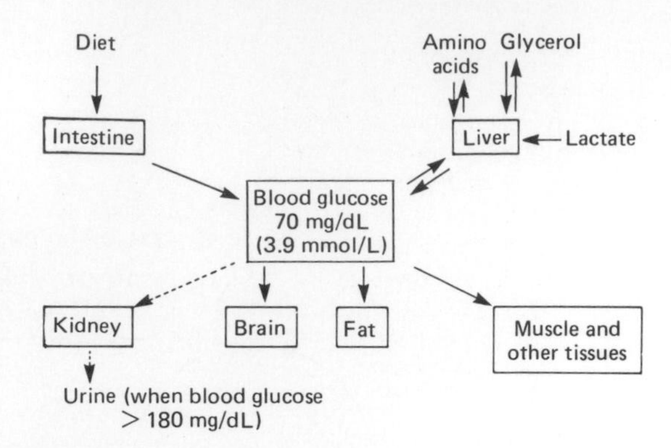

Figure 17–14. Blood glucose homeostasis, illustrating the glucostatic function of the liver.

initially rises with increased hepatic glycogenolysis but may fall with strenuous, prolonged exercise. There is an increase in gluconeogenesis (Fig 17–15). Plasma insulin falls, and plasma glucagon rises.

After exercise, liver glycogen is replenished by additional gluconeogenesis and a decrease in hepatic glucose output. Insulin levels rise sharply, especially in hepatic portal blood. The insulin entering the liver presumably promotes glycogen storage (see Chapter 19).

Metabolism of Hexoses Other Than Glucose

Other hexoses that are absorbed from the intestine include galactose, which is fiberated by the digestion of lactose and converted to glucose in the body; and

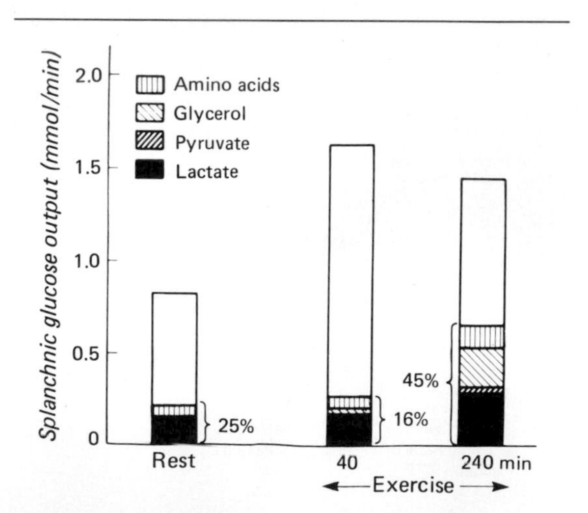

Figure 17–15. Splanchnic (hepatic) glucose output, showing output due to glycogenolysis (open bars) and output presumably due to gluconeogenesis (in brackets). The values for gluconeogenesis are measured values for splanchnic uptake of the various gluconeogenetic precursors. (Reproduced, with permission, from Felig P, Wahren J: Fuel homeostasis in exercise. *N Engl J Med* 1975; 293:1078.)

fructose, part of which is ingested and part produced by hydrolysis of sucrose. After phosphorylation, galactose is converted to uridine diphosphogalactose. The uridine diphosphogalactose is converted to uridine diphosphoglucose, which functions in glycogen synthesis (Fig 17–12). The latter reaction is reversible, and this is the way the galactose necessary for formation of glycolipids and mucoproteins is formed when dietary galactose intake is inadequate. The utilization of galactose, like that of glucose, is dependent upon insulin (see Chapter 19). In the inborn error of metabolism known as **galactosemia,** there is a congenital deficiency of phosphogalactose uridyl transferase, the enzyme responsible for the reaction between galactose 1-phosphate and uridine diphosphoglucose, so that ingested galactose accumulates in the circulation. Serious disturbances of growth and development result. Treatment with galactose-free diets improves this condition without leading to galactose deficiency because the enzymes necessary for the formation of uridine diphosphogalactose from uridine diphosphoglucose are present.

Fructose is converted in part to fructose 6-phosphate and then metabolized via fructose 1,6-diphosphate (Fig 17–9). The enzyme catalyzing the formation of fructose 6-phosphate is hexokinase, the same enzyme that catalyzes the conversion of glucose to glucose 6-phosphate. However, much more fructose is converted to fructose 1-phosphate in a reaction catalyzed by fructokinase. Most of the fructose 1-phosphate is then split into dihydroxyacetone phosphate and glyceraldehyde. The glyceraldehyde is phosphorylated, and it and the dihydroxyacetone phosphate enter the pathways for glucose metabolism. Since the reactions proceeding through phosphorylation of fructose in the 1 position can occur at a normal rate in the absence of insulin, it has been recommended that fructose be given to diabetics to replenish their carbohydrate stores. However, most of the fructose is metabolized in the intestines and liver, so its value in replenishing carbohydrate elsewhere in the body is limited.

Fructose 6-phosphate can also be phosphorylated in the 2 position, forming fructose 2,6-diphosphate. This compound is an important regulator of hepatic gluconeogenesis. When the fructose 2,6-diphosphate level is high, conversion of fructose 6-phosphate to fructose 1,6-diphosphate is facilitated, and thus breakdown of glucose to pyruvate is increased. A decreased level of fructose 2,6-diphosphate facilitates the reverse reaction and consequently aids gluconeogenesis. One of the actions of the protein kinase produced by the action of glucagon is to decrease hepatic fructose 2,6-diphosphate (see Chapter 19).

PROTEIN METABOLISM

Proteins

Proteins are made up of large numbers of amino acids (Fig 17–16) linked into chains by **peptide bonds** joining the amino group of one amino acid to the carboxyl group of the next. In addition, some proteins contain carbohydrates (glycoproteins) and lipids (lipoproteins). Smaller chains of amino acids are called **peptides** or **polypeptides.** The boundaries between peptides, polypeptides, and proteins are not well defined, but in this book, chains containing 2–10 amino acid residues are called peptides, chains containing more than 10 but less than 100 amino acid residues are called polypeptides, and chains containing 100 or more amino acid residues are called proteins. The term oligopeptide, which is employed by others to refer to small peptides, is not used.

The order of the amino acids in the peptide chains is called the **primary structure** of a protein. The chains are twisted and folded in complex ways, and the term **secondary structure** of a protein refers to the spatial arrangement produced by the twisting and folding. The most common secondary structure is a regular coil with 3.7 amino acid residues per turn (α-helix). The **tertiary structure** of a protein is the arrangement of the twisted chains into layers, crystals, or fibers. Some protein molecules are made of subunits (eg, hemoglobin; see Chapter 27), and the term **quaternary structure** is used to refer to the arrangement of the subunits.

Amino Acids

The amino acids that are found in proteins are shown in Table 17–2. These amino acids are identified by the 3-letter abbreviations or, less frequently,

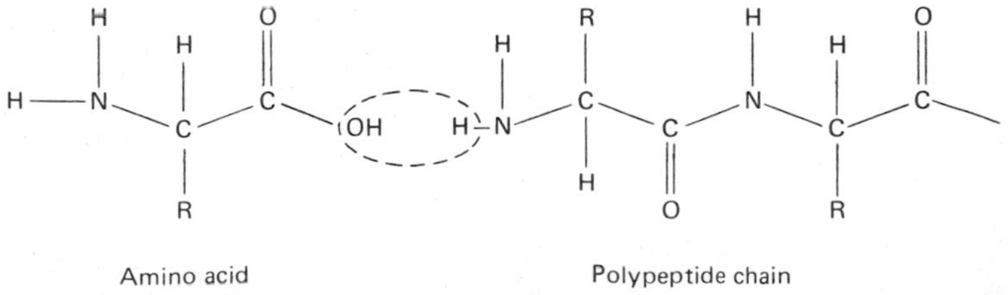

Amino acid

Polypeptide chain

Figure 17–16. Amino acid structure and formation of peptide bonds. The dotted lines show how the peptide bonds are formed, with the production of H_2O. R, remainder of amino acid. For example, in glycine, R = H; in glutamate, R = $-(CH_2)_2-COOH$.

Table 17–2. Amino acids found in proteins. Those in bold type are the nutritionally essential amino acids. The generally accepted 3-letter and 1-letter abbreviations for the amino acids are shown in parentheses.

Neutral Amino Acids	Acidic Amino Acids (Monoamino acids containing 2 or more carboxyl groups)
Amino acids with unsubstituted chains	Aspartic acid (Asp, D)
Glycine (Gly, G)	Asparagine (Asn, N)
Alanine (Ala, A)	Glutamine (Gln, Q)
Valine (Val, V)	Glutamic acid (Glu, E)
Leucine (Leu, L)	γ-Carboxyglutamic acid† (Gla)
Isoleucine (Ile, I)	Basic Amino Acids (Diaminomonocarboxyllc acids)
Hydroxyl-substituted amino acids	**Arginine*** (Arg, R)
Serine (Ser, S)	**Lysine** (Lys, K)
Threonine (Thr, T)	Hydroxylysine† (Hyl)
Sulfur-containing amino acids	**Histidine*** (His, H)
Cysteine (Cys, C)	Imino Acids (contain imino group but no amino group)
Methionine (Met, M)	Proline (Pro, P)
Aromatic amino acids	4-Hydroxyproline† (Hyp)
Phenylalanine (Phe, F)	3-Hydroxyproline†
Tyrosine (Tyr, Y)	
Tryptophan (Trp, W)	

*Arginine and histidine are sometimes called "semi-essential"; they are not necessary for maintenance of nitrogen balance but are needed for normal growth.
† There are no tRNAs for these 4 amino acids; they are formed by posttranslational modification of the corresponding unmodified amino acid in peptide linkage. There are tRNAs for the remaining 20 amino acids, and they are incorporated into peptides and proteins under direct genetic control.

the single-letter abbreviations shown in the table. Various other important amino acids such as ornithine, 5-hydroxytryptophan, L-dopa, taurine, and thyroxine (T_4) occur in the body but are not found in proteins. In higher animals, the L isomers of the amino acids are the only naturally occurring forms. The L isomers of hormones such as thyroxine are much more active than the D isomers. The amino acids are acidic, neutral, or basic in reaction, depending upon the relative proportions of free acidic (–COOH) or basic (–NH_2) groups in the molecule.

Some of the amino acids are **nutritionally essential amino acids,** which must be obtained in the diet, whereas others can be synthesized in vivo at rates sufficient to meet metabolic needs (see below).

The Amino Acid Pool

Small amounts of proteins are absorbed from the gastrointestinal tract in infants, and even smaller amounts in adults, but most ingested proteins are digested and their constituent amino acids absorbed. The body's own proteins are being continuously hydrolyzed to amino acids and resynthesized. The turnover rate of endogenous proteins averages 80–100 g/d, being highest in the intestinal mucosa and practically nil in collagen. The amino acids formed by endogenous protein breakdown are in no way different from those derived from ingested protein. With the latter, they form a common **amino acid pool** that supplies the needs of the body (Fig 17–17). In the kidney, most of the filtered amino acids are reabsorbed. During growth, the equilibrium between amino acids and body proteins shifts toward the latter, so that synthesis exceeds breakdown. At all ages, a small amount of protein is lost as hair. Some small proteins are lost in the urine, and there are unreabsorbed protein digestive secretions in the stools. These losses are made up by synthesis from the amino acid pool.

Specific Metabolic Functions of Amino Acids

Thyroxine (T_4), catecholamines, histamine, serotonin, melatonin, and intermediates in the urea cycle are formed from specific amino acids. Methionine, cystine, and cysteine provide the sulfur contained in proteins, CoA, taurine, and other biologically important compounds. Methionine is converted into S-adenosylmethionine, which is the active methylating agent in the synthesis of compounds such as epinephrine, acetylcholine, and creatine. It is a major donor of biologically labile methyl groups, but methyl groups can also be synthesized from a derivative of formic acid bound to folic acid derivatives if the diet contains adequate amounts of folic acid and cyanocobalamin.

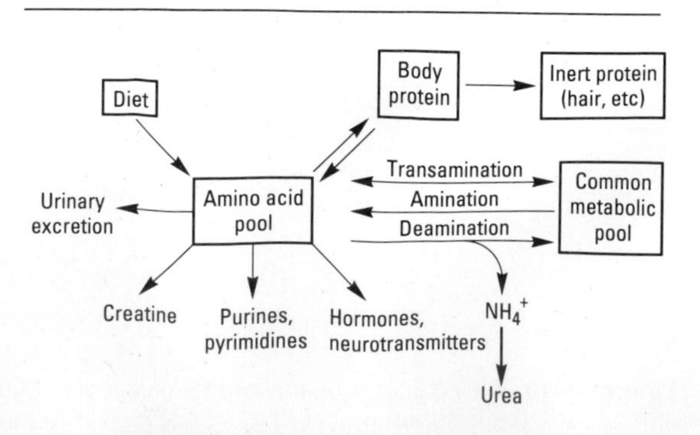

Figure 17–17. Amino acid metabolism.

Urinary Sulfates

Sulfur-containing amino acids are the source of the sulfates in the urine. A few unoxidized sulfur-containing compounds are excreted (urinary **neutral sulfur**), but most of the urinary excretion is in the form of **sulfate** (SO_4^{2-}) accompanied by corresponding amounts of cation (Na^+, K^+, NH_4^+, or H^+). The **ethereal sulfates** in the urine are organic sulfate esters ($R\text{–}O\text{–}SO_3H$) formed in the liver from endogenous and exogenous phenols, including estrogens and other steroids, indoles, and drugs.

Deamination, Amination, & Transamination

Interconversions between amino acids and the products of carbohydrate and fat catabolism at the level of the common metabolic pool and the citric acid cycle involve transfer, removal, or formation of amino groups. **Transamination** reactions, conversion of one amino acid to the corresponding keto acid with simultaneous conversion of another keto acid to an amino acid, occur in many tissues.

$$\text{Alanine} + \alpha\text{-Ketoglutarate} \rightleftharpoons \text{Pyruvate} + \text{Glutamate}$$

The **transaminases** involved are also present in the circulation. When damage to many active cells occurs as a result of a pathologic process, serum transaminase levels rise. An example is the rise in **serum glutamic-oxaloacetic transaminase (SGOT)** following myocardial infarction.

Oxidative deamination of amino acids occurs in the liver. An imino acid is formed by dehydrogenation, and this compound is hydrolyzed to the corresponding keto acid, with production of NH_4^+.

$$\text{Amino acid} + NAD^+ \rightarrow \text{Imino acid} + NADH$$

$$\text{Imino acid} + H_2O \rightarrow \text{Keto acid} + NH_4^+$$

NH_4^+ is in equilibrium with NH_3. Amino acids can also take up NH_4^+, forming the corresponding amide. An example is the binding of NH_4^+ in the brain by glutamine (Fig 17–18). The reverse reaction occurs in the kidney, with conversion of NH_4^+ to NH_3 and secretion of NH_3 into the urine. The NH_3 reacts with H^+ in the urine to form NH_4^+, thus permitting more H^+ to be secreted into the urine (see Chapter 38).

Interconversions between the amino acid pool and the common metabolic pool are summarized in Fig 17–19. Leucine, isoleucine, phenylalanine, and tyrosine are said to be **ketogenic** because they are converted to the ketone body acetoacetate (see below). Alanine and many other amino acids are **glucogenic** or **gluconeogenic,** ie, they give rise to compounds that can readily be converted to glucose.

Urea Formation

Most of the NH_4^+ formed by deamination of amino acids in the liver is converted to urea, and the urea is excreted in the urine. Except for the brain, the liver is probably the only site of urea formation, and in severe hepatic disease, the blood urea nitrogen level falls and the blood NH_3 level rises. The synthesis of urea via the **urea cycle** (Fig 17–20) involves conversion of the amino acid ornithine to citrulline and then to arginine, following which urea is split off and ornithine is regenerated.

Creatine & Creatinine

Creatine is synthesized in the liver from methionine, glycine, and arginine. In skeletal muscle, it is phosphorylated to form phosphorylcreatine (Fig 17–21), which is an important energy store for ATP synthesis (see Chapter 3). The ATP formed by glycolysis and oxidative phosphorylation reacts with creatine to form ADP and large amounts of phosphorylcreatine. During exercise, the reaction is reversed, maintaining the supply of ATP, which is the immediate source of the energy for muscle contraction.

The creatinine in the urine is formed from phosphorylcreatine. Creatine is not converted directly to creatinine. The rate of creatinine excretion is relatively constant from day to day. Indeed, creatinine output is sometimes measured as a check on the accuracy of the urine collections in metabolic studies; an average daily creatinine output is calculated, and the values for the daily output of other substances are corrected to what they would have been at this creatinine output.

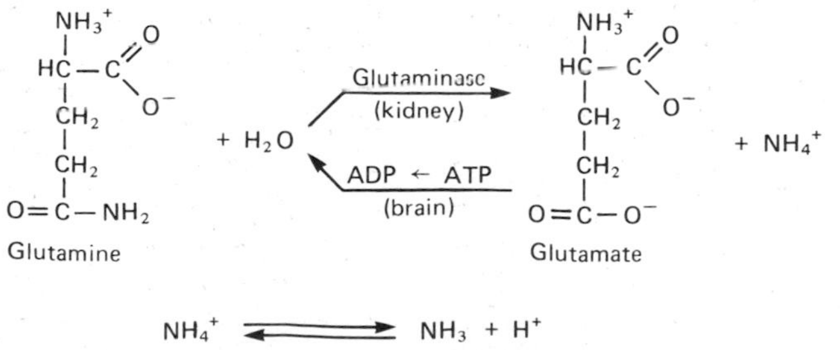

Glutamine + H_2O ⇄ Glutamate + NH_4^+ (Glutaminase (kidney) →; ← ADP ← ATP (brain))

$$NH_4^+ \rightleftharpoons NH_3 + H^+$$

Figure 17–18. Release and uptake of NH_4^+ by interconversion of glutamine and glutamate. NH_4^+ is in equilibrium with NH_3. The reaction goes predominantly to the right in the kidney, and NH_3 is secreted into the urine. In the brain, the reaction goes predominantly to the left, removing NH_3, which is toxic to nerve cells.

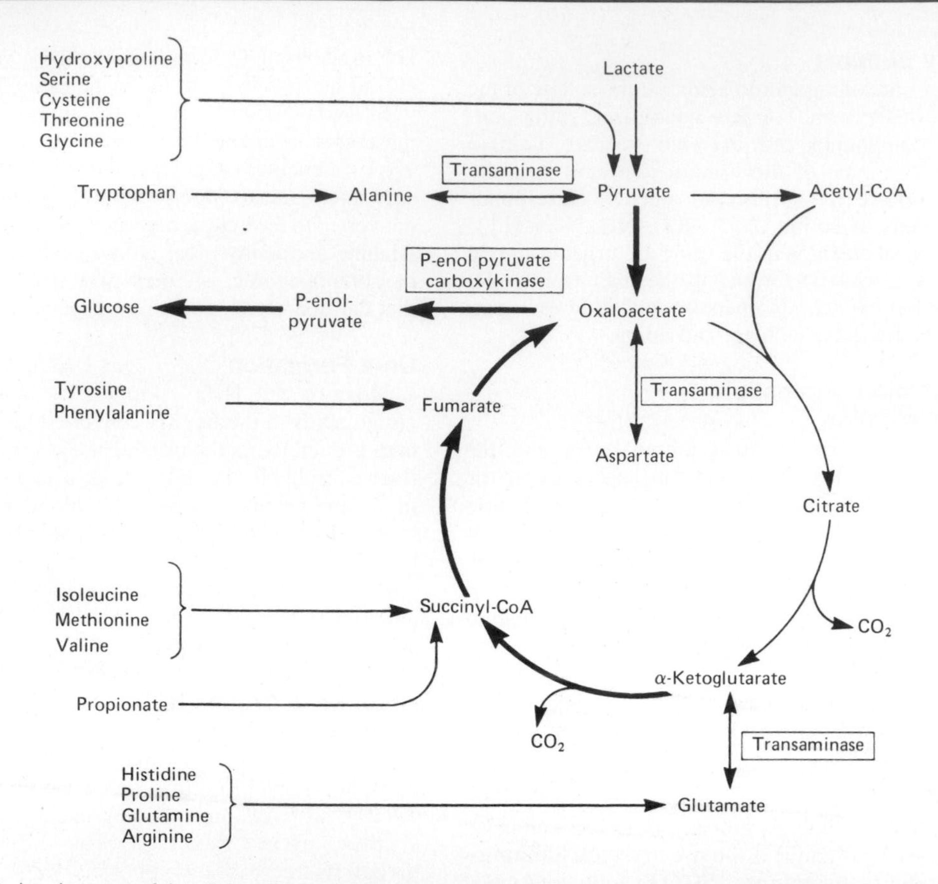

Figure 17–19. Involvement of the citric acid cycle in transamination and gluconeogenesis. The bold arrows indicate the main pathway of gluconeogenesis. (Reproduced, with permission, from Murray RK et al: *Harper's Biochemistry,* 21st ed. Appleton & Lange, 1988.)

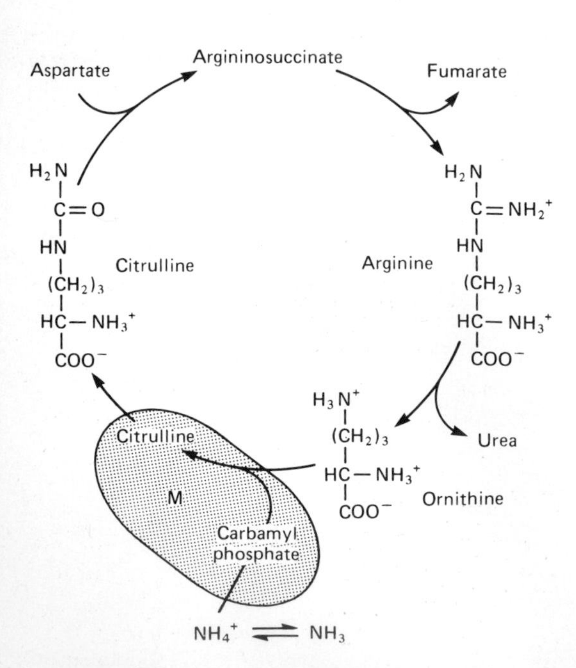

Figure 17–20. Urea cycle. M, mitochondrion.

Creatinuria occurs normally in children, in women during and after pregnancy, and occasionally in nonpregnant women. There is very little, if any, creatine in the urine of normal men, but appreciable quantities are excreted in any condition associated with extensive muscle breakdown. Thus, creatinuria occurs in starvation, thyrotoxicosis, poorly controlled diabetes mellitus, and the various primary and secondary diseases of muscle (**myopathies**).

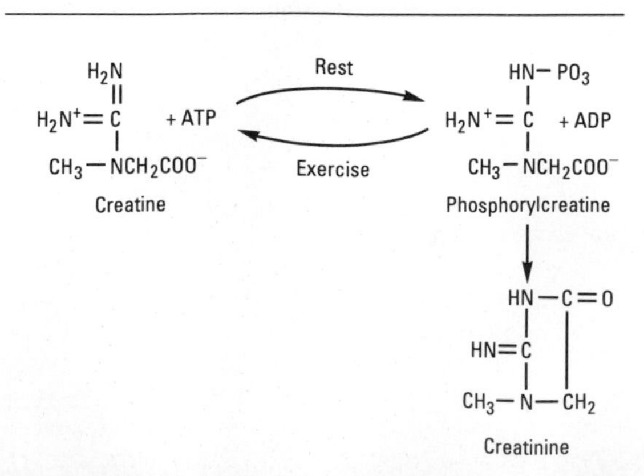

Figure 17–21. Creatine, phosphorylcreatine, and creatinine.

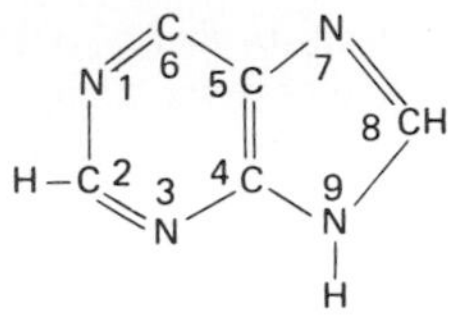

Purine nucleus

Adenine: 6-Aminopurine
Guanine: 1-Amino-6-oxypurine
Hypoxanthine: 6-Oxypurine
Xanthine: 2,6-Dioxypurine

Pyrimidine nucleus

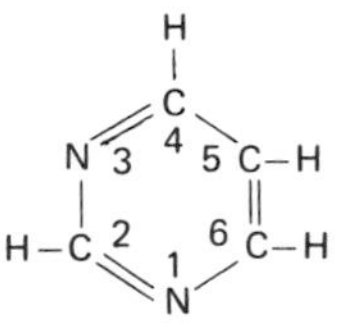

Cytosine: 4-Amino-2-oxypyrimidine
Uracil: 2,4-Dioxypyrimidine
Thymine: 5-Methyl-2,4-dioxypyrimidine

Figure 17–22. Principal physiologically important purines and pyrimidines. Oxypurines and oxypyrimidines may form enol derivatives (hydroxypurines and hydroxypyrimidines) by migration of hydrogen to the oxygen substituents.

Purines & Pyrimidines

The physiologically important **purines** and **pyrimidines** are shown in Fig 17–22. **Nucleosides**—purines or pyrimidines combined with ribose—are components not only of a variety of coenzymes and related substances (NAD^+, $NADP^+$, ATP, UDPG, etc) but of RNA and DNA as well (Table 17–3). The structure and function of DNA and RNA and their role in protein synthesis are discussed in Chapter 1.

Nucleic acids in the diet are digested and their constituent purines and pyrimidines absorbed, but most of the purines and pyrimidines are synthesized from amino acids, principally in the liver. The nucleotides and RNA and DNA are then synthesized. RNA is in dynamic equilibrium with the amino acid pool, but DNA, once formed, is metabolically stable throughout life.

The purines and pyrimidines released by the breakdown of nucleotides may be reused or catabolized. Minor amounts are excreted unchanged in the urine. The pyrimidines are catabolized to Co_2 and NH_3, and the purines are converted to uric acid.

Protein Degradation

Like protein synthesis, protein degradation is a carefully regulated, complex process. The rates at which individual proteins are metabolized vary, and the body appears to have mechanisms by which abnormal proteins are recognized and degraded more rapidly than normal body constituents. For example, abnormal hemoglobins are metabolized rapidly in individuals with congenital hemoglobinopathies (see Chapter 27). The rate of protein degradation is decreased during hypertrophy in exercised skeletal muscle and increased during atrophy in denervated or unused skeletal muscle. In addition, the rate of protein degradation is a factor in the determination of organ size (eg, the rate of degradation of liver protein is markedly reduced during the compensatory hypertrophy that follows partial hepatectomy).

Table 17–3. Purine- and pyrimidine-containing compounds.

Purine or pyrimidine + Ribose or 2-deoxyribose = Nucleoside
Nucleoside + Phosphoric acid residue = Nucleotide (mononucleotide)
Many nucleotides forming double helical structure of 2 polynucleotide chains = Nucleic acid
Nucleic acid + 1⁺ Simple basic protein = Nucleoprotein
Ribonucleic acids (RNA) contain ribose
Deoxyribonucleic acids (DNA) contain 2-deoxyribose

Uric Acid

Uric acid is formed by the breakdown of purines and by direct synthesis from 5-phosphoribosyl pyrophosphate (5-PRPP) and glutamine (Fig 17–23). In humans, uric acid is excreted in the urine; but in other mammals, uric acid is further oxidized to allantoin before excretion. The normal blood uric acid level in humans is approximately 4 mg/dL (0.24 mmol/L). In the kidney, uric acid is filtered, reabsorbed, and secreted. Normally, 98% of the filtered uric acid is reabsorbed, and the remaining 2% makes up approximately 20% of the amount excreted. The remaining 80% comes from tubular secretion. The uric acid excretion on a purine-free diet is about 0.5 g/24 h and on a regular diet about 1 g/24 h.

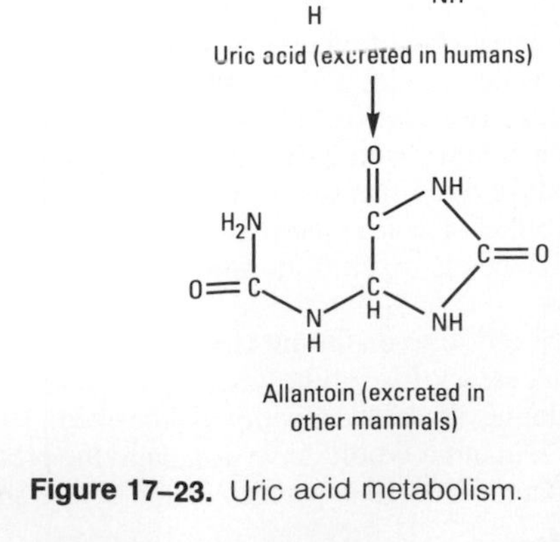

Other purines → Xanthine → (Xanthine oxidase) → Uric acid (excreted in humans) ← 5-PRPP + Glutamine

Uric acid (excreted in humans) → Allantoin (excreted in other mammals)

Figure 17–23. Uric acid metabolism.

"Primary" & "Secondary" Gout

Gout is a disease characterized by recurrent attacks of arthritis; urate deposits in the joints, kidneys, and other tissues; and elevated blood and urine uric acid levels. The joint most commonly affected initially is the metatarsophalangeal joint of the great toe. There are 2 forms of "primary" gout: one in which, because of various enzyme abnormalities, uric acid production is increased and another in which there is a selective deficit in renal tubular transport of uric acid. In "secondary" gout, the uric acid levels in the body fluids are elevated as a result of decreased excretion or increased production secondary to some other disease process. For example, excretion is decreased in patients treated with thiazide diuretics (see Chapter 38) and those with renal disease. Production is increased in leukemia and pneumonia because of increased breakdown of uric acid-rich white blood cells.

The treatment of gout is aimed at relieving the acute arthritis with drugs such as colchicine or nonsteroidal anti-inflammatory agents and decreasing the uric acid level in the blood. Colchicine does not affect uric acid metabolism, and it apparently relieves gouty attacks by inhibiting the phagocytosis of uric acid crystals by leukocytes, a process that in some way produces the joint symptoms. Phenylbutazone and probenecid decrease uric acid reabsorption in the renal tubules. Allopurinol, which inhibits xanthine oxidase (Fig 17–23), is one of the drugs used to decrease uric acid production.

Nitrogen Balance

A moderate daily protein intake is necessary to replace protein and amino acid losses. This requirement is not for protein itself but for the constituent amino acids, and the need can be met by feeding pure amino acids. Loss of protein and its derivatives in the stools is normally very small. Consequently, the amount of nitrogen in the urine is a reliable indicator of the amount of irreversible protein and amino acid breakdown. When the amount of nitrogen in the urine is equal to the nitrogen content of the protein in the diet, the individual is said to be in **nitrogen balance.** If protein intake is increased in a normal individual, the extra amino acids are deaminated and urea excretion increases, maintaining nitrogen balance. However, in conditions in which secretion of the catabolic hormones of the adrenal cortex is elevated or that of insulin is decreased, and during starvation and forced immobilization, nitrogen losses exceed intake and the nitrogen balance is **negative.** During growth or recovery from severe illnesses, or following administration of anabolic steroids such as testosterone, nitrogen intake exceeds excretion and nitrogen balance is **positive.**

When any one of the nutritionally essential amino acids necessary for synthesis of a particular protein is unavailable, the protein is not synthesized. The other amino acids that would have gone into the protein are deaminated, like other excess amino acids, and their nitrogen is excreted as urea. This is probably why nitrogen balance becomes negative whenever a single essential amino acid is omitted from the diet.

Response to Starvation

When an individual eats a diet that is low in protein but calorically adequate, excretion of urea and inorganic and ethereal sulfates declines. Uric acid excretion falls by 50%. Creatinine excretion is not affected. The creatinine and about half of the uric acid in the urine must therefore be the result of "wear and tear" processes that are unaffected by the protein intake. Total nitrogen excretion fails to fall below 3.6 g/d during protein starvation when the diet is calorically adequate because of the negative nitrogen balance produced by essential amino acid deficiencies.

On a diet that is inadequate in calories as well, urea nitrogen excretion averages about 10 g/d as proteins are catabolized for energy. Small amounts of glucose counteract this catabolism to a marked degree (**protein-sparing** effect of glucose). This protein-sparing effect is probably due for the most part to the increased insulin secretion produced by the glucose. The insulin in turn inhibits the breakdown of protein in muscle. Intravenous injection of relatively small amounts of amino acids also exerts a considerable protein-sparing effect.

Fats also spare nitrogen. During prolonged starvation, keto acids derived from fats (see below) are used by the brain and other tissues. These substances share cofactors for metabolism in muscle with 3 branched-chain amino acids, leucine, isoleucine, and valine, and to the extent that the fat-derived keto acids are utilized, these amino acids are apparently spared. Infusion of the non-nitrogen-containing analogs of these amino acids produces protein sparing and decreases urea and ammonia formation in patients with renal and hepatic failure.

Most of the protein burned during total starvation comes from the liver, spleen, and muscles and relatively little from the heart and brain. The blood glucose falls somewhat after liver glycogen is depleted (see above) but is maintained above levels that produce hypoglycemic symptoms by gluconeogenesis. Ketosis is present, and neutral fat is rapidly catabolized. When fat stores are used up, protein catabolism increases even further, and death soon follows. An average 70-kg man has 0.1 kg of glycogen in his liver, 0.4 kg of glycogen in his muscles, and 12 kg of fat. The glycogen is enough fuel for about 1 day of starvation. In hospitalized obese patients given nothing except water and vitamins, weight loss was observed to be about 1kg/d for the first 10 days. It then declined and stabilized at about 0.3 kg/d. The patients did quite well for a time, although postural hypotension and attacks of acute gouty arthritis were troublesome complications in some instances. In the Irish prisoners who starved themselves to death several years ago, the average time from the start of the fast to death was about 60 days.

There has been considerable interest in reducing diets in which the sole source of calories is liquid protein or amino acid mixtures. Such diets have been hypothesized to promote ketosis, which in turn decreases appetite while sparing body protein. However, it is as yet uncertain whether these diets are more successful than others in reducing body weight, and there have been a relatively large number of unexplained sudden deaths in patients on these diets. Until the cause of these deaths is elucidated, it seems unwise to use the diets.

FAT METABOLISM

Lipids

The biologically important lipids are the neutral fats (triglycerides), the phospholipids and related compounds, and the sterols. The triglycerides are made up of 3 fatty acids bound to glycerol (Table 17-4). Naturally occurring fatty acids contain an even number of carbon atoms. They may be saturated (no double bonds) or unsaturated (dehydrogenated, with various numbers of double bonds). The phospholipids are constituents of cell membranes. The sterols include the various steroid hormones and cholesterol.

Table 17–4. Lipids.

Typical fatty acids:

Palmitic acid: $CH_3(CH_2)_{14}—\overset{\overset{\displaystyle O}{\|}}{C}—OH$

Stearic acid: $CH_3(CH_2)_{16}—\overset{\overset{\displaystyle O}{\|}}{C}—OH$

Oleic acid: $CH_3(CH_2)_7CH{=}CH(CH_2)_7—\overset{\overset{\displaystyle O}{\|}}{C}—OH$
(Unsaturated)

Triglycerides (triacylglycerols): Esters of glycerol and 3 fatty acids.

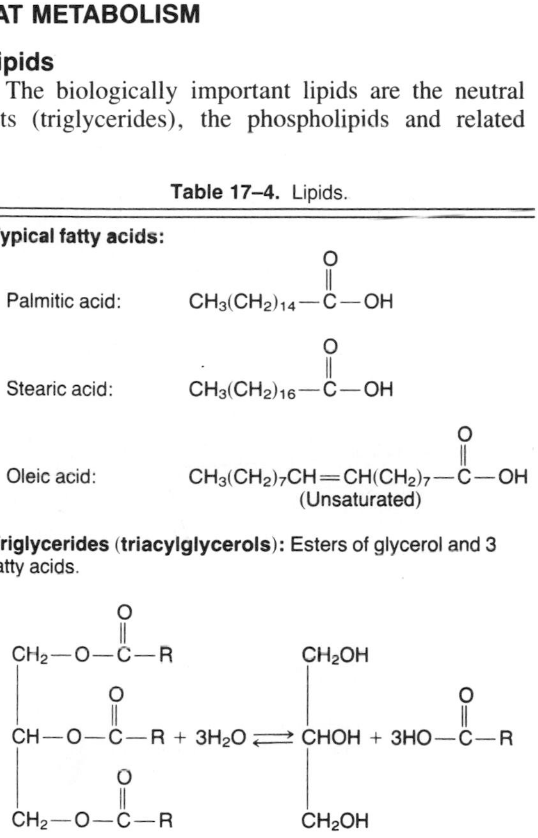

$$
\begin{array}{l}
CH_2—O—\overset{\overset{\displaystyle O}{\|}}{C}—R \\
|\\
CH—O—\overset{\overset{\displaystyle O}{\|}}{C}—R \\
|\\
CH_2—O—\overset{\overset{\displaystyle O}{\|}}{C}—R
\end{array}
+ 3H_2O \rightleftharpoons
\begin{array}{l}
CH_2OH \\
|\\
CHOH \\
|\\
CH_2OH
\end{array}
+ 3HO—\overset{\overset{\displaystyle O}{\|}}{C}—R
$$

Triglyceride   Glycerol

R = Aliphatic chain of various lengths and degrees of saturation.

Phospholipids:

A. Esters of glycerol, 2 fatty acids, and
   1. Phosphate = phosphatidic acid
   2. Phosphate plus inositol = phosphatidylinositol
   3. Phosphate plus choline = phosphatidylcholine (lecithin)
   4. Phosphate plus ethanolamine = phosphatidylethanolamine (cephalin)
   5. Phosphate plus serine = phosphatidylserine

B. Other phosphate-containing derivatives of glycerol

C. Sphingomyelins: Esters of fatty acid, phosphate, choline, and the amino alcohol sphingosine.

Cerebrosides: Compounds containing galactose, fatty acid, and sphingosine.

Sterols: Cholesterol and its derivatives, including steroid hormones, bile acids, and various vitamins.

Fatty Acid Oxidation & Synthesis

In the body, fatty acids are broken down to acetyl-CoA, which enters the citric acid cycle. Fatty acid oxidation begins with activation of the fatty acid (Fig 17–24), and this reaction occurs both inside and outside the mitochondria. Active long-chain fatty acids formed outside the mitochondria must be linked to **carnitine,** a lysine derivative, to cross the mitochondrial membrane; thus, carnitine stimulates fat oxidation. The rest of the oxidation steps occur in the mitochondria, where 2-carbon fragments are serially split off the fatty acid (**beta-oxidation**). The energy yield of this process is large. For example, catabolism of 1 mol of a 6-carbon fatty acid through the citric acid cycle to Co_2 and H_2O generates 44 mol of ATP, compared to the 38 mol generated by catabolism of 1 mol of the 6-carbon carbohydrate glucose.

Many tissues can synthesize fatty acids from acetyl-CoA. Some synthesis of long-chain fatty acids from short-chain fatty acids occurs in the mitochondria by simple reversal of the reactions shown in Fig 17–24. There is in addition a similar chain elongation system outside the mitochondria. However, most of the synthesis of fatty acids occurs de novo from acetyl-CoA via a different pathway located principally outside the mitochondria, in the microsomes. The steps in this pathway are summarized in Fig 17–25.

For unknown reasons, fatty acid synthesis stops in practically all cells when the chain is 16 carbon atoms long. Only small amounts of 12- and 14-carbon fatty acids are formed, and none with more than 16 carbons. Particularly in fat depots, the fatty acids are combined with glycerol to form neutral fats. This combination takes place in the mitochondria.

Ketone Bodies

In many tissues, acetyl-CoA units condense to form acetoacetyl-CoA (Fig 17–26). In the liver, which (unlike other tissues) contains a deacylase, free acetoacetate is formed. This β-keto acid is converted to β-hydroxybutyrate and acetone, and because these compounds are metabolized with difficulty in the liver, they diffuse into the circulation. Acetoacetate is also formed in the liver via the formation of β-hydroxy-β-methylglutaryl-CoA (Fig 17–26), and this pathway is quantitatively more important than deacylation. Acetoacetate, β-hydroxybutyrate, and acetone are called **keone bodies.** Tissues other than liver transfer CoA from succinyl-CoA to acetoacetate and metabolize the "active" acetoacetate to CO_2 and H_2O via the citric acid cycle. There are also other pathways whereby ketone bodies are metabolized.

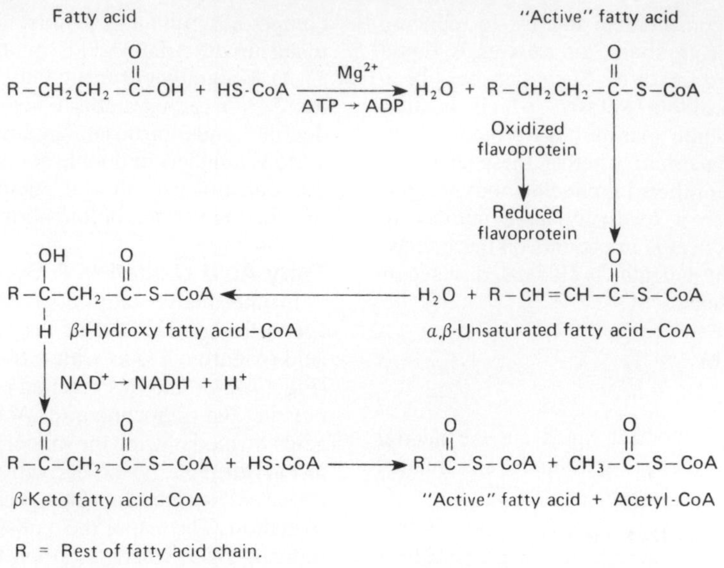

Fatty acid → "Active" fatty acid

$$R{-}CH_2CH_2{-}\overset{O}{\overset{\|}{C}}{-}OH + HS{\cdot}CoA \xrightarrow[ATP \rightarrow ADP]{Mg^{2+}} H_2O + R{-}CH_2CH_2{-}\overset{O}{\overset{\|}{C}}{-}S{-}CoA$$

Oxidized flavoprotein → Reduced flavoprotein

$$H_2O + R{-}CH{=}CH{-}\overset{O}{\overset{\|}{C}}{-}S{-}CoA \longrightarrow R{-}\underset{H}{\overset{OH}{C}}{-}CH_2{-}\overset{O}{\overset{\|}{C}}{-}S{-}CoA$$

α,β-Unsaturated fatty acid-CoA → β-Hydroxy fatty acid-CoA

$NAD^+ \rightarrow NADH + H^+$

$$R{-}\overset{O}{\overset{\|}{C}}{-}CH_2{-}\overset{O}{\overset{\|}{C}}{-}S{-}CoA + HS{\cdot}CoA \longrightarrow R{-}\overset{O}{\overset{\|}{C}}{-}S{-}CoA + CH_3{-}\overset{O}{\overset{\|}{C}}{-}S{-}CoA$$

β-Keto fatty acid-CoA → "Active" fatty acid + Acetyl-CoA

R = Rest of fatty acid chain.

Figure 17–24. Fatty acid oxidation. This process, splitting off 2 carbon fragments at a time, is repeated to the end of the chain.

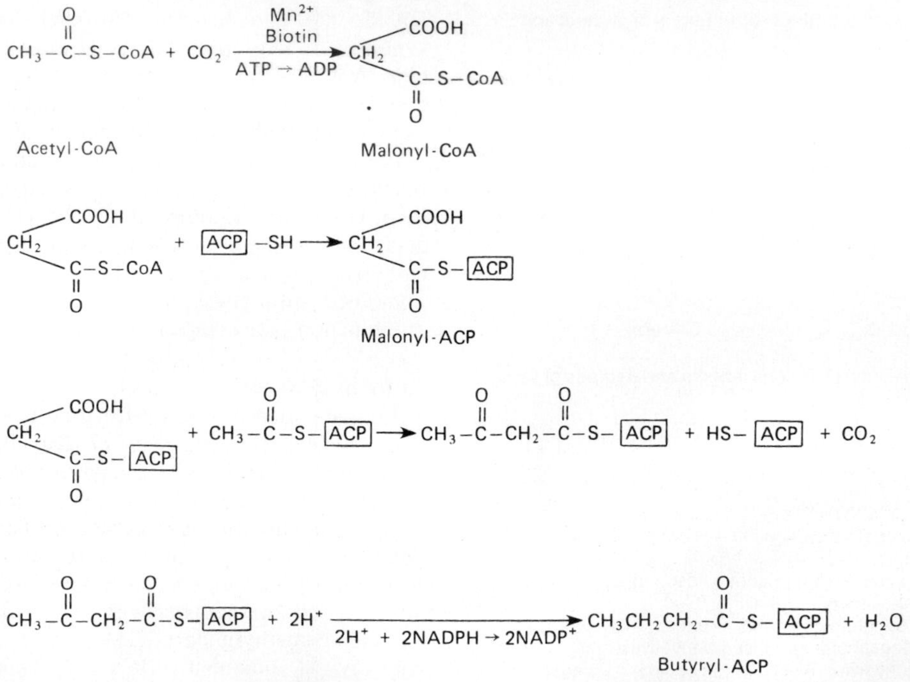

$$CH_3{-}\overset{O}{\overset{\|}{C}}{-}S{-}CoA + CO_2 \xrightarrow[ATP \rightarrow ADP]{Mn^{2+},\ Biotin} HOOC{-}CH_2{-}\overset{O}{\overset{\|}{C}}{-}S{-}CoA$$

Acetyl-CoA → Malonyl-CoA

$$HOOC{-}CH_2{-}\overset{O}{\overset{\|}{C}}{-}S{-}CoA + \boxed{ACP}{-}SH \longrightarrow HOOC{-}CH_2{-}\overset{O}{\overset{\|}{C}}{-}S{-}\boxed{ACP}$$

Malonyl-ACP

$$HOOC{-}CH_2{-}\overset{O}{\overset{\|}{C}}{-}S{-}\boxed{ACP} + CH_3{-}\overset{O}{\overset{\|}{C}}{-}S{-}\boxed{ACP} \longrightarrow CH_3{-}\overset{O}{\overset{\|}{C}}{-}CH_2{-}\overset{O}{\overset{\|}{C}}{-}S{-}\boxed{ACP} + HS{-}\boxed{ACP} + CO_2$$

$$CH_3{-}\overset{O}{\overset{\|}{C}}{-}CH_2{-}\overset{O}{\overset{\|}{C}}{-}S{-}\boxed{ACP} + 2H^+ \xrightarrow[2H^+ + 2NADPH \rightarrow 2NADP^+]{} CH_3CH_2CH_2{-}\overset{O}{\overset{\|}{C}}{-}S{-}\boxed{ACP} + H_2O$$

Butyryl-ACP

Figure 17–25. Fatty acid synthesis via pathway found in microsomes. ACP = acyl carrier protein, a protein that is part of an enzyme complex to which the acyl residues are attached as fatty acid synthesis proceeds. The process shown here repeats itself, adding 1 acetyl-ACP unit to the butyryl-ACP to form a 6-carbon fatty acid-ACP derivative, then another acetyl-ACP to form an 8-carbon unit, etc.

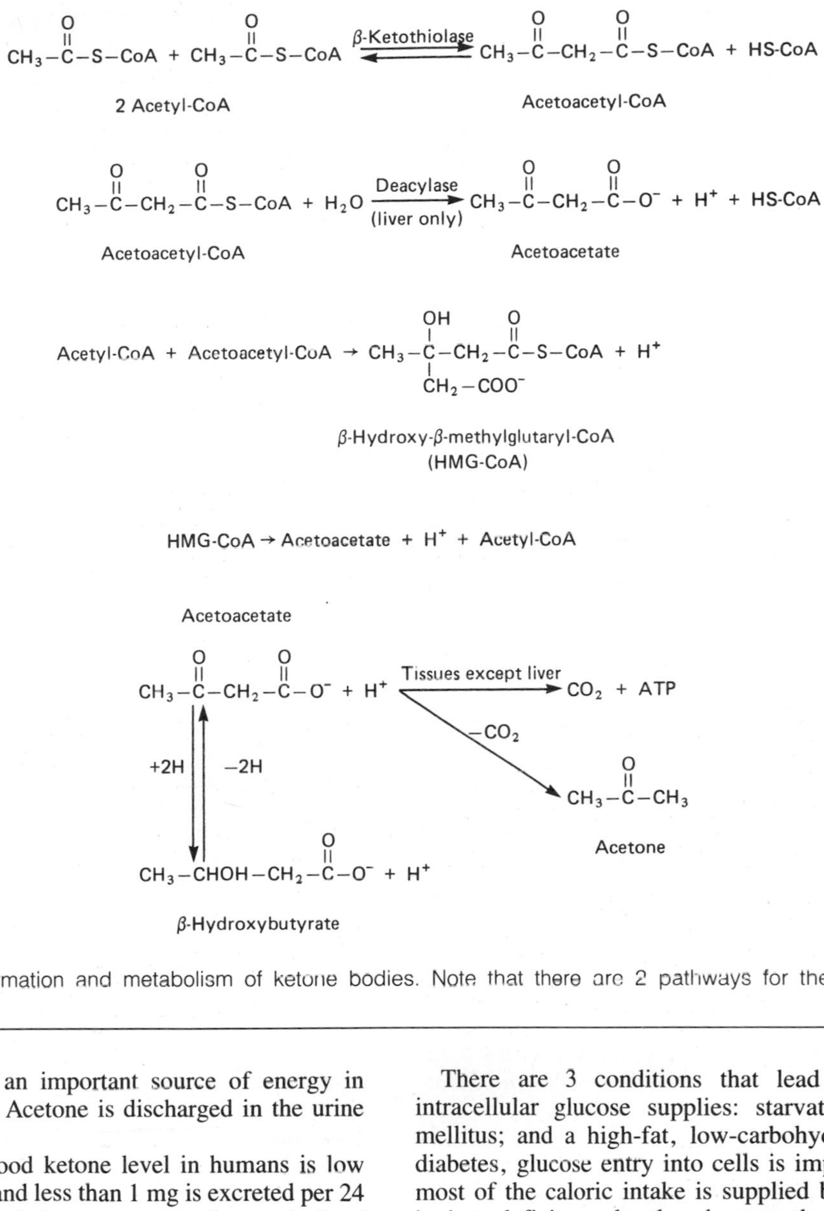

$$CH_3-\overset{O}{\overset{\|}{C}}-S-CoA + CH_3-\overset{O}{\overset{\|}{C}}-S-CoA \xrightleftharpoons{\beta\text{-Ketothiolase}} CH_3-\overset{O}{\overset{\|}{C}}-CH_2-\overset{O}{\overset{\|}{C}}-S-CoA + HS\text{-}CoA$$

2 Acetyl-CoA → Acetoacetyl-CoA

$$CH_3-\overset{O}{\overset{\|}{C}}-CH_2-\overset{O}{\overset{\|}{C}}-S-CoA + H_2O \xrightarrow[\text{(liver only)}]{\text{Deacylase}} CH_3-\overset{O}{\overset{\|}{C}}-CH_2-\overset{O}{\overset{\|}{C}}-O^- + H^+ + HS\text{-}CoA$$

Acetoacetyl-CoA → Acetoacetate

$$\text{Acetyl-CoA} + \text{Acetoacetyl-CoA} \rightarrow CH_3-\underset{CH_2-COO^-}{\overset{OH}{C}}-CH_2-\overset{O}{\overset{\|}{C}}-S-CoA + H^+$$

β-Hydroxy-β-methylglutaryl-CoA
(HMG-CoA)

$$\text{HMG-CoA} \rightarrow \text{Acetoacetate} + H^+ + \text{Acetyl-CoA}$$

Acetoacetate

$$CH_3-\overset{O}{\overset{\|}{C}}-CH_2-\overset{O}{\overset{\|}{C}}-O^- + H^+ \xrightarrow{\text{Tissues except liver}} CO_2 + ATP$$

$$CH_3-\overset{O}{\overset{\|}{C}}-CH_2-\overset{O}{\overset{\|}{C}}-O^- + H^+ \xrightarrow{-CO_2} CH_3-\overset{O}{\overset{\|}{C}}-CH_3 \quad \text{(Acetone)}$$

+2H ⇅ −2H

$$CH_3-CHOH-CH_2-\overset{O}{\overset{\|}{C}}-O^- + H^+$$

β-Hydroxybutyrate

Figure 17–26. Formation and metabolism of ketone bodies. Note that there are 2 pathways for the formation of acetoacetate.

The ketones are an important source of energy in some conditions. Acetone is discharged in the urine and expired air.

The normal blood ketone level in humans is low (about 1 mg/dL) and less than 1 mg is excreted per 24 hours, because the ketones are normally metabolized as rapidly as they are formed. However, if the entry of acetyl-CoA into the citric acid cycle is depressed because of a decreased supply of the products of glucose metabolism, or if the entry does not increase when the supply of acetyl-CoA increases, acetyl-CoA accumulates, the rate of condensation to acetoacetyl-CoA increases, and more acetoacetate is formed in the liver. The ability of the tissues to oxidize the ketones is soon exceeded, and they accumulate in the bloodstream (**ketosis**). Two of the 3 ketone bodies, acetoacetate and β-hydroxybutyrate, are anions of the moderately strong acids acetoacetic acid and β-hydroxybutyric acid, and they are buffered, reducing the decline in pH that would otherwise occur. However, the buffering capacity can be exceeded, and the metabolic acidosis that develops in conditions such as diabetic ketosis can be severe and even fatal.

There are 3 conditions that lead to deficient intracellular glucose supplies: starvation; diabetes mellitus; and a high-fat, low-carbohydrate diet. In diabetes, glucose entry into cells is impaired. When most of the caloric intake is supplied by fat, carbohydrate deficiency develops because there is no major pathway for converting fat to carbohydrate. The liver cells also become filled with fat, which damages them and displaces any glycogen that is formed. In all of these conditions, ketosis develops primarily because the supply of ketones is overabundant.

The acetone odor on the breath of children who have been vomiting is due to the ketosis of starvation. Parenteral administration of relatively small amounts of glucose abolishes the ketosis, and it is for this reason that carbohydrate is said to be **antiketogenic.**

Cellular Lipids

The lipids in cells are of 2 main types: **structural lipids,** which are an inherent part of the membranes and other parts of cells; and **neutral fat,** stored in the adipose cells of the fat depots. Depot fat is mobilized during starvation, but structural lipid is preserved.

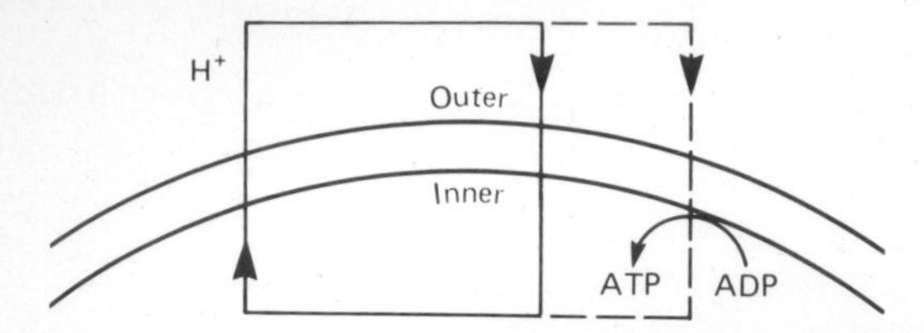

Figure 17–27. Proton transport across mitochondrial membrane in brown fat. Protons are transported outward by the electron transport system, as in other mitochondria, but in addition to the inward proton movement that generates ATP, there is an inward proton "leak" that does not generate ATP. Consequently; metabolism of fat and generation of ATP are uncoupled.

The size of the fat depots obviously varies, but in nonobese individuals, the depots make up about 10% of body weight. They are not the inert lumps they were once thought to be, but active dynamic tissues undergoing continuous breakdown and resynthesis. In the depots, glucose is metabolized to fatty acids, and neutral fats are synthesized. Neutral fat is also broken down, and free fatty acids are released into the circulation.

Brown Fat

A third, special type of lipid is **brown fat,** which makes up a small percentage of total body fat. Brown fat, which is somewhat more abundant in infants but is present in adult animals and humans as well, is located between the scapulas, at the nape of the neck, along the great vessels in the thorax and abdomen, and in other scattered locations in the body. In the depots made up of white fat, only the blood vessels have a sympathetic innervation, but in brown fat depots, the fat cells themselves as well as the blood vessels are extensively innervated with sympathetic nerve fibers. In addition, ordinary lipocytes have only a single large droplet of white fat, whereas brown fat cells contain several small droplets of fat. Brown fat cells also contain many mitochondria. In these mitochondria, there is the usual inward proton conductance that generates ATP (oxidative phosphorylation; see above), but there is in addition a second proton conductance that does not generate ATP. This "short-circuit" conductance, which is associated with a polypeptide of molecular weight 32,000 in the membrane, causes uncoupling of metabolism and generation of ATP, so that more heat is produced (Fig 17–27). Stimulation of the sympathetic innervation to brown fat releases norepinephrine, which acts via β-adrenergic receptors to increase lipolysis, and increased fatty acid oxidation in the mitochondria increases heat production. Thus, variations in the activity in nerves to brown fat produce variations in the efficiency with which food is utilized and energy produced; ie, they provide a mechanism for varying the weight gained per unit of food ingested.

There is evidence that brown fat functions in this way in 2 situations: (1) In animals and presumably in humans adapted to cold, the rate of heat production in brown fat is increased, and there is in addition a marked increase in blood flow. (2) Nerve discharge to brown fat is also increased after eating, so that heat production is increased. Note that there are 2 components to the heat production after eating: the prompt specific dynamic action (SDA; see above) due to assimilation of foodstuff, and a second, somewhat slower increase in heat produced by brown fat. The possible role of abnormalities in brown fat metabolism in the genesis of obesity is discussed below, and the relation of brown fat to food intake is discussed in Chapter 14.

Plasma Lipids & Lipid Transport

The major lipids in plasma do not circulate in the free form. **Free fatty acids** (variously called FFA, UFA, or NEFA) are bound to albumin, whereas cholesterol, triglycerides, and phospholipids are transported in the form of lipoprotein complexes. There are 6 families of lipoproteins (Table 17–5), which are graded in size and lipid content. The density of these lipoproteins (and consequently the speed at which

Table 17–5. The principal lipoproteins. The plasma lipids include these components, free fatty acids from adipose tissue, which circulate bound to albumin, and chylomicron remnants.

	Size (nm)	Composition (%)					Origin
		Protein	Free Cholesterol	Cholesteryl Esters	Triglyceride	Phospholipid	
Chylomicrons	75–1000	2	2	3	90	3	Intestine
Chylomicron remnants	30–80	...	...	...	...	...	Capillaries
Very low density lipoproteins (VLDL)	30–80	8	4	16	55	17	Liver and intestine
Intermediate-density lipoproteins (IDL)	25–40	10	5	25	40	20	VLDL
Low-density lipoproteins (LDL)	20	20	7	46	6	21	IDL
High-density lipoproteins (HDL)	7.5–10	50	4	16	5	25	Liver and intestine

they sediment in the ultracentrifuge) is inversely proportionate to their lipid content. In general, the lipoproteins consist of a hydrophobic core of triglycerides and cholesteryl esters surrounded by phospholipids and protein (Fig 17–28). Their interrelations and their roles in lipid transport are summarized in Fig 17–29.

The protein constituents of the lipoproteins are called **apoproteins.** The major apoproteins are called E, C, and B (Fig 17–29). There are 2 forms of apoprotein B, a low-molecular-weight form called B-48, which is characteristic of the exogenous system that transports exogenous ingested lipids, and a high-molecular-weight form called B-100, which is characteristic of the endogenous system that transports lipids that come from the liver.

Chylomicrons are formed in the intestinal mucosa during the absorption of the products of fat digestion (see Chapter 25). They are very large lipoprotein complexes that enter the circulation via the lymphatic ducts. After meals, there are so many of these particles in the blood that the plasma may have a milky appearance (**lipemia**). The chylomicrons are cleared from the circulation by the action of **lipoprotein lipase,** which is located on the surface of the endothelium of the capillaries. This enzyme catalyzes the breakdown of the triglyceride in the chylomicrons to FFA and glycerol, which then enter

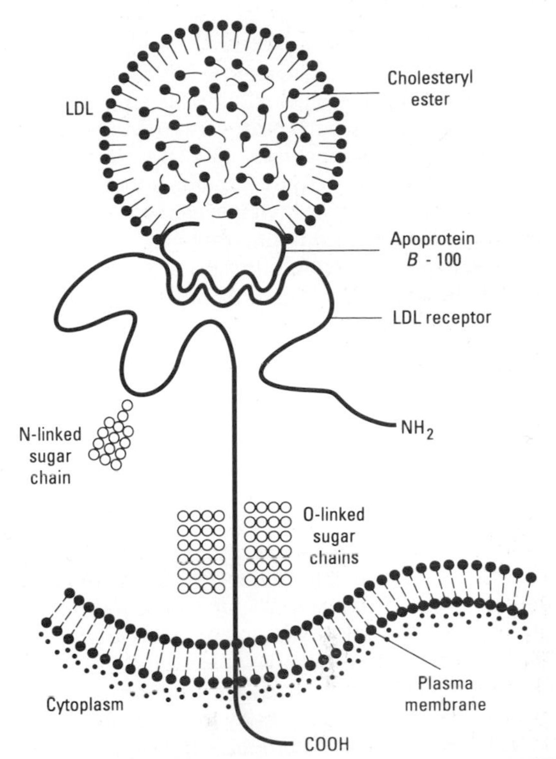

Figure 17–28. Diagrammatic representation of the structure of LDL, the LDL receptor, and the binding of the LDL to the receptor via apoprotein B-100 (Courtesy of National Institutes of Health).

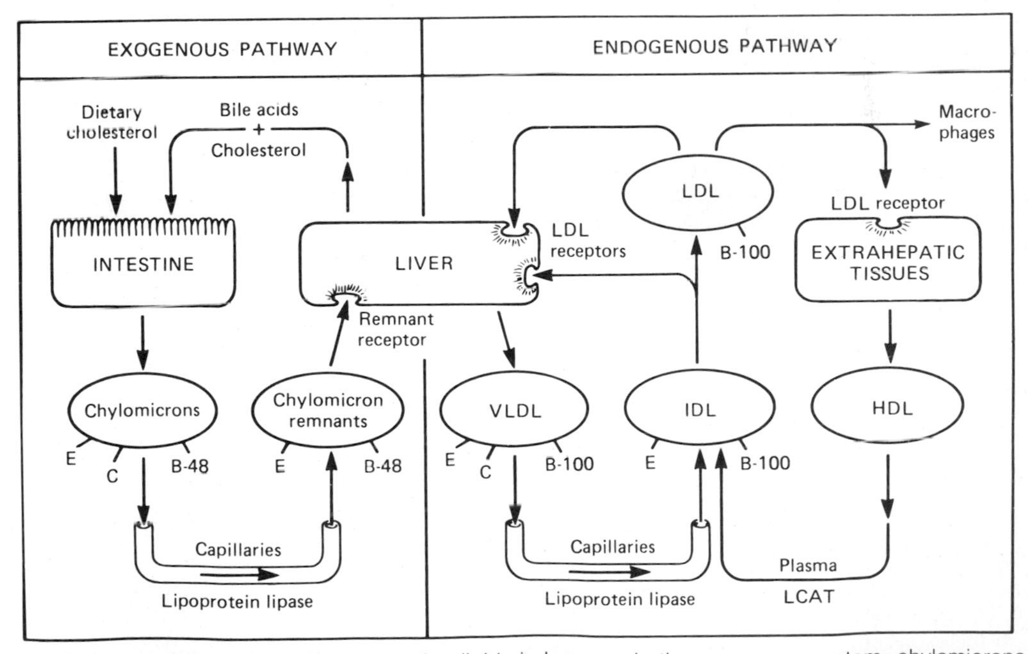

Figure 17–29. Lipoprotein systems for transporting lipids in humans. In the exogenous system, chylomicrons rich in triglycerides of dietary origin are converted to chylomicron remnants rich in cholesteryl esters by the action of lipoprotein lipase. In the endogenous system, VLDL rich in triglycerides are secreted by the liver and converted to IDL, then to LDL rich in cholesteryl esters. LCAT, lechithin-cholesterol acyltransferase. The letters on chylomicrons, chylomicron remnants, VLDL, IDL, and LDL identify the principal apoproteins found in them. One-third of the LDL is taken up by macrophages and other cells by alternative, receptor-independent mechanisms. (Slightly modified and reproduced, with permission, from Goldstein JL, Kita T, Brown MS: Defective lipoprotein receptors and atherosclerosis. *N Engl J Med* 1983;**309**:288.)

adipose cells and are reesterified. Alternatively, the FFA remain in the circulation bound to albumin. Lipoprotein lipase, which required heparin as a cofactor, also removes triglycerides from circulating **very low density lipoproteins (VLDL;** see below). Chylomicrons and VLDL contain apoprotein C, a complex of proteins, which leaves them in the capillaries. One component of the complex, apolipoprotein C II, activates lipoprotein lipase.

Chylomicrons depleted of their triglyceride remain in the circulation as cholesterol-rich lipoproteins called **chylomicron remnants,** which are 30–80 nm in diameter. The remnants are carried to the liver, where they bind to receptors, are immediately internalized by receptor-mediated endocytosis (see Chapter 1), and are degraded in lysosomes.

The chylomicrons and their remnants constitute a transport system for ingested exogenous lipids (Fig 17–29). There is also an endogenous system made up of VLDL, **intermediate-density lipoproteins (IDL), low-density lipoproteins (LDL),** and **high-density lipoproteins (HDL),** which transports triglycerides and cholesterol throughout the body. VLDL are formed in the liver and transport triglycerides formed from fatty acids and carbohydrates in the liver to extrahepatic tissues. After their triglyceride is largely removed by the action of lipoprotein lipase, they become IDL. The IDL give up phospholipids and, through the action of the plasma enzyme lecithin-cholesterol acyltransferase (LCAT; Fig 17–29), pick up cholesteryl esters formed from cholesterol in the HDL. Some IDL are taken up by the liver. The remaining IDL then lose more triglyceride and protein, probably in the sinusoids of the liver, and become LDL. During this conversion, they lose apoprotein E, but apoprotein B-100 remains.

LDL provide cholesterol to the tissues. The cholesterol is an essential constituent in cell membranes and is used by gland cells to make steroid hormones. In the liver and most extrahepatic tissues, LDL are taken up by receptor-mediated endocytosis in coated pits (see Chapter 1). The receptors recognize the apoprotein B-100 component of the LDL (Fig 17–28).

The structure of the human LDL receptor is now known. It is a large, complex molecule made up of a cysteine-rich region of 292 amino acid residues that binds LDL; a region of about 400 amino acid residues that is homologous to the precursor for epidermal growth factor; a 58-amino-acid region that is rich in serine and threonine and is the site of glycosylation; a stretch of 22 hydrophobic amino acid residues that spans the cell membrane; and a portion of 50 amino acid residues that projects into the cytoplasm (Fig 17–28). The gene for this protein contains 18 exons, and 13 of the exons encode protein sequences homologous to sequences in other proteins. Thus, it appears that the LDL receptor is a mosaic protein formed by exons which code for parts of other proteins.

In the process of receptor-mediated endocytosis, each coated pit is pinched off to form a coated vesicle and then an endosome. Protein pumps in the membranes of the endosomes lower the pH in this organelle, and this triggers release of the LDL receptors, which recycle to the cell membrane (Fig 17–30). The endosome then fuses with a lysosome, where cholesterol formed from the cholesteryl esters by the acid lipase in the lysosomes becomes available to meet the cell's needs (Fig 17–30). The cholesterol in the cells also inhibits intracellular synthesis of cholesterol by inhibiting HMG-CoA reductase (see below), stimulates esterification of any excess cholesterol that is released, and inhibits synthesis of new LDL receptors. All of these reactions provide feedback control of the amount of cholesterol in the cell.

LDL are also taken up by a lower-affinity system in the macrophages and some other cells. When overloaded by a high plasma level of LDL, the macrophages become full of cholesteryl esters and make up the "foam cells" frequently seen in atherosclerotic lesions.

It is apparent that in the steady state, cholesterol leaves as well as enters cells. The cholesterol leaving the cells is absorbed into HDL, lipoproteins that are synthesized in the liver and the intestine. The primary function of the HDL is in cholesterol exchange and esterification; as noted above, HDL provides through LCAT the cholesteryl esters that are transferred to IDL and thence back again to LDL.

Free Fatty Acid Metabolism

Free fatty acids (FFA) are provided to fat cells and other tissues by chylomicrons and VLDL (see above). They are also synthesized in the fat depots in which they are stored. They circulate bound to albumin and are a major source of energy for many organs. They are used extensively in the heart, but probably all tissues, including the brain, can oxidize FFA to Co_2 and H_2O.

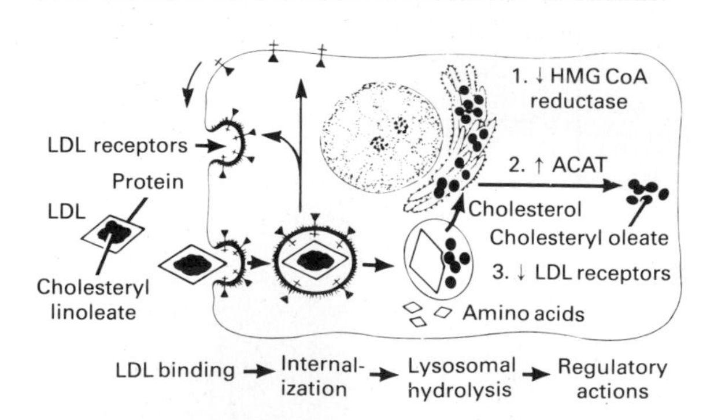

Figure 17–30. Cellular uptake and metabolism of cholesterol. LDL bind to receptors and are internalized by receptor-mediated endocytosis. Receptors are freed and recycle to the membrane. The cholesteryl esters enter lysosomes, where free cholesterol is released and is used for cellular processes. Cholesterol also (1) inhibits HMG-CoA reductase, (2) is processed in part to other cholesteryl esters by the enzyme acetyl-CoA: cholesterol acyltransferase (ACAT), and (3) inhibits the formation of LDL receptors. (Courtesy of MS Brown.)

The supply of FFA to the tissues is regulated by 2 lipases. As noted above, lipoprotein lipase on the surface of the endothelium of the capillaries hydrolyzes the triglycerides in chylomicrons and VLDL, providing FFA and glycerol, which are reassembled into new triglycerides in the fat cells. The intracellular **hormone-sensitive lipase** of adipose tissue catalyzes the breakdown of stored triglycerides into glycerol and fatty acids, with the latter entering the circulation.

The hormone-sensitive lipase is converted from an inactive to an active form by cyclic AMP via protein kinase A (Fig 17–31). The adenylate cyclase in adipose cells is in turn activated by glucagon. It is also activated by the catecholamines norepinephrine and epinephrine via a β_1-adrenergic receptor. ACTH, TSH, LH, serotonin, and vasopressin increase lipolysis via cyclic AMP, but the physiologic role of these substances in the regulation of lipolysis is uncertain. Growth hormone, glucocorticoids, and thyroid hormone also increase the activity of the hormone-sensitive lipase, but they do it by a slower process that requires synthesis of new protein. Growth hormone appears to produce a protein that increases the ability of catecholamines to activate cyclic AMP, whereas cortisol produces a protein that increases the action of cyclic AMP. On the other hand, insulin and prostaglandin E decrease the activity of the hormone-sensitive lipase, possibly by inhibiting the formation of cyclic AMP.

Given the hormonal effects described in the preceding paragraph, it is not surprising that the activity of the hormone-sensitive lipase is increased by fasting and stress and decreased by feeding and insulin. Conversely, feeding increases and fasting and stress decrease the activity of lipoprotein lipase.

Cholesterol Metabolism

Cholesterol is the precursor of the steroid hormones and bile acids and is an essential constituent of cell membranes (see Chapter 1). It is found only in animals. Related sterols occur in plants, but plant sterols are not normally absorbed from the gastrointestinal tract. Most of the dietary cholesterol is contained in egg yolks and animal fat.

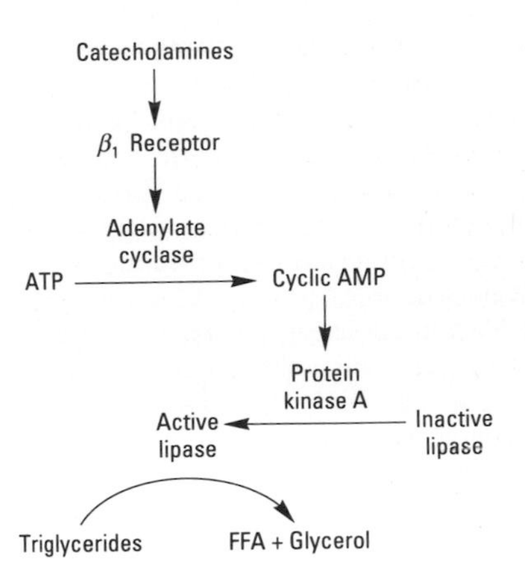

Figure 17–31. Mechanism by which catecholamines increase the activity of the hormone-sensitive lipase in adipose tissue.

Cholesterol is absorbed from the intestine and incorporated into the chylomicrons formed in the mucosa. After the chylomicrons discharge their triglyceride in adipose tissue, the chylomicron remnants bring cholesterol to the liver. The liver and other tissues also synthesize cholesterol. Some of the cholesterol in the liver is excreted in the bile, both in the free form and as bile acids. Some of the biliary cholesterol is reabsorbed from the intestine. Most of the cholesterol in the liver is incorporated into VLDL, and all of it circulates in lipoprotein complexes (see above).

The biosynthesis of cholesterol from acetate is summarized in Fig 17–32. Cholesterol feeds back to

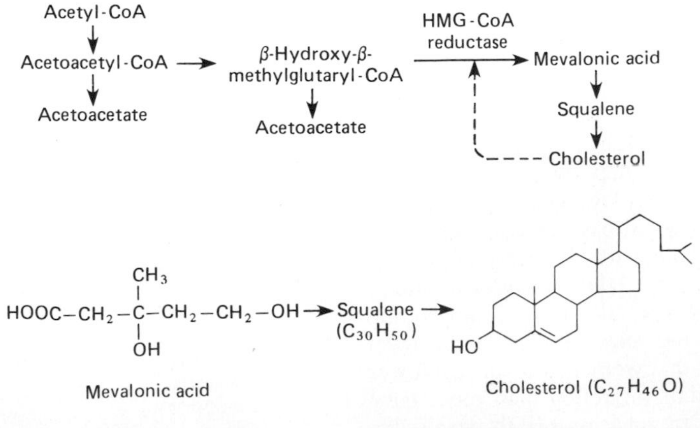

Figure 17–32. Biosynthesis of cholesterol. Six mevalonic acid molecules condense to form squalene, which is then hydroxylated and converted to cholesterol. The dashed arrow indicates feedback inhibition by cholesterol of HMG-CoA reductase, the enzyme that catalyzes mevalonic acid formation.

inhibit its own synthesis by inhibiting **HMG-CoA reductase,** the enzyme that converts β-hydroxy-β-methylglutaryl-CoA to mevalonic acid. Thus, when dietary cholesterol intake is high, hepatic cholesterol synthesis is decreased, and vice versa. However, the feedback compensation is incomplete, because a diet that is low in cholesterol and saturated fats leads to a modest decline in circulating blood cholesterol.

The plasma cholesterol level is decreased by thyroid hormones, which increase LDL receptors, and by estrogens, which lower LDL and increase HDL. Plasma cholesterol is elevated by biliary obstruction and in untreated diabetes mellitus. If bile acid reabsorption in the intestine is decreased by resins such as cholestipol, more cholesterol is diverted to bile acid formation. However, the drop in plasma cholesterol is relatively small because there is a compensatory increase in cholesterol synthesis. Big doses of the vitamin niacin decrease LDL and increase HDL. Compactin, mevinolin, and their derivatives inhibit HMG-CoA reductase and show considerable promise for clinical use in lowering plasma cholesterol.

Relation to Atherosclerosis

A subject of great interest is the role of cholesterol in the etiology and course of **atherosclerosis.** This extremely widespread disease predisposes to myocardial infarction, cerebral thrombosis, and other serious illnesses. It is characterized by infiltration of cholesterol and appearance of foam cells in certain lesions of the arterial walls, distorting the vessels and making them rigid. In individuals with elevated plasma cholesterol levels, there is an increased incidence of atherosclerosis and its complications. Of the cholesterol in the body, 93% is in cells and only 7% in the plasma, but it is the increases in the 7% in plasma that predispose to atherosclerosis.

Familial hypercholesterolemia can be produced by a variety of mutations in the LDL receptor. In individuals with this condition and in experimental animals, lowering plasma cholesterol slows the progression of atherosclerosis. In addition, it now seems clear that in humans with plasma cholesterol levels in the upper part of the normal range, lowering plasma cholesterol by diet, supplemented, if necessary, with drugs, decreases the incidence of atherosclerosis, myocardial infarctions, and strokes. The plasma concentrations of LDL and HDL as well as of cholesterol need to be considered when evaluating patients. Individuals with elevated LDL have a higher than normal incidence of the disease and its complications, whereas individuals with elevated HDL have a lower incidence. Thus, the HDL cholesterol ratio is probably a better index of the risk of developing atherosclerosis than the plasma cholesterol level alone. It is interesting that women, who have a lower incidence of myocardial infarction than men, have higher HDL levels. In addition, HDL levels are increased in individuals who exercise and those who drink 1–2 alcoholic drinks per day, whereas they are decreased in individuals who smoke, are obese, or live sedentary lives. Exercise and moderate drinking decrease the incidence of myocardial infarction, and obesity and smoking are known risk factors that increase it. There is evidence that elevated levels of IDL and chylomicron remnants also predispose to atherosclerosis, whereas elevated chylomicron and VLDL levels do not.

Essential Fatty Acids

Animals fed a fat-free diet fail to grow, develop skin and kidney lesions, and become infertile. Adding linolenic acid to the diet restores growth to normal, and linoleic and arachidonic acids cure all the deficiency symptoms. These 3 acids are polyunsaturated fatty acids and because of their action are called **essential fatty acids.** Similar deficiency symptoms have not been unequivocally demonstrated in humans, but there is reason to believe that some unsaturated fats are essential dietary constituents, especially in children. Dehydrogenation of fats is known to occur in the body, but there does not appear to be any synthesis of carbon chains with the arrangement of double bonds found in the essential fatty acids.

Prostaglandins

One of the reasons, and possibly the only reason, that essential fatty acids are necessary for health is that they are the precursors of prostaglandins and related compounds (Fig 17–33). The **prostaglandins** are a series of 20-carbon unsaturated fatty acids containing a cyclopentane ring. They were first isolated from semen but have now been shown to be synthesized in most and possibly in all organs in the body. The structures of some of them are shown in Fig 17–34. The prostaglandins are divided into groups—PGE and PGF, for example—on the basis of

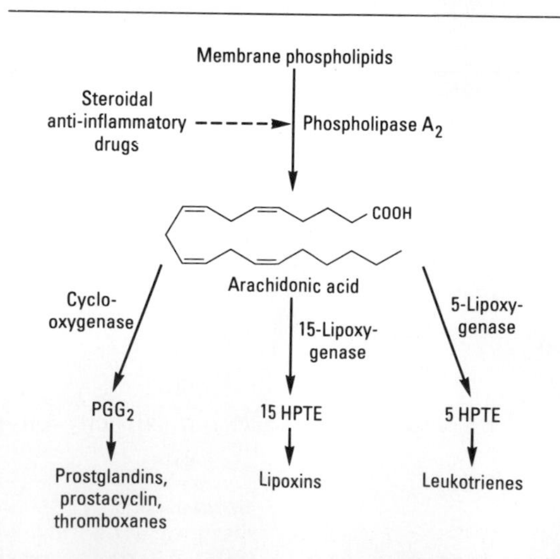

Figure 17–33. Principal biologically active substances produced from arachidonic acid.

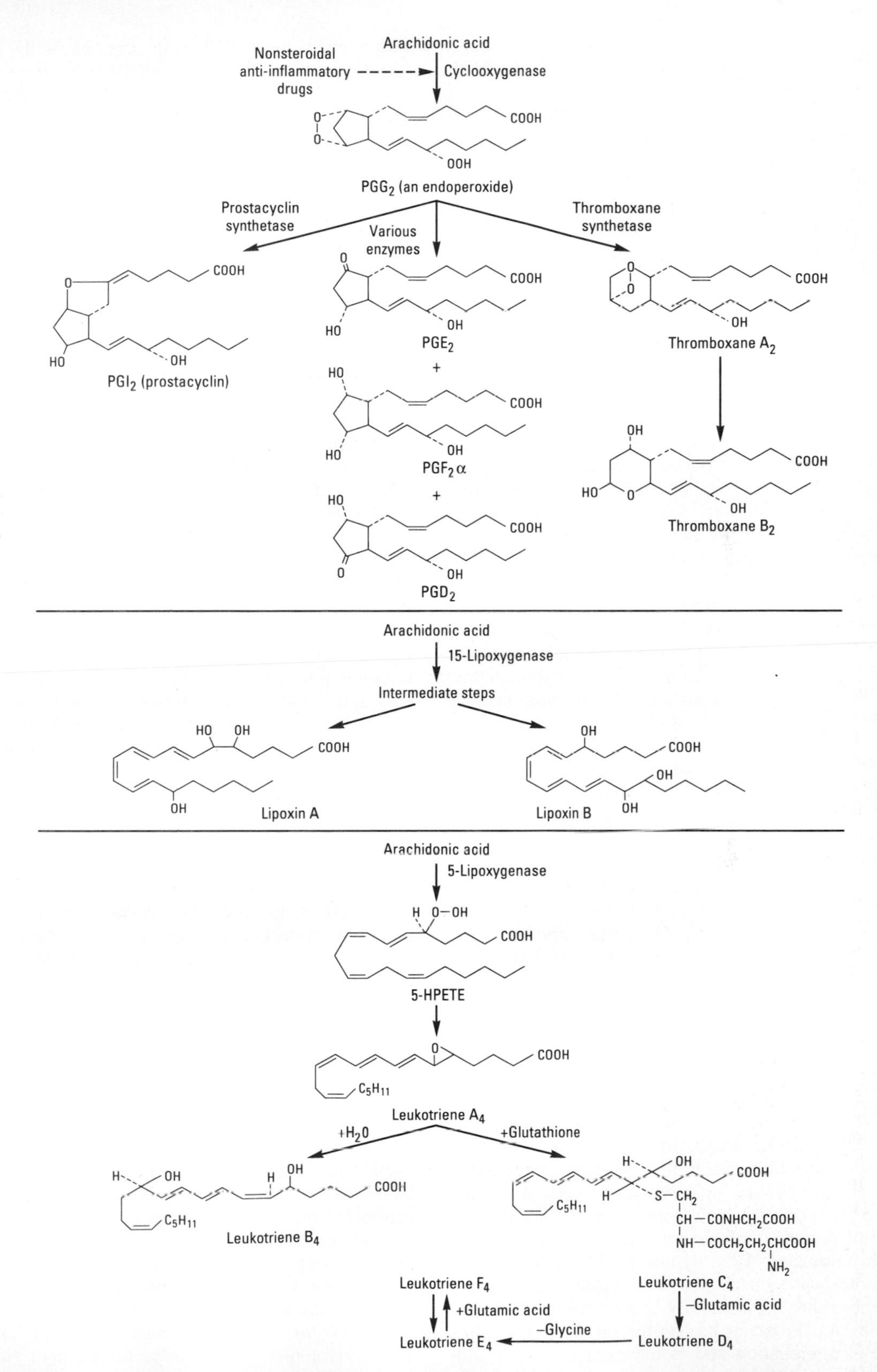

Figure 17–34. Biosynthesis of prostaglandins and thromboxanes (*Top*), lipoxins (*Middle*), and leukotrienes (*Bottom*), from arachidonic acid. Leukotriene C_4 (LTC_4) is converted to leukotriene D_4 by removal of glutamic acid, and leukotriene D_4 is converted to leukotriene E_4 by further removal of glycine. Leukotriene E_4 can take up glutamic acid to form leukotriene F_4. 5-HPETE, 5-hydroperoxyeicosate-traenoic acid.

the configuration of the cyclopentane ring. The number of double bonds in the side chains is indicated by subscript numbers; for example, the E series prostaglandin shown in Fig 17–34 is PGE_2.

Prostaglandins are synthesized via endoperoxides from arachidonic acid and other essential fatty acids. These unsaturated fatty acids become incorporated into the phospholipids of cell membranes and can be released by the action of phospholipase A_2. In humans, the main precursor of prostaglandins is arachidonic acid (Fig 17–34). This substance is converted to the endoperoxide PGG_2 and then, via 3 different pathways, to prostacyclin (PGI_2); prostaglandins PGE_2, PGF_2, and PGD_2; and thromboxanes A_2 and B_2. Arachidonic acid is also converted to 5-hydroperoxy-eicosatetraenoic acid (5-HPETE), which is then converted to the **leukotrienes.** Four of the leukotrienes are **aminolipids** that contain amino acids; leukotriene C_4 (LTC_4) contains the tripeptide glutathione, LTD_4 contains glycine and cysteine, LTE_4 contains cysteine, and LTF_4 contains cysteine and glutamic acid. In addition, arachidonic acid is converted to **lipoxins** (Figs 17–33 and 17–34) and a few other compounds.

Anti-inflammatory steroids such as cortisol inhibit the release of arachidonic acid from phospholipid stores and thus block formation of leukotrienes as well as prostacyclin, PGE_2 and related prostaglandins, and thromboxanes. Nonsteroidal anti-inflammatory drugs such as aspirin and indomethacin inhibit cyclooxygenase, leaving the leukotriene pathway intact. Aspirin irreversibly inactivates cyclooxygenase by acetylating it, and the effect of aspirin only wears off in 5–7 days as new enzyme is synthesized by the cells. Much effort is being devoted to a search for drugs that will produce even more selective alterations by acting farther along the biosynthetic pathways.

The leukotrienes, thromboxanes, lipoxins, and prostaglandins have been called local hormones. They have short half-lives and are inactivated in many different tissues. They undoubtedly act mainly in the tissues and sites in which they are produced.

The leukotrienes are probably mediators of allergic responses and inflammation. Their release is provoked when specific allergens combine with IgE antibodies on the surfaces of mast cells. They produce bronchoconstriction, constrict arterioles, increase vascular permeability, and attract neutrophils and eosinophils to inflammatory sites. They are found in relatively high concentrations in the joint fluid of patients with rheumatoid arthritis.

Thromboxane A_2 is synthesized by platelets and promotes vasoconstriction and platelet aggregation; thromboxane B_2 is mainly a metabolite of thromboxane A_2. Prostacyclin, on the other hand, inhibits platelet aggregation and is a vasodilator. It is produced by endothelial and smooth muscle cells in the walls of blood vessels. Release of thromboxane A_2 by platelets at the immediate site of injury to a blood vessel promotes clot formation, whereas prostacyclin formation in the neighboring portions of the blood vessel wall keeps the clot localized and maintains the patency of the rest of the vessel. Obviously, the effects of inhibiting cyclooxygenase in the platelets and the vessel walls would cancel each other. However, small doses of aspirin inhibit platelet cyclooxygenase without inhibiting vessel wall cyclooxygenase, and doses in the range of 300–600 mg/d have been used with some success to prevent formation of clots in cerebral and myocardial vessels.

Lipoxin A dilates the microvasculature, and lipoxin A and lipoxin B both inhibit the cytotoxic effects of natural killer cells (see Chapter 27). However, their physiologic role is still uncertain.

The prostaglandins have attracted great attention because they have profound effects when administered in very small doses. In addition, they have a bewildering array of actions. Prostaglandins may play a role in regulating the capacity of red blood cells to undergo deformation in passing through capillaries. Some of them decrease gastric acid secretion and inhibit the formation of peptic ulcers in experimental animals. They cause luteolysis and may play a role in the regulation of the female reproductive cycle (see Chapter 23). They also have the capacity to modify pituitary responses to hypothalamic hormones. They induce abortion when injected intra-amniotically in midtrimester pregnancy, and they induce labor when injected near term. They can mimic the effects of TSH, ACTH, and other hormones. They stimulate renin secretion. Some of them are antilipolytic and may play a role along with glucagon, catecholamines, and other hormones in the regulation of FFA release. They appear to be released from inflamed tissue and cause fever when injected into the third ventricle (see Chapter 14). They are found in the eye and produce miosis. They exist in the brain and may alter the release or effects of neurotransmitters.

It is difficult to find a common theme in these multifarious actions, but it may be significant that cyclic AMP is involved in most if not all of them. It may be that the various prostaglandins act to adjust the generation of cyclic AMP in response to various stimuli.

Obesity

Obesity in humans is a health problem of appreciable magnitude, since it is associated with an increased incidence of various diseases, including cardiovascular and gallbladder disease and diabetes. In addition, obese individuals have an increased mortality rate. According to a commonly used definition, obesity is said to be present when more than 20% of body weight is due to fat in men and more than 25% in women. Normal values of fat are 12–18% for men and 18–24% for women. Normal height and weight tables are also used extensively, but a value that correlates better to body fat is the **Quetelet index,** which is the body weight divided by the height squared. The normal value for this index is

Table 17–6. Recommended daily dietary allowances.[1] (Revised 1980.) Designed for the maintenance of good nutrition of practically all healthy people in the USA.

	Age (years)	Weight (kg)	Weight (lb)	Height (cm)	Height (in)	Protein (g)	Fat-Soluble Vitamins: Vitamin A (μg RE)[2]	Vitamin D (μg)[3]	Vitamin E (mg α-TE)[4]	Water-Soluble Vitamins: Vitamin C (mg)	Thiamine (mg)	Riboflavin (mg)	Niacin (mg NE)[5]	Vitamin B_6 (mg)	Folacin[6] (μg)	Vitamin B_{12} (μg)	Minerals: Calcium (mg)	Phosphorus (mg)	Magnesium (mg)	Iron (mg)	Zinc (mg)	Iodine (μg)
Infants	0.0–0.5	6	13	60	24	kg × 2.2	420	10	3	35	0.3	0.4	6	0.3	30	0.5[7]	360	240	50	10	3	40
	0.5–1.0	9	20	71	28	kg × 2.0	400	10	4	35	0.5	0.6	8	0.6	45	1.5	540	360	70	15	5	50
Children	1–3	13	29	90	35	23	400	10	5	45	0.7	0.8	9	0.9	100	2.0	800	800	150	15	10	70
	4–6	20	44	112	44	30	500	10	6	45	0.9	1.0	11	1.3	200	2.5	800	800	200	10	10	90
	7–10	28	62	132	52	34	700	10	7	45	1.2	1.4	16	1.6	300	3.0	800	800	250	10	10	120
Males	11–14	45	99	157	62	45	1000	10	8	50	1.4	1.6	18	1.8	400	3.0	1200	1200	350	18	15	150
	15–18	66	145	176	69	56	1000	10	10	60	1.4	1.7	18	2.0	400	3.0	1200	1200	400	18	15	150
	19–22	70	154	177	70	56	1000	7.5	10	60	1.5	1.7	19	2.2	400	3.0	800	800	350	10	15	150
	23–50	70	154	178	70	56	1000	5	10	60	1.4	1.6	18	2.2	400	3.0	800	800	350	10	15	150
	51+	70	154	178	70	56	1000	5	10	60	1.2	1.4	16	2.2	400	3.0	800	800	350	10	15	150
Females	11–14	46	101	157	62	46	800	10	8	50	1.1	1.3	15	1.8	400	3.0	1200	1200	300	18	15	150
	15–18	55	120	163	64	46	800	10	8	60	1.1	1.3	14	2.0	400	3.0	1200	1200	300	18	15	150
	19–22	55	120	163	64	44	800	7.5	8	60	1.1	1.3	14	2.0	400	3.0	800	800	300	18	15	150
	23–50	55	120	163	64	44	800	5	8	60	1.0	1.2	13	2.0	400	3.0	800	800	300	18	15	150
	51+	55	120	163	64	44	800	5	8	60	1.0	1.2	13	2.0	400	3.0	800	800	300	10	15	150
Pregnant						+30	+200	+5	+2	−20	+0.4	+0.3	+2	+0.6	+400	+1.0	+400	+400	+150	[8]	+5	+25
Lactating						+20	+400	+5	+3	+40	+0.5	+0.5	+5	+0.5	+100	+1.0	+400	+400	+150	[8]	+10	+50

Reference: *Recommended Dietary Allowances,* 9th ed. Food and Nutrition Board, National Research Council–National Academy of Sciences, 1980.

[1] The allowances are intended to provide for individual variations among most normal persons as they live in the United States under usual environmental stresses. Diets should be based on a variety of common foods in order to provide other nutrients for which human requirements have been less well defined.

[2] Retinol equivalents. 1 retinol equivalent = 1 μg of retinol or 6 μg of β-carotene.

[3] As cholecalciferol. 10 μg cholecalciferol = 400 IU of vitamin D.

[4] α-Tocopherol equivalents. 1 mg of α-tocopherol = 1 α-TE.

[5] 1 NE (niacin equivalent) is equal to 1 mg of niacin or 60 mg of dietary tryptophan.

[6] The folacin allowances refer to dietary sources as determined by *Lactobacillus casei* assay after treatment with enzymes (conjugases) to make polyglutamyl forms of the vitamin available to the test organism.

[7] The recommended dietary allowance for vitamin B_{12} in infants is based on average concentration of the vitamin in human milk. The allowances after weaning are based on energy intake (as recommended by the American Academy of Pediatrics) and consideration of other factors, such as intestinal absorption.

[8] The increased requirement during pregnancy cannot be met by the iron content of habitual American diets or by the existing iron stores of many women; therefore the use of 30–60 mg of supplemental iron is recommended. Iron needs during lactation are not substantially different from those of nonpregnant women, but continued supplementation of the mother for 2–3 months after parturition is advisable in order to replenish stores depleted by pregnancy.

20–25 Kg/m^2. Lean body mass rises to a plateau in the third decade of life, then in men declines at an accelerating rate with advancing age. In women, the decline is small until age 50–55 but rapid thereafter. Consequently, if food intake is not reduced with advancing age, obesity will result. In addition, the basal metabolic rate declines with age.

There is a strong genetic component to obesity in humans, but environmental factors contribute as well. For instance, in the USA, there is a significantly lower incidence of obesity in women from higher socioeconomic groups than in those from lower ones. The relation of obesity to diabetes is discussed in Chapter 19.

Obese individuals often complain that they eat very little and still get fat. It now appears that there may be some truth to this claim. In certain strains of obese mice, the response of brown fat to norepinephrine is impaired, so that they cannot uncouple oxidative phosphorylation and they use their food more efficiently than normal individuals. In other cases in experimental animals, the mechanism in the brain that increases sympathetic discharge to brown fat is defective. There is some indirect evidence that similar defects may occur in humans.

NUTRITION

The aim of the science of nutrition is the determination of the kinds and amounts of foods that promote health and well-being. This includes not only the problems of undernutrition but those of overnutrition, taste, and availability. However, certain substances are essential constituents of any human diet. Many of these compounds have been mentioned in previous sections of this chapter, and a brief summary of the essential and desirable dietary components is presented below.

Essential Dietary Components

An optimal diet includes, in addition to sufficient water (see Chapter 38), adequate calories, protein, fat, minerals, and vitamins.

Caloric Intake & Distribution

As noted above, the caloric value of the dietary intake must be approximately equal to the energy expended as heat and work if body weight is to be maintained. When the caloric intake is insufficient, body stores of protein and fat are catabolized, and when the intake is excessive, obesity results. In addition to the 2000 kcal/d necessary to meet basal needs, 500–2500 or more kcal/d are required to meet the energy demands of daily activities.

The distribution of the calories among carbohydrate, protein, and fat foodstuffs is determined partly by physiologic factors and partly by taste and economic considerations. A daily protein intake of at least 1 g/kg body weight to supply the 8 nutritionally essential amino acids and other amino acids is now regarded as desirable. The source of the protein is also important. **Grade I proteins,** the animal proteins of meat, fish, and eggs, contain amino acids in approximately the proportions required for protein synthesis and other uses. Some of the plant proteins are also grade I, but most are **grade II** because they supply different proportions of amino acids, and some lack one or more of the essential amino acids. Protein needs can be met with a mixture of grade II proteins, but the intake must be large because of the amino acid wastage.

Fat is the most compact form of food, since it supplies 9.3 kcal/g. However, it is also the most expensive. Indeed, there is a reasonably good positive correlation between fat intake and standard of living, and in the past, western diets have contained moderately large amounts (100 g/d or more). The evidence suggesting that a high unsaturated/saturated fat ratio in the diet is of value in the prevention of arteriosclerosis and the current interest in preventing obesity may change this. In Central and South American Indian communities where corn (carbohydrate) is the dietary staple, adults live without ill effects for years on a very small fat intake. Therefore, provided the needs for essential fatty acids are met, a low fat intake does not seem to be harmful, and a diet low in saturated fats may be desirable.

Carbohydrate is the cheapest source of calories and provides 50% or more of the calories in most diets. In the average middle-class American diet, approximately 50% of the calories come from carbohydrate, 15% from protein, and 35% from fat. When calculating dietary needs, it is usual to meet the protein requirement first and then split the remaining calories between fat and carbohydrate, depending upon taste, income, and other factors. For example, a 70-kg man who is moderately active needs about 2800 kcal/d. He should eat at least 65 g of protein daily, supplying 267 (65 × 4.1) kcal. Some of this should be grade I protein. Fat intake depends upon taste, but a reasonable figure is 60–70 g. The rest of the caloric requirement can be met by supplying carbohydrate.

Mineral Requirements

A number of minerals must be ingested daily for the maintenance of health. Besides those for which recommended daily dietary allowances have been set (Table 17–6), a variety of different trace elements should be included. Trace elements are defined as elements found in tissues in minute amounts. Those believed to be essential for life, at least in experimental animals, are listed in Table 17–7. In humans, iron

Table 17–7. Trace elements believed essential for life.*

Arsenic	Manganese
Chromium	Molybdenum
Cobalt	Nickel
Copper	Selenium
Fluorine	Silicon
Iodine	Vanadium
Iron	Zinc

*Data from Mertz W: The essential trace elements. *Science* 1981;**213**:1332.

Table 17–8. Vitamins essential or probably essential to human nutrition. Choline is not listed because it is synthesized in the body in adequate amounts except in special circumstances.

Vitamin	Action	Deficiency Symptoms	Sources	Chemistry
$A(A_1, A_2)$	Constituents of visual pigments (see Chapter 8); maintain epithelia.	Night blindness, dry skin	Yellow vegetables and fruit	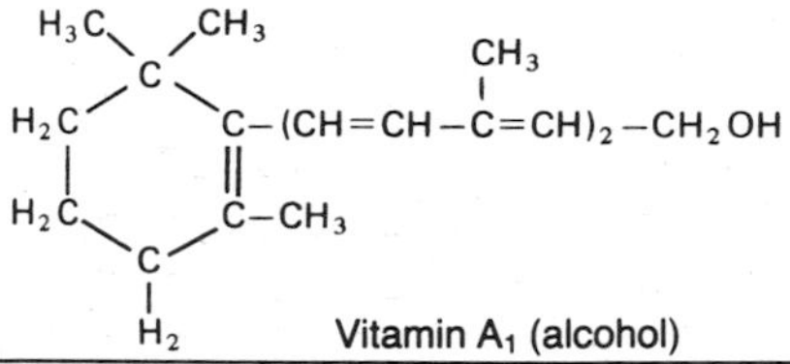 Vitamin A_1 (alcohol)
B complex Thiamine (vitamin B_1)	Cofactor in decarboxylations.	Beriberi, neuritis	Liver, unrefined cereal grains	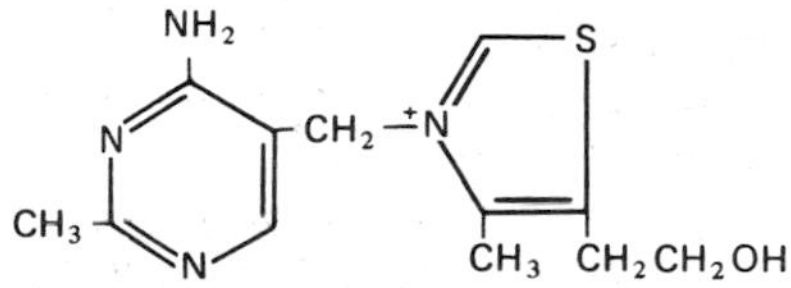
Riboflavin (vitamin B_2)	Constituent of flavoproteins.	Glossitis, cheilosis	Liver, milk	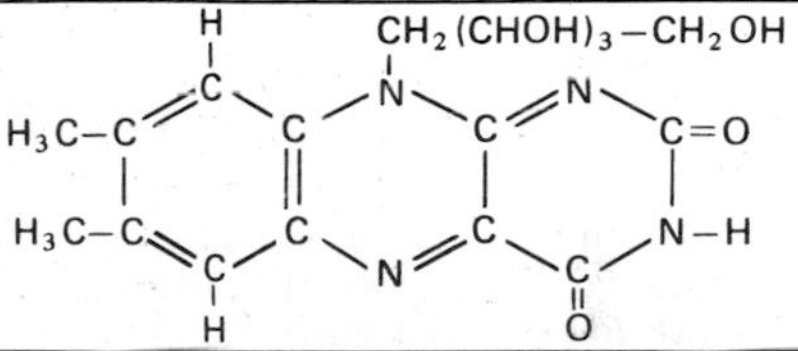
Niacin	Constituent of NAD^+, $NADP^+$	Pellagra	Yeast, lean meat, liver	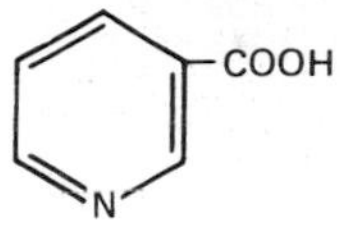 Can be synthesized in body from tryptophan.
Pyridoxine (vitamin B_6)	Forms prosthetic group of certain decarboxylases and transaminases. Converted in body into pyridoxal phosphate and pyridoxamine phosphate.	Convulsions, hyperirritability	Yeast, wheat, corn, liver	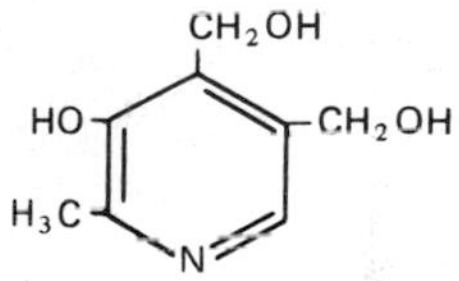
Pantothenic acid	Constituent of CoA.	Dermatitis, enteritis, alopecia, adrenal insufficiency	Eggs, liver, yeast	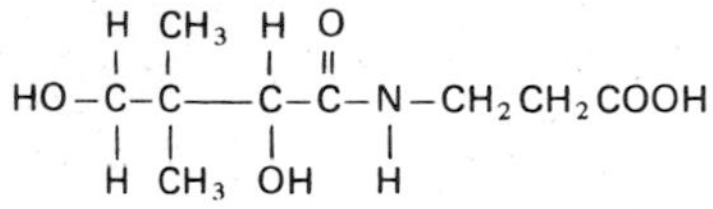
Biotin	Catalyzes CO_2 "fixation" (in fatty acid synthesis, etc).	Dermatitis, enteritis	Egg yolk, liver, tomatoes	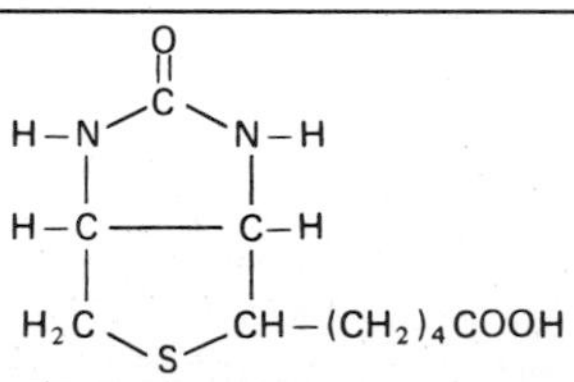
Folates (folic acid) and related compounds	Coenzymes for "1-carbon" transfer; involved in methylating reactions.	Sprue, anemia	Leafy green vegetables	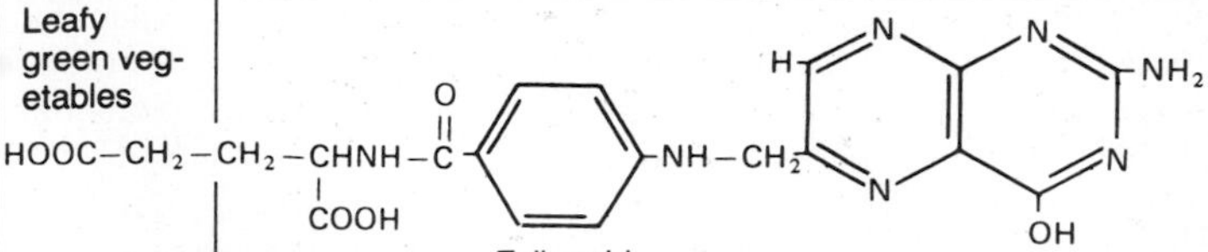 Folic acid
Cyanocobalamin (vitamin B_{12})	Coenzyme in amino acid metabolism. Stimulates erythropoiesis.	Pernicious anemia (see Chapter 26)	Liver, meat, eggs, milk	Complex of 4 substituted pyrrole rings around a cobalt atom (see Chapter 26).

Table 17–8 (cont'd).

Vitamin	Action	Deficiency Symptoms	Sources	Chemistry
C	Necessary for hydroxylation of proline and lysine in collagen synthesis.	Scurvy	Citrus fruits, leafy green vegetables	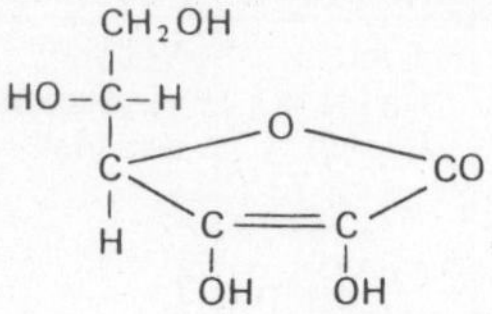 Ascorbic acid (synthesized in most mammals except guinea pigs and primates, including humans).
D group	Increase intestinal absorption of calcium and phosphate (see Chapter 21).	Rickets	Fish liver	Family of sterols (see Chapter 21).
E group	Antioxidants; cofactors in electron transport in cytochrome chain?	Muscular dystrophy and fetal death in animals	Milk, eggs, meat, leafy vegetables	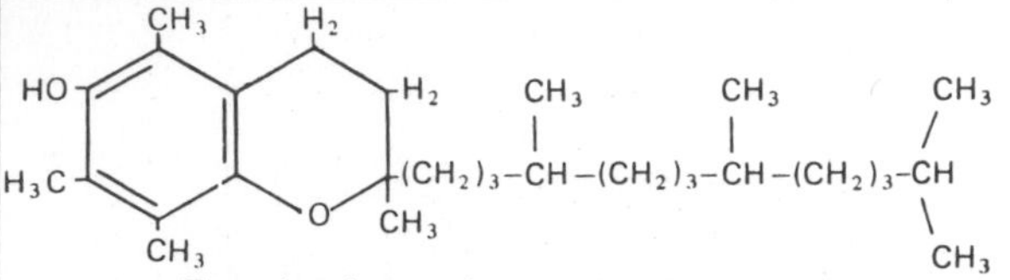 α-Tocopherol; β- and γ-tocopherol also active.
K group	Catalyze γ carboxylation of glutamic acid residues on various proteins concerned with blood clotting.	Hemorrhagic phenomena	Leafy green vegetables	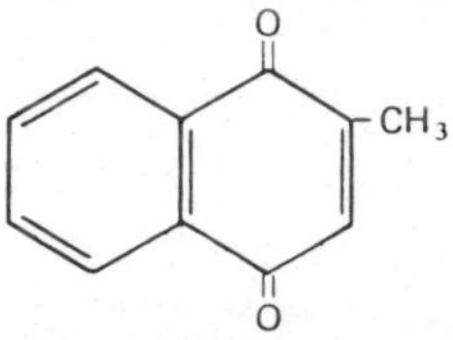 Vitamin K_3; a large number of similar compounds have biologic acitivity.

deficiency causes anemia (see Chapter 26). Cobalt is part of the vitamin B_{12} molecule, and vitamin B_{12} deficiency leads to megaloblastic anemia (see Chapter 26). Iodine deficiency causes thyroid disorders (see Chapter 18). Zinc deficiency causes skin ulcers, depressed immune responses, and hypogonadal dwarfism. Copper deficiency causes anemia, changes in ossification, and possibly elevated plasma cholesterol. Chromium deficiency causes insulin resistance. Fluorine deficiency increases the incidence of dental caries. Trace element deficiencies are rare, because any diet that is adequate in other respects easily supplies the needed minerals.

Conversely, some minerals can be toxic when excess amounts are present in the body. For example, iron overload causes hemochromatosis (see Chapter 25), copper excess causes brain damage (Wilson's disease; see Chapter 12), and aluminum poisoning in patients with renal failure who are receiving dialysis treatment causes a rapidly progressive dementia that resembles Alzheimer's disease (see Chapter 16).

Sodium and potassium are also essential minerals, but listing them is academic, because it is very difficult to prepare a sodium-free or potassium-free diet. A low-salt diet is well tolerated for prolonged periods because of the compensatory mechanisms that conserve sodium, and salt depletion is a problem only when sodium is lost in excessive amounts in the stools, sweat, or urine.

Vitamins

Vitamins were discovered when it was observed that diets adequate in calories, essential amino acids, fats, and minerals failed to maintain health. The term **vitamin** has now come to refer to any organic dietary constituent necessary for life, health, and growth that does not function by supplying energy.

Because there are minor differences in metabolism between mammalian species, some substances are vitamins in one species and not in another. The sources and functions of the major vitamins in humans are listed in Table 17–8 and the recommended daily dietary allowances in Table 17–6. Most vitamins have important functions in intermediary metabolism or the special metabolism of the various organ systems. Those that are water-soluble (vitamin B complex, vitamin C) are easily absorbed, but the fat-soluble vitamins (vitamins A, D, E, and K) are poorly absorbed in the absence of bile or pancreatic lipase. Some dietary fat intake is necessary for their absorption, and in obstructive jaundice or in pancreatic disease, deficiencies of the fat-soluble vitamins can develop even if their intake is adequate (see Chapter 26).

The diseases caused by deficiency of each of these vitamins are listed in Table 17–8. It is worth remembering, however, particularly in view of the advertising campaigns for vitamin pills and supplements, that very large doses of the fat-soluble vitamins are definitely toxic. **Hypervitaminosis A** is characterized by anorexia, headache, hepatosplenomegaly, irritability, scaly dermatitis, patchy loss of hair, bone pain, and hyperostosis. Acute vitamin A intoxication was first described by Arctic explorers, who developed headache, diarrhea, and dizziness after eating polar bear liver. The liver of this animal is particularly rich in vitamin A. **Hypervitaminosis D** is associated with weight loss and calcification of many soft tissues, and excessive vitamin D intake eventually causes renal failure. **Hypervitaminosis K** is characterized by gastrointestinal disturbances and anemia. Large doses of water-soluble vitamins have been thought to be less likely to cause problems because they can be rapidly cleared from the body. However, it has now been demonstrated that ingestion of megadoses of pyridoxine (vitamin B_6) can produce peripheral neuropathy.

18 The Thyroid Gland

INTRODUCTION

The thyroid gland maintains the level of metabolism in the tissues that is optimal for their normal function. Thyroid hormone stimulates the O_2 consumption of most of the cells in the body, helps regulate lipid and carbohydrate metabolism, and is necessary for normal growth and maturation. The thyroid gland is not essential for life, but in its absence, there is poor resistance to cold, mental, and physical slowing, and, in children, mental retardation and dwarfism. Conversely, excess thyroid secretion leads to body wasting, nervousness, tachycardia, tremor, and excess heat production. Thyroid function is controlled by the thyroid-stimulating hormone (TSH, thyrotropin) of the anterior pituitary. The secretion of this tropic hormone is in turn regulated in part by a direct inhibitory feedback of high circulating thyroid hormone levels on the pituitary and in part via neural mechanisms operating through the hypothalamus. In this way, changes in the internal and external environment bring about appropriate adjustments in the rate of thyroid secretion.

In mammals, the thyroid gland also secretes calcitonin, a calcium-lowering hormone. This hormone is discussed in Chapter 21.

ANATOMIC CONSIDERATIONS

Thyroid tissue is present in all vertebrates. In mammals, the thyroid originates from an evagination of the floor of the pharynx, and a **thyroglossal duct** marking the path of the thyroid from the tongue to the neck sometimes persists in the adult. The 2 lobes of the human thyroid are connected by a bridge of tissue, the **thyroid isthmus,** and there is sometimes a **pyramidal lobe** arising from the isthmus in front of the larynx (Fig 18–1). The gland is well vascularized, and the thyroid has one of the highest rates of blood flow per gram of tissue of any organ in the body.

The thyroid is made up of multiple **acini (follicles)**. Each spherical follicle is surrounded by a single layer of cells and filled with pink-staining proteinaceous material called **colloid.** When the gland is inactive, the colloid is abundant, the follicles are large, and the cells lining them are flat. When the gland is active, the follicles are small, the cells are cuboid or columnar, and the edge of the colloid is scalloped, forming many small "reabsorption lacunae" (Fig 18–2).

Microvilli project into the colloid from the apexes of the thyroid cells, and canaliculi extend into them.

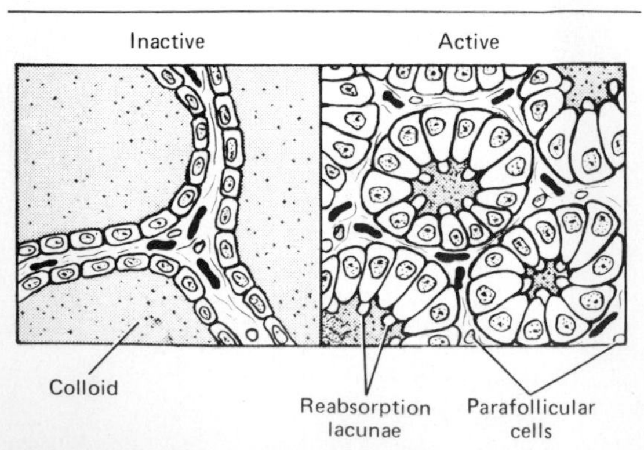

Figure 18–2. Thyroid histology. Note the small, punched out "reabsorption lacunae" in the colloid next to the cells in the active gland.

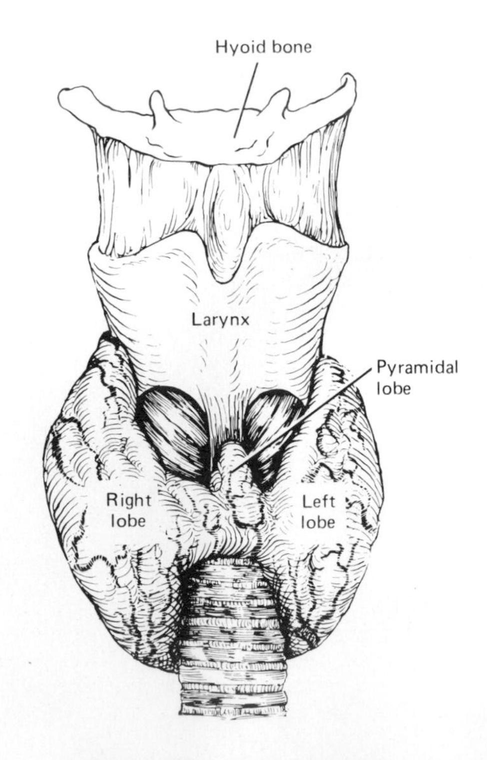

Figure 18–1. The human thyroid.

There is a prominent endoplasmic reticulum, a feature common to most glandular cells, and secretory droplets of thyroglobulin are seen (Fig 18–3). The individual thyroid cells rest on a basal lamina that separates them from the adjacent capillaries. The endothelial cells are attenuated at places, forming gaps (fenestrations; see Chapter 30) in the walls of the capillaries (Fig 18–3). This fenestration of the capillary walls occurs in most endocrine organs.

FORMATION & SECRETION OF THYROID HORMONES

Chemistry

The principal hormones secreted by the thyroid are **thyroxine (T_4)** and **triiodothyronine (T_3).** T_3 is also formed in the peripheral tissues by deiodination of T_4 (see below). Both hormones are iodine-containing amino acids (Fig 18–4). Small amounts of reverse triiodothyronine (3,3′,5′-triiodothyronine, RT_3), monoiodotyrosine, and other compounds are also found in thyroid venous blood. T_3 is more active than T_4, whereas RT_3 is inactive. The naturally occurring forms of T_4 and its congeners with an asymmetric carbon atom are the L isomers. D-Thyroxine has only a small fraction of the activity of the L form.

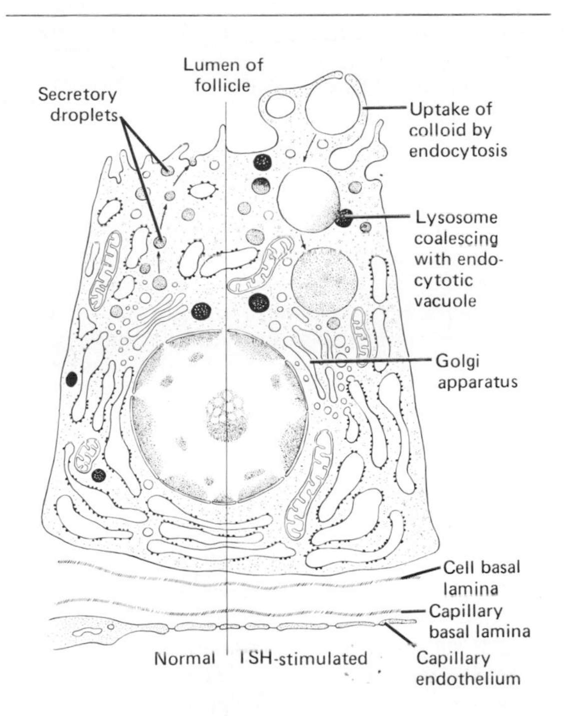

Figure 18–3. Thyroid cell. *Left:* Normal pattern. *Right:* After marked TSH stimulation. The arrows on the left show the secretion of thyroglobulin into colloid. On the right, endocytosis of the colloid and merging of a colloid-containing vacuole with a lysosome are shown. The cell rests on a capillary with gaps (fenestrations) in the endothelial wall. (Modified and reproduced, with permission, from Fawcett DW, Long JA, Jones AL: The ultrastructure of endocrine glands. *Recent Prog Horm Res* 1969;**25**:315.)

Thyroglobulin

T_4 and T_3 are synthesized in the colloid by iodination and condensation of tyrosine molecules bound in peptide linkage in **thyroglobulin.** This glycoprotein is made up of 2 subunits and has a molecular weight of 660,000. It contains 10% carbohydrate by weight. It also contains 123 tyrosine residues, but only 4–8 of these are normally incorporated into thyroid hormones. Thyroglobulin is synthesized in the thyroid cells and secreted into the colloid by exocytosis of granules that also contain thyroid peroxidase (see below). The hormones remain bound to thyroglobulin until secreted. When they are secreted, colloid is ingested by the thyroid cells, the peptide bonds are hydrolyzed, and free T_4 and T_3 are discharged into the capillaries. The thyroid cells thus have 3 functions: They collect and transport iodine; they synthesize thyroglobulin and secrete it into the colloid; and they remove the thyroid hormones from thyroglobulin and secrete them into the circulation.

Thyroglobulin enters the blood as well as the colloid. The normal serum thyroglobulin concentration in humans is about 6 ng/mL, and this level is increased in hyperthyroidism and some forms of thyroid cancer. However, the function, if any, of circulating thyroglobulin is unknown.

Iodine Metabolism

Iodine is a raw material essential for thyroid hormone synthesis. Ingested iodine is converted to iodide and absorbed. The fate of the absorbed I^- is summarized in Fig 18–5. The minimum daily iodine intake that will maintain normal thyroid function is 150 μg in adults (Table 17–6), but in the USA the average dietary intake is approximately 500 μg/d.

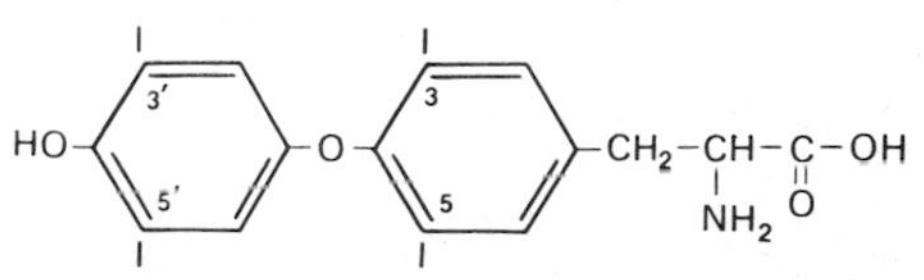

3,5,3′,5′-Tetraiodothyronine (thyroxine, T_4)

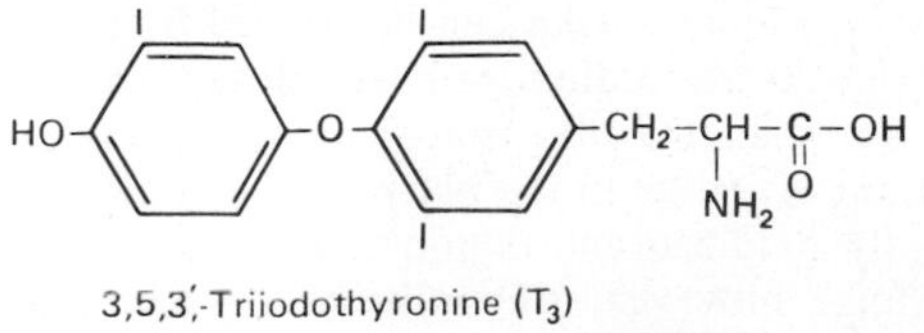

3,5,3′-Triiodothyronine (T_3)

Figure 18–4. Thyroid hormones. The numbers in the rings in the T_4 formula indicate the numbering of positions in the molecule. RT_3 is 3,3′,5′-triiodothyronine.

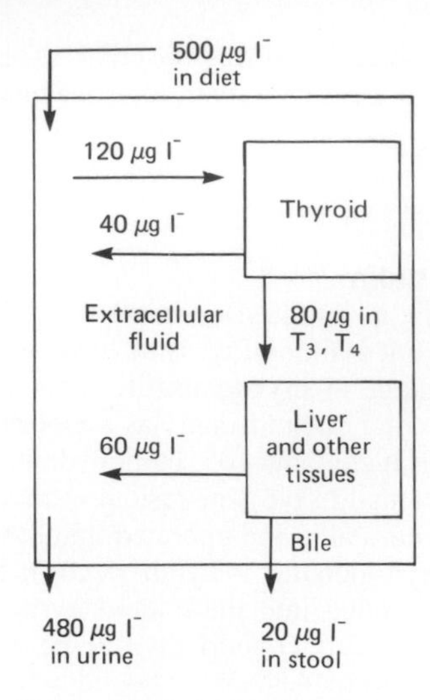

Figure 18–5. Iodine metabolism.

The normal plasma I^- level is about 0.3 μg/dL, and I^- is distributed in a "space" of approximately 25 L (35% of body weight). The principal organs that take up the I^- are the thyroid, which uses it to make thyroid hormones, and the kidneys, which excrete it in the urine. About 120 μg/d enter the thyroid at normal rates of thyroid hormone synthesis and secretion. The thyroid secretes 80 μg/d as iodine in T_3 and T_4. Forty micrograms of I^- per day diffuse into the ECF. The secreted T_3 and T_4 are metabolized in the liver and other tissues, with the release of 60 μg of I^- into the ECF. Some thyroid hormone derivatives are excreted in the bile, and some of the iodine in them is reabsorbed (enterohepatic circulation), but there is a net loss of I^- in the stool of approximately 20 μg/d. The total amount of I^- entering the ECF is thus 500 + 40 + 60, or 600 μg/d; 20% of this I^- enters the thyroid, whereas 80% is excreted in the urine.

Iodide Trapping

The thyroid concentrates iodide by actively transporting it from the circulation to the colloid. The transport mechanism is frequently called the "iodide trapping mechanism" or "iodide pump." The thyroid cell is about 50 mV negative relative to the interstitial area and the colloid, ie, it has a resting membrane potential of −50 mV. Iodide is pumped into the cell at its base against this electrical gradient, then diffuses down the electrical gradient into the colloid. Iodine uptake can be studied by administering radioactive iodine in **tracer doses,** amounts that are so small they do not significantly increase the amount of iodide in the body. In the gland, iodide is rapidly oxidized and bound to tyrosine. Despite the binding, however, the ratio of thyroid to plasma serum free iodide **(T/S ratio)** is normally greater than 1, and if binding to tyrosine residues is blocked by antithyroid drugs such as propylthiouracil (see below), iodide accumulates in the thyroid, and the T/S ratio is markedly increased. Perchlorate and a number of other anions decrease iodide transport by competitive inhibition. The active transport mechanism of iodide is stimulated by TSH. It also depends on Na^+-K^+ ATPase and consequently is inhibited by ouabain (see Chapter 1).

The salivary glands, the gastric mucosa, the placenta, the ciliary body of the eye, the choroid plexus, and the mammary glands also transport iodide against a concentration gradient, but their uptake is not affected by TSH. The mammary glands also bind the iodine; diiodotyrosine is formed in mammary tissue, but T_4 and T_3 are not. The physiologic significance of all these extrathyroidal iodide-concentrating mechanisms is obscure. It has been claimed that traces of T_4 may be formed by nonthyroidal tissues, but if such formation does occur, the amount is insufficient to prevent the development of the full picture of hypothyroidism after surgical thyroidectomy.

Thyroid Hormone Synthesis

In the thyroid gland, iodide is oxidized to iodine and bound in a matter of seconds to the 3 position of tyrosine molecules attached to thyroglobulin (Fig 18–6). The enzyme responsible for the oxidation and binding of iodide is **thyroid peroxidase,** with hydrogen peroxide accepting the electrons. Monoiodotyrosine (MIT) is next iodinated in the 5 position to form diiodotyrosine (DIT). Two DIT molecules then undergo an oxidative condensation to form T_4 with the elimination of the alanine side chain from the molecule that forms the outer ring. There are 2 theories of how this **coupling reaction** occurs. One holds that the coupling occurs with both DIT molecules attached to thyroglobulin (intramolecular coupling). The other holds that the DIT that forms the outer ring is first detached from thyroglobulin (intermolecular coupling). In any case, thyroid peroxidase is probably involved in coupling as well as iodination. T_3 is probably formed by condensation of MIT with DIT. A small amount of RT_3 is also formed, probably by condensation of DIT with MIT. In the normal human thyroid, the average distribution of iodinated compounds is 23% MIT, 33% DIT, 35% T_4, and 7% T_3. Only traces of RT_3 and other components are present.

MIT and DIT residues that are not coupled are deiodinated in the thyroid, and deiodination of these residues normally provides about twice as much iodide for hormone synthesis as the iodide pump. In patients with congenital absence of the thyroid deiodinase, MIT and DIT appear in the urine and there are symptoms of iodine deficiency (see below).

Secretion

The human thyroid secretes about 80 μg (103 nmol) of T_4, 4 μg (7 nmol) of T_3, and 2 μg (3.5 nmol) of RT_3, per day (Fig 18–7). The thyroid cells ingest colloid by endocytosis (see Chapter 1). This chewing away at the edge of the colloid produces the reabsorption lacunae seen in active glands (Fig 18–2).

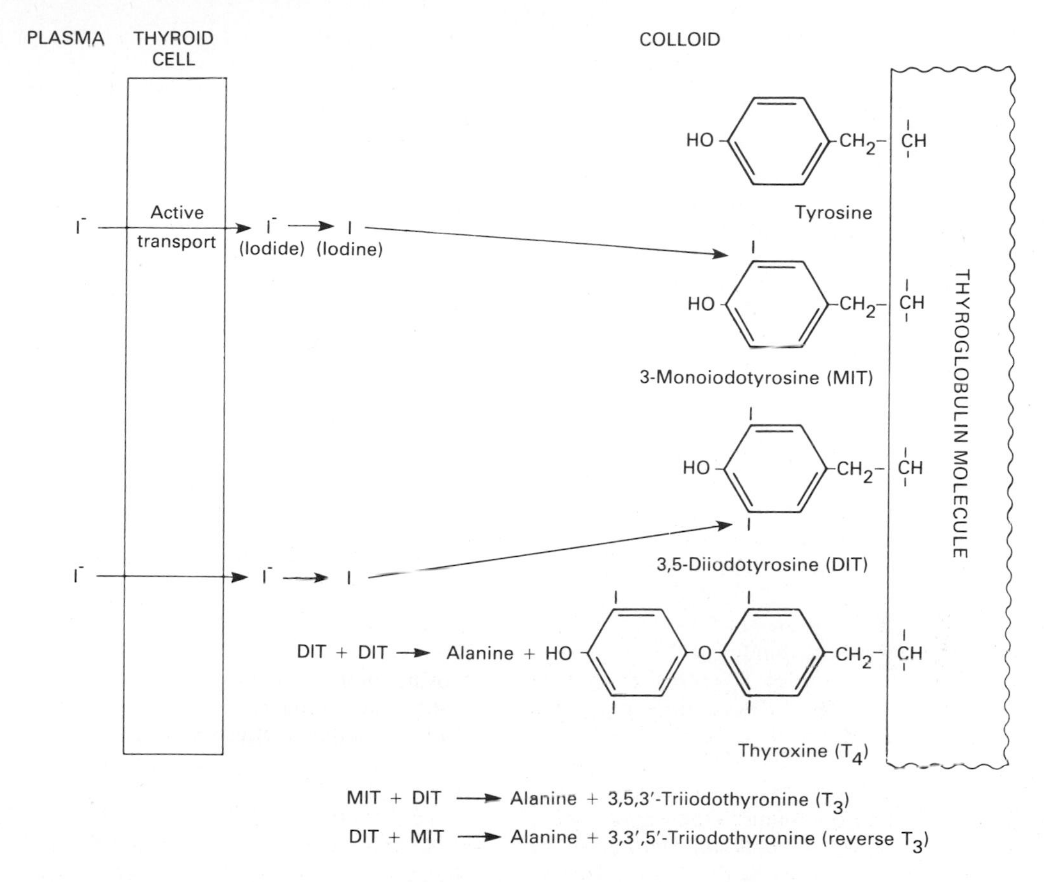

Figure 18–6. Outline of thyroid hormone biosynthesis. Iodination of tyrosine and the condensation reaction take place at the apical border of the thyroid cells while the molecules are bound in peptide linkage in thyroglobulin.

In the cells, the globules of colloid merge with lysosomes (Fig 18–3). The peptide bonds between the iodinated residues and the thyroglobulin are broken by the proteases in the lysosomes, and T_4, T_3, diiodotyrosine, and monoiodotyrosine are liberated into the cytoplasm. The iodinated tyrosines are deiodinated by a microsomal **iodotyrosine dehalogenase,** but this enzyme does not attack iodinated thyronines, and T_4 and T_3 pass on the circulation. The iodine liberated by deiodination of the tyrosines is reutilized.

TRANSPORT & METABOLISM OF THYROID HORMONES

Protein Binding

The normal total plasma T_4 level is approximately 8 μg/dL (103 nmol/L), and that of T_3 is approximately 0.15 μg/dL (2.3 nmol/L). Both are bound to plasma proteins. Both are now measured by radioimmunoassay, and their measurement has largely replaced use of **protein-bound iodine (PBI)** as an index of the circulating level of thyroid hormones. The normal PBI level is approximately 6 μg/dL.

Capacity & Affinity of Plasma Proteins for Thyroid Hormones

The plasma proteins that bind thyroid hormones are **albumin;** a prealbumin called **thyroxine-binding prealbumin (TBPA);** and a globulin with an electrophoretic mobility between α_1- and α_2-globulin,

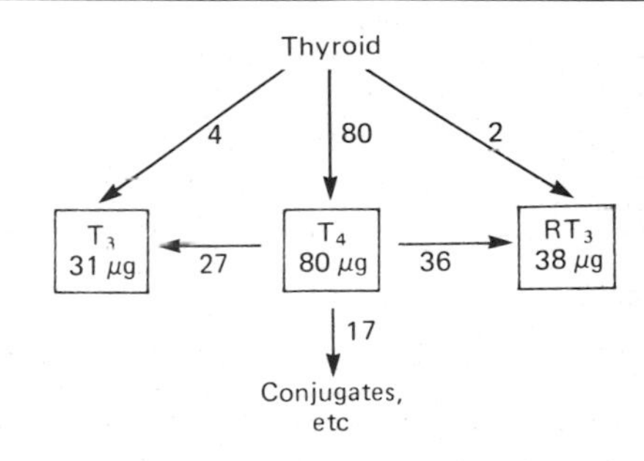

Figure 18–7. Secretion and interconversion of thyroid hormones in normal adult humans. Figures are in micrograms per day. Note that most of the T_3 and RT_3 are formed from T_4 deiodination in the tissues, and only small amounts are secreted by the thyroid. (Data from Cavalieri RR: Peripheral metabolism of thyroid hormones. *Thyroid Today* [Nov 1980;3[7].)

thyroxine-binding globulin (TBG). Of the 3 proteins, albumin has the largest **capacity** to bind T_4—ie, it can bind the most T_4 before becoming saturated—and TBG the smallest. However, the **affinities** of the proteins for T_4—ie, the avidity with which they bind T_4 under physiologic conditions—are such that most of the circulating T_4 is bound to TBG (Table 18–1), with over a third of the binding sites on the protein occupied. Smaller amounts of T_4 are bound to TBPA and albumin. The half-life of TBPA is 2 days, that of TBG is 5 days, and that of albumin 13 days.

Normally, 99.98% of the T_4 in plasma is bound; the free T_4 level is only about 2 ng/dL. There is very little T_4 in the urine. Its biologic half-life is long (about 6–7 days), and its volume of distribution is less than that of ECF (10 L, or about 15% of body weight). All of these properties are characteristic of a substance that is strongly bound to protein.

T_3 is not bound to quite as great an extent; of the 0.15 μg/dL normally found in plasma, 0.2% (0.3 ng/dL) is free. The remaining 99.8% is protein-bound, 46% to TBG and most of the remainder to albumin, with very little binding to TBPA (Table 18–1). The lesser binding of T_3 correlates with the facts that T_3 has a shorter half-life than T_4 and that its action on the issues is much more rapid. RT_3 also binds to TBG.

The free thyroid hormones in plasma are in equilibrium with the protein-bound thyroid hormones in the tissues (Fig 18–8). Free thyroid hormones are added to the circulating pool by the thyroid. It is the free thyroid hormones in plasma that are physiologically active, and it is this fraction that inhibits the pituitary secretion of TSH (see below).

Free T_4 and T_3 can be measured by equilibrium dialysis, but this is a laborious technique. Tissue uptake of the free hormones is proportionate to their concentration in plasma, and consequently, red blood cell uptake is sometimes measured as an index of their level. Similar results are obtained when a resin is used to take up the free thyroid hormones. For example, radioactive T_4 is added to plasma, and after equilibration, resin is added and the percentage of the radioactive T_4 taken up by the resin is determined. Multiplication of the resin uptake by the total T_4 concentration provides the **free T_4 index** (FT_4I), an index of the amount of free T_4 in the specimen. In a similar fashion, a **free T_3 index** (FT_3I) can be calculated. It is important to emphasize that the FT_4I and FT_3I are not direct measures of free T_4 and T_3 but indices of their concentration.

Table 18–1. Binding of thyroid hormones to plasma proteins in normal adult humans.

Protein	Plasma Concentration (mg/dL)	Amount of Circulating Hormone Bound (%) T_4	T_3
Thyroxine-binding globulin (TBG)	2	67	46
Thyroxine-binding prealbumin (TBPA)	15	20	1
Albumin	3500	13	53

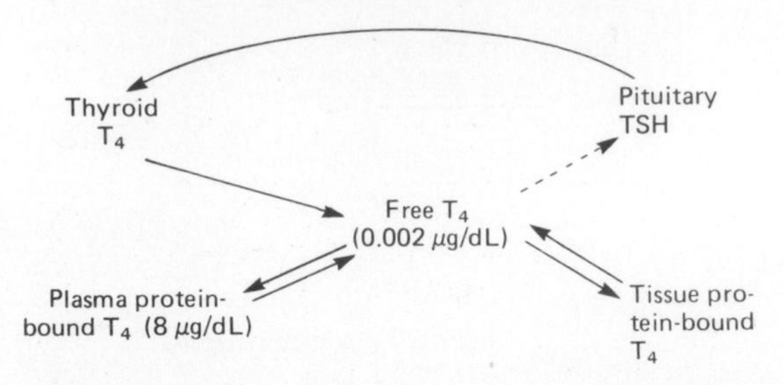

Figure 18–8. Distribution of T_4 in the body. The distribution of T_3 is similar. The dashed arrow indicates inhibition of TSH secretion by increases in the free T_4 level of ECF. Approximate concentrations in human blood are shown in parentheses.

Fluctuations in Binding

When there is a sudden, sustained increase in the concentration of thyroid-binding proteins in the plasma, the concentration of free thyroid hormones falls. This change is temporary, however, because there is a concomitant decrease in the rate of entry of free thyroid hormones into tissue, leaving more free T_3 and T_4 in the circulation. In addition, the decrease in the concentration of free thyroid hormones in the circulation stimulates TSH secretion, which in turn causes an increase in the production of free thyroid hormones. A new equilibrium is eventually reached at which the total quantity of thyroid hormones in the blood is elevated but the concentration of free hormones, the rate of their metabolism, and the rate of TSH secretion are normal. Corresponding changes in the opposite direction occur when the concentration of thyroid-binding proteins is reduced. Consequently, patients with elevated or decreased concentrations of binding proteins, particularly TBG, are neither hyper- nor hypothyroid, ie, they are **euthyroid.**

TBG levels are elevated in estrogen-treated patients and during pregnancy, as well as following treatment with various drugs (Table 18–2). They are depressed by glucocorticoids, androgens, the antiestrogen danazol, and the cancer chemotherapeutic agent L-asparaginase. A number of other drugs, including salicylates, the anticonvulsant phenytoin, and the cancer chemotherapeutic agents mitotane (*o,p′*-DDD) and 5-fluorouracil, inhibit binding of T_4 and T_3 to TBG and consequently produce changes similar to those produced by a decrease in TBG concentration. Changes in total plasma T_4 and T_3 can also be produced by changes in plasma concentrations of albumin and prealbumin.

Metabolism of Thyroid Hormones

T_4 and T_3 are deiodinated in the liver, the kidneys, and many other tissues. One-third of the circulating T_4 is normally converted to T_3 in adult humans, and

Table 18–2. Effect of variations in the concentrations of thyroid hormone-binding proteins in the plasma on various parameters of thyroid function after equilibrium has been reached.

	Concentrations of Binding Proteins	Total Plasma T_4, T_3, RT_3	Free Plasma T_4, T_3, RT_3	Plasma TSH	Clinical State
Hyperthyroidism	Normal	High	High	Low	Hyperthyroid
Hypothyroidism	Normal	Low	Low	High	Hypothyroid
Estrogens, methadone, heroin, major tranquilizers, clofibrate	High	High	Normal	Normal	Euthyroid
Glucocorticoids, androgens, danazol, L-asparaginase	Low	Low	Normal	Normal	Euthyroid

45% is converted to RT_3. As shown in Fig. 18–7, only about 13% of the circulating T_3 is secreted by the thyroid and 87% is formed by deiodination of T_4, similarly, only 5% of the circulating RT_3 is secreted by the thyroid and 95% is formed by deiodination of T_4. Two different enzymes are involved, 5′-deiodinase catalyzing the formation of T_3 and 5-deiodinase catalyzing the formation of RT_3 (Fig 18–9). T_3 and RT_3 are then converted to various diiodothyronines. 5′-Deiodinase catalyzes the conversion of RT_3 to 3,3′-diiodothyronine, whereas 5-deiodinase catalyzes the conversion of T_3 to 3,3′-diiodothyronine. 3,5-Diiodothyronine and 3′,5′-diiodothyronine are also formed. Much more RT_3 and much less T_3 are formed during fetal life, and the ratio shifts to that of adults about 6 weeks after birth.

Because T_3 acts more rapidly than T_4 and is 3–5 times as potent on a molar basis, T_4 is believed to be metabolically inert until it is converted to T_3. Additional evidence in favor of this view is that T_3 binds to nuclear receptors to a much greater extent than T_4. It is difficult to prove that T_4 has no metabolic activity of its own, but T_4 certainly qualifies as a prohormone in the sense that it is the major precursor of T_3.

In the liver, T_4 and T_3 are conjugated to form sulfates and glucuronides. These conjugates enter the bile and pass into the intestine. The thyroid conjugates are hydrolyzed, and some are reabsorbed (enterohepatic circulation), but some are excreted in the stool. The iodide lost in this way amounts to about 4% of the total daily iodide loss.

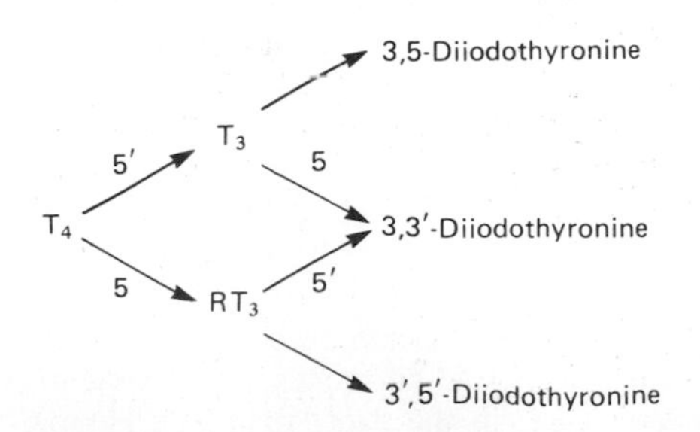

Figure 18–9. Conversion of T_4 to T_3 and RT_3 and thence to diiodothyronines in liver, kidney, and other tissues. 5′ = 5′-deiodinase; 5 = 5-deiodinase.

Fluctuations in Deiodination

Various drugs inhibit the 5′-deiodinase that converts T_4 to T_3 and RT_3 to 3,3′-diiodothyronine, producing a fall in plasma T_3 and a rise in plasma RT_3. A wide variety of nonthyroidal illnesses also depress 5′-deiodinase. These include burns, trauma, advanced cancer, cirrhosis, renal failure, myocardial infarction, and febrile states. The low T_3 state produced by these conditions disappears with recovery. It is difficult to decide whether individuals with the low T_3 state produced by drugs and illness have mild hypothyroidism, and thyroid hormone measurements are difficult to interpret if thyroid disease is suspected, partly because many of the drugs have other effects on thyroid hormones and partly because the drugs and the illnesses produce other changes that make clinical hypothyroidism hard to diagnose.

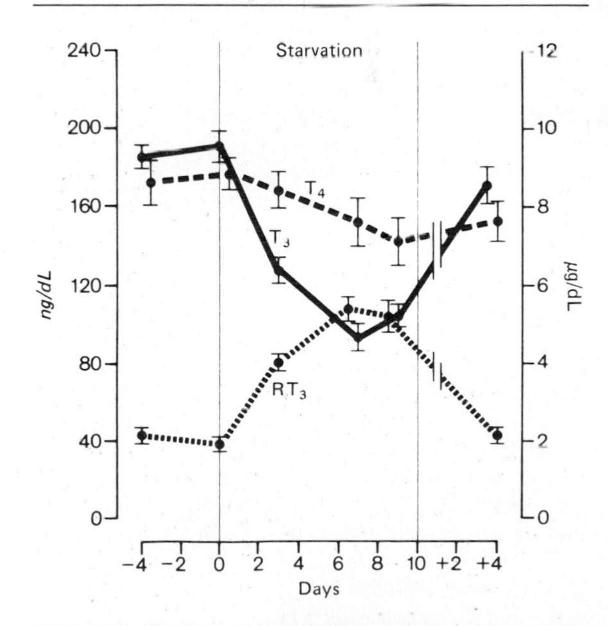

Figure 18–10. Effect of starvation on plasma levels of T_4, T_3, and RT_3 in humans. Similar changes occur in wasting diseases. The scale for T_3 and RT_3 is on the left and the scale for T_4 on the right. (Reproduced, with permission, from Burger AG: New aspects of the peripheral action of thyroid hormones. *Triangle, Sandoz J Med Sci* 1983; 22:175. Copyright Sandoz Ltd., Basel, Switzerland.)

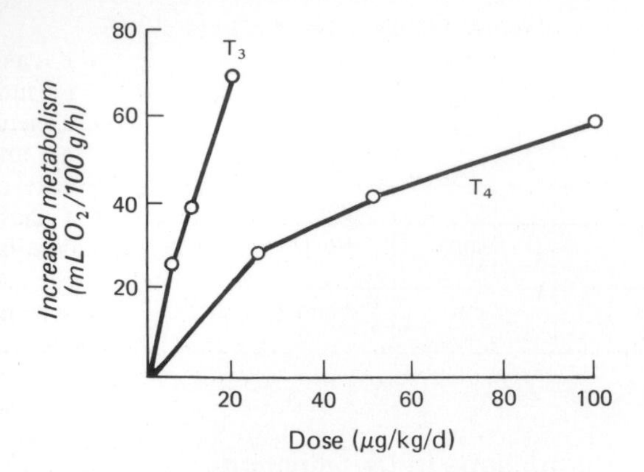

Figure 18–11. Calorigenic responses of thyroidectomized rats to subcutaneous injections of T_4 and T_3. (Redrawn and reproduced, with permission, from Barker SB: Peripheral actions of thyroid hormones. *Fed Proc* 1962;21:635.)

Diet also has a clear-cut effect on conversion of T_4 to T_3. In fasted individuals, plasma T_3 is reduced 10–20% in 24 hours and about 50% in 3–7 days, with a corresponding rise in RT_3 (Fig 18–10). Free and bound T_4 levels remain normal. During more prolonged starvation, RT_3 returns to normal but T_3 remains depressed. At the same time, the BMR falls and urinary nitrogen excretion, an index of protein breakdown, is decreased. Thus, the decline in T_3 conserves calories and protein. Conversely, overfeeding increases T_3 and reduces RT_3.

EFFECTS OF THYROID HORMONES

Most of the widespread effects of thyroid hormones in the body are secondary to stimulation of O_2 consumption **(calorigenic action),** although the hormones also affect growth and maturation in mammals, help regulate lipid metabolism, and increase the absorption of carbohydrates from the intestine. They also increase the dissociation of oxygen from hemoglobin by increasing red cell 2,3-diphosphoglycerate (DPG) (see Chapter 35). In these actions, T_3 is 3–5 times more potent than T_4 (Fig 18–11), whereas RT_3 is inert. It is of interest that the iodine atoms in thyroid hormones are not necessary for biologic activity; several active thyroxine analogs that do not contain iodine have been synthesized.

Calorigenic Action

T_4 and T_3 increase the O_2 consumption of almost all metabolically active tissues. The exceptions are the adult brain, testes, uterus, lymph nodes, spleen, and anterior pituitary. T_4 actually depresses the O_2 consumption of the anterior pituitary, presumably because it inhibits TSH secretion. The increase in metabolic rate produced by a single dose of T_4 becomes measurable after a latent period of several hours and lasts 6 days or more. The magnitude of the calorigenic effect depends upon the level of catecholamine secretion and on the metabolic rate before injection. If the initial rate is low, the rise is great; but if it is high, the rise is small. This is true not only in euthyroid subjects but in athyreotic T_4-treated patients as well. The cause of the decreased effect at higher metabolic rates is obscure.

Effects Secondary to Calorigenesis

When the metabolic rate is increased by T_4 and T_3 in adults, nitrogen excretion is increased; if food intake is not increased, endogenous protein and fat stores are catabolized, and weight is lost. In hypothyroid children, small doses of thyroid hormones cause a positive nitrogen balance because they stimulate growth, but large doses cause protein catabolism similar to that produced in the adult. The potassium liberated during protein catabolism appears in the urine, and there is an increase in urinary hexosamine and uric acid excretion.

The skin normally contains a variety of proteins combined with polysaccharides, hyaluronic acid, and chondroitin sulfuric acid. In hypothyroidism, these complexes accumulate, promoting water retention and the characteristic puffiness of the skin (myxedema). When thyroid hormones are administered, the proteins are mobilized, and diuresis continues until the myxedema is cleared.

Large doses of thyroid hormones cause enough extra heat production to cause a slight rise in body temperature (see Chapter 14), which in turn activates heat-dissipating mechanisms. Peripheral resistance decreases because of cutaneous vasodilation, but cardiac output is increased by the combined action of thyroid hormones and catecholamines on the heart, so that pulse pressure and cardiac rate are increased and circulation time is shortened.

In the absence of thyroid hormones, a moderate anemia occurs as a result of decreased bone marrow metabolism and poor absorption of cyanocobalamin (vitamin B_{12}) from the intestine. T_4 and T_3 correct these defects.

When the metabolic rate is increased, the need for all vitamins is increased, and vitamin deficiency syndromes may be precipitated. Thyroid hormones are necessary for hepatic conversion of carotene to vitamin A, and the accumulation of carotene in the bloodstream **(carotenemia)** in hypothyroidism is responsible for the yellowish tint of the skin. Carotenemia can be distinguished from jaundice because in the former condition the scleras are not yellow.

Milk secretion is decreased in hypothyroidism and stimulated by thyroid hormones, a fact sometimes put to practical use in the dairy industry. Thyroid hormones do not stimulate the metabolism of the uterus but are essential for normal menstrual cycles and fertility.

Effects on the Nervous System

In hypothyroidism, mentation is slow and the CSF protein level elevated. Thyroid hormones reverse these changes, and large doses cause rapid mentation, irritability, and restlessness. This is somewhat surprising, because it is generally agreed that cerebral blood flow and the glucose and O_2 consumption of the brain are normal and in adult hypo- and hyperthyroidism. However, it has now been proved that thyroid hormones enter the brain in adults and are found in gray matter in numerous different locations. In addition, the brain converts T_4 to T_3, and there is a sharp increase in brain 5′-deiodinase activity after thyroidectomy that is reversed within 4 hours by a single intravenous dose of T_3. Some of the effects of thyroid hormones on the brain are probably secondary to increased responsiveness to catecholamines, with consequent increased activation of the reticular activating system (see Chapter 11). There is disagreement about whether thyroid hormones affect brain O_2 consumption in young animals. It is clear, however, that thyroid hormones have marked effects on brain development. In hypothyroid infants, synapses develop abnormally, myelination is defective, and mental development is seriously retarded. The mental changes are irreversible if replacement therapy is not begun soon after birth.

Thyroid hormones also exert effects on the peripheral nervous system. The reaction time of stretch reflexes (see Chapter 6) is shortened in hyperthyrodism and prolonged in hypothyroidism. Measurement of the reaction time of the ankle jerk (Achilles reflex) has attracted considerable attention as a clinical test for evaluating thyroid function, but the reaction time is also affected by certain other diseases.

Effects on Skeletal Muscle

Muscle weakness occurs in most patients with hyperthyroidism **(thyrotoxic myopathy),** and when the hyperthyroidism is severe and prolonged, the myopathy may be severe. The muscle weakness may be due in part to increased protein catabolism, but there may also be changes in the myosin similar to those that occur in the heart (see below). On the other hand, hypothyroidism is also associated with muscle weakness, cramps, and stiffness.

Effects on the Heart

Thyroid hormones increase the number and affinity of β-adrenergic receptors in the heart and consequently increase its sensitivity to the inotropic and chronotropic effects of catecholamines. They also affect the type of myosin found in cardiac muscle. In rats, there are 2 myosin heavy chain (MHC) isoforms in the heart (see Chapter 3). Alpha-MHC predominates in the ventricles in adults, and its level is increased by treatment with thyroid hormone. However, expression of the α-MHC gene is depressed and that of the β-MHC gene is enhanced in hypothyroidism. The β-MHC has less myosin ATPase activity than the α-MHC.

Relation to Catecholamines

The actions of thyroid hormones and the catecholamines norepinephrine and epinephrine are intimately interrelated. Epinephrine increases the metabolic rate, stimulates the nervous system, and produces cardiovascular effects similar to those of thyroid hormones, although the duration of these actions is brief. Norepinephrine has generally similar actions. Thyroid hormones increase the number and affinity of β-adrenergic receptors in the heart and possibly in some other tissues, and the effects of thyroid hormones on the heart resemble those of β-adrenergic stimulation. The toxicity of the catecholamines is markedly increased in rats treated with T_4. Although plasma catecholamines are normal in hyperthyroidism, the cardiovascular effects, tremulousness, and sweating produced by thyroid hormones can be reduced or abolished by sympathectomy. They can also be reduced by drugs such as propranolol that block β-adrenergic receptors. Indeed, propranolol and other β blockers are used extensively in the treatment of thyrotoxicosis and in the treatment of the severe exacerbations of hyperthyroidism called **thyroid storms.** It is worth pointing out, however, that although β blockers are weak inhibitors of extrathyroidal conversion of T_4 to T_3, and consequently may produce a small fall in plasma T_3, they do not prevent or ameliorate the other effects of thyroid hormones.

Effects on Carbohydrate Metabolism

Thyroid hormones increase the rate of absorption of carbohydrate from the gastrointestinal tract, an action that is probably independent of their calorigenic action. In hyperthyroidism, therefore, the blood glucose level rises rapidly after a carbohydrate meal, sometimes exceeding the renal threshold. However, it falls again at a rapid rate.

Effects on Cholesterol Metabolism

Thyroid hormones lower circulating cholesterol levels. The plasma cholesterol level drops before the metabolic rate rises, which indicates that his action is independent of the stimulation of O_2 consumption. As noted in Chapter 17, the decrease in plasma cholesterol concentration is due to increased formation of LDL receptors.

Effects on Growth & Development

Thyroid hormones are essential for normal growth and skeletal maturation (see Chapter 22). In hypothyroid children, bone growth is slowed and epiphyseal closure delayed. In the absence of thyroid hormones, growth hormone secretion may also be depressed, and thyroid hormones potentiate the effect of growth hormone on the tissues.

Another example of the role of thyroid hormones in growth and maturation is their effect on amphibian metamorphosis. Tadpoles treated with T_4 and

T_3 metamorphose early into dwarf frogs, whereas hypothyroid tadpoles never become frogs. The effects of T_4 on metamorphosis are probably independent of its effects on O_2 consumption, even though it raises the O_2 consumption of metamorphosing tadpole skin in vitro. The hormone also exerts a calorigenic effect on tadpoles in vivo.

MECHANISM OF ACTION OF THYROID HORMONES

Thyroid hormones enter cells, and T_3 binds to receptors in the nuclei (Fig 1–27). T_4 can also bind, but not as avidly, and in many organs, much of the T_4 is converted to T_3 in the cytoplasm. The T_3-receptor complex then binds to DNA in a manner that appears to be analogous to the steroid receptor binding (see Chapter 1). The expression of specific genes is increased, with the induction of mRNAs that in turn alter cell function. It seems clear that a variety of different enzymes are produced, since the effects of thyroid hormones are multiple and varied.

Thyroid hormones increase the activity of the membrane-bound Na^+-K^+ ATPase in many tissues, and it has been argued that it is the increase in energy consumption associated with the increase in Na^+ transport which is responsible for the increase in metabolic rate. However, inhibition of the increase in Na^+-K^+ ATPase activity with ouabain does not completely abolish the calorigenic effect of thyroid hormones are multiple and varied.

Thyroid hormones increase the activity of the membrane-bound Na^+-K^+ ATPase in many tissues, and it has been argued that it is the increase in energy consumption associated with the increase in Na^+ transport which is responsible for the increase in metabolic rate. However, inhibition of the increase in Na^+-K^+ ATPase activity with ouabain does not completely abolish the calorigenic effect of thyroid hormones. Mitochondrial protein synthesis is increased, but the role of the mitochondria in the response is obscure. The old theory that thyroid hormones increased energy consumption by uncoupling oxidation of substrate from phosphorylation in the mitochondria has now been largely abandoned because such uncoupling could not be demonstrated in mitochondria from hyperthyroid rats. In addition, drugs such as 2,4-dinitrophenol uncouple oxidative phosphorylation without reproducing the full metabolic or other effects of thyroid hormones.

REGULATION OF THYROID SECRETION

Thyroid function is regulated primarily by variations in the circulating level of pituitary TSH. TSH secretion is increased by the hypophyseotropic hormone TRH (see Chapter 14) and inhibited in a negative feedback fashion by circulating free T_4 and T_3. TSH secretion is also inhibited by stress, and in experimental animals, it is increased by cold and decreased by warmth.

Chemistry & Metabolism of TSH

Human TSH is a glycoprotein that contains 211 amino acid residues, plus hexoses, hexosamines, and sialic acid. It is made up of 2 subunits, designated α and β. These subunits are encoded by genes on separate chromosomes, and the α and β subunits become noncovalently linked in the thyrotropes. TSH-α is identical in structure to the α subunit of LH and FSH and differs only slightly from hCG-α (see Chapters 22 and 23). The functional specificity of TSH is conferred by the β unit. The structure of TSH varies from species to species, but other mammalian TSHs are biologically active in humans.

Recent research has done much to clarify the mechanisms involved in the glycosylation of TSH. Glycosylation of both subunits begins with attachment in the rough endoplasmic reticulum of a 14-sugar unit to asparagine residues with modification of these carbohydrate side chains in the Golgi apparatus. The initial glycosylation is necessary for proper combination of the α and the β subunits, and the final pattern of carbohydrate side chains is necessary for full biologic activity. In addition, deglycosylated TSH is removed more rapidly from the circulation; thus, its effectiveness is decreased. An interesting new finding is that TRH increases the biologic activity of TSH by altering its glycosylation. In patients with TRH deficiency, there is a drop in circulating TSH bioactivity, but in up to 40% of the cases, there is little if any decrease in circulating TSH immunoactivity.

The biologic half-life of human TSH is about 60 minutes. TSH is degraded for the most part in the kidneys and to a lesser extent in the liver. Secretion is pulsatile, and mean output starts to rise at about 9:00 PM, peaks at midnight, and then declines during the day. The normal secretion rate is about 110 μg/d. The average plasma level is about 2 μU/mL (Fig 18–12).

Thyroid-stimulating activity can be extracted from the placenta, and benign and malignant tumors of the placenta may contain enough of this material to cause hyperthyroidism. However, they also contain and secrete hCG, and hCG has a modest amount of TSH activity. As noted above, hCG-α is very similar to TSH-α, and hCG can displace TSH from TSH receptors. Consequently, it seems likely that the so-called placental TSH is actually hCG.

Effects of TSH on the Thyroid

When the pituitary is removed, thyroid function is depressed and the gland atrophies; when TSH is administered, thyroid function is stimulated. Within a few minutes after the injection of TSH, there are increases in iodide binding; synthesis of T_3, T_4, and iodotyrosines; secretion of thyroglobulin into the colloid; and endocytosis of colloid. Iodide trapping is increased in a few hours; blood flow increases; and,

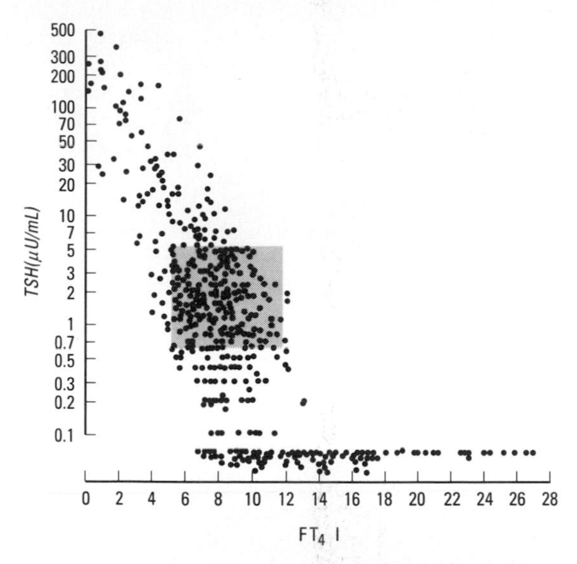

Figure 18–12. Relation between plasma TSH, measured by a new, highly sensitive immunoenzymatic assay, and the free T_4 index (FT_4I). The shaded area identifies the normal range. (Reproduced, with permission, from Spencer CA, Clinical utility of sensitive TSH assays. *Thyroid Today* (Apr) 1986;9:1. The Boots Co [USA] Inc.)

with chronic TSH treatment, the cells hypertrophy and the weight of the gland increases. In order for the growth response to occur, growth hormone, corticosteroids, and insulin must be present; TSH alone does not produce thyroid hypertrophy in hypophysectomized animals.

Whenever TSH stimulation is prolonged, the thyroid becomes detectably enlarged. Enlargement of the thyroid is called **goiter.**

TSH Receptors

TSH receptors have been studied in considerable detail with monoclonal antibodies. The receptor in the thyroid cell membranes are made up of a glycoprotein bound to a ganglioside. When TSH binds to the receptor, it activates adenylate cyclase, and the resultant increase in intracellular cyclic AMP brings about most of the effects of TSH. However, it also stimulates the metabolism of phospholipids in the membrane, and it is the alteration in phospholipid metabolism that mediates the thyroid hypertrophy.

Control Mechanisms

The mechanisms regulating thyroid secretion are summarized in Fig 18–13. The negative feedback effect of thyroid hormones on TSH secretion is exerted in part at the hypothalamic level, but it is also due in large part to an action on the pituitary, since T_4 and T_3 block the increase in TSH secretion produced by TRH. The relation between plasma TSH and the FT_4I index is shown in Fig 18–12. Infusion of T_4 as well as T_3 reduces TSH, and there is a measurable decline in the level of TSH within 1 hour. In experimental animals, there is an initial rise in pituitary TSH content before the decline, indicating that thyroid hormones inhibit secretion before they inhibit synthesis. The effects on secretion and synthesis of TSH both appear to depend on protein synthesis, even though the former is relatively rapid.

The day-to-day maintenance of thyroid secretion depends on the feedback interplay of thyroid hormones with TSH and TRH (Fig 18–13). The adjustments that appear to be mediated via TRH include the increased secretion of thyroid hormones produced by cold and, presumably, the decrease produced by heat. It is worth noting that although cold produces clear-cut increases in circulating TSH in experimental animals and human infants, the rise produced by cold in adult humans is negligible. Consequently, in adults, increased heat production due to increased thyroid hormone secretion **(thyroid hormone thermogenesis)** plays little if any role in the response to cold. The inhibiting effect of stress on TSH secretion is probably due to an inhibiting effect of glucocorticoids on TRH secretion, but the effect of stress is a minor one. Dopamine and somatostatin act at the pituitary level to inhibit TSH secretion, but it is not known whether they play a physiologic role in the regulation of TSH secretion.

The amount of thyroid hormone necessary to maintain normal cellular function in thyroidectomized individuals used to be defined as the amount necessary to normalize the BMR, but it is now usually defined as the amount necessary to return plasma TSH to normal. This averages about 112 μg of levothyroxine by mouth per day in adults. Eighty percent of this dose is absorbed from the gastrointestinal tract. It produces a slightly greater than

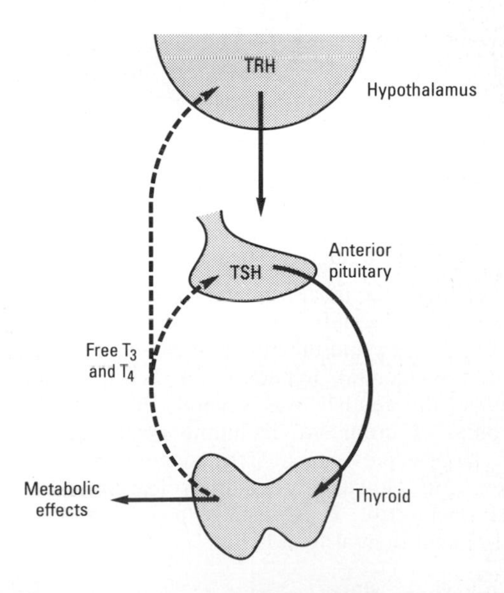

Figure 18–13. Feedback control of thyroid secretion. The dashed arrows indicate inhibitory effects and the solid arrows stimulatory effects. Compare with Figs 20–22, 22–9, 23–21, and 23–35.

normal FT_4I but a normal FT_3I, indicating that in humans, unlike some experimental animals, it is circulating T_3 rather than T_4 that is the principal feedback regulator of TSH secretion.

CLINICAL CORRELATES

Hypothyroidism

The signs, symptoms, and complications of hypothyroidism and hyperthyroidism in humans are predictable consequences of the physiologic effects of thyroid hormones discussed above. The syndrome of adult hypothyroidism is generally called **myxedema,** although the term myxedema is also used to refer specifically to the skin changes in this syndrome. Hypothyroidism may be the end result of a number of diseases of the thyroid gland, or it may be secondary to pituitary failure (''pituitary hypothyroidism'') or hypothalamic failure (''hypothalamic hypothyroidism''). In the latter 2 conditions, unlike the first, the thyroid responds to a test dose of TSH, and at least in theory, hypothalamic hypothyroidism can be distinguished from pituitary hypothyroidism by the presence in the former of a rise in plasma TSH following a test dose of TRH. The TSH response to TRH is usually normal in hypothalamic hypothyroidism, while it is increased in hypothyroidism due to thyroid disease and decreased in hyperthyroidism due to the feedback of thyroid hormones on the pituitary gland.

In completely athyreotic humans, the BMR falls to about -40. The hair is coarse and sparse, the skin is dry and yellowish (carotenemia), and cold is poorly tolerated. The voice is husky and slow, the basis of the aphorism that ''myxedema is the one disease that can be diagnosed over the telephone.'' Mentation is slow and memory is poor, and in some patients there are severe mental symptoms (''myxedema madness'').

Cretinism

Children who are hypothyroid from birth are called cretins. They are dwarfed and mentally retarded and have enlarged, protruding tongues and pot bellies (Fig 18–14). Before the use of iodized salt became widespread, the most common cause of cretinism was maternal iodine deficiency. Various congenital abnormalities of thyroid function that cause goiter can also cause congenital hypothyroidism with cretinism. Unless the mother was severely hypothyroid, the stigmas of cretinism, including mental deficiency, are largely preventable if treatment is started soon after birth. However, once the typical clinical picture has developed, it is usually too late to prevent permanent mental retardation.

Hyperthyroidism

Hyperthyroidism (thyrotoxicosis) is characterized by nervousness; weight loss; hyperphagia; heat intolerance; increased pulse pressure; a fine tremor of the outstretched fingers; a warm, soft skin; sweating; and a basal metabolic rate from $+10$ to as high as $+100$. It may be caused by a variety of thyroid disorders, including, in rare instances, benign and malignant tumors. Cases due to TSH-secreting pituitary tumors have also been reported. However, the most common form is **Graves' disease (exophthalmic goiter).** In Graves' disease, the thyroid is diffusely enlarged and hyperplastic, and there is a protrusion of the eyeballs called **exophthalmos** (Fig 18–15). It now seems clear that Graves' disease is an autoimmune disease (see Chapter 27) in which T lymphocytes activated by antigens in the thyroid gland stimulate B lymphocytes to produce circulating antibodies against the antigens. Some of the antibodies damage the thyroid, producing **Hashimoto's thyroiditis,** and this condition can progress to hypothyroidism. However, other antibodies are formed against components of the TSH receptors. These antibodies, which are called **thyroid-stimulating immunoglobulins (TSI),** activate the receptors, producing hyperthyroidism. One theory of the genesis of Graves' disease holds that there is a defect in suppressor T lymphocytes that permits helper T lymphocytes to stimulate B lymphocytes to produce the thyroid autoantibodies. However, the

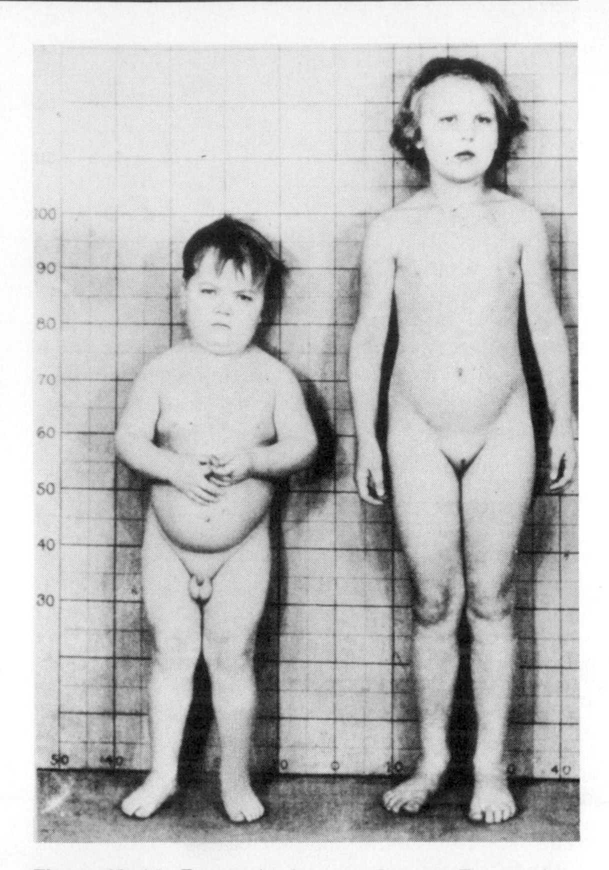

Figure 18–14. Fraternal twins, age 8 years. The boy has congenital hypothyroidism. (Reproduced, with permission, from Wilkins L in: *Clinical Endocrinology I.* Astwood EB, Cassidy CE [editors]. Grune & Stratton, 1960.)

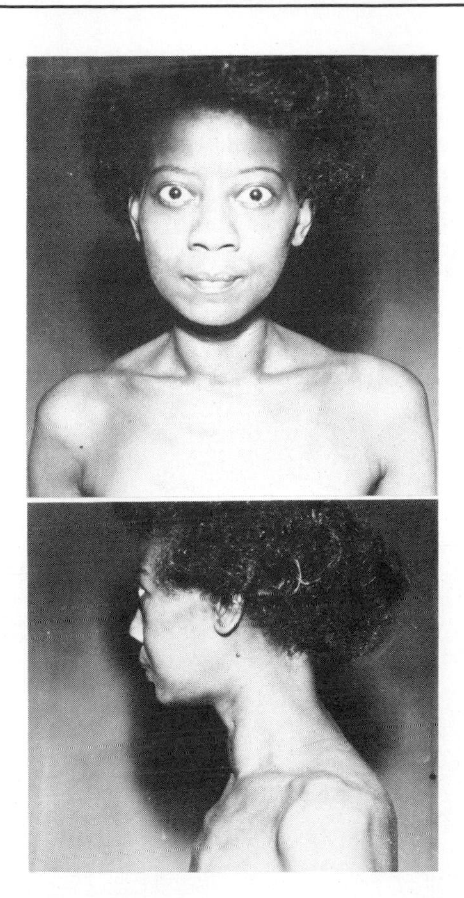

Figure 18–15. Graves' disease. Note the goiter and the exophthalmos. (Courtesy of PH Forsham.)

mechanism that initiates the immunologic events is still unknown. There is marked stimulation of the secretion of thyroid hormones, and the high circulating T_4 and T_3 levels inhibit TSH secretion (Fig 18–16), so the circulating TSH level is depressed. Increasing levels of thyroid hormones tend to cause increased formation of the antibodies, whereas decreased levels produced by treatment of the disease tend to lower but not completely prevent formation of the antibodies.

The exophthalmos in Graves' disease is due to swelling of the tissue, particularly the extraocular muscles, within the rigid bony walls of the orbits. This pushes the eyeballs forward. Thyroid hormones are not directly responsible for the exophthalmos, and there has been much debate about its etiology. Although exophthalmos is usually associated with hyperthyroidism, it may occur before, during, or after the thyrotoxic state, and in a significant proportion of patients with thyrotoxicosis, it is made worse by thyroidectomy. In addition, it is occasionally seen in euthyroid patients with Hashimoto's thyroiditis, in patients with myxedema, and in patients with no known thyroid disease. The exophthalmos may be due to deposition of thyroglobulin-antithyroglobulin and other thyroid immune complexes in the extraocular muscles, with the production of an immune complex inflammatory reaction. Alternatively, it may be an autoimmune disease of the orbital muscles.

Thyrotoxicosis places a considerable load on the cardiovascular system, and in some patients with hyperthyroidism, most or even all of the symptoms are cardiovascular. Thyrotoxic heart disease is a curable form of heart disease, but the diagnosis is often missed. It is important to remember that heart failure develops whenever the cardiac output, even though it is elevated, is inadequate to maintain tissue perfusion. The drop in peripheral resistance caused by cutaneous vasodilation in thyrotoxicosis is equivalent to opening a large arteriovenous fistula; if the compensatory increase in cardiac output is not great enough, "high-output failure" results.

Iodine Deficiency

When the dietary iodine intake falls below 10 μg/d, thyroid hormone synthesis is inadequate, and secretion declines. As a result of increased TSH secretion, the thyroid hypertrophies, producing an **iodine deficiency goiter** that may become very large. Such "endemic goiters" have been known since ancient times. Before the practice of adding iodide to table salt became widespread, they were very common in Central Europe and the area around the Great Lakes in the USA, the inland "goiter belts" where iodine has been leached out of the soil by rainwater so that food grown in the soil is iodine-deficient.

Radioactive Iodine Uptake

Iodine uptake is an index of thyroid function that can be easily measured by using tracer doses of radioactive isotopes of iodine that have no known deleterious effect on the thyroid. The tracer is administered orally and the thyroid uptake determined by

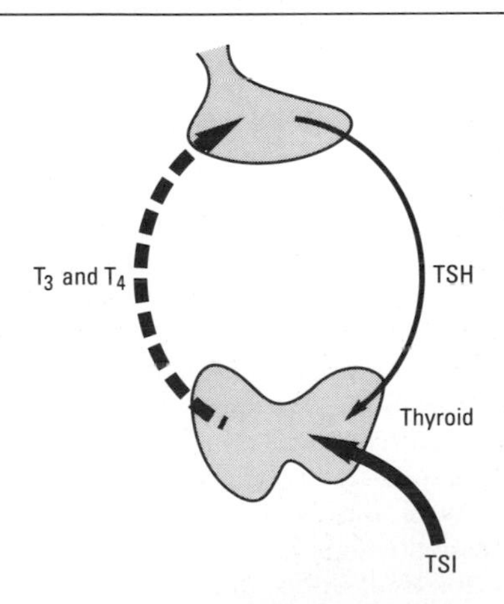

Figure 18–16. Thyroid and pituitary function in Graves' disease. Thyroid-stimulating immunoglobulins (TSI) act on the thyroid to produce a marked increase in the secretion of T_4 and T_3 and secretion of TSH from the pituitary is markedly inhibited.

placing a gamma ray counter over the neck. An area such as the thigh is also counted, and counts in this region are subtracted from the neck counts to correct for nonthyroidal radioactivity in the neck. The isotope of iodine that is most commonly used is ^{123}I because it has a half-life of only 0.55 day, compared with ^{131}I, which has a half-life of 8.1 days, and ^{125}I, which has a half-life of 60 days. Clinical use of radioactive iodine uptake has decreased, in part because of the general availability of methods for measuring T_4, T_3, and TSH in plasma and in part because of the widespread use of iodized salt. This causes uptake to be low, because the iodide pool is so large that the tracer is excessively diluted. However, an analysis of radioactive iodine uptake is helpful in understanding the physiology of the thyroid gland. The uptake in a normal subject is plotted in Fig 18–17. In hyperthyroidism, iodide is rapidly incorporated into T_4 and T_3, and these hormones are released at an accelerated rate. Therefore, the amount of radioactivity in the thyroid rises sharply, but it then levels off and may start to decline within 24 hours, at a time when the uptake in normal subjects is still rising. In hypothyroidism, the uptake is low.

Large amounts of radioactive iodine destroy thyroid tissue because the radiation kills the cells. Radioiodine therapy is useful in some cases of thyroid cancer and can be used to treat benign thyroid diseases, although, especially in young patients, the dangers of possible radiation carcinogenesis and germ cell mutations must be kept in mind.

Radioactive isotopes of iodine are major products of nuclear fission, and if fission products are released into the atmosphere as a result of an accident at a nuclear power plant or explosion of a nuclear bomb, the isotopes spread for considerable distances in the atmosphere because they are more volatile than the other products. Treatment with potassium iodide is regularly instituted in fallout areas to enlarge the iodide pool and depress thyroid uptake to low levels.

Antithyroid Drugs

Most of the drugs that inhibit thyroid function act either by interfering with the iodide-trapping mechanism or by blocking the organic binding of iodine. In either case, TSH secretion is stimulated by the decline in circulating thyroid hormones, and goiter is produced. A number of monovalent anions compete with iodide for active transport into the thyroid and therefore inhibit uptake to the point where the T/S ratio approaches unity; this inhibition of iodide transport can be overcome by administration of extra iodide. The anions include chlorate, pertechnetate, periodate, biodate, nitrate, and perchlorate. Thiocyanate, another monovalent anion, inhibits iodide transport but is not itself concentrated within the gland. The activity of perchlorate is about 10 times that of thiocyanate.

The **thiocarbamides,** a group of compounds related to thiourea, inhibit the iodination of monoiodotyrosine (organic binding of iodide) and block the coupling reaction. The 2 used clinically are propylthiouracil and methimazole (Fig 18–18). Iodination of tyrosine is inhibited because propylthiouracil and methimazole compete with tyrosine residues for iodine and become iodinated. In addition, propylthiouracil but not methimazole inhibits the conversion of T_4 to T_3 in extrathyroidal tissues. Both drugs may also ameliorate thyrotoxicosis by suppressing the immune sys-

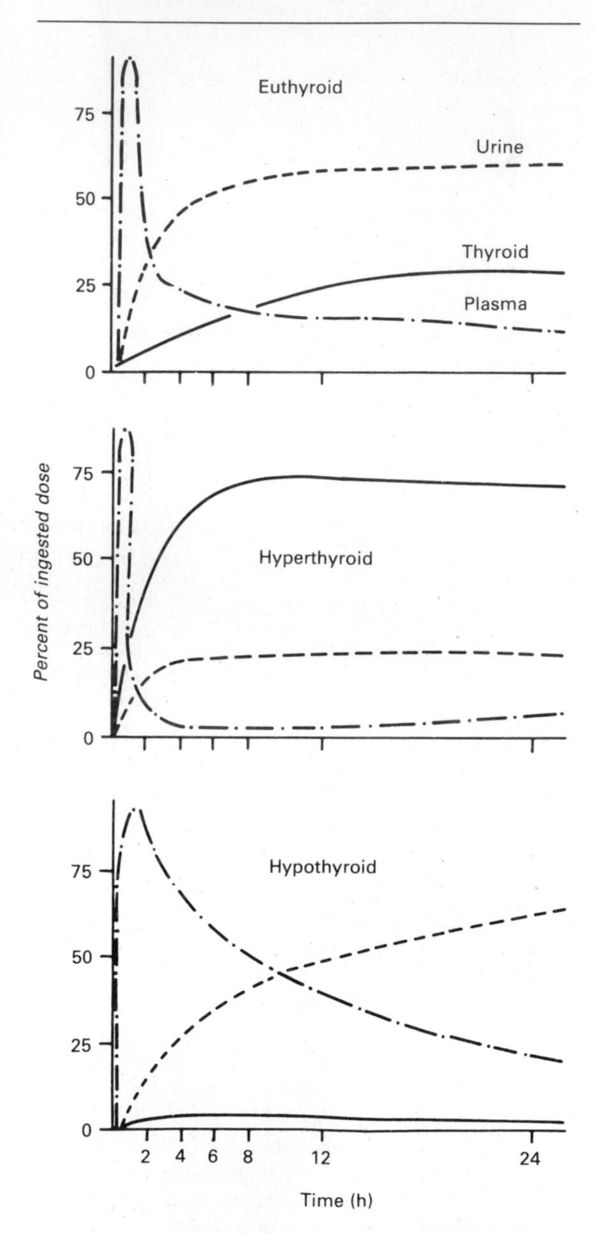

Figure 18–17. Radioactive iodine uptake by individuals on a relatively low-iodine diet. Percentages are plotted against time after an oral dose of radioactive iodine for plasma, urine, and the thyroid gland. In hyperthyroidism, plasma radioactivity falls rapidly, then rises again as a result of release of labeled T_4 and T_3 from the thyroid. (Modified from Ingbar SH, Woeber KA in: *Textbook of Endocrinology,* 4th ed. Williams RH [editor]. Saunders, 1968.)

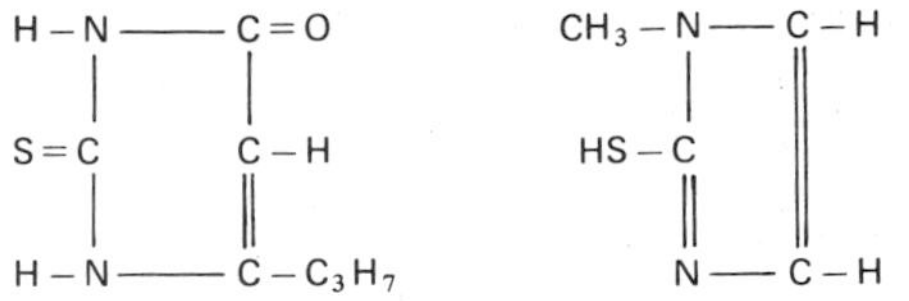

Propylthiouracil

1-Methyl-2-mercapto-imidazole (methimazole; Tapazole)

Figure 18–18. Antithyroid thiocarbamides.

tem and thereby depressing TSI formation. They may also inhibit the biosynthesis or alter the structure of thyroglobulin.

The effect of thiocarbamides on coupling is produced by lower doses than the doses that affect iodination. The thiocarbamides do not block iodide trapping. Because of the increase in TSH secretion, the initial uptake of radioactive iodine is actually increased during thiocarbamide therapy, and the T/S ratio may reach 250. However, since binding is inhibited, the iodine is not retained, and 24 hours after administration of the isotope, uptake is subnormal.

Another substance that inhibits thyroid function under certain conditions is iodide itself. The position of iodide in thyroid physiology is thus unique in that while some iodide is needed for normal thyroid function, too little iodide and too much both cause abnormal thyroid function. In normal individuals, large doses of iodides act directly on the thyroid to produce a mild and transient inhibition of organic binding of iodide and hence of hormone synthesis. This inhibition is known as the Wolff-Chaikoff effect. The Wolff-Chaikoff effect is greater and more prolonged when iodide transport is increased, and this is why patients with thyrotoxicosis are more responsive to iodide than normal individuals. Susceptibility to the effect is also increased when there is a defect in the organic binding mechanism, and consequently, the inhibition is increased in individuals with partially destroyed or removed thyroids or thyroiditis. There are at least 2 additional mechanisms by which excess I^- inhibits thyroid function. It reduces the effect of TSH on the gland by reducing the cyclic AMP response to this hormone, and it inhibits proteolysis of thyroglobulin. There is no direct effect on the I^- trapping mechanism, but the total I^- uptake is low because of the inhibition of organic binding and, to a lesser extent, because the amount of circulating iodide is so great that added tracer is immensely diluted. In thyrotoxicosis, iodides cause colloid to accumulate, and the vascularity of the hyperplastic gland is decreased, making iodide treatment of considerable value in preparing thyrotoxic patients for surgery.

Naturally Occurring Goitrogens

Thiocyanates are sometimes ingested with food, and there are relatively large amounts of naturally occurring goitrogens in some foods. Vegetables of the Brassicaceae family, particularly rutabagas, cabbage. and turnips, contain **progoitrin** and a substance that converts this compound into **goitrin,** an active antithyroid agent (Fig 18–19). The progoitrin activator in vegetables is heat-labile, but because there are activators in the intestine (presumably of bacterial origin), goitrin is formed even if the vegetables are cooked. The goitrin intake on a normal mixed diet is usually not great enough to be harmful, but in vegetarians and food faddists, "cabbage goiters" do occur. Others as yet unidentified plant goitrogens probably exist and may be responsible for the occasional small "goiter epidemics" reported from various parts of the world.

Use of Thyroid Hormones in Nonthyroidal Diseases

When the pituitary-thyroid axis is normal, doses of exogenous thyroid hormone that provide less than the amount secreted endogenously have no significant effect on metabolism, because there is a compensatory decline in endogenous secretion resulting from inhibition of TSH secretion. In euthyroid humans, the oral dose of T_4 that merely suppresses thyroid function is 200–500 μg/d. Suppression of TSH secretion due to larger doses of T_4 or pituitary disease leads eventually to thyroid atrophy. An atrophic gland initially responds sluggishly to TSH, and if the TSH suppression has been prolonged, it may take some time for normal thyroid responsiveness to return. The adrenal cortex and some other endocrine glands respond in an analogous fashion; when they are deprived of the support of their tropic hormones for some time, they become atrophic and only sluggishly responsive to their tropic hormone until the hormone has had some time to act on the gland.

In patients with so-called metabolic insufficiency—fatigued individuals who feel weak and have a slightly low BMR with normal plasma T_4 and T_3 levels—thyroid therapy has been proved to be of no more value than placebo medication. Use of thyroid to promote weight loss can be of value only if the patient pays the price of some nervousness and heat intolerance and curbs the appetite so that there is no compensatory increase in caloric intake.

Progoitrin

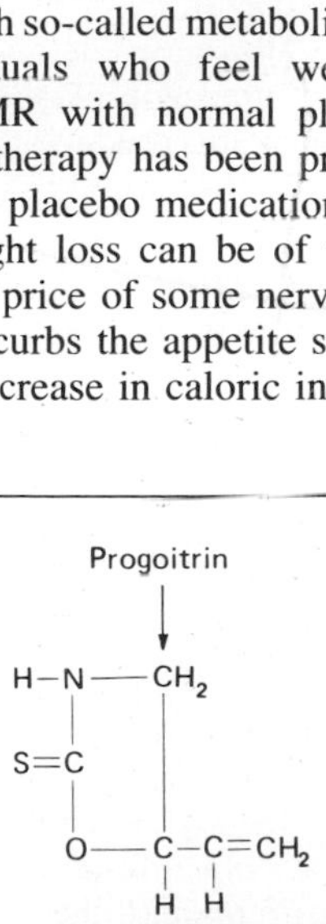

Goitrin
(L-5-Vinyl-2-thiooxazolidone)

Figure 18–19. The naturally occurring goitrogen in vegetables of the family Brassicaceae.

19 Endocrine Functions of the Pancreas & the Regulation of Carbohydrate Metabolism

INTRODUCTION

At least 4 peptides with hormonal activity are secreted by the islets of Langerhans in the pancreas. Two of these hormones, **insulin** and **glucagon,** have important functions in the regulation of the intermediary metabolism of carbohydrates, proteins, and fats. The third hormone, **somatostatin,** may play a role in the regulation of islet cell secretion, and the physiologic function of the fourth, **pancreatic polypeptide,** is unsettled. Glucagon, somatostatin, and possibly pancreatic polypeptide are also secreted by cells in the mucosa of the gastrointestinal tract.

Insulin is anabolic, increasing the storage of glucose, fatty acids, and amino acids. Glucagon is catabolic, mobilizing glucose, fatty acids, and amino acids from stores into the bloodstream. The 2 hormones are thus reciprocal in their overall action and are reciprocally secreted in most circumstances. Insulin excess causes hypoglycemia, which leads to convulsions and coma. Insulin deficiency, either absolute or relative, causes diabetes mellitus, a complex and debilitating disease that if untreated is eventually fatal. Glucagon deficiency can cause hypoglycemia, and glucagon excess makes diabetes worse. Excess pancreatic production of somatostatin produces hyperglycemia and other manifestations of diabetes.

A variety of other hormones also have important roles in the regulation of carbohydrate metabolism.

ISLET CELL STRUCTURE

The islets of Langerhans (Fig 19–1) are ovoid, 76 × 175 μm collections of cells scattered throughout the pancreas, although they are more plentiful in the tail than in the body and head. They make up 1–2% of the weight of the pancreas. In humans, there are 1–2 million islets. Each has a copious blood supply; and blood from the islets, like that from the gastrointestinal tract but unlike that from any other endocrine organs, drains into the portal vein.

The cells in the islets can be divided into types on the basis of their staining properties and morphology. There are at least 4 distinct cell types in humans: A, B, D, and F cells. A, B, and D cells are also called α, β, and δ cells. However, this leads to confusion in view of the use of Greek letters to refer to other structures in the body, particularly adrenergic receptors (see Chapter 4). The A cells secrete glucagon, the B cells secrete insulin, the D cells secrete somatostatin, and the F cells secrete pancreatic polypeptide. The B cells, which are the most common and account for 60–75% of the cells in the islets, are generally located in the center of each islet. They tend to be surrounded by the A cells, which make up 20% of the total, and the less common D and F cells. The islets in the tail, the body, and the anterior and superior part of the head of the human pancreas have many A cells and few if any F cells in the outer rim, whereas in rats and probably in humans, the islets in the posterior part of the head of the pancreas have a relatively large number of F cells and few A cells. The A cell-rich (glucagon-rich) islets arise embryologically from the dorsal pancreatic bud and the F cell-rich (pancreatic polypeptide-rich) islets arise from the ventral pancreatic bud. These buds arise separately from the duodenum.

The B cell granules are packets of insulin in the cell cytoplasm. Each packet is contained in a membrane-

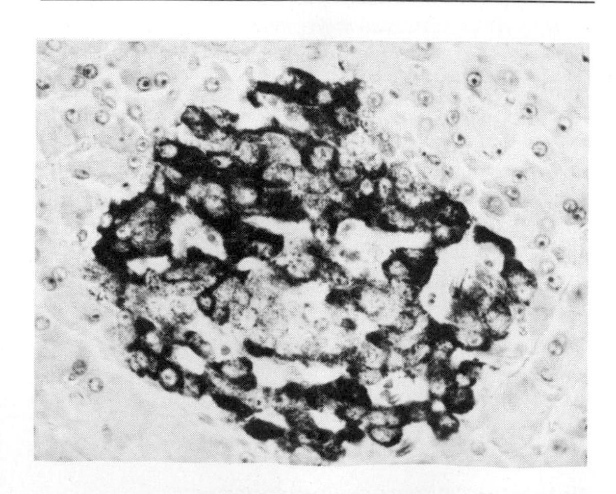

Figure 19–1. Islet of Langerhans, rat pancreas. Darkly stained cells are B cells. Surrounding pancreatic acinar tissue is light-colored. (× 400; courtesy of LL Bennett.)

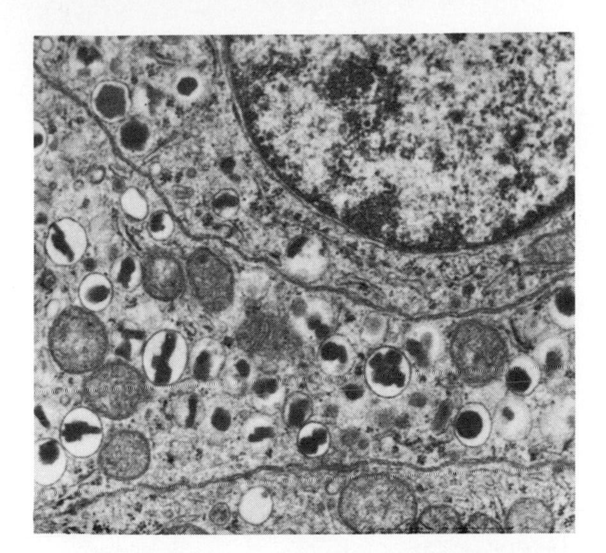

Figure 19–2. Electron micrograph of 2 adjoining B cells in human pancreas. The B granules are the membrane-lined vesicles containing crystals that vary in shape from rhombic to round. (× 26,000; courtesy of A Like. Reproduced, with permission, from Fawcett DW: *Bloom and Fawcett, A Textbook of Histology,* 11th ed. Saunders, 1986.)

lined vesicle (Fig 19–2), and, characteristically, there is a clear space (halo) between the wall of the vesicle and the packet. The shape of the packets varies from species to species; in humans, some are round while others are rectangular. In the B cells, the insulin molecule forms polymers and also complexes with zinc. The differences in shape of the packets are probably due to differences in the size of polymers or zinc aggregates of insulin. The A granules, which contain glucagon, are relatively uniform from species to species (Fig 19–3). The D cells also contain large numbers of relatively homogeneous granules.

STRUCTURE, BIOSYNTHESIS & SECRETION OF INSULIN

Structure & Species Specificity

Insulin is a polypeptide containing 2 chains of amino acids linked by disulfide bridges (Table 19–1). There are minor differences in the amino acid composition of the molecule from species to species. The differences are generally not sufficient to affect the biologic activity of a particular insulin in heterologous species but are sufficient to make the insulin antigenic. If the insulin of one species is injected for a prolonged period into another species, the anti-insulin antibodies formed inhibit the injected insulin. Almost all humans who have received commercial beef insulin for more than 2 months have antibodies against beef insulin, but the titer is usually low and presents no clinical problem. Patients who develop high titers and become resistant to beef insulin on this basis are usually responsive to insulins from other species. Pork insulin differs from human insulin by only one amino acid residue and has low antigenicity. Human insulin produced in bacteria by recombinant DNA technology is now available.

Biosynthesis & Secretion

Insulin is synthesized in the endoplasmic reticulum of the B cells (Fig 19–4). It is then transported to the Golgi apparatus, where it is packaged in membrane-bound granules. These granules move to the cell wall by a process that appears to involve microtubules,

Table 19–1. Structure of human insulin (molecular weight 5808) and (below) variations in this structure in other mammalian species. In the rat, the islet cells secrete 2 slightly different insulins, and in certain fish 4 different chains are found.

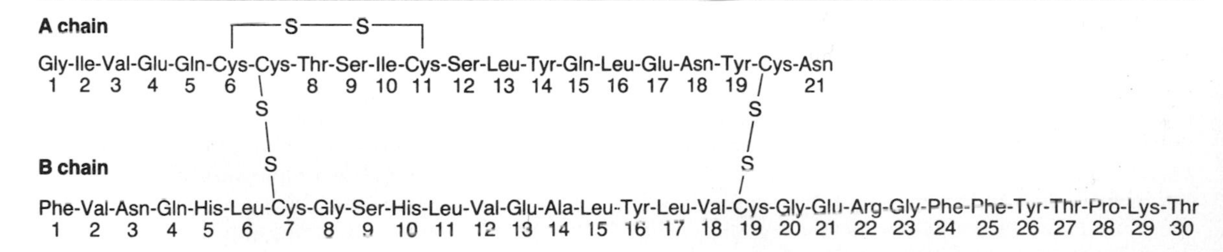

Species	Variations From Human Amino Acid Sequence: A Chain Position 8 9 10	B Chain Position 30
Pig, dog, sperm whale	Thr-Ser-Ile	Ala
Rabbit	Thr-Ser-Ile	Ser
Cattle, goat	Ala-Ser-Val	Ala
Sheep	Ala-Gly-Val	Ala
Horse	Thr-Gly-Ile	Ala
Sei whale	Ala-Ser-Thr	Ala

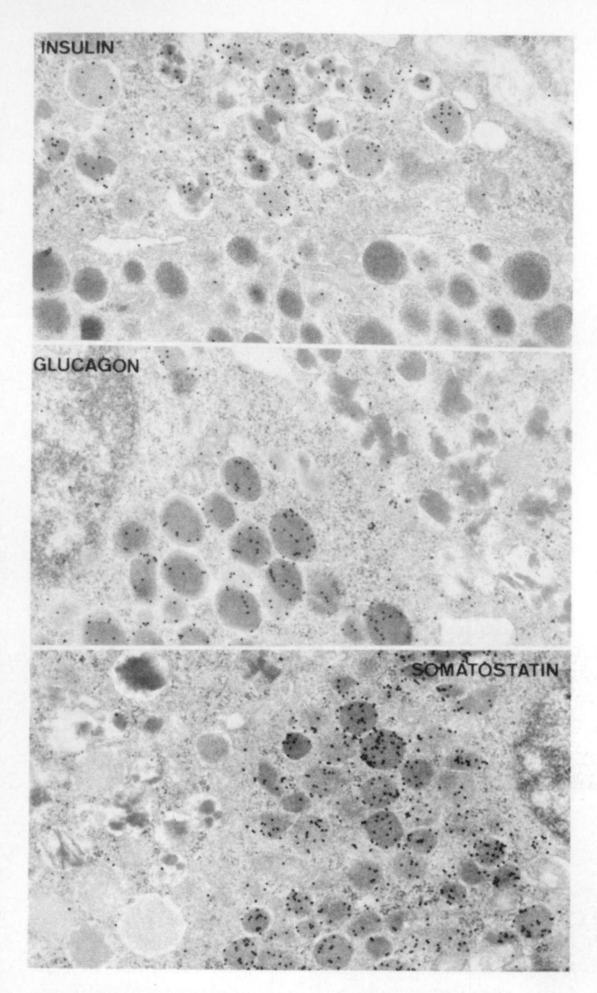

Figure 19–3. Sections of human pancreas, showing the location of individual hormones. The black spots identify insulin in granules in a B cell (top, × 28,000); glucagon in granules in an A cell (middle, × 31,500); and somatostatin in granules in a D cell (bottom, × 27,000). The hormones are stained by specific antisera, using the protein A–gold technique. (Courtesy of L Orci, Geneva, Switzerland.)

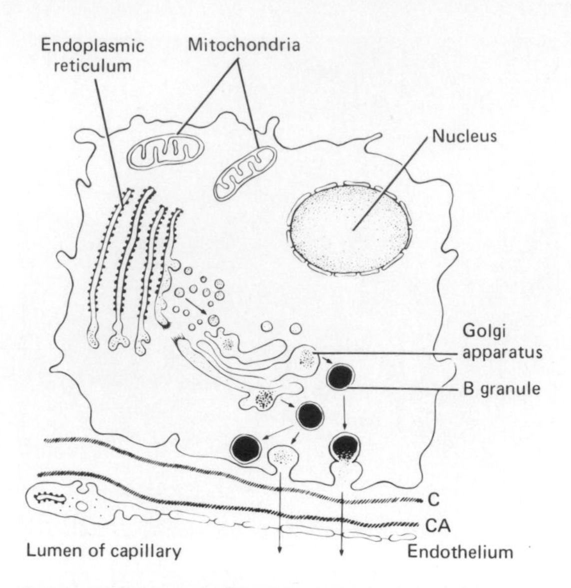

Figure 19–4. Schematic representation of insulin biosynthesis and secretion. Insulin is synthesized in the rough endoplasmic reticulum and translocated to the Golgi apparatus, where the B granules are formed. The granules fuse to the cell membranes, insulin is released by exocytosis, and the hormone passes through the basal lamina of the B cell (C) and the basal lamina (CA) plus the fenestrated endothelium of a nearby capillary on its way to the bloodstream.

and their membranes fuse with the membrane of the cell, expelling the insulin to the exterior by exocytosis (see Chapter 1). The insulin then crosses the basal laminas of the B cell and a neighboring capillary and the fenestrated endothelium of the capillary (Fig 19–4) to reach the bloodstream. The fenestrations are discussed in detail in Chapter 30.

Like other polypeptide hormones and related proteins that enter the endoplasmic reticulum, insulin is synthesized as part of a larger preprohormone (see Chapter 1). The gene for insulin is located on the short arm of chromosome 11 in humans. It has 2 introns and 3 exons (Fig 19–5). **Preproinsulin** has a 23-amino-acid signal peptide (leader sequence) removed as it enters the endoplasmic reticulum. The remainder of the molecule is then folded, and the disulfide bonds are formed to make **proinsulin.** Proinsulin is secreted after prolonged stimulation and by some islet cell tumors, but the peptide segment connecting the A and B chains is normally detached in the granules before secretion. Without the connection, the proper folding of the molecule for the formation of the disulfide bridges would be difficult. The polypeptide that remains in addition to insulin after the connection is severed is called the **connecting peptide (C peptide).** It contains 31 amino acid residues and has about 10% of the biologic activity of insulin. It enters the bloodstream along with the insulin when the granule contents are extruded by exocytosis. It can be measured by radioimmunoassay, and its level provides an index of B cell function in patients receiving exogenous insulin.

Tissue kallikrein (see Chapter 31) may play a role in the conversion of proinsulin to insulin. This endopeptidase is found in pancreatic islets, and its distribution parallels that of insulin.

FATE OF SECRETED INSULIN

Insulin & Insulinlike Activity in Blood

Plasma contains a variety of substances with insulinlike activity in addition to insulin (Table 19–2). Indeed, if insulinlike activity is measured by deter-

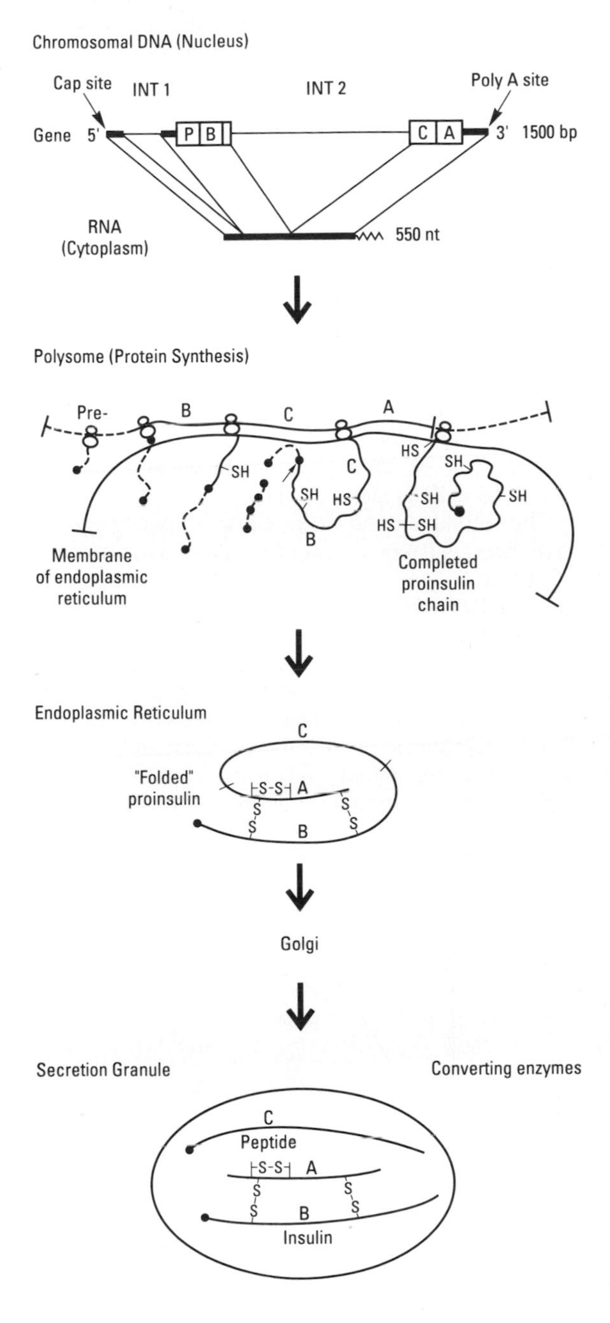

Figure 19–5. Biosynthesis of insulin. The 3 exons of the insulin gene (top) are separated by 2 introns (INT 1 and INT 2). Exons 1 and 2 code for an untranslated part of the mRNA, exon 2 codes for the signal peptide (P) and the B chain (B), exons 2 and 3 code for the C peptide (C), and exon 3 codes for the A chain (A) plus an untranslated part of the mRNA. bp, base pairs; nt, nucleotides. The signal peptide guides the polypeptide chain into the endoplasmic reticulum and is then removed. The molecule is next folded, with formation of the disulfide bonds. The C peptide is separated by one or more converting enzymes in the secretory granule.

Table 19–2. Substances with insulinlike activity in human plasma.

Insulin
Proinsulin
Nonsuppressible insulinlike activity (NSILA)
Low-molecular-weight fraction
IGF-I (somatomedin C)
IGF-II (multiplication-stimulating activity)
High-molecular-weight fraction

mining the glucose uptake and gas exchange in adipose tissue, only 7% of the plasma insulin activity is suppressed by anti-insulin antibodies, and the remaining 93%, which has been called **nonsuppressible insulinlike activity (NSILA),** is unaffected by these antibodies. Much of the NSILA persists after pancreatectomy. About 5% of the NSILA is due to polypeptide growth factors (somatomedins) that have insulinlike activity (see Chapter 22). These include 2 insulinlike growth factors, IGF-I and IGF-II. The structures of IGF-I and IGF-II resemble the structure of insulin. Human IGF-I contains 70 amino acid residues and is formed from a preprohormone containing 130 amino acid residues. Human IGF-II contains 67 amino acid residues and is formed from a preprohormone containing 180 amino acid residues. The properties of IGF-I and IGF-II are compared with those of insulin in Table 22–3. The remaining NSILA is high-molecular-weight material sometimes called **nonsuppressible insulinlike protein (NSILP),** the origin and significance of which are uncertain.

One may well ask why pancreatectomy causes diabetes mellitus (see below) when it only removes the source of insulin, with much of the NSILA persisting in the plasma. Of course the bioassay assesses only one of the actions of insulin in vivo, but the main point seems to be that even in the presence of NSILA, the additional action of insulin is required to maintain glucose metabolism in the normal range.

Metabolism

The half-life of insulin in the circulation in humans is about 5 minutes. Almost all tissues have the ability to metabolize insulin, but over 80% of secreted insulin is normally degraded in the liver and kidneys. Three insulin-inactivating systems have been described. Two break the disulfide linkages in the molecule—one enzymatically and one nonenzymatically—and one cleaves the peptide chains. The enzyme involved in the enzymatic disruption of the disulfide linkages is **hepatic glutathione insulin transhydrogenase,** which breaks the insulin molecule into A and B chains. Glutathione is a sulfur-containing tripeptide that in this case is acting as a coenzyme for the transhydrogenase. The enzymes responsible for the inactivation of insulin used to be grouped together under the term "insulinase." Variations in the rate of insulin inactivation in various tissues have been reported, but their significance is uncertain.

CONSEQUENCES OF INSULIN DEFICIENCY & ACTIONS OF INSULIN

The biologic effects of insulin are so far-reaching and complex that they are best illustrated by a consideration of the consequences of insulin deficiency.

Diabetes Mellitus

In humans, insulin deficiency is a common and serious pathologic condition. In animals, it can be produced by pancreatectomy; by the administration of alloxan, streptozocin, or other toxins that in appropriate doses cause selective destruction of the B cells of the pancreatic islets; by drugs that inhibit insulin secretion; and by administration of anti-insulin antibodies. Strains of mice, rats, hamsters, guinea pigs, miniature swine, and monkeys that have a high incidence of spontaneous diabetes mellitus have also been described.

The constellation of abnormalities caused by insulin deficiency is called **diabetes mellitus.** Greek and Roman physicians used the term ''diabetes'' to refer to conditions in which the cardinal finding was a large urine volume, and 2 types were distinguished: ''diabetes mellitus,'' in which the urine tasted sweet; and ''diabetes insipidus,'' in which the urine was tasteless. Today, the term diabetes insipidus is reserved for the condition produced by lesions of the supraoptic-posterior pituitary system (see Chapter 14), and the unmodified word diabetes is generally used as a synonym for diabetes mellitus.

Diabetes is characterized by polyuria, polydipsia, weight loss in spite of polyphagia (increased appetite), hyperglycemia, glycosuria, ketosis, acidosis, and coma. There are widespread biochemical abnormalities, but the fundamental defects to which most of the abnormalities can be traced are (1) reduced entry of glucose into various ''peripheral'' tissues and (2) increased liberation of glucose into the circulation from the liver (increased **hepatic glucogenesis).** There is therefore an extracellular glucose excess and, in many cells, an intracellular glucose deficiency, a situation that has been called ''starvation in the midst of plenty.'' There is also a decrease in the entry of amino acids into muscle and an increase in lipolysis.

It is now clear that there is an absolute or relative hypersecretion of glucagon in diabetes. This is true even when the pancreas is removed, because glucagon is secreted by the gastrointestinal tract as well as the pancreas. Somatostatin inhibits the secretion of insulin and glucagon (see Chapter 14), and when it is infused in pancreatectomized animals, the blood glucose falls toward normal. Consequently, it appears that the extracellular glucose excess in diabetes is due in part to hyperglucagonemia. However, some hyperglycemia persists even when glucagon secretion is reduced to zero.

Glucose Tolerance

In diabetes, glucose piles up in the bloodstream, especially after meals. If a glucose load is given to a diabetic, the blood glucose rises higher and returns to the baseline more slowly than it normally does. The response to a standard oral test dose of glucose, the **oral glucose tolerance test,** is used in the clinical diagnosis of diabetes (Fig 19–6).

Impaired glucose tolerance in diabetes is due in part to reduced entry of glucose into cells **(decreased peripheral utilization).** In the absence of insulin, the entry of glucose into skeletal muscle, cardiac and smooth muscle, and other tissues is decreased. Glucose uptake by the liver is also reduced, but the effect is indirect (see below). Intestinal absorption of glucose is unaffected, as is its reabsorption from the urine by the cells of the proximal tubules of the kidney. Glucose uptake by most of the brain and the red blood cells is also normal.

The second and the major cause of hyperglycemia in diabetes is derangement of the glucostatic function of the liver (see Chapter 17). The liver takes up glucose from the bloodstream and stores it as glycogen, but because the liver contains glucose 6-phosphatase it also discharges glucose into the bloodstream. Indeed, Claude Bernard spoke of the liver as an endocrine gland that secreted glucose. Insulin facilitates glycogen synthesis and inhibits hepatic glucose output. When the blood glucose is high, insulin secretion is normally increased and hepatic glucogenesis is decreased. After an overnight fast, 50% of glucose utilization is accounted for by

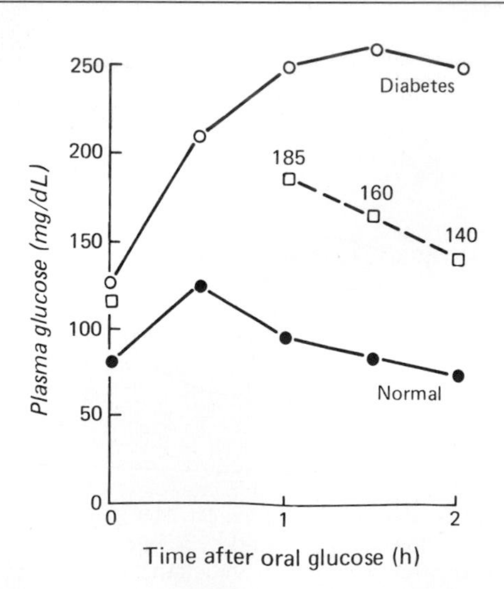

Figure 19–6. Glucose tolerance curves following oral administration of glucose, 40 g/m^2 body surface. According to American Diabetes Association standards (*Diabetes* 1969;**18**:299), chemical diabetes is present if the fasting glucose is greater than 115 mg/dL or if the 1-, 1.5-and 2-hour postglucose values are greater than 185, 165, and 140 mg/dL, respectively, with all 3 values abnormal.

the brain and 25% is accounted for by other tissues that do not respond to insulin; only 25% of utilization is affected by insulin. Under these circumstances, the main function of insulin is clearly inhibition of glucose output by the liver.

In diabetes, the glucose output remains elevated (Fig 19–7), in part because of the hyperglucagonemia. Glucagon exerts its effect on glucose output via a protein kinase (Fig 19–8) that inhibits glycogen synthesis, facilitates glycogenolysis, and inhibits conversion of phosphoenolpyruvate to pyruvate. It also decreases the concentration of fructose 2,6-diphosphate, and this in turn inhibits the conversion of fructose 6-phosphate to fructose, 1,6-diphosphate. The resultant buildup of glucose 6-phosphate leads to increased release of glucose. Catecholamines, cortisol, and growth hormone also cause an increase in glucose output when the stress of illness is severe enough to increase circulating levels of these hormones.

Hypoglycemic Action of Insulin

In muscles, fat, and a variety of other tissues (Table 19–3), insulin facilitates the entry of glucose into the cells by an action on the cell membranes. The rate of phosphorylation of the glucose, once it has entered the cells, is apparently regulated by other hormones. Growth hormone and cortisol both have been reported to inhibit phosphorylation in certain tissues. However, the process is normally so rapid that it is a rate-limiting step in glucose metabolism only when the rate of glucose entry is high.

Because of the rapid phosphorylation, the intracellular free glucose concentration is normally low, and the concentration gradient for glucose is directed inward. Some glucose moves down this gradient into cells in the absence of insulin, but insulin greatly increases the rate of glucose transport. The transport shows the saturation kinetics of a system in which there is limited availability of a specific carrier. It is thus an example of **facilitated diffusion.** The process is highly specific, being limited for the most part to glucose and a few other sugars with the same configuration in the first 3 carbon atoms; furthermore, these sugars compete with one another for transport.

The carrier involved in glucose transport across the cell membrane is the **glucose transporter.** This is a large protein, which appears to span the membrane 12 times. Insulin causes insertion of more glucose transporters into the cell membrane. Blood glucose starts to fall immediately, and following a single intravenous dose of insulin, the decline is maximal in about 30 minutes.

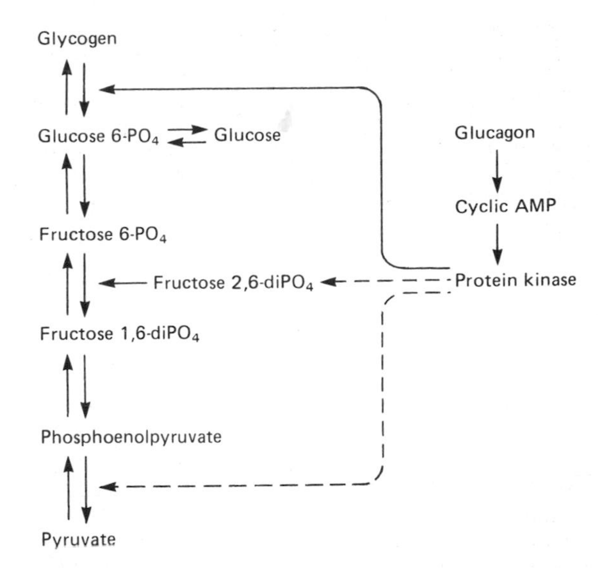

Figure 19–8. Mechanisms by which glucagon increases glucose output from the liver. Dashed arrows indicate inhibition. (Modified from Foster DW, McGarry JD: The metabolic derangements and treatment of diabetic ketoacidosis. *N Engl J Med* 1983;**309**:159.)

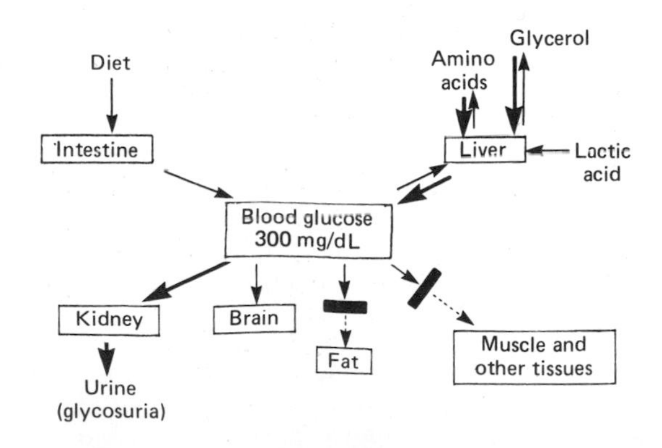

Figure 19–7. Disordered blood glucose homeostasis in insulin deficiency. Compare with Fig 17–14. The heavy arrows indicate reactions that are accentuated. The rectangles across arrows indicate reactions that are blocked.

Table 19–3. Effect of insulin on glucose uptake in tissues in which it has been investigated.

Tissues in which insulin facilitates glucose uptake
Skeletal muscle
Cardiac muscle
Smooth muscle
Adipose tissue
Leukocytes
Crystalline lens of the eye
Pituitary
Fibroblasts
Mammary gland
Aorta
A cells of pancreatic islets
Tissues in which insulin does not facilitate glucose uptake
Brain (except probably part of hypothalamus)
Kidney tubules
Intestinal mucosa
Red blood cells

Insulin does not affect the movement of glucose across the membranes of hepatic cells directly, but it brings about a facilitation of glycogen synthesis and a net decrease in glucose output. Insulin suppresses the synthesis of key glyconeogenic enzymes and induces the synthesis of key glycolytic enzymes such as glucokinase. Glycogen synthase activity is also increased. However, it has not been proved that these increases are directly due to insulin.

Insulin Preparations

As noted above, the maximal decline in blood glucose occurs 30 min after intravenous injection of crystalline insulin. After subcutaneous administration, the maximal fall occurs in 2–3 hours (Table 19–4). A wide variety of insulin preparations are now available commercially. These include insulins that have been complexed with protamine and other polypeptides to delay absorption. In general, however, they fall into 3 categories: rapid-, intermediate-, and long-acting. The main intermediate-acting insulins are NPH and lente insulins (Table 19–4), and the main long-acting insulin is protamine zinc insulin.

Distribution of Endogenous & Exogenous Insulin

The action of insulin on the liver assumes added significance in view of the fact that endogenously secreted insulin enters the portal vein, so that the liver is normally exposed to concentrations of insulin which are 3–10 times greater than those in peripheral tissues. The liver binds about half a dose of insulin injected into the portal vein but only 25% of a peripherally injected dose. In addition, the liver is more sensitive to insulin than are peripheral tissues. Exogenous insulin injected in the treatment of diabetes enters the peripheral rather than the portal circulation, although there is debate about whether the resultant unphysiologic distribution of insulin has any important consequences in terms of control of the disease or prevention of its complications.

Table 19–4. Characteristics of the blood-glucose-lowering effects of crystalline zinc insulin and several modified insulins.

	Hours After Subcutaneous Administration	
Type of Insulin	**Peak Action**	**Duration of Action**
Crystalline zinc insulin (CZI)	2–3	5–7
NPH insulin (neutral protamine-Hagedorn insulin)	10–20	24–28
Lente insulin	10–20	24–28
Protamine zinc insulin (PZI)	16–24	36+

Effects of Hyperglycemia

Hyperglycemia by itself can cause symptoms resulting from the hyperosmolality of the blood. In addition, there is glycosuria because the renal capacity for glucose reabsorption is exceeded. Excretion of the osmotically active glucose molecules entails the loss of large amounts of water (osmotic diuresis; see Chapter 38). The resultant dehydration activates the mechanisms regulating water intake, leading to polydipsia. There is an appreciable urinary loss of Na^+ and K^+ as well. For every gram of glucose excreted, 4.1 kcal are lost from the body. Increasing oral caloric intake to cover this loss simply raises the blood glucose further and increases the glycosuria, so mobilization of endogenous protein and fat stores and weight loss are not prevented.

Relation to Potassium

Insulin causes K^+ to enter cells, with a resultant lowering of the extracellular K^+ concentration. Infusions of insulin and glucose significantly lower the plasma K^+ level in normal individuals and are very effective for the temporary relief of hyperkalemia in patients with renal failure. Hypokalemia often develops when patients with diabetic acidosis are treated with insulin. The reason for the intracellular migration of K^+ is still uncertain. However, insulin increases the activity of Na^+-K^+ ATPase in cell membranes, so that more K^+ is pumped into cells. This activation may be secondary to activation of the transport system that moves H^+ out of cells in exchange for Na^+. The activity of the Na^+-K^+ ATPase is increased by a rise in intracellular Na^+, and insulin has been reported to increase intracellular pH.

K^+ depletion decreases insulin secretion, and K^+-depleted patients, eg, patients with primary hyperaldosteronism (see Chapter 20), develop diabetic glucose tolerance curves. These curves are restored to normal by K^+ repletion. The thiazide diuretics, which cause loss of K^+ as well as Na^+ in the urine (Table 38–10), decrease glucose tolerance and make diabetes worse. They apparently exert this effect primarily because of their K^+-depleting effects, although some of them also cause pancreatic islet cell damage.

Exercise

The entry of glucose into skeletal muscle is increased during exercise in the absence of insulin. The increased muscle uptake may be due in part to relative O_2 deficiency, because glucose entry into cells is increased under anaerobic conditions. However, exercise also increases the affinity of the insulin receptors in muscle. Exercise can precipitate hypoglycemia in diabetics taking insulin not only because of the effect on insulin receptors and the production of relative O_2 deficiency but also because absorption of injected insulin is more rapid during exercise. Patients with diabetes should take in extra calories or reduce their insulin dosage when they exercise.

Effects of Intracellular Glucose Deficiency

The plethora of glucose outside the cells in diabetes contrasts with the intracellular deficit. Glucose catabolism is normally a major source of energy for cellular processes, and in diabetes, energy requirements can be met only by drawing on protein and fat reserves. Mechanisms are activated that greatly increase the catabolism of protein and fat, and one of the consequences of increased fat catabolism is ketosis.

Deficient glucose utilization in the cells of the hypothalamic ventromedial nuclei is probably the cause of the hyperphagia in diabetes. When the activity of the satiety center is decreased in response to decreased glucose utilization in its cells, the lateral appetite center operates unopposed, and food intake is increased (see Chapter 14).

Glycogen depletion is a common consequence of intracellular glucose deficit, and the glycogen content of liver and skeletal muscle in diabetic animals is usually reduced. Insulin regularly increases the glycogen content of skeletal muscle, and it increases liver glycogen unless it produces sufficient hypoglycemia to also activate glycogenolytic mechanisms.

Changes in Protein Metabolism

In diabetes, the rate at which amino acids are catabolized to CO_2 and H_2O is increased. In addition, more amino acids are converted to glucose in the liver. This shift is indicated in Fig 19–9, which also shows the other principal abnormalities of intermediary metabolism in the liver.

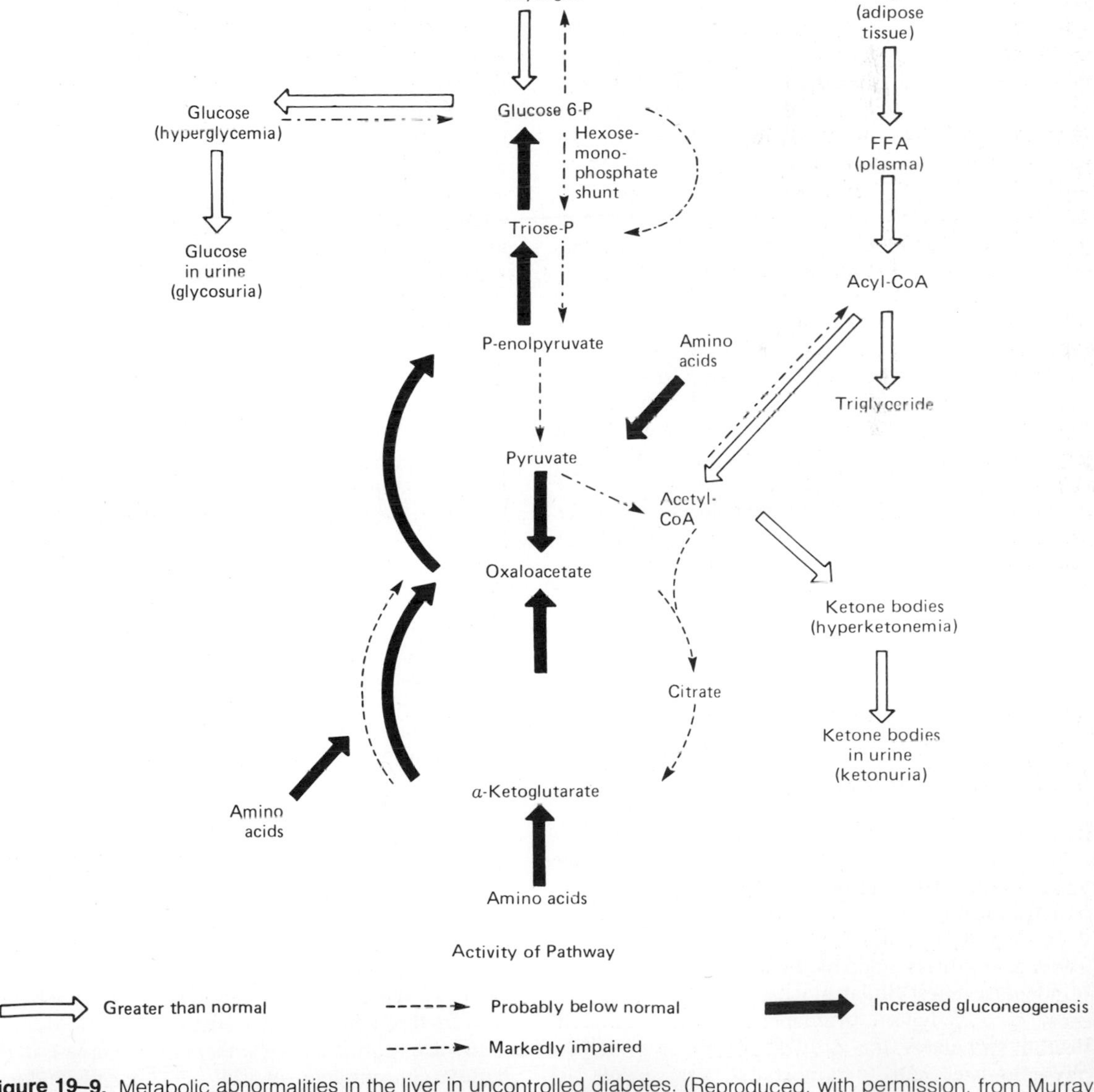

Figure 19–9. Metabolic abnormalities in the liver in uncontrolled diabetes. (Reproduced, with permission, from Murray RK, Granner DK, Mayes PA, Rodwell VW: *Harper's Review of Biochemistry,* 21st ed. Lange, 1988.)

Some idea of the rate of gluconeogenesis in fasting diabetic animals is obtained by measuring the ratio of glucose (dextrose) to nitrogen in the urine **(D/N ratio).** In fasting animals, liver glycogen is depleted and glycerol is converted to glucose at a very limited rate, so that the only important source of plasma glucose is protein (see Chapter 17). It can be calculated that the amount of carbon in the protein represented by 1 g of urinary nitrogen is sufficient to form 8.3 g of glucose. Consequently, the D/N ratio of approximately 3 seen in diabetes indicates the conversion to glucose of about 33% of the carbon of the protein metabolized.

The causes of the increased gluconeogenesis are multiple. Glucagon stimulates gluconeogenesis, and hyperglucagonemia is generally present in diabetes. Adrenal glucocorticoids also contribute to increased gluconeogenesis when they are elevated in severely ill diabetics. There is an increased supply of amino acids for gluconeogenesis because, in the absence of insulin, less protein synthesis occurs in muscle, and blood amino acid levels rise. Alanine is particularly easily converted to glucose. In addition, the activity of the enzymes that catalyze the conversion of pyruvate and other 2-carbon metabolic fragments to glucose is increased. These include phosphoenolpyruvate carboxykinase, which facilitates conversion of oxaloacetate to phosphoenolpyruvate (see Chapter 17). They also include fructose 1,6-diphosphatase, which catalyzes the conversion of fructose diphosphate to fructose 6-phosphate, and glucose 6-phosphatase, which controls the entry of glucose into the circulation from the liver. Increased acetyl-CoA increases pyruvate carboxylase activity, and insulin deficiency increases the supply of acetyl-CoA because lipogenesis is decreased. Pyruvate carboxylase catalyzes the conversion of pyruvate to oxaloacetate (Fig 17–9).

When plasma glucose is episodically elevated over a period of time, small amounts of hemoglobin A are nonenzymatically glycosylated to form **HbA$_{IC}$** (see Chapter 27). Careful control of the diabetes with insulin reduces this level, and consequently HbAIC concentration is measured clinically as an integrated index of diabetic control for the 4- to 6-week period before the measurement. Albumin is also glycosylated, and glycosylated albumin turns over more rapidly, so it provides an index of more rapid fluctuations in diabetic control.

Insulin & Growth

Not only is protein catabolism accelerated in the absence of insulin, but protein synthesis is depressed and insulin stimulates protein formation by increasing the transport of amino acids into cells. This anabolic effect of insulin is aided by the protein-sparing action of adequate intracellular glucose supplies. Failure to grow is a symptom of diabetes in children, and insulin stimulates the growth of immature hypophysectomized rats to almost the same degree as growth hormone. Maximum insulin-induced growth is present, however, only when the protein-sparing action of glucose is fostered by feeding a high-carbohydrate diet.

Negative Nitrogen Balance

In diabetes, the net effect of accelerated protein catabolism to CO_2 and H_2O and to glucose, plus diminished protein synthesis, is a markedly negative nitrogen balance, protein depletion, and wasting. Protein depletion from any cause is associated with poor "resistance" to infections, and the sugar-rich body fluids are undoubtedly good culture media for microorganisms. This is probably why diabetics are particularly prone to bacterial infections.

Fat Metabolism in Diabetes

The principal abnormalities of fat metabolism in diabetes are acceleration of lipid catabolism, with increased formation of ketone bodies, and decreased synthesis of fatty acids and triglycerides. The manifestations of the disordered lipid metabolism are so prominent that diabetes has been called "more a disease of lipid than of carbohydrate metabolism."

Fifty percent of an ingested glucose load is normally burned to CO_2 and H_2O; 5% is converted to glycogen; and 30–40% is converted to fat in the fat depots. In diabetes, less than 5% is converted to fat even though the amount burned to CO_2 and H_2O is also decreased and the amount converted to glycogen is not increased. Therefore, glucose accumulates in the bloodstream and spills over into the urine.

The role of lipoprotein lipase and hormone-sensitive lipase in the regulation of the metabolism of fat depots is discussed in Chapter 17. In diabetes, there is decreased conversion of glucose to fatty acids in the depots because of the intracellular glucose deficiency. Insulin inhibits the hormone-sensitive lipase in adipose tissue, and, in the absence of this hormone, the plasma level of **free fatty acids** (NEFA, UFA, FFA) is more than doubled. The increased glucagon also contributes to the mobilization of FFA. Thus, the FFA level parallels the blood glucose level in diabetes and in some ways is a better indicator of the severity of the diabetic state. In the liver and other tissues, the fatty acids are catabolized to acetyl-CoA. Some of the acetyl-CoA is burned along with amino acid residues to yield CO_2 and H_2O in the citric acid cycle. However, the supply exceeds the capacity of the tissues to catabolize the acetyl-CoA.

The events occurring in the liver in diabetes are summarized in Fig 19–9. In addition to the previously mentioned increase in gluconeogenesis and marked outpouring of glucose into the circulation, there is a marked impairment of the conversion of acetyl-CoA to malonyl-Coa and thence to fatty acids. This is due to a deficiency of acetyl carboxylase, the enzyme that catalyzes the conversion. The excess acetyl-CoA is converted to ketone bodies (see below).

In uncontrolled diabetes, there is an increase in the plasma concentration of triglycerides and chylomicrons as well as FFA, and the plasma is often lipemic.

The rise in these constituents is due mainly to decreased removal of triglycerides into the fat depots. The decreased activity of lipoprotein lipase contributes to this decreased removal.

Ketosis

When there is excess acetyl-CoA in the body, some of it is converted to acetoacetyl-CoA and then, in the liver, to acetoacetate. Acetoacetate and its derivatives, acetone and β-hydroxybutyrate, enter the circulation in large quantities (see Chapter 17).

These circulating ketone bodies are an important source of energy in fasting. Half of the metabolic rate in fasted normal dogs is said to be due to metabolism of ketones. The rate of ketone utilization in diabetics is also appreciable. It has been calculated that the maximal rate at which fat can be catabolized without significant ketosis is 2.5 g/kg body weight/d in diabetic humans. In untreated diabetes, production is much greater than this, and ketone bodies pile up in the bloodstream. There is some evidence that in severe diabetes the rate of ketone utilization may also decline, making the ketosis worse, and insulin is said to increase ketone uptake in muscle.

Acidosis.

Most of the hydrogen ions liberated from acetoacetate and β-hydroxybutyrate are buffered, but severe metabolic acidosis still develops. The low plasma pH stimulates the respiratory center, producing the rapid, deep respiration described by Kussmaul as ''air hunger'' and named, for him, **Kussmaul breathing.** The urine becomes acidic. However, when the ability of the kidneys to replace the plasma cations accompanying the organic anions with H^+ and NH_4^+ is exceeded, Na^+ and K^+ are lost in the urine. The electrolyte and water losses lead to dehydration, hypovolemia, and hypotension. Finally, the acidosis and dehydration depress consciousness to the point of coma. Diabetic acidosis is a medical emergency. Now that the infections which used to complicate the disease can be controlled with antibiotics, acidosis is the commonest cause of early death in clinical diabetes.

In severe acidosis, total body sodium is markedly depleted, and when sodium loss exceeds water loss, plasma Na^+ is not infrequently low. Total body potassium is also low, but the plasma K^+ is usually normal, partly because ECF volume is reduced and partly because K^+ moves from cells to ECF when the ECF H^+ concentration is high. Another factor tending to maintain the plasma K^+ is the lack of insulin-induced entry of K^+ into cells. This must be kept in mind in treating acidosis; total body stores of this ion are low, and severe and even fatal hypokalemia may develop when insulin is given (see above).

The degree to which ketoacidosis complicates experimental diabetes varies in different species. The size of the body fat stores is also a factor conditioning the response to diabetes. Thin, wasted pancreatectomized dogs rarely develop ketoacidosis, whereas well-fed, plump animals do so readily. It is pertinent that before the isolation of insulin by Banting and Best in 1921 the principal treatment of human diabetes was a starvation diet **(Allen regiment).** This not only lowered the blood glucose level but also reduced depot fat stores to the point where there was little fat to mobilize.

Coma

Coma in diabetes can be due to acidosis and dehydration. However, the blood glucose can be elevated to such a degree that independent of plasma pH, the hyperosmolarity of the plasma causes unconsciousness **(hyperosmolar coma).** Accumulation of lactate in the blood **(lactic acidosis)** may also complicate diabetic ketoacidosis if the tissues become hypoxic (see Chapter 33), and lactic acidosis may itself cause coma. Brain edema is seen in a significant number of patients with diabetic acidosis, and it can cause coma. The reason for the brain edema is unknown, but it is a serious complication with a bad prognosis.

Cholesterol Metabolism

In diabetes, the plasma cholesterol level is usually elevated, and this plays a role in the accelerated development of the arteriosclerotic vascular disease that is a major long-term complication of diabetes in humans. The rise in plasma cholesterol level is due to an increase in the plasma concentration of VLDL and LDL (see Chapter 17), which may be due to increased hepatic production of VLDL or decreased removal of VLDL and LDL from the circulation.

Summary

Because of the complexities of the metabolic abnormalities in diabetes, a summary is in order. One of the key features of insulin deficiency (Fig 19–10) is decreased entry of glucose into many tissues (decreased peripheral utilization). There is also increased net release of glucose from the liver (increased production), due in part to glucagon excess.

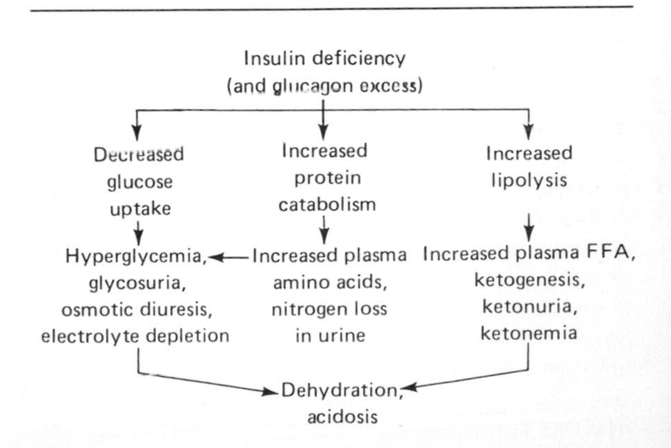

Figure 19–10. Effects of insulin deficiency. (Courtesy of RJ Havel.)

The resultant hyperglycemia leads to glycosuria and a dehydrating osmotic diuresis. Dehydration leads to polydipsia. In the face of intracellular glucose deficiency, appetite is stimulated, glucose is formed from protein (gluconeogenesis), and energy supplies are maintained by metabolism of proteins and fats. Weight loss, debilitating protein deficiency, and inanition are the result.

Fat catabolism is increased, and the system is flooded with triglycerides and FFA. Fat synthesis is inhibited, and the overloaded catabolic pathways cannot handle the excess acetyl-CoA formed. In the liver, the acetyl-CoA is converted to ketone bodies. The ketones are for the most part organic acids, which accumulate in the circulation (ketosis) because their production rate exceeds the ability of the body to utilize them. Therefore, metabolic acidosis develops as the ketones accumulate. Na^+ and K^+ depletion are added to the dehydration because these plasma cations are excreted with the organic anions not covered by the H^+ and $NH_4{}^+$ secreted by the kidney. Finally, the acidotic, hypovolemic, hypotensive, depleted animal or patient becomes comatose because of the toxic effects of acidosis, dehydration, and hyperosmolarity on the nervous system and dies if treatment is not instituted.

All of these abnormalities are corrected by administration of insulin. Although emergency treatment of acidosis also includes administration of alkali to combat the acidosis and parenteral water, Na^+, and K^+ to replenish body stores, only insulin repairs the fundamental defects in a way that permits a return to normal.

Table 19–5. Principal actions of insulin.

Adipose tissue
1. Increased glucose entry
2. Increased fatty acid synthesis
3. Increased glycerol phosphate synthesis
4. Increased triglyceride deposition
5. Activation of lipoprotein lipase
6. Inhibition of hormone-sensitive lipase
7. Increased K^+ uptake
Muscle
1. Increased glucose entry
2. Increased glycogen synthesis
3. Increased amino acid uptake
4. Increased protein synthesis in ribosomes
5. Decreased protein catabolism
6. Decreased release of gluconeogenic amino acids
7. Increased ketone uptake
8. Increased K^+ uptake
Liver
1. Decreased ketogenesis
2. Increased protein synthesis
3. Increased lipid synthesis
4. Decreased glucose output due to decreased gluconeogenesis and increased glycogen synthesis
General
1. Increased cell growth

MECHANISM OF ACTION OF INSULIN

The principal actions of insulin are summarized in Table 19–5. It is secreted after meals (see below), and its net effect is to promote the storage of carbohydrate, protein, and fat. It is therefore appropriately called the "hormone of abundance."

Insulin binds to insulin receptors in the cell membrane, and this triggers its actions. There is an appealing parsimony to the concept that insulin has a single action that underlies all its diverse effects on metabolism. One of its major actions is to increase the number of glucose transporters in the membrane. Another is to increase the activity of glycogen synthase. However, it also has many other actions (Table 19–5). Some of these occur in a matter of seconds or minutes, whereas others such as cell growth and synthesis of proteins and lipids require hours to develop. Thus, it appears that after insulin binds to its receptor, it triggers many different intracellular events.

Insulin Receptors

An important advance in endocrine physiology has been the characterization and cloning of the insulin receptor and the exploration of its actions. This receptor is a complex protein with a molecular weight of approximately 340,000. Insulin receptors are found on many different cells in the body, including cells in which insulin does not increase glucose uptake. The receptor is a tetramer made up of 2 α and 2 β glycoprotein subunits. The subunits are linked to each other and to the β subunits by disulfide bonds (Fig 19–11). The α subunits bind insulin and are extracellular, whereas the β subunits span the membrane. The intracellular ends of the β subunits have tyrosine kinase activity. The α and β subunits are both glycosylated, with sugar residues extending into the interstitial fluid. Binding of insulin triggers the tyrosine kinase activity of the β subunits, producing autophosphorylation of the β subunits on tyrosine residues. This autophosphorylation is necessary for insulin to exert its biologic effects. At present, it is uncertain how phosphorylation of the insulin receptors triggers the actions of the hormone, but presumably the phosphorylated receptor acts by phosphorylating and dephosphorylating other proteins inside the cell.

It is interesting to compare the insulin receptor with other related receptors. It is very similar to the receptor for IGF-I but different from the receptor for IGF-II, which has recently been cloned (Fig 19–11). Several other receptors for growth factors and receptors for various oncogenes are also tyrosine kinases. However, the amino acid composition of these receptors is quite different.

When insulin binds to its receptors, they aggregate in patches and are taken into the cell by receptor-mediated endocytosis (see Chapter 1). Eventually, the insulin-receptor complexes enter lysosomes, where the receptors are presumably broken down or recycled. The half-life of the insulin receptor is about

7 hours. There has been speculation that the internalized insulin has physiologic effects inside the cell, but this has not been proved.

The number or the affinity, or both, of insulin receptors is affected by insulin and other hormones, exercise, food, and other factors. Exposure to increased amounts of insulin decreases receptor concentration (down regulation), and exposure to decreased insulin increases the affinity of the receptors. The number of receptors per cell is increased in starvation and decreased in obesity and acromegaly. The affinity of the receptors is increased in adrenal insufficiency and decreased by excess glucocorticoids (see below).

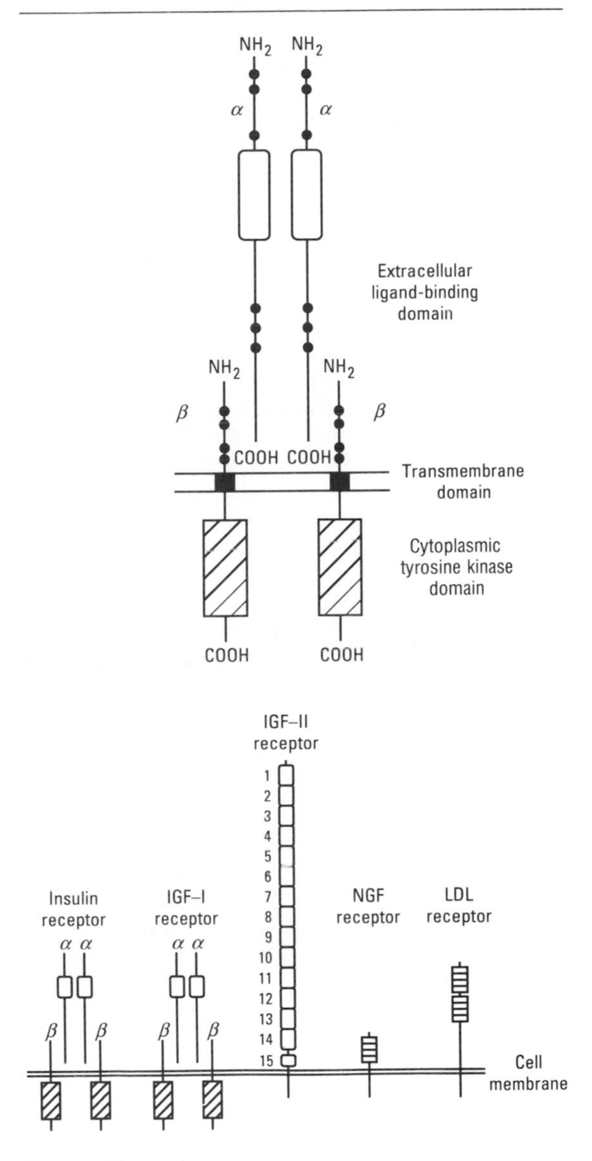

Figure 19-11. Structure of insulin receptor (top) and comparison of its structure with that of the IGF-I, IGF-II, nerve growth factor (NGF), and LDL receptors. The open boxes identify extracellular cysteine-rich regions and repeat sequences. The dots on the α and β chains in the top diagram identify single cysteine residues that may be involved in the formation of disulfide bonds. The striped boxes identify intracellular tyrosine kinase domains. Note the marked similarity between the insulin receptor and the IGF-I receptor; also note the 15 repeat sequences in the extracellular portion of the IGF-II receptor. (Modified from Ulrich A et al: Human insulin receptor and its relationship to the tyrosine kinase family of oncogenes *Nature* 1985;**313**:756; and Morgan DA et al: Insulinlike growth factor II receptor as a multifunctional binding protein. *Nature* 1987;**329**:301.)

INSULIN EXCESS

Symptoms

All the known consequences of insulin excess are manifestations, directly or indirectly, of the effects of hypoglycemia on the nervous system. Except in individuals who have been fasting for some time, glucose is the only fuel used in appreciable quantities by the brain. The carbohydrate reserves in neural tissue are very limited, and normal function depends upon a continuous glucose supply. As the blood glucose level falls, the cortex and the other brain areas with high metabolic rates are affected first, followed by the more slowly respiring vegetative centers in the diencephalon and hindbrain (see Chapter 32). Thus, early cortical symptoms of confusion, weakness, dizziness, and hunger are followed by convulsions and coma. If the hypoglycemia is prolonged, irreversible changes develop in the same cortical-diencephalic-medullary sequence, and death results from depression of the respiratory center. Therefore, prompt treatment with glucose is in order. Although a dramatic disappearance of symptoms is the usual response, abnormalities ranging from some intellectual dulling to permanent coma may persist if the hypoglycemia was severe or prolonged. Hypoglycemia is a potent stimulus to sympathetic discharge and increased secretion of catecholamines, particularly epinephrine. The tremors, palpitations, and nervousness in hypoglycemia are probably due to sympathetic overactivity.

Hypoglycemia is sometimes said to be present when the blood glucose concentration is below 45 mg/dL, but the concentration at which symptoms of hypoglycemia appear is variable. It has been argued that some patients with insulin-secreting tumors adapt to blood glucose levels as low as 20 mg/dL, whereas diabetics adapted to chronic hyperglycemia may have definite symptoms of hypoglycemia with blood glucose values of 100 mg/dL. However, there is considerable debate about whether such adaptation really occurs.

Compensatory Mechanisms

Hypoglycemia triggers increased secretion of 5 counterregulatory hormones that antagonize the fall in the blood glucose level: epinephrine, norepinephrine, glucagon, growth hormone, and cortisol. Epinephrine, norepinephrine, glucagon, and cortisol increase hepatic output of glucose, the first 3 by increasing glycogenolysis. Growth hormone decreases the utilization of glucose in various peripheral tissues, and cortisol has a similar action (see below). The keys to counterregulation appear to be epinephrine and glucagon: if the plasma concentration of either increases, the decline in the blood glucose level is reversed; but if both fail to increase, there is little if any compensatory rise in the blood glucose level. The actions of the other hormones are supplementary.

REGULATION OF INSULIN SECRETION

Insulin secretion is now known to be influenced by a variety of stimulatory and inhibitory factors (Table 19–6). However, many of these factors are either substances related to glucose metabolism or agents that affect cyclic AMP. Normal secretion requires the presence of adequate quantities of Ca^{2+} and K^+, and there is evidence that increased intracellular Ca^{2+} triggers increased insulin secretion in a fashion similar to Ca^{2+} stimulation of exocytosis in neurons and other endocrine cells.

As noted above, only 7% of the insulinlike activity in plasma is due to insulin per se. The remaining bioassayable but not immunoassayable material is made up of a variety of growth factors and high-molecular-weight material. The normal concentration of insulin measured by radioimmunoassay in the peripheral venous plasma of fasting normal humans is 0–70 μU/mL (0–502 pmol/L). The amount of insulin secreted per day in a normal human has been calculated to be about 40 units (287 nmol).

Effect of the Blood Glucose Level

The major control of insulin secretion is exerted by a feedback effect of the blood glucose level directly on the pancreas. Glucose penetrates the B cells readily, and its rate of entrance is unaffected by insulin. When the level of glucose in the blood perfusing the pancreas is elevated (> 110 mg/dL in rats), insulin secretion in the pancreatic venous blood is increased; when the level is normal or low, the rate of insulin secretion is low (Fig 19–12). Mannose also stimulates insulin secretion. Fructose has a moderate stimulatory effect, but it is converted intracellularly into glucose. Neither galactose nor a variety of other sugars, citric acid cycle intermediates, and non-metabolizable sugars stimulate insulin secretion. However, the stimulatory effect of glucose depends on its metabolism, since 2-deoxyglucose and mannoheptulose, agents that prevent glucose from being metabolized, also inhibit insulin secretion.

In experimental animals (Fig 19–13) and humans, the action of glucose on insulin secretion is biphasic; there is a rapid increase in secretion followed by a more slowly developing prolonged response. The initial response is due to release of preformed insulin and is triggered by increased cytoplasmic Ca^{2+}. Glucose partially depolarizes the B cell, and this opens voltage-gated Ca^{2+} channels, with a resultant influx of extracellular Ca^{2+}. The prolonged response is due to release of newly synthesized insulin, since it is blocked by inhibitors of protein synthesis. If glucose perfusion is resumed after a rest period, the response to glucose is increased.

The feedback control of blood glucose on insulin secretion normally operates with great precision, so that blood glucose and blood insulin levels parallel each other with remarkable consistency.

Protein & Fat Derivatives

Arginine, leucine, and certain other amino acids stimulate insulin secretion, and so do β-keto acids

Table 19–6. Factors affecting insulin secretion.

Stimulators	Inhibitors
Glucose	Somatostatin
Mannose	2-Deoxyglucose
Amino acids (leucine, arginine, others)	Mannoheptulose
Intestinal hormones (GIP, gastrin, secretin, CCK, glucagon, others?)	α-Adrenergic stimulating agents (norepinephrine, epinephrine)
β-Keto acids	β-Adrenergic blocking agents (propranolol)
Acetylcholine	Diazoxide
Glucagon	Thiazide diuretics
Cyclic AMP and various cyclic AMP-generating substances	K^+ depletion
β-Adrenergic stimulating agents	Phenytoin
Theophylline	Alloxan
Sulfonylureas	Microtubule inhibitors
	Insulin

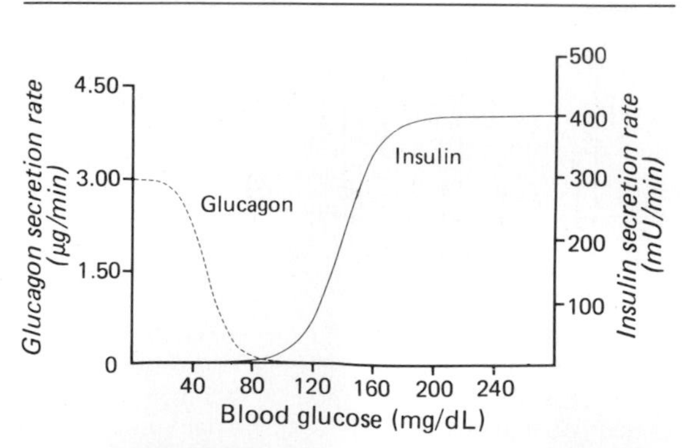

Figure 19–12. Mean rates of insulin and glucagon delivery from an artificial pancreas at various blood glucose levels. The device was programmed to establish and maintain normal blood glucose in insulin-requiring diabetic humans, and the values for hormone output approximate the output of the normal human pancreas. The shape of the insulin curve also resembles the insulin response of incubated B cells to graded concentrations of glucose. (Reproduced, with permission, from Marliss EB et al: Normalization of glycemia in diabetics during meals with insulin and glucagon delivery by the artificial pancreas. *Diabetes* 1977;26:663.)

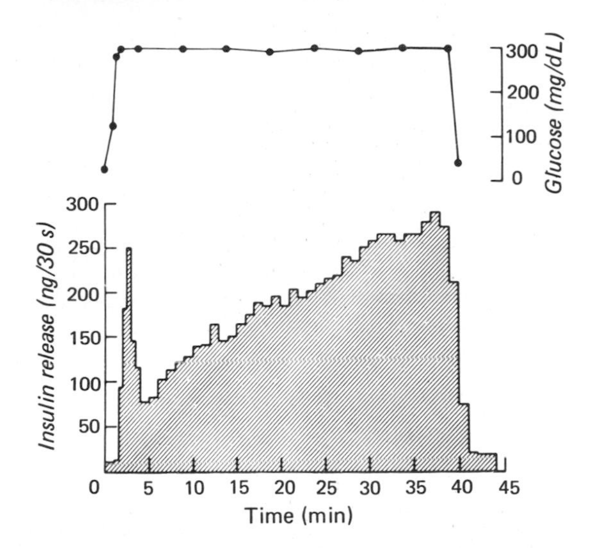

Figure 19–13. Insulin secretion from perfused rat pancreas in response to sustained glucose infusion. Values are means of 3 preparations. The top record shows the glucose concentration in the effluent perfusion mixture. (Reproduced, with permission, from Curry DL, Bennett LL, Grodsky GM: Dynamics of insulin secretion by the perfused rat pancreas. *Endocrinology* 1968;**83**:572.)

such as acetoacetate. The mechanisms by which these acids bring about the stimulation are unknown. However, it is worth noting that insulin stimulates the incorporation of amino acids into proteins and combats the fat catabolism that produces the β-keto acids.

Cyclic AMP & Insulin Secretion

Stimuli that increase cyclic AMP in B cells increase insulin secretion, probably by increasing intracellular Ca^{2+}. These include β-adrenergic agonists, glucagon, and phosphodiesterase inhibitors such as theophylline.

Catecholamines have a dual effect on insulin secretion; they inhibit insulin secretion via α_2-adrenergic receptors and stimulate insulin secretion via β-adrenergic receptors. The net effect of epinephrine and norepinephrine is usually inhibition. However, if catecholamines are infused after administration of α-adrenergic blocking drugs, the inhibition is converted to stimulation.

Effect of Autonomic Nerves

Branches of the right vagus nerve innervate the pancreatic islets, and stimulation of the right vagus causes increased insulin secretion. Atropine blocks the response, and acetylcholine stimulates insulin secretion. The effect of acetylcholine, like that of glucose, is due to increased cytoplasmic Ca^{2+}, but acetylcholine activates phospholipase C, with the released IP_3 releasing the Ca^{2+} from the endoplasmic reticulum.

Stimulation of the sympathetic nerves to the pancreas inhibits insulin secretion. The inhibition is produced via release of norepinephrine, and the inhibitory response is converted to an excitatory one by infusion of α-adrenergic blockers. Thus, the autonomic innervation of the pancreas is involved in the regulation of insulin secretion. The effects of glucose do not require intact innervation, since they occur in the transplanted pancreas, but there is some evidence that the nerve fibers maintain normal islet sensitivity to glucose.

Intestinal Hormones

Orally administered glucose exerts a greater insulin-stimulating effect than intravenously administered glucose, and orally administered amino acids also produce a greater insulin response than intravenous amino acids. These observations led to exploration of the possibility that a substance secreted by the gastrointestinal mucosa stimulated insulin secretion. Glucagon, secretin, CCK, gastrin, and gastric inhibitory peptite (GIP) all have such an action (see Chapter 26), and CCK potentiates the insulin-stimulating effects of amino acids. However, it appears that GIP is the physiologic "gut factor" that normally stimulates insulin secretion, since this factor produces stimulation when administered in doses that produce blood GIP levels comparable to those produced by oral glucose. GIP probably acts by increasing cyclic AMP in the B cells. The possible role of pancreatic somatostatin and glucagon in the regulation of insulin secretion is discussed below.

Oral Hypoglycemic Agents

Tolbutamide and other sulfonylurea derivatives such as chlorpropamide and the newer compounds glipizide and glyburide lower blood glucose, initially by stimulating insulin secretion but subsequently by enhancing the effect of endogenous insulin at a receptor or postreceptor level. Thus, they are of value in diabetics who have some residual pancreatic function. There is some evidence that some of these agents cause premature deaths from cardiovascular disease, but this evidence has been questioned.

Phenformin and other biguanides that have been used as oral hypoglycemic agents do not affect insulin secretion but increase glucose utilization, apparently because they inhibit oxidative metabolism of glucose and consequently increase anaerobic glycolysis within the cells. They also decrease glucose absorption from the gastrointestinal tract. However, they cause lactic acidosis in a significant number of patients, and because of the seriousness of this complication, they have been removed from the market in the USA.

Drugs That Inhibit Insulin Secretion

The drug diazoxide, which was originally developed for the treatment of hypertension, has been found to be diabetogenic, and it exerts this effect by inhibiting insulin secretion. It is sometimes used in the treatment of hyperinsulinism due to diffuse islet cell hyperplasia. A number of other drugs that inhibit insulin secretion have been reported, and in some individuals the commonly used thiazide diuretics may exert an inhibitory effect.

Long-Term Changes in B Cell Responses

The magnitude of the insulin response to a given stimulus is determined in part by the secretory history of the B cells. Individuals fed a high-carbohydrate diet for several weeks not only have higher fasting plasma insulin levels but also show a greater secretory response to a glucose load than individuals fed an isocaloric low-carbohydrate diet.

Although the B cells respond to stimulation with hypertrophy like other endocrine cells, they become exhausted and stop secreting **(B cell exhaustion)** when the stimulation is marked or prolonged. With such stimulation, they become vacuolated and hyalinized (hydropic and hyaline degeneration). If the stimulation stops soon after the cells stop secreting, they can recover; but if it is continued, the cells eventually die and disappear. In spite of some claims to the contrary, there is no convincing evidence that such "exhaustion atrophy" occurs in any other endocrine gland.

The pancreatic reserve is large, and it is difficult to produce B cell exhaustion in normal animals; but if the pancreatic reserve is reduced by partial pancreatectomy or small doses of alloxan, exhaustion of the remaining B cells can be produced by any procedure that chronically raises the blood glucose level. This is the cause of the diabetes produced in animals with limited pancreatic reserves by anterior pituitary extracts, growth hormone, thyroid hormones, or the prolonged continuous infusion of glucose alone. The diabetes precipitated by hormones in animals is at first reversible, but with prolonged treatment it becomes permanent. The diabetes is usually named for the agent producing it, eg, "hypophyseal diabetes," "thyroid diabetes." Permanent diabetes persisting after treatment has been discontinued is indicated by the prefix meta-, eg, "metahypophyseal diabetes" or "metathyroid diabetes." When insulin is administered along with the diabetogenic hormones, the B cells are protected, probably because the blood glucose is lowered, and diabetes does not develop. Adrenal glucocorticoids raise blood glucose, but it has proved difficult to produce permanent diabetes in animals with adrenal steroid treatment alone.

Effect of Exogenous Insulin

When insulin is administered to normal individuals, the rate at which they secrete their own insulin drops; this rate can be measured by following the circulating level of C peptide or the endogenous insulin level by radioimmunoassay with an antibody that does not cross-react with the administered insulin. The drop in insulin secretion is due in large part to the decrease in blood glucose concentration. However, insulin also feeds back to inhibit its own secretion when the blood glucose level is held constant, and this feedback effect may be reduced in obesity.

It is worth noting that in normal individuals, insulin secretion is low between meals and increases with each meal. Duplication of this pattern of secretion can be achieved in diabetes by islet cell transplants. It can also be accomplished with portable or implantable electronic insulin pumps programmed to inject increased amounts of insulin through an implanted catheter automatically or on demand.

GLUCAGON

Chemistry

Human glucagon, a linear polypeptide with a molecular weight of 3485, is produced by the A cells of the pancreatic islets. It contains 29 amino acid residues (Table 26–1). All mammalian glucagons appear to have the same structure. Glucagon and glicentin, a larger molecule with some glucagon activity, are also found in the gastrointestinal mucosa (see Chapter 26). The C-terminal portion of glicentin consists of glucagon extended at its C terminal by an octapeptide. Human preproglucagon is a 179-amino-acid polypeptide that contains glicentin near its N terminal, followed by glucagon and then by 2 glucagonlike peptides of unknown function.

It has been demonstrated that glicentin as well as glucagon occurs in the A cells of the pancreatic islets, with glicentin in the peripheral portion of the A cell granules and glucagon in the center. However, the relation between these 2 molecules in the pancreas is still unknown.

Action

Glucagon is glycogenolytic, gluconeogenic, lipolytic, and ketogenic (Fig 19–14). When it binds to receptors on liver cells, it acts via G_s to activate adenylate cyclase and increase intracellular cyclic AMP. This leads to the activation of phosphorylase and therefore to increased breakdown of glycogen and an increase in blood glucose. The steps involved in this classic example of mediation of a hormone effect via cyclic AMP are discussed in Chapter 17. However, it now appears that glucagon also acts on different glucagon receptors located on the same hepatic cells to activate lipase C and that the resulting increase in cytoplasmic Ca^{2+} also stimulates glycogenolysis. The protein kinase activated by cyclic AMP also decreases the metabolism of glucose 6-phosphate (Fig 19–8). Glucagon does not cause glycogenolysis in muscle. It increases gluconeogenesis from available amino acids in the liver and elevates the metabolic rate. It increases ketone body formation by decreasing malonyl-CoA in the liver (see Chapter 17). Its lipolytic activity, which leads in turn to increased ketogenesis, is discussed in Chapter 17. The calorigenic action of glucagon is not due to

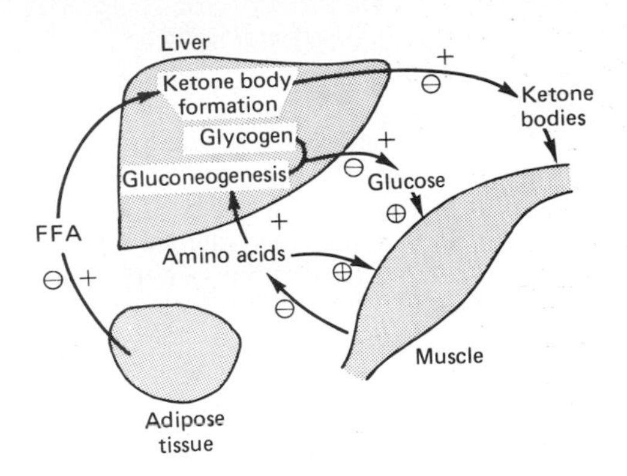

Figure 19–14. Bihormonal control of human substrate metabolism. ⊕, increased by insulin; ⊖, decreased by insulin; +, increased by glucagon. (Modified from Gerich J et al: Prevention of human diabetic ketoacidosis by somatostatin. *N Engl J Med* 1975;**292**:985.)

the hyperglycemia per se but probably to the increased hepatic deamination of amino acids.

Large doses of exogenous glucagon exert a positively inotropic effect on the heart (see Chapter 29) without producing increased myocardial excitability, presumably because they increase myocardial cyclic AMP. Use of the hormone in the treatment of heart disease has been advocated, but there is no evidence for a physiologic role of glucagon in the regulation of cardiac function. Glucagon also stimulates the secretion of growth hormone, insulin, and pancreatic somatostatin.

Glucagon receptors have been identified in a variety of tissues and partially purified. The molecular weight of the receptor protein is approximately 190,000.

Metabolism

Glucagon has a half-life in the circulation of 5–10 minutes. It is degraded by many tissues but particularly by the liver. Since glucagon is secreted into the portal vein and reaches the liver before it reaches the peripheral circulation, peripheral blood levels are relatively low. The rise in peripheral blood glucagon levels produced by excitatory stimuli (see below) is exaggerated in patients with cirrhosis, presumably because of decreased hepatic degradation of the hormone.

Regulation of Secretion

The principal factors known to affect glucagon secretion are summarized in Table 19–7. Secretion is decreased by a rise in plasma glucose, and this inhibitory effect requires the presence of insulin; thus, the A cells appear to be an insulin-dependent tissue. Secretion is increased by stimulation of the sympathetic nerves to the pancreas, and this sympathetic effect is mediated via β-adrenergic receptors and cyclic AMP. It appears that the A cells are like the B cells in that stimulation of β-adrenergic receptors increases secretion and stimulation of α-adrenergic receptors inhibits secretion (see above). However, the pancreatic response to sympathetic stimulation in the absence of blocking drugs is increased secretion of glucagon, so the effect of β-receptors predominates in the glucagon-secreting cells. The stimulatory effects of various stresses and possibly of exercise and infection are mediated at least in part via the sympathetic nervous system. Vagal stimulation also increases glucagon secretion.

A protein meal and infusion of various amino acids increase glucagon secretion. It seems appropriate that the glucogenic amino acids are particularly potent in this regard, since these are the amino acids that are converted to glucose in the liver under the influence of glucagon. The increase in glucagon secretion following a protein meal is also valuable, since the amino acids stimulate insulin secretion, and the secreted glucagon prevents the development of hypoglycemia while the insulin promotes storage of the absorbed carbohydrates, fat, and lipids. Glucagon secretion increases during starvation. It reaches a peak on the third day of a fast, at the time of maximal gluconeogenesis. Thereafter, the plasma glucagon level declines as fatty acids and ketones become the major sources of energy.

During exercise, there is a balanced increase in glucose utilization and glucose production. Plasma glucagon increases and plasma insulin falls. The increased glucose uptake is due to hypoxia, increased affinity of insulin receptors, increased muscle blood flow permitting insulin to gain access to the muscles, and other factors that in effect amplify the action of insulin. The increase in glucose production by the liver is made possible by the decrease in plasma insulin and is brought about primarily by the increase in plasma glucagon.

The glucagon response to oral administration of amino acids is greater than the response to intravenous infusion of amino acids, suggesting that a glucagon-stimulating factor is secreted from the gastrointestinal mucosa. Cholecystokinin-pancreozymin (CCK) and gastrin increase glucagon secretion, whereas secretin inhibits it. Since CCK and gastrin secretion are both increased by a protein meal, either

Table 19–7. Factors affecting glucagon secretion.

Stimulators	Inhibitors
Amino acids (particularly the glucogenic amino acids: alanine, serine, glycine, cysteine, and threonine)	Glucose
CCK, gastrin	Somatostatin
Cortisol	Secretin
Exercise	FFA
Infections	Ketones
Other stresses	Insulin
β-Adrenergic stimulators	Phenytoin
Theophylline	α-Adrenergic stimulators
Acetylcholine	

hormone could be the gastrointestinal mediator of the glucagon response. The inhibition produced by somatostatin is discussed below.

Glucagon secretion is also inhibitied by FFA and ketones. However, this inhibition can apparently be overridden, since plasma glucagon levels are high in diabetic ketoacidosis.

Insulin-Glucagon Molar Ratios

As noted above, insulin is glycogenic, antigluconeogenetic, antilipolytic, and antiketotic in its actions. It thus favors storage of absorbed nutrients and is a ''hormone of energy storage.'' Glucagon, on the other hand, is glycogenolytic, gluconeogenetic, lipolytic, and ketogenic. It mobilizes energy stores and is a ''hormone of energy release.'' Because of their opposite effects, the blood levels of both hormones must be considered in any given situation. A good example is exercise, as discussed above. It is convenient to think in terms of the molar ratios of these hormones. The insulin-glucagon molar ratio in any given blood specimen can readily be calculated from the blood levels of the hormones, as determined by immunoassays, and their molecular weights.

It has been shown that the insulin-glucagon molar ratios fluctuate markedly because the secretion of glucagon and insulin are both modified by the conditions that preceded the application of any given stimulus (Table 19–8). Thus, for example, the insulin-glucagon molar ratio on a balanced diet is approximately 2.3. An infusion of arginine increases the secretion of both hormones and raises the ratio to 3.0. After 3 days of starvation, the ratio falls to 0.4, and an infusion of arginine in this state lowers the ratio to 0.3. Conversely, the ratio is 25 in individuals receiving a constant infusion of glucose, and ingestion of a protein meal during the infusion causes the ratio to rise to 170. The rise is due to the fact that insulin secretion rises sharply, while the usual glucagon response to a protein meal is abolished. Thus, when energy is needed during starvation, the insulin-glucagon molar ratio is low, favoring glycogen breakdown and gluconeogenesis; conversely, when the need for energy mobilization is low, the ratio is high, favoring the deposition of glycogen, protein, and fat.

Table 19–8. Insulin-glucagon molar ratios (I/G) in blood in various conditions. 1+ to 4+ indicate relative magnitude. (Courtesy of RH Unger.)

Condition	Hepatic Glucose Storage (S) or Production (P)	I/G
Glucose availability		
Large carbohydrate meal	4+ (S)	70
IV glucose	2+ (S)	25
Small meal	1+ (S)	7
Glucose need		
Overnight fast	1+ (P)	2.3
Low-carbohydrate diet	2+ (P)	1.8
Starvation	4+ (P)	0.4

OTHER ISLET CELL HORMONES

In addition to insulin and glucagon, the pancreatic islets secrete somatostatin and pancreatic polypeptide. These substances are released from the pancreas. In addition, somatostatin may be involved in regulatory processes within the islets that adjust the pattern of hormones secreted in response to various stimuli.

Somatostatin

Somatostatin 14 and its N-terminal-extended form somatostatin 28 are found in the D cells of pancreatic islets. Both forms inhibit the secretion of insulin, glucagon, and pancreatic polypeptide and may act locally within the pancreatic islets in a paracrine fashion. Furthermore, patients with somatostatin-secreting pancreatic tumors develop hyperglycemia and other manifestations of diabetes that disappear when the tumor is removed. They also develop dyspepsia due to slow gastric emptying and decreased gastric acid secretion, and gallstones, which are probably precipitated by decreased gallbladder contraction. The secretion of pancreatic somatostatin is increased by several of the same stimuli that increase insulin secretion, ie, glucose and amino acids, particularly arginine and leucine. Its secretion is also increased by CCK. Somatostatin is released from the pancreas and the gastrointestinal tract into the peripheral blood.

Pancreatic Polypeptide

Human pancreatic polypeptide is a linear peptide containing 36 amino acid residues. It can be measured by radioimmunoassay in peripheral blood. At least in part, its secretion is under cholinergic control; plasma levels fall after administration of atropine. Its secretion is increased by a meal containing protein and by fasting, exercise, and acute hypoglycemia. Its secretion is decreased by somatostatin and intravenous glucose. Infusions of leucine, arginine, and alanine do not affect it, so the stimulatory effect of a protein meal may be mediated indirectly. A closely related but different polypeptide, neuropeptide Y, is found in the brain (see Chapter 4). Pancreatic polypeptide slows the absorption of food in humans, and it may smooth out the peaks and valleys of absorption. However, its exact physiologic function is unknown.

Organization of the Pancreatic Islets

The presence in the pancreatic islets of hormones that affect the secretion of other islet hormones raises the possibility that the islets function as secretory units in the regulation of nutrient homeostasis. Somatostatin inhibits the secretion of insulin, glucagon, and pancreatic polypeptide (Fig 19–15), whereas insulin inhibits the secretion of glucagon, and glucagon stimulates the secretion of insulin and somatostatin. As noted above, A and D cells and pancreatic polypeptide-secreting cells are generally

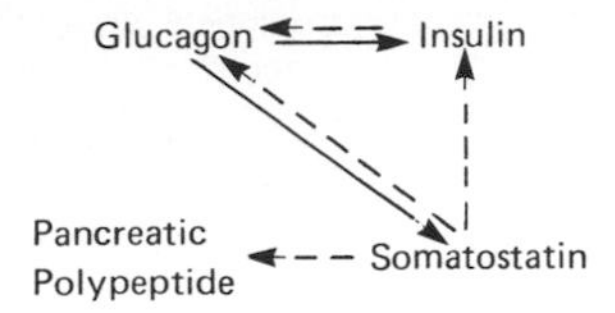

Figure 19–15. Effects of islet cell hormones on the secretion of other islet cell hormones. Solid arrows indicate stimulation; dashed arrows indicate inhibition.

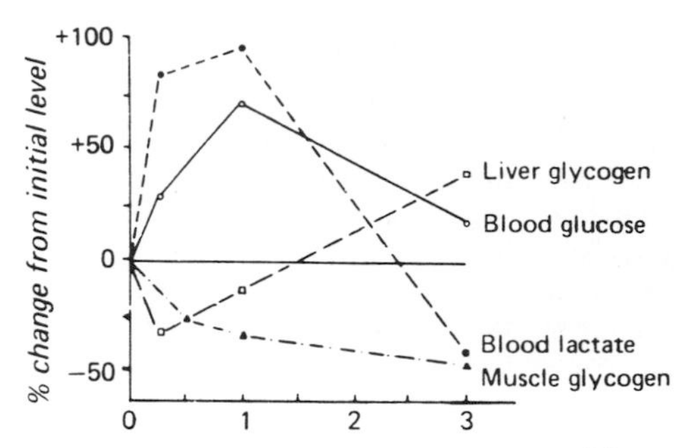

Figure 19–16. Effect of epinephrine on tissue glycogen, blood glucose, and blood lactate levels in fed rats. (Reproduced, with permission, from Ruch TC, Patton HD [editors]: *Physiology and Biophysics,* 20th ed. Vol. 3. Saunders, 1973.)

located around the periphery of the islets, with the B cells in the center. There are clearly 2 types of islets, glucagon-rich islets and pancreatic polypeptide-rich islets, but the functional significance of this separation is not known. It is possible that the islet cell hormones released into the ECF diffuse to other islet cells and influence their function (paracrine communication; see Chapter 1). However, this is difficult to prove. It has been demonstrated that there are gap junctions between A, B, and D cells, and these permit the passage of ions and other molecules from one cell to another, which could coordinate their secretory functions.

ENDOCRINE REGULATION OF CARBOHYDRATE METABOLISM

Many hormones in addition to insulin, IGF-I, IGF-II, glucagon, and somatostatin have important roles in the regulation of carbohydrate metabolism. They include epinephrine, thyroid hormones, glucocorticoids, and growth hormone. The other functions of these hormones are considered elsewhere, but it seems wise to summarize their effects on carbohydrate metabolism in the context of the present chapter. The "directional flow valves" in metabolism at which these hormones act are discussed in Chapter 17.

Catecholamines

The activation of phosphorylase in liver by catecholamines is discussed in Chapters 17 and 20. Activation occurs via β-adrenergic receptors, which increase intracellular cyclic AMP, and α-adrenergic receptors, which increase intracellular Ca^{2+}. Hepatic glucose output is increased, producing hyperglycemia. In muscle, the phosphorylase is also activated via cyclic AMP and presumably via Ca^{2+}, but the glucose 6-phosphate formed can only be catabolized to pyruvate because of the absence of glucose 6-phosphatase. For reasons that are not entirely clear, large amounts of pyruvate are converted to lactate, which diffuses from the muscle into the circulation (Fig 19–16). The lactate is oxidized in the liver to pyruvate and converted to glycogen. Therefore, the response to an injection of epinephrine is an initial glycogenolysis followed by a rise in hepatic glycogen content. Lactate oxidation may be responsible for the calorigenic effect of epinephrine (see Chapter 20). Epinephrine and norepinephrine also liberate FFA into the circulation, and epinephrine decreases peripheral utilization of glucose.

Adrenal medullary tumors (pheochromocytomas) that secrete epinephrine cause hyperglycemia, glycosuria, and an elevated metabolic rate. However, the hyperglycemic effect of epinephrine is generally too transient to produce permanent diabetes, and the metabolic abnormalities disappear as soon as the tumors are removed.

Thyroid Hormones

Thyroid hormones make experimental diabetes worse; thyrotoxicosis aggravates clinical diabetes; and metathyroid diabetes can be produced in animals with decreased pancreatic reserve. The principal diabetogenic effect of thyroid hormones is to increase absorption of glucose from the intestine, but the hormones also cause (probably by potentiating the effects of catecholamines) some degree of hepatic glycogen depletion. Glycogen-depleted liver cells are easily damaged. When the liver is damaged, the glucose tolerance curve is diabetic because the liver takes up less of the absorbed glucose. Thyroid hormones may also accelerate the degradation of insulin. All these actions have a hyperglycemic effect and, if the pancreatic reserve is low, may lead to B cell exhaustion.

Adrenal Glucocorticoids

Glucocorticoids from the adrenal cortex (see Chapter 20) elevate blood glucose and produce a diabetic type of glucose tolerance curve. In humans, this effect may occur only in individuals with a genetic predisposition to diabetes. Glucose tolerance is reduced in 80% of patients with Cushing's syndrome (see Chapter 20), and 20% of these patients have frank diabetes. The glucocorticoids are necessary for glucagon to exert its gluconeogenetic action during fasting. They are gluconeogenetic themselves, but their role is mainly permissive. In adrenal insufficiency, the blood glucose is normal as long as food intake is maintained, but fasting precipitates hypoglycemia and collapse. The blood-glucose-lowering effect of insulin is greatly enhanced in patients with adrenal insufficiency. In animals with experimental diabetes, adrenalectomy markedly ameliorates the diabetes.

The complex actions of the glucocorticoids on intermediary metabolism are incompletely understood. Known or suspected actions that affect carbohydrate metabolism are summarized in Table 19–9. The major diabetogenic effects are an increase in protein catabolism with increased gluconeogenesis in the liver; increased hepatic glycogenesis and ketogenesis; and a decrease in peripheral glucose utilization relative to the blood insulin level that may be due to inhibition of glucose phosphorylation (see below).

Table 19–9. Actions of glucocorticoids that affect carbohydrate metabolism. Items 1–6 represent reactions leading to an elevated blood glucose level due to increased gluconeogenesis.

1. Increased protein catabolism in periphery, resulting in increased concentration of amino acids in plasma
2. Increased hepatic uptake ("trapping") of amino acids
3. Increased deamination and transamination of amino acids
4. Increased hepatic conversion of oxaloacetate to phosphopyruvate
5. Increased hepatic fructose diphosphatase activity, facilitating dephosphorylation of fructose 1,6-diphosphate
6. Increased hepatic glucose 6-phosphatase activity, releasing more glucose into the circulation
7. Decrease in glucose utilization peripherally and in the liver, possibly due to inhibition of phosphorylation
8. Increased blood lactate and pyruvate
9. Decreased hepatic lipogenesis
10. Increased plasma FFA levels and increased ketone formation (when pancreatic reserve is low)
11. Increased formation of the active form of glycogen synthase

Growth Hormone

The diabetogenic effect of anterior pituitary extracts is due in part to the ACTH and TSH they contain, but it is also due to growth hormone. There is some species variation in the diabetogenicity of growth hormone; but human growth hormone makes clinical diabetes worse, and 25% of patients with growth-hormone-secreting tumors of the anterior pituitary have diabetes. Hypophysectomy ameliorates diabetes and increases sensitivity to insulin even more than adrenalectomy, whereas growth hormone treatment decreases insulin responsiveness.

Growth hormone mobilizes FFA from adipose tissue, thus favoring ketogenesis. It decreases glucose uptake into some tissues ("anti-insulin action"), increases hepatic glucose output, and may decrease tissue binding of insulin. Indeed, it has been suggested that the ketosis and decreased glucose tolerance produced by starvation are due to hypersecretion of growth hormone. Growth hormone does not stimulate insulin secretion directly, but the hyperglycemia it produces secondarily stimulates the pancreas and may eventually exhaust the B cells.

There is evidence that growth hormone decreases the number of insulin receptors and glucocorticoids decrease their affinity. However, the decrease in glucose utilization produced by these hormones is due more to inhibition of glucose phosphorylation than to inhibition of glucose entry into cells. Entry into cells is usually the rate-limiting step in glucose metabolism, but when adequate insulin is present, a decline in phosphorylation can decrease utilization.

HYPOGLYCEMIA & DIABETES MELLITUS IN HUMANS

Hypoglycemia

"Insulin reactions" are common in juvenile diabetics, and in most diabetics, occasional hypoglycemic episodes are the price of good diabetic control. The increase in glucose uptake of skeletal muscle and the increased absorption of injected insulin during exercise have been mentioned above. Diabetics taking insulin should be taught to anticipate this effect by adjusting their diet or insulin dose when they exercise.

Symptomatic hypoglycemia also occurs in nondiabetics, and a review of some of the more important causes serves to emphasize the variables affecting blood glucose to homeostasis. Chronic mild hypoglycemia can cause incoordination and slurred speech, and the condition can be mistaken for drunkenness. Mental aberrations and convulsions in the absence of frank coma also occur. When the level of insulin secretion is chronically elevated by an insulin-secreting tumor **(insulinoma)** or as a result of hyperplasia of the B cells, symptoms are most common in the morning. This is because a night of fasting has depleted hepatic glycogen reserves. However, symptoms can develop at any time, and in such patients, the diagnosis may be missed. Some cases of insulinoma have been erroneously diagnosed as epilepsy or psychosis. Hypoglycemia also occurs in some patients with large malignant tumors that do not involve the pancreatic islets, and the hypoglycemia in these cases is apparently due to excess secretion of IGF-II or some of the other insulinlike growth factors (see above), or both. The affinity of the insulin receptors for these factors is low enough that they do not normally cause a significant decline in blood glucose concentration, but such a decline may occur when plasma concentrations of these factors are elevated. A third cause of hypoglycemia is an autoimmune disorder in which spontaneously produced antibodies to insulin bind appreciable amounts of the hormone, then release free insulin at irregular intervals.

In liver disease, the glucose tolerance curve is diabetic but the fasting blood glucose level is low (Fig 19–17). In **functional hypoglycemia,** the blood glucose rise is normal after a test dose of glucose, but the subsequent fall overshoots to hypoglycemic levels, producing symptoms 3–4 hours after meals. This pattern is sometimes seen in individuals who later develop diabetes. Most patients with this syndrome are tense, conscientious people with symptoms of autonomic overactivity. It has been postulated that the overshoot of the blood glucose is due to insulin secretion stimulated by impulses in the right vagus, but cholinergic blocking agents do not routinely correct the abnormality. In some thyrotoxic patients and in patients who have had gastrectomies or other operations that speed the passage of food into the intestine, glucose absorption is abnormally rapid. The blood glucose rises to a high, early peak, but it then falls rapidly to hypoglycemic levels because the wave of hyperglycemia evokes a greater than normal rise in insulin secretion. Symptoms characteristically occur about 2 hours after meals.

Diabetes Mellitus

Diabetes mellitus is common in humans, although it also occurs in other species. In the USA, it has been calculated that almost 5% of the population has the disease. Not only does it sometimes cause acidosis and coma, but it is associated over long periods with serious complications. These include proliferative scarring of the retina (**diabetic retinopathy**), renal disease (**diabetic nephropathy**), loss of nerve function, particularly in the autonomic nervous system (**diabetic neuropathy**), and accelerated atherosclerosis. The atherosclerosis leads to circulatory insufficiency in the legs, with chronic ulceration and gangrene, and to an increased incidence of stroke and myocardial infarction.

The cause of clinical diabetes is always a deficiency of the action of insulin at the tissue level, but the deficiency may be due to a variety of abnormalities. Thus, diabetes, like hypertension (see Chapter 33), is a syndrome with multiple causes. There are 2 common forms, type I and type II diabetes (Table 19–10), and a collection of other less frequent forms due to conditions such as excess secretion of diabetogenic hormones or congenitally defective insulin receptors. **Type II diabetes** usually develops after age 40 and is not associated with total loss of the ability to secrete insulin. Consequently, it is also called **maturity-onset diabetes** and **NIDDM.** It has an insidious onset, is rarely associated with ketosis, and is usually associated with normal B cell morphology and insulin content if the B cells have not become exhausted. Plasma insulin levels are often normal or elevated, with an exaggerated, prolonged insulin response to glucose. Almost all patients with this form of diabetes are obese, and their glucose tolerance is restored to or toward normal if they reduce. They have a reduced number of insulin receptors on their adipose tissue cells, and the number of insulin receptors increases with weight reduction. Similar changes in receptor numbers are seen in obese nondiabetics, and volunteers who are force-fed have a reduction in the number of insulin receptors per adipose cell as they gain weight, with a return to the normal number when the weight is subsequently lost. Therefore, it has been hypothesized that individuals who develop maturity-onset diabetes overeat, with consequent increased stimulation of insulin secretion, and the elevated plasma insulin levels cause down regulation of the receptors. The compensatory response to the resultant decrease in insulin sensitivity is a further increase in insulin secretion. Normal individuals show the same response, but frank diabetes occurs when the pancreatic reserve is exceeded. There is also a strong genetic, or at least familial, component in this form of the disease. When one identical twin develops maturity-onset diabetes, the other almost invariably does as well—ie, the **concordance rate** is close to 100%.

In **type I diabetes,** on the other hand, the concordance rate is about 50%, indicating that an environmental factor as well as a genetic predisposition is involved. This disease usually develops before age 40, although it can occur at any age, and it is characterized by loss of B cells with eventual absence of insulin in the circulation. Consequently, it is also called **juvenile diabetes** and **IDDM.** It is not associ-

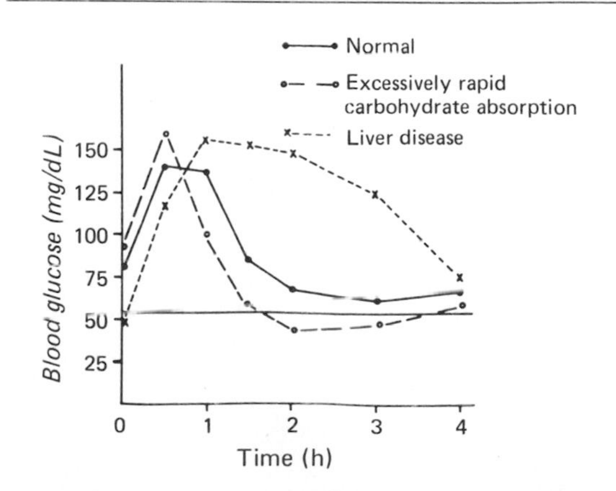

Figure 19–17. Typical glucose tolerance curves after an oral glucose load in liver disease and in conditions causing excessively rapid absorption of glucose from the intestine. Horizontal line is approximate blood glucose level at which hypoglycemic symptoms may appear.

Table 19–10. Types of human diabetes mellitus.

Type	Other Names
Type I	Insulin-dependent diabetes (IDDM). Juvenile diabetes. Ketosis-prone diabetes.
Type II	Non-insulin-dependent diabetes (NIDDM). Maturity-onset diabetes. Ketosis-resistant diabetes.
Diabetes associated with other conditions	Examples include diabetes due to pancreatoectomy or pancreatic disease; diabetes due to defective forms of insulin or insulin receptors; and diabetes in patients with Cushing's syndrome, acromogaly, or other endocrine diseases.

ated with obesity and is commonly complicated by ketosis and acidosis. A family history of diabetes is less common than in type II diabetes, but over 80% of type I diabetics have a pattern of histocompatibility antigens (see Chapter 27) characterized by the presence of HLA-DR3, HLA-DR4, or both. In the general population, the incidence of these 2 antigens is about 40%. Another gene not linked to HLA also seems to be involved. In addition, it is clear that type I diabetes is an autoimmune disease in which antibodies against islet tissue cause lymphoid infiltration and eventual destruction of the islets (''insulitis''). It now appears that in individuals who have the genetic predisposition indicated by the HLA pattern, a precipitating event that is environmental initiates formation of antibodies against islet cells, and the resulting insulitis leads over a period of years to loss of B cells and frank diabetes. The nature of the precipitating event is unknown. It is significant, however, that immunosuppression with drugs such as cyclosporin produces amelioration of type I diabetes if given early in the course of the disease before all the B cells are lost.

Insulin Requirements

The dose of insulin required to control diabetes mellitus varies from patient to patient and from time to time in the same patient. As discussed in previous sections of this chapter, the factors affecting insulin requirements are those that influence glucose tolerance. Thus, insulin requirements rise when patients gain weight, when they secrete or receive increased amounts of glucocorticoids and other hormones that are diabetogenic, during pregnancy, and in the presence of infection and fever. They fall when patients lose weight and during exercise.

Some degree of insulin resistance is found in almost all forms of diabetes mellitus. The causes are multiple and can be classified as prereceptor, receptor, and postreceptor in nature. Examples of prereceptor abnormalities include the presence of circulating antibodies to insulin and, possibly, increased degradation of insulin. The receptors themselves may be down-regulated or congenitally defective. At the postreceptor level, the pool of glucose transporters available for insertion into the membrane may be reduced in type I diabetes. The pool increases with insulin treatment, and this is one reason for the frequently observed transient fall in insulin requirements when type I diabetes is first treated. Indeed, in about 3% of newly diagnosed juvenile diabetics, the diabetes ''disappears'' after initial treatment and the remissions last from 2 weeks to 1 year. However, the diabetes reappears later on.

The Adrenal Medulla & Adrenal Cortex

20

INTRODUCTION

There are 2 endocrine organs in the adrenal gland, one surrounding the other. The main secretions of the inner **adrenal medulla** (Fig 20–1) are the catecholamines **epinephrine, norepinephrine,** and **dopamine;** the outer **adrenal cortex** secretes steroid hormones.

The adrenal medulla is in effect a sympathetic ganglion in which the postganglionic neurons have lost their axons and become secretory cells. The cells secrete when stimulated by the preganglionic nerve fibers that reach the gland via the splanchnic nerves. Adrenal medullary hormones are not essential for life, but they help to prepare the individual to deal with emergencies.

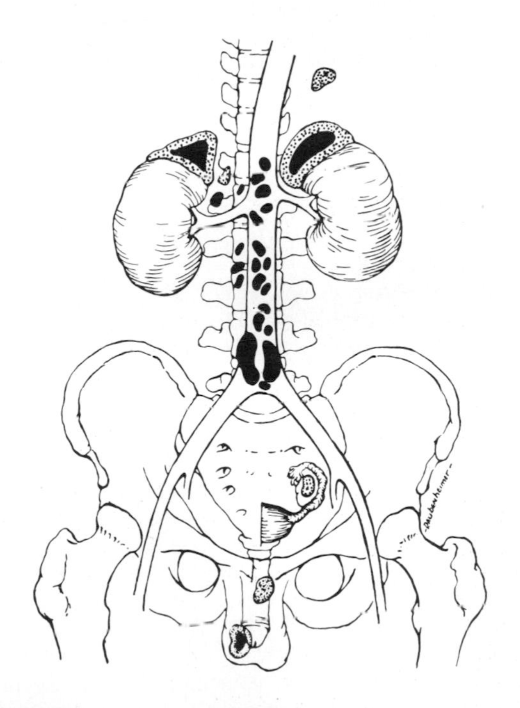

Figure 20–1. Human adrenal glands. Adrenocortical tissues is stippled; adrenal medullary tissue is black. Note location of adrenals at superior pole of each kidney. Also shown are extra-adrenal sites at which cortical and medullary tissue is sometimes found. (Reproduced, with permission, from Forsham PH: The adrenal cortex. In: *Textbook of Endocrinology,* 4th ed. Williams RH [editor]. Saunders, 1968.)

On the other hand, the adrenal cortex is essential for life. It secretes **glucocorticoids,** steroids with widespread effects on the metabolism of carbohydrate and protein; **a mineralocorticoid** essential to the maintenance of sodium balance and ECF volume; and **sex hormones** that exert minor effects on reproductive function. Of these, the mineralocorticoids and the glucocorticoids are necessary for survival. Adrenocortical secretion is controlled primarily by ACTH from the anterior pituitary, but mineralocorticoid secretion is also subject to independent control by circulating factors, of which the most important is angiotensin II, a peptide formed in the bloodstream. The formation of angiotensin is in turn dependent on renin, which is secreted by the kidney.

ADRENAL MORPHOLOGY

The adrenal medulla, which constitutes 28% of the mass of the adrenal gland, is made up of interlacing cords of densely innervated granule-containing cells that abut on venous sinuses. Two cell types can be distinguished morphologically: an epinephrine-secreting type that has larger, less dense granules; and a norepinephrine-secreting type in which smaller, very dense granules fail to fill the vesicles in which they are contained (Fig 20–2). Ninety percent of the cells are the epinephrine-secreting type, and 10% are the norepinephrine-secreting type. The type of cell that secretes dopamine is unknown. **Paraganglia,** small groups of cells resembling those in the adrenal medulla, are found near the thoracic and abdominal sympathetic ganglia (Fig 20–1).

In adult mammals, the adrenal cortex is divided into 3 zones of variable distinctness (Fig 20–3). The outer **zona glomerulosa** is made up of whorls of cells that are continuous with the columns of cells which form the **zona fasciculata.** These columns are separated by venous sinuses. The inner portion of the zona fasciculata merges into the **zona reticularis,** where

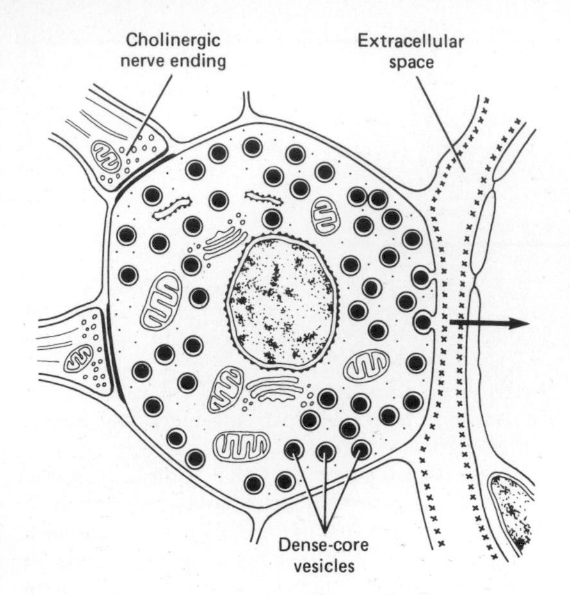

Figure 20–2. Norepinephrine-secreting adrenal medullary cell. The granules are released by exocytosis and the granule contents enter the bloodstream (arrow). (Modified from Poirier J, Dumas JLR: *Review of Medical Histology.* Saunders, 1977.)

the cell columns become interlaced in a network. The zona glomerulosa makes up 15% of the mass of the adrenal gland, the zona fasciculata 50%, and the zona reticularis 7%. The adrenal cells contain abundant lipid, especially in the outer portion of the zona fasciculata. All 3 cortical zones secrete corticosterone (see below), but the enzymatic mechanism for aldosterone biosynthesis is limited to the zona glomerulosa, whereas the enzymatic mechanism for forming cortisol and sex hormones is found in the 2 inner zones.

Arterial blood reaches the adrenal from many small branches of the phrenic and renal arteries and the aorta. From a plexus in the capsule, blood flows to the sinusoids of the medulla. The medulla is also supplied by a few arterioles that pass directly to it from the capsule. In most species, including humans, there is a single large adrenal vein. The blood flow through the adrenal is large, as it is in most endocrine glands.

During fetal life, the human adrenal is large and under pituitary control, but the 3 zones of the permanent cortex represent only 20% of the gland. The remaining 80% is the large **fetal adrenal cortex,** which undergoes rapid degeneration at the time of birth. A major function of this fetal adrenal is secretion of sulfate conjugates of androgens that are converted in the placenta to estrogens (see Chapter 23). There is no structure comparable to the human fetal adrenal in laboratory animals.

An important function of the zona glomerulosa, in addition to aldosterone biosynthesis, is the formation of new cortical cells. Like other tissues of neural origin, the adrenal medulla does not regenerate; but when the inner 2 zones of the cortex are removed, a new zona fasciculata and zona reticularis regenerate from glomerular cells attached to the capsule. Small capsular remnants will regrow large pieces of adrenocortical tissue. Immediately after hypophysectomy, the zona fasciculata and zona reticularis begin to atrophy, whereas the zona glomerulosa is unchanged (Fig 20–3), because of the action of angiotensin II on this zone. The ability to secrete aldosterone and conserve Na^+ is normal for some time, but in long-standing hypopituitarism, aldosterone deficiency may develop, apparently because of the absence of a pituitary factor that maintains the responsiveness of the zona glomerulosa (see below). Injections of ACTH and stimuli that cause endogenous ACTH secretion produce hypertrophy of the zona fasciculata and zona reticularis but do not increase the size of the zona glomerulosa.

The cells of the adrenal cortex contain large amounts of smooth endoplasmic reticulum, which seems to be involved in the steroid-forming process. Other steps in steroid biosynthesis occur in the mitochondria. The structure of steroid-secreting cells is very similar throughout the body. The typical features of such cells are shown in Fig 20–4.

ADRENAL MEDULLA

STRUCTURE & FUNCTION OF MEDULLARY HORMONES

Catecholamines

Norepinephrine, epinephrine, and dopamine are secreted by the adrenal medulla. Cats and some other species secrete mainly norepinephrine, but in dogs and humans, most of the catecholamine output in the adrenal vein is epinephrine. Norepinephrine also enters the circulation from noradrenergic nerve endings.

The structures of norepinephrine, epinephrine, and dopamine and the pathways for their biosynthesis and metabolism are shown in Figs 4–18, 4–19, and 4–20. Norepinephrine is formed by hydroxylation and decarboxylation of tyrosine, and epinephrine by methylation of norepinephrine. Phenylethanolamine-N-methyltransferase (PNMT), the enzyme that catalyzes the formation of epinephrine from norepinephrine, is found in appreciable quantities only in the brain and the adrenal medulla. Adrenal medullary PNMT is induced by glucocorticoids, and although relatively large amounts are required, the glucocorticoid concentration is high in the blood draining from the cortex to the medulla. After hypophysectomy, the glucocorticoid concentration of this blood falls and epinephrine synthesis is decreased.

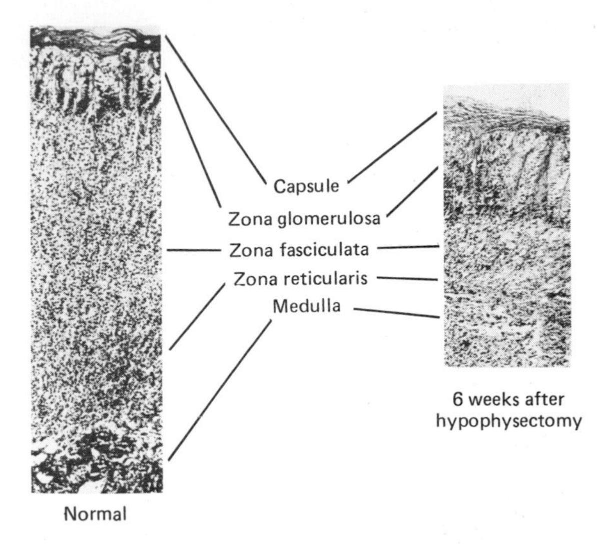

Figure 20–3. Effect of hypophysectomy on the morphology of the adrenal cortex of the dog. Note that the atrophy does not involve the zona glomerulosa. The morphology of the human adrenal is similar.

In plasma, about 95% of the dopamine and 70% of the norepinephrine and epinephrine are conjugated to sulfate. Sulfate conjugates are inactive, and their function is unsettled. In recumbent humans, the normal plasma level of free norepinephrine is about 300 pg/mL (1.8 nmol/L). There is a 50–100% increase upon standing (Fig 20–5). Plasma norepinephrine is generally unchanged after adrenalectomy, but the free epinephrine level, which is normally about 30 pg/mL (0.16 nmol/L), falls to essentially zero. The epinephrine found in tissues other than the adrenal medulla and the brain is for the most part absorbed from the bloodstream rather than synthesized in situ. The plasma free dopamine level is about 35 pg/mL (0.23 nmol/L), and there are appreciable quantities of dopamine in the urine. Half the plasma dopamine comes from the adrenal medulla, whereas the remaining half presumably comes from the sympathetic ganglia or other components of the autonomic nervous system.

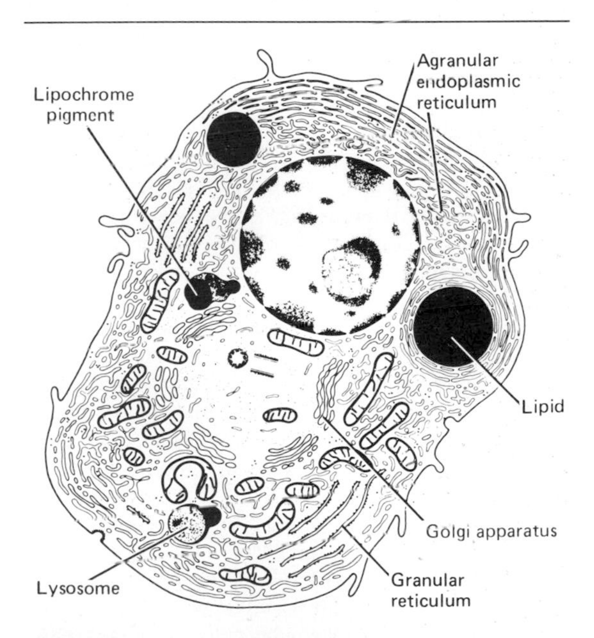

Figure 20–4. Diagrammatic representation of the cytologic features of steroid-secreting cells. Note the abundant agranular endoplasmic reticulum, the pleomorphic mitochondria, and the lipid droplets. (Reproduced, with permission, from Fawcett DW, Long JA, Jones AL: The ultrastructure of endocrine glands. *Recent Prog Horm Res* 1969;**25**:315.)

The catecholamines have a half-life of about 2 minutes in the circulation. For the most part, they are methoxylated and then oxidized to 3-methoxyl-4-hydroxymandelic acid (vanillylmandelic acid, VMA). About 50% of the secreted catecholamines appear in the urine as free or conjugated metanephrine and normetanephrine, and 35% as VMA. Only small amounts of free norepinephrine and epinephrine are excreted. In normal humans, about 30 μg of norepinephrine, 6 μg of epinephrine, and 700 μg of VMA are excreted per day.

Other Substances Secreted by the Adrenal Medulla

In the medulla, norepinephrine and epinephrine are stored in granules with ATP. The granules also contain chromogranins (see Chapter 4). Secretion is initiated by acetylcholine released from the preganglionic neurons that innervate the secretory cells. The acetylcholine increases the permeability of the cells, and the Ca^{2+} that enters the cells from the ECF triggers exocytosis (see Chapter 1). In this fashion, the catecholamines, ATP, and proteins in the granules are all released together.

Epinephrine-containing cells of the medulla also contain and secrete opioid peptides (see Chapter 4). The precursor molecule is preproenkephalin (Table 4–4), a protein containing a leader sequence of 25 amino acid residues and 242 additional amino acid residues that include 4 met-enkephalins, a heptapeptide which contains met-enkephalin, an octapeptide which contains met-enkephalin, and one leu-enkephalin (Fig 1–19). Most of the circulating met-enkephalin comes from the adrenal medulla. However, the circulating opioid peptides do not cross the blood-brain barrier to any degree, and their function in the blood is presently unknown.

Effects of Epinephrine & Norepinephrine

In addition to mimicking the effects of noradrenergic nervous discharge, norepinephrine and epinephrine stimulate the nervous system and exert metabolic effects that include glycogenolysis in liver and skeletal muscle, mobilization of FFA, increased plasma lactate, and stimulation of the metabolic rate. The effects of norepinephrine and epinephrine are

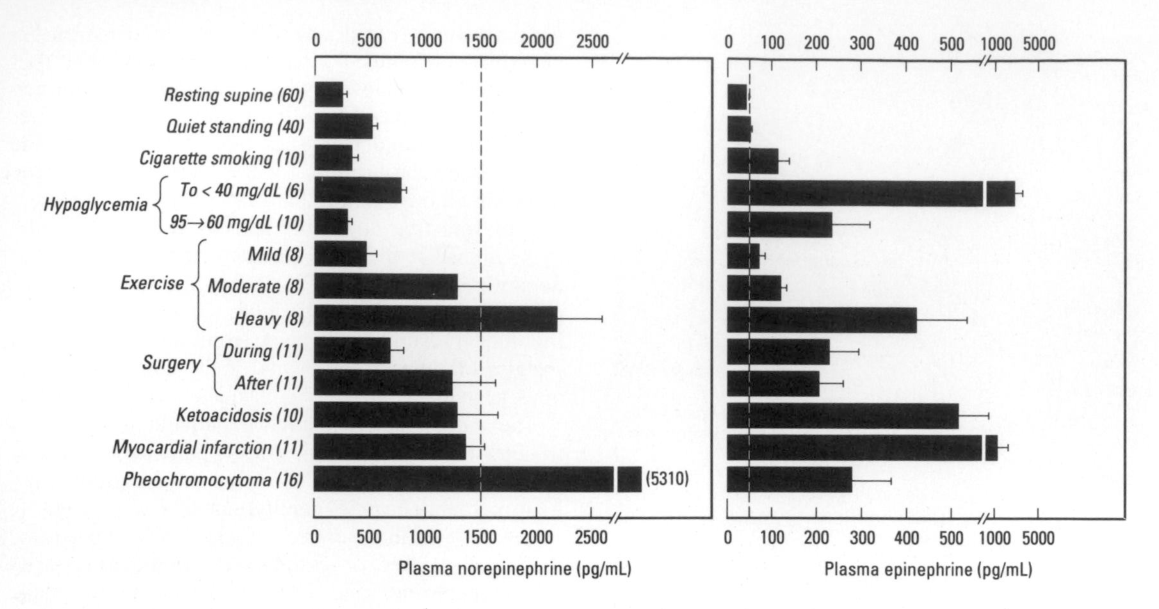

Figure 20–5. Norepinephrine and epinephrine levels in human venous blood in various physiologic and pathologic states. Note that the horizontal scales are different. The numbers in parentheses are the numbers of subjects tested. In each case, the vertical dashed line identifies the threshold plasma concentration at which detectable physiologic changes are observed. (Modified and reproduced, with permission, from Cryer PE: Physiology and pathophysiology of the human sympathoadrenal neuroendocrine system. *N Eng J Med* 1980;**303**:436.)

brought about by actions on 2 classes of receptors, α- and β-adrenergic receptors. Alpha receptors are subdivided into α_1 and α_2 receptors, and β receptors into β_1 and β_2 receptors, as outlined in Chapter 4. Table 13–2 lists the receptors in each of the visceral organs.

Norepinephrine and epinephrine both increase the force and rate of contraction of the isolated heart. These receptors are mediated by β_1 receptors. The catecholamines also increase myocardial excitability, causing extrasystoles and, occasionally, more serious cardiac arrhythmias. Norepinephrine produces vasoconstriction in most if not all organs via α_1 receptors, but epinephrine dilates the blood vessels in skeletal muscle and the liver via β_2 receptors. This usually overbalances the vasoconstriction epinephrine produces elsewhere, and the total peripheral resistance drops. When norepinephrine is infused slowly in normal animals or humans, the systolic and diastolic blood pressures rise. The hypertension stimulates the carotid and aortic baroreceptors, producing reflex bradycardia that overrides the direct cardioacceleratory effect of norepinephrine. Consequently, cardiac output per minute falls. Epinephrine causes a widening of the pulse pressure, but because baroreceptor stimulation is insufficient to obscure the direct effect of the hormone on the heart, cardiac rate and output increase. These changes are summarized in Fig 20–6.

Catecholamines increase alertness (see Chapter 11). Epinephrine and norepinephrine are equally potent in this regard, although in humans epinephrine usually evokes more anxiety and fear.

The catecholamines have several different actions that affect blood glucose. Epinephrine and norepinephrine both cause glycogenolysis. They produce this effect via β-adrenergic receptors that increase cyclic AMP, with activation of phosphorylase, and via α-adrenergic receptors that increase intracellular Ca^{2+} (see Chapter 17). In addition, the catecholamines increase the secretion of insulin and glucagon via β-adrenergic mechanisms and inhibit the secretion of these hormones via α-adrenergic mechanisms.

Norepinephrine and epinephrine also produce a prompt rise in the metabolic rate that is independent of the liver and a smaller, delayed rise that is abolished by hepatectomy and coincides with the rise in blood lactate concentration. The calorigenic action does not occur in the absence of the thyroid and the adrenal cortex. The cause of the initial rise in metabolic rate is not clearly understood. It may be due to cutaneous vasoconstriction, which decreases heat loss and leads to a rise in body temperature, or to increased muscular activity, or to both. The second rise is probably due to oxidation of lactate in the liver.

When injected, epinephrine and norepinephrine cause an initial rise in plasma K^+ because of release of K^+ from the liver, then a prolonged fall in plasma K^+ because of an increased entry of K^+ into skeletal muscle that is mediated by β_2-adrenergic receptors. There is some evidence that activation of α receptors opposes this effect. Thus, the catecholamines may play a significant role in regulating the ratio between extracellular and intracellular K^+.

The increase in plasma norepinephrine and epinephrine that are needed to produce the various effects listed above have been determined by infusion of catecholamines in resting humans. In general, the

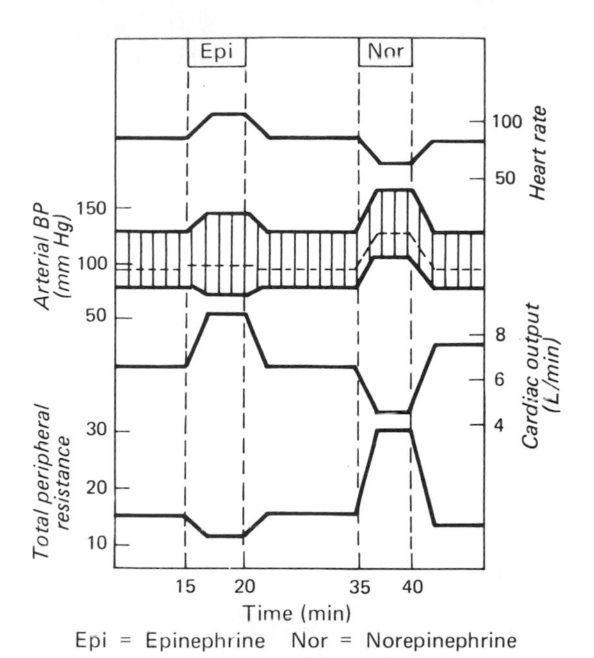

Figure 20–6. Circulatory changes produced in humans by the slow intravenous infusion of epinephrine and norepinephrine. (Modified and reproduced, with permission, from Barcroft H, Swan HJC: *Sympathetic Control of Human Blood Vessels.* Arnold, 1953.)

threshold for the cardiovascular and the metabolic effects of norepinephrine is about 1500 pg/mL, ie, about 5 times the resting value. Epinephrine, on the other hand, produces tachycardia when the plasma level is about 50 pg/mL, ie, about twice the resting value. The threshold for increased systolic blood pressure and lipolysis is about 75 pg/mL; the threshold for hyperglycemia, increased plasma lactate, and decreased diastolic blood pressure is about 150 pg/mL; and the threshold for the α-mediated decrease in insulin secretion is about 400 pg/mL. As shown in Fig 20–5, plasma epinephrine often exceeds those thresholds. On the other hand, plasma norepinephrine rarely exceeds the threshold for its cardiovascular and metabolic effects, and most of its effects are due to its local release from postganglionic sympathetic neurons. Most adrenal medullary tumors (**pheochromocytomas**) secrete norepinephrine, and episodic or sustained hypertension is the most prominent finding. Pheochromocytomas that secrete epineprine cause less hypertension and frequently produce episodic hyperglycemia, glycosuria, and other metabolic effects.

Effects of Dopamine

The physiologic function of the dopamine in the circulation is unknown. However, injected dopamine produces renal vasodilation, probably by acting on a specific dopaminergic receptor. It also produces vasodilation in the mesentery. Elsewhere, it produces vasoconstriction, probably by releasing norepinephrine, and it has a positively inotropic effect on the heart by an action on β_1-adrenergic receptors. The net effect of moderate doses of dopamine is an increase in systolic pressure and no change in diastolic pressure. Because of these actions, dopamine is useful in the treatment of traumatic and cardiogenic shock (see Chapter 33).

REGULATION OF ADRENAL MEDULLARY SECRETION

Neural Control

Certain drugs act directly on the adrenal medulla, but physiologic stimuli affect medullary secretion through the nervous system. Catecholamine secretion is low in basal states, but the secretion of epinephrine and, to a lesser extent, that of norepinephrine is reduced even further during sleep.

Increased adrenal medullary secretion is part of the diffuse sympathetic discharge provoked in emergency situations, which Cannon called the "emergency function of the sympathoadrenal system." The ways in which this discharge prepares the individual for flight or fight are described in Chapter 13, and the increases in plasma catecholamines under various conditions are shown in Fig 20–5. Some of the secreted norepinephrine is taken up by noradrenergic neurons, providing them with more transmitter and thus reinforcing their function.

The metabolic effects of circulating catecholamines are probably important, especially in certain situations. The calorigenic action of catecholamines in animals exposed to cold is an example. Animals with denervated adrenal glands shiver sooner and more vigorously than normal controls when exposed to cold. The glycogenolysis produced by epinephrine and norepinephrine in hypoglycemic animals is another example. Hypoglycemia is a potent stimulus to catecholamine secretion. As noted in Chapter 19, glucagon can substitute for catecholamines as a counterregulatory hormone, and vice versa; but if the secretion of both is blocked, insulin tolerance is markedly reduced.

Selective Secretion

When adrenal medullary secretion is increased, the ratio of epinephrine to norepinephrine in the adrenal effluent is generally unchanged or elevated. However, asphyxia and hypoxia increase the ratio of norepinephrine to epinephrine. The fact that the output of norepinephrine can be increased selectively has unfortunately led to the teleologic speculation that the adrenal medulla secretes epinephrine or norepinephrine depending upon which hormone best equips the animal to meet the emergency it faces. The fallacy of such speculation is illustrated by the response to hemorrhage, in which the predominant catecholamine secreted is not norepinephrine but epinephrine, which lowers the peripheral resistance. Norepinephrine secretion is increased by emotional stresses with which the individual is familiar, whereas epinephrine secretion rises in situations in which the individual does not know what to expect.

ADRENAL CORTEX

STRUCTURE & BIOSYNTHESIS OF ADRENOCORTICAL HORMONES

Classification & Structure

The hormones of the adrenal cortex are derivatives of cholesterol. Like cholesterol, bile acids, vitamin D, and ovarian and testicular steroids, they contain the **cyclopentanoperhydrophenanthrene nucleus** (Fig 20–7). The adrenocortical steroids are of 2 structural types: those that have a 2-carbon side chain attached at position 17 of the D ring and contain 21 carbon atoms ("C_{21} steroids"), and those that have a keto or hydroxyl group at position 17 and contain 19 carbon atoms ("C_{19} steroids"). Most of the C_{19} steroids have a keto group at position 17 and are therefore called **17-ketosteroids.** The C_{21} steroids that have a hydroxyl group at the 17 position in addition to the side chain are often called 17-hydroxycorticoids or 17-hydroxycorticosteroids.

The C_{19} steroids have androgenic activity. The C_{21} steroids are classified, using Selye's terminology, as mineralocorticoids or glucocorticoids. All secreted C_{21} steroids have both mineralocorticoid and glucocorticoid activity; **mineralocorticoids** are those in which effects on Na^+ and K^+ excretion predominate, and **glucocorticoids** those in which effects on glucose and protein metabolism predominate.

Steroid Nomenclature & Isomerism

For the sake of simplicity, the steroid names used here and in Chapter 23 are the most commonly used trivial names. A few common synonyms are shown in Table 20–1. The details of steroid nomenclature and isomerism can be found in the texts listed in the references at the end of this section, but it is pertinent to mention that the Greek letter Δ indicates a double bond and that the groups which lie above the plane of each of the steroid rings are indicated by the Greek letter β and a solid line (–OH), whereas those which lie below are indicated by α and a dotted line (---OH). Thus, the C_{21} steroids secreted by the adrenal have a Δ^4-3-keto configuration in the A ring. In most naturally occurring adrenal steroids, 17-hydroxy groups are in the α configuration, while 3-, 11-, and 21-hydroxy groups are in the β configuration. The 18-aldehyde configuration on naturally occurring aldosterone is the D form. L-Aldosterone is physiologically inactive.

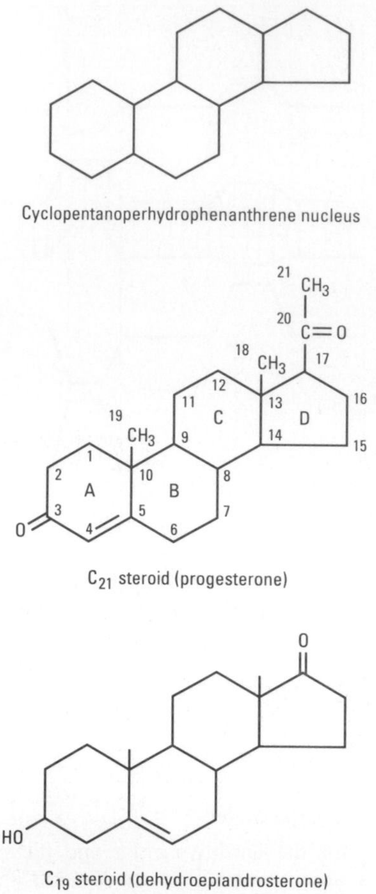

Cyclopentanoperhydrophenanthrene nucleus

C_{21} steroid (progesterone)

C_{19} steroid (dehydroepiandrosterone)

Figure 20–7. Structure of adrenocortical steroids. The letters in the formula for progesterone identify the A, B, C, and D rings; the numbers show the positions in the basic C_{21} steroid structure. The angular methyl groups (positions 18 and 19) are usually indicated simply by straight lines, as in the bottom formula. Dehydroepiandrosterone is a "17-ketosteroid" formed by cleavage of the side chain of the C_{21} steroid 17-hydroxypregnenolone and its replacement by an O atom. Other C_{21} steroids are converted to 17-ketosteroids in a similar way.

Secreted Steroids

Innumerable steroids have been isolated from adrenal tissue, but the only steroids normally secreted in physiologically significant amounts are the mineralocorticoid **aldosterone,** the glucocorticoids **cortisol** and **corticosterone,** and the androgens **dehydroepiandrosterone** and **androstenedione.** The structures of these steroids are shown in Figs 20–8 and 20–9. **Deoxycorticosterone** is a mineralocorticoid that is normally secreted in about the same amount as aldosterone (Table 20–1) but has only 3% of the mineralocorticoid activity of aldosterone. Its effect on mineral metabolism is usually negligible, but in diseases in which its secretion is increased, its effects can be appreciable. The adrenals may also secrete small amounts of estrogen, although most of the estrogens that are not formed in the ovaries are

Table 20–1. Principal adrenocortical hormones in adult humans.*

Name	Synonyms	Average Plasma Concentration (Free and Bound) (μg/dL)	Average Amount Secreted (mg/24 h)
Cortisol	Compound F, hydrocortisone	13.9	20
Corticosterone	Compound B	0.4	3
Aldosterone		0.006	0.15
Deoxycorticosterone	DOC	0.006	0.20
Dehydroepiandrosterone	DEA, DHEA	175.0	20

* All plasma concentration values except DEA are morning values after overnight recumbency.

produced in the circulation from adrenal androstenedione. Almost all the dehydroepiandrosterone is secreted conjugated with sulfate, although most if not all of the other steroids are secreted in the free, unconjugated form.

The secretion rate for individual steroids can be determined by injecting a very small dose of isotopically labeled steroid and determining the degree to which the radioactive steroid excreted in the urine is diluted by unlabeled secreted hormone. This technique is used to measure the output of many different hormones.

Species Differences

In all species from amphibia to humans, the major C_{21} steroid hormones secreted by adrenocortical tissue appear to be aldosterone, cortisol, and corticosterone, although the ratio of cortisol to corticosterone varies. Birds, mice, and rats secrete corticosterone almost exclusively; dogs secrete approximately equal amounts of the 2 glucocorticoids; and cats, sheep, monkeys, and humans secrete predominantly cortisol. In humans, the ratio of secreted cortisol to corticosterone is approximately 7.

Synthetic Steroids

The glucocorticoids are among the large number of naturally occurring substances that chemists have been able to improve upon. A number of synthetic steroids are now available that have many times the activity of cortisol. The relative glucocorticoid and mineralocorticoid potencies of the natural steroids are compared to those of the synthetic steroids 9α-fluorocortisol, prednisolone, and dexamethasone in Table 20–2. The potency of dexamethasone is due to its high affinity for glucocorticoid receptors and its long half-life (see below). Prednisolone also has a long half-life.

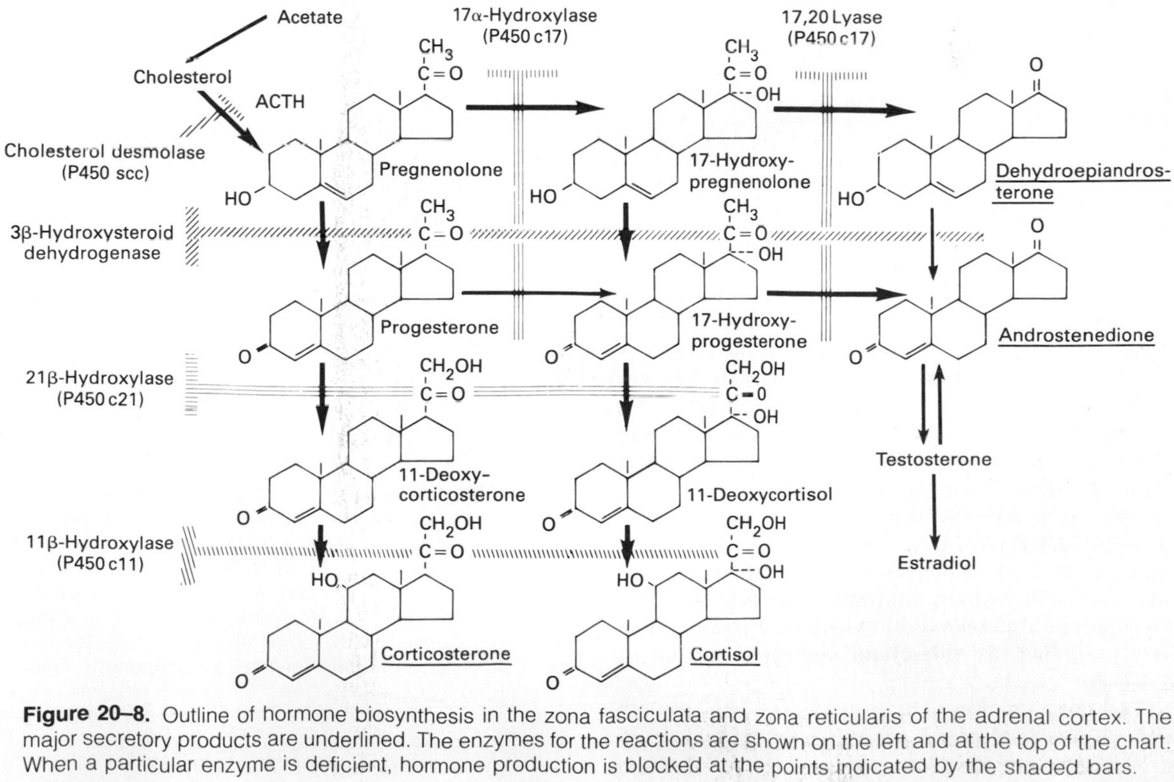

Figure 20–8. Outline of hormone biosynthesis in the zona fasciculata and zona reticularis of the adrenal cortex. The major secretory products are underlined. The enzymes for the reactions are shown on the left and at the top of the chart. When a particular enzyme is deficient, hormone production is blocked at the points indicated by the shaded bars.

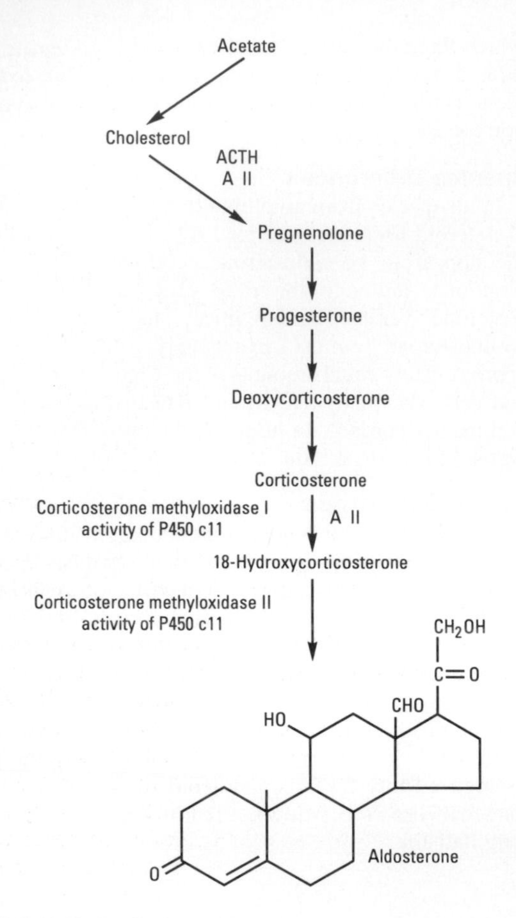

Figure 20–9. Outline of hormone synthesis in the zona glomerulosa. The steps from acetate to corticosterone are the same as in the zona fasciculata and zona reticularis. However, the zona glomerulosa lacks 17α-hydroxylase activity, and only the zona glomerulosa can convert 18-hydroxycorticosterone to aldosterone. A II, angiotensin II.

Table 20–2. Relative potencies of corticosteroids compared to cortisol. Values are approximations based on liver glycogen deposition or anti-inflammatory assays for glucocorticoid activity, and effect on urinary Na^+/K^+ or maintenance of adrenalectomized animals for mineralocorticoid activity. The last 3 steroids listed are synthetic compounds that do not occur naturally. (Data from various sources.)

	Glucocorticoid Activity	Mineralocorticoid Activity
Cortisol	1.0	1.0
Corticosterone	0.3	15
Aldosterone	0.3	3000
Deoxycorticosterone	0.2	100
Cortisone	0.7	1.0
Prednisolone	4	0.8
9α-Fluorocortisol	10	125
Dexamethasone	25	~0

Steroid Biosynthesis

The major paths by which the naturally occurring adrenocortical hormones are synthesized in the body are summarized in Figs 20–8 and 20–9. The precursor of all steroids is cholesterol. Some of the cholesterol is synthesized from acetate, but most of it is taken up from LDL in the circulation (see Chapter 17). LDL receptors are especially abundant in adrenocortical cells. The cholesterol is esterified and stored in lipid droplets. **Cholesterol ester hydrolase** catalyzes the formation of free cholesterol in the lipid droplets (Fig 20–10). The cholesterol is transported to mitochondria by a sterol carrier protein. In the mitochondria, it is converted to pregnenolone in a reaction catalyzed by **cholesterol desmolase.** This enzyme is a mitochondrial cytochrome P450, which is also known as **side-chain cleavage enzyme** or **P450scc.**

The pregnenolone then moves to the smooth endoplasmic reticulum, where some of it is dehydrogenated to form progesterone in a reaction catalyzed by **3β-hydroxysteroid dehydrogenase.** This enzyme has a molecular weight of 46,000 and is not a P450. Some of the pregnenolone and some of the progesterone are hydroxylated in the smooth endoplasmic reticulum to 17α-hydroxypregnenolone. The enzyme that catalyzes these reactions, **17α-hydroxylase,** is another P450, **P450c17.** Some of the 17α-hydroxypregnenolone is converted to 17α-hydroxyprogesterone. P450c17 is also the **17,20 lyase** that catalyzes the cleavage of the 17,20 bond to convert 17α-hydroxypregnenolone to dehydroepinandrosterone and 17α-hydroxyprogesterone to andro-

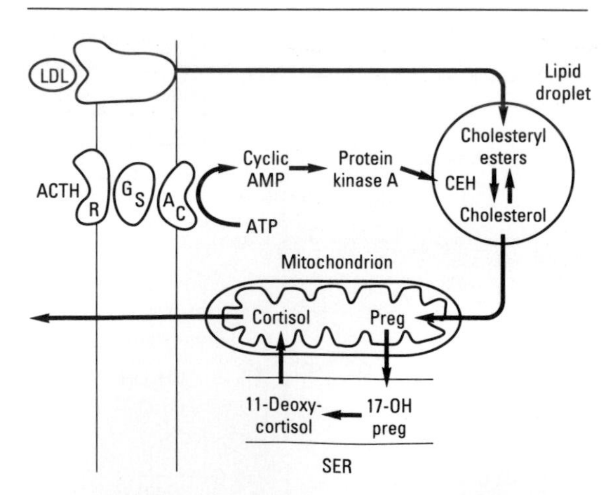

Figure 20–10. Mechanism of action of ACTH on cortisol-secreting cells in the inner 2 zones of the adrenal cortex. When ACTH binds to its receptor (R), adenylate cyclase (AC) is activated via G_S. The resulting increase in cyclic AMP activates protein kinase A, and the kinase phosphorylates cholesterol ester hydrolase (CEH), increasing its activity. Consequently, more free cholesterol is formed and converted to pregnenolone in the mitochondria. Note that in the subsequent steps in steroid biosynthesis, products are shuttled between the mitochondria and the smooth endoplasmic reticulum (SER). Additional corticosterone is also synthesized and secreted.

stenedione (Fig 20–8). This lyase activity of P450c17 is normally not very great in the adrenal unless 21-hydroxylation (see below) is deficient, but it is prominent in the gonads (see Chapter 23).

Hydroxylation of progesterone and 17α-hydroxyprogesterone also occurs in the smooth endoplasmic reticulum, where it is catalyzed by **21β-hydroxylase,** a mitochondrial cytochrome which is also called **P450c21.** The end products, 11-deoxycorticosterone and 11-deoxycortisol, move back to the mitochondria, where they are 11-hydroxylated to form corticosterone and cortisol. The enzyme that catalyzes these reactions, **11β-hydroxylase,** is a fourth cytochrome, **P450c11.** In the zona fasciculata and zona reticularis, the corticosterone and the cortisol then diffuse into the circulation.

P450c11 also has **18-hydroxylase activity (corticosterone methyl oxidase I activity).** In the zona glomerulosa, the product of this activity is 18-hydroxycorticosterone, and in the inner 2 zones, it is 18-hydroxydeoxycorticosterone. 18-Hydroxycorticosterone is converted to aldosterone by the **corticosterone methyl oxidase II** activity of P450c11. This activity is unique to the zona glomerulosa, which also lacks 17α-hydroxylase. This is why the zona glomerulosa makes aldosterone but fails to make 17-hydroxysteroids or sex hormones.

Much of the dehydroepiandrosterone formed in the inner 2 zones is converted to dehydroepiandrosterone sulfate by **adrenal sulfokinase.** Androsterone is converted to testosterone and estradiol.

Action of ACTH

ACTH binds to high-affinity receptors on the plasma membrane of adrenocortical cells. This activates adenylate cyclase via G_s (see Chapter 1), and the resultant increase in intracellular cyclic AMP activates protein kinase A. Protein kinase A phosphorylates cholesterol ester hydrolase, increasing its activity, and conversion of cholesterol esters to free cholesterol is increased (Fig 20–10). This, in turn, leads to a prompt increase in the formation of pregnenolone and its derivatives. Over longer periods, ACTH also increases the synthesis of all 5 P450s involved in the formation of adrenocortical hormones.

Action of Angiotensin II

Angiotensin II binds to receptors in the zona glomerulosa which act via G_p to activate phospholipase C (see Chapter 1). The resulting increase in protein kinase C fosters the conversion of cholesterol to pregnenolone and also facilitates the formation of 18-hydroxycorticosterone, which, in turn, facilitates the production of aldosterone.

Enzyme Deficiencies

The consequences of inhibiting any of the enzyme systems involved in steroid biosynthesis can be predicted from Figs 20–8 and 20–9. Single enzyme defects can occur as congenital "inborn errors of metabolism" or can be produced by drugs. Congenital defects in any of the 5 enzymes lead to deficient cortisol secretion and the syndrome of **congenital adrenal hyperplasia**. The hyperplasia is due to increased ACTH secretion. The disease is severe when cholesterol desmolase is deficient, so that conversion of cholesterol to pregnenolone is compromised. The gene coding for this enzyme is on chromosome 15. The defect, which fortunately is rare, causes diffuse adrenal insufficiency and, in most cases, death soon after birth. Since androgens are not formed, female genitalia develop regardless of genetic sex (see Chapter 23). In 3β-hydroxysteroid deficiency, another rare condition, DHEA secretion is increased. This steroid is a weak androgen that can cause some masculinization in females with the disease, but it is not adequate to produce full masculinization of the genitalia in genetic males. Consequently, hypospadias is common. In 17α-hydroxylase deficiency, a third rare condition due to deficiency of an enzyme encoded by a gene on chromosome 10, no sex hormones are produced, so there are female external genitalia. However, the pathway leading to corticosterone and aldosterone is intact, and elevated levels of 11-deoxycorticosterone and other mineralocorticods produce hypertension and hypokalemia.

Unlike the 3 defects discussed in the preceding paragraph, 21β-hydroxylase and 11β-hydroxylase deficiencies are common; the former accounts for 90–95% of the cases of congenital adrenal hyperplasia, and the latter accounts for all but 1% of the remaining cases. Both are characterized by **virilization,** because the increase in ACTH secretion causes steroids to pile up behind the blockades and to be shunted to the production of androgens. The characteristic pattern that develops in females in the absence of treatment is the **adrenogenital syndrome** (Fig 20–11). P450c21 is on the short arm of chromosome 6, closely linked to the HLA major histocompatibility complex (see Chapter 19), and various defects can produce degrees of 21β-hydroxylase deficiency that range from mild to severe. Enough glucocorticoids and mineralocorticoids are usually formed to sustain life. In severe cases, the genitalia of genetic females are masculinized (**female pseudohermaphroditism;** see Chapter 23). However, masculinization may not be marked until later in life, and mild cases can be detected only by laboratory tests. About 30% of the patients with 21β-hydroxylase deficiency lose sodium to an excessive degree (**salt-losing form** of congenital virilizing adrenal hyperplasia). The sodium loss is due both to the mineralocorticoid deficiency and to the aldosterone-antagonizing effect of some of the steroids being secreted in excess amounts. In 11β-hydroxylase deficiency, there is excess secretion of 11-deoxycortisol and deoxycorticosterone; since the latter is an active mineralocorticoid, salt and water retention and hypertension result (**hypertensive form** of congenital virilizing adrenal hyperplasia). The gene coding for 11β-hydroxylase (P450c11) is located on chromosome 8, but molecular defects leading to

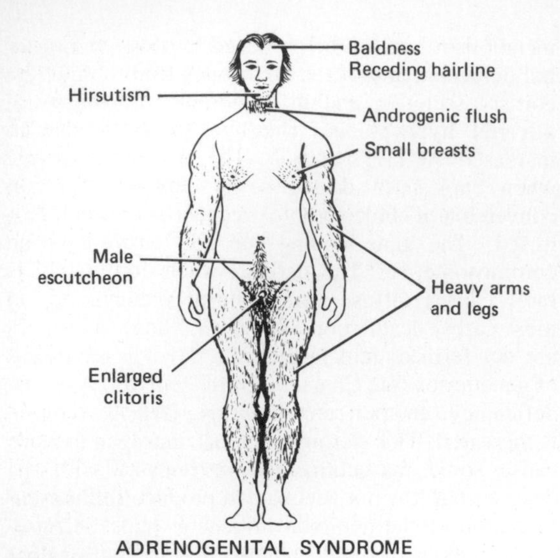

Figure 20–11. Typical findings in the adrenogenital syndrome in a post pubertal woman. (Reproduced, with permission, from Forsham PH, Di Raimondo VC: *Traumatic Medicine and Surgery for the Attorney.* Butterworth, 1960.)

11β-hydroxylase deficiency have not yet been worked out in detail.

Glucocorticoid treatment is indicated in all of these forms of congenital adrenal hyperplasia because it repairs the glucocorticoid deficit and inhibits ACTH secretion (see below), thus preventing the abnormal secretion of androgens and other steroids.

Enzyme Inhibition With Drugs

Metyrapone (methopyrapone, SU-4885, Metopirone) inhibits adrenal 11β-hydroxylase when administered in proper doses. It is used clinically to test pituitary reserve by producing a transient cortisol deficiency. The deficiency stimulates ACTH secretion, and the resultant increase in 11-deoxycortisol secretion is proportionate to the ability of the pituitary to respond. **Amphenone** and the *o,p′* isomer of DDD (mitotane, Lysodren) block the secretion of all steroids. The latter compound is a derivative of DDT. Amphenone causes adrenocortical hypertrophy, but ***o,p′*-DDD** produces necrosis of cortical cells and is of value in the treatment of adrenocortical cancer.

TRANSPORT, METABOLISM, & EXCRETION OF ADRENOCORTICAL HORMONES

Glucocorticoid Binding

Cortisol is bound in the circulation to an α globulin called **transcortin** or **corticosteroid-binding globulin (CBG).** There is also a minor degree of binding to albumin. Corticosterone is similarly bound, but to a lesser degree. The half-life of cortisol in the circulation is therefore longer (about 60–90 minutes) than that of corticosterone (50 minutes). Bound steroids appear to be physiologically inactive. Because of protein binding, there is relatively little free cortisol and corticosterone in the urine.

The equilibrium between cortisol and its binding protein and the implications of binding in terms of tissue supplies and ACTH secretion are summarized in Fig 20–12. The bound cortisol functions as a circulating reservoir of hormone that keeps a supply of free cortisol available to the tissues. The relationship is similar to that of T_4 and its binding protein (Fig 18–8). The usual methods for determining unconjugated "free 17-hydroxycorticoids" in the plasma measure both unbound and bound cortisol. At normal levels of total plasma cortisol (13.5 μg/dL, or 375 nmol/L), there is very little free cortisol in the plasma, but the binding sites on CBG become saturated when the total plasma cortisol exceeds 20 μg/dL. At higher plasma levels, there is some increased binding to albumin but the main increase is in the unbound fraction.

CBG is synthesized in the liver, and its production is increased by estrogen. CBG levels are elevated during pregnancy and depressed in cirrhosis, nephrosis, and multiple myeloma. When the CBG level rises, more cortisol is bound, and initially there is a drop in the free cortisol level. This stimulates ACTH secretion, and more cortisol is secreted until a new equilibrium is reached at which the free cortisol level returns to normal. The bound cortisol level remains elevated, but ACTH secretion returns to normal. Changes in the opposite direction occur when the CBG level falls. This explains why pregnant women have high total plasma 17-hydroxycorticoid levels without symptoms of glucocorticoid excess; conversely, it also explains why some patients with nephrosis have low plasma 17-hydroxycorticoids without adrenal insufficiency.

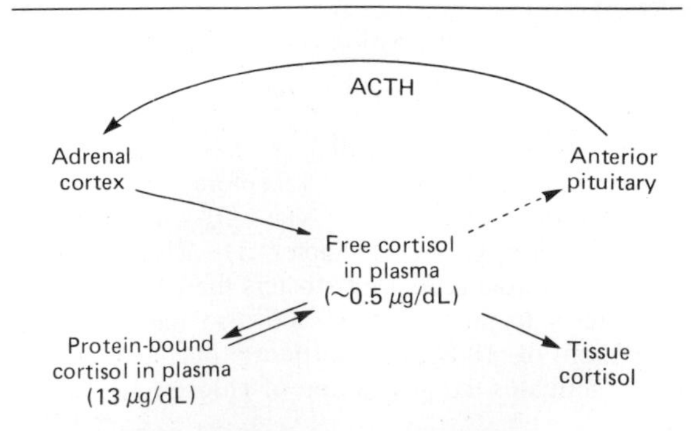

Figure 20–12. The interrelationships of free and bound cortisol. The dashed arrow indicates that cortisol inhibits ACTH secretion. The value for free cortisol is an approximation; in most studies, it is calculated by subtracting the protein-bound cortisol from the total plasma cortisol.

Metabolism & Excretion of Glucocorticoids

Cortisol is metabolized in the liver, which is the principal site of glucocorticoid catabolism. Most of the cortisol is reduced to dihydrocortisol and then to tetrahydrocortisol, which is conjugated to glucuronic acid (Fig 20–13). The glucuronyl transferase system responsible for this conversion also catalyzes the formation of the glucuronides of bilirubin (see Chapter 26) and a number of hormones and drugs. There is competitive inhibition between these substrates for the enzyme system.

Some of the cortisol in the liver is converted to cortisone. It should be emphasized that cortisone and other steroids with an 11-keto group are metabolites of the secreted glucocorticoids. Cortisone is an active glucocorticoid and is well known because of its extensive use in medicine, but it is not secreted in appreciable quantities by the adrenal glands. Little if any of the cortisone formed in the liver enters the circulation, because it is promptly reduced and conjugated to form tetrahydrocortisone glucuronide. The tetrahydroglucuronide derivatives ("conjugates") of cortisol and corticosterone are freely soluble. They enter the circulation, where they do not become bound to protein. They are rapidly excreted in the urine, in part by tubular secretion.

About 10% of the secreted cortisol is converted in the liver to the 17-ketosteroid derivatives of cortisol and cortisone. The ketosteroids are conjugated for the most part to sulfate and then excreted in the urine. Other metabolites, including 20-hydroxy derivatives, are formed. There is an enterohepatic circulation of glucocorticoids, and about 15% of the secreted cortisol is excreted in the stool. The distribution of cortisol and its derivatives in the plasma and urine is summarized in Table 20–3. The metabolism of corticosterone is similar to that of cortisol, except that it does not form a 17-ketosteroid derivative.

Variations in the Rate of Hepatic Metabolism

The rate of hepatic inactivation of glucocorticoids is depressed in liver disease and, interestingly, during surgery and other stresses. Thus, in stressed humans, the plasma free cortisol level rises higher than it does with maximal ACTH stimulation in the absence of stress. Glucocorticoids are not degraded in the process of exerting their physiologic effects.

Aldosterone

Aldosterone is bound to protein to only a slight extent, and its half-life is short (about 20 minutes). The amount secreted is small (Table 20–1), and the normal plasma aldosterone level in humans is about 0.006 μg/dL (0.17 nmol/L), compared with a cortisol level (bound and free) of about 13.5 μg/dL (375 nmol/L). Much of the aldosterone is converted in the

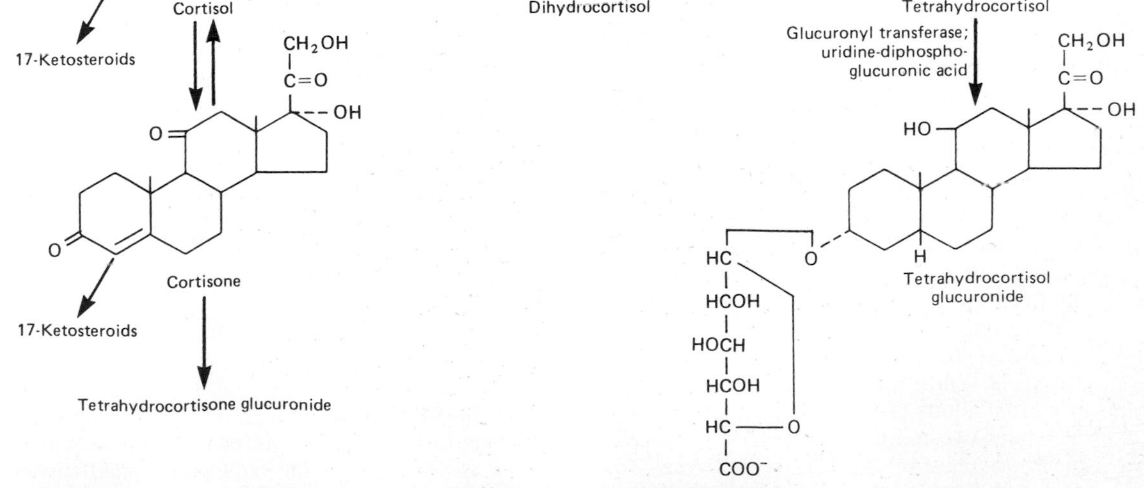

Figure 20–13. Outline of hepatic metabolism of cortisol.

Table 20–3. Distribution of cortisol and its principal derivatives in human plasma and urine. Note that the derivatives of corticosterone and the 17-ketosteroids from other sources are not included. The "plasma 17-hydroxycorticoids" include, in addition to cortisol and its derivatives, small amounts of other 17 α-hydroxylated steroids.

Plasma: Average concentration at 8:00 AM

	μg/dL	
Free cortisol	0.5	90% of "plasma 17-hydroxy-corticoids"
Protein-bound cortisol	13	
Unconjugated dihydro- and tetrahydro-cortisol and cortisone	Traces	
Tetrahydrocortisol glucuronide	10	"Conjugates"
Tetrahydrocortisone glucuronide	6	
17-Ketosteroids derived from cortisol and cortisone (mostly sulfate conjugates)	Actual plasma concentrations have not been measured	

Urine: Average amount excreted per 24 hours

	mg
Free cortisol	0.03
Tetrahydrocortisol glucoronide	5
Tetrahydrocortisone glucuronide	3
20-Hydroxy derivatives of tetrahydroglucuronides	6
17-Ketosteroids derived from cortisol and cortisone (mostly sulfate conjugates)	1
Unidentified metabolites	7

liver to the tetrahydroglucuronide derivative, but some is changed in the liver and in the kidneys to an 18-glucuronide. This glucuronide, which is unlike the breakdown products of other steroids, is converted to free aldosterone by hydrolysis at pH 1.0, and it is therefore often referred to as the "acid-labile conjugate." Less than 1% of the secreted aldosterone appears in the urine in the free form. Another 5% is in the form of the acid-labile conjugate, and up to 40% is in the form of the tetrahydroglucuronide.

17-Ketosteroids

The major adrenal androgen is the 17-ketosteroid dehydroepiandrosterone, although androstenedione is also secreted. The 11-hydroxy derivative of androstenedione and the 17-ketosteroids formed from cortisol and cortisone by side chain cleavage in the liver are the only 17-ketosteroids that have an =O or an −OH group in the 11 position ("11-oxy-17-ketosteroids"). Testosterone is also converted to a 17-ketosteroid. Since the daily 17-ketosteroid excretion in normal adults is 15 mg in men and 10 mg in women, about two-thirds of the urinary ketosteroids in men are secreted by the adrenal or formed from cortisol in the liver, and about one-third is of testicular origin.

Etiocholanolone, one of the metabolites of the adrenal androgens and testosterone (Fig 23–18), can cause fever when it is unconjugated (see Chapter 14). Certain individuals have episodic bouts of fever due to periodic accumulation in the blood of unconjugated etiocholanolone ("etiocholanolone fever").

EFFECTS OF ADRENAL ANDROGENS & ESTROGENS

Androgens

Androgens are the hormones that exert masculinizing effects, and they promote protein anabolism and growth (see Chapter 23). Testosterone from the testes is the most active androgen, and the adrenal androgens have less than 20% of its activity. Secretion of the adrenal androgens is controlled by ACTH and possibly by a pituitary adrenal androgen-stimulating hormone (see Chapter 23) but not by gonadotropins. The normal plasma concentration of dehydroepiandrosterone in adult men and women is about 300 μg/dL, and all but 0.1% of this is conjugated to sulfate. After the menopause, the plasma concentration in women falls to about 90 μg/dL. The secretion of adrenal androgens is as great in castrated males and females as it is in normal males, so it is clear that these hormones exert very little masculinizing effect when secreted in normal amounts. However, they can produce appreciable masculinization when secreted in excessive amounts. In adult males, excess adrenal androgens merely accentuate existing characteristics; but in prepuberal boys, they can cause precocious development of the secondary sex characteristics without testicular growth **(precocious pseudopuberty).** In females they cause female pseudohermaphroditism and the adrenogenital syndrome (Fig 20–11).

Estrogens

The adrenal androgen androstenedione is converted to estrogens (aromatized) in the circulation, and the adrenal may also secrete some estrogens. The amount of estrogen formed from adrenal precursors is normally too small to have any apparent physiologic effect. However, feminizing estrogen-secreting tumors of the adrenal cortex have been described, and ovariectomized patients with estrogen-dependent breast cancer improve when their adrenals are removed or when they are treated with glucocorticoids in doses sufficient to inhibit ACTH secretion.

PHYSIOLOGIC EFFECTS OF GLUCOCORTICOIDS

Adrenal Insufficiency

In untreated adrenalectomized animals and humans, there is Na^+ loss with circulatory insufficiency, hypotension, and, eventually, fatal shock due to the mineralocorticoid deficiency. Because glucocorticoids are absent, water, carbohydrate, protein, and fat metabolism are abnormal, and collapse and death follow exposure to even minor noxious stimuli. Small amounts of glucocorticoids correct the metabolic abnormalities, in part directly and in part by permitting other reactions to occur. It is important to separate these physiologic actions of glucocorticoids from the quite different effects produced by large amounts of the hormones.

Mechanism of Action

The multiple effects of glucocorticoids are triggered by binding to a glucocorticoid receptor, which promotes the transcription of certain segments of DNA (see Chapter 1). This, in turn, leads via the appropriate mRNAs (Fig 1–28) to synthesis of enzymes that alter cell function.

Effects on Intermediary Metabolism

The actions of glucocorticoids on the intermediary metabolism of carbohydrate, protein, and fat are discussed in Chapter 19. They include increased protein catabolism and increased hepatic glycogenesis and gluconeogenesis (Fig 20–14). Glucose 6-phosphatase activity is increased, and the blood glucose level rises. Glucocorticoids exert an anti-insulin action in peripheral tissues and make diabetes worse. However, the brain and the heart are spared, and the extra glucose ensures a supply of glucose to these vital organs. In diabetics, glucocorticoids raise plasma lipid levels and increase ketone body formation, but in normal individuals, the increase in insulin secretion provoked by the rise in blood glucose obscures these actions. In adrenal insufficiency, the blood glucose level is normal as long as an adequate caloric intake is maintained, but fasting causes hypoglycemia that can be fatal. The adrenal cortex is not essential for the ketogenic response to fasting.

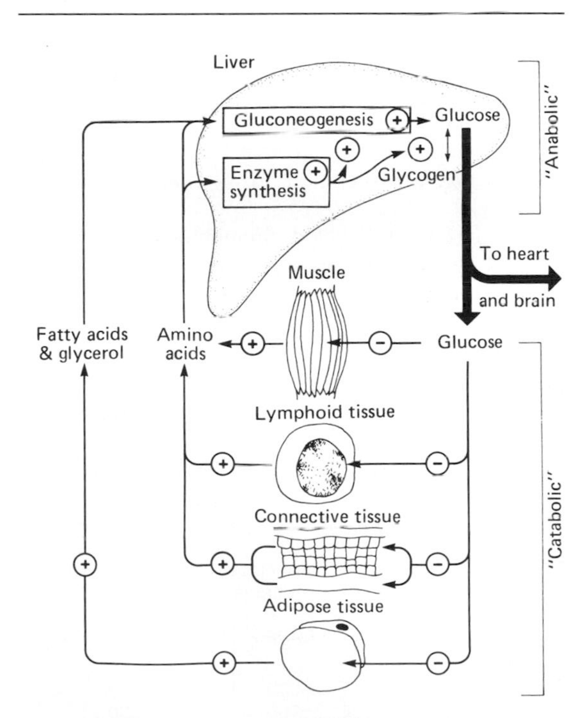

Figure 20–14. Actions of glucocorticoids in promoting gluconeogenesis, −, inhibition; +, stimulation. (Modified and reproduced, with permission, from Baxter JD, Forsham PH: Tissue effects of glucocorticoids. *Am J Med* 1972;**52**:573.)

Permissive Action

Small amounts of glucocorticoids must be present for a number of metabolic reactions to occur, although the glucocorticoids do not produce the reactions by themselves. This effect is called their **permissive action.** Permissive effects include the requirement for glucocorticoid to be present for glucagon and catecholamines to exert their calorigenic effects (see above and Chapter 19), for catecholamines to exert their lipolytic effects, and for catecholamines to produce pressor responses and bronchodilation.

Effects on ACTH Secretion

Glucocorticoids inhibit ACTH secretion, and ACTH secretion is increased in adrenalectomized animals. The consequences of the feedback action of cortisol on ACTH secretion are discussed below in the section on regulation of glucocorticoid secretion.

Vascular Reactivity

In adrenally insufficient animals, vascular smooth muscle becomes unresponsive to norepinephrine and epinephrine. The capillaries dilate and, terminally, become permeable to colloidal dyes. Failure to respond to the norepinephrine liberated at noradrenergic nerve endings probably impairs vascular compensation for the hypovolemia of adrenal insufficiency and promotes vascular collapse. Glucocorticoids restore vascular reactivity.

Effects on the Nervous System

Changes in the nervous system in adrenal insufficiency that are reversed only by glucocorticoids include the appearance of electroencephalographic waves slower than the normal α rhythm, personality changes, and increased sensitivity to olfactory and gustatory stimuli. The personality changes, which are mild, include irritability, apprehension, and inability to concentrate.

Effects on Water Metabolism

Adrenal insufficiency is characterized by inability to excrete a water load (Fig 20–15), and only glucocorticoids repair this deficit. The load is eventually excreted, but the excretion is so slow that there is danger of water intoxication. In patients with adrenal insufficiency who have not received glucocorticoids, glucose infusions may cause high fever (''glucose fever'') followed by collapse and death.

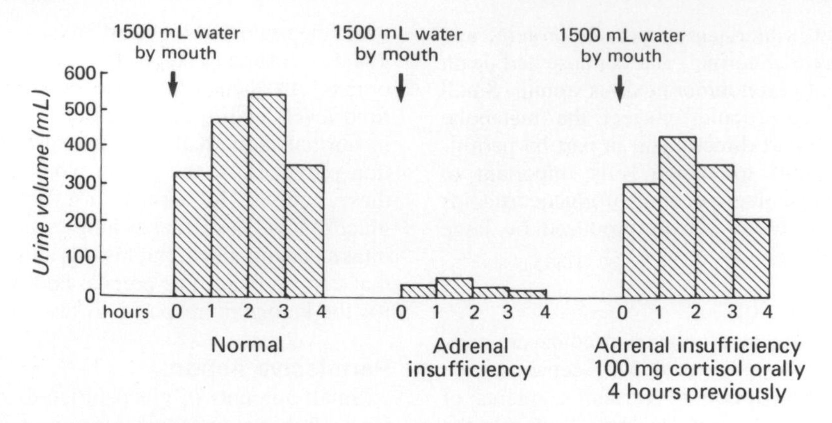

Figure 20–15. Response to a 1500-mL water load in a normal subject and a patient with adrenal insufficiency before and after cortisol treatment.

Presumably, the glucose is metabolized, the water dilutes the plasma, and the resultant osmotic gradient between the plasma and the cells causes the cells of the thermoregulatory centers in the hypothalamus to swell to such an extent that their function is disrupted. It should be noted that water intoxication can occur in adrenal insufficiency even without an absolute excess of water in the body; the important factor is retention of water in excess of Na^+.

The cause of defective water excretion in adrenal insufficiency is unsettled. Plasma vasopressin and neurophysin levels are elevated in adrenal insufficiency and reduced by glucocorticoid treatment. The glomerular filtration rate is low, and this probably contributes to the deficiency in water excretion. The selective effect of glucocorticoids on the abnormal water excretion is consistent with this possibility, because even though the mineralocorticoids improve filtration by restoring plasma volume, the glucocorticoids raise the glomerular filtration rate to a much greater degree.

Effects on the Blood Cells & Lymphathic Organs

Glucocorticoids decrease the number of circulating eosinophils by increasing their sequestration in the spleen and lungs. Changes in the eosinophil level have been used as an index of changes in ACTH secretion, but they are an unreliable indicator because some stresses cause eosinopenia in the absence of the adrenals. Glucocorticoids also lower the number of basophils in the circulation and increase the number of neutrophils, platelets, and red blood cells (Table 20–4).

Glucocorticoids decrease the circulating lymphocyte count and the size of the lymph nodes and thymus by inhibiting lymphocyte mitotic activity. It appears that their primary action is to inhibit production of interleukin-2 by T lymphocytes, thus effectively stopping lymphocyte proliferation (see Chapter 27).

Resistance to "Stress"

When an animal or human is exposed to any of an immense variety of noxious or potentially noxious stimuli, there is an increased secretion of ACTH and, consequently, a rise in the circulating glucocorticoid level. This rise is essential for survival. Hypophysectomized animals, or adrenalectomized animals treated with maintenance doses of glucocorticoids, die when exposed to the same noxious stimuli.

Selye defined noxious stimuli that increase ACTH secretion as "stressors," and it is fashionable today to lump these stimuli together under the term "stress." This word is a short, emotionally charged word for something that otherwise takes many words to say, and it is a convenient term to use as long as it is understood that, in the context of this book, it denotes only those stimuli that have been proved to increase ACTH secretion in normal animals and humans.

The reason an elevated circulating glucocorticoid level is essential for resisting stress remains for the most part unknown. Most of the stressful stimuli that increase ACTH secretion also activate the sympathoadrenal medullary system, and part of the function of circulating glucocorticoids may be maintenance of vascular reactivity to catecholamines. Glucocorticoids are also necessary for the catecholamines to exert their full FFA-mobilizing action,

Table 20–4. Typical effects of cortisol on the white and red blood cell counts in humans (cells/μL).

	Normal	Cortisol-Treated
White blood cells		
Total	9000	10,000
PMNs	5760	8330
Lymphocytes	2370	1080
Eosinophils	270	20
Basophils	60	30
Monocytes	450	540
Red blood cells	5 million	5.2 million

and the FFAs are an important emergency energy supply. However, sympathectomized animals tolerate a variety of stresses with relative impunity. Another theory holds that glucocorticoids prevent other stress-induced changes from becoming excessive. At present, all that can be said is that stress causes increases in plasma glucocorticoids to high "pharmacologic" levels that in the short run are life-saving but in the long run are definitely harmful and disruptive.

PHARMACOLOGIC & PATHOLOGIC EFFECTS OF GLUCOCORTICOIDS

Cushing's Syndrome

The clinical picture produced by prolonged increases in plasma glucocorticoids is called **Cushing's syndrome** (Fig 20–16). It may be caused not only by the administration of large amounts of exogenous hormones but also by glucocorticoid-producing adrenocortical tumors and by hypersecretion of ACTH. Most patients with hypersecretion of ACTH from the pituitary have small pituitary tumors **(microadenomas)** that are often so small that the sella turcica is not enlarged. However, some cases of excess ACTH secretion may be due to hypersecretion of corticotropin-releasing hormone (CRH). Cushing's syndrome is also produced by tumors of nonendocrine tissues that secrete substances with CRH activity or, more commonly, ACTH **(ectopic ACTH syndrome).** Tumors of nonendocrine tissues have also been reported to secrete substances with actions like those of almost every known hormone.

Patients with Cushing's syndrome are protein-depleted as a result of excess protein catabolism. The skin and subcutaneous tissues are therefore thin, and the muscles are poorly developed. Wounds heal poorly, and minor injuries cause bruises and ecchymoses. The hair is thin and scraggly. Many patients with the disease have some increase in facial hair and acne, but this is caused by the increased secretion of adrenal androgens that often accompanies the increase in glucocorticoid secretion.

Body fat is redistributed in a characteristic way. The extremities are thin, but fat collects in the abdominal wall, face, and upper back, where it produces a "buffalo hump." As the thin skin of the abdomen is stretched by the increased subcutaneous fat depots, the subdermal tissues rupture to form prominent reddish-purple **striae.** These scars are seen normally whenever there is rapid stretching of skin (eg, around the breasts of girls at puberty or in the abdominal skin during pregnancy), but in normal individuals, the striae are usually inconspicuous and lack the intense purplish color.

Many of the amino acids liberated from catabolized proteins are converted into glucose in the liver,

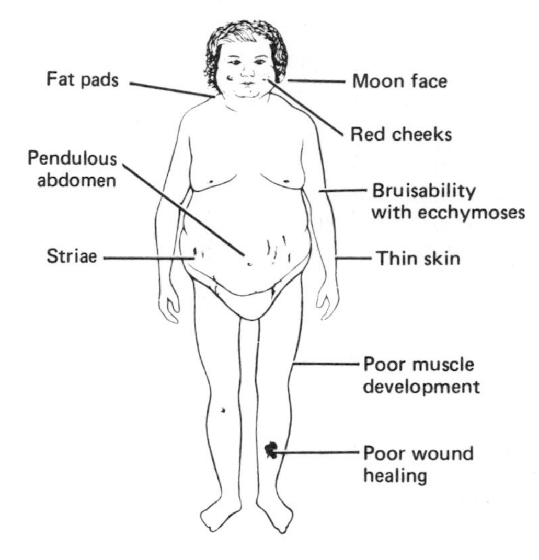

Figure 20–16. Typical findings in Cushing's syndrome. (Reproduced, with permission, from Forsham PH, Di Raimondo VC: *Traumatic Medicine and Surgery for the Attorney.* Butterworth, 1960.)

and the resultant hyperglycemia and decreased peripheral utilization of glucose may be sufficient to precipitate insulin-resistant diabetes mellitus, especially in patients genetically predisposed to diabetes. Hyperlipemia and ketosis are associated with the diabetes, but acidosis is usually not severe.

The glucocorticoids are present in such large amounts in Cushing's syndrome that they exert a significant mineralocorticoid action. Deoxycorticosterone secretion is also elevated in cases due to ACTH hypersecretion. The salt and water retention plus the facial obesity cause the characteristic plethoric, rounded "moon-faced" appearance, and there may be significant K^+ depletion and weakness. About 85% of patients with Cushing's syndrome are hypertensive. The hypertension may be due to increased deoxycorticosterone secretion, increased secretion of angiotensinogen (see Chapter 24), or a direct glucocorticoid effect on blood vessels.

Excess glucocorticoids lead to bone dissolution by decreasing bone formation and increasing bone resorption. This lead to **osteoporosis,** a loss of bone mass that leads eventually to collapse of vertebral bodies and other fractures. The mechanisms by which glucocorticoids produce their effects on bone are discussed in Chapter 21.

Glucocorticoids in excess accelerate the basic electroencephalographic rhythms and produce mental aberrations ranging from increased appetite, insomnia, and euphoria to frank toxic psychoses. As noted above, glucocorticoid deficiency is also associated with mental symptoms, but the symptoms produced by glucocorticoid excess are more severe.

Anti-inflammatory & Antiallergic Effects of Glucocorticoids

In large doses, glucocorticoids inhibit the inflammatory response to tissue injury. This reaction is discussed in Chapter 27. The glucocorticoids also suppress manifestations of allergic disease that are due to the release of histamine from tissues. Both of these effects require high levels of circulating glucocorticoids and cannot be produced without producing the other manifestations of glucocorticoid excess. Furthermore, large doses of exogenous glucocorticoids inhibit ACTH secretion to the point that several adrenal insufficiency can be a dangerous problem when therapy is stopped.

Glucocorticoids inhibit fibroblastic activity, decrease local swelling, and block the systemic effects of bacterial toxins. The decreased local inflammatory reaction is due, at least in part, to inhibition of phospholipase A_2, with a consequent reduction in the release of arachidonic acid from tissue phospholipids. Therefore, the formation of leukotrienes, thromboxanes, prostaglandins, and prostacyclin (Fig 17–33) is reduced. The leukotrienes are powerful mediators of inflammation, and high concentrations of leukotriene B_4 have been found in synovial fluid in rheumatoid arthritis. By stabilizing lysosomal membranes, glucocorticoids inhibit the breakdown of lysosomes that takes place in inflamed tissue. Despite their protein-catabolic effects, they slow the degrading effect of collagenase on joint tissues in rheumatoid arthritis, and this is the basis of their effectiveness following intra-articular administration. The glucocorticoids also inhibit the release of interleukin-1 (endogenous pyrogen) from granulocytes (see Chapter 14). Large doses of glucocorticoids in humans initially elevate and later depress antibody levels. The inhibition of fibroblastic activity prevents the walling off of infections and also prevents such phenomena as keloid formation and the development of adhesions after abdominal surgery.

The actions of glucocorticoids in patients with bacterial infections are dramatic but dangerous. For example, in pneumococcal pneumonia or active tuberculosis, the febrile reaction, the toxicity, and the lung symptoms disappear; but unless antibiotics are given at the same time, the bacteria spread throughout the body. It is important to remember that the symptoms are the warning that disease is present; when these symptoms are masked by treatment with glucocorticoids, there may be serious and even fatal delays in diagnosis and the institution of treatment with antimicrobial drugs.

When certain types of antibodies combine with their antigens, they provoke the release of histamine from mast cells in various tissues, and this in turn causes many of the symptoms of allergy. Glucocorticoids do not affect the combination of antigen with antibody and have no influence on the effects of histamine once it is released, but they prevent histamine release. Glucocorticoids therefore relieve the symptoms of asthma and delayed hypersensitivity reactions and may be of benefit in such immunoglobulin-mediated diseases as serum sickness.

Other Effects

Large doses of glucocorticoids inhibit growth, decrease growth hormone secretion (see Chapter 22), induce PNMT, and decrease TSH secretion. During fetal life, glucocorticoids accelerate the maturation of surfactant in the lungs (see Chapter 34).

REGULATION OF GLUCOCORTICOID SECRETION

Role of ACTH

Both basal secretion of glucocorticoids and the increased secretion provoked by stress are dependent upon ACTH from the anterior pituitary. Angiotensin II also stimulates the adrenal cortex, but its effect is mainly on aldosterone secretion. Large doses of a number of other naturally occurring substances, including vasopressin, serotonin, and VIP, are capable of stimulating the adrenal directly, but there is no evidence that these agents play any role in the physiologic regulation of glucocorticoid secretion.

Chemistry & Metabolism of ACTH

ACTH is a single-chain polypeptide containing 39 amino acids. Its origin from pro-opiomelanocortin (POMC) in the pituitary is discussed in Chapter 22. The first 23 amino acids in the chain, which are the same in all species that have been examined, constitute the active "core" of the molecule, and a synthetic peptide containing these 23 amino acids has been shown to have the full activity of the 39-amino-acid peptide. Amino acids 24–39 are therefore a "tail" that apparently stabilizes the molecule and varies slightly in amino acid composition from species to species (Fig 20–17). Each of the ACTHs that have been isolated is active in all species but generally antigenic in heterologous species.

ACTH is inactivated in blood in vitro at a slower rate than it is in vivo; its half-life in the circulation in humans is about 10 minutes. A large part of an injected dose of ACTH is found in the kidneys, but neither nephrectomy nor evisceration appreciably enhances its in vivo activity, and the site of its inactivation is not known.

Effect of ACTH on the Adrenal

After hypophysectomy, glucocorticoid synthesis and output decline within 1 hour to very low levels, although some hormone is still secreted. Within a short time after an injection of ACTH (in dogs, less than 2 minutes), glucocorticoid output rises (Fig 20–18). With low doses of ACTH, there is a linear relationship between the log of the dose and the increase in glucocorticoid secretion. However, the maximal rate at which glucocorticoids can be secreted is rapidly reached; and in dogs, doses larger than 10

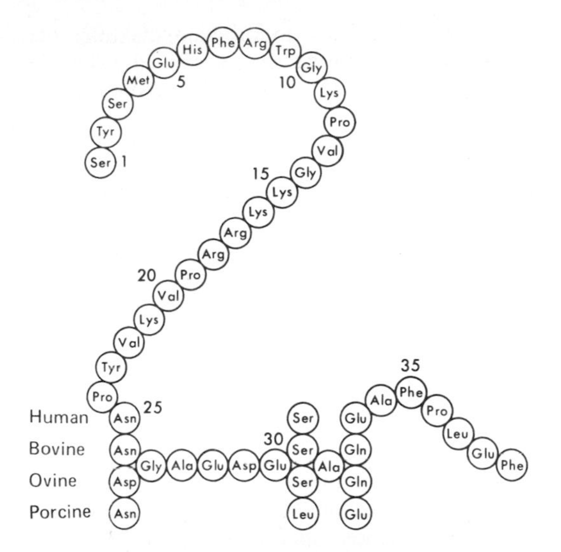

Figure 20–17. Structure of ACTH in various species. The amino acid composition varies only at positions 25, 31, and 33. Thus, human ACTH has Asn in position 25, Ser in position 31, and Glu in position 33; bovine ACTH has Asn in position 25, Ser in position 31, and Gln in position 33; etc. (Reproduced, with permission, from Li CH: Adrenocorticotropin 45: Revised amino acid sequences for sheep and bovine hormones. *Biochem Biophys Res Commun* 1972;**49**:835.)

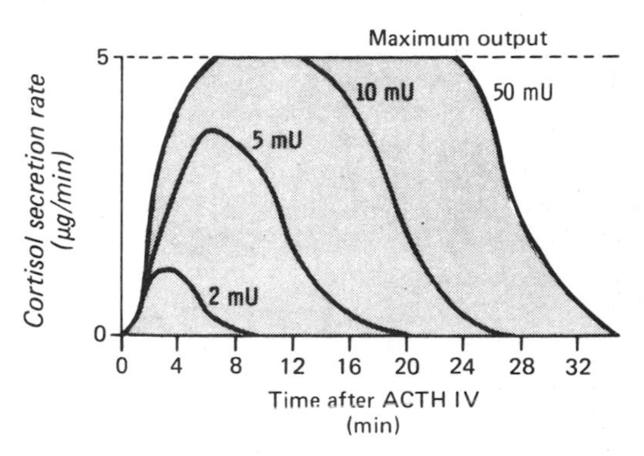

Figure 20–18. Changes in cortisol output in the hypophysectomized dog following the administration of various doses of ACTH. (Reproduced, with permission, from Ganong WF: The central nervous system and the synthesis and release of ACTH. In: *Advances in Neuroendocrinology*. Nalbandov A [editor]. Univ. of Illinois Press, 1963.)

milliunits (mU) only prolong the period of maximal secretion. A similar "ceiling on output" exists in rats and humans. The effects of ACTH on adrenal morphology and the mechanism by which it increases steroid secretion are discussed above.

Adrenal Responsiveness

ACTH not only produces prompt increases in glucocorticoid secretion but also increases the sensitivity of the adrenal to subsequent doses of ACTH. Conversely, single doses of ACTH do not increase glucocorticoid secretion in chronically hypohysectomized animals and patients with hypopituitarism, and repeated injections or prolonged infusions of ACTH are necessary to restore normal adrenal responses to ACTH. Decreased responsiveness is also produced by doses of glucocorticoids that inhibit ACTH secretion. The decreased adrenal responsiveness to ACTH is detectable within 24 hours after hypophysectomy and increases progressively with time (Fig 20–19). It is marked when the adrenal is atrophic but develops before there are visible changes in adrenal size or morphology. There is evidence that the sensitivity of the inner 2 zones and the zona glomerulosa to ACTH is increased by a portion of the N-terminal region of the POMC molecule (see Chapter 22) that is secreted along with ACTH.

Circadian Rhythm

ACTH is secreted in irregular bursts throughout the day, and plasma cortisol tends to rise and fall in response to these bursts (Fig 20–20). In humans, the bursts are most frequent in the early morning and least frequent in the evening. This **diurnal (circadian) rhythm** in ACTH secretion is present in patients with adrenal insufficiency receiving constant doses of glucocorticoids. It is not due to the stress of getting up in the morning, traumatic as that may be, because the increased ACTH secretion occurs before waking up. If the "day" is lengthened experimentally to more than 24 hours—ie, if the individual is isolated

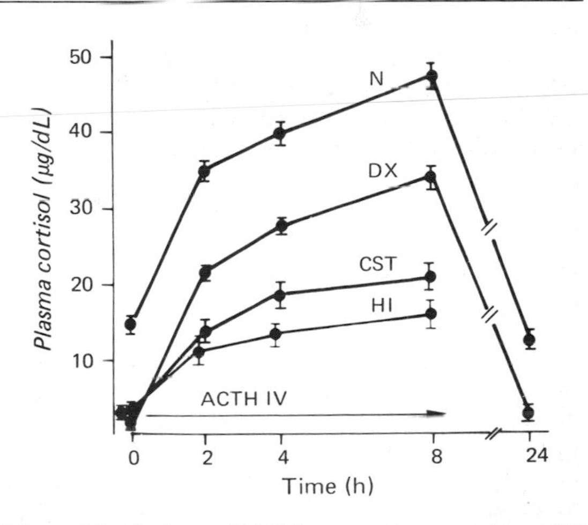

Figure 20–19. Loss of ACTH responsiveness when ACTH secretion is decreased in humans. The 1–24 amino acid sequence of ACTH was infused intravenously in a dose of 250 µg over 8 h. N, normal subjects; DX, dexamethasone 0.75 mg every 8 h for 3 days; CST, long-term corticosteroid therapy; HI, anterior pituitary insufficiency. (Reproduced, with permission, from Kolanowski J et al: Adrenocortical response upon repeated stimulation with corticotrophin in patients lacking endogenous corticotrophin secretion. *Acta Endocrinol* [*Kbh*] 1977;**85**:595.)

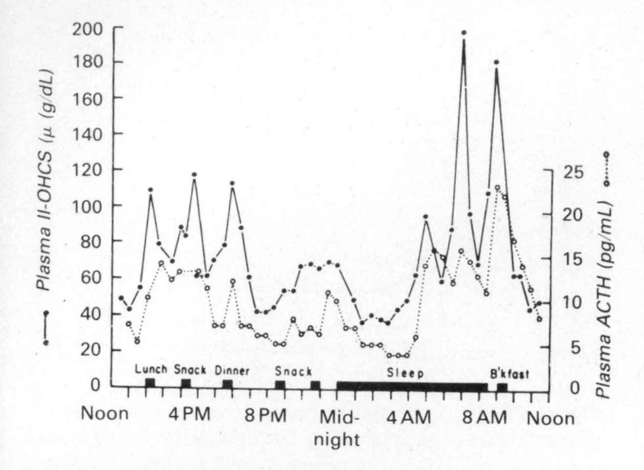

Figure 20–20. Fluctuations in plasma ACTH and glucocorticoids throughout the day in a normal girl (age 16). The ACTH was measured by immunoassay and the glucocorticoids as 11-oxysteroids (11-OHCS). Note the greater ACTH and glucocorticoid rises in the morning, before awakening from sleep. (Reproduced, with permission, from Krieger DT et al: Characterization of the normal temporal pattern of plasma corticosteroid levels. *J Clin Endocrinol Metab* 1971;**32**:266.)

and the day's activities are spread over more than 24 hours—the adrenal cycle also lengthens, but the increase in ACTH secretion still occurs during the period of sleep. The biologic clock responsible for the diurnal ACTH rhythm is located in the suprachiasmatic nuclei of the hypothalamus (see Chapter 14).

The Response to Stress

The morning plasma ACTH concentration in a healthy resting human is about 25 pg/mL (5.5 pmol/L). ACTH and cortisol values in various abnormal conditions are summarized in Fig 20–21. During severe stress, the amount of ACTH secreted exceeds the amount necessary to produce maximal glucocorticoid output. However, prolonged exposure to ACTH in conditions such as the ectopic ACTH syndrome increases the adrenal maximum.

Increases in ACTH secretion to meet emergency situations are mediated almost exclusively through the hypothalamus via release of CRH. This polypeptide is secreted in the median eminence and transported in the portal-hypophyseal vessels to the anterior pituitary, where it stimulates ACTH secretion (see Chapter 14). If the median eminence is destroyed, some basal glucocorticoid secretion continues and the adrenals do not atrophy, but increased secretion in response to many different stresses is blocked. Afferent nerve pathways from many parts of the brain converge on the median eminence. Fibers from the amygdaloid nuclei mediate responses to emotional stresses, and fear, anxiety, and apprehension cause marked increases in ACTH secretion. Input from the suprachiasmatic nuclei also provides the drive for the diurnal rhythm. Impulses ascending to the hypothalamus via the nociceptive pathways and the reticular formation trigger ACTH secretion in response to injury (Fig 20–22). There is an inhibitory input from the baroreceptors via the nucleus of the tractus solitarius. Contrary to previously proposed theories, circulating epinephrine and norepinephrine do not increase ACTH secretion in humans, and adrenocortical and adrenal medullary secretion are independently regulated from the hypothalamus.

Glucocorticoid Feedback

High circulating levels of free glucocorticoids inhibit ACTH secretion, and the degree of pituitary inhibition is apparently a linear function of the circulating glucocorticoid level. The inhibitory effect is exerted at both the pituitary and the hypothalamic level. The inhibition is due to an action on DNA, and maximal inhibition takes several hours to develop, although there is some evidence for an additional, more rapid "fast-feedback." The ACTH-inhibiting

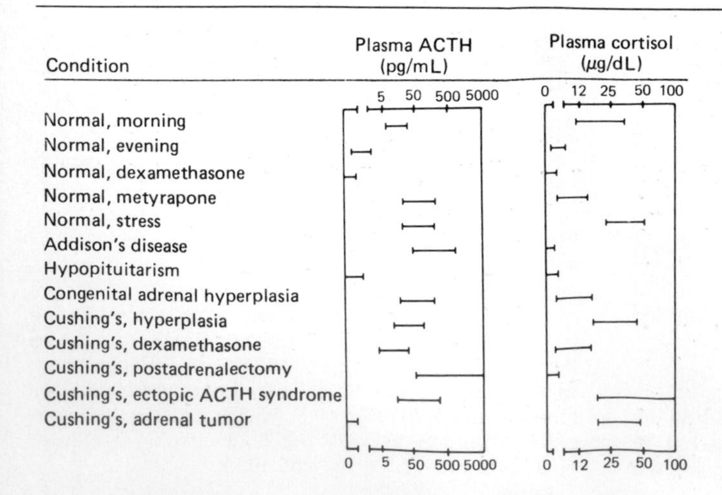

Figure 20–21. Plasma concentrations of ACTH and cortisol in various clinical states. (Reproduced, with permission, from Liddle G: The adrenal cortex. In: *Textbook of Endocrinology*, 5th ed. Williams RH [editor]. Saunders, 1974.)

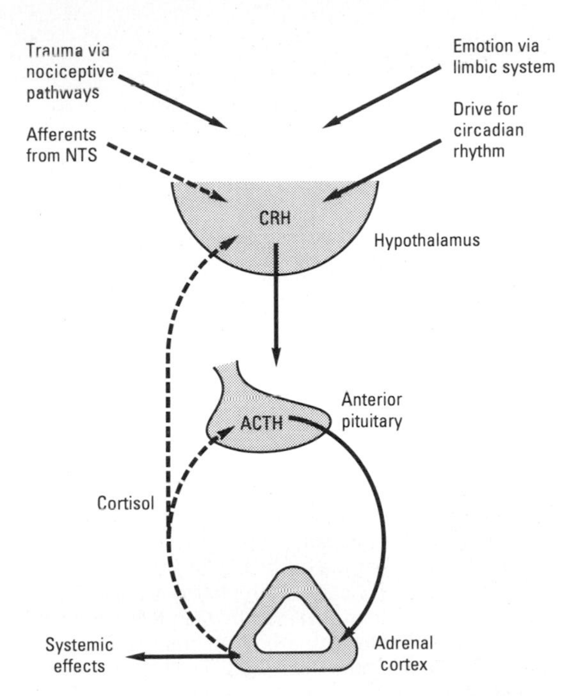

Figure 20–22. Feedback control of the secretion of cortisol and other glucocorticoids. The dashed arrows indicate inhibitory effects, and the solid arrows indicate stimulatory effects. Compare with Fig 18–13, 22–9, 23–21 and 23–35.

activity of the various steroids parallels their glucocorticoid potency. A drop in resting corticoid levels stimulates ACTH secretion, and in chronic adrenal insufficiency, there is a marked increase in the rate of ACTH synthesis and secretion.

It was once thought that stress stimulated ACTH secretion by initially lowering the plasma corticoid level, but this has been shown to be incorrect. In fact, an untreated adrenalectomized animal whose circulating corticoid level is zero responds to stress with a much greater than normal rise in plasma ACTH. Thus, it seems clear that the rate of ACTH secretion is determined by 2 opposing forces: the sum of the neural and possibly other stimuli converging through the median eminence to increase ACTH secretion, and the magnitude of the braking action of glucocorticoids on ACTH secretion, which is proportionate to their level in the circulating blood (Fig 20–22).

The dangers involved when prolonged treatment with anti-inflammatory doses of glucocorticoids is stopped deserve emphasis. Not only is the adrenal atrophic and unresponsive after such treatment; even if its responsiveness is restored by injecting ACTH, the pituitary may be unable to secrete normal amounts of ACTH for as long as a month. The cause of the deficiency is presumably diminished ACTH synthesis. Thereafter, there is a slow rise in ACTH to supranormal levels. These in turn stimulate the adrenal, and glucocorticoid output rises, with feedback inhibition gradually reducing the elevated ACTH levels to normal (Fig 20–23). The complications of sudden cessation of steroid therapy can usually be avoided by slowly decreasing the steroid dose over a long period.

EFFECTS OF MINERALOCORTICOIDS

Actions

Aldosterone and other steroids with mineralocorticoid activity increase the reabsorption of Na^+ from the urine, sweat, saliva, and gastric juice. Thus, they cause retention of Na^+ in the ECF. In the kidney, they act on the epithelium of the distal tubule and collecting duct. They may also increase the K^+ and decrease the Na^+ in muscle and brain cells. Under the influence of aldosterone, increased amounts of Na^+ are in effect exchanged for K^+ and H^+ in the renal tubules, producing a K^+ diuresis (Fig 20–24) and an increase in urine acidity.

Mechanism of Action

Aldosterone, like other steroids, acts by stimulating DNA-dependent mRNA synthesis (see Chapter 1). Sodium ions diffuse out of the urine (or saliva, sweat, or gastric juice) into the surrounding epithelial cells and are actively transported from these cells into the interstitial fluid. The amount of Na^+ removed from these fluids is proportionate to the rate of active transport of Na^+. The energy for the active transport is supplied by ATP, and ATP synthesis depends for the most part on the oxidation of substrate to CO_2 and H_2O via the citric acid cycle. Aldosterone binds to the mineralocorticoid receptor (see Chapter 1), and like other steroids, the steroid-receptor complex acts in

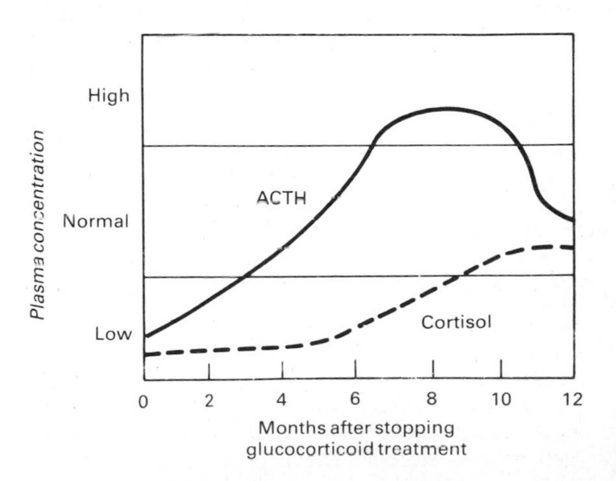

Figure 20–23. Pattern of plasma ACTH and cortisol values in patients recovering from prior long-term daily treatment with large doses of glucocorticoids. (Courtesy of R Ney.)

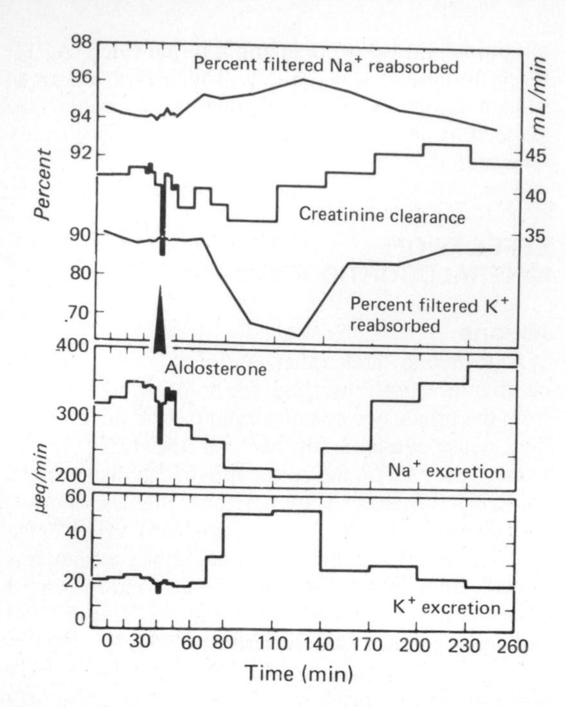

Figure 20–24. Effect of aldosterone (5 μg as a single dose injected into the aorta) on electrolyte excretion in an adrenalectomized dog.

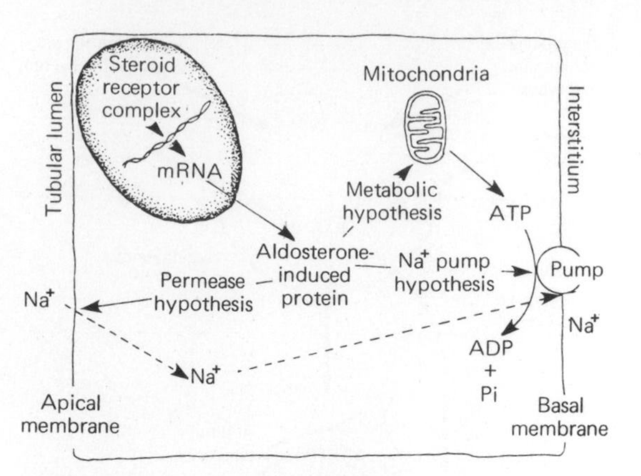

Figure 20–25. Mechanism of action of aldosterone. The steroid induces the formation of one or more proteins that in turn may increase the permeability of the apical (luminal) membrane to Na^+, increase the active transport of Na^+ out of the cell across the basal and lateral membranes to the interstitium, or increase the energy available to the pump. (Modified from Edelman IS: Candidate mediators in the action of aldosterone on Na^+ transport. In: *Membrane Transport Processes*. Vol I. Hoffman JF [editor]. Raven, 1978.)

the nucleus to increase transcription of DNA (Fig 20–25). The induced mRNA stimulates protein synthesis at the ribosomal level. Aldosterone fails to exert any effect on Na^+ excretion for 10–30 minutes or more even when injected directly into the renal artery. This latent period represents the time needed to increase protein synthesis. The function of the aldosterone-induced protein or proteins remains unsettled. One hypothesis holds (Fig 20–25) that the protein increases the passive permeability of the cell to Na^+ from the tubular lumen (permease hypothesis). Another holds that the protein increases the oxidation of substrate to provide ATP (metabolic hypothesis). A third holds that the hormone increases the synthesis of Na^+-K^+ ATPase, and when these additional pump molecules are inserted in the membrane, Na^+ extrusion is increased (Na^+ pump hypothesis). Evidence is now accumulating that this third hypothesis is correct. In any case, the effect is increased active transport of Na^+ from the tubular lumen to the interstitium and thence to the bloodstream.

Aldosterone is the principal mineralocorticoid secreted by the adrenal, although corticosterone is secreted in sufficient amounts to exert a minor mineralocorticoid effect (Tables 20–1 and 20–2). Deoxycorticosterone, which is secreted in appreciable amounts only in abnormal situations, has about 3% of the activity of aldosterone. It is used clinically as a mineralocorticoid because its synthetic acetate **(desoxycorticosterone acetate, DOCA)** is cheaper and more readily available than aldosterone. Large amounts of progesterone and some other steroids cause natriuresis, but there is little evidence that they play any normal role in the control of Na^+ excretion.

Effect of Adrenalectomy

In adrenal insufficiency, Na^+ is lost in the urine. K^+ is retained, and the plasma K^+ rises. When adrenal insufficiency develops rapidly, the decline in ECF Na^+ exceeds the amount excreted in the urine, indicating that Na^+ also must be entering the cells. When the posterior pituitary is intact, salt loss exceeds water loss, and the plasma Na^+ falls (Table 20–5). However, plasma volume also is reduced, resulting in hypotension, circulatory insufficiency, and, eventually, fatal shock. These changes can be prevented to a degree by increasing the dietary salt intake. Rats survive indefinitely on extra salt alone, but in dogs and most humans, the amount of supplementary salt needed is so large that it is almost impossible to prevent eventual collapse and death unless mineralocorticoid treatment is also instituted.

Secondary Effects of Excess Mineralocorticoids

K^+ depletion due to prolonged K^+ diuresis is a prominent feature of prolonged mineralocorticoid treatment (Table 20–5). When the K^+ loss is marked, intracellular K^+ is replaced by Na^+. This shift can be prevented by the administration of supplemental K^+ along with the mineralocorticoid. There is no net increase in body sodium if salt intake is restricted to the level of output, but on a normal diet, total body

Table 20–5. Typical plasma electrolyte levels in normal humans and patients with adrenocortical diseases.

	Plasma Electrolytes (meq/L)			
	Na^+	K^+	Cl^-	HCO_3^-
Normal	142	4.5	105	25
Adrenal insufficiency	120	6.7	85	25
Primary hyper-aldosteronism	145	2.4	96	41

sodium rises. However, the plasma Na^+ is elevated only slightly if at all, because water is retained with the osmotically active sodium ions. Consequently, ECF volume is expanded and the blood pressure rises. When the ECF expansion passes a certain point, Na^+ excretion is usually increased in spite of the continued action of mineralocorticoids on the renal tubules. This "escape phenomenon" (Fig 20–26) is probably due to increased secretion of ANP (see Chapter 24). Because of increased excretion of Na^+ when the ECF volume is expanded, mineralocorticoids do not produce edema in normal individuals and patients with hyperaldosteronism. However, escape may not occur in certain disease states, and in these situations, continued expansion of ECF volume leads to edema (see Chapter 38).

Primary Hyperaldosteronism

The effects of chronic mineralocorticoid excess are seen in patients with aldosterone-secreting tumors of the adrenal cortex that produce the syndrome of **primary hyperaldosteronism** (Conn's syndrome). These patients are severely K^+-depleted. They are hypertensive, and their ECF volume is expanded, but they are not edematous or markedly hypernatremic, because the escape phenomenon discussed in the preceding paragraph causes the excretion of extra ingested salt.

Prolonged K^+ depletion damages the kidney, resulting in a loss of concentrating ability and polyuria **(hypokalemic nephropathy).** The K^+ depletion also causes muscle weakness and metabolic alkalosis (see Chapter 38), and the alkalosis lowers the plasma Ca^{2+} level to the point where latent or even frank tetany is present (see Chapter 21). The K^+ depletion causes a minor but detectable decrease in glucose tolerance that is corrected by K^+ repletion (see Chapter 19).

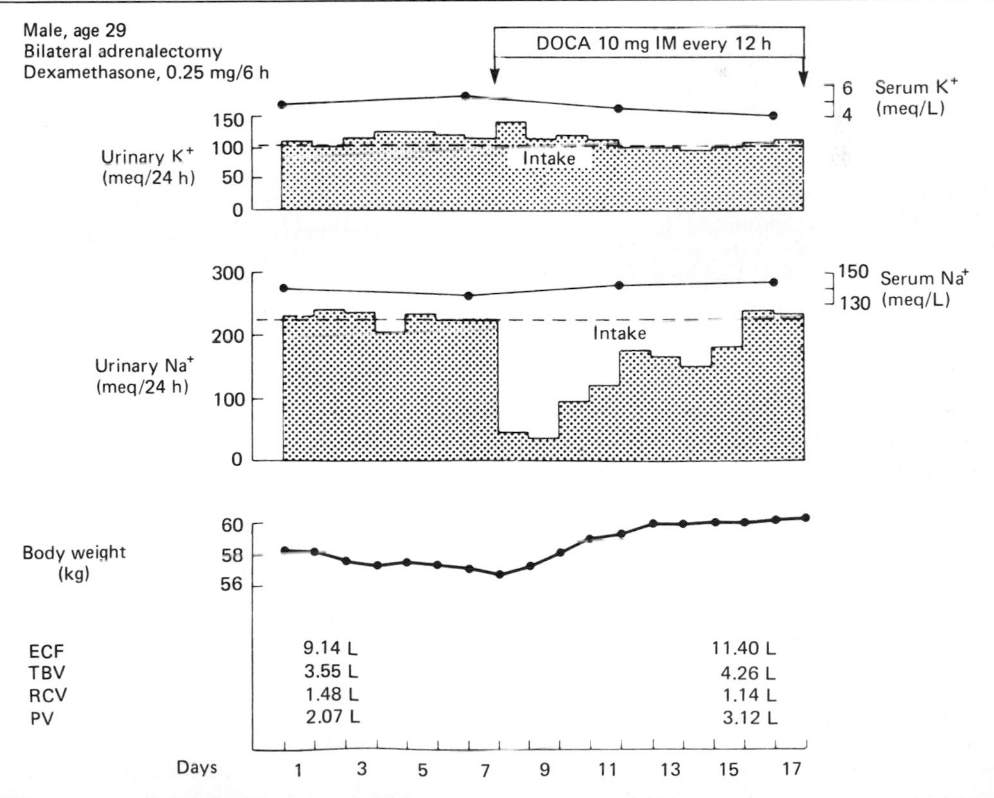

Figure 20–26. "Escape" from the sodium-retaining effect of desoxycorticosterone acetate (DOCA) in an adrenalectomized patient. ECF, extracellular fluid volume; TBV total blood volume; RCV, red cell volume; PV, plasma volume. (Courtesy of EG Biglieri.)

REGULATION OF ALDOSTERONE SECRETION

Stimuli

The principal stimuli that increase aldosterone secretion are summarized in Table 20–6. Some of them also increase glucocorticoid secretion; others selectively affect the output of aldosterone. The primary regulatory factors involved are ACTH from the pituitary, renin from the kidney via angiotensin II, and a direct stimulatory effect of a rise in plasma K^+ concentration—or a drop in plasma Na^+ concentration—on the adrenal cortex. As noted above, the effects of ACTH are mediated by cyclic AMP and protein kinase A, whereas the effects of angiotensin II are mediated by diacylglycerol and protein kinase C.

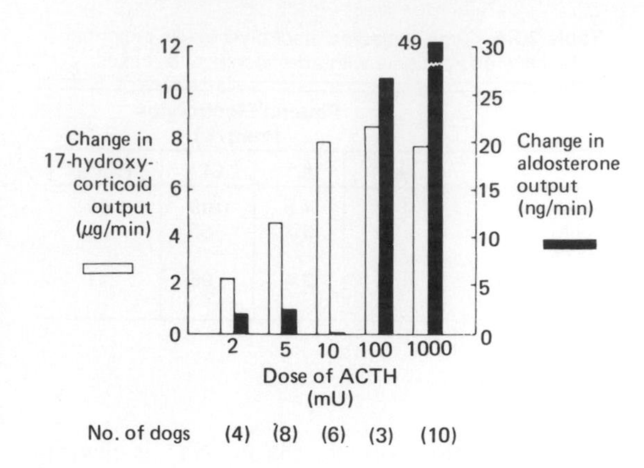

Figure 20–27. Changes in adrenal venous output of steroids produced by ACTH in nephrectomized hypophysectomized dogs.

Effect of ACTH

When first administered, ACTH stimulates the output of aldosterone as well as that of glucocorticoids and sex hormones. Although the amount of ACTH required to increase aldosterone output is somewhat greater than the amount that stimulates maximal glucocorticoid secretion (Fig 20–27), it is well within the range of endogenous ACTH secretion. The effect is transient, and even if ACTH secretion remains elevated, aldosterone output begins to decline in 1 or 2 days. On the other hand, the output of the mineralocorticoid deoxycorticosterone remains elevated. The decline in aldosterone output is partly due to decreased renin secretion secondary to hypervolemia (see below), but it is possible that some other factor also decreases the conversion of corticosterone to aldosterone. After hypophysectomy, the basal rate of aldosterone secretion is normal. The increase normally produced by surgical and other stresses is absent, but the increase produced by dietary salt restriction is unaffected for some time. The atrophy of the zona glomerulosa that complicates the picture in long-standing hypopituitarism has been discussed above.

Effects of Angiotensin II & Renin

The octapeptide angiotensin II is formed in the body from angiotensin I, which is liberated by the action of renin on a circulating α_2 globulin (see Chapter 24). Injections of angiotensin II stimulate adrenocortical secretion and, in small doses, affect primarily the secretion of aldosterone (Fig 20–28). At least in experimental animals, the sites of action of angiotensin II are both early and late in the steroid biosynthetic pathway. The early action is on the conversion of cholesterol to pregnenolone, and the late action is on the conversion of corticosterone to aldosterone. In humans, angiotensin II does not increase the secretion of deoxycorticosterone, which is controlled by ACTH.

Renin is secreted from the juxtaglomerular cells that surround the renal afferent arterioles as they enter the glomeruli (see Chapter 24). Aldosterone secretion is regulated via the renin-angiotensin system in a feedback fashion (Fig 20–29). A drop in ECF volume or intra-arterial vascular volume leads to a reflex increase in renal nerve discharge and decreases renal arterial pressure. Both changes increase renin secre-

Table 20–6. Stimuli that increase aldosterone secretion.

Glucocorticoid secretion also increased
Surgery
Anxiety
Physical trauma
Hemorrhage
Glucocorticoid secretion unaffected
High potassium intake
Low sodium intake
Constriction of inferior vena cava in thorax
Standing
Secondary hyperaldosteronism (in some cases of congestive heart failure, cirrhosis, and nephrosis)

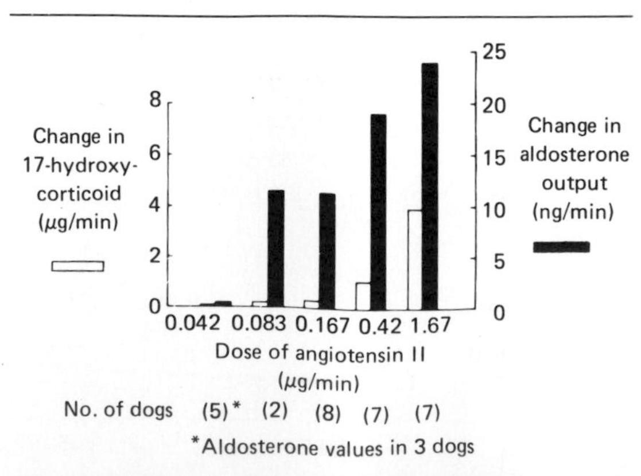

Figure 20–28. Changes in adrenal venous output of steroids produced by angiotensin II in nephrectomized hypophysectomized dogs.

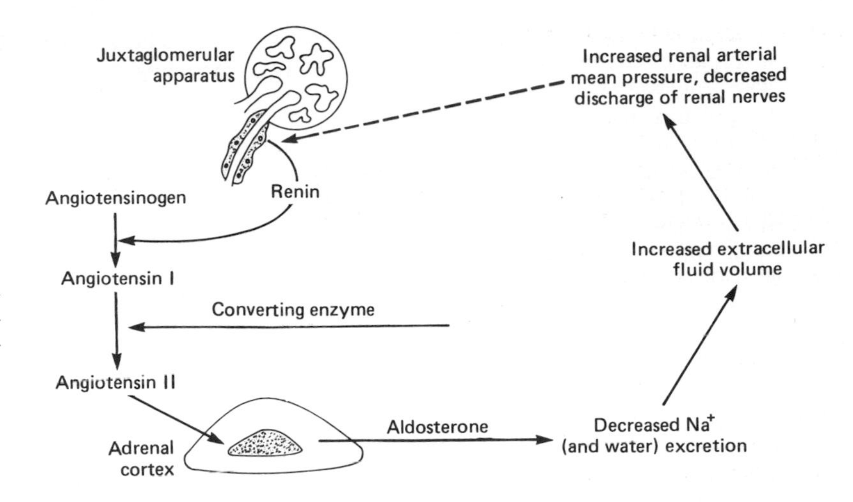

Figure 20–29. Feedback mechanism regulating aldosterone secretion. The dashed arrow indicates inhibition.

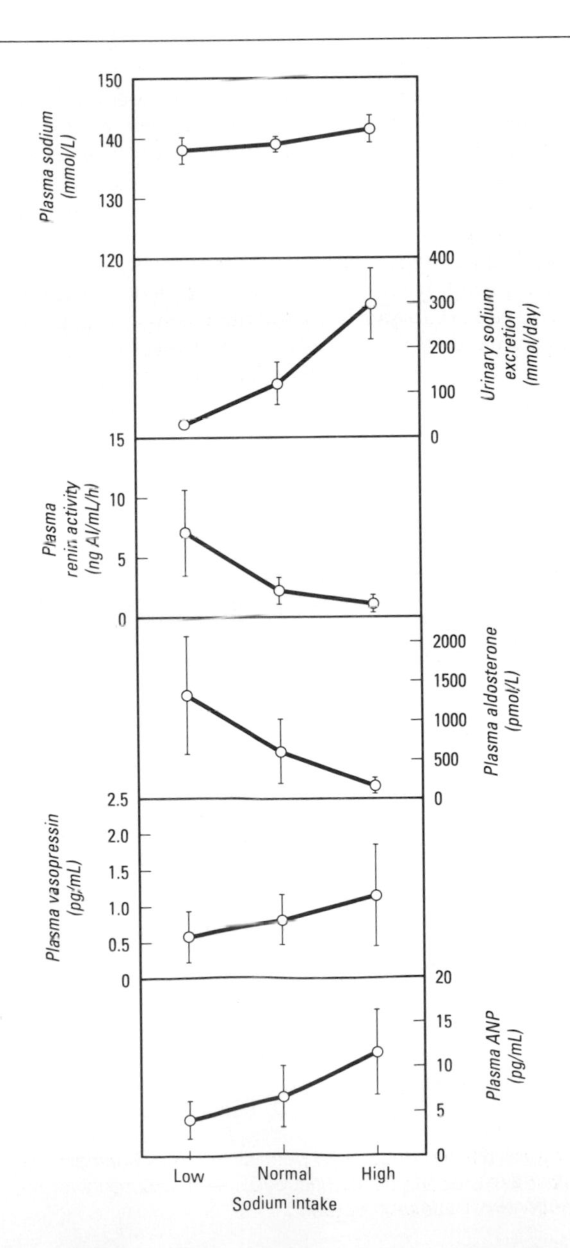

tion, and the angiotensin II formed by the action of the renin increases the rate of secretion of aldosterone. The aldosterone causes Na^+ and water retention, expanding ECF volume and shutting off the stimulus that initiated increased renin secretion.

Hemorrhage stimulates ACTH secretion, but it also increases aldosterone secretion in the absence of the pituitary. Like hemorrhage, standing and constriction of the thoracic inferior vena cava decrease intra-arterial vascular volume. Dietary sodium restriction also increases aldosterone secretion via the renin-angiotensin system (Fig 20–30). Such restriction reduces ECF volume, but aldosterone and renin secretion are increased before there is any consistent decrease in blood pressure. Consequently, the initial increase in renin secretion produced by dietary sodium restriction is probaby due to reflex increases in the activity of the renal nerves. Sodium depletion increases the binding of angiotension II to the adrenal cortex, first by increasing the affinity of angiotensin II receptors and then by increasing their number. At the same time, there is a decrease in angiotensin II binding to blood vessels.

Effects of Electrolytes

It seems unlikely that changes in plasma Na^+ are a major factor in the regulation of aldosterone secretion, since, at least when it occurs acutely as an isolated change, the plasma Na^+ concentration must drop about 20 meq/L to produce stimulation. Changes

Figure 20–30. Effect of low-, normal-, and high-sodium diets on sodium metabolism and plasma renin activity, aldosterone, vasopressin, and ANP in normal humans. (Data from Sagnella GA et al: Plasma atrial natriuretic peptide: its relationship to changes in sodium intake, plasma renin activity, and aldosterone in man. *Clin Sci* 1987;72:25.)

of this magnitude are unusual. However, the plasma K^+ level need increase only 1 meq/L or less to stimulate aldosterone secretion, and transient increases of this magnitude may occur after a meal, particularly if it is rich in K^+. Like angiotensin II, K^+ stimulates the conversion of cholesterol to pregnenolone and the conversion of corticosterone to aldosterone. The sensitivity of the zona glomerulosa to angiotensin II and consequently to a low-sodium diet is decreased by a low-potassium diet. However, in most clinical situations, K^+ depletion is produced by an aldosterone-producing tumor or is secondary to hyperaldosteronism from some other cause.

In normal individuals, there is an increase in plasma aldosterone concentration during the portion of the day that the individual is carrying on activities in the upright position. This increase is due to a decrease in the rate of removal of aldosterone from the circulation by the liver and an increase in aldosterone secretion due to a postural increase in renin secretion. Individuals who are confined to bed show a circadian rhythm of aldosterone and renin secretion, with the highest values in the early morning before awakening from sleep.

Other Factors

From time to time, it has been claimed that factors other than renin, ACTH, and plasma electrolytes play a role in the regulation of aldosterone secretion, and there is evidence that a pituitary hormone other than ACTH helps maintain the responsiveness of the zona glomerulosa. However, there is as yet no definite proof for the existence of other physiologically important regulatory factors.

Clinical Correlates

Circulating renin and angiotensin II levels are elevated in many patients with congestive heart failure, nephrosis, or cirrhosis of the liver, and this in turn increases aldosterone secretion **(secondary hyperaldosteronism).** Increased renin secretion is also found in individuals with the salt-losing form of the adrenogenital syndrome (see above), presumably because their ECF volume is low. In patients with elevated renin secretion due to renal artery constriction, aldosterone secretion is increased; in those in whom renin secretion is not elevated, aldosterone secretion is normal. The relationship of aldosterone to hypertension is discussed in Chapter 33.

ROLE OF MINERALOCORTICOIDS IN THE REGULATION OF SALT BALANCE

Variation in aldosterone secretion is only one of many factors affecting Na^+ excretion. Other major factors include the glomerular filtration rate, ANP, the presence or absence of osmotic diuresis, and changes in tubular reabsorption of Na^+ independent of aldosterone, It takes some time for aldosterone to act. Thus, when one rises from the supine to the standing position, aldosterone secretion is increased and there is Na^+ retention; but the decrease in Na^+ excretion develops too rapidly to be explained solely by increased aldosterone secretion. The primary function of the aldosterone-secreting mechanism is probably the defense of intravascular volume, but it is only one of the homeostatic mechanisms involved.

SUMMARY OF THE EFFECTS OF ADRENOCORTICAL HYPER- & HYPOFUNCTION IN HUMANS

Recapitulating the manifestations of excess and deficiency of the adrenocortical hormones in humans is a convenient way to summarize the multiple and complex actions of these steroids. A characteristic clinical syndrome is associated with excess secretion of each of the types of hormones. Excess androgen secretion causes masculinization **(adrenogenital syndrome)** and precocious pseudopuberty or female pseudohermaphroditism. Feminizing estrogen-secreting adrenal tumors are sometimes seen. Excess glucocorticoid secretion produces a moon-faced, plethoric appearance, with trunk obesity, purple abdominal striae, hypertension, osteoporosis, protein depletion, mental abnormalities, and, frequently, diabetes mellitus **(Cushing's syndrome).** Excess mineralocorticoid secretion leads to K^+ depletion and Na^+ retention, usually without edema but with weakness, hypertension, tetany, polyuria, and hypokalemic alkalosis **(Conn's syndrome).**

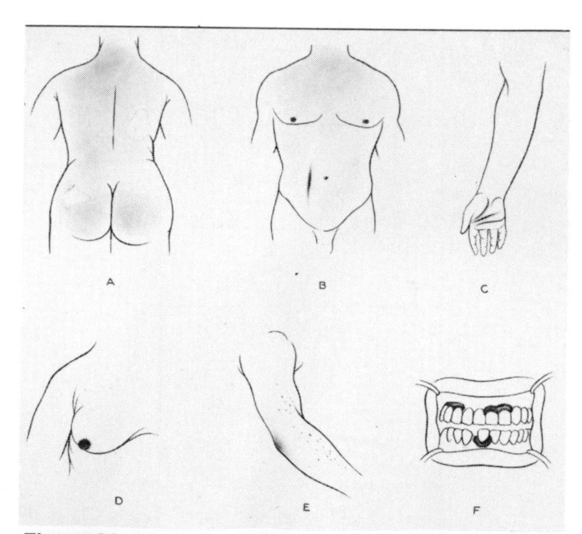

Figure 20–31. Pigmentation in Addison's disease. *A,* tan and vitiligo; *B,* pigmentation of scars from lesions that occurred after the development of the disease; *C,* pigmentation of skin creases; *D,* darkening of areolas; *E,* pigmentation of pressure points; *F,* pigmentation of the gums. (Reproduced, with permission, from Forsham PH, Di Raimondo VC: *Traumatic Medicine and Surgery for the Attorney.* Butterworth, 1960.)

If hormone treatment is stopped in a surgically adrenalectomized patient, Na^+ excretion is increased and Na^+ enters the cells. The plasma volume declines, the blood pressure falls, and fatal shock develops in a few days if treatment is not reinstituted. These changes can be prevented by mineralocorticoids. Many adrenalectomized patients, provided they eat regularly, can be maintained for prolonged periods on mineralocorticoid replacement alone, but their existence is a precarious one. Fasting causes fatal hypoglycemia, and any stress causes collapse. Water is poorly excreted, and there is always the danger of water intoxication. Circulating ACTH levels are elevated. The diffuse tanning of the skin and the spotty pigmentation characteristic of chronic glucocorticoid deficiency (Fig 20–31) are due, at least in part, to the MSH activity of the ACTH in the blood (see Chapter 22). Minor menstrual abnormalities occur in women, but the deficiency of adrenal sex hormones usually has little effect in the presence of normal testes or ovaries.

Adrenocortical insufficiency in cases of adrenocortical atrophy due to autoimmune disease or destruction of the adrenal glands by diseases such as tuberculosis and cancer is called **Addison's disease.** Total adrenal insufficiency is rapidly fatal, but in Addison's disease, destruction of the adrenals is usually incomplete. Therefore, patients with this disease have time to develop marked pigmentation and a decrease in cardiac size that is apparently secondary to chronic hypotension and a decrease in cardiac work. They often get along fairly well until some minor stress precipitates a collapse **(addisonian crisis)** requiring emergency medical treatment.

Cases of isolated aldosterone deficiency have also been reported in patients with renal disease and a low circulating renin level. These patients have marked hyperkalemia and may develop metabolic acidosis.

21 Hormonal Control of Calcium Metabolism & the Physiology of Bone

INTRODUCTION

Three hormones are primarily concerned with the regulation of calcium metabolism. 1,25-Dihydroxycholecalciferol is a steroid hormone formed from vitamin D by successive hydroxylations in the liver and kidneys. Its primary action is to increase calcium absorption from the intestine. Parathyroid hormone, which is secreted by the parathyroid glands, mobilizes calcium from bone and increases urinary phosphate excretion. Calcitonin, a calcium-lowering hormone that in mammals is primarily secreted by cells in the thyroid gland, inhibits bone resorption. Although the role of calcitonin seems to be relatively minor, all 3 hormones probably operate in concert to maintain the constancy of the Ca^{2+} level in the body fluids. Glucocorticoids, growth hormone, estrogens, and various growth factors also affect calcium metabolism.

CALCIUM & PHOSPHORUS METABOLISM

Calcium

The adult human body contains about 1100 g (27.5 mol) of calcium (1.5% of body weight). Ninety-nine percent of the calcium is in the skeleton. The plasma calcium, normally about 10 mg/dL (5 meq/L, 2.5 mmol/L), is partly bound to protein and partly diffusible (Table 21–1). The distribution of calcium inside cells is discussed in Chapter 1.

It is the free, ionized calcium in the body fluids that is a vital second messenger (see Chapter 1) and is necessary for blood coagulation, muscle contraction, and nerve function. A decrease in extracellular Ca^{2+} at the myoneural junction inhibits transmission (see Chapter 4), but this effect is overbalanced by the excitatory effect of a low Ca^{2+} level on the nerve and muscle cells (see Chapter 2). The result is **hypocalcemic tetany,** which is due to increased activity of the motor nerve fibers. This condition is characterized by extensive spasms of skeletal muscle, involving especially the muscles of the extremities and the larynx. Laryngospasm becomes so severe that the airway is obstructed, and fatal asphyxia is produced. Ca^{2+} plays an important role in clotting (see Chapter 27) and many different aspects of cell function (see Chapter 1); in vivo, however, the level of plasma Ca^{2+} at which fatal tetany occurs is still above the level at which clotting defects would occur.

Since the extent of Ca^{2+} binding by plasma proteins is proportionate to the plasma protein level, it is important to know plasma protein level when evaluating the total plasma calcium. Plasma ionized calcium can be measured by use of a calcium-sensitive electrode. Other electrolytes and pH affect the Ca^{2+} level. Thus, for example, symptoms of tetany appear at much higher total calcium levels if the patient hyperventilates, increasing plasma pH. Plasma proteins are more ionized when the pH is high, providing more protein anion to bind with Ca^{2+}.

The calcium in bone is of 2 types: a readily exchangeable reservoir and a much larger pool of stable calcium that is only slowly exchangeable. It is worth noting that there are 2 independent but interacting homeostatic systems affecting the calcium in bone. One is the system that regulates plasma Ca^{2+}, and in the operation of this system, about 500 mmol of Ca^{2+} per day moves into and out of the readily exchangeable pool in the bone (Fig 21–1). The other system is the one concerned with bone remodeling by the constant interplay of bone resorption and deposition (see below), which, in the adult, accounts for 95% of bone formation. However, the Ca^{2+} interchange between plasma and this stable pool of bone calcium is only about 7.5 mmol/d.

A large amount of calcium is filtered in the kidneys, but 98–99% of the filtered calcium is reabsorbed. About 60% of the reabsorption occurs in the proximal tubules and the remainder in the ascending limb of the loop of Henle and the distal tubule. Distal tubular reabsorption is regulated by parathyroid hormone.

Table 21–1. Distribution (mmol/L) of calcium in normal human plasma.

Diffusible		1.34
Ionized (Ca^{2+})	1.18	
Complexed to HCO_3^-, citrate, etc	0.16	
Nondiffusible (protein-bound)		1.16
Bound to albumin	0.92	
Bound to globulin	0.24	
Total plasma calcium		2.50

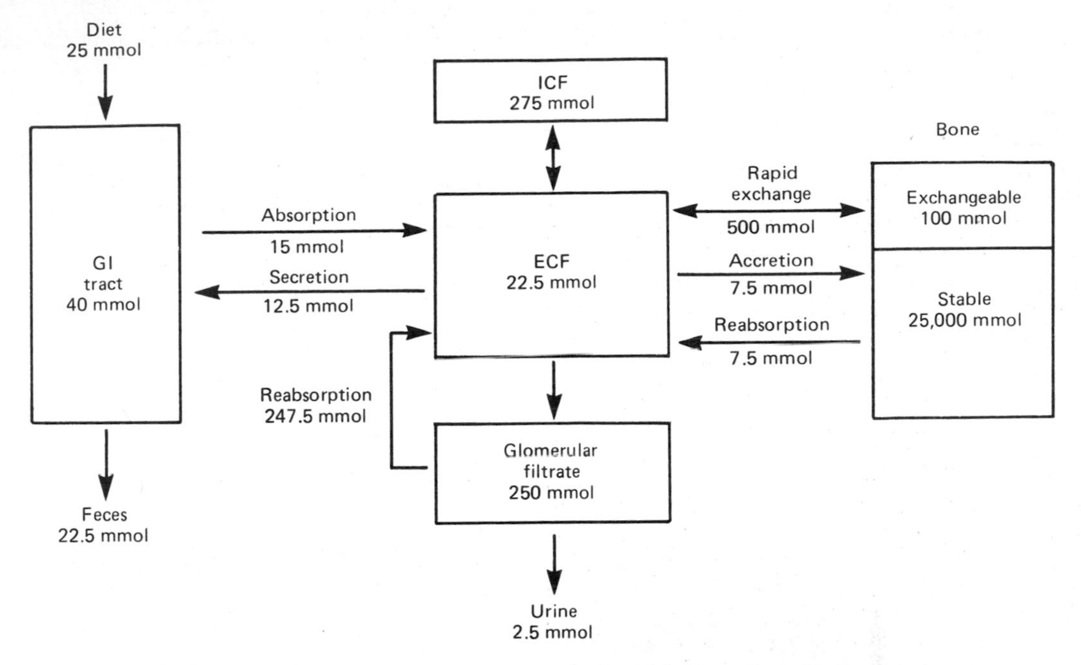

Figure 21–1. Calcium metabolism in an adult human ingesting 25 mmol (1000 mg) of calcium per day.

The absorption of Ca^{2+} from the gastrointestinal tract is discussed in Chapter 25. Ca^{2+} is actively transported out of the intestine by a system in the brush border of the epithelial cells that involves a calcium-dependent ATPase, and this process is regulated by 1,25-dihydroxycholecalciferol (see below). There is also some absorption by passive diffusion. When Ca^{2+} intake is high, the active transport mechanism becomes saturated. In addition, 1,25-dihydroxycholecalciferol levels fall in the presence of increased plasma Ca^{2+}. Consequently, Ca^{2+} absorption undergoes adaption ie, it is high when the calcium intake is low and decreased when the calcium intake is high. Calcium absorption is also decreased by substances that form insoluble salts with Ca^{2+} (eg, phosphates and oxalates) or by alkalis, which favor formation of insoluble calcium soaps. A high-protein diet increases absorption in adults.

Phosphorus

Phosphorus metabolism is not as finely regulated as calcium metabolism, but phosphorus is found in ATP, cyclic AMP, 2,3-diphosphoglycerate, and other vitally important compounds in the body. Total body phosphorus is 500–800 g (16.1–25.8 mol), 85–90% of which is in the skeleton. Total plasma phosphorus is about 12 mg/dL, with two-thirds of this total in organic compounds and the remaining inorganic phosphorus (Pi) mostly in PO_4^{3-}, HPO_4^{2-}, and $H_2PO_4^{-}$.

The amount of phosphorus normally entering bone is about 3 mg (97 μmol)/kg/d, with an equal amount leaving via reabsorption.

The Pi in the plasma is filtered in the glomeruli, and 85–90% of the filtered Pi is reabsorbed. Active transport in the proximal tubule accounts for most of the reabsorption, and this active transport process is powerfully inhibited by parathyroid hormone (see below).

Pi is absorbed in the duodenum and small intestine by both active transport and passive diffusion. However, unlike the absorption of Ca^{2+}, the absorption of Pi is linearly proportionate to dietary intake. Many stimuli that increase Ca^{2+} absorption, including 1,25-dihydroxycholecalciferol, also increases Pi absorption.

BONE PHYSIOLOGY

Structure of Bone

Bone is a living tissue with a collagenous matrix that has been impregnated with mineral salts, especially phosphates of calcium. It supports the body, provides a storehouse of Ca^{2+} and other minerals that aid in maintaining mineral homeostasis, and aids the lungs and kidneys in the maintenance of acid-base balance by providing additional phosphate and carbonate buffers. The protein in bone matrix is mostly type I collagen, which is also the major structural protein in tendons and skin. This collagen, which weight for weight is as strong as steel, is made up of a triple helix of 3 polypeptides bound tightly together. Two of these are identical α_1 polypeptides encoded by one gene, and one is an α_2 polypeptide encoded by a different gene.

Adequate amounts of both protein and minerals must be available for the maintenance of normal bone structure. Mineral in bone is mostly in the form of

hydroxyapatites, which have the general formula $Ca^{2+}{}_{10-x}(H_3O^+)_{2x} \times (PO_4^{3-})_6(OH^-)_2$. These salts form crystals that measure 20 by 3–7 nm. Sodium and small amounts of magnesium and carbonate are also present in bone.

Bone is cellular and well vascularized; the total bone blood flow in humans has been estimated to be 200–400 mL/min. Throughout life, the mineral in the skeleton is being actively turned over, and bone is constantly resorbed and re-formed. The calcium in bone turns over at a rate of 100% per year in infants and 18% per year in adults. There is continuous remodeling of bone throughout life, with continuous local cycles of bone resorption followed by bone formation. This remodeling is related in part to the stresses and strains imposed on the skeleton by gravity and involves local as well as systemic factors, whose exact nature is unsettled. These factors include various growth factors, including those that regulate hematopoietic function (see Chapter 27), IGF-I, and other hormones.

Most bones are made up of an outer layer of compact bone surrounding trabecular bone and, in many instances, a bone marrow cavity (Fig 21–2). **Trabecular** or **spongy bone** is made up of bone spicules separated by spaces. **Compact bone** is much denser and is less active metabolically. In spongy bone, nutrients diffuse from bone ECF into the trabeculae, but in compact bone, nutrients are provided via **Haversian canals,** which contain blood vessels (Fig 21–3). Around each Haversian canal, collagen is arranged in concentric layers, forming cylinders called **osteons** or **Haversian systems.**

The cells in bone that are primarily concerned with bone formation and resorption are osteoblasts, osteocytes, and osteoclasts. **Osteoblasts** are the bone-forming cells that secrete collagen, forming a matrix around themselves which then calcifies. When surrounded by calcified matrix, they are called **osteocytes.** They send processes into the canaliculi that ramify throughout the bone (Fig 21–3). **Osteoclasts** are multinuclear cells that erode and resorb previously formed bone. They are derived from hematopoietic stem cells via monocytes (see Chapter 27). They phagocytose bone, digesting it in their cytoplasm; this is why bone around an active osteoclast has a characteristic ruffled or chewed-out edge. Osteoblasts, on the other hand, arise from **osteoprogenitor cells** that are of mesenchymal origin. Osteoblasts remain in contact with one another via tight junctions between long protoplasmic processes that run through canaliculi in the bone. Osteoblasts and osteoclasts are both affected by hormones; the actions of these hormones are summarized in Fig 21–3 and discussed in more detail below.

Bone Formation & Resorption

The osteoblasts in bone appear to form at least a partial membrane that separates bone fluid (the fluid in most immediate contact with hydroxyapatites) from the ECF of the rest of the body. In this way, the Ca^{2+} and Pi concentration in bone fluid can be carefully regulated. The details of the process responsible for the calcification of newly formed bone matrix are uncertain despite intensive investigation. Whether calcium phosphate precipitates out of a solution depends upon the product of concentrations of Ca^{2+} and PO_4^{3-}. At a certain value for this product (the **solubility product),** the solution is saturated. Whenever $[Ca^{2+}] \times [PO_4^{3-}]$ exceeds the solubility product, calcium phosphate precipitates. Associated with the osteoblasts in new bone is an alkaline phosphatase that hydrolyzes phosphate esters. The phosphates liberated by ester hydrolysis increase the concentration of phosphate in the vicinity of the osteoblasts to a point where the solubility product is exceeded and calcium phosphate precipitates. This mechanism may be involved in the calcification process, but it is almost certainly only part of the story. Bone contains a protein with a large number of γ-carboxyglutamic acid residues, and these residues bind Ca^{2+}. Furthermore, a high content of this protein correlates with ongoing calcification. However, γ-carboxylation is catalyzed by vitamin K (see Chapter 17), and vitamin K deficiency only causes skeletal abnormalities in the fetus. On the other hand, bone resorption is brought about by the osteoclasts. Osteoclasts and osteoblasts both increase their Ca^{2+} permeability in response to parathyroid hormone.

Bone Growth

The bones of the skull are formed by ossification of membranes (**intramembranous** bone formation). The

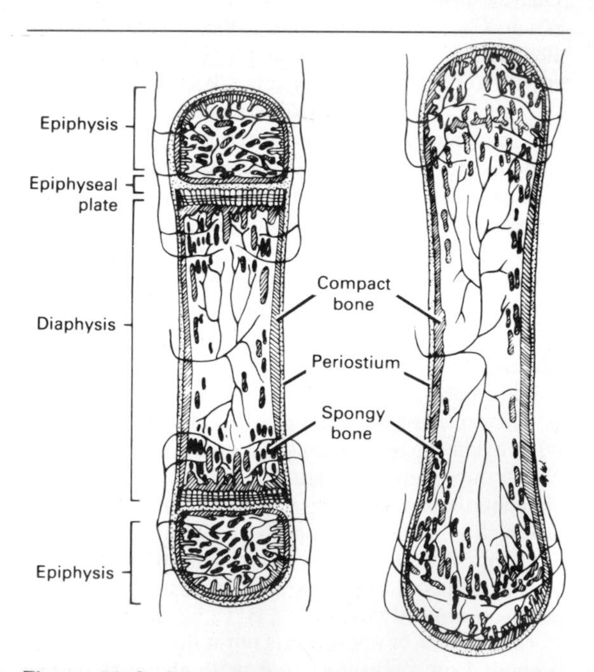

Figure 21–2. Structure of a typical long bone before (left) and after (right) epiphyseal closure. (Modified from Junqueira LC, Carneiro J, Kelley RO: *Basic Histology,* 6th ed. Appleton & Lange, 1989.)

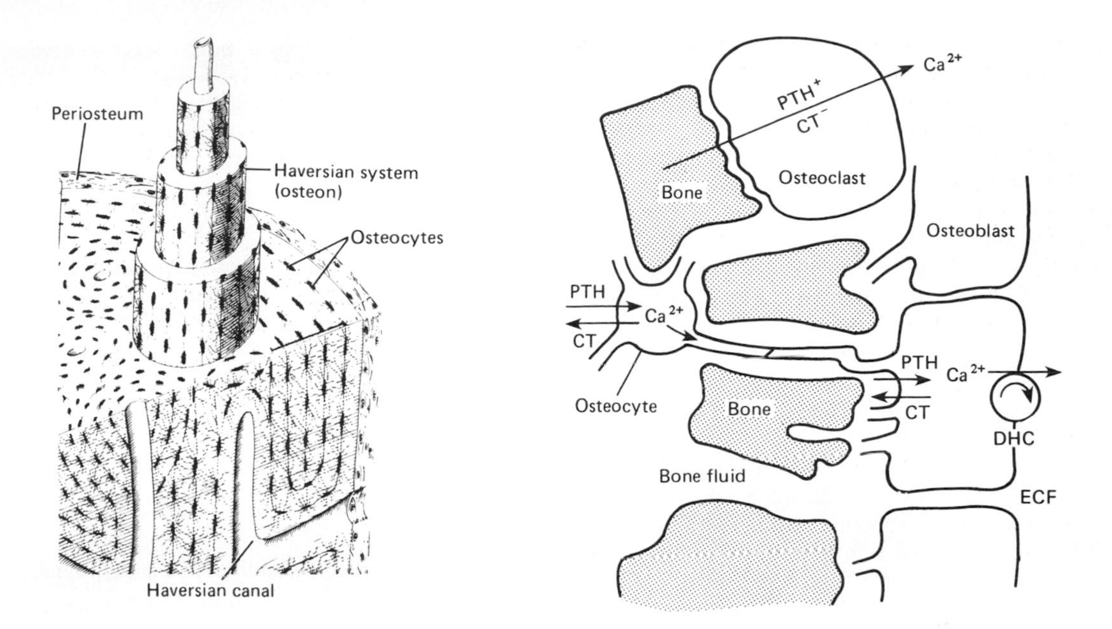

Figure 21–3. *Left:* Schematic drawing of compact bone. (Modified from Junqueira LC, Carneiro J, Kelley, RO: *Basic Histology,* 6th ed. Lange, 1989.). *Right:* Proposed organization of bone cells. Note that the osteoblasts form a partial membrane which separates bone fluid from EFC and that their cytoplasmic processes connect via tight junctions to the cytoplasmic processes of osteocytes deep in the bone. Parathyroid hormone (PTH) increases and calcitonin (CT) decreases the permeability of bone cells to Ca^{2+}, whereas 1,25-dihydroxycholecalciferol (DCH) may act by facilitating the active transport of Ca^{2+} from osteoblasts into ECF. (Courtesy of H Heath III and CD Arnaud.)

long bones are first modeled in cartilage and then transformed into bone by ossification that begins in the shaft and in the ends of the bone **(enchondral** bone formation). Osteoblasts form a network of collagen fibers. This matrix then calcifies. In the meantime, the cartilage in the center of the shaft is invaded by osteoclasts that erode it away.

During growth, specialized areas at the ends of each long bone **(epiphyses)** are separated from the shaft of the bone by a plate of actively proliferating cartilage, the **epiphyseal plate** (Fig 21–2). Growth in bone length occurs as this plate lays down new bone on the end of the shaft. The width of the epiphyseal plate is proportionate to the rate of growth. Its width is affected by a number of hormones but most markedly by the pituitary growth hormone and IGF-I (see Chapter 22). Changes in the width of the tibial epiphyseal plate are used as the end point in the ''tibia test,'' a bioassay that has been used to measure growth hormone.

Linear bone growth can occur as long as the epiphyses are separated from the shaft of the bone, but such growth ceases after the epiphyses unite with the shaft **(epiphyseal closure).** The epiphyses of the various bones close in an orderly temporal sequence, the last epiphyses closing after puberty. The normal age at which each of the epiphyses closes is known, and the ''bone age'' of an individual can be determined by x-raying the skeleton and noting which epiphyses are open and which are closed.

Uptake of Other Minerals

Lead and some other toxic elements are taken up and released by bone in a manner similar to that in which calcium is turned over. The rapid bone uptake of these elements is sometimes called a ''detoxifying mechanism'' because it serves to remove them from the body fluids, thus ameliorating the toxic manifestations. The radioactive elements radium and plutonium (and the radioactive isotope of strontium that is a by-product of atomic bomb explosions) are also taken up by bone. In this situation, the uptake is definitely harmful, because radiation from these elements may cause malignant degeneration of bone cells and the formation of osteogenic sarcomas. Fluoride taken up by bone causes new bone to form. It is also incorporated into the enamel of the teeth, and small amounts of fluoride increase resistance to dental caries. However, large amounts cause discoloration of the enamel **(mottled enamel).**

Bone Disease

The diseases produced by selective abnormalities of the cells and processes discussed above illustrate the interplay of factors that maintain normal bone function. Various mutations in the genes that code for collagen make bones brittle, causing **osteogenesis imperfecta** (brittle bones disease). In **osteopetrosis,** another rare and often severe disease, the osteoclasts are defective and are unable to resorb bone in their usual fashion. The result is a steady increase in bone density, neurological defects due to narrowing and distortion of foramina through which nerves normally pass, and hematological abnormalities due to crowding out of the marrow cavities. The condition in which the amount of calcium accretion per unit bone matrix is deficient is called **rickets** in children and **osteomalacia** in adults. In **osteoporosis,** matrix and mineral are both lost, and there is a loss of bone mass (Fig 21–4). This disease, in a sense, is the opposite of osteopetrosis in that it is characterized over time by a net excess of bone resorption over bone formation. It has multiple causes but is associated mostly with advancing age. Involutional osteoporosis is a common disease and is becoming a major public health problem in the USA and Europe as the number of elderly people in the population increases.

All normal humans gain bone early in life, during growth. After a plateau, they begin to lose bone as they grow older (Fig 21–5). When this loss is accelerated or exaggerated, as it is in osteoporosis, it leads to an increased incidence of fractures, particularly fractures of the distal forearm (Colles' fracture), vertebral bodies, and hip. All these areas have a high content of trabecular bone, and since trabecular bone is more active metabolically, it is lost more rapidly. Fractures of the vertebras with compression cause kyphosis, with the production of the typical widow's hump that is common in elderly women with osteoporosis. Fractures of the hip in elderly individuals are associated with a mortality rate of 12–20%, and half of those who survive require prolonged, expensive care.

Adult women have less bone mass than adult men, and after menopause, they lose it more rapidly than men do. Consequently, they are more prone to development of serious osteoporosis. The cause of the bone loss after menopause is estrogen deficiency, but why estrogen deficiency leads to bone loss is a matter of considerable debate. Recent evidence indicates that contrary to previous views, human osteoblasts have estrogen-binding sites and that the hormone may act directly on these bone cells. Osteoporosis can be prevented by small doses of estrogens, but the problem is that even these small doses may increase the incidence of cancer of the endometrium. Increased intake of calcium, particularly from natural sources such as milk, and moderate exercise may also help prevent or slow the progress of the disease, although their effects are not great. Fluoride stimulates osteoblasts and may be beneficial in the treatment of the disease.

In patients who are immobilized for any reason, and during space flight (see Chapter 33), bone resorption exceeds bone formation and disuse osteoporosis develops. The plasma calcium level is not markedly elevated, but plasma concentrations of parathyroid hormone and 1,25-dihydroxycholecalciferol fall, and large amounts of calcium are lost in the urine. Osteoporosis also occurs in patients with excess glucocorticoid secretion (Cushing's syndrome; see Chapter 20).

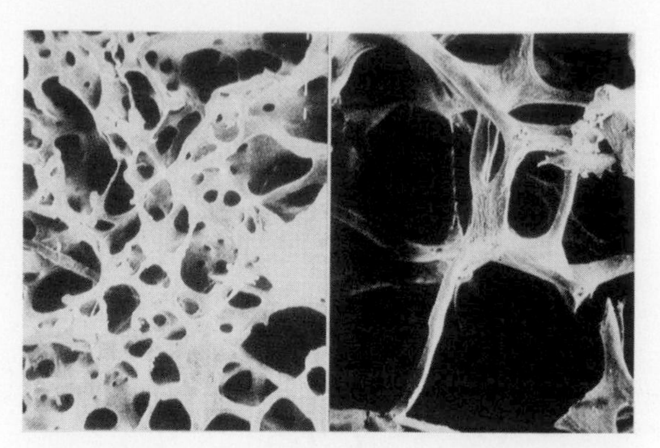

Figure 21–4. Normal trabecular bone (left) compared with trabecular bone from a patient with osteoporosis.

VITAMIN D & THE HYDROXYCHOLECALCIFEROLS

Chemistry

The active transport of Ca^{2+} and PO_4^{3-} from the intestine is increased by a metabolite of vitamin D. The term vitamin D is used to refer to a group of closely related sterols produced by the action of ultraviolet light on certain provitamins (Fig 21–6). Vitamin D_3, which is also called cholecalciferol, is produced in the skin of mammals from 7-dehy-

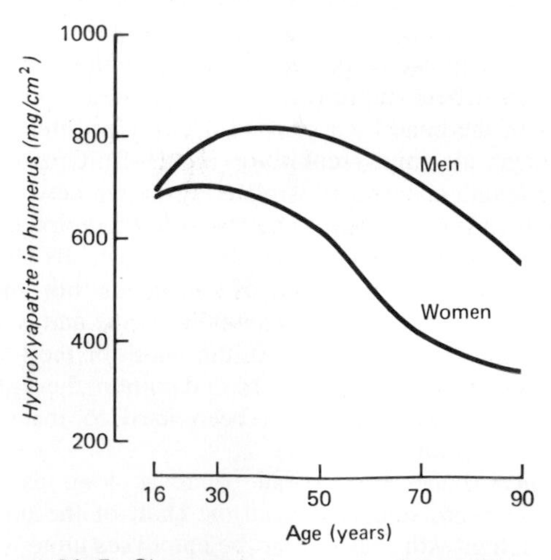

Figure 21–5. Changes in bone mass from adolescence to old age in normal men and women. Curves are drawn from values of hydroxyapatite (mg/cm of the radius) determined by radiodensitometry. (Data from Gordon GS, Genant HK: Aging of bone in two sexes. In: *Steroid Modulation of Neuroendocrine Function and Sterols, Steroids, and Bone Metabolism.* Martini L, Gordon GS, Sciara F [editors]. Elsevier, 1984.)

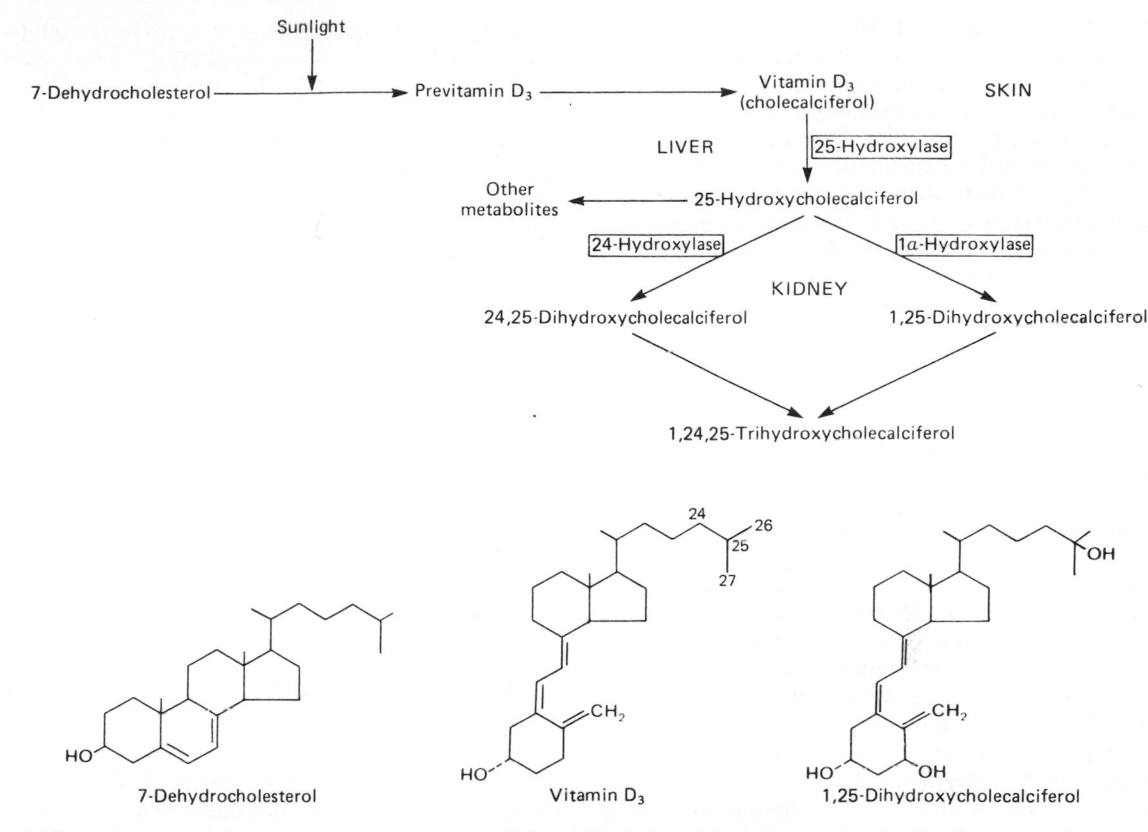

Figure 21–6. Formation and hydroxylation of vitamin D_3. 25-Hydroxylation takes place in the liver, and the other hydroxylations occur primarily in the kidneys. The formulas of 7-dehydrocholestérol, vitamin D_3, and 1,25-dihydroxycholecalciferol are also shown.

drocholesterol by the action of sunlight. The reaction involves the rapid formation of previtamin D_3, which is then converted more slowly to vitamin D_3 (cholecalciferol). Vitamin D_3 and its hydroxylated derivatives are transported in the plasma bound to a specific globulin vitamin-D-binding protein (DBP). The affinity of previtamin D_3 for DBP is low, but the affinity of vitamin D_3 is high, so DBP moves vitamin D_3 from the skin into the circulation. Vitamin D_3 is also ingested in the diet.

In the liver, vitamin D_3 is converted to 25-hydroxycholecalciferol (calcidiol, 25-OHD_3). The 25-hydroxycholecalciferol is converted in the proximal tubules of the kidneys to the more active metabolite 1,25-dehydroxycholecalciferol, which is also called calcitriol or 1,25-$(OH)_2D_3$. The enzyme involved is mitochondrial 1α-hydroxylase, and it is located in the cells of the proximal tubules. In healthy humans, this is the only site of formation of 1,25-dihydroxycholecalciferol, but in patients with sarcoidosis, pulmonary alveolar and other macrophages also produce 1,25-dihydroxycholecalciferol, apparently upon stimulation by γ-interferon. The normal plasma level of 25-hydroxycholecalciferol is about 30 ng/mL, and that of 1,25-dihydroxycholecalciferol is about 0.03 ng/mL (approximately 100 pmol/L). The less active metabolite 24,25-dihydroxycholecalciferol is also formed in the kidney, along with several other relatively inactive metabolites (Fig 21–6).

Actions

Because 1,25-dihydroxycholecalciferol is produced in the body and transported in the bloodstream to affect Ca^{2+} transport at a distance from its site of production, it is properly called a hormone. In intestinal epithelial cells, it binds to a cytoplasmic receptor, and the complex is then translocated to the nucleus, where it increases the formation of mRNA. The mRNA dictates the formation of a calcium-binding protein, but the relation of this protein to facilitation of Ca^{2+} absorption is unsettled. 1,25-Dihydroxycholecalciferol also increases the pumping of Ca^{2+} out of the basolateral membranes of intestinal epithelial cells and increases phosphatidylcholine and phosphatidylethanolamine metabolism in their cell membranes, but the relationship of these actions to Ca^{2+} absorption is not clear.

The calcium-binding proteins induced by 1,25-dihydroxycholecalciferol have been isolated and their structure determined. One is induced in the intestine of mammals and has a molecular weight of 9700. The other is found in other mammalian tissues and has a molecular weight of 28,000. 1,25-Dihydroxycholecalciferol receptors have been identified in many tissues in addition to the intestine, including bone, brain, kidney, various endocrine glands, and monocytes. Many of the cells with receptors contain one or the other of the 2 calcium-binding proteins. However, the functions of the receptors and the calcium-binding proteins in these cells are largely unknown.

Additional well-documented actions of 1,25-dihydroxycholecalciferol include facilitation of Ca^{2+} reabsorption in the kidney. It also acts on bone, where it mobilizes Ca^{2+} and PO_4^{3-}, probably by increasing the active transport of Ca^{2+} out of osteoblasts into ECF (Fig 21–3).

The poor intestinal absorption of Ca^{2+} in vitamin D deficiency often leads to hypocalcemia, and because of the calcium deficiency, the protein of new bone fails to mineralize. The result in children and young animals is the disease called **rickets.** Rickets used to be most commonly due to inadequate intake of the provitamins that are converted to vitamin D_3 by sunlight or inadequate exposure to sun, but it can also be due to failure of the liver to produce 25-hydroxycholecalciferol, failure of the kidney to make 1,25-dihydroxycholecalciferol, or failure of target cells to bind 1,25-dihydroxycholecalciferol.

Regulation of Synthesis

The formation of 25-hydroxycholecalciferol does not appear to be stringently regulated, but the formation of 1,25-dihydroxycholecalciferol in the kidneys, which is catalyzed by 1 α-hydroxylase, is regulated in a feedback fashion by plasma Ca^{2+} and PO_4^{3-} (Fig 21–7). The hormone acts on intestine and bone to increase plasma Ca^{2+} and PO_4^{3-} levels. Its formation is facilitated by parathyroid hormone, and when the plasma Ca^{2+} level is low, parathyroid hormone secretion is increased. When the plasma Ca^{2+} level is high, little 1,25-dihydroxycholecalciferol is produced, and the kidney produces the relatively inactive metabolite 24,25-dihydroxycholecalciferol instead. This effect of Ca^{2+} on production of 1,25-dihydroxycholecalciferol is the mechanism that brings about adaptation of Ca^{2+} absorption from the intestine (see above). The production of 1,25-dihydroxycholecalciferol is also increased by low and inhibited by high plasma PO_4^{3-} levels, by a direct inhibitory effect of PO_4^{3-} on 1 α-hydroxylase. Additional control of 1,25-dihydroxycholecalciferol formation is exerted by a direct negative feedback effect of the metabolite on 1 α-hydroxylase and a positive feedback action on the formation of 24,25-dihydroxycholecalciferol. Estrogens and prolactin increase the activity of 1 α-hydroxylase, and circulating 1,25-dihydroxycholecalciferol is increased during lactation. 1,25-dihydroxycholecalciferol production is depressed by metabolic acidosis and insulin deficiency.

THE PARATHYROID GLANDS

Anatomy

In humans, there are usually 4 parathyroid glands: 2 embedded in the superior poles of the thyroid and 2 in its inferior poles (Fig 21–8). However, the location of the individual parathyroids and their number vary considerably. Parathyroid tissue is sometimes found in the mediastinum.

Each parathyroid gland is a richly vascularized disk, about 3 × 6 × 2 mm, containing 2 distinct types of cells. The abundant **chief cells,** which have a clear cytoplasm, secrete parathyroid hormone. The less abundant and larger **oxyphil cells,** which have oxyphil granules in their cytoplasm (Fig 21–9), contain large numbers of mitochondria. The function of the oxyphil cells is unknown.

Effects of Parathyroidectomy

Parathyroid hormone (parathormone, PTH) is essential for life. After parathyroidectomy, there is a steady decline in the plasma calcium level. Signs of neuromuscular hyperexcitability appear, followed by full-blown hypocalcemic tetany (see above). Plasma phosphate levels usually rise as the plasma calcium level falls after parathyroidectomy, but the rise does not always occur.

In humans, tetany is most often due to inadvertent parathyroidectomy during thyroid surgery. Symptoms usually develop 2–3 days postoperatively but

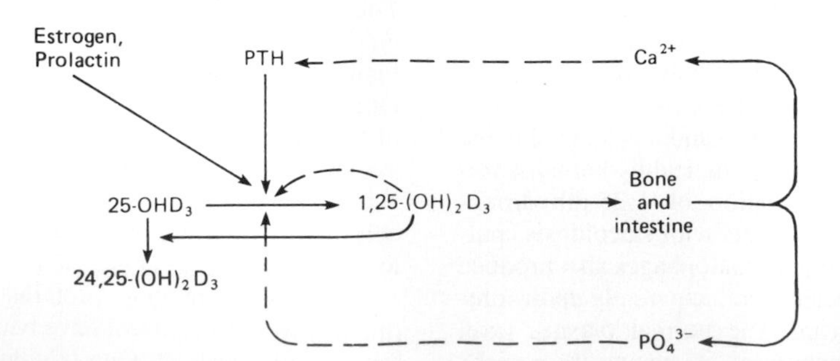

Figure 21–7. Feedback control of the formation of 1,25-dihydroxycholecalciferol (1,25-[OH]$_2$D$_3$) from 25-hydroxycholecalciferol (25-OHD$_3$) in the kidney. Solid arrows indicate stimulation, dashed arrows inhibition.

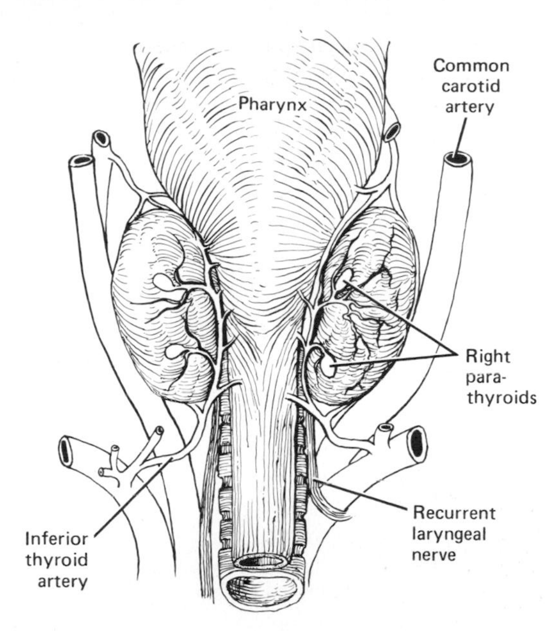

Figure 21–8. The human parathyroid glands, viewed from behind.

may not appear for several weeks or more. In rats on a low-calcium diet, tetany develops much more rapidly, and death occurs 6–10 hours after parathyroidectomy. Injections of parathyroid hormone correct the chemical abnormalities, and the symptoms disappear. Injections of calcium salts give temporary relief, and 1-hydroxylated derivatives of vitamin D restore plasma Ca^{2+} to normal. 25-hydroxycholecalciferol has a direct stimulatory effect on Ca^{2+} absorption from the intestine when administered in large doses, and it is sometimes used in the treatment of hypoparathyroidism even though there is little 1-hydroxylation of vitamin D metabolites in the absence of parathyroid hormone.

The signs of tetany in humans include **Chvostek's sign,** a quick contraction of the ipsilateral facial muscles elicited by tapping over the facial nerve at the angle of the jaw; and **Trousseau's sign,** a spasm of the muscles of the upper extremity that causes flexion of the wrist and thumb with extension of the fingers (Fig 21–10). In individuals with mild tetany in whom spasm is not evident, Trousseau's sign can sometimes be produced by occluding the circulation for a few minutes with a blood pressure cuff.

Parathyroid Hormone Excess

Hyperparathyroidism due to injections of large doses of parathyroid extract in animals or hypersecretion of a functioning parathyroid tumor in humans is characterized by hypercalcemia, hypophosphatemia, demineralization of the bones, hypercalciuria, and the formation of calcium-containing kidney stones. The bone disease caused by hyperparathyroidism **(osteitis fibrosa cystica)** is characterized by multiple bone cysts.

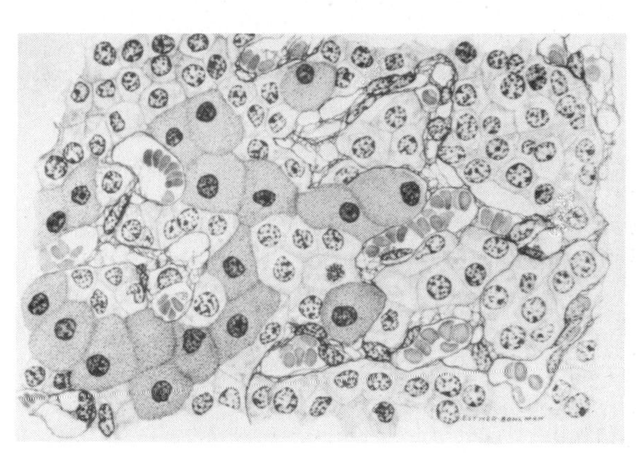

Figure 21–9. Section of human parathyroid. (Reduced 50% from × 960.) Small cells are chief cells; large cells with granules in cytoplasm (especially prominent in lower left of picture) are oxyphil cells. (Reproduced, with permission, from Fawcett DW: *Bloom and Fawcett, A Textbook of Histology.* 11th ed. Saunders, 1986.)

Synthesis & Metabolism of PTH

Human parathyroid hormone is a linear polypeptide with a molecular weight of 9500 that contains 84 amino acid residues (Fig 21–11). Its structure is very similar to that of bovine and porcine PTH. It is synthesized as part of a larger molecule containing 115 amino acid residues **(prepro-PTH).** Upon entry of prepro-PTH into the endoplasmic reticulum, a leader sequence containing 25 amino acid residues is removed from the N terminal to form the 90-amino-acid polypeptide **pro-PTH.** Six additional amino acid residues are removed from the N terminal of pro-PTH in the Golgi apparatus, and the 84-amino-acid polypeptide PTH is the main secretory product of the chief cells.

The half-life of PTH is less than 20 minutes, and the secreted polypeptide is rapidly cleaved by the Kupffer cells in the liver into 2 polypeptides, a biologically inactive C-terminal fragment with a

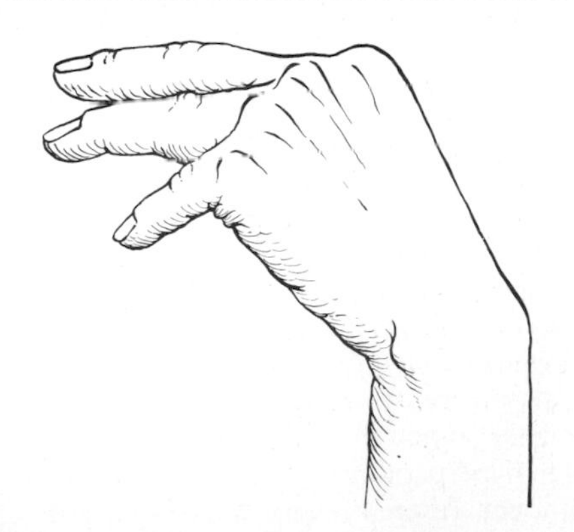

Figure 21–10. Position of the hand in hypocalcemic tetany (Trousseau's sign).

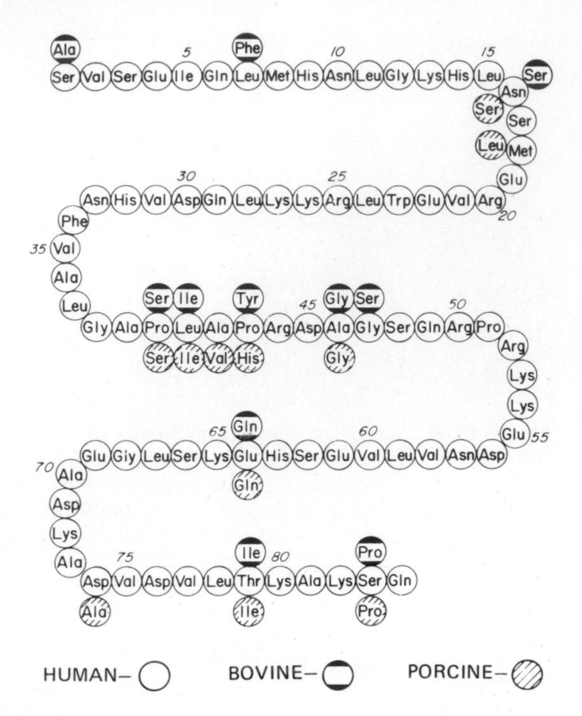

Figure 21–11. Parathyroid hormone. The structure of human PTH is shown. The symbols above and below the human structure show where amino acid residues are different in bovine and porcine PTH. (Reproduced, with permission, from Keutmann HT et al: Complete amino acid sequence of human parathyroid hormone. *Biochemistry* 1978;17:5723. Copyright by the American Chemical Society.)

molecular weight of 7000 and a biologically active N-terminal fragment with a molecular weight of 2500. It is interesting in this regard that a synthetic polypeptide containing amino acid residues 1–34 of bovine parathyroid hormone has all the known biologic effects of the full molecule. Further metabolism occurs at a slower rate. There is some evidence that the 2500- and 7000-molecular-weight fragments are also formed in the chief cells, with secretion of small amounts of each. The fragments may also be produced in the kidneys.

Actions

Parathyroid hormone acts directly on bone to increase bone resorption and mobilize Ca^{2+}. In addition to increasing the plasma Ca^{2+} and depressing the plasma phosphate, parathyroid hormone increases phosphate excretion in the urine. This **phosphaturic action** is due to a decrease in reabsorption of phosphate in the proximal tubules. Parathyroid hormone also increases reabsorption of Ca^{2+} in the distal tubules, although Ca^{2+} excretion is usually increased in hyperparathyroidism because the increase in the amount filtered overwhelms the effect on reabsorption. Parathyroid hormone also increases the formation of 1,25-dihydroxycholecalciferol, the physiologically active metabolite of vitamin D (see above). It increases Ca^{2+} absorption from the intestine, but this action is apparently due solely to the stimulation of 1,25-dihydroxycholecalciferol production.

Mechanism of Action

The actions of parathyroid hormone on the bones and the kidneys involve activation of adenylate cyclase by way of a membrane receptor and G_s, with consequent increased formation of cyclic AMP in the affected cells. How cyclic AMP affects calcium in bone is unsettled. The hormone increases the permeability of osteoclasts and osteoblasts to the Ca^{2+} in bone fluid and, presumably, the osteoblasts then pump the Ca^{2+} into the ECF (Fig 21–3). The pump is stimulated by 1,25-dihydroxycholecalciferol. This is why the action of parathyroid hormone is facilitated by 1,25-dihydroxycholecalciferol without any effect of the vitamin derivative on cyclic AMP. Starting about 12 hours after injection, parathyroid hormone also promotes increased osteoclastic activity and triggers the formation of more osteoclasts while inhibiting the formation of osteoblasts.

In the disease called **pseudohypoparathyroidism,** the signs and symptoms of hypoparathyroidism develop but the circulating level of parathyroid hormone is normal or elevated. Since the tissues fail to respond to the hormone, the disease is sometimes classified as a receptor disease. There are 2 forms, both associated with an abnormal response to parathyroid hormone. In the most common form, there is a congenital 50% reduction of the activity of G_s, the nucleotide regulatory protein that is associated with adenylate cyclase in cell membranes (Fig 1–33), and parathyroid hormone fails to produce a normal increase in cyclic AMP concentration. In a different, less common form, the cyclic AMP response is normal but the phosphaturic action of the hormone is defective.

Regulation of Secretion

It has been proved by perfusion experiments that the circulating level of ionized calcium acts directly on the parathyroid glands in a feedback fashion to regulate the secretion of parathyroid hormone. When the plasma Ca^{2+} level is high, secretion is inhibited, and the calcium is deposited in the bones. When it is low, secretion is increased, and Ca^{2+} is mobilized from the bones. Magnesium also appears to have a direct effect, with an acute decrease in plasma Mg^{2+} concentration stimulating parathyroid secretion. In addition, 1,25-dihydroxycholecalciferol acts directly on the parathyroid glands to decrease preproparathyroid hormone mRNA. Secretion of PTH is also increased by β-adrenergic discharge and cyclic AMP.

Another factor regulating plasma Ca^{2+} is the readily exchangeable bone calcium pool. However, this pool can only maintain the total plasma calcium at about 7 mg/dL in the absence of parathyroid hormone, and maintenance of the normal level of 10 mg/dL is due to the activity of the parathyroid glands.

In conditions such as chronic renal disease and rickets, in which the plasma Ca^{2+} level is chronically low, feedback stimulation of the parathyroid glands causes compensatory parathyroid hypertrophy and **secondary hyperparathyroidism.** The plasma C^{2+} level is low in chronic renal disease primarily because the diseased kidneys lose the ability to form 1,25-dihydroxycholecalciferol.

CALCITONIN

Origin

In dogs, perfusion of the thyroproparathyroid region with solutions containing high concentrations of Ca^{2+} leads to a fall in peripheral plasma calcium, and after damage to this region, Ca^{2+} infusions cause a greater increase in plasma Ca^{2+} than they do in control animals. These and other observations led to the discovery that a calcium-lowering as well as a calcium-elevating hormone was secreted by structures in the neck. The calcium-lowering hormone has been named **calcitonin.** In nonmammalian vertebrates, the source of calcitonin is the **ultimobranchial bodies,** a pair of glands derived embryologically from the fifth branchial arches. In mammals, these bodies have for the most part become incorporated into the thyroid gland, where the ultimobranchial tissue is distributed around the follicles as the **parafollicular cells** (Figs 18–2 and 21–12). Since these cells (also known as clear cells or C cells) secrete the hormone, it is sometimes called **thyrocalcitonin.** However, total thyroidectomy does not reduce the circulating level of the hormone to zero, and the hormone has been found by radioimmunoassay in cerebrospinal fluid and in the brain, pituitary, thymus, lung, gut, liver, bladder, and other tissues. The significance of extrathyroidal calcitonin is unsettled, but at least it seems appropriate to refer to it by its original name, calcitonin.

Structure

Human calcitonin has a molecular weight of 3500 and contains 32 amino acid residues (Fig 21–13). Preprocalcitonin is the source not only of calcitonin but also of an additional circulating peptide, **katacalcin,** the physiologic significance of which is

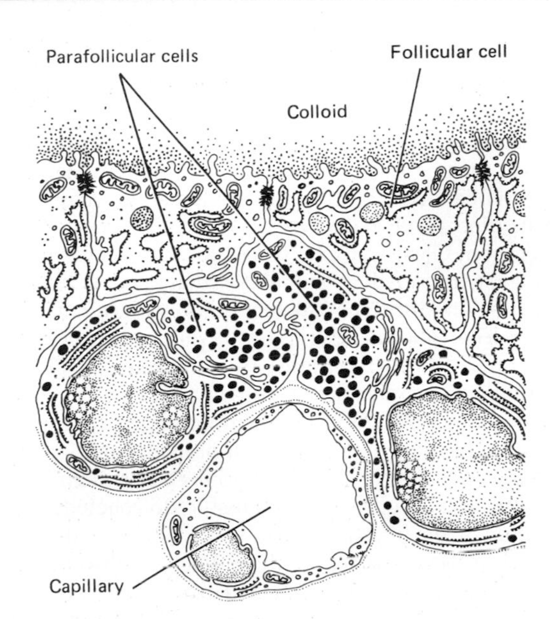

Figure 21–12. Parafollicular cells in thyroid. (Modified from Poirier J, Dumas JLR: *Review of Medical Histology.* Saunders, 1977.)

unsettled. Much of the mRNA transcribed from the calcitonin gene is processed to a different mRNA in the brain, so that calcitonin-gene-related peptide (CGRP) is formed rather than calcitonin (see Chapter 4). CGRP is also found in the intermediate lobe of the pituitary, the adrenal medulla, the pancreas, and the circulating blood. Some is also found in the thyroid. When injected systemically, CGRP produces vasodilation, but its physiologic function is unknown. It is now known that 2 CGRPs that differ by only one amino acid residue, CGRPα and CGRPβ, are formed in humans. These polypeptides are encoded by 2 different genes (see Chapter 4). However, it is not known whether their effects differ.

The calcitonins of the other species that have been studied also contain 32 amino acid residues, but the amino acid composition varies considerably. Salmon calcitonin is of interest because it is more than 20 times more active in humans than human calcitonin.

Secretion & Metabolism

Secretion of calcitonin is increased when the thyroid gland is perfused with solutions containing a

S ——— S (disulfide bridge between Cys 1 and Cys 7)

Cys-Gly-Asn-Leu-Ser-Thr-Cys-Met-Leu-Gly-Thr-Tyr-Thr-Gln-Asp-Phe-Asn-
1 2 3 4 5 6 7 8 9 10 11 12 13 14 15 16 17

Lys-Phe-His-Thr-Phe-Pro-Gln-Thr-Ala-Ile-Gly-Val-Gly-Ala-Pro-NH_2
18 19 20 21 22 23 24 25 26 27 28 29 30 31 32

Figure 21–13. Human calcitonin.

high Ca^{2+} concentration. Measurement of circulating calcitonin by immunoassay indicates that it is not secreted until the plasma calcium level reaches approximately 9.5 mg/dL and that above this calcium level, plasma calcitonin is directly proportionate to plasma calcium (Fig 21–14). β-Adrenergic agonists, dopamine, and estrogens also stimulate calcitonin secretion. Gastrin, CCK, glucagon, and secretin have all been reported to stimulate calcitonin secretion, gastrin being the most potent stimulus (see Chapter 26). The plasma calcitonin level is elevated in Zollinger-Ellison syndrome (see Chapter 26) and in pernicious anemia, in which the plasma gastrin level is also elevated. However, the dose of gastrin needed to stimulate calcitonin secretion produces an increase in plasma gastrin concentration greater than that produced by food, so it is premature to conclude that calcium in the intestine initiates secretion of a calcium-lowering hormone before the calcium is absorbed.

Human calcitonin has a half-life of less than 10 minutes.

Actions

Calcitonin receptors are found in bones and the kidneys. Calcitonin lowers the circulating calcium and phosphate levels. It exerts its calcium-lowering effect by inhibiting bone resorption. This action is direct and is apparently due to inhibition of the Ca^{2+} permeability of osteoclasts as well as osteoblasts (Fig 21–3). It also increases Ca^{2+} excretion in the urine.

The exact physiologic role of calcitonin is uncertain. The calcitonin content of the human thyroid is low, and after thyroidectomy, bone density and plasma Ca^{2+} level are normal as long as the parathyroid glands are intact. Indeed, there are only transient abnormalities of calcium metabolism when a calcium load or parathyroid extract is injected. This may be explained in part by secretion of calcitonin from tissues other than the thyroid. However, there is general agreement that the hormone has little long-term effect on the plasma Ca^{2+} level in adult animals and humans. In addition, patients with medullary carcinoma of the thyroid have a very high circulating calcitonin level but no symptoms directly attributable to the hormone, and their bones are essentially normal. No syndrome due to calcitonin deficiency has been described. The hormone is more active in young individuals and may play a role in skeletal development. It has been suggested that it protects against postprandial hypercalcemia (see above). In addition, it may protect the bones of the mother from excess calcium loss during pregnancy. Bone formation in the infant and lactation are major drains on calcium stores, and plasma concentrations of 1,25-dihydroxycholecalciferol are elevated in pregnancy. This would cause bone loss in the mother if bone resorption were not simultaneously inhibited by an increase in the plasma calcitonin level.

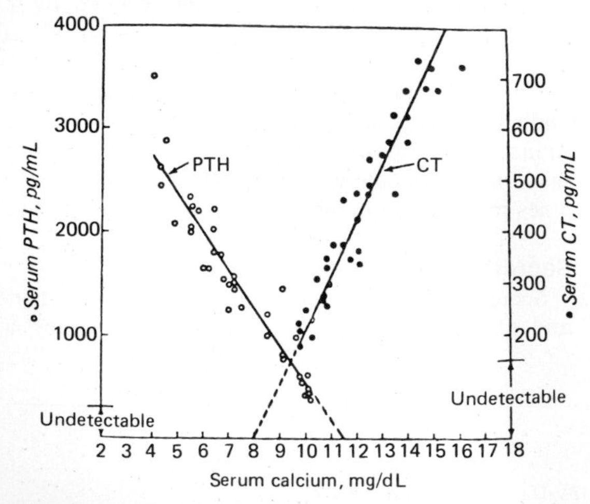

Figure 21–14. Concentration of calcitonin (CT) and parathyroid hormone (PTH) as a function of serum calcium level in pigs given EDTA or calcium infusions to decrease or increase serum calcium. (Reproduced, with permission, from Arnaud CD et al, in: Taylor S [editor: *Calcitonin: Proceedings of the Second International Symposium.* Heinemann, 1969.)

Clinical Correlates

Calcitonin is useful in the treatment of Paget's disease, a condition in which increased osteoclastic activity triggers compensatory formation of disorganized new bone. It also has the beneficial effects in severe hypercalcemia, but the hormone must be injected and its effect generally wears off. It is of questionable value in the treatment of osteoporosis.

Summary

The actions of the 3 principal hormones that regulate the plasma concentration of Ca^{2+} are summarized in Fig 21–15. Parathyroid hormone increases plasma Ca^{2+} by mobilizing this ion from bone. It increases Ca^{2+} reabsorption in the kidney, but this is offset by the increase in filtered Ca^{2+}. It also increases the formation of 1,25-dihydroxycholecalciferol. 1,25-Dihydroxycholecalciferol increases Ca^{2+} absorption from the intestine, mobilizes the ion from bone, and increases Ca^{2+} reabsorption in the kidneys. Calcitonin inhibits bone resorption and increases the amount of Ca^{2+} in the urine.

EFFECTS OF OTHER HORMONES & HUMORAL AGENTS ON CALCIUM METABOLISM

Calcium metabolism is affected by various hormones and growth factors in addition to 1,25-dihydroxycholecalciferol, parathyroid hormone, and calcitonin (Table 21–2). The glucocorticoids lower plasma Ca^{2+} levels by inhibiting osteoclast formation and activity, but over long periods, they cause osteoporosis by decreasing bone formation and increasing bone resorption (see Chapter 20). They

Table 21–2. Factors that affect bone formation and calcium metabolism.

Parathyroid hormone
1,25-Dihydroxycholecalciferol
Calcitonin
Glucocorticoids
Growth hormone and somatomedins
Thyroid hormones
Estrogens
Insulin
IGF-I
Epidermal growth factor
Fibroblast growth factor
Platelet-derived growth factor
Prostaglandin E_2
Osteoclast activating factor

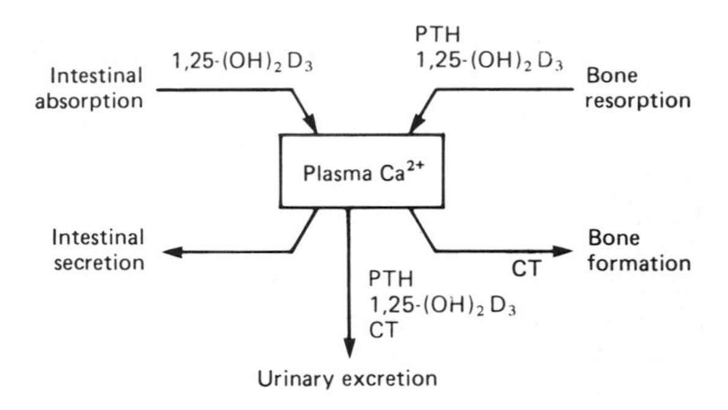

Figure 21–15. Hormonal control of plasma Ca^{2+}. PTH, parathyroid hormone; 1,25-$(OH)_2D_3$, 1,25-dihydroxycholecalciferol; CT, calcitonin. See text.

decrease bone formation by inhibiting cellular replication and protein synthesis in bone, and they inhibit the function of osteoblasts. They also decrease the absorption of Ca^{2+} and PO_4^{3-} from the intestine by an anti-vitamin D action and increase the renal excretion of these ions. This is why they depress the hypercalcemia of vitamin D intoxication. The decrease in plasma Ca^{2+} concentration increases the secretion of parathyroid hormone, and bone resorption is facilitated. Growth hormone increases calcium excretion in the urine, but it also increases intestinal absorption of calcium, and this effect may be greater than the effect on excretion, with a resultant positive calcium balance. In addition, IGF-I generated by the action of growth hormone stimulates protein synthesis in bone. Thyroid hormones may cause hypercalcemia, hypercalciuria, and, in some instances, osteoporosis, but the mechanism by which they produce these effects is uncertain. Estrogens prevent osteoporosis, possibly by a direct effect on osteoblasts (see above). Insulin increases bone formation, and there is significant bone loss in untreated diabetes. Additional local factors must be involved in the constant remodeling of bone in response to stress, but their nature is unknown.

Hypercalcemia is a common metabolic complication of cancer. In patients with bone metastases, local erosion of the bone contributes to the hypercalcemia. There is evidence that this erosion is produced by prostaglandins such as PGE from the tumor. Hypercalcemia is also seen in the absence of bone metastases in patients with squamous cell carcinoma of the lung, and a calcium-mobilizing factor has been isolated from them. The structure of this protein resembles that of PTH, but it does not cross-react with antibodies to PTH and is encoded by a different gene. An osteoclast-activating factor is secreted by T and B lymphocytes and appears to be the cause of the hypercalcemia in patients with hematologic cancers.

22 The Pituitary Gland

INTRODUCTION

The anterior, intermediate, and posterior lobes of the pituitary gland are actually 3 more or less separate endocrine organs that, at least in some species, contain 14 or more hormonally active substances. The 6 established hormones that are secreted by the anterior pituitary are **thyroid-stimulating hormone (TSH, thyrotropin), adrenocorticotropic hormone (ACTH), luteinizing hormone (LH), follicle-stimulating hormone (FSH), prolactin,** and **growth hormone** (Table 22–1). ACTH, prolactin, and growth hormone are simple polypeptides or proteins, whereas TSH, LH, and FSH are glycoproteins. Prolactin acts on the breast (Fig 14–6). The remaining 5 are, at least in part, **tropic hormones;** ie, they stimulate secretion of hormonally active substances by other endocrine glands or, in the case of growth hormone, the liver (see below). In addition, the anterior lobe of the pituitary secretes **β-lipotropin (β-LPH).** This linear polypeptide contains 91 amino acid residues, and although its physiologic role is uncertain, it contains the amino acid sequences of endorphins and enkephalins, peptides that bind to opiate receptors (see Chapter 4). In addition, the anterior and intermediate lobes contain other hormonally active derivatives of the pro-opiomelanocortin molecule (see below). The hormones tropic to a particular endocrine gland are discussed in the chapter on that gland: TSH in Chapter 18; ACTH in Chapter 20; and the gonadotropins FSH and LH in Chapter 23, along with prolactin. The hormones secreted by the posterior pituitary in mammals **(oxytocin** and **vasopressin)** and the neural regulation of anterior and posterior pituitary secretion are discussed in Chapter 14. Growth hormone, β-lipotropin, and the melanocyte-stimulating hormones of the intermediate lobe of the pituitary, **α-MSH** and **β-MSH,** are the subject of this chapter, along with a number of general considerations about the pituitary.

MORPHOLOGY

Gross Anatomy

The anatomy of the pituitary gland is summarized in Fig 22–1 and discussed in detail in Chapter 14. The posterior pituitary is largely made up of the endings on blood vessels of axons from the supraoptic and paraventricular nuclei of the hypothalamus, whereas the anterior pituitary has a special vascular connection with the brain, the portal hypophyseal vessels. The intermediate lobe is formed in the embryo from the dorsal half of **Rathke's pouch,** an evagination of the roof of the pharynx, but is closely adherent to the posterior lobe in the adult. It is separated from the anterior lobe by the remains of Rathke's pouch, the **residual cleft.**

Histology

In the posterior lobe, the endings of the supraoptic and paraventricular axons can be observed in close relation to blood vessels. In some species, the endings are rodlike palisaded structures. There are also neuroglial cells and **pituicytes,** stellate cells containing fat globules that were once thought to secrete posterior lobe hormones but are now believed to be modified astroglia.

The intermediate lobe is rudimentary in humans and a few other mammalian species. In these species, most of its cells are agranular, although there are often a few basophilic elements that resemble anterior lobe cells. Along the residual cleft are small thyroid-like follicles, some containing a little colloid. The function of the colloid, if any, is unknown.

The anterior pituitary is made up of interlacing cell cords and an extensive network of sinusoidal capillaries. The endothelium of the capillaries is fenestrated, like that in other endocrine organs. The cells contain granules of stored hormone that are extruded from the cells by exocytosis. The granules have not been seen entering the capillaries, and they presumably break down in the pericapillary space.

Cell Types in the Anterior Pituitary

Human anterior pituitary cells have traditionally been divided on the basis of their staining reactions into agranular chromophobes and granular chromophils. The chromophilic cells are subdivided into acidophils (about 40% of the cells), which stain with acidic dyes, and basophils (about 10% of the cells), which stain with basic dyes. The cells secreting growth hormone and prolactin are acidophilic, whereas those secreting the glycoprotein hormones TSH, LH, and FSH are basophilic and those secreting ACTH are variously classified as basophilic or chromophobic. With more modern techniques of immunocytochemistry and electron microscopy, it is possible to distinguish 5 types of cells (Fig 22–2): somatotropes, which secrete growth hormone; lactotropes (also called mammotropes), which secrete

Table 22–1. Pituitary hormones in mammals.*

Name and Source	Principal Actions
Anterior lobe	
Thyroid-stimulating hormone (TSH, thyrotropin)	Stimulates thyroid secretion and growth.
Adrenocorticotropic hormone (ACTH, corticotropin)	Stimulates adrenocortical secretion and growth.
Growth hormone (GH, somatotropin, STH)	Accelerates body growth; stimulates secretion of IGF-I.
Follicle-stimulating hormone (FSH)	Stimulates ovarian follicle growth in female and spermatogenesis in male.
Luteinizing hormone (LH, interstitial cell-stimulating hormone, ICSH)	Stimulates ovulation and luteinization of ovarian follicles in female and testosterone secretion in male.
Prolactin (luteotropic hormone, LTH, luteotropin, lactogenic hormone, mammotropin)	Stimulates secretion of milk and maternal behavior. Maintains corpus luteum in female rodents but apparently not in other species.
β-Lipotropin (β-LPH)	?
γ-Melanocyte-stimulating hormone (γ-MSH)	May maintain adrenal sensitivity.
Intermediate lobe	
α- and β-Melanocyte-stimulating hormones (α- and β-MSH; referred to collectively as melanotropin or intermedin)	Expands melanophores in fish, amphibians, and reptiles; stimulates melanin synthesis in melanocytes in humans.
γ-Lipotropin (γ-LPH), corticotropinlike intermediate lobe peptide (CLIP), other fragments of pro-opiomelanocortin	?
Posterior lobe	
Vasopressin (antidiuretic hormone, ADH)	Promotes water retention.
Oxytocin	Causes milk ejection.

*In addition, a variety of gastrointestinal and other polypeptides are found in one or more lobes of the pituitary gland. These include CCK, gastrin, renin, angiotensin II, and calcitonin-gene-related peptide (CGRP).

prolactin; thyrotropes, which secrete TSH; gonadotropes, which appear in most instances to secrete both LH and FSH; and corticotropes, which secrete both ACTH and β-LPH. Note that in 2 instances, 2 or more polypeptide hormones are secreted by the same cell.

Two-Unit Structure of FSH, LH, & TSH

The 3 pituitary glycoprotein hormones, FSH, LH, and TSH, are each made up of 2 subunits. The subunits, which have been designated α and β, have some activity but must be combined for maximal physiologic activity. In addition, the placental glycoprotein gonadotropin human chorionic gonadotropin (hCG) has an α and a β subunit (see Chapter 23). All of the α subunits of these hormones are similar to one another, and they are products of a single gene. The amino acid composition of the α subunits of LH, FSH, and TSH is identical although the carbohydrate residues may differ, and the amino acid composition

Third ventricle
Third ventricle
Pars tuberalis
Posterior lobe
Rathke's pouch
Anterior lobe
Intermediate lobe

Figure 22–1. Diagrammatic outline of the formation of the pituitary and the various parts of the organ in the adult.

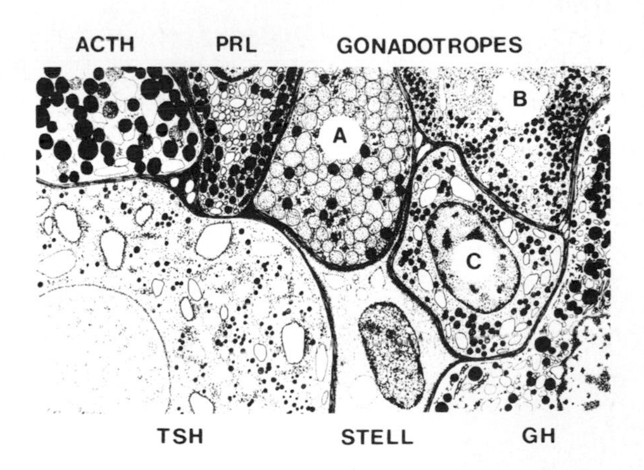

Figure 22–2. Ultrastructure of the anterior pituitary, showing the characteristics of 5 types of secretory cells. Note that 3 morphologically different subtypes of gonadotropes (A, B, and C) can be identified. (Reproduced, with permission, from Steger RW, Peluso JJ: Anterior pituitary. In: *Handbook of Endocrinology.* Gass GH, Kaplan HM [editors]. CRC Press, 1982. Copyright CRC Press, Inc., Boca Raton, FL.)

of the α subunit of hCG differs only slightly from the others. The β subunits, which are produced by separate genes and differ in structure, confer hormonal specificity. The α units are remarkably interchangeable, and the hybrid molecule made up of TSH-α and LH-β, for example, has much more gonadotropic activity than LH-β alone. However, the physiologic and evolutionary significance of the unique 2-unit structure of these glycoprotein hormones remains to be determined.

INTERMEDIATE LOBE HORMONES

Pro-opiomelanocortin

Intermediate lobe cells and corticotropes of the anterior lobe both synthesize a large precursor protein that is cleaved to form a family of hormones. After removal of the signal peptide, this prohormone is known as pro-opiomelanocortin (POMC). This molecule is also synthesized in the hypothalamus and other parts of the nervous system, the lungs, the gastrointestinal tract, and the placenta (see Chapter 4). Its structure is shown in Fig 22–3. In the corticotropes, it is hydrolyzed to ACTH and β-LPH plus a small amount of β-endorphin, and these substances are secreted. It also appears that γ-MSH is secreted. In the intermediate lobe, POMC is further hydrolyzed to α-MSH, corticotropinlike intermediate lobe peptide (CLIP), γ-LPH, and appreciable quantities of β-endorphin, and these substances are presumably secreted. The functions, if any, of CLIP and γ-LPH are unknown, whereas β-endorphin is an opioid peptide (see Chapter 4) that has the 5 amino acid residues of met-enkephalin at its N-terminal end. The intermediate lobe in humans is rudimentary, and in adults, it appears that neither α-MSH nor β-MSH is secreted.

Melanocyte-Stimulating Hormones

In the skins of fish, reptiles, and amphibians, there are cells called **melanophores.** These contain melanin granules. Melanins are synthesized from tyrosine via dopa (see Chapter 4) and dopaquinone. They are of 2 types: eumelanins, which are responsible for black or brown pigmentation, and pheomelanins, which contain more sulfhydryl groups and are responsible for yellow and red pigmentation. Melanins are synthesized in cellular organelles called **melanosomes.** Other cells called **iridophores** contain reflecting platelets. When the granules aggregate around the nuclei of melanophores and the reflecting platelets disperse to the periphery of the iridophores, the color of the skin lightens; when the melanin granules disperse and the reflecting platelets aggregate, the skin darkens. Birds and mammals do not have pigment cells of this type, but they have cells that contain melanin, and their pituitaries contain 3 polypeptides called **melanotropins** or **melanocyte-stimulating hormones (MSHs),** which when injected into lower vertebrates disperse melanophore granules and aggregate iridophore platelets.

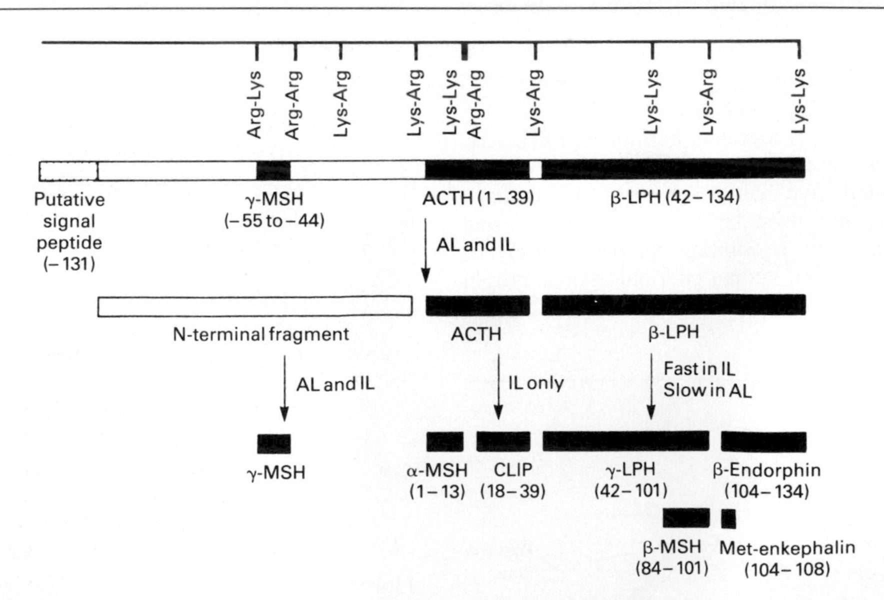

Figure 22–3. Schematic representation of the prepro-opiomelanocortin molecule formed in pituitary cells, neurons, and other tissues. The numbers in parentheses identify the amino acid sequences in each of the polypeptide fragments. For convenience, the amino acid sequences are numbered from the N terminal of ACTH and read toward the C-terminal portion of the parent molecule, whereas the amino acid sequences in the other portion of the molecule read to the left to −131, the N terminal of the parent molecule. The locations of Lys-Arg and other pairs of basic amino acids residues are also indicated; these are the sites of proteolytic cleavage in the formation of the smaller fragments of the parent molecule. AL, anterior lobe; IL, intermediate lobe.

α-MSH is made up of amino acid residues 1–13 of the ACTH molecule (Fig 22–3). β-MSH is made up of the 17 amino acid residues at the C-terminal end of γ-LPH. It contains a 7-amino-acid sequence that is identical to amino acid residues 4–10 in α-MSH and ACTH. Consequently, it is not surprising that ACTH has considerable MSH activity. The effects of γ-MSH resemble those of the other MSHs. It is present in highest concentration in the intermediate lobe, but it is also present in the anterior lobe, and its blood level is increased in patients with adrenal insufficiency and decreased in patients with high levels of circulating glucocorticoids.

Control of Skin Coloration

Changes in skin coloration in fish, reptiles, and amphibians are probably mediated in part via neuroendocrine reflex mechanisms. The receptors for these reflexes are in the retina, and the changes are brought about by MSH. On a dark background, MSH secretion is not inhibited and the animal is dark. On a white background, release of MSH is inhibited and the animal's color lightens.

In mammals, there are no melanophores containing pigment granules that disperse and aggregate as they do in fish, reptiles, and birds, but there are **melanocytes,** which contain melanosomes and synthesize melanins. The melanocytes then transfer the melanosomes to skin cells (keratinocytes) in hair follicles and other portions of the skin. This accounts for the pigmentation of hair and skin. Treatment with MSHs accelerates melanin synthesis and causes readily detectable darkening of the skin of humans in 24 hours. However, the physiologic function of MSHs in humans is unknown.

Pigment Abnormalities in Humans

The pigmentary changes in several endocrine diseases are probably due to changes in circulating ACTH, since ACTH has MSH activity. For example, abnormal pallor is a hallmark of hypopituitarism. Hyperpigmentation occurs in patients with adrenal insufficiency due to primary adrenal disease. Indeed, the presence of hyperpigmentation in association with adrenal insufficiency rules out the possibility that the insufficiency is secondary to pituitary disease, because the pituitary must be intact for pigmentation to occur.

Albinos have a congenital inability to synthesize melanin. Albinism occurs in humans and many other mammalian species. It can be due to a variety of different genetic defects in the pathways for melanin synthesis. **Piebaldism** is characterized by patches of skin that lack melanin owing to congenital defects in the migration of pigment cell precursors from the neural crest during embryonic development. Not only the condition but also the precise pattern of the loss is passed from one generation to the next. **Vitiligo** is due to a similar patchy loss of melanin, but the loss develops after birth and is progressive.

GROWTH HORMONE

Chemistry & Species Specificity

Growth hormone, like other protein hormones, varies considerably in structure from species to species. The principal growth hormone in humans has a molecular weight of approximately 22,000 (22K), 2 disulfide bridges, and 191 amino acid residues (Fig 22–4). It accounts for 90% of the growth hormone in the pituitary gland. The remaining 10% has a molecular weight of about 20,000 (20K), lacks amino acid residues 32–46, and is formed from the same gene as the 22K growth hormone. Both the 22K and 20K growth hormones are secreted, and both are biologically active. The physiologic significance of the existence of 2 forms is unknown.

Human growth hormone (hGH) bears a marked structural resemblance to prolactin and a placental hormone, human chorionic somatomammotropin (hCS), that has growth-promoting activity (see Chapter 23), and it seems likely that they all evolved from a common single hormone. Porcine and simian growth hormones have only a transient effect in the guinea pig, probably because they stimulate the rapid formation of anti-growth-hormone antibodies. In monkeys and humans, bovine and porcine growth hormones do not even have a significant transient effect on growth, although monkey and human growth hormones are fully active in both monkeys and humans (Table 22–2). Human growth hormone has intrinsic lactogenic activity.

Plasma Levels & Metabolism

The basal growth hormone level measured by radioimmunoassay in adult humans is normally less than 3 ng/mL. Growth hormone is metabolized rapidly, probably at least in part in the liver. The half-life of circulating growth hormone in humans is 6–20 minutes, and the daily growth hormone output has been calculated to be 0.2–1.0 mg/d in adults.

Effects on Growth

Most of the effects of growth hormone are due at least in part to its interaction with somatomedins. The place of growth hormone and somatomedins in the complex of factors that promote growth is discussed below. In young animals in which the epiphyses have not yet fused to the long bones (see Chapter 21), growth is inhibited by hypophysectomy (Fig 22–5) and stimulated by growth hormone. Chondrogenesis is accelerated, and as the cartilaginous epiphyseal plates widen, they lay down more matrix at the ends of long bones (Fig 22–6). In this way, stature is increased, and prolonged treatment leads to gigantism. The increase in width of the tibial epiphyseal plate in hypophysectomized rats has been used as a bioassay for growth hormone ("tibia test").

When the epiphyses are closed, linear growth is no longer possible, and growth hormone produces the

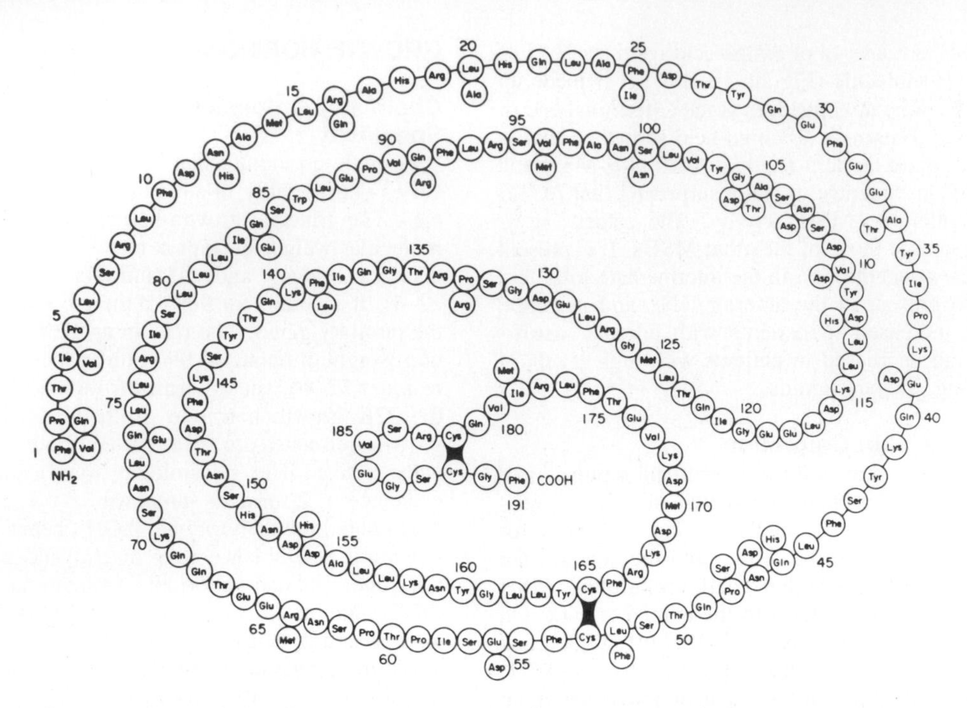

Figure 22–4. Structure of the principal human growth hormone (continuous chain). The black bars indicate disulfide bridges. The 29 residues alongside the chain identify those residues that differ in human chorionic somatomammotropin (hCS; see Chapter 23). All the other residues in hCS are the same, and hCS also has 191 amino acid residues. (Reproduced, with permission, from Parsons JA [editor]: *Peptide Hormones.* University Park Press, 1976.)

pattern of bone and soft tissue deformities known in humans as **acromegaly** (Fig 22–7). The size of most of the viscera is increased. Endocrine organs may also be affected by growth hormone, and the hormone is said to synergize with ACTH in increasing adrenal size and with androgens in increasing the size of accessory reproductive organs. The protein content of the body is increased and the fat content is decreased.

Table 22–2. Activities of growth hormones in various species. (+), active; (−), inactive.

Growth Hormone From	Stimulates Growth In				
	Fish	Birds	Rats	Monkeys	Humans
Fish	+		−		
Reptiles			+		
Amphibia			+		
Birds		+	+		
Cows	+	−*	+	−	−†
Sheep	+		+		−
Pigs	+	−*	+	−	−†
Whales			+		−
Monkeys	+		+	+	+
Humans	+		+	+	+

*Probable diabetogenic effect.
†Slight diabetogenic effect.

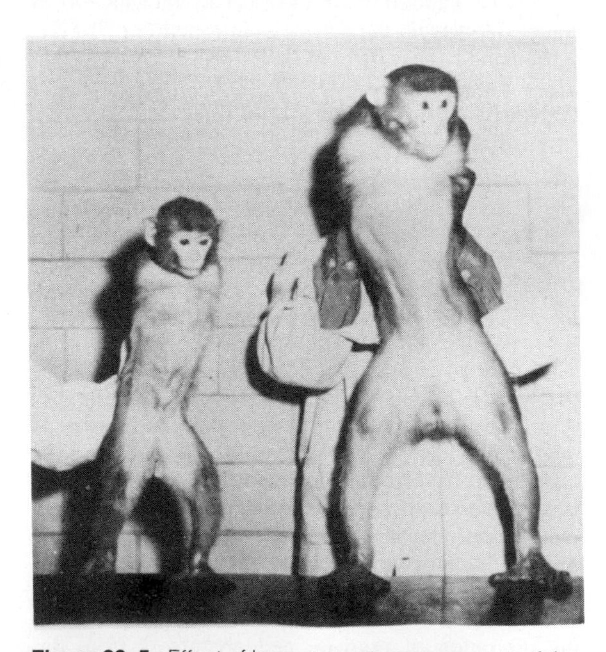

Figure 22–5. Effect of hypophysectomy on growth of the immature rhesus monkey. Both monkeys were the same size and weight 2 years previously, when the one on the left was hypophysectomized. (Reproduced, with permission, from Knobil E in: *Growth in Living Systems.* Zarrow MX [editor]. Basic Books, 1961.)

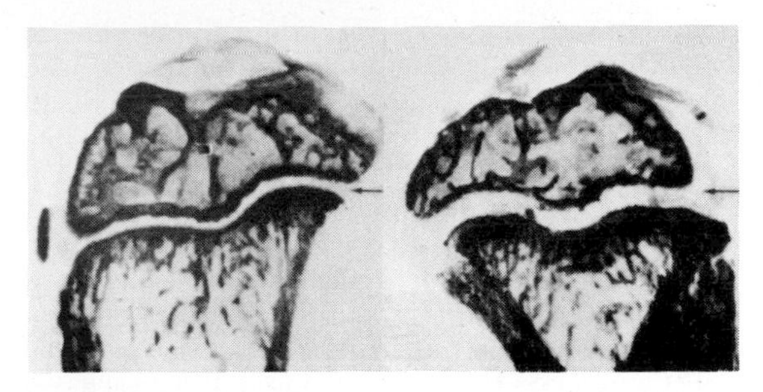

Figure 22–6. Effect of growth hormone treatment for 4 days on the proximal tibial epiphysis of the hypophysectomized rat. Note the increased width of the unstained cartilage plate in the tibia of the treated animal on the right, compared with the control on the left. (Reproduced, with permission, from Evans HM et al: Bioassay of pituitary growth hormone. *Endocrinology* 1943;**32**:14.)

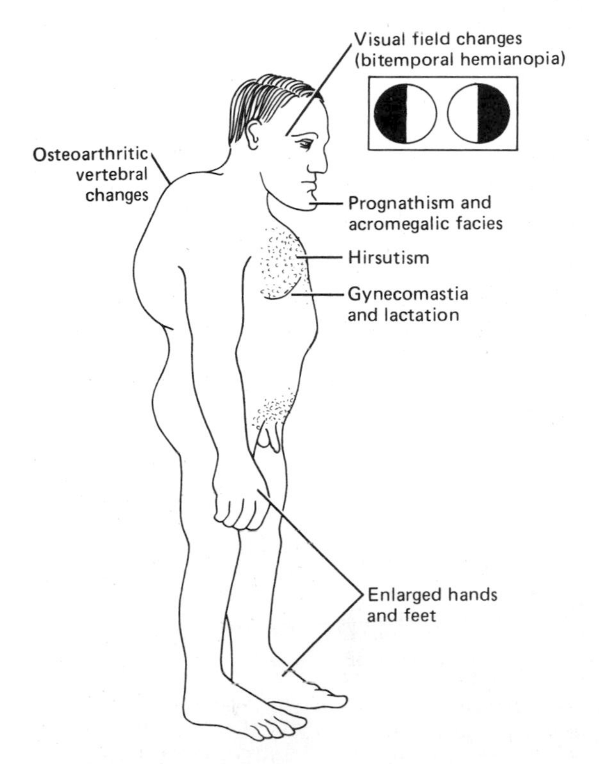

Figure 22–7. Typical findings in acromegaly.

Effects on Protein & Electrolyte Metabolism

Growth hormone is a protein anabolic hormone and produces a positive nitrogen and phosphorus balance, a rise in plasma phosphorus, and a fall in the blood urea nitrogen and amino acid levels. It also stimulates erythropoiesis. Gastrointestinal absorption of Ca^{2+} is increased. Na^+ and K^+ excretion are reduced by an action independent of the adrenal glands, probably because these electrolytes are diverted from the kidney to the growing tissues. Excretion of the amino acid 4-hydroxyproline is increased during growth and in acromegaly, but it is also increased in a number of other diseases. Much of the excreted hydroxyproline comes from collagen, and hydroxyproline excretion is increased in diseases associated with increased collagen destruction. However, it is also increased when synthesis of soluble collagen is increased, and growth hormone stimulates the synthesis of soluble collagen.

Effects on Carbohydrate & Fat Metabolism

The actions of growth hormone on carbohydrate metabolism are discussed in Chapter 19. Growth hormone is diabetogenic because it increases hepatic glucose output and exerts an anti-insulin effect in muscle. It is ketogenic because it increases circulating FFA levels. The increase in plasma FFA, which takes several hours to develop, provides a ready source of energy for the tissues during hypoglycemia, fasting, and stressful stimuli. Growth hormone does not stimulate B cells of the pancreas directly, but it increases the ability of the pancreas to respond to insulinogenic stimuli such as arginine and glucose. This is an additional way growth hormone promotes growth, since insulin has a protein anabolic effect (see Chapter 19).

Somatomedins

The effects of growth hormone on growth, cartilage, and protein metabolism depend on an interaction between growth hormone and **somatomedins,** which are polypeptide growth factors secreted by the liver and other tissues in response to stimulation by growth hormone. The first of these factors isolated was called sulfation factor because it stimulated the incorporation of sulfate into cartilage. However, it also stimulated collagen formation, and its name was changed to somatomedin. It then became clear that there are a variety of different somatomedins and that they are related to an increasingly long list of growth factors that affect many different tissues and organs. These include nerve growth factor (NGF; see Chapter 2), epidermal growth factor (EGF), ovarian growth factor (OGF), fibroblast growth factor (FGF), and platelet-derived growth factor (PDGF). Several of the growth factors are related to the proteins produced by oncogenes (see Chapter 1).

The principal (and in humans probably the only) circulating somatomedins are insulinlike growth factor I (IGF-I, somatomedin C) and insulinlike growth factor II (IGF-II). These factors and the hormone relaxin, which relaxes the pubic ligaments at the time of delivery (see Chapter 3), are all related to insulin (Fig 22–8). The mRNAs for IGF-I and IGF-II are found not only in the liver but also in cartilage and many other tissues, indicating that they are synthesized in these tissues.

The properties of IGF-I, IGF-II, and insulin are compared in Table 22–3. Both are bound to proteins in the plasma, and at least in the case of IGF-I, this prolongs the IGF half-life in the circulation. The

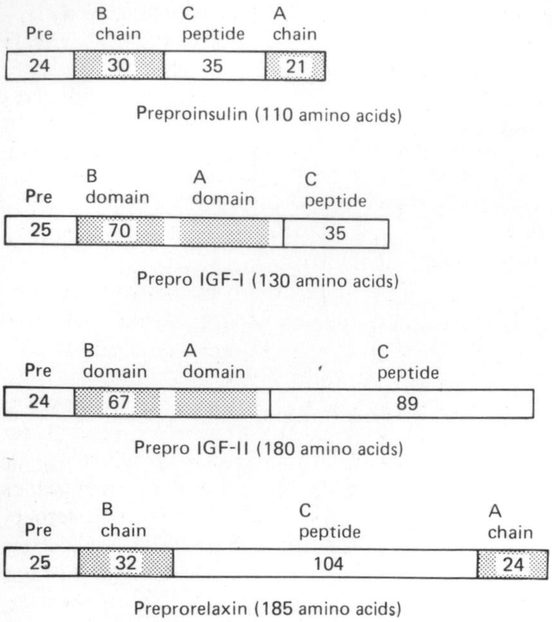

Pre B chain C peptide A chain
24 30 35 21
Preproinsulin (110 amino acids)

Pre B domain A domain C peptide
25 70 35
Prepro IGF-I (130 amino acids)

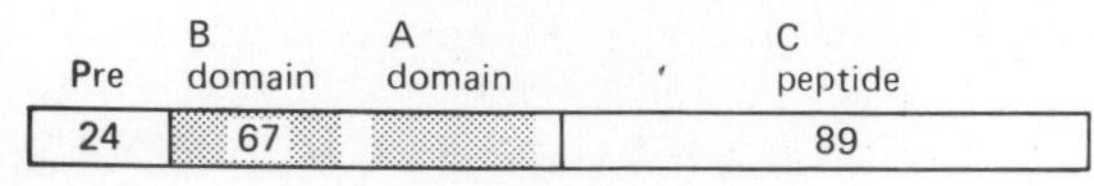

Prepro IGF-II (180 amino acids)

Pre B chain C peptide A chain
25 32 104 24
Preprorelaxin (185 amino acids)

Figure 22–8. Formation of insulin and insulinlike hormones from their preprohormones. The shaded areas identify homologous chains or domains. In insulin and relaxin, the C peptides are eliminated and the A and B chains joined by disulfide bridges. In IGF-I and IGF-II, the A and B domains remain connected in a single peptide. The vertical lines identify sites where the preprohormones are cleaved, and the numbers indicate the number of amino acid residues in each segment. Pre, signal sequence. Note that the A domains in the IGFs are extended at their C terminals by 8 amino acid residues (IGF-I) and 6 amino acid residues (IGF-I), respectively. (Modified from Bell GI et al: Sequence of a cDNA clone encoding human preproinsulin-like growth factor II. *Nature* 1984;310:776.)

contributions of the IGFs to the insulinlike activity in blood is discussed in Chapter 19. Secretion of IGF-I is stimulated by growth hormone, and it has pronounced growth-stimulating activity. IGF-II is much less affected by growth hormone and may play a role in the growth of the fetus before birth. It is also present in the brain, where its function is unknown.

The secretion of somatomedins is affected by various factors in addition to growth hormone. Glucocorticoids and protein deficiency reduce plasma somatomedin activity, and large doses of estrogens inhibit the production of IGF-I. The secretion of somatomedins is reduced in untreated diabetes and restored to normal by insulin treatment.

Mechanism of Action of Growth Hormone

Ideas about the mechanism of action of growth hormone have undergone a series of changes as new information has become available. The growth hormone receptor has now been cloned and shown to be a linear protein that spans the cell membrane one time. Growth hormone was originally thought to produce growth by a direct action on tissues, and later, it was believed to act solely through somatomedins. However, if growth hormone is injected into one proximal tibial epiphysis, a unilateral increase in cartilage width is produced, and it has now been shown that cartilage, like other tissues, makes IGF-I. A current hypothesis to explain these results holds that growth hormone acts on cartilage to convert stem cells into cells that respond to IGF-I and then that locally produced and circulating IGF-I make the cartilage grow. However, the role of circulating IGF-I remains important, since infusion of IGF-I to hypophysectomized rats restores bone and body growth.

Presumably, growth hormone combines with circulating and locally produced IGF-I in various proportions to produce the other effects of growth hormone, but the details of these interactions remain to be determined.

Hypothalamic Control of Growth Hormone Secretion

The secretion of growth hormone is controlled via the hypothalamus. The hypothalamus secretes growth hormone-releasing hormone (GRH) and the growth hormone-inhibiting hormone somatostatin into the portal hypophyseal blood, and hypothalamic lesions or section of the pituitary stalk inhibits growth hormone secretion. If growth hormone were secreted in a constant and continuous fashion and only in children, the hypothalamic control of its secretion might be surprising; however, there are large quantities of growth hormone in the pituitaries of adults as

Table 22–3. Comparison of insulin and the insulinlike growth factors.*

	Insulin	IGF-I	IGF-II
Other names	. . .	Somatomedin C	Multiplication-stimulating activity (MSA)
Number of amino acids	51	70	67
Source	Pancreatic B cells	Liver and other tissues	Diverse tissues
Level regulated by	Glucose	Growth hormone, nutritional status	Unknown
Plasma levels	0.3–2 ng/mL	ng/mL range	ng/mL range
Plasma binding protein	No	Yes	Yes
Major physiologic role	Control of metabolism	Skeletal and cartilage growth	Unknown; probably induced in fetal development

*Modified and reproduced, with permission, from Murray RK et al: *Harper's Biochemistry*, 21st ed. Appleton & Lange, 1987.

well as children, and both in children and in adults, the rate of growth hormone secretion undergoes marked and rapid fluctuations in response to a variety of stimuli.

Growth hormone secretion is under feedback control, like the secretion of other anterior pituitary hormones. Growth hormone increases circulating IGF-I, and IGF-I in turn exerts a direct inhibitory action on growth hormone secretion from the pituitary. It also stimulates somatostatin secretion (Fig 22–9).

Stimuli Affecting Growth Hormone Secretion

As noted above, the basal plasma growth hormone concentration ranges from 0 to 3 ng/mL in normal adults. The values in newborn infants are higher, but resting plasma growth hormone levels throughout the rest of childhood are not significantly greater than they are in adults. On the other hand, growth hormone is secreted in irregular "spikes" throughout the day, and there is some evidence that these spikes increase in amplitude during puberty.

The stimuli that increase growth hormone secretion are summarized in Table 22–4. Most of them fall into 3 general categories: (1) conditions such as hypoglycemia and fasting, in which there is an actual or threatened decrease in the substrate for energy production in the cells; (2) conditions in which there are increased amounts of certain amino acids in the plasma; and (3) stressful stimuli. It has been claimed that the response to glucagon is useful as a test of the growth-hormone-secreting mechanism in patients with endocrine diseases. Although the growth hormone responses to the other stimuli are relatively reproducible, they may vary from individual to individual. A spike in growth hormone secretion occurs with considerable regularity upon going to sleep, but the significance of the association between growth hormone and sleep is an enigma. Growth hormone secretion is increased in subjects deprived of REM sleep (see Chapter 11) and inhibited during normal REM sleep.

Glucose infusions lower plasma growth hormone levels and inhibit the response to exercise. The increase produced by 2-deoxyglucose is presumably due to intracellular glucose deficiency, since this compound blocks the catabolism of glucose 6-phosphate. Sex hormones, particularly estrogens, increase growth hormone responses to provocative stimuli such as arginine and insulin. Growth hormone secretion is inhibited by cortisol, FFA, and medroxyprogesterone.

Growth hormone secretion is increased by L-dopa, which increases the release of dopamine and norepinephrine in the brain, and by the dopamine receptor agonist apomorphine. Central release of norepinephrine is known to increase GRH secretion, but it is uncertain whether the effect mediated by dopamine receptors is central or peripheral.

PHYSIOLOGY OF GROWTH

Growth is a complex phenomenon that is affected not only by growth hormone and somatomedins but also by thyroid hormones, androgens, estrogens, glucocorticoids, and insulin. It is also affected, of course, by genetic factors and it depends on adequate

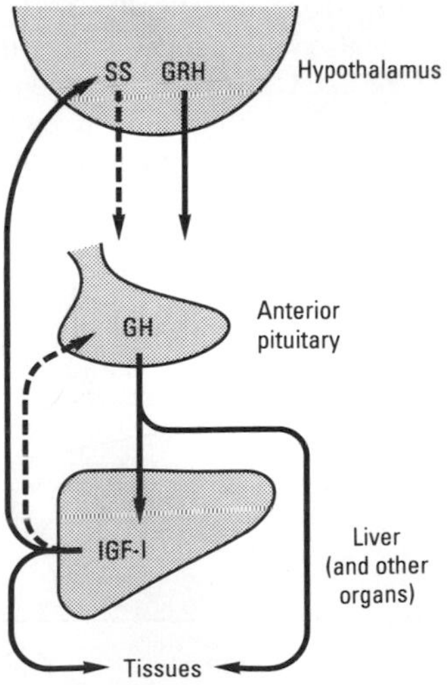

Figure 22–9. Feedback control of growth hormone secretion. The dashed arrows indicate inhibitory effects and the solid arrows stimulatory effects. Note that IGF-I stimulates the secretion of somatostatin (SS) from the hypothalamus and acts directly on the pituitary to inhibit growth hormone (GH) secretion. Compare with Fig 18–13, 20–22, 23–21 and 23–35.

Table 22–4. Stimuli that affect growth hormone secretion in humans.

Stimuli that increase secretion
Deficiency of energy substrate
Hypoglycemia
2-Deoxyglucose
Exercise
Fasting
Increase in circulating levels of certain amino acids
Protein meal
Infusion of arginine and some other amino acids
Glucagon
Stressful stimuli
Pyrogen
Lysine vasopressin
Various psychologic stresses
Going to sleep
L-Dopa and α-adrenergic agonists that penetrate the brain
Apomorphine and other dopamine receptor agonists
Estrogens and androgens
Stimuli that decrease secretion
REM sleep
Glucose
Cortisol
FFA
Medroxyprogesterone
Growth hormone

nutrition. It is normally accompanied by an orderly sequence of maturational changes, and it involves accretion of protein and increase in length and size, not just an increase in weight, which may be due to the formation of fat or retention of salt and water.

Role of Nutrition

The food supply is the most important extrinsic factor affecting growth. The diet must be adequate not only in protein content but also in essential vitamins and minerals (see Chapter 17) and in calories, so that ingested protein is not burned for energy. However, the age at which a dietary deficiency occurs appears to be an important consideration. For example, once the pubertal growth spurt has commenced, considerable linear growth continues even if caloric intake is reduced. Injury and disease stunt growth because they increase protein catabolism.

Growth Periods

Patterns of growth vary somewhat from species to species. Rats continue to grow, although at a declining rate, throughout life. In humans, there are 2 periods of rapid growth (Fig 22–10), the first in infancy and the second in late puberty just before growth stops. The first period of accelerated growth is partly a continuation of the fetal growth period. The second growth spurt at the time of puberty and the subsequent cessation of growth are due in large part to the action of androgens and estrogens. Since girls mature earlier than boys, this growth spurt appears earlier in girls. Of course in both sexes, the rate of growth of individual tissues varies (Fig 22–11).

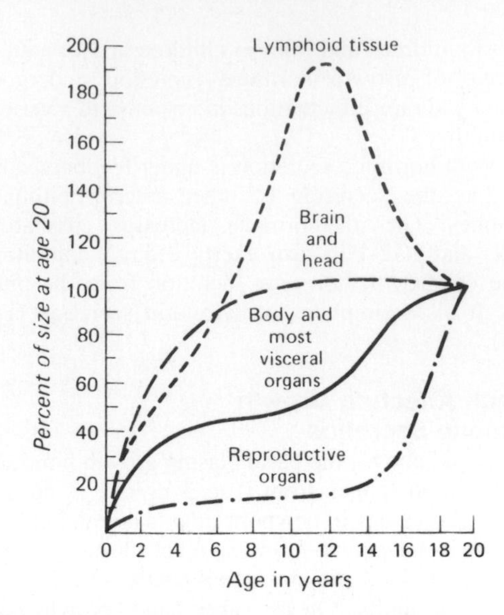

Figure 22–11. Growth of different tissues at various ages as percentage of size at age 20. The curves are composites that include both boys and girls.

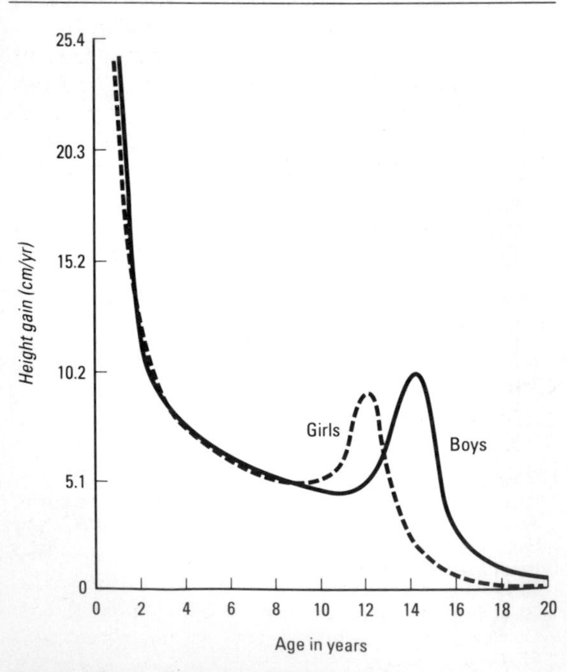

Figure 22–10. Rate of growth in boys and girls from birth to age 20.

Following illnesses in children, there is a period of "catch-up growth" during which the growth rate may be as much as 400% above normal. The acceleration continues until the previous growth curve is reached, at which point growth slows to normal. The mechanisms that bring about and control "catch-up growth" are not known.

Hormonal Effects

The contributions of hormones to growth after birth are shown diagrammatically in Fig 22–12. In laboratory animals and in humans, growth in utero is independent of fetal growth hormone. Rats hypophysectomized at birth continue to grow for about 30 days before they stop. Interestingly, the brain continues to grow in these rats after growth of the

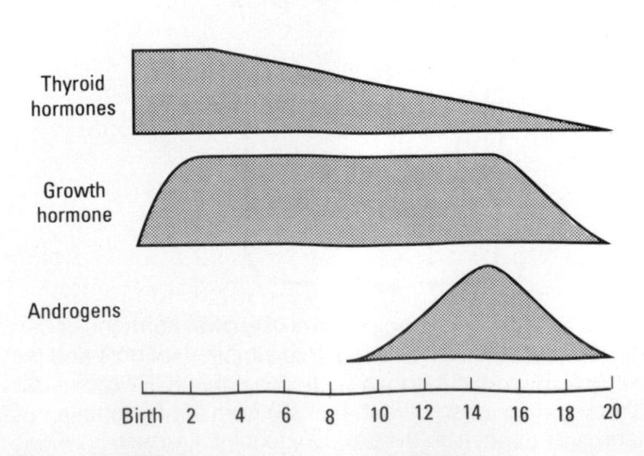

Figure 22–12. Relative importance of hormones in human growth at various ages. (Courtesy of DA Fisher.)

skull stops, and the resulting compression of the brain may lead to death.

As noted above, plasma growth hormone is elevated in newborns but subsequently falls to adult levels, even though the spikes of growth hormone secretion may be larger at the time of puberty. One of the factors stimulating IGF-I secretion is growth hormone, and plasma IGF-I levels rise during childhood, reaching a peak at 13–17 years of age. In contrast, IGF-II levels are constant throughout postnatal growth.

The growth spurt that occurs at the time of puberty (Fig 22–10) is due in part to the protein anabolic effect of androgens, and the secretion of adrenal androgens increases at this time in both sexes (Fig 23–10). However, it is also due to an interaction between sex steroids, growth hormone, and IGF-I. Treatment with estrogens and androgens increases the growth hormone responses to stimuli such as insulin and arginine. Sex steroids also increase plasma IGF-I but fail to produce this increase in individuals with growth hormone deficiency. Thus, it appears that the sex hormones produce enough of an increase in amplitude of the spikes in growth hormone secretion to increase IGF-I secretion, and this in turn causes growth.

Although androgens and estrogens initially stimulate growth, they ultimately terminate growth by causing the epiphyses to fuse to the long bones (epiphyseal closure). Once the epiphyses have closed, linear growth ceases (see Chapter 21). This is why pituitary dwarfs treated with testosterone first grow a few inches and then stop. It is also why patients with sexual precocity are apt to be dwarfed. On the other hand, men who were castrated before puberty tend to be tall because the epiphyses remain open and some growth continues past the normal age of puberty.

When growth hormone is administered to hypophysectomized animals, the animals do not grow as rapidly as they do when treated with growth hormone plus thyroid hormones. Thyroid hormones alone have no effect on growth in this situation. Their action is therefore permissive to that of growth hormone, possibly via potentiation of the actions of somatomedins. Thyroid hormones also appear to be necessary for a completely normal rate of growth hormone secretion; basal growth hormone levels are normal in hypothyroidism, but the response to hypoglycemia is frequently subnormal in hypothyroid children. Thyroid hormones have widespread effects on the ossification of cartilage, the growth of teeth, the contours of the face, and the proportions of the body. Cretins are therefore dwarfed and have infantile features (Fig 22–13). Patients who are dwarfed because of panhypopituitarism have features consistent with their chronologic age until puberty, but since they do not mature sexually, they have juvenile features in adulthood.

The effect of insulin on growth is discussed in Chapter 19. Diabetic animals fail to grow, and insulin causes growth in hypophysectomized animals. However, the growth is appreciable only when large amounts of carbohydrate and protein are supplied with the insulin.

Adrenocortical hormones other than androgens exert a permissive action on growth in the sense that adrenalectomized animals fail to grow unless their blood pressures and circulations are maintained by replacement therapy. On the other hand, glucocorticoids are potent inhibitors of growth because of their direct action on cells, and treatment of children with pharmacologic doses of steroids slows or stops growth for as long as the treatment is continued.

Dwarfism

Short stature can be due to GRH deficiency, growth hormone deficiency, deficient secretion of IGF-I, or other causes. Isolated growth hormone deficiency is often due to GRH deficiency, and in these instances, the growth hormone response to GRH is normal. However, some patients with isolated growth hormone deficiency have abnormalities of the growth hormone-secreting cells. In another group of dwarfed children, the plasma growth hormone concentration is normal or elevated but there is end-organ unresponsiveness to growth hormone and consequently a deficiency of circulating growth factors. This syndrome was described by Laron, and the condition is known as **Laron dwarfism.** African pygmies have normal plasma growth hormone levels and normal plasma concentrations of IGF-I and IGF-II before puberty. However, there is no IGF-I increase at the time of puberty and no pubertal growth spurt. Thus, it seems likely that in individuals who are not dwarfed, IGF-I is essential for the pubertal growth spurt in that sex hormones fail to have a marked effect on growth in the absence of IGF-I.

As noted above, short stature is characteristic of cretinism and occurs in patients with precocious puberty. It is also part of the syndrome of **gonadal dysgenesis** seen in patients who have an XO chromosomal pattern instead of an XX or XY pattern (see Chapter 23). Various bone and metabolic diseases also cause stunted growth, and in many cases there is no known cause (''constitutional delayed growth'').

PITUITARY INSUFFICIENCY

Changes in Other Endocrines

The widespread changes that develop when the pituitary is removed surgically or destroyed by disease in humans or animals are predictable in terms of the known hormonal functions of the gland. The adrenal cortex atrophies, and the secretion of adrenal glucocorticoids and sex hormones falls to low levels, although some secretion persists. Stress-induced rises in aldosterone secretion are absent, but basal aldosterone secretion and increases induced by salt depletion are normal, at least for some time. Since there is no mineralocorticoid deficiency, salt loss and

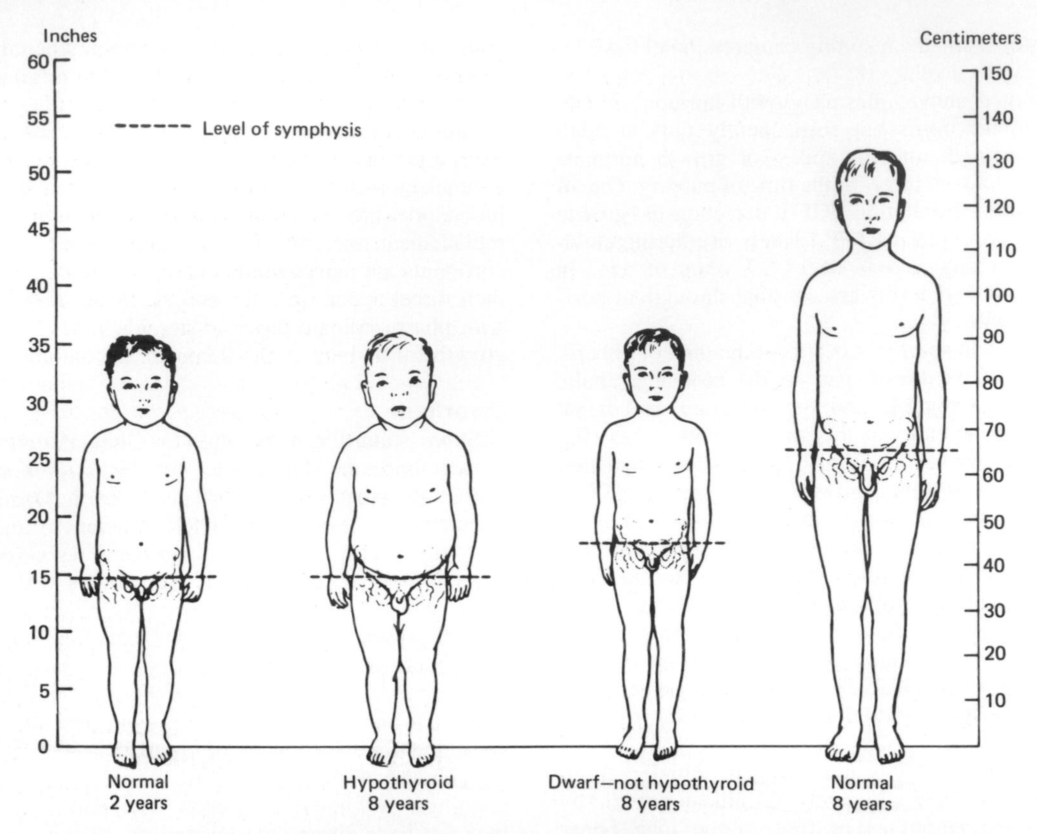

Figure 22–13. Normal and abnormal growth. Hypothyroid dwarfs retain their infantile proportions, whereas dwarfs of the constitutional type and, to a lesser extent, of the hypopituitary type have proportions characteristic of their chronologic age. (Reproduced, with permission, from Wilkins L: *The Diagnosis and Treatment of Endocrine Disorders in Childhood and Adolescence,* 3rd ed. Thomas, 1966.)

hypovolemic shock do not develop, but the inability to increase glucocorticoid secretion makes patients with pituitary insufficiency sensitive to stress. Growth is inhibited (see above). Thyroid function is depressed to low levels, and cold is tolerated poorly. The gonads atrophy, sexual cycles stop, and some of the secondary sex characteristics disappear.

Insulin Sensitivity

Hypophysectomized animals have a tendency to become hypoglycemic, especially when fasted. In some species, but not in humans, fatal hypoglycemic reactions are fairly common. Hypophysectomy ameliorates diabetes mellitus (see Chapter 19) and markedly increases the hypoglycemic effect of insulin. This is due in part to the deficiency of adrenocortical hormones, but hypophysectomized animals are more sensitive to insulin than adrenalectomized animals because they also lack the anti-insulin effect of growth hormone.

Water Metabolism

Although selective destruction of the supraoptic-posterior pituitary mechanism causes diabetes insipidus (see Chapter 14), removal of both the anterior and posterior pituitary usually causes no more than a transient polyuria. In the past, there was speculation that the anterior pituitary secreted a "diuretic hormone," but the amelioration of the diabetes insipidus is probably explained on the basis of a decrease in the osmotic load presented for excretion. Osmotically active particles hold water in the renal tubules (see Chapter 38). Because of the ACTH deficiency, the rate of protein catabolism is decreased in hypophysectomized animals. Because of the TSH deficiency, the metabolic rate is low. Consequently, fewer osmotically active products of catabolism are filtered and urine volume declines, even in the absence of vasopressin. Growth hormone deficiency contributes to the depression of the glomerular filtration rate in hypophysectomized animals, and growth hormone increases the glomerular filtration rate and renal plasma flow in humans. Finally, because of the glucocorticoid deficiency, there is the same defective excretion of a water load that is seen in adrenalectomized animals. The "diuretic" activity of the anterior pituitary can thus be explained in terms of the actions of ACTH, TSH, and growth hormone.

Other Defects

The deficiency of ACTH and other pituitary hormones with MSH activity may be responsible for the pallor of the skin in patients with hypopituitarism. There may be some loss of protein in adults, but it should be emphasized that wasting is not a feature of hypopituitarism in humans, and most patients with pituitary insufficiency are well nourished (Fig 22–14). It used to be thought that cachexia was part of the clinical picture, but it is now generally accepted that emaciated patients described in the older literature had anorexia nervosa rather than hypopituitarism.

Causes of Pituitary Insufficiency in Humans

Tumors of the anterior pituitary are generally classified on the basis of staining characteristics as chromophobe, acidophil, or basophil tumors. Some of the apparently nonfunctioning tumors are chromophobe tumors, and they produce hypopituitarism by destroying normal pituitary tissue. Up to 70% of patients with chromophobe tumors have high circulating levels of prolactin. This indicates that the tumors secrete prolactin, although pressure on the hypothalamus or pituitary stalk interrupting the transport of dopamine from the hypothalamus to the anterior pituitary could also contribute. Suprasellar cysts, remnants of Rathke's pouch that enlarge and compress the pituitary, are another cause of hypopituitarism. In women who have an episode of shock in the course of pregnancy, the pituitary may become infarcted, with the subsequent development of postpartum necrosis. The blood supply to the anterior lobe is vulnerable because it descends on the pituitary stalk through the rigid diaphragma sellae, and during pregnancy, the pituitary is enlarged. Pituitary infarction is extremely rare in men, but it was fairly common in soldiers who contracted hemorrhagic fever in Korea. In hemorrhagic fever, there was a diffuse vasculitis that probably caused pituitary enlargement due to edema, and the patients who developed pituitary infarction were those who went into shock in the course of their disease.

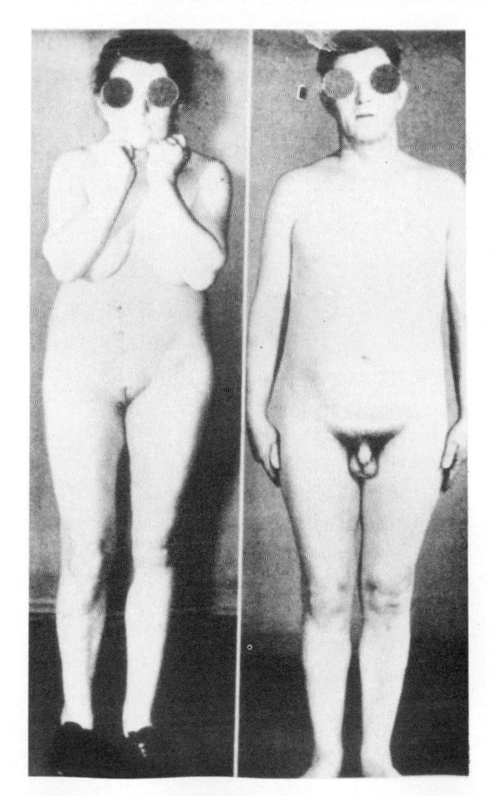

Figure 22–14. Typical picture of hypopituitarism in adults. Note well-nourished appearance and pallor. (Reproduced, with permission, from Daughaday WH: The adenohypophysis. In: *Textbook of Endocrinology,* 5th ed. Williams RH [editor]. Saunders, 1974.)

Partial Pituitary Insufficiency

The anterior pituitary has a large reserve, and much of it can be destroyed without producing detectable endocrine abnormalities. It now appears that with progressive loss of pituitary tissue, growth hormone secretion is the first function to be impaired. In dogs and rats, there is a depression of gonadotropin secretion when 70–90% of the anterior pituitary is removed or destroyed. When 90–95% is destroyed, thyroid function is also impaired, but almost 100% of the anterior pituitary must be destroyed before a marked degree of adrenal insufficiency is seen (Table 22–5). The same relationships hold in humans. Isolated pituitary tropin deficiencies occur, but they are often due to hypothalamic rather than pituitary disorders. For example, an appreciable number of patients with isolated TSH deficiency show an increase in TSH secretion when TRH is injected. Similarly, some patients with gonadotropin deficiency respond to LHRH.

In rats and in some other species, there appears to be some localization within the pituitary of the cells secreting the various hormones. It is not known if similar localization occurs in humans.

Table 22–5. Effects of removing various amounts of pituitary tissue on endocrine function in male dogs.* (+) indicates that the finding was present and (0) that it was absent.

Number of Dogs	Gonadal Atrophy	Thyroid Atrophy	Adrenal Atrophy	Mean % Anterior Pituitary Remaining
6	0	0	0	27
3	+	0	0	11
2	+	+	0	5
19	+	+	+	2

*Data from Ganong WF, Hume DM: The effect of graded hypophysectomy on thyroid, gonadal, and adrenocortical function in the dog. *Endocrinology* 1956;**59**:293.

PITUITARY HYPERFUNCTION IN HUMANS

Acromegaly

Tumors of the somatotropes of the anterior pituitary secrete large amounts of growth hormone, leading in children to **gigantism** and in adults to **acromegaly.** The principal findings are those related to the local effects of the tumor (enlargement of the sella turcica, headache, visual disturbances) and those due to growth hormone secretion. In adults, there is enlargement of the hands and feet (**acral** parts; hence the term acromegaly) and a protrusion of the lower jaw called **prognathism** (Fig 22–7). Overgrowth of the malar, frontal, and basal bones combines with prognathism to produce the coarse facial features called **acromegalic facies.** Body hair is increased in amount. The skeletal changes predispose to osteoarthritis. About 25% of patients have abnormal glucose tolerance tests, and 4% develop lactation in the absence of pregnancy.

Cushing's Syndrome

The clinical picture of Cushing's syndrome and its various causes are described in Chapter 20. Many patients with bilaterally hyperplastic adrenals have small ACTH-secreting pituitary tumors (microadenomas) that are difficult to detect. However, a significant percentage of the patients who have bilaterally hyperplastic adrenals removed develop rapidly growing ACTH-secreting pituitary tumors (**Nelson's syndrome).** These tumors cause hyperpigmentation of the skin and neurologic signs due to pressure on structures in the sellar region. Most of them are made up of chromophobic cells rather than basophils, and some are malignant. Blood ACTH levels are extremely high, and the intrinsic MSH activity of the ACTH probably accounts for the cutaneous pigmentation. It is difficult to say whether these patients had undetected tumors to start with or developed neoplastic changes in the pituitary when the feedback check on ACTH secretion was removed.

Other Hormone-Secreting Tumors

Animals sometimes develop TSH-secreting tumors after thyroidectomy and gonadotropin-secreting tumors after gonadectomy. However, TSH-secreting tumors are rare in humans, and gonadotropin-secreting tumors have not been reported. On the other hand, prolactin-secreting tumors are common (see Chapter 23).

The Gonads: Development & Function of the Reproductive System

23

INTRODUCTION

Modern genetics and experimental embryology make it clear that, in most species of mammals, the multiple differences between the male and the female depend primarily on a single chromosome (the Y chromosome) and a single pair of endocrine structures, the testes in the male and the ovaries in the female. The differentiation of the primitive gonads into testes or ovaries in utero is genetically determined in humans, but the formation of male genitalia depends upon the presence of a functional, secreting testis. There is evidence that male sexual behavior and, in some species, the male pattern of gonadotropin secretion are due to the action of male hormones on the brain in early development. After birth, the gonads remain quiescent until adolescence, when they are activated by gonadotropins from the anterior pituitary. Hormones secreted by the gonads at this time cause the appearance of features typical of the adult male or female and the onset of the sexual cycle in the female. In males, the gonads remain more or less active from puberty onward. In human females, ovarian function regresses after a period of time, and sexual cycles cease (the menopause).

In both sexes, the gonads have a dual function: the production of germ cells **(gametogenesis)** and the secretion of **sex hormones.** The **androgens** are the steroid sex hormones that are masculinizing in their action; the **estrogens** are those that are feminizing. Both types of hormones are normally secreted in both sexes. The testes secrete large amounts of androgens, principally **testosterone,** but they also secrete small amounts of estrogens. The ovaries secrete large amounts of estrogens and small amounts of androgens. Androgens and, probably, small amounts of estrogens are secreted from the adrenal cortex in both sexes. The ovaries also secrete **progesterone,** a steroid that has special functions in preparing the uterus for pregnancy. During pregnancy, the ovaries secrete the peptide hormone **relaxin,** which loosens the ligaments of the pubic symphysis and softens the cervix, facilitating delivery of the fetus. In both sexes, the gonads secrete other polypeptides, including **inhibin,** a polypeptide that inhibits FSH secretion.

The secretory and gametogenic functions of the gonads are both dependent upon the secretion of the anterior pituitary gonadotropins, FSH and LH. The sex hormones and inhibin feed back to inhibit gonadotropin secretion. In males, gonadotropin secretion is noncyclic; but in postpuberal females an orderly, sequential secretion of gonadotropins is necessary for the occurrence of menstruation, pregnancy, and lactation.

SEX DIFFERENTIATION & DEVELOPMENT

CHROMOSOMAL SEX

The Sex Chromosomes

Sex is determined genetically by 2 chromosomes, called the **sex chromosomes** to distinguish them from the **somatic chromosomes (autosomes).** In humans and many other mammals, the sex chromosomes are called X and Y chromosomes. The Y chromosome is necessary and sufficient for the production of testes, and the testis-determining gene product is a testis-determining factor (TDF) encoded by a gene on the short arm of the Y chromosome. Male cells with the diploid number of chromosomes contain an X and a Y chromosome (XY pattern), whereas female cells contain two X chromosomes (XX pattern). As a consequence of meiosis during gametogenesis, each normal ovum contains a single X chromosome, but half the normal sperms contain an X chromosome and half a Y chromosome (Fig 23–1). When a sperm containing a Y chromosome fertilizes an ovum, an XY pattern results, and the zygote develops into a **genetic male.** When fertilization occurs with an X-containing sperm, an XX pattern and a **genetic female** result. Cell division and the chemical nature of chromosomes are discussed in Chapter 1.

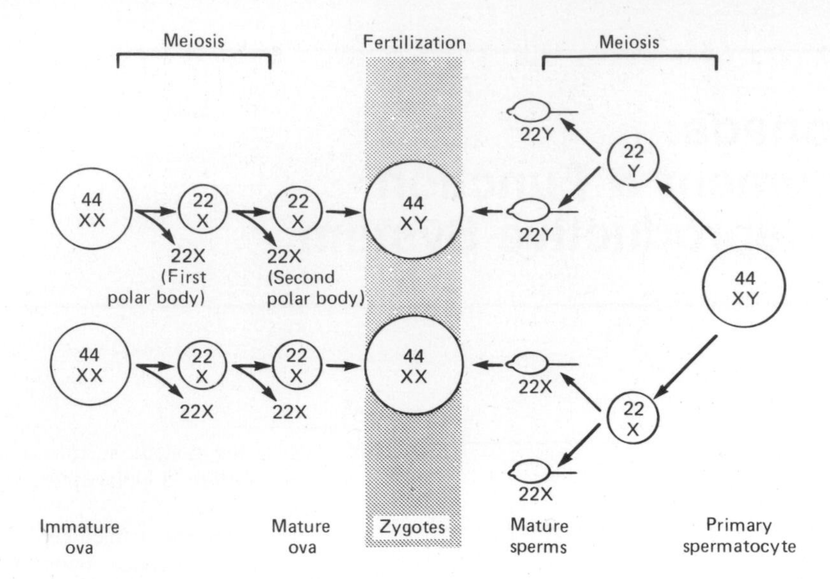

Figure 23–1. Basis of genetic sex determination. In the case of the 2-stage meiotic division in the female, only one body survives as the mature ovum. In the case of the male, the meiotic division results in the formation of 4 sperms, 2 containing the X and 2 the Y chromosome. Fertilization thus produces a 44XY (male) zygote or a 44XX (female) zygote.

Human Chromosomes

Human chromosomes can be studied in detail. Human cells are grown in tissue culture; treated with the drug colchicine, which arrests mitosis at the metaphase; exposed to a hypotonic solution that makes the chromosomes swell and disperse; and then "squashed" onto slides. Fluorescent and other staining techniques make it possible to identify the individual chromosomes and study them in detail (Fig 23–2). There are 46 chromosomes: in males, 22 pairs of autosomes plus a large X chromosome and a small Y chromosome; in females, 22 pairs of autosomes plus two X chromosomes. The individual chromosomes are usually arranged in an arbitrary pattern **(karyotype).** The individual autosome pairs are identified by the numbers 1–22 on the basis of their morphologic characteristics. Because the human Y chromosome is smaller than the X chromosome, it has been hypothesized that sperms containing the Y chromosome are lighter and able to "swim" faster up the female genital tract, thus reaching the ovum more rapidly. This supposedly accounts for the fact that the number of males born is slightly greater than the number of females.

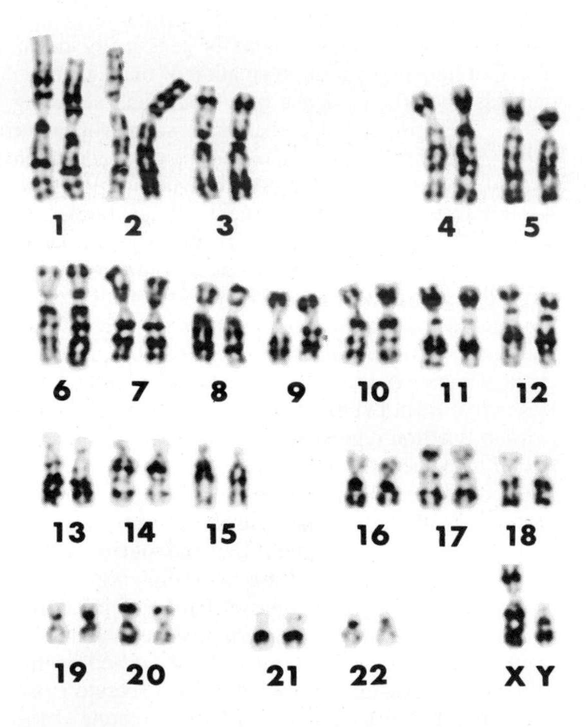

Figure 23–2. Karyotype of chromosomes from a normal male. The chromosomes have been stained with Giemsa's stain, which produces a characteristic banding pattern. This banding, and the patterns produced by other techniques, make it possible to identify individual chromosomes with great accuracy. (Reproduced, with permission, from Conte FA, Grumbach MM: Abnormalities of sexual differentiation. In: *Basic & Clinical Endocrinology,* 3rd ed. FS Greenspan, PH Forsham [editors]. Appleton & Lange, 1989.)

Sex Chromatin

Soon after cell division has started during embryonic development, one or the other of the two X chromosomes of the somatic cells in normal females becomes functionally inactive. The choice of which X chromosome becomes inactive in any given cell is apparently random, so one X chromosome is inactivated in approximately half of the cells and the other X chromosome in the other half. The selection persists through subsequent divisions of these cells,

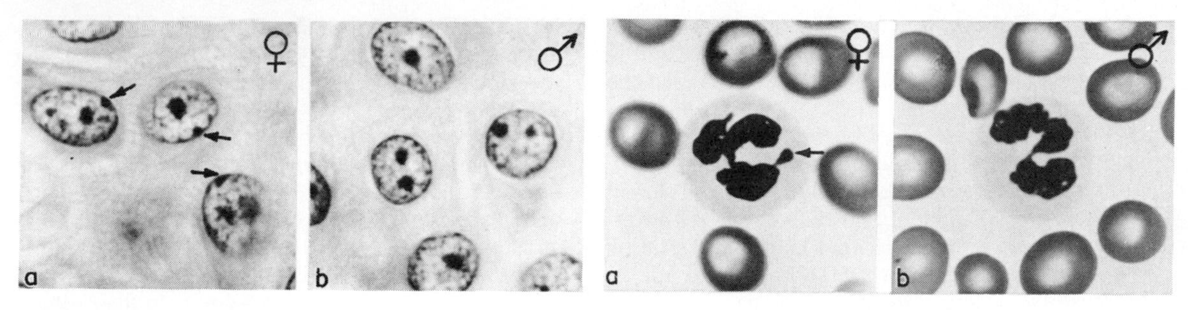

Figure 23–3. Barr body (arrows) in epidermal spinous cell layer (*left*) and nuclear appendage ("drumstick") in white blood cells (*right*). (Reproduced, with permission, from Grumbach MM, Barr ML: Cytologic tests of chromosomal sex in relation to sex anomalies in man. *Recent Prog Horm Res* 1958;**14**:255.)

and consequently some of the somatic cells in adult females contain an active X chromosome of paternal origin and some contain an active X chromosome of maternal origin.

The inactive X chromosome condenses and can be seen in various types of cells, usually near the nuclear membrane, as the **Barr body** (Fig 23–3). Thus, there is a Barr body for each X chromosome in excess of one in the cell. The inactive X chromosome is also visible as a small "drumstick" of chromatin projecting from the nuclei of 1–15% of the polymorphonuclear leukocytes in females but not in males (Fig 23–3).

The Barr body has also been called sex chromatin. Another form of sex chromatin is the **F body,** a portion of the Y chromosome that can be stained with modern fluorescence techniques. Thus, the number of Y chromosomes in a cell can be estimated by the number of F bodies it contains.

EMBRYOLOGY OF THE HUMAN REPRODUCTIVE SYSTEM

Development of the Gonads

On each side of the embryo, a primitive gonad arises from the genital ridge, a condensation of tissue near the adrenal gland. The gonad develops a **cortex** and a **medulla** (Fig 23–4). Until the sixth week of

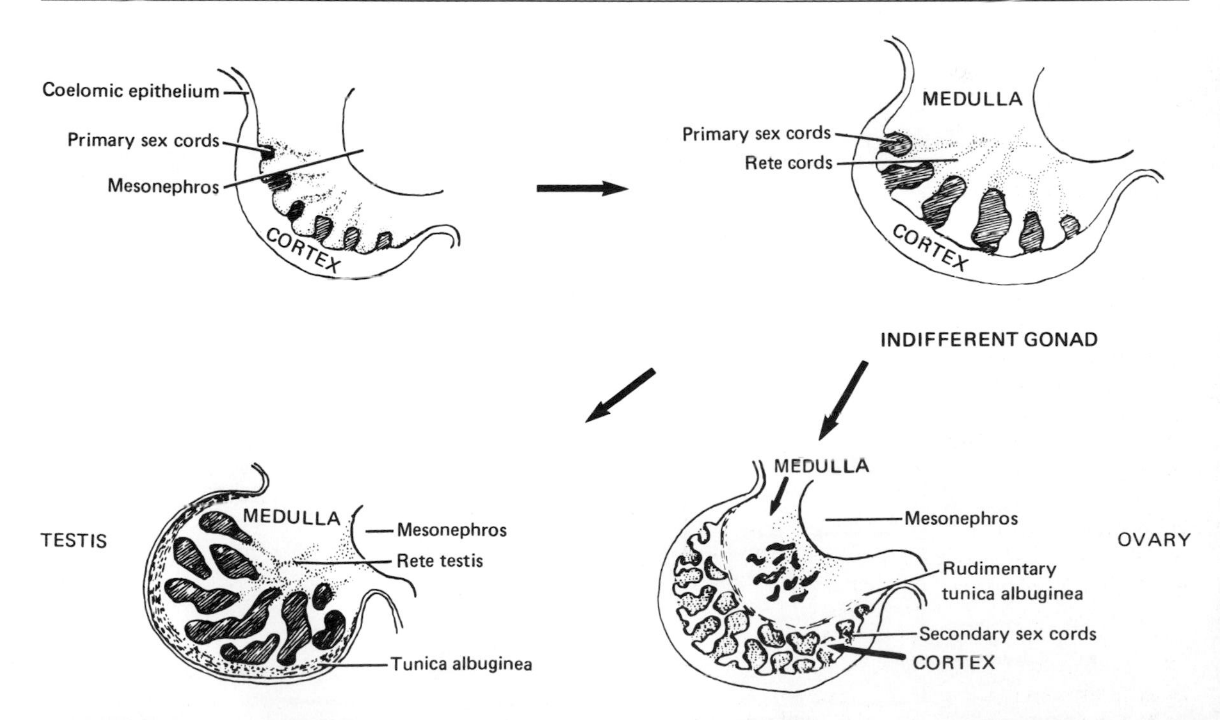

Figure 23–4. Diagrammatic representation of the development of an ovary from the cortex or a testis from the medulla of a bipotential primordial human gonad. (Reproduced, with permission, from Grumbach M in: *Clinical Endocrinology I.* Astwood EB [editor]. Grune & Stratton, 1960.)

development, these structures are identical in both sexes. In genetic males, the medulla develops during the seventh and eighth weeks into a testis, and the cortex progresses (Fig 23–5). This development depends on the presence of testis-determining factor (TDF; see above). The mechanism of action of TDF is unknown. It used to be thought that a different protein, the H-Y antigen, was responsible for development of testes, but it is now known that TDF is produced by a gene on the short arm of the Y chromosome, whereas the H-Y antigen is produced by a gene on the long arm of the Y chromosome. Leydig cells appear next, and testosterone and müllerian inhibiting substance (see below) are secreted. In genetic females, the cortex develops into an ovary, and the medulla regresses. The embryonic ovary does not secrete hormones. Hormonal treatment of the mother has no effect on gonadal (as opposed to ductal and genital) differentiation in humans, although it does in some experimental animals.

Embryology of the Genitalia

In the seventh week of gestation, the embryo has both male and female primordial genital ducts (Fig 23–6). In a normal female fetus, the müllerian duct system then develops into uterine tubes (oviducts) and a uterus. In the normal male fetus, the wolffian duct system on each side develops into the epididymis and vas deferens. The external genitalia are similarly bipotential until the eighth week (Fig 23–7). Thereafter, the urogenital slit disappears and male genitalia form, or, alternatively, it remains open and female genitalia form.

When there are functional testes in the embryo, male internal and external genitalia develop. The Leydig cells of the fetal testis secrete testosterone, and the Sertoli cells secrete **müllerian inhibiting substance** (MIS; also called müllerian regression factor, or MRF). MIS is a 535-amino-acid protein that is related to inhibin (see below). In their effects on the internal as opposed to the external genitalia, MIS and testosterone act unilaterally. MIS causes regression of the müllerian ducts on the side on which it is secreted, and testosterone fosters the development of the vas deferens and related structures from the wolffian ducts. Testosterone induces the formation of male external genitalia (Fig 23–5). MIS is also produced by the granulosa cells of the ovarian follicles, where its function is unknown. It now appears that MIS requires a high androgen:estrogen ratio to exert its effects on the müllerian ducts, and this is present in the male but not in the female fetus, so the ducts persist in the female despite the MIS in the ovaries.

Development of the Brain

At least in some species, the development of the brain as well as the external genitalia is affected by androgens early in life. In rats, a brief exposure to

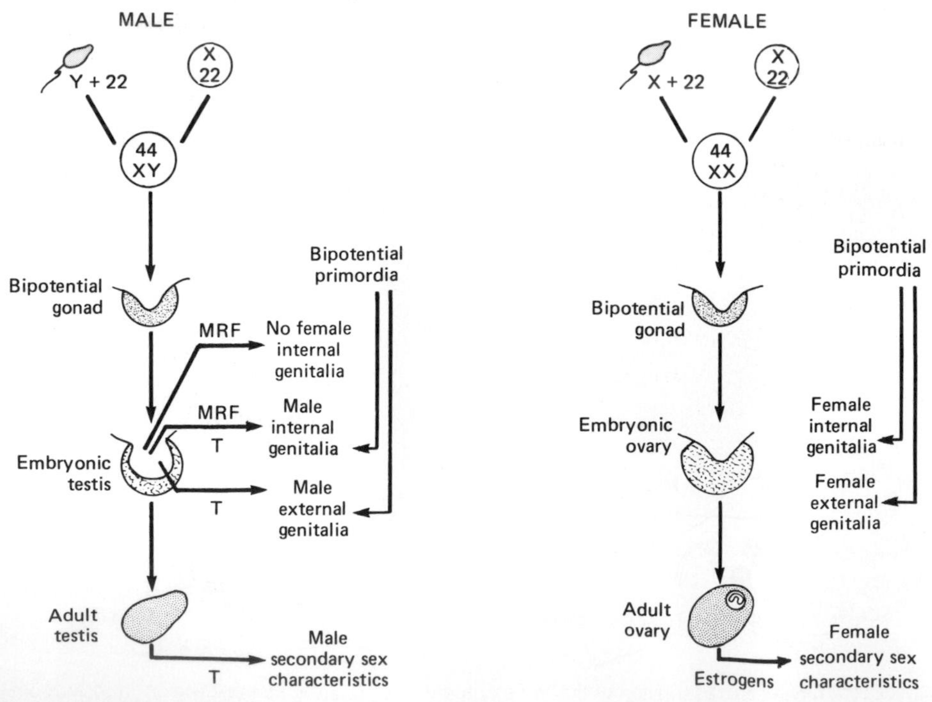

Figure 23–5. Diagrammatic summary of normal sex determination, differentiation, and development in humans. MIS, müllerian inhibiting substance; T, testosterone or other androgen.

androgens during the first few days of life causes the male pattern of sexual behavior and the male pattern of hypothalamic control of gonadotropin secretion to develop after puberty. In the absence of androgens, female patterns develop (see Chapter 15). In monkeys, similar effects on sexual behavior are produced by exposure to androgens in utero, but the pattern of gonadotropin secretion remains cyclic. Early exposure of female human fetuses to androgens also appears to cause subtle but significant masculinizing

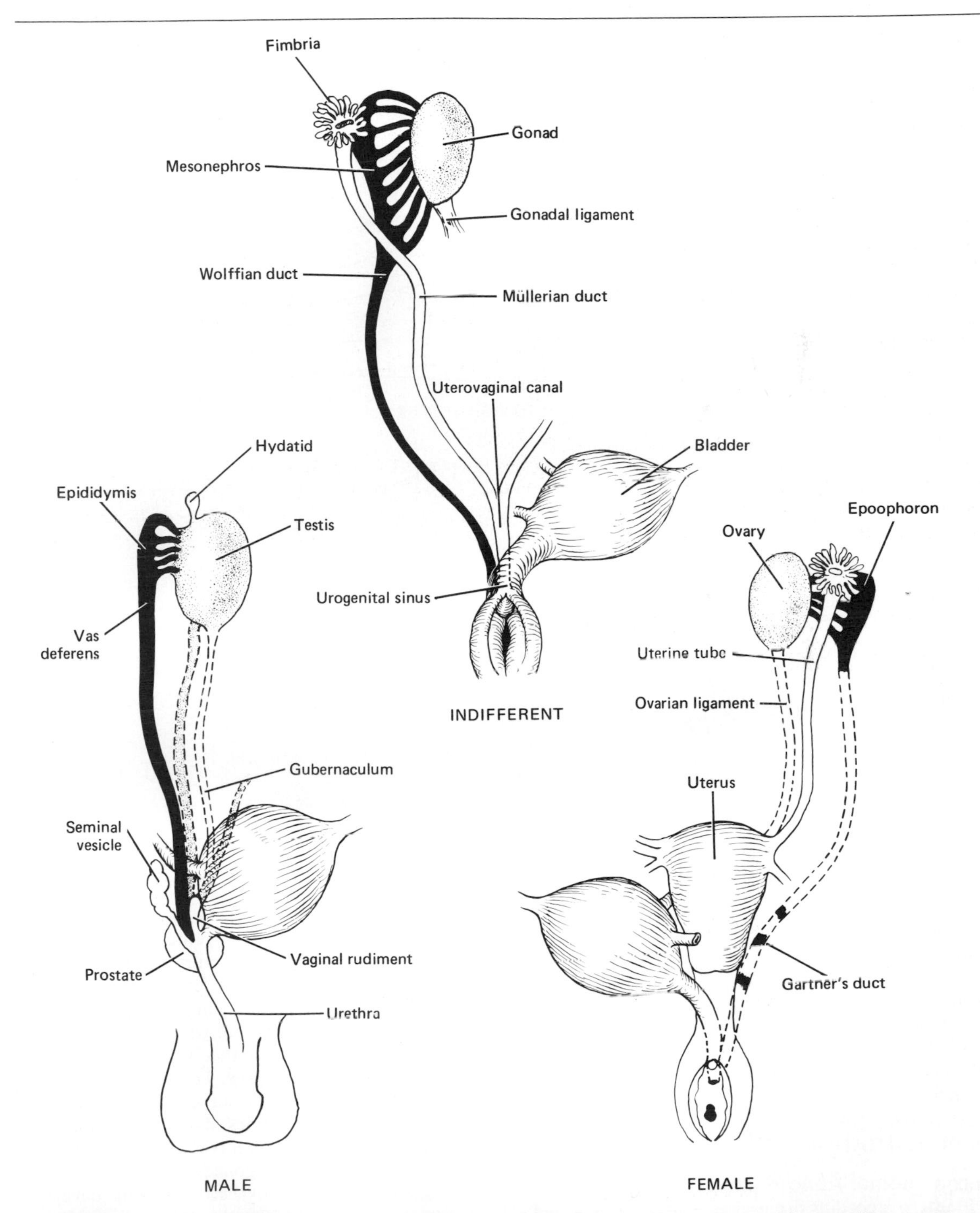

Figure 23–6. Embryonic differentiation of male and female internal genitalia (genital ducts) from wolffian (male) and müllerian (female) primordia. (After Corning HK, Wilkins L. Redrawn and reproduced, with permission, from Van Wyk J, Grumbach M in: *Textbook of Endocrinology*. 5th ed. Williams RH [editor]. Saunders, 1974.)

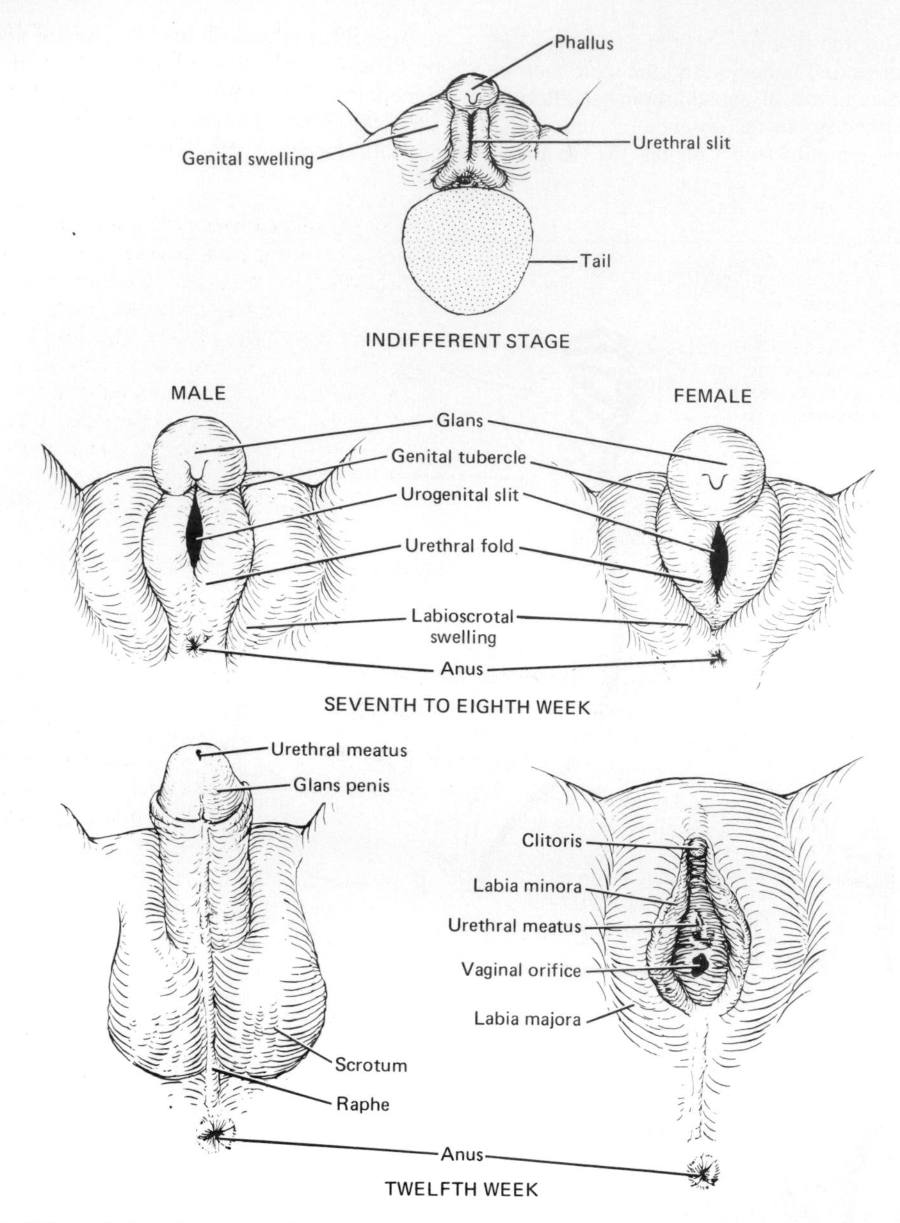

Figure 23–7. Differentiation of male and female external genitalia from indifferent primordial structures in the embryo.

effects on behavior. However, women with adrenogenital syndrome due to congenital adrenocortical enzyme deficiency (see Chapter 20) develop normal menstrual cycles when treated with cortisol. Thus, the human, like the monkey, appears to retain the cyclic pattern of gonadotropin secretion despite the exposure to androgens in utero.

ABERRANT SEXUAL DIFFERENTIATION

Chromosomal Abnormalities

From the preceding discussion, it might be expected that abnormalities of sexual development could be caused by genetic or hormonal abnormalities as well as by other nonspecific teratogenic influences, and this is indeed the case. The major classes of abnormalities are listed in Table 23–1.

An established defect in gametogenesis is **nondisjunction,** a phenomenon in which a pair of chromosomes fail to separate, so that both go to one of the daughter cells during meiosis. Four of the abnormal zygotes that can form as a result of nondisjunction of one of the X chromosomes during oogenesis are shown in Fig 23–8. In individuals with the XO chromosomal pattern, the gonads are rudimentary or absent, so that female external genitalia develop. Stature is short, other congenital abnormalities are often present, and no maturation occurs at puberty. This syndrome is now called **gonadal dysgenesis** or,

Table 23–1. Classification of the major disorders of sex differentiation in humans. Many of these syndromes can have great variation in degree and, consequently, in manifestations.

Chromosomal disorders
Gonadal dysgenesis (XO and variants)
"Superfemales" (XXX)
Seminiferous tubule dysgenesis (XXY and variants)
True hermaphroditism
Developmental disorders
Female pseudohermaphroditism
Congenital virilizing adrenal hyperplasia of fetus
Maternal androgen excess
Virilizing ovarian tumor
Iatrogenic: Treatment with androgens or certain synthetic progestational drugs
Male pseudohermaphroditism
Androgen resistance
Defective testicular development
Congenital 17α-hydroxylase deficiency
Congenital adrenal hyperplasia due to blockade of pregnenolone formation
Various nonhormonal anomalies

alternatively, **ovarian agenesis** or **Turner's syndrome.** Individuals with the XXY pattern, the most common sex chromosome disorder, have the genitalia of a normal male. Testosterone secretion at puberty is often great enough for the development of male characteristics. However, the seminiferous tubules are abnormal, and there is a higher than normal incidence of mental retardation. This syndrome is known as **seminiferous tubule dysgenesis** or **Klinefelter's syndrome.** The XXX ("superfemale") pattern is second in frequency only to the XXY pattern and may be even more common in the general population, since it does not seem to be associated with any characteristic abnormalities. The YO combination is probably lethal.

The XYY pattern has also been observed in males who tend to be tall and have severe acne. This karyotype has attracted considerable attention because its incidence was initially reported to be high in prison populations. However, it is also relatively common in the general population, and there is little solid evidence that individuals with the XYY pattern are predisposed to aggressive behavior.

Meiosis is a 2-stage process, and although nondisjunction usually occurs during the first meiotic division, it can occur in the second, producing more complex chromosomal abnormalities. In addition, nondisjunction or simple loss of a sex chromosome can occur during the early mitotic divisions after fertilization. The result of faulty mitoses in the early zygote is the production of a mosaic, an individual with 2 or more populations of cells with different chromosome complements. **True hermaphroditism,** the condition in which the individual has both ovaries and testes, is probably due to XX/XY mosaicism and related mosaic patterns, although other genetic aberrations are possible.

Chromosomal abnormalities also include transposition of parts of chromosomes to other chromosomes. Rarely, genetic males are found to have the XX karyotype because the short arm of their father's Y chromosome was transposed to their father's X chromosome during meiosis and they received that X chromosome along with their mother's. Similarly, deletion of the small portion of the Y chromosome coding for TDF produces females with the XY karyotype.

Sex chromosome abnormalities are of course not the only abnormalities associated with disease states; nondisjunction of several different autosomal chromosomes is known to occur. For example, nondisjunction of chromosome 21 (Fig 23–2) produces **trisomy 21,** the chromosomal abnormality associated with Down's syndrome (mongolism).

There are many other chromosomal abnormalities as well as numerous diseases due to defects in single genes. Over 2300 different inherited disorders in humans have been catalogued in the literature.

Hormonal Abnormalities

Development of the male external genitalia occurs normally in genetic males in response to androgen secreted by the embryonic testes, but male genital development may also occur in genetic females exposed to androgens from some other source during the eighth to the thirteenth weeks of gestation. The syndrome that results in **female pseudohermaphroditism.** A pseudohermaphrodite is an individual with the genetic constitution and gonads of one sex

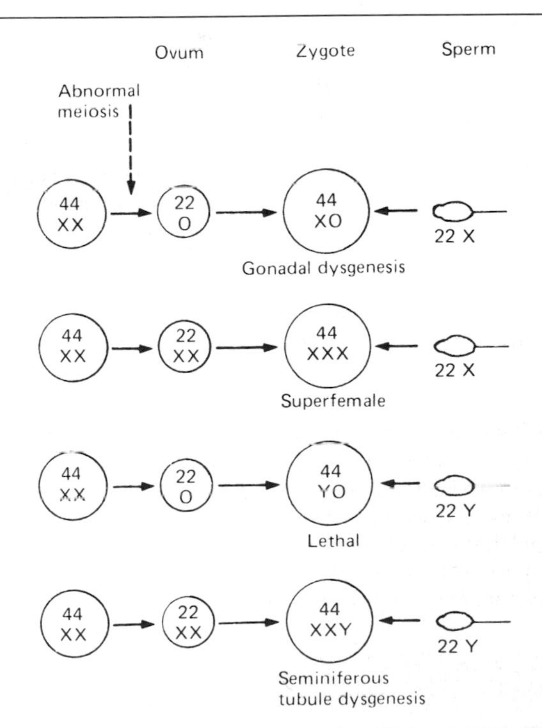

Figure 23–8. Summary of 4 possible defects produced by maternal nondisjunction of the sex chromosomes at the time of meiosis. The YO combination is believed to be lethal, and the fetus dies in utero.

and the genitalia of the other. After the thirteenth week, the genitalia are fully formed, but exposure to androgens can cause hypertrophy of the clitoris. Female pseudohermaphroditism may be due to congenital virilizing adrenal hyperplasia (see Chapter 20), or it may be caused by androgens administered to the mother. Conversely, female external genitalia development in genetic males **(male pseudohermaphroditism)** occurs when the embryonic testes are defective. Because the testes also secrete MIS, genetic males with defective testes have female internal genitalia.

Another cause of male pseudohermaphroditism is **androgen resistance,** in which, as a result of various congenital abnormalities, male hormones cannot exert their full effects on the tissues. One form of androgen resistance is a **5α-reductase deficiency,** in which the enzyme responsible for the formation of dihydrotestosterone, the active form of testosterone, is decreased. The consequences of this deficiency are discussed in the section on the male reproductive system. In other forms of androgen resistance, androgen receptors are absent or abnormal. When the loss of receptor function is complete, the **testicular feminizing syndrome** results. In this condition, MIS is present and testosterone is secreted at normal or even elevated rates. The external genitalia are female, but the vagina ends blindly because there are no female internal genitalia. Individuals with this syndrome develop enlarged breasts at puberty and usually are considered to be normal women until they are diagnosed when they seek medical advice because of lack of menstruation. Androgen resistance can also occur in the presence of normal androgen receptors when there are congenital defects in the molecular events that occur after the receptor binding **(receptor positive androgen resistance).**

It is worth noting that genetic males with congenital adrenal hyperplasia due to blockade of the formation of pregnenolone are pseudohermaphrodites because testicular as well as adrenal androgens are normally formed from pregnenolone. Male pseudohermaphroditism also occurs when there is a congenital deficiency of 17α-hydroxylase (see Chapter 20).

PUBERTY

After birth, the androgen-secreting Leydig cells in the fetal testes become quiescent. There follows in all mammals a period in which the gonads of both sexes are quiescent until they are activated by gonadotropins from the pituitary to bring about the final maturation of the reproductive system. This period of final maturation is known as **adolescence.** It is often also called **puberty,** although puberty, strictly defined, is the period when the endocrine and gametogenic functions of the gonads have first developed to the point where reproduction is possible. In girls, the first event is **thelarche,** the development of breasts, followed by **pubarche,** the development of axillary and pubic hair, and then by **menarche,** the first menstrual period. The initial periods are generally anovalatory, and regular ovulation appears about a year later. In contrast to the situation in adulthood, removal of the gonads during the period from birth to puberty causes little or no increase in gonadotropin secretion, so gonadotropin secretion is not being held in check by gonadal hormones. In children between the ages of 7 and 10, a slow increase in estrogen and androgen secretion precedes the more rapid rise in the early teens (Fig 23–9). The age at the time of puberty is variable. In Europe and the USA, it has been declining at the rate of 1–3 months per decade for more than 175 years. In the USA in recent years, puberty generally occurs between the ages of 8 and 13 in girls and 9 and 14 in boys.

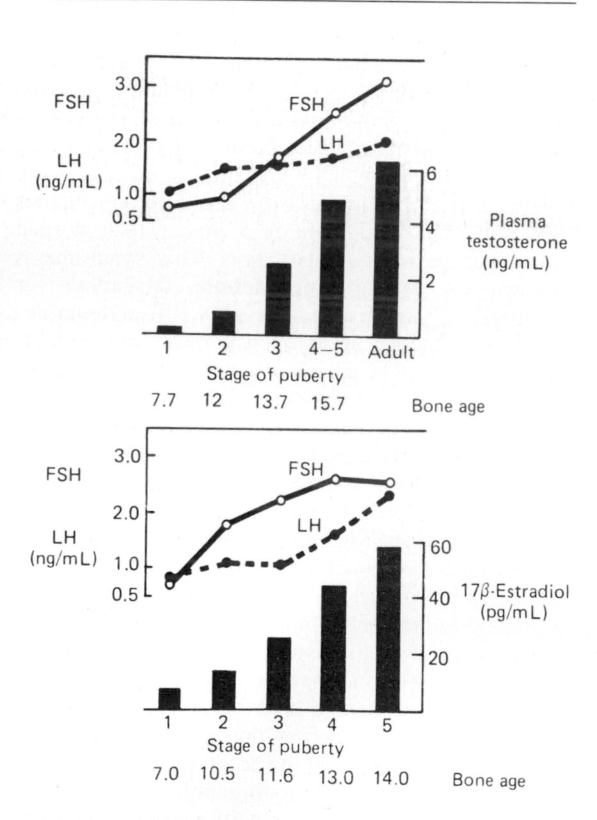

Figure 23–9. Changes in plasma hormone concentrations during puberty in boys (*top*) and girls (*bottom*). Stage 1 of puberty is preadolescence in both sexes. In boys, stage 2 is characterized by beginning enlargement of the testes, stage 3 by penile enlargement, stage 4 by growth of the glans penis, and stage 5 by adult genitalia. In girls, stage 2 is characterized by breast buds, stage 3 by elevation and enlargement of the breasts, stage 4 by projection of the areolas, and stage 5 by adult breasts. (Reproduced, with permission, from Grumbach MM: Onset of puberty. In: *Puberty: Biologic and Psychosocial Components*. Berenberg SR [editor]. HE Stenfoert Kroese BV, 1975.)

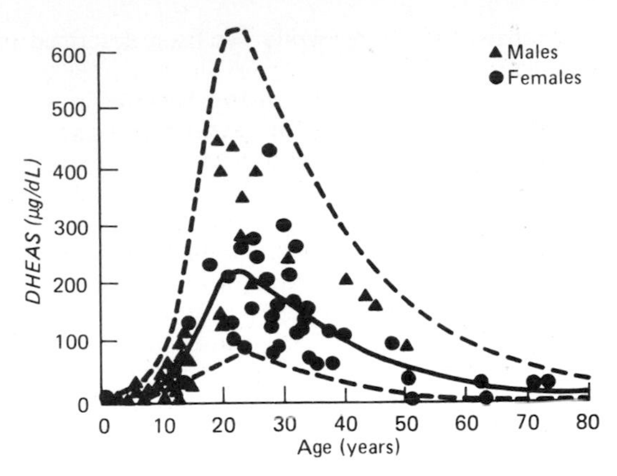

Figure 23–10. Change in serum dehydroepiandrosterone sulfate (DHEAS) with age. There are no significant differences between males and females. The middle line is the mean, and the dashed lines identify ±1.96 standard deviations. (Reproduced, with permission, from Smith MR et al: A radioimmunoassay for the estimation of serum dehydroepiandrosterone sulfate in normal and pathological sera. *Clin Chim Acta* 1975;**65**:5.)

Another event that occurs in humans at the time of puberty is an increase in the secretion of adrenal androgens (Fig 23–10). The onset of this increase is called **adrenarche.** It occurs at age 8–10 years, without any change in the secretion of cortisol or ACTH. It may be due to a change in the enzyme systems in the adrenal so that more pregnenolone is diverted to the androgen pathway (see Chapter 20). On the other hand, there is some evidence that it is due to increased secretion of an as yet unisolated **adrenal androgen-stimulating hormone (AASH)** from the pituitary gland.

Control of the Onset of Puberty

The gonads of children can be stimulated by gonadotropins; their pituitaries contain gonadotropins; and their hypothalami contain LHRH (see Chapter 14). However, their gonadotropins are not secreted. In immature monkeys, normal menstrual cycles can be brought on by pulsatile injection of LHRH, and they persist as long as the pulsatile injection is continued. In addition, LHRH is secreted in a pulsatile fashion in utero. Thus, it seems clear that during the period from birth to puberty, a neural mechanism is operating to prevent the normal pulsatile release of LHRH. It is interesting in this regard that in experimental animals and in humans, lesions in the ventral hypothalamus near the infundibulum cause precocious puberty. However, it is possible that the effect of lesions is due instead to chronic stimulation originating in irritative foci in tissue around the lesion, and at present, the nature of the mechanism inhibiting the LHRH pulse generator is unknown.

PRECOCIOUS & DELAYED PUBERTY

Sexual Precocity

The major causes of precocious sexual development in humans are listed in Table 23–2. Early development of secondary sexual characteristics without gametogenesis is caused by abnormal exposure of immature males to androgen or females to estrogen. This syndrome should be called **precocious pseudopuberty** to distinguish it from **true precocious puberty** due to an early but otherwise normal pubertal pattern of gonadotropin secretion from the pituitary (Fig 23–11). In one large series of cases, precocious puberty was the most frequent endocrine symptom of hypothalamic disease (see Chapter 14). Pineal tumors are sometimes associated with precocious puberty, but there is evidence that these tumors are associated with precocious puberty only when there is secondary damage to the hypothalamus. Precocity due to this and other forms of hypothalamic damage probably occurs with equal frequency in both sexes, although the constitutional form of precocious puberty is more common in girls. In addition, it has now been proved that precocious gametogenesis and steroidogenesis can occur without the pubertal pattern of gonadotropin secretion (gonadotropin-independent precocity).

Delayed or Absent Puberty

The normal variation in the age at which adolescent changes occur is so wide that puberty cannot be considered to be pathologically delayed until the menarche has failed to occur by the age of 17 or testicular development by the age of 20. Failure of maturation due to panhypopituitarism is associated with dwarfing and evidence of other endocrine abnormalities. Patients with the XO chromosomal pattern and gonadal dysgenesis are also dwarfed. In some individuals, puberty is delayed even though the gonads are present and other endocrine functions are

Table 23–2. Classification of the causes of precocious sexual development in humans.

True precocious puberty
Constitutional
Cerebral: Disorders involving posterior hypothalamus
Tumors
Infections
Developmental abnormalities
Gonadotropin-independent precocity
Precocious pseudopuberty (no spermatogenesis or ovarian development)
Adrenal
Congenital virilizing adrenal hyperplasia (without treatment in males; following cortisone treatment in females)
Androgen-secreting tumors (in males)
Estrogen-secreting tumors (in females)
Gonadal
Interstitial cell tumors of testis
Granulosa cell tumors of ovary
Miscellaneous

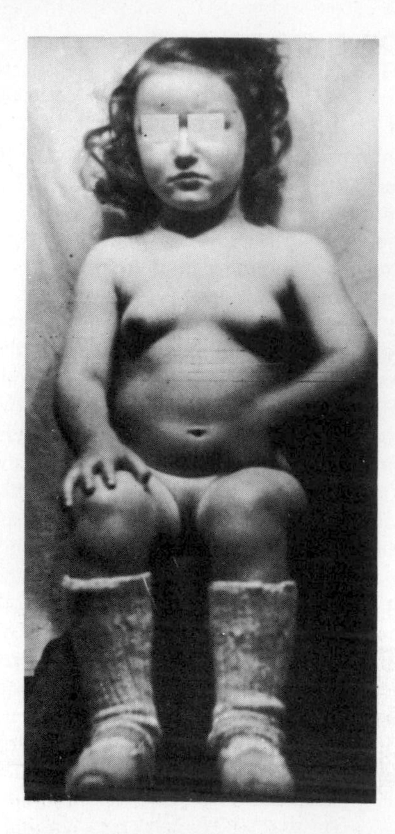

Figure 23–11. Constitutional precocious puberty in a 3½-year-old girl. The patient developed pubic hair and started to menstruate at the age of 17 months. (Reproduced, with permission, from Jolly H: *Sexual Precocity*. Thomas, 1955.)

normal. In males, this clinical picture is called **eunuchoidism.** In females, it is called **primary amenorrhea** (see below).

MENOPAUSE

The human ovary becomes unresponsive to gonadotropins with advancing age, and its function declines, so that sexual cycles disappear **(menopause).** This unresponsiveness is associated with and probably caused by a decline in the number of primordial follicles, which becomes precipitous at the time of menopause (Fig 23–12). The ovaries no longer secrete progesterone and 17β-estradiol in appreciable quantities, and estrogen is only formed in small amounts by aromatization of androstenedione in the circulation (see Chapter 20). The uterus and the vagina gradually become atrophic. As the negative feedback effect of estrogens and progesterone is reduced, secretion of FSH and LH is increased, and plasma FSH and LH increase to high levels. Old female mice and rats have long periods of diestrus and increased levels of gonadotropin secretion, but a clear-cut "menopause" has apparently not been described in animals.

In women, the menses usually become irregular and cease between the ages of 45 and 55. The average age at onset of the menopause has been increasing since the turn of the century and is currently 52 years.

Sensations of warmth spreading from the trunk to the face ("hot flashes") and various psychic symptoms are common after ovarian function has ceased. The hot flashes are prevented by estrogen treatment. They are not peculiar to the menopause; they also occur in premenopausal women and men whose gonads are removed surgically or destroyed by disease. Their cause is unknown. However, it has been demonstrated that they coincide with surges of LH secretion. LH is secreted in episodic bursts at intervals of 30–60 minutes or more **(circhoral secretion),** and in the absence of gonadal hormones, these bursts are large. Each hot flash begins with the start of a burst. However, LH itself is not responsible for the symptoms, because they can continue after removal of the pituitary. Instead, it appears that some event in the hypothalamus initiates both the release of LH and the episode of flushing.

Although the function of the testes tends to decline slowly with advancing age, the evidence is clear that there is no "male menopause" **(climacteric)** similar to that occurring in women.

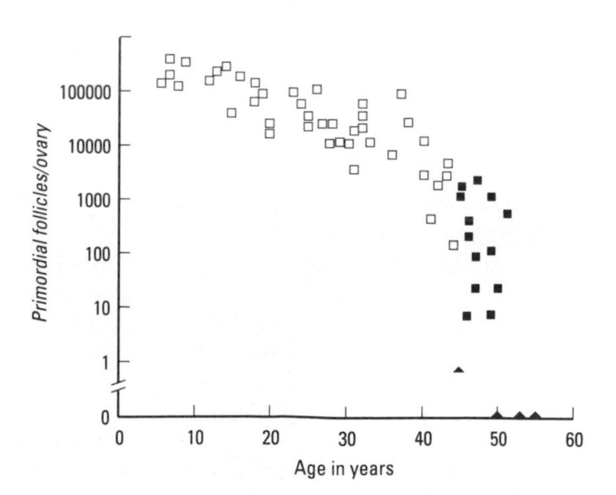

Figure 23–12. Number of primordial follicles per ovary in women at various ages. □, premenopausal women (regular menses); ■, perimenopausal women (irregular menses for at least 1 year); ▲, postmenopausal women (no menses for at least 1 year). Note that the vertical scale is a log scale and that the values are from one rather than 2 ovaries. (Redrawn by PM Wise and reproduced, with permission, from Richardson SJ, Senikas V, Nelson JF: Follicular depletion during the menopausal transition: evidence for accelerated loss and ultimate exhaustion. *J Clin Endocrinol Metab* 1987;**65**:1231.)

PITUITARY GONADOTROPINS & PROLACTIN

Actions

The testes and ovaries become atrophic when the pituitary is removed or destroyed. The actions of prolactin and the gonadotropins FSH and LH, as well as those of the gonadotropin secreted by the placenta, are described in detail in succeeding sections of this chapter. In brief, FSH helps maintain the spermatogenic epithelium and Sertoli cells in the male and is responsible for the early growth of ovarian follicles in the female. LH is tropic to the Leydig cells, and testosterone output in the spermatic vein is increased within minutes after its injection in dogs. In females, LH is responsible for the final maturation of the ovarian follicles and estrogen secretion from them. It is also responsible for ovulation and the initial formation of the corpus luteum and secretion of progesterone. Prolactin causes milk secretion from the breast after estrogen and progesterone priming. Its effect on the breast involves increased action of mRNA and increased production of casein and lactalbumin. However, the action of the hormone is not exerted on the cell nucleus and is prevented by inhibitors of microtubules. Prolactin also inhibits the effects of gonadotropins, possibly by an action at the level of the ovary. Its role in preventing ovulation in lactating women is discussed below. The function of prolactin in males (if any) is unknown. In rodents, prolactin maintains the corpus luteum, but it is not luteotropic in humans, and the corpus luteum is maintained by LH. An action of prolactin that has been used as the basis for bioassay of this hormone is stimulation of the growth and "secretion" of the crop sacs in pigeons and other birds. The paired crop sacs are outpouchings of the esophagus which form, by desquamation of their inner cell layers, a nutritious material ("milk") that the birds feed to their young. However, prolactin, FSH, and LH are now regularly measured by radioimmunoassay.

Chemistry

FSH and LH are each made up of an α and a β subunit the nature of which is discussed in Chapter 22. They are glycoproteins that contain the hexoses mannose and galactose, the hexosamines N-acetylgalactosamine and N-acetylglycosamine, and the methyl pentose fucose. They also contain sialic acid. The carbohydrate in the gonadotropin molecules increases their potency by markedly slowing their metabolism. The half-life of human FSH is about 170 minutes; the half-life of LH is about 60 minutes.

Human prolactin contains 199 amino acid residues and 3 disulfide bridges (Fig 23–13) and has considerable structural similarity to human growth hormone and hCS.

Regulation of Prolactin Secretion

The normal plasma prolactin concentration is approximately 5 ng/mL in men and 8 ng/mL in women. Secretion is tonically inhibited by the hypothalamus, and section of the pituitary stalk leads to an increase in circulating prolactin. Thus, the effect of hypothalamic prolactin-inhibiting hormone (PIH) usually overbalances the effect of the putative prolactin-releasing hormone (see Chapter 14). It now seems clear that PIH is dopamine secreted by the tubero-infundibular dopaminergic neurons into the portal hypophyseal vessels. In humans, prolactin secretion is increased by exercise, surgical and psychologic stresses, and stimulation of the nipple (Table 23–3). The plasma prolactin level rises during sleep, the rise starting after the onset of sleep and persisting throughout the sleep period. Secretion is increased during pregnancy, reaching a peak at the time of parturition. After delivery, plasma concentration falls to nonpregnant levels in about 8 days. Suckling produces a prompt increase in secretion, but the magnitude of this rise gradually declines after a woman has been nursing for more than 3 months. With prolonged lactation, milk secretion occurs with prolactin levels that are in the normal range.

L-Dopa decreases prolactin secretion by increasing the formation of dopamine, and bromocriptine and other dopamine agonists inhibit secretion because they stimulate dopamine receptors. Chlorpromazine and related drugs that block dopamine receptors increase prolactin secretion. TRH stimulates the secretion of prolactin in addition to TSH, but it seems likely that there is an additional prolactin-releasing hormone (PRH) in hypothalamic tissue that is different from TRH. Estrogens also produce a slowly developing increase in prolactin secretion.

It has now been established that prolactin acts to facilitates the secretion of dopamine in the median eminence. Thus, prolactin acts in the hypothalamus in a negative feedback fashion to inhibit its own secretion.

Hyperprolactinemia

Up to 70% of the patients with chromophobe adenomas of the anterior pituitary have elevated plasma prolactin levels. In some instances, the elevation may be due to damage to the pituitary stalk, but in most cases, the tumor cells are actually secreting the hormone. The hyperprolactinemia may cause galactorrhea, but in many individuals there are no demonstrable abnormalities. Indeed, most women with galactorrhea have normal prolactin levels; definite elevations are found in less than a third of patients with this condition.

Another interesting observation is that 15–20% of women with secondary amenorrhea have elevated prolactin levels and that when prolactin secretion is reduced, normal menstrual cycles and fertility return. It appears that the prolactin may produce amenorrhea by blocking the action of gonadotropins on the ovaries, but definitive proof of this hypothesis must

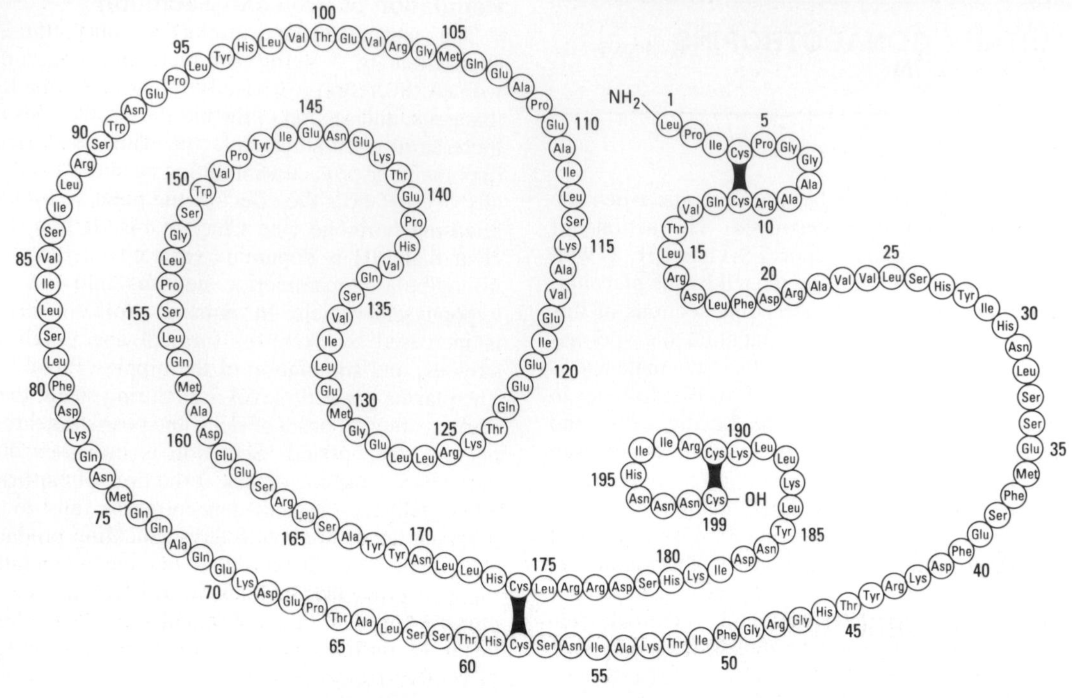

Figure 23–13. Structure of human prolactin.

await further research. In men, hyperprolactinemia is associated with impotence that disappears when prolactin secretion is reduced.

THE MALE REPRODUCTIVE SYSTEM

STRUCTURE

The testes are made up of loops of convoluted **seminiferous tubules,** along the walls of which the spermatozoa are formed from the primitive germ cells **(spermatogenesis).** Both ends of each loop drain into a network of ducts in the head of the **epididymis.** From there, spermatozoa pass through the tail of the epididymis into the **vas deferens.** They enter through the **ejaculatory ducts** into the urethra in the body of the **prostate** at the time of ejaculation (Fig 23–14). Between the tubules in the testes are nests of cells containing lipid granules, the **interstitial cells of Leydig,** which secrete testosterone into the bloodstream (Figs 23–15 and 23–16). The spermatic arteries to the testes are tortuous, and blood in them runs parallel but in the opposite direction to blood in the pampiniform plexus of spermatic veins. This anatomic arrangement may permit countercurrent exchange of heat and testosterone. The principles of countercurrent exchange are considered in detail in relation to the kidney in Chapter 38.

GAMETOGENESIS & EJACULATION

Blood-Testis Barrier

The walls of the seminiferous tubules are lined by primitive germ cells (see below) and by **Sertoli cells,** large, complex glycogen-containing cells that stretch from the basal lamina of the tubule to the lumen (Fig 23–16). Tight junctions between adjacent Sertoli cells near the basal lamina form a blood-testis barrier that prevents proteins and other large molecules from passing from the interstitial tissue and the part of the tubule near the basal lamina (basal compartment) to the region near the tubular lumen (adluminal compartment) and the lumen. However, steroids penetrate this barrier with ease. In addition, maturing germ cells must pass through the barrier as they move to the lumen. This appears to occur without disruption of the barrier by progressive breakdown of the tight junctions above the germ cells with concomitant formation of new tight junctions below them.

The fluid of the lumen of the seminiferous tubules is quite different from plasma; it contains very little protein and glucose, but it is rich in androgens, estrogens, K^+, inositol, and glutamic and aspartic acids. Maintenance of its composition presumably depends on the blood-testis barrier. The barrier also protects the germ cells from blood-borne noxious agents, prevents antigenic products of germ cell division and maturation from entering the circulation and generating an autoimmune response, and may help establish an osmotic gradient that facilitates movement of fluid into the tubular lumen.

Table 23–3. Factors affecting the secretion of human prolactin and growth hormone. I, moderate increase; I+, marked increase; I++, very marked increase; N, no change; D, moderate decrease; D+, marked decrease.*

Factor	Prolactin	Growth Hormone
Sleep	I+	I+
Nursing	I++	N
Breast stimulation in nonlactating women	I	N
Stress	I+	I+
Hypoglycemia	I	I+
Strenuous exercise	I	I
Sexual intercourse in women	I	N
Pregnancy	I++	N
Estrogens	I	I
Hypothyroidism	I	N
TRH	I+	N
Phenothiazines, butyrophenones	I+	N
Opiates	I	I
Glucose	N	D
Somatostatin	N	D+
L-Dopa	D+	I+
Apomorphine	D+	I+
Bromocriptine and related ergot derivatives	D+	I

*Modified from Frantz A: Prolactin. *N Engl J Med* 1978;**298:** 201.

Spermatogenesis

The **spermatogonia,** the primitive germ cells next to the basal lamina of the seminiferous tubules, mature into **primary spermatocytes** (Fig 23–16). This process begins during adolescence. The primary spermatocytes undergo meiotic division, reducing the number of chromosomes. In this 2-stage process, they divide into **secondary spermatocytes** and then into **spermatids,** which contain the haploid number of 23 chromosomes. The spermatids mature into **spermatozoa (sperms).** As a single spermatogonium divides and matures, its descendants remain tied together by cytoplasmic bridges until the late spermatid stage. This apparently ensures synchrony of the differentiation of each clone of germ cells. The estimated number of spermatids formed from a single spermatogonium is 512. In humans, it takes an average of 74 days to form a mature sperm from a primitive germ cell by this orderly process of spermatogenesis. Each sperm is an intricate motile cell, rich in DNA, with a head that is made up mostly of chromosomal material (Fig 23–17).

The spermatids mature into spermatozoa in deep folds of the cytoplasm of the Sertoli cells (Fig 23–16). Mature spermatozoa are released from the Sertoli cells and become free in the lumens of the tubules. The Sertoli cells may secrete estrogens, and they secrete **androgen-binding protein (ABP),** a protein that probably functions to maintain a high, stable supply of androgen in the tubular fluid. In addition, the Sertoli cells secrete **inhibin,** a polypeptide that inhibits FSH secretion. The development of the Sertoli cells is stimulated by FSH. During fetal life, the Sertoli cells secrete MIS (see above).

FSH and androgens maintain the gametogenic function of the testis. After hypophysectomy, injection of LH produces a high local concentration of androgen in the testes, and this maintains spermatogenesis. There is some evidence that testosterone passes from the spermatic veins to the spermatic arteries as they parallel each other in the scrotum, thus helping to maintain a high androgen concentration in the testes. Less androgen is required if FSH is present. The stages from spermatogonia to spermatids appear to be androgen-independent. However, the maturation from spermatids to spermatozoa depends on androgen acting on the Sertoli cells in which the developing spermatozoa are embedded. The role of FSH in spermatogenesis is uncertain, although it also appears to facilitate the last stages of spermatid maturation via an action on the Sertoli

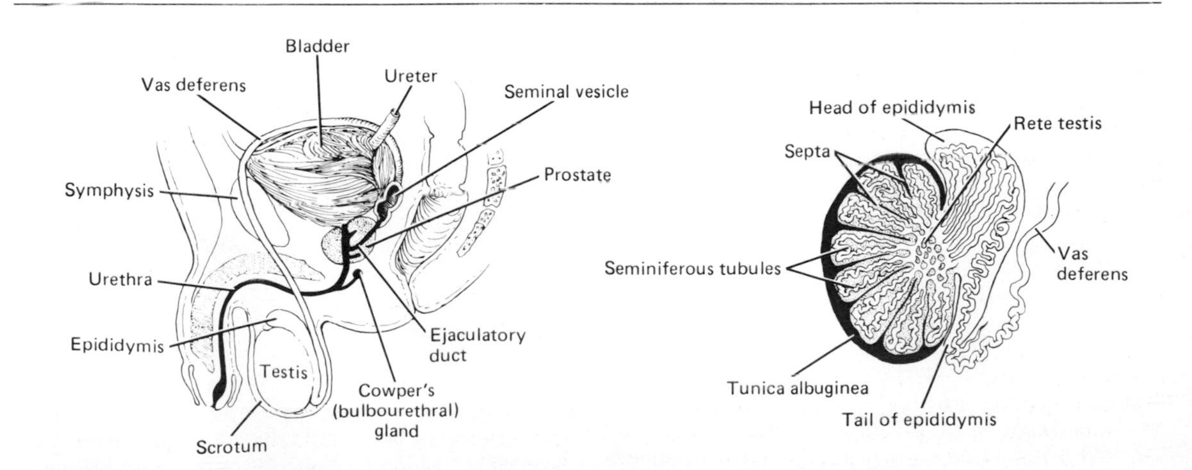

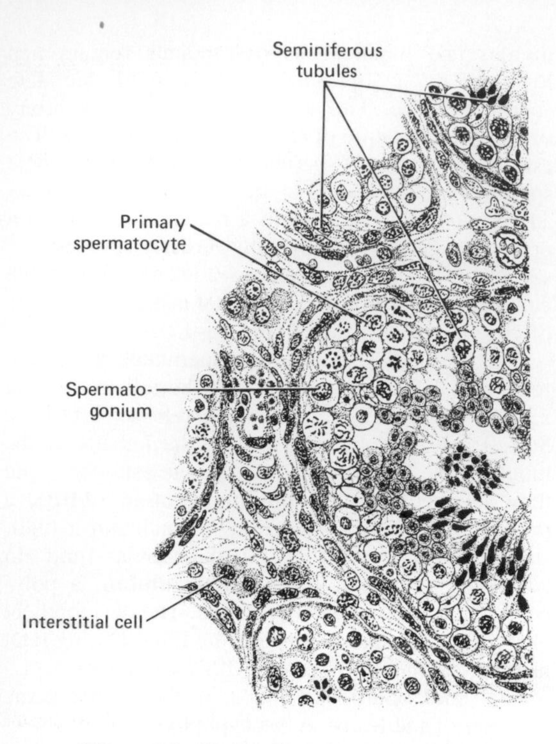

Figure 23–15. Section of human testis.

cells. In addition, it promotes the production of ABP.

Spermatozoa leaving the testes are not fully mobile. They continue their maturation and acquire motility during their passage through the epididymis. Their capacity to produce fertilization is further enhanced if they spend some additional time in the female reproductive tract. The initial process occurring in the female reproductive tract is called **capacitation.** It makes the spermatozoa better able to adhere to the ovum. Capacitation can also be achieved by incubating spermatozoa with tissue fluids. The subsequent events that occur when sperms attach to the ovum are discussed below, in the section on fertilization.

Effect of Temperature

Spermatogenesis requires a temperature considerably lower than that of the interior of the body. The testes are normally maintained at a temperature of about 32 °C. They are kept cool by air circulating around the scrotum and probably by heat exchange in a countercurrent fashion between the spermatic arteries and veins. When the testes are retained in the abdomen or when, in experimental animals, they are held close to the body by tight cloth binders, degeneration of the tubular walls and sterility result. Hot baths (43–45 °C for 30 minutes per day) and insulated athletic supporters reduce sperm count in humans,

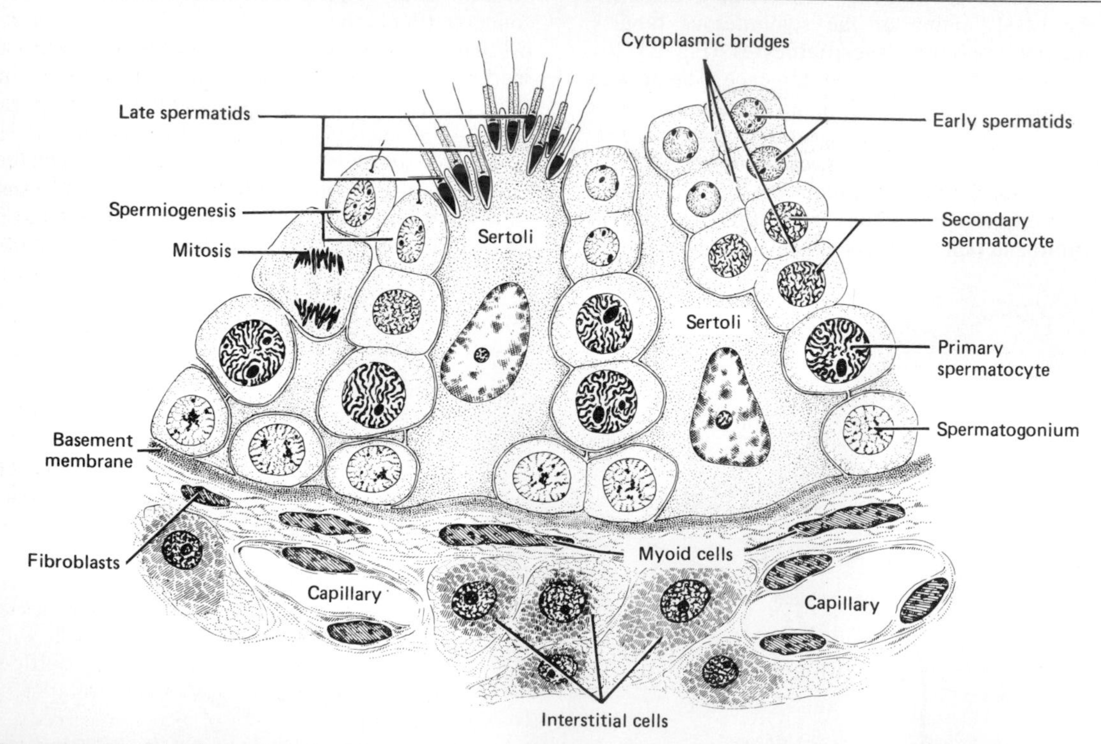

Figure 23–16. Seminiferous epithelium. Note that maturing germ cells remain connected by cytoplasmic bridges through the early spermatid stage, and that these cells are closely invested by Sertoli cell cytoplasm as they move from the basal lamina to the lumen. (Reproduced, with permission, from Junqueira LC, Carneiro J, Kelley RO: *Basic Histology*, 6th ed. Appleton & Lange, 1989.)

but the reductions produced in this fashion have not been large enough or consistent enough to make the procedures reliable forms of male contraception.

Semen

The fluid that is ejaculated at the time of orgasm, the **semen,** contains sperms and the secretions of the seminal vesicles, prostate, Cowper's glands, and, probably, the urethral glands (Table 23–4). An average volume per ejaculate is 2.5–3.5 mL after several days of continence. Volume of semen and sperm count decrease rapidly with repeated ejaculation. Even though it takes only one sperm to fertilize the ovum, there are normally about 100 million sperms per milliliter of semen. Fifty percent of men with counts of 20–40 million/mL and essentially all of those with counts under 20 million/mL are sterile. The **prostaglandins** in semen, which actually come from the seminal vesicles, are in high concentration, but the function of these fatty acid derivatives in semen is not known. Their structure and their multiple actions in other parts of the body are discussed in Chapter 17.

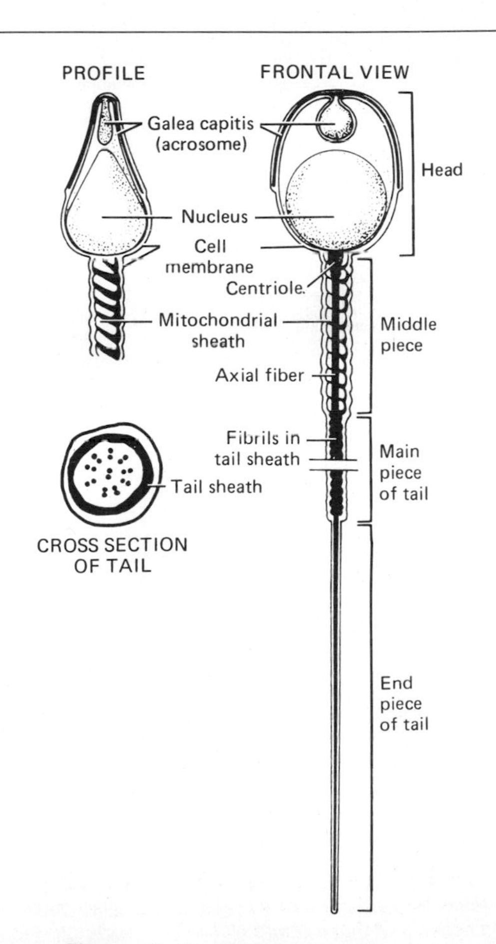

Figure 23–17. Human spermatozoan.

Table 23–4. Composition of human semen.

Component	Source
Color: White, opalescent	
Specific gravity: 1.028	
pH: 7.35–7.50	
Sperm count: Average about 100 million/mL, with fewer than 20% abnormal forms	
Other components:	
Fructose (1.5–6.5 mg/mL)	From seminal vesicles (contribute 60% of total volume)
Phosphorylcholine	
Ergothioneine	
Ascorbic acid	
Flavins	
Prostaglandins	
Spermine	From prostate (contributes 20% of total volume)
Citric acid	
Cholesterol, phospholipids	
Fibrinolysin, fibrinogenase	
Zinc	
Acid phosphatase	
Phosphate	Buffers
Bicarbonate	
Hyaluronidase	

Human sperms move at a speed of about 3 mm/min through the female genital tract. Sperms reach the uterine tubes 30–60 minutes after copulation. In some species, contractions of the female organs facilitate the transport of the sperms to the uterine tubes, but it is not known if such contractions occur in humans.

Ejaculation

Ejaculation is a 2-part spinal reflex that involves **emission,** the movement of the semen into the urethra; and **ejaculation** proper, the propulsion of the semen out of the urethra at the time of orgasm. The afferent pathways are mostly fibers from touch receptors in the glans penis that reach the spinal cord through the internal pudendal nerves. Emission is a sympathetic response, integrated in the upper lumbar segments of the spinal cord and effected by contraction of the smooth muscle of the vasa deferentia and seminal vesicles in response to stimuli in the hypogastric nerves. The semen is propelled out of the urethra by contraction of the bulbocavernosus muscle, a skeletal muscle. The spinal reflex centers for this part of the reflex are in the upper sacral and lowest lumbar segments of the spinal cord, and the motor pathways traverse the first to third sacral roots and the internal pudendal nerves.

Erection

Erection is initiated by dilation of the arterioles of the penis. As the erectile tissue of the penis fills with blood, the veins are compressed, blocking outflow and adding to the turgor of the organ. The integrating centers in the lumbar segments of the spinal cord are activated by impulses in afferents from the genitalia and descending tracts that mediate erection in response to erotic psychic stimuli. The efferent parasympathetic fibers are in the pelvic splanchnic nerves **(nervi**

erigentes). The fibers presumably contain acetylcholine and VIP as cotransmitters (see Chapter 4), and release of both produces the vasodilation; in any case, local injection of VIP produces erection. Sympathetic vasoconstrictor impulses to the arterioles terminate the erection.

Vasectomy

Bilateral ligation of the vas deferens (vasectomy) has proved to be a relatively safe and convenient contraceptive procedure. However, it has proved difficult to restore the patency of the vas in those wishing to restore fertility, and the current success rate for such operations, as measured by the subsequent production of pregnancy, is about 50%. Half of the men who have been vasectomized develop antibodies against spermatozoa, and in monkeys, the presence of such antibodies is associated with a higher incidence of infertility after restoration of the patency of the vas. However, there do not appear to be any other adverse effects of the antisperm antibodies.

ENDOCRINE FUNCTION OF THE TESTES

Chemistry & Biosynthesis of Testosterone

Testosterone, the principal hormone of the testes, is a C_{19} steroid (see Chapter 20) with an $-OH$ group in the 17 position (Fig 23–18). It is synthesized from cholesterol in the Leydig cells. According to current concepts, the biosynthetic pathways in all endocrine organs that form steroid hormones are similar, the organs differing from one another only in the enzyme systems they contain. In the Leydig cells, the 11- and 21-hydroxylases found in the adrenal cortex (Fig 20–8) are absent, but 17α-hydroxylase is present. Pregnenolone is therefore hydroxylated in the 17 position, then subjected to side chain cleavage to form 17-ketosteroids. These in turn are converted to testosterone. Testosterone is also formed via progesterone and 17-hydroxyprogesterone, but this pathway is less prominent in humans. The secretion of testosterone is under the control of LH, and the

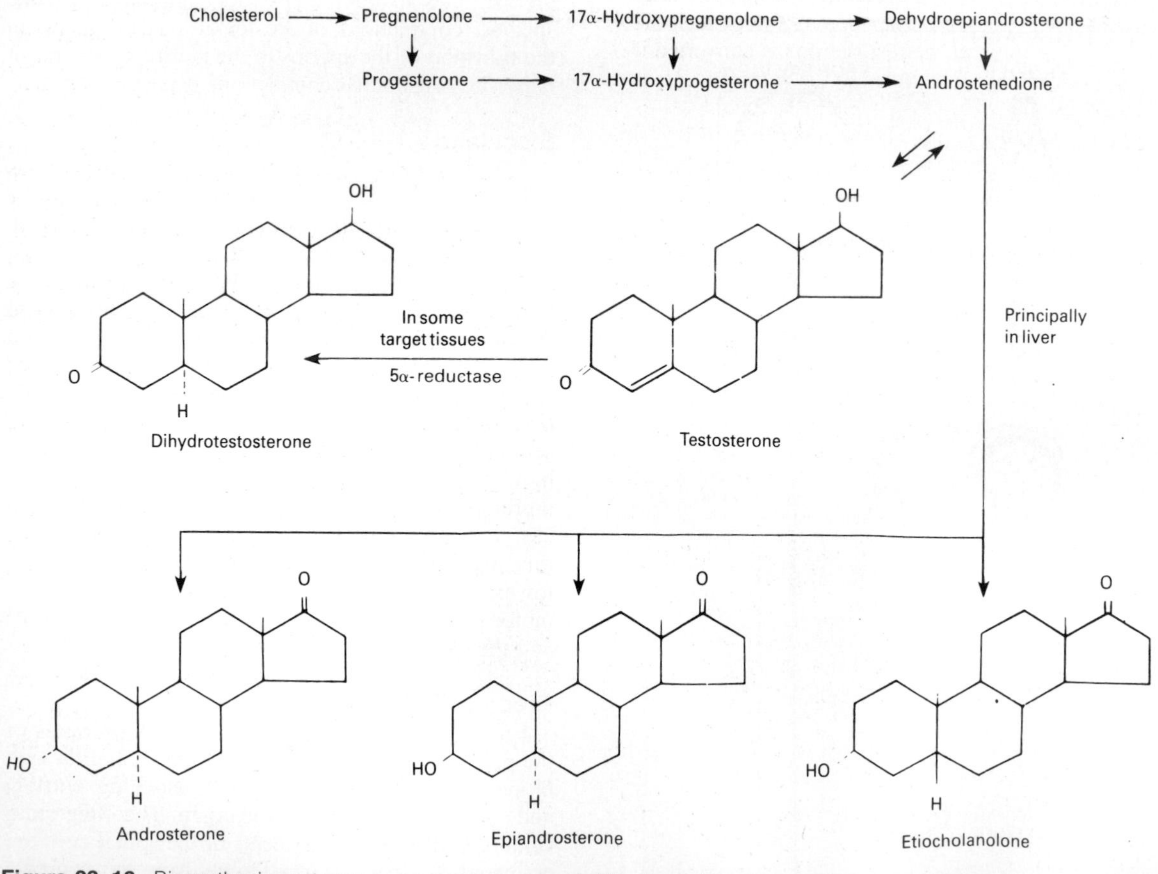

Figure 23–18. Biosynthesis and metabolism of testosterone. The formulas of the precursor steroids are shown in Fig 20–8. Although the main secretory product of the Leydig cells is testosterone, some of the precursors also enter the circulation. Note that dihydrotestosterone is formed in some target tissues and that the three 17-ketosteroids produced from testosterone in the liver are isomers of one another.

mechanism by which LH stimulates the Leydig cells involves increased formation of cyclic AMP (see Chapter 17). Cyclic AMP increases the formation of cholesterol from cholesteryl esters and the conversion of cholesterol to pregnenolone via the activation of protein kinase A. Testosterone is also formed in the adrenal cortex.

Secretion

The testosterone secretion rate is 4–9 mg/d (13.9–31.2 nmol/d) in normal adult males. Small amounts of testosterone are also secreted in females, probably from the ovary but possibly from the adrenal as well.

Transport & Metabolism

Ninety-seven percent of the testosterone in plasma is bound to protein: 40% is bound to a β-globulin called **gonadal steroid-binding globulin (GBG)** or **sex steroid-binding globulin,** 40% to albumin, and 17% to other proteins. GBG also binds estradiol. The plasma testosterone level (free and bound) is approximately 0.65 μg/dL (22.5 nmol/L) in adult men and 0.03 μg/dL (1.0 nmol/L) in adult women. It declines somewhat with age in males.

A small amount of circulating testosterone is converted to estrogen somewhere in the body (see below), but most of the testosterone is converted into 17-ketosteroids and excreted in the urine (Fig 23–18). About two-thirds of the urinary 17-ketosteroids are of adrenal origin, and one-third are of testicular origin (Table 23–5). Although most of the 17-ketosteroids are weak androgens (they have 20% or less the potency of testosterone), it is worth emphasizing that not all 17-ketosteroids are androgens and not all androgens are 17-ketosteroids. Etiocholanolone, for example, has no androgenic activity, and testosterone itself is not a 17-ketosteroid.

Actions

In addition to their actions during development (see above), testosterone and other androgens exert an inhibitory feedback effect on pituitary LH secretion; develop and maintain the male secondary sex characteristics; and exert an important protein anabolic, growth-promoting effect. Along with FSH, testosterone is responsible for the maintenance of gametogenesis (see above).

Secondary Sex Characteristics

The widespread changes in hair distribution, body configuration, and genital size that develop in boys at puberty—the male **secondary sex characteristics**—are summarized in Table 23–6. Not only do the prostate and seminal vesicles enlarge, but the seminal vesicles begin to secrete fructose. This sugar appears to function as the main nutritional supply for the spermatozoa. The psychic effects of testosterone are difficult to define in humans, but in experimental animals, androgens provoke boisterous and aggressive play. The effects of androgens and estrogens on sexual behavior are considered in detail in Chapter 15. Although body hair is increased by androgens, scalp hair is decreased. Hereditary baldness often fails to develop unless androgens are present (Fig 23–19).

Anabolic Effects

Androgens increase the synthesis and decrease the breakdown of protein, leading to an increase in the rate of growth. They also cause the epiphyses to fuse to the long bones, thus eventually stopping growth. Their role in the adolescent growth spurt is discussed in Chapter 22. Secondary to their anabolic effect, they cause moderate sodium, potassium, water, calcium, sulfate, and phosphate retention; and they also increase the size of the kidneys. Doses of exogenous testosterone that exert significant anabolic effects are also masculinizing and increase libido, which limits the usefulness of the hormone as an anabolic agent in patients with wasting diseases. Attempts to develop synthetic steroids in which the anabolic action is divorced from the androgenic action have not been particularly successful.

Mechanism of Action

Like other steroids (see Chapter 1), testosterone binds to an intracellular receptor, and the receptor-steroid complex then binds to DNA in the nucleus, facilitating transcription of various genes. In addition, testosterone is converted to dihydrotestosterone

Table 23–5. Origin of principal plasma and urinary 17-ketosteroids. The boldface compounds and their derivatives are 17-ketosteroids with an −O or −OH group in position 11 ("11-oxy-17-ketosteroids").

	Adrenal Androgens and Their Metabolites	Hepatic Metabolite of		Metabolite of Testosterone
		Cortisol	Cortisone	
Dehydroepiandrosterone	X			
Androstenedione	X			
11β-Hydroxyandrostenedione	X	X		X
Androsterone	X			X
Epiandrosterone	X			X
Etiocholanolone	X			X
Adrenosterone			X	

Table 23–6. Body changes at puberty in boys (male secondary sex characteristics).

External genitalia: Penis increases in length and width. Scrotum becomes pigmented and rugose.
Internal genitalia: Seminal vesicles enlarge and secrete and begin to form fructose. Prostate and bulbourethral glands enlarge and secrete.
Voice: Larynx enlarges, vocal cords increase in length and thickness, and voice becomes deeper.
Hair growth: Beard appears. Hairline on scalp recedes anterolaterally. Pubic hair grows with male (triangle with apex up) pattern. Hair appears in axillas, on chest, and around anus; general body hair increases.
Mental: More aggressive, active attitude. Interest in opposite sex develops.
Body conformation: Shoulders broaden, muscles enlarge.
Skin: Sebaceous gland secretion thickens and increases (predisposing to acne).

(DHT) by 5α-reductase in many target cells (Fig 23–18), and DHT binds to the same intracellular receptor as testosterone. DHT also circulates, with a plasma level that is about 10% of the testosterone level. Testosterone-receptor complexes are less stable than DHT-receptor complexes in target cells, and they transform less well to the DNA-binding state. Thus, DHT formation is a way of amplifying the action of testosterone in target tissues.

Testosterone-receptor complexes are responsible for the maturation of wolffian duct structures and consequently for the formation of male internal genitalia during development, but DHT-receptor complexes are needed to form male external genitalia. DHT-receptor complexes are also primarily responsible for enlargement of the prostate and probably of the penis at the time of puberty as well as the facial hair, the acne, and temporal recession of the hairline. On the other hand, the increase in muscle mass and the development of male sex drive and libido depend primarily on testosterone rather than DHT.

Congenital 5α-reductase deficiency, which is common in certain parts of the Dominican Republic, produces an interesting form of male pseudohermaphroditism. Individuals with this syndrome are born with male internal genitalia including testes, but they have female external genitalia and are usually raised as girls. However, when they reach puberty, LH secretion and circulating testosterone levels are increased. Consequently, they develop male body contours and male libido. At this point, they usually change their gender identities and "become boys." They develop little facial hair or baldness, but their clitorises enlarge ("penis-at-12 syndrome") to the point that some of them have intercourse with women. This enlargement probably occurs because with the high LH, there is enough testosterone to overcome the need for DHT amplification in the genitalia, and possibly because high levels of testosterone induce an increase in 5α-reductase.

Testicular Production of Estrogens

Seventy percent of the estradiol in the plasma of adult men is formed by aromatization of circulating testosterone and androstenedione. Small amounts may be secreted by the adrenal cortex, but most of the remaining 30% is secreted by the testes. Some of the estradiol in testicular venous blood comes from the Leydig cells, but some may also come from the Sertoli cells. In men, the plasma estradiol level is approximately 2 ng/dL (70 pmol/L), and the total production rate is 0.05 mg/d (0.18 μmol/d). In contrast to the situation in women, there is a moderate increase in estrogen production with advancing age in men.

CONTROL OF TESTICULAR FUNCTION

FSH is tropic to the Sertoli cells, and FSH and androgens maintain the gametogenic function of the testes. FSH also stimulates the secretion of androgen-binding protein and inhibin. Inhibin feeds back to inhibit FSH secretion. LH is tropic to the Leydig cells and stimulates secretion of testosterone, which in turn feeds back to inhibit LH secretion. Hypothalamic

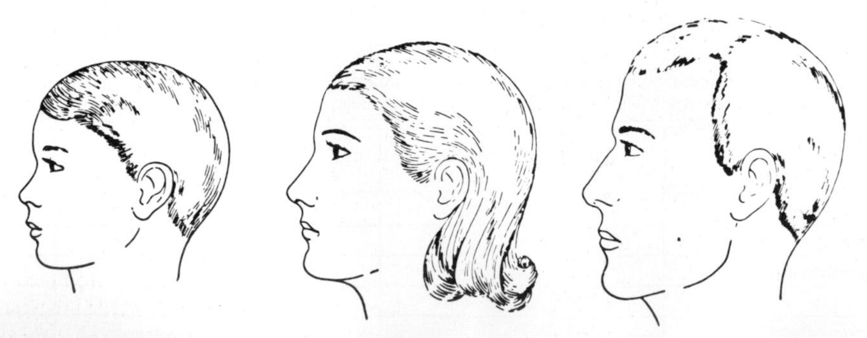

Figure 23–19. Hairline in children and adults. The hairline of the woman is like that of the child, whereas that of the man is indented in the lateral frontal region. (Reproduced, with permission, from Greulich WW et al: Somatic and endocrine studies of pubertal and adolescent boys. *Monogr Soc Res Child Dev* 1942;7:1.)

lesions in animals and hypothalamic disease in humans lead to atrophy of the testes and loss of their function.

Inhibin

Testosterone reduces plasma LH, but except in large doses, it has no effect on plasma FSH. Plasma FSH is elevated in patients who have atrophy of the seminiferous tubules but normal levels of testosterone and LH secretion. These observations led to the search for **inhibin,** a factor of testicular origin that regulates FSH secretion. There is FSH-inhibiting activity in extracts of testes and, in women, in antral fluid from ovarian follicles. Follicular fluid contains 3 polypeptide subunits: an α subunit with a molecular weight of 18,000, and 2 β subunits, β_A and β_B, each with a molecular weight of 14,000. The α subunit combines with β_A to form a heterodimer and with β_B to form another heterodimer, with the subunits linked by disulfide bonds (Fig 23–20). Both $\alpha\beta_A$ (inhibin A) and $\alpha\beta_B$ (inhibin B) inhibit FSH secretion by a direct action on the pituitary. However, heterodimer $\beta_A\beta_B$ and the homodimer $\beta_A\beta_A$ also form, and they both stimulate FSH secretion. Consequently, they have been called **activins.** Furthermore, the β dimers have marked homology to transforming growth factor-β (TGFβ), a dimeric polypeptide with a molecular weight of 25,000; and TGFβ also stimulates FSH secretion. The physiologic roles of these various stimulators are not yet known.

Steroid Feedback

A current "working hypothesis" of the way the functions of the testes are regulated is shown in Fig 23–21. Castration is followed by a rise in the pituitary content and secretion of FSH and LH, and hypothalamic lesions prevent this rise. Testosterone inhibits LH secretion by acting directly on the anterior pituitary and by inhibiting the secretion of LHRH from the hypothalamus. Inhibin acts directly on the anterior pituitary to inhibit FSH secretion.

In response to LH, some of the testosterone secreted from the Leydig cells bathes the seminiferous epithelium and provides the high local concentration of androgen to the Sertoli cells that is necessary for normal spermatogenesis. Systemically administered testosterone does not raise the androgen level in the testes to as great a degree, and it inhibits LH secretion. Consequently, the net effect of systemically administered testosterone is generally a decrease in sperm count. Testosterone therapy has been suggested as a means of male contraception. However, the dose of testosterone needed to suppress spermatogenesis causes sodium and water retention. The possible use of inhibin as a male contraceptive is now being explored.

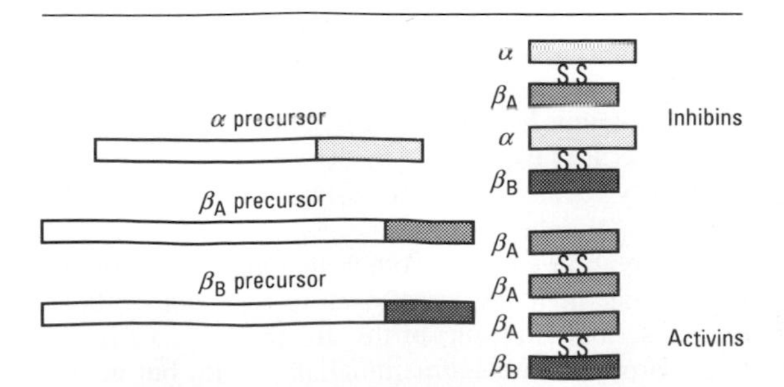

Figure 23–20. Inhibin precursors and the various inhibins and activins that are formed from the C-terminal regions of these precursors.

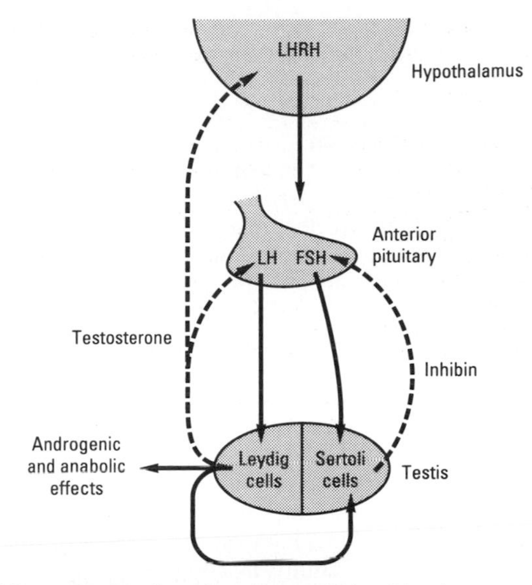

Figure 23–21. Postulated interrelationships between the hypothalamus, anterior pituitary, and testes. Solid arrows indicate excitatory effects; dashed arrows indicate inhibitory effects. Compare with Figs 18–13, 20–22, 22–9, and 23–35.

ABNORMALITIES OF TESTICULAR FUNCTION

Cryptorchidism

The testes develop in the abdominal cavity and normally migrate to the scrotum during fetal development. Testicular descent is incomplete on one or, less commonly, both sides in 10% of newborn males, the testes remaining in the abdominal cavity or inguinal canal. Spontaneous descent of these testes is the rule, however, and the proportion of boys with undescended testes **(cryptorchidism)** falls to 2% at age 1 year and 0.3% after puberty. Gonadotropic hormone treatment speeds descent in some cases, or the defect can be corrected surgically. Treatment should be instituted, probably before puberty, because there is a higher incidence of malignant tumors in undescended than in scrotal testes and because after puberty the higher temperature in the abdomen eventually causes irreversible damage to the spermatogenic epithelium.

Male Hypogonadism

The clinical picture of male hypogonadism depends upon whether testicular deficiency develops before or after puberty and whether the gametogenic or the endocrine function is compromised. The causes of testicular deficiency include hypothalamic and pituitary disease as well as a variety of primary testicular and chromosomal disorders. Loss or failure of maturation of the gametogenic function causes sterility. If the endocrine function is lost in adulthood, the secondary sex characteristics regress slowly because it takes very little androgen to maintain them once they are established. The growth of the larynx during adolescence is permanent, and the voice remains deep. Men castrated in adulthood suffer some loss of libido, although the ability to copulate persists for some time. They occasionally have hot flashes and are generally more irritable, passive, and depressed than men with intact testes. When the Leydig cell deficiency dates from childhood, the clinical picture is that of **eunuchoidism.** Eunuchoid individuals over the age of 20 are characteristically tall, although not so tall as hyperpituitary giants, because their epiphyses remain open and some growth continues past the normal age of puberty. They have narrow shoulders and small muscles, a body configuration resembling that of the adult female. The genitalia are small and the voice high-pitched. Pubic and axillary hair do appear, because of adrenocortical androgen secretion; but the hair is sparse, and the pubic hair has the female ''triangle with the base up'' distribution rather than the ''triangle with the base down'' pattern (male escutcheon) seen in normal males.

Androgen-Secreting Tumors

''Hyperfunction'' of the testes in the absence of tumor formation is not a recognized entity. Androgen-secreting testicular tumors are rare and cause detectable endocrine symptoms only in prepuberal boys, who develop precocious pseudopuberty (Table 23–2).

THE FEMALE REPRODUCTIVE SYSTEM

THE MENSTRUAL CYCLE

The reproductive system of the female (Fig 23–22), unlike that of the male, shows regular cyclic changes that teleologically may be regarded as periodic preparations for fertilization and pregnancy. In primates, the cycle is a **menstrual** cycle, and its most conspicuous feature is the periodic vaginal bleeding that occurs with the shedding of the uterine mucosa **(menstruation).** The length of the cycle is notoriously variable in women, but an average figure is 28 days from the start of one menstrual period to the start of the next. By common usage, the days of the cycle are identified by number, starting with the first day of menstruation.

Ovarian Cycle

From the time of birth, there are many **primordial follicles** under the ovarian capsule. Each contains an immature ovum (Fig 23–23). At the start of each cycle, several of these follicles enlarge, and a cavity forms around the ovum **(antrum formation).** In humans, one of the follicles in one ovary starts to grow rapidly on about the sixth day, while the others regress. It is not known how one follicle is singled out for development during this **follicular phase** of the menstrual cycle. When women are given highly purified human pituitary gonadotropin preparations by injection, many follicles develop simultaneously.

The structure of a maturing ovarian **(graafian)** follicle is shown in Fig 23–23. The cells of the **theca interna** of the follicle are the primary source of circulating estrogens. However, the follicular fluid has a high estrogen content, and much of this estrogen appears to come from the granulosa cells.

At about the 14th day of the cycle, the distended follicle ruptures, and the ovum is extruded into the abdominal cavity. This is the process of **ovulation.** The ovum is picked up by the fimbriated ends of the uterine tubes (oviducts). It is transported to the uterus and, unless fertilization occurs, on out through the vagina. Follicles that enlarge but fail to ovulate degenerate, forming **atretic follicles** (Fig 23–23).

The follicle that ruptures at the time of ovulation promptly fills with blood, forming what is sometimes called a **corpus hemorrhagicum.** Minor bleeding from the follicle into the abdominal cavity may cause peritoneal irritation and fleeting lower abdominal pain (''mittelschmerz''.) The granulosa and theca cells of the follicle lining promptly begin to proliferate, and the clotted blood is rapidly replaced with yellowish, lipid-rich **luteal** cells, forming the **corpus luteum.** This is the **luteal phase** of the menstrual cycle, during which the luteal cells secrete estrogens and progesterone. If pregnancy occurs, the corpus luteum persists, and there are usually no more periods until after delivery. If there is no pregnancy, the corpus luteum begins to degenerate about 4 days before the next menses (24th day of the cycle) and is eventually replaced by scar tissue, forming a **corpus albicans.**

The ovarian cycle in other mammals is similar, except that in many species more than one follicle ovulates, and multiple births are the rule. Corpora lutea form in some submammalian species but not in others.

In humans, no new ova are formed after birth. During fetal development, the ovaries contain over 7

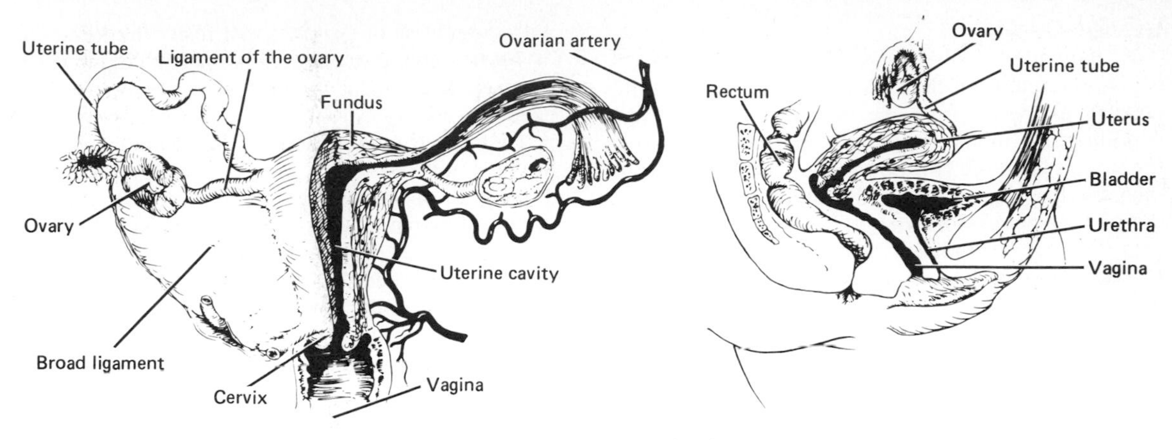

Figure 23–22. The female reproductive system.

million primordial follicles. However, many undergo atresia (involution) before birth, and others are lost after birth. At the time of birth, there are 2 million ova, but 50% of these are atretic. The million that are normal undergo the first part of the first meiotic division at about this time and enter a stage of arrest in which those that survive persist until adulthood. The arrest is probably due to the presence of an inhibitory substance in the primordial follicles. There is continuing atresia during development, and the number of ova in both of the ovaries at the time of puberty is less than 300,000 (Fig 23–12). Only one of these ova per cycle (or about 500 in the course of a normal reproductive life) is stimulated to mature; the remainder degenerate. Just before ovulation, the first meiotic division is completed. One of the daughter cells, the **secondary oocyte,** receives most of the cytoplasm, while the other, the **first polar body,** fragments and disappears. The secondary oocyte immediately begins the second meiotic division, but this division stops at metaphase and is completed only when a sperm penetrates the oocyte. At that time, the **second polar body** is cast off and the fertilized ovum proceeds to form a new individual.

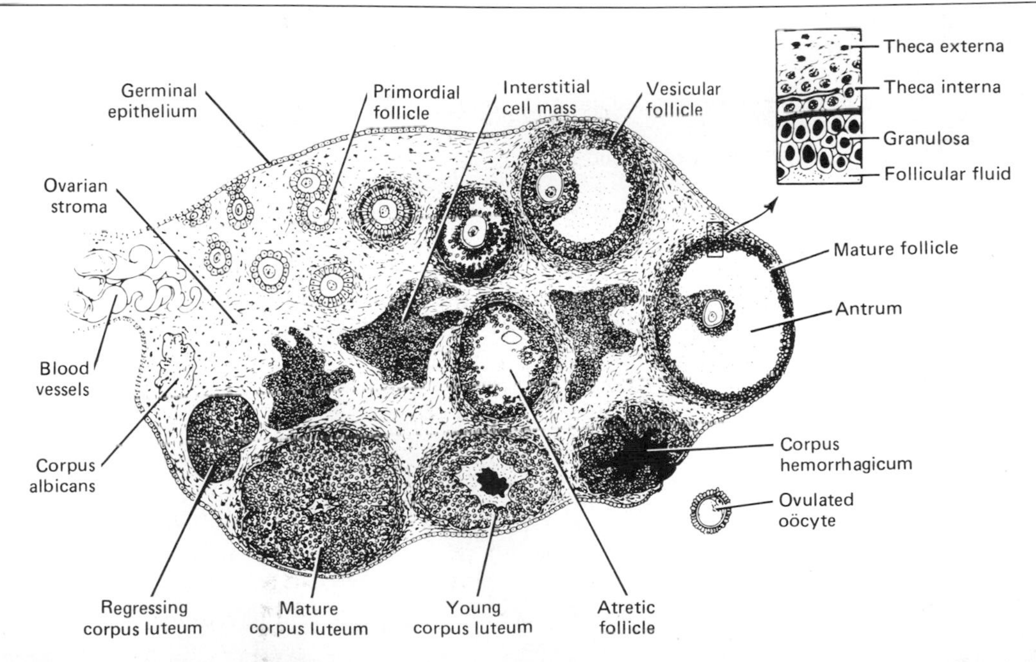

Figure 23–23. Diagram of a mammalian ovary, showing the sequential development of a follicle, formation of a corpus luteum, and, in the center, follicular atresia. A section of the wall of a mature follicle is enlarged at the upper right. The interstitial cell mass is not prominent in primates. (Reproduced, with permission, from Gorbman A, Bern H: *Textbook of Comparative Endocrinology.* Wiley, 1962.)

Uterine Cycle

At the end of menstruation, all but the deep layers of the endometrium have sloughed. Under the influence of estrogens from the developing follicle, the endometrium increases rapidly in thickness during the period from the fifth to the fourteenth days of the menstrual cycle. As the thickness increases, the uterine glands are drawn out so that they lengthen (Fig 23–24), but they do not become convoluted or secrete to any degree. These endometrial changes are called proliferative, and this part of the menstrual cycle is sometimes called the **proliferative phase.** It is also called the preovulatory or follicular phase of the cycle. After ovulation, the endometrium becomes more highly vascularized and slightly edematous under the influence of estrogen and progesterone from the corpus luteum. The glands become coiled and tortuous (Fig 23–24), and they begin to secrete a clear fluid. Consequently, this phase of the cycle is called the **secretory** or **luteal phase.**

The endometrium is supplied by 2 types of arteries. The superficial two-thirds of the endometrium that is shed during menstruation, the **stratum functionale,** is supplied by long, coiled **spiral arteries** (Fig 23–25), whereas the deep layer that is not shed, the **stratum basale,** is supplied by short, straight **basilar arteries.**

When the corpus luteum regresses, hormonal support for the endometrium is withdrawn. The endometrium becomes thinner, which adds to the coiling of the spiral arteries. Foci of necrosis appear in the endometrium, and these coalesce. There is in addition a necrosis of the walls of the spiral arteries, leading to spotty hemorrhages that become confluent and produce the menstrual flow.

The cause of the vascular necrosis is unknown, but it is associated with spasm of the blood vessels, which may be produced by locally released prostaglandins. There are large quantities of prostaglandins in the secretory endometrium and in menstrual blood, and infusions of $PGF_{2\alpha}$ produce endometrial necrosis and bleeding. One theory of the onset of menstruation holds that in necrotic endometrial cells, lysosomal membranes break down, with the release of enzymes that foster the formation of prostaglandins from cellular phospholipids. The prostaglandins then produce vasospasm, vascular necrosis, and menstrual flow. After menstruation, a new endometrium regenerates from cells that remain in the stratum basale.

From the point of view of endometrial function, the proliferative phase of the menstrual cycle represents restoration of the epithelium from the preceding menstruation, and the secretory phase represents preparation of the uterus for implantation of the fertilized ovum. The length of the secretory phase is remarkably constant at about 14 days, and the variations seen in the length of the menstrual cycle are due for the most part to variations in the length of the proliferative phase. When fertilization fails to occur during the secretory phase, the endometrium is shed and a new cycle starts.

Normal Menstruation

Menstrual blood is predominantly arterial, with only 25% of the blood being of venous origin. It contains tissue debris, prostaglandins, and relatively large amounts of fibrinolysin from endometrial tissue. The fibrinolysin lyses clots, so that menstrual blood does not normally contain clots unless the flow is excessive.

The usual duration of the menstrual flow is 3–5 days, but flows as short as 1 day and as long as 8 days can occur in normal women. The amount of blood lost may range normally from slight spotting to 80 mL; the average amount lost is 30 mL. Loss of more than 80 mL is abnormal. Obviously, the amount of flow can be affected by various factors, including the thickness of the endometrium, medication, and diseases that affect the clotting mechanism.

Anovulatory Cycles

In some instances, ovulation fails to occur during the menstrual cycle. Such anovulatory cycles are common for the first 12–18 months after menarche and again before the onset of the menopause. When ovulation does not occur, no corpus luteum is formed and the effects of progesterone on the endometrium

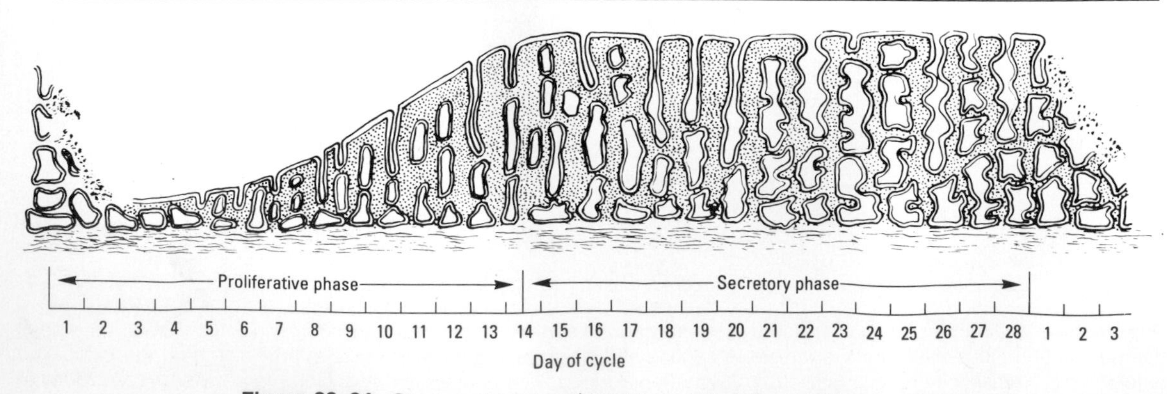

Figure 23–24. Changes in the endometrium during the menstrual cycle.

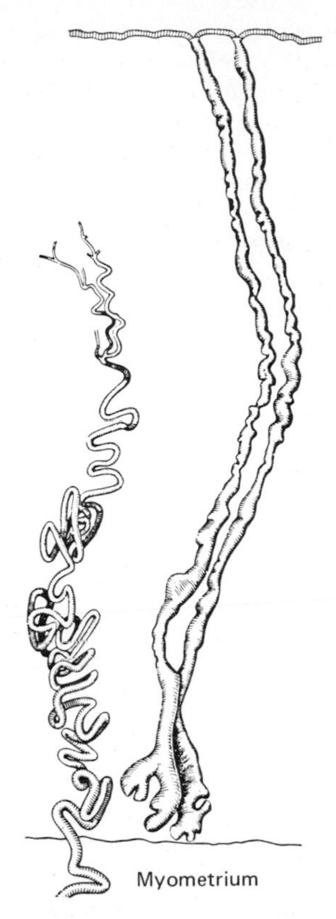

Figure 23–25. Spiral artery of endometrium. Drawing of a spiral artery (*left*) and 2 uterine glands (*right*) from the endometrium of a rhesus monkey; early progestational phase. The uterine cavity is at the top. (Reproduced, with permission, from Daron GH: The arterial pattern of the tunica mucosa of the uterus in the *Macacus rhesus. Am J Anat* 1936;**58**:349.)

are absent. Estrogens continue to cause growth, however, and the proliferative endometrium becomes thick enough to break down and begins to slough. The time it takes for bleeding to occur is variable, but it usually occurs in less than 28 days from the last menstrual period. The flow is also variable and ranges from scanty to relatively profuse.

Cyclic Changes in the Uterine Cervix

Although it is continuous with the body of the uterus, the cervix of the uterus is different in a number of ways. The mucosa of the uterine cervix does not undergo cyclic desquamation, but there are regular changes in the cervical mucus. Estrogen makes the mucus thinner and more alkaline, changes that promote the survival and transport of sperm. Progesterone makes it thick, tenacious, and cellular. The mucus is thinnest at the time of ovulation, and its elasticity, or **spinnbarkeit,** increases so that by midcycle, a drop can be stretched into a long, thin thread that may be 8–12 cm or more in length. In addition, it dries in an arborizing, fernlike pattern (Fig 23–26) when a thin layer is spread on a slide. After ovulation and during pregnancy, it becomes thick and fails to form the fern pattern.

Vaginal Cycle

Under the influence of estrogens, the vaginal epithelium becomes cornified, and cornified epithelial cells can be identified in the vaginal smear. Under the influence of progesterone, a thick mucus is secreted, and the epithelium proliferates and becomes infiltrated with leukocytes. The cyclic change in the vaginal smear in rats are particularly well known (Fig 23–27). The changes in humans and other species are similar but unfortunately not so clear-cut.

Cyclic Changes in the Breasts

Although lactation normally does not occur until the end of pregnancy, there are cyclic changes in the breasts during the menstrual cycle. Estrogens cause proliferation of mammary ducts, whereas progesterone causes growth of lobules and alveoli. The breast swelling, tenderness, and pain experienced by many women during the 10 days preceding menstruation are probably due to distention of the ducts, hyperemia, and edema of the interstitial tissue of the breast. All these changes regress, along with the symptoms, during menstruation.

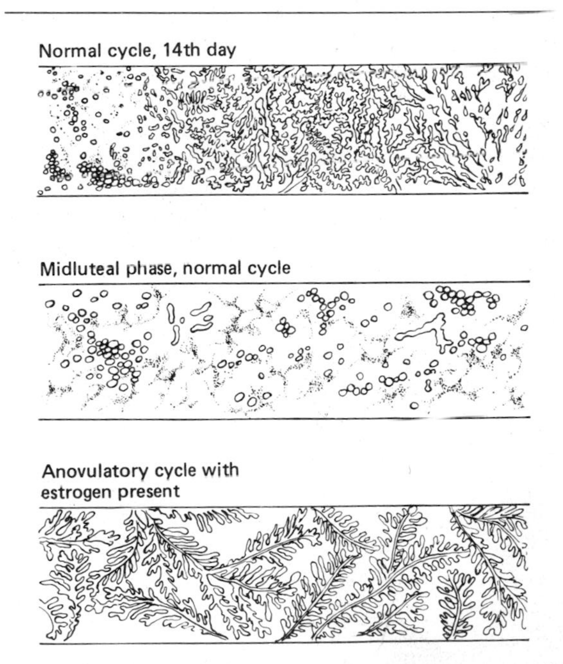

Figure 23–26. Patterns formed when cervical mucus is smeared on a slide, permitted to dry, and examined under the microscope. Progesterone makes the mucus thick and cellular. In the smear from a patient who failed to ovulate (*bottom*), there is no progesterone to inhibit the estrogen-induced fern pattern.

Changes During Intercourse

During sexual excitement in women, the vaginal walls become moist as a result of transudation of fluid through the mucous membrane. A lubricating mucus is secreted by the vestibular glands. The upper part of the vagina is sensitive to stretch, while tactile stimulation from the labia minora and clitoris adds to the sexual excitement. These stimuli are reinforced by tactile stimuli from the breasts and, as in men, by visual, auditory, and olfactory stimuli, which may build to the crescendo known as orgasm. During orgasm, there are autonomically mediated rhythmic contractions of the vaginal walls. Impulses also travel via the pudendal nerves and produce rhythmic contraction of the bulbocavernosus and ischiocavernosus muscles. The vaginal contractions may aid sperm transport but are not essential for it, since fertilization of the ovum is not dependent on orgasm.

Indicators of Ovulation

It is often important in clinical practice to know that ovulation has occurred and to know when during the cycle it occurred. The finding of a secretory pattern in a biopsy of the endometrium indicates that a functioning corpus luteum is present. Less reliably, finding thick, cellular cervical mucus that does not form a fern pattern in a woman who has regular menses is evidence of the same thing. A convenient and reasonably reliable indicator of the time of ovulation is a change—usually a rise—in the basal body temperature (Fig 23–28). Women interested in obtaining an accurate temperature chart should use a thermometer with wide gradations and take their temperatures (oral or rectal) in the morning before getting out of bed. The cause of the temperature change at the time of ovulation is unsettled. However, it is probably due to the increase in progesterone secretion, since progesterone is thermogenic (see Chapter 14).

The ovum lives for approximately 72 hours after it is extruded from the follicle, but it is probably fertilizable for less than half this time. Sperms apparently survive in the female genital tract for no more than 48 hours. Consequently, the "fertile period" during a 28-day cycle is no longer than 120 hours, and it is probably much shorter. Unfortunately for those interested in the "rhythm method" of contraception, the time of ovulation is rather variable even from one menstrual cycle to another in the same woman. Before the ninth and after the twentieth day, there is little chance of conception; but there are documented cases of pregnancy resulting from isolated coitus on every day of the cycle.

The Estrous Cycle

Mammals other than primates do not menstruate, and their sexual cycle is called an **estrous cycle.** It is named for the conspicuous period of "heat" **(estrus)** at the time of ovulation, normally the only time during which the sexual interest of the female is aroused (see Chapter 15). In spontaneously ovulating species with estrous cycles, such as the rat, there is no episodic vaginal bleeding but the underlying endocrine events are essentially the same as those in the menstrual cycle. In other species, ovulation is produced by copulation (reflex ovulation).

OVARIAN HORMONES

Chemistry, Biosynthesis & Metabolism of Estrogens

The naturally occuring estrogens are steroids that do not have an angular methyl group attached to the 10 position or a Δ^4-3-keto configuration in the A ring. They are secreted by the theca interna and granulosa cells of the ovarian follicles, by the corpus luteum, by the placenta, and, in small amounts, by the adrenal cortex and the testis. The biosynthetic pathway (Fig 23–29) involves their formation from androgens. They are also formed by aromatization of androstenedione in the circulation. Aromatase is the enzyme that catalyzes the conversion of androstenedione to estrone (Fig 23–29). It also catalyzes the conversion of testosterone to estradiol.

Theca interna cells have many LH receptors, and LH acts via cyclic AMP to increase conversion of cholesterol to androstenedione. Some of the androstenedione is converted to estradiol, which enters the circulation. The theca interna cells also supply androstenedione to the granulosa cells. The granulosa cells only make estradiol when provided with androgens (Fig 23–30), and in primates, it appears that the estradiol they form is secreted into the follicular fluid. Granulosa cells have many FSH receptors, and FSH facilitates their secretion of estradiol by acting via cyclic AMP to increase aromatase activity in the cells.

The stromal tissues of the ovary also have the potential to make androgens and estrogens. However, they probably do so in insignificant amounts in normal premenopausal women. **17β-Estradiol,** the major secreted estrogen, is in equilibrium in the circulation with **estrone.** Estrone is further metabolized to **estriol,** probably for the most part in the liver. Estradiol is the most potent estrogen of the 3, and estriol the least.

Three percent of the circulating estradiol is free, and the remainder is bound to protein: 60% to albumin and 37% to the same gonadal steroid-binding globulin (GBG) that binds testosterone.

In the liver, estrogens are oxidized or converted to glucuronide and sulfate conjugates. Appreciable amounts are secreted in the bile and reabsorbed into the bloodstream (enterohepatic circulation). There are at least 10 different metabolites of estradiol in human urine.

Secretion

The concentration of estradiol in the plasma during the menstrual cycle is shown in Fig 23–28. Almost all of this estrogen comes from the ovary, and there are

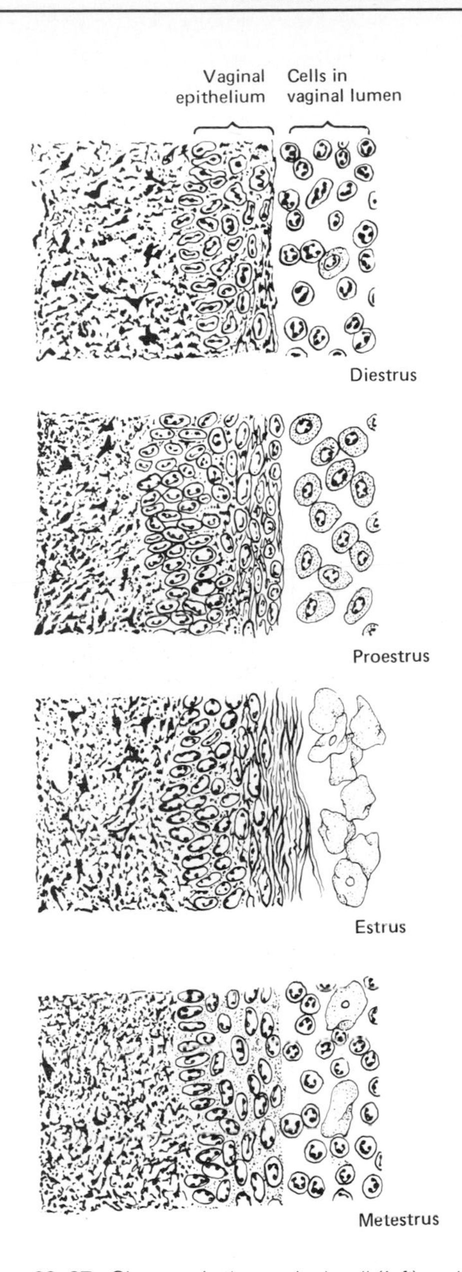

Figure 23–27. Changes in the vaginal wall (*left*) and type of cells found in the vaginal smear (*right*) during the estrous cycle in rats. The vaginal smear pattern in women is similar but not so clear-cut. (Redrawn and reproduced, with permission, from Turner CD, Bagnara JT: *General Endocrinology,* 6th ed. Saunders, 1976.)

2 peaks of secretion: one just before ovulation and one during the midluteal phase. The estradiol secretion rate is 0.07 mg/d (0.26 μmol/d) in the early follicular phase, 0.6 mg/d just before ovulation, and 0.25 mg/d during the midluteal phase. After menopause, estrogen secretion declines to low levels.

As noted above, the estradiol production rate in men is 0.05 mg/d (0.18 μmol/d).

Effects on the Female Genitalia

Estrogens facilitate the growth of the ovarian follicles and increase the motility of the uterine tubes. Their role in the cyclic changes in the endometrium, cervix, and vagina is discussed above. They increase uterine blood flow and have important effects on the smooth muscle of the uterus. In immature and castrate females, the uterus is small and the myometrium atrophic and inactive. Estrogens increase the amount of uterine muscle and its content of contractile proteins. Under the influence of estrogens, the muscle becomes more active and excitable, and action potentials in the individual fibers become more frequent (see Chapter 3). The "estrogen-dominated" uterus is also more sensitive to oxytocin.

Chronic treatment with estrogens causes the endometrium to hypertrophy. When estrogen therapy is discontinued, there is sloughing with **withdrawal bleeding.** Some "breakthrough" bleeding may occur during treatment when estrogens are given for long periods.

Effects on Endocrine Organs

Estrogens decrease FSH secretion. Under some circumstances, they inhibit LH secretion (negative feedback); in other circumstances, they increase LH secretion (positive feedback; see below). Estrogens also increase the size of the pituitary. Women are sometimes given large doses of estrogens for 4–6 days to prevent conception after coitus during the fertile period (postcoital or "morning after" contraception). However, in this instance, pregnancy is probably prevented by interference with implantation of the fertilized ovum rather than changes in gonadotropin secretion.

Estrogens cause increased secretion of angiotensinogen (see Chapter 24) and thyroid-binding globulin (see Chapter 18). They exert an important protein anabolic effect in chickens and cattle, possibly by stimulating the secretion of androgens from the adrenal, and estrogen treatment has been used commercially to increase the weight of domestic animals. Estrogens have been reported to exert anabolic effects and to cause epiphyseal closure in humans.

Behavioral Effects

The estrogens are responsible for estrous behavior in animals, and they increase libido in humans. They apparently exert this action by a direct effect on certain neurons in the hypothalamus (Fig 23–31). The relation of estrogens, progesterone, and androgens to sexual behavior is discussed in Chapter 15.

Effects on the Breasts

Estrogens produce duct growth in the breasts and are largely responsible for breast enlargement at puberty in girls. Breast enlargement that occurs when estrogen-containing skin creams are applied locally is

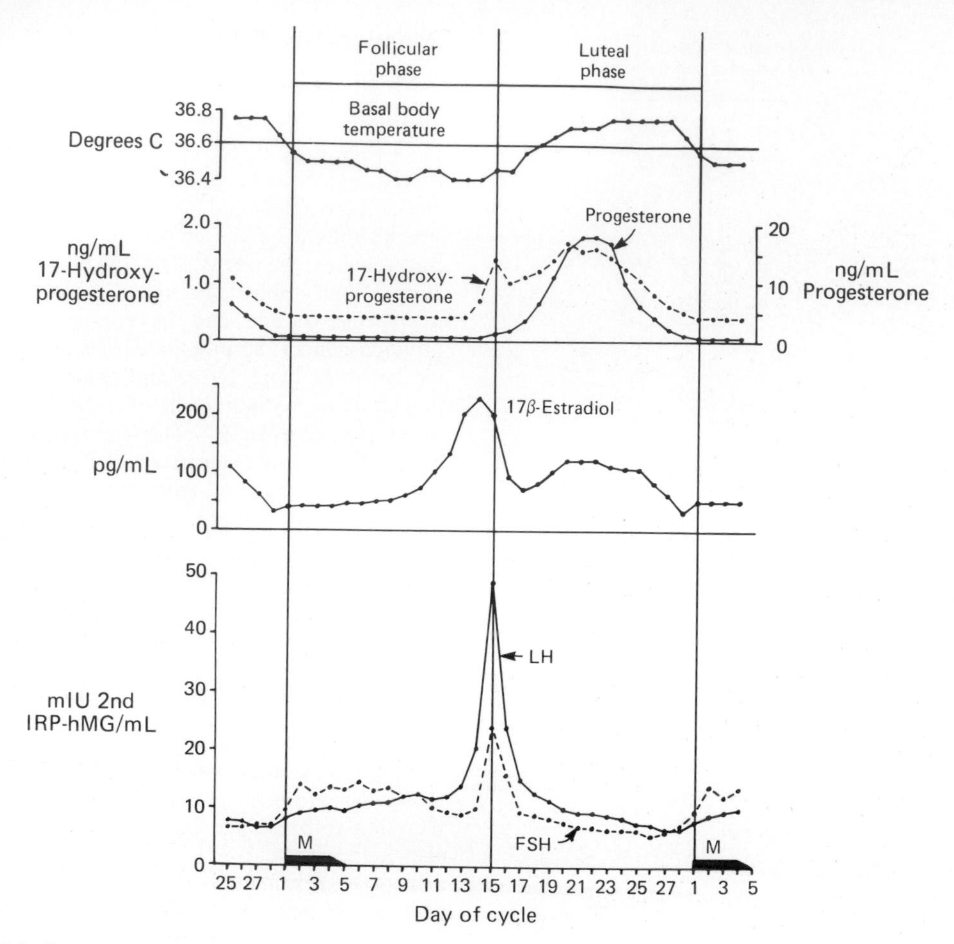

Figure 23–28. Typical basal body temperature and plasma hormone concentrations during a normal 28-day human menstrual cycle. M, menstruation; 2nd IRP-hMG, second international reference standard for gonadotropins. (Reproduced, with permission, from Midgley AR in: *Human Reproduction.* Hafez ESE, Evans TN [editors]. Harper & Row, 1973.)

due primarily to systemic absorption of the estrogen, although a slight local effect is also produced. Estrogens are responsible for the pigmentation of the areolas, although pigmentation usually becomes more intense during the first pregnancy than it does at puberty. The role of the estrogens in the overall control of breast growth and lactation is discussed below.

Female Secondary Sex Characteristics

The body changes that develop in girls at puberty—in addition to enlargement of breasts, uterus, and vagina—are due in part to estrogens, which are the "feminizing hormones," and in part simply to the absence of testicular androgens. Women have narrow shoulders and broad hips, thighs that converge, and arms that diverge (wide **carrying angle**). This body configuration, plus the female distribution of fat in the breasts and buttocks, is seen also in castrate males. In women, the larynx retains its prepuberal proportions and the voice stays high-pitched. There is less body hair and more scalp hair, and the pubic hair generally has a characteristic flat-topped pattern (female escutcheon). Growth of pubic and axillary hair in the female is due primarily to androgens rather than estrogens, although estrogen treatment may cause some hair growth. The androgens come from the adrenal cortex and, to a lesser extent, from the ovaries.

Other Actions

Normal women retain salt and water and gain weight just before menstruation. Estrogens cause some degree of salt and water retention. However, it is also possible that increased vasopressin secretion contributes to the premenstrual fluid retention.

Estrogens are said to make sebaceous gland secretions more fluid and thus to counter the effect of testosterone and inhibit formation of **comedones** ("blackheads") and acne. The liver palms, spider angiomas, and slight breast enlargement seen in advanced liver disease are due to increased circulating estrogens. The increase is due not only to

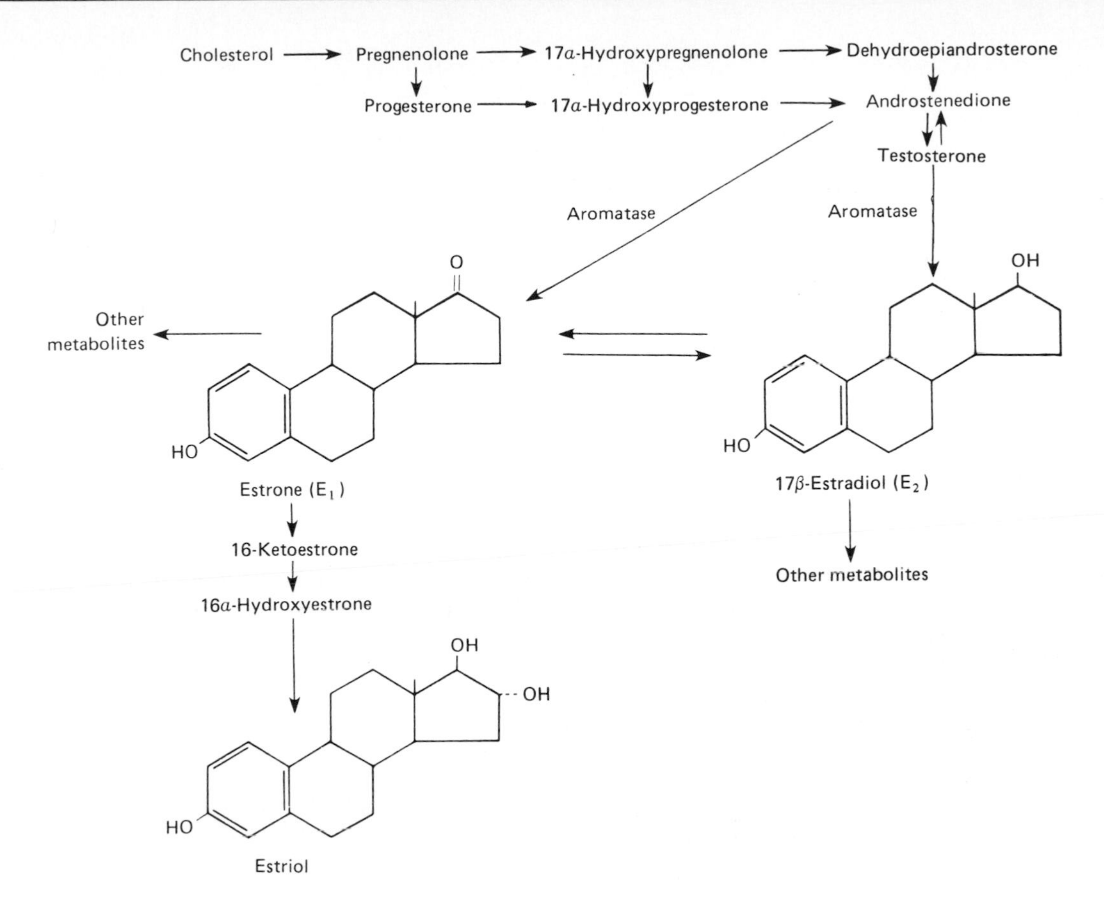

Figure 23–29. Biosynthesis and metabolism of estrogens. The formulas of the precursor steroids are shown in Fig 20–8.

complex alterations in the hepatic metabolism of estrogens but also to increased conversion of androgens to estrogens.

Estrogens have a significant plasma cholesterol-lowering action (see Chapter 17), and they inhibit atherogenesis. They may contribute to the low incidence of myocardial infarction and other complications of arteriosclerotic vascular disease in premenopausal women. However, pharmacologic doses of orally active estrogens appear to promote thrombosis by acting on the liver to alter its production of clotting factors. Thus, estrogen treatment is of little practical value in the prevention of coronary artery disease.

Mechanism of Action

Like other steroids (see Chapter 1), estrogens combine with an intracellular protein receptor, and the complex binds to DNA, promoting formation of mRNAs that in turn direct the formation of new proteins which modify cell function. It seems probable that almost all estrogen actions are produced in this fashion, although it is possible that other mechanisms are also involved.

Synthetic Estrogens

The ethinyl derivative of estradiol (Fig 23–32) is a potent estrogen and—unlike the naturally occurring estrogens—is relatively active when given by mouth, because it is resistant to hepatic metabolism. The activity of the naturally occurring hormones is low when they are administered by mouth because the portal venous drainage of the intestine carries them to the liver, where they are inactivated before they can reach the general circulation. Some nonsteroidal substances and a few compounds found in plants have estrogenic activity. The plant estrogens are rarely a problem in human nutrition, but they may cause undesirable effects in farm animals. Diethylstilbestrol (Fig 23–32) and a number of related compounds are estrogenic, possibly because they are converted to a steroidlike ring structure in the body.

Chemistry, Biosynthesis, & Metabolism of Progesterone

Progesterone is a C_{21} steroid (Fig 23–33) secreted by the corpus luteum and the placenta. It is an important intermediate in steroid biosynthesis in all

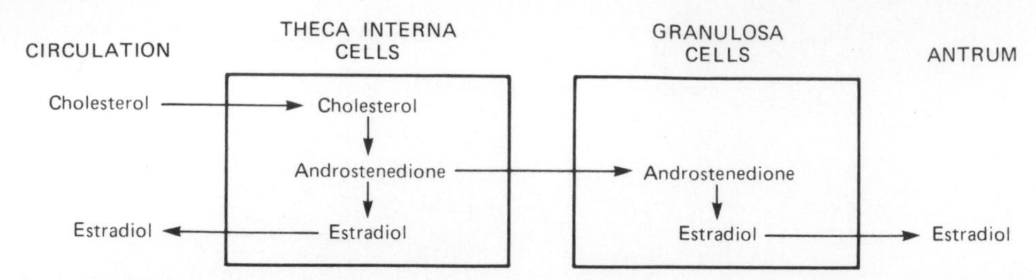

Figure 23–30. Interactions between theca and granulosa cells in estradiol synthesis and secretion.

tissues that secrete steroid hormones, and small amounts apparently enter the circulation from the testes and adrenal cortex. 17α-Hydroxyprogesterone is apparently secreted along with estrogens from the ovarian follicle, and its secretion parallels that of 17β-estradiol. About 45% of the progesterone in the circulation is bound to transcortin (see Chapter 20) and 50% of albumin. About 1–2% is free. Progesterone has a short half-life and is converted in the liver to pregnanediol, which is conjugated to glucuronic acid and excreted in the urine (Fig 23–33).

Secretion

In men, the plasma progesterone level is approximately 0.3 ng/mL (1 nmol/L). In women, the level is approximately 0.9 ng/mL (3 nmol/L) during the follicular phase of the menstrual cycle, the difference being due to secretion of small amounts of progesterone by cells in the ovarian follicles. During the luteal phase, the corpus luteum produces large quantities of progesterone, and ovarian secretion increases about 20-fold (Fig 23–28). The result is an increase in plasma progesterone to a peak value of approximately 18 ng/mL (60 nmol/L).

The stimulating effect of LH on progesterone secretion by the corpus luteum has been shown to be accompanied by increased formation of cyclic AMP. The increase in progesterone secretion produced by LH or exogenous cyclic AMP is reduced by puromycin, which indicates that it depends upon the synthesis of new protein (see Chapter 1). However, the increase in the cyclic AMP content of the corpus luteum produced by LH is not blocked. The data suggest that LH activates adenylate cyclase in the corpus luteum and that in a series of events analogous to that triggered by ACTH in the adrenal, the increased cyclic AMP initiates a reaction which involves protein synthesis and facilitates steroid secretion.

Actions

The principal target organs of progesterone are the uterus, the breasts, and the brain. Progesterone is responsible for the progestational changes in the endometrium and the cyclic changes in the cervix and vagina described above. It has an antiestrogenic effect on the myometrial cells, decreasing their excitability, their sensitivity to oxytocin, and their spontaneous electrical activity while increasing their mem-

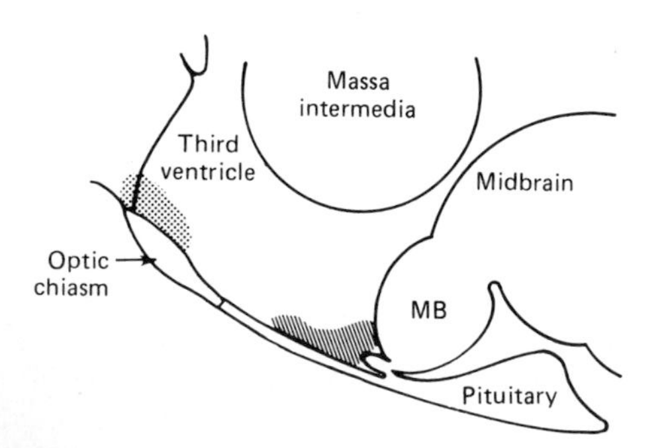

Figure 23–31. Loci where implantations of estrogen in the hypothalamus affect ovarian weight and sexual behavior in rats, projected on a sagittal section of the hypothalamus. The implants that stimulate sex behavior are located in the suprachiasmatic area above the optic chiasm (dotted area), whereas ovarian atrophy is produced by implants in the arcuate nucleus and surrounding ventral hypothalamus (striped area). MB, mamillary body.

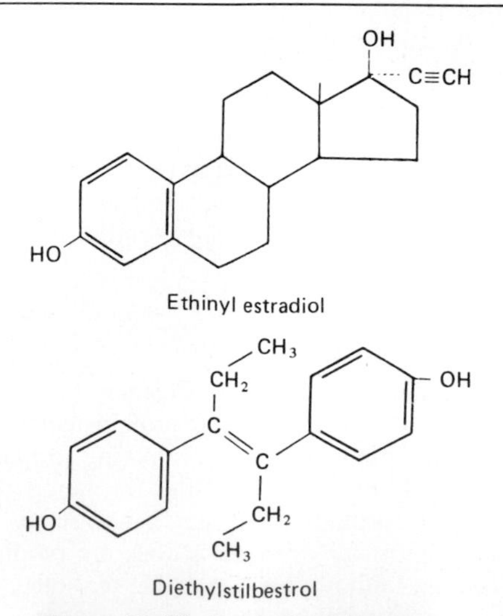

Figure 23–32. Synthetic estrogens.

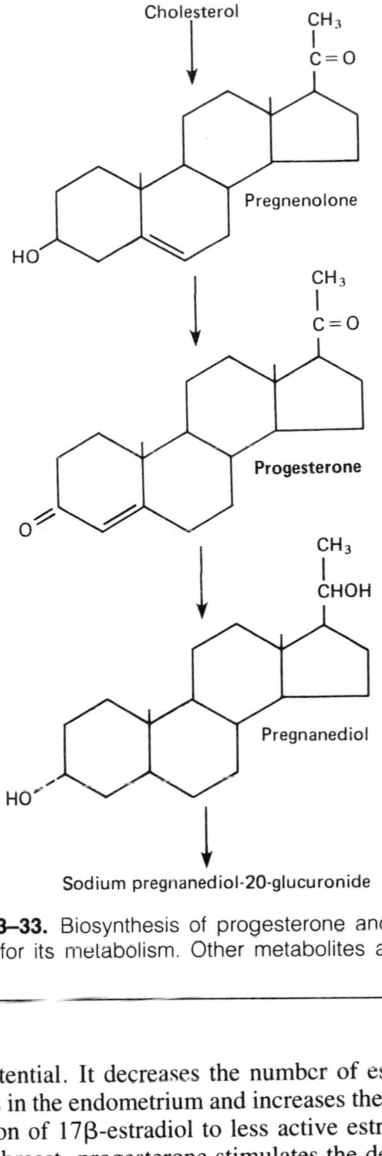

Figure 23–33. Biosynthesis of progesterone and major pathway for its metabolism. Other metabolites are also formed.

brane potential. It decreases the number of estrogen receptors in the endometrium and increases the rate of conversion of 17β-estradiol to less active estrogens.

In the breast, progesterone stimulates the development of lobules and alveoli. It induces differentiation of estrogen-prepared ductal tissue and supports the secretory function of the breast during lactation. The feedback effects of progesterone are complex and are exerted at both the hypothalamic and the pituitary levels. Large doses of progesterone inhibit LH secretion and potentiate the inhibitory effect of estrogens. Progesterone injections can prevent ovulation in humans. Progesterone does not induce heat in castrate animals, except possibly in huge doses, but in some species it lowers the dose of estrogen necessary to produce estrous behavior.

Progesterone is thermogenic and is probably responsible for the rise in basal body temperature at the time of ovulation. It stimulates respiration, and the alveolar P_{CO_2} (P_{ACO_2}; see Chapter 34) in women during the luteal phase of the menstrual cycle is lower than that in men. In pregnancy, the P_{ACO2} falls as progesterone secretion rises.

Large doses of progesterone produce natriuresis, probably by blocking the action of aldosterone on the kidney. The hormone does not have a significant anabolic effect.

The effects of progesterone, like those of other steroids, are brought about primarily by an action on DNA to initiate synthesis of new mRNA.

Substances that mimic the action of progesterone are sometimes called **progestational agents, gestagens,** or **progestins.** They are used along with synthetic estrogens as oral contraceptive agents (see below).

Relaxin

Relaxin is a polypeptide hormone that relaxes the pubic symphysis and other pelvic joints and softens and dilates the uterine cervix during pregnancy. Thus, it facilitates delivery. It also inhibits uterine contractions and may play a role in the development of the mammary glands. Human relaxin resembles human insulin, with 2 polypeptide chains and 3 disulfide bridges located in the same relative positions as those in insulin (Fig 22–8). However, it has a 24-amino-acid A chain and a 32-amino-acid B chain, with many amino acid residues that differ from those in insulin. In pregnant women, it is found in the corpus luteum and in the placenta, where it is found in the chorionic cytotrophoblast and the cells of the placental base plate. In nonpregnant women, it is found in the corpus luteum and the endometrium during the secretory but not the proliferative phase of the menstrual cycle.

CONTROL OF OVARIAN FUNCTION

FSH from the pituitary is responsible for the early maturation of the ovarian follicles, and FSH and LH together are responsible for their final maturation. A burst of LH secretion (Fig 23–28) is responsible for ovulation and the initial formation of the corpus luteum. There is also a smaller midcycle burst of FSH secretion the significance of which is uncertain. LH stimulates the secretion of estrogen and progesterone from the corpus luteum. A late rise in FSH just before menstruation primes the follicles for their subsequent maturation during the next cycle.

Hypothalamic Components

The hypothalamus occupies a key position in the control of gonadotropin secretion. Hypothalamic control is exerted by LHRH secreted into the portal hypophyseal vessels (see Chapter 14). LHRH stimulates the secretion of FSH as well as LH, and it is uncertain whether there is an additional separate FRH.

LHRH is normally secreted in episodic bursts, and these bursts produce the circhoral peaks of LH secretion (Fig 23–34). They are essential for normal secretion of gonadotropins. If LHRH is administered

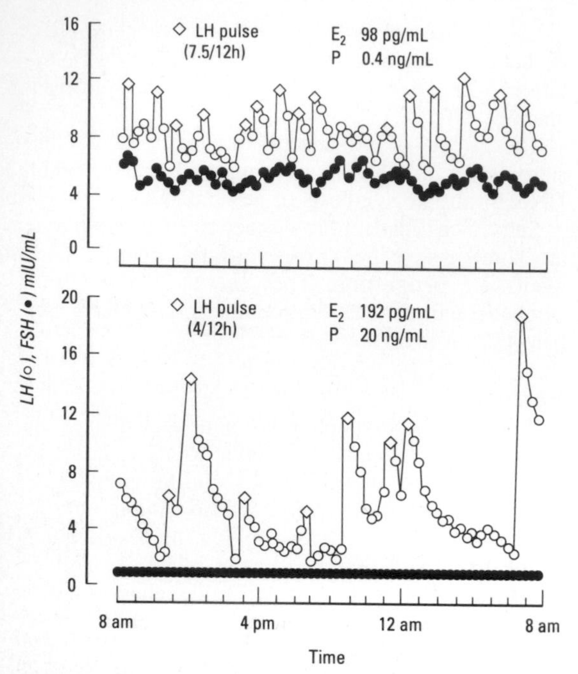

Figure 23–34. Episodic secretion of LH (○) and FSH (•) du;ring the follicular stage (**top**) and the luteal stage (**bottom**) of the menstrual cycle. The numbers above each graph indicate the numbers of LH pulses per 12h and the plasma estradiol (E_2) and progesterone (P) concentrations at these 2 times of the cycle. (Reproduced, with permission, from Marshall JC, Kelch RO: Gonadotropin-releasing hormone: role of pulsatile secretion in the regulation of reproduction. *N Engl J Med* 1986;**315**:1459.)

by constant infusion, the LHRH receptors in the anterior pituitary down-regulate (see Chapter 1) and LH secretion declines to zero. However, if LHRH is administered episodically at a rate of one pulse per hour, LH secretion is stimulated. This is true even when endogenous LHRH secretion has been prevented by a lesion of the ventral hypothalamus.

It is now clear not only that episodic secretion of LHRH is a general phenomenon but also that fluctuations in the frequency and amplitude of the LHRH bursts are important in generating the other hormonal changes that are responsible for the menstrual cycle. Frequency is increased by estrogens and decreased by progesterone and testosterone. The frequency increases late in the follicular phase of the cycle, culminating in the LH surge. During the proliferative phase, the frequency decreases as a result of the action of progesterone (Fig 23–34), but when estrogen and progesterone secretion decrease at the end of the cycle, the frequency once again increases.

The nature and the exact location of the LHRH pulse generator in the hypothalamus are still unsettled. However, it is known in a general way that norepinephrine and possibly epinephrine in the hypothalamus increase LHRH pulse frequencies. Conversely, opioid peptides such as the enkephalins and β-endorphin reduce the frequency of LHRH pulses.

The down-regulation of pituitary receptors and the consequent decrease in LH secretion produced by constantly elevated levels of LHRH has led to the suggestion that this hormone or some of its long-acting analogs could be used as effective contraceptive agents, and preliminary results have been promising in this regard in both men and women.

Feedback Effects

Estrogens inhibit FSH and LH secretion during the early part of the follicular phase of the cycle (Fig 23–35), and inhibin from the ovarian follicles inhibits FSH secretion. The rise in circulating estrogens 24 hours before ovulation initiates the burst of LH secretion (LH surge) that produces ovulation. Ovulation occurs about 9 hours after the LH peak. The secretion of FSH and LH is again inhibited by the high circulating estrogen and progesterone levels during the luteal phase of the menstrual cycle. Thus, a moderate, constant level of circulating estrogen exerts a negative feedback effect on LH secretion, whereas an elevated estrogen level exerts a positive feedback effect and stimulates LH secretion. It has been demonstrated that in monkeys, there is also a minimum time that estrogens must be elevated to produce positive feedback. When circulating estro-

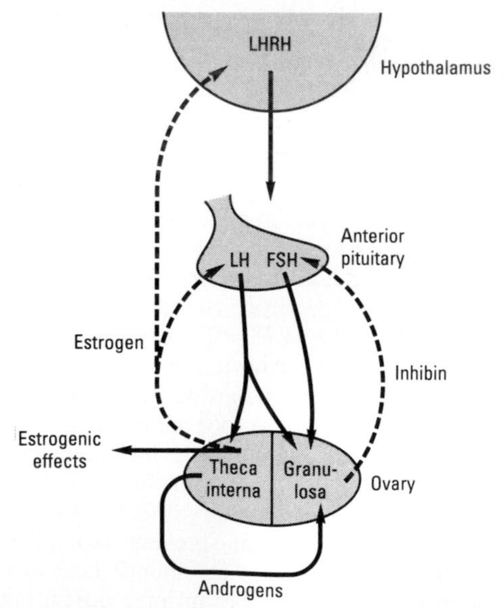

Figure 23–35. Feedback regulation of ovarian function. The cells of the theca interna provide androgens to the granulosa cells, and theca cells also produce the circulating estrogens that inhibit the secretion of LHRH, LH, and FSH. Inhibin from the granulosa cells inhibits FSH secretion. LH regulates the thecal cells, whereas the granulosa cells are regulated by both LH and FSH. The dashed arrows indicate inhibitory effects and the solid arrows stimulatory effects. Compare with Fig 18–13, 20–22, 22–9, and 23–21.

gen was increased about 300% for 24 hours, only negative feedback was seen; but when it was increased about 300% for 36 hours or more, a brief decline in secretion was followed by a burst of LH secretion that resembled the midcycle surge. When circulating levels of progesterone were high, the positive feedback effect of estrogen was inhibited.

Control of the Cycle

In an important sense, regression of the corpus luteum **(luteolysis)** is the key to the menstrual cycle. There is evidence that prostaglandins may play a role in this process, possibly by inhibiting the effect of LH on cyclic AMP, but their exact function is unknown. In some domestic animals, oxytocin appears to be secreted by the corpus luteum and to exert a local luteolytic effect, possibly by causing release of prostaglandins. Once luteolysis begins, the estrogen and progesterone levels fall and the secretion of FSH and LH increases. A new follicle develops and matures as a result of the action of FSH and LH. Near midcycle, there is a rise in estrogen secretion from the follicle. This rise augments the responsiveness of the pituitary to LHRH and triggers a burst of LH secretion. The resulting ovulation is followed by formation of a corpus luteum. There is a drop in estrogen secretion, but progesterone and estrogen levels then rise together. The elevated estrogen and progesterone levels inhibit FSH and LH secretion for a while, but luteolysis again occurs and a new cycle starts.

Reflex Ovulation

Female cats, rabbits, mink, and some other animals have long periods of estrus, during which they ovulate only after copulation. Such **reflex ovulation** is brought about by afferent impulses from the genitalia and the eyes, ears, and nose that converge on the ventral hypothalamus and provoke an ovulation-inducing release of LH from the pituitary. In species such as rats, monkeys, and humans, ovulation is a spontaneous periodic phenomenon, but neural mechanisms are also involved. Ovulation can be prevented in rats by administering pentobarbital or various other neurally active drugs 12 hours before the expected time of ovulation. In women, menstrual cycles may be markedly influenced by emotional stimuli.

Effects of Intrauterine Foreign Bodies

Implantation of foreign bodies in the uterus causes changes in the duration of the sexual cycle in a number of mammalian species. In humans, such foreign bodies do not alter the menstrual cycle, but they act as effective contraceptive devices. Intrauterine implantation of pieces of metal or plastic **(intrauterine devices, IUDs)** has been used in programs aimed at controlling population growth. The mechanism by which these devices exert their contraceptive effect is unsettled, but there is some evidence that they speed the passage of the fertilized ovum through the uterus, preventing its implantation in the endometrium. They also disturb the orderly sequential changes in the endometrium during the menstrual cycle, and this may be a factor. Their usefulness is limited by their tendency to cause intrauterine infections.

Contraceptive Steroids

Women treated over a long period with relatively large doses of estrogen do not ovulate, probably because they have depressed FSH levels and multiple irregular bursts of LH secretion rather than a single midcycle peak. Women treated with similar doses of estrogen plus a progestational agent do not ovulate because the secretion of both gonadotropins is suppressed. In addition, the progestin makes the cervical mucus thick and unfavorable to sperm migration, and it may also interfere with implantation. For contraception, an orally active estrogen such as ethinyl estradiol (Fig 23–32) is often combined with a synthetic progestin such as norethindrone (Fig 23–36). The pills are administered for 21 days, then withdrawn for 5–7 days to permit menstrual flow, and started again. Norethindrone has an ethinyl group on position 17 of the steroid nucleus, so it is resistant to hepatic metabolism and consequently effective by mouth. In addition to being a progestin, it is partly metabolized to ethinyl estradiol, and for this reason it also has estrogenic activity. In sequential therapy, which has been used as an alternative to combination therapy, an estrogen is administered and a progestin is added for the last 5 days of each cycle. The treatments are effective, but there are complications including thromboses in some women. Orally active estrogens, as opposed to naturally produced estradiol, reach the liver in relatively high concentrations and alter its production of clotting factors. This appears to be the cause of the thromboses. Sequential therapy has been discontinued in the USA because of the possibility that it increases the incidence of carcinoma of the endometrium.

Implants made up primarily of progestins are now seeing increased use in some parts of the world. These are inserted under the skin and can prevent pregnancy for up to 5 years. They often produce amenorrhea, but otherwise they appear to be effective and well tolerated.

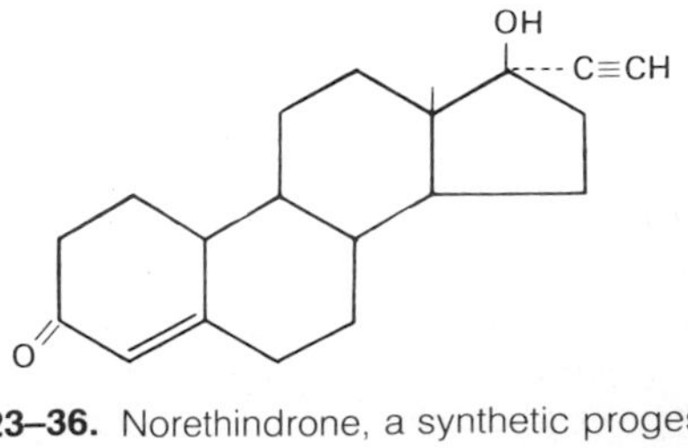

Figure 23–36. Norethindrone, a synthetic progestational agent.

ABNORMALITIES OF OVARIAN FUNCTION

Menstrual Abnormalities

Some women who are infertile have **anovulatory cycles;** they fail to ovulate but have menstrual periods at fairly regular intervals. Such anovulatory cycles are the rule for the first 1–2 years after menarche and again before the menopause. **Amenorrhea** is the absence of menstrual periods. If menstrual bleeding has never occurred, the condition is called **primary amenorrhea.** Some women with primary amenorrhea have small breasts and other signs of failure to mature sexually. Cessation of cycles in a woman with previously normal periods is called **secondary amenorrhea.** The commonest cause of secondary amenorrhea is pregnancy, and the old clinical maxim that "secondary amenorrhea should be considered to be due to pregnancy until proven otherwise" has considerable merit. Other causes of amenorrhea include emotional stimuli and changes in the environment, hypothalamic diseases, pituitary disorders, primary ovarian disorders, and various systemic diseases. There is evidence that in some women with hypothalamic amenorrhea, the frequency of LHRH pulses is slowed as a result of excess opioid activity in the hypothalamus, and in encouraging preliminary studies, the frequency of the LHRH pulses has been increased by administration of the orally active opioid blocker naltrexone.

The terms **oligomenorrhea** and **menorrhagia** refer to scanty and abnormally profuse flow, respectively, during regular periods. **Metrorrhagia** is bleeding from the uterus between periods. **Dysmenorrhea** is painful menstruation. The severe menstrual cramps that are common in young women quite often disappear after the first pregnancy. Most of the symptoms of dysmenorrhea are due to accumulation of prostaglandins in the uterus, and symptomatic relief has been obtained by treatment with inhibitors of prostaglandin synthesis (see Chapter 17).

Some women develop symptoms such as irritability, bloating, edema, emotional lability, decreased ability to concentrate, headache, and constipation during the last 7–10 days of their menstrual cycles. These symptoms of the **premenstrual syndrome** have been attributed to salt and water retention produced by estrogens, but this seems unlikely because the peak of estrogen secretion occurs before the symptoms develop. Other explanations have been advanced, but at present, the cause of the premenstrual syndrome must be listed as unknown.

The Polycystic Ovary Syndrome

An interesting cause of infertility and amenorrhea is the **polycystic ovary syndrome** (Stein-Leventhal syndrome), a condition characterized by thickening of the ovarian capsule and the formation of multiple follicular cysts, usually in both ovaries. Plasma testosterone, estradiol, and LH are elevated in this syndrome, whereas plasma FSH is low. It has been suggested that this condition is due to an abnormality of the hypothalamic pulse generator in which LHRH pulses are too frequent. This would favor LH secretion and diminish FSH secretion (see above).

Ovarian Tumors

Androgen-secreting ovarian tumors can cause masculinization, and estrogen-secreting ovarian tumors in childhood can cause precocious sexual development (Table 23–2).

PREGNANCY

Fertilization & Implantation

In humans, **fertilization** of the ovum by the sperm usually occurs in the mid portion of the uterine tube. Many sperms attach to the **zona pellucida,** a membranous structure that surrounds the ovum. The sperms then bind to the zona pellucida by a reaction between sperm receptors in the zona and a specific egg-binding protein on the sperm plasma membrane. Binding is followed by the **acrosomal reaction,** which appears to be triggered by the sperm receptor. This reaction is the breakdown of the acrosome, a lysosomelike organelle on the head of the sperm (Fig 23–17), with the release of various enzymes including the trypsinlike protease **acrosin.** Acrosin facilitates the penetration of the sperm through the zona pellucida. When one sperm reaches the membrane of the ovum, it fuses to it, setting off a reduction in the membrane potential of the ovum that prevents polyspermy, the fertilization of the ovum by more than one sperm. This transient potential change is followed by a structural change in the zona pellucida that provides protection against polyspermy on a more long-term basis. Fusion of the cell membranes of the sperm and the ovum activates the cells, and embryonic development begins.

The developing embryo, now called a **blastocyst,** moves down the tube into the uterus. Once in contact with the endometrium, the blastocyst becomes surrounded by an outer layer of **syncytiotrophoblast,** a multinucleate mass with no discernible cell boundaries, and an inner layer of **cytotrophoblast** made up of individual cells. The syncytiotrophoblast erodes the endometrium, and the blastocyst burrows into it **(implantation).** The blastocyst spends about 3 days in the uterine tube and 3 more days in the fluids of the uterus before it implants. The implantation site is usually on the dorsal wall of the uterus. A placenta then develops, and the trophoblast remains associated with it.

It should be noted that the fetus and the mother are 2 genetically distinct individuals, and the fetus is in effect a transplant of foreign tissue in the mother. However, the transplant is tolerated, and the rejection reaction that is characteristically produced when other foreign tissues are transplanted (see Chapter 27) fails to occur. The reason the "fetal graft" is not rejected is unknown, although it may be that the major histocompatibility antigens are not expressed by the trophoblast in areas where it is in contact with maternal tissue.

Endocrine Changes

In all mammals, the corpus luteum in the ovary at the time of fertilization fails to regress and instead enlarges in response to stimulation by gonadotropic hormones secreted by the placenta. The placental gonadotropin in humans is called **human chorionic gonadotropin (hCG).** The enlarged **corpus luteum of pregnancy** secretes estrogens, progesterone, and relaxin. In most species, removal of the ovaries at any time during pregnancy precipitates abortion. In humans, however, the placenta produces sufficient estrogen and progesterone from maternal and fetal precursors to take over the function of the corpus luteum after the sixth week of pregnancy. Ovariectomy before the sixth week leads to abortion but thereafter has no effect on the pregnancy. The function of the corpus luteum begins to decline after 8 weeks of pregnancy, but it persists throughout pregnancy. hCG secretion decreases after an initial marked rise, but estrogen and progesterone secretion increase until just before parturition (Fig 23–37).

hCG

hCG is a glycoprotein that contains lactose and hexosamine. It is produced by the syncytiotrophoblast. Like the pituitary glycoprotein hormones, it is made up of α and β subunits. hCG-α is very similar to the α subunit of LH, FSH, and TSH, differing only in having 2 amino acid residues inverted and 3 amino acid residues deleted at the N terminal. The molecular weight of hCG-α is 18,000, and that of hCG-β is 28,000. hCG is primarily luteinizing and luteotropic and has little FSH activity. It can be measured by radioimmunoassay and detected in the blood as early as 6 days after conception. Its presence in the urine in early pregnancy is the basis of the various laboratory tests for pregnancy, and it can sometimes be detected in the urine as early as 14 days after conception.

hCG is not absolutely specific for pregnancy. Small amounts are secreted by a variety of gastrointestinal and other tumors in both sexes, and hCG has been measured in individuals with suspected tumors as a "tumor marker." It also appears that the fetal liver and kidney normally produce small amounts of hCG.

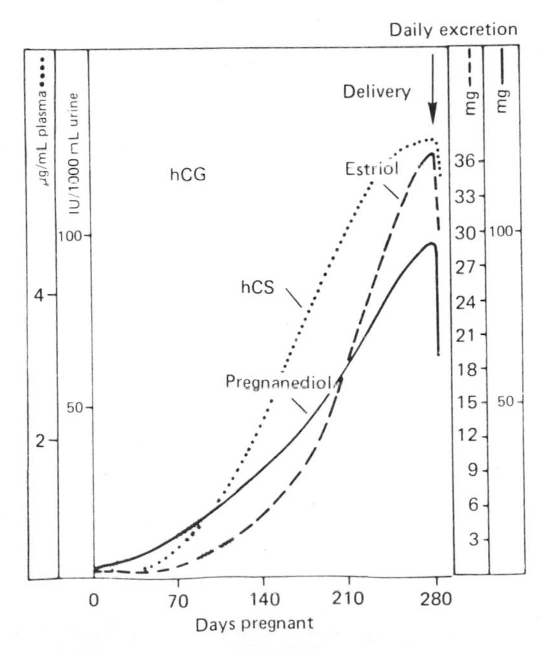

Figure 23–37. Hormone levels during normal pregnancy. (Data from various authors.)

hCS

The syncytiotrophoblast also secretes large amounts of a protein hormone that is lactogenic and has a small amount of growth-stimulating activity. This hormone has been called **chorionic growth hormone–prolactin** (CGP) and **human placental lactogen** (hPL), but it is now generally called **human chorionic somatomammotropin** (hCS). The structure of hCS is very similar to that of human growth hormone (Fig 22–4), and it appears that these 2 hormones and prolactin evolved from a common progenitor hormone. Large quantities of hCS are found in maternal blood, but very little reaches the fetus. Like the plasma hCS level, the plasma prolactin level increases progressively throughout pregnancy. On the other hand, secretion of growth hormone from the maternal pituitary is not increased during pregnancy and may actually be decreased by hCS. However, hCS has most of the actions of growth hormone and apparently functions as a "maternal growth hormone of pregnancy" to bring about the nitrogen, potassium, and calcium retention and decreased glucose utilization seen in this state. The amount of hCS secreted is proportionate to the size of the placenta, which normally weighs about one-sixth as much as the fetus, and low hCS levels are a sign of placental insufficiency.

Other Placental Hormones

In addition to hCG, hCS, progesterone, and estrogens, the human placenta apparently secretes relaxin. Placental extracts have TSH activity, but this activity is probably due to hCG, which has intrinsic TSH activity (see Chapter 18). The placenta contains 2 β-endorphin-like materials the significance of which is presently unknown. The cytotrophoblast of the human chorion contains prorenin (see Chapter 24), a material that appears to be identical to LHRH, and inhibin. There are large amounts of prorenin in amniotic fluid, but its function in this location is unknown. The placental LHRH and inhibin are secreted by human placental cells in tissue culture, and LHRH stimulates and inhibin inhibits hCG secretion. Thus, locally produced LHRH and inhibin may act in a paracrine fashion to regulate hCG secretion.

Fetoplacental Unit

The fetus and the placenta interact in the formation of steroid hormones. The placenta synthesizes pregnenolone and progesterone from cholesterol. Some of the progesterone enters the fetal circulation and provides the substrate for the formation of cortisol and corticosterone in the fetal adrenal glands (Fig 23–38). Some of the pregnenolone enters the fetus, and, along with pregnenolone synthesized in the fetal liver, it is the substrate for the formation of dehydroepiandrosterone sulfate (DHEAS) and 16-hydroxydehydroepiandrosterone sulfate (16-OHDHEAS) in the fetal adrenal. Some 16-hydroxylation also occurs in the fetal liver. DHEAS and 16-OHDHEAS are transported back to the placenta, where DHEAS forms estradiol and 16-OHDHEAS forms estriol. The principal estrogen formed is estriol, and since fetal 16-OHDHEA sulfate is the principal substrate for the estrogens, the urinary estriol excretion of the mother can be followed as an index of the state of the fetus.

Parturition

The duration of pregnancy in humans averages 270 days from fertilization (284 days from the first day of the menstrual period preceding conception). Irregular uterine contractions increase in frequency in the last month of pregnancy.

The difference between the body of the uterus and the cervix becomes evident at the time of delivery. The cervix, which is firm in the nonpregnant state and throughout pregnancy until near the time of delivery, softens and dilates, while the body of the uterus contracts and expels the fetus.

The number of oxytocin receptors in the myometrium and the decidua (the endometrium of pregnancy) increases more than 100-fold during pregnancy and reaches a peak during early labor. Estrogens increase the number of oxytocin receptors, and uterine distention late in pregnancy may also increase their formation. In early labor, the oxytocin concentration in maternal plasma is not elevated from the prelabor value of about 25 pg/mL, but the marked increase in oxytocin receptors may cause the uterus to respond to normal plasma oxytocin concentrations. Once labor is started, the uterine contractions dilate the cervix, and this dilation in turn sets up signals in afferent nerves that increase oxytocin secretion (Fig 23–39). The plasma oxytocin level rises, and more oxytocin becomes available to act on the uterus. Thus, a positive feedback loop is established that aids delivery and terminates on expulsion of the products of conception. Oxytocin increases uterine contractions in 2 ways: (1) It acts directly on uterine smooth muscle cells to make them contract. (2) It stimulates the formation of prostaglandins in the decidua, and the prostaglandins enhance the oxytocin-induced contractions. There is evidence that both processes are needed for normal labor to occur; labor is prolonged by inhibitors of prostaglandin synthesis and by inhibition of oxytocin secretion. Some women with diabetes insipidus have been reported to have normal deliveries, but this is probably because some oxytocin-secreting neurons were spared by the disease process that caused the defect in vasopressin secretion.

During labor, spinal reflexes and voluntary contractions of the abdominal muscles (''bearing down'') also aid in delivery. However, it appears that delivery can occur without bearing down and without a reflex increase in secretion of oxytocin from the posterior pituitary gland, since paraplegic women can go into labor and deliver.

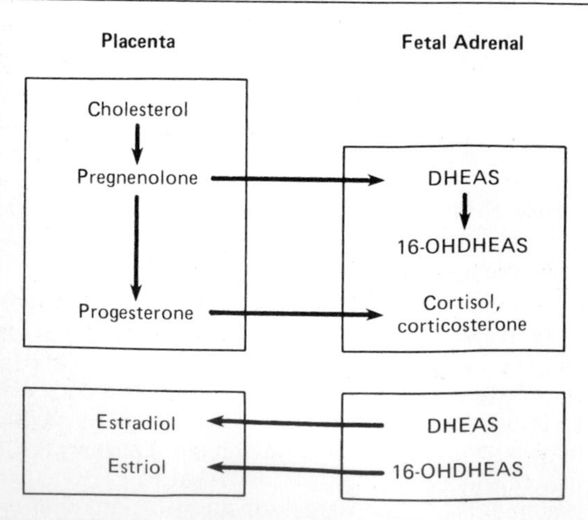

Figure 23–38. Interactions between the placenta and the fetal adrenal cortex in the production of steroids.

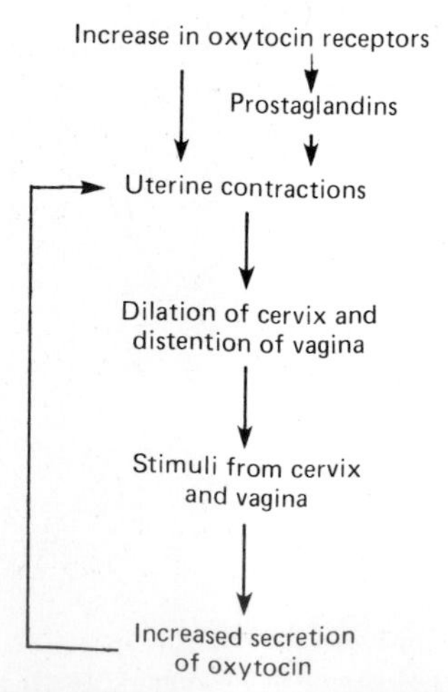

Figure 23–39. Role of oxytocin in parturition.

The fetus may also play a role in triggering the onset of labor. It is interesting in this regard that the plasma oxytocin concentration in the human fetus rises in very early labor, before the maternal concentration rises. However, it is uncertain whether fetal oxytocin can reach the decidua and the myometrium. In some species, an increase in fetal plasma glucocorticoid concentration may be involved in the initiation of labor, but it is uncertain whether such a mechanism plays any role in humans.

Pseudopregnancy

In rats, mice, and various other species, sterile coitus, stimulation of the cervix with a glass rod, or suckling a foster litter initiates prolonged secretion of prolactin and retention of the corpus luteum, with a consequent delay in the return of normal cycles for some time. This state is called **pseudopregnancy.** The release of prolactin that maintains pseudopregnancy is clearly a neuroendocrine reflex response.

The type of pseudopregnancy seen in rodents does not occur in women, but women do sometimes imagine themselves pregnant. In false pregnancy **(pseudocyesis),** there may be amenorrhea, abdominal enlargement, breast changes, and morning sickness. The fact that these changes can occur in the absence of pregnancy emphasizes the degree to which endocrine secretion can be affected by emotional states.

LACTATION

Development of the Breasts

Many hormones are necessary for full mammary development. In general, estrogens are primarily responsible for proliferation of the mammary ducts and progesterone for the development of the lobules. In rats, some prolactin is also needed for development of the glands at puberty, but it is not known if prolactin is necessary in humans. In hypophysectomized rats, glucocorticoids, insulin, and growth hormone are necessary for mammary development in response to other hormones, but they do not by themselves cause growth of the breasts (Fig 23–40). During pregnancy, prolactin levels increase steadily until term, and under the influence of this hormone plus the high levels of estrogens and progesterone, full lobulo-alveolar development of the breasts takes place.

Secretion & Ejection of Milk

The composition of human and cows' milk is shown in Table 23–7. In estrogen- and progesterone-primed rodents, injections of prolactin cause the formation of milk droplets and their secretion into the ducts. Oxytocin causes contraction of the myoepithelial cells lining the duct walls, with consequent ejection of the milk through the nipple (Fig 23–40). The reflex release of oxytocin initiated by touching the nipples and areolas (milk ejection reflex) is discussed in Chapter 14. Oxytocin is not essential for milk ejection in some species, but it is in humans.

The other hormonal relations in humans are generally similar to those in rats, although normal breast growth and lactation can occur in dwarfs with congenital growth hormone deficiency.

Initiation of Lactation After Delivery

The breasts enlarge during pregnancy in response to high circulating levels of estrogens, progesterone, prolactin, and possibly hCG. Some milk is secreted into the ducts as early as the fifth month, but the amounts are small compared with the surge of milk secretion that follows delivery. A similar increase in milk secretion follows abortions after the fourth month, so expulsion of the uterine contents in some way stimulates mild secretion. In most animals, milk is secreted within an hour after delivery, but in women it takes 1–3 days for the milk to "come in."

After expulsion of the placenta at parturition, there is an abrupt decline in circulating estrogens and progesterone. The drop in circulating estrogen initiates lactation. Prolactin and estrogen synergize in producing breast growth, but estrogen antagonizes the milk-producing effect of prolactin on the breast. Indeed, in women who do not wish to nurse their babies, estrogens may be administered to stop lactation.

Suckling not only evokes reflex oxytocin release and milk ejection; it also maintains and augments the secretion of milk because of the stimulation of prolactin secretion produced by suckling (see above).

Effect of Lactation on Menstrual Cycles

Women who do not nurse their infants usually have their first menstrual period 6 weeks after delivery. Nursing stimulates prolactin secretion, and there is evidence that prolactin inhibits LHRH secretion, inhibits the action of LHRH on the pituitary, and antagonizes the action of gonadotropins on the ovaries. Ovulation is inhibited, and the ovaries are inactive, so estrogen and progesterone output fall to low levels. Fifty percent of nursing mothers do not ovulate until the child is weaned.

Chiari-Frommel Syndrome

An interesting although rare condition is persistence of lactation **(galactorrhea)** and amenorrhea in women who do not nurse after delivery. This condition, called the **Chiari-Frommel syndrome,** may be associated with some genital atrophy and is due to persistent prolactin secretion without the secretion of the FSH and LH necessary to produce maturation of new follicles and ovulation. A similar pattern of galactorrhea and amenorrhea with high circulating prolactin levels is seen in nonpregnant women with

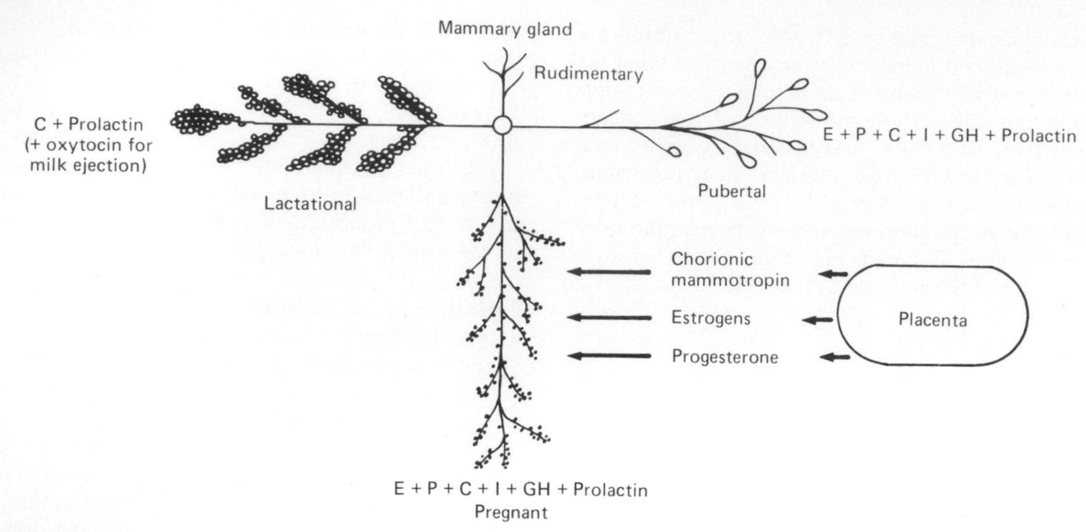

Figure 23–40. Hormonal control of breast development and lactation in rats. Estrogens (E): plus some progesterone (P) and some prolactin in the presence of glucocorticoids (C), insulin (I), and growth hormone (GH) cause duct proliferation and growth at puberty (*right*). During pregnancy, all of these hormones bring about full alveolar development and some milk secretion (*below*). After delivery, increased secretion of prolactin and a decline in estrogen and progesterone levels bring about copious secretion and, in the presence of oxytocin, ejection of milk (*left*). Chorionic mammotropin is the lactogenic hormone presumably secreted by the placenta in rats and is analogous to hCS. It supplements the action of prolactin.

chromophobe pituitary tumors and in women in whom the pituitary stalk has been sectioned in treatment of cancer.

Gynecomastia

Breast development in the male is called **gynecomastia.** It may be unilateral but is more commonly bilateral. It is seen in mild, transient form in 70% of normal boys at the time of puberty. It is a complication of estrogen therapy and occurs in patients with estrogen-secreting tumors. It also occurs in a wide variety of seemingly unrelated conditions, including eunuchoidism due to primary testicular disease, hypothyroidism, hyperthyroidism, and cirrhosis of the liver. Digitalis can produce it, apparently because cardiac glycosides are weakly estrogenic. It has been seen in malnourished prisoners of war, but only after they were liberated and eating an adequate diet. A feature common to many cases of gynecomastia is an increase in the plasma estrogen-androgen ratio. However, there are many other theories about the causes of gynecomastia, and the causes are probably multiple.

Hormones & Cancer

About 35% of carcinomas of the breast in women of childbearing age are **estrogen-dependent;** their continued growth depends upon the presence of estrogens in the circulation. The tumors are not cured by decreasing estrogen secretion, but symptoms are dramatically relieved, and the tumor regresses for months or years before recurring. Women with estrogen-dependent tumors often have a remission when their ovaries are removed. The incidence of a favorable response is greater when the tumor contains estrogen receptors and greatest when it contains both estrogen and progesterone receptors, because estrogen stimulates the formation of progesterone receptors and their presence indicates that estrogen is not

Table 23–7. Composition of colostrum and milk.* (Units are weight per deciliter.)

	Human Colostrum	Human Milk	Cows' Milk
Water, g	. . .	88	88
Lactose, g	5.3	6.8	5.0
Protein, g	2.7	1.2	3.3
Casein:lactalbumin ratio	. . .	1:2	3:1
Fat, g	2.9	3.8	3.7
Linoleic acid	. . .	8.3% of fat	1.6% of fat
Sodium, mg	92	15	58
Potassium, mg	55	55	138
Chloride, mg	117	43	103
Calcium, mg	31	33	125
Magnesium, mg	4	4	12
Phosphorus, mg	14	15	100
Iron, mg	†0.09	†0.15	†0.10
Vit A, μg	89	53	34
Vit D, μg	. . .	†0.03	†0.06
Thiamine, μg	15	16	42
Riboflavin, μg	30	43	157
Nicotinic acid, μg	75	172	85
Ascorbic acid, mg	‡4.4	‡4.3	†1.6

*Reproduced, with permission, from Findlay ALR: Lactation. *Res Reproduction* (Nov) 1974; **6(6).**
†Poor source.
‡Just adequate.

only binding to but acting on the tumor cells. However, a few women with neither type of receptor still respond to this type of endocrine therapy. When the disease recurs, another remission follows bilateral adrenalectomy. Since ovarian and adrenal estrogen secretion are both inhibited by hypophysectomy, this operation has been performed in cancer patients. The number of remissions produced by hypophysectomy is at least as great as the number produced by castration. There also is some evidence that growth hormone and prolactin stimulate the growth of breast carcinomas, and hypophysectomy removes these stimuli. Hypophysectomy induces a significant incidence of remissions in carcinoma of the male breast, an uncommon but serious disease.

Some carcinomas of the prostate are **androgen-dependent** and regress temporarily after the removal of the testes or treatment with LHRH agonists in doses that are sufficient to produce down-regulation of the LHRH receptors on gonadotropes and decrease LH secretion. The formation of pituitary tumors after removal of the target endocrine glands controlled by pituitary tropic hormones is discussed in Chapter 22.

24 Other Endocrine Organs

INTRODUCTION

The organs with endocrine functions include a number of structures in addition to the posterior, intermediate, and anterior lobes of the pituitary; the thyroid; the parathyroids; the pancreas; the adrenal cortex; the adrenal medulla; and the gonads. Hormones that stimulate or inhibit the secretion of anterior pituitary hormones are secreted by the hypothalamus (see Chapter 14), and a number of hormones are secreted by the mucosa of the gastrointestinal tract (see Chapter 26). The thymus also secretes one or more hormones that affect lymphocytes (see Chapter 27). The kidney produces 3 hormones: 1,25-dihydroxycholecalciferol (see Chapter 21), renin, and erythropoietin. Atrial natriuretic peptide, a substance secreted by the heart, increases excretion of sodium by the kidneys, and there may be an additional natriuretic hormone. The pineal gland secretes melatonin, and this indole may have an endocrine function. The endocrine functions of the kidneys, heart, and pineal gland are considered in this chapter.

THE RENIN-ANGIOTENSIN SYSTEM

Renin & Angiotensin

The rise in blood pressure produced by injection of kidney extracts is due to **renin,** an acid protease secreted by the kidney into the bloodstream. This glycoprotein hormone has a molecular weight of 37,326 in humans. The molecule is made up of 2 lobes, or domains, between which the active site of the enzyme is located in a deep cleft. Two aspartic acid residues, one at position 32 and one at position 215, are juxtaposed at the mouth of the cleft and are essential for activity. Thus, renin is an aspartyl protease.

Like other hormones, renin is synthesized as a large preprohormone. Human **preprorenin** contains 406 amino acid residues. The **prorenin** contains 406 amino acid residues. The **prorenin** that remains after removal of a leader sequence of 23 amino acid residues from the N terminal contains 383 amino acid residues, and after removal of the pro sequence from the N terminal of prorenin, active **renin** contains 340 amino acid residues. Prorenin has relatively little biologic activity. The renin in mouse salivary glands (see below) undergoes an additional cleavage near the C-terminal end to produce 2 separate peptide chains connected by a disulfide bond, but this cleavage does not occur in human renal renin.

Some prorenin is converted to renin in the kidneys, and some is secreted and possibly converted to renin in the circulation. Prorenin is converted to renin by tissue kallikrein (see Chapter 31), but the details of the process by which renin is formed from prorenin in vivo are still unsettled.

Renin has a half-life in the circulation of 80 minutes or less. It splits the decapeptide **angiotensin I** from the N terminal of a glycoprotein in the α_2 globulin fraction of the proteins in the circulating plasma (Fig 24–1). The α_2 globulin, which contains about 13% carbohydrate, is made up of 453 amino acid residues. It is synthesized in the liver and is called **angiotensinogen** or **renin substrate.** Its circulating level is increased by glucocorticoid hormones and estrogens. **Angiotensin converting enzyme (ACE)** is a dipeptidyl-carboxypeptidase that splits off histidyl-leucine from the physiologically inactive angiotensin I, forming the octapeptide **angiotensin II** (Fig 24–2). Most of the converting enzyme that forms angiotensin II in the circulation is located in endothelial cells. Much of the conversion occurs as the blood passes through the lungs, but there is also conversion in many other parts of the body.

Angiotensin II is destroyed rapidly, its half-life in humans being 1-2 minutes. The enzymes that destroy angiotensin II are lumped together under the term **angiotensinase.** They include an aminopeptidase that removes the Asp residue from the N terminal of the

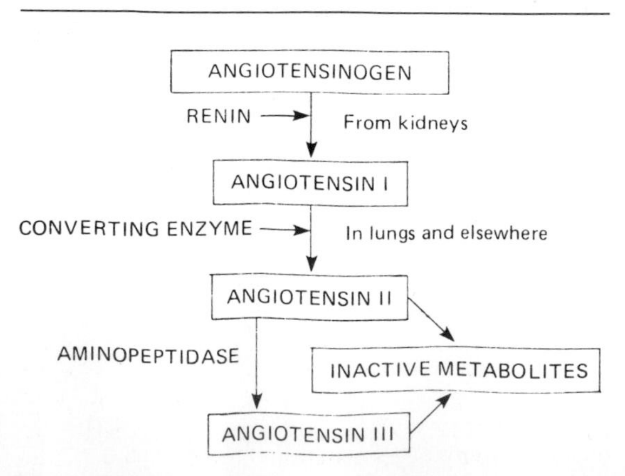

Figure 24–1. Formation and metabolism of circulating angiotensins.

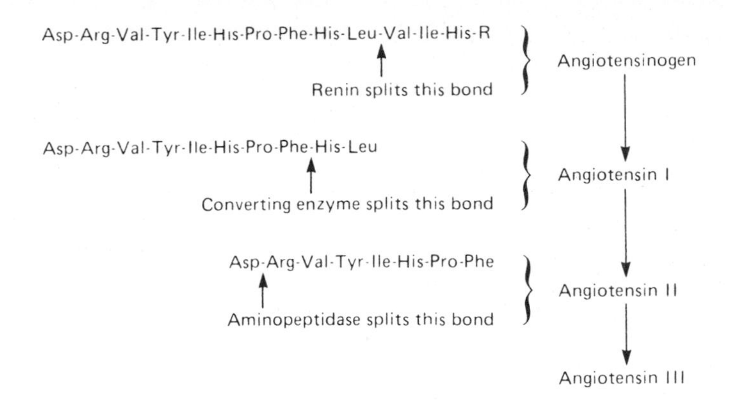

Figure 24–2. Structure of the N-terminal end of angiotensinogen and angiotensins I, II, and III in humans. R, remainder of protein. After removal of a 24-amino-acid leader sequence, rat angiotensinogen contains 453 amino acid residues; human angiotensinogen probably has a comparable structure. The structure of angiotensin II in dogs, rats, and many other mammals is the same as that in humans. Bovine and ovine angiotensin II have valine instead of isoleucine in position 5.

peptide. The resulting heptapeptide, unlike the other peptide fragments, has physiologic activity and is sometimes called **angiotensin III** (see below). In addition aminopeptidase can act on angiotensin I to produce (des-Asp[1]) angiotensin I, and this compound can be converted directly to angiotensin III by the action of converting enzyme. Angiotensinase activity is found in red blood cells and many tissues. In addition, angiotensin II appears to be removed from the circulation by some sort of trapping mechanism in the vascular beds of tissues other than the lungs.

Renin is usually measured by incubating the sample to be assayed and measuring by immunoassay the amount of angiotensin I generated. This measures the **plasma renin activity (PRA)** of the sample. Angiotensinogen as well as renin deficiency can cause low PRA values, and to avoid this problem, exogenous angiotensinogen is often added, so that **plasma renin concentration (PRC)** rather than PRA is measured. The normal PRA in supine subjects eating a normal amount of sodium is approximatley 1 ng of angiotensin I generated per milliliter per hour. The plasma angiotensin II concentration in such subjects is about 25 pg/mL (approximately 25 pmol/L).

Actions of Angiotensins

Angiotensin I appears to function solely as the precursor of angiotensin II and does not have any other established action.

Angiotensin II—previously called hypertensin or angiotonin—produces arteriolar constriction and a rise in systolic and diastolic blood pressure (Fig 39–2). It is one of the most potent vasoconstrictors known, being 4-8 times as active as norepinephrine on a weight basis in normal individuals. However, its pressor activity is decreased in sodium-depleted individuals and in patients with cirrhosis and some other diseases. In these conditions, circulating angiotensin II is increased, and this down-regulates the angiotensin receptors in vascular smooth muscle. Consequently, there is less response to injected angiotensin II.

Angiotensin II also acts directly on the adrenal cortex to increase the secretion of aldosterone, and the renin-angiotensin system is a major regulator of aldosterone secretion (see Chapter 20). Additional actions of angiotensin II include facilitation of the release of norepinephrine by a direct action on postganglionic sympathetic neurons, contraction of mesangial cells with a resultant decrease in glomerular filtration rate (see Chapter 38), and possibly, additional direct effects on the kidneys.

Angiotensin II also acts on the brain to increase blood pressure, increase water intake (see Chapter 14), and increase the secretion of vasopressin and ACTH. It does not penetrate the blood-brain barrier, but it triggers these responses by acting on the circumventricular organs, 4 small structures in the brain that are outside the blood-brain barrier (see Chapter 32). One of these structures, the area postrema, is primarily responsible for the pressor effect, whereas 2 of the others, the subfornical organ (SFO) and the organum vasculosum of the lamina terminalis (OVLT), are responsible for the increase in water intake (dipsogenic effect). It is not certain which of the circumventricular organs are responsible for the increases in vasopressin and ACTH secretion.

Angiotensin III ([des-Asp[1]] angiotensin II) has about 40% of the pressor activity of angiotensin II but 100% of the aldosterone-stimulating activity. It has been suggested that angiotensin III is the natural aldosterone-stimulating peptide whereas angiotensin II is the blood-pressure-regulating peptide. However, this appears not to be the case, and instead angiotensin III is simply a breakdown product with some biologic activity.

Angiotensin II Receptors

Receptors for angiotensin II have been found in many different organs, and progress has been made in isolating and purifying the angiotensin II receptor. In most instances, binding of angiotensin II to its receptor produces intracellular effects by activation of phospholipase C with a resultant increase in cytosolic free Ca^{2+} (see Chapter 1). The receptors in the arterioles and the adrenal cortex are regulated in opposite ways: an excess of angiotensin II down-regulates the vascular receptors, but it up-regulates the adrenocortical receptors, making the gland more sensitive to the aldosterone-stimulating effect of the peptide. In addition, there is some evidence for the presence of 2 different kinds of angiotensin II receptors in the kidneys.

Other Angiotensin-Generating Enzymes

There are other acid proteases in the body in addition to renal renin that are capable of splitting angiotensin I from angiotensinogen. Enzymes that are very similar to renin have been extracted from or identified immunocytochemically in the uterus, placenta, fetal membranes, amniotic fluid, adrenal cortex, anterior and intermediate lobes of the pituitary, pineal, testes, ovaries, blood vessel walls, and brain. In mice, an angiotensin-generating enzyme is found in the submaxillary glands. The mRNA for renin is present in many of these organs, but the enzymes differ in their immunologic properties, so they probably represent different posttranscriptional modifications of the product of the renin gene. Their physiologic roles are uncertain; they contribute very little to the circulating renin pool, since plasma renin activity drops to near zero when the kidneys are removed.

The Juxtaglomerular Apparatus

The source of the renin in kidney extracts and the and the bloodstream is the **juxtaglomerular cells (JG cells).** These epithelioid cells are located in the media of the afferent arterioles as they enter the glomeruli. The membrane-lined secretory granules in them have been shown to contain renin. Renin is also found in agranular **lacis cells** that are located in the junction between the afferent and efferent arterioles, but its significance in this location is unknown.

At the point where the afferent arteriole enters the glomerulus and the efferent arteriole leaves it, the tubule of the nephron touches the arterioles of the glomerulus from which it arose. At this point, which marks the start of the distal convolution, there is a modified region of tubular epithelium called the macula densa (Fig 24–3). The macula densa is in close proximity to the JG cells. The lacis cells, the JG cells, and the macula densa constitute the **juxtaglomerular apparatus.**

Regulation of Renin Secretion

Renin secretion is increased by stimuli that decrease ECF volume and blood pressure or increase sympathetic output (Table 24–1). At least 5 different regulatory factors appear to be involved (Fig 24–4). One is an intrarenal baroreceptor mechanism that causes renin secretion to increase when the intra-arteriolar pressure at the level of JG cells is decreased and to decrease when this pressure is increased. Another sensor involved in the regulation of renin secretion is the macula densa. There is some debate about what is being sensed, but the bulk of the evidence supports the view that renin secretion is inversely proportion-

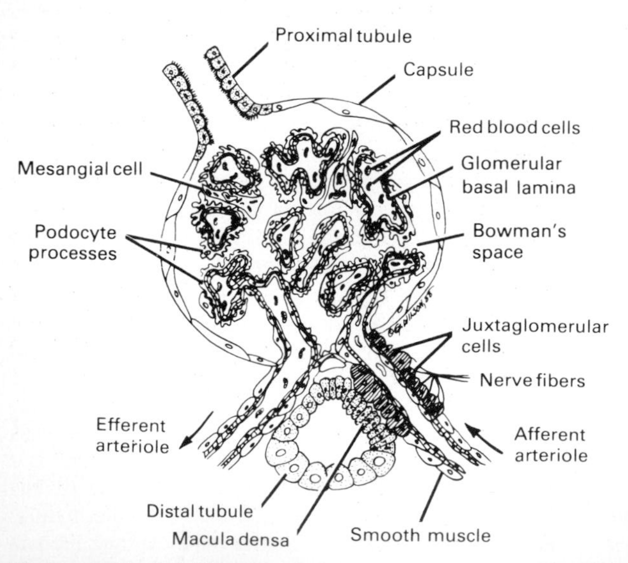

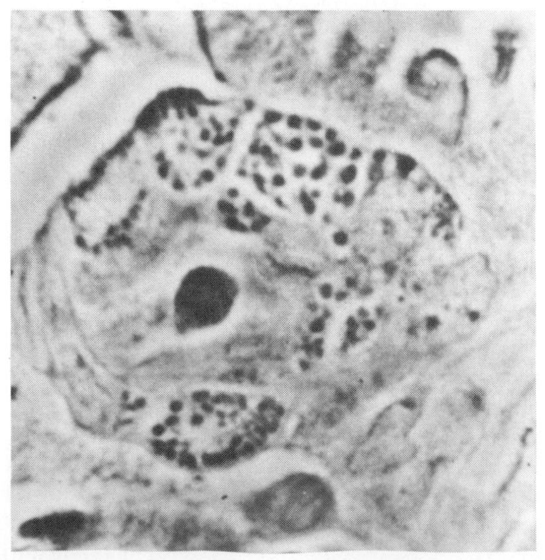

Figure 24–3. *Left:* Diagram of glomerulus, showing the juxtaglomerular apparatus. (Modified from Ham AW: *Histology,* 6th ed. Lippincott, 1969.) *Right:* Phase contrast photomicrograph of afferent arteriole in unstained, freeze-dried preparation of the kidney of a mouse. Note the red blood cell in the lumen of the arteriole and the granulated juxtaglomerular cells in the wall. (Courtesy of C Peil.)

Table 24–1. Stimuli that increase renin secretion.

Stimuli
Sodium depletion
Diuretics
Hypotension
Hemorrhage
Upright posture
Dehydration
Constriction of renal artery or aorta
Cardiac failure
Cirrhosis
Various psychological stimuli

ate to the rate of transport of Cl^- or possibly Na^+ across this portion of the distal tubule. The rate of anisms in the macula densa cells but also on the amount of electrolyte reaching the macula densa. Therefore, decreased delivery of Na^+ and Cl^- to the distal tubules is associated with increased renin secretion. Prostaglandins, especially prostacyclin (see Chapter 17), stimulate renin secretion, apparently by a direct action on the juxtaglomerular cells. There is some evidence that the effects of the macula densa on renin secretion are mediated by prostaglandins generated in the renal cortex. Renin secretion also varies inversely with the plasma K^+ level, but the effect of K^+ appears to be mediated by the changes it produces in Na^+ and Cl^- delivery to the macula densa.

Angiotensin II feeds back to inhibit renin secretion by a direct action on the JG cells, and vasopressin also inhibits renin secretion.

Finally, increased activity of the sympathetic nervous system increases renin secretion. The increase is mediated both by way of increased circulating catecholamines and by way of the renal sympathetic nerves. The sympathetic effects on renin secretion are mostly mediated via β_1-adrenergic receptors and cyclic AMP generated in the JG cells. Renin secretion at any given time is apparently due to the combined activity of these various regulators.

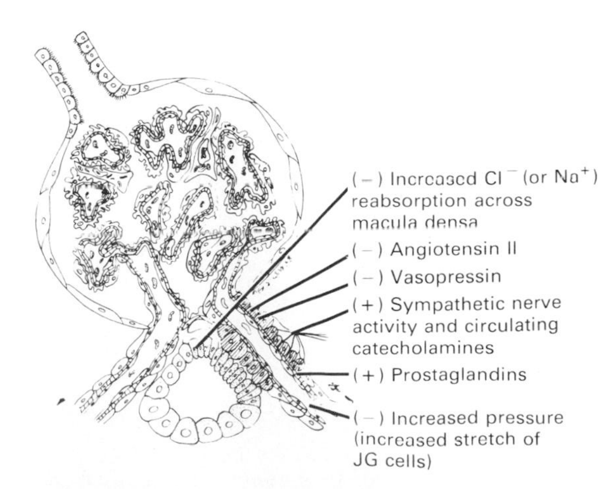

Figure 24–4. Factors that inhibit (−) and stimulate (+) renin secretion.

Pharmacologic Manipulation of the Renin-Angiotensin System

It is now possible to inhibit the secretion or the effects of renin in a variety of ways. Inhibitors of prostaglandin synthesis such as **indomethacin** and β-adrenergic blocking drugs such as **propranolol** reduce renin secretion. The peptide **pepstatin** and some of its derivatives prevent renin from generating angiotensin I and angiotensin-converting enzyme inhibitors (ACE inhibitors) such as **captopril** prevent conversion of angiotensin I to angiotensin II. **Saralasin** and several other analogs of angiotensin II are competitive inhibitors of the action of angiotensin II on its receptors. These pharmacologic agents have been important tools in exploring the physiology of the renin-angiotensin system, and the converting enzyme inhibitors are used clinically in the treatment of hypertension.

Role of Renin in Hypertension

Constriction of one renal artery causes the development of sustained hypertension (**renal** or **Goldblatt hypertension**). Since Goldblatt first demonstrated this effect of renal artery constriction, it has often been assumed that the hypertension is due to increased renin secretion. Removal of the ischemic kidney cures the hypertension if it has not persisted for too long. Some patients with unilateral renal artery stenosis have high circulating renin and angiotensin levels, hypokalemia, and high aldosterone secretion rates (see Chapter 20). However, most animals and humans with this from of renal hypertension do not have high levels of aldosterone secretion and their circulating levels of renin and angiotensin are not elevated. These and other data indicate that after renal artery constriction, the increase in renin secretion is transient and that—1 day to several weeks later—renin secretion returns to normal. The explanation of the continued hypertension is unknown. It could be due to secretion of another renal pressor agent or decreased secretion of a renal depressor agent. However, many patients with hypertension and normal or low plasma renin activity have a fall in blood pressure when they are treated with captopril or other converting enzyme inhibitors. It is possible that these patients are abnormally sensitive to angiotensin II, or the converting enzyme inhibitors could be acting on the renin-angiotensin system in blood vessel walls. However, the decline in blood pressure could also be due to a rise in the plasma bradykinin level, since converting enzyme also catalyzes the breakdown of this depressor peptide.

An interesting syndrome occurs in patients with idiopathic hypertrophy and hyperplasia of their juxtaglomerular apparatuses **(Bartter's syndrome).** These patients have persistent hypokalemia, elevated aldosterone secretion, and high circulating levels of angiotensin II. However, their blood pressure is normal. The role of renin in a feedback mechanism that helps maintain the constancy of ECF volume through regulation of aldosterone secretion has been described in Chapter 20. A high level of renin secretion is responsible for the elevated aldosterone secretion **(secondary hyperaldosteronism)** seen in some normotensive patients with cirrhosis and nephrosis.

ERYTHROPOIETIN

Structure and Function

When an individual is bled or becomes hypoxic, hemoglobin synthesis is enhanced, and production and release of red blood cells from the bone marrow **(erythropoiesis)** are increased (see Chapter 27). Conversely, when the red cell volume is increased above normal by transfusion, the erythropoietic activity of the bone marrow decreases. These adjustments are brought about by changes in the circulating level of **erythropoietin,** a circulating glycoprotein that contains 166 amino acid residues and has a molecular weight of 34,000. Forty percent of this weight is due to the carbohydrate residues attached to the molecule. The gene for this hormone has been cloned, and recombinant erythropoietin produced in cultures of animal cells is now available. This recombinant erythropoietin has been shown to relieve the anemia produced by renal failure (see below).

Erythropoietin causes certain stem cells in the bone marrow to be converted to proerythroblasts (Fig 27–2). Although they resemble other stem cells, they are somewhat more differentiated and are referred to as **erythropoietin-sensitive stem cells.** The action of erythropoietin is apparently mediated by stimulation of mRNA synthesis.

The principal site of inactivation of erythropoietin is the liver, and the hormone has a half-life in the circulation of about 5 hours. However, the increase in circulating red cells that it triggers takes 2-3 days to appear, since red-cell maturation is a relatively slow process. Loss of even a small portion of the sialic acid residues in the carbohydrate moieties that are part of the erythropoietin molecule shortens its half-life to 5 min, making it biologically ineffective.

Sources

In adults, about 85% of the erythropoietin comes from the kidneys and 15% from the liver. Both these organs contain the mRNA for erythropoietin. Erythropoietin can also be extracted from the spleen and salivary glands, but these tissues do not contain the mRNA and consequently do not appear to manufacture the hormone. During fetal and neonatal life, the major site of erythropoiesis is the liver, and it is also the major site of erythropoietin production before erythropoiesis is taken over by the bone marrow and erythropoietin production by the kidneys. When renal mass is reduced in adults by renal disease or nephrectomy, the liver cannot compensate and anemia develops.

The mesangial cells in the renal glomeruli (see Chapter 38) appear to be responsible for the production of some of the erythropoietin that is secreted by the kidneys, since these cells produce the hormone when they are cultured in vitro. However, much of the erythropoietin mRNA in the kidneys appears to be in the renal tubules. In the liver, both Kupfer cells and hepatocytes have been claimed to produce erythropoietin.

Regulation of Secretion

The usual stimulus for erythropoietin secretion is hypoxia, but secretion of the hormone can also be stimulated by cobalt salts, by androgens and possibly by other hormones. When cultured mesangial cells are exposed to hypoxia, their production of erythropoietin is increased. Secrection of hormone is facilitated by the alkalosis that develops at high altitudes. Like renin secretion, erythropoietin secretion is facilitated by catecholamines via a β-adrenergic mechanism, although the renin angiotensin system is totally separate from the erythropoietin system. It has been postulated that hypoxia causes prostaglandins to be formed in the glomeruli, and these in turn activate adenylate cyclase, which triggers erythropoietin secretion. The mechanism that triggers release of extrarenal erythropoietin is not known.

THE ENDOCRINE FUNCTION OF THE HEART: ATRIAL NATRIURETIC PEPTIDE

Structure

The existence of various **natriuretic hormones** has been postulated for some time. One of these is secreted by the heart. The muscle cells in the atria contain secretory granules (Fig 24–5) that increase in number when sodium chloride intake is increased and extracellular fluid expanded, and extracts of atrial

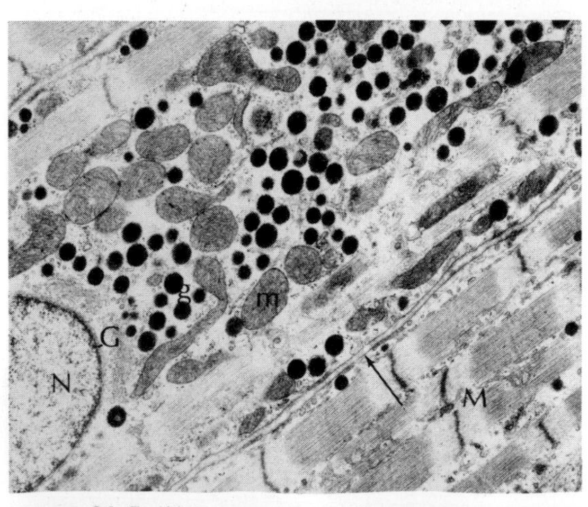

Figure 24–5. ANP granules (g) interspersed between mitochondria (m) in rat atrial muscle cell. G, golgi complex; M, myofilaments; N, nucleus. The arrow identifies the cell membrane. The granules in human atrial cells are similar. (x17,640. Courtesy of M. Cantin.)

Ser-Ser-Cys-Phe-Gly-Gly-Arg- Ile -Asp-Arg-Ile-Gly-Ala-Gln-Ser-Gly-Leu-Gly-Cys-Asn-Ser-Phe-Arg-Tyr

Ser-Leu-Arg-Arg-Ser-Ser-Cys-Phe-Gly-Gly-Arg-Met-Asp-Arg-Ile-Gly-Ala-Gln-Ser-Gly-Leu-Gly-Cys-Asn-Ser-Phe-Arg-Tyr

Figure 24–6. *Top:* Structure of one of the atrial natriuretic peptides in the rat (atriopeptin III). ***Bottom:*** Structure of the circulating form of human atrial natriuretic peptide (α-hANP). Note that the ring structure is identical in the 2 peptides except that in α-hANP one isoleucine has been replaced by a methionine residue.

tissue cause natriuresis. Several related natriuretic polypeptides have been isolated from the atria of various species. They all contain a similar ring formed by one disulfide bond but differ in the number of amino acid residues at the C- and N-terminal ends. The structure of the principal circulating form of **atrial natriuretic peptide (ANP)** in humans (α-hANP) is shown in Fig 24–6. This 28-amino-acid form and related human polypeptides are formed from a large precursor molecule that contains 151 amino acid residues, including a 24-amino-acid signal peptide.

Actions

ANP causes natriuresis, which may be due to an increase in glomerular filtration rate. There are ANP receptors on the mesangial cells in the glomeruli, and the relaxation of these cells produced by ANP presumably increases the effective surface area available for filtration (see Chapter 38). Alternatively, ANP could act on the tubules to promote sodium excretion. However, the exact mechanism of action of ANP on the kidneys is currently unsettled, and additional research is needed. ANP also lowers blood pressure, decreases the responsiveness of vascular smooth muscle to many vasoconstrictor substances, decreases the responsiveness of the zona glomerulosa to stimuli that normally increase aldosterone secretion, and inhibits the secretion of vasopressin. It is worth noting that these actions are generally the opposite of those of angiotensin II, and, indeed, the actions of ANP are most readily demonstrated against a background of stimulation by angiotensin II. In addition, ANP inhibits renin secretion and consequently lowers circulating angiotensin II levels.

Another link between ANP and angiotensin II is that both are found in the anterior pituitary gland and the brain as well as the circulation. The form of ANP in the brain contains 24–25 amino acid residues. An ANP-containing neural pathway projects from the anterior, medial part of the hypothalamus to the areas in the lower brain stem that are concerned with neural regulation of the cardiovascular system. However, the function of ANP in the pituitary and the brain has not been determined.

Secretion

The concentration of ANP in plasma is about 5 fmol/mL in normal humans ingesting moderate amounts of sodium. ANP secretion is increased when the ECF volume is increased by infusion of isotonic saline or ingestion of a high-sodium diet. It is also increased by immersion in water up to the neck (Fig 24–7), a procedure that counteracts the effect of

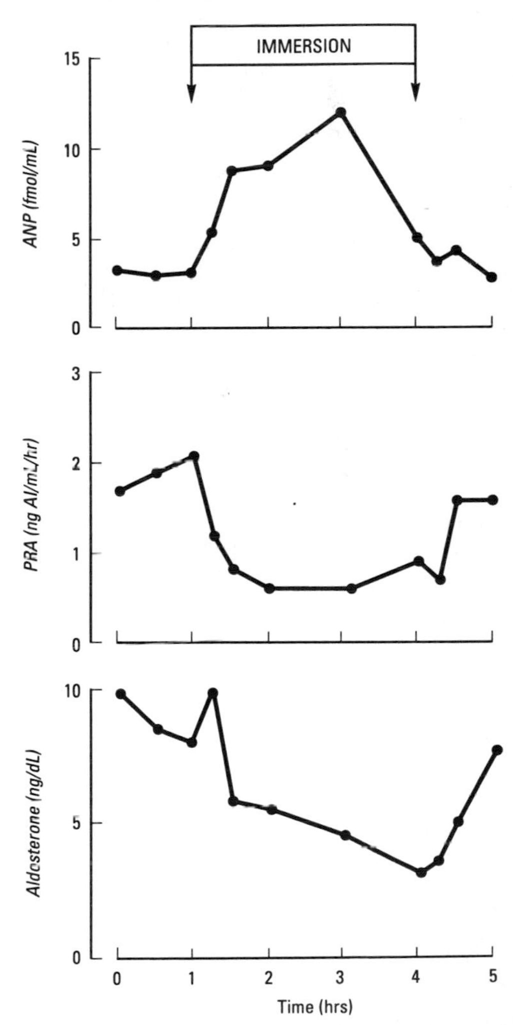

Figure 24–7. Effect of immersion in water up to the neck for 3 hours on plasma concentrations of ANP, PRA, and aldosterone. (Modified and reproduced, with permission, from Epstein M et al: Increases in circulating atrial natriuretic factor during immersion-induced central hypervolaemia in normal humans. *Hypertension* 1986; 4(suppl 2):593.)

gravity on the circulation and increases central venous and consequently atrial pressure. Note that immersion also decreases the secretion of renin and aldosterone. Conversely, there is a small but measurable decrease in plasma ANP in association with a decrease in central venous pressure on rising from the supine to the standing position. Strips of atrial muscle release ANP when stretched in vitro. Thus, it appears likely that the atria respond directly to stretch in vivo and that the rate of ANP secretion is proportionate to the degree to which the atria are stretched by increases in central venous pressure.

Na^+-K^+ ATPase-Inhibiting Factor

There is considerable evidence for the existence of another natriuretic factor in the blood. This factor produces natriuresis by inhibiting Na^+-K^+ ATPase and has been postulated to be a steroid that cross-reacts with antibodies to digitalis glycosides. However, its source in the body and its physiologic significance are not yet known.

PINEAL

The **pineal** (epiphysis), believed by Descartes to be the seat of the soul, has at one time or another been regarded as having a wide variety of functions. It is now known to secrete melatonin, and there is speculation that it may function as a timing device to keep internal events synchronized with the light-dark cycle in the environment.

Anatomy

The pineal arises from the roof of the third ventricle under the posterior end of the corpus callosum and is connected by a stalk to the posterior commissure and habenular commissure. There are nerve fibers in the stalk, but they apparently do not reach the gland. The pineal stroma contains neuroglia and parenchymal cells with features suggesting they have a secretory function (Fig 24–8). Like other endocrine glands, the pineal has highly permeable fenestrated capillaries. In young animals and infants, the pineal is large, and the cells tend to be arranged in alveoli. It begins to involute before puberty, and, in humans, small concretions of calcium phosphate and carbonate **(pineal sand)** appear in the tissue. Because the concretions are radiopaque, the normal pineal is often visible on x–ray films of the skull in adults. Displacement of a calcified pineal from its normal position indicates the presence of a space-occupying lesion such as a tumor in the brain.

Melatonin

The amphibian pineal contains an indole, N-acetyl-5-methroxytryptamine, named **melatonin** because it lightens the skin of tadpoles by an action on melanophores. However, it does not appear to play a physiologic role in the regulation of skin color. Melatonin and the enzymes responsible for its synthesis from serotonin by N-acetylation and O-methylation are also present in mammalian pineal tissue (Fig 24–9), although some melatonin is synthesized in other parts of the body as well. Melatonin is

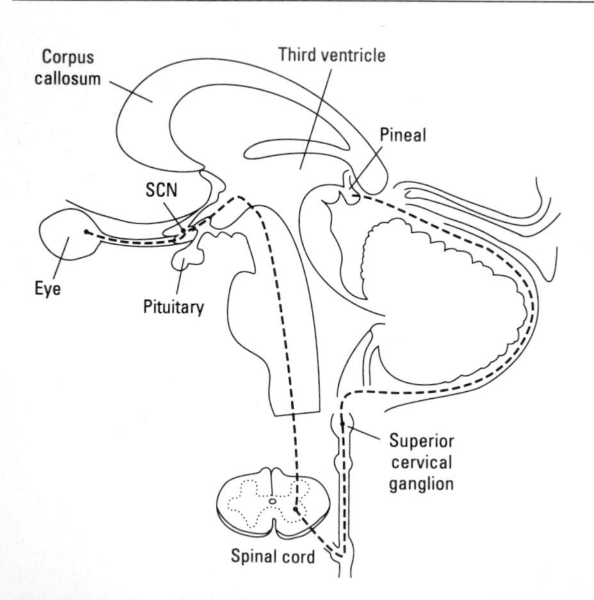

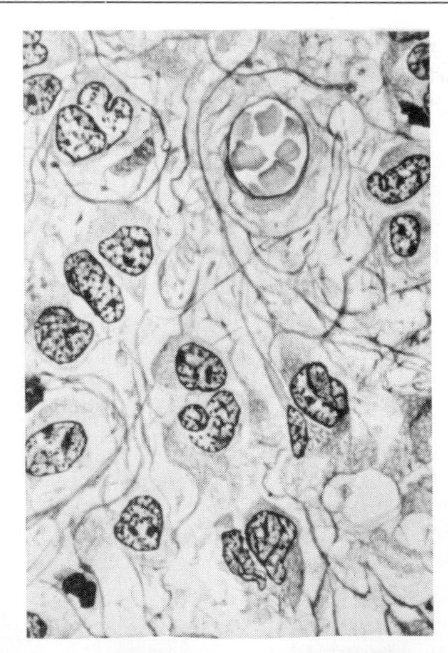

Figure 24–8. *Left:* Sagittal section of human brain stem showing the pineal and its innervation. Retino-hypothalamic fibers synapse in the suprachiasmatic nuclei (SCN), and there are connections from the SCN to the intermediolateral gray column in the spinal cord. Preganglionic neurons pass from the spinal cord to the superior cervical ganglion and the postganglionic neurons project from this ganglion to the pineal in the nervi conarii. *Right:* Histology of pineal. Drawing of hematoxylin and eosin-stained section. (Reproduced, with permission, from Fawcett DW: Bloom and Fawcett, *A Textbook of Histology,* 11th ed. Saunders, 1986.)

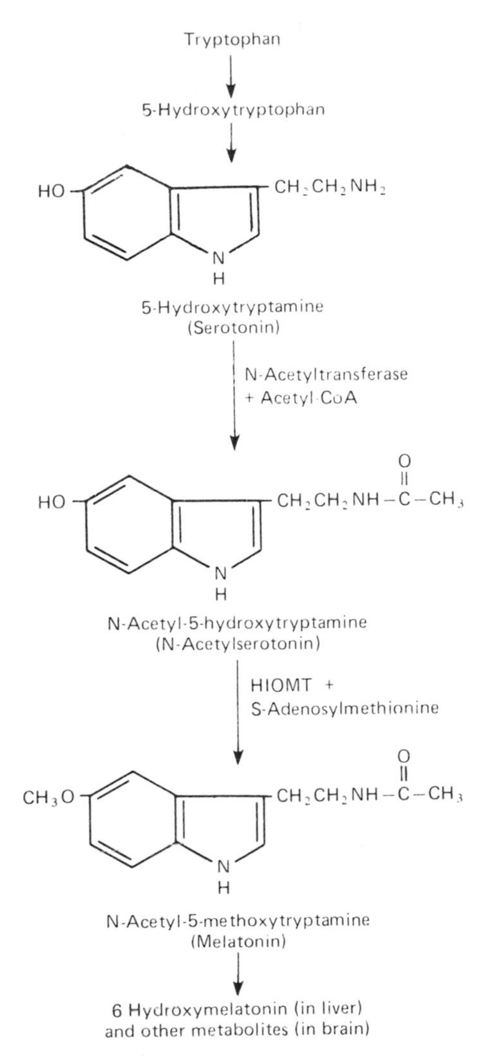

Figure 24–9. Formation and metabolism of melatonin. HIOMT, hydroxyindole-O-methyltransferase. For details of the synthesis of serotonin, see Fig 4-21.

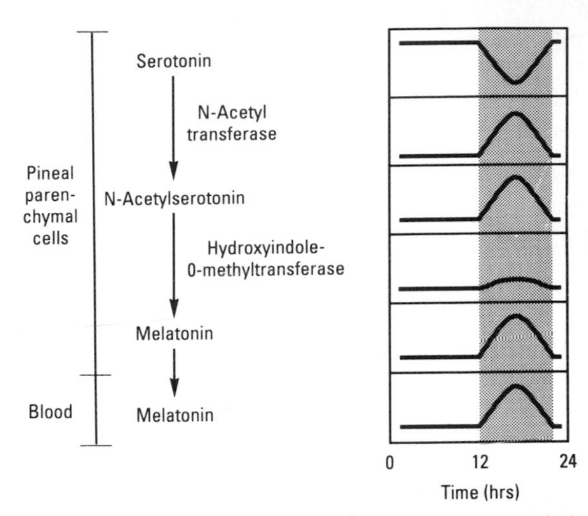

Figure 24–10. Diurnal rhythms of various compounds in the pineal and melatonin in blood. The shaded area represents the hour of darkness during the 24-hour day. (Modified from Reiter RJ: The pineal gland: an important link to the environment. *News Physiol Sci* 1986;**1**:20.)

synthesized by pineal parenchymal cells and secreted by them into the blood and the cerebrospinal fluid.

In humans and all other species studied to date, melatonin synthesis and secretion are increased during the dark period of the day and maintained at a low level during the daylight hours (Fig 24–10). This remarkable diurnal variation in secretion is brought about by norepinephrine secreted by the postganglionic sympathetic nerves (nervi conarii) that innervate the pineal (Fig 24–8). The norepinephrine acts via β-adrenergic receptors in the pineal to increase intracellular cyclic AMP, and the cyclic AMP in turn produces a marked increase in N-acetyltransferase activity. This results in increased melatonin synthesis and secretion.

The discharge of the sympathetic nerves to the pineal is entrained to the light-dark cycle in the environment via the retinohypothalamic nerve fibers and the suprachiasmatic nuclei (Fig 24–8). From the hypothalamus, descending pathways converge on the intermediolateral gray column of the thoracic spinal cord and end on the preganglionic sympathetic neurons that in turn innervate the superior cervical ganglion, the site of origin of the postganglionic neurons to the pineal.

Circulating melatonin is rapidly metabolized in the liver by 6-hydroxylation followed by conjugation, and over 90% of the melatonin that appears in the urine is in the form of 6-hydroxy conjugate. The pathway by which the brain metabolizes melatonin is unsettled but may involve cleavage of the indole nucleus.

Function of the Pineal

Injected melatonin has effects on the gonads, but these effects vary markedly from species to species and also depend on the time of injection. In some situations, melatonin inhibits gonadal function, and in others, its effect is facilitatory. This variability has led to the hypothesis that it is not the melatonin per se but the diurnal change in melatonin secretion that functions as some sort of timing signal that coordinates internal events with the light-dark cycle in the environment. This may be correct, especially in seasonally breeding animals, which respond to changes in day length. However, the events that are regulated in humans have not been identified.

It has been argued that the pineal normally inhibits the onset of puberty in humans, because pineal tumors are sometimes associated with sexual precocity. However, as noted in Chapter 23, there is evidence that pineal tumors produce precocity only when they produce hypothalamic damage, and except for a small rise during breast development in girls, there are no significant changes in melatonin excretion during puberty. Thus, the function of melatonin and the pineal in humans remains obscure.

REFERENCES
Section IV: Endocrinology, Metabolism, & Reproductive Function

Bauer C, Kurtz A: Erythropoietin production in the kidney. *News Physiol Sci* 1987;**2:**69.

Bender AE, Brookes LJ (editors): *Body Weight Control: The Physiology, Clinical Treatment and Prevention of Obesity.* Churchill Livingston, 1987.

Bilzekian JP, Loeb JN: The influence of hyperthyroidism and hypothyroidism on α- and β-adrenergic receptor systems and adrenergic responsiveness. *Endocr Rev* 1983;**4:**378.

Blundell TL, Humbel RE: Hormone families: Pancreatic hormones and homologous growth factors. *Nature* 1980; **287:**781.

Boss GR, Seegmiller JE: Hyperuricemia and gout. *N Engl J Med* 1979;**300:**1459.

Bravo EM, Gifford RW Jr: Pheochromocytoma: Diagnosis, localization, and management. *N Engl J Med* 1984; **311:**1298.

Brown MS, Goldstein JL: A receptor-mediated pathway for cholesterol homeostasis. *Science* 1986;**232:**34.

Bryant-Greenwood GD: Relaxin as a new hormone. *Endocr Rev* 1982;**3:**62.

Burman KD, Baker JR: Immune mechanisms in Graves' disease. *Endocr Rev* 1985;**6:**183.

Cantin M, Genest J: The heart and the atrial natriuretic factor. *Endocr Rev* 1985;**6:**107.

Carlson HE: Gynecomastia. *N Engl J Med* 1980;**303:**795.

Carmichael SW, Winkler H: The adrenal chromaffin cell. *Sci AM* (Aug) 1985;**253:**40.

Cavalieri RR, Pitt-Rivers R: The effects of drugs on the distribution and metabolism of thyroid hormones. *Pharmacol Rev* 1981;**33:**55.

Cohen MP:*Diabetes and Protein Glycosylation.* Springer-Verlag, 1986.

Cooper DS: Antithyroid drugs. *N Engl J Med* 1984; **311:1353.**

Cryer PE, Gerich JE: Glucose counter-regulation, hypoglycemia, and intensive insulin therapy in diabetes mellitus. *N Engl J Med* 1985;**313:**232.

Davis JO, Freeman RH: Mechanisms regulating renin release. *Physiol Rev* 1976;**56:1.**

Deluca H: The vitamin D story: a collaborative effort of basic science and clinical medicine. *FASEB J* 1988;**2:**224.

Eggo MC, Burrow GN (editors): *Thryoglobulin: The Prothyroid Hormone,* Raven, 1985.

Eisenbarth GS: Type I diabetes mellitus: A chronic autoimmune disease. *N Engl J Med* 1986;**314:**1360.

Erickson EF: Normal and pathophysiological remodelling of human trabecular bone: three dimensional reconstruction of the remodelling in normals and in metabolic bone disease. *Endocr Rev* 1986;**7:**379.

Falkner F, Tanner JM (editors): *Human Growth,* 2nd ed. 3 vols. Plenum, 1986.

Felig P et al (editors): *Endrocrinology and Metabolism,* 2nd ed. McGraw Hill, 1987.

Fieser LF, Fieser M: *Steroids.* Reinhold, 1959.

Foster DW, McGarry JD: The metabolic derangements and treatment of diabetic ketoacidosis. *N Engl J Med* 1983;**309:**159.

Fuchs F, Klopper AI (editors): *Endocrinology of Pregnancy,* 2nd ed. Harper, 1977.

Goodman DS: Vitamin A and retinoids in health and disease. *N Engl J Med* 1984;**310:**1023.

Gosden RG: *Biology of Menopause.* Academic Press, 1985.

Griffin JE, Wilson JD: The syndromes of androgen resistance. *N Engl J Med* 1980;**302:**198.

Guillemin R et al: Growth hormone-releasing factor: Chemistry and physiology. *Proc Soc Exp Biol Med* 1984; **175:**407.

Habener JF, Rosenblatt M, Potts JT Jr: Parathyroid hormone: Biochemical aspects of biosynthesis, secretion, action, and metabolism. *Physiol Rev* 1984;**64:985.**

Himms-Hagen J: Thermogenesis in brown adipose tissue as an energy buffer. *N Engl J Med* 1984;**311:**1549.

Howell RR, Morriss FH Jr, Pickering LK (editors): *Human Milk in Infant Nutrition and Health.* Thomas, 1986.

Ingbar SH, Braverman IE (editors): *Werner's The Thyroid,* 5th ed. Lippincott, 1986.

Izuma S, Nadal-Girard B, Mahdavi V: All members of the MHC multigene family respond to thyroid hormone in a highly tissue-specific manner. *Science* 1986;**231:**597.

Jaffe RB: *Prolactin.* Elsevier, 1981.

Kelch RP: Management of precocious puberty. *N Engl J Med* 1985;**312:**1057.

Knobil E: A hypothalamic pulse generator governs mammalian reproduction. *News Physiol Sci* 1987;**2:**42.

Knobil E, Neill JD (editors): *The Physiology of Reproduction.* 2 vols. Raven, 1987.

Kols A et al: Oral contraceptives in the 1980s. *Population Rep* 1982;**10:**A-189.

Larsen PR: Thyroid-pituitary interaction. *N Engl J Med* 1982;**396:**23.

Levine BS, Coburn JW: Magnesium, the mimic/antagonist of calcium. *N Engl J Med* 1984;**310:**1253.

Mahley RW: Apolipoprotein E: cholesterol transport protein with expanding role in cell biology. *Science* 1988; **240:**622.

Manacada S (editor): Prostacyclin, thromboxane, and leukotrienes. *Br Med Bull* 1983;**39:**209.

Martin JB, Reichlin S: *Clinical Neuroendocrinology,* 2nd ed. Davis, 1987.

Martini L, Ganong WF (editors): *Frontiers in Neuroendocrinology,* vol 10, Raven, 1988.

Mertz W: The essential trace elements. *Science* 1981; **213:**1332.

Morris BJ et al: A structural analysis of human renin. *Clin Exp Pharmacol Physiol* 1985;**12:**299.

Naftolin F, Butz E (editors): Sexual dimorphism. *Science* 1981;**211:**1263.

Needleman P et al: Atriopeptins as cardiac hormones. *Hypertension* 1985;**7:**469.

Norman AW, Litwack G:*Hormones.* Academic, 1987.

Ondetti MA, Cushman DW: Enzymes of the renin-angiotensin system and their inhibitors. *Annu Rev Biochem* 1982;**51:**283.

Orci L, Vassalli, J-D, Perrelot A: The insulin factory. *Sci Am* (Sept) 1988;**259:**85.

Pawelek JM, Korner AM: The biosynthesis of mammalian melanin. *Am Sci* 1982;**70:**136.

Rechler MM, Nissley SP, Roth J: Hormonal regulation of human growth. *N Eng J Med* 1987;**316:**941.

Reid IA, Morris BJ, Ganong WF: The renin-angiotensin system. *Annu Rev Physiol* 1978;**40:**377.

Riggs BL, Melton LJ III: Involutional osteoporosis. *N Engl J Med* 1986;**314:**1676.

Robbins DC, Toger HS, Rubenstein AH: Biologic and clinical importance of proinsulin. *N Engl J Med* 1984; **310:**1165.

Robbins J: Iodine deficiency, iodine excess and the use of iodine for protection against radioactive iodine. *Thyroid Today* (Dec) 1980;**(8).**

Rodger JC, Drake BL: The enigma of the fetal graft. *Am Sci* 1987;**75:**51.

Rosen O: After insulin binds. Science 1987;**237:**1452.

Samuelsson B et al: Leukotrienes and lipoxins: structures, biosynthesis, and biological effects. *Science* 1987; **237:**1171.

Seibel MM: A new era in reproductive technology: In vitro fertilization, gamete intrafallopian transfer, and donated gametes and embryos. *N Eng J Med* 1988;**318:**828.

Selden RF et al: Regulation of insulin-gene expression. Implications for gene therapy. *N Eng J Med* 1987; **317:**1067.

Shils ME, Young VR (editors): *Modern Nutrition in Health and Disease*, 7th ed. Lea & Febiger, 1987.

Stoner HB: Metabolism after trauma and in sepsis. *Circ Shock* 1986;**19:**75.

Sutherland DER: Pancreas and islet transplantation. *Diabetes* 1981;**20:**161.

Tamarkin L, Baird CJ, Almcida OF: Melatonin: A coordinating signal for mammalian reproduction. *Science* 1985;**227:**714.

Therman E: *Human Chromosomes: Structure, Behavior, Effects*, 2nd ed. Springer-Verlag, 1986.

Ungar A, Phillips JH: Regulation of adrenal medulla. *Physiol Rev* 1983;**63:**787.

Unger RH, Orci L: Glucagon and the A cell: Physiology and pathophysiology. *N Engl J Med* 1981;**304:**1518.

Vaitukaitis JL: Premenstrual syndrome. *N Engl J Med.* 1984;**311:**1371.

Van Middlesworth L: Nuclear reactor accidents and the thyroid. *Thyroid Today* (Apr) 1987;**10:**1.

Vranic M et al: Hormonal interaction on control of metabolism during exercise in physiology and diabetes. In: *Diabetes Mellitus: Theory and Practice*, 3rd ed. Ilenberg HE, Ritkin H (editors). Medical Examination Publishing Co., 1983.

Wasserman PM. The biology and chemistry of fertilization. *Science* 1987;**235:**553.

Weintraub BD, Gesundheit N: Thyroid-stimulating hormone synthesis and glycosylation: Clinical implication. *Thyroid Today* (Jan) 1987;**10:**1.

White PC, New MI, DuPont B: Congenital adrenal hyperplasia. *N Engl J Med* 1987;**316:**1519.

Williams JA: Electrical correlates of secretion in endocrine and exocrine cells. *Fed Proc* 1981;**40:**128.

Wood SC, Porte D Jr: Ncural control of the endocrine pancreas. *Physiol Rev* 1974;**54:**596.

Wortman J: Vasectomy: What are the problems? *Population Rep* 1975;**2:**D-25.

Section V. Gastrointestinal Function

25 Digestion & Absorption

INTRODUCTION

The gastrointestinal system is the portal through which nutritive substances, vitamins, minerals, and fluids enter the body. Proteins, fats, and complex carbohydrates are broken down into absorbable units **(digested),** principally in the small intestine. The products of digestion and the vitamins, minerals, and water cross the mucosa and enter the lymph or the blood **(absorption).** The digestive and absorptive processes are the subject of this chapter. The details of the functions of the various parts of the gastrointestinal system are considered in Chapter 26.

Digestion of the major foodstuffs is an orderly process involving the action of a large number of **digestive enzymes** (Table 25–1). Some of these enzymes are found in the secretions of the salivary glands, the stomach, and the exocrine portion of the pancreas. Other enzymes are found in the luminal membranes and the cytoplasm of the cells that line the small intestine. The action of the enzymes is aided by the hydrochloric acid secreted by the stomach and the bile secreted by the liver.

The mucosal cells in the small intestine have a **brush border** made up of numerous microvilli lining their apical surface (Fig 26–25). This border is rich in enzymes. It is lined on its luminal side by a layer that is rich in neutral and amino sugars, the **glycocalyx.** The membranes of the mucosal cells contain glycoprotein enzymes that hydrolyze carbohydrates and peptides, and the glycocalyx is made up in part of the carbohydrate portions of these glycoproteins that extend into the intestinal lumen. Next to the brush border and glycocalyx is an **unstirred layer** similar to the layer adjacent to other biologic membranes (see Chapter 1). Solutes must diffuse across this layer to reach the mucosal cells. The mucous coat overlying the cells also constitutes a significant barrier to diffusion.

Substances pass from the lumen of the gastrointestinal tract to the extracellular fluid and thence to the lymph and blood by diffusion, facilitated diffusion, solvent drag, active transport, secondary active transport (coupled transport), and endocytosis. Most substances must pass from the intestinal lumen into the mucosal cells and then out of the mucosal cells to the extracellular fluid, and the processes responsible for movement across the luminal cell membrane are often quite different from those responsible for movement across the basal and lateral cell membranes to the extracellular fluid. The dynamics of transport in all parts of the body are considered in Chapter 1.

CARBOHYDRATES

Digestion

The principal dietary carbohydrates are polysaccharides, disaccharides, and monosaccharides. Starches (glucose polymers) and their derivatives are the only polysaccharides that are digested to any degree in the human gastrointestinal tract. In glycogen, the glucose molecules are mostly in long chains (glucose molecules in 1,4α linkage), but there is some chain branching (produced by 1,6α linkages; see Fig 17–12). Amylopectin, which constitutes 80–90% of dietary starch, is similar but less branched, whereas amylose is a straight chain with only 1,4α linkages. Glycogen is found in animals, whereas amylose and amylopectin are of plant origin. The disaccharides **lactose** (milk sugar) and **sucrose** (table sugar) are also ingested, along with the monosaccharides fructose and glucose.

Starch is attacked by ptyalin, the α-amylase in the saliva. However, the optimal pH for this enzyme is 6.7, and its action is inhibited by the acid gastric juice when food enters the stomach. In the small intestine, the potent pancreatic α-amylase also acts on the ingested polysaccharides. Both the salivary and the pancreatic α-amylases hydrolyze 1,4α linkages but spare 1,6α linkages, terminal 1,4α linkages, and the 1,4α linkages next to branching points. Consequently, the end products of α-amylase digestion are oligosaccharides: the disaccharide **maltose,** the trisaccharide **maltotriose,** some slightly larger polymers with glucose in 1,4α linkage, and **α-limit dextrins,** branched polymers containing an average of about 8 glucose molecules (Fig 25–1).

Table 25–1. Principal digestive enzymes.
The corresponding proenzymes are shown in parentheses.

Source	Enzyme	Activator	Substrate	Catalytic Function or Products
Salivary glands	Salivary α-amylase	Cl^-	Starch	Hydrolyzes 1,4α linkages, producing α-limit dextrins, maltotriose, and maltose
Lingual glands	Lingual lipase		Triglycerides	Fatty acids plus 1,2-diocylglycerols
Stomach	Pepsins (pepsinogens)	HCl	Proteins and polypeptides	Cleave peptide bonds adjacent to aromatic amino acids
	Gastric lipase		Triglycerides	Fatty acids and glycerols
Exocrine pancreas	Trypsin (trypsinogen)	Enteropeptidase	Proteins and polypeptides	Cleave peptide bonds adjacent to arginine or lysine
	Chymotrypsins (chymotrypsinogens)	Trypsin	Proteins and polypeptides	Cleave peptide bonds adjacent to aromatic amino acids
	Elastase (proelastase)	Trypsin	Elastin, some other proteins	Cleaves bonds adjacent to aliphatic amino acids
	Carboxypeptidase A (procarboxypeptidase A)	Trypsin	Proteins and polypeptides	Cleaves carboxy terminal amino acids that have aromatic or branched aliphatic side chains
	Carboxypeptidase B (procarboxypeptidase B)	Trypsin	Proteins and polypeptides	Cleaves carboxy terminal amino acids that have basic side chains
	Colipase (procolipase)	Trypsin	Fat droplets	Binds to bile salt-triglyceride-water interface, making anchor for lipase
	Pancreatic lipase	. . .	Triglycerides	Monoglycerides and fatty acids
	Cholesteryl ester hydrolase	. . .	Cholesteryl esters	Cholesterol
	Pancreatic α-amylase	Cl^-	Starch	Same as salivary α-amylase
	Ribonuclease	. . .	RNA	Nucleotides
	Deoxyribonuclease	. . .	DNA	Nucleotides
	Phospholipase, A_2 (prophospholipase A_2)	Trypsin	Phospholipids	Fatty acids, lysophospholipids
Intestinal mucosa	Enteropeptidase	. . .	Trypsinogen	Trypsin
	Aminopeptidases	. . .	Polypeptides	Cleave N-terminal amino acid from peptide
	Dipeptidases	. . .	Dipeptides	Two amino acids
	Glucoamylase	. . .	Maltose, maltotriose	Glucose
	Lactase	. . .	Lactose	Galactose and glucose
	Sucrase*	. . .	Sucrose	Fructose and glucose
	α-Limit dextrinase*	. . .	α-Limit dextrins	Glucose
	Nuclease and related enzymes	. . .	Nucleic acids	Pentoses and purine and pyrimidine bases
Cytoplasm of mucosal cells	Various peptidases	. . .	Di, tri-, and tetrapeptides	Amino acids

*Sucrase and α-limit dextrinase are separate polypeptide chains that are subunits of a single protein.

The oligosaccharides responsible for the further digestion of the starch derivatives are located in the outer portion of the brush border, the membrane of the microvilli of the small intestine. α-Limit dextrinase hydrolyzes the α-limit dextrins, and glucomylase splits glucose from maltose, maltotriose, and other polymers of glucose in 1,4α linkage. Most of the glucose molecules that are formed enter adjacent mucosal cells, although some remain in the intestinal lumen and are absorbed farther along. Ingested disaccharides are hydrolyzed by lactase or sucrase on the luminal surface of mucosal cells (Fig 25–2). Deficiency of one or more of these disaccharidases may cause diarrhea, bloating, and flatulence after ingestion of sugar. The diarrhea is due to the increased number of osmotically active oligosaccharide molecules that remain in the intestinal lumen, causing the volume of the intestinal contents to increase. In the colon, bacteria break down some of the oligosaccharides, further increasing the number of osmotically active particles. The bloating and flatulence are due to the production of gas (CO_2 and H_2) from disaccharide residues in the lower small intestine and colon. The problem of milk intolerance

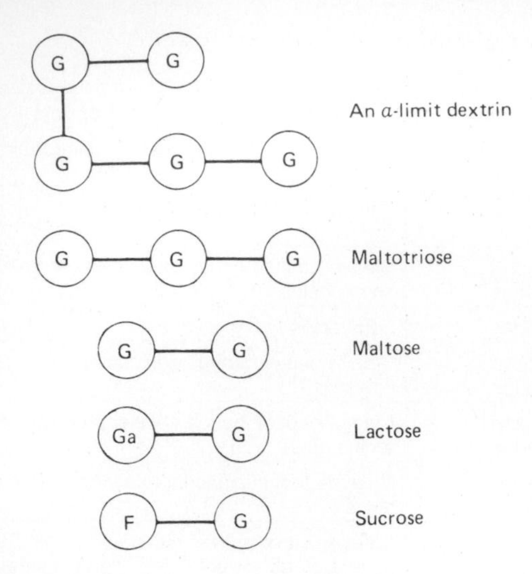

Figure 25–1. Principal end products of carbohydrate digestion in the intestinal lumen. Each circle represents a hexose molecule. G, glucose; F, fructose; Ga, galactose.

can be relieved by administration of commercial lactase preparations, but this is expensive. Yogurt is better tolerated than milk in intolerant individuals because it contains its own bacterial lactase.

Lactase is of interest because, in most mammals and in many races of humans, intestinal lactase activity is high at birth, declines to low levels during childhood, and remains low in adulthood. The low lactase levels are associated with intolerance to milk (lactose intolerance). Most Europeans and their American descendants retain their intestinal lactase activity in adulthood; the incidence of lactase deficiency in northern and western Europeans is only about 15%. However, the incidence in blacks, American Indians, Orientals, and Mediterranean populations is 70–90%.

Absorption

Hexoses and pentoses are rapidly absorbed across the wall of the small intestine (Table 25–2). Essentially all of the hexoses are removed before the remains of a meal reach the terminal part of the ileum. The sugar molecules pass from the mucosal cells to the blood in the capillaries draining into the portal vein.

The transport of some sugars is uniquely affected by the amount of Na^+ in the intestinal lumen; a high concentration of Na^+ on the mucosal surface of the cells facilitates and a low concentration inhibits sugar influx into the epithelial cells. This is because glucose and Na^+ share the same **symport.** Intracellular Na^+ is low, and Na^+ moves into the cell along its concentration gradient. Glucose moves with the Na^+ and is released in the cell (Fig 25–3). The Na^+ is transported into the lateral intercellular spaces, and the glucose diffuses into the interstitium and thence to the capillaries. Thus, glucose transport is an example of secondary active transport (see Chapter 1); the energy for glucose transport is provided indirectly, by the active transport of Na^+ out of the cell. This maintains the concentration gradient across the lumi-

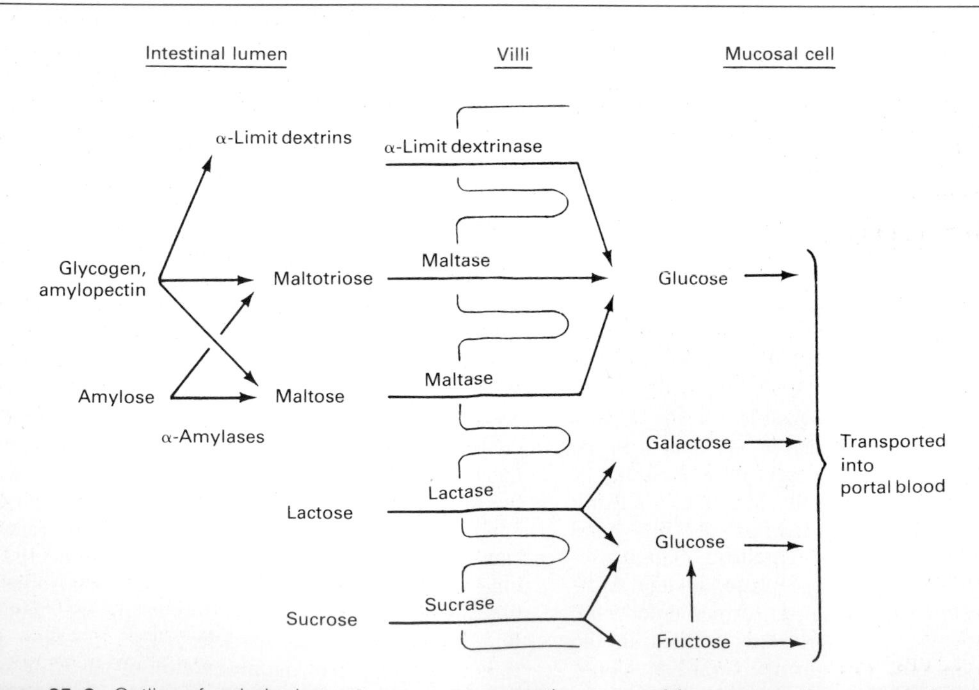

Figure 25–2. Outline of carbohydrate digestion and absorption. Some of the monosaccharides are also released into the intestinal lumen. (Modified from Gray GM: Carbohydrate digestion and absorption. *N Engl J Med* 1975;**292**:1225.)

Table 25–2. Normal transport of substances by the intestine and location of maximum absorption or secretion.*

	Small Intestine			
	Upper†	Mid	Lower	Colon
Absorption				
Sugars (glucose, galactose, etc)	+ +	+ + +	+ +	0
Amino acids	+ +	+ + +	+ +	0
Water-soluble vitamins	+ + +	+ +	0	0
Betaine, dimethylglycine, sarcosine	+	+ +	+ +	?
Antibodies in newborns	+	+ +	+ + +	?
Pyrimidines (thymine and uracil)	+	+	?	?
Fatty acid absorption and conversion to triglyceride	+ + +	+ +	+	0
Bile salts	+	+	+ + +	
Vitamin B_{12}	0	+	+ + +	0
Na^+	+ + +	+ +	+ + +	+ + +
K^+	+	+	+	sec
Ca^{2+}	+ + +	+ +	+	?
Fe^{2+}	+ + +	+ +	+	?
Cl^-	+ + +	+ +	+	+
SO_4^{2-}	+ +	+	0	?

*Amount of absorption is graded + to + + +. Sec: secreted when luminal $K^+ < 25$ mm
†Upper small intestine refers primarily to jejunum, although the duodenum is similar in most cases studied (with the notable exception that the duodenum secretes HCO_3^- and shows little net absorption or secretion of NaCl).

nal border of the cell, so that more Na^+ and consequently more glucose enter. The glucose mechanism also transports galactose. Fructose utilizes a different carrier, and its absorption is independent of Na^+ or the transport of glucose and galactose; it is transported instead by facilitated diffusion. Some fructose is converted to glucose in the mucosal cells. Pentoses are absorbed by simple diffusion.

Insulin has little effect on intestinal transport of sugars. In this respect, intestinal absorption resembles glucose reabsorption in the proximal convoluted tubules of the kidneys (see Chapter 38); neither process requires phosphorylation, and both are essentially normal in diabetes but depressed by the drug phlorhizin. The maximal rate of glucose absorption from the intestine is about 120 g/h.

PROTEINS & NUCLEIC ACIDS

Protein Digestion

Protein digestion begins in the stomach, where pepsins cleave some of the peptide linkages. Like many of the other enzymes concerned with protein digestion, pepsins are secreted in the form of inactive precursors **(proenzymes)** and activated in the gastrointestinal tract. The pepsin precursors are called pepsinogens and are activated by gastric hydrochloric acid. Human gastric mucosa contains a number of related pepsinogens, which can be divided into 2 immunohistochemically distinct groups, pepsinogen I and pepsinogen II. Pepsinogen I is found only in acid-secreting regions, whereas pepsinogen II is also found in the pyloric region. Maximal acid secretion correlates with pepsinogen I levels, and patients with congenitally elevated circulating pepsinogen I levels have a 5-fold greater incidence of peptic ulcers than individuals with normal levels (see Chapter 26).

Pepsins hydrolyze the bonds between aromatic amino acids such as phenylalanine or tyrosine and a second amino acid, so the products of peptic digestion are polypeptides of very diverse sizes. A **gelatinase** that liquefies gelatin is also found in the stomach. **Chymosin,** a milk-clotting gastric enzyme also known as **rennin,** is found in the stomachs of young animals but is probably absent in humans.

Because pepsins have a pH optimum of 1.6–3.2, their action is terminated when the gastric contents are mixed with the alkaline pancreatic juice in the duodenum and jejunum. The pH of the intestinal contents in the duodenal cap is 2.0–4.0, but in the rest of the duodenum, it is about 6.5.

In the small intestine, the polypeptides formed by digestion in the stomach are further digested by the powerful proteolytic enzymes of the pancreas and intestinal mucosa. Trypsin, the chymotrypsins, and elastase act at interior peptide bonds in the peptide molecules and are called **endopeptidases.** The formation of the active endopeptidases from their inactive precursors is discussed in Chapter 26. The carboxypeptidases of the pancreas and the aminopeptidases of the brush border are **exopeptidases** that hydrolyze the amino acids at the carboxy and amino ends of the polypeptides. Some free amino acids are liberated in the intestinal lumen, but others are liberated at the cell surface by the aminopeptidases and dipeptidases in the brush border of the mucosal cells. Some di- and tripeptides are actively transported into the intestinal cells and hydrolyzed by intracellular peptidases, with the amino acids entering the bloodstream. Thus, the final digestion to amino acids occurs in 3 locations: the intestinal

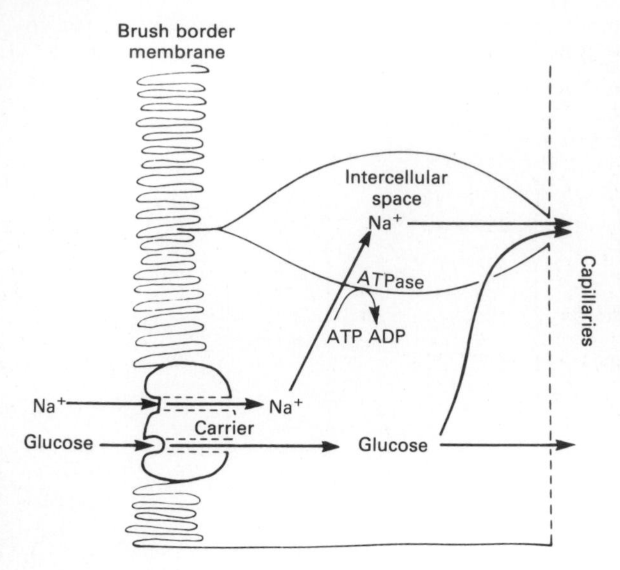

Figure 25–3. Mechanism for glucose transport across intestinal epithelium. Glucose transport into the intestinal cell is coupled to Na^+ transport, utilizing a common carrier protein. Na^+ is then actively transported out of the cell, and glucose diffuses into the capillaries. (Reproduced, with permission, from Gray GM: Carbohydrate digestion and absorption. *N Engl J Med* 1975;**292**:1225.)

lumen, the brush border, and the cytoplasm of the mucosal cells.

Absorption

After ingestion of a protein meal, there is a sharp transient rise in the amino nitrogen content of the portal blood. L Amino acids are absorbed more rapidly than the corresponding D isomers. The D amino acids are apparently absorbed solely by passive diffusion, whereas most L amino acids are actively transported out of the intestinal lumen. There are at least 3 separate transport systems: one that transports neutral amino acids; one that transports basic amino acids; and one that transports imino acids such as proline and hydroxyproline. A separate system transports di- and tripeptides into the mucosal cells. Absorption of amino acids by the neutral and imino acid systems is coupled to Na^+ transport, like glucose transport (Fig 25–3), whereas transport by the basic system is not Na^+ dependent. The transported amino acids and those produced by intracellular hydrolysis of di- and tripeptides accumulate in the mucosal cells, and from these cells they apparently diffuse passively into the blood. The only small peptides known to enter portal blood are those from gelatin that contain proline and hydroxyproline and those from certain meats that contain carnosine and anserine.

Absorption of amino acids is rapid in the duodenum and jejunum but slow in the ileum. Approximately 50% of the digested protein comes from ingested food, 25% from proteins in digestive juices, and 25% from desquamated mucosal cells. Only 2–5% of the protein in the small intestine escapes digestion and absorption. Some of the ingested protein enters the colon and is eventually digested by bacterial action. The protein in the stools is not of dietary origin but comes from bacteria and cellular debris. There is evidence that the peptidase activities of the brush border and the mucosal cell cytoplasm are increased by resection of part of the ileum and that they are independently altered in starvation. Thus, these enzymes appear to be subject to homeostatic regulation. In humans, a congenital defect in the mechanism that transports neutral amino acids in the intestine and renal tubules causes **Hartnup disease.** A congenital defect in the transport of basic amino acids causes **cystinuria.**

In infants, moderate amounts of undigested proteins are also absorbed. The protein antibodies in maternal colostrum that contribute to passive immunity against infections enter the circulation from the intestine, although this transfer is relatively minor in humans. Absorption is by endocytosis and subsequent exocytosis. Protein absorption declines with age, but adults still absorb small quantities. Foreign proteins that enter the circulation provoke the formation of antibodies, and the antigen-antibody reaction occurring upon subsequent entry of more of the same protein may cause allergic symptoms. Thus, absorption of proteins from the intestine may explain the occurrence of allergic symptoms after eating certain foods. However, true food allergies are probably rare, and the diagnosis is difficult to establish.

Absorption of protein antigens, particularly bacterial and viral proteins, takes place in large **microfold cells** or **M cells,** specialized intestinal epithelial cells that overlie aggregates of lymphoid tissue (Peyer's patches). These cells pass the antigens to the lymphoid cells, and lymphoblasts are activated. The activated lymphoblasts enter the circulation, but they later return to the intestinal mucosa and other epithelia, where they secrete IgA in response to subsequent exposures to the same antigen. This **secretory immunity** is an important defense mechanism; it is discussed in more detail in Chapter 27.

Nucleic Acids

Nucleic acids are split into nucleotides in the intestine by the pancreatic nucleases, and the nucleosides are split into the nucleosides and phosphoric acid by enzymes that appear to be located on the luminal surfaces of the mucosal cells. The nucleosides are then split into their constituent sugars and purine and pyrimidine bases. The bases are absorbed by active transport.

LIPIDS

Fat Digestion

A lingual lipase is secreted by Ebner's glands on the dorsal surface of the tongue, and the stomach also secretes a lipase (Table 25–1). The gastric lipase is of little importance except in pancreatic insufficiency, but lingual lipase is active in the stomach and can digest as much as 30% of dietary triglyceride.

Most fat digestion begins in the duodenum, pancreatic lipase being the most important enzyme involved. This enzyme hydrolyzes the 1- and 3-bonds of the triglycerides with relative ease but acts on the 2-bonds at a very low rate, so the principal products of its action are free fatty acids and 2-monoglycerides. It acts on fats that have been emulsified. However, it cannot act on fat droplets covered by emulsifying agents without **colipase,** a protein with a molecular weight of about 11,000 that binds to the surface of the fat droplets, displacing the emulsifying agents and anchoring lipase to the droplet. Colipase is secreted into the pancreatic juice in an inactive proform (Table 25–1) and is activated in the intestinal lumen by trypsin.

Most of the dietary cholesterol is in the form of cholesteryl esters, and cholesteryl ester hydrolase hydrolyzes these esters in the intestinal lumen.

Fats are finely emulsified in the small intestine by the detergent action of bile salts, lecithin, and monoglycerides; bile salts alone are relatively poor emulsifying agents, but in the presence of phospholipids and monoglycerides, particles 200–5000 nm in diameter are formed. The structure of the bile salts is discussed in Chapter 26.

When the concentration of bile salts in the intestine is high, as it is after contraction of the gallbladder, lipids and bile salts interact spontaneously to form **micelles** (Fig 25–4). These cylindrical aggregates are 3–10 nm in diameter, with the polar hydroxy, peptide bond, and carboxyl or sulfonic acid portions of the bile salts facing outward and the nonpolar steroid nuclear portions facing inward. Although their lipid concentration varies, they generally contain fatty acids, monoglycerides, and cholesterol in their hydrophobic centers. Micellar formation further solubilizes the lipids and provides a mechanism for their transport to the mucosal cells. Thus, the micelles move down their concentration gradient through the unstirred layer to the brush border of the mucosal cells. The lipids diffuse out of the micelles, and a saturated aqueous solution of the lipids is maintained in contact with the brush border of the mucosal cells (Fig 25–4). The lipids enter the cells by passive diffusion and are rapidly esterified inside the cells, maintaining a favorable concentration gradient from the lumen into the cells. Unlike the ileal mucosa, the rate of uptake of bile salts by the jejunal mucosa is low, and for the most part, the bile salts remain in the intestinal lumen, where they are available for the formation of new micelles. Thus, the bile salt micelles solubilize lipids, transport them across the unstirred layer, and keep a saturated solution of lipids in contact with the mucosal cells.

Pancreatectomized animals and patients with diseases that destroy the exocrine portion of the pancreas have fatty, bulky, clay-colored stools **(steatorrhea)** because of the impaired digestion and absorption of fat. The steatorrhea is due mostly to the lipase deficiency, but the micelle formation is also depressed in pancreatic insufficiency because, in the absence of the bicarbonate that is secreted from the pancreas, the relatively acid milieu in the duodenum inhibits the incorporation of fatty acids in the micelles. Acid also inhibits pancreatic lipase and precipitates some bile salts. This is why patients with excess secretion of gastric acid due to a gastrin-secreting tumor (gastrinoma; see Chapter 26) and a consequent low duodenal pH may develop steatorrhea. Another cause of steatorrhea is defective reabsorption of bile salts in the distal ileum (see Chapter 26).

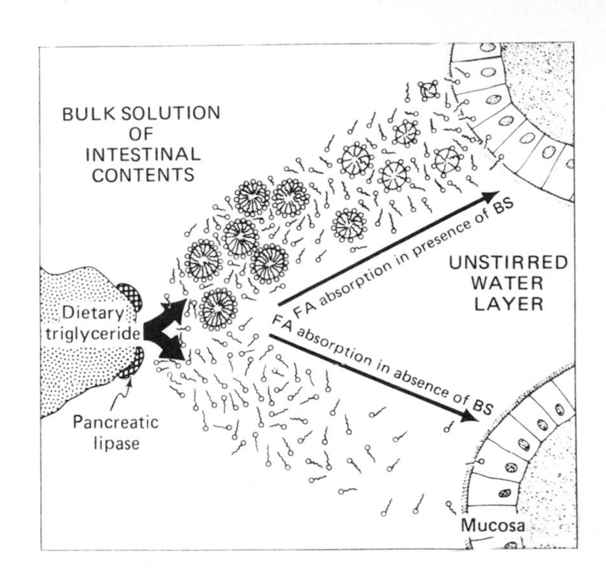

Figure 25–4. Lipid digestion and passage to intestinal mucosa. Fatty acids (FA) are liberated by the action of pancreatic lipase on dietary triglycerides and, in the presence of bile salts (BS), form micelles (the circular structures), which diffuse through the unstirred layer to the mucosal surface. (Reproduced, with permission, from Thomson ABR: Intestinal absorption of lipids: Influence of the unstirred water layer and bile acid micelle. In: *Disturbances in Lipid and Lipoprotein Metabolism.* Dietschy JM, Gotto AM Jr, Ontko JA [editors]. American Physiological Society, 1978.)

Fat Absorption

Monoglycerides, cholesterol, and fatty acids from the micelles enter the mucosal cells by passive diffusion. The subsequent fate of the fatty acids depends on their size. Fatty acids containing less than 10–12 carbon atoms pass from the mucosal cells directly into the portal blood, where they are transported as free (unesterified) fatty acids. The fatty acids containing more than 10–12 carbon atoms are reesterified to triglycerides in the mucosal cells. In addition, some of the absorbed cholesterol is esterified. The triglycerides and cholesteryl esters are then coated with a layer of protein, cholesterol, and phospholipid to form chylomicrons, which leave the cell and enter the lymphatics (Fig 25–5).

In mucosal cells, most of the triglyceride is formed by the acylation of the absorbed 2-monoglycerides, primarily in the smooth endoplasmic reticulum. However, some of the triglyceride is formed from glycerophosphate, which in turn is a product of glucose catabolism. Glycerophosphate is also converted into glycerophospholipids that participate in chylomicron formation. The acylation of glycerophosphate and the formation of lipoproteins occur in the rough endoplasmic reticulum. Carbohydrate moieties are added to the proteins in the Golgi apparatus, and the finished chylomicrons are extruded by exocytosis from the basal or lateral aspects of the cell.

Fat absorption is greatest in the upper parts of the small intestine, but appreciable amounts are also absorbed in the ileum (Fig 25–6). On a moderate fat intake, 95% or more of the ingested fat is absorbed. The stools contain 5% fat, but much of the fecal fat is probably derived from cellular debris and microorganisms rather than from the diet. The processes involved in fat absorption are not fully mature at birth, and infants fail to absorb 10–15% of ingested fat. Thus, they are more susceptible to the ill effects of disease processes that reduce fat absorption.

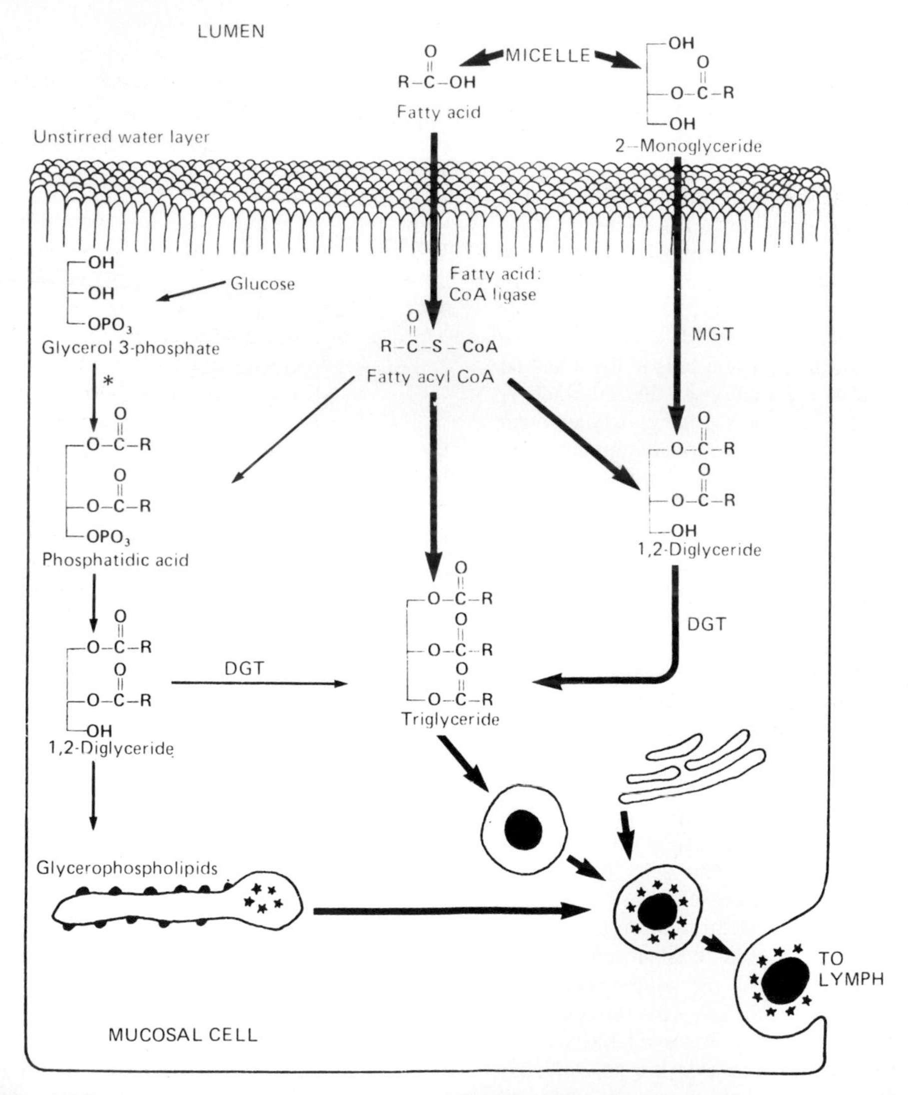

Figure 25–5. Lipid absorption. Triglycerides are formed in the mucosal cells from monoglycerides and fatty acids. Some of the glycerides also come from glucose via phosphatidic acid. The triglycerides are then converted to chylomicrons and released by exocytosis. From the extracellular space, they enter the lymph. ★, reaction inhibited by monoglyceride; MGT, monoacylglycerol acyltransferase; DGT, diacylglycerol acyltransferase. (Modified and reproduced, with permission, from Johnston JM: Esterification reactions in the intestinal mucosa and lipid absorption. In: *Disturbances in Lipid and Lipoprotein Metabolism.* Dietschy JM, Gotto AM Jr, Ontko JA [editors]. American Physiological Society, 1978.)

Absorption of Cholesterol & Other Sterols

Cholesterol is readily absorbed from the small intestine if bile, fatty acids, and pancreatic juice are present. Closely related sterols of plant origin are poorly absorbed. Almost all the absorbed cholesterol is incorporated into chylomicrons that enter the circulation via the lymphatics, as noted above. Nonabsorbable plant sterols such as those found in soybeans reduce the absorption of cholesterol, probably by competing with cholesterol for esterification with fatty acids.

ABSORPTION OF WATER & ELECTROLYTES

Water, Sodium, & Potassium

Overall water balance in the gastrointestinal tract is summarized in Table 25–3. The intestines are presented each day with about 2000 mL of ingested fluid plus 7000 mL of secretions from the mucosa of the gastrointestinal tract and associated glands, Ninety-eight percent of this fluid is reabsored, with a daily fluid loss of only 200 mL in the stools.

It should be noted that the figures for intestinal reabsorption are net rather than gross. Only small amounts of water move across the gastric mucosa, but water moves in both directions across the mucosa of the small and large intestines in response to osmotic gradients. Some Na^+ diffuses into or out of the small intestine depending on the concentration gradient. In addition, Na^+ is actively transported out of the lumen in the small intestine and colon by pumps that appear to be located on the basal and lateral walls of the cells. In the ileum and jejunum, Na^+ transport from intestine to blood is facilitated by aldosterone.

In the small intestine, active transport of Na^+ is important in bringing about absorption of glucose, some amino acids, and other substances (see above). Conversely, the presence of glucose in the intestinal lumen facilitates the reabsorption of Na^+. This is the physiologic basis for the treatment of Na^+ and water loss in diarrhea by oral administration of solutions containing NaCl and glucose. This type of treatment has even proved to be beneficial in the treatment of cholera, a disease associated with very severe and, if untreated, frequently fatal diarrhea. The cholera vibrio stays in the intestinal lumen, but it produces a toxin that activates adenylate cyclase, causing a marked increase in intracellular cyclic AMP. Some strains of diarrhea-producing *Escherichia coli* produce a similar toxin. The accumulation of cyclic AMP increases Cl^- secretion from the intestinal glands and inhibits the function of the mucosal carrier for Na^+, reducing NaCl absorption. The resultant increase in electrolyte and water content of the intestinal contents causes the diarrhea. However, the Na^+ pump and the common carrier for glucose and Na^+ are unaffected, so coupled reabsorption of glucose and Na^+ bypasses the defect.

Water moves into or out of the intestine until the osmotic pressure of the intestinal contents equals that of the plasma. The osmolality of the duodenal contents may be hypertonic or hypotonic, depending on the meal ingested, but by the time the meal enters the jejunum, its osmolality is close to that of plasma. This osmolality is maintained throughout the rest of the small intestine; the osmotically active particles produced by digestion are removed by absorption, and water moves passively out of the gut along the osmotic gradient thus generated. In the colon, Na^+ is pumped out and water moves passively with it, again along the osmotic gradient. **Saline cathartics** such as magnesium sulfate are poorly absorbed salts that retain their osmotic equivalent of water in the intes-

Table 25–3. Daily net water turnover (mL) in the gastrointestinal tract.*

Ingested		2000
Endogenous secretions		7000
Salivary glands	1500	
Stomach	2500	
Bile	500	
Pancreas	1500	
Intestine	1000	
	7000	
Total input		9000
Reabsorbed		8800
Jejunum	5500	
Ileum	2000	
Colon	1300	
	8800	
Balance in stool		200

*Data from Moore EW: *Physiology of Intestinal Water and Electrolyte Absorption.* American Gastroenterological Society, 1976.

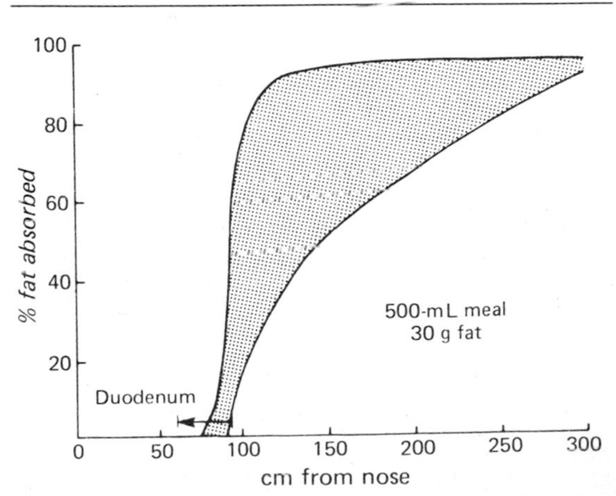

Figure 25–6. Fat absorption, based on measurement after a fat meal in humans. (Redrawn and reproduced, with permission, from Davenport HW: *Physiology of the Digestive Tract,* 2nd ed. Year Book, 1966.)

tine, thus increasing intestinal volume and consequently exerting a laxative effect.

There is some secretion of K^+ into the intestinal lumen, especially as a component of mucus, but for the most part, the movement of K^+ across the gastrointestinal mucosa is due to diffusion. The mucosal cells of the colon behave like renal tubular cells in regard to K^+. On a normal diet, K^+ is pumped from the interstitial fluid into the cells by the Na^+-K^+ ATPase on the basolateral membranes of the cells and then diffuses into the colon. The transmembrane potential on the luminal side of the cells is less than on the basolateral side, and this fosters the movement of K^+ into the colon. This is why the loss of ileal or colonic fluids in chronic diarrhea can lead to severe hypokalemia.

When the dietary intake of K^+ is high for a period of time, aldosterone secretion is increased and more K^+ is secreted into the colon. This is due in part to the appearance of more Na^+-K^+ ATPase pumps in the basolateral membranes of the cells, with a consequent increase in intracellular K^+ and K^+ diffusion across the luminal membranes of the cells.

Chloride & Bicarbonate

In the ileum and the colon, it appears that Cl^- is actively reabsorbed in a one-for-one exchange for HCO_3^-. This tends to make the intestinal contents more alkaline. However, the physiologic significance of this exchange is uncertain.

ABSORPTION OF VITAMINS & MINERALS

Vitamins

Absorption of water-soluble vitamins is rapid, but absorption of the fat-soluble vitamins A,D,E, and K is deficient if fat absorption is depressed because of lack of pancreatic enzymes or if bile is excluded from the intestine by obstruction of the bile duct. Most vitamins are absorbed in the upper small intestine, but vitamin B_{12} is absorbed in the ileum. This vitamin binds to intrinsic factor, a protein secreted by the stomach, and the complex is absorbed across the ileal mucosa (see Chapter 26).

Calcium

Thirty to 80% of ingested calcium is absorbed. Active transport of Ca^{2+} out of the intestinal lumen occurs primarily in the upper small intestine, and there is also some absorption by passive diffusion. Active transport is facilitated by 1,25-dihydroxycholecalciferol, the metabolite of vitamin D that is produced in the kidney. The metabolite induces the synthesis of a Ca^{2+}-binding protein in the mucosal cells (see Chapter 21). The rate of production of 1,25-dihydroxycholecalciferol is increased when the plasma calcium level is decreased and reduced when the plasma calcium level is elevated (see Chapter 21). Consequently, Ca^{2+} absorption is adjusted to body needs; absorption is increased in the presence of Ca^{2+} deficiency and decreased in the presence of Ca^{2+} excess. Ca^{2+} absorption is also facilitated by protein. It is inhibited by phosphates and oxalates because these anions form insoluble salts with Ca^{2+} in the intestine. Magnesium absorption is facilitated by protein.

Iron

In adults, the amount of iron lost from the body is relatively small. The losses are generally unregulated, and total body stores of iron are regulated by changes in the rate at which it is absorbed from the intestine. Men lose about 0.6 mg/d, whereas women have a variable, larger loss averaging about twice this value because of the additional iron lost in the blood shed during menstruation. The average daily iron intake in the USA and Europe is about 20 mg, but the amount absorbed is equal only to the losses; if it were any greater, iron overload would develop. Thus, the amount of iron absorbed ranges normally from about 3 to 6% of the amount ingested.

Iron is more readily absorbed in the ferrous state (Fe^{2+}), but most of the dietary iron is in the ferric form (Fe^{3+}). No more than a trace of iron is absorbed in the stomach, but the gastric secretions dissolve the iron and permit it to form soluble complexes with ascorbic acid and other substances that aid its reduction to the Fe^{2+} form. The importance of this function in humans is indicated by the fact that iron deficiency anemia is a troublesome and relatively frequent complication of partial gastrectomy. Heme is also absorbed, and the Fe^{2+} that it contains is released in the mucosal cells. Various dietary factors affect the availability of iron for absorption; for example, the phytic acid found in cereals reacts with iron to form insoluble compounds in the intestine. So do phosphates and oxalates. Pancreatic juice inhibits iron absorption.

Most of the iron is absorbed in the upper part of the small intestine (Table 25–2). Other mucosal cells can transport iron, but the duodenum and adjacent jejunum contain most of the iron suitable for absorption. The mucosal cells contain an intracellular iron carrier. Some iron is supplied to the mitochondria from the carrier, but the remainder is partitioned between **apoferritin** in the mucosal cells and **tranferrin,** the iron-transporting polypeptide in the plasma (Fig 25–7). Apoferritin, which is also found in many other tissues, combines with iron to form **ferritin.** The iron bound to ferritin in intestinal cells is lost with the cells when they are shed into the intestinal lumen at the end of their life cycle and passed in the stool.

Apoferritin is a globular protein made up of 24 subunits. Iron forms a micelle of ferric hydroxyphosphate, and in ferritin, the subunits surround this micelle. The ferritin molecule can contain as many as 4500 atoms of iron. Ferritin is readily visible under the electron microscope and has been used as a tracer in studies of phagocytosis and related phenomena. It is the principal storage form of iron in tissues. Ferritin molecules in lysosomal membranes may aggregate in deposits that contain as much as 50% iron. These deposits are called **hemosiderin.** Seventy percent of the iron in the body is in hemoglobin, 3% in myoglobin, and the remainder in ferritin. Ferritin is also found in plasma, but most iron is transported bound to transferrin. This polypeptide has 2 iron-binding sites. Normally, transferrin is about 35% saturated with iron, and the normal plasma iron level is about 130 μg/dL (23 μmol/L) in men and 110 μg/dL (19 μmol/L) in women.

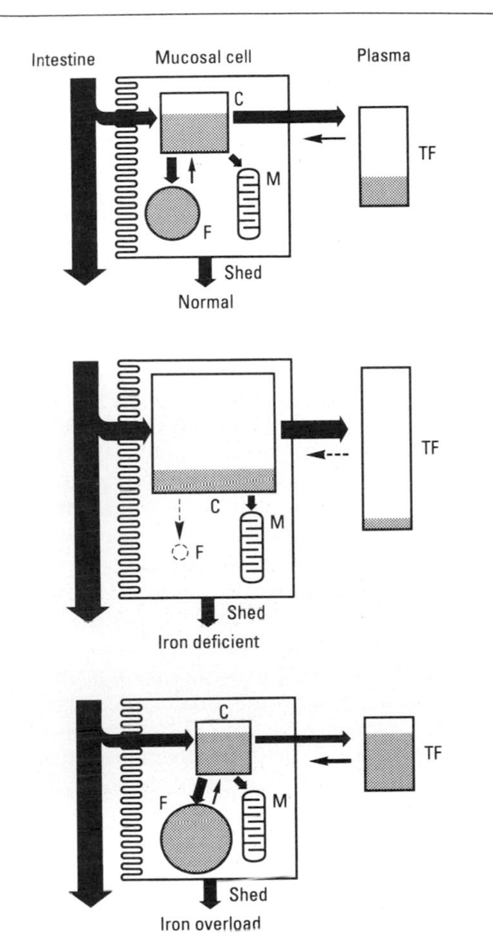

Figure 25–7. Diagrammatic summary of the control of iron absorption by the intestinal mucosa. C, intracellular iron carrier; F, ferritin; M, mitochondria; TF, plasma transferrin. The width of the black arrows is proportionate to the movement of iron from one compartment to another. In C and TF, the shedding is proportionate to the percent saturation with iron. Mucosal cells are shed into the intestinal lumen, and the more ferritin they contain, the more they contribute to iron loss in the stools. Note that the amount of TF in the plasma is inversely proportionate to the amount of iron in the body and that the greater the saturation of TF, the more iron goes into ferritin in the mucosal cells. (Modified from Stanbury JB et al: *The Metabolic Basis of Inherited Disease,* 5th ed. McGraw Hill, 1983.)

Iron absorption is increased when body iron stores are depleted or when erythropoiesis is increased, and decreased under the opposite conditions. In iron deficiency, the amount of transferrin in the plasma is increased and its percent of saturation with iron is decreased. Consequently, more iron moves from the intracellular iron carrier to transferrin and less binds to apoferritin (Fig 25–7). Ferritin stores in the mucosa are decreased, and consequently, less iron is lost when the mucosal cells are shed. Conversely, in the presence of iron overload, the amount of circulating transferrin is decreased and its saturation is increased, so more iron is diverted to apoferritin, ferritin stores increase, and more iron is lost when the mucosal cells are shed. Recent evidence indicates that in addition, iron fosters apoferritin synthesis by binding to a sequence at the 5′ untranslated portion of the ferritin mRNA, making the mRNA more active. This is one of the few instances of regulation of this type at the level of translation rather than transcription.

The normal operation of the factors that maintain iron balance is essential for health. Iron deficiency causes anemia. If more iron is absorbed than is excreted, iron overload results. However, normal individuals can maintain a normal rate of absorption even when the ingested load is 5 or 10 times more than needed. Hemosiderin accumulates in the tissues when the overload is prolonged and severe, producing **hemosiderosis.** Large amounts of hemosiderin can damage the tissues, producing **hemochromatosis.** This syndrome is characterized by pigmentation of the skin, pancreatic damage with diabetes ("bronze diabetes"), cirrhosis of the liver, a high incidence of hepatic carcinoma, and gonadal atrophy. Hemochromatosis can be produced not only by prolonged excessive iron intake but also by a number of other conditions. In idiopathic hemochromatosis, a congenital disorder due to an autosomal recessive gene, the mucosal regulatory mechanism behaves as if the iron deficiency were present and absorbs iron at a high rate in the face of elevated rather than depleted body iron stores.

26 Regulation of Gastrointestinal Function

INTRODUCTION

The digestive and absorptive functions of the gastrointestinal system outlined in the previous chapter depend upon a variety of mechanisms that soften the food, propel it through the gastrointestinal tract, and mix it with bile from the gallbladder and digestive enzymes secreted by the salivary glands and pancreas. Some of these mechanisms depend upon intrinsic properties of the intestinal smooth muscle. Others involve the operation of visceral reflexes or the actions of **gastrointestinal hormones.** The hormones are humoral agents secreted by parts of the mucosa and transported in the circulation to influence the functions of the stomach, the intestines, the pancreas, and the gallbladder.

ANATOMIC CONSIDERATIONS

Organization

The organization of the structures that make up the wall of the gastrointestinal tract from the posterior pharynx to the anus is shown in Fig 26–1. There is some local variation, but in general there are 3 layers of smooth muscle, 2 longitudinal and 1 circular. The wall is lined by mucosa and, except in the case of the esophagus, is covered by serosa. The serosa continues onto the mesentery, which contains the nerves, lymphatics, and blood vessels supplying the tract.

Innervation of the Gastrointestinal Tract

There are 2 major networks of nerve fibers that are intrinsic to the gastrointestinal tract: the **myenteric nerve plexus** (Auerbach's plexus), between the outer longitudinal and middle circular muscle layers; and the **submucous plexus** (Meissner's plexus), between the middle circular layer and the mucosa (Fig 26–1). The plexuses are interconnected, and they contain nerve cells with processes that originate in receptors in the wall of the gut or the mucosa. They also contain many interneurons. The mucosal receptors are mechanoreceptors sensitive to stretch of the intestinal wall, and some may also be chemoreceptors that sense the composition of the intestinal contents. The nerve cells innervate hormone-secreting cells and all the muscle layers in the mucosa. The nerves in the intestinal wall constitute a complex **enteric nervous system** that some investigators add to the sympathetic and parasympathetic divisions as a third division of the autonomic nervous system. There are about 1 million neurons in this system. They contain not only acetylcholine and norepinephrine but serotonin (5-hydroxytryptamine), vasoactive intestinal polypeptide (VIP), substance P, somatostatin, enkephalins, cholecystokinin, gastrin-releasing peptide (GRP), neurotensin, and, possibly, angiotensin II. Most of these peptides and serotonin are also found in gland cells in the mucosa. The way the peptide-secreting and other neurons interact to regulate gastrointestinal function is still unsettled. However, it is clear that the plexuses are responsible for peristaltic and other contractions, and coordinated motor activity occurs in the total absence of extrinsic innervation.

The intestine receives a dual extrinsic innervation from the autonomic nervous system, with parasympathetic cholinergic activity generally increasing the activity of intestinal smooth muscle and sympathetic

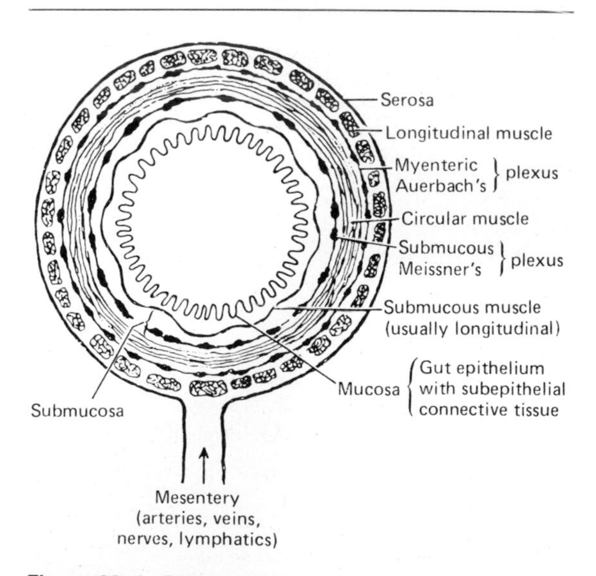

Figure 26–1. Diagrammatic representation of the layers of the wall of the stomach, small intestine, and colon. The structure of the esophagus and the distal rectum is similar except that they have no mesentery. (Reproduced, with permission, from Bell GH, Emslie-Smith D, Paterson CR: *Textbook of Physiology and Biochemistry*, 9th ed. Churchill Livingstone, 1976.)

noradrenergic activity generally decreasing it while causing sphincters to contract. The preganglionic parasympathetic fibers consist of about 2000 vagal efferents and other efferents in the sacral nerves. They generally end on cholinergic nerve cells of the myenteric and submucous plexuses. The sympathetic fibers are postganglionic, but many of them end on postganglionic cholinergic neurons, where they inhibit acetylcholine secretion. Others innervate blood vessels, where they produce vasoconstriction, and some appear to end directly on intestinal smooth muscle cells. The electrical properties of intestinal smooth muscle are discussed in Chapter 3.

Gastrointestinal Circulation

The blood flow to the stomach, intestines, pancreas, and liver is arranged in a series of parallel circuits, with all the blood from the intestines and pancreas draining via the portal vein to the liver. The physiology of this important portion of the circulation is discussed in Chapter 32.

GASTROINTESTINAL HORMONES

Many different hormonally active polypeptides have been isolated from the gastrointestinal mucosa. Experiments with them and measurement of their concentrations in blood by radioimmunoassay have identified the roles these **gastrointestinal hormones** play in the regulation of gastrointestinal secretion and motility. When large doses of the hormones are given, their actions overlap. However, their physiologic effects appear to be relatively discrete. On the basis of structural similarity (Table 26–1) and, to a degree, similarity of function, many of the hormones fall into one of 2 families: the gastrin family, the primary members of which are gastrin and cholecystokinin (CCK); and the secretin family, the primary members of which are secretin, glucagon, glicentin (GLI), VIP, and gastric inhibitory peptide (GIP).

Gastrin

Gastrin is produced by cells called G cells in the lateral walls of the glands in the antral portion of the gastric mucosa (Fig 26–2). G cells are flask shaped, with a broad base containing many gastrin granules and a narrow apex that reaches the mucosal surface. Microvilli project from the apical end into the lumen. Receptors mediating gastrin responses to changes in gastric contents are present on the microvilli. Like the other cells in the gastrointestinal tract that secrete hormones, these cells contain amines related to norepinephrine or serotonin and appear to be of neural crest origin. Because they take up amine precursors and decarboxylate them, they are sometimes called APUD cells (for *a*mine *p*recursor *u*ptake and *d*ecarboxylation). There are APUD cells in many different parts of the body in addition to the gastrointestinal tract. A second type of gastrin-producing cell, the TG cell, is found throughout the stomach and small intestine. It contains G 34 and the C-terminal tetrapeptide of gastrin but lacks G 17.

Gastrin is also found in the pancreatic islets in fetal life. Gastrin-secreting tumors called **gastrinomas** occur in the pancreas, but it is uncertain whether any gastrin is found in the anterior and intermediate lobes of the pituitary gland, in the hypothalamus and medulla oblongata, and in the vagus and sciatic nerves.

Gastrin is typical of a number of polypeptide hormones in that it shows both **macroheterogeneity** and **microheterogeneity.** Macroheterogeneity refers to the occurrence in tissues and body fluids of peptide chains of varying length; microheterogeneity refers to differences in molecular structure due to derivatization of single amino acid residues. Preprogastrin is processed into fragments of varying size. Three main forms of gastrin contains 34, 17, and 14 amino acid residues, respectively. All have the same C-terminal configuration (Table 26–1). These forms are also known as G 34, G 17, and G 14 gastrins. Another form is the C-terminal tetrapeptide, and there is also a large form that is extended at the N terminal and contains more than 45 amino acid residues. One form of derivatization is sulfation of the tyrosine that is the sixth amino acid residue from the C terminal. There are approximately equal amounts of nonsulfated and sulfated forms in blood and tissues, and they are equally active. Another derivatization is amidation of the C-terminal phenylalanine.

What is the physiologic significance of this marked heterogeneity? There are some differences in activity between the various components, and the proportions of the components also differ in the various tissues in which gastrin is found. This suggests that different forms are tailored for different actions. However, all that can be concluded at present is that G 17 is the principal form with respect to gastric acid secretion. The C-terminal tetrapeptide has all the activities of gastrin but only 10% of the strength of G 17.

G 14 and G 17 have half-lives of 2–3 minutes in the circulation, whereas G 34 has a half-life of 15 minutes. Gastrins are inactivated primarily in the kidney and small intestine.

In large doses, gastrin has a variety of actions, but its principal physiologic actions are stimulation of gastric acid and pepsin secretion and stimulation of the growth of the gastric mucosa. Stimulation of gastric motility is probably a physiologic action as well. Gastrin also causes contraction of the musculature that closes the gastroesophageal junction (see below), but this effect is of questionable physiologic significance. It stimulates insulin and glucagon secretion, but only after a protein meal and not after a carbohydrate meal does circulating endogenous gastrin reach the level necessary to stimulate the B cells. The functions of gastrin in the pituitary gland, brain, and peripheral nerves are unknown.

Table 26–1. Structure of hormonally active polypeptides secreted by cells in the human gastrointestinal tract. Homologous amino acid residues are enclosed by the lines that generally cross from one polypeptide to another. Arrows indicate points of cleavage to form smaller variants. Tys, tyrosine sulfate. All gastrins occur in unsulfated (gastrin I) and sulfated (gastrin II) forms. Glicentin, an additional member of the secretin family which is not shown, is a C terminally extended relative of glucagon.

Gastrin Family		Secretin Family				Other Polypeptides			
CCK 39	Gastrin 34	GIP	Glucagon	Secretin	VIP	Motilin	Substance P	Gastrin-Releasing Peptide	Somatostatin 14
Tyr		Tyr	His	His	His	Phe	Arg	Val	Ala
Ile		Ala	Ser	Ser	Ser	Val	Pro	Pro	Gly
Gln		Glu	Gln	Asp	Asp	Pro	Lys	Leu	Cys
Gln		Gly	Gly	Gly	Ala	Ile	Pro	Pro	Lys
Ala		Thr	Thr	Thr	Val	Phe	Gln	Ala	Asn
Arg	(pyro)Glu	Phe	Phe	Phe	Phe	Thr	Gln	Gly	Phe
→Lys	Leu	Ile	Thr	Thr	Thr	Tyr	Phe	Gly	Phe
Ala	Gly	Ser	Ser	Ser	Asp	Gly	Phe	Gly	Trp
Pro	Pro	Asp	Asp	Glu	Asn	Glu	Gly	Thr	Lys
Ser	Gln	Tyr	Tyr	Leu	Tyr	Leu	Leu	Val	Thr
Gly	Gly	Ser	Ser	Ser	Thr	Gln	Met-NH$_2$	Leu	Phe
Arg	Pro	Ile	Lys	Arg	Arg	Arg		Thr	Thr
Met	Pro	Ala	Tyr	Leu	Leu	Met		Lys	Ser
Ser	His	Met	Leu	Arg	Arg	Gln		Met	Cys
Ile	Leu	Asp	Asp	Glu	Lys	Glu		Tyr	
Val	Val	Lys	Ser	Gly	Gln	Lys		Pro	
Lys	Ala	Ile	Arg	Ala	Met	Glu		Arg	
Asn	Asp	His	Arg	Arg	Ala	Arg		Gly	
Leu	Pro	Gln	Ala	Leu	Val	Asn		Asn	
Gln	Ser	Gln	Gln	Gln	Lys	Lys		His	
Asn	Lys	Asp	Asp	Arg	Lys	Gly		Trp	
Leu	Lys	Phe	Phe	Leu	Tyr	Gln		Ala	
Asp	→Gln	Val	Val	Leu	Leu			Val	
Pro	Gly	Asn	Gln	Gln	Asn			Gly	
Ser	Pro	Trp	Trp	Gly	Ser			His	
His	→Trp	Leu	Leu	Leu	Ile			Leu	
Arg	Leu	Leu	Met	Val-NH$_2$	Leu			Met-NH$_2$	
→Ile	Glu	Ala	Asn		Asn-NH$_2$				
Ser	Glu	Glu	Thr						
Asp	Glu	Lys							
Arg	Glu	Gly							
→Asp	Glu	Lys							
Tys	Ala	Lys							
Met	Tys	Asn							
Gly	Gly	Asp							
→Trp	→Trp	Trp							
Met	Met	Lys							
Asp	Asp	His							
Phe-NH$_2$	Phe-NH$_2$	Asn							
		Ile							
		Thr							
		Gln							

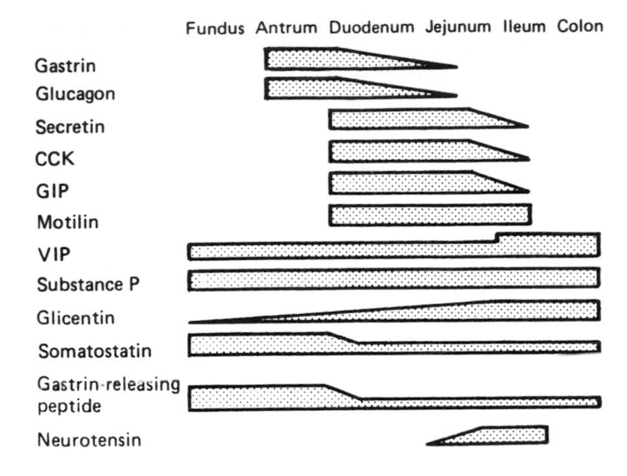

Figure 26–2. Distribution of gastrointestinal peptides along the gastrointestinal tract. The thickness of each bar is proportionate to the concentration of the peptide in the mucosa.

Gastrin secretion is affected by the contents of the stomach, the rate of discharge of the vagus nerves, and blood-borne factors (Table 26–2). Atropine does not inhibit the gastrin response to a test meal in humans because in all probability the transmitter secreted by the vagal fibers is gastrin-releasing peptide (GRP; see below) rather than acetylcholine. Gastrin secretion is also increased by the presence of the products of protein digestion in the stomach, particularly amino acids, which act directly on the G cells. Phenylalanine and tryptophan are particularly effective.

Acid in the antrum inhibits gastrin secretion. The effect of acid is the basis of a negative feedback loop regulating gastrin secretion. Increased secretion of the hormone increases acid secretion, but the acid then feeds back to inhibit further gastrin secretion.

The role of gastrin in the pathophysiology of duodenal ulcers is discussed below. In conditions such as pernicious anemia in which the acid-secreting cells of the stomach are damaged, gastrin secretion is chronically elevated.

Table 26–2. Stimuli that affect gastrin secretion.

Stimuli that affect gastrin secretion
Stimuli that increase gastrin secretion
Luminal
Peptides and amino acids
Distention
Neural
Increased vagal discharge, probably mediated by GRP
Blood-borne
Calcium
Epinephrine
Stimuli that inhibit gastrin secretion
Luminal
Acid
Blood-borne
Secretin, GIP, VIP, glucagon, calcitonin

Cholecystokinin-Pancreozymin

It was formerly thought that a hormone called cholecystokinin produced contraction of the gallbladder whereas a separate hormone called pancreozymin increased the secretion of pancreatic juice rich in enzymes. It is now clear that a single hormone secreted by cells in the mucosa of the upper small intestine has both activities, and the hormone has therefore been named **cholecystokinin-pancreozymin.** It is also called **CCK-PZ** or, most commonly, **CCK.**

Like gastrin, CCK shows both macroheterogeneity and microheterogeneity. Prepro-CCK is processed into many fragments. A large CCK contains 58 amino acid residues (CCK 58). There are, in addition, CCK peptides containing 39 amino acid residues (CCK 39) and 33 amino acid residues (CCK 33), several forms that have 12 (CCK 12) or slightly more amino acid residues, and a form containing 8 amino acid residues (CCK 8). All of these forms have the same 5 amino acids at the C terminal as gastrin (Table 26–1). The C-terminal tetrapeptide (CCK 4) also exists in tissues. The C terminal is amidated, and the tyrosine that is the seventh amino acid residue from the C terminal is sulfated. Unlike gastrin, the nonsulfated form of CCK has not been found in tissues. However, it is known that derivatization of other amino acid residues in CCK can occur. The half-life of circulating CCK is about 5 minutes, but little is known about its metabolism.

In addition to its occurrence in endocrine cells in the upper intestine, CCK is found in nerves in the distal ileum and colon. It is also found in neurons in the brain, especially the cerebral cortex, and in nerves in many parts of the body (see Chapter 4). The CCK secreted in the duodenum and jejunum is probably mostly CCK 8 and CCK 12, although CCK 58 is also present in the intestine and circulating blood in some species. The enteric and pancreatic nerves contain primarily CCK 4. CCK 58 and CCK 8 are found in the brain.

In addition to causing contraction of the gallbladder and secretion of a pancreatic juice rich in enzymes, CCK augments the action of secretin in producing secretion of an alkaline pancreatic juice. It also inhibits gastric emptying, exerts a trophic effect on the pancreas, increases the secretion of enterokinase, and may enhance the motility of the small intestine and colon. There is some evidence that, along with secretin, it augments the contraction of the pyloric sphincter, thus preventing the reflux of duodenal contents into the stomach. Gastrin and CCK stimulate glucagon secretion, and since the secretion of both gastrointestinal hormones is increased by a protein meal, either or both may be the "gut factor" that stimulates glucagon secretion (see Chapter 19). The action of CCK on pancreatic acinar cells is produced by activation of phospholipase C, with consequent increases in intracellular IP_3 and DAG.

The secretion of CCK is increased by contact of the intestinal mucosa with the products of digestion, particularly peptides and amino acids, and also by the presence in the duodenum of fatty acids containing more than 10 carbon atoms. Since the bile and pancreatic juice that enter the duodenum in response to CCK further the digestion of protein and fat and the products of this digestion stimulate further CCK secretion, a sort of positive feedback operates in the control of the secretion of this hormone. The positive feedback is terminated when the products of digestion move on to the lower portions of the gastrointestinal tract.

Secretin

Secretin occupies a unique position in the history of physiology. In 1902, Bayliss and Starling first demonstrated that the excitatory effect of duodenal stimulation on pancreatic secretion was due to a blood-borne factor. Their research led to the identification of secretin. They also suggested that many chemical agents might be secreted by cells in the body and passed in the circulation to affect organs some distance away, and Starling introduced the term **hormone** to categorize such "chemical messengers." Modern endocrinology is the proof of the correctness of this hypothesis.

Secretin is secreted by cells that are located deep in the glands of the mucosa of the upper portion of the small intestine. The structure of secretin (Table 26–1) is different from that of CCK and gastrin but very similar to that of glucagon, GLI, VIP, and GIP. Only one form of secretin has been isolated, and the fragments of the molecule that have been tested to date are inactive. Its half-life is about 5 minutes, but little is known about its metabolism.

Secretin increases the secretion of bicarbonate by the duct cells of the pancreas and biliary tract. It thus causes the secretion of a watery, alkaline pancreatic juice. Its action on pancreatic duct cells is mediated via cyclic AMP. It also augments the action of CCK in producing pancreatic secretion of digestive enzymes. It decreases gastric acid secretion and may cause contraction of the pyloric sphincter.

The secretion of secretin is increased by the products of protein digestion and by acid bathing the mucosa of the upper small intestine. The release of secretin by acid is another example of feedback control: Secretin causes alkaline pancreatic juice to flood into the duodenum, neutralizing the acid from the stomach and thus stopping further secretion of the hormone.

GIP

GIP contains 43 amino acid residues (Table 26–1) and is found in the mucosa of the duodenum and jejunum. Its secretion is stimulated by glucose and fat in the duodenum, and it inhibits gastric secretion and motility. It also stimulates insulin secretion. Gastrin, CCK, secretin and glucagon also have this effect, but the evidence summarized in Chapter 19 indicates that GIP is the physiologic B-cell-stimulating hormone of the gastrointestinal tract. For this reason, it is sometimes called glucose-dependent insulinotropic peptide.

The integrated action of gastrin, CCK, secretin, and GIP in facilitating digestion and utilization of absorbed nutrients is summarized in Fig 26–3.

VIP

VIP contains 28 amino acid residues (Table 26–1). It is found in nerves in the gastrointestinal tract. Prepro-VIP contains both VIP and a closely related polypeptide **(PHM-27** in humans, PHI-27 in other species). VIP is also found in blood, in which it has a half-life of about 2 minutes. It markedly stimulates intestinal secretion of electrolytes and hence of water. Its other actions include dilation of peripheral blood vessels and inhibition of gastric acid secretion. It is also found in the brain and many autonomic nerves (see Chapter 4), where it occurs in the same neurons as acetylcholine. It potentiates the action of acetylcholine in salivary glands. VIP-secreting tumors (VIPomas) have been described in patients with severe diarrhea. The relation of GIP and VIP to **enterogastrone,** a putative hormone that inhibits gastric acid secretion and motility, is unknown.

Other Gastrointestinal Hormones

Motilin, a polypeptide containing 22 amino acid residues (Table 26–1), has been extracted from duodenal mucosa. It causes contraction of intestinal smooth muscle and appears to be a regulator of interdigestive motility, preparing the intestine for the next meal. **Substance P** (Table 26–1) is found in endocrine cells in the gastrointestinal tract, but it has

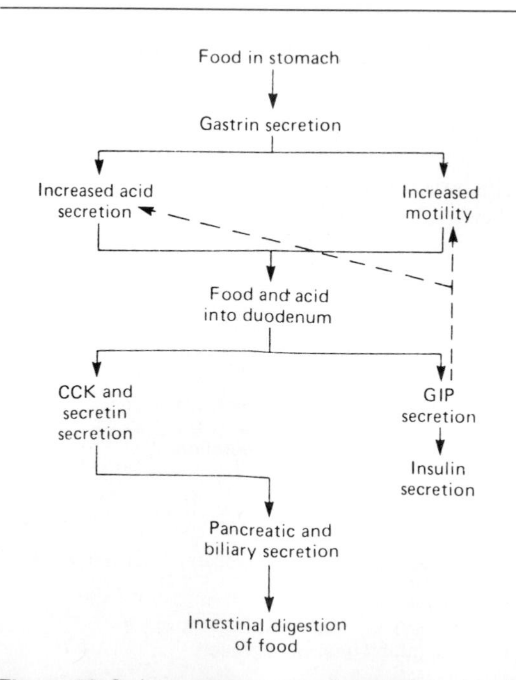

Figure 26–3. Integrated action of gastrointestinal hormones in regulating digestion and utilization of absorbed nutrients. The dashed arrows indicate inhibition.

not been proved to enter the circulation. It increases the motility of the small intestine. **Gastrin-releasing peptide (GRP)** contains 27 amino acid residues, and the 10 amino acid residues at its C terminal are almost identical to amphibian bombesin. It is present in the vagal nerve endings that terminate on G cells and is probably the neurotransmitter producing vagally mediated increases in gastrin secretion. **Somatostatin,** the growth-hormone-inhibiting hormone originally isolated from the hypothalamus, is secreted into the circulation by D cells in the pancreatic islets (see Chapter 19) and by similar D cells in the gastrointestinal mucosa. As noted in Chapter 14, it exists in tissues in 2 forms, somatostatin 14 (Table 26–1) and somatostatin 28, and both are secreted. In addition, SS28 (1–12) is found in tissues. Somatostatin inhibits the secretion of gastrin, VIP, GIP, secretin, and motilin. Like several other gastrointestinal hormones, somatostatin is secreted in larger amounts into the gastric lumen than into the bloodstream. Its secretion is stimulated by acid in the lumen, and it may act in a "paracrine" fashion via the gastric juice to mediate the inhibition of gastrin secretion produced by acid. It also inhibits pancreatic exocrine secretion; gastric acid secretion and motility; gallbladder contraction; and the absorption of glucose, amino acids, and triglycerides. **Glucagon** is secreted by the A cells in the mucosa of the stomach and duodenum in addition to the A cells in the pancreatic islets. As noted in Chapter 19, glucagon from the gastrointestinal tract appears to play a role in the hyperglycemia of diabetes. The intestine also contains **glicentin (glucagonlike immunoreactivity, GLI),** which has a different distribution that glucagon in the intestinal mucosa (Fig 26–2). Glicentin has some glucagon activity, but it is larger (molecular weight approximately 11,000 versus 3485 for glucagon), and the C-terminal portion of glicentin is made up of glucagon extended at its C terminal by an octapeptide. Its function is unknown. Glicentin is found along with glucagon in the granules of A cells in the pancreatic islets (see Chapter 19).

Enkephalins, serotonin, and neurotensin may also be gastrointestinal hormones. As noted above, serotonin, substance P, VIP, somatostatin, enkephalins, CCK, and neurotensin are present in gland cells and nerve fibers in the gastrointestinal tract. All these substances, along with gastrin and glucagon, are also found in the brain (see Chapter 4). The opioid peptide dynorphin is found in the posterior pituitary and the duodenum. TRH and ACTH are found in the gastrointestinal tract, but their functions in this location are unknown. TRH is also found in the pancreatic islets, where it is probably located in B cells. A substance called **urogastrone,** which was originally isolated from urine and found to aid the healing of ulcers, is now known to be epidermal growth factor (see Chapter 22).

The possible role of **pancreatic polypeptide** in the regulation of gastrointestinal function is discussed in Chapter 19.

Caerulein

An interesting decapeptide isolated from the skin of the Australian frog *Hyla caerulea* has the same five C-terminal and two N-terminal amino acid residues as gastrin and also has a sulfated tyrosyl residue. This material, which has been named **caerulein,** has all the properties of gastrin and CCK in mammals and has been used in humans to contract the gallbladder during cholecystography (see below). The skin of various species of frogs also contains TRH and many other biologically active peptides (see Chapter 14).

MOUTH & ESOPHAGUS

In the mouth, food is mixed with saliva and propelled into the esophagus. Peristaltic waves in the esophagus move the food into the stomach.

Mastication

Chewing **(mastication)** breaks up large food particles and mixes the food with the secretions of the salivary glands. This wetting and homogenizing action aids subsequent digestion. Large food particles can be digested, but they cause strong and often painful contractions of the esophageal musculature. Edentulous patients are generally restricted to a soft diet and have considerable difficulty eating dry food.

Salivary Glands & Saliva

In the salivary glands, the secretory **(zymogen)** granules containing the salivary enzymes are discharged from the acinar cells into the ducts (Fig 26–4). The characteristics of each of the 3 pairs of salivary glands in humans are summarized in Table 26–3.

Saliva contains 2 digestive enzymes: **lingual lipase,** secreted by glands on the tongue, and **ptyalin (salivary α-amylase),** secreted by the salivary glands. The functions of these enzymes are discussed in Chapter 25. Saliva also contains **mucin,** a glycoprotein that lubricates the food. About 1500 mL of saliva is secreted per day. The pH of saliva is about 7.0. It performs a number of important functions. It facilitates swallowing, keeps the mouth moist, serves as a solvent for the molecules that stimulate the taste buds, aids speech by facilitating movements of the lips and tongue, and keeps the mouth and teeth clean. The saliva may also have some antibacterial action, and patients with deficient salivation **(xerostomia)** have a higher than normal incidence of dental caries. The buffers in saliva help maintain the oral pH at about 7.0. They also help neutralize gastric acid and relieve heartburn when gastric juice is regurgitated into the esophagus.

The concentrations of Na^+, Cl^-, and HCO_3^- in saliva increase as the amount of saliva that is secreted increases, whereas the concentration of K^+ decreases. It appears that the salivary acini elaborate a primary secretion containing these ions, with the concentrations of K^+ and HCO_3^- greater than those in plasma

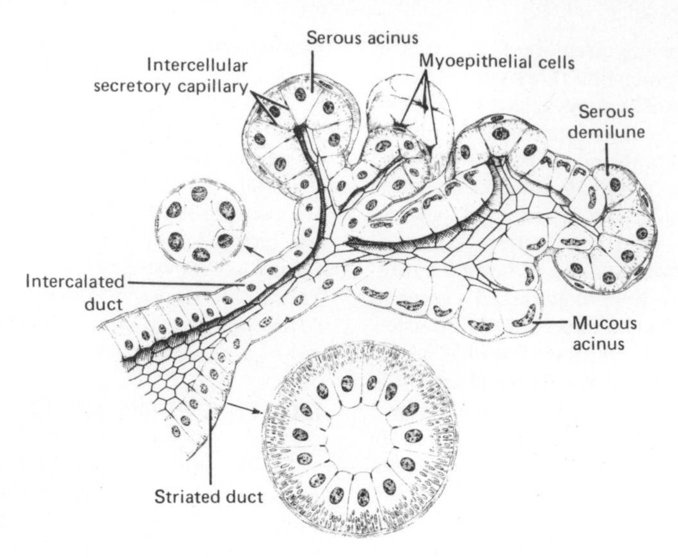

Figure 26–4. Structure of submandibular gland (also known as submaxillary gland). Note that the cells in the mucous acini have flattened basal nuclei whereas the cells in the serous acini have round nuclei, abundant rough endoplasmic reticulum, and collections of zymogen secretory granules at their apexes. The intercalated ducts drain into the striated ducts, where the cells are specialized for ion transport. (Modified and reproduced, with permission, from Junqueira LC, Carneiro J, Kelley RO: *Basic Histology,* 6th ed. Appleton & Lange, 1989.)

because of active transport. Cl^-, HCO_3^-, and Na^+ are reabsorbed in the ducts, whereas K^+ is secreted. When salivary flow rates are high, there is little time for reabsorption, so saliva contains more of the normally reabsorbed and less of the normally secreted ions. Aldosterone increases the K^+ concentration and reduces the Na^+ concentration of saliva in an action analogous to its action on the kidney (see Chapters 20 and 38), and a high salivary Na^+/K^+ ratio is seen in Addison's disease.

Control of Salivary Secretion

Salivary secretion is under neural control. Stimulation of the parasympathetic nerve supply causes profuse secretion of watery saliva with a relatively low content of organic material. Associated with this secretion is a pronounced vasodilation in the gland, which appears to be due to the local release of VIP. This polypeptide is a cotransmitter with acetylcholine in some of the postganglionic parasympathetic neurons. Atropine and other cholinergic blocking agents reduce salivary secretion. Stimulation of the sympathetic nerve supply causes vasoconstriction and, in humans, secretion of small amounts of saliva rich in organic constitutents from the submaxillary glands.

Food in the mouth causes reflex secretion of saliva, and so does stimulation of vagal afferent fibers at the gastric end of the esophagus. Salivary secretion is easily conditioned, as shown in Pavlov's original experiments (see Chapter 16). In humans, the sight, smell, and even thought of food causes salivary secretion (''makes the mouth water'').

Like the thyroid gland, the salivary glands and the gastric mucosa concentrate iodide from the plasma, the salivary/plasma iodide ratio sometimes reaching 60. The physiologic significance of this iodide-trapping phenomenon is uncertain. The salivary glands have also been reported to contain somatostatin, glucagon, and, at least in some species, renin and various growth factors. The functions of most of these factors in the salivary glands are unknown, but it has been suggested that glucagon secreted from the salivary glands contributes to the hyperglycemia in pancreatectomized animals (see Chapter 19).

Table 26–3. Nerve supply, histologic type, and relative contribution to total salivary output of each of the pairs of salivary glands in humans. Serous cells secrete ptyalin; mucous cells secrete mucin.

Gland	Parasympathetic Nerve Supply Via	Histologic Type	Percentage of Total Salivary Secretion in Humans (1.5 L/d)
Parotid	Glossopharyngeal	Serous	25
Submandibular (submaxillary)	Facial	Mixed	70
Sublingual	Facial	Mucous	5

Deglutition

Swallowing (deglutition) is a reflex response that is triggered by afferent impulses in the trigeminal, glossopharyngeal, and vagus nerves. These impulses are integrated in the nucleus of the tractus solitarius and the nucleus ambiguus. The efferent fibers pass to the pharyngeal musculature and the tongue via the trigeminal, facial, and hypoglossal nerves. Swallowing is initiated by the voluntary action of collecting the oral contents on the tongue and propelling them backward into the pharynx. This starts a wave of involuntary contraction in the pharyngeal muscles that pushes the material into the esophagus. Inhibition of respiration and glottic closure are part of the reflex response. Swallowing is difficult if not impossible when the mouth is open, as anyone who has spent time in the dentist's chair feeling saliva collect in the throat is well aware. A normal adult swallows at a rapid rate while eating, but swallowing also continues between meals. The total number of swallows per day is about 2400.

At the pharyngoesophageal junction, there is a 3-cm segment of esophagus in which the resting wall tension is high. This segment relaxes reflexly upon swallowing, permitting the swallowed material to enter the body of the esophagus. A ring contraction of the esophageal muscle forms behind the material, which is then swept down the esophagus by a peristaltic wave at a speed of approximately 4 cm/s. When humans are in an upright position, liquids and semisolid foods generally fall by gravity to the lower esophagus ahead of the peristaltic wave. Unlike the rest of the esophagus, the musculature of the gastroesophageal junction **(lower esophageal sphincter, LES)** is tonically active but relaxes upon swallowing. The relaxation may be mediated via neurons that secrete VIP. The tonic activity of the lower esophageal sphincter between meals prevents reflux of gastric contents into the esophagus. Large doses of gastrin increase the tone of the lower esophageal sphincter, but this effect is not produced by doses of gastrin that produce circulating gastrin levels comparable to those that occur after a meal.

Motor Disorders of the Esophagus

Achalasia is a condition in which food accumulates in the esophagus and the organ becomes massively dilated. It is due to increased resting lower esophageal sphincter tension, incomplete relaxation of this sphincter on swallowing, and weak esophageal peristalsis. There are lesions in the vagus nerves and the nerve plexuses in the esophageal wall and lower esophageal sphincter. The condition can be treated by pneumatic dilation of the sphincter or incision of the esophageal muscle (myotomy).

The opposite condition is lower esophageal sphincter incompetence, which permits reflux of acid gastric contents into the esophagus. This causes heartburn and esophagitis and can lead to ulceration and stricture due to scarring. The condition can be treated by avoiding irritating foods or surgically, by making a fold of gastric tissue **(fundoplication).**

Aerophagia & Intestinal Gas

Nervous persons who hyperventilate sometimes swallow large amounts of air, and some air is unavoidably swallowed in the process of eating and drinking **(aerophagia).** Some of the swallowed air is regurgitated (belching), and some of the gases it contains are absorbed, but much of it passes on to the colon. Here, some of the oxygen is absorbed, and hydrogen, hydrogen sulfide, carbon dioxide, and methane formed by the colonic bacteria are added to it. It is then expelled as **flatus.** The volume of gas normally found in the human gastrointestinal tract is about 200 mL. In some individuals, gas in the intestines causes cramps, **borborygmi** (rumbling noises), and abdominal discomfort.

STOMACH

Food is stored in the stomach; mixed with acid, mucus, and pepsin; and released at a controlled, steady rate into the duodenum.

Anatomic Considerations

The gross anatomy of the stomach is shown in Fig 26–5. The gastric mucosa contains many deep glands. In the pyloric and cardiac regions, the glands secrete mucus. In the body of the stomach, including the fundus, the glands contain **parietal (oxyntic) cells,** which secrete hydrochloric acid and intrinsic factor, and **chief (zymogen, peptic) cells,** which secrete pepsinogens (Fig 26–6). These secretions mix with mucus secreted by the cells in the necks of the glands.

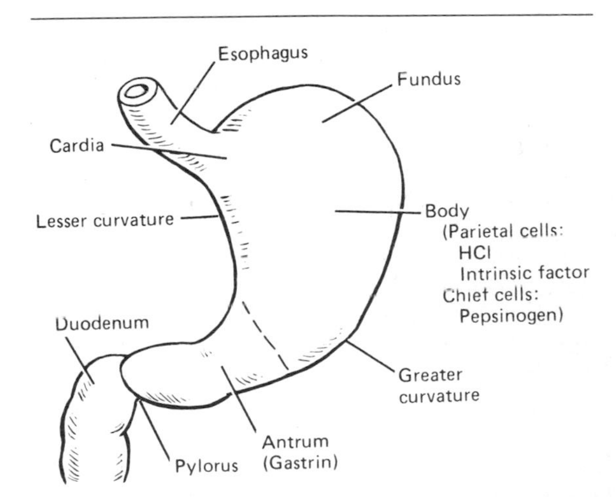

Figure 26–5. Anatomy of the stomach. The principal secretions are listed in parentheses under the labels indicating the locations where they are produced. In addition, mucus is secreted in all parts of the stomach. The dashed line marks the border between the body and the antrum.

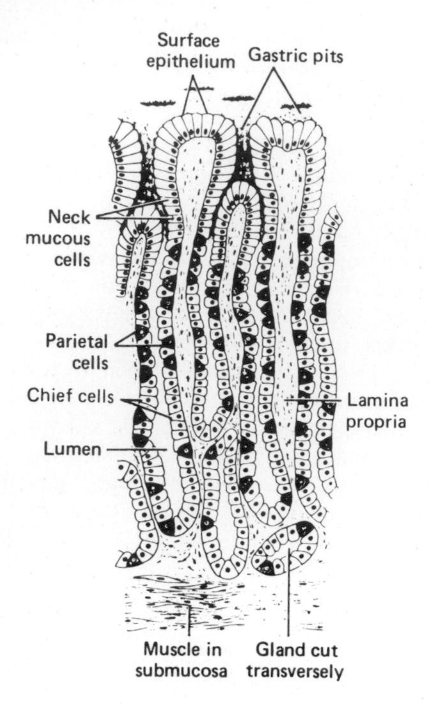

Figure 26–6. Diagram of glands in the mucosa of the body of the human stomach. (Reproduced, with permission, from Bell GH, Davidson N, Scarborough G: *Textbook of Physiology and Biochemistry,* 6th ed. Livingstone, 1965.)

Several of the glands open on a common chamber **(gastric pit)** that opens in turn on the surface of the mucosa.

The stomach has a very rich blood and lymphatic supply. Its parasympathetic nerve supply comes from the vagi and its sympathetic supply from the celiac plexus.

Gastric Secretion

The cells of the gastric glands secrete about 2500 mL of **gastric juice** daily. This juice contains a variety of substances (Table 26–4). The gastric enzymes are discussed in Chapter 25. The **hydrochloric acid** secreted by the glands in the body of the stomach kills many ingested bacteria, aids protein digestion, provides the necessary pH for pepsin to start protein digestion, and stimulates the flow of bile and pancreatic juice. It is concentrated enough to cause tissue damage, but in normal individuals the gastric mucosa does not become irritated or digested, in part because the gastric juice also contains **mucus.** Mucus, which is secreted by the neck and surface mucous cells in the body and fundus and similar cells elsewhere in the stomach, is made up of glycoproteins. Each mucus glycoprotein contains 4 subunits joined by disulfide bridges. The mucus forms a flexible gel that coats the mucosa.

Table 26–4. Contents of normal gastric juice (fasting state).

Cations: Na^+, K^+, Mg^{2+}, H^+ (pH approximately 1.0)
Anions: Cl^-, HPO_4^{2-}, SO_4^{2-}
Pepsins I–III
Gelatinase
Mucus
Intrinsic factor
Water

The gastric mucosa also secretes bicarbonate; the bicarbonate and the mucus form an unstirred layer that has a pH of about 7.0. This unstirred layer plus the surface membranes of the mucosal cells and the tight junctions between them constitute the mucosal bicarbonate barrier that protects the mucosal cells from damage by gastric acid.

Substances that tend to disrupt the barrier and cause gastric irritation include ethanol, vinegar, bile salts, and aspirin and other nonsteroidal anti-inflammatory drugs. Prostaglandins stimulate mucus secretion, and aspirin and related drugs inhibit prostaglandin synthesis.

The electrolyte content of the gastric juice varies with the rate of secretion. At low secretory rates, the Na^+ concentration is high and the H^+ concentration is low, but as acid secretion increases, the Na^+ concentration falls.

Pepsinogen Secretion

The chief cells that secrete pepsinogens, the precursors of the pepsins in gastric juice (see Chapter 25), contain zymogen granules. The secretory process is similar to that involved in the secretion by ptyalin by the salivary glands and trypsinogen and the other pancreatic enzymes by the pancreas. Pepsinogen activity can be detected in the plasma and in the urine, where it is called **uropepsinogen.**

Hydrochloric Acid Secretion

The parietal cells contain numerous tubulovesicular structures in their cytoplasm (Fig 26–7). When the cells are stimulated to secrete acid, many of the tubulovesicles move to the mucosal membrane of the cell and fuse with it, so its area increases and forms many microvilli. These microvilli project into canaliculi that project from the lumen deep into the cell. Transport of H^+ from the cytoplasm to the lumen of the canaliculi is brought about by the H^+-K^+ ATPase of the parietal cells, and there is evidence that the ATPase is synthesized in the tubulovesicles at rest and inserted with the tubulovesicle membranes into the mucosal membrane during stimulation.

It is difficult to obtain the products of parietal cell secretion free of contamination by other gastric secretions, but the purest specimens that have been analyzed are essentially isotonic. Their H^+ concentration is equivalent to approximately 0.17 N HCl, with pHs as low as 0.87. Therefore, parietal cell secretion may well be an isotonic solution of essentially pure

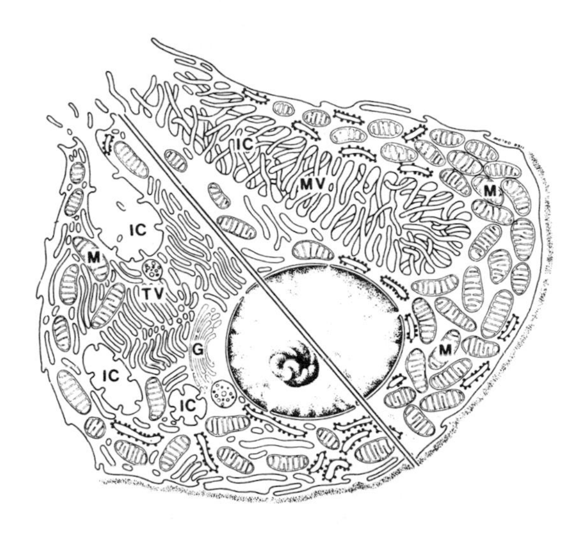

Figure 26–7. Composite diagram of a parietal cell, showing the resting state (lower left) and the active state (upper right). The cell has intracellular canaliculi (IC), and when the cell is active, the tubulovesicles (TV) that are prominent at rest form multiple microvilli (MV) projecting into the intracellular canaliculi. M, mitochondrion; G, Golgi apparatus. (Reproduced, with permission, from Junqueira LC, Carneiro J, Kelley RO: *Basic Histology,* 6th ed. Appleton & Lange, 1989.)

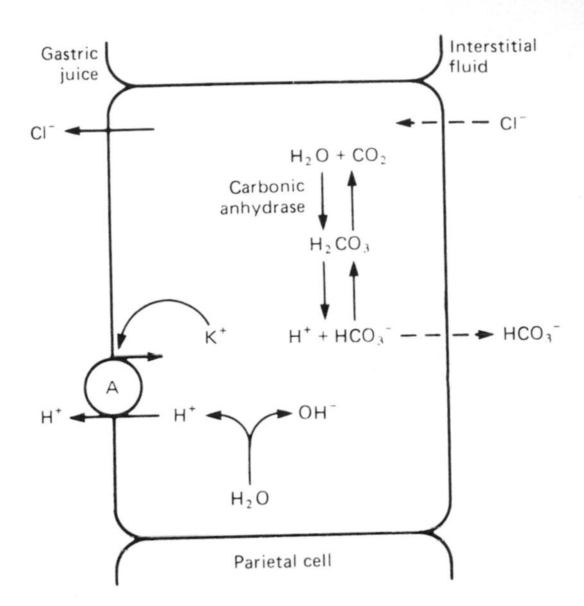

Figure 26–8. Postulated mechanisms responsible for HCl secretion by parietal cells in the stomach. Compare to Fig 38–19. H^+ is actively transported into the gastric lumen in exchange for K^+ by H^+-K^+ ATPase (A). The K^+ may recycle through the membrane. Cl^- is also actively transported into the gastric lumen. HCO_3^- diffuses into the interstitial fluid, and Cl^- diffuses into the parietal cell from the interstitial fluid. Solid arrows crossing cell membranes indicate active transport, and dashed arrows indicate diffusion.

HCl that contains 150 meq of Cl^- and 150 meq of H^+ per liter. Yet the pH of the cytoplasm of the parietal cells, like that of other cells, is 7.0–7.2, and the comparable concentrations per liter of plasma are about 100 meq of Cl^- and 0.00004 meq of H^+.

It is the H^+-K^+ ATPase in the mucosal membrane of the parietal cells that transports H^+ against a concentration gradient of this magnitude. The primary source of the secreted H^+ is the ionization of water. The H^+ formed in this fashion is promptly secreted by active transport into the gastric juice in exchange for K^+ by H^+-K^+ ATPase (Fig 26–8). However, external K^+ is not required for secretion, and it has been suggested that the K^+ that exchanges for H^+ recycles from the membrane rather than from the exterior. In the stomach, the H^+-K^+ ATPase is almost entirely limited to the parietal cells. It provides the energy for H^+ transport by hydrolyzing ATP, so a ready supply of ATP is needed for secretion. Inhibition of the generation of ATP prevents H^+ secretion.

Cl^- is also actively transported into the gastric juice. For each H^+ secreted, an OH^- remains in the cell (Fig 26–8). The OH^- is neutralized by an H^+ formed by the dissociation of carbonic acid (carbonic acid buffer system; see Chapters 35 and 39). The HCO_3^- formed by this dissociation enters the interstitial fluid, being replaced by Cl^- that diffuses from the interstitial fluid into the parietal cell. There is an HCO_3^--Cl^- exchange mechanism in the membrane facing the interstitial fluid, but Cl^- also enters with Na^+. The H_2CO_3 supply is replenished by the hydration of CO_2, which comes from the interstitial fluid or is produced by cellular metabolism. The gastric mucosa contains an abundance of carbonic anhydrase, the enzyme that catalyzes the hydration of CO_2, and the stomach has a negative respiratory quotient (RQ)—ie, the amount of CO_2 in arterial blood is greater than the amount in gastric venous blood. Blood coming from the stomach is alkaline and has a high HCO_3^- content. When gastric acid secretion is elevated after a meal, sufficient H^+ may be secreted to raise the pH of systemic blood and make the urine alkaline. This is the probable explanation of the high pH of the urine excreted after a meal, the so-called **postprandial alkaline tide.**

Acid secretion is stimulated by histamine via H_2 receptors, by acetylcholine via M_1 muscarinic receptors, and by gastrin via gastrin receptors in the membranes of the parietal cells (Fig 26–9). The H_2 receptors increase intracellular cyclic AMP, whereas the muscarinic receptors and the gastrin receptors exert their effects by increasing intracellular free Ca^{2+}. The intracellular events interact so that activation of one receptor type potentiates the response of another to stimulation. The histamine comes from cells in the mucosa that resemble mast cells, the acetylcholine comes from the endings of postganglionic cholinergic neurons innervating the parietal cells, and the gastrin reaches the cells via the circulation.

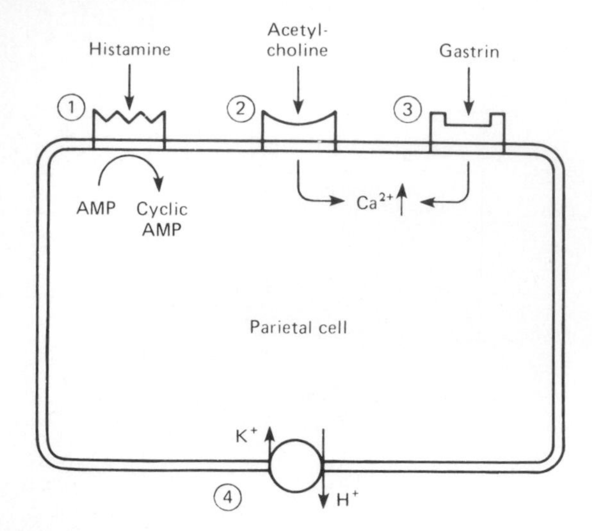

Figure 26–9. Regulation of gastric acid secretion. Acid secretion is increased by histamine acting on H_2 receptors (1), acetylcholine acting on M_1 muscarinic receptors (2), and gastrin acting on gastrin receptors (3). All increase the transport of H^+ into the gastric lumen by H^+-K^+ ATPase (4).

Gastric Motility & Emptying

When food enters the stomach, the organ relaxes by the reflex process of **receptive relaxation.** This relaxation of the gastric musculature is triggered by movement of the pharynx and esophagus. It is followed by peristaltic contractions that mix the food and squirt food into the duodenum at a controlled rate. The peristaltic waves are most marked in the distal half of the stomach. When well developed, they occur at a rate of 3/min.

The pyloric sphincter has a limited function in the control of gastric emptying. Gastric emptying is normal if the pylorus is held open or even if it is surgically resected. The antrum, pylorus, and upper duodenum apparently function as a unit. Contraction of the antrum is followed by sequential contraction of the pyloric region and the duodenum. In the antrum, contraction ahead of the advancing gastric contents prevents solid masses from entering the duodenum. The gastric contents are thus squirted a bit at a time into the small intestine. Normally, regurgitation from the duodenum does not occur, because the contraction of the pyloric segment tends to persist slightly longer than that of the duodenum. The prevention of regurgitation may also be due to the stimulating action of CCK and secretin on the pyloric sphincter.

Gastric Slow Wave

Peristaltic contractions in the stomach are coordinated by the **gastric slow wave,** a wave of depolarization of smooth muscle cells proceeding from the circular muscle of the fundus of the stomach to the pylorus approximately every 20 seconds. This wave is also called the **basic electric rhythm (BER).** It is the pacemaker for antral peristalsis; the wave and, consequently, peristalsis become irregular and chaotic after vagotomy or transection of the stomach wall. Thus, the slow wave plays a major role in the control of gastric emptying. A similarly aborally directed slow wave coordinates smooth muscle contractions in the small intestine and the colon (see below).

Hunger Contractions

The musculature of the stomach is rarely inactive. Soon after the stomach is emptied, mild peristaltic contractions begin. They gradually increase in intensity over a period of hours. The more intense contractions can be felt and may even be mildly painful. These **hunger contractions** are associated with the sensation of hunger and were once thought to be an important regulator of appetite. However, it has now been demonstrated that food intake is normal in animals after denervation of the stomach and intestines (see Chapter 14).

REGULATION OF GASTRIC SECRETION & MOTILITY

Gastric motility and secretion are regulated by neural and humoral mechanisms. The neural components are local autonomic reflexes, involving cholinergic neurons, and impulses from the central nervous system by way of the vagus nerves. The humoral components are the hormones discussed above. Vagal stimulation increases gastrin secretion by release of gastrin-releasing peptide (see above). Other vagal fibers release acetylcholine, which acts directly on the cells in the glands in the body and the fundus to increase acid and pepsin secretion. Stimulation of the vagus nerve in the chest or neck increases acid and pepsin secretion, but vagotomy does not abolish the secretory response to local stimuli.

For convenience, the physiologic regulation of gastric secretion is usually discussed in terms of cephalic, gastric, and intestinal influences, although these overlap. The **cephalic** influences are vagally mediated responses induced by activity in the central nervous system. The **gastric** influences are primarily local reflex responses and responses to gastrin. The **intestinal** influences are the reflex and hormonal feedback effects on gastric secretion initiated from the mucosa of the small intestine.

Cephalic Influences

The presence of food in the mouth reflexly stimulate gastric secretion. The efferent fibers for this reflex are in the vagus nerves. Vagally mediated increases in gastric secretion are easily conditioned. In humans, for example, the sight, smell, and thought of food increases gastric secretion. These increases are due to alimentary conditioned reflexes that become established early in life.

The neural mechanisms involved in the formation of conditioned reflexes are discussed in Chapter 16. The impulses descending in the vagi originate in part in the diencephalon and limbic system. Stimulation of the anterior hypothalamus and parts of the adjacent orbital frontal cortex increases vagal efferent activity and gastric secretion.

Emotional Responses

Psychic states have effects on gastric secretion and motility that are principally mediated via the vagi. Among his famous observations on Alexis St. Martin, the Canadian with a permanent gastric fistula resulting from a gunshot wound, William Beaumont noted that anger and hostility were associated with turgor, hyperemia, and hypersecretion of the gastric mucosa. Similar observations have been made on other patients with gastric fistulas. Fear and depression decrease gastric secretion and blood flow and inhibit gastric motility.

Gastric Influences

Food in the stomach accelerates the increase in gastric secretion produced by the sight and smell of food and the presence of food in the mouth (Fig 26–10). Receptors in the wall of the stomach and the mucosa respond to stretch and chemical stimuli, mainly amino acids and related products of digestion. The fibers from the receptors enter Meissner's plexus, where the cell bodies of the receptor neurons are located. They synapse on postganglionic parasympathetic neurons that end on parietal cells and stimulate acid secretion. Thus, the acid responses are produced by local reflexes in which the reflex arc is totally within the wall of the stomach. The postganglionic neurons in the local reflex arc are the same ones innervated by the descending vagal preganglionic neurons from the brain that mediate the cephalic phase of secretion. The products of protein digestion also bring about increased secretion of gastrin, and this augments the flow of acid.

Intestinal Influences

Although there are gastrin-containing cells in the mucosa of the small intestine as well as in the stomach, instillation of amino acids directly into the duodenum does not increase circulating gastrin. Fats, carbohydrates, and acid in the duodenum inhibit gastric acid and pepsin secretion and gastric motility, probably via GIP and secretin and possibly via other hormones. Gastric acid secretion is increased following removal of large parts of the small intestine. The hypersecretion, which is roughly proportionate in degree to the amount of intestine removed, may be due in part to removal of the source of hormones that inhibit acid secretion.

Other Influences

Hypoglycemia acts via the brain and vagal efferents to stimulate acid and pepsin secretion. Other stimulants include **alcohol** and **caffeine,** both of which act directly on the mucosa. The beneficial effects of moderate amounts of alcohol on appetite and digestion, due to this stimulatory effect on gastric secretion, have been known since ancient times.

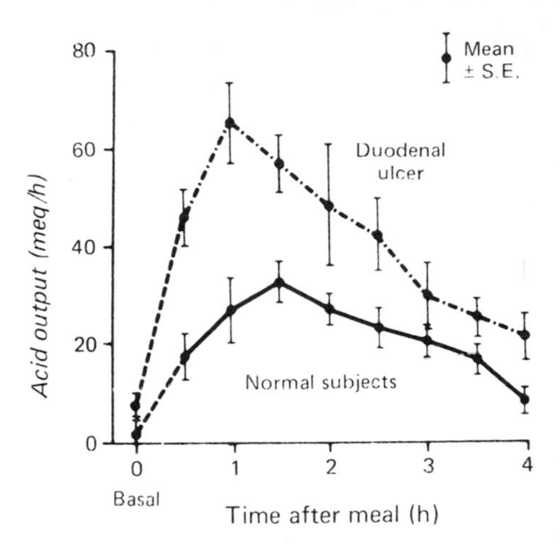

Figure 26–10. Human gastric acid secretion after a steak meal. (Reproduced, with permission, from Brooks FP: Integrative lecture: Response of the GI tract to a meal. *Undergraduate Teaching Project.* American Gastroenterological Association, 1974.)

Regulation of Gastric Motility & Emptying

The rate at which the stomach empties into the duodenum depends on the type of food ingested. Food rich in carbohydrate leaves the stomach in a few hours. Protein-rich food leaves more slowly, and emptying is slowest after a meal containing fat (Fig 26–11). The rate of emptying also depends on the osmotic pressure of the material entering the duodenum. Hyperosmolality of the duodenal contents is sensed by "duodenal osmoreceptors" that initiate a decrease in gastric emptying which is probably neural in origin.

Products of protein digestion and hydrogen ions bathing the duodenal mucosa initiate a neurally mediated decrease in gastric motility, the **enterogastric reflex.** Distention of the duodenum also initiates this reflex. GIP and other hormones inhibit gastric motility and secretion. Cutting the vagus nerves slows gastric emptying. In humans, vagotomy may cause relatively severe gastric atony and distention. Excitement is said to hasten gastric emptying, and fear to slow it.

Since fats are particularly effective in inhibiting gastric emptying, some people drink milk, cream, or even olive oil before a cocktail party. The fat keeps the alcohol in the stomach for a long time, where its absorption is slower than in the small intestine, and the intoxicant enters the small intestine in a slow, steady stream so that—theoretically, at least—a sudden rise of the blood alcohol to a high level and consequent embarrassing intoxication are avoided.

Peptic Ulcer

Gastric and duodenal ulceration in humans is related in some poorly understood way to a breakdown of the barrier that normally prevents irritation and autodigestion of the mucosa by the gastric secretions. In patients with ulcers of the duodenum and prepyloric portion of the stomach, gastric acid is always present ("no acid, no ulcer") and its secretion is often elevated (Fig 26–10). As noted in Chapter 25, there is a correlation between pepsinogen I levels, maximum acid secretion, and the incidence of peptic ulcer. Resting gastrin levels do not seem to be elevated in most patients with gastric and duodenal ulcers, but their gastrin responses to feeding are greater than normal and their parietal cells are hyperresponsive to gastrin. Another illustration of the importance of acid hypersecretion in the genesis of duodenal and prepyloric ulcers is provided by the **Zollinger-Ellison syndrome.** This syndrome is seen in patients with gastrinomas, tumors that secrete gastrin. These tumors can occur in the stomach and duodenum, but most of them are found in the pancreas. The gastrin causes prolonged hypersecretion of acid and severe ulcers are produced.

More extensive discussions of the problem of peptic ulceration can be found in textbooks of pathophysiology. However, it is pertinent to note that the procedures used in the treatment of ulcers are aimed at inhibiting acid secretion and enhancing mucosal resistance to acid. A variety of different antacids, most of which contain aluminum hydroxide, magnesium hydroxide, or calcium carbonate, are available. Blockade of the gastric H_2 receptors with the H_2 receptor-blocking drugs cimetidine, ranitidine, and famotidine produces effective reduction of acid responses because it removes the potentiating effect of stimulation of this receptor on the responses to other stimuli (see above); indeed, on the basis of its ability to inhibit acid secretion, cimetidine was at one point the most commonly prescribed drug in the USA. M_1 muscarinic receptors can be blocked with atropine or the new, more specific anticholinergic drug pirenzepine. Gastric receptors are sometimes blocked with proglumide. The gastric H^+-K^+ ATPase can be inhibited by omeprazole, and this drug has proved to be effective over long periods. Sucralfate, a basic aluminum salt of sucrose octasulfate, increases the resistance of the mucosa to acid by forming adherent protein complexes at the ulcer site, and it also shows promise in the treatment of ulcers.

Severe duodenal and prepyloric ulcers are sometimes treated by vagotomy combined with removal of the gastrin-secreting antral mucosa. Vagotomy is incomplete in about 20% of patients, but removal eliminates the potentiating action of gastrin and the effects of a few remaining vagal fibers on acid secretion are too small to be a problem.

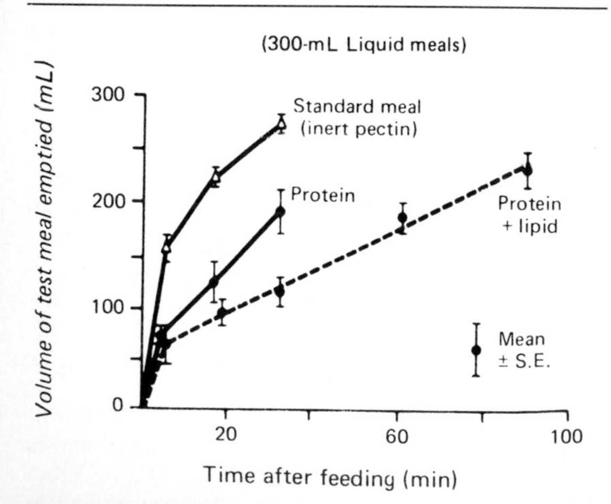

Figure 26–11. Effect of protein and fat on the rate of emptying of the human stomach. Subjects were fed 300-mL liquid meals. (Reproduced, with permission, from Brooks FP: Integrative lecture: Response of the GI tract to a meal. *Undergraduate Teaching Project.* American Gastroenterological Association, 1974.)

OTHER FUNCTIONS OF THE STOMACH

The stomach has a number of functions in addition to the storage of food and the control of its release into the duodenum. The hydrochloric acid kills many of the ingested bacteria. The parietal cells in the gastric mucosa also secrete **intrinsic factor,** a substance necessary for the absorption of **cyanocobalamin** (vitamin B_{12}) from the small intestine.

Cyanocobalamin (Fig 26–12) is a complex cobalt-containing vitamin that is necessary for normal erythropoiesis. Inadequate absorption of this vitamin causes an anemia characterized by the appearance in the bloodstream of large primitive red cell precursors called **megaloblasts.** A complete remission of the deficiency syndrome occurs when cyanocobalamin is injected parenterally but not when it is administered by mouth unless the intrinsic factor secreted by the gastric mucosa is present. Anemia due to an inadequate dietary intake of cyanocobalamin is very rare, apparently because the minimum daily requirements are quite low and the vitamin is found in most foods of animal origin. The deficiency states seen clinically are therefore due primarily to defective cyanocobalamin absorption. Inadequate absorption may be due to primary intestinal diseases such as sprue, or it may be caused by intrinsic factor deficiency following gastrectomy. In patients with the specific megaloblastic anemia called **pernicious anemia,** there is intrinsic factor deficiency due to an idiopathic atrophy of much of the gastric mucosa.

Intrinsic factor is a glycoprotein with a molecular weight of 45,000. Cyanocobalamin becomes firmly bound to it in the intestine. The intrinsic factor-cyanocobalamin complex then becomes bound to specific receptors in the ileum, and the cyanocobalamin is transferred across the intestinal epithelium. Trypsin is required for this process to be

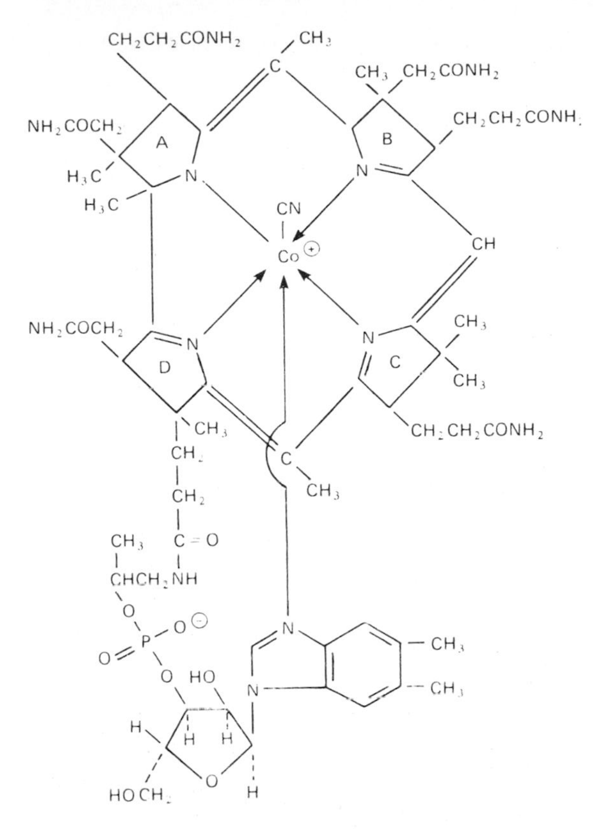

Figure 26–12. Cyanocobalamin (vitamin B_{12}). Empirical formula: $C_{63}H_{88}O_{14}N_{14}PCo$.

efficient, and absorption is sometimes decreased in patients with pancreatic insufficiency. The details of the absorptive process are uncertain, but it has been suggested that the role of intrinsic factor in the process is to stimulate endocytosis. This would explain how such a large molecule can be absorbed. A syndrome has been described in which there is congenital absence of the ileal receptors for the cyanocobalamin-intrinsic factor complex.

In the bloodstream, vitamin B_{12} is bound to a protein, **transcobalamin II,** which is necessary for the efficient distribution of the vitamin to the tissues. Two other binding proteins, transcobalamin I and transcobalamin III, are also found in the bloodstream. These 2 proteins are called R binder proteins because of their rapid electrophoretic mobility. The vitamin B_{12} metabolites that are bound to R binder proteins are removed from the circulation by hepatocytes and subsequently excreted in the bile. Individuals in whom the R binder proteins are congenitally deficient develop neurologic symptoms such as loss of position sense and ataxia, and it has been suggested that the normal function of these proteins is to remove harmful metabolites of vitamin B_{12} from the brain.

In totally gastrectomized patients, the intrinsic factor deficiency must be circumvented by parenteral injection of cyanocobalamin. Protein digestion is normal in the absence of pepsin, and nutrition can be maintained. However, these patients are prone to develop iron deficiency anemia (see Chapter 25) and other abnormalities, and they must eat frequent small meals. Because of rapid absorption of glucose from the intestine and the resultant hyperglycemia and abrupt rise in insulin secretion, gastrectomized patients sometimes develop hypoglycemic symptoms about 2 hours after meals (see Chapter 19). Weakness, dizziness, and sweating after meals, due in part to hypoglycemia, are part of the picture of the "dumping syndrome," a distressing syndrome that develops in patients in whom portions of the stomach have been removed or the jejunum has been anastomosed to the stomach. Another cause of the symptoms is rapid entry of hypertonic meals into the intestine; this provokes the movement of so much water into the gut that significant hypovolemia and hypotension are produced.

EXOCRINE PORTION OF THE PANCREAS

The pancreatic juice contains enzymes that are of major importance in the digestion (Table 25–1). Its secretion is controlled in part by a reflex mechanism and in part by the gastrointestinal hormones secretin and CCK.

Anatomic Considerations

The portion of the pancreas that secretes pancreatic juice is a compound alveolar gland resembling the salivary glands. Granules containing the digestive enzymes **(zymogen granules)** are formed in the cell and discharged by exocytosis (see Chapter 1) from the apexes of the cells into the lumens of the pancreatic ducts (Fig 26–13). The small duct radicles coalesce into a single duct (duct of Wirsung), which usually joins the common bile duct to form the ampulla of Vater (Fig 26–14). The ampulla opens through the duodenal papilla, and its orifice is encir-

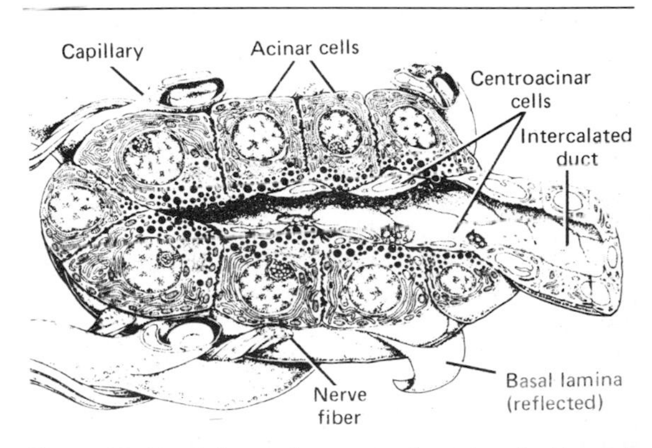

Figure 26–13. Acinar cells surrounding a terminal branch of the pancreatic ducts. Note the abundant rough endoplasmic reticulum and the zymogen granules concentrated at the apexes of the cells. (Reproduced, with permission, from Krstić RV: *Die Gewebe des Menschen und der Säugetiere.* Springer-Verlag, 1978.)

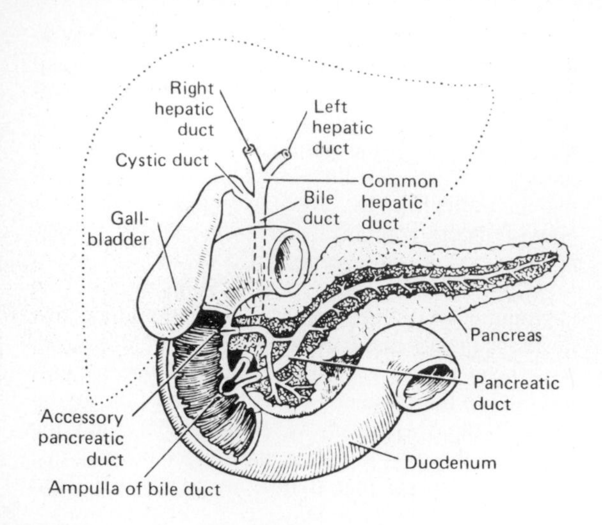

Figure 26–14. Connections of the ducts of the gallbladder, liver, and pancreas. (Reproduced, with permission, from Bell GH, Emslie-Smith D, Paterson CR: *Textbook of Physiology and Biochemistry,* 9th ed. Churchill Livingstone, 1976.)

cled by the sphincter of Oddi. In some individuals, there is an accessory pancreatic duct (duct of Santorini) that enters the duodenum more proximally. There is said to be a high incidence of nonpatency of the accessory pancreatic duct in patients with gastric acid hypersecretion and duodenal ulcer. It is possible that failure of the alkaline pancreatic secretion to enter the upper part of the duodenum via this duct contributes to the development of the ulcer.

Composition of Pancreatic Juice

The pancreatic juice is alkaline (Table 26–5) and has a high bicarbonate content (approximately, 113 meq/L versus 24 meq/L in plasma). About 1500 mL of pancreatic juice is secreted per day. Bile and intestinal juices are also neutral or alkaline, and these 3 secretions neutralize the gastric acid, raising the pH of the duodenal contents to 6.0–7.0. By the time the chyme reaches the jejunum, its reaction is nearly neutral, but the intestinal contents are rarely alkaline.

The powerful protein-splitting enzymes of the pancreatic juice are secreted as inactive proenzymes. Trypsinogen is converted to the active enzyme trypsin by the brush border enzyme **enteropeptidase (enterokinase)** when the pancreatic juice enters the duodenum. The secretion of enteropeptidase is increased by CCK. Enteropeptidase contains 41% polysaccharide, and this high polysaccharide content apparently prevents it from being digested itself before it can exert its effect. Trypsin converts chymotrypsinogens into chymotrypsins and other proenzymes into active enzymes (Fig 26–15). Trypsin can also activate trypsinogen; therefore, once some trypsin is formed, there is an autocatalytic chain reaction. Enteropeptidase deficiency occurs as a congenital abnormality and leads to protein malnutrition.

Table 26–5. Composition of normal human pancreatic juice.

Cations: Na^+, K^+, Ca^{2+}, Mg^{2+} (pH approximately 8.0)
Anions: HCO_3^-, Cl^-, SO_4^{2-}, HPO_4^{2-}
Digestive enzymes (see Table 25–1)
Albumin and globulin

The potential danger of the release into the pancreas of a small amount of trypsin is apparent; the resulting chain reaction would produce active enzymes that could digest the pancreas. It is therefore not surprising that the pancreas normally contains a trypsin inhibitor.

Another enzyme activated by trypsin is phospholipase A. This enzyme splits a fatty acid off lecithin, forming lysolecithin. Lysolecithin damages cell membranes. It has been hypothesized that in **acute pancreatitis,** a severe and sometimes fatal disease, phospholipase A is activated in the pancreatic ducts, with the formation of lysolecithin from the lecithin that is a normal constituent of bile. This causes disruption of pancreatic tissue and necrosis of surrounding fat. Another theory of the cause of acute pancreatitis holds that release of the pancreatic proenzymes by exocytosis from acinar cells is inhibited and that the secretory granules fuse with lysosomes in the cells **(crinophagy).** This process, which is known to occur in other cells, usually does not cause tissue damage, but in pancreatic acinar cells, the lysosomal hydrolases activate the digestive enzymes, digesting the cells.

A small amount of pancreatic α-amylase normally leaks into the circulation, but in acute pancreatitis, the circulating level of this enzyme rises markedly. Measurement of plasma amylase is therefore of value in diagnosing the disease.

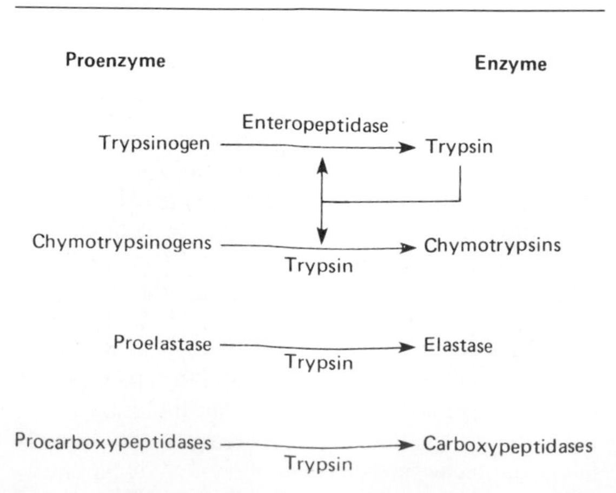

Figure 26–15. Activation of the pancreatic proteases in the duodenal lumen.

Regulation of the Secretion of Pancreatic Juice

Secretion of pancreatic juice is primarily under hormonal control. Secretin acts on the pancreatic ducts to cause copious secretion of a very alkaline pancreatic juice that is rich in HCO_3^- and poor in enzymes. Secretin also stimulates bile secretion. CCK acts on the acinar cells to cause release of zymogen granules and production of pancreatic juice rich in enzymes.

The response to intravenous secretin is shown in Fig 26–16. Note that as the volume of pancreatic secretion increases, its Cl^- concentration falls and its HCO_3^- concentration increases. Although HCO_3^- is secreted in the small ducts, it is reabsorbed in the large ducts in exchange for Cl^-. The magnitude of the exchange is inversely proportionate to the rate of flow.

Like CCK, acetylcholine acts on acinar cells to cause discharge of zymogen granules, and stimulation of the vagi causes secretion of a small amount of pancreatic juice rich in enzymes. This effect is blocked by atropine and by denervation of the pancreas, whereas the effects of secretin and CCK are not. There is evidence for vagally mediated conditioned reflex secretion of pancreatic juice in response to the sight or smell of food. Acetylcholine and CCK produce their effects on acinar cells by activating phospholipase C (see Chapter 1), whereas secretin acts on the duct cells by increasing intracellular cyclic AMP.

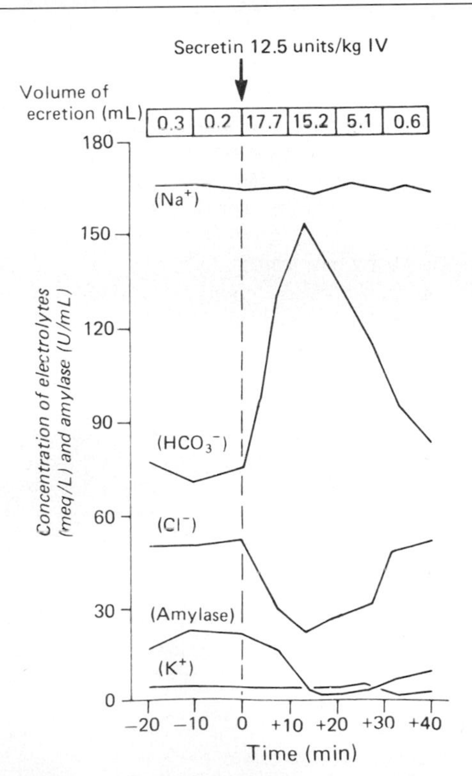

Figure 26–16. Effect of a single dose of secretin on the composition and volume of the pancreatic juice in humans. (Redrawn and reproduced, with permission, from Janowitz HD: Pancreatic secretion. *Physiol Physicians* [Nov] 1964; 2:[11].)

LIVER & BILIARY SYSTEM

Bile is secreted by the cells of the liver into the bile duct, which drains into the duodenum. Between meals, the duodenal orifice of this duct is closed and bile flows into the gallbladder, where it is stored. When food enters the mouth, the sphincter around the orifice relaxes; when the gastric contents enter the duodenum, the hormone CCK from the intestinal mucosa causes the gallbladder to contract.

Anatomic Considerations

The liver is organized in lobules, within which blood flows past hepatic cells via sinusoids from branches of the portal vein to the central vein of each lobule (Figs 26–17 and 26–18). The endothelium of the sinusoids has large fenestrations, and plasma is in

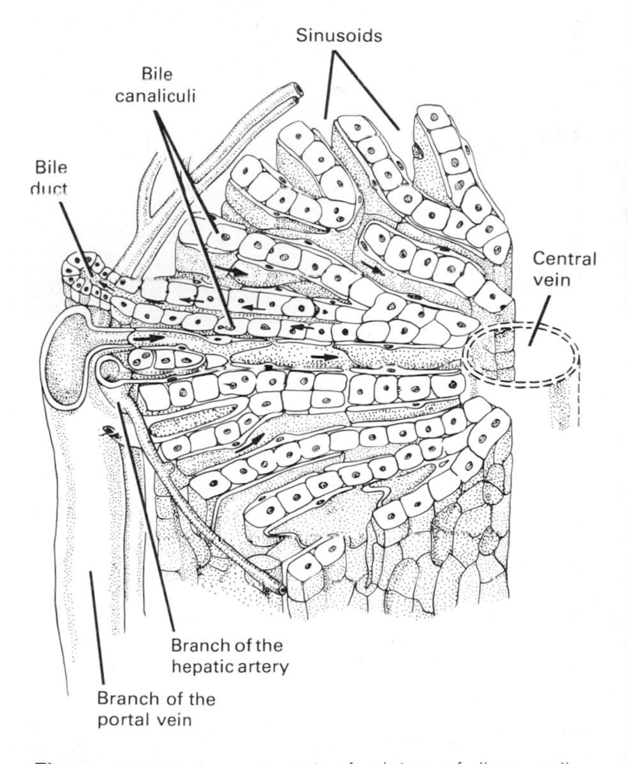

Figure 26–17. Arrangement of plates of liver cells, sinusoids, and bile ducts in a liver lobule, showing centripetal flow of blood in sinusoids to central vein and centrifugal flow of bile in bile canaliculi to bile ducts. (Reproduced, with permission, from Fawcett DW: *Bloom and Fawcett, A Textbook of Histology,* 11th ed. Saunders, 1986.)

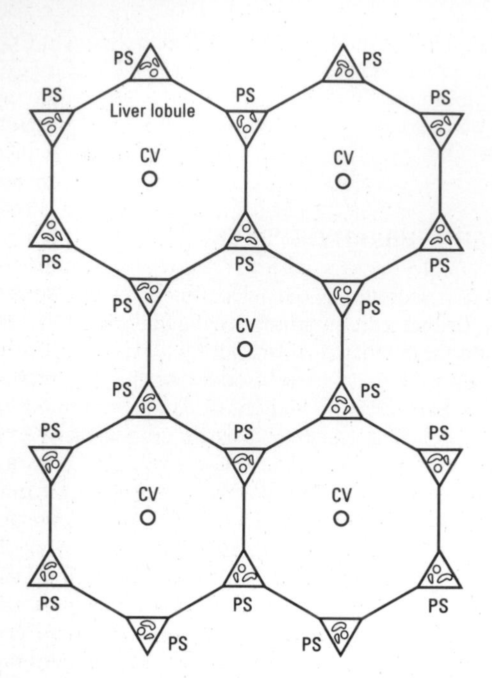

Figure 26–18. Organization of hepatic lobules. CV, central vein; PS, portal space containing branches of portal vein and hepatic artery.

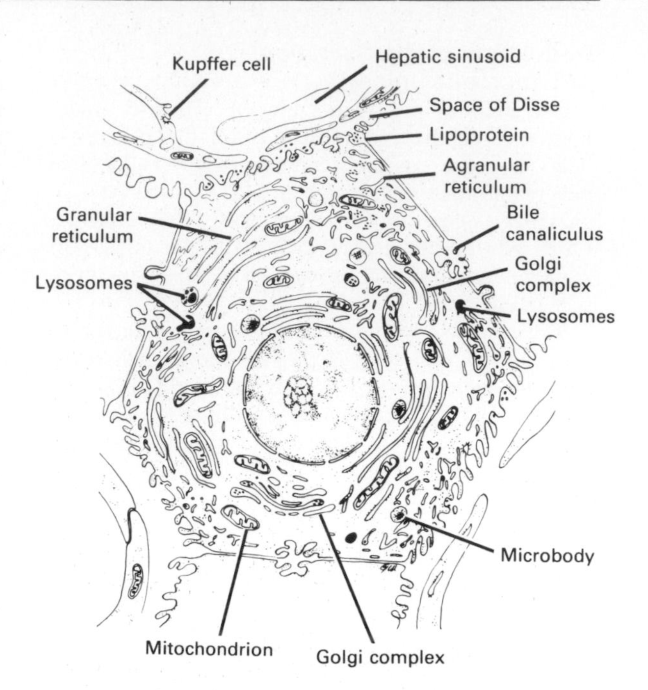

Figure 26–19. Liver cell, showing relation of hepatocytes to bile canaliculi and sinusoids. Note the wide openings between the endothelial cells next to the hepatocyte. (Reproduced, with permission, from Fawcett DW: *Bloom and Fawcett, A Textbook of Histology,* 11th ed. Saunders, 1986.)

close contact with liver cells (Fig 26–19). There is usually only one layer of hepatocytes between sinusoids, so the total area of contact between liver cells and plasma is very large. Hepatic artery blood also enters the sinusoids. The central veins coalesce to form the hepatic veins, which drain into the inferior vena cava. Hepatic blood flow is discussed in more detail in Chapter 32. The average transit time for blood across the liver lobule from the portal venule to the central hepatic vein is about 8.4 seconds. Numerous macrophages **(Kupffer cells)** are anchored to the endothelium of the sinusoids and project into the lumen. The functions of these phagocytic cells are discussed in Chapter 27.

Each liver cell is also apposed to several bile canaliculi (Fig 26–17). The canaliculi drain into intralobular bile ducts, and these coalesce via interlobular bile ducts to form the right and left hepatic ducts. These ducts join outside the liver to form the common hepatic duct. The cystic duct drains the gallbladder. The hepatic duct unites with the cystic duct to form the common bile duct. The common bile duct enters the duodenum at the duodenal papilla. Its orifice is surrounded by the sphincter of Oddi, and it usually unites with the main pancreatic duct just before entering the duodenum (Fig 26–14).

The walls of the extrahepatic biliary ducts and the gallbladder contain fibrous tissue and smooth muscle. The mucous membrane contains mucous glands and is lined by a layer of columnar cells. In the gallbladder, the mucous membrane is extensively folded; this increases its surface area and gives the interior of the gallbladder a honeycombed appearance. In primates, the mucous membranes of the cystic duct are also folded to form the so-called spiral valves.

Functions of the Liver

The liver, the largest gland in the body, has many complex functions, including those listed in Table 26–6. Discussion of these functions in one place requires that they be taken out of their proper context at considerable cost in terms of clarity and integration of function. Therefore, in this book they are discussed individually in the chapters on the systems of which each is a part. References to the relevant pages can be found in the index under the heading ''liver functions.''

Table 26–6. Principal functions of the liver.

Formation of bile
Carbohydrate storage and release
Formation of urea
Manufacture of plasma proteins
Many functions related to metabolism of fat
Inactivation of some polypeptide hormones
Reduction and conjugation of adrenocortical and gonadal steroid hormones
Synthesis of 25-hydroxycholecalciferol
Detoxification of many drugs and toxins

Table 26–7. Composition of human hepatic duct bile.

Water	97.0%
Bile salts	0.7%
Bile pigments	0.2%
Cholesterol	0.06%
Inorganic salts	0.7%
Fatty acids	0.15%
Lecithin	0.1%
Fat	0.1%
Alkaline phosphatase	. . .

Composition of Bile

Bile is made up of bile salts, bile pigments, and other substances dissolved in an alkaline electrolyte solution that resembles pancreatic juice (Table 26–7). About 500 mL is secreted per day. Some of the components of the bile are reabsorbed in the intestine and then excreted again by the liver **(enterohepatic circulation).**

The glucuronides of the **bile pigments,** biliverdin and bilirubin, are responsible for the golden yellow color of bile. The formation of these breakdown products of hemoglobin is discussed in detail in Chapter 27, and their excretion is discussed below.

The **bile salts** are sodium and potassium salts of **bile acids** conjugated to glycine or taurine, a derivative of cystine. Four bile acids found in humans are listed in Fig 26–20. In common with vitamin D, cholesterol, a variety of steroid hormones, and the digitalis glycosides, the bile acids contain the cyclopentanoperhydrophenanthrene nucleus (see Chapter 20). The 2 principal (primary) bile acids formed in the liver are cholic acid and chenodeoxycholic acid. In the colon, bacteria convert cholic acid to deoxycholic acid and chenodeoxycholic acid to lithocholic acid. Since they are formed by bacterial action, deoxycholic acid and lithocholic acid are called secondary bile acids. Conjugation of the acids to glycine or taurine occurs in the liver, and the conjugates—eg, **glycocholic acid** and **taurocholic acid**—form sodium and potassium salts in the alkaline hepatic bile. The bile salts have a number of important actions. They combine with lipids to form micelles, water-soluble complexes from which the lipids can be more easily absorbed. This action is called their **hydrotropic** effect. They reduce surface tension and, in conjunction with phospholipids and monoglycerides, are responsible for the solubilization of fat preparatory to its digestion and absorption in the small intestine (see Chapter 25).

Ninety to 95% of the bile salts are absorbed from the small intestine. Some are absorbed by nonionic diffusion, but most are absorbed from the terminal ileum by an extremely efficient active transport process (Fig 26–21). The remaining 5% enter the colon and are converted to the salts of deoxycholic acid and lithocholic acid. Lithocholate is relatively insoluble and is mostly excreted in the stools; only 1% is absorbed. However, deoxycholate is absorbed. The absorbed bile salts are transported back to the liver in the portal vein and reexcreted in the bile (enterohepatic circulation). The normal rate of bile salt synthesis is 0.2–0.4 g/d. The total bile salt pool of approximately 3.5 g recycles repeatedly via the

Cholic acid

	Group at Position 3	7	12	Percent in Human Bile
Cholic acid	OH	OH	OH	50
Chenodeoxycholic acid	OH	OH	H	30
Deoxycholic acid	OH	H	OH	15
Lithocholic acid	OH	H	H	5

Figure 26–20. Human bile acids. The numbers in the formula for cholic acid refer to the positions in the steroid ring.

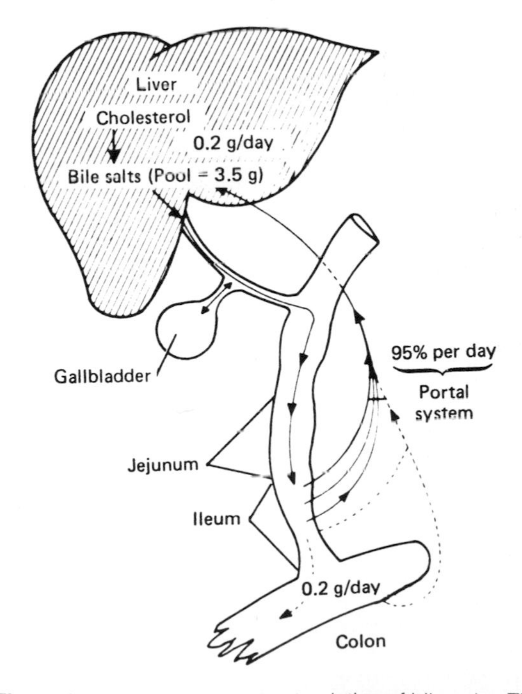

Figure 26–21. Enterohepatic circulation of bile salts. The solid lines entering the portal system represent bile salts of hepatic origin, whereas the dashed lines represent bile salts resulting from bacterial action. (Courtesy of M Tyor. Modified from Way LW [editor]: *Current Surgical Diagnosis & Treatment,* 8th ed. Appleton & Lange, 1988.)

enterohepatic circulation; it has been calculated that the entire pool recycles twice per meal and 6–8 times per day. When bile is excluded from the intestine, up to 50% of ingested fat appears in the feces. There is also severe malabsorption of fat-soluble vitamins. When bile salt reabsorption is prevented by resection of the terminal ileum or by disease in this portion of the small intestine, fatty stools are also produced, because when the enterohepatic circulation is interrupted, the liver cannot increase the rate of bile salt production to a sufficient degree to compensate for the loss, and fat digestion and absorption are compromised.

Bilirubin Metabolism & Excretion

Most of the bilirubin in the body is formed in the tissues by the breakdown of hemoglobin (see Chapter 27). The bilirubin is bound to albumin in the circulation. Some of it is tightly bound, but most of it can dissociate in the liver, and free bilirubin enters liver cells, where it is bound to cytoplasmic proteins (Fig 26–22). It is next conjugated to glucuronic acid in a reaction catalyzed by the enzyme **glucuronyl transferase.** This enzyme is located primarily in the smooth endoplasmic reticulum. Each bilirubin molecule reacts with 2 uridine diphosphoglucuronic acid (UDPGA) molecules to form bilirubin diglucuronide. This glucuronide, which is more water-soluble than the free bilirubin, is then transported against a concentration gradient by a presumably active process into the bile canaliculi. A small amount of the bilirubin glucuronide escapes into the blood, where it is bound less tightly to albumin than free bilirubin, and is excreted in the urine. Thus, the total plasma bilirubin normally includes free bilirubin plus a small amount of conjugated bilirubin. Most of the bilirubin glucuronide passes via the bile ducts to the intestine.

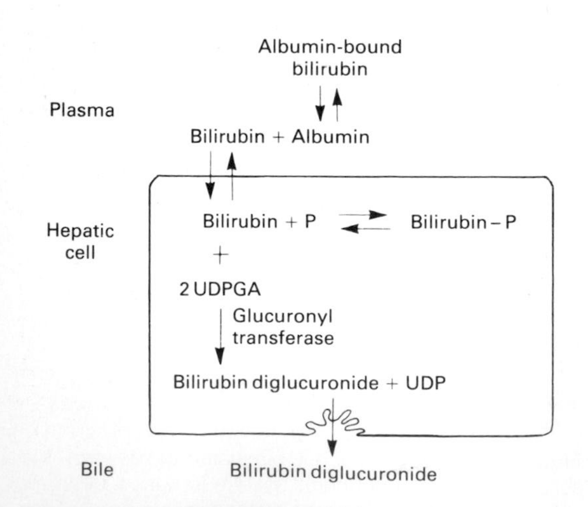

Figure 26–22. Metabolism of bilirubin in the liver. P, intracellular binding proteins; UDPGA, uridine diphosphoglucuronic acid; UDP, uridine diphosphate.

The intestinal mucosa is relatively impermeable to conjugated bilirubin but is permeable to unconjugated bilirubin and to urobilinogens, a series of colorless derivatives of bilirubin formed by the action of bacteria in the intestine. Consequently, some of the bile pigments and urobilinogens are reabsorbed in the portal circulation. Some of the reabsorbed substances are again excreted by the liver (enterohepatic circulation), but small amounts of urobilinogens enter the general circulation and are excreted in the urine.

Jaundice

When free or conjugated bilirubin accumulates in the blood, the skin, scleras, and mucous membranes turn yellow. This yellowness is known as **jaundice** (icterus) and is usually detectable when the total plasma bilirubin is greater than 2 mg/dL (34 μmol/L). Hyperbilirubinemia may be due to (1) excess production (hemolytic anemia, etc); (2) decreased uptake of bilirubin into hepatic cells; (3) disturbed intracellular protein binding or conjugation; (4) disturbed secretion of bilirubin into the bile canaliculi; or (5) intrahepatic or extrahepatic bile duct obstruction. When it is due to processes (1)–(4), the free bilirubin rises. When it is due to bile duct obstruction, bilirubin glucuronide regurgitates into the blood, and it is predominantly the conjugated bilirubin in the plasma that is elevated.

Other Substances Conjugated by Glucuronyl Transferase

The glucuronyl transferase system in the smooth endoplasmic reticulum catalyzes the formation of the glucuronides of a variety of substances in addition to bilirubin. The list includes steroids (see Chapters 20 and 23) and various drugs. These other compounds can compete with bilirubin for the enzyme system when they are present in appreciable amounts. In addition, several barbiturates, antihistamines, anticonvulsants, and other compounds have been found to cause marked proliferation of the smooth endoplasmic reticulum in the hepatic cells, with a concurrent increase in hepatic glucuronyl transferase activity. Phenobarbital has been used successfully for the treatment of a congenital disease in which there is a relative deficiency of glucuronyl transferase (type 2 UDP glucuronyl transferase deficiency.)

Other Substances Excreted in the Bile

Cholesterol and alkaline phosphatase are excreted in the bile. In patients with jaundice due to intra- or extrahepatic obstruction of the bile duct, the blood levels of these 2 substances usually rise; a much smaller rise is generally seen when the jaundice is due to nonobstructive hepatocellular disease. Adrenocortical and other steroid hormones and a number of drugs are excreted in the bile and subsequently reabsorbed (enterohepatic circulation). The dye sulfobromophthalein (Bromsulphalein, BSP) is removed from the bloodstream by the liver cells and excreted in the bile. Its rate of removal from the circulation has been used as a test of hepatic function. The rate depends not only on the functional capacity of the liver cells but also on the hepatic blood flow.

Functions of the Gallbladder

In normal individuals, bile flows into the gallbladder when the sphincter of Oddi is closed. In the gallbladder, the bile is concentrated by absorption of water. The degree of this concentration is shown by the increase in the concentration of solids (Table 26–8); liver bile is 97% water, whereas the average water content of gallbladder bile is 89%. When the bile duct and cystic duct are clamped, the intrabiliary pressure rises to about 320 mm of bile in 30 minutes, and bile secretion stops. However, when the bile duct is clamped and the cystic duct is left open, water is reabsorbed in the gallbladder, and the intrabiliary pressure rises only to about 100 mm of bile in several hours. Acidification of the bile is another function of the gallbladder (Table 26–8).

Regulation of Biliary Secretion

When food enters the mouth, the resistance of the sphincter of Oddi decreases. Fatty acids and amino acids in the duodenum release CCK, which causes gallbladder contraction. Acid and Ca^{2+} also stimulate the secretion of CCK. Substances that cause contraction of the gallbladder are called **cholagogues.**

The production of bile is increased by stimulation of the vagus nerves and by the hormone secretin, which increases the water and HCO_3^- content of bile. Substances that increase the secretion of bile are known as **choleretics.** Bile salts themselves are among the most important physiologic choleretics. The bile salts reabsorbed from the intestine actually inhibit the synthesis of new bile acids, but they themselves are promptly secreted, and they markedly increase bile flow.

Effects of Cholecystectomy

The periodic discharge of bile from the gallbladder aids digestion but is not essential for it. Cholecystectomized patients maintain good health and nutrition with a constant slow discharge of bile into the duodenum, although eventually the bile duct becomes somewhat dilated, and more bile tends to enter the duodenum after eating than at other times. Cholecystectomized patients can even tolerate fried foods, although they generally must avoid foods that are particularly high in fat content.

Table 26–8. Comparison of human hepatic duct bile and gallbladder bile.

	Hepatic Duct Bile	Gallbladder Bile
Percent of solids	2–4	10–12
Bile salts (mmol/L)	10–20	50–200
pH	7.8–8.6	7.0–7.4

Cholecystography

Certain radiopaque iodine-containing dyes such as tetraiodophenolphthalein are excreted in the bile and concentrated with the bile in the gallbladder. When the gallbladder contains an adequate concentration of such a dye, it can be visualized by x-ray **(cholecystography).** To determine if the gallbladder is capable of normal contraction, a fatty meal can be fed. CCK is released and the gallbladder contracts, the size of its x-ray shadow normally decreasing to less than one-third of its original size within 30 minutes. When administered intravenously, technetium 99-labeled derivatives of iminidoacetic acid are excreted in the bile and provide excellent gamma camera images of the gallbladder and bile ducts **(radionuclide scan).** Additional visualization techniques include **percutaneous transhepatic cholangiography,** in which a needle is inserted through the skin and liver into dilated bile ducts and a radiopaque dye injected, and **endoscopic retrograde cholangiography,** in which a dye is injected through an endoscopically placed cannula in the sphincter of Oddi.

Gallstones

Cholelithiasis, ie, the presence of gallstones, is a common condition. Its incidence increases with age, so that in the USA, for example, 20% of the women and 5% of the men between the ages of 50 and 65 have gallstones. The stones are of 2 types: calcium bilirubinate stones and cholesterol stones. In the USA and Europe, 85% of the stones are cholesterol stones.

Three factors appear to be involved in the formation of cholesterol stones. One is bile stasis; stones form in the bile that is sequestrated in the gallbladder rather than the bile that is flowing in the bile ducts. A second is supersaturation of the bile with cholesterol. Cholesterol is very insoluble in bile, and it is maintained in solution in micelles only at certain concentrations of bile salts and lecithin (Fig 26–23). At concentrations above line ABC in Fig 26–23, the bile is supersaturated and contains small crystals of cholesterol in addition to micelles. However, many normal individuals who do not develop gallstones also have supersaturated bile. The third factor is a mix of nucleation factors that favors formation of stones from the supersaturated bile. Outside the body, bile from patients with cholelithiasis forms stones in 2–3 days, whereas it takes more than 2 weeks for stones to form in bile from normal individuals. The exact nature of nucleation factors is unsettled, although glycoproteins in gallbladder mucus have been implicated. In addition, it is unsettled whether stones form as a result of excess production of components that favor nucleation or decreased production of antinucleation components that prevent stones from forming in normal individuals.

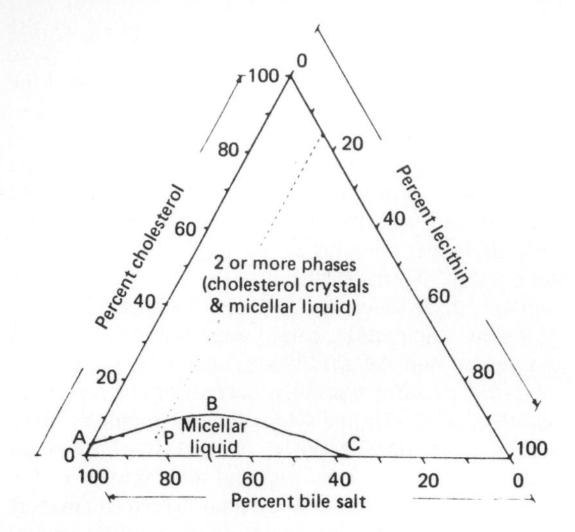

Figure 26–23. Cholesterol solubility in bile as a function of the proportions of lecithin, bile salts, and cholesterol. In bile that has a composition described by any point below line ABC (eg, point P), cholesterol is solely in micellar solution; points above line ABC describe bile in which there are cholesterol crystals as well. (Reproduced, with permission, from Small DM: Gallstones. *N Engl J Med* 1968;**279**:588.)

SMALL INTESTINE

In the small intestine, the intestinal contents are mixed with the secretions of the mucosal cells and with pancreatic juice and bile. Digestion, which begins in the mouth and stomach, is completed in the lumen and mucosal cells of the small intestine, and the products of digestion are absorbed, along with most of the vitamins and fluid. The small intestine is presented with about 9 L of fluid per day—2 L from dietary sources and 7 L of gastrointestinal secretions (Table 25–3); however, only 1–2 L passes into the colon.

Anatomic Considerations

The general arrangement of the muscular layers, nerve plexuses, and mucosa in the small intestine is shown in Fig 26–1. The first portion of the duodenum is sometimes called the duodenal cap or bulb. It is the region struck by the acid gastric contents squirted through the pylorus and is a common site of peptic ulceration. At the ligament of Treitz, the duodenum becomes the jejunum. Arbitrarily, the upper 40% of the small intestine below the duodenum is called the jejunum and the lower 60% the ileum, although there is no sharp anatomic boundary between the two. The ileocecal valve marks the point where the ileum ends in the colon.

The small intestine is shorter during life than it is in cadavers; it relaxes and elongates after death. In living humans, the distance from the pylorus to the ileocecal valve has been reported to be 285 cm (Table 26–9), but at autopsy, this distance is about 700 cm.

Table 26–9. Mean lengths of various segments of the gastrointestinal tract as measured by intubation in living humans.*

Segment	Length (cm)
Pharynx, esophagus, and stomach	65
Duodenum	25
Jejunum and ileum	260
Colon	110

*Data from Hirsch JE, Ahrens EH Jr, Blankenhorn DH: Measurement of human intestinal length in vivo and some causes of variation. *Gastroenterology* 1956;**31**:274.

The mucosa of the small intestine contains **solitary lymph nodules** and, especially in the ileum, **aggregated lymphatic nodules** (Peyer's patches) along the antimesenteric border. Throughout the small intestine there are simple tubular **intestinal glands** (crypts of Lieberkühn). In the duodenum there are, in addition, the small, coiled acinotubular **duodenal glands** (Brunner's glands). The **enterochromaffin cells** in the mucosa of the intestine secrete serotonin (see Chapter 4). These cells are often located deep in the intestinal glands. There are many valvelike folds **(valvulae conniventes)** in the mucous membrane.

Throughout the length of the small intestine, the mucous membrane is covered by **villi** (Fig 26–24). There are 20–40 villi per square millimeter of mucosa. Each intestinal villus is a fingerlike projection, 0.5–1 mm long, covered by a single layer of columnar

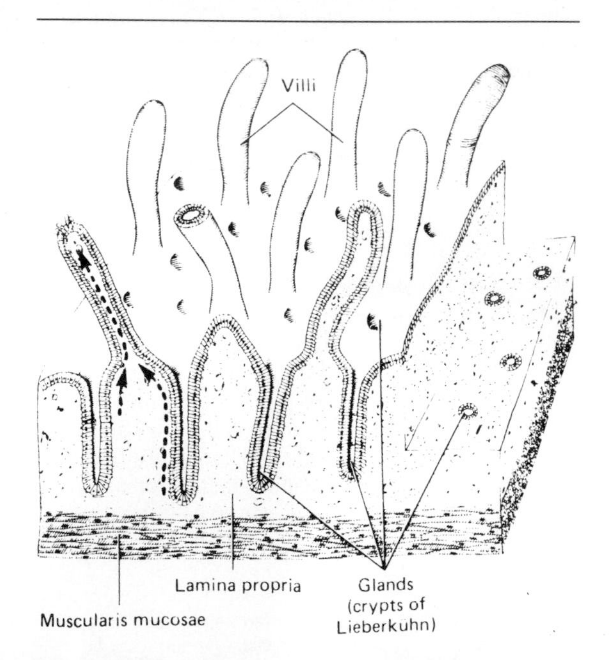

Figure 26–24. Structure of small intestine. The epithelium lining the glands continues over the villi. Mucosal cells are formed in the glands and migrate up to the tips of the villi, where they are shed into the intestinal lumen. (Modified from Junqueira LC, Carneiro J, Kelley RO: *Basic Histology,* 6th ed. Appleton & Lange, 1989.)

epithelium and containing a network of capillaries and a lymphatic vessel **(lacteal).** Fine extensions of the smooth muscle of the submucosa run longitudinally up each villus to its tip. The free edges of the cells of the epithelium of the villi are divided into minute **microvilli** (Fig 26–25), which form a **brush border.** The cells are connected to each other by tight junctions. The outer layer of the cell membrane of the mucosal cells contains many of the enzymes involved in the digestive processes initiated by salivary, gastric, and pancreatic enzymes. Enzymes found in this membrane include various disaccharidases, peptidases, and enzymes involved in the breakdown of nucleic acids (see Chapter 25). The brush border is lined on its luminal side by an amorphous layer called the glycocalyx, which is rich in neutral and amino sugars and may serve a protective function.

The absorptive surface of the small intestine is increased about 600-fold by the valvulae conniventes, villi, and microvilli. It has been estimated that the inner surface area of a mucosal cylinder the size of the small intestine would be about 3300 cm^2, that the valvulae increase the surface area to 10,000 cm^2, that the villi increase it to 100,000 cm^2, and that the microvilli increase it to 2 million cm^2.

The mucosal cells in the small intestine are formed from mitotically active undifferentiated cells in the crypts of Lieberkühn. They migrate up to the tips of the villi, where they are sloughed into the intestinal lumen in large numbers (Fig 26–24). The average life of mucosal cells is 2–5 days, depending on the species. The number of cells shed per day has been calculated to be about 17 billion in humans, and the amount of protein "secreted" in this fashion is about 30 g/d. Mucosal cells are also rapidly sloughed and replaced in the stomach. The crypts are also the site of cyclic AMP-mediated secretion of water and electrolytes (see Chapter 25).

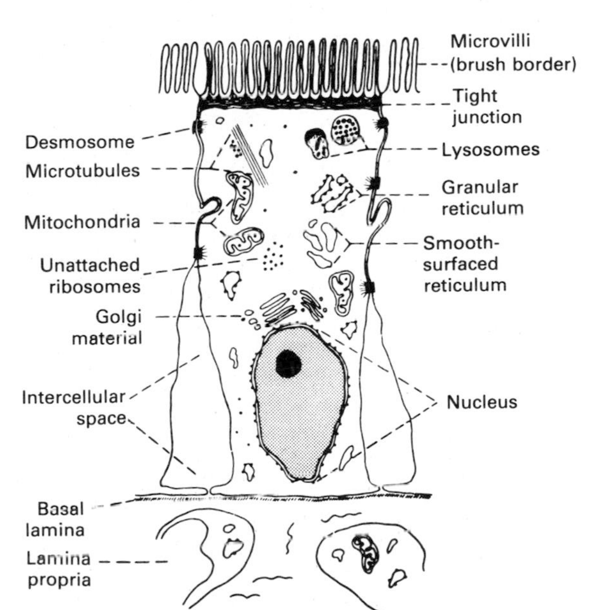

Figure 26–25. Diagram of mucosal cells from human small intestine. Note microvilli, tight connections of cells at mucosal edge (desmosomes), and space between cells at base (intercellular space). (Reproduced, with permission, from Trier JS: Structure of the mucosa of the small intestine as it relates to intestinal function. *Fed Proc* 1967;**26**:1391.)

Intestinal Motility

The contractions of the small intestine are coordinated by the **small bowel slow wave,** a wave of smooth muscle depolarization that moves caudally from the circular smooth muscle of the duodenum. The frequency of the slow waves decreases from about 12/min in the jejunum to about 9/min in the ileum.

The movements of the small intestine mix and churn the intestinal contents **(chyme)** and propel it toward the large intestine. There are 2 types of movement, segmentation contractions and peristaltic waves. Both occur in the absence of extrinsic innervation but require an intact myenteric nerve plexus. The innervation of the intestine is discussed above. **Segmentation contractions** are ringlike contractions that appear at fairly regular intervals along the gut, then disappear and are replaced by another set of ring contractions in the segments between the previous contractions (Fig 26–26). They move the chyme to and fro and increase its exposure to the mucosal surface.

Peristaltic waves move the chyme along the intestine. When the intestinal wall is stretched, a deep circular contraction (peristaltic wave) forms behind the point of stimulation and passes along the intestine toward the rectum at rates varying from 2 to 25 cm/s. This response to stretch is called the **myenteric reflex.** It was originally believed that each peristaltic wave was preceded by a wave of smooth muscle relaxation, but such relaxation is often absent. Peristaltic waves differ in intensity and in the distance they travel. Very intense peristaltic waves called **peristaltic rushes** are not seen in normal individuals

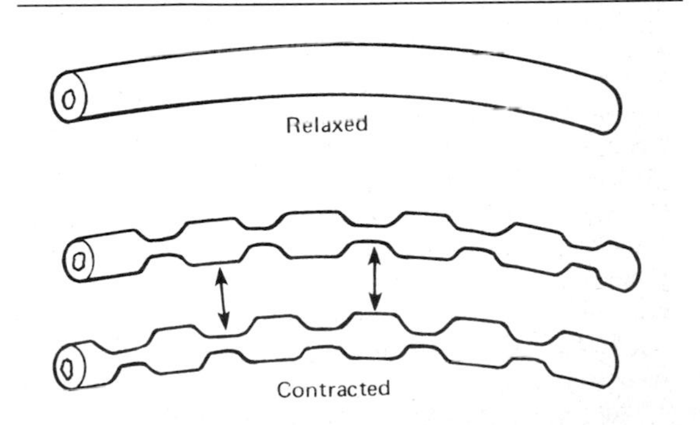

Figure 26–26. Diagram of segmentation contractions of the intestine. Arrows indicate how areas of relaxation become areas of constriction and vice versa.

but do occur when the intestine is obstructed. Weak antiperistalsis is sometimes seen in the colon, but most waves pass regularly in an oral-caudal direction. Removal and resuture of a segment of intestine in its original position does not block progression, and waves will even cross a small gap where the intestine has been replaced by a plastic tube. However, if an intestinal segment is reversed and sewed back into place, peristalsis stops at the reversed segment.

Regulation of Intestinal Secretion

Brunner's glands in the duodenum secrete a thick alkaline mucus that probably helps protect the duodenal mucosa from the gastric acid. There is also an appreciable secretion of HCO_3^- that is independent of Brunner's glands. Decreased duodenal HCO_3^- secretion may play a role in the genesis of duodenal ulcers.

The intestinal glands secrete an isotonic fluid. Most of the enzymes usually found in this secretion are in desquamated mucosal cells; cell-free intestinal juice probably contains few if any enzymes. Gastrointestinal hormones such as VIP (see above) appear to stimulate the secretion of intestinal juice. Vagal stimulation increases the secretion of Brunner's glands but probably has no effect on that of the intestinal glands.

The Malabsorption Syndrome

The digestive and absorptive functions of the small intestine are essential for life. Removal of short segments of the jejunum or ileum generally does not cause severe symptoms, and there is compensatory hypertrophy and hyperplasia of the remaining mucosa, with gradual return of the absorptive function toward normal **(intestinal adaptation).** This adaptation is partly due to a direct effect of nutrients in the intestinal lumen on the mucosa and partly due to circulating factors, probably gastrointestinal hormones. However, when more than 50% of the small intestine is resected or bypassed, the absorption of nutrients and vitamins is so compromised that it is very difficult to prevent malnutrition and wasting. Various disease processes also impair absorption (Table 26–10). The pattern of deficiencies that results is sometimes called the **malabsorption syndrome.** This pattern varies somewhat with the cause, but it can include deficient absorption of amino acids, with marked body wasting and, eventually, hypoproteinemia and edema. Carbohydrate and fat absorption is also depressed. Because of the defective fat absorption, the fat-soluble vitamins (vitamins A, D, E, and K) are not absorbed in adequate amounts. The amount of fat and protein in the stools is increased, and the stools become bulky, pale, foul-smelling, and greasy **(steatorrhea).**

The increased gastric acid secretion produced by intestinal resection has been mentioned above. Resection of the ileum prevents the absorption of bile acids, and this leads in turn to deficient fat absorption. It also causes diarrhea because the unabsorbed bile salts enter the colon, where they inhibit the absorption of Na^+ and water. For this reason, and because the capacity of the jejunum to adapt is lower than that of the ileum, distal small bowel resection causes a greater degree of malabsorption than removal of a comparable length of proximal small bowel. Other complications of intestinal resection or bypass include hypocalcemia, arthritis, hyperuricemia, and possibly fatty infiltration of the liver, followed by cirrhosis. Operations in which segments of the small intestine are bypassed have been recommended for the treatment of obesity, but in view of their complications and dangers, they should not be undertaken lightly. Another surgical approach to obesity is reduction of the volume the stomach can hold, but procedures to accomplish this also have complications, including vomiting and unexplained neurologic disorders.

Table 26–10. Disease processes associated with malabsorption.

Abnormalities of digestion in the intestinal lumen
Inadequate lipolysis (e.g., due pancreatic insufficiency or excess secretion of gastric acid)
Decreased conjugated bile salts (eg, due to ileal resection or bacterial overgrowth)
Abnormalities of mucosal cell transport
Nonspecific, due to tropical sprue, celiac disease, etc
Specific, due to deficiency of various disaccharidases, etc
Abnormalities of fat transport in intestinal lymphatics

The defective intestinal function in tropical sprue (Table 26–10) is apparently due in part to folic acid deficiency. However, the intestinal changes in experimentally produced folic acid deficiency are less intense than those in tropical sprue, and patients with tropical sprue respond to treatment with antibiotics such as tetracycline. In celiac disease, the absorption defect is caused by congenital absence of the enzyme gluten hydrolase from the mucosal cells. This results in the formation of the toxic polypeptide gliadin from gluten, the principal protein in wheat, and gliadin disrupts the formation of microvilli. When wheat products are omitted from the diet, bowel function is generally restored to normal.

Adynamic Ileus

When the intestines are traumatized, there is a decrease in intestinal motility due to direct inhibition of smooth muscle. When the peritoneum is irritated, there is reflex inhibition due to increased discharge of noradrenergic fibers in the splanchnic nerves. Both types of inhibition operate to cause **paralytic (adynamic) ileus** after abdominal operations. Because of the diffuse decrease in peristaltic activity in the small intestine, its contents are not propelled into the colon, and it becomes irregularly distended by pockets of gas and fluid. Intestinal peristalsis returns in 6–8 hours, followed by gastric peristalsis, but colonic activity takes 2–3 days to return. Adynamic ileus can be relieved by passing a tube through the nose down to the small intestine and aspirating the fluid and gas for a few days until peristalsis returns.

Mechanical Obstruction of the Small Intestine

Localized mechanical obstruction of the small intestine causes severe cramping pain **(intestinal colic),** whereas adynamic ileus is often painless. The segment of intestine above the point of mechanical obstruction dilates and becomes filled with fluid and gas. The pressure in the segment rises, and the blood vessels in its wall are compressed, causing local ischemia. Activity in visceral afferent nerve fibers from the distended segment causes sweating, a drop in blood pressure, and severe vomiting, with resultant metabolic alkalosis and dehydration. If the obstruction is not relieved, the condition is fatal.

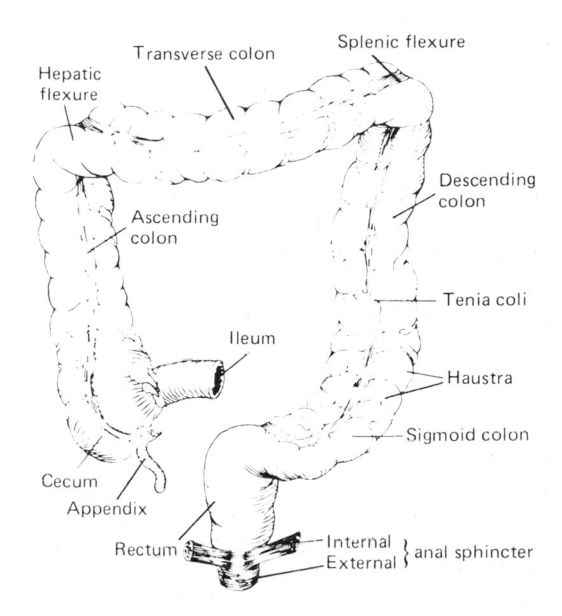

Figure 26–27. The human colon.

COLON

The main function of the colon is absorption of water, Na^+, and other minerals. By removal of about 90% of the fluid, it converts the 1000–2000 mL of isotonic chyme that enters it each day from the ileum to about 200–250 mL of semisolid feces. Certain vitamins are also absorbed, and some of them are synthesized by the large numbers of bacteria that grow in the colon.

Anatomic Considerations

The diameter of the colon is greater than that of the small intestine. Its length is about 100 cm in living adults and about 150 cm at autopsy. The fibers of its external muscular layer are collected into 3 longitudinal bands, the **teniae coli.** Because these bands are shorter than the rest of the colon, the wall of the colon forms outpouchings **(haustra)** between the teniae (Fig 26–27). There are no villi on the mucosa. The colonic glands are short inward projections of the mucosa that secrete mucus. Solitary lymph follicles are present, especially in the cecum and appendix.

Motility & Secretion of the Colon

The portion of the ileum containing the ileocecal valve projects slightly into the cecum, so that increases in colonic pressure squeeze it shut whereas increases in ileal pressure open it. Therefore, it effectively prevents reflux of colonic contents into the ileum. It is normally closed. Each time a peristaltic wave reaches it, it opens briefly, permitting some of the ileal chyme to squirt into the cecum. If the valve is resected in experimental animals, the chyme enters the colon so rapidly that absorption in the small intestine is reduced; however, a significant reduction does not occur in humans. When food leaves the stomach, the cecum relaxes and the passage of chyme through the ileocecal valve increases **(gastroileal reflex).** This is presumably a vagal reflex, although there is some argument about whether vagal stimulation affects the ileocecal valve. Sympathetic stimulation increases the tonal contraction of the valve.

The movements of the colon include segmentation contractions and peristaltic waves like those occurring in the small intestine. Segmentation contractions mix the contents of the colon and, by exposing more of the contents to the mucosa, facilitate absorption. Peristaltic waves propel the contents toward the rectum, although weak antiperistalsis is sometimes seen. A third type of contraction that occurs only in the colon is the **mass action contraction,** in which there is simultaneous contraction of the smooth muscle over large confluent areas. These contractions move material from one portion of the colon to another. They also move material into the rectum, and rectal distension initiates the defecation reflex (see below).

The movements of the colon are propagated in the circular smooth muscle by a slow wave of the colon. The frequency of this wave, unlike the wave in the small intestine, increases along the colon, from about 2/min at the ileocecal valve to 6/min at the sigmoid.

Mucus secretion by the colonic glands is stimulated by contact between the gland cells and the colonic contents. No hormonal or neural link is involved in the basic secretory response, although some additional secretion may be produced by local reflex responses via the pelvic and splanchnic nerves. No digestive enzymes are secreted in the colon.

Transit Time in the Small Intestine & Colon

The first part of a test meal reaches the cecum in about 4 hours, and all of the undigested portions have entered the colon in 8 or 9 hours. On the average, the first remnants of the meal reach the hepatic flexure in 6 hours, the splenic flexure in 9 hours, and the pelvic colon in 12 hours. From the pelvic colon to the anus, transport is much slower. As much as 25% of the residue of a test meal may still be in the rectum in 72 hours. When small colored beads are fed with a meal, an average of 70% of them are recovered in the stool in 72 hours, but total recovery requires more than a week.

Absorption in the Colon

The absorptive capacity of the mucosa of the large intestine is great. Na^+ is actively transported out of the colon, and water follows along the osmotic gradient thus generated. Normally, there is net secretion of K^+ and HCO_3^- into the colon (see Chapter 25). The absorptive capacity of the colon makes rectal instillation a practical route for drug administration, especially in children. Many compounds, including anesthetics, sedatives, tranquilizers, and steroids, are absorbed rapidly by this route. Some of the water in an enema is absorbed, and if the volume of an enema is large, absorption may be rapid enough to cause water intoxication. Coma and death due to water intoxication have been reported following tap-water enemas in children with megacolon.

Feces

The stools contain inorganic material, undigested plant fibers, bacteria, and water. Their composition (Table 26–11) is relatively unaffected by variations in diet because a large fraction of the fecal mass is of nondietary origin. This is why appreciable amounts of feces continue to be passed during prolonged starvation.

Intestinal Bacteria

The chyme in the jejunum normally contains few if any bacteria. There are more microorganisms in the ileum, but it is only the colon that regularly contains large numbers of bacteria. The reason for the relative sterility of the jejunal contents is unsettled, although gastric acid and the comparatively rapid transit of the chyme through this region may inhibit bacterial growth.

Table 26–11. Approximate composition of feces on an average diet.

	Percentage of Total Weight
Water	75
Solids	25
	Percentage of Total Solids
Cellulose and other indigestible fiber	Variable
Bacteria	30
Inorganic material (mostly calcium and phosphates)	15
Fat and fat derivatives	5
Also desquamated mucosal cells, mucus, and small amounts of digestive enzymes	

The microorganisms present in the colon include not only bacilli such as *Escherichia coli* and *Enterobacter aerogenes* but also pleomorphic organisms such as *Bacteroides fragilis,* cocci of various types, and organisms such as gas gangrene bacilli, which can cause serious disease in tissues outside the colon. Great masses of bacteria are passed in the stool. At birth, the colon is sterile, but the intestinal bacterial flora becomes established early in life.

The effects of the intestinal bacteria on their host are complex, some being definitely beneficial and others possibly harmful.

Antibiotics improve growth rates in a variety of species, including humans; and small amounts of antibiotics are frequently added to the diets of domestic animals. Animals raised under sanitary but not germ-free conditions grow faster than controls. They assimilate food better and do not require certain amino acids that are essential dietary constituents in other animals. They also have larger litters and a lower neonatal death rate.

The reason for the improved growth is unsettled. Nutritionally important substances such as ascorbic acid, cyanocobalamin, and choline are utilized by some intestinal bacteria. On the other hand, some enteric microorganisms synthesize vitamin K and a number of B complex vitamins, and the folic acid produced by bacteria can be shown to be absorbed in significant amounts.

The brown color of the stools is due to pigments formed from the bile pigments by the intestinal bacteria. When bile fails to enter the intestine, the stools become white **(acholic stools).** Bacteria produce some of the gases in the flatus. Organic acids formed from carbohydrates by bacteria are responsible for the slightly acid reaction of the stools (pH 5.0–7.0). Intestinal bacteria also appear to play a role in cholesterol metabolism, since the poorly absorbed antibiotic neomycin that modifies the intestinal flora lowers LDL and the plasma cholesterol level.

A number of amines, including such potentially toxic substances as histamine and tyramine, are formed in the colon by bacterial enzymes that decarboxylate amino acids. At one time, the symptoms seen in constipated patients were blamed on absorption of these amines, but this theory of "autointoxication" has been discredited. Other amines formed by the intestinal bacteria, especially indole and skatole, are largely responsible for the odor of the feces.

Ammonia is also produced in the colon and absorbed. When the liver is diseased, ammonia is not removed from the blood, and hyperammonemia can cause neurologic symptoms **(hepatic encephalopathy).** Elevated blood ammonia is also one of the causes of hepatic encephalopathy that occurs in patients in whom the portal vein has been anastomosed to the vena cava in order to reduce portal pressure and stop bleeding from dilated esophageal varices. There is improvement in encephalopathy if the colon is removed or isolated by anastomosing the ileum to the sigmoid. However, the improvement is transient because colonic bacteria soon colonize the ileum, and the procedure is not used therapeutically.

Secretory Immunity in the Intestine

It is now well established that cells in the gut-associated lymphoid tissue are stimulated to proliferate and differentiate by intestinal antigens that reach them via the M cells. The lymphoid cells enter the circulation but return to the intestinal wall, where they differentiate further into plasma cells that secrete immunoglobulins, particularly IgA and IgM (see Chapter 27). These antibodies coat the mucosa and do much to prevent pathogenic organisms from penetrating the intestinal wall.

When normal animals with the usual intestinal bacterial flora are exposed to ionizing radiation, the body defenses that prevent intestinal bacteria from invading the rest of the body break down, and a major cause of death in **radiation poisoning** is overwhelming sepsis. Germ-free animals have extremely hypoplastic lymphoid tissue and poorly developed immune mechanisms, probably because these mechanisms have never been challenged. However, they are much more resistant to radiation than animals with the usual intestinal flora because they have no intestinal bacteria to cause sepsis.

Blind Loop Syndrome

Overgrowth of bacteria within the intestinal lumen can cause definite harmful effects. Such overgrowth occurs when there is stasis of the contents of the small intestine, and it causes macrocytic anemia, malabsorption of cyanocobalamin (vitamin B_{12}), and steatorrhea. Because the condition is prominent in patients with surgically created blind intestinal loops, it has acquired the name **blind loop syndrome;** however, it can occur in any condition that promotes massive bacterial contamination of the small intestine. The cause of the anemia appears to be the cyanocobalamin deficiency, and the deficiency in cyanocobalamin absorption is probably due to uptake of ingested cyanocobalamin by the bacteria. The steatorrhea is probably due to excessive hydrolysis of conjugated bile salts by the bacteria. The important role of bile salts in fat digestion is discussed in Chapter 25.

Dietary Fiber

Adequate nutrition in herbivorous animals depends upon the action of gastrointestinal microorganisms that break down cellulose and related plant carbohydrates. In humans, there is no appreciable digestion of these vegetable products. Cellulose, hemicellulose, and lignin in the diet are important components of the **dietary fiber,** which by definition is all ingested food that reaches the large intestine in an essentially unchanged state. Various gums, algal polysaccharides, and pectic substances also contribute to dietary fiber.

If the amount of dietary fiber is low, the diet is said to lack **bulk.** Since the amount of material in the colon is small, the colon is inactive and bowel movements are infrequent. So-called bulk laxatives work by providing a larger volume of indigestible material to the colon. It has been claimed that some individuals with chronic constipation have a greater than normal capacity to break down cellulose and related products, thus reducing the residue in their colons.

There has been a recent upsurge of interest in dietary fiber because of epidemiologic evidence indicating that groups of people who live on a diet which contains large amounts of vegetable fiber have a low incidence of diverticulitis, cancer of the colon, diabetes mellitus, and coronary artery disease. However, the relationship between dietary fiber and the incidence of disease is still unsettled and needs further study.

Defecation

Distention of the rectum with feces initiates reflex contractions of its musculature and the desire to defecate. In humans, the sympathetic nerve supply to the internal (involuntary) anal sphincter is excitatory, whereas the parasympathetic supply is inhibitory. This sphincter relaxes when the rectum is distended. The nerve supply to the striate muscle of the external anal sphincter comes from the pudendal nerve. The sphincter is maintained in a state of tonic contraction, and moderate distention of the rectum increases the force of its contraction (Fig 26–28). The urge to defecate first occurs when rectal pressure increases to about 18 mm Hg. When this pressure reaches 55 mm Hg, the external as well as the internal sphincter relaxes and the contents of the rectum are expelled. This is why there is reflex evacuation of the rectum in chronic spinal animals and humans. Before the pressure that relaxes the external anal sphincter is reached, voluntary defecation can be initiated by voluntarily relaxing the external sphincter and contracting the abdominal muscles (straining), thus aiding the reflex emptying of the distended rectum. Defecation is therefore a spinal reflex that can be voluntarily inhibited by keeping the external sphincter contracted or facilitated by relaxing the sphincter and contracting the abdominal muscles.

Distension of the stomach by food initiates contractions of the rectum and, frequently, a desire to defecate. The response is called the **gastrocolic reflex,** although there is some evidence that it is due to an action of gastrin on the colon and is not naturally mediated. Because of the response, defecation after meals is the rule in children. In adults, habit and cultural factors play a large role in determining when defecation occurs.

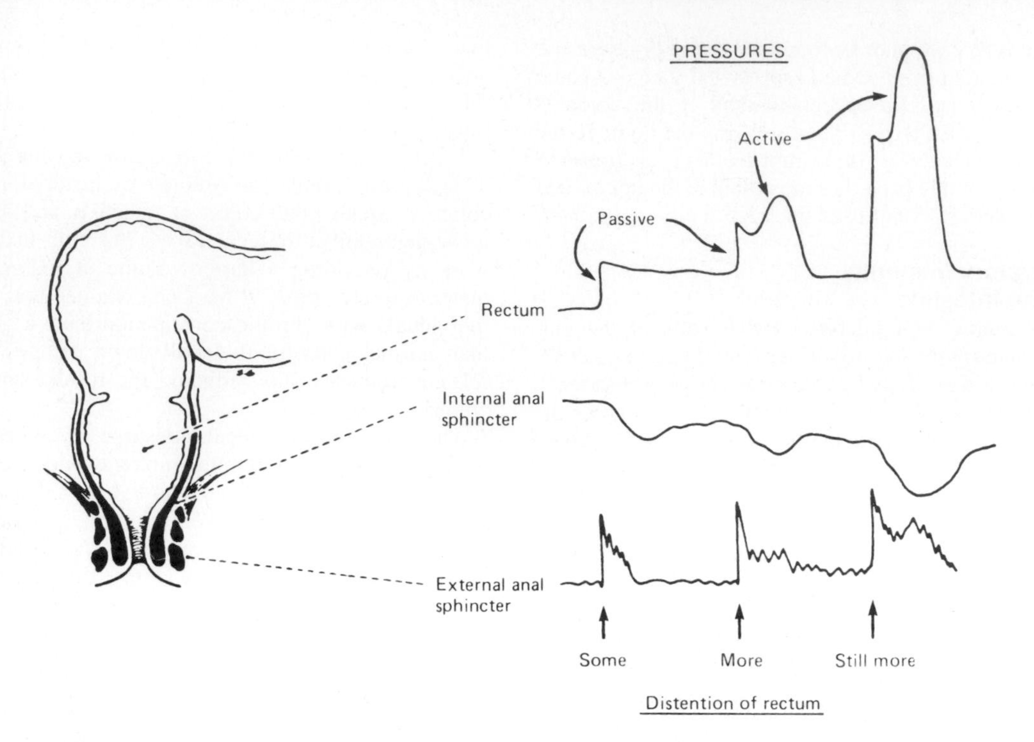

Figure 26–28. Responses to distention of the rectum by pressures less than 55 mm Hg. Distention produces passive tension due to stretching of the wall of the rectum, and additional active tension when the smooth muscle in the wall contracts. (Reproduced, with permission, from Davenport HW: *A Digest of Digestion,* 2nd ed. Year Book, 1978.)

Effects of Colectomy

Humans can survive after total removal of the colon if fluid and electrolyte balance is maintained. When total colectomy is performed, the ileum is brought out through the abdominal wall **(ileostomy)** and the chyme expelled from the ileum is collected in a plastic bag fastened around the opening. If the diet is carefully regulated, the volume of ileal discharge decreases and its consistency increases over a period of time. Care of an ileostomy used to be a time-consuming, difficult job, but when ileostomies are constructed by modern techniques, they can be relatively trouble-free and patients with them can lead essentially normal lives.

Constipation

In bowel-conscious America, the amount of misinformation and undue apprehension about constipation probably exceeds that about any other health topic. Patients with persistent constipation, and particularly those with a recent change in bowel habits, should of course be examined carefully to rule out underlying organic disease. However, many normal humans defecate only once every 2–3 days, even though others defecate once a day and some as often as 3 times a day. Furthermore, the only symptoms caused by constipation are slight anorexia and mild abdominal discomfort and distention. These symptoms are not due to absorption of "toxic substances," because they are promptly relieved by evacuating the rectum and can be reproduced by distending the rectum with inert material. Other symptoms attributed by the lay public to constipation are due to anxiety or other causes.

Megacolon

The lack of harmful effects of infrequent bowel movements is emphasized by the relative absence for months and even years of symptoms other than abdominal distention, anorexia, and lassitude in children with **aganglionic megacolon** (Hirschsprung's disease). This disease is due to congenital absence of the ganglion cells in both the myenteric and submucous plexuses of a segment of the distal colon. The substance P content of the segment is low. Feces pass the aganglionic region with difficulty, and children with the disease may defecate as infrequently as once every 3 weeks. The condition can be cured if the aganglionic region is resected and the portion of the colon above it anastomosed to the rectum.

Diarrhea

Severe diarrhea is debilitating and can be fatal, especially in infants. Large amounts of Na^+, K^+, and water are washed out of the colon and the small intestine in the diarrheal stools, causing dehydration, hypovolemia, and, eventually, shock and cardiovascular collapse. A more insidious complication of chronic diarrhea, if fluid balance is maintained, is severe hypokalemia.

REFERENCES
Section V: Gastrointestinal Function

Brooks FP (editor): *Peptic Ulcer Disease*. Churchill Livingstone, 1985.

Castell DO, Johnson LF (editors): *Esophageal Function in Health and Disease*. Elsevier, 1983.

Cohen S, Soloway RD (editors): *Gallstones*. Churchill Livingstone, 1985.

Davenport HW: *A Digest of Digestion*, 2nd ed. Year Book, 1978.

Davenport HW: *Physiology of the Digestive Tract*, 5th ed. Year Book, 1982.

Diamond JM, Karasov WH: Trophic control of the intestinal mucosa. *Nature* 1983;**304:**18.

Dobbins WO: Gut immunophysiology: A gastroenterologist's view with emphasis on pathophysiology. *Am J Physiol* 1982;**242:**G1.

Field M, Fordtran JS, Schultz SG (editors): *Secretory Diarrhea*. American Physiological Society, 1980.

Finch CA, Huebers H: Perspectives in iron metabolism. *N Engl J Med* 1982;**306:**1520.

Forte JG, Machen TE, Obrink KJ: Mechanisms of gastric H^+ and Cl^- transport. *Annu Rev Physiol* 1980;**42:**111.

Fraser CL, Arieff AI: Hepatic encephalopathy. *N Engl J Med* 1985;**313:**865.

Furness JB, Costa M: Types of nerves in the enteric nervous system. *Neuroscience* 1980;**5:**1.

Hirsch DJ, Hayslett JP: Adaptation to potassium. *News Physiol Sci* 1986;**1:**54.

Holt KM, Isenberg JI. Peptic ulcer disease: Physiology and pathophysiology. *Hosp Pract* (Jan) 1985;**20:**89.

Howat HT, Searles H (editors): *The Exocrine Pancreas*. Saunders, 1979.

Johnson LR: Regulation of the mucosal gastrin receptor. *Proc Soc Exp Biol Med* 1983;**173:**167.

Johnson LR (editor): *Gastrointestinal Physiology*. Mosby, 1977.

Jones AL et al: The architecture of bile secretion. *Dig Dis Sci* 1980;**25:**609.

Klaassen CD, Watkins JB III: Mechanisms of bile formation, hepatic uptake and biliary excretion. *Pharmacol Rev* 1984;**361:**1.

Kretchmer N: Lactose and lactase. *Sci Am* (Oct) 1972; **227:**70.

Levitt MD, Bond JH: Volume, composition and source of intestinal gas. *Gastroenterology* 1970;**59:**921.

Mackowiak PA: The normal microbial flora. *N Engl J Med* 1982;**307:**83.

McArthur KE, Jensen RT, Gardner JD: Treatment of acid-peptic diseases by inhibition of gastric H^+, K^+-ATPase. *Annu Rev Med* 1986;**37:**47.

Mendeloff AI: Dietary fiber and human health. *N Engl J Med* 1977;**297:**811.

Miller AJ: Deglutition. *Physiol Rev* 1982;**62:**129.

Rehfeld JF: Four basic characteristics of the gastrin-cholecystokinin system. *Am J Physiol* 1981;**240:**G255.

Reichlin S: Somatostatin. *N Engl J Med* 1983;**309:**18.

Shepherd AP, Granger DN (editors): *Physiology of the Intestinal Circulation*. Raven, 1984.

Sleisenger MH, Brandborg LL: *Malabsorption*. Saunders, 1977.

Thompson JC et al (editors): *Gastrointestinal Endocrinology*. McGraw-Hill, 1987.

Weisbrodt NW: Gastrointestinal motility. *Annu Rev Physiol* 1981;**43:**7.

Williams JA: Regulatory mechanisms in pancreas and salivary acini. *Annu Rev Physiol* 1984;**46:**361.

Williamson RCN: Intestinal adaptation. *N Engl J Med* 1978;**298:**1393.

Wolfe MM, Jensen RT: Zollinger-Ellison syndrome: Current concepts in diagnosis and management. *N Engl J Med* 1987;**317:**1200.

Young JA, van Lennep EW: *The Morphology of Salivary Glands*. Academic Press, 1978.

Zakim D, Boyer TD (editors): *Hepatology: A Textbook of Liver Disease*. Saunders, 1982.

Section VI. Circulation

27 Circulating Body Fluids

INTRODUCTION

The **circulatory system** is the transport system that supplies O_2 and substances absorbed from the gastrointestinal tract to the tissues, returns CO_2 to the lungs and other products of metabolism to the kidneys, functions in the regulation of body temperature, and distributes hormones and other agents that regulate cell function. The blood, the carrier of these substances, is pumped through a closed system of blood vessels by the heart, which in mammals is really 2 pumps in series with each other. From the left ventricle, blood is pumped through the arteries and arterioles to the capillaries, where the blood equilibrates with the interstitial fluid. The capillaries drain through venules into the veins and back to the right atrium. This is the **major (systemic) circulation.** From the right atrium, blood flows to the right ventricle, which pumps it through the vessels of the lungs—the **lesser (pulmonary) circulation**—and the left atrium to the left ventricle. In the pulmonary capillaries, the blood equilibrates with the O_2 and CO_2 in the alveolar air. Some tissue fluids enter another system of closed vessels, the lymphatics, which drain lymph via the thoracic duct and the right lymphatic duct into the venous system (the **lymphatic circulation).** The circulation is controlled by multiple regulatory systems that function in general to maintain adequate capillary blood flow—when possible, in all organs, but particularly in the heart and brain. This chapter is concerned with blood and lymph.

BLOOD

The cellular elements of the blood—white blood cells, red blood cells, and platelets—are suspended in the plasma. The normal total circulating blood volume is about 8% of the body weight (5600 mL in a 70-kg man). About 55% of this volume is plasma.

BONE MARROW

In the adult, red blood cells, many white blood cells, and platelets are formed in the bone marrow. In the fetus, blood cells are also formed in the liver and spleen, and in adults such **extramedullary hematopoiesis** may occur in diseases in which the bone marrow becomes destroyed or fibrosed. In children, blood cells are actively produced in the marrow cavities of all the bones. By age 20, the marrow in the cavities of the long bones, except for the upper humerus and femur, has become inactive (Fig 27–1). Active cellular marrow is called **red marrow;** inactive marrow that is infiltrated with fat is called **yellow marrow.**

The bone marrow is actually one of the largest organs in the body, approaching the size and weight of the liver. It is also one of the most active. Normally, 75% of the cells in the marrow belong to the white blood cell-producing myeloid series, and only 25% are maturing red cells, even though there are over 500 times as many red cells in the circulation as there are white cells. This difference in the marrow

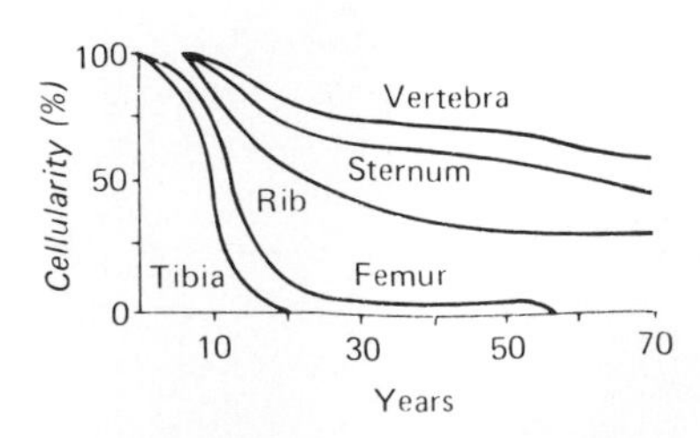

Figure 27–1. Changes in red bone marrow cellularity with age. 100% equals degree of cellularity at birth. (Reproduced, with permission, from Whitby LEH, Britton CJC: *Disorders of the Blood,* 10th ed. Churchill Livingstone, 1969.)

probably reflects the fact that the average life span of white cells is short, whereas that of red cells is long.

The bone marrow contains pluripotent uncommitted stem cells and unipotent committed stem cells. The former differentiate only into one of the differentiated cell types found in marrow and blood (Fig 27–2). It appears likely that the bone marrow contains pools of committed stem cells for megakaryocytes, lymphocytes, erythrocytes, eosinophils, and basophils, whereas neutrophils and monocytes arise from a common precursor.

WHITE BLOOD CELLS

There are normally 4000–11,000 white blood cells per microliter of human blood (Table 27–1). Of these, the **granulocytes (polymorphonuclear leukocytes, PMNs)** are the most numerous. Young granulocytes have horseshoe-shaped nuclei that become multilobed as the cells grow older (Fig 27–2). Most of them contain neutrophilic granules **(neutrophils),** but a few contain granules that stain with acid dyes **(eosinophils),** and some have basophilic granules **(basophils).** The other 2 cell types found normally in peripheral blood are **lymphocytes,** cells with large, round nuclei and scanty cytoplasm; and **monocytes,** cells with abundant agranular cytoplasm and kidney-shaped nuclei (Fig 27–2). Acting together, these cells provide the body with powerful defenses against tumors and viral, bacterial, and parasitic infections.

Functions & Life Cycle of Granulocytes

All granulocytes contain the enzyme **myeloperoxidase.** This enzyme, which has a molecular weight of about 150,000, catalyzes the formation of ClO^- and other hypohalite ions that aid in killing ingested bacteria. The basophils, which resemble but are not identical to mast cells, contain histamine and heparin. They release histamine and other inflammatory mediators when activated by a histamine-releasing factor secreted by T cells and are essential for immediate-type hypersensitivity reactions. These range from mild urticaria and rhinitis to severe anaphylactic shock. The eosinophils attack some parasites, and they inactivate mediators released from mast cells during allergic reactions. The circulating eosinophil level is often elevated in patients with allergic diseases. The neutrophils seek out, ingest, and kill bacteria and have been called the body's first line of defense against bacterial infections.

Table 27–1. Normal values for the cellular elements in human blood.

	Cells/μL (average)	Approximate Normal Range	Percentage of Total White Cells
Total WBC	9000	4000–11,000	. . .
Granulocytes			
Neutrophils	5400	3000–6000	50–70
Eosinophils	275	150–300	1–4
Basophils	35	0–100	0.4
Lymphocytes	2750	1500–4000	20–40
Monocytes	540	300–600	2–8
Erythrocytes			
Females	4.8×10^6	. . .	. . .
Males	5.4×10^6	. . .	. . .
Platelets	300,000	200,000–500,000	. . .

The average half-life of a neutrophil in the circulation is 6 hours. To maintain the normal circulating blood level, it is therefore necessary to produce over 100 billion neutrophils per day. Many of the neutrophils enter the tissues; they first adhere to the endothelium and then insinuate themselves through the walls of the capillaries between endothelial cells by a process called **diapedesis.** Many of those that leave the circulation enter the gastrointestinal tract and are lost from the body.

Invasion of the body by bacteria triggers the **inflammatory response.** The bone marrow is stimulated to produce and release large numbers of neutrophils. Bacterial products interact with plasma factors and cells to produce agents that attract neutrophils to the infected area **(chemotaxis).** The chemotactic agents include a component of the complement system (C5a), leukotrienes, and polypeptides from lymphocytes, mast cells, and basophils. Other plasma factors act on the bacteria to make them "tasty" to the phagocytes **(opsonization).** The principal opsonins that coat the bacteria are immunoglobulins of a particular class (IgG) and complement proteins (see below). The neutrophils then actively ingest the bacteria **(phagocytosis).** The steps involved in the phagocytosis of bacteria or other antigens are summarized in Table 27–2. The phagocytic vesicles formed by endocytosis of the antigen-antibody complex fuse with lysosomes. Since lysosomes are the granules of the neutrophils, this step is called **degranulation.** Associated with the degranulation is a sharp increase in O_2 uptake and metabolism (the **respiratory burst**), with increased activity in the hexosemonophosphate shunt and the production of hydrogen peroxide (H_2O_2) and superoxide anions ($O_2^{\dot{-}}$). $O_2^{\dot{-}}$ is a **free radical** formed by the loss of one electron from O_2, and like other free radicals, it is highly reactive chemically. $O_2^{\dot{-}}$, H_2O_2, and other O_2 derivatives formed in the granulocytes kill the bacteria, and the lytic enzymes in the granules digest them. The movements of the cell in phagocytosis as well as migration to the site of infection involve microtubules and microfilaments (see Chapter 1). Proper function of the microfilaments involves the interaction of the actin they contain with myosin and an actin-binding protein on the inside of the cell membrane. Fusion of lysosomes with endocytotic pits before they are pinched off to form phagocytic vesicles can occur if there are many antigens in the vicinity of the cell or the antigen is too large to be ingested. When this occurs, lysosomal enzymes enter the extracellular fluid, producing the inflammatory response. Neutrophils also release thromboxanes that are vasoconstrictors and platelet-aggregating agents, leukotrienes that increase vascular permeability and attract other neutrophils to the site, and other prostaglandins that exert a moderate anti-inflammatory effect.

Monocytes

The monocytes, like neutrophilic leukocytes, are actively phagocytic and contain peroxidase and

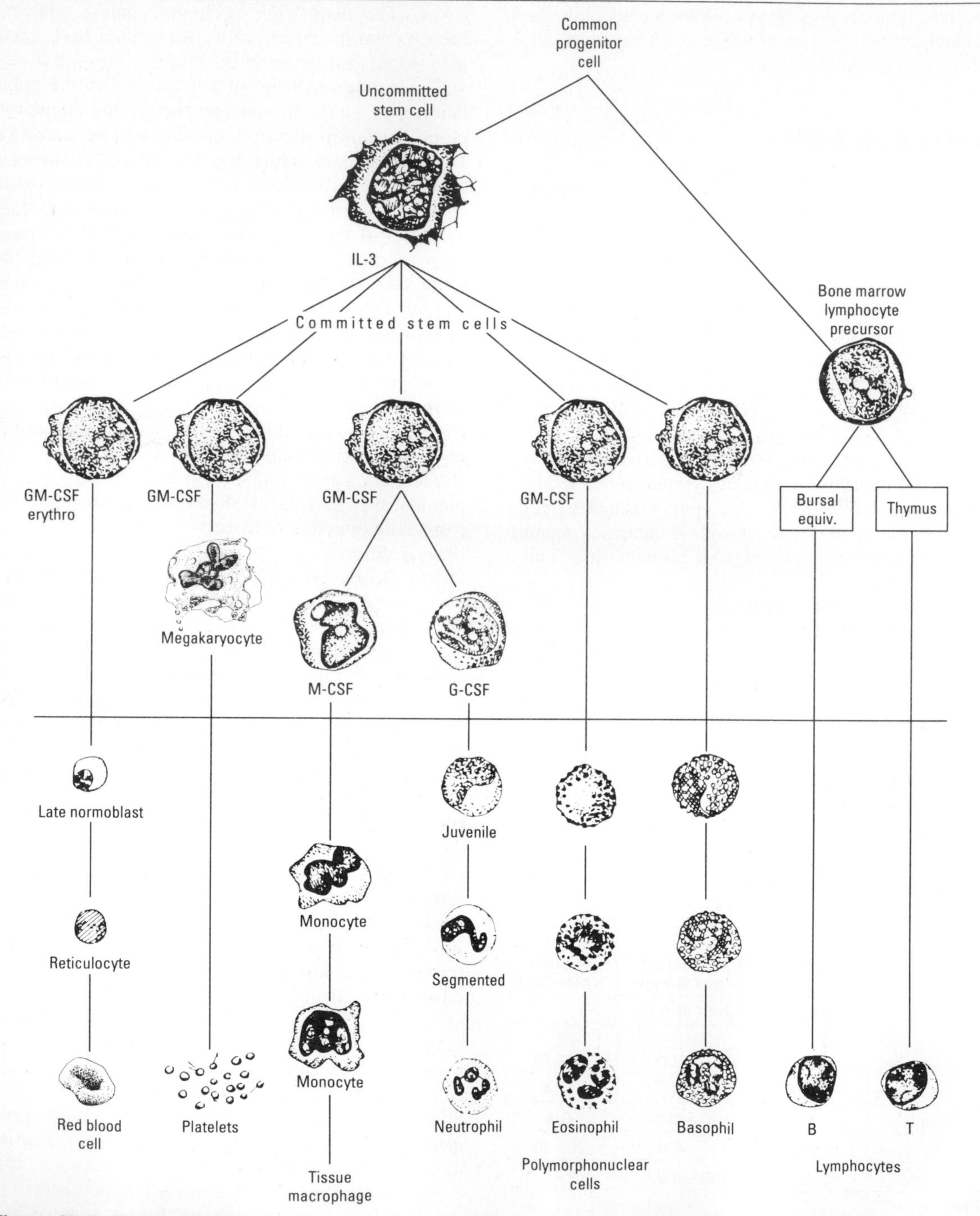

Figure 27–2. The development of various formed elements of the blood from bone marrow cells. Cells below the horizontal line are found in normal peripheral blood. The principal sites of action of erythropoietin (erythro) and the various colony-stimulating factors (CSF) that stimulate the differentiation of the components are indicated. G, granulocyte; M, macrophage; IL-3, interleukin-3.

Table 27–2. Events occurring during phagocytosis.

1. Binding of antibody-coated antigen to receptors on membrane of phagocytic cell.
2. Hyperpolarization of cell membrane, followed by depolarization and slow repolarization (in 5–10 seconds). Increased free Ca^{2+} in cytoplasm.
3. Endocytosis of antigen-antibody complex, with release of superoxide ions (in 30 seconds).
4. Fusion of phagocytic vacuole with lysosome.
5. Release of thromboxanes, prostaglandins, and leukotrienes.

lysosomal enzymes. They are mobilized along with neutrophils as part of the inflammatory response and constitute a second line of defense against bacterial infections. They enter the circulation from the bone marrow, but after about 24 hours they enter the tissues to become **tissue macrophages** (Fig 27–3). All of the tissue macrophages, including the Kupffer cells of the liver and the alveolar macrophages in the lung, come from circulating monocytes. The tissue macrophage system has generally been called the **reticuloendothelial system.** The macrophages migrate in response to chemotactic stimuli and engulf and kill bacteria by processes generally similar to those occurring in neutrophils. They play a key role in immunity (see below). They also secrete many substances that affect lymphocytes and other cells, along with prostaglandins of the E series and clot-promoting factors.

Granulocyte & Macrophage Colony-Stimulating Factors

The production of red and white blood cells is regulated with great precision in healthy individuals, and the production of granulocytes is rapidly and dramatically increased in infections. The regulatory factors include a number of glycoprotein factors or hormones that cause cells in one or more of the committed cell lines in the bone marrow to proliferate (Fig 27–2, Table 27–3). The regulation of erythrocyte production by **erythropoietin** is discussed below and in Chapter 24. To date, 4 additional glycoprotein factors that regulate the production of white blood cells have been identified and characterized. These factors are called **colony-stimulating factors** because they cause appropriate single stem cells to proliferate in tissue culture, forming colonies in the culture medium. The 4 factors are: **granulocyte-macrophage colony-stimulating factor** (GM-CSF), **granulocyte colony-stimulating factor** (G-CSF), **multipotential colony-stimulating factor** (multi-CSF, also known as IL-3), and **macrophage colony-stimulating factor** (M-CSF). Their actions are complex. They increase the biologic activities of different types of white blood cells in addition to promoting cell growth and differentiation. In general, however, their principal actions are at the points shown in Fig 27–2. These hormonal factors are produced by various types of cells (Table 27–3). In general, the mRNAs for the factors are difficult to detect in these cells at rest, but after activation of the cells by mitogens, the mRNAs for the CSFs are readily detectable.

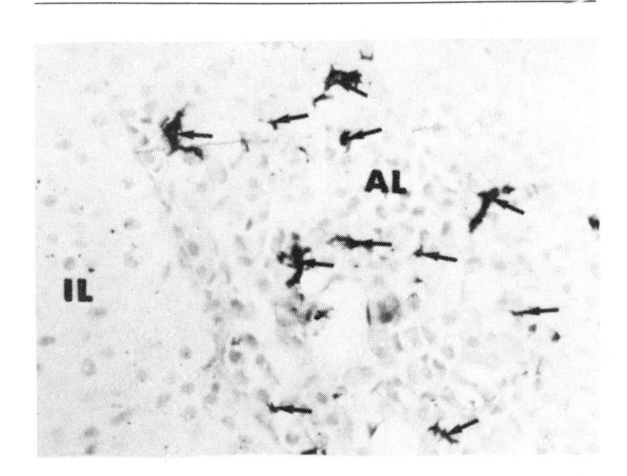

Figure 27–3. Tissue macrophages in the pituitary gland. The macrophages, which are identified by arrows, have been stained immunocytochemically using a monoclonal antibody that is specific for these cells. AL, anterior lobe; IL, intermediate lobe. (Courtesy of S Gordon.)

Disorders of Phagocytic Function

More than 15 primary defects in neutrophil function have been described, along with at least 30 other conditions in which there is a secondary depression of the function of the neutrophils. Patients with these diseases are prone to infections that are relatively mild when only the neutrophil system is involved but severe when the monocyte-tissue macrophage system is also involved. In one syndrome (neutrophil hypomotility), actin in the neutrophils does not polymerize normally, and the neutrophils move slowly. In another, more serious disease (chronic granulomatous disease), there is a failure to generate $O_2^{\cdot-}$ in both the neutrophils and monocytes and consequent inability

Table 27–3. Factors regulating hematopoiesis.

Name	Cellular sources	Cell types produced in increased numbers*
Erythropoietin	Kidney cells, other	rbc
G-CSF	Monocytes, fibroblasts	n
M-CSF	Monocytes, fibroblasts, endothelial cells	m
GM-CSF	T cells, endothelial cells, fibroblasts	n, m, e, meg, rbc
IL-3	T cells	n, m, e, b, meg, rbc

*n, neutrophils; m, monocytes; e, eosinophils; b, basophils; meg, megakaryocytes; rbc, red blood cells.

to kill many phagocytosed bacteria. In severe congenital glucose 6-phosphate dehydrogenase deficiency, there are multiple infections because of failure to generate the NADPH necessary for $O_2^{\dot{-}}$ production. In congenital myeloperoxidase deficiency, microbial killing power is reduced because hypohalite ions are not formed, but it is not absent, because other bactericidal mechanisms are intact.

Lymphocytes

After birth, some lymphocytes are formed in the bone marrow, but most are formed in the lymph nodes, thymus, and spleen from precursor cells that originally came from the bone marrow. Lymphocytes enter the bloodstream for the most part via the lymphatics. It has been calculated that in humans, 3.5×10^{10} lymphocytes per day enter the circulation via the thoracic duct alone; however, this count includes cells that reenter the lymphatics and thus traverse the thoracic duct more than once. The effects of adrenocortical hormones on the lymphoid organs, the circulating lymphocytes, and the granulocytes are discussed in Chapter 20.

IMMUNE MECHANISMS

Lymphocytes are key constituents of the immune system. In mammals, this system has the remarkable ability to produce antibodies against many millions of different foreign agents that may invade the body. In addition, the immune system "remembers," and a second exposure to a foreign substance produces a more rapid and greater response. There are 2 types of immune defense systems: humoral and cellular. Both react to antigens—usually proteins that are foreign to the body, such as bacteria or foreign tissue. **Humoral immunity** is immunity due to circulating antibodies in the γ globulin fraction of the plasma proteins. It is a major defense against bacterial infections. **Cellular immunity** is responsible for delayed allergic reactions and rejection of transplants of foreign tissue. It constitutes a major defense against infections due to viruses, fungi, and a few bacteria such as the tubercle bacillus. There have been spectacular advances in immunology in recent years, and the field is now large and complex. Only the fundamentals are presented in this book.

Development of the Immune System

During fetal development, lymphocyte precursors come from the bone marrow. Those that populate the thymus (Fig 27–4) become transformed by the environment in this organ into the lymphocytes responsible for cellular immunity **(T lymphocytes).** In birds, the precursors that populate the bursa of Fabricius, a lymphoid structure near the cloaca, become transformed into the lymphocytes responsible for humoral immunity **(B lymphocytes).** There is no bursa in mammals, and it appears that the transformation to B lymphocytes occurs in the fetal liver and possibly the fetal spleen. After residence in the thymus or liver and spleen, many of the T and B lymphocytes migrate to the lymph nodes and bone marrow. T and B lymphocytes are morphologically indistinguishable but can be identified with special techniques. B cells differentiate into **plasma cells** and **memory B cells.** Four different varieties of T cells have been identified: **helper/inducer T cells** (Fig 27–4), **suppressor T cells, cytotoxic T cells** (which are also known as **effector T cells** or **killer cells),** and **memory T cells.** The first 2 types are involved in the regulation of antibody production by B cell derivatives, whereas the cytotoxic T cells destroy transplanted and other foreign cells. Cytotoxic and suppressor T cells generally have on their surface a glycoprotein marker called T_8 that can be detected by monoclonal antibodies, so they are frequently called T_8 cells. Helper/inducer T cells generally have on their surface a glycoprotein marker called T_4 and are called T_4 cells.

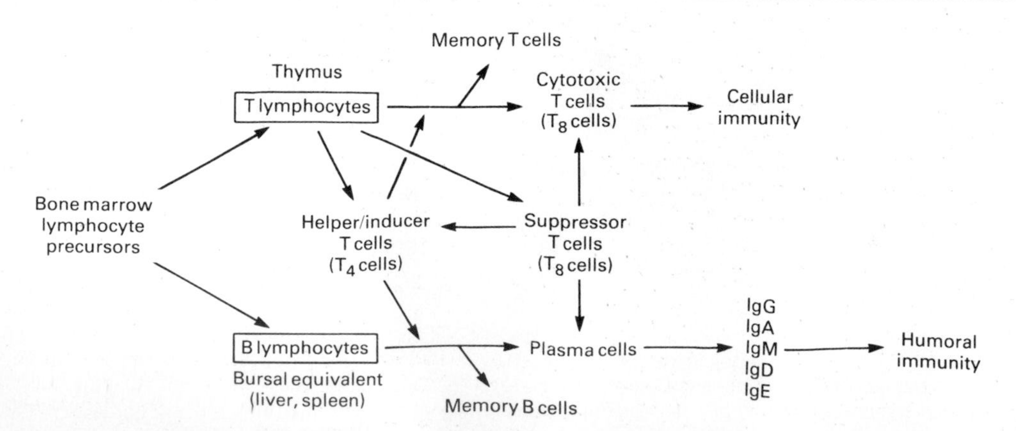

Figure 27–4. Development of the immune system.

There is evidence that maturation of B and T lymphocytes is affected by hormonal factors that are analogous to the colony-stimulating factors which regulate the development of the other blood cells. Maturation of T lymphocytes in the thymus appears to be promoted by humoral factors of thymic origin. The maturation-promoting activity of thymic extracts has been called **thymosin,** but it is now clear that there are a number of different thymic polypeptides with thymosin activity.

Humoral Immunity

When viruses, bacteria, or other foreign proteins and related substances enter the body, they are ingested by macrophages. The macrophages expose part of the ingested antigen plus proteins of the **major histocompatibility complex (MHC)** on their surfaces (Fig 27–5). The macrophages then contact lymphocytes. T_4 cells are activated when they bind simultaneously to the antigen and a class II MHC protein on the surface of the macrophage. The T_4 cells then contact B cells, activating them and causing them to proliferate and transform into **memory B cells** and **plasma cells.** The plasma cells secrete large quantities of antibodies into the general circulation. The antibodies circulate in the globulin fraction of the plasma (see below) and, like antibodies elsewhere, are called **immunoglobulins.** B cells can also bind free antigens in the blood and lymph, but they require contact with helper/inducer T_4 cells to mature and differentiate.

Antigens can also be processed and presented to T_4 cells by several types of cells in the body in addition to macrophages. The other types of **antigen-presenting cells** include the Langerhans cells of the skin and cells in the lymph nodes and spleen.

The number of different antigens recognized by lymphocytes in the body is extremely large. How can so many different substances be recognized? It is now clear that the ability is innate and develops without exposure to the antigen. Stem cells differentiate into more than a million different T and B lymphocytes, each with the ability to respond to a particular antigen. When the antigen first enters the body, it is processed by antigen-binding cells and then binds to the appropriate lymphocytes by the process described above. These cells are stimulated to divide, forming **clones** of cells that respond to this antigen **(clonal selection).**

Immunoglobulins

Five general types of immunoglobulin antibodies are produced by the lymphocyte–plasma cell system (Table 27–4). The basic component of each is a symmetric unit containing 4 polypeptide chains (Fig 27–6). The 2 long chains are called **heavy chains,** whereas the 2 short chains are called **light chains.** There are 2 types of light chains, κ and λ, and 5 types of heavy chains (Table 27–4). The chains are joined by disulfide bridges that permit mobility, and there are intrachain disulfide bridges as well. In addition, the heavy chains are flexible in a region called the hinge. Each heavy chain has a variable (V) segment in which the amino acid sequence is highly variable, a diversity (D) segment in which the amino acid segment is also highly variable, a joining (J) segment in which it is moderately variable, and a constant (C) segment in which the sequence is constant. Each light chain has a V, a J, and a C segment. The antigen binding sites are made up of the V segments, but the shape of the sites is highly variable because of the variability in the amino acid composition of the V, D, and J segments. The sites where binding occurs to effectors such as complement that mediate reactions initiated by antibodies are in the Fc portion of the C segments of the molecule (Fig. 27-6).

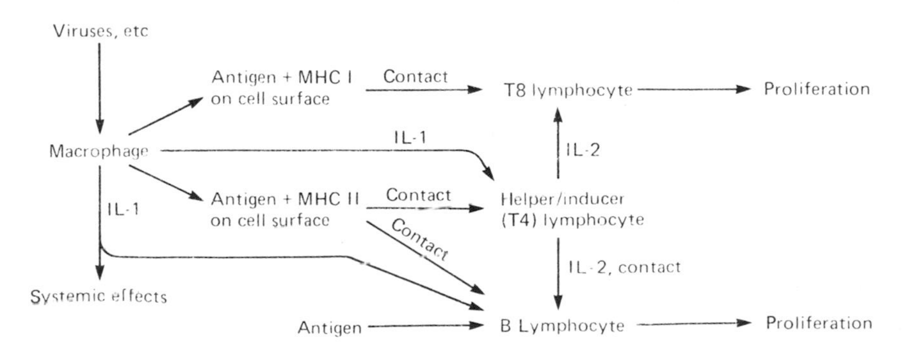

Figure 27–5. Activation of the immune system by viruses and other foreign substances. Macrophages and other antigen-presenting cells ingest the foreign material and express part of it (antigen) plus class I and II proteins of the major histocompatibility complex (MHC) on their surfaces. When antigen and MHC-II proteins are presented to T_4 lymphocytes, the T_4 lymphocytes in turn contact B lymphocytes that have been exposed to antigen and MHC-II proteins, and the B cells proliferate and secrete immunoglobulins. Antigen and MHC-I proteins activate T_8 lymphocytes to proliferate and form cytotoxic and suppressor T cells. Macrophages that have ingested foreign material secrete interleukin-1 (IL-1), which also stimulates T_4 cells and produces systemic effects. Activated T_4 cells produce interleukin-2 (IL-2), which causes proliferation of T_8 cells and B cells.

Table 27–4. Human immunoglobulins. In all instances, the light chains are κ or λ.

Immuno-globulin	Function	Heavy Chain	Additional Chain	Structure	Plasma Concentration (μg/mL)
IgG	Complement fixation	γ_1, γ_2, γ_3, γ_4		Monomer	12,100
IgA	Localized protection in external secretions (tears, intestinal secretions, etc)	α_1, α_2	J, SC	Monomer; dimer with J or SC chain; trimer with J chain	2600
IgM	Complement fixation	μ	J	Pentamer with J chain	930
IgD	Antigen recognition by B cells	δ		Monomer	23
IgE	Reagin activity; releases histamine from basophils and mast cells	ϵ		Monomer	0.5

Two of the classes of immunoglobulins contain additional polypeptide components (Table 27–4). In IgMs, 5 of the basic immunoglobulin units join around a polypeptide called the J chain to form a pentamer. In IgAs, the **secretory immunoglobulins,** the immunoglobulin units form dimers and trimers around a J chain and a polypeptide that comes from epithelial cells, the secretory component (SC).

In the intestine, bacterial and viral antigens are taken up by M cells (see Chapter 25) and passed on to underlying aggregates of lymphoid tissue, where they stimulate lymphoblasts. These lymphoblasts then enter the circulation via the lymphatic ducts, but after "maturing" in the circulation, they move to diffuse lymphoid tissue underlying the intestinal mucosa and epithelia in the lung, breast, genitourinary tract, and female reproductive tract. There they secrete large amounts of IgAs when exposed again to the antigen that initially stimulated them. The epithelial cells produce the SC, which acts as a receptor for and binds the IgA. The resulting secretory immunoglobulin passes through the epithelial cell and is secreted by exocytosis. This system of **secretory immunity** is an important and effective defense mechanism.

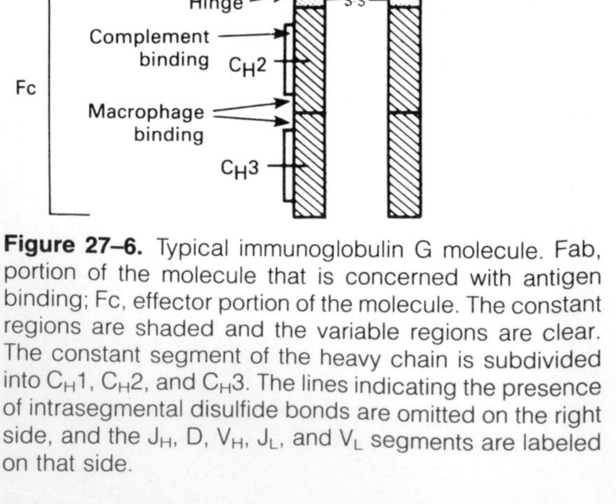

Figure 27–6. Typical immunoglobulin G molecule. Fab, portion of the molecule that is concerned with antigen binding; Fc, effector portion of the molecule. The constant regions are shaded and the variable regions are clear. The constant segment of the heavy chain is subdivided into C_H1, C_H2, and C_H3. The lines indicating the presence of intrasegmental disulfide bonds are omitted on the right side, and the J_H, D, V_H, J_L, and V_L segments are labeled on that side.

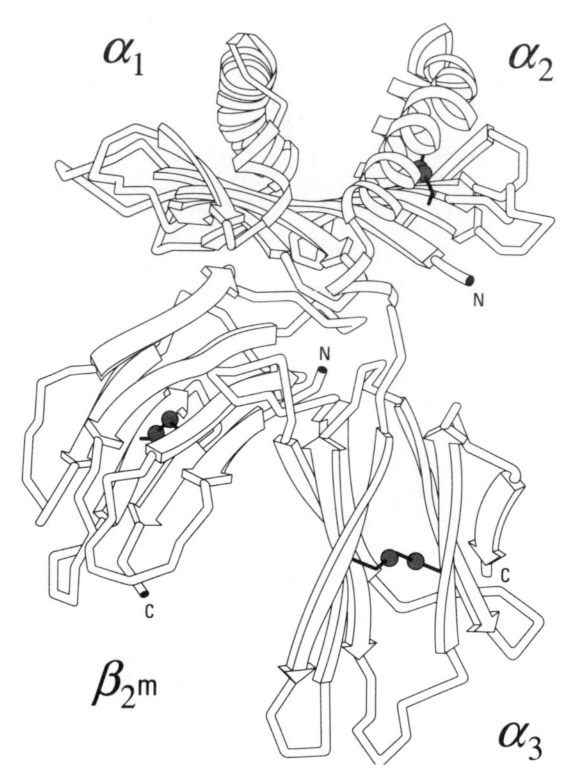

Figure 27–7. Structure of human histocompatibility antigen HLA-A2. The antigen-binding pocket is at the top formed by the α_1 and the α_2 parts of the molecule. The α_3 portion and the associated β_2 microglobulin are close to the membrane, and an extension of the C terminal from α_3 that provides the transmembrane domain and the small cytoplasmic portion of the molecule has been removed prior to x-ray crystallography. (Reproduced, with permission, from Bjorkman PJ et al: Structure of the human histocompatibility antigen HLA-A2. *Nature* 1987;**329**:506.)

Major Histocompatibility Complex

The genes of the MHC, which are located on the short arm of chromosome 6, encode glycoproteins that are located on the surface of all cells and function in antibody processing and distinguishing self from nonself. They are divided into 2 classes on the basis of tissue distribution and function. Class I antigens are composed of a 45-kilodalton heavy chain associated noncovalently with β_2 microglobulin encoded by a gene outside the MHC (Fig 27–7). They are found on all nucleated cells and, as noted above, they must be presented with antigen to activate T_8 cells. The known class I loci in humans are called HLA-A, B, and C. Class II antigens are heterodimers made up of a 29-34K α chain associated noncovalently with a 25-28K β chain. They are found on macrophages, B cells, and activated T cells. As noted above, they must be presented with antigen to activate T_4 cells. Three families of loci code for class II antigens: HLA-DR, DQ, and DP.

T Cell Receptors

It appears that the receptors on T cells must recognize MHC antigens combined with parts of a wide variety of different antigens. T cell receptors have now been shown to be made up of 2 polypeptide subunits associated with a collection of invariant proteins called CD_3. The subunits, which are α and β subunits in most but apparently not all T cells, form heterodimers and contain variable regions. Thus, like the immunoglobulins, their structure is highly variable.

Genetic Basis of Diversity in the Immune System

The genetic mechanism for the production of the immensely large number of different configurations of immunoglobulins in the body store of lymphocytes is a fascinating biologic problem. Diversity is brought about in part by the fact that there are 2 kinds of light chains and 5 kinds of heavy chains. As noted above, the variable portion of the heavy chains consists of a variable (V) segment. In the gene family responsible for this portion, there are several hundred different coding regions for the V segment, about 20 for the D segment, and 4 for the J segment. During B cell development, one V coding region, one D coding region, and one J coding region are selected at random and recombined to form the gene that produces that particular variable portion. There is similar variable recombination in the coding regions responsible for the 2 variable segments (V and J) in the light chain. In addition, the J segments are variable because the gene segments join in an imprecise and variable fashion (junctional site diversity) and nucleotides are sometimes added (junctional insertion diversity). Finally, somatic mutation adds to the variability. It has been calculated that these mechanisms permit the production of more than 18 billion different immunoglobulin molecules.

Similar gene rearrangement and joining mechanisms operate to produce the diversity in T cell receptors. In humans, the α subunit has a V region encoded by one of about 50 different genes and a J region encoded by another 50 different genes. The β subunits have a V region encoded by one of about 50 genes, a D region encoded by 1 or 2 genes, and a J region encoded by one of 13 genes.

Monoclonal Antibodies

It is now possible to obtain large quantities of the immunoglobulin produced by a single plasma cell by fusing the cell with a tumor cell, producing an antibody "factory." In practice, animals are immunized with a particular antigen or cell preparation. They are then sacrificed, and the antibody-producing cells are extracted from their spleens and fused to myeloma cells. Myelomas are B lymphocyte tumors that readily fuse with plasma cells to form antibody-producing **hybridomas** that grow and reproduce very well. The fused cells are separated by standard techniques, and each starts a line, a **clone,** of cells descended from a single cell. The hybridomas have the potential to produce the heavy and light chain of the spleen cell and the heavy and light chain of the myeloma cell, but hybridomas tend to lose chromosomes, and clones can be selected that form only the antibody of the spleen cell (monoclonal antibody). Alternatively, the spleen cells can be fused to mutant myeloma cells that do not produce an immunoglobulin. The value of this technology is very great not only in terms of producing large amounts of pure antibodies for research use but in terms of using the antibodies to treat disease.

The Complement System

When antigens combine with circulating antibodies, the antigens are frequently neutralized and prevented from exerting their effects. In addition, cells are lysed, bacteria are opsonized, leukocytes are attracted to the antigen, and histamine is released from elements in the blood. These effects are mediated in part by a system of plasma enzymes called the **complement system.** The enzymes are identified by the numbers C1 to C9. C1 is made up of 3 subunits, C1q, C1r, and C1s, so that there are actually 11 proteins in the system. C1 binds to immunoglobulins that have bound antigen, and this triggers a sequence of events that activates other components of the system; activation of complement in this fashion is called activation via the **classical pathway.** One consequence of activating the system is insertion of pore-forming molecules in the membranes of antibody-sensitized cells. Ions move through these pores, and the cells become lysed. Another consequence of activation is formation of the fragments C3a and C5a from C3 and C5, respectively. These fragments release histamine from granulocytes, mast cells, and platelets. The histamine dilates blood vessels and increases capillary permeability. C5a and a complex formed by C5b, C6, and C7 are chemotactic and attract leukocytes to the site of the antigen-antibody reaction. C3b is the complement component responsible for opsonization of bacteria, the step that precedes phagocytosis by neutrophils. IgG can also serve as an opsonin.

Complement-mediated cell lysis can also occur in the absence of antibodies via the **alternative,** or **properdin pathway.** The key to this pathway is a circulating protein, **factor I,** that recognizes repetitive sugar structures such as polyglucose or polyfructose in cell membranes. These sequences are found in bacteria and viruses but not in mammalian cells. Interaction of factor I with the surface of invading cells triggers reactions that activate C3 and C5. **Properdin** is another circulating protein that stabilizes the activating enzyme complex.

Cellular Immunity

Cellular immunity is mediated by T_8 cells. These cells are activated when they are presented with antigens and MHC-I proteins on the surfaces of antigen-presenting cells and when also exposed to interleukin-2 (see below), they proliferate and differentiate into memory T cells, cytotoxic T cells, and suppressor T cells (Fig 27–4). The cytotoxic T cells attack and destroy cells that have the antigen which activated them. They probably kill by inserting pore-forming molecules in the membranes of their target cells in the same fashion as the complement system does. However, they may also act by inducing an as yet undefined change within the cells that leads to death. Suppressor T cells, which develop more slowly than cytotoxic T cells, help terminate the immune response by dampening the immune responses of T and B cells. This includes turning off the helper/inducer cells. The memory T cells, along with the memory B cells, persist for long periods and are responsible for the accelerated response to a second exposure to the same antigens.

Recognition of Self

An intriguing question is why T and B cells do not form antibodies against and destroy the cells and organs of the individual in which they develop. One theory holds that during their development, lymphocytes go through a transient period during which exposure to an antigen kills them, and since they are continuously exposed to self antigens, as opposed to foreign antigens, self-destructive clones are killed off. Another theory holds that suppressor T cells keep full-blown responses to self antigens in check. However, the exact mechanism that generally protects self is unknown.

Natural Killer Cells

Another type of cytotoxic lymphocyte is the **natural killer cell (NK cell).** These cells are large lymphocytes that are not of thymic origin. They kill virus-infected cells, transplanted bone marrow cells, and malignant tumor cells without interacting with other lymphocytes or recognizing antigens. Thus, they appear to be part of the body's defense against cancer as well as against infection. Furthermore, they produce and are activated by interferon (see below).

Lymphokines & Cytokines

Lymphocytes, macrophages, and in some instances endothelial cells, neurons, glial cells, and other types of cells secrete many hormonelike chemical messengers that affect other cells. The messengers secreted by lymphocytes are often called lymphokines. However, since they are produced by other cells as well, it seems more appropriate to call them cytokines. This field is growing very rapidly, and most of the cytokines are initially named for other actions, eg, B cell-differentiating factor, tumor necrosis factor. There is a convention that once the amino acid sequence of a factor in humans is known, its name is changed to **interleukin.** Thus, for example, the name of B cell-stimulating factor was recently changed to interleukin-4. However, nomenclature in this field remains somewhat confused and uncertain. In a field that is moving this fast, it is very difficult to construct a comprehensive list of factors, but many of the principal cytokines are listed in Table 27-5.

Interleukin-1 (IL-1) exists in 2 forms (α and β) with little sequence homology but the same receptor and identical activities. Their molecular weights are about 17,000. IL-1 effects are very widespread. They include activation of T_4 and B lymphocytes during activation of the immune system (Fig 27–5). In addition, IL-1 exerts effects on other blood cells, the brain, the liver, and the vascular wall (Table 27–6). Cachectin has overlapping, similar systemic effects. Thus, many of the effects are those that are seen in humans at the start of an infection.

Interleukin-2 (IL-2), a polypeptide with a molecular weight of approximately 15,000, is a growth factor produced by T_4 cells that increases the synthesis and maturation of T_8 and B cells (Fig 27–5).

α and β interferon have antiviral effects, and α interferon fosters the appearance of MCH class I antigens. α and β interferon are quite different from γ interferon, which is a potent activator of macrophages and an inducer of MHC class II antigens. Interferon also increases the activity of NK cells. This action has led to the hope that α interferon may be of value in the treatment of cancer.

Tissue Transplantation

It is the T lymphocyte system that is responsible for the rejection of transplanted tissue. When tissues such as skin and kidneys are transplanted from a donor to a recipient of the same species, the transplants ''take'' and function for a while but then become necrotic and ''rejected'' because the recipient develops an immune response to the transplanted tissue. This is generally true even if the donor and recipient are close relatives, and the only transplants that are never rejected are those from an identical twin.

Table 27–5. Principal cytokines.

Cytokine	Sources	Principal Actions
Interleukin-1α, Interleukin-1β	Macrophages, T lymphocytes, keratinocytes, glia cells, etc	Multiple; see Table 27–6; α and β act on same receptor.
Interleukin-2	T_4 lymphocytes	Proliferation of T_8 and B lymphocytes.
Interleukin-3, G-CSF, M-CSF, GM-CSF	See Table 27–3	See Table 27–3.
Interleukin-4 (B-cell differentiating factor)	T_4 lymphocytes	Differentiation of B lymphocytes.
Interleukin-6 (B-cell stimulatory factor 2, interferon-β2)	Fibroblasts, tumor cells	Increased synthesis and secretion of immunoglobulins by B lymphocytes.
Tumor necrosis factor (cachectin)	Macrophages	Multiple (include fever, lysis of bone, hemorrhagic necrosis in tumors).
Tumor necrosis factor β (lymphotoxin)	T cells	Generally similar to those of cachexin; acts on same receptor as cachexin.
α Interferon	Leukocytes	Antiviral; antiproliferative; induces MHC class I antigens on lymphocytes; increases NK cell activity.
β Interferon	Fibroblasts	Antiviral; antiproliferative.
γ Interferon	T lymphocytes, NK cells	Activates macrophages; induces MHC class II antigens on macrophages
Platelet-derived factor (PDGF)	Platelets, macrophages, endothelial cells	Mitogen for vascular smooth muscle; fosters wound healing.
Platelet-activating factor (PAF)	Neutrophils, monocytes, platelets	Platelet aggregation; neutrophil chemotoxis; bronchoconstriction; increased capillary permeability.

A number of treatments have been developed to overcome the rejection of transplanted organs in humans. The goal of treatment is to stop rejection without leaving the patient vulnerable to massive infections. One approach is to kill T lymphocytes by killing all rapidly dividing cells with drugs such as azathioprine, a purine antimetabolite, but this makes patients susceptible to infections and cancer. Another is to administer glucocorticoids, which inhibit cytotoxic T cell proliferation by inhibiting production of IL-2 by T_4 cells but cause osteoporosis, mental changes, and the other stigmas of Cushing's syndrome (see Chapter 20). A third is to suppress T cells with antibodies against these cells, using **antilymphocyte globulin.** A fourth approach is treatment with **cyclosporin A.** This cyclic peptide extracted from fungi does not affect B lymphocytes but has the unique ability to prevent activated T lymphocytes from dividing. It inhibits IL-2 synthesis by preventing transcription of the IL-2 gene. It may also be of value in the treatment of autoimmune disease, and it kills the parasites that cause malaria and schistosomiasis. However, it also causes renal damage.

Other Clinical Correlates

As knowledge about immune mechanisms has increased, diseases due to defects at various stages in the development of the immune system have now been identified. Failure of the lymphocyte precursors to develop or to migrate to the thymus, liver, and spleen causes absence of cellular and humoral immunity, with marked susceptibility to infections. In individuals with congenital absence of the thymus (DiGeorge's syndrome), cellular immunity is absent but humoral immunity is present. Conversely, in Bruton's X-linked agammaglobulinemia, the B lym-

Table 27–6. Systemic effects of recombinant IL-1.*

Central nervous system effects
- Fever
- Increased slow-wave sleep
- Increased secretion of CRH
- Anorexia

Metabolic effects
- Increased synthesis of hepatic proteins
- Increased sodium excretion
- Decreased plasma zinc and iron
- Decreased cytochrome P-450
- Lactic acidosis

Hematologic effects
- Increased circulating neutrophils
- Decreased circulating lymphocytes
- Inhibition of lipoprotein lipase
- Increased secretion of colony stimulating factors
- Increased nonspecific resistance

Vascular wall effects
- Increased leukocyte adherence
- Increased prostaglandin synthesis
- Increased release of platelet-activating factor
- Increased capillary permeability
- Hypertension

*Modified from Dinarello CA: Biology of interleukin-1. *FASEB J* 1988; **2:**108.

phocytes fail to develop. Patients with this condition have many bacterial infections but are relatively resistant to viral and fungal diseases. Abnormalities of the helper and suppressor T cells may also cause disease. Deficiency of helper T cells could cause decreased production of humoral antibodies, and deficiency of suppressor T cells could cause some forms of **autoimmune disease.** Other forms of autoimmune disease may be due to increased T_4 cell function, and still others may be due to the formation of antibodies against antibodies (anti-idiotope antibodies) to receptors. These antibodies bind not only to the immunoglobulins but also to the receptors and may activate them. One example of a disease in which antibodies stimulate receptors is Graves' disease (see Chapter 18).

Malignant transformation can also occur at various stages of development. Most if not all cases of chronic lymphocytic leukemia are due to uncontrolled proliferation of B lymphocytes, whereas multiple myeloma is due to malignant proliferation of clones of mature plasma cells. Some cases of acute lymphocytic leukemia are T lymphocyte cancers.

Acquired immune deficiency syndrome (AIDS), a disease that is currently attracting a good deal of attention, is unique in that **HIV (human immunodeficiency virus),** the retrovirus that causes it, produces a decrease in the number of circulating T_4 cells. The loss of helper lymphocytes leads in turn to failure of proliferation of T_8 and B cells, with eventual loss of immune function and death from infections due to normally nonpathogenic bacteria or cancer.

RED BLOOD CELLS

The red blood cells **(erythrocytes)** carry hemoglobin in the circulation. They are biconcave disks (Fig 27–8) that are manufactured in the bone marrow. In mammals, they lose their nuclei before entering the circulation. In humans, they survive in the circulation for an average of 120 days. The average normal red blood cell count is 5.4 million/μL in men and 4.8 million/μL in women. Each human red blood cell is about 7.5 μm in diameter and 2 μm thick, and each contains approximately 29 pg of hemoglobin (Table 27–7). There are thus about 3×10^{13} red blood cells and about 900 g of hemoglobin in the circulating blood of an adult man (Fig 27–9).

Erythropoiesis

The formation of red blood cells **(erythropoiesis)** is subject to a feedback control. It is inhibited by a rise in the circulating red cell level to supernormal values and stimulated by anemia. It is also stimulated by hypoxia, and an increase in the number of circulating red cells is a prominent feature of acclimatization to altitude (see Chapter 37). Erythropoiesis is controlled by a circulating glycoprotein hormone called **erythropoietin,** which is secreted primarily by the kidneys (see Chapter 24). This hormone promotes the differentiation of committed stem cells (erythropoietin-sensitive stem cells) into proerythroblasts (see above).

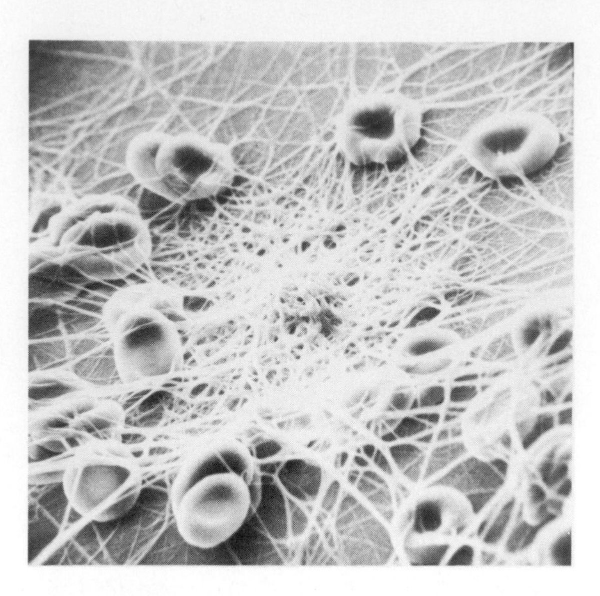

Figure 27–8. Human red blood cells and fibrin fibrils. Blood was placed on a polyvinyl chloride surface, then fixed and photographed, using a scanning electron microscope. Reduced from × 2590. (Courtesy of NF Rodman.)

Red Cell Fragility

Red blood cells, like other cells, shrink in solutions with an osmotic pressure greater than that of normal plasma. In solutions with a lower osmotic pressure, they swell, becoming spherical rather than disk-shaped, and eventually lose their hemoglobin **(hemolysis).** The hemoglobin of hemolyzed red cells dissolves in the plasma, coloring it red. A 0.9% sodium chloride solution is isotonic with plasma. When **osmotic fragility** is normal, red cells begin to hemolyze when suspended in 0.48% saline, and hemolysis is complete in 0.33% saline. In **hereditary spherocytosis** (congenital hemolytic icterus), the cells are spherocytic in normal plasma and hemolyze more readily than normal cells in hypotonic sodium chloride solutions **(abnormal red cell fragility).** Spherocytosis is apparently produced by abnormalities of the protein lattice that maintains cell shape and flexibility.

Red cells can also be lysed by drugs and infections. The susceptibility of red cells to hemolysis by these agents is increased by deficiency of the enzyme glucose 6-phosphate dehydrogenase (G6PD), which catalyzes the initial step in the oxidation of glucose via the hexosemonophosphate pathway (see Chapter 17). This pathway generates NADPH, which is needed in some way for the maintenance of normal red cell fragility. Congenital deficiency of G6PD activity in the red cells as a result of the presence of enzyme variants is common; indeed, G6PD defi-

Table 27–7. Characteristics of human red cells.* Cells with MCVs > 95 are called macrocytes; cells with MCVs < 80 are called microcytes; cells with MCHs < 25 are called hypochromic.

		Male	Female
Hematocrit (Hct) (%)		47	42
Red blood cells (RBC) (millions/μL)		5.4	4.8
Hemoglobin (Hb) (g/dL)		16	14
Mean corpuscular volume (MCV) (fL)	$= \frac{\text{Hct} \times 10}{\text{RBC}\ (10^6/\mu\text{L})}$	87	87
Mean corpuscular hemoglobin (MCH) (pg)	$= \frac{\text{Hb} \times 10}{\text{RBC}\ (10^6/\mu\text{L})}$	29	29
Mean corpuscular hemoglobin concentration (MCHC) (g/dL)	$= \frac{\text{Hb} \times 100}{\text{Hct}}$	34	34
Mean cell diameter (MCD) (μm)	= Mean diameter of 500 cells in smear	7.5	7.5

*Values are from Wintrobe M: *Clinical Hematology*, 6th ed. Lea & Febiger, 1967.

ciency is the commonest known genetically determined human enzyme abnormality. More than 80 genetic variants of G6PD have been described; 40 of these do not cause appreciable decreases in enzyme activity, but the others produce increased activity, increased sensitivity to hemolytic agents, and hemolytic anemia. Severe G6PD deficiency also inhibits the killing of bacteria by granulocytes and predisposes to severe infections (see above).

Role of the Spleen

The spleen is an important blood filter that removes spherocytes and other abnormal red cells. It also plays a significant role in the immune system. The circulation of the spleen has 2 components: a fast component, mainly nutritive in function, in which blood stays within blood vessels; and a slow component in which blood leaves arterioles and percolates through large numbers of phagocytes and lymphocytes before entering the splenic sinuses and passing back into the general circulation. The phagocytes remove bacteria and initiate immune responses. Abnormal red cells are removed if they are not as flexible as normal red cells and consequently are unable to squeeze through the slits between the endothelial cells that line the splenic sinuses. In the absence of the spleen, bacterial infections are more common and more severe. In addition, malaria results in a higher mortality rate because the deformed red cells that contain the malaria parasite are not removed.

CIRCULATION
3×10^{13} Red blood cells
900 g Hemoglobin
1×10^{10} RBC
0.3 g Hemoglobin
per hour
1×10^{10} RBC
0.3 g Hemoglobin
per hour
BONE MARROW
IRON
RETICULO-ENDOTHELIAL SYSTEM
DIET
AMINO ACIDS
Bile pigments in stool, urine
Small amount of iron

Figure 27–9. Red cell formation and destruction.

Hemoglobin

The red, oxygen-carrying pigment in the red blood cells of vertebrates is **hemoglobin,** a protein with a molecular weight of 64,450. Hemoglobin is a globular molecule made up of 4 subunits (Fig 27–10). Each subunit contains a **heme** moiety conjugated to a

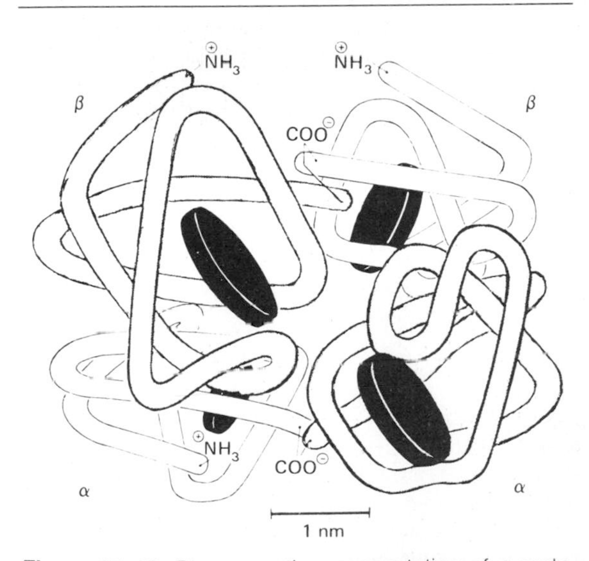

Figure 27–10. Diagrammatic representation of a molecule of hemoglobin A, showing the 4 subunits. There are two α and two β polypeptide chains, each containing a heme moiety. These moieties are represented by the disks. (Reproduced, with permission, from Harper HA et al: *Physiologische Chemie.* Springer-Verlag, 1975.)

polypeptide. Heme is an iron-containing porphyrin derivative (Fig 27–11). The polypeptides are referred to collectively as the **globin** portion of the hemoglobin molecule. There are 2 pairs of polypeptides in each hemoglobin molecule, 2 of the subunits containing one type of polypeptide and 2 containing another. In normal adult human hemoglobin **(hemoglobin A),** the 2 types of polypeptide are called the α chains, each of which contains 141 amino acid residues, and the β chains, each of which contains 146 amino acid residues. Thus, hemoglobin A is designated $\alpha_2\beta_2$. Not all the hemoglobin in the blood of normal adults is hemoglobin A. About 2.5% of the hemoglobin is hemoglobin A_2, in which β chains are replaced by δ chains ($\alpha_2\delta_2$). The δ chains also contain 146 amino acid residues, but 10 individual residues differ from those in the β chains.

There are small amounts of 3 hemoglobin A derivatives closely associated with hemoglobin A that probably represent glycosylated hemoglobins. One of these, **hemoglobin A_{I_c} (HbA_{I_c}),** has a glucose attached to the terminal valine in each β chain and is of special interest because the quantity in the blood increases in poorly controlled diabetes mellitus (see Chapter 19).

Reactions of Hemoglobin

Hemoglobin binds O_2 to form **oxyhemoglobin,** O_2 attaching to the Fe^{2+} in the heme. The affinity of hemoglobin for O_2 is affected by pH, temperature, and the concentration in the red cells of 2,3-diphosphoglycerate (2,3-DPG). 2,3-DPG and H^+ compete with O_2 for binding to deoxygenated hemoglobin, decreasing the affinity of hemoglobin for O_2 by shifting the positions of the 4 peptide chains (quaternary structure). The details of the oxygenation and deoxygenation of hemoglobin and the physiologic role of these reactions in O_2 transport are discussed in Chapter 35.

When blood is exposed to various drugs and other oxidizing agents in vitro or in vivo, the ferrous iron (Fe^{2+}) in the molecule is converted to ferric iron (Fe^{3+}, forming **methemoglobin.** Methemoglobin is dark-colored, and when it is present in large quantities in the circulation, it causes a dusky discoloration of the skin resembling cyanosis (see Chapter 37). Some oxidation of hemoglobin to methemoglobin occurs normally, but an enzyme system in the red cells, the NADH-methemoglobin reductase system, converts methemoglobin back to hemoglobin. Congenital absence of this system is one cause of hereditary methemoglobinemia.

Carbon monoxide reacts with hemoglobin to form **carbonmonoxyhemoglobin (carboxyhemoglobin).** The affinity of hemoglobin for O_2 is much lower than its affinity for carbon monoxide, which consequently displaces O_2 on hemoglobin, reducing the oxygen-carrying capacity of blood (see Chapter 37).

Heme is also part of the structure of **myoglobin,** an oxygen-binding pigment found in red (slow) muscles (see Chapter 3) and in the respiratory enzyme **cytochrome c** (see Chapter 17). Porphyrins other than that found in heme play a role in the pathogenesis of a number of metabolic diseases (congenital and acquired porphyria, etc.).

Hemoglobin in the Fetus

The blood of the human fetus normally contains **fetal hemoglobin (hemoglobin F).** Its structure is similar to that of hemoglobin A except that the β chains are replaced by γ chains; ie, hemoglobin F is $\alpha_2\gamma_2$. The γ chains also contain 146 amino acid residues but have 37 that differ from those in the β chain. Fetal hemoglobin is normally replaced by

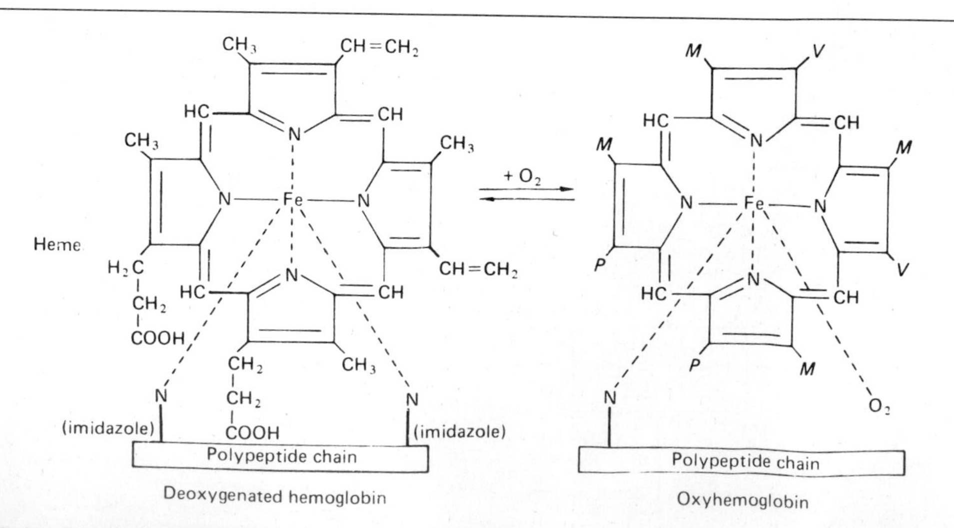

Figure 27–11. Reaction of heme with O_2. The abbreviations M, V, and P stand for the groups shown on the molecule on the left.

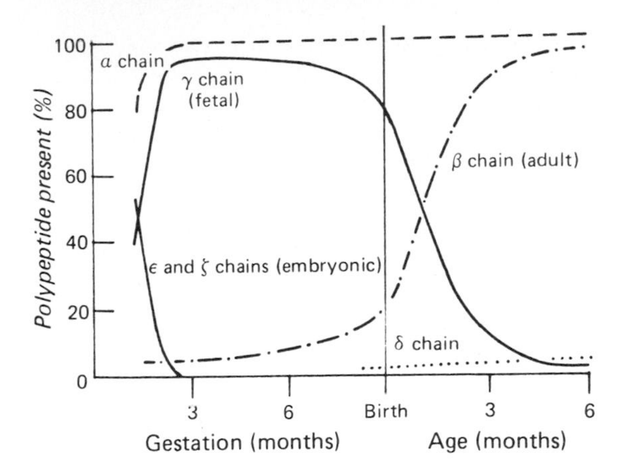

Figure 27–12. Development of human hemoglobin chains.

adult hemoglobin soon after birth (Fig 27–12). In certain individuals, it fails to disappear and persists throughout life. In the body, its O_2 content at a given P_{O_2} is greater than that of adult hemoglobin because it binds 2,3-DPG less avidly. This facilitates movement of O_2 from the maternal to the fetal circulation (see Chapter 32). In young embryos there are, in addition, ζ and ε chains, forming Gower I hemoglobin ($\zeta_2\epsilon_2$) and Gower II hemoglobin ($\zeta_2\gamma_2$).

Abnormalities of Hemoglobin Production

The amino acid sequences in the polypeptide chains of hemoglobin are determined by globin genes.

There are 2 major types of inherited disorders of hemoglobin in humans: the **hemoglobinopathies,** in which abnormal polypeptide chains are produced, and the **thalassemias** and related disorders, in which the chains are normal in structure but produced in decreased amounts because globin genes have been deleted or rendered nonfunctional. The α and β thalassemias are defined by decreased or absent α and β polypeptides, respectively. Normal individuals have 4 globin genes. If the genes for both α and β chains are missing, only Barts hemoglobin (4 γ chains) is produced, and there is severe anemia, hydrops (edema) and death in utero.

Mutant genes that cause the production of abnormal hemoglobins are widespread, and many abnormal hemoglobins have been described in humans. They are usually identified by letter—hemoglobin C, E, I, J, S, etc. In most instances, the abnormal hemoglobins differ from normal hemoglobin A in the structure of the polypeptide chains. For example, in hemoglobin S, the α chains are normal but the β chains are abnormal, because among the 146 amino acid residues in each β polypeptide chain, one glutamic acid residue has been replaced by a valine residue (Table 27–8).

When an abnormal gene inherited from one parent dictates formation of an abnormal hemoglobin—ie, when the individual is heterozygous—half the circulating hemoglobin is abnormal and half is normal. When identical abnormal genes are inherited from both parents, the individual is homozygous and all of the hemoglobin is abnormal. It is theoretically possible to inherit 2 different abnormal hemoglobins, one from the father and one from the mother. Studies of the inheritance and geographic distribution of abnormal hemoglobins have made it possible in some cases to decide where the mutant gene originated and approximately how long ago the mutation occurred. In general, harmful mutations tend to die out, but mutant genes that confer traits with survival value persist and spread in the population.

Many of the abnormal hemoglobins are harmless. However, some have abnormal O_2 equilibriums. Others cause anemia. For example, hemoglobin S is very insoluble at low O_2 tensions, and this causes the red cells to become sickle-shaped. These abnormal sickle cells hemolyze, producing the severe anemia known as **sickle cell anemia.** Heterozygous individuals have the **sickle cell trait** and rarely have severe symptoms, but homozygous ones develop the full-blown disease. The sickle cell gene is an example of a gene that has persisted and spread in the population. It originated in the black population in Africa, and it confers resistance to one type of malaria. This is an important benefit in Africa, and in some parts of

Table 27–8. Partial amino acid composition of normal human β chain, and some hemoglobins with abnormal β chains. Other hemoglobins have abnormal α chains. Abnormal hemoglobins that are very similar electrophoretically but differ slightly in composition are indicated by the same letter and a subscript indicating the geographic location where they were first discovered; hence, $M_{Saskatoon}$ and $M_{Milwaukee}$.

	Positions on β Polypeptide Chain of Hemoglobin									
Hemoglobin	1	2	3	6	7	26	63	67	121	146
A (normal)	Val-	His-	Leu	Glu-	Glu	Glu	His	Val	Glu	His
S (sickle cell)				Val						
C				Lys						
$G_{San\ Jose}$					Gly					
E						Lys				
$M_{Saskatoon}$							Tyr			
$M_{Milwaukee}$								Glu		
O_{Arabia}									Lys	

Africa, 40% of the population have the sickle cell trait. In the American black population, its incidence is about 10%.

Synthesis of Hemoglobin

The average normal hemoglobin content of blood is 16 g/dL in men and 14 g/dL in women, all of it in red cells. In the body of a 70-kg man, there is about 900 g of hemoglobin, and 0.3 g of hemoglobin is destroyed and 0.3 g synthesized every hour (Fig 27–9). The heme portion of the hemoglobin molecule is synthesized from glycine and syccinyl-CoA.

Catabolism of Hemoglobin

When old red blood cells are destroyed in the reticuloendothelial system, the globin portion of the hemoglobin molecule is split off, and the heme is converted to **biliverdin.** In humans, most of the biliverdin is converted to **bilirubin** (Fig 27–13) and excreted in the bile (see Chapter 26). The iron from the heme is reused for hemoglobin synthesis. Iron is essential for hemoglobin synthesis; if blood is lost from the body and the iron deficiency is not corrected, **iron deficiency anemia** results. The metabolism of iron is discussed in Chapter 25.

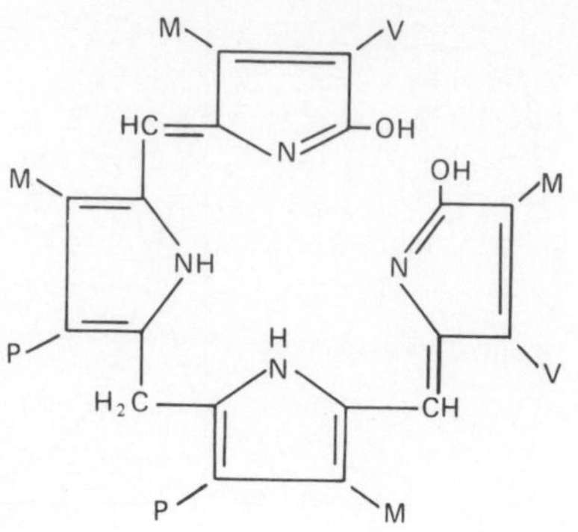

Figure 27–13. Bilirubin. The abbreviations M, V, and P stand for the groups shown on the molecule on the left in Fig 27–11.

BLOOD TYPES

The membranes of human red cells contain a variety of antigens called **agglutinogens.** The most important and best known of these are the A and B agglutinogens, and individuals are divided into 4 major **blood types,** types A, B, AB, and O, on the basis of the agglutinogens present in their red cells. There are A and B antigens in many tissues other than blood. They have been found in salivary glands, saliva, pancreas, kidney, liver, lungs, testes, semen, and amniotic fluid. The A and B agglutinogens are glycoproteins that differ in composition by only one sugar residue. Type A individuals have an enzyme (glycosyltransferase) that puts acetylgalactosamine on the glycoprotein skeleton, whereas type B individuals have an enzyme that puts galactose on the skeleton. Individuals with type AB blood have both enzymes.

Antibodies against agglutinogens are called **agglutinins.** They may occur naturally (ie, be inherited), or they may be produced by exposure to the red cells of another individual. This exposure may occur via a transfusion or during pregnancy, when fetal red cells cross the placenta and enter the circulation of the mother. Agglutinins against A and B agglutinogens are inherited, whereas antibodies against the Rh group (see below) and other agglutinogens are produced by exposure to foreign red cells. Individuals with type A blood (ie, those who have agglutinogen A on their red cells) always have an appreciable titer of an antibody against agglutinogen B called the β or anti-B agglutinin. When their plasma is mixed with type B cells, the agglutinins and the B cell agglutinogens react, causing the type B cells to become clumped (agglutinated) and subsequently hemolyzed (Fig 27–14). Similarly, individuals with type B blood have a circulating titer of an α (anti-A) agglutinin. Individuals with type O blood have circulating anti-A and anti-B agglutinins, and those with type AB blood have no circulating agglutinins. **Blood typing** is performed by mixing an individual's red cells with appropriate antisera on a slide and seeing if agglutination occurs.

Some individuals with A agglutinogen have an additional agglutinogen called A_1. Thus, the A group is subdivided into types A_1 (those with both A agglutinogens) and A_2 (those with only the A agglutinogen). Therefore, there are 6 ABO groups instead of 4: B, O, A_1, A_2, A_1B, and A_2B. Individuals who lack the A_1 agglutinogen normally have little if any circulating anti-A agglutinin, but appreciable titers are sometimes found.

Transfusion Reactions

Dangerous **hemolytic transfusion reactions** occur when blood is transfused into an individual with an incompatible blood type, ie, an individual who has agglutinins against the red cells in the transfusion. The plasma in the transfusion is usually so diluted in the recipient that it rarely causes agglutination even when the titer of agglutinins against the recipient's cells is high. However, when the recipient's plasma

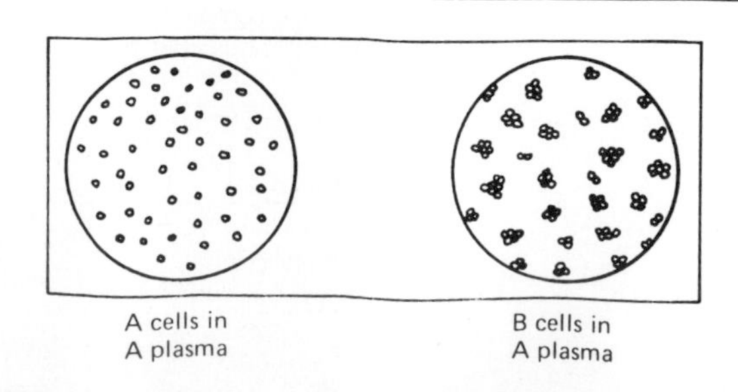

Figure 27–14. Red cell agglutination in incompatible plasma.

has agglutinins against the donor's red cells, the cells agglutinate and hemolyze. Free hemoglobin is liberated into the plasma. The severity of the resulting transfusion reaction may vary from an asymptomatic minor rise in the plasma bilirubin level to severe jaundice and renal tubular damage (caused in some way by the products liberated from hemolyzed cells), with anuria and death.

Incompatibilities in the ABO blood group system are summarized in Table 27–9. Persons with type AB blood are "universal recipients" because they have no circulating agglutinins and can be given blood of any type without developing a transfusion reaction due to ABO incompatibility. Type O individuals are "universal donors" because there are no regular anti-O agglutinins, and type O blood can be given to anyone without producing a transfusion reaction due to ABO incompatibility. This does not mean, however, that blood should ever be transfused without being cross-matched except in the most extreme emergencies, since the possibility of reactions or sensitization due to incompatibilities other than ABO incompatibilities always exists. There are also individuals with appreciable titers of the anti-A_1 agglutinin, and transfusion reactions due to this agglutinin can be prevented if blood is cross-matched. In cross-matching, donor red cells are mixed with recipient plasma on a slide and checked for agglutination (Fig 27–14). It is advisable to check the action of the donor's plasma on the recipient cells even though, as noted above, this is rarely a source of trouble.

A procedure that has recently become popular is to withdraw the patient's own blood in advance of elective surgery and then infuse this blood back if a transfusion is needed during the surgery. With iron treatment, 1000–1500 mL can be withdrawn over a 3-week period. The popularity of banking one's own blood is due primarily to fear of transmission of AIDS by heterologous transfusions, but of course another advantage is elimination of the risk of transfusion reactions.

Inheritance of A & B Antigens

The A_1, A_2, and B antigens are inherited as mendelian allelomorphs, A_1, A_2, and B being dominants. For example, an individual with type B blood may have inherited a B antigen from each parent or a B antigen from one parent and an O from the other; thus, an individual whose **phenotype** is B may have the **genotype** BB **(homozygous)** or the **genotype** BO **(heterozygous).**

When the blood types of the parents are known, the possible genotypes of their children can be stated. When both parents are type B, they could have children with genotype BB (B antigen from both parents), BO (B antigen from one parent, O from the other heterozygous parent), or OO (O antigen from both parents, both being heterozygous). When the blood types of a mother and her child are known, typing can prove that a man could not be the father, although it cannot prove that he is the father. The predictive value is increased if the blood typing of the parties concerned includes identification of antigens other than the ABO agglutinogens. With the addition of HLA typing and other molecular biological techniques, the exclusion rate rises to more than 92%.

Other Agglutinogens

In addition to the 6 antigens of the ABO system in human red cells, there are other agglutinogen systems containing many individual antigens in red cells (Table 27–10). There are over 500 billion possible known blood group phenotypes, and because undiscovered antigens undoubtedly exist, it has been calculated that the number of phenotypes is actually in the trillions.

The number of blood groups in animals is as large as it is in humans. An interesting question is why this degree of polymorphism developed and has persisted through evolution. Certain diseases are more common in individuals with one blood type or another, but the differences are not great. It seems likely that the actual function of the blood group antigens is cell recognition, but the exact significance of a recognition code of this complexity is unknown.

The Rh Group

Aside from the antigens of the ABO system, those of the Rh system are of the greatest clinical importance. The "Rh factor," named for the rhesus monkey because it was first studied using the blood of this animal, is a system composed of many antigens (Table 27–9). D is by far the most antigenic, and the

Table 27–9. Summary of ABO blood system.*

Blood Type	Agglutinins in Plasma	Frequency in USA (%)	Plasma Agglutinates Red Cells of Type
O	Anti-A, anti-B	45	A_1, A_1B, B, A_2B
A_1	Anti-B	41	B, A_1B, A_2B
A_2			B, A_1B, A_2B
B	Anti-A	10	A_1, A_1B, A_2, A_2B
A_1B	None	4	None
A_2B			

*Some type A_2 and A_2B individuals may also have sufficient anti-A_1 agglutinin to agglutinate A_1 and A_1B red cells.

Table 27–10. Human blood group antigens.*

System	Antigens Detected by: Positive Reactions With Specific Antibody	Antigens Detected by: Positive Reaction With One Antibody, Negative With Another†
A_1A_2BO	A_1, B, ‡H	A_2, A_3, A_x, and other A and B variants
MNSs	M, N, S, s, U, M^g, M_1, M′, Tm, Sj, Hu, He, Mi^a, Vw (Gr), Mur, Hil, Hut, M^v, Vr, Ri^a, St^a, Mt^a, Cl^a, Ny^a, Sul, Far	M_2, N_2, M^c, M^a, N^a, M^r, M^z, S_2
P	P_1, P^k, ‡Luke	P_2
Rh	D, C, c, C^w, C^x, E, e, e^s (VS), E^w, G, ce(f), ce^s(V), Ce, CE, cE, D^w, E^T, Go^a, hr^s, hr^H, hr^B, $\bar{\bar{R}}^N$, Rh33, Rh35, Be^a, ‡LW	D^u, C^u, E^u, and many other variant forms of D, C, and e
Lutheran	Lu^a, Lu^b, Lu^aLu^b (Lu3), Lu6, Lu9, §Lu4, Lu5, Lu7, Lu8, Lu10–17	
Kell	K, k, Kp^a, Kp^b, Ku, Js^a, Js^b, Ul^a, Wk^a, K11, §KL, K12–16	
Lewis	Le^a, Le^b, Le^c, Le^d, Le^x	
Duffy	Fy^a, Fy^b, Fy3, Fy4	
Kidd	Jk^a, Jk^b, Jk^aJk^b (Jk3)	
Diego	Di^a, Di^b	
Yt	Yt^a, Yt^b	
Auberger	Au^a	
Dombrock	Do^a, Do^b	
Colton	Co^a, Co^b, Co^aCo^b	
Sid	Sd^a	
Scianna	Sc1, Sc2 (Bu^a)	
Very frequent antigens	Vel, Ge, Lan, Gy^a, At^a, En^a, Wr^b, Jr^a, Kn^a; El, Dp, Gn^a, Jo^a, and many unpublished examples	
Very infrequent antigens	An^a, By, Bi, Bp^a, Bx^a, Chr^a, Evans, Good, Gf, Heibel, Hey, Hov, Ht^a, Je^a, Jn^a, Levay, Ls^a, Mo^a, Or, Pt^a, Rl^a, Rd, Re^a, Sw^a, To^a, Tr^a, Ts, Wb, Wr^a, Wu, Zd, and many unpublished examples	
Other antigens	I, i, Bg (HL-A), Chido, Cs^a, Yk^a	
Xg	Xg^a	

*Reproduced, with permission, from Race RR, Sanger R: *Blood Groups in Man*, 6th ed. Blackwell, 1975.
†Recognizable only in favorable genotypes.
‡A genetically independent part of the system.
§Place in system not yet genetically clear.

term "Rh-positive" as it is generally used means that the individual has agglutinogen D. The "Rh-negative" individual has no D antigen and forms the anti-D agglutinin when injected with D-positive cells. The Rh typing serum used in routine blood typing is anti-D serum. Eighty-five percent of Caucasians are D-positive and 15% are D-negative; over 99% of Orientals are D-positive. D-negative individuals who have received a transfusion of D-positive blood (even years previously) can have appreciable anti-D titers and thus may develop transfusion reactions when transfused again with D-positive blood.

Hemolytic Disease of the Newborn

Another complication due to "Rh incompatibility" arises when an Rh-negative mother carries an Rh-positive fetus. Small amounts of fetal blood leak into the maternal circulation at the time of delivery, and some mothers develop significant titers of anti-Rh agglutinins during the postpartum period. During the next pregnancy, the mother's agglutinins cross the placenta to the fetus. In addition, there are some cases of fetal-maternal hemorrhage during pregnancy, and sensitization can occur during pregnancy. In any case, when anti-Rh agglutinins cross the placenta to an Rh-positive fetus, they can cause hemolysis and various forms of **hemolytic disease of the newborn (erythroblastosis fetalis).** If hemolysis in the fetus is severe, the infant may die in utero or may develop anemia, severe jaundice, and edema **(hydrops fetalis). Kernicterus,** a neurologic syndrome in which the bile pigments are deposited in the basal ganglia, may also develop, especially if birth is complicated by a period of hypoxia. Bile pigments do not enter the brain in the adult, but in the fetus and newborn infant the blood-brain barrier is not developed (see Chapter 32).

About 50% of Rh-negative individuals are sensitized (develop an anti-Rh titer) by transfusion of Rh-positive blood. Since sensitization of Rh-negative mothers by carrying an Rh-positive fetus generally occurs at birth, the first child is usually normal. However, hemolytic disease occurs in about 17% of the Rh-positive fetuses born to Rh-negative mothers who have previously been pregnant one or more times with Rh-positive fetuses. Fortunately, it is usually possible to prevent sensitization from occurring the first time by administering a single dose of anti-Rh antibodies in the form of Rh immune globulin **(Rh_oD) immune globulin)** during the postpartum period. Such passive immunization does not harm the mother and has been demonstrated to prevent active antibody formation by the mother. In obstetric clinics, the institution of such treatment on a routine basis to unsensitized Rh-negative women who have delivered an Rh-positive baby has reduced the overall incidence of hemolytic disease by more than 90%. Treatment with a small dose during pregnancy will also prevent sensitization due to fetal-maternal hemorrhage before delivery.

PLASMA

The fluid portion of the blood, the **plasma,** is a remarkable solution containing an immense number of ions, inorganic molecules, and organic molecules that are in transit to various parts of the body or aid in the transport of other substances. The normal plasma volume is about 5% of body weight, or roughly 3500 mL in a 70-kg man. Plasma clots on standing, remaining fluid only if an anticoagulant is added. If whole blood is allowed to clot and the clot is removed, the remaining fluid is called **serum.** Serum has essentially the same composition as plasma except that its fibrinogen and clotting factors II, V, and VIII (Table 27–11) have been removed, and it has a higher serotin content because of the breakdown of platelets during clotting. The normal plasma levels of various substances are discussed in the chapters on the systems with which the substances are concerned and summarized on the inside back cover.

Plasma Proteins

The plasma proteins consist of **albumin, globulin,** and **fibrinogen** fractions. The globulin fraction is subdivided into numerous components. One classification divides it into α_1, α_2, β_1, β_2, and γ globulins and fibrinogen. Protein fractions can be separated and characterized by their relative speed of sedimentation in the ultracentrifuge or their rate of migration in an electrical field (electrophoresis). The molecular weights and configurations of several of the plasma proteins are shown in Fig 27–15.

Table 27–11. System for naming blood clotting factors. Factor VI is not a separate entity and has been dropped.

Factor	
I	Fibrinogen
II	Prothrombin
III	Thromboplastin
IV	Calcium
V	Proaccelerin, labile factor, accelerator globulin
VII	Proconvertin, SPCA, stable factor
VIII	Antihemophilic factor (AHF), antihemophilic factor A, antihemophilic globulin (AHG)
IX	Plasma thromboplastic component (PTC), Christmas factor, antihemophilic factor B
X	Stuart-Prower factor
XI	Plasma thromboplastin antecedent (PTA), antihemophilic factor C
XII	Hageman factor, glass factor
XIII	Fibrin-stabilizing factor, Laki-Lorand factor
HMW-K	High-molecular-weight kininogen, Fitzgerald factor
Pre-K	Prekallikrein, Fletcher factor
Ka	Kallikrein
PL	Platelet phospholipid

The capillary walls are relatively impermeable to the proteins in plasma, and the proteins therefore exert an osmotic force of about 25 mm Hg across the capillary wall **(oncotic pressure;** see Chapter 1) that tends to pull water into the blood. The plasma proteins are also responsible for 15% of the buffering capacity of the blood (see Chapter 35), because of the weak ionization of their substituent –COOH and $-NH_2$ groups. At the normal plasma pH of 7.40, the proteins are mostly in the anionic form (see Chapter 1) and constitute a significant part of the anionic complement of plasma. Plasma proteins such as antibodies and the proteins concerned with blood clotting have specific functions. Some of the proteins function in the transport of thyroid, adrenocortical, and gonadal hormones. Binding keeps these hormones from being rapidly filtered through the glomeruli and provides a stable reservoir of hormones on which the tissues can draw. In addition, albumin serves as a carrier for metals, ions, fatty acids, amino acids, bilirubin, enzymes, and drugs.

Origin of Plasma Proteins

Circulating antibodies in the γ globulin fraction of the plasma proteins are manufactured in the plasma cells (see above). The albumin fraction and the proteins concerned with blood clotting (fibrinogen, prothrombin) are manufactured in the liver.

Data on the turnover of albumin provide an indication of the role played by synthesis in the maintenance of normal albumin levels. In normal adult humans, the plasma albumin level is 3.5–5.0 g/dL, and the total exchangeable albumin pool is 4.0–5.0 g/kg body weight; 38–45% of this albumin is

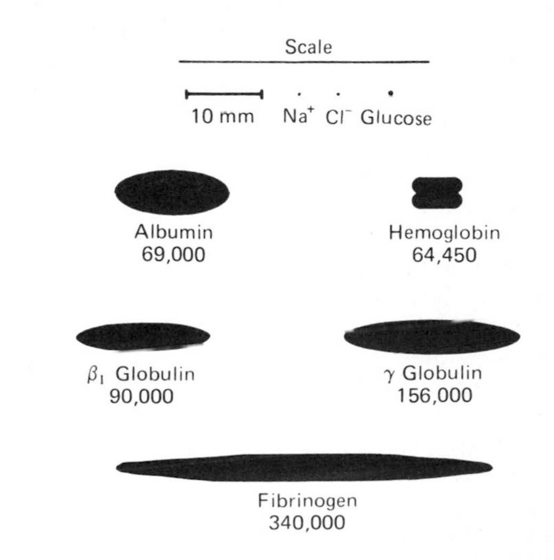

Figure 27–15. Relative dimensions and molecular weights of some of the protein molecules in the blood. (Modified and reproduced, with permission, from Murray RK et al: *Harper's Biochemistry,* 21st ed. Appleton & Lange, 1988.)

intravascular, and much of the rest of it is in the skin. Six to 10% of the exchangeable pool is degraded per day, and the degraded albumin is replaced by hepatic synthesis of 200–400 mg/kg/d. The albumin is probably transported to the extravascular areas by vesicular transport across the walls of the capillaries (see Chapter 1). Albumin synthesis is carefully regulated. It is decreased during fasting and increased in conditions such as nephrosis in which there is excessive albumin loss.

Hypoproteinemia

Plasma protein levels are maintained during starvation until body protein stores are markedly depleted; but in prolonged starvation and in the malabsorption syndrome due to intestinal diseases such as sprue, plasma protein levels are low **(hypoproteinemia).** They are also low in liver disease, because hepatic protein synthesis is depressed; and in nephrosis, because large amounts of albumin are lost in the urine. Because of the decrease in the plasma oncotic pressure, edema tends to develop (see Chapter 30). Rarely, there may be congenital absence of one or another plasma protein fraction. Examples of congenital protein deficiencies are **agammaglobulinemia,** a condition in which there is markedly lowered resistance to infections due to the absence of circulating antibodies; and the congenital form of **afibrinogenemia,** characterized by defective blood clotting.

PLATELETS

The platelets are small, granulated bodies 2–4 μm in diameter (Fig 27–2). There are about 300,000 per microliter of circulating blood, and they normally have a half-life of about 4 days. The **megakaryocytes,** giant cells in the bone marrow, form platelets by pinching off bits of cytoplasm and extruding them into the circulation. Platelet production is regulated by the colony-stimulating factors that control the production of megakaryocytes (Table 27–3).

Platelets have a ring of microtubules around their periphery and contain actin and myosin. They also contain glycogen, lysosomes, and 2 types of granules: dense granules, which contain the nonprotein substances that are secreted in response to platelet activation, including serotonin and ADP and other adenine nucleotides; and α-granules, which contain secreted proteins other than the hydrolases in lysosomes. These proteins include clotting factors and **platelet-derived growth factor** (PDGF). PDGF is also produced by macrophages and endothelial cells. It is a dimer made up of A and B subunit polypeptides. Homodimers (AA and BB) as well as the heterdimer (AB) are produced. PDGF stimulates wound healing and is a potent mitogen for vascular smooth muscle.

The platelets have an extensively invaginated membrane with an intricate canalicular system in contact with the ECF. When suitably activated, they collect at the site of injury **(platelet aggregation),** change shape, put out pseudopodia, and discharge the contents of their granules via the canaliculi **(platelet release).**

Platelet activation is caused by collagen fibers, ADP, and thrombin. Aggregation is also fostered by platelet-activating factor (PAF), a cytokine secreted by neutrophils and monocytes as well as platelets (Table 27–5). This compound is an ether phospholipid, 1-alkyl-2-acetylglyceryl-3-phosphorylcholine. When blood vessel walls are injured, receptors on the surface of the platelets interact with exposed collagen and adhere to the site of injury. The collagen causes hydrolysis of phosphatidylinositol on the inner surface of the platelet membrane, and diacylglycerol, one of the products of the hydrolysis, causes the platelets to release the contents of their granules. The ADP from the granules triggers release by other platelets. ADP acts by causing an increase in intraplatelet Ca^{2+} that is not secondary to release of diacylglycerol. The Ca^{2+} and the diacylglycerol activate phospholipase A_2, which causes release of arachidonic acid from membrane phospholipids, and the arachidonic acid is converted to thromboxane A_2 (see Chapter 17). Thromboxane A_2 facilitates further Ca^{2+} influx and phosphatidylinositol breakdown. Thrombin also acts by fostering Ca^{2+} influx and hydrolysis of phosphatidylinositol. Aspirin causes a moderate inhibition of thromboxane formation, and there is evidence that when taken in low doses, it is of value in reducing the incidence of recurrences in patients with some myocardial infarctions and some forms of stroke. In addition, there is preliminary evidence that it decreases the incidence of initial myocardial infarctions.

Von Willebrand factor, a protein produced by endothelial cells and platelets, is also necessary for normal adherence of platelets at a site of injury. This factor has an additional role in blood clotting (see below).

Although thromboxane A_2 is a major product of arachidonic acid metabolism in platelets, blood vessel walls contain abundant prostacyclin synthetase (see Chapter 17), and relatively large quantities of prostacyclin are formed. Prostacyclin inhibits platelet aggregation and consequently inhibits thrombus formation. Evidence now indicates that prostacyclin is produced in the intima, whereas the action of thromboxane A_2 is relatively unopposed in the deeper layers of blood vessel walls. Thus, there is a gradient across the wall of a cut blood vessel, with antiaggregation predominating in the intima and aggregation predominating in the adventitia. Therefore, the clot does not extend into the blood vessel and blood flow is not compromised. Presumably, aggregation predominates close to the injury, but some of the endoperoxides formed from the platelets are converted to prostacyclin in the neighboring vessel wall, keeping the platelet plug localized and preventing excess aggregation. Other mechanisms that prevent extension of the clot are discussed below.

HEMOSTASIS

When a small blood vessel is transected or damaged, the injury initiates a series of events (Fig 27–16) that leads to the formation of a clot **(hemostasis).** This seals off the damaged region and prevents further blood loss. The initial event is constriction of the vessel and formation of a temporary hemostatic plug of platelets, followed by conversion of the plug into the definitive clot.

The in vivo action of the clotting mechanism is balanced by limiting reactions that normally prevent clots from developing in uninjured vessels and maintain the blood in a fluid state. It is worth emphasizing that a balance between many complex, interrelated systems must be maintained to prevent hemorrhage while at the same time preventing intravascular coagulation. The factors involved include the endothelium of the blood vessels and the collagen underlying it, vascular tone, the platelets, the clotting and the fibrinolytic systems, and the flow of characteristics of blood within the blood vessels.

The Temporary Hemostatic Plug

When a blood vessel is damaged, the endothelium is disrupted and an underlying layer of collagen is exposed. Collagen attracts platelets, which adhere to it and liberate serotonin and ADP. The ADP in turn rapidly attracts other platelets, and a loose plug of aggregated platelets is formed.

Local Vasoconstriction

The constriction of an injured arteriole or small artery may be so marked that its lumen is obliterated. The vasoconstriction is due to serotonin and other vasoconstrictors liberated from platelets that adhere to the walls of the damaged vessels. It is claimed that at least for a time after being divided transversely, arteries as large as the radial artery constrict and stop bleeding; however, this is no excuse for delay in ligating the damaged vessel. Furthermore, arterial walls cut longitudinally or irregularly do not constrict in such a way that the lumen of the artery is occluded, and bleeding continues.

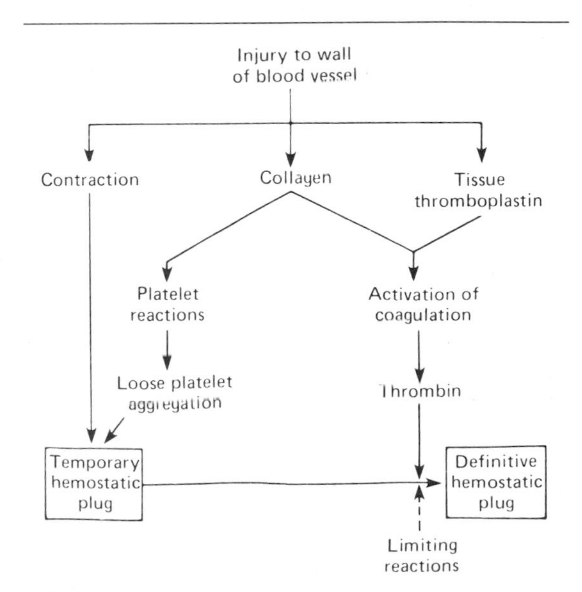

Figure 27–16. Summary of reactions involved in hemostasis. The dashed arrow indicates inhibition. (Modified and reproduced, with permission, from Deykin D: Thrombogenesis. *N Engl J Med* 1967;**276**:622.)

The Clotting Mechanism

The loose aggregation of platelets in the temporary plug is bound together and converted into the definitive clot by **fibrin.** The clotting mechanism responsible for the formation of fibrin involves a cascade of reactions in which inactive enzymes are activated, and the activated enzymes in turn activate other inactive enzymes. The complexity of the system has in the past been compounded by variations in nomenclature, but acceptance of a numbering system for most of the various clotting factors (Table 27–11) has simplified the situation.

The fundamental reaction in the clotting of blood is conversion of the soluble plasma protein fibrinogen to insoluble fibrin (Fig 27–17). The process involves the release of 2 pairs of polypeptides from each fibrinogen molecule. The remaining portion, **fibrin monomer,** then polymerizes with other monomer molecules to form **fibrin.** The fibrin is initially a loose mesh of interlacing strands. It is converted by the formation of covalent cross-linkages to a dense, tight aggregate. This latter reaction is catalyzed by factor XIII, the fibrin-stabilizing factor, and requires Ca^{2+}.

The conversion of fibrinogen to fibrin is catalyzed by thrombin. Thrombin is a serine protease that is formed from its circulating precursor, prothrombin, by the action of activated factor X. Factor X can be activated by reactions that proceed along either of 2 systems, an intrinsic and an extrinsic system (Fig 27–17).

The initial reaction in the **intrinsic system** is conversion of inactive factor XII to active factor XII (XIIa). This activation can be brought about in vitro by exposing the blood to electronegatively charged wettable surfaces such as glass and collagen fibers. Activation in vivo occurs when blood is exposed to the collagen fibers underlying the endothelium in the blood vessels. Active factor XII then activates factor XI, and active factor XI activates factor IX. Activated factor IX forms a complex with factor VIII, activating factor X. Phospholipid from aggregated platelets (PL) and Ca^{2+} are necessary for full activation of factor X. The **extrinsic system** is triggered by the release of tissue thromboplastin, a protein-phospholipid mixture that activates factor VII. The tissue thromboplastin and factor VII activate factors IX and X. In the presence of PL, Ca^{2+}, and factor V, activated factor X catalyzes the conversion of prothrombin to thrombin.

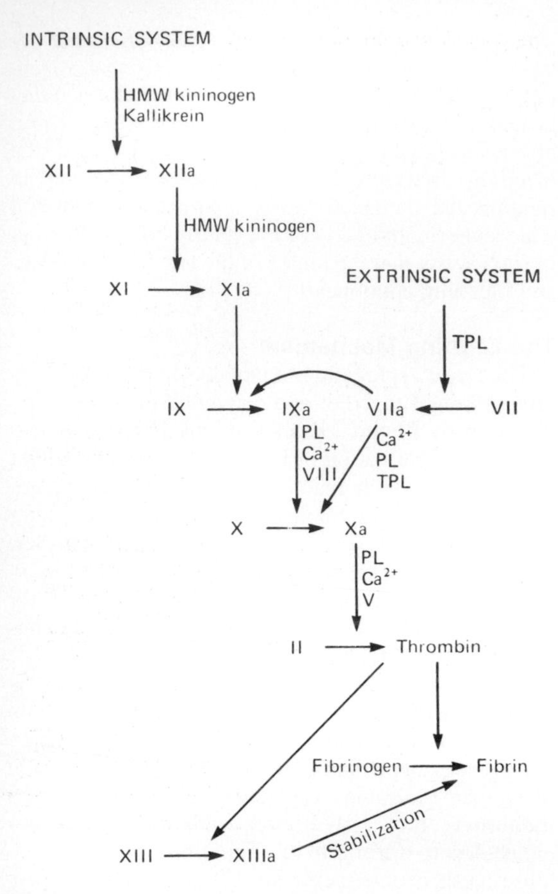

Figure 27–17. The clotting mechanism. a, active form of clotting factor; HMW, high-molecular-weight; PL, platelet phospholipid; TPL, tissue thromboplastin. (Reproduced, with permission, from Wessler S, Gitel SN: Warfarin: From bedside to bench. (Reproduced, with permission of the *New England Journal of Medicine,* 1984;**311**:645.)

Anticlotting Mechanisms

The tendency of blood to clot is balanced in vivo by limiting reactions that tend to prevent clotting inside the blood vessels and to break down any clots that do form. These reactions include removal of some activated clotting factors from the circulation by the liver and reduction in the supply of clotting factors to the degree that they are used up during clotting. Another is the interaction between the platelet-aggregating effect of thromboxane A_2 and the antiaggregating effect of prostacyclin, which causes clots to form in the walls of the injured blood vessels but keeps the vessel lumens free of clot.

Antithrombin III is a circulating protease inhibitor that binds to the serine proteases in the coagulation system, blocking their activity as clotting factors. This binding is facilitated by **heparin,** a naturally occurring anticoagulant that is a mixture of sulfated polysaccharides with molecular weights averaging 15,000–18,000. The clotting factors that are inhibited are the active forms of factors IX, X, XI, and XII. Note that "antithrombin" is a misnomer; it is actually a different substance, heparin cofactor II, that inhibits thrombin.

The endothelium of the blood vessels also plays an active role in preventing the extension of clots to normal blood vessels. All endothelial cells except those in the cerebral microcirculation produce **thrombomodulin,** a thrombin-binding protein that converts thrombin into **protein C activator.** This activates protein C (Fig 27–18), a naturally occurring anticoagulant protein that inactivates factors V and VIII and inactivates an inhibitor of plasminogen activator, increasing the formation of plasmin.

Plasmin (fibrinolysin) is the active component of the **fibrinolytic system** (Fig 27–18). This enzyme lyses fibrin and fibrinogen, with the production of fibrinogen degradation products (FDP) that inhibit thrombin. Plasmin is formed from its active precursor, plasminogen, by the action of thrombin and a **tissue plasminogen activator (+-PA).** Human +-PA is now produced by recombinant DNA techniques and is available for clinical use. It shows considerable promise in lysing clots in the coronary arteries if given to patients soon after the onset of myocardial infarction. Streptokinase, an enzyme from cultured human fetal cells, are also fibrinolytic and are used in the treatment of early myocardial infarction to lyse clots.

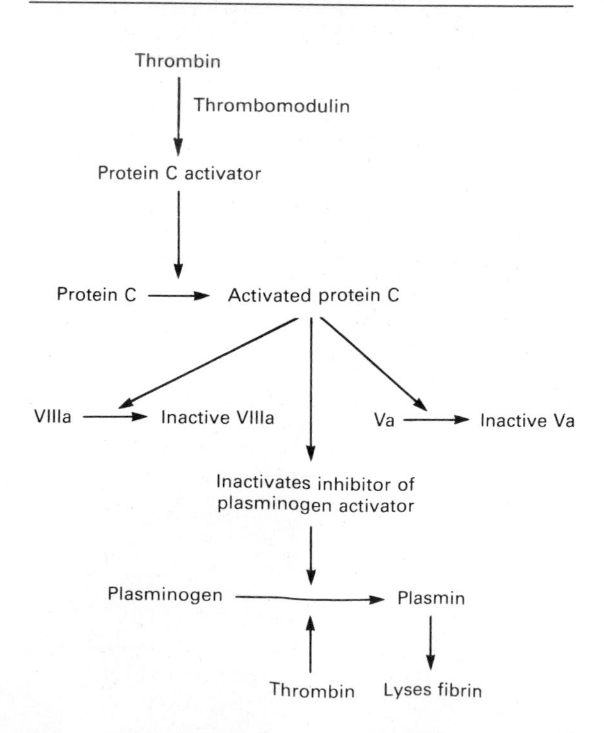

Figure 27–18. The fibrinolytic system and its regulation by protein C. (Modified from Wessler S, Gitel SN: Warfarin: From bedside to bench. *N Engl J Med* 1984;**311**:645.)

Anticoagulants

As noted above, heparin is a naturally occurring anticoagulant that facilitates the action of antithrombin III. It is also a cofactor for the lipoprotein lipase (clearing factor; see Chapter 17). The highly basic protein protamine forms an irreversible complex with heparin and is used clinically to neutralize heparin.

Heparin and histamine are found in the granules of the **mast cells,** wandering cells found in large numbers in tissues that are rich in connective tissue. In rodents but not in humans, mast cells also contain serotonin. Histamine is also found in the granules of circulating basophils, along with heparinlike polysulfated chondroitins. Mast cells have IgE receptors on their surfaces and discharge the contents of their granules when IgE-coated antigens bind to the receptors. The cells also bring about the secretion of leukotrienes from other cells (see Chapter 17) and mediate some aspects of allergic and inflammatory reactions. In addition to mediating allergic reactions, the mast cells may aid in defense against parasitic infections.

Clotting can be prevented in vitro if Ca^{2+} is removed from the blood by the addition of substances such as oxalates, which form insoluble salts with Ca^{2+}, or citrate or other **chelating agents,** which bind Ca^{2+}. Coumarin derivatives such as **dicumarol** and **warfarin** are also effective anticoagulants. They inhibit the action of vitamin K, and this vitamin catalyzes the conversion of glutamic acid residues to γ-carboxyglutamic acid residues. Six of the proteins involved in clotting—factor II (prothrombin); factors VII, IX, and X; protein C; and protein S, which facilitates the inactivation of factors Va and VIIIa by facilitating the action of activated protein C—are unique in that they each require conversion of a number of glutamic acid residues to carboxyglutamic acid residues before being released into the circulation. Hence, all 6 are vitamin K-dependent.

Abnormalities of Hemostasis

In vivo, a plasma Ca^{2+} level low enough to interfere with blood clotting is incompatible with life, but hemorrhagic diseases due to selective deficiencies of almost all the other clotting factors have been described (Table 27–12). In some genetic conditions, normal clotting factors are replaced by abnormal factors. Hemophilia A is of interest because it is relatively common. The disease has been treated with factor VIII-rich preparations made from plasma, but unfortunately this has led to transmission of the AIDS virus to a significant number of patients. However, factor VIII has been produced by recombinant DNA techniques and should be available for general clinical use in the near future. In addition to promoting platelet adherence (see above), von Willebrand factor forms a complex with factor VIII and regulates the plasma levels of the latter. Congenital deficiency of von Willebrand factor also causes a bleeding disorder (von Willebrand's disease).

Table 27–12. Examples of diseases due to deficiency of clotting factors.

Deficiency of Factor	Clinical Syndrome	Cause
I	Afibrinogenemia	Depletion during pregnancy with premature separation of placenta; also congenital (rare).
II	Hypoprothrombinemia (hemorrhagic tendency in liver disease)	Decreased hepatic synthesis, usually secondary to vitamin K deficiency.
V	Parahemophilia	Congenital.
VII	Hypoconvertinemia	Congenital.
VIII	Hemophilia A (classical hemophilia)	Congenital defect due to various abnormalities of the gene on X chromosome that codes for factor VIII; disease is therefore inherited as sex-linked characteristic.
IX	Hemophilia B (Christmas disease)	Congenital.
X	Stuart-Prower factor deficiency	Congenital.
XI	PTA deficiency	Congenital.
XII	Hageman trait	Congenital.

The absorption of the K vitamins, along with that of the other fat-soluble vitamins, is depressed in obstructive jaundice because of the lack of bile in the intestine and consequent depression of fat absorption (see Chapter 26). The resulting clotting factor deficiencies may cause the development of a significant bleeding tendency. Aspirin inhibits platelet aggregation by inhibiting the cyclooxygenase that catalyzes the formation of prostaglandins and thromboxanes, but it rarely causes abnormal bleeding, probably because it does not inhibit aggregation to a sufficient degree.

When the platelet count is low, clot retraction is deficient and there is poor constriction or ruptured vessels. The resulting clinical syndrome **(thrombocytopenic purpura)** is characterized by easy bruisability and multiple subcutaneous hemorrhages. Purpura may also occur when the platelet count is normal, and in some of these cases, the circulating platelets are abnormal **(thrombasthenic purpura).**

Formation of clots inside blood vessels is called **thrombosis** to distinguish it from the normal extravascular clotting of blood. Thromboses are a major medical problem. They are particularly prone to occur where blood flow is sluggish, eg, in the veins of the legs after operations and delivery, because the slow flow permits activated clotting factors to accumulate instead of being washed away. They also occur in vessels such as the coronary and cerebral arteries at sites where the intima is damaged by arteriosclerotic plaques, and over areas of damage to the endocardium. They frequently occlude the arterial supply to the organs in which they form, and bits of thrombus **(emboli)** sometimes break off and travel in the bloodstream to distant sites, damaging other organs. Examples are obstruction of the pulmonary artery or its branches **(pulmonary embolism)** by thrombi from the leg veins, and embolism of cerebral or leg vessels by bits of clot breaking off from a thrombus in the left ventricle **(mural thrombus)** overlying a myocardial infarct.

Congenital absence of protein C leads to uncontrolled intravascular coagulation and, in general, death in infancy. If this condition is diagnosed and treatment is instituted with blood concentrates rich in protein C, the coagulation defect disappears. It has also been found that protein C deficiency occurs in *E coli*-induced shock (see Chapter 33). Correction of the deficiency lowers the mortality rate, but the mechanism by which protein C exerts this protective effect is unknown.

LYMPH

Lymph is tissue fluid that enters the lymphatic vessels. It drains into the venous blood via the thoracic and right lymphatic ducts. It contains clotting factors and clots on standing in vitro. In most locations, it also contains proteins that traverse capillary walls and return to the blood via the lymph. Its protein content is generally lower than that of plasma, which contains about 7 g/dL, but lymph protein varies with the region from which the lymph drains (Table 27–13). Water-insoluble fats are absorbed from the intestine into the lymphatics, and the lymph in the thoracic duct after a meal is milky because of its high fat content (see Chapter 25). Lymphocytes enter the circulation principally through the lymphatics, and there are appreciable numbers of lymphocytes in thoracic duct lymph.

Table 27–13. Probable approximate protein content of lymph in humans.*

Lymph From	Protein Content (g/dL)
Choroid plexus	0
Ciliary body	0
Skeletal muscle	2
Skin	2
Lung	4
Gastrointestinal tract	4.1
Heart	4.4
Liver	6.2

* Data largely from JN Diana.

Origin of the Heartbeat & the Electrical Activity of the Heart

28

INTRODUCTION

The parts of the heart normally beat in an orderly sequence: Contraction of the atria (**atrial systole**) is followed by contraction of the ventricles (**ventricular systole**), and during **diastole** all 4 chambers are relaxed. The heartbeat originates in a specialized **cardiac conduction system** and spreads via this system to all parts of the myocardium. The structures that make up the conduction system (Fig 28–1) are the **sinoatrial node** (SA node), the **internodal atrial pathways,** the **atrioventricular node** (AV node), the **bundle of His** and its branches, and the **Purkinje system.** The various parts of the conduction system and, under abnormal conditions, parts of the myocardium are capable of spontaneous discharge. However, the SA node normally discharges most rapidly, depolarization spreading from it to the other regions before they discharge spontaneously. The SA node is therefore the normal **cardiac pacemaker,** its rate of discharge determining the rate at which the heart beats. Impulses generated in the SA node pass through the atrial pathways to the AV node, through this node to the bundle of His, and through the branches of the bundle of His via the Purkinje system to the ventricular muscle.

ORIGIN & SPREAD OF CARDIAC EXCITATION

Anatomic Considerations

In 4-chambered mammalian hearts, the SA node is located at the junction of the superior vena cava with the right atrium. The AV node is located in the right posterior portion of the interatrial septum (Fig 28–1).

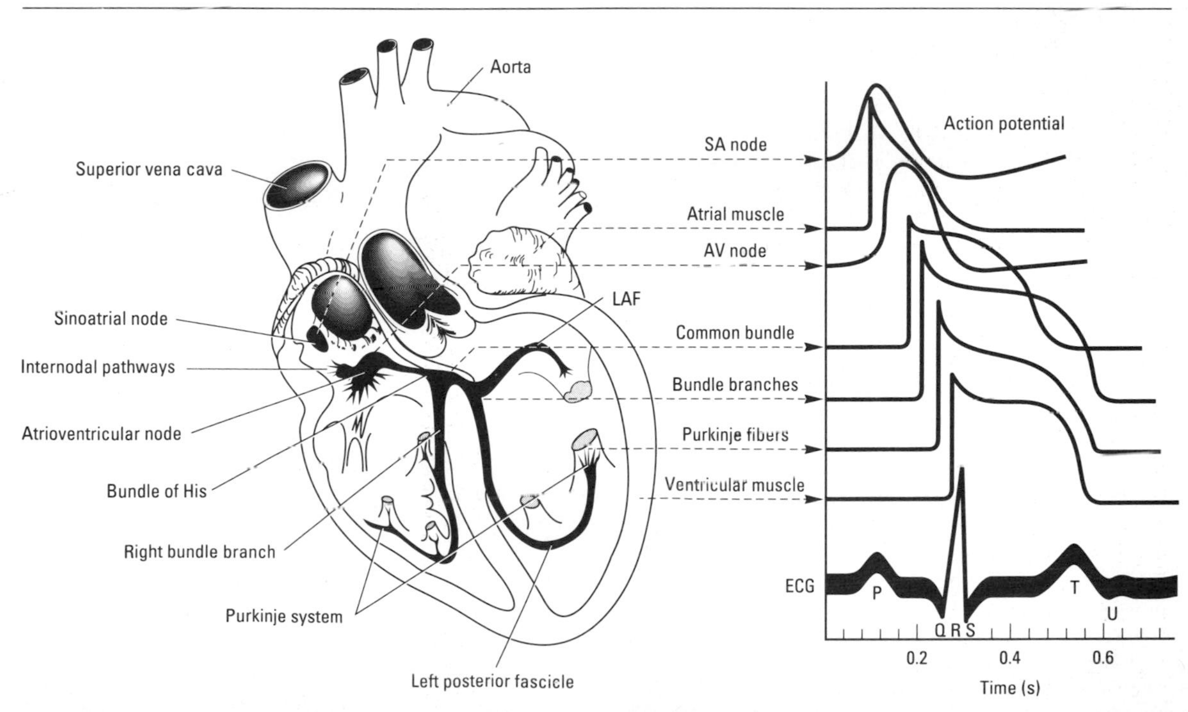

Figure 28–1. Conducting system of the heart. Typical transmembrane action potentials for the SA and AV node, other parts of the conduction system, and the atrial and ventricular muscle are shown along with the correlation to the extracellularly recorded electrical activity, ie, the electrocardiogram (ECG). The action potentials and ECG are plotted on the same time axis but with different zero points on the vertical scale.

There are 3 bundles of atrial fibers that contain Purkinje type fibers and conduct impulses from the SA node to the AV node: the anterior internodal tract of Bachman, the middle internodal tract of Wenckebach, and the posterior internodal tract of Thorel. These fibers converge and interdigitate with the fibers in the AV node. The AV node is continuous with the bundle of His, which gives off a left bundle branch at the top of the interventricular septum and continues as the right bundle branch. The left bundle branch divides into an anterior fascicle and a posterior fascicle. The branches and fascicles run subendocardially down either side of the septum and come into contact with the Purkinje system, the fibers of which spread to all parts of the ventricular myocardium.

The histology of cardiac muscle is described in Chapter 3. The conduction system is composed of modified cardiac muscle that is striate but has indistinct boundaries. It is richer in glycogen and has more sarcoplasm than the rest of the cardiac muscle fibers. The atrial muscle fibers are separated from those of the ventricles by a fibrous tissue ring, and normally the only conducting tissue between the atria and ventricles is the bundle of His.

The SA node develops from structures on the right side of the embryo and the AV node from structures on the left. This is why in the adult the right vagus is distributed mainly to the SA node and the left vagus mainly to the AV node. Both areas receive noradrenergic nerves from the cervical sympathetic ganglia via the cardiac nerves. Noradrenergic fibers are distributed to the atrial and ventricular myocardium as well; the vagal fibers are probably distributed only to the nodal tissue and the atrial musculature.

Properties of Cardiac Muscle

The electrical responses of cardiac muscle and nodal tissue and the ionic fluxes that underlie them are discussed in detail in Chapter 3. As shown in Fig 3–14, myocardial fibers have a resting membrane potential of approximately −90 mV. The relation of the action potential to the contractile response is shown in Fig 3–13. The individual fibers are separated from one another by membranes, but depolarization spreads radially through them as if they were a synctium, because of the presence of gap junctions. The transmembrane action potential of single cardiac muscle cells is characterized by rapid depolarization, a plateau, and a slow repolarization process (Fig 3–14). The initial depolarization is due to an increase in Na^+ permeability (increased conductance in fast Na^+ channels in the cell membrane). This is followed by a slower increase in Ca^{2+} permeability (increased conductance in slow Ca^{2+} channels), which produces the plateau. Repolarization after the plateau is due to a delayed increase in K^+ permeability. Recorded extracellularly, the summed electrical activity of all the cardiac muscle fibers is the ECG. The timing of the discharge of the individual units relative to the ECG is shown in Fig 28–1.

Rhythmically discharging cells have an unstable membrane potential that after each impulse declines to the firing level (**prepotential** or **pacemaker potential;** see Chapter 3). This triggers the next impulse. The prepotential is due to a steady decrease in K^+ permeability. The speed with which the membrane potential is reduced to the firing level determines the rate at which the tissue discharges. Prepotentials are normally prominent only in the SA and AV nodes (Fig 28–1), but there are "latent pacemakers" in other portions of the conduction system that can take over when the SA and AV nodes are depressed or conduction from them is blocked. Atrial and ventricular muscle fibers do not have prepotentials, and they discharge spontaneously only under abnormal conditions.

When the cholinergic vagal fibers to nodal tissue are stimulated, the membrane becomes slightly hyperpolarized and the slope of the prepotentials is decreased, because the acetylcholine liberated at the nerve endings increases the permeability of nodal tissue to K^+. This action is mediated by muscarinic receptors and is due to the opening of a special set of K^+ channels. The result is a decrease in the rate of firing. In addition, acetylcholine decreases conductance in the Ca^{2+} channels via muscarinic receptors. Strong vagal stimulation may abolish spontaneous discharge for some time. Conversely, stimulation of the sympathetic cardiac nerves makes the membrane potential fall more rapidly, and the rate of spontaneous discharge increases (Fig 28–2). This is due to the noradrenergic mediator norepinephrine, which, via β-adrenergic receptors, produces an increase in the rate at which K^+ permeability declines between action potentials. In addition, norepinephrine acts via β-adrenergic receptors to increase conductance in Ca^{2+} channels, thus increasing the strength of each cardiac contraction.

The rate of discharge of the SA node and other nodal tissue is influenced by temperature and by drugs. The discharge frequency is increased when the temperature rises, and this may contribute to the tachycardia associated with fever. Digitalis depresses

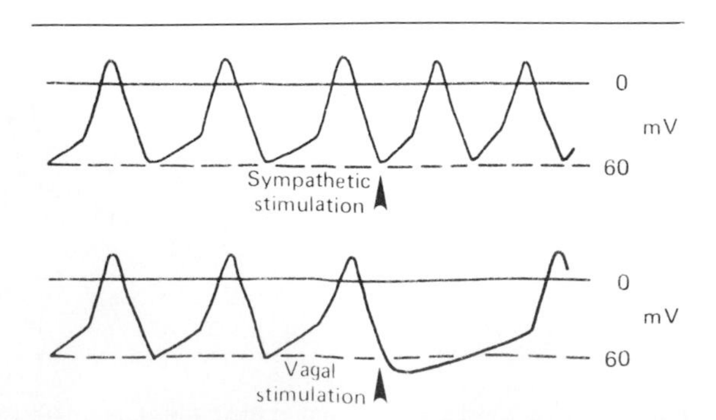

Figure 28–2. Effect of sympathetic (noradrenergic) and vagal (cholinergic) stimulation on the membrane potential of the SA node.

nodal tissue and exerts an effect like that of vagal stimulation, particularly on the AV node. The speeds of conduction in the various types of cardiac tissue are shown in Table 28–1.

Spread of Cardiac Excitation

Depolarization initiated in the SA node spreads radially through the atria, then converges on the AV node. Atrial depolarization is complete in about 0.1 second. Because conduction in the AV node is slow, there is a delay of about 0.1 second (**AV nodal delay**) before excitation spreads to the ventricles. This delay is shortened by stimulation of the sympathetic nerves to the heart and lengthened by stimulation of the vagi. From the top of the septum, the wave of depolarization spreads in the rapidly conducting Purkinje fibers to all parts of the ventricles in 0.08–0.1 second. In humans, depolarization of the ventricular muscle starts at the left side of the interventricular septum and moves first to the right across the mid portion of the septum. The wave of depolarization then spreads down the septum to the apex of the heart. It returns along the ventricular walls to the AV groove, proceeding from the endocardial to the epicardial surface (Fig 28–3). The last parts of the heart to be depolarized are the posterobasal portion of the left ventricle, the pulmonary conus, and the uppermost portion of the septum.

THE ELECTROCARDIOGRAM

Because the body fluids are good conductors (ie, because the body is a **volume conductor**), fluctuations in potential that represent the algebraic sum of the action potentials of myocardial fibers can be recorded extracellularly. The record of these potential fluctuations during the cardiac cycle is the **electrocardiogram (ECG).** Most electrocardiograph machines record these fluctuations on a moving strip of paper.

The ECG may be recorded using an **active** or **exploring** electrode connected to an indifferent electrode at zero potential (**unipolar** recording), or between 2 active electrodes (**bipolar** recording). In a volume conductor, the sum of the potentials at the points of an equilateral triangle with a current source in the center is zero at all times. A triangle with the heart at its center (**Einthoven's triangle**) can be

Table 28–1. Conduction speeds in cardiac tissue.

Tissue	Conduction Rate (m/s)
SA node	0.05
Atrial pathways	1
AV node	0.05
Bundle of His	1
Purkinje system	4
Ventricular muscle	1

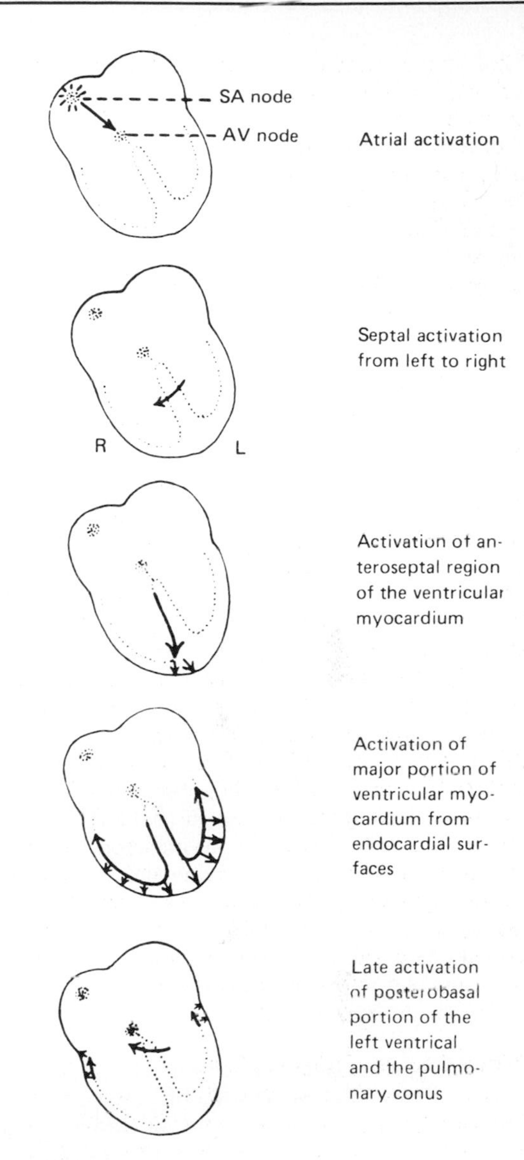

Figure 28–3. Normal spread of electrical activity in heart. (Reproduced, with permission, from Goldman MJ: *Principles of Clinical Electrocardiography,* 12th ed. Lange, 1986.)

approximated by placing electrodes on both arms and on the left leg. These are 3 **standard limb leads** used in electrocardiography. If these electrodes are connected to a common terminal, an indifferent electrode that stays near zero potential is obtained. Depolarization moving toward an active electrode in a volume conductor produces a positive deflection, whereas depolarization moving in the opposite direction produces a negative deflection.

The names of the various waves of the ECG and their timing in humans are shown in Fig 28–4. By convention, an upward deflection is written when the active electrode becomes positive relative to the indifferent electrode; and when the active electrode

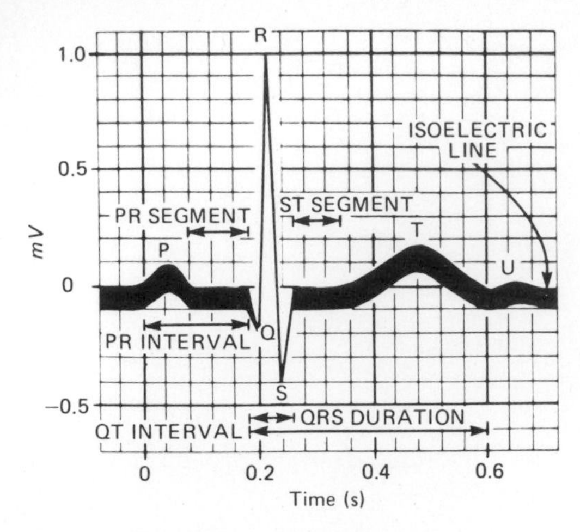

Figure 28–4. Waves of the ECG.

becomes negative, a downward deflection is written. The P wave is produced by atrial depolarization, the QRS complex by ventricular depolarization, and the ST segment and T wave by ventricular repolarization. The manifestations of atrial repolarization are not normally seen because they are obscured by the QRS complex. However, a depression of the PR segment (Ta wave) is sometimes seen in association with sinus tachycardia in normal individuals or when the atria are hypertrophied. The U wave is an inconstant finding, believed to be due to slow repolarization of the papillary muscles. The intervals between the various waves of the ECG and the events in the heart that occur during these intervals are shown in Table 28–2.

The magnitude and configuration of the individual waves of the ECG vary with the location of the electrodes. When recorded from the surface of the body, all the waves are small compared with the transmembrane potentials of single fibers, because the ECG is being recorded at a considerable distance from the heart.

Bipolar Leads

Bipolar leads were used before unipolar leads were developed. The **standard limb leads,** leads I, II, and III, are records of the differences in potential between 2 limbs. Since current flows only in the body fluids, the records obtained are those that would be obtained if the electrodes were at the points of attachment of the limbs, no matter where on the limbs the electrodes are placed. In lead I, the electrodes are connected so that an upward deflection is inscribed when the left arm becomes positive relative to the right (left arm positive). In lead II, the electrodes are on the right arm and left leg, with the leg positive; and in lead III, the electrodes are on the left arm and left leg, with the leg positive.

Unipolar (V) Leads

An additional 9 unipolar leads, ie, leads that record the potential difference between an exploring electrode and an indifferent electrode, are commonly used in clinical electrocardiography. There are 6 unipolar chest leads (precordial leads) designated V_1–V_6 (Fig 28–5) and 3 unipolar limb leads: VR (right arm), VL (left arm), and VF (left foot). **Augmented** limb leads, designated by the letter a (aVR, aVL, aVF) are generally used. The augmented limb leads are recordings between one limb and the other 2 limbs. This increases the size of the potentials by 50% without any change in configuration from the nonaugmented record.

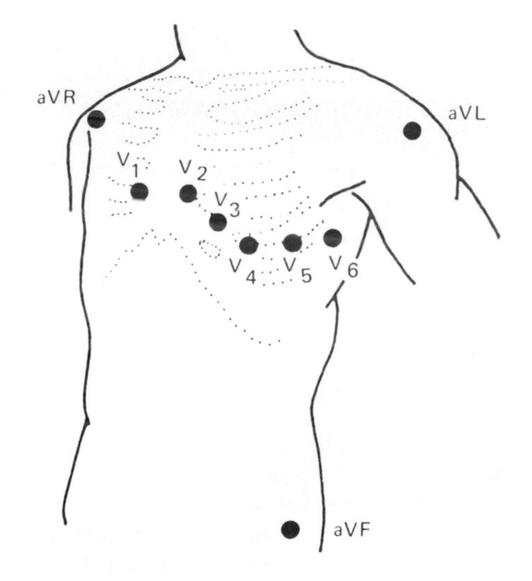

Figure 28–5. Unipolar electrocardiographic leads.

Table 28–2. ECG intervals.

	Normal Duration (s)		
	Average	**Range**	**Events in Heart During Interval**
PR interval*	0.18†	0.12–0.20	Atrial depolarization and conduction through AV node
QRS duration	0.08	to 0.10	Ventricular depolarization and atrial repolarization
QT interval	0.40	to 0.43	Ventricular depolarization plus ventricular repolarization
ST interval (QT minus QRS)	0.32	. . .	Ventricular repolarization

* Measured from the beginning of the P wave to the beginning of the QRS complex.
† Shortens as heart rate increases from average of 0.18 at rate of 70 to 0.14 at rate of 130.

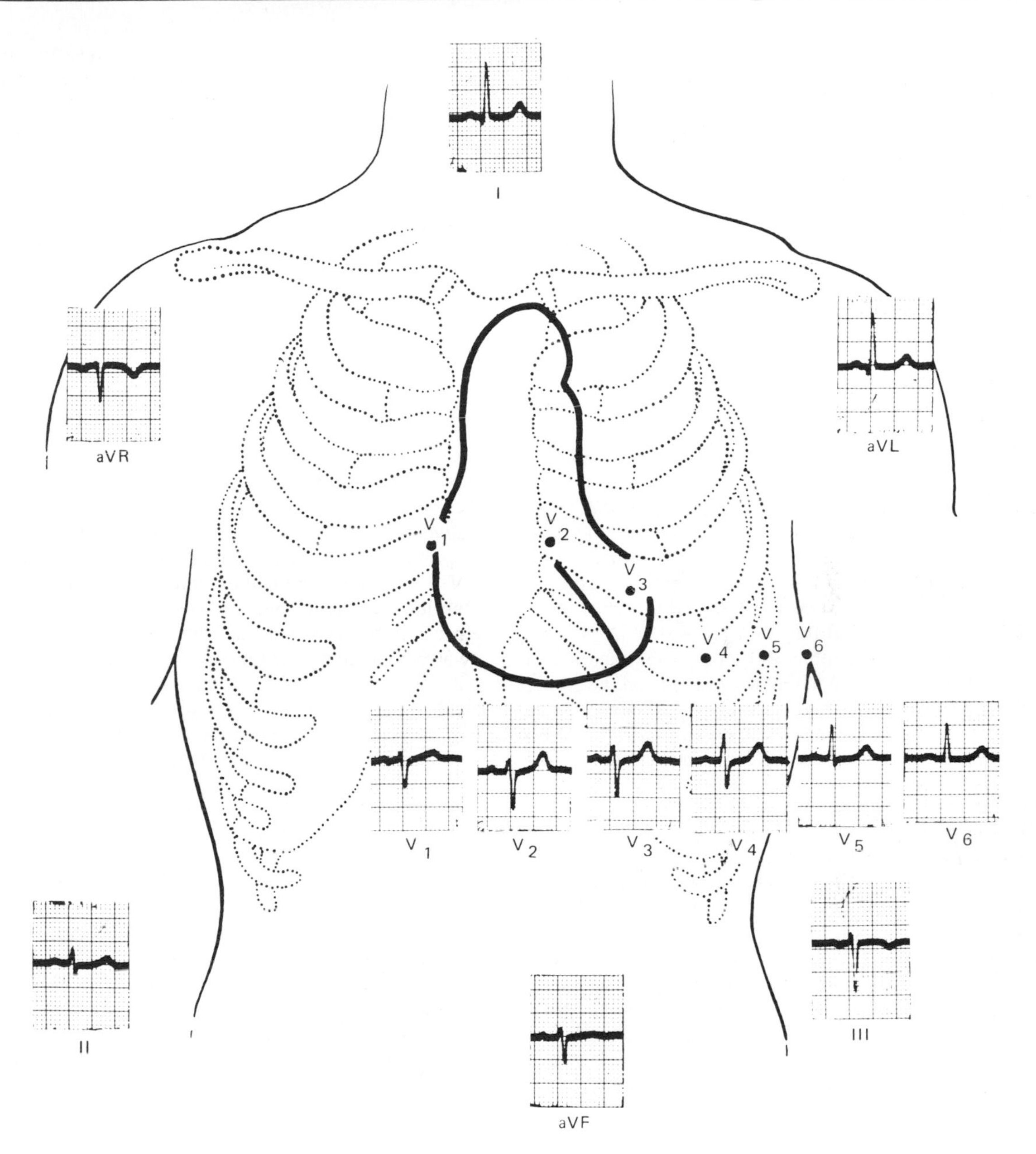

Figure 28–6. Normal ECG. (Reproduced, with permission, from Goldman MJ: *Principles of Clinical Electrocardiography*, 12th ed. Lange, 1986.)

An esophageal electrode is sometimes used to study atrial activity. The electrode is inserted in a catheter and swallowed, and each esophageal lead is identified by the letter E followed by the number of centimeters from the teeth to the electrode tip. Thus, for example, lead E_{35} is a unipolar record in which the exploring electrode is 35 cm down the esophagus. Unipolar leads can also be placed at the tips of catheters and inserted into the heart.

Normal ECG

The ECG of a normal individual is shown in Fig 28–6. The sequence in which the parts of the heart are depolarized (Fig 28–3) and the position of the heart relative to the electrodes are the important considerations in interpreting the configurations of the waves in each lead. The atria are located posteriorly in the chest. The ventricles form the base and anterior surface of the heart, and the right ventricle is anterolateral to the left (Fig 28–7). Thus, a VR "looks at" the cavities of the ventricles. Atrial depolarization, ventricular depolarization, and ventricular repolarization move away from the exploring electrode, and the P wave, QRS complex, and T wave are therefore all negative (downward) deflections; a VL and a VF look at the ventricles, and the deflections are therefore predominantly positive or biphasic. There is no Q wave in V_1 and V_2, and the initial portion of the QRS complex is a small upward deflection because ventricular depolarization first moves across the mid portion of the septum from left to right toward the exploring electrode. The wave of excitation then moves down the septum and into the left ventricle away from the electrode, producing a large S wave. Finally, it moves back along the ventricular wall toward the electrode, producing the return to the isoelectric line. Conversely, in the left ventricular leads (V_{4-6}), there may be an initial small Q wave (left to right septal depolarization), and there is a large R wave (septal and left ventricular depolarization) followed in V_4 and V_5 by a moderate S wave (late depolarization of the ventricular walls moving back toward the AV junction).

There is considerable variation in the position of the normal heart, since the heart rotates on the 2 axes shown in Fig 28–7 plus the transverse axis (rotation forward and backward). These position changes affect the configuration of the electrocardiographic complexes in the various leads.

Bipolar Limb Leads & the Cardiac Vector

Because the standard limb leads are records of the potential differences between 2 points, the deflection in each lead at any instant indicates the magnitude and direction in the axis of the lead of the electromotive force generated in the heart (**cardiac vector** or **axis**). The vector at any given moment in the 2 dimensions of the frontal plane can be calculated from any 2 standard limb leads (Fig 28–8) if it is assumed that the 3 electrode locations form the points of an equilateral triangle (Einthoven's triangle) and that the heart lies in the center of the triangle. These assumptions are not completely warranted, but calculated vectors are useful approximations. An approximate **mean QRS vector** ("electrical axis of the heart") is often plotted using the average QRS deflection in each lead as shown in Fig 28–8. This is a **mean** vector as opposed to an **instantaneous** vector, and the average QRS deflections should be measured by integrating the QRS complexes. However, they can be approximated by measuring the net differences between the positive and negative peaks of the QRS. The normal direction of the mean QRS vector is generally said to be −30 to + 110 degrees on the coordinate system shown in Fig 28–8. **Left** or **right axis deviation** is said to be present if the calculated axis falls to the left of −30 degrees or to the right of +110 degrees. Right axis deviation suggests right ventricular hypertrophy, and left axis deviation may be due to left ventricular hypertrophy. However, simple differences in heart position may produce axis values in the "abnormal" range, and there are better and more reliable electrocardiographic criteria for ventricular hypertrophy.

Vectorcardiography

If the tops of the arrows representing all of the **instantaneous** cardiac vectors in the frontal plane during the cardiac cycle are connected, from first to last, the line connecting them forms a series of 3 loops: one for the P wave, one for the QRS complex, and one for the T wave. This can be done electronically and the loops, called **vectorcardiograms,** projected on the face of a cathode ray oscilloscope.

His Bundle Electrogram

In patients with heart block, the electrical events in the AV node, bundle of His, and Purkinje system are frequently studied with a catheter containing ring electrodes at its tip that is passed through a vein to the right side of the heart and manipulated into a position close to the tricuspid valve. Three or more standard electrocardiographic leads are recorded simultaneously. The record of the electrical activity obtained with the catheter (Fig 28–9) is the **His bundle electrogram (HBE).** It normally shows an A deflection when the AV node is activated, an H spike during transmission through the His bundle, and a V deflection during ventricular depolarization. With the HBE and the standard electrocardiographic leads, it is possible to accurately time 3 intervals: (1) the PA interval, the time from the first appearance of atrial depolarization to the A wave in the HBE, which represents conduction time from the SA node to the AV node; (2) the AH interval, from the A wave to the start of the H spike, which represents the AV nodal conduction time; and (3) the HV interval, the time from the start of the H spike to the start of the QRS deflection in the ECG, which represents conduction in the bundle of His and the bundle branches. The approximate normal values for these intervals in adults are PA, 27 ms; AH, 92 ms; and HV, 43 ms. These values illustrate the relative slowness of conduction in the AV node (Table 28–1).

Monitoring

The ECG is often recorded continuously in hospital coronary care units, with alarms arranged to sound at the onset of life-threatening arrhythmias. Using a small portable tape recorder (**Holter monitor**), it is also possible to record the ECG in ambulatory individuals as they go about their normal activities. The recording is later played back at high speed and analyzed. Long-term continuous records can be obtained, and transtelephonic ECG devices are also available so that records can be forwarded to a doctor's office or a laboratory by telephone. Long-term recordings have proved valuable in the diagnosis of arrythmias and in planning treatment of patients recovering from myocardial infarctions.

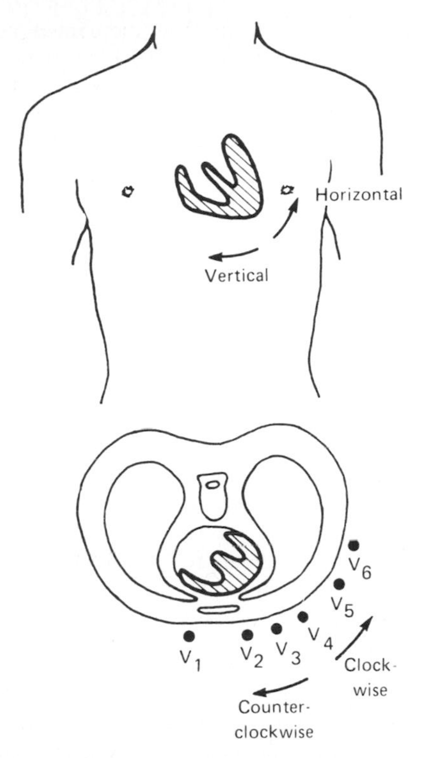

Figure 28–7. Diagram of position of heart in chest. Arrows indicate directions of cardiac rotation with normal variations in cardiac position. By convention, the direction of rotation in the horizontal plane is defined by its appearance as viewed from the inferior surface of the heart looking upward from the diaphragm.

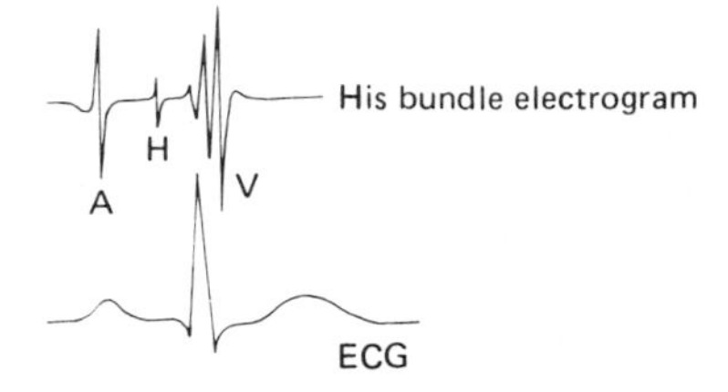

Figure 28–9. Normal His bundle electrogram (HBE) with simultaneously recorded ECG.

CARDIAC ARRHYTHMIAS

Normal Cardiac Rate

In the normal human heart, each beat originates in the SA node (**normal sinus rhythm, NSR**). The heart beats about 70 times a minute at rest. The rate is slowed (**bradycardia**) during sleep and accelerated (**tachycardia**) by emotion, exercise, fever, and many other stimuli. The control of heart rate is discussed in Chapter 31. In healthy young individuals breathing at a normal rate, the heart rate varies with the phases of respiration: it accelerates during inspiration and decelerates during expiration, especially if the depth of breathing is increased. This **sinus arrhythmia** (Fig 28–10) is a normal phenomenon. Table 28–3 shows that it is due primarily to fluctuations in parasympathetic output to the heart. In the normal men studied in the work reported in this table, the time interval between R waves in the ECG (R-R interval) and the variation in this interval over time were not affected

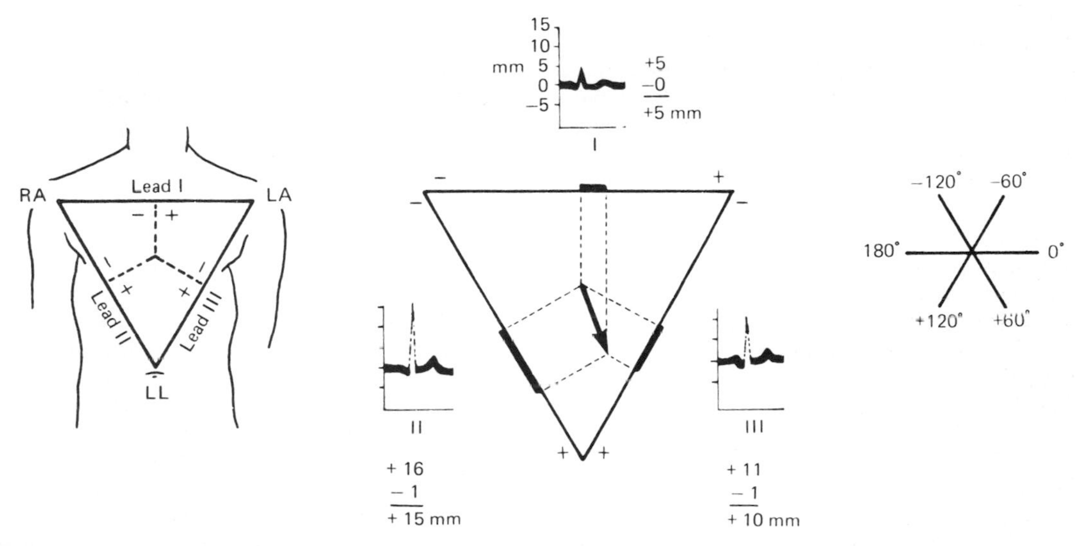

Figure 28–8. Cardiac vector. *Left:* Einthoven's triangle. Perpendiculars dropped from the midpoints of the sides of the equilateral triangle intersect at the center of electrical activity. RA, right arm; LA, left arm; LL, left leg. *Center:* Calculation of mean QRS vector. In each lead, distances equal to the height of the R wave minus the height of the largest negative deflection in the QRS complex are measured off from the midpoint of the side of the triangle representing that lead. An arrow drawn from the center of electrical activity to the point where perpendiculars extended from the distances measured off on the sides intersect represents the magnitude and direction of the mean QRS vector. *Right:* Reference axes for determining the direction of the vector.

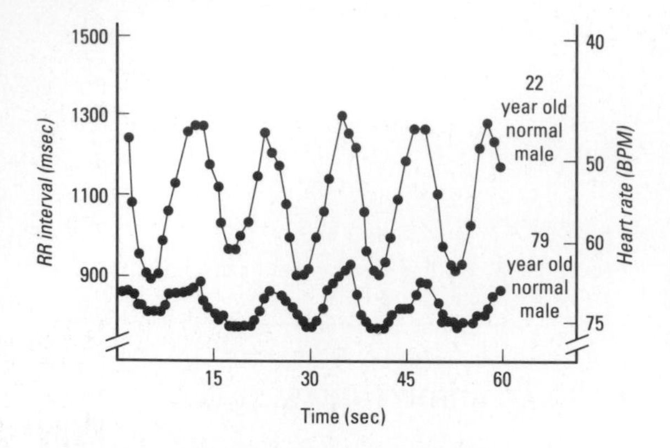

Figure 28–10. Sinus arrythmia in a young man and an old man. The subjects breathed 5 times per minute, and with each inspiration, the R-R interval declined, indicating an increase in heart rate. Note the marked reduction in the magnitude of the arrythmia in the older individual. These records were obtained after β-adrenergic blockade but would have been generally similar in its absence. (Reproduced, with permission, from Pfiefer MA et al: Differential changes of autonomic nervous system function with age in man. *Am J Med* 1983;*75:*49.)

to a significant degree by blocking the sympathetic input to the heart with propranolol, but blocking the parasympathetic input increased the heart rate (shortened the R-R interval) and reduced the variation. During inspiration, impulses in the vagi from the stretch receptors in the lungs inhibit the cardioinhibitory area in the medulla oblongata. The tonic vagal discharge that keeps the heart rate slow decreases, and the heart rate rises.

Disease processes affecting the sinus node lead to marked bradycardia accompanied by dizziness and syncope (**sick sinus syndrome**). When the condition causes severe symptoms, the treatment is implantation of a pacemaker, and sinus node dysfunction accounts for over half of the pacemaker implants in the USA.

Abnormal Pacemakers

The AV node and other portions of the conduction system can in abnormal situations become the cardiac pacemaker. In addition, diseased atrial and ventricular muscle fibers can have their membrane potentials reduced and discharge repetitively.

A simple but enlightening experiment that demonstrates the hierarchy of pacemakers in the heart can be performed in cold-blooded animals such as frogs and turtles. In these animals, the heartbeat originates in a separate cardiac chamber, the **sinus venosus,** instead of in an SA node. There is no coronary circulation, the heart receiving O_2 by diffusion from the blood in the cardiac chambers. If a ligature (Stannius ligature I) is tied around the junction of the sinus venosus and right atrium, conduction from the sinus to the rest of the heart is prevented. The atria and ventricles stop for a moment, then resume beating at a slower rate as a focus in the atrium becomes the pacemaker for the portion of the heart below the ligature. If a second ligature (Stannius ligature II) is tied between the atria and ventricles, the ventricles stop, then resume beating at an even slower rate than the atria. With both ligatures in place, there are 3 separate regions of the heart beating at 3 different rates.

It is not possible to perform the Stannius experiment in mammals because the SA node is embedded in the atrial wall, and tying a ligature between the atria and the ventricles interrupts the coronary circulation, usually causing immediate ventricular fibrillation. However, experiments on laboratory mammals involving cooling or crushing of the SA and AV nodes and the "experiments of nature" in humans with diseased nodal tissue show that the same pacemaker hierarchy prevails.

When conduction from the atria to the ventricles is completely interrupted, complete (third-degree) heart block is said to be present, and the ventricles beat at a low rate (**idioventricular rhythm**) independently of the atria (Fig 28–11). The block may be due to disease in the AV node (**AV nodal block**) or in the conducting system below the block (**infranodal block**). In patients with AV nodal block, the remaining nodal tissue becomes the pacemaker, and the rate of the idioventricular rhythm is approximately 45 beats/min. In patients with infranodal block due to disease in the bundle of His, the ventricular pacemaker is located more peripherally in the conduction system, and the ventricular rate is lower; it averages 35 beats/min, but in individual cases it can be as low as 15 beats/min. In such individuals, there may also be periods of asystole lasting a minute or more. The resultant cerebral ischemia causes dizziness and fainting (**Stokes-Adams syndrome**). Causes of third-degree heart block include septal myocardial infarction and damage to the bundle of His during surgical correction of congenital interventricular septal defects. In infranodal block with Stokes-Adams syndrome, implantation of a permanent cardiac pacemaker is indicated in order to drive the ventricles at a more rapid, regular rate.

When conduction between the atria and venticles is slowed but not completely interrupted, **incomplete heart block** is present. In the form called **first-degree heart block,** all the atrial impulses reach the

Table 28–3. Effect of β-adrenergic blockade with propranolol and muscarinic cholinergic blockade with atropine on sinus arrythmia in healthy men.*

Treatment	R-R Interval (ms)	Variation (ms)
Saline	1026 ± 57	82 ± 11
Propranolol	1148 ± 51	83 ± 14
Atropine	761 ± 24	17 ± 1

*Values are means ± standard errors. (Data from Pfeifer MA et al: Quantitative evaluation of cardiac parasympathetic activity in normal and diabetic mean. *Diabetes* 1982; **31**:339.)

ventricles, but the PR interval is abnormally long. In the form called **second-degree heart block,** not all atrial impulses are conducted to the ventricles. There may be, for example, a ventricular beat following every second or every third atrial beat (2:1 block, 3:1 block, etc). In another form of incomplete heart block, there are repeated sequences of beats in which the PR interval lengthens progressively until a ventricular beat is dropped (**Wenckebach phenomenon**). The PR interval of the cardiac cycle that follows each dropped beat is usually normal or only slightly prolonged (Fig 28–11).

Sometimes one branch of the bundle of His is interrupted, causing **right** or **left bundle branch block.** In bundle branch block, excitation passes normally down the bundle on the intact side and then sweeps back through the muscle to activate the ventricle on the blocked side. The ventricular rate is therefore normal, but the QRS complexes are prolonged and deformed (Fig 28–11). Block can also occur in the anterior or posterior fascicle of the left bundle branch, producing the condition called **hemiblock** or **fascicular block.** Left anterior hemiblock produces abnormal left axis deviation in the ECG, whereas left posterior hemiblock produces abnormal right axis deviation. It is not uncommon to find combinations of fascicle and branch blocks. The His bundle electrogram permits detailed analysis of the site of block when there is a defect in the conduction system.

Ectopic Foci of Excitation

Normally, myocardial cells do not discharge spontaneously, and the possibility of spontaneous discharge of the His bundle and Purkinje system is low because the normal pacemaker discharge of the SA node is more rapid than their rate of spontaneous discharge. However, in abnormal conditions, the His-Purkinje fibers may discharge spontaneously or myocardial fibers may develop oscillating membrane potentials and discharge spontaneously. In these conditions, **increased automaticity** of the heart is said to be present. If an irritable **ectopic focus** discharges once, the result is a beat that occurs before the expected next normal beat and transiently interrupts the cardiac rhythm (atrial, nodal, or ventricular **extrasystole** or **premature beat**). If the focus discharges repetitively at a rate higher than that of the SA node, it produces rapid, regular tachycardia (atrial, ventricular, or nodal **paroxysmal tachycardia** or **atrial flutter**).

Reentry

A more common cause of paroxysmal arrhythmias is a defect in conduction that permits a wave of excitation to propagate continuously within a closed circuit (**circus movement**). The mechanism that initiates repeated reentry of the impulse into a given part of the heart is illustrated diagrammatically in Fig 28–12. If the ring of tissue is in the AV node, the reentrant activity depolarizes the atrium, and the

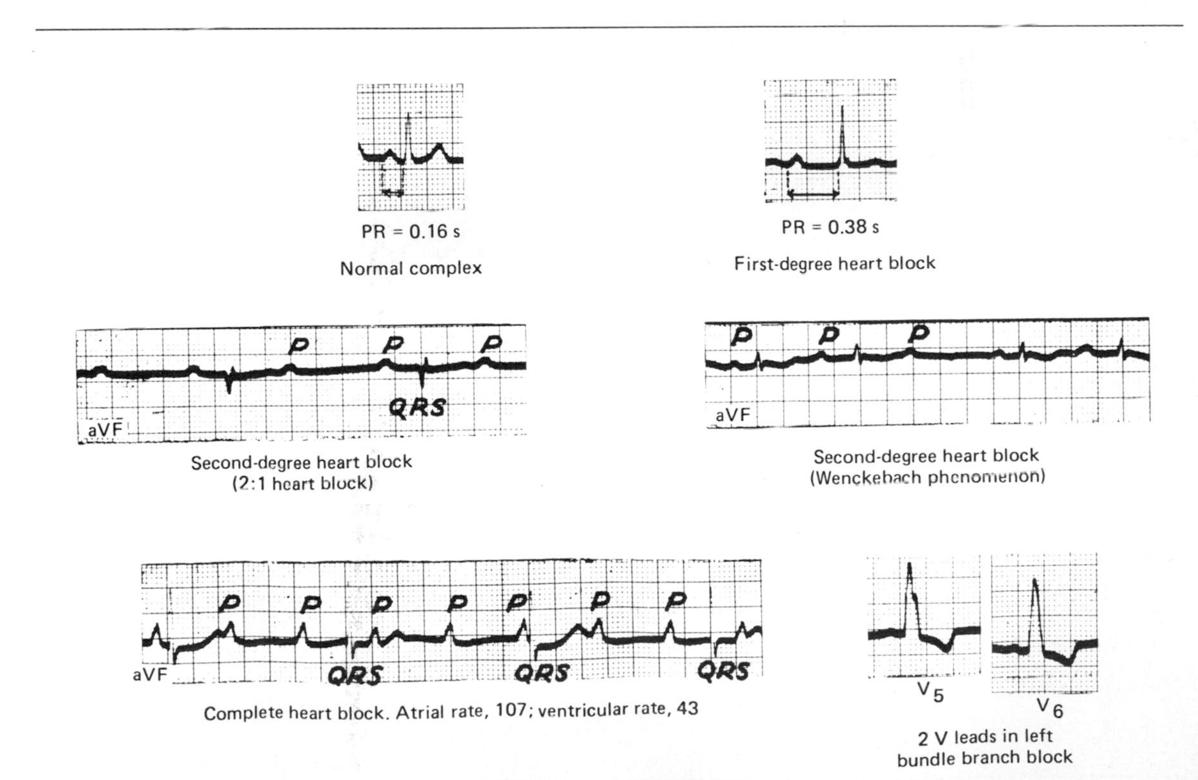

Figure 28–11. Heart block.

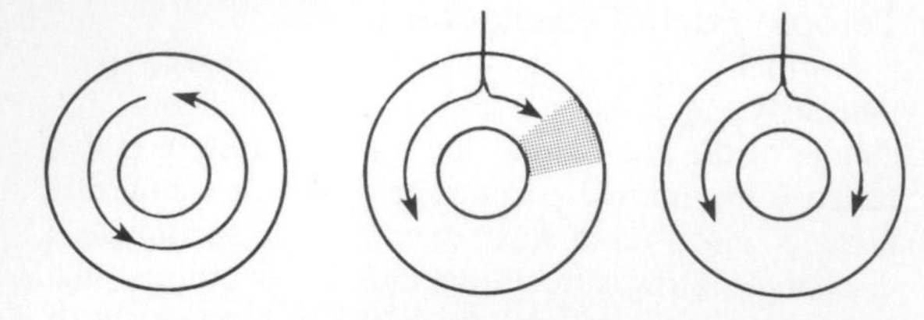

Figure 28–12. Depolarization of a ring of cardiac muscle. Normally, the impulse spreads in both directions in the ring (*left*) and the tissue immediately behind each branch of the impulse is refractory. When there is a transient block on one side (shaded area), the impulse goes around the ring, and if the transient block has now worn off (*right*), the impulse passes this area and continues to circle indefinitely (circus movement).

resulting atrial beat is called an echo beat. In addition, the reentrant activity in the node propagates back down to the ventricle, producing paroxysmal nodal tachycardia. Circus movements can also become established in the atria or ventricles. In individuals with an abnormal bundle of muscle or nodal tissue connecting the atria to the ventricles (bundle of Kent), the circus activity can pass in one direction through the AV node and in the other direction through the bundle, thus involving both the atria and the ventricles (see below).

Atrial Arrhythmias

Excitation spreading from an independently discharging focus in the atria stimulates the AV node and is conducted to the ventricles. The P waves of atrial extrasystoles are abnormal, but the QRST configurations are normal. The excitation depolarizes the SA node, which must repolarize and then depolarize to the firing level before it can initiate the next normal beat. Consequently, there is a pause between the extrasystole and the next normal beat that is usually equal in length to the interval between the normal beats preceding the extrasystole.

Atrial tachycardia occurs when an atrial focus discharges regularly or there is reentrant activity producing atrial rates of 150–220/min (Table 28–4). Sometimes, especially in digitalized patients, some degree of atrioventricular block is associated with the tachycardia (**paroxysmal atrial tachycardia with**

Table 28–4. Atrial arrhythmias. The tracings are portions of the continuous esophageal (E_{35}) ECG of a patient who manifested all the rhythms illustrated within a single 5-minute period.*

Rate of Discharge of Atrial Ectopic Focus	Arrhythmia
Occasional discharge at a rate slower than the basic sinus rhythm Atrial premature contractions (at arrows)	E_{35}
About 160 to about 220 Atrial tachycardia (with 1:1 conduction)	E_{35}
About 220 to about 350 Atrial flutter	E_{35}
Over 350 Atrial fibrillation	E_{35}

* Reproduced, with permission, from Goldman MJ: *Principles of Clinical Electrocardiography,* 12th ed. Lange, 1986.

block). When the atrial rate is 200–350/min, the condition is called atrial flutter. Atrial flutter is almost always associated with 2:1 or 3:1 AV block, because the normal AV node has a long refractory period and, in adults, it cannot conduct more than about 230 impulses per minute.

In **atrial fibrillation,** the atria beat very rapidly (300–500/min) in a completely irregular and disorganized fashion. The cause of atrial fibrillation is still a matter of debate, but it is probably due to multiple, concurrently circulating reentrant excitation wave fronts in the atrial muscle. Because the AV node discharges at irregular intervals, the ventricles beat at a completely irregular rate, usually 80–160/min (Table 28–4).

Consequences of Atrial Arrhythmias

Occasional atrial extrasystoles occur from time to time in most normal humans and have no pathologic significance. In paroxysmal atrial tachycardia and flutter, the ventricular rate may be so high that diastole is too short for adequate filling of the ventricles with blood between contractions. Consequently, cardiac output is reduced and symptoms of heart failure appear. The relationship between cardiac rate and cardiac output is discussed in detail in Chapter 29. Heart failure may also complicate atrial fibrillation when the ventricular rate is high. Acetylcholine liberated at vagal endings depresses conduction in the atrial musculature and AV node. This is why stimulating reflex vagal discharge by pressing on the eyeball (**"oculocardiac reflex"**) or massaging the carotid sinus often converts tachycardia and sometimes converts atrial flutter to normal sinus rhythm. Alternatively, vagal stimulation increases the degree of AV block, abruptly lowering the ventricular rate. Digitalis also depresses AV conduction and is used to lower a rapid ventricular rate in atrial fibrillation.

Ventricular Arrhythmias

Premature beats that originate in an ectopic ventricular focus usually have bizarrely shaped prolonged QRS complexes (Fig 28–13) owing to the slow spread of the impulse from the focus through the ventricular muscle to the rest of the ventricle. They are usually incapable of exciting the bundle of His, and retrograde conduction to the atrium therefore does not occur. In the meantime, the next succeeding normal SA nodal impulse depolarizes the atria. The P wave is usually buried in the QRS of the extrasystole. If the normal impulse reaches the ventricles, they are still in the refractory period following depolarization from the ectopic focus. However, the second succeeding impulse from the SA node produces a normal beat. Thus, ventricular premature beats are followed by a **compensatory pause** that is longer than the pause after an atrial extrasystole. Furthermore, ventricular premature beats do not interrupt the regular discharge of the SA node, whereas atrial premature beats interrupt and "reset" the normal rhythm. These differences make it possible to distinguish between atrial and ventricular premature beats by palpation of the pulse or auscultation of the heart at the bedside. If one foot is tapped in time with the regular normal rhythm, the beats after an atrial premature beat occur early with respect to the previous regular beats. However, the first normal beat after a ventricular premature beat occurs on the second foot tap after the premature beat, because the duration of a ventricular premature beat plus the preceding normal beat is equal to the duration of 2 normal beats (Fig 28–13).

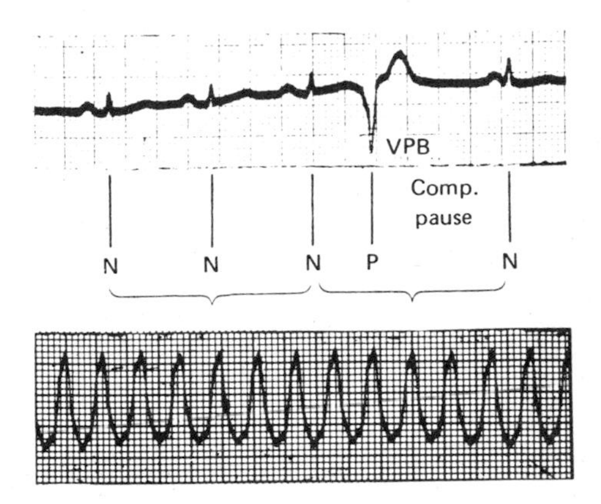

Figure 28–13. *Top:* Ventricular premature beats (VPB). The lines under the tracing illustrate the compensatory pause and show that the duration of the premature beat plus the preceding normal beat is equal to the duration of 2 normal beats. *Bottom:* Ventricular tachycardia.

Atrial and ventricular premature beats are not strong enough to produce a pulse at the wrist if they occur early in diastole, when the ventricles have not had time to fill with blood and the ventricular musculature is still in its relatively refractory period. They may not even open the aortic and pulmonary valves, in which case there is, in addition, no second heart sound (see Chapter 29).

Paroxysmal ventricular tachycardia (Fig 28–13) is in effect a series of rapid, regular ventricular depolarizations usually due to a circus movement involving the ventricles. Tachycardias originating above the ventricles (supraventricular tachycardias such as paroxysmal nodal tachycardia) can be distinguished from paroxysmal ventricular tachycardia by use of the HBE; in supraventricular tachycardias, there is a His bundle (H) deflection, whereas in ventricular tachycardias, there is none. Ventricular premature beats are common, and, in the absence of ischemic heart disease, they are usually benign. Ventricular tachycardia is more serious because cardiac output is decreased, and ventricular fibrillation is an occasional complication of ventricular tachycardia.

In **ventricular fibrillation** (Fig 28–14), the ventricular muscle fibers contract in a totally irregular and ineffective way because of the very rapid dis-

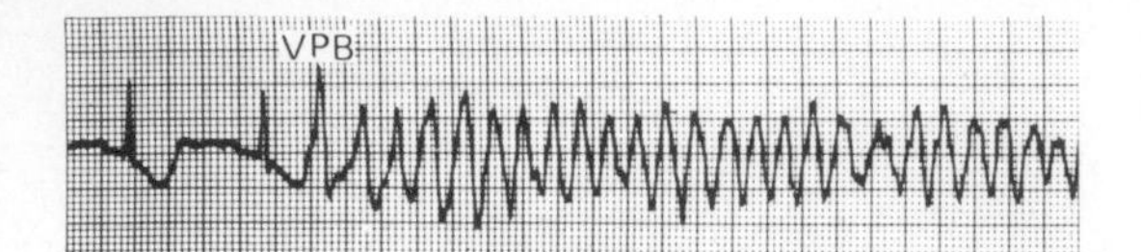

Figure 28–14. Ventricular fibrillation triggered by a ventricular premature beat (VPB) in the vulnerable period in a patient with a myocardial infarct. The patient was immediately defibrillated electrically and made an uneventful recovery.

charge of multiple ventricular ectopic foci or a circus movement. The fibrillating ventricles, like the fibrillating atria, look like a quivering ''bag of worms.'' Ventricular fibrillation can be produced by an electric shock or an extrasystole during a critical interval, the **vulnerable period.** The vulnerable period coincides in time with the mid portion of the T wave—ie, it occurs at a time when some of the ventricular myocardium is depolarized, some is incompletely repolarized, and some is completely repolarized. These are excellent conditions in which to establish reentry and a circus movement. The fibrillating ventricles cannot pump blood effectively, and circulation of the blood stops. Therefore, in the absence of emergency treatment, ventricular fibrillation that lasts more than a few minutes is fatal. The most frequent cause of sudden death in patients with myocardial infarcts is ventricular fibrillation.

Although ventricular fibrillation may be produced by electrocution, it can often be stopped and converted to normal sinus rhythm by means of electrical shocks. Electronic defibrillators are now available not only in hospitals but also in emergency vehicles and should be used as rapidly as possible. In addition, defibrillators can now be implanted surgically in patients who are at high risk for attacks of ventricular fibrillation.

In patients who are fibrillating or whose hearts have stopped, cardiac output and perfusion of the coronaries can be maintained by **cardiac massage.** Effective massage can be carried out without opening the chest. The person conducting external massage places the heel of one hand on the lower sternum above the xiphoid process and the heel of the other hand on top of the first (Fig 28–15). Pressure is applied straight down, depressing the sternum 4 or 5 cm toward the spine. This procedure is repeated 60 times per minute. Manually squeezing the ventricles is also effective if the chest is already open, but emergency thoracotomies should not be performed. If breathing has also stopped, cardiac massage should alternate with mouth-to-mouth resuscitation (see Chapter 37).

Accelerated AV Conduction

An interesting condition seen in some otherwise normal individuals who are prone to attacks of paroxysmal atrial arrhythmias is **accelerated AV**

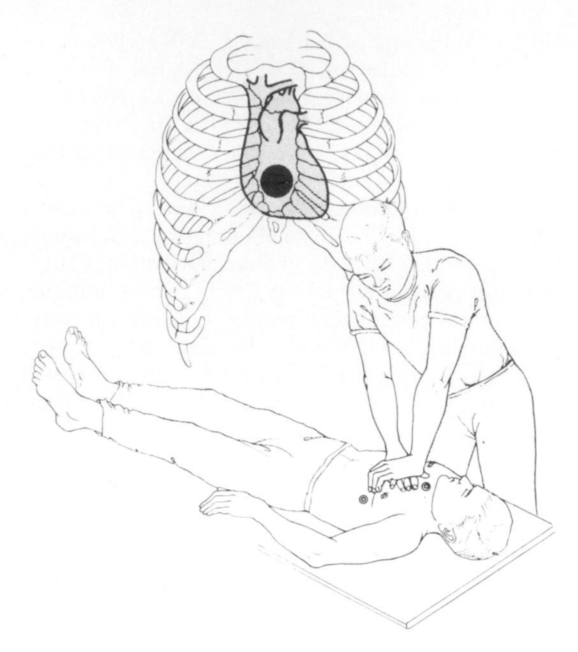

Figure 28–15. Technique of external, closed chest cardiac massage. The black circle on the diagram of the heart shows the area where force should be applied. Circles on the supine figure at the apex of the heart and just to the right of the upper portion of the sternum show where electrodes should be applied for defibrillation. (Reproduced, with permission, from Krupp MA, Schroeder SA, Tierney LM Jr [editors]: *Current Medical Diagnosis & Treatment*, 1989. Appleton & Lange, 1989.)

conduction (Wolff-Parkinson-White syndrome). Normally, the only conducting pathway between the atria and the ventricles is the AV node. Individuals with Wolff-Parkinson-White syndrome have an additional aberrant muscular or nodal tissue connection (**bundle of Kent**) between the atria and ventricles. This conducts more rapidly than the slow-conducting AV node, and one ventricle is excited early. The manifestations of its activation merge with the normal QRS pattern, producing a short PR interval and a prolonged QRS deflection slurred on the upstroke (Fig 28–16), with a normal interval between the start of the P wave and the end of the QRS complex (''PJ interval''). The paroxysmal atrial tachycardias seen in this syndrome often follow an atrial premature beat. This beat conducts normally down the AV node but finds the aberrant bundle refractory, since the bundle has a longer refractory period than the AV node. However, when ventricular activation spreads to the aberrant bundle, it is no longer refractory, and the impulse is transmitted retrograde to the atrium. A circus movement is thus established (Fig 28–17). Less commonly, an atrial premature beat finds the AV node refractory but reaches the ventricles via the bundle of Kent, setting up a circus movement in which the impulse passes from the ventricles to the atria via the AV node.

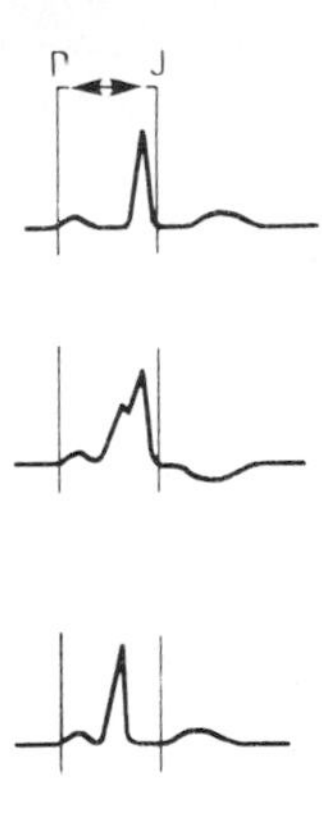

Figure 28–16. Accelerated AV conduction. *Top:* Normal sinus beat. *Middle:* Short PR interval; wide, slurred QRS complex; normal PJ interval (Wolff-Parkinson-White syndrome). *Bottom:* Short PR interval, normal QRS complex (Lown-Ganong-Levine syndrome). (Reproduced, with permission, from Goldman MJ: *Principles of Clinical Electrocardiography,* 12th ed. Lange, 1986.)

Attacks of paroxysmal supraventricular tachycardia, usually nodal tachycardia, are seen in individuals with short PR intervals and normal QRS complexes (**Lown-Ganong-Levine syndrome**). In this condition, depolarization presumably passes from the atria to the ventricles via an aberrant bundle that bypasses the AV node but enters the intraventricular conducting system distal to the node.

Slow Channel Ca^{2+} Blockers

Two Ca^{2+} channel blockers, verapamil and diltiazem, are of considerable value in the treatment of supraventricular tachycardias. Both drugs depress Ca^{2+} entry into conduction tissue and cardiac muscle. The resulting depression of conduction interrupts circus movements. These drugs plus a third, nifedipine, are also of value in the treatment of angina pectoris and myocardial infarction because they inhibit contraction of vascular smooth muscle, producing vasodilation and decreasing the afterload on the heart by lowering blood pressure (see Chapter 29).

ELECTROCARDIOGRAPHIC FINDINGS IN OTHER CARDIAC & SYSTEMIC DISEASES

Myocardial Infarction

When the blood supply to part of the myocardium is interrupted, there are prompt, profound changes in the myocardium that lead to irreversible changes and death of muscle cells (**myocardial infarction;** see Chapter 32). The ECG is very useful for diagnosing and locating areas of infarction. The underlying electrical events and the resulting electrocardiographic changes are complex, and only a brief review can be presented here.

The 3 major abnormalities that cause electrocardiographic changes in acute myocardial infarction are summarized in Table 28–5. The first change, abnormally rapid repolarization of the infarcted muscle fibers due to accelerated opening of K^+ channels, develops seconds after occlusion of a coronary artery in experimental animals. It lasts only a few minutes, but before it is over the resting membrane potential of the infarcted fibers declines because of the loss of intracellular K^+. Starting about 30 minutes later, the infarcted fibers also begin to depolarize more slowly than the surrounding normal fibers.

All 3 of these changes cause current flow that produces elevation of the ST segment in electrocardiographic leads recorded with electrodes over the infarcted area (Fig 28–18). Because of the rapid repolarization in the infarct, the membrane potential of the area is greater than it is in the normal area during the latter part of repolarization, making the normal region negative relative to the infarct. Extracellularly, current therefore flows out of the infarct into the normal area (since, by convention, current flow is from positive to negative). This current flows toward electrodes over the injured area, causing increased positivity between the S and T waves of the ECG. Similarly, the delayed depolarization of the infarcted cells causes the infarcted area to be positive relative to the healthy tissue (Table 28–5) during the early part of repolarization, and the result is also ST segment elevation. The remaining change, the decline in resting membrane potential during diastole, causes a current flow into the infarct during ventricular diastole. The result of this current flow is a depression of the TQ segment of the ECG. However, the electronic arrangement in electrocardiographic recorders is such that a TQ segment depression is recorded as an ST segment elevation. Thus, the hallmark of acute myocardial infarction is elevation of the ST segments in the leads overlying the area of infarction (Fig 28–18). Leads on the opposite side of the heart show ST segment depression.

After some days or weeks, the ST segment abnormalities subside. The dead muscle and scar tissue become electrically silent. The infarcted area is therefore negative relative to the normal myocardium during systole, and it fails to contribute its share of positivity to the electrocardiographic complexes. The manifestations of this negativity are multiple and subtle. Common changes include the appearance of a Q wave in some of the leads in which it was not previously present and an increase in the size of the normal Q wave in some of the other leads, although so-called non-Q-wave infarcts are also seen. These infarcts tend to be less severe, but there is a high incidence of subsequent reinfarction. Another finding in infarction of the anterior left ventricle is "failure of progression of the R wave"; the R wave fails to become successively larger in the precordial leads as the electrode is moved from right to left over the left ventricle. If the septum is infarcted, the conduction system may be damaged, causing bundle branch block or other forms of heart block.

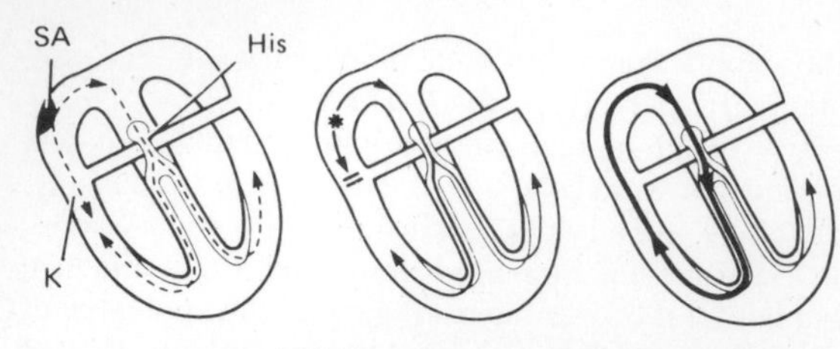

Figure 28–17. Initiation of circus movement causing paroxysmal tachycardia in patients with Wolff-Parkinson-White syndrome. K, bundle of Kent; His, AV node and bundle of His; SA, SA node. *Left:* Conduction pattern during normal sinus rhythm. *Center:* Atrial premature beat finds bundle of Kent refractory but reaches ventricles via AV node. *Right:* Impulse returns to atria via bundle of Kent, setting up circus movement. (Reproduced, with permission of the American Heart Association, Inc., from Wellens HJJ: Wolff-Parkinson-White syndrome. *Mod Concepts Cardiovasc Dis* 1983;**52**:53.)

Myocardial infarctions are often complicated by serious ventricular arrhythmias, with the threat of ventricular fibrillation and death. In experimental animals and presumably in humans, ventricular arrhythmias occur during 3 periods. During the first 30 minutes of an infarct, arrhythmias due to reentry are common. There follows a period relatively free from arrhythmias, but, starting 12 hours after infarction, arrhythmias occur as a result of increased automaticity. Arrhythmias occurring 3 days to several weeks after infarction are once again usually due to reentry.

Effects of Changes in the Ionic Composition of the Blood

Changes in ECF Na^+ and K^+ concentration would be expected to affect the potentials of the myocardial fibers, because the electrical activity of the heart depends upon the distribution of these ions across the muscle cell membranes. Clinically, a fall in the plasma level of Na^+ may be associated with low-voltage electrocardiographic complexes, but changes in the plasma K^+ level produce severe cardiac abnormalities. Hyperkalemia is a very dangerous and potentially lethal condition because of its effects on the heart. As the plasma K^+ level rises, the first change in the ECG is the appearance of tall peaked T waves, a manifestation of altered repolarization (Fig 28–19). At higher K^+ levels, paralysis of the atria and prolongation of the QRS complexes occur. Ventricular arrhythmias may develop. The resting membrane potential of the muscle fibers decreases as the extracellular K^+ concentration increases. The fibers eventually become unexcitable, and the heart stops in diastole. Conversely, a decrease in the plasma K^+ level causes prolongation of the PR interval, prominent U waves, and, occasionally, late T wave inversion in the precordial leads. If the T waves and U waves merge, the apparent QT interval is often

Table 28–5. Summary of the 3 major abnormalities of membrane polarization associated with acute myocardial infarction and the electrocardiographic abnormalities they produce. Because of the characteristics of their input capacitors, ECG records show a TQ segment depression as an ST segment elevation.

Abnormality	Cause	Resultant Extracellular Current Flow (Positive to Negative)	Resultant ECG Change in Leads Over Infarct
During repolarization − − + + − − + + | − − | + + + + | − − | + + − − + + − −	Rapid repolarization of infarcted cells	Out of infarct	ST segment elevation
At rest + + − − + + − − | + + | − − − − | + + | − − + + − − + +	Decreased resting membrane potential of infarcted cells	Into infarct	TQ segment depression (manifest as ST segment elevation)
During depolarization − − + + − − + + | − − | + + + + | − − | + + − − + + − −	Delayed depolarization of infarcted cells	Out of infarct	ST segment elevation

Legend:

Extracellular fluid		
Normal muscle	Infarcted muscle	Normal muscle
Extracellular fluid		

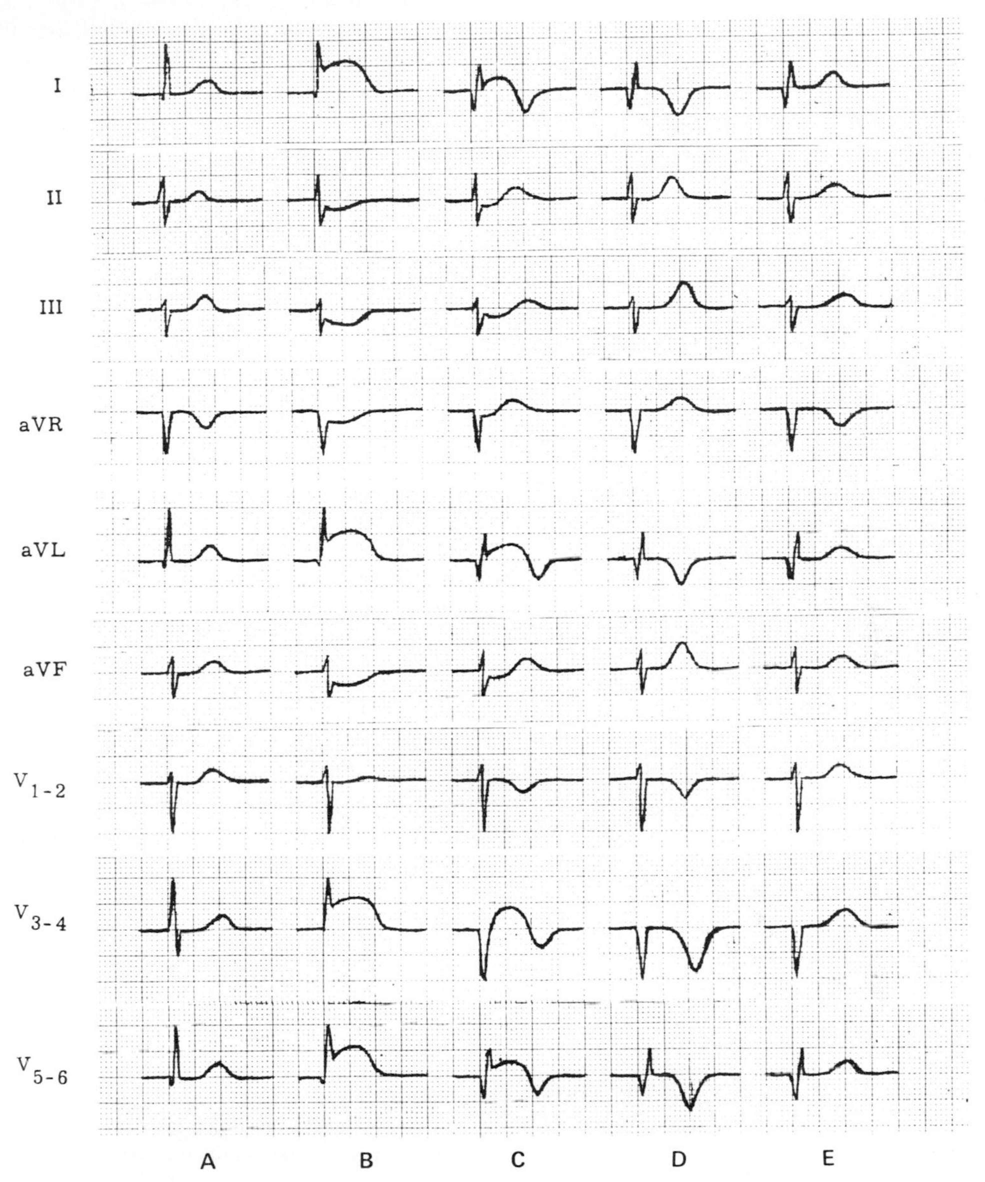

Figure 28–18. Diagrammatic illustration of serial electrocardiographic patterns in anterior infarction. *A:* Normal tracing. *B:* Very early pattern (hours after infarction): ST segment elevation in I, aVL, and V_{3-6}; reciprocal ST depression in II, III, and aVF. *C:* Later pattern (many hours to a few days): Q waves have appeared in I, aVL, and V_{5-6}. QS complexes are present in V_{3-4}. This indicates that the major transmural infarction is underlying the area recorded by V_{3-4}; ST segment changes persist but are of lesser degree, and the T waves are beginning to invert in those leads in which the ST segments are elevated. *D:* Late established pattern (many days to weeks): The Q waves and QS complexes persist; the ST segments are isoelectric; the T waves are symmetric and deeply inverted in leads that had ST elevation and tall in leads that had ST depression. This pattern may persist for the remainder of the patient's life. *E:* Very late pattern: This may occur many months to years after the infarct. The abnormal Q waves and QS complexes persist. The T waves have gradually returned to normal. (Reproduced, with permission, from Goldman MJ: *Principles of Clinical Electrocardiography,* 12th ed. Lange, 1986.)

prolonged; if the T and U waves are separated, the true QT interval is seen to be of normal duration. Hypokalemia is a serious condition, but it is not as rapidly fatal as hyperkalemia.

Increases in extracellular Ca^{2+} concentration enhance myocardial contractility. When large amounts of Ca^{2+} are infused into experimental animals, the heart relaxes less during diastole and eventually stops in systole (**calcium rigor**). However, in clinical conditions associated with hypercalcemia, the plasma calcium level is rarely if ever high enough to affect the heart. Hypocalcemia causes

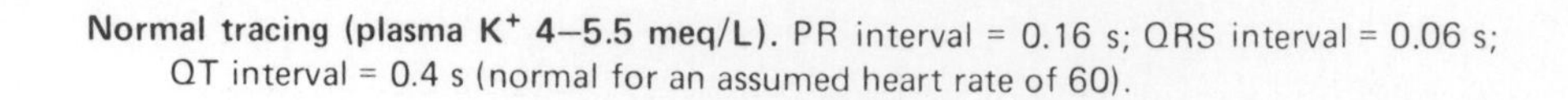

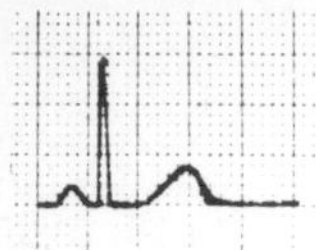

Normal tracing (plasma K⁺ 4–5.5 meq/L). PR interval = 0.16 s; QRS interval = 0.06 s; QT interval = 0.4 s (normal for an assumed heart rate of 60).

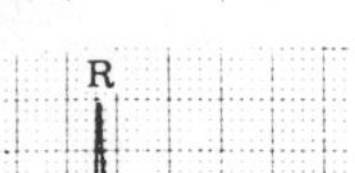

Hypokalemia (plasma K⁺ ±3.5 meq/L). PR interval = 0.2 s; QRS interval = 0.06 s; ST segment depression. A prominent U wave is now present immediately following the T. The actual QT interval remains 0.4 s. If the U wave is erroneously considered a part of the T, a falsely prolonged QT interval of 0.6 s will be measured.

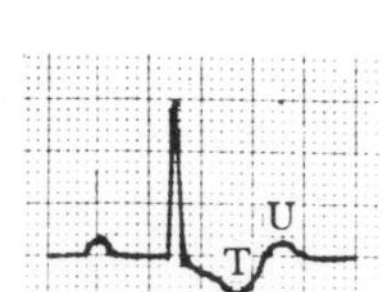

Hypokalemia (plasma K⁺ ±2.5 meq/L). The PR interval is lengthened to 0.32 s; the ST segment is depressed; the T wave is inverted; a prominent U wave is seen. The true QT interval remains normal.

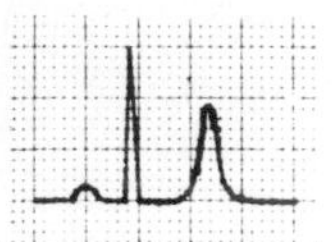

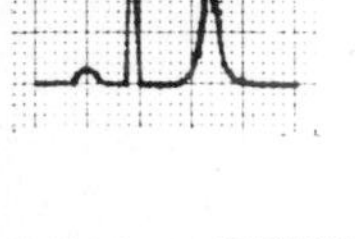

Hyperkalemia (plasma K⁺ ±7.0 meq/L). The PR and QRS intervals are within normal limits. Very tall, slender peaked T waves are now present.

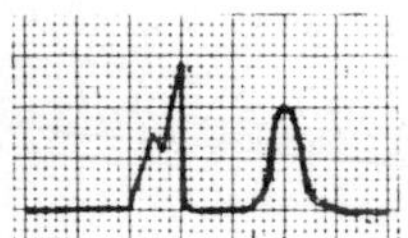

Hyperkalemia (plasma K⁺ ±8.5 meq/L). There is no evidence of atrial activity; the QRS complex is broad and slurred and the QRS interval has widened to 0.2 s. The T waves remain tall and slender. Further elevation of the plasma K⁺ level may result in ventricular tachycardia and ventricular fibrillation.

Figure 28–19. Correlation of plasma K⁺ level and the ECG, assuming plasma calcium level is normal. The diagrammed complexes are left ventricular epicardial leads. (Reproduced, with permission, from Goldman MJ: *Principles of Clinical Electrocardiography*, 12th ed. Lange, 1986.)

prolongation of the ST segment and consequently of the QT interval, a change that is also produced by phenothiazines and tricyclic antidepressant drugs and by various diseases of the central nervous system.

Changes in plasma calcium and potassium levels produce relatively marked changes in the sensitivity of the heart to digitalis. Hypercalcemia enhances digitalis toxicity, and hyperkalemia counteracts it. Mg^{2+}, which depresses the myocardium, also counteracts digitalis toxicity.

In experimental animals, acidosis increases the duration of diastole and decreases the strength of the heartbeat, but no electrocardiographic abnormalities specifically due to the changes in H^+ concentration of the body fluids in patients with acidosis or alkalosis have been demonstrated.

The Heart as a Pump

29

INTRODUCTION

The orderly depolarization process described in the previous chapter triggers a wave of contraction that spreads through the myocardium. In single muscle fibers, contraction starts just after depolarization and lasts until about 50 ms after repolarization is completed (Fig 3–13). Atrial systole starts after the P wave of the ECG; ventricular systole starts near the end of the R wave and ends just after the T wave. The contraction produces sequential changes in pressures and flows in the heart chambers and blood vessels. It should be noted that the term **systolic pressure** in the vascular system refers to the peak pressure reached during systole, not the mean pressure; similarly, the **diastolic pressure** refers to the lowest pressure during diastole.

The heart is separated from the rest of the thoracic viscera by the pericardium. The myocardium itself is covered by the fibrous epicardium. The pericardial sac normally contains 5–30 mL of clear fluid, which lubricates the heart and permits it to contract with minimal friction.

MECHANICAL EVENTS OF THE CARDIAC CYCLE

Events in Late Diastole

Late in diastole, the mitral and tricuspid valves between the atria and ventricles are open and the aortic and pulmonary valves are closed. Blood flows into the heart throughout diastole, filling the atria and ventricles. The rate of filling declines as the ventricles become distended, and—especially when the heart rate is low—the cusps of the atrioventricular (AV) valves drift toward the closed position (Fig 29–1). The pressure in the ventricles remains low.

Atrial Systole

Contraction of the atria propels some additional blood into the ventricles, but about 70% of the ventricular filling occurs passively during diastole. Contraction of the atrial muscle that surrounds the orifices of the venae cavae and pulmonary veins narrows their orifices, and the inertia of the blood moving toward the heart tends to keep blood in it; however, there is some regurgitation of blood into the veins during atrial systole.

Ventricular Systole

At the start of ventricular systole, the mitral and tricuspid (AV) valves close. Ventricular muscle initially shortens relatively little, but intraventricular pressure rises sharply as the myocardium presses on the blood in the ventricle (Fig 29–2). This period of **isovolumetric (isovolumic, isometric) ventricular contraction** lasts about 0.05 second, until the pressures in the left and right ventricles exceed the pressures in the aorta (80 mm Hg; 10.6 kPa) and pulmonary artery (10 mm Hg) and the aortic and pulmonary valves open. During isovolumetric contraction, the AV valves bulge into the atria, causing a small but sharp rise in atrial pressure (Fig 29–3).

When the aortic and pulmonary valves open, the phase of **ventricular ejection** begins. Ejection is

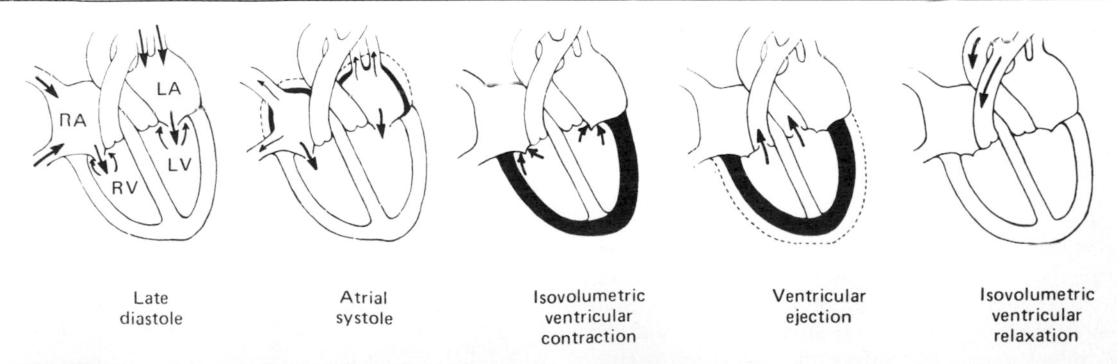

Figure 29–1. Blood flow in the heart and great vessels during the cardiac cycle. The portions of the heart contracting in each phase are indicated in black. RA and LA, right and left atria; RV and LV, right and left ventricles.

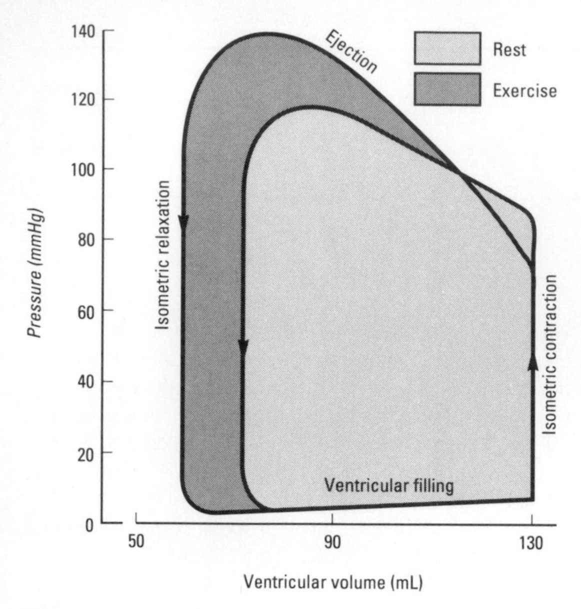

Figure 29–2. Pressure-volume loop of the cardiac ventricles at rest and during exercise.

rapid at first, slowing down as systole progresses. The intraventricular pressure rises to a maximum and then declines somewhat before ventricular systole ends. Peak left ventricular pressure is about 120 mm Hg, and peak right ventricular pressure is 25 mm Hg or less. Late in systolc, the aortic pressure actually exceeds the ventricular, but for a short period momentum keeps the blood moving forward. The AV valves are pulled down by the contractions of the ventricular muscle, and atrial pressure drops. The amount of blood ejected by each ventricle per stroke at rest is 70–90 mL. The **end-diastolic ventricular volume** is about 130 mL. Thus, the **ejection fraction,** the percentage of the ventricular volume ejected with each stroke, is about 65%, and about 50 mL of blood remains in each ventricle at the end of systole (**end-systolic ventricular volume**). The ejection fraction is a relatively good index of ventricular function, and as such is calculated in a variety of clinical situations.

Early Diastole

Once the ventricular muscle is fully contracted, the already falling ventricular pressures drop more rapidly. This is the period of **protodiastole.** It lasts about 0.04 second. It ends when the momentum of the ejected blood is overcome and the aortic and pulmonary valves close, setting up transient vibrations in the blood and blood vessel walls. After the valves are closed, pressure continues to drop rapidly during the period of **isovolumetric ventricular relaxation.** Isovolumetric relaxation ends when the ventricular pressure falls below atrial pressure and the AV valves open, permitting the ventricles to fill. Filling is rapid at first, then slows as the next cardiac contraction approaches. Atrial pressure continues to rise after the end of ventricular systole until the AV valves open, then drops and slowly rises again until the next atrial systole.

Timing

Although events on the 2 sides of the heart are similar, they are somewhat asynchronous. Right atrial systole precedes left atrial systole, and contraction of the right ventricle starts after that of the left (see Chapter 28). However, since pulmonary arterial pressure is lower than aortic pressure, right ventricular ejection begins before left ventricular ejection. During expiration, the pulmonary and aortic valves close at the same time; but during inspiration, the aortic valve closes slightly before the pulmonary. The slower closure of the pulmonary valve is due to a decrease in the impedance of the pulmonary vascular tree, with more prolonged ejection plus an increase in systemic venous return, which also prolongs ejection. When measured over a period of minutes, the output of the 2 ventricles is, of course, equal, but transient differences in output during the respiratory cycle occur in normal individuals.

Length of Systole & Diastole

Cardiac muscle has the unique property of contracting and repolarizing faster when the heart rate is high (see Chapter 3), and the duration of systole decreases from 0.3 second at a heart rate of 65 to 0.16 second at a rate of 200 (Table 29–1). The shortening is due mainly to a decrease in the duration of systolic ejection. However, the duration of systole is much more fixed than that of diastole, and when the heart rate is increased, diastole is shortened to a much greater degree. For example, at a heart rate of 65, the duration of diastole is 0.62 second, whereas at a heart rate of 200, it is only 0.14 second. This fact has important physiologic and clinical implications. It is during diastole that the heart muscle rests, and coronary blood flow to the subendocardial portions of the left ventricle occurs only during diastole (see

Table 29–1. Variation in length of action potential and associated phenomena with cardiac rate. All values are in seconds. (Courtesy of AC Barger and GS Richardson.)

	Heart Rate 75/min	Heart Rate 200/min	Skeletal Muscle
Duration, each cardiac cycle	0.80	0.30	. . .
Duration of systole	0.27	0.16	. . .
Duration of action potential	0.25	0.15	0.005
Duration of absolute refractory period	0.20	0.13	0.004
Duration of relative refractory period	0.05	0.02	0.003
Duration of diastole	0.53	0.14	. . .

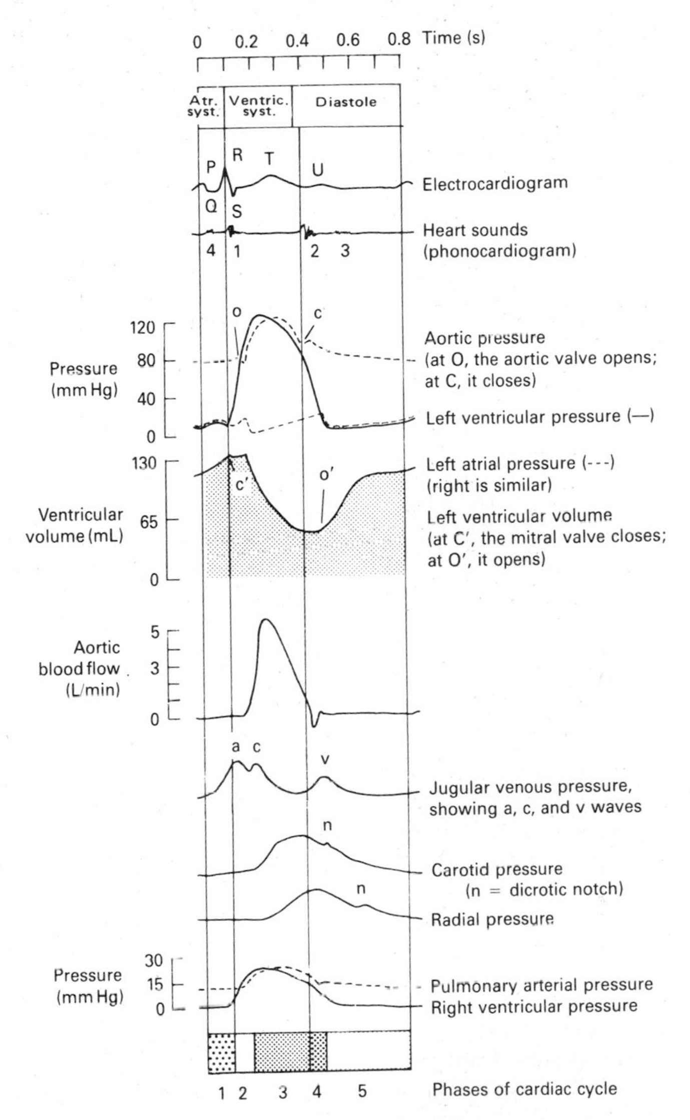

Figure 29–3. Events of the cardiac cycle at a heart rate of 75 beats/min. The phases of the cardiac cycle identified by the numbers at the bottom are as follows: 1, atrial systole; 2, isovolumetric ventricular contraction; 3, ventricular ejection; 4, isovolumetric ventricular relaxation; 5, ventricular filling. Note that late in systole, aortic pressure actually exceeds left ventricular pressure. However, the momentum of the blood keeps it flowing out of the ventricle for a short period. The pressure relationships in the right ventricle and pulmonary artery are similar. Atr syst, atrial systole; Ventric syst, ventricular systole.

Chapter 32). Furthermore, most of the ventricular filling occurs in diastole. At heart rates up to about 180 beats/min, filling is adequate as long as there is ample venous return, and cardiac output per minute is increased by an increase in rate. However, at very high heart rates, filling may be compromised to such a degree that cardiac output falls and symptoms of heart failure develop.

Because it has a prolonged action potential, cardiac muscle is in its refractory period and will not contract in response to a second stimulus until near the end of the initial contraction (Fig 3–13; Table 29–1). Therefore, cardiac muscle cannot be tetanized like skeletal muscle. The highest rate at which the ventricles can contract is theoretically about 400/min, but in adults the AV node will not conduct more than about 230 impulses/min because of its long refractory period. A ventricular rate of more than 230 is seen only in paroxysmal ventricular tachycardia (see Chapter 28).

Exact measurement of the duration of isometric ventricular contraction is difficult in clinical situations, but it is relatively easy to measure the duration of **total electromechanical systole (QS_2)**, the **preejection period (PIP)**, and the **left ventricular ejection time (LVET)** by recording the ECG, phonocardiogram, and carotid pulse simultaneously.

QS_2 is the period from the onset of the QRS complex to the closure of the aortic valves, as determined by the onset of the second heart sound. LVET is the period from the beginning of the carotid pressure rise to the dicrotic notch (see below). PEP is the difference between QS_2 and LVET and represents the time for the electrical as well as the mechanical events that precede systolic ejection. The ratio PEP/LVET is normally about 0.35, and it increases without a change in QS_2 as left ventricular performance is compromised in a variety of cardiac diseases.

Arterial Pulse

The blood forced into the aorta during systole not only moves the blood in the vessels forward but also sets up a pressure wave that travels along the arteries. The pressure wave expands the arterial walls as it travels, and the expansion is palpable as the **pulse.** The rate at which the wave travels, which is independent of and much higher than the velocity of blood flow, is about 4 m/s in the aorta, 8 m/s in the large arteries, and 16 m/s in the small arteries of young adults. Consequently, the pulse is felt in the radial artery at the wrist about 0.1 second after the peak of systolic ejection into the aorta (Fig 29–3). With advancing age, the arteries become more rigid, and the pulse wave moves faster.

The strength of the pulse is determined by the pulse pressure and bears little relation to the mean pressure. The pulse is weak (''thready'') in shock. It is strong when stroke volume is large, eg, during exercise or after the administration of histamine. When the pulse pressure is high, the pulse waves may be large enough to be felt or even heard by the individual (palpitation, ''pounding heart''). When the aortic valve is incompetent (aortic insufficiency), the pulse is particularly strong, and the force of systolic ejection may be sufficient to make the head nod with each heartbeat. The pulse in aortic insufficiency is called a **collapsing, Corrigan,** or **water-hammer pulse.** A water hammer is an evacuated glass tube half-filled with water that was a popular toy in the 19th century. When held in the hand and inverted, it delivers a short, hard knock.

The **dicrotic notch,** a small oscillation on the falling phase of the pulse wave caused by vibrations set up when the aortic valve snaps shut (Fig 29–3), is visible if the pressure wave is recorded but is not palpable at the wrist. There is also a dicrotic notch on the pulmonary artery pressure curve due to the closure of the pulmonary valves.

Atrial Pressure Changes & the Jugular Pulse

Atrial pressure rises during atrial systole and continues to rise during isovolumetric ventricular contraction when the AV valves bulge into the atria. When the AV valves are pulled down by the contracting ventricular muscle, pressure falls rapidly and then rises as blood flows into the atria until the AV valves open early in diastole. The return of the AV valves to their relaxed position also contributes to this pressure rise by reducing atrial capacity. The atrial pressure changes are transmitted to the great veins, producing 3 characteristic waves in the jugular pulse (Figs 29–3 and 29–4). The **a wave** is due to atrial systole. As noted above, some blood regurgitates into the great veins when the atria contract, even though the orifices of the great veins are constricted. In addition, venous inflow stops, and the resultant rise in venous pressure contributes to the a wave. The **c wave** is the transmitted manifestation of the rise in atrial pressure produced by the bulging of the tricuspid valve into the atria during isometric ventricular contraction. the **v wave** mirrors the rise in atrial pressure before the tricuspid valve opens during diastole. The jugular pulse waves are superimposed on the respiratory fluctuations in venous pressure. Venous pressure falls during inspiration as a result of the increased negative intrathoracic pressure and rises again during expiration.

Careful bedside inspection of the pulsations of the jugular veins may give clinical information of some importance. For example, in tricuspid insufficiency there is a giant c wave with each ventricular systole. In complete heart block, when the atria and ventricles are beating at different rates, the a waves that are not synchronous with the radial pulse can be made out, and there is a giant a wave (''cannon wave'') whenever the atria contract while the tricuspid valve is closed. It may also be possible to distinguish atrial from ventricular extrasystoles by inspection of the jugular pulse, because atrial premature beats produce an a wave, whereas ventricular premature beats do not.

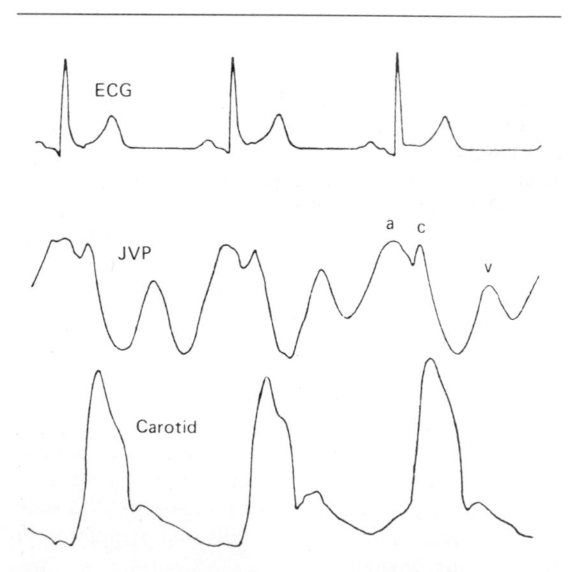

Figure 29–4. Jugular venous pressure record (JVP), showing *a*, *c*, and *v* waves, compared with the simultaneously recorded ECG and carotid pressure record. (Redrawn and reproduced, with permission, from Wood PW: *Diseases of the Heart*, Lippincott, 1956.)

Heart Sounds

Two sounds are normally heard through a stethoscope during each cardiac cycle. The first is a low, slightly prolonged "lub" (**first sound**), caused by vibrations set up by the sudden closure of the mitral and tricuspid valves at the start of ventricular systole. The second is a shorter, high-pitched "dup" (**second sound**), caused by vibrations associated with closure of the aortic and pulmonary valves just after the end of ventricular systole. A soft, low-pitched **third sound** is heard about one-third of the way through diastole in many normal young individuals. It coincides with the period of rapid ventricular filling and is probably due to vibrations set up by the inrush of blood. A **fourth sound** can sometimes be heard immediately before the first sound when atrial pressure is high or the ventricle is stiff in conditions such as ventricular hypertrophy. It is due to ventricular filling and is rarely heard in normal adults.

The first sound has a duration of about 0.15 second and a frequency of 25–45 Hz. It is soft when the heart rate is slow, because the ventricles are well filled with blood and the leaflets of the AV valves float together before systole. The second sound lasts about 0.12 second, with a frequency of 50 Hz. It is loud and sharp when the diastolic pressure in the aorta or pulmonary artery is elevated, causing the respective valves to shut briskly at the end of systole. The interval between aortic and pulmonary valve closure during inspiration is frequently long enough for the second sound to be reduplicated (physiologic splitting of the second sound). Splitting also occurs in various diseases. The third sound has a duration of 0.1 second.

Murmurs

Murmurs and **bruits** are abnormal sounds heard in various parts of the vascular system. As discussed in detail in Chapter 30, blood flow is laminar and nonturbulent up to a critical velocity; above this velocity, and beyond an obstruction, blood flow is turbulent. Laminar flow is silent, but turbulent flow creates sounds. Blood flow speeds up when an artery or a heart valve is narrowed.

Examples of vascular sounds outside the heart are the bruit heard over a large, highly vascular goiter and the murmurs heard over an aneurysmal dilation of one of the large arteries, an arteriovenous (A-V) fistula, or a patent ductus arteriosus.

The major but certainly not the only cause of cardiac murmurs is disease of the heart valves. When a valve is narrowed (**stenosis**), blood flow through it in the normal direction is accelerated and turbulent. When a valve is incompetent, blood flows backwards through it (**regurgitation** or **insufficiency**), again through a narrow orifice that accelerates flow. The timing (systolic or diastolic) of a murmur due to stenosis or insufficiency of any particular valve (Table 29–2) can be predicted from a knowledge of the mechanical events of the cardiac cycle. Murmurs due to disease of a particular valve can generally be heard best when the stethoscope is over that particular valve, and there are other aspects of the duration, character, accentuation, and transmission of the sound that help to locate its origin in one valve or the other. One of the loudest murmurs is that produced when blood flows backward in diastole through a hole in a cusp of the aortic valve. Most murmurs can be heard only with the aid of the stethoscope, but this high-pitched musical diastolic murmur is sometimes audible to the unaided ear several feet from the patient.

In patients with congenital interventricular septal defects, flow from the left to the right ventricle causes a systolic murmur. Soft murmurs may also be heard in patients with interatrial septal defects, although they are not a constant finding.

It should be emphasized that soft systolic murmurs are common in individuals, especially children, who have no cardiac disease. Systolic murmurs are also common in anemic patients as a result of the low viscosity of the blood and the rapid flow (see Chapter 30).

Echocardiography

Ventricular output and other aspects of cardiac function can be evaluated by **echocardiography,** a noninvasive technique that does not involve injections or insertion of a catheter. In echocardiography, pulses of ultrasonic waves, commonly at a frequency of 2.25 MHz, are emitted from a transducer that also functions as a receiver to detect waves reflected back from various parts of the heart. Reflections occur wherever acoustic impedance changes, and a recording of the echoes displayed against time on an oscilloscope provides a record of the movements of the ventricular wall, septum, and valves during the cardiac cycle. This technique has considerable clinical usefulness, particularly in evaluating and planning therapy in patients with valvular lesions.

CARDIAC OUTPUT

Methods of Measurement

In experimental animals, cardiac output can be measured with an electromagnetic flow meter placed on the ascending aorta. Two methods of measuring output that are applicable to humans are the **direct Fick method** and the **indicator dilution method.**

The **Fick principle** states that the amount of a substance taken up by an organ (or by the whole

Table 29 – 2. Heart murmurs.

Valve	Abnormality	Timing of Murmur
Aortic or pulmonary	Stenosis	Systolic
	Insufficiency	Diastolic
Mitral or tricuspid	Stenosis	Diastolic
	Insufficiency	Systolic

body) per unit of time is equal to the arterial level of the substance minus the venous level (**A-V difference**) times the blood flow. This principle can be applied, of course, only in situations in which the arterial blood is the sole source of the substance taken up. The principle can be used to determine cardiac output by measuring the amount of O_2 consumed by the body in a period of time and dividing this value by the A-V difference across the lungs. Because arterial blood has the same content in all parts of the body, the arterial O_2 content can be measured in a sample obtained from any convenient artery. A sample of venous blood in the pulmonary artery is obtained by means of a cardiac catheter. Right atrial blood has been used in the past, but mixing of this blood may be incomplete, so that the sample is not representative of the whole body. An example of the calculation of cardiac output using a typical set of values is as follows:

$$\textbf{Output of left ventricle} = \frac{\textbf{O}_2\textbf{ consumption (mL/min)}}{[\textbf{A}_{\textbf{O}_2}] - [\textbf{V}_{\textbf{O}_2}]}$$

$$= \frac{\textbf{250 mL/min}}{\textbf{190 mL/L arterial blood} - \textbf{140 mL/L venous blood in pulmonary artery}}$$

$$= \frac{\textbf{250 mL/min}}{\textbf{50 mL/L}}$$

$$= \textbf{5 L/min}$$

It has now become commonplace to insert a long catheter through a forearm vein and to guide its tip into the heart with the aid of a fluoroscope. The technique was initially developed by Forssmann, who catheterized himself but was summarily fired from his job when he sought permission to explore the use of the catheter in others for diagnostic purposes. However, the procedure is now known to be essentially harmless. Catheters can be inserted not only into the right atrium but also through the atrium and the right ventricle into the small branches of the pulmonary artery. Catheters can also be inserted in peripheral arteries and guided in a retrograde direction to the heart and into coronary or other arteries.

In the indicator dilution technique, a known amount of a dye such as indocyanine green or a radioactive isotope is injected into an arm vein, and the concentration of the indicator in serial samples of arterial blood is determined. The output of the heart is equal to the amount of indicator injected divided by its average concentration in arterial blood after a single circulation through the heart (Fig 29–5). The indicator must, of course, be a substance that stays in the bloodstream during the test and that has no harmful or hemodynamic effects. In practice, the log of the indicator concentration in the serial arterial samples is plotted against time as the concentration rises, falls, and then rises again as the indicator recirculates. The initial decline in concentration, linear on a semilog plot, is extrapolated to the abscissa, giving the time for first passage of the indicator through the circulation. The cardiac output for that period is calculated (Fig 29–5) and then converted to output per minute. Similar estimates of ventricular output have been made by injecting a suitable radioactive substance (radionuclide) and monitoring radioactivity over the heart during the first pass of the radionuclide through it.

One of the most popular indicator dilution techniques in recent years has been **thermodilution** in which the indicator used is cold saline. The saline is injected into the right atrium through one side of a double-lumen catheter, and the temperature change in

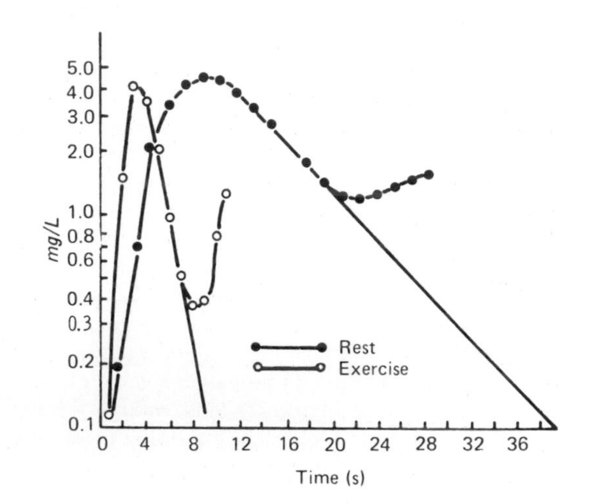

$$F = \frac{E}{\int_0^{\infty} C\,dt}$$

F = flow
E = amount of indicator injected
C = instantaneous concentration of indicator in arterial blood

In the **rest** example above,

$$\text{Flow in 39 s (time of first passage)} = \frac{\text{5 mg injection}}{\text{1.6 mg/L (avg concentration)}}$$

$$\text{Flow} = \text{3.1 L in 39 s}$$

$$\text{Flow (cardiac output)/min} = 3.1 \times \frac{60}{39} = 4.7\text{ L}$$

For the **exercise** example,

$$\text{Flow in 9 s} = \frac{\text{5 mg}}{\text{1.51 mg/L}} = 3.3\text{ L}$$

$$\text{Flow/min} = 3.3 \times \frac{60}{9} = 22.0\text{ L}$$

Figure 29–5. Determination of cardiac output by indicator (dye) dilution. (Data and graph from Asmussen E, Nielsen M: The cardiac output in rest and work determined by the acetylene and the dye injection methods. *Acta Physiol Scand* 1952:**27**:217.)

the blood is recorded in the pulmonary artery, using a thermistor in the other, longer side of the catheter. The temperature change is inversely proportionate to the amount of blood flowing through the pulmonary artery, ie, to the extent that the cold saline is diluted by blood. This technique has 2 important advantages: (1) the cold is dissipated in the tissues, so recirculation is not a problem; and (2) since the saline is completely innocuous, it is easy to make repeated determinations.

Cardiac Output in Various Conditions

The amount of blood pumped out of each ventricle per beat, the **stroke volume,** is about 80 mL in a resting man of average size in the supine position (80 mL from the left ventricle and 80 mL from the right, with the 2 ventricular pumps in series). The output of the heart per unit time is the **cardiac output.** In a resting, supine man, it averages about 5.5 L/min (80 mL x 69 beats/min). There is a correlation between resting cardiac output and body surface area. The output per minute per square meter of body surface (the **cardiac index**) averages 3.2 L. The effects of various conditions on cardiac output are summarized in Table 29–3.

Factors Controlling Cardiac Output

Variations in cardiac output can be produced by changes in cardiac rate or stroke volume (Fig 29–6). The cardiac rate is controlled primarily by the cardiac innervation, sympathetic stimulation increasing the rate and parasympathetic stimulation decreasing it (see Chapter 28). The stroke volume is also determined in part by neural input, sympathetic stimuli making the myocardial muscle fibers contract with greater strength at any given length and parasympathetic stimuli having the opposite effect. When the strength of contraction increases without an increase in fiber length, more of the blood that normally remains in the ventricles is expelled; ie, the ejection fraction increases and the end-systolic ventricular blood volume falls. The cardiac accelerator action of the catecholamines liberated by sympathetic stimulation is referred to as their **chronotropic action,** whereas their effect on the strength of cardiac contraction is called their **inotropic action.** Factors that increase the strength of cardiac contraction are said to be positively inotropic; those that decrease it are said to be negatively inotropic.

The force of contraction of cardiac muscle is dependent upon its preloading and its afterloading. These factors are illustrated in Fig 29–7, in which a muscle strip is stretched by a load (the **preload**) that rests on a platform. The initial phase of the contraction is isometric; the elastic component in series with the contractile element is stretched, and tension increases until it is sufficient to lift the load. The tension at which the load is lifted is the **afterload.** The muscle then contracts isotonically without developing further tension. In vivo, the preload is the degree to which the myocardium is stretched before it contracts and the afterload is the resistance against which blood is expelled.

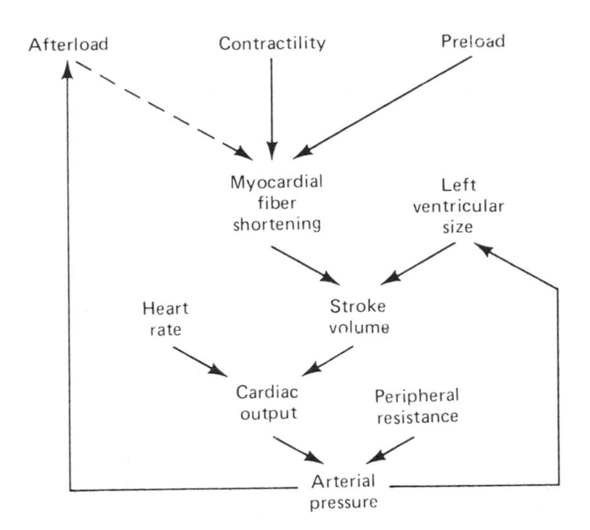

Figure 29–6. Interactions between the components that regulate cardiac output and arterial pressure. Solid lines indicate increases, and the dashed line indicates a decrease. (Modified from Braunwald E: Regulation of the circulation, *N Engl J Med* 1974:**290**:1124.)

Table 29 – 3. Effect of various conditions on cardiac output. Approximate percentage changes are shown in parentheses.

	Condition or Factor
No change	Sleep Moderate changes in environmental temperature
Increase	Anxiety and excitement (50 – 100%) Eating (30%) Exercise (up to 700%) High environmental temperature Pregnancy Epinephrine Histamine
Decrease	Sitting or standing from lying position (20 – 30%) Rapid arrhythmias Heart disease

Relation of Tension to Length in Cardiac Muscle

The length-tension relationship in cardiac muscle (Fig 3–15) is similar to that in skeletal muscle (Fig 3–9); as the muscle is stretched, the developed tension increases to a maximum and then declines as stretch becomes more extreme. Starling pointed this out when he stated that the "energy of contraction is proportional to the initial length of the cardiac muscle fiber." This pronouncement has come to be known as **Starling's law of the heart** or the **Frank-Starling law.** For the heart, the length of the muscle fibers (ie, the extent of the preload) is proportionate to the

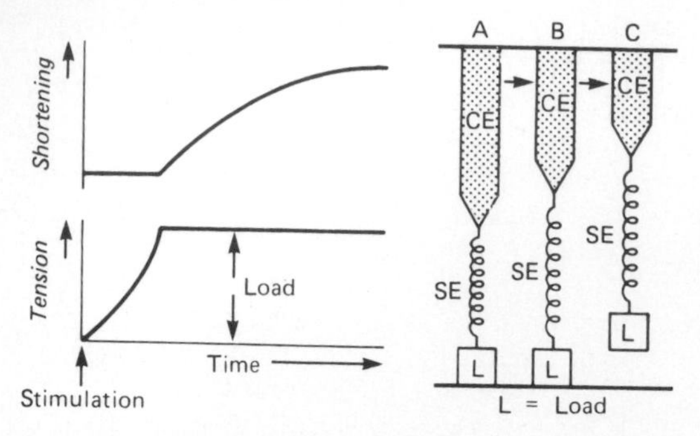

Figure 29–7. Model for isotonic contraction of afterloaded muscles. *A:* Rest. *B:* Partial contraction of the contractile element of the muscle (CE), with stretching of the series elastic element (SE) but no shortening. *C:* Complete contraction, with shortening. (Reproduced, with permission, from Sonnenblick EH in: *The Myocardial Cell: Structure, Function and Modification.* Briller SA, Conn HL [editors]. Univ of Pennsylvania Press, 1966.)

end-diastolic volume. The relation between ventricular stroke volume and end-diastolic volume is called the Frank-Starling curve.

Regulation of cardiac output as a result of changes in cardiac muscle fiber length is sometimes called **heterometric regulation,** whereas regulation due to changes in contractility independent of length is sometimes called **homometric regulation.**

Factors Affecting End-Diastolic Volume

The factors that normally operate to regulate end-diastolic volume, ie, the degree to which cardiac muscle is stretched, are listed in Table 29–4. An increase in intrapericardial pressure limits the extent to which the ventricle can fill. So does a decrease in ventricular compliance, ie, an increase in ventricular stiffness produced by myocardial infarction, infiltrative disease, and other abnormalities. Atrial contractions aid ventricular filling. The other factors affect the amount of blood returning to the heart and hence the degree of cardiac filling during diastole. An increase in total blood volume increases venous return. Constriction of the veins reduces the size of the venous reservoirs, decreasing venous pooling and thus increasing venous return. An increase in the normal negative intrathoracic pressure increases the pressure gradient along which blood flows to the heart, whereas a decrease impedes venous return. Standing decreases venous return, and muscular activity increases it as a result of the pumping action of skeletal muscle.

Table 29–4. Factors that normally increase or decrease the length of ventricular cardiac muscle fibers.

Increase
Stronger atrial contractions
Increased total blood volume
Increased venous tone
Increased pumping action of skeletal muscle
Increased negative intrathoracic pressure
Decrease
Standing
Increased intrapericardial pressure
Decreased ventricular compliance

Effect of Changes in Aortic Impedance

As noted above, the strength of cardiac contractions is also determined by the resistance (afterload, impedance) against which the ventricles pump blood. This impedance is low in the pulmonary artery, but the aortic impedance is high. It is proportionate to the resistance to flow through the aortic valve and the systemic blood pressure.

The effect of changes in aortic impedance can be demonstrated in the **heart-lung preparation** (Fig 29–8). In this preparation, the heart and lungs of an anesthetized experimental animal, usually a dog, are cannulated in such a way that blood flows from the aorta through a system of tubing and reservoirs to the right atrium, and from there through the animal's heart and lungs back to the aorta. Deprived of its blood supply, the rest of the animal dies, so that the heart is functionally denervated and the heart rate varies little if at all. By decreasing the caliber of the outflow tubing, the resistance against which the heart pumps (**"peripheral resistance"**) can be increased. When the peripheral resistance is increased, the heart

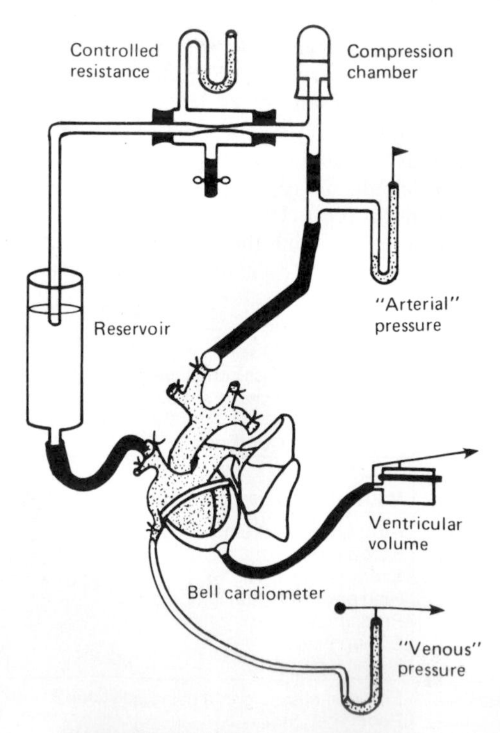

Figure 29–8. Heart-lung preparation.

puts out less blood than it receives for several beats. Blood accumulates in the ventricles, and the size of the heart increases. The distended heart beats more forcefully, and output returns to its previous level.

The effect of variations in venous return can also be demonstrated in the heart-lung preparation. When the reservoir emptying into the right atrium in Fig 29–8 is raised, venous pressure and hence venous return are increased. The myocardial fibers are stretched, and cardiac output is increased.

Myocardial Contractility

The contractility of the myocardium exerts a major influence on stroke volume. When the sympathetic nerves to the heart are stimulated, the whole length-tension curve shifts upward and to the left (Fig 29–9). The positively inotropic effect of the norepinephrine liberated at the nerve endings is augmented by circulating norepinephrine, and epinephrine has a similar effect. There is a negatively inotropic effect of vagal stimulation on the atrial muscle and a small negative inotropic effect on the ventricular muscle.

Changes in cardiac rate and rhythm also affect myocardial contractility (force-frequency relation, Fig 29–9). Ventricular extrasystoles condition the myocardium in such a way that the next succeeding contraction is stronger than the preceding normal contraction. This **postextrasystolic potentiation** is independent of ventricular filling, since it occurs in isolated cardiac muscle, and is due to increased availability of intracellular Ca^{2+}. A sustained increment in contractility can be produced by delivering paired electrical stimuli to the heart in such a way that the second stimulus is delivered shortly after the refractory period of the first. It has also been shown that myocardial contractility increases as the heart rate increases, although this effect is relatively small.

The catecholamines exert their inotropic effect via an action on cardiac β_1-adrenergic receptors and G_s with resultant activation of adenylate cyclase and increased intracellular cyclic AMP (see Chapter 3). Xanthines such as caffeine and theophylline that inhibit the breakdown of cyclic AMP are positively inotropic. Glucagon, which increases the formation of cyclic AMP, is positively inotropic, and it has been recommended for use in the treatment of some heart diseases. The positively inotropic effect of digitalis and related drugs (Fig 29–9) is due to their inhibitory effect on the Na^+-K^+ ATPase in the myocardium. The inhibition causes an increase in intracellular Na^+, which in turn increases the availability of Ca^{2+} in the cell. The key role of Ca^{2+} in the initiation of contraction in skeletal and cardiac muscle is discussed in Chapter 3. Hypercapnia, hypoxia, acidosis, and drugs such as quinidine, procainamide, and barbiturates depress myocardial contractility. The contractility of the myocardium is also reduced in heart failure (intrinsic depression). This depression is associated with a reduction in the catecholamine content of the myocardium, but its exact cause is not known. If part of the myocardium becomes fibrotic and nonfunctional as the result of a myocardial infarction, total ventricular performance is reduced.

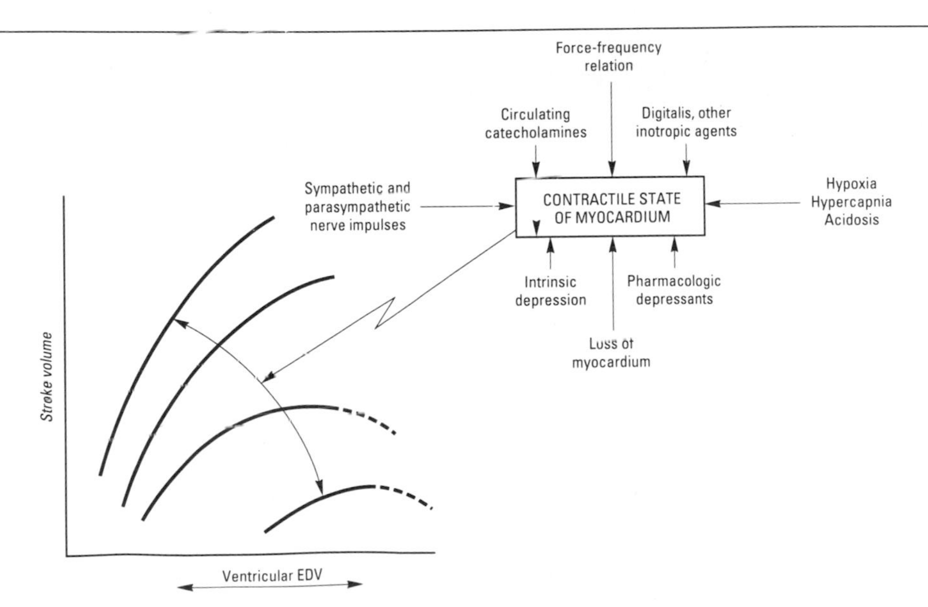

Figure 29–9. Effect of changes in myocardial contractility on the Frank-Starling curve. The major factors influencing contractility are summarized on the right. The dashed lines indicate portions of the ventricular function curves where maximum contractility has been exceeded, ie, they identify points on the "falling limb" of the Frank-Starling curve. EDV, end-diastolic volume. (Reproduced, with permission, from Braunwald E, Ross J, Sonnenblick EH: Mechanisms of contraction of the normal and failing heart. *N Engl J Med* 1967;**277**:794. Courtesy of Little, Brown, Inc.)

Control of Cardiac Output in Vivo

In intact animals and humans, the mechanisms listed above operate in an integrated way to maintain cardiac output. During muscular exercise, there is increased sympathetic discharge, so that myocardial contractility is increased and the heart rate rises. The increase in heart rate is particularly prominent in normal individuals, and there is only a modest increase in stroke volume (Table 29–5). However, patients with transplanted hearts are able to increase their cardiac output during exercise in the absence of cardiac innervation through the operation of the Frank-Starling mechanism (Fig 29–10). Circulating catecholamines also contribute. The increase seen in these patients is not as rapid, and their maximal increase is less than in normal individuals, but the increase is appreciable. It is important to note in this regard that the heart is the servant, rather than the master; it normally adjusts its function so that it expels all the blood that returns in the veins. If venous return increases and there is no change in sympathetic tone, venous pressure rises, diastolic inflow is greater, ventricular end-diastolic pressure increases, and the heart muscle contracts more forcefully. During muscular exercise, venous return is increased by the pumping action of the muscles and the increase in respiration (see Chapter 33). In addition, because of vasodilation in the contracting muscles, peripheral resistance, and consequently afterload are decreased. The end result in both normal and transplanted hearts is thus a prompt and marked increase in cardiac output.

One of the differences between untrained individuals and trained athletes is that the athletes have lower heart rates, greater end-systolic ventricular volumes, and greater stroke volumes at rest. Therefore, they can potentially achieve a given increase in cardiac output by without increasing their heart rate to as great a degree as an untrained individual.

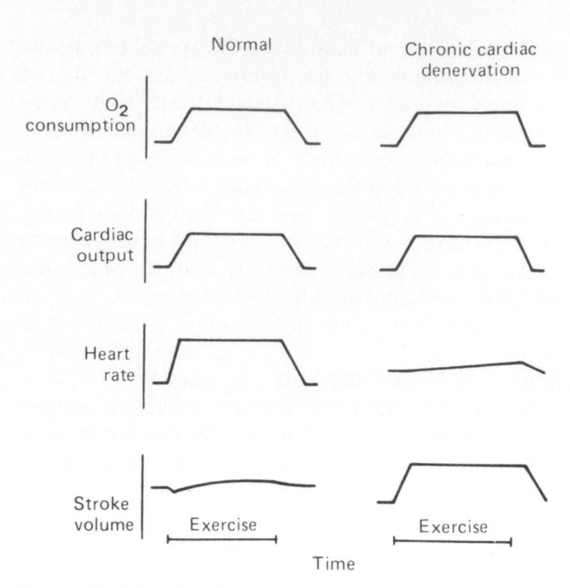

Figure 29–10. Cardiac responses to moderate supine exercise in normal humans and patients with cardiac transplants. (Reproduced, with permission, from Kent KM, Cooper T: The denervated heart. *N Engl J Med* 1974;**291**:1017.)

Oxygen Consumption of the Heart

The basal O_2 consumption of the myocardium, which can be determined by stopping the heart while artificially maintaining the coronary circulation, is about 2 mL/100g/min. This value is considerably higher than that of resting skeletal muscle. O_2 consumption of the beating heart is about 9mL/100g/min at rest. Increases occur during exercise and in a number of different states. Cardiac venous O_2 tension is low, and little additional O_2 can be extracted from the blood in the coronaries, so increases in O_2 consumption require increases in coronary blood flow. The regulation of coronary flow is discussed in Chapter 32.

The O_2 consumption of the heart is determined primarily by the intramyocardial tension, the contractile state of the myocardium, and the heart rate. Ventricular work per beat correlates with O_2 consumption. The work is the product of stroke volume and mean arterial pressure in the pulmonary artery (for the right ventricle) or the aorta (for the left ventricle). Since aortic pressure is 7 times greater than pulmonary artery pressure, the stroke work of the left ventricle is approximately 7 times the stroke

Table 29–5. Changes in cardiac function with exercise. Note that stroke volume levels off, then falls somewhat (as a result of the shortening of diastole) when the heart rate rises to high values.*

Work (kg-m/min)	O_2 Usage (mL/min)	Pulse Rate (per min)	Cardiac Output (L/min)	Stroke Volume (mL)	A-V O_2 Difference (mL/dL)
Rest	267	64	6.4	100	4.3
288	910	104	13.1	126	7.0
540	1430	122	15.2	125	9.4
900	2143	161	17.8	110	12.3
1260	3007	173	20.9	120	14.5

* Reproduced, with permission, from Asmussen E, Nielsen M: The cardiac output in rest and work determined by the acetylene and the dye injection methods. *Acta Physiol Scand* 1952;**27**:217.

work of the right ventricle. In theory, a 25% increase in stroke volume without a change in arterial pressure should produce the same increase in O_2 consumption as a 25% increase in arterial pressure without a change in stroke volume. However, for reasons that are incompletely understood, pressure work produces a greater increase in O_2 consumption than volume work. In other words, an increase in afterload causes a greater increase in cardiac O_2 consumption than an increase in preload. This is why angina pectoris due to deficient delivery of O_2 to the myocardium is more common in aortic stenosis than in aortic insufficiency. In aortic stenosis, intraventricular pressure must be increased to force blood through the stenotic valve, whereas in aortic insufficiency, regurgitation of blood produces an increase in stroke volume with little change in aortic impedance.

It is worth noting that the increase in O_2 consumption produced by increased stroke volume when the myocardial fibers are stretched is an example of the operation of the law of Laplace. This law, which is discussed in detail in Chapter 30, states that the tension developed in the wall of a hollow viscus is proportionate, among other things, to the radius of the viscus, and the radius of a dilated heart is increased. O_2 consumption per unit of time increases when heart rate is increased by sympathetic stimulation because of the increased number of beats and the increased velocity and strength of each contraction. However, this is somewhat offset by the decrease in end-systolic volume and hence in the radius of the heart.

30 Dynamics of Blood & Lymph Flow

INTRODUCTION

The blood vessels are a closed system of conduits that carry blood from the heart to the tissues and back to the heart. Some of the interstitial fluid enters the lymphatics and passes via these vessels to the vascular system. Blood flows through the vessels primarily because of the forward motion imparted to it by the pumping of the heart, although in the case of the systemic circulation, diastolic recoil of the walls of the arteries, compression of the veins by skeletal muscles during exercise, and the negative pressure in the thorax during inspiration also move the blood forward. The resistance to flow depends to a minor degree upon the viscosity of the blood but mostly upon the diameter of the vessels, principally the arterioles. The blood flow to each tissue is regulated by local chemical and general neural and humoral mechanisms that dilate or constrict the vessels of the tissue. All of the blood flows through the lungs, but the systemic circulation is made up of numerous different circuits in parallel (Fig 30–1), an arrangement that permits wide variations in regional blood flow without changing total systemic flow.

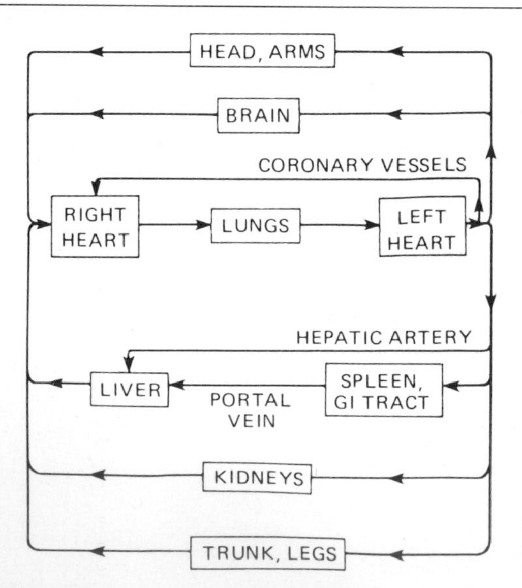

Figure 30–1. Diagram of the circulation in the adult.

This chapter is concerned with the general principles that apply to all parts of the circulation and with pressure and flow in the systemic circulation. The homeostatic mechanisms operating to adjust flow are the subject of Chapter 31. The special characteristics of pulmonary and renal circulation are discussed in Chapters 34 and 38 and the unique features of the circulation to other organs in Chapter 32.

ANATOMIC CONSIDERATIONS

Arteries & Arterioles

The characteristics of the various types of blood vessels are shown in Table 30–1. The walls of the aorta and other arteries of large diameter contain a relatively large amount of elastic tissue. They are stretched during systole and recoil on the blood during diastole. The walls of the arterioles contain less elastic tissue but much more smooth muscle. The muscle is innervated by noradrenergic nerve fibers, which are constrictor in function, and in some instances by cholinergic fibers, which dilate the vessels. The arterioles are the major site of the resistance to blood flow, and small changes in their caliber cause large changes in the total peripheral resistance.

Capillaries

The arterioles divide into smaller muscle-walled vessels, sometimes called **metarterioles,** and these in turn feed into capillaries (Fig 30–2). In some of the vascular beds that have been studied in detail, a metarteriole is connected directly with a venule by a capillary **thoroughfare vessel,** and the true capillaries are an anastomosing network of side branches of this thoroughfare vessel. The openings of the true capillaries are surrounded on the upstream side by minute smooth muscle **precapillary sphincters.** It is unsettled whether the metarterioles are innervated, and it appears that the precapillary sphincters are not. However, they can of course respond to local or circulating vasoconstrictor substances. The true capillaries are about 5 μm in diameter at the arterial end and 9 μm in diameter at the venous end. When the sphincters are dilated, the diameter of the capillaries is just sufficient to permit red blood cells to squeeze through in "single file." As they pass through the

Table 30–1. Characteristics of various types of blood vessels in humans.

	Lumen Diameter	Wall Thickness	All Vessels of Each Type: Approximate Total Cross-Sectional Area (cm^2)	All Vessels of Each Type: Percentage of Blood Volume Contained*
Aorta	2.5 cm	2 mm	4.5	2
Artery	0.4 cm	1 mm	20	8
Arteriole	30 μm	20 μm	400	1
Capillary	5 μm	1 μm	4500	5
Venule	20 μm	2 μm	4000	
Vein	0.5 cm	0.5 mm	40	54
Vena cava	3 cm	1.5 mm	18	

*In systemic vessels. There is an additional 12% in the heart and 18% in the pulmonary circulation.

capillaries, the red cells become thimble- or parachute-shaped, with the concavity pointing in the direction of flow. This configuration appears to be due simply to the pressure in the center of the vessel whether or not the edges of the red blood cell are in contact with the capillary walls.

The total area of all the capillary walls in the body exceeds 6300 m^2 in the adult. The walls, which are about 1 μm thick, are made up of a single layer of endothelial cells. The structure of the walls varies from organ to organ. In many beds, including those in skeletal, cardiac, and smooth muscle, the junctions between the endothelial cells (Fig 30–3) permit the passage of molecules up to 10 nm in diameter. It also appears that plasma and its dissolved proteins are taken up by endocytosis, transported across the endothelial cells, and discharged by exocytosis (**vesicular transport;** see Chapter 1). However, this process can account for only a small portion of the transport across the endothelium. In the brain, the capillaries resemble the capillaries in muscle, but the junctions between endothelial cells are tighter; they permit the passage of small molecules only. In most endocrine glands, the intestinal villi, and parts of the kidney, the cytoplasm of the endothelial cells is attenuated to form gaps called **fenestrations.** These fenestrations are 20–100 nm in diameter. They permit the passage of relatively large molecules and make the capillaries porous. Except in the renal glomeruli, they appear to be closed by a thin membrane. However, in a number of different tissues, the membrane can be shown by a rapid freeze-fracture technique to be discontinuous, consisting of a central hub joined by spokes of membrane to the edges of the fenestration (Fig 30–4). In the liver, where the sinusoidal capillaries are extremely porous, the endothelium is discontinuous and there are large gaps between endothelial cells that are not closed by membranes (Fig 26–19). Some of the gaps are 600 nm in diameter, and others may be as large as 3000 nm. The permeabilities of capillaries in various parts of the body, expressed in terms of their hydraulic conductivity, are summarized in Table 30–2.

Lymphatics

The lymphatics drain from the lungs and from the rest of the body tissues via a system of vessels that coalesce and eventually enter the right and left subclavian veins at their junctions with the respective internal jugular veins. The lymph vessels contain valves and regularly traverse lymph nodes along their course. The ultrastructure of the small lymph vessels

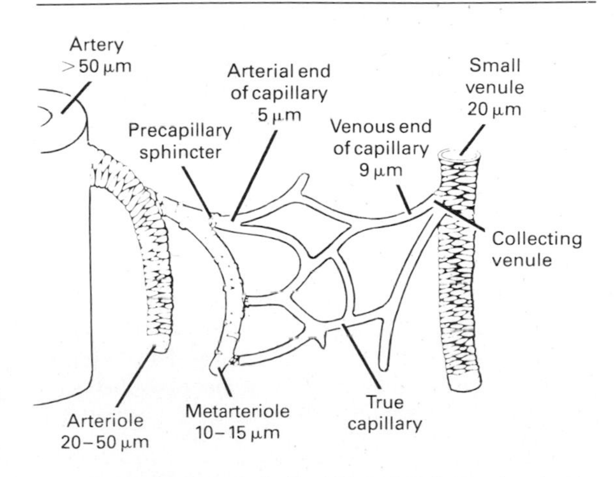

Figure 30–2. The microcirculation. Arterioles give rise to metarterioles, which give rise to capillaries. The capillaries drain via short collecting venules to the venules. The walls of the arteries, arterioles, and small venules contain relatively large amounts of smooth muscle. There are scattered smooth muscle cells in the walls of the metarterioles, and the openings of the capillaries are guarded by muscular precapillary sphincters. The diameters of the various vessels are also shown. (Courtesy of JN Diana.)

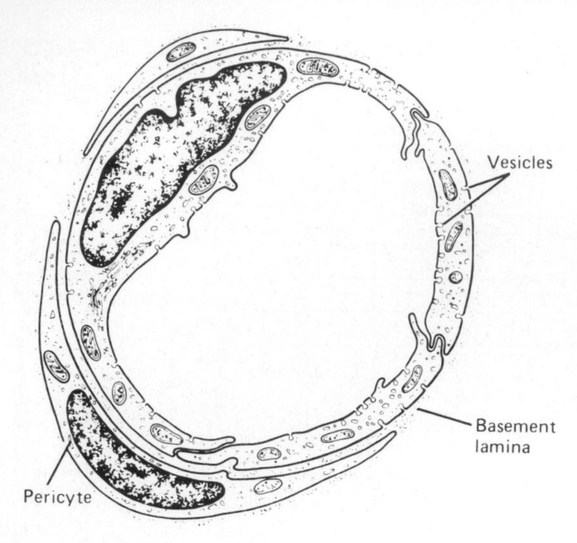

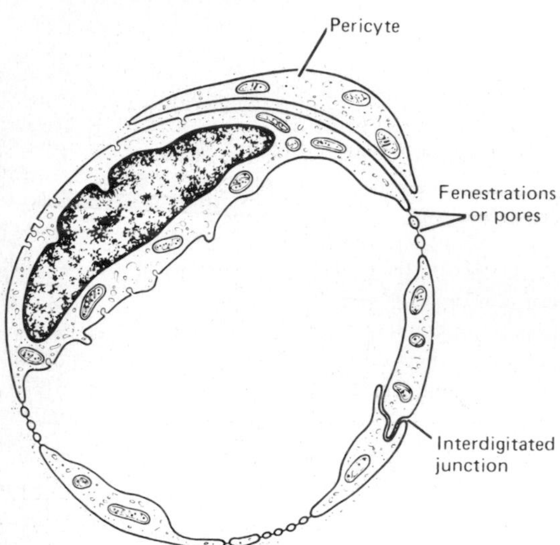

Figure 30–3. Capillaries. *Top:* Continuous type of capillary found in muscle. *Bottom:* Fenestrated type of capillary. (Reproduced, with permission, from Fawcett DW: *Bloom and Fawcett—A Textbook of Histology,* 11th ed. Saunders, 1986.)

differs from that of the capillaries in several details: There are no visible fenestrations in the lymphatic endothelium; there is very little if any basal lamina under the endothelium; and the junctions between endothelial cells are open, with no tight intercellular connections.

Arteriovenous Anastomoses

In the fingers, palms, and ear lobes of humans and the paws, ears, and some other tissues of animals, there are short channels that connect arterioles to venules, bypassing the capillaries. These **arteriovenous (A-V) anastomoses,** or **shunts,** have thick, muscular walls and are abundantly innervated, presumably by vaso-constrictor nerve fibers.

Venules & Veins

The walls of the venules are only slightly thicker than those of the capillaries. The walls of the veins are also thin and easily distended. They contain relatively little smooth muscle, but considerable venoconstriction is produced by activity in the noradrenergic nerves to the veins and by chemical agents such as norepinephrine. Anyone who has had trouble making venipunctures has observed the marked local venospasm produced in superficial forearm veins by injury. Variations in venous tone are important in circulatory adjustments.

The intima of the limb veins is folded at intervals to form **venous valves** that prevent retrograde flow. The way these valves function was first demonstrated by William Harvey (Fig 30–5). There are no valves in the very small veins, the great veins, or the veins from the brain and viscera.

Angiogenesis

When tissues grow, blood vessels must proliferate if the tissue is to maintain a normal blood supply. Eight different factors that promote the formation of new blood vessels have been isolated and characterized: the acidic and basic forms of fibroblast growth factor (FGF), the α and β forms of transforming growth factor (TGF-α and TGF-β), angiogenin, and tumor necrosis factor (TNF-α), all of which are polypeptides, plus nicotinamide and a more complex molecule that contains nicotinamide. These factors act in many different poorly understood ways to cause more blood vessels to form in growing tissues and in tumors. Tumors develop a blood supply as they grow, and if this blood supply fails to develop, the tumors do not grow. Consequently, efforts are being made to develop substances that inhibit angiogenesis in tumors. Heparin (see Chapter 27) promotes angiogenesis, and protamine inhibits it.

BIOPHYSICAL CONSIDERATIONS

Flow, Pressure, & Resistance

Blood always flows, of course, from areas of high pressure to areas of low pressure, except in certain situations when momentum transiently sustains flow (Fig 29–3). The relationship between mean flow, mean pressure, and resistance in the blood vessels is analogous in a general way to the relationship between the current, electromotive force, and resistance in an electrical circuit expressed in Ohm's law:

$$\text{Current (I)} = \frac{\text{Electromotive force (E)}}{\text{Resistance (R)}}$$

$$\text{Flow (F)} = \frac{\text{Pressure (P)}}{\text{Resistance (R)}}$$

Flow in any portion of the vascular system is equal to the **effective perfusion pressure** in that portion divided by the **resistance.** The effective perfusion

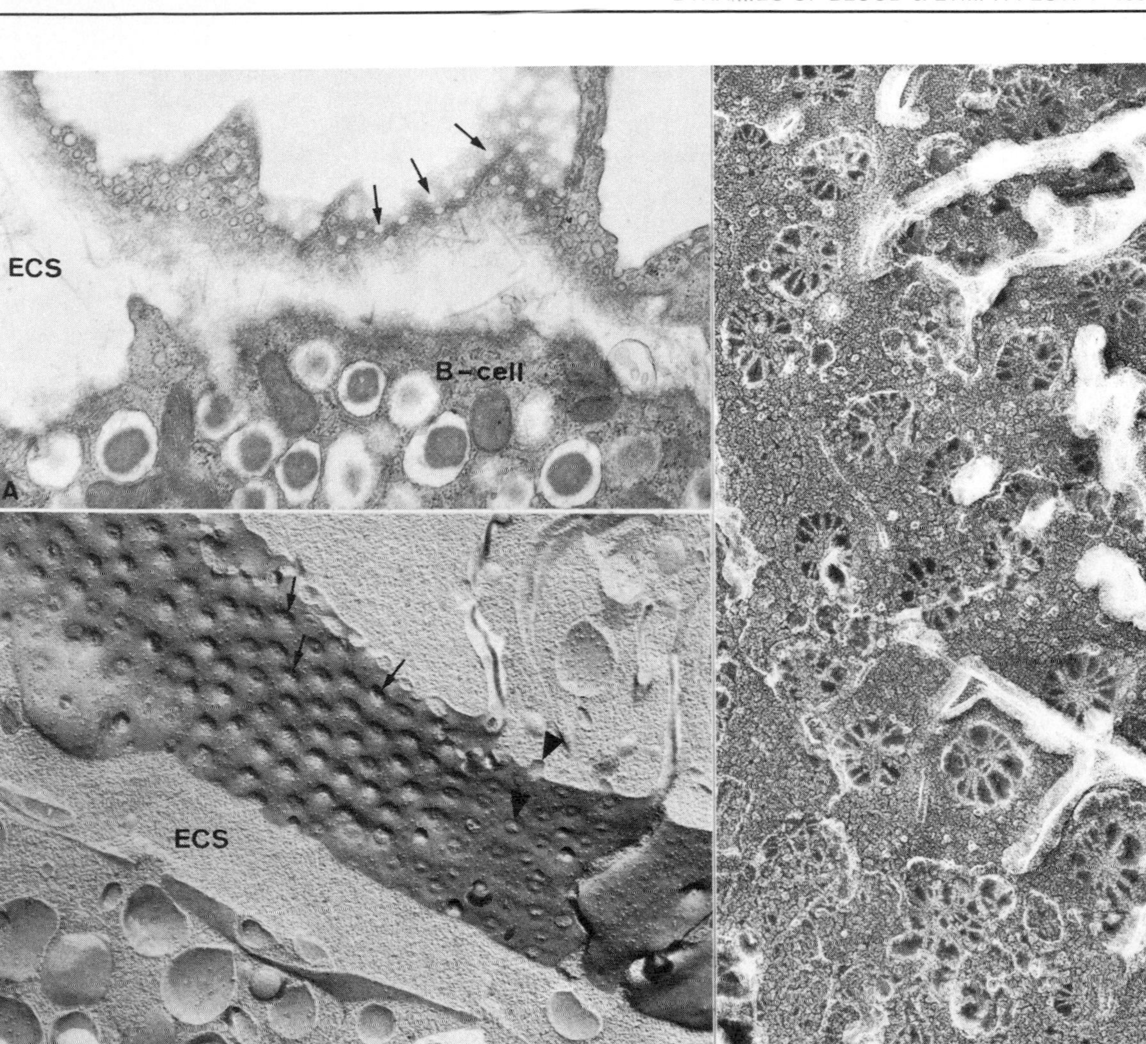

Figure 30–4. Fenestrations in capillaries in pancreatic islets. In *A,* arrows identify fenestrations in a grazing section of the endothelium (x 15,000). ECS, extracellular space. In *B,* arrows identify fenestrations in a freeze-fracture preparation (x 31,500). The arrowheads identify the smaller pits due to endocytosis. In *C,* fenestrations are seen in a quick-frozen, deep-etched face view of the capillary endothelium (x 64,000). Note the central material and the spokes attaching it to the rim of each fenestration. (Reproduced, with permission, from Orci L: The insulin cell: Its cellular environment and how it processes (pro)-insulin. *Diabetes/Metabolism Reviews* 1986;2:71.)

pressure is the mean intraluminal pressure at the arterial end minus the mean pressure at the venous end. The units of resistance (pressure divided by flow) are dyne-s-cm^{-5}. To avoid dealing with such complex units, resistance in the cardiovascular system is sometimes expressed in **R units** by dividing pressure in mm Hg by flow in mL/s (see also Table 32–1). Thus, for example, when the mean aortic pressure is 90 mm Hg and the left ventricular output is 90 mL/s, the total peripheral resistance is

$$\frac{90 \text{ mm Hg}}{90 \text{ mL/s}} = 1 \text{ R unit}$$

Methods for Measuring Blood Flow

Blood flow can be measured by cannulating a blood vessel, but this has obvious limitations. Various devices have been developed to measure flow in a blood vessel without opening it. **Electromagnetic flow meters** depend on the principle that a voltage is generated in a conductor moving through a magnetic field and that the magnitude of the voltage is proportionate to the speed of movement. Since blood is a conductor, a magnet is placed around the vessel, and the voltage, which is proportionate to the volume flow, is measured with an appropriately placed elec-

Table 30–2. Hydraulic conductivity of capillaries in various parts of the body.*

Organ	Conductivity†	Type of Endothelium
Brain (excluding circumventricular organs)	3	Continuous
Skin	100	
Skeletal muscle	250	
Lung	340	
Heart	860	
Gastrointestinal tract (intestinal mucosa)	13,000	Fenestrated
Glomerulus in kidney	15,000	

* Data courtesy of JN Diana.
† Units of conductivity are cm^3 sec^{-1} $dyne^{-1}$ $\times 10^{-13}$.

trode on the surface of the vessel. Blood flow velocity can be measured with **Doppler flow meters.** Ultrasonic waves are sent into a vessel diagonally from one crystal, and the waves reflected from the red and white blood cells are picked up by a second, downstream crystal. The frequency of the reflected waves is higher by an amount that is proportionate to the rate of flow toward the second crystal because of the Doppler effect.

Indirect methods for measuring the blood flow of various organs in humans include adaptations of the Fick and indicator dilution techniques described in Chapter 29. An ingenious example of the use of the Fick principle to measure flow in an organ is the Kety N_2O method for measuring cerebral blood flow (see Chapter 32). Another example is determination of the renal blood flow by measuring the clearance of paraaminohippuric acid (see Chapter 38). A considerable amount of data on blood flow in the extremities has been obtained by **plethysmography** (Fig 30–6). The forearm, for example, is sealed in a watertight chamber (**plethysmograph**). Changes in the volume of the forearm, reflecting changes in the amount of blood and interstitial fluid it contains, displace the water, and this displacement is measured with a volume recorder. When the venous drainage of the forearm is occluded, the rate of increase in the volume of the forearm is a function of the arterial blood flow (**venous occlusion plethysmography**).

Applicability of Physical Principles to Flow in Blood Vessels

Physical principles and equations that are applicable to the description of the behavior of perfect fluids in rigid tubes have often been used indiscriminately to explain the behavior of blood in blood vessels. Blood vessels are not rigid tubes, and the blood is not a perfect fluid but a 2-phase system of liquid and cells. Therefore, the behavior of the circulation deviates, sometimes markedly, from that predicted by these principles. However, the physical principles are of value when used as an aid to understanding what goes on in the body rather than as an end in themselves or as a test of the memorizing ability of students.

Laminar Flow

The flow of blood in the blood vessels, like the flow of liquids in narrow rigid tubes, is normally **laminar (streamline).** Within the blood vessels, an infinitely thin layer of blood in contact with the wall of the vessel does not move. The next layer within the vessel has a low velocity, the next a higher velocity, and so forth, velocity being greatest in the center of the stream (Fig 30–7). Laminar flow occurs at velocities up to a certain **critical velocity.** At or above this velocity, flow is turbulent. Streamline flow is silent, but turbulent flow creates sounds.

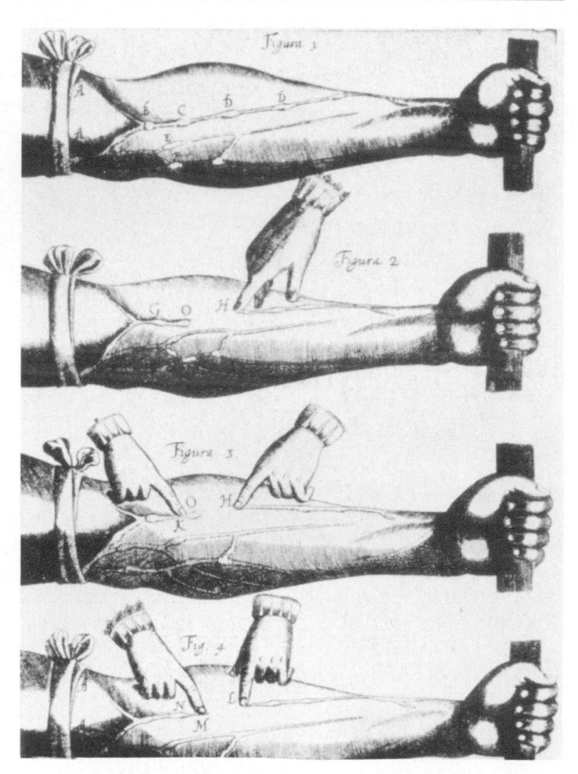

Figure 30–5. Illustration showing the function of the venous valves in the human forearm, from William Harvey's book *Exercitatio Anatomica de Motu Cordis et Sanguinis in Animalibus,* which was published in 1628. *A* is a tourniquet above the elbow, obstructing the venous return. If the vein (Fig 2) is milked from *G* to *H,* it fills from above only to the valve, *O.* If it is milked from *K* to *O* with another finger (Fig 3), it bulges above *O,* but *O* to H remains empty. If the vein is occluded at *L* (Fig 4) and milked with another finger (*M*) from *L* to the valve *N* , it remains empty from *L* to *N.* If the finger at *L* is removed, the vein fills from below.

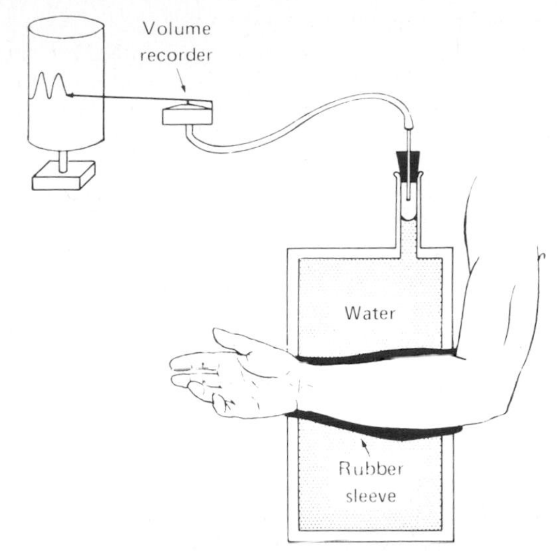

Figure 30–6. Plethysmograph.

The probability of turbulence is also related to the diameter of the vessel and the viscosity of the blood. This probability can be expressed by the ratio of inertial to viscous forces as follows:

$$R = \frac{\rho DV}{\eta}$$

where R is the Reynolds number, named for the man who described the relationship; ρ is the density of the fluid; D is the diameter of the tube under consideration; V is the velocity of the flow; and η is the viscosity of the fluid. The higher the value of R, the greater the probability of turbulence. Constriction of an artery increases the velocity of blood flow through the constriction, producing turbulence beyond the constriction (Fig 30–8). As noted in Chapter 29, streamline flow is silent, whereas turbulence creates sound. Examples are bruits heard over arteries constricted by atherosclerotic plaques and the sounds of Korotkow heard when measuring blood pressure (see below).

In humans, critical velocity is sometimes exceeded in the ascending aorta at the peak of systolic ejection, but it is usually exceeded only when an artery is constricted. Turbulence occurs more frequently in anemia because the viscosity of the blood is lower. This may be the explanation of the systolic murmurs that are common in anemia.

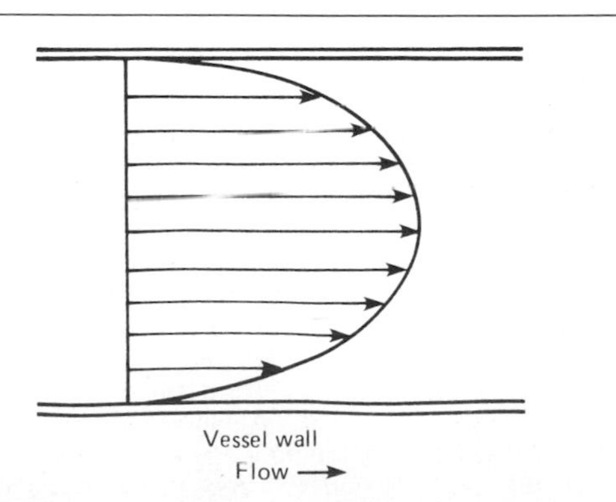

Figure 30–7. Diagram of the velocities of concentric laminas of a viscous fluid flowing in a tube, illustrating the parabolic distribution of velocities (streamline flow).

Average Velocity

When considering flow in a system of tubes, it is important to distinguish between velocity, which is displacement per unit time (eg, cm/s), and flow, which is volume per unit time (eg, cm^3/s). Velocity (V) is proportionate to flow (Q) divided by the area of the conduit (A):

$$V = \frac{Q}{A}$$

The average velocity of fluid movement at any point in a system of tubes is inversely proportionate to the *total* cross-sectional area at that point. Therefore, the average velocity of the blood is high in the aorta, declines steadily in the smaller vessels, and is lowest in the capillaries, which have 1000 times the *total* cross-sectional area of the aorta (Table 30–1). The average velocity of blood flow increases again as the blood enters the veins and is relatively high in the vena cava, although not so high as in the aorta. Clinically, the velocity of the circulation can be measured by injecting a bile salt preparation into an arm vein and timing the first appearance of the bitter taste it produces (Fig 30–9). The average normal arm to tongue **circulation time** is 15 seconds.

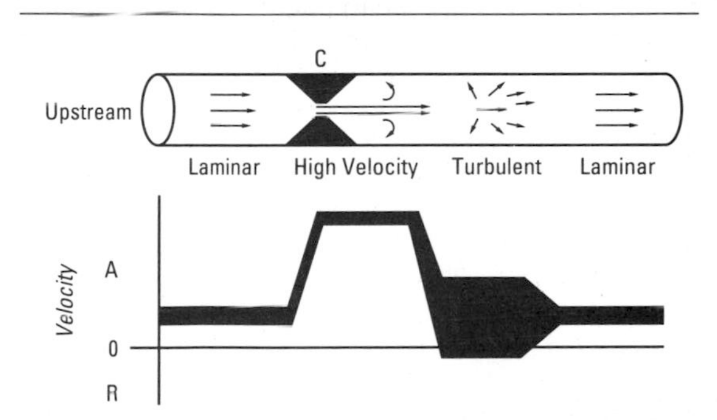

Figure 30–8. Effect of constriction (C) on the profile of velocities in a blood vessel. The arrows indicate direction of velocity components, and their length is proportionate to their magnitude. The graph below the vessel shows the range of velocities at each point along the vessel. Note that in the area of turbulence, there are many different anterograde (A) velocities, and some in the retrograde (R) direction. (Modified and reproduced, with permission, from Richards KE: Doppler echocardiography in diagnosis and quantification of vascular disease. *Mod Concepts Cardiovasc. Dis* 1987;**56**:43. By permission of the American Heart Association, Inc.)

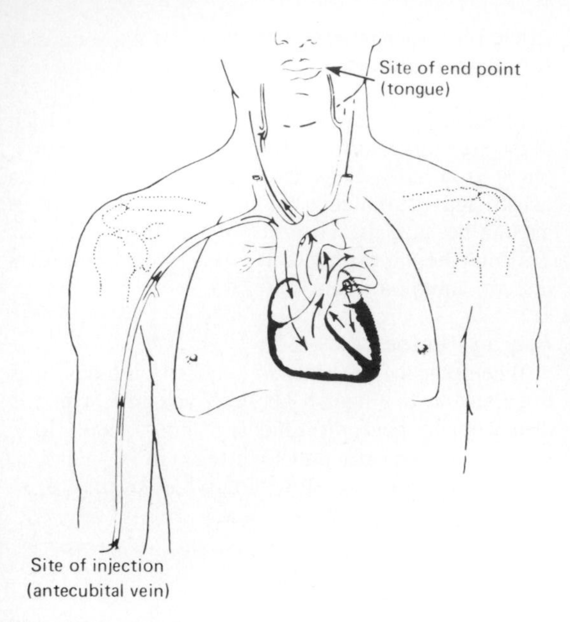

Figure 30–9. Arm to tongue circulation time.

Poiseuille-Hagen Formula

The relation between the flow in a long narrow tube, the viscosity of the fluid, and the radius of the tube is expressed mathematically in the **Poiseuille-Hagen formula:**

$$F = (P_A - P_B) \times \left(\frac{\pi}{8}\right) \times \left(\frac{1}{\eta}\right) \times \left(\frac{r^4}{L}\right)$$

where

F = flow
$P_A - P_B$ = pressure difference between the 2 ends of the tube
η = viscosity
r = radius of tube
L = length of tube

Since flow is equal to pressure difference divided by resistance (R),

$$R = \frac{8\eta L}{\pi r^4}$$

Since flow varies directly and resistance inversely with the fourth power of the radius, blood flow and resistance in vivo are markedly affected by small changes in the caliber of the vessels. Thus, for example, flow through a vessel is doubled by an increase of only 19% in its radius; and when the radius is doubled, resistance is reduced to 6% of its previous value. This is why organ blood flow is so effectively regulated by small changes in the caliber of the arterioles and why variations in arteriolar diameter have such a pronounced effect on systemic arterial pressure.

Viscosity & Resistance

The resistance to blood flow is determined not only by the radius of the blood vessels **(vascular hindrance)** but by the viscosity of the blood. The effect of viscosity in vivo deviates from that predicted by the Poiseuille-Hagen formula. Viscosity depends for the most part on the **hematocrit,** ie, the percentage of the volume of blood occupied by red blood cells. In large vessels, increases in hematocrit cause appreciable increases in viscosity. However, in vessels smaller than 100 μm in diameter, ie, in arterioles, capillaries, and venules, the viscosity change per unit change in hematocrit is much less than it is in large-bore vessels. This is due to a difference in the nature of flow through the small vessels. Therefore, the net change in viscosity per unit change in hematocrit is considerably less in the body than it is in vitro (Fig 30–10). This is why hematocrit changes have relatively little effect on the peripheral resistance except when the changes are large. In severe polycythemia, the increase in resistance does increase the work of the heart. Conversely, in severe anemia, the peripheral resistance is decreased, although the decreased resistance is only partly due to the decrease in viscosity. Cardiac output is increased, and as a result, the work of the heart is also increased in anemia.

Viscosity is also affected by the composition of the plasma and the resistance of the cells to deformation. Clinically significant increases in viscosity are seen in diseases in which plasma proteins such as the immunoglobulins are markedly elevated and in diseases such as hereditary spherocytosis, in which the red blood cells are abnormally rigid.

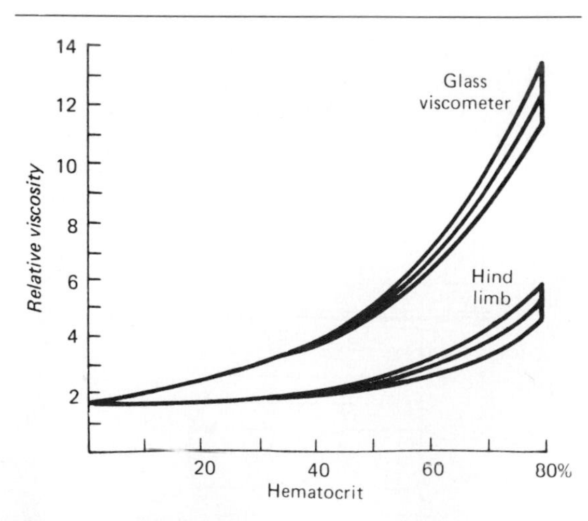

Figure 30–10. Effect of changes in hematocrit on the relative viscosity of blood measured in a glass viscometer and in the hind leg of a dog. In each case, the middle line represents the mean and the upper and lower lines the standard deviation. (Reproduced, with permission, from Whittaker, SRF, Winton FR: The apparent viscosity of blood flowing in the isolated hind limb of the dog, and its variation with corpuscular concentration. *J Physiol [Lond]* 1933;**78**:338.)

In the vessels, red cells tend to accumulate in the center of the flowing stream. Consequently, the blood along the side of the vessels has a low hematocrit, and branches leaving a large vessel at right angles may receive a disproportionate amount of red-cell-poor blood. This phenomenon, which has been called **plasma skimming,** may be the reason the hematocrit of capillary blood is regularly about 25% lower than the whole body hematocrit.

Critical Closing Pressure

In rigid tubes, the relation between pressure and flow of homogeneous fluids is linear, but in blood vessels in vivo it is not. When the pressure in a small blood vessel is reduced, a point is reached at which there is no flow of blood even though the pressure is not zero (Fig 30–11). This is in part a manifestation of the fact that it takes some pressure to force red cells through capillaries which have diameters less than the red cells. Also, the vessels are surrounded by tissues that exert a small but definite pressure on the vessels, and when the intraluminal pressure falls below the tissue pressure, the vessels collapse. In inactive tissues, for example, the pressure in many capillaries is low because the precapillary sphincters and met-arterioles are constricted, and many of these capillaries are collapsed. The pressure at which flow ceases is called the **critical closing pressure.**

Law of Laplace

The relation between distending pressure and tension is shown diagrammatically in Fig 30–12. It is perhaps surprising that structures as thin-walled and delicate as the capillaries are not more prone to rupture. The principal reason for their relative invulnerability is their small diameter. The protective effect of small size in this case is an example of the operation of the **law of Laplace,** an important physical principle with several other applications in physiology. This law states that the distending pressure (P) in a distensible hollow object is equal at equilibrium to the tension in the wall (T) divided by the 2 principal radii of curvature of the object (R_1 and R_2):

$$P = T(1/R_1 + 1/R_2)$$

In the equation, P is actually the **transmural pressure,** the pressure on one side of the wall minus that on the other. T is expressed in dynes/cm and R_1 and R_2 in cm, so P is expressed in dynes/cm^2. In a sphere, $R_1 = R_2$ so

$$P = 2T/R$$

In a cylinder such as a blood vessel, one radius is infinite, so

$$P = T/R$$

Consequently, the smaller the radius of a blood vessel, the lower the tension in the wall necessary to balance the distending pressure. In the human aorta, for example, the tension at normal pressures is about 170,000 dynes/cm, and in the vena cava it is about 21,000 dynes/cm; but in the capillaries, it is approximately 16 dynes/cm.

The law of Laplace also makes clear a disadvantage faced by dilated hearts. When the radius of a cardiac chamber is increased, a greater tension must be developed in the myocardium to produce any given pressure; consequently, a dilated heart must do more work than a nondilated heart. In the lungs, the

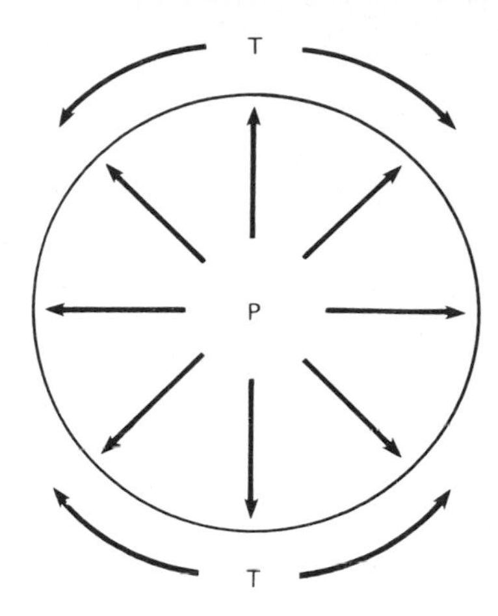

Figure 30–12. Relation between distending pressure (*P*) and wall tension (*T*) in a hollow viscus.

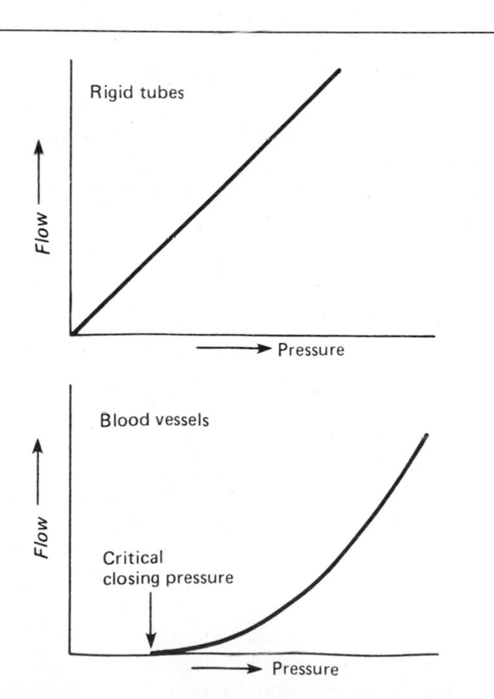

Figure 30–11. Relation of pressure to flow in a rigid-walled system (*top*) and the vascular system (*bottom*).

radii of curvature of the alveoli becomes smaller during expiration, and these structures would tend to collapse because of the pull of surface tension if the tension were not reduced by the surface-tension-lowering agent, surfactant (see Chapter 34). Another example of the operation of this law is seen in the urinary bladder (see Chapter 38).

Resistance & Capacitance Vessels

When a segment of aorta is filled and then distended with increasing volumes of fluid, the pressure in the segment initially rises in a linear fashion. When the same experiment is carried out on a segment of the vena cava or another large distensible vein, the pressure does not rise rapidly until large volumes of fluid are injected. In vivo, the veins are an important blood reservoir. Normally they are partially collapsed and oval in cross section. A large amount of blood can be added to the venous system before the veins become distended to the point where further increments in volume produce a large rise in venous pressure. The veins are therefore called **capacitance vessels.** The small arteries and arterioles are referred to as **resistance vessels** because they are the principal site of the peripheral resistance (see below).

At rest, at least 50% of the circulating blood volume is in the systemic veins. Twelve percent is in the heart cavities, and 18% is in the low-pressure pulmonary circulation. Only 2% is in the aorta, 8% in the arteries, 1% in the arterioles, and 5% in the capillaries (Table 30–1). When extra blood is administered by transfusion, less than 1% of it is distributed in the arterial system (the **"high-pressure system"),** and all the rest is found in the systemic veins, pulmonary circulation, and heart chambers other than the left ventricle (the **"low-pressure system").**

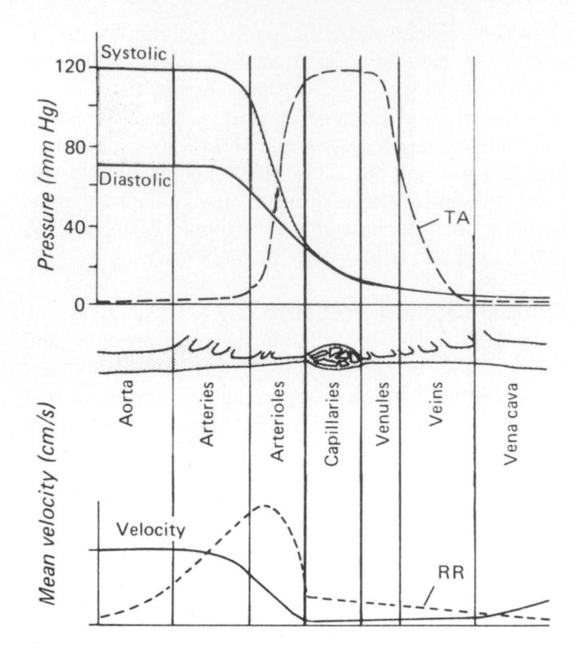

Figure 30–13. Diagram of the changes in pressure and velocity as blood flows through the systemic circulation. TA, total cross-sectional area of the vessels, which increases from 4.5 cm^2 in the aorta to 4500 cm^2 in the capillaries (Table 30–1). RR, relative resistance, which is highest in the arterioles.

ARTERIAL & ARTERIOLAR CIRCULATION

The pressure and velocities of the blood in the various parts of the systemic circulation are summarized in Fig 30-13. The general relationships in the pulmonary circulation are similar, but the pressure in the pulmonary artery is 25/10 mm Hg or less.

Velocity & Flow of Blood

Although the mean velocity of the blood in the proximal portion of the aorta is 40 cm/s, the flow is phasic, and velocity ranges from 120 cm/s during systole to a negative value at the time of the transient backflow before the aortic valves close in diastole. In the distal portions of the aorta and in the large arteries, velocity is also greater in systole than it is in diastole. However, the vessels are elastic, and forward flow is continuous because of the recoil during diastole of the vessel walls that have been stretched during systole (Fig 30–14). This recoil effect is sometimes called the **Windkessel effect,** and the vessels are called Windkessel vessels; Windkessel is the German word for an elastic reservoir. Pulsatile flow appears in some poorly understood way to maintain optimal function of the tissues. If an organ is perfused with a pump that delivers a nonpulsatile flow, there is a gradual rise in vascular resistance, and tissue perfusion fails.

Arterial Pressure

The pressure in the aorta and in the brachial and other large arteries in a young adult human rises to a peak value (**systolic pressure**) of about 120 mm Hg during each heart cycle and falls to a minimum value (**diastolic pressure**) of about 70 mm Hg. The arterial pressure is conventionally written as systolic pressure over diastolic pressure—eg, 120/70 mm Hg. One mm Hg equals 0.133 kPa, so in SI units this value is 16.0/9.3 kPa. The **pulse pressure,** the difference between the systolic and diastolic pressures, is normally about 50 mm Hg. The **mean pressure** is the average pressure throughout the cardiac cycle. Because systole is shorter than diastole, the mean pressure is slightly less than the value halfway between systolic and diastolic pressure. It can actually be determined only by integrating the area of the pressure curve (Fig 30–15); however, as an approximation, the diastolic pressure plus one-third of the pulse pressure is reasonably accurate.

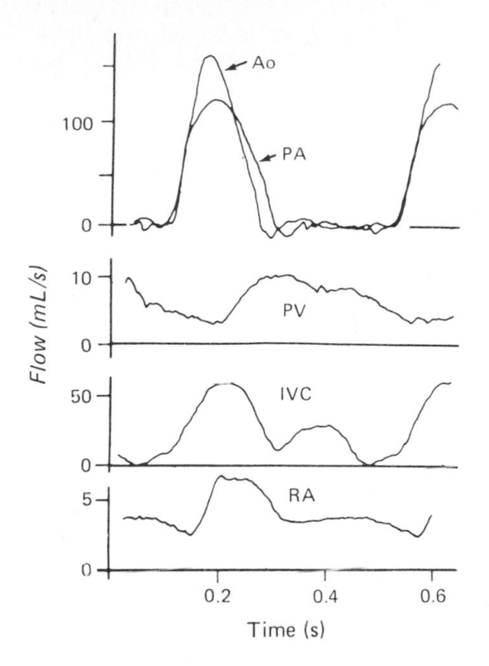

Figure 30–14. Changes in blood flow during the cardiac cycle in the dog. Systole at 0.2 and 0.6 second. Flow patterns in humans are similar. Ao, aorta; PA, pulmonary artery; PV, pulmonary vein; IVC, inferior vena cava; RA, renal artery. (Reproduced, with permission, from Milnor WR: Pulsatile blood flow. *N Engl J Med* 1972;**287**:27.)

The pressure falls very slightly in the large and medium-sized arteries because their resistance to flow is small, but it falls rapidly in the small arteries and arterioles, which are the main sites of the peripheral resistance against which the heart pumps. The mean pressure at the end of the arterioles is 30–38 mm Hg. Pulse pressure also declines rapidly to about 5 mm Hg at the ends of the arterioles (Fig 30–13). The magnitude of the pressure drop along the arterioles varies considerably depending upon whether they are constricted or dilated.

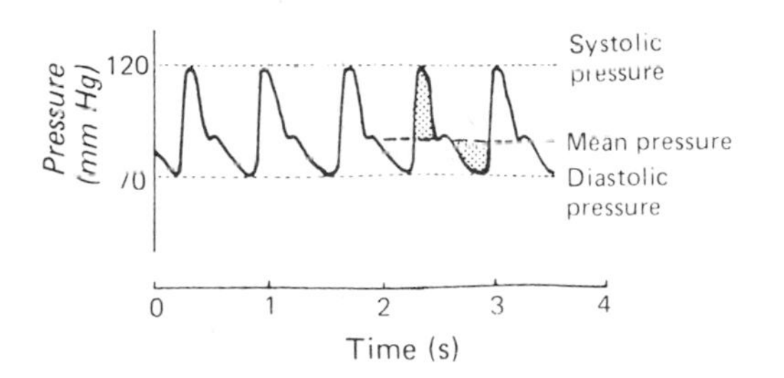

Figure 30–15. Brachial artery pressure curve of normal young human, showing the relation of systolic and diastolic pressure to mean pressure. The shaded area above the mean pressure line is equal to the shaded area below it.

Effect of Gravity

The pressures in Fig 30–13 are those in blood vessels at heart level. The pressure in any vessel below heart level is increased and that in any vessel above heart level is decreased by the effect of gravity. The magnitude of the gravitational effect—the product of the density of the blood, the acceleration due to gravity (980 cm/s/s), and the vertical distance above or below the heart—is 0.77 mm Hg/cm at the density of normal blood. Thus, in the upright position, when the mean arterial pressure at heart level is 100 mm Hg, the mean pressure in a large artery in the head (50 cm above the heart) is 62 mm Hg (100 − [0.77 x 50]) and the pressure in a large artery in the foot (105 cm below the heart) is 180 mm Hg (100 + [0.77 x 105]) (Fig 30–16). The effect of gravity on venous pressure is similar (see below).

Methods of Measuring Blood Pressure

If a cannula is inserted into an artery, the arterial pressure can be measured directly with a mercury manometer or a suitably calibrated strain gauge and

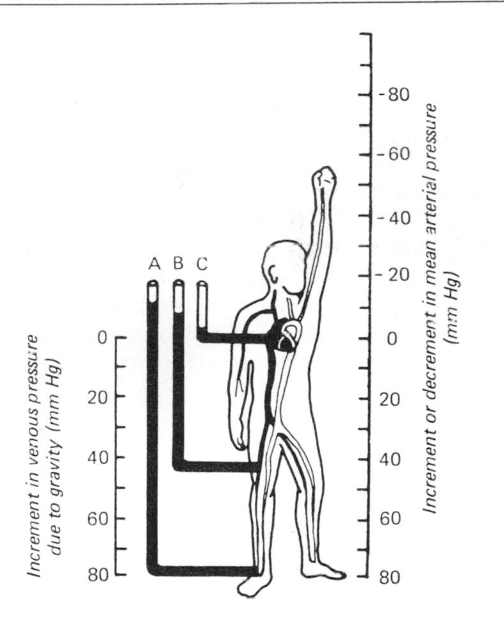

Figure 30–16. Effects of gravity on arterial and venous pressure. The scale on the right indicates the increment (or decrement) in mean pressure in a large artery at each level. The mean pressure in all large arteries is approximately 100 mm Hg when they are at the level of the left ventricle. The scale on the left indicates the increment in venous pressure at each level due to gravity. The manometers on the left of the figure indicate the height to which a column of blood in a tube would rise if connected to an ankle vein (*A*), the femoral vein (*B*), or the right atrium (*C*), with the subject in the standing position. The approximate pressures in these locations in the recumbent position—ie, when the ankle, thigh, and right atrium are at the same level—are (*A*), 10 mm Hg; (*B*), 7.5 mm Hg; and (*C*), 4.6 mm Hg.

an oscillograph arranged to write directly on a moving strip of paper. When an artery is tied off beyond the point at which the cannula is inserted, an **end pressure** is recorded. Flow in the artery is interrupted, and all the kinetic energy of flow is converted into pressure energy. If, alternatively, a T tube is inserted into a vessel and the pressure is measured in the side arm of the tube, under conditions where pressure drop due to resistance is negligible, the recorded **side pressure** is less than the end pressure by the kinetic energy of flow. This is because in a tube or a blood vessel the total energy—the sum of the kinetic energy of flow and the pressure energy—is constant **(Bernoulli's principle).**

It is worth noting that the pressure drop in any segment of the arterial system is due both to resistance and to conversion of potential into kinetic energy. The pressure drop due to energy lost in overcoming resistance is irreversible, since the energy is dissipated as heat; but the pressure drop due to conversion of potential to kinetic energy as a vessel narrows is reversed when the vessel widens out again (Fig 30–17).

Bernoulli's principle also has a significant application in pathophysiology. According to the principle, the greater the velocity of flow in a vessel, the less the lateral pressure distending its walls. When a vessel is narrowed, the velocity of flow in the narrowed portion increases and the distending pressure decreases. Therefore, when a vessel is narrowed by a pathologic process such as an arteriosclerotic plaque, the lateral pressure at the constriction is decreased and the narrowing tends to maintain itself.

Auscultatory Method

The arterial blood pressure in humans is routinely measured by the **auscultatory method.** An inflatable cuff **(Riva-Rocci cuff)** attached to a mercury manometer **(sphygmomanometer)** is wrapped around the arm and a stethoscope is placed over the brachial artery at the elbow (Fig 30–18). The cuff is rapidly inflated until the pressure in it is well above the expected systolic pressure in the brachial artery. The artery is occluded by the cuff, and no sound is heard with the stethoscope. The pressure in the cuff is then lowered slowly. At the point at which systolic pressure in the artery just exceeds the cuff pressure, a spurt of blood passes through with each heartbeat and, synchronously with each beat, a tapping sound is heard below the cuff. The cuff pressure at which the sounds are first heard is the systolic pressure. As the cuff pressure is lowered further, the sounds become louder, then dull and muffled, and finally, in most individuals, they disappear. These are the **sounds of Korotkow.** When direct and indirect blood pressure measurements are made simultaneously, the diastolic pressure in resting adults correlates best with the pressure at which the sound disappears. However, in adults after exercise and in children, the diastolic pressure correlates best with the pressure at which the sounds become muffled. This is also true in diseases such as hyperthyroidism and aortic insufficiency.

The sounds of Korotkow are produced by turbulent flow in the brachial artery. The streamline flow in the unconstricted artery is silent, but when the artery is narrowed, the velocity of flow through the constriction exceeds the **critical velocity** and turbulent flow results (Fig 30–8). At cuff pressures just below the systolic pressure, flow through the artery occurs only at the peak of systole, and the intermittent turbulence produces a tapping sound. As long as the pressure in the cuff is above the diastolic pressure in the artery, flow is interrupted at least during part of diastole, and the intermittent sounds have a staccato quality. When the cuff pressure is just below the arterial diastolic

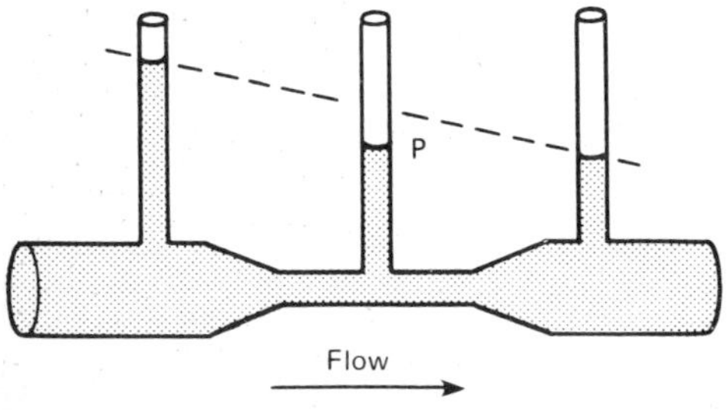

Figure 30–17. Bernoulli's principle. When fluid flows through the narrow portion of the tube, the kinetic energy of flow is increased as the velocity increases, and the pressure energy is reduced. Consequently, the measured pressure (P) is less than it would have been at that point if the tube had not been narrowed. The dashed line indicates what the pressure drop due to frictional forces would have been if the tube had been of uniform diameter.

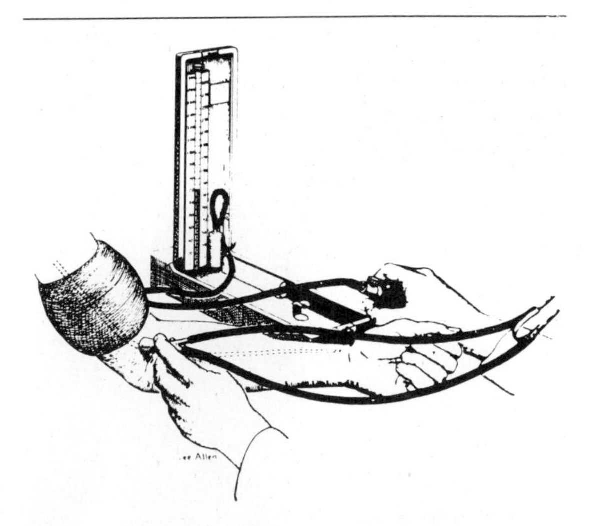

Figure 30–18. Determination of blood pressure by the auscultatory method. (Reproduced, with permission, from Schottelius BA, Schottelius D: *Textbook of Physiology,* 18th ed. Mosby, 1978.)

pressure, the vessel is still constricted, but the turbulent flow is continuous. Continuous sounds have a muffled rather than a staccato quality.

The auscultatory method is accurate when used properly, but a number of precautions must be observed. The cuff must be at heart level to obtain a pressure that is uninfluenced by gravity. The blood pressure in the thighs can be measured with the cuff around the thigh and the stethoscope over the popliteal artery, but there is more tissue between the cuff and the artery in the leg than there is in the arm, and some of the cuff pressure is dissipated. Therefore, pressures obtained using the standard arm cuff are falsely high. The same thing is true when brachial arterial pressures are measured in individuals with obese arms, because the blanket of fat dissipates some of the cuff pressure. In both situations, accurate pressures can be obtained by using a cuff that is wider than the standard arm cuff. If the cuff is left inflated for some time, the discomfort may cause generalized reflex vasoconstriction, raising the blood pressure. It is always wise to compare the blood pressure in both arms when examining an individual for the first time. Persistent major differences between the pressure on the 2 sides indicate the presence of vascular obstruction.

Palpation Method

The systolic pressure can be determined by inflating an arm cuff and then letting the pressure fall and determining the pressure at which the radial pulse first becomes palpable. Because of the difficulty in determining exactly when the first beat is felt, pressures obtained by this **palpation method** are usually 2–5 mm Hg lower than those measured by the auscultatory method.

It is wise to form a habit of palpating the radial pulse while inflating the blood pressure cuff during measurement of the blood pressure by the auscultatory method. When the cuff pressure is lowered, the sounds of Korotkow sometimes disappear at pressures well above diastolic pressure, then reappear at lower pressures (''auscultatory gap''). If the cuff is initially inflated until the radial pulse disappears, the examiner can be sure that the cuff pressure is above systolic pressure, and falsely low pressure values will be avoided.

Normal Arterial Blood Pressure

The blood pressure in the brachial artery in young adults in the sitting or lying position at rest is approximately 120/70 mm Hg. Since the arterial pressure is the product of the cardiac output and the peripheral resistance, it is affected by conditions that affect either or both of these factors. Emotion, for example, increases the cardiac output, and it may be difficult to obtain a truly resting blood pressure in an excited or tense individual. In general, increases in cardiac output increase the systolic pressure, whereas increases in peripheral resistance increase the diastolic pressure. There is a good deal of controversy about where to draw the line between normal and elevated blood pressure levels (**hypertension**), particularly in older patients. However, the evidence seems incontrovertible than in apparently healthy humans both the systolic and the diastolic pressure rise with age (Fig 30–19). The systolic pressure increase is greater than the diastolic. An important cause of the systolic pressure rise is decreased distensibility of the arteries as their walls become increasingly rigid. At the same level of cardiac output, the systolic pressure is higher in old subjects than in young ones because there is less increase in the volume of the arterial system during systole to accommodate the same amount of blood.

CAPILLARY CIRCULATION

At any one time, only 5% of the circulating blood is in the capillaries, but this 5% is in a sense the most important part of the blood volume because it is across the systemic capillary walls that O_2 and nutrients enter the interstitial fluid and CO_2 and waste

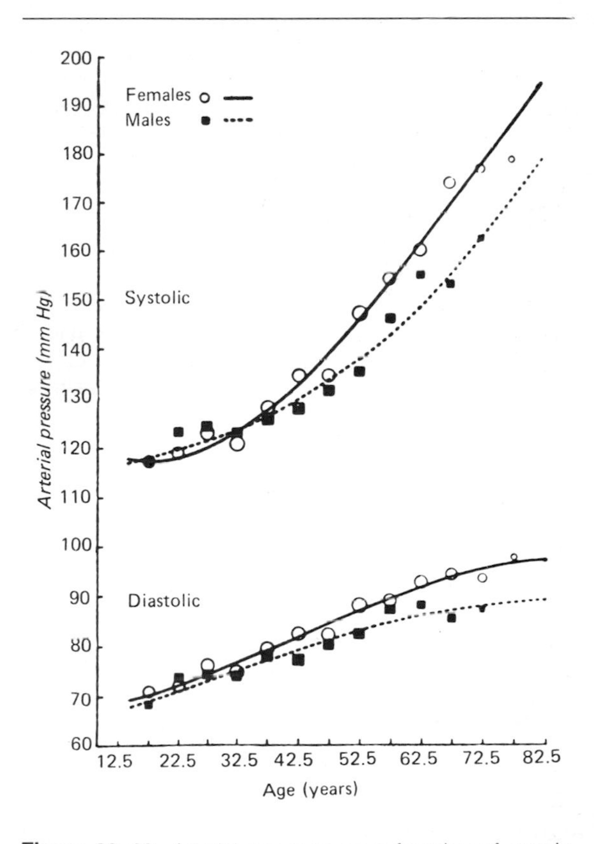

Figure 30–19. Arterial pressure as a function of age in the general population. The size of the squares and the circles is proportionate to the number of subjects studied in that age group. (Reproduced, with permission, from Hamilton M et al: The aetiology of essential hypertension. 1. The arterial pressure in the general population. *Clin Sci* 1954;13:11.)

products enter the bloodstream. The exchange across the capillary walls is essential to the survival of the tissues.

Methods of Study

It is difficult to obtain accurate measurements of capillary pressures and flows. The capillaries in the mesentery of experimental animals and the fingernail beds of humans are readily visible under the dissecting microscope, and observations on flow patterns under various conditions have been made in these and in some other tissues. Capillary pressure has been estimated by determining the amount of external pressure necessary to occlude the capillaries or the amount of pressure necessary to make saline start to flow through a micropipette inserted so that its tip faces the arteriolar end of the capillary.

Capillary Pressure & Flow

Capillary pressures vary considerably, but typical values in human nail bed capillaries are 32 mm Hg at the arteriolar end and 15 mm Hg at the venous end. The pulse pressure is approximately 5 mm Hg at the arteriolar end and zero at the venous end. The capillaries are short, but blood moves slowly (about 0.07 cm/s) because the total cross-sectional area of the capillary bed is large. Transit time from the arteriolar to the venular end of an average-sized capillary is 1–2 seconds.

Equilibration With Interstitial Fluid

As noted above, the capillary wall is a thin membrane made up of endothelial cells. Substances pass through the junctions between endothelial cells, and some also pass through the cells by vesicular transport or, in the case of lipid-soluble substances, by diffusion.

The factors other than vesicular transport that are responsible for transport across the capillary wall are diffusion and filtration (see Chapter 1). Diffusion is quantitatively much more important in terms of the exchange of nutrients and waste materials between blood and tissue. O_2 and glucose are in higher concentration in the bloodstream than in the interstitial fluid and diffuse into the interstitial fluid, whereas CO_2 diffuses in the opposite direction. Lipid-soluble substances diffuse across the capillary walls with greater ease than lipid-insoluble substances, probably by passing directly through the endothelial cells.

The rate of filtration at any point along a capillary depends upon a balance of forces sometimes called the **Starling forces** after the physiologist who first described their operation in detail. One of these forces is the **hydrostatic pressure gradient** (the hydrostatic pressure in the capillary minus the hydrostatic pressure of the interstitial fluid) at that point. The interstitial fluid pressure varies from one organ to another, and there is considerable evidence that it is subatmospheric—about −2 mm Hg—in subcutaneous tissue. It is positive in the liver and kidney and is as high as +6 mm Hg in the brain. The other force is the **osmotic pressure gradient** across the capillary wall (colloid osmotic pressure of plasma minus colloid osmotic pressure of interstitial fluid). This component is directed inward.

Thus:

$$\text{Fluid movement} = k[(P_C + \pi_i) - (P_i + \pi_C)]$$

where

k = capillary filtration coefficient
P_C = capillary hydrostatic pressure
P_i = interstitial hydrostatic pressure
π_C = capillary colloid osmotic pressure
π_i = interstitial colloid osmotic pressure

π_I is usually negligible, so the osmotic pressure gradient($\pi_C - \pi_I$) usually equals the oncotic pressure. The capillary filtration coefficient takes into account, and is proportionate to, the permeability of the capillary wall and the area available for filtration. The magnitude of the Starling forces along a typical muscle capillary is shown in Fig 30–20. Fluid moves into the interstitial space at the arteriolar end of the capillary, where the filtration pressure across its wall exceeds the oncotic pressure; and into the capillary at the venular end, where the oncotic pressure exceeds the filtration pressure. In other capillaries, the balance of Starling forces is different and, for example, fluid moves out of almost the entire length of the capillaries in the renal glomeruli. On the other hand, fluid moves into the capillaries through almost their entire length in the intestines.

It is worth noting that small molecules often equilibrate with the tissues near the arteriolar end of each capillary. In this situation, total diffusion can be increased by increasing blood flow; ie, exchange is **flow-limited** (Fig 30–21). Conversely, transfer of substances that do not reach equilibrium with the tissues during their passage through the capillaries is said to be **diffusion-limited.**

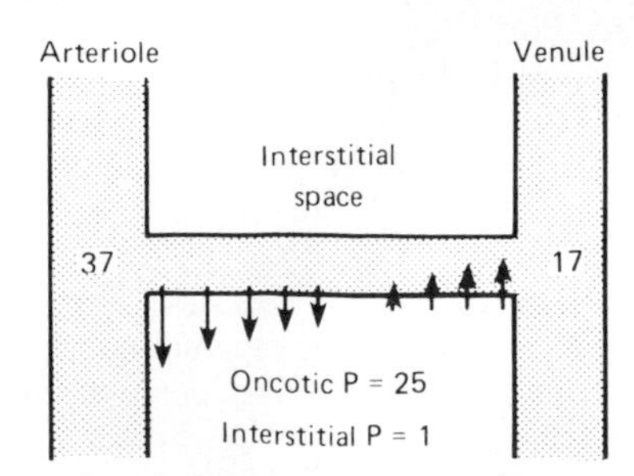

Figure 30–20. Schematic representation of pressure gradients across the wall of a muscle capillary. The numbers at the arteriolar and venular ends of the capillary are the hydrostatic pressures in mm Hg in these locations. The arrows indicate the approximate magnitude and direction of fluid movement. In this example, the pressure differential at the arteriolar end of the capillary is 11 mm Hg ([37–1]−25) outward; at the opposite end, it is 9 mm Hg(25–[17–1]) inward.

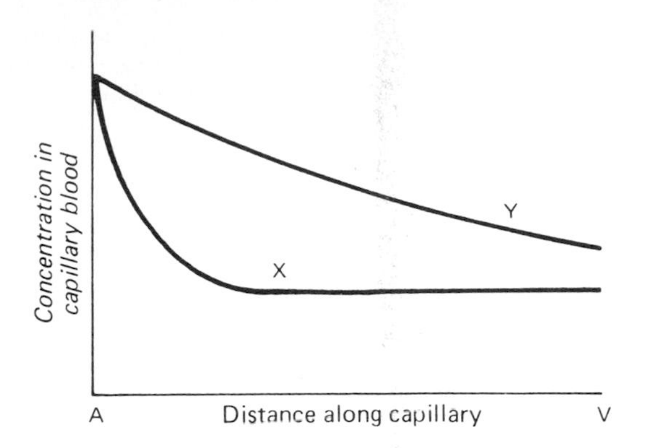

Figure 30–21. Flow-limited and diffusion-limited exchange across capillary walls. A and V indicate the arteriolar and venular ends of the capillary. Substance X equilibrates with the tissues (movement into the tissues equals movement out) well before the blood leaves the capillary, whereas substance Y does not equilibrate. If other factors stay constant, the amount of X entering the tissues can only be increased by increasing blood flow, ie, it is flow-limited. The movement of Y is diffusion-limited.

It has been estimated that about 24 L of fluid is filtered through the capillaries per day. This is about 0.3% of the cardiac output. About 85% of the filtered fluid is reabsorbed into the capillaries, and the remainder returns to the circulation via the lymphatics.

Active & Inactive Capillaries

In resting tissues, most of the capillaries are collapsed, and blood flows for the most part through the thoroughfare vessels from the arterioles to the venules. In active tissues, the metarterioles and the precapillary sphincters dilate. The intracapillary pressure rises, overcoming the critical closing pressure of the vessels, and blood flows through all the capillaries. Relaxation of the smooth muscle of the metarterioles and precapillary sphincters is due to the action of vasodilator metabolites formed in active tissue (see Chapter 31) and possibly also to a decrease in the activity of the sympathetic vasoconstrictor nerves that innervate the smooth muscle.

After noxious stimulation, substance P released by the axon reflex (see Chapter 32) increases capillary permeability. Bradykinin and histamine also increase capillary permeability. When capillaries are stimulated mechanically, they empty (white reaction; see Chapter 32), but this is probably due to contraction of the precapillary sphincters.

LYMPHATIC CIRCULATION & INTERSTITIAL FLUID VOLUME

Lymphatic Circulation

Fluid efflux normally exceeds influx across the capillary walls, but the extra fluid enters the lymphatics and drains through them back into the blood. This keeps the interstitial fluid pressure from rising and promotes the turnover of tissue fluid. The normal 24-hour lymph flow is 2–4L. The composition of lymph is discussed in Chapter 27.

Lymph flow is due to movements of skeletal muscle, the negative intrathoracic pressure during inspiration, the suction effect of high-velocity flow of blood in the veins in which the lymphatics terminate, and rhythmic contractions of the walls of the large lymph ducts. Since lymph vessels have valves that prevent backflow, skeletal muscle contractions push the lymph toward the heart. Pulsations of arteries near lymphatics may have a similar effect. However, the contractions of the walls of the lymphatic ducts are important, and the rate of these contractions increases in direct proportion to the volume of lymph in the vessels. There is evidence that the contractions are the principal factor propelling the lymph.

Agents that increase lymph flow are called **lymphagogues.** They include a variety of agents that increase capillary permeability. Agents that cause contraction of smooth muscle also increase lymph flow from the intestines.

Other Functions of the Lymphatic System

Appreciable quantities of protein enter the interstitial fluid in the liver and intestine, and smaller quantities enter from the blood in other tissues. The walls of the lymphatics are permeable to macromolecules, and the proteins are returned to the bloodstream via the lymphatics. The amount of protein returned in this fashion in 1 day is equal to 25–50% of the total circulating plasma protein. In the kidneys, formation of a maximally concentrated urine depends upon an intact lymphatic circulation; removal of reabsorbed water from the medullary pyramids is essential for the efficient operation of the countercurrent mechanism (see Chapter 38), and water enters the vasa recta only if an appreciable osmotic gradient is maintained between the medullary interstitium and the vasa recta blood by drainage of protein-containing interstitial fluid into the renal lymphatics. Some large enzymes—notably histamines and lipase—may reach the circulation largely or even exclusively via the lymphatics after their secretion from cells into the interstitial fluid. The transport of absorbed long-chain fatty acids and cholesterol from the intestine via the lymphatics has been discussed in Chapter 25.

Interstitial Fluid Volume

The amount of fluid in the interstitial spaces depends upon the capillary pressure, the interstitial fluid pressure, the oncotic pressure, the capillary filtration coefficient, the number of active capillaries, the lymph flow, and the total ECF volume. The ratio of precapillary to postcapillary venular resistance is also important. Precapillary constriction lowers filtration pressure, whereas postcapillary constriction raises it. Changes in any of these variables lead to changes in the volume of interstitial fluid. Factors promoting an increase in this volume are summarized in Table 30–3. **Edema** is the accumulation of interstitial fluid in abnormally large amounts.

In active tissues, capillary pressure rises, often to the point where it exceeds the oncotic pressure throughout the length of the capillary. In addition, osmotically active metabolites may temporarily accumulate in the interstitial fluid because they cannot be washed away as rapidly as they are formed. To the extent that they accumulate, they exert an osmotic effect that decreases the magnitude of the osmotic gradient due to the oncotic pressure. The amount of fluid leaving the capillaries is therefore markedly increased and the amount entering them reduced. Lymph flow is increased, decreasing the degree to which the fluid would otherwise accumulate, but exercising muscle, for example, still increases in volume by as much as 25%.

Interstitial fluid tends to accumulate in dependent parts because of the effect of gravity. In the upright position, the capillaries in the legs are protected from the high arterial pressure by the arterioles, but the high venous pressure is transmitted to them through the venules. Skeletal muscle contractions keep the venous pressure low by pumping blood toward the heart (see below) when the individual moves about; but if one stands still for long periods, fluid accumulates and edema eventually develops. The ankles also swell during long trips when travelers sit for prolonged periods with their feet in a dependent position. Venous obstruction may contribute to the edema in these situations.

Whenever there is abnormal retention of salt in the body, water is also retained. The salt and water are distributed throughout the ECF, and since the interstitial fluid volume is therefore increased, there is a predisposition to edema. Salt and water retention is a factor in the edema seen in heart failure, nephrosis, and cirrhosis, but there are also variations in the mechanisms that govern fluid movement across the capillary walls in these diseases. In congestive heart failure, for example, there is usually an elevation in venous pressure, and capillary pressure is consequently elevated. In cirrhosis of the liver, oncotic pressure is low because hepatic synthesis of plasma proteins is depressed; and in nephrosis, oncotic pressure is low because large amounts of protein are lost in the urine.

Table 30–3. Causes of increased interstitial fluid volume and edema.

Increased fiiltration pressure
Arteriolar dilation
Venular constriction
Increased venous pressure (heart failure, incompetent valves, venous obstruction, increased total ECF volume, effect of gravity, etc)
Decreased osmotic pressure gradient across capillary
Decreased plasma protein level
Accumulation of osmotically active substances in interstitial space
Increased capillary permeability
Substance P
Histamine and related substances
Kinins, etc
Inadequate lymph flow

Another cause of edema is inadequate lymphatic drainage. A complication of **radical mastectomy,** an operation for cancer of the breast in which the axillary lymph nodes are also removed, is edema of the arm due to interruption of its lymph drainage. In filariasis, parasitic worms migrate into the lymphatics and obstruct them. Fluid accumulation plus tissue reaction lead in time to massive swelling, usually of the legs or scrotum (**elephantiasis**). The extent of the reaction is perhaps most graphically illustrated by the remarkable account of the man with elephantiasis whose scrotum was so edematous that he had to place it in a wheelbarrow and wheel it along with him when he walked.

VENOUS CIRCULATION

Blood flows through the blood vessels, including the veins, primarily because of the pumping action of the heart, although venous flow is aided by the heartbeat, the increase in the negative intrathoracic pressure during each inspiration, and contractions of skeletal muscles that compress the veins (**muscle pump).**

Venous Pressure & Flow

The pressure in the venules is 12–18 mm Hg. It falls steadily in the larger veins to about 5.5 mm Hg in the great veins outside the thorax. The pressure in the great veins at their entrance into the right atrium (**central venous pressure**) averages 4.6 mm Hg but fluctuates with respiration and heart action.

Peripheral venous pressure, like arterial pressure, is affected by gravity. It is increased by 0.77 mm Hg for each cm below the right atrium and decreased a like amount for each cm above the right atrium the pressure is measured (Fig 30–16).

When blood flows from the venules to the large veins, its average velocity increases as the total cross-sectional area of the vessels decreases. In the great veins, the velocity of blood is about one-fourth as great as that in the aorta, averaging about 10 cm/s.

Thoracic Pump

During inspiration, the intrapleural pressure falls from −2.5 mm Hg to −6 mm Hg. This negative pressure is transmitted to the great veins and, to a lesser extent, the atria, so that central venous pressure fluctuates from about 6 mm Hg during expiration to approximately 2 mm Hg during quiet inspiration. The drop in venous pressure during inspiration aids venous return. When the diaphragm descends during inspiration, intra-abdominal pressure rises, and this also squeezes blood toward the heart because backflow into the leg veins is prevented by the venous valves.

Effects of Heartbeat

The variations in atrial pressure are transmitted to the great veins to produce the **a, c,** and **v waves** of the venous pressure–pulse curve (see Chapter 29). Atrial pressure drops sharply during the ejection phase of ventricular systole because the atrioventricular valves are pulled downward, increasing the capacity of the atria. This action sucks blood into the atria from the great veins. The sucking of the blood into the atria during systole contributes appreciably to the venous return, especially at rapid heart rates.

Close to the heart, venous flow becomes pulsatile. When the heart rate is slow, 2 periods of peak flow are detectable, one during ventricular systole, due to pulling down of the atrioventricular valves, and one in early diastole, during the period of rapid ventricular filling (Fig 30–14).

Muscle Pump

In the limbs, the veins are surrounded by skeletal muscles, and contraction of these muscles during activity compresses the veins. Pulsations of nearby arteries may also compress veins. Since the venous valves prevent reverse flow, the blood moves toward the heart. During quiet standing, when the full effect of gravity is manifest, venous pressure at the ankle is 85–90 mm Hg (Fig 30–16). Pooling of blood in the leg veins reduces venous return, with the result that cardiac output is reduced, sometimes to the point where fainting occurs. Rhythmic contractions of the leg muscles while the person is standing serve to lower the venous pressure in the legs to less than 30 mm Hg by propelling blood toward the heart. This heartward movement of the blood is decreased in patients with **varicose veins,** whose valves are incompetent, and such patients may have venous stasis and ankle edema. However, even when the valves are incompetent, muscle contractions will continue to produce a basic heartward movement of the blood because the resistance of the larger veins in the direction of the heart is less than the resistance of the small vessels away from the heart.

Venous Pressure in the Head

In the upright position, the venous pressure in the parts of the body above the heart is decreased by the force of gravity. The neck veins collapse above the point where the venous pressure is close to zero, and the pressure all along the collapsed segments is close to zero rather than subatmospheric. However, the dural sinuses have rigid walls and cannot collapse. The pressure in them in the standing or sitting position is therefore subatmospheric. The magnitude of the negative pressure is proportionate to the vertical distance above the top of the collapsed neck veins and in the superior sagittal sinus may be as much as − 10 mm Hg. This fact must be kept in mind by neurosurgeons. Neurosurgical procedures are sometimes performed with the patient seated. If one of the sinuses is opened during such a procedure, it sucks air, causing **air embolism.**

Air Embolism

Because air, unlike fluid, is compressible, its presence in the circulation has serious consequences. The forward movement of the blood depends upon the fact that blood is incompressible. Large amounts of air fill the heart and effectively stop the circulation, causing sudden death, because most of the air is compressed by the contracting ventricles rather than propelled into the arteries. Small amounts of air are swept through the heart with the blood, but the bubbles lodge in the small blood vessels. The surface capillarity of the bubbles markedly increases the resistance to blood flow, and flow is reduced or abolished. Blockage of small vessels in the brain leads to serious and even fatal neurologic abnormalities. In experimental animals, the amount of air that produces fatal air embolism varies considerably, depending in part upon the rate at which it enters the veins. Sometimes as much as 100 mL can be injected without ill effects, whereas at other times as little as 5 mL is lethal.

Measuring Venous Pressure

Central venous pressure can be measured directly by inserting a catheter into the thoracic great veins. **Peripheral venous pressure** correlates well with central venous pressure in most conditions. To measure peripheral venous pressure, a needle attached to a manometer containing sterile saline is inserted into an arm vein (Fig 30–22). The peripheral vein should be at the level of the right atrium (a point 10 cm or

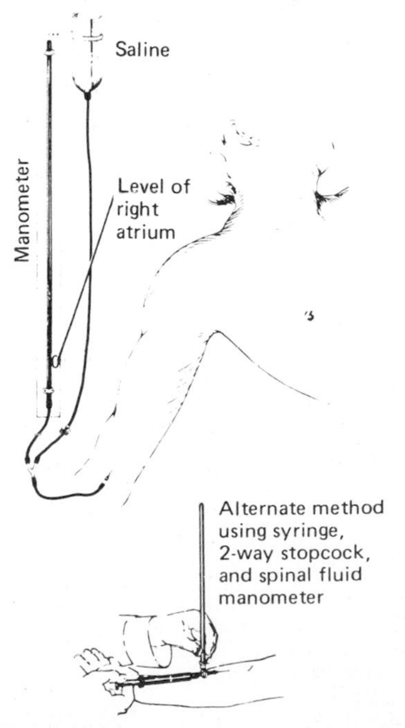

Figure 30–22. Two techniques of measuring peripheral venous pressure.

half the chest diameter from the back in the supine position). The values obtained in millimeters of saline can be converted into millimeters of mercury by dividing by 13.6 (the density of mercury). The amount by which peripheral venous pressure exceeds central venous pressure increases with the distance from the heart along the veins. The mean pressure in the antecubital vein is normally 7.1 mm Hg, compared with a mean pressure of 4.6 mm Hg in the central veins.

A fairly accurate estimate of central venous pressure can be made without any equipment by simply noting the height to which the external jugular veins are distended when the subject lies with the head slightly above the heart. The vertical distance between the right atrium and the place the vein collapses (the place where the pressure in it is zero) is the venous pressure in millimeters of blood.

Central venous pressure is decreased during negative pressure breathing and shock. It is increased by positive pressure breathing, straining, expansion of the blood volume, and heart failure. In advanced congestive heart failure or obstruction of the superior vena cava, the pressure in the antecubital vein may reach values of 20 mm Hg or more.

Cardiovascular Regulatory Mechanisms

31

INTRODUCTION

In humans and other mammals, multiple cardiovascular regulatory mechanisms have evolved. These mechanisms increase the blood supply to active tissues and increase or decrease heat loss from the body by redistributing the blood. In the face of challenges such as hemorrhage, they maintain the blood flow to the heart and brain. When the challenge faced is severe, flow to these vital organs is maintained at the expense of the circulation to the rest of the body.

Circulatory adjustments are effected by altering the output of the pump (the heart), changing the diameter of the resistance vessels (primarily the arterioles), or altering the amount of blood pooled in the capacitance vessels (the veins). Regulation of cardiac output is discussed in Chapter 29. The caliber of the arterioles is increased in active tissues by locally produced vasodilator metabolites and is regulated systemically by circulating vasoactive substances and the nerves that innervate the arterioles. The caliber of the capacitance vessels is also affected by circulating vasoactive substances and by vasomotor nerves. The systemic regulatory mechanisms synergize with the local mechanisms and adjust vascular responses throughout the body.

The terms **vasoconstriction** and **vasodilation** are generally used to refer to constriction and dilation of the resistance vessels. Changes in the caliber of the veins are referred to specifically as **venoconstriction** or **venodilation.**

LOCAL REGULATORY MECHANISMS

Autoregulation

The capacity of tissues to regulate their own blood flow is referred to as **autoregulation.** Most vascular beds have an intrinsic capacity to compensate for moderate changes in perfusion pressure by changes in vascular resistance, so that blood flow remains relatively constant. This capacity is well developed in the kidney (see Chapter 38), but it has also been observed in the mesentery, skeletal muscle, brain, liver, and myocardium. It is probably due in part to the intrinsic contractile response of smooth muscle to stretch (**myogenic theory of autoregulation**). As the pressure rises, the blood vessels are distended, and the vascular smooth muscle fibers that surround the vessels contract. If it is postulated that the muscle responds to the tension in the vessel wall, this theory could explain the greater degree of contraction at higher pressures; the wall tension is proportionate to the distending pressure times the radius of the vessel (law of Laplace; see Chapter 30), and the maintenance of a given wall tension as the pressure rises would require a decrease in radius. Vasodilator substances tend to accumulate in active tissues, and these "metabolites" also contribute to autoregulation (**metabolic theory of autoregulation**). When blood flow decreases, they accumulate and the vessels dilate; whereas when blood flow increases, they tend to be washed away. It has also been argued that as blood flow increases, the accumulation of interstitial fluid compresses the capillaries and venules; but the bulk of the available evidence is against this **tissue pressure hypothesis of autoregulation.**

"Vasodilator Metabolites"

The metabolic changes that produce vasodilation include, in most tissues, decreases in O_2 tension and pH. These changes cause relaxation of the arterioles and precapillary sphincters. Increases in CO_2 tension and osmolality also dilate the vessels. The direct dilator action of CO_2 is most pronounced in the skin and brain. The neurally mediated vasoconstrictor effects of systemic as opposed to local hypoxia and hypercapnia are discussed below. A rise in temperature exerts a direct vasodilator effect, and the temperature rise in active tissues (due to the heat of metabolism) may contribute to the vasodilation. K^+ is another substance that accumulates locally, has demonstrated dilator activity, and probably plays a role in the dilation that occurs in skeletal muscle. Lactate may also contribute to the dilation. In injured tissues, histamine released from damaged cells increases capillary permeability. Thus, it is probably responsible for some of the swelling in areas of inflammation. Adenosine may play a vasodilator role in cardiac muscle but not in skeletal muscle.

Local Vasoconstrictors

Injured arteries and arterioles constrict strongly. The constriction appears to be due in part to the local liberation of serotonin from platelets that stick to the vessel wall in the injured area (see Chapter 27).

A drop in tissue temperature causes vasoconstriction, and this local response to cold plays a part in temperature regulation (see Chapter 14).

Endothelial Effects on Vascular Tone

A polypeptide called **endothelin**, which contains 21 amino acid residues and 2 disulfide bridges, has recently been isolated from endothelial cells. This substance is one of the most potent vasoconstrictors known and may act as an endogenous agonist for dihydropyridine-sensitive Ca^{2+} channels in vascular smooth muscle. Its exact physiologic role remains to be determined.

Several years ago, a chance observation led to the discovery that the endothelium also plays a key role in vasodilation. Many different stimuli act on the endothelial cells to produce **endothelium-derived relaxing factor (EDRF),** a substance that has now been identified as nitric oxide (NO). Adenosine, ANP, and histamine via H_2 receptors produce relaxation of vascular smooth muscle that is independent of the endothelium. However, bradykinin, VIP, substance P, and some other polypeptides act via the endothelium, and various vasoconstrictors that act directly on vascular smooth muscle would produce much greater constriction if they did not simultaneously cause the release of EDRF. A good example is acetylcholine, which, when applied to blood vessels with damaged endothelium, produces vasoconstriction but which, when applied to vessels with intact endothelium, produces vasodilation. Other vasoconstrictors that also cause the release of some EDRF include norepinephrine, serotonin, vasopressin, and angiotesin II.

SYSTEMIC REGULATORY MECHANISMS

Vasoactive Substances in the Circulation

Systemic regulation is brought about by circulating substances and by the vasomotor nerves. Substances in the circulation that bring about vasodilation include kinins, VIP, and ANP. Circulating vasoconstrictors include vasopressin, norepinephrine, epinephrine, and angiotensin II.

Kinins

Two related vasodilator peptides called **kinins** are found in the body. The nonapeptide **bradykinin** is formed in the plasma and the decapeptide **lysylbradykinin** is formed in tissues (Fig 31–1) from substrates called **kininogens** by the action of proteolytic enzymes called **kallikreins.** Tissue kallikrein is found in the kidneys, sweat glands, pancreas, salivary glands, and intestine.

Plasma kallikrein is formed from an inactive precursor, prekallikrein (Fig 31–2). The prekallikrein is converted to kallikrein in the presence of prekallikrein activators, which are proteolytic fragments of the active form of clotting factor XII. The formation of prekallikrein activators from active factor XII is catalyzed by plasmin (see Chapter 27) and in a positive feedback fashion by plasma kallikrein itself. Tissue kallikrein converts prorenin to renin, and it has been suggested that plasma kallikrein has the same effect (see Chapter 24).

Kinins are converted to inactive peptides by kininase I, a carboxypeptidase that removes the C-terminal amino acid residue, and kininase II (angiotensin converting enzyme), the same dipeptidycarboxypeptidase that converts angiotensin I to angiotensin II (see Chapter 24). The latter enzyme is found in high concentration in the lungs, and the lungs are particularly active in removing kinins from the circulation.

The actions of the kinins resemble those of histamine. They cause contraction of visceral smooth muscle, but they relax vascular smooth muscle via EDRF, lowering blood pressure. They also increase capillary permeability, attract leukocytes, and cause pain upon injection under the skin. They appear to be formed during active secretion in sweat glands, salivary glands, and the exocrine portion of the pancreas (see Chapter 26), and they are probably responsible for the increase in blood flow when these tissues are actively secreting their products.

Atrial Natriuretic Peptide

The atrial natriuretic peptide (ANP) secreted by the heart (see Chapter 24) antagonizes the action of various vasoconstrictor agents and lowers blood pressure, but its exact role in the regulation of the circulation is still unknown.

Circulating Vasoconstrictors

Vasopressin is a potent vasoconstrictor, but when it is injected in normal individuals, there is a compensating decrease in cardiac output, so that there is little

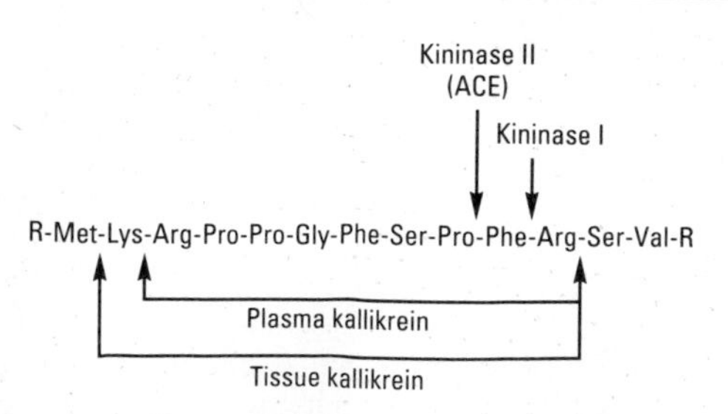

Figure 31–1. Formation of kinins from kininogen, and their metabolism. R, rest of kininogen molecule. Plasma kallikrein hydrolyzes the Lys-Arg and Arg-Ser bonds, forming the nonapeptide bradykinin, whereas tissue kallikrein hydrolyzes the Met-Lys and Arg-Ser bond, forming the decapeptide lysylbradykinin. The peptides are inactivated by kininase I, a carboxypeptidase, and kininase II, the dipeptidylcarboxypeptidase also known as angiotensin converting enzyme (ACE).

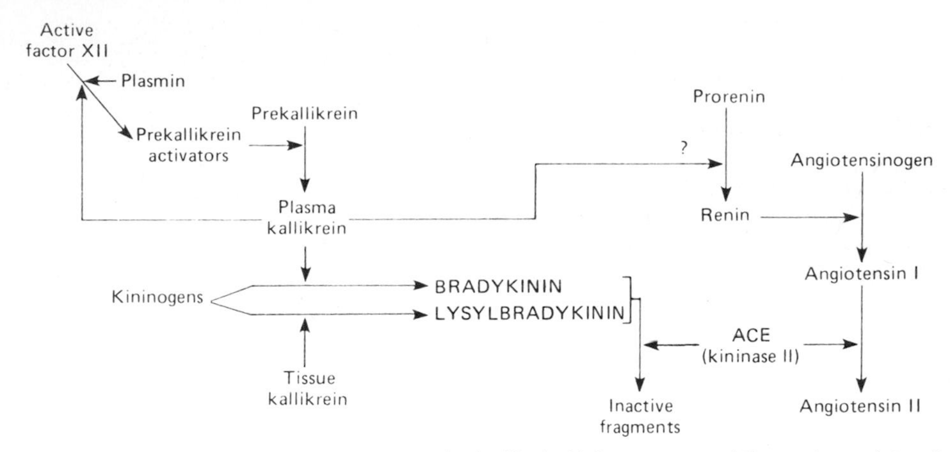

Figure 31–2. Interrelations between clotting factors, the kallikrein-kinin system, and the renin-angiotensin system.

change in blood pressure. Its role in blood pressure regulation is discussed in Chapter 14.

Norepinephrine has a generalized vasoconstrictor action, whereas epinephrine dilates the vessels in skeletal muscle and the liver. The relative unimportance of circulating norepinephrine, as opposed to norepinephrine released from vasomotor nerves, is pointed out in Chapter 20, in which the cardiovascular actions of catecholamines are discussed in detail.

The octapeptide angiotensin II has a generalized vasoconstrictor action. It is formed from angiotensin I liberated by the action of renin from the kidney on circulating angiotensinogen (see Chapter 24). Its formation is increased because renin secretion is increased when the blood pressure falls or ECF volume is reduced, and it helps maintain blood pressure. Angiotensin II also increases water intake and stimulates aldosterone secretion, and increased formation of angiotensin II is part of a homeostatic mechanism that operates to maintain ECF volume (see Chapter 20).

Neural Regulatory Mechanisms

Although the arterioles and the other resistance vessels are most densely innervated, all blood vessels except capillaries and venules contain smooth muscle and receive motor nerve fibers from the sympathetic division of the autonomic nervous system. The fibers to the resistance vessels regulate tissue blood flow and arterial pressure. The fibers to the venous capacitance vessels vary the volume of blood "stored" in the veins. The innervation of most veins is sparse, but the splanchnic veins are well innervated. Venoconstriction is produced by stimuli that also activate the vasoconstrictor nerves to the arterioles. The resultant decrease in venous capacity increases venous return, shifting blood to the arterial side of the circulation.

Innervation of the Blood Vessels

Noradrenergic fibers end on vessels in all parts of the body (Fig 31–3). The noradrenergic fibers are vasoconstrictor in function. In addition to their vasoconstrictor innervation, the resistance vessels of the skeletal muscles are innervated by vasodilator fibers that, although they travel with the sympathetic nerves, are cholinergic (the **sympathetic vasodilator system**). There is some evidence that blood vessels in the heart, lungs, kidney, and uterus also receive a cholinergic innervation. Bundles of smooth noradrenergic and cholinergic fibers form a plexus on the

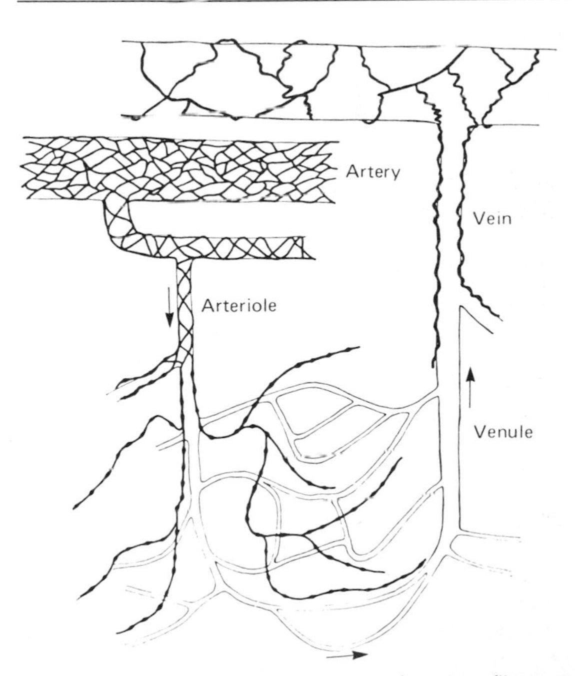

Figure 31–3. Relation of noradrenergic nerve fibers to blood vessels in the mesentery of the rat. Arrows indicate direction of flow. (Reproduced, with permission, from Furness JB, Marshall JM: Correlation of the directly observed responses of mesenteric vessels of the rat to nerve stimulation and noradrenaline with the distribution of adrenergic nerves. *J Physiol* 1974:**75**:239.)

adventitia of the arterioles. Fibers with multiple varicosities extend from this plexus to the media and end primarily on the outer surface of the smooth muscle of the media without penetrating it. Transmitter reaches the inner portions of the media by diffusion, and current spreads from one smooth muscle cell to another via tight junctions.

There is no tonic discharge in the vasodilator fibers, but the vasoconstrictor fibers to most vascular beds have some tonic activity. When the sympathetic nerves are cut (**sympathectomy**), the blood vessels dilate. In most tissues, vasodilation is produced by decreasing the rate of tonic discharge in the vasoconstrictor nerves, although in skeletal muscles it can also be produced by activating the sympathetic vasodilator system (Table 31–1).

Nerves containing peptides are found on many blood vessels. The cholinergic nerves also contain VIP, which produces vasodilation; and postganglionic sympathetic nerves also contain neuropeptide Y, which is a vasoconstrictor. Substance P and CGRPα, which also produce vasodilation, are found in sensory nerves near blood vessels.

Afferent impulses in sensory nerves from the skin are relayed antidromically down branches of the sensory nerves that innervate blood vessels, and these impulses cause release of substance P from the nerve endings. Substance P causes vasodilation and increased capillary permeability. This local neural mechanism is called the **axon reflex** (Fig 32–16). Other cardiovascular reflexes are integrated in the central nervous system.

Cardiac Innervation

Impulses in the noradrenergic sympathetic nerves to the heart increase the cardiac rate (chronotropic effect) and the force of cardiac contraction (inotropic effect). Impulses in the cholinergic vagal cardiac fibers decrease heart rate. There is a moderate amount of tonic discharge in the cardiac sympathetic nerves at rest, but there is a good deal of tonic vagal discharge (**vagal tone**) in humans and other large animals. When the vagi are cut in experimental animals, the heart rate rises, and after the administration of parasympatholytic drugs such as atropine, the cardiac rate in humans increases from its normal resting value of 70 to 150–180 beats per minute because the sympathetic tone is unopposed. In humans in whom both noradrenergic and cholinergic systems are blocked, the heart rate is approximately 100.

Table 31–1. Summary of factors affecting the caliber of the arterioles.

Constriction	Dilation
Increased noradrenergic discharge	Decreased noradrenergic discharge
Circulating catecholamines (except epinephrine in skeletal muscle and liver)	Circulating epinephrine in skeletal muscle and liver
Circulating angiotensin II	Activation of cholinergic dilators in skeletal muscle
Locally released serotonin	Histamine
Decreased local temperature	Kinins
Endothelin	Substance P (axon reflex)
Neuropeptide Y	CGRPα
	VIP
	EDRF
	Decreased O_2 tension
	Increased CO_2 tension
	Decreased pH
	Lactate, K^+, adenosine, etc
	Increased local temperature

Vasomotor Control

The sympathetic nerves that constrict arterioles and veins and increase heart rate and stroke volume discharge in a tonic fashion, and blood pressure is adjusted by variations in the rate of this tonic discharge. Spinal reflex activity affects blood pressure, but the main control of blood pressure is exerted by groups of neurons in the medulla that are sometimes called collectively the **vasomotor area** or **vasomotor center.** Excitatory neurons with their cell bodies in the ventrolateral medulla, including the so-called C1 area, project directly to sympathetic preganglionic neurons in the intermediolateral gray column of the spinal cord (Fig 31–4). Inhibitory pathways from the medulla also converge on the preganglionic neurons. Impulses reaching the medulla also affect the heart rate via vagal discharge to the heart. The neurons from which the vagal fibers arise are in the dorsal motor nucleus of the vagus, the nucleus of the tractus solitarius, and the nucleus ambiguus.

When vasoconstrictor tone is increased, there is increased arteriolar constriction and a rise in blood pressure. Venoconstriction and a decrease in the stores of blood in the venous reservoirs usually accompany these changes, although changes in the capacitance vessels do not always parallel changes in the resistance vessels. Heart rate and stroke volume are increased because of activity in the sympathetic nerves to the heart, and cardiac output is increased. There is usually an associated decrease in the tonic activity of vagal fibers to the heart. Conversely, a decrease in the discharge rate of vasoconstrictor fibers causes vasodilation, a fall in blood pressure, and an increase in the storage of blood in the venous reservoirs. There is usually a concomitant decrease in heart rate, but this is mostly due to stimulation of the vagal innervation of the heart.

Afferents to the Vasomotor Area

The afferents that converge on the vasomotor area are summarized in Table 31–2. They include not only the very important fibers from arterial and venous baroreceptors but fibers from other parts of the nervous system and from the carotid and aortic chemoreceptors as well. In addition, some stimuli act directly on the vasomotor area.

There are descending tracts to the vasomotor area from the cerebral cortex (particularly the limbic cortex) that relay in the hypothalamus and possibly also in the mesencephalon. These fibers are responsible for the blood pressure rise and tachycardia produced by emotions such as sexual excitement and

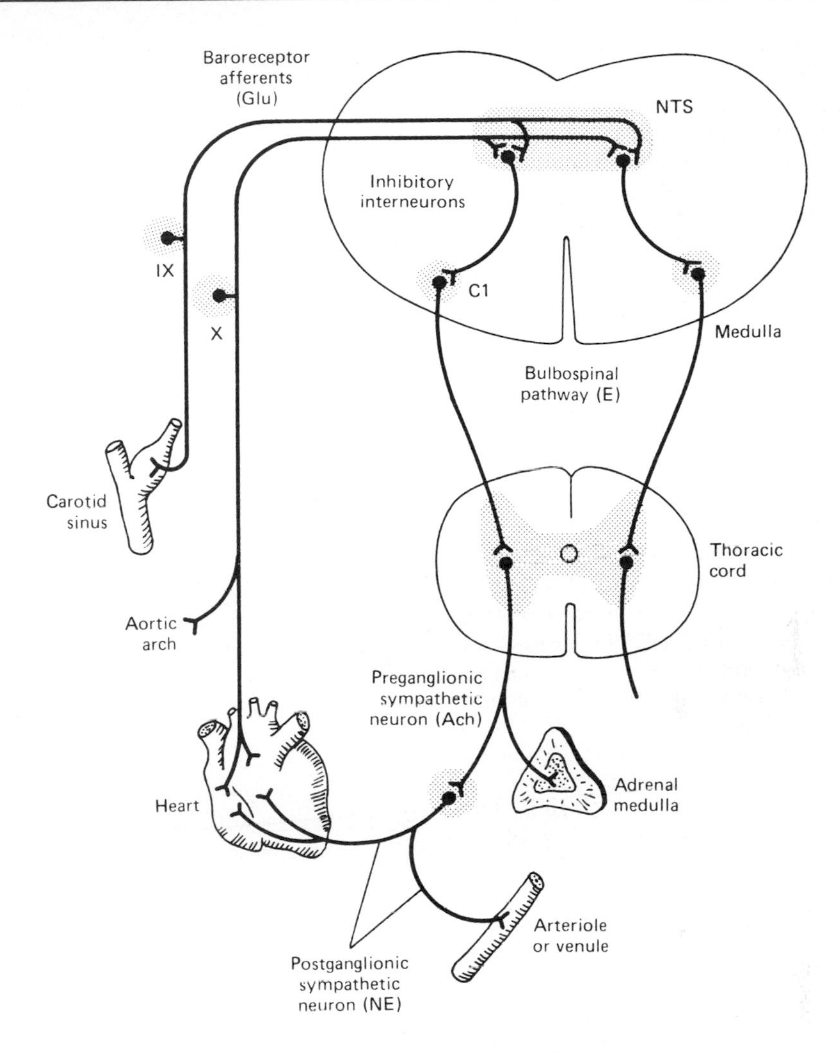

Figure 31–4. Basic pathways involved in the medullary control of blood pressure. The vagal efferent pathways that slow the heart are not shown. There is evidence that the synaptic transmitter released by baroreceptor afferents in the nucleus of the tractus solitarius (NTS) is glutamate (Glu). The transmitter secreted by the inhibitory interneurons from the NTS to the vasomotor region (C1) is unknown. The transmitter secreted by the bulbospinal neurons that excite the preganglionic sympathetic neurons in the intermediolateral gray column (ML) of the spinal cord may be epinephrine (E) or substance P. The preganglionic neurons secrete acetylcholine (Ach), and the postganglionic neurons secrete norepinephrine (NE). (Modified and reproduced, with permission, from Reis DJ et al: Role of adrenaline neurons of ventrolateral medulla [the C1 group] in the tonic and phasic control of arterial pressure. *Clin Exp Hypertens* [A] 1984;**6**:221. By courtesy of Marcel Dekker, Inc.)

Table 31–2. Factors affecting the activity of the vasomotor area in the medulla.

Factors
Direct stimulation
CO_2
Hypoxia
Excitatory inputs
From cortex via hypothalamus
From pain pathways
From carotid and aortic chemoreceptors
Inhibitory inputs
From cortex via hypothalamus
From lungs
From carotid, aortic, and cardiopulmonary baroreceptors

anger. The connections between the hypothalamus and the vasomotor area are reciprocal, with afferents from the brain stem closing the loop.

Inflation of the lungs causes vasodilation and a decrease in blood pressure. This response is mediated via vagal afferents from the lungs that inhibit vasomotor discharge. Pain usually causes a rise in blood pressure, presumably as a result of afferent impulses in the reticular formation converging on the vasomotor area. However, prolonged severe pain may cause vasodilation and fainting.

Baroreceptors

The **baroreceptors** are stretch receptors in the walls of the heart and blood vessels. The **carotid sinus** and **aortic arch** receptors monitor the arterial circulation. Receptors are also located in the walls of the right and left atria at the entrance of the superior and inferior venae cavae and the pulmonary veins, in the wall of the left ventricle, and in the pulmonary circulation. These receptors in the low-pressure part of the circulation and the left ventricle are referred to collectively as the cardiopulmonary receptors. The baroreceptors are stimulated by distention of the structures in which they are located, and so they discharge at an increased rate when the pressure in these structures rises. Their afferent fibers pass via the glossopharyngeal and vagus nerves to the medulla. Most of them end in the nucleus of the tractus solitarius (NTS) on each side. From the NTS, inhibitory interneurons project to the vasomotor area in the ventrolateral medulla. It is important to remember that impulses generated in the baroreceptors *inhibit* the tonic discharge of the vasoconstrictor nerves and *excite* the vagal innervation of the heart, producing vasodilation, venodilation, a drop in blood pressure, bradycardia, and a decrease in cardiac output.

One hypothesis about the synaptic transmitters involved in this basic vasomotor reflex pathway is shown in Fig 31–4. However, there is considerable debate about whether the transmitter in the descending bulbospinal pathway is adrenergic, and there is some evidence that epinephrine inhibits rather than excites preganglionic sympathetic neurons. In addition, the transmitter released by baroreceptor afferents may be substance P rather than glutamate.

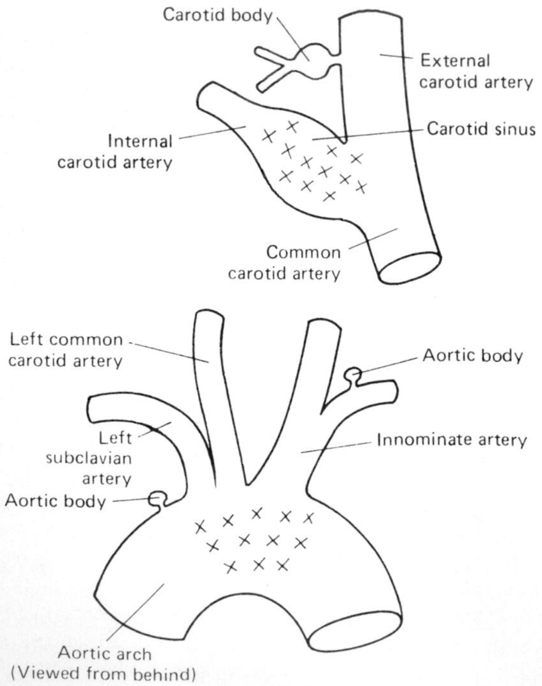

Carotid Sinus & Aortic Arch

The carotid sinus is a small dilation of the internal carotid artery just above the bifurcation of the common carotid into external and internal carotid branches (Fig 31–5). Baroreceptors are located in this dilation and are also found in the wall of the arch of the aorta. The receptors are located in the adventitia of the vessels. They are extensively branched, knobby, coiled, and intertwined ends of myelinated nerve fibers that resemble Golgi tendon organs (Fig 6–5). Similar receptors have been found in various other parts of the large arteries of the thorax and neck in some species. The afferent nerve fibers from the carotid sinus and carotid body form a distinct branch of the glossopharyngeal nerve, the **carotid sinus nerve,** but the fibers from the aortic arch form a separate distinct branch of the vagus only in the rabbit. The carotid sinus nerves and vagal fibers from the aortic arch are commonly called the **buffer nerves.**

Buffer Nerve Activity

At normal blood pressure levels, the fibers of the buffer nerves discharge at a slow rate (Fig 31–6). When the pressure in the sinus and aortic arch rises, the discharge rate increases; and when the pressure

Carotid body
External carotid artery
Internal carotid artery
Carotid sinus
Common carotid artery
Left common carotid artery
Aortic body
Left subclavian artery
Innominate artery
Aortic body
Aortic arch (Viewed from behind)

Figure 31–5. Baroreceptor areas in carotid sinus and aortic arch. The Xs identify sites where receptors are located.

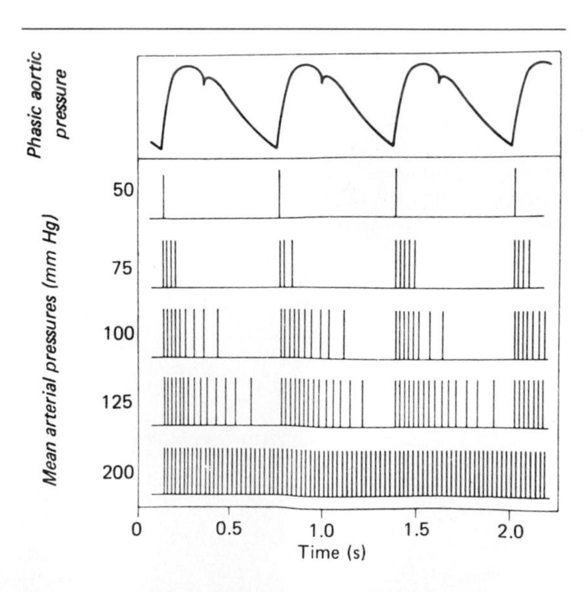

Figure 31–6. Discharges (vertical lines) in a single afferent nerve fiber from the carotid sinus at various arterial pressures, plotted against changes in aortic pressure with time. (Reproduced, with permission, from Berne RM, Levy MN: *Cardiovascular Physiology,* 3rd ed. Mosby, 1977.)

falls, the rate declines. The compensatory response produced by increased discharge is a fall in blood pressure, because activity in the baroreceptor afferents inhibits the tonic discharge in the vasoconstrictor nerves.

When one carotid sinus of a monkey is isolated and perfused and the other baroreceptors are denervated, there is no discharge in the afferent fibers from the perfused sinus and no drop in the animal's arterial pressure or heart rate when the perfusion pressure is below 30 mm Hg. At perfusion pressures of 70–110 mm Hg, there is an essentially linear relation between the perfusion pressure and the fall in blood pressure and heart rate produced in the monkey. At perfusion pressures above 150 mm Hg there is no further increase in response (Fig 31–7), presumably because the rate of baroreceptor discharge and the degree of inhibition of the vasomotor center are maximal.

The carotid receptors respond both to sustained pressure and to pulse pressure. A decline in carotid pulse pressure without any change in mean pressure decreases the rate of baroreceptor discharge and provokes a rise in blood pressure and tachycardia. The receptors also respond to changes in pressure as well as steady pressure; when the pressure is fluctuating, they sometimes discharge during the rises and are silent during the falls (Fig 31–6) at mean pressures at which if there were no fluctuations, there would be a steady discharge.

The aortic receptors have not been studied in such great detail, but there is no reason to believe that their responses differ significantly from those of the receptors in the carotid sinus.

From the foregoing discussion, it is apparent that the baroreceptors on the arterial side of the circulation, their afferent connections to the vasomotor and cardioinhibitory areas, and the efferent pathways from these areas constitute a reflex feedback mechanism that operates to stabilize the blood pressure and heart rate. Any drop in systemic arterial pressure decreases the inhibitory discharge in the buffer nerves, and there is a compensatory rise in blood pressure and cardiac output. Any rise in pressure produces dilation of the arterioles and decreases cardiac output until the blood pressure returns to its previous normal level.

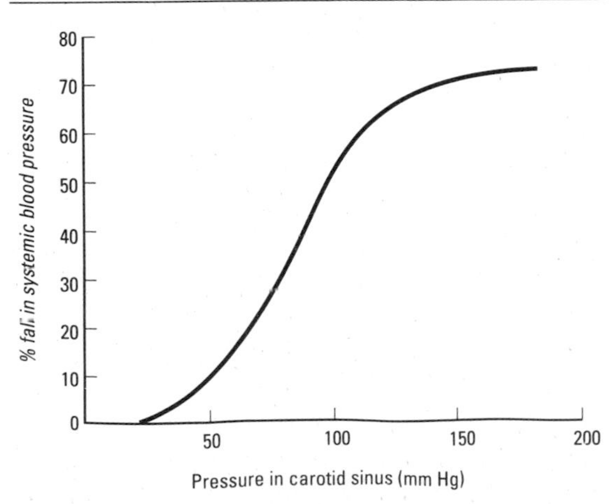

Figure 31–7. Fall in systemic blood pressure by raising the pressure in the isolated carotid sinus of a monkey to various values.

In chronic hypertension, the baroreceptor reflex mechanism is "reset" to maintain an elevated rather than a normal blood pressure. In perfusion studies on hypertensive experimental animals, raising the pressure in the isolated carotid sinus lowers the elevated systemic pressure, and decreasing the perfusion pressure raises the systemic pressure. Little is known about how and why this occurs, but resetting occurs rapidly in experimental animals. It is also rapidly reversible, both in experimental animals and in clinical situations.

If norepinephrine is painted on the carotid sinus, baroreceptor discharge increases and blood pressure falls. According to some investigators, stimulation of the sympathetic nerves to the sinus has a similar effect. The increased discharge is due to the fact that stretch receptors are in series with contractile elements in the media. However, there is considerable evidence that variations in the rate of sympathetic discharge and catecholamine secretion in intact animals have little if any effect on the sensitivity of the carotid sinus mechanism.

Effect of Carotid Clamping & Buffer Nerve Section

Bilateral clamping of the carotid arteries proximal to the carotid sinuses elevates the blood pressure and heart rate because the procedure lowers the pressure in the sinuses. Cutting the carotid sinus nerves on each side has the same effect. The pressor response following these 2 procedures is moderate, because the aortic baroreceptors are still functioning normally, and they buffer the rise. If baroreceptor afferents in the vagi are also interrupted, blood pressure rises to 300/200 mm Hg or higher. Bilateral lesions of the nucleus of the tractus solitarius, the site of termination of the baroreceptor afferents, cause severe hypertension that can be fatal. These forms of experimental hypertension are called **"neurogenic hypertension."**

Atrial Stretch Receptors

The stretch receptors in the atria are of 2 types: those that discharge primarily during atrial systole (type A), and those that discharge primarily late in diastole, at the time of peak atrial filling (type B). The discharge of type B baroreceptors is increased when venous return is increased and decreased by positive pressure breathing, indicating that they respond primarily to distention of the atrial walls. The reflex circulatory adjustments initiated by increased discharge from most if not all of these receptors include vasodilation and a fall in blood pressure. However, the heart rate is increased rather than decreased. The atrial baroreceptors are probably part of a reflex mechanism that combats excessive rises in central venous pressure and venous return.

Bainbridge Reflex

Rapid infusion of blood or saline in anesthetized animals sometimes produces a rise in heart rate if the initial heart rate is low. This effect was described by Bainbridge in 1915, and since then it has been known as the **Bainbridge reflex.** It appears to be a true reflex rather than a response to local stretch, since it is abolished by bilateral vagotomy, and infusion of fluids in animals with transplanted hearts increases the rate of the recipient's atrial remnant but fails to affect the rate of the transplanted heart. The receptors may be the tachycardia-producing atrial receptors mentioned above. The reflex competes with the baroreceptor-mediated decrease in heart rate produced by volume expansion and is diminished or absent when the initial heart rate is high. There has been much debate about its significance, and its physiologic role remains unsettled.

Left Ventricular Receptors

When the left ventricle is distended in experimental animals, there is a fall in systemic arterial pressure and heart rate. It takes considerable ventricular distention to produce this response, and its physiologic significance is uncertain. However, left ventricular stretch receptors may play a role in the maintenance of the vagal tone that keeps the heart rate low at rest.

In experimental animals, injections of the drug **veratridine** into the branches of the coronary arteries that supply the left ventricle cause apnea, hypotension, and bradycardia (the **coronary chemoreflex** or **Bezold-Jarisch reflex**). The response is prevented by vagotomy. Injections into the coronary arteries supplying the right ventricle and atria are ineffective. Nicotine produces a similar response when it is injected into the arterial supply of the left ventricle or applied to the surface of the left ventricle near the apex of the heart on pieces of filter paper. The response to nicotine by both routes is absent if procaine is first injected into the pericardial sac. The coronary chemo-reflex might be triggered by chemical stimulation of the stretch receptors in the ventricular wall, or it might be due to stimulation of as yet unidentified chemoreceptors in the myocardium. There has been speculation that in patients with myocardial infarcts, substances released from the infarcted tissue stimulate ventricular receptors, contributing to the hypotension that is not infrequently a stubborn complication of this disease.

Pulmonary Receptors

Distention of the pulmonary vascular bed causes reflex bradycardia and systemic hypotension. The location of the receptors involved is not settled.

Injections of veratridine, phenyl biguanide, and serotonin into the pulmonary artery produce apnea, hypotension, and bradycardia (**pulmonary chemoreflex**). The response, which is blocked by vagotomy, is essentially the same as that produced by injection of veratridine into the arterial supply of the left ventricle, but it occurs too rapidly to be caused by the drugs reaching the left ventricular receptors. The receptors are probably in the pulmonary veins, but their exact location has not been determined, and it is not now known whether they are different from the receptors for the response to distention of the pulmonary circulation.

Mesenteric Baroreceptors

There is some evidence that the pacinian corpuscles in the mesentery function as baroreceptors. They probably initiate reflexes that control the local blood flow in the viscera.

Other Effects of Baroreceptor Stimulation

Increased activity in baroreceptor afferents inhibits respiration, but this effect is slight and of little physiologic importance.

Impulses initiated in atrial stretch receptors are relayed from the nucleus of the tractus solitarius to the hypothalamus, where they inhibit the secretion of ACTH and vasopressin (see Chapter 14). The resultant diuresis helps reduce the venous distention that caused the stimulation. Atrial distention also tends to lower renin secretion, but the effect on aldosterone secretion is slight. A drop in systemic arterial pressure stimulates ACTH and vasopressin secretion. The increase in vasopressin secretion is inhibited if the carotid sinus nerves and the vagi are cut, but not by vagotomy alone.

Clinical Testing & Stimulation

The changes in pulse rate and blood pressure that occur in humans on standing up or lying down (see Chapter 33) are due for the most part to baroreceptor reflexes. The function of the receptors can be tested by monitoring changes in heart rate as a function of increasing arterial pressure during infusion of the α-adrenergic agonist phenylephrine. A normal response is shown in Fig 31–8; from a systolic pressure of about 120–150 mm Hg, there is a linear relation between pressure and lowering of the heart rate (greater R-R interval).

The function of the receptors can also be tested by monitoring the changes in pulse and blood pressure that occur in response to brief periods of straining (forced expiration against a closed glottis: the **Valsalva maneuver**). The blood pressure rises at the onset of straining (Fig 31–9) because the increase in intrathoracic pressure is added to the pressure of the blood in the aorta. It then falls because the high intrathoracic pressure compresses the veins, decreasing venous return and cardiac output. The decreases in arterial pressure and pulse pressure inhibit the baroreceptors, causing tachycardia and a rise in peripheral resistance. When the glottis is opened and the intrathoracic pressure returns to normal, cardiac output is restored but the peripheral vessels are constricted. The blood pressure therefore rises above normal, and this stimulates the baroreceptors, causing bradycardia and a drop in pressure to normal levels.

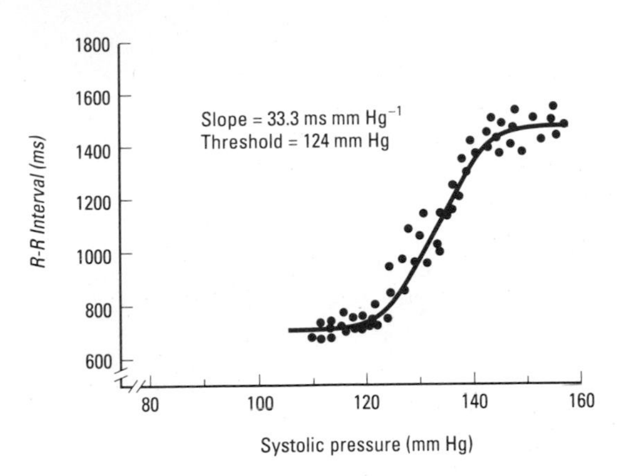

Figure 31–8. Baroreflex-mediated lowering of the heart rate during infusion of phenylephrine in a human subject. Note that the values for the R-R interval of the EKG, which are plotted on the vertical axis, are inversely proportionate to the heart rate. (Reproduced, with permission, from Kotrly K et al: Effects of fentanyl-diazepam-nitrous oxide anaesthesia on arterial baroreflex control of heart rate in man *Br J Anaesth* 1986;**58**:406.)

In sympathectomized patients, heart rate changes still occur because the baroreceptors and the vagi are intact. However, in patients with autonomic insufficiency, a disease of unknown cause in which there is widespread disruption of autonomic function, the heart rate changes are absent. For reasons that are still obscure, patients with primary hyperaldosteronism also fail to show the heart rate changes and the blood pressure rise when the intrathoracic pressure returns to normal. Their response to the Valsalva maneuver returns to normal after removal of the aldosterone-secreting tumor.

Effects of Chemoreceptor Stimulation on the Vasomotor Area

Afferents from the chemoreceptors in the carotid and aortic bodies exert their main effect on respiration, and their function is discussed in Chapter 36. However, they also converge on the vasomotor area. The cardiovascular response to chemoreceptor stimulation is peripheral vasoconstriction and bradycardia. However, hypoxia also produces hyperpnea and increased catecholamine secretion from the adrenal medulla, both of which produce tachycardia and an increase in cardiac output. Hemorrhage that produces hypotension leads to chemoreceptor stimulation. This is due to decreased blood flow to the chemoreceptors and consequent stagnant anoxia of these organs (see Chapter 37). In hypotensive animals, baroreceptor discharge is low (see below), and section of the glossopharyngeal and vagus nerves leads to a fall rather than a rise in blood pressure, because the chemoreceptor drive to the vasomotor area is removed. Chemoreceptor discharge may also contribute to the production of **Mayer waves.** These should not be confused with **Traube-Hering waves,** which are fluctuations in blood pressure synchronized with respiration. The Mayer waves are slow regular oscillations in arterial pressure that occur at the rate of about one per 20–40 seconds during hypotension. Under these conditions, hypoxia stimulates the chemoreceptors. The stimulation raises the blood pressure, which improves the blood flow in the receptor organs and eliminates the stimulus to the chemoreceptors, so that the pressure falls and a new cycle is initiated. However, Mayer waves are reduced but not abolished by chemoreceptor denervation and are sometimes present in spinal animals, so oscillation in spinal vasopressor reflexes is also involved.

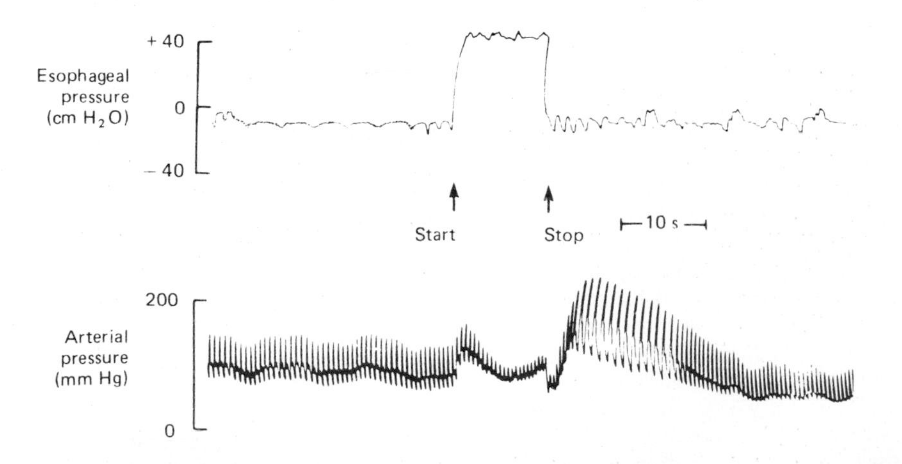

Figure 31–9. Diagram of the response to straining (the Valsalva maneuver) in a normal man, recorded with a needle in the brachial artery. (Courtesy of M McIlroy.)

Direct Effects on the Vasomotor Area

Hypoxia and hypercapnia both stimulate the vasomotor area directly, although the direct effect of hypoxia is small. When intracranial pressure is increased, the blood supply to the vasomotor area is compromised, and the local hypoxia and hypercapnia increase its discharge. The resultant rise in systemic arterial pressure (**Cushing reflex**) tends to restore the blood flow to the medulla. The rise in blood pressure causes a reflex decrease in heart rate via the arterial baroreceptors (see below), and this is why bradycardia rather than tachycardia is characteristically seen in patients with increased intracranial pressure.

A rise in arterial P_{CO_2} stimulates the vasomotor area, but the direct peripheral effect of hypercapnia is vasodilation. Therefore, the peripheral and central actions tend to cancel each other. Moderate hyperventilation, which significantly lowers the CO_2 tension of the blood, causes cutaneous and cerebral vasoconstriction in humans, but there is little change in blood pressure. Exposure to high concentrations of CO_2 is associated with marked cutaneous and cerebral vasodilation, but there is vasoconstriction elsewhere and usually a slow rise in blood pressure.

Sympathetic Vasodilator System

The cholinergic sympathetic vasodilator fibers are part of a regulatory system that originates in the cerebral cortex, relays in the hypothalamus and mesencephalon, and passes through the medulla without interruption to the intermediolateral gray column of the spinal cord (Fig 31–10). The preganglionic neurons which are part of this system activate postganglionic neurons to blood vessels in skeletal muscle that are anatomically sympathetic but secrete acetylcholine. Stimulation of this system produces vasodilation in skeletal muscle, but the resultant increase in blood flow is associated with a decrease rather than an increase in muscle O_2 consumption. This suggests that the blood is being diverted through thoroughfare channels rather than capillaries. Adrenal medullary secretion of norepinephrine and epinephrine is apparently increased when this system is stimulated, the epinephrine probably reinforcing the dilation of muscle blood vessels. In cats and dogs, the system has been shown to discharge in response to emotional stimuli such as fear, apprehension, and rage. Its role in humans is uncertain, but it has been suggested that the sympathetic vasodilator system is responsible for fainting in emotional situations. It has also been argued that this system is responsible for the increase in muscle blood flow that occurs at or even before the start of muscular exercise (see Chapter 33). However, it now appears that the vasodilation before the start of exercise is not a constant or marked phenomenon.

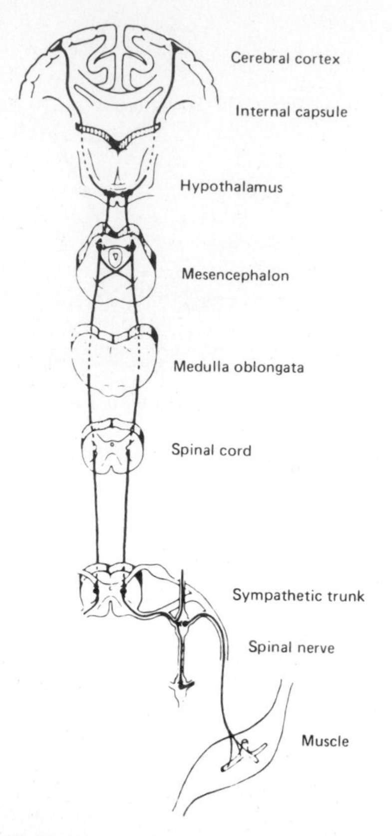

Figure 31–10. The sympathetic vasodilator pathways. (Reproduced, with permission, from Lindgren P: The mesencephalic and vascular system. *Acta Physiol Scand [Suppl]* 1955;**35**:121.)

Control of Heart Rate

The sympathetic and parasympathetic nerves to the heart and baroreceptor-mediated reflex changes in

Table 31–3. Factors affecting heart rate. Norepinephrine has a direct chronotropic effect on the heart, but in the intact animal its pressor action stimulates the baroreceptors, leading to enough reflex increase in vagal tone to overcome the direct effect and produce bradycardia.

Factors affecting heart rate
Heart rate **accelerated** by:
Decreased activity of baroreceptors in the arteries, left ventricle, and pulmonary circulation
Inspiration
Excitement
Anger
Most painful stimuli
Hypoxia
Exercise
Norepinephrine
Epinephrine
Thyroid hormones
Fever
Bainbridge reflex
Heart rate **slowed** by:
Increased activity of baroreceptors in the arteries, left ventricle, and pulmonary circulation
Expiration
Fear
Grief
Stimulation of pain fibers in trigeminal nerve
Increased intracranial pressure

heart rate have been considered in detail in preceding sections of this chapter. However, Table 31–3 is a convenient summary of conditions that affect the heart rate. In general, stimuli that increase the heart rate also increase blood pressure, whereas those that decrease the heart rate lower the blood pressure. Thus, for example, anger and excitement are associated with tachycardia and a rise in blood pressure, whereas fear and grief are usually associated with bradycardia and hypotension. One exception is the production of hypotension and tachycardia by stimulation of atrial stretch receptors (see above). Another is the production of hypertension and bradycardia by increased intracranial pressure. The 2 occur together in this latter condition because, as noted above, there is hypercapnic stimulation of the vasomotor center and reflex bradycardia. When the body temperature rises, the heart rate is increased, but the cutaneous vessels dilate and blood pressure is unchanged or lowered. When the sinoatrial node is warmed, its rate of discharge increases, and the effect of fever on the heart rate is probably due in part to the rise in cardiac temperature. Thyroid hormones increase the pulse pressure and accelerate the heart rate by potentiating the action of catecholamines. Epinephrine and norepinephrine both act directly on the heart to increase its rate, but the marked pressor response produced by norepinephrine stimulates the arterial baroreceptors, and the reflex bradycardia obscures the cardioacceleratory action (see Chapter 20).

32 Circulation Through Special Regions

INTRODUCTION

The distribution of the cardiac output to various parts of the body at rest in a normal man is shown in Table 32–1. The general principles described in preceding chapters apply to the circulation of all these regions, but the vascular supplies of most organs have additional special features. The portal circulation of the anterior pituitary is discussed in Chapter 14, the renal circulation in Chapter 38, and the pulmonary circulation in Chapter 34. The circulation of skeletal muscle is discussed with the physiology of exercise in Chapter 33. This chapter is concerned with the circulation of the brain, the heart, the splanchnic area, the skin, the placenta, and the fetus.

CEREBRAL CIRCULATION

ANATOMIC CONSIDERATIONS

Vessels

The principal arterial inflow to the brain in humans is via 4 arteries: 2 internal carotids and 2 vertebrals. The vertebral arteries unite to form the basilar artery; and the circle of Willis, formed by the carotids and the basilar artery, is the orgin of the 6 large vessels supplying the cerebral cortex. In some animals the vertebrals are large and the internal carotids small, but in humans a relatively small fraction of the total arterial flow is carried by the vertebral arteries. Substances injected into one carotid artery are distributed almost exclusively to the cerebral hemisphere on that side. There is normally no crossing over, probably because the pressure is equal on both sides. Even when it is not, the anastomotic channels in the circle do not permit a very large flow. Occlusion of one carotid artery, particularly in older patients, often causes serious symptoms of cerebral ischemia. There are precapillary anastomoses between the cerebral arterioles in humans and some other species, but flow through these channels is generally insufficient to maintain the circulation and prevent infarction when a cerebral artery is occluded.

Venous drainage from the brain by way of the deep veins and dural sinuses empties principally into the internal jugular veins in humans, although a small amount of venous blood drains through the ophthalmic and pterygoid venous plexuses, through emissary veins to the scalp, and down the system of paravertebral veins in the spinal canal. In other species, the internal jugular veins are small, and the venous blood from the brain mixes with blood from other structures.

Table 32–1. Resting blood flow and O_2 consumption of various organs in a 63-kg adult human with a mean arterial blood pressure of 90 mm Hg and an O_2 consumption of 250 mL/min. R units are pressure (mm Hg) divided by blood flow (mL/s).*

Region	Mass (kg)	Blood Flow		Arteriovenous Oxygen Difference (mL/L)	Oxygen Consumption		Resistance in R units		Percentage of Total	
		mL/min	mL/100 g/min		mL/min	mL/100 g/min	Absolute	per kg	Cardiac Output	Oxygen Consumption
Liver	2.6	1500	57.7	34	51	2.0	3.6	9.4	27.8	20.4
Kidneys	0.3	1260	420.0	14	18	6.0	4.3	1.3	23.3	7.2
Brain	1.4	750	54.0	62	46	3.3	7.2	10.1	13.9	18.4
Skin	3.6	462	12.8	25	12	0.3	11.7	42.1	8.6	4.8
Skeletal muscle	31.0	840	2.7	60	50	0.2	6.4	198.4	15.6	20.0
Heart muscle	0.3	250	84.0	114	29	9.7	21.4	6.4	4.7	11.6
Rest of body	23.8	336	1.4	129	44	0.2	16.1	383.2	6.2	17.6
Whole body	63.0	5400	8.6	46	250	0.4	1.0	63.0	100.0	100.0

*Reproduced, with permission, from Bard P (editor): *Medical Physiology*, 11th ed. Mosby, 1961.

The cerebral vessels have a number of unique anatomic features. In the choroid plexuses there are gaps between the endothelial cells of the capillary wall, but the choroid epithelial cells are densely intermeshed and interlocked. The capillaries in the brain substance resemble nonfenestrated capillaries in muscle and other parts of the body (see Chapter 30). However, there are tight junctions between the endothelial cells that do not permit the passage of substances which pass through the junctions between endothelial cells in other tissues. In addition, there are relatively few vesicles in the endothelial cytoplasm, and presumably there is little vesicular transport. The brain capillaries are surrounded by the end-feet of astrocytes (Fig 32–1). These end-feet are closely applied to the basal lamina of the capillaries, but they do not cover the entire capillary wall, and there are gaps of about 20nm between end-feet. The protoplasma of astrocytes is also found around synapses, where it appears to isolate the synapses in the brain from one another.

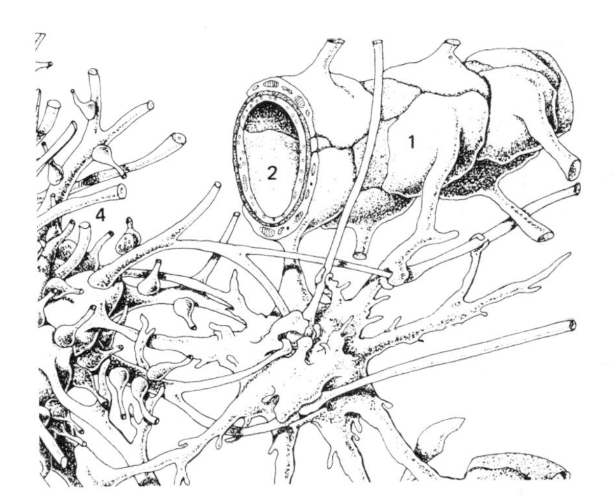

Figure 32–1. Membrane of end-feet (1) of fibrous astrocytes around a cerebral capillary (2). On the left, the processes of an astrocyte (3) also envelop parts of the dendrites of neighboring neurons (4). (Reproduced, with permission, from Krstić RV: *Die Gewebe des Menschen und der Säugetiere.* Springer-Verlag, 1978.)

Innervation

Three systems of nerves innervate the cerebral blood vessels. Postganglionic sympathetic neurons have their cell bodies in the superior cervical ganglia, and their endings contain norepinephrine and neuropeptide Y. Cholinergic neurons that probably originate in the sphenopalatine ganglia also innervate the cerebral vessels, and the postganglionic cholinergic neurons on the blood vessels contain acetylcholine, VIP, and PHM-27 (see Chapter 26). Sensory nerves on the blood vessels have their cell bodies in the trigeminal ganglia and contain substance P, a polypeptide closely related to substance P and produced by the same gene, and CGRP. Substance P, CGRP, VIP, and PHM-27 cause vasodilation, whereas neuropeptide Y is a vasoconstrictor. Touching or pulling on the cerebral vessels causes pain.

CEREBROSPINAL FLUID

Formation & Absorption

About 50% of the cerebrospinal fluid (CSF) that fills the cerebral ventricles and subarachnoid space is formed in the choroid plexuses; the remaining 50% is formed around the cerebral vessels and along the ventricular walls. In humans, the CSF turns over about 4 times per day. The composition of the CSF depends on filtration and diffusion from the blood, along with facilitated diffusion and active transport, much of it across the choroid plexus. The composition (Table 32–2) is essentially the same as that of brain ECF. There appears to be free communication between the brain interstitial fluid and CSF, although the diffusion distances from parts of the brain to the CSF are appreciable. Consequently, equilibration may take some time to occur, and local areas of the brain may have extracellular microenvironments that are transiently different from CSF. The CSF in the ventricles flows through the foramens of Magendie and Luschka to the subarachnoid space, and is absorbed through the arachnoid villi into the cerebral venous sinuses. Bulk flow via the villi is 500 mL/d in humans. Substances also leave the CSF by diffusion across adjacent membranes, and there is facilitated diffusion of glucose and active transport of cations and organic acids out of the CSF.

Lumbar CSF pressure is normally 70–180 mm CFS. Up to pressures well above this range, the rate of CSF formation is independent of intraventricular pressure. However, absorption, which takes place largely by bulk flow, is proportionate to the pressure (Fig 32–2). At a pressure of 112 mm CSF, which is the average normal CSF pressure, filtration and absorption are equal. Below a pressure of approximately 68 mm CSF, absorption stops. Large amounts of fluid accumulate when the reabsorptive capacity of the arachnoid villi is decreased **(external hydrocephalus, communicating hydrocephalus).** Fluid also accumulates proximal to the block and distends the ventricles when the foramens of Luschka and Magendie are blocked or there is obstruction within the ventricular system **(internal hydrocephalus, noncommunicating hydrocephalus).**

Brain Extracellular Space

There has been controversy about the size of the brain extracellular space. The Na^+ space in brain tissue is about 35% of brain volume and the Cl^- space is about 30%, but both ions are known to be present inside as well as outside of brain cells. Inulin and ferrocyanide, which do not enter cells, distribute in about 15% of brain volume. Under the electron microscope, brain cells appear very close together,

Table 32–2. Concentration of various substances in human CSF and plasma.

Substance		CSF	Plasma	Ratio CSF/Plasma
Na^+	(meq/kg H_2O)	147.0	150.0	0.98
K^+	(meq/kg H_2O)	2.9	4.6	0.62
Mg^{2+}	(meq/kg H_2O)	2.2	1.6	1.39
Ca^{2+}	(meq/kg H_2O)	2.3	4.7	0.49
Cl^-	(meq/kg H_2O)	113.0	99.0	1.14
HCO_3^-	(meq/L)	25.1	24.8	1.01
P_{CO_2}	(mm Hg)	50.2	39.5	1.28
pH		7.33	7.40	...
Osmolality	(mosm/kg H_2O)	289.0	289.0	1.00
Protein	(mg/dL)	20.0	6000.0	0.003
Glucose	(mg/dL)	64.0	100.0	0.64
Inorganic P	(mg/dL)	3.4	4.7	0.73
Urea	(mg/dL)	12.0	15.0	0.80
Creatinine	(mg/dL)	1.5	1.2	1.25
Uric acid	(mg/dL)	1.5	5.0	0.30
Lactic acid	(mg/dL)	18.0	21.0	0.86
Cholesterol	(mg/dL)	0.2	175.0	0.001

but the brain extracellular space shrinks to 4% or less of brain volume when the brain is exposed to approximately the same amount of asphyxia as brain tissue prepared for electron photomicrography. Thus, the extracellular space in living humans occupies about 15% of brain volume.

Protective Function

The meninges and the CSF protect the brain. The dura is attached firmly to bone. There is normally no "subdural space," the arachnoid being held to the dura by the surface tension of the thin layer of fluid between the 2 membranes. As shown in Figure 32–3, the brain itself is supported within the arachnoid by the blood vessels and nerve roots and by the multiple, fine fibrous **arachnoid trabeculae.** The brain weighs about 1400 g in air, but in its "water bath" of CSF it has a net weight of only 50 g. The buoyancy of the brain in the CSF permits its relatively flimsy attachments to suspend it very effectively. When the head receives a blow, the arachnoid slides on the dura and the brain moves, but its motion is gently checked by the CSF cushion and by the arachnoid trabeculae.

The pain produced by spinal fluid deficiency illustrates the importance of spinal fluid in supporting the brain. As a diagnostic procedure in patients suspected of having brain tumors, spinal fluid is removed and replaced by air to make the outline of the ventricles visible by x-ray **(pneumoencephalography).** This procedure causes a severe headache after the fluid is removed, because the brain hangs on the vessels and nerve roots, and traction on them stimulates pain fibers. The pain can be relieved by intrathecal injection of sterile isotonic saline.

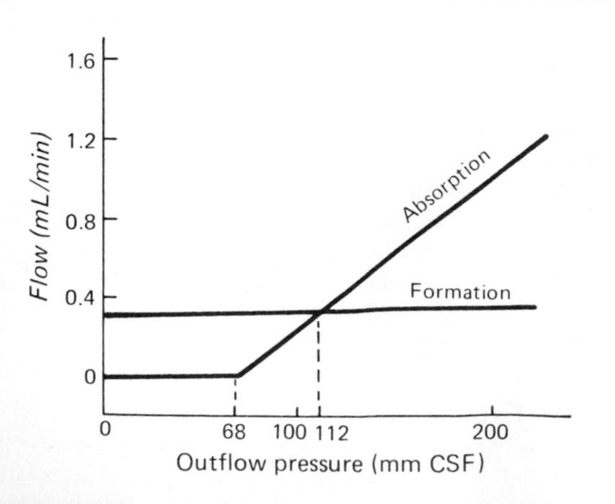

Figure 32–2. CSF formation and absorption in humans at various CSF pressures. Note that at 112 mm CSF, formation and absorption are equal, and at 68 mm CSF, absorption is zero. (Modified and reproduced, with permission, from Cutler RWP et al: Formation and absorption of cerebrospinal fluid in man. *Brain* 1968;**91**:707.)

Head Injuries

Without the protection of the spinal fluid and the meninges, the brain would probably be unable to withstand even the minor traumas of everyday living; but with the protection afforded, it takes a fairly severe blow to produce cerebral damage. The brain is damaged most commonly when the skull is fractured and bone is driven into neural tissue (depressed skull fracture), when the brain moves far enough to tear the

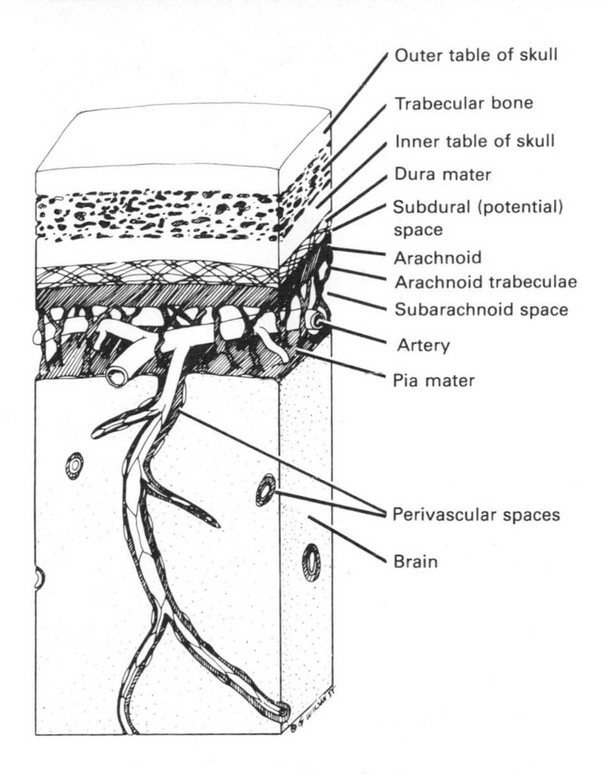

Figure 32–3. Investing membranes of the brain, showing their relation to the skull and to brain tissue. (Reproduced, with permission, from Wheater PR et al, *Functional Histology.* Churchill Livingstone, 1979.)

delicate bridging veins from the cortex to the bone, or when the brain is accelerated by a blow on the head and is driven against the skull or the tentorium at a point opposite where the blow was struck **(contrecoup injury).**

THE BLOOD-BRAIN BARRIER

Over 70 years ago, it was demonstrated that when acidic dyes such as trypan blue were injected into living animals, all the tissues were stained except most of the brain and spinal cord. To explain the failure of the neural tissue to stain, the existence of a **blood-brain barrier** was postulated. Subsequent research has established the fact that only water, CO_2, and O_2 cross the cerebral capillaries with ease, and the exchange of other substances is slow.

The general features of exchange across capillary walls between the plasma and the interstitial fluid are described in Chapter 30. There is considerable variation in capillary permeability from organ to organ in the body. However, the exchange across the cerebral vessels is so different from that in other capillary beds—and the rate of exchange of many physiologically important substances is so low—that it seems justifiable to speak specifically of a blood-brain barrier.

Penetration of Substances Into Brain

As noted above, brain ECF is essentially identical to CSF, and substances enter brain ECF and CSF by filtration, diffusion, facilitated diffusion, and active transport. There is an H^+ gradient between brain ECF and blood; the pH of brain ECF is 7.33, whereas that of blood is 7.40.

In general, the rapidity with which substances penetrate brain tissue is inversely related to their molecular size and directly related to their lipid solubility; water-soluble polar compounds generally cross slowly. There are bidirectional transport systems in the endothelium of brain capillaries for K^+, organic acids, amino acids, and glucose. Water, CO_2, and O_2 cross the blood-brain barrier readily, whereas glucose crosses more slowly. Na^+, K^+, Mg^{2+}, Cl^-, HCO_3^-, and HPO_4^{2-} in plasma require 3–30 times as long to equilibrate with spinal fluid as they do with other portions of the interstitial fluid. The relatively slow penetration of urea into the brain and the CSF is illustrated in Fig 32–4. Bile salts and catecholamines do not enter the adult brain in more than minute amounts. Proteins cross the barrier to a very limited extent, and it is because they are bound to protein that acidic dyes fail to stain neural tissue. It is worth noting that no substance is completely excluded from the brain and that the important consideration is the rate of transfer of the substance. Certain compounds cross the blood-brain barrier slowly, whereas closely related compounds enter rapidly. For example, the amines dopamine and serotonin penetrate to a very limited degree, but their corresponding acids, L-dopa and 5-hydroxytryptophan, enter with relative ease (see Chapter 15).

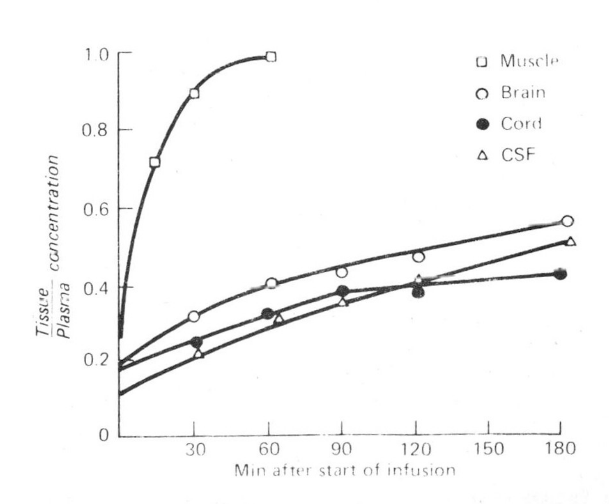

Figure 32–4. Penetration of urea into muscle, brain, spinal cord, and CSF. Urea administered by constant influsion. (Modified and reproduced, with permission, from Kleeman CR, Davson H, Levin E: Urea transport in the central nervous system. *Am J Physiol* 1962;**203**:739.)

Development of the Blood-Brain Barrier

The cerebral capillaries are much more permeable at birth than in adulthood, and the blood-brain barrier develops during the early years of life. In severely jaundiced infants, bile pigments penetrate into the nervous system and, in the presence of asphyxia, damage the basal ganglia (kernicterus). However, in jaundiced adults, the nervous system is unstained and not directly affected.

Circumventricular Organs

When an acidic dye is injected into an animal, 4 small areas in or near the brain stain like the tissues outside the brain. These areas are (1) the **posterior pituitary** (neurohypophysis) and the adjacent ventral part of the median eminence of the hypothalamus, (2) the **area postrema,** (3) the **organum vasculosum of the lamina terminalis** (OVLT, supraoptic crest), and (4) the **subfornical organ** (intercolumnar tubercle).

These areas are referred to collectively as the **circumventricular organs** (Fig 32–5). All have fenestrated capillaries, and because of their permeability they are said to be "outside the blood-brain barrier." Some of them function as **neurohemal organs,** ie, areas in which substances secreted by neurons enter the circulation; for example, oxytocin and vasopressin enter the general circulation in the posterior pituitary, and hypothalamic hypophyseotropic hormones enter the portal hypophyseal circulation in the median eminence. Other circumventricular organs function as chemoreceptor zones, ie, areas in which substances in the circulating blood can act to trigger changes in brain function without penetrating the blood-brain barrier. The area postrema is a chemoreceptor zone that initiates vomiting in response to chemical changes in the plasma (Chapter 14). It is also concerned with cardiovascular control, and in some species, circulating angiotensin II acts on the area postrema to trigger an increase in blood pressure. Angiotensin II also acts on the subfornical organ and possibly on the OVLT to increase water intake.

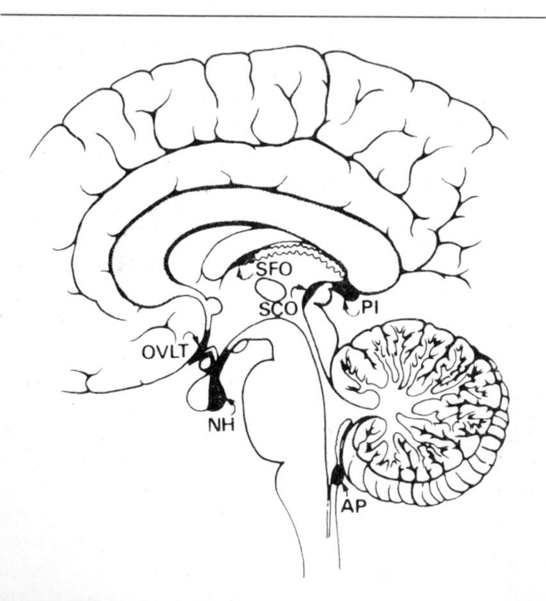

Figure 32–5. Circumventricular organs. The neurohypophysis (NH), organum vasculosum of the lamina terminalis (OVLT, supraoptic crest), subfornical organ (SFO), and area postrema (AP) are shown projected on a sagittal section of the human brain. SCO, subcommissural organ; PI, pineal.

The subcommissural organ (Fig 32–5) is closely associated with the pineal and histologically resembles the circumventricular organs. However, it does not have fenestrated capillaries, is less permeable, and has no established function. The pineal and the anterior pituitary have fenestrated capillaries and are outside the blood-brain barrier, but both are endocrine glands and are not part of the brain.

Function of the Blood-Brain Barrier

The blood-brain barrier probably maintains the constancy of the environment of the neurons in the central nervous system. These neurons are so dependent upon the concentration of K^+, Ca^{2+}, Mg^{2+}, H^+, and other ions in the fluid bathing them that even minor variations have far-reaching consequences. The constancy of the composition of the ECF in all parts of the body is maintained by multiple homeostatic mechanisms (see Chapter 1 and 39), but because of the sensitivity of the cortical neurons to ionic change it is not surprising that an additional defense has evolved to protect them. Similarly, a blood-testis barrier protects the composition of the fluid surrounding the germinal epithelium and a placental barrier protects the composition of the body fluids of the fetus.

Other suggested functions for the blood-brain barrier are protection of the brain from endogenous and exogenous toxins in the blood and prevention of the escape of neurotransmitters into the general circulation.

Clinical Implications

The physician must know the permeability of the blood-brain barrier to drugs in order to treat diseases of the nervous system intelligently. For example, among the antibiotics, penicillin and chlortetracycline enter the brain to a very limited degree. Sulfadiazine and erythromycin, on the other hand, enter quite readily.

Another important clinical consideration is the fact that the blood-brain barrier tends to break down in areas of the brain that are irradiated, infected, or the site of tumors. The breakdown helps in identifying the location of tumors; substances such as radioactive iodine-labeled albumin penetrate normal brain tissue very slowly, but they enter tumor tissue, making the tumor stand out as an island of radioactivity in the surrounding normal brain. The blood-brain barrier can also be temporarily disrupted by sudden marked increases in blood pressure or by intravenous injection of hypertonic fluids.

CEREBRAL BLOOD FLOW

Kety Method

According to the **Fick principle** (see Chapter 29), the blood flow of any organ can be measured by determining the amount of a given substance (Q_x) removed from the bloodstream by the organ per unit of time and dividing that value by the difference between the concentration of the substance in arterial blood and the concentration in the venous blood from the organ ($[A_x]) - [V_x]$). Thus:

$$\text{Cerebral blood flow (CBF)} = \frac{Q_x}{[A_x] - [V_x]}$$

When an individual inhales small subanesthetic amounts of nitrous oxide (N_2O), this gas is taken up by the brain, and the brain N_2O equilibrates with the N_2O in blood in 9–11 minutes. After equilibration, the N_2O concentration in cerebral venous blood is equal to the concentration in the brain because the brain-blood partition coefficient for N_2O is 1. Therefore, the level in cerebral venous blood after equilibration divided by the mean arteriovenous N_2O difference during equilibration equals the cerebral blood flow (CBF) per unit of brain:

$$\text{CBF (mL/100 g brain/min)} = \frac{100\ V_uS}{\int_0^u (A - V)dt}$$

where

V = Cerebral venous N_2O concentration (vol/100 g)
u = Time of equilibrium (minutes)
S = Partition coefficient of N_2O in blood and brain (= 1)
A = Arterial N_2O concentration (vol/100 g)

This is the **Kety method** for measuring cerebral blood flow. The mean arteriovenous difference can also be estimated from a plot of the differences at various times during equilibration (Fig 32–6). The average cerebral blood flow in young adults is 54 mL/100 g/min. The average adult brain weighs about 1400 g, so the flow for the whole brain is about 756 mL/min. Note that the Kety method provides an average value for perfused areas of brain during the 10-minute equilibration period; that it gives no information about regional differences in blood flow; and that since it depends on N_2O uptake, it measures flow to perfused parts of the brain only. If the blood flow to a portion of the brain is occluded, there is no change in the measured flow, because the nonperfused area does not take up any N_2O.

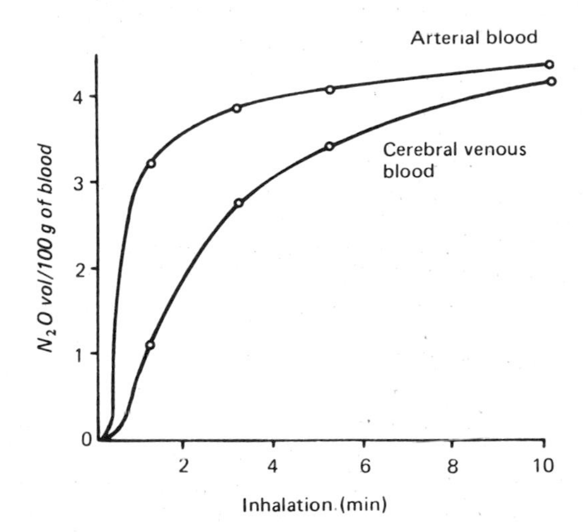

Figure 32–6. Arterial and cerebral venous blood N_2O levels while inhaling N_2O. (Redrawn and reproduced, with permission, from Kety S in: *Methods in Medical Research.* Vol 1. Potter VR [editor]. Year Book, 1948.)

Blood Flow in Various Parts of the Brain

Regional blood flow in the brain has been determined by measuring the distribution of an inert radioactive gas in frozen sections of the brain and comparing the values obtained with the level of the gas in the blood. It can also be determined in living animals and humans by monitoring the distribution of ^{133}Xe, ^{123}I-labeled iodoamphetamine (^{123}I-IMP), or other tracers. The arrival and clearance of the tracer are monitored by a battery of 254 scintillation detectors placed over the head. The output from the detectors is processed in a computer and displayed on a color television screen in such a way that the color corresponding to the location of each detector is proportionate to the flow it is detecting. Resolution can be improved with computerized tomographic reconstruction of the type used for computer-assisted tomography (CAT) and nuclear magnetic resonance imaging (NMRI) scanning (see appendix); this technique is called single photo emission computed tomography (SPECT).

Since blood flow is tightly coupled to brain metabolism, local uptake of 2-deoxyglucose is also a good index of blood flow (see below and Chapter 8). If the 2-deoxyglucose is labeled with short-half-life positron emitters such as ^{18}F, ^{11}O, and ^{15}O, its concentration in any part of the brain can be monitored by **positron emission tomography (PET)** scanning through the intact skull in living subjects.

Values for regional blood flow using the frozen section method are shown in Table 32–3. Blood flow in the cerebral cortex and cerebellar cortex is large, but the part of the brain with the largest blood flow per gram is the inferior colliculus. The blood flow in gray matter is about 6 times that in white matter. Using the ^{133}Xe method, the average hemispheric

Table 32–3. Blood flow of representative areas of the brain of unanesthetized cats.*

Area	Mean Blood Flow (mL/g/min)
Inferior colliculus	1.80
Sensorimotor cortex	1.38
Auditory cortex	1.30
Visual cortex	1.25
Medial geniculate body	1.22
Lateral geniculate body	1.21
Superior colliculus	1.15
Caudate nucleus	1.10
Thalamus	1.03
Association cortex	0.88
Cerebellar nuclei	0.87
Cerebellar white matter	0.24
Cerebral white matter	0.23
Spinal cord white matter	0.14

*Data from Landau WM et al: The local circulation of the living brain: Values in the unanesthetized cat. *Trans Am Neurol Assoc* 1955;**80:**125.

blood flow in resting humans is 48 mL/100 g/min, with the flow in gray matter averaging 69 mL/100 g/min compared to 28 mL/100 g/min in white matter.

The striking feature of cerebral function is the marked fluctuation in regional blood flow with fluctuation in activity. In subjects who are awake but at rest, blood flow is greatest in the premotor and frontal regions. This is the part of the brain that is believed to be concerned with decoding and analyzing afferent input and with intellectual activity. During voluntary clenching of the right hand, flow is increased in the hand area of the left motor cortex and the corresponding sensory areas in the postcentral gyrus. Especially when the movements being performed are sequential, the flow is also increased in the supplementary motor area. When subjects talk, there is a bilateral increase in blood flow in the face, tongue, and mouth-sensory and motor areas and the upper premotor cortex in the categorical (usually the left) hemisphere. When the speech is stereotyped, Broca's and Wernicke's area do not show increased flow, but when the speech is creative, ie, when it involves ideas, there are flow increases in both these areas. Reading produces widespread increases in blood flow. Problem solving, reasoning, and motor ideation without movement produce increases in selected areas of the premotor and frontal cortex (Fig 32–7). In right-handed individuals, blood flow to the left hemisphere is greater when a verbal task is being performed and blood flow to the right hemisphere is greater when a spatial task is being performed.

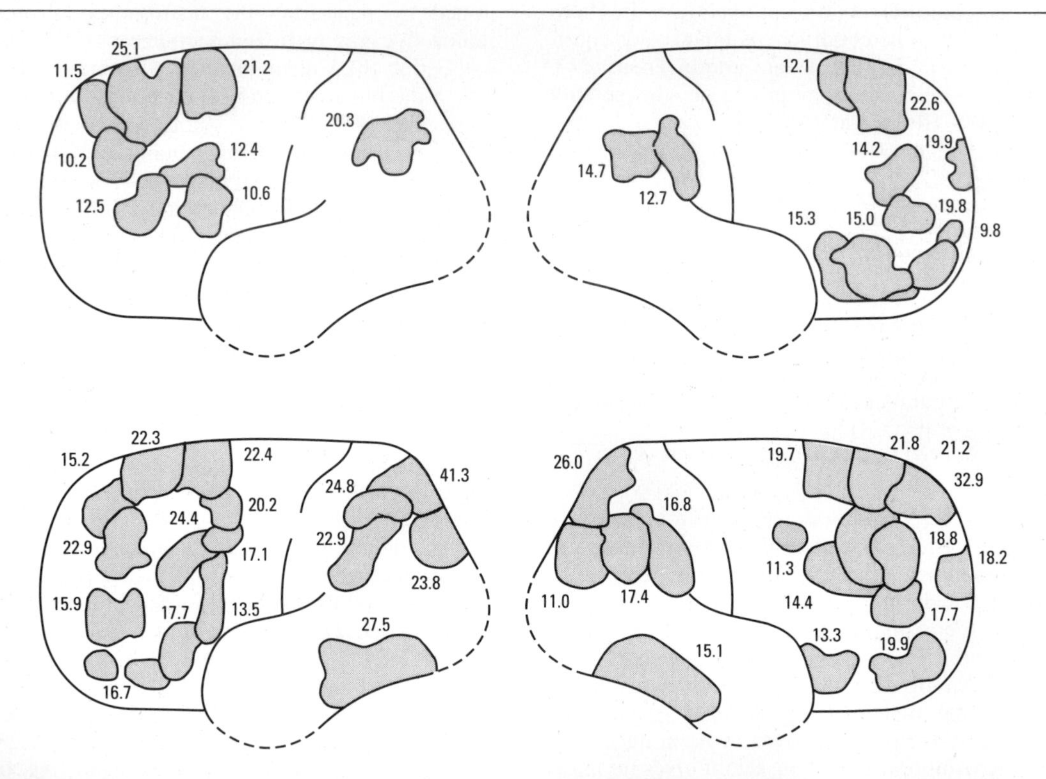

Figure 32–7. Areas showing increases in regional blood flow in left and right cerebral hemispheres of humans carrying out 2 kinds of thinking. *Top:* Subjects performing mental subtraction task. *Bottom:* Subjects imagining that they are walking a specific route in familiar surroundings. The numbers beside each area indicate the percent increase. The left hemisphere was analyzed in 6 subjects and the right in 5. Note that the areas activated are different for the 2 kinds of thinking. However, the areas activated during each kind of thinking are remarkably similar in all subjects. (Redrawn and reproduced, with permission, from Roland PE: Change in brain blood and oxidase metabolism during mental activity. *News Physiol Sci* 1987;**2**;120).

SPECT and PET scanning have been applied to the study of various diseases. Epileptic foci are hyperemic during seizures, whereas flow is reduced in other parts of the brain. Between seizures, flow is sometimes reduced in the foci that generate the seizures. There is decreased temporal lobe flow in patients with memory deficits and decreased parieto-occipital flow in patients with symptoms of agnosia (see Chapter 16). In senile dementia, there are patchy decreases in flow, particularly in the temporoparietal cortex, and the magnitude of the decreases is proportionate to the degee of dementia. In Huntington's disease, there is a bilatcrial reduction in the blood flow to the caudate nucleus, and this alteration in flow occurs so early in the disease that it is probably of potential value in detecting asymptomatic individuals at risk for the disease. In chronic schizophrenia, there is decreased blood flow at rest; and in manic depressives but not in patients with unipolar depression, there is a general decrease in cortical blood flow when the patients are depressed.

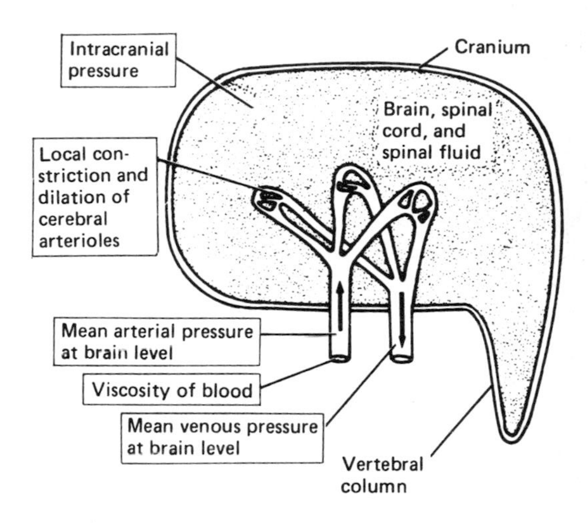

Figure 32–8. Diagrammatic summary of the factors affecting cerebral blood flow.

Cerebral Vascular Resistance

The cerebral vascular resistance (CVR) is equal to the cerebral perfusion pressure divided by the cerebral blood flow. The CVR in the supine position can be calculated on the basis of the mean brachial artery pressure (ignoring the relatively low cerebral venous pressure) without introducing a very large error. Calculated in this way, the normal CVR is approximately 10 "R units" per kilogram of brain (7.2 R units for the whole brain) (Table 32–1).

REGULATION OF CEREBRAL CIRCULATION

Normal Flow

The cerebral circulation is regulated in such a way that a constant total cerebral blood flow is generally maintained under varying conditions. For example, despite extensive shifts in the pattern of flow, total cerebral blood flow is not increased by strenuous mental activity. The factors affecting the total cerebral blood flow are the arterial pressure at brain level, the venous pressure at brain level, the intracranial pressure, the viscosity of the blood, and the degree of active constriction or dilation of the cerebral arterioles (Fig 32–8). The caliber of the arterioles is controlled by local vasodilator metabolites that include the products of metabolism and other factors, and by autoregulation, circulating peptides such as angiotensin II, and vasomotor nerves.

Role of Intracranial Pressure

In adults, the brain, spinal cord, and spinal fluid are encased, along with the cerebral vessels, in a rigid bony enclosure. The cranial cavity normally contains a brain weighing approximately 1400 g, 75 mL of blood, and 75 mL of spinal fluid. Because brain tissue and spinal fluid are essentially incompressible, the volume of blood, spinal fluid, and brain in the cranium at any time must be relatively constant **(Monro-Kellie doctrine).** More importantly, the cerebral vessels are compressed whenever the intracranial pressure rises. Any change in venous pressure promptly causes a similar change in intracranial pressure. Thus, a rise in venous pressure decreases cerebral blood flow both by decreasing the effective perfusion pressure and by compressing the cerebral vessels. This relationship helps to compensate for changes in arterial blood pressure at the level of the head. For example, if the body is accelerated upward (positive *g*"), blood moves toward the feet, and arterial pressure at the level of the head decreases. However, venous pressure also falls and intracranial pressure falls, so that the pressure on the vessels decreases and blood flow is much less severely compromised than it would otherwise be. Conversely, during acceleration downward, force acting toward the head (negative *g*") increases arterial pressure at head level, but intracranial pressure also rises, so that the vessels are supported and do not rupture. The cerebral vessels are protected during the straining associated with defecation or delivery in the same way.

Effect of Intracranial Pressure Changes on Systemic Blood Pressure

When intracranial pressure is elevated to more than 450 mm of water (33 mm Hg) over a short period, cerebral blood flow is significantly reduced. The resultant ischemia stimulates the vasomotor area (see Chapter 31), and systemic blood pressure rises. Stimulation of vagal outflow produces bradycardia, and respiration is slowed. The blood pressure rise, which was described by Cushing and is sometimes called the **Cushing reflex,** helps to maintain the cerebral blood flow. Over a considerable range, the rise in systemic blood pressure is proportionate to the rise in intracranial pressure, although eventually a point is reached where the intracranial pressure exceeds the arterial pressure and cerebral circulation ceases.

Effects of Brain Metabolism on Cerebral Vessels

As noted above, there are marked local fluctuations in blood flow with alterations in neuronal activity. Increased neuronal activity produces a local increase in ECF K^+, and the K^+ produces vasodilation when the O_2 supply falls behind demand. There is a decline in local ECF pH, and increased H^+ causes vasodilation. So does adenosine released from active cells. The arterioles in the brain, like those in other parts of the body, are directly affected by local changes in CO_2 and O_2 tension. A rise in P_{CO2}, which increases the local H^+ concentration, exerts a particularly potent dilator effect on the cerebral vessels. A fall in P_{CO2} has a constrictor effect, and cerebral vasoconstriction is an important factor in the production of the cerebral symptoms seen when the arterial P_{CO2} falls during hyperventilation. Changes in local O_2 tension also affect the cerebral arterioles; a low P_{O2} is associated with vasodilation and a high P_{o2} with mild vasoconstriction.

Autoregulation

Autoregulation is prominent in the brain (Fig 32–9). This process, by which the flow to many tissues is maintained at relatively constant levels despite variations in perfusion pressure, is discussed in Chapter 31. As in other tissues, cerebral autoregulation may depend upon an inherent capacity of vascular smooth muscle to contract when it is stretched or upon the washing away of CO_2 and other vasodilator metabolites when the perfusion pressure and hence the blood flow is increased–or upon both mechanisms.

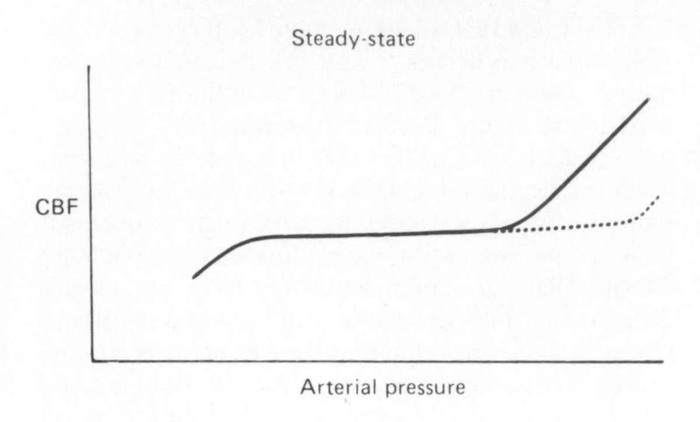

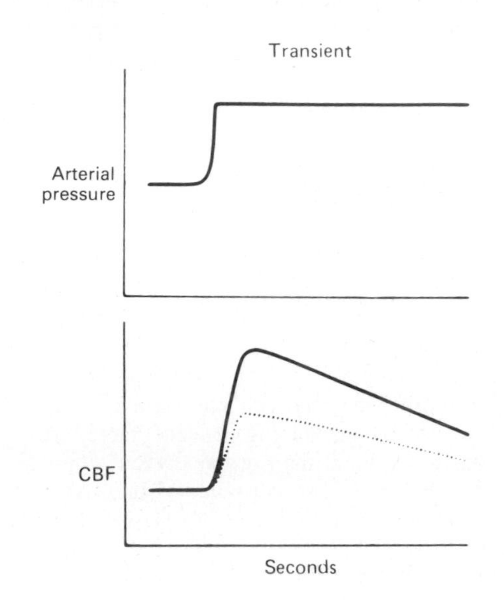

Figure 32–9. Autoregulation of cerebral blood flow (CBF) during steady-state conditions (***top***) and during a sudden increase in perfusion pressure (***bottom***). In each case, the dotted line shows the alteration produced by sympathetic stimulation during autoregulation. (Reproduced, with permission, from Heistad DD, Kontos H: *Handbook of Physiology.* Section 2, Vol 3. American Physiological Society, 1983; and by permission of the American Heart Association, Inc., from Busija DW, Heistad DD, Marcus ML: Effects of sympathetic nerves on cerebral blood vessels during acute moderate increases in arterial pressure in cats and dogs. *Circ Res* 1980;**46**:696.)

Role of Vasomotor Nerves

The innervation of the cerebral blood vessels by postganglionic sympathetic, postganglionic parasympathetic, and sensory nerves is described above. The role of these nerves remains a matter of debate. Stimulation of the cervical sympathetic nerves does cause some constriction of pial vessels in animals, and it has been argued that noradrenergic discharge occurs when the blood pressure is markedly elevated. This reduces the resultant passive increase in blood flow and helps protect the blood-brain barrier from the disruption that could otherwise occur (see above).

Vasomotor discharges also affect autoregulation. With sympathetic stimulation, the constant-flow, or plateau, part of the pressure-flow curve is extended to the right (Fig 32–9); ie, greater increases in pressure can occur without an increase in flow. On the other hand, the vasodilator hydralazine and the ACE inhibitor captopril reduce the length of the plateau.

Integrated Operation of Regulatory Mechanisms

Several examples of the way the regulatory mechanisms discussed above maintain cerebral blood flow are discussed in Chapter 33. It is worth remembering that these mechanisms operate together to overcome such formidable challenges as the effect of gravity on the cerebral blood flow, not only in humans but also, for example, in giraffes when these animals stoop to drink water and when they raise their heads to nibble leaves from the tops of trees.

BRAIN METABOLISM & OXYGEN REQUIREMENTS

Uptake & Release of Substances by the Brain

If the cerebral blood flow is known, it is possible to calculate the consumption or production by the brain of O_2, CO_2, glucose, or any other substance present in the bloodstream by multiplying the cerebral blood flow by the difference between the concentration of the substance in arterial blood and its concentration in cerebral venous blood (Table 32–4). When calculated in this fashion, a negative value indicates that the brain is producing the substance.

Oxygen Consumption

The O_2 consumption of human brain **(cerebral metabolic rate for O_2, $CMRO_2$)** averages about 3.5 mL/100 g brain/min (49 mL/min for the whole brain) in an adult. This figure represents approximately 20% of the total resting O_2 consumption (Table 32–1). The brain is extremely sensitive to hypoxia, and occlusion of its blood supply produces unconsciousness in as short a period as 10 seconds. The vegetative structures in the brain stem are more resistant to hypoxia than the cerebral cortex, and patients may recover from accidents such as cardiac arrest and other conditions causing fairly prolonged hypoxia with normal vegetative functions but severe, permanent intellectual deficiencies. The basal ganglia use O_2 at a very high rate, and symptoms of Parkinson's disease as well as intellectual deficits can be produced by chronic hypoxia. The thalamus and the inferior colliculus are also very suspectible to hypoxic damage.

Energy Sources

Glucose is the major ultimate source of energy for the brain under normal conditions. It is taken up from the blood in large amounts, and the RQ (respiratory quotient; see Chapter 17) of cerebral tissue is 0.95–0.99 in normal individuals. This does not mean that the total source of energy is always glucose. During prolonged starvation, there is appreciable utilization of other substances, Indeed, there is evidence that as much as 30% of the glucose taken up under normal conditions is converted to amino acids, lipids, and proteins and that substances other than glucose are metabolized for energy during convulsions. There may also be some utilization of amino acids from the circulation even though the amino acid arteriovenous difference across the brain is normally minute. Insulin is not required for most cerebral cells to utilize glucose.

Glucose uptake is increased in active neurons, and 2-deoxyglucose is taken up like glucose but is not metabolized. Therefore, the uptake of radioactive 2-deoxyglucose can be used to map the activity of neurons on a fine scale in experimental animals and in living humans by PET scanning. For example, illumination of one eye in monkeys causes an increase in the amount of 2-deoxyglucose, as determined by radioautography, in columns of cells in the visual cortex that are less than 1 mm apart (see Chapter 8). In general, glucose utilization at rest parallels blood flow (Table 32–3) and O_2 consumption. There is an average decrease of 30% in the uptake of all areas during slow-wave sleep.

Table 32–4. Utilization and production of substances by adult human brain in vivo.*

	Uptake (+) or output (−) per 100 g brain/min	Total/min
Substances utilized		
Oxygen	+3.5 mL	+49 mL
Glucose	+5.5 mg	+77 mg
Glutamate	+0.4 mg	+ 5.6 mg
Substances produced		
Carbon dioxide	−3.5 mL	−49 mL
Glutamine	−0.6 mg	− 8.4 mg

Substances not utilized or produced in the fed state: lactate, pyruvate, total ketones, α-ketoglutarate.

*From data compiled by Sokoloff L in: *Handbook of Physiology.* Field J, Magoun HW (editors). Washington: The American Physiological Society, 1960. Section 1, pp. 1843–1865.

Hypoglycemia

The symptoms of hypoglycemia are mental changes, ataxia, confusion, sweating, coma, and convulsions (see Chapter 19). The total glycogen content of the brain is about 1.6 mg/g in fasted animals, but the available glycogen and glucose are used up in 2 minutes if the blood supply is totally occluded. Thus, the brain can withstand hypoglycemia for somewhat longer periods than it can withstand hypoxia, but glucose and O_2 are both needed for survival. The cortical regions are more sensitive to hypoglycemia than the vegetative centers in the brain stem, and sublethal exposures to hypoglycemia, like similar exposures to hypoxia, may cause irreversible cortical changes. In diabetic patients with chronic hyperglycemia, maximum glucose transport capacity across the blood-brain barrier is reduced, and these patients may develop symptoms of hypoglycemia when their blood glucose is reduced to values that are still in the normal rather than the hypoglycemic range for normal individuals.

Glutamate & Ammonia Removal

The brain uptake of glutamate is approximately balanced by its output of glutamine. Glutamate entering the brain takes up ammonia and leaves as glutamine (see Chapter 17). The glutamate-glutamine conversion in the brain—the opposite of the reaction in the kidney that produces some of the ammonia entering the tubules—probably serves as a detoxifying mechanism to keep the brain free of ammonia.

Ammonia is very toxic to nerve cells, and ammonia intoxication is believed to be a major cause of the bizarre neurologic symptoms in hepatic coma.

CORONARY CIRCULATION

Anatomic Considerations

The 2 coronary arteries that supply the myocardium arise from the sinuses behind the cusps of the aortic valve at the root of the aorta (Fig 32–10). Eddy currents keep the valves away from the orifices of the arteries, and they are patent throughout the cardiac cycle. The right coronary artery has a greater flow in 50% of individuals, the left has a greater flow in 20%, and the flow is equal in 30%. There are 2 venous drainage systems: a superficial system, ending in the coronary sinus and anterior cardiac veins, that drains the left ventricle; and a deep system that drains the rest of the heart (Fig 32–11). The deep system is made up largely of the **arteriosinusoidal vessels,** sinusoidal capillarylike vessels that empty directly into the heart chambers. There are also direct connections to the atria and ventricles from the coronary arterioles **(arterioluminal vessels)** and the veins **(thebesian veins).** There are a few anastomoses between the coronary arterioles and extracardiac arterioles, especially around the mouths of the great veins. Anastomoses between coronary arterioles in humans only pass particles less than 40 μm in diameter, but there is evidence that these channels enlarge and increase in number in patients with coronary artery disease.

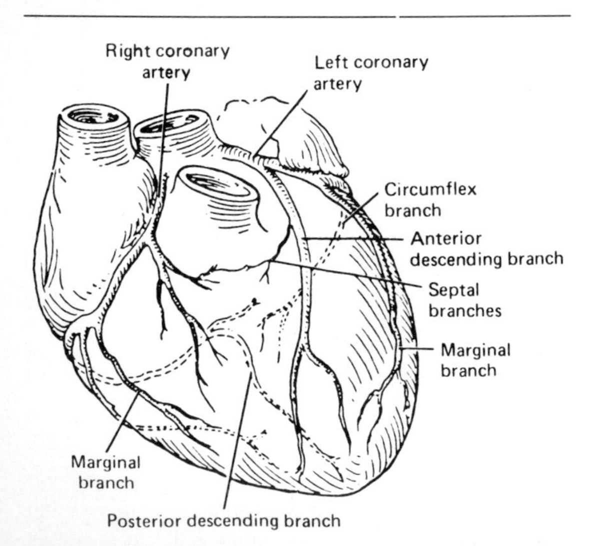

Figure 32–10. Coronary arteries and their principal branches in humans. (Reproduced, with permission, from Ross G: The cardiovascular system. In Ross G [editor]: *Essentials of Human Physiology.* Copyright © 1978 by Year Book Medical Publishers, Inc., Chicago.)

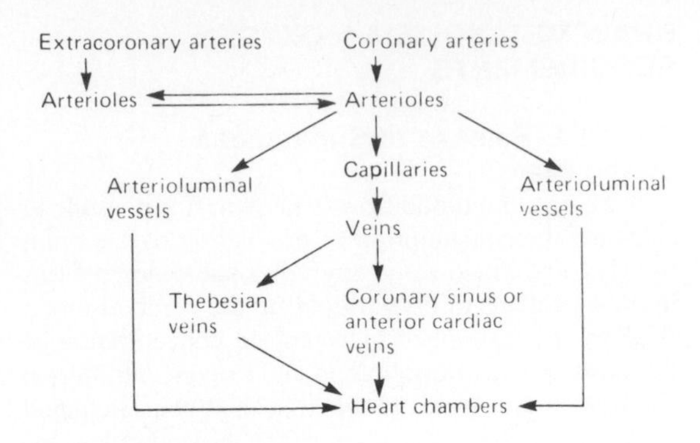

Figure 32–11. Diagram of the coronary circulation.

Pressure Gradients & Flow in the Coronary Vessels

The heart is a muscle that, like skeletal muscle, compresses its blood vessels when it contracts. The pressure inside the left ventricle is slightly higher than in the aorta during systole (Table 32–5). Consequently, flow occurs in the arteries supplying the subendocardial portion of the left ventricle only during diastole, although the force is sufficiently dissipated in the more superficial portions of the left ventricular myocardium to permit some flow in this region throughout the cardiac cycle. Since diastole is shorter when the heart rate is high, left ventricular coronary flow is reduced during tachycardia. On the other hand, the pressure differential between the aorta and the right ventricle, and the differential between the aorta and the atria, are somewhat greater during systole than during diastole. Consequently, coronary flow in those parts of the heart is not appreciably reduced during systole. Flow in the right and left coronary arteries is shown in Fig 32–12. Because there is no blood flow during systole in the subendocardial portion of the left ventricle, this region is prone to ischemic damage and is the most common site of myocardial infarction. Blood flow to the left ventricle is decreased in patients with stenotic aortic valves because in aortic stenosis the pressure in the left ventricle must be much higher than that in the aorta to eject the blood. Consequently, the coronary vessels are severely compressed during systole. Patients with this disease are particularly prone to

Table 32–5. Pressures in aorta and left and right ventricles in systole and diastole.

	Pressure (mm Hg) in			Pressure Differential (mm Hg) Between Aorta and	
	Aorta	Left Vent	Right Vent	Left Vent	Right Vent
Systole	120	121	25	−1	95
Diastole	80	0	0	80	80

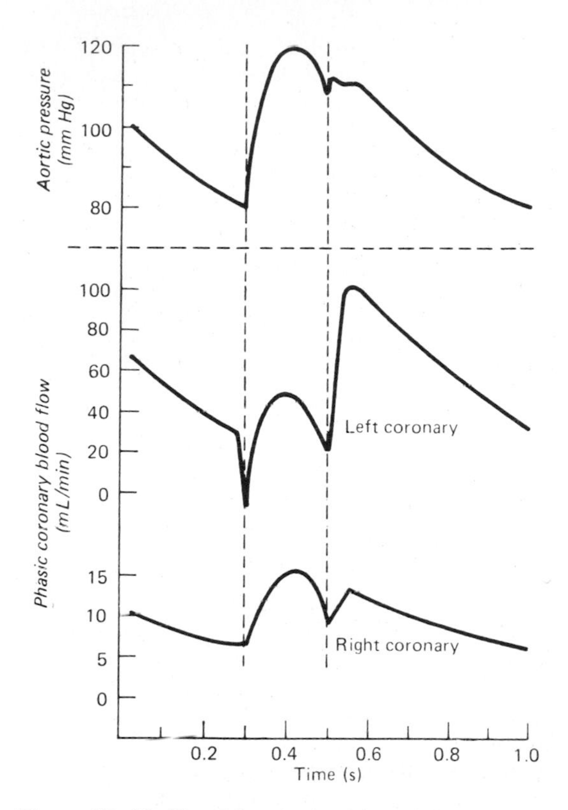

Figure 32–12. Blood flow in the left and right coronary arteries during various phases of the cardiac cycle. Systole occurs between the 2 vertical dashed lines. (Reproduced, with permission, from Berne RM, Levy MN: *Physiology.* The C.V. Mosby Co., St. Louis, 1983.)

develop symptoms of myocardial ischemia, in part because of this compression and in part because the myocardium requires more O_2 to expel blood through the stenotic aortic valve. Coronary flow is also decreased when the aortic diastolic pressure is low. The rise in venous pressure in conditions such as congestive heart failure reduces coronary flow because it decreases effective coronary perfusion pressure.

Coronary blood flow has been measured by inserting a catheter into the coronary sinus and applying the Kety method to the heart on the assumption that the N_2O content of coronary venous blood is typical of the entire myocardial effluent. Coronary flow at rest in humans is about 250 mL/min (5% of the cardiac output). A number of techniques utilizing **radionuclides,** radioactive tracers that can be detected with γ scintillation cameras over the chest, have been used to study regional blood flow in the heart and to detect areas of ischemia and infarction as well as to evaluate ventricular function. Radionuclides such as thallium 201 (^{201}Tl) are pumped into cardiac muscle cells by Na^+-K^+ ATPase and equilibrate with the intracellular K^+ pool. For the first 10–15 minutes after intravenous injection, ^{201}Tl distribution is directly proportionate to myocardial blood flow, and areas of ischemia can be detected by their low uptake. The uptake of this isotope is often determined soon after exercise and again several hours later to bring out areas in which exertion leads to compromised flow. Conversely, radiopharmaceuticals such as technetium 99m stannous pyrophosphate (^{99m}Tc-PYP) are selectively taken up by infarcted tissue by an incompletely understood mechanism and make infarcts stand out as "hot spots" on scintiscans of the chest. Coronary angiography can be combined with measurement of ^{133}Xe washout (see above) to provide detailed analysis of coronary blood flow. Radiopaque contrast medium is first injected into the coronary arteries, x-rays being used to outline their distribution. The angiographic camera is then replaced with a multiple-crystal scintillation camera and ^{133}Xe washout is measured. An example of normal flow distribution after injection in a left coronary artery is shown in Fig 32–13.

Variations in Coronary Flow

At rest, the heart extracts 70–80% of the O_2 from each unit of blood delivered to it (Table 32–1). O_2 consumption can be increased significantly only by increasing blood flow. Therefore, it is not surprising

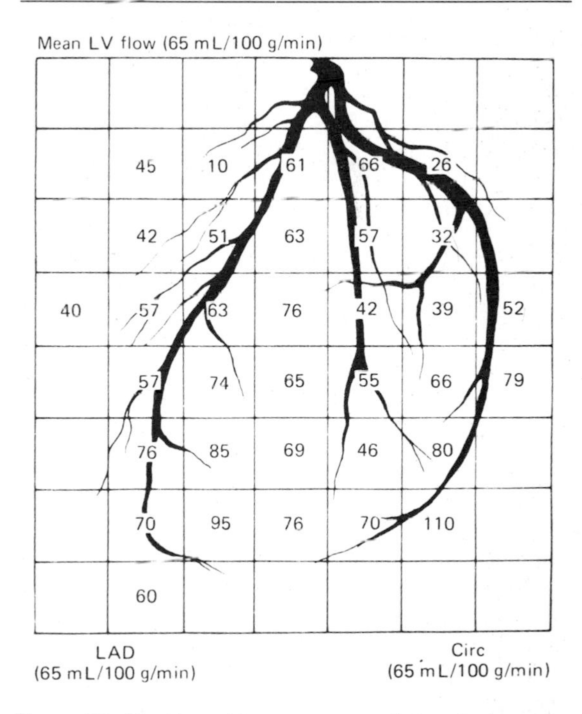

Figure 32–13. Normal human myocardial perfusion pattern following injection of ^{133}Xe into the left main coronary artery. The branches of the artery are shown in black, and the numbers in the squares are the flow values (in mL/100 g/min) for the regions under each scintillation detector. Circ, circumflex artery; LAD, left anterior descending artery; LV, left ventricle. (Reproduced, with permission, from Cannon PJ et al: Evaluation of myocardial circulation with radionuclides. *Cardiovasc Med* 1978;2:371.)

that blood flow increases when the metabolism of the myocardium is increased. The caliber of the coronary vessels, and consequently the rate of coronary blood flow, is influenced not only by pressure changes in the aorta but also by chemical and neural factors. The coronary circulation shows considerable autoregulation.

Chemical Factors

The close relationship between coronary blood flow and myocardial O_2 consumption indicates that one or more of the products of metabolism cause coronary vasodilation. Factors suspected of playing this role include O_2 lack and increased local concentrations of CO_2, H^+, K^+, lactate, prostaglandins, adenine nucleotides, and adenosine. More than one of these vasodilator metabolites could be involved. Asphyxia, hypoxia, and intracoronary injections of cyanide all increase coronary blood flow 200–300% in denervated as well as intact hearts, and the feature common to these 3 stimuli is hypoxia of the myocardial fibers. A similar increase in flow is produced in the area supplied by a coronary artery if the artery is occluded and then released. This **reactive hyperemia** is similar to that seen in the skin (see below).

Neural Factors

The coronary arterioles contain α-adrenergic receptors, which mediate vasoconstriction, and β-adrenergic receptors, which mediate vasodilation. Activity in the noradrenergic nerves to the heart and injections of norepinephrine cause coronary vasodilation. However, norepinephrine increases the heart rate and the force of cardiac contraction, and the vasodilation is due to production of vasodilator metabolites in the myocardium secondary to the increase in its activity. When the inotropic and chronotropic effects of noradrenergic discharge are blocked by a β-adrenergic blocking drug, stimulation of the noradrenergic nerves or injection of norepinephrine in unanesthetized animals elicits coronary vasoconstriction. Thus, the direct effect of noradrenergic stimulation is constriction rather than dilation of the coronary vessels. Stimulation of vagal fibers to the heart dilates the coronaries.

When the systemic blood pressure falls, the overall effect of the reflex increase in noradrenergic discharge is increased coronary blood flow secondary to the metabolic changes in the myocardium at a time when the cutaneous, renal, and splanchnic vessels are constricted. In this way the circulation of the heart, like that of the brain, is preserved when flow to other organs is compromised.

Coronary Artery Disease

When flow through a coronary artery is reduced to the point that the myocardium it supplies becomes hypoxic, "P factor" accumulates and **angina pectoris** develops (see Chapter 7). If the myocardial ischemia is severe and prolonged, irreversible changes occur in the muscle, and the result is **myocardial infarction.** The cause of myocardial infarction is usually obstruction of at least 75% of the lumen of a coronary artery by a thrombus in a region narrowed by atherosclerotic plaques. Myocardial infarction is a common cause of death, particularly in developed countries. Other events that precipitate a thrombosis include spasm of a coronary artery, transient platelet aggregation in a severely sclerotic vessel, or rupture of or hemorrhage into an arteriosclerotic plaque.

The electrical events produced by myocardial ischemia and infarction and the resulting changes produced in the ECG are discussed in Chapter 28. It is interesting that before the ischemic muscle cells die, they cease contracting, and this cessation of contraction has the homeostatic effect of prolonging the period before the onset of irreversible changes. The cause of the loss of contractile activity is uncertain. In dogs in which blood flow to a portion of myocardium has been interrupted, necrosis starts in 20 minutes but is not complete for 3–6 hours.

Damaged cells leak enzymes into the circulation, and the rises in the serum levels of enzymes and isoenzymes produced by myocardial infarction play an important ancillary role in the diagnosis of this disease. The enzymes most commonly measured are creatine kinase (CK) and lactate dehydrogenase (LDH). The MB isoenzyme of CK and fraction 1 of LDH are found in higher concentrations in heart muscle than in many other organs, and measurement of the serum concentration of the particular isoenzymes helps improve the diagnostic specificity of the enzyme measurement.

Nitrates such as nitroglycerin often produce prompt relief of the pain in angina pectoris. These substances dilate normal arterial vessels, and even though atherosclerotic coronary arteries are generally too rigid to dilate to any degree, large doses produce dilation of coronary arteries with eccentric stenosos. However, the main effect of the nitrates is to decrease venous return to the heart. This reduces stroke volume, and consequently myocardial O_2 consumption is reduced. The decreased venous return is secondary to dilation of peripheral veins, while pooling of blood in the periphery.

Discrete areas of narrowing in the coronary arteries can be detected by arteriography and can be bypassed by implantation of a graft (aortocoronary artery bypass graft). This procedure often produces marked and sustained relief of intractable angina, but it is unsettled whether it protects against myocardial infarction or produces significant prolongation of life when compared with modern medical treatment of coronary artery disease. In the treatment of myocardial infarction, emphasis is now placed on removing the obstruction and/or lysing the clot as soon as possible after the onset of pain in order to establish reperfusion and prevent irreversible changes. One technique is to insert a catheter, manipulate it into the affected coronary artery, and then use it to dilate the area of constriction above where the clot has formed **(coronary angioplasty).** Another is lysis of the thrombus by intravenous or intracoronary injection of **streptokinase,** a protein from hemolytic streptococci that converts plasminogen to plasmin (see Chapter 27). An even better lytic agent that converts plasminogen to plasmin in the thrombus is **tissue plasminogen activator (t-PA),** a naturally occurring protease that is produced by means of recombinant DNA techniques. If injection is carried out within the first few hours after the onset of pain in developing myocardial infarction, coronary perfusion can be improved in a significant number of patients and the increased patency can be maintained with antiplatelet drugs, calcium channel blockers, and other medical therapy.

SPLANCHNIC CIRCULATION

The blood from the intestines, pancreas, and spleen drains via the portal vein to the liver and from the liver via the hepatic veins to the inferior vena cava. The viscera and the liver receive about 30% of the cardiac output via the celiac, superior mesenteric, and inferior mesenteric arteries (Fig 32–14). The liver receives about 1000 mL/min from the portal vein and 500 mL/min from the hepatic artery.

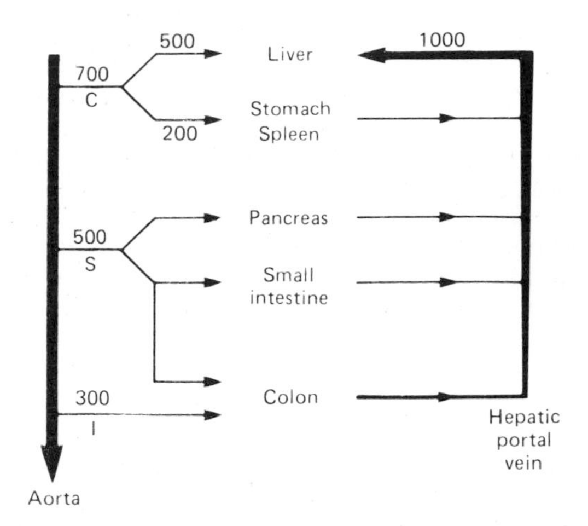

Figure 32–14. Splanchnic circulation. Note that most of the viscera are supplied by a series of parallel circuits, whereas the liver receives blood from the hepatic artery and the portal vein. The figures are average blood flows (mL/min). C, celiac axis; S, superior mesenteric artery; I, inferior mesenteric artery. (Modified from Sernka T, Jacobson E: *Gastrointestinal Physiology—The Essentials.* Williams & Wilkins, 1979.)

Intestinal Circulation

The intestines are supplied by a series of parallel circulations via the branches of the superior and inferior mesenteric arteries (Fig 32–14). There are extensive anastomoses between these vessels, but blockage of a large intestinal artery still leads to infarction of the bowel. The blood flow to the mucosa is greater than that to the rest of the intestinal wall, and it responds to changes in metabolic activity. Thus, blood flow to the small intestine (and hence blood flow in the portal vein) doubles after a meal, and the increase lasts up to 3 hours. The intestinal circulation is capable of extensive autoregulation.

Hepatic Circulation

There are large fenestrations in the walls of hepatic sinusoids, and they are highly permeable. The way the intrahepatic branches of the hepatic artery and portal vein converge on the sinusoids and drain into the central lobular veins of the liver is shown in Figs 26–17 and 32–15. The functional unit of the liver is the acinus. Each acinus is at the end of a vascular stalk containing terminal branches of portal veins, hepatic arteries, and bile ducts. Blood flows from the center of this functional unit to the terminal branches of the hepatic veins at the periphery (Fig 32–15). This is why the central portion of the acinus, sometimes called zone 1, is well oxygenated, the intermediate zone (zone 2) is moderately well oxygenated, and the peripheral zone (zone 3) is least well oxygenated and most susceptible to anoxic injury. The hepatic veins drain into the inferior vena cava. The acini have been likened to grapes or berries, each on a vascular stem. There are about 100,000 acini in the human liver.

Portal venous pressure is normally about 10 mm Hg in humans, and hepatic venous pressure approximately 5 mm Hg. The mean pressure in the hepatic artery branches that converge on the sinusoids is

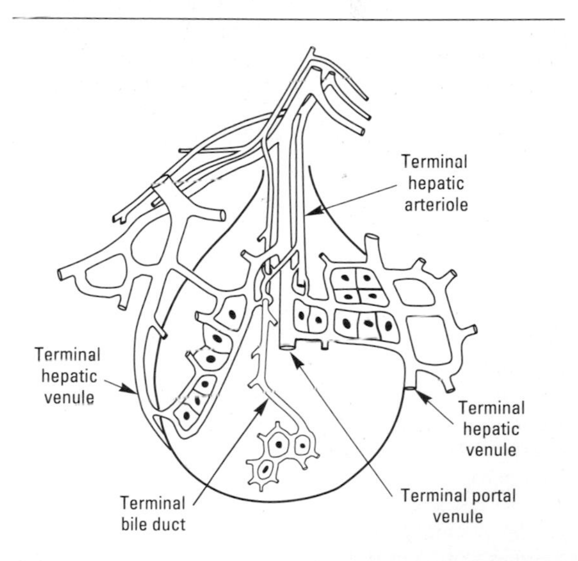

Figure 32–15. Concept of the acinus as the functional unit of the liver. In each acinus, blood in the portal venule and hepatic arteriole enters the center of the acinus and flows outward to the hepatic venule. (Reproduced, with permission, from Lautt WW, Greenway CV: Conceptual review of the hepatic vascular bed. *Hepatology* 1987; 7:952 © by Am. Assoc. for the Study of Liver Diseases.)

about 90 mm Hg, but the pressure in the sinusoids is lower than the portal venous pressure, so there is a marked pressure drop along the hepatic arterioles. This pressure drop is adjusted so that there is an inverse relationship between hepatic arterial and portal venous blood flow. This inverse relationship may be maintained in part by the rate at which adenosine is removed from the region around the arterioles. According to this hypothesis, adenosine is produced by metabolism at a constant rate. When portal flow is reduced, it is washed away at a lower rate, and the local accumulation of adenosine dilates the terminal arterioles.

The intrahepatic portal vein radicles have smooth muscle in their walls that is innervated by noradrenergic vasoconstrictor nerve fibers reaching the liver via the third to eleventh thoracic ventral roots and the splanchnic nerves. The vasoconstrictor innervation of the hepatic artery comes from the hepatic sympathetic plexus. There are no known vasodilator fibers reaching the liver. At rest, circulation in the peripheral portions of the liver is sluggish, and only a portion of the organ is actively perfused. When systemic venous pressures rises, the portal vein radicles are dilated passively and the amount of blood in the liver increases. In congestive heart failure, this hepatic venous congestion may be extreme. Conversely, when there is diffuse noradrenergic discharge in response to a drop in systemic blood pressure, the intrahepatic portal radicles constrict, portal pressure rises, and blood flow through the liver is brisk, bypassing most of the organ. Most of the blood in the liver enters the systemic circulation. Constriction of the hepatic arterioles diverts blood from the liver, and constriction of the mesenteric arterioles reduces portal inflow. In severe shock, hepatic blood flow may be reduced to such a degree that there is patchy necrosis of the liver.

Reservoir Function of the Splanchnic Circulation

In dogs and other carnivores, there is a large amount of smooth muscle in the capsule of the spleen. The spleen traps blood, and rhythmic contractions of its capsule pump plasma into the lympatics. The spleen therefore contains a reservoir of blood rich in cells. Noradrenergic nerve discharge and epinephrine make the spleen contract strongly, discharging the blood into the circulation. This function of the spleen is quantitatively unimportant in humans. However, the reservoir function of the whole visceral circulation is important. For example, 25–30% of the volume of the liver is accounted for by blood. Contraction of the capacitance vessels in the viscera can pump a liter of blood into the arterial circulation in less than a minute.

Other blood reservoirs that contain a large volume of blood at rest are the skin and lungs. During severe exercise, constriction of the vessels in these organs and decreased blood ''storage'' in the liver and other portions of the splanchnic bed, the skin, and the lungs may increase the volume of actively circulating blood perfusing the muscles by as much as 30%.

CIRCULATION OF THE SKIN

The amount of the heat lost from the body is regulated to a large extent by varying the amount of blood flowing through the skin (see Chapter 14). The fingers, toes, palms, and earlobes contain well-innervated anastomotic connections between arterioles and venules (arteriovenous anastomoses; see Chapter 30). Blood flow in response to thermoregulatory stimuli can vary from 1 to as much as 150 mL/100 g of skin per minute, and it has been postulated that these variations are possible because blood can be shunted through the anastomoses. The subdermal capillary and venous plexus is a blood reservoir of some importance, and the skin is one of the few places where the reactions of blood vessels can be observed visually.

White Reaction

When a pointed object is drawn lightly over the skin, the stroke lines become pale **(white reaction).** The mechanical stimulus apparently initiates contraction of the precapillary sphincters, and blood drains out of the capillaries and small veins. The response appears in about 15 seconds.

Triple Response

When the skin is stroked more firmly with a pointed instrument, instead of the white reaction there is reddening at the site that appears in about 10 seconds **(red reaction).** This is followed in a few minutes by local swelling and diffuse, mottled reddening around the injury. The initial redness is due to capillary dilation, a direct response of the capillaries to pressure. The swelling **(wheal)** is local edema due to increased permeability of the capillaries and postcapillary venules, with consequent extravasation of fluid. The redness spreading out from the injury **(flare)** is due to arteriolar dilation. This three-part response, the red reaction, wheal, and flare, is called the **triple response** and is part of the normal reaction to injury (see Chapter 20). It is present after total sympathectomy. The flare is absent in locally anesthetized skin and in denervated skin after the sensory nerves have degenerated, but it is present immediately after nerve block or section above the site of the injury. This plus other evidence indicates that it is due to an **axon reflex,** a response in which impulses initiated in sensory nerves by the injury are relayed antidromically down other branches of the sensory nerve fibers (Fig 32–16). This is the one situation in the body where there is substantial evidence for a physiologic effect due to antidromic conduction. The

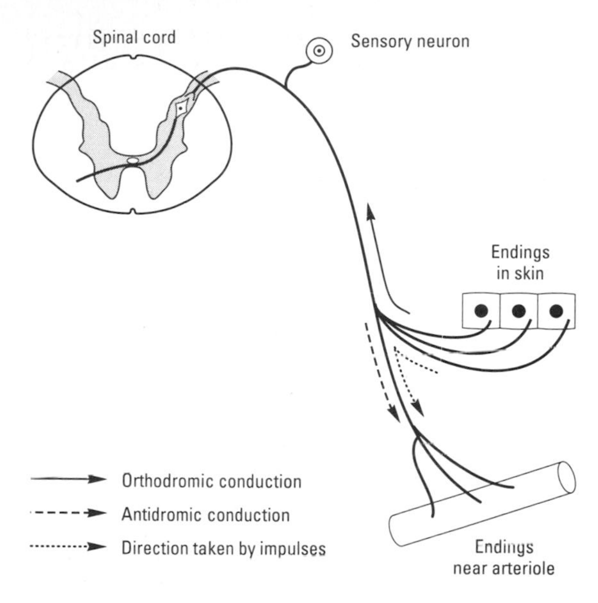

Figure 32–16. Axon reflex. Hypothetical pathway involved in the flare response.

increase in vascular permeability responsible for wheal formation is produced in part by histamine or a histaminelike substance released from local mast cells and mediated via H_1 receptors. Serotonin released from mast cells also contributes to the swelling in rats, but human mast cells do not contain serotonin. The transmitter released at the central termination of the sensory C fiber neurons is substance P (see Chapter 4), and substance P and CGRP are present in all parts of the neurons. Both dilate arterioles and, in addition, substance P causes extravasation of fluid. Furthermore, substance P antagonists reduce the extravasation. Thus, there is a neurogenic as well as a histaminergic and serotonergic component in the wheal.

Reactive Hyperemia

A response of the blood vessels that occurs in many organs but is visible in the skin is **reactive hyperemia,** an increase in the amount of blood in a region when its circulation is reestablished after a period of occlusion. When the blood supply to a limb is occluded, the cutaneous arterioles below the occlusion dilate; and when the circulation is reestablished, blood flowing into the dilated vessels makes the skin become fiery red. O_2 diffuses a short distance through the skin, and reactive hyperemia is prevented if the circulation of the limb is occluded in an atmosphere of 100% O_2. Therefore, the arteriolar dilation is apparently due to a local effect of hypoxia, and it is believed to be effected by the release of a chemical substance.

Generalized Responses

Noradrenergic nerve stimulation and circulating epinephrine and norepinephrine constrict cutaneous blood vessels. There are no known vasodilator nerve fibers to the cutaneous vessels, and vasodilation is brought about by a decrease in constrictor tone as well as the local production of bradykinin in sweat glands and vasodilator metabolites. Skin color and temperature also depend on the state of the capillaries and venules. A cold blue or gray skin is one in which the arterioles are constricted and the capillaries dilated; a warm red skin is one in which both are dilated.

Because painful stimuli cause diffuse noradrenergic discharge, a painful injury causes generalized cutaneous vasoconstriction in addition to the local triple response. When the body temperature rises during exercise, the cutaneous blood vessels dilate in spite of continuing noradrenergic discharge in other parts of the body. Dilation of cutaneous vessels in response to a rise in hypothalamic temperature (see Chapter 14) is a prepotent reflex response that overcomes other reflex activity. Cold causes cutaneous vasoconstriction; however, with severe cold, superficial vasodilation may supervene. This vasodilation is the cause of the ruddy complexion seen on a cold day.

Shock is more profound in patients with elevated temperatures because of the cutaneous vasodilation, and patients in shock should not be warmed to the point that their body temperature rises. This is sometimes a problem because well-meaning laymen have read in first-aid books that "injured patients should be kept warm," and they pile blankets on accident victims who are in shock.

PLACENTAL & FETAL CIRCULATION

Uterine Circulation

The blood flow of the uterus parallels the metabolic activity of the myometrium and endometrium and undergoes cyclic fluctuations that correlate well with the menstrual cycle in the nonpregnant woman. The function of the spiral and basal arteries of the endometrium in menstruation is discussed in Chapter 23. During pregnancy, blood flow increases rapidly as the uterus increases in size (Fig 32–17). Metabolites with a vasodilator action are undoubtedly produced in the uterus, as they are in other active tissues; but in early pregnancy the arteriovenous O_2 difference across the uterus is small, and it has been suggested that estrogens act on the blood vessels to increase uterine blood flow in excess of tissue O_2 needs. However, even though uterine blood flow increases 20-fold during pregnancy, the size of the conceptus increases much more, changing from a single cell to a fetus plus a placenta that weigh 4–5 kg at term in humans. Consequently, more O_2 is extracted from the uterine blood during the latter part of pregnancy, and the O_2 saturation of uterine blood

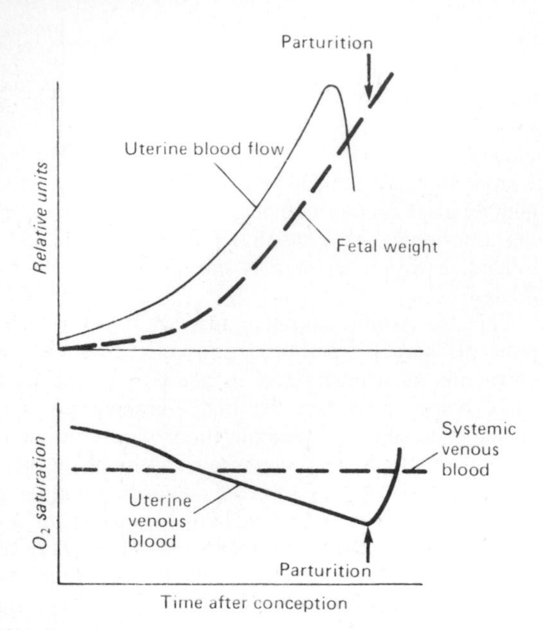

Figure 32–17. Changes in uterine blood flow and the amount of O_2 in uterine venous blood during pregnancy. (After Barcroft H. Modified and redrawn, with permission, from Keele CA, Neil E: *Samson Wright's Applied Physiology,* 12th ed. Oxford Univ Press, 1971.)

falls. Just before parturition there is a sharp decline in uterine blood flow, but the significance of this is not clear.

Placenta

The placenta is the "fetal lung." Its maternal portion is in effect a large blood sinus. Into this "lake" project the villi of the fetal portion containing the small branches of the fetal umbilical arteries and vein (Fig 32–18). O_2 is taken up by the fetal blood and CO_2 is discharged into the maternal circulation across the walls of the villi in a fashion analogous to O_2 and CO_2 exchange in the lungs (see Chapter 34), but the cellular layers covering the villi are thicker and less permeable than the alveolar membranes in the lungs, and exchange is much less efficient. The placenta is also the route by which all nutritive materials enter the fetus and by which fetal wastes are discharged to the maternal blood.

Fetal Circulation

The arrangement of the circulation in the fetus is shown diagrammatically in Fig 32–19. Fifty-five percent of the fetal cardiac output goes through the placenta. The blood in the umbilical vein in humans is believed to be about 80% saturated with O_2, compared with 98% saturation in the arterial circulation of the adult. The **ductus venosus** (Fig 32–20) diverts some of this blood directly to the inferior vena cava, and the remainder mixes with the portal blood of the fetus. The portal and systemic venous blood of the fetus is only 26% saturated, and the saturation of the mixed blood in the inferior vena cava is approximately 67%. Most of the blood entering the heart through the inferior vena cava is diverted directly to the left atrium via the patent foramen ovale. Most of the blood from the superior vena cava enters the right ventricle and is expelled into the pulmonary artery. The resistance of the collapsed lungs is high, and the pressure in the pulmonary artery is several mm Hg higher than it is in the aorta, so that most of the blood in the pulmonary artery passes through the **ductus arteriosus** to the aorta. In this fashion, the relatively unsaturated blood from the right ventricle is diverted to the trunk and lower body of the fetus, while the head of the fetus receives the better-oxygenated blood from the left ventricle. From the aorta, some of the blood is pumped into the umbilical arteries and back to the placenta. The O_2 saturation of the blood in the lower aorta and umbilical arteries of the fetus is approximately 60%.

Fetal Respiration

The tissues of fetal and newborn mammals have a remarkable but poorly understood resistance to hypoxia. However, the O_2 saturation of the maternal blood in the placenta is so low that the fetus might suffer hypoxic damage if fetal red cells did not have

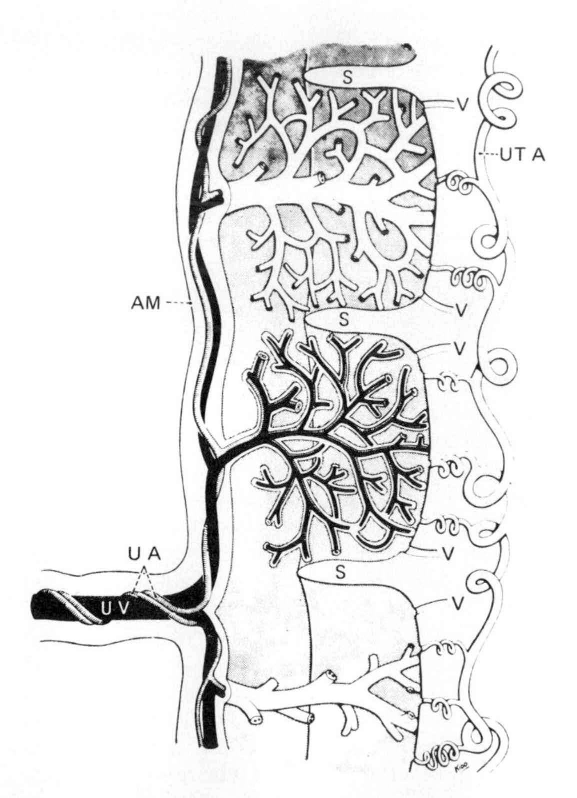

Figure 32–18. Diagram of a section through the human placenta. AM, amnion; S, septa; UA, umbilical arteries; UV, umbilical vein; UT A, uterine artery; V, veins. (Reproduced, with permission, from Harrison RG. *Textbook of Human Embryology,* 2nd ed. Blackwell, 1964.)

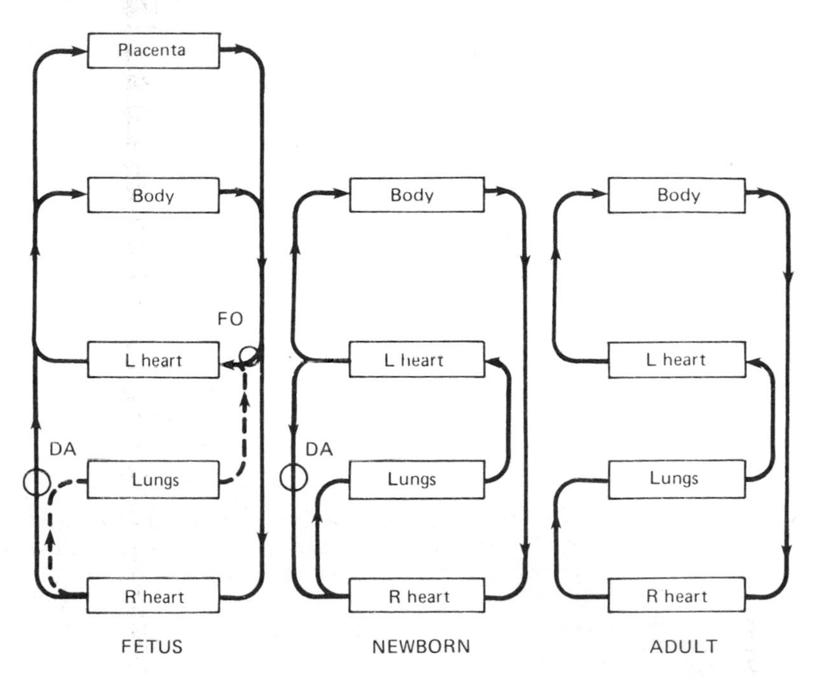

Figure 32–19. Diagram of the circulation in the fetus, the newborn infant, and the adult. DA, ductus arteriosus; FO, foramen ovale. (Redrawn and reproduced, with permission, from Born GVR et al: Changes in the heart and lungs at birth. *Cold Spring Harbor Symp Quant Biol* 1954; **19**:102.)

a greater O_2 affinity than adult red cells (Fig 32–21). The fetal red cells contain fetal hemoglobin (hemoglobin F), while the adult cells contain adult hemoglobin (hemoglobin A). The cause of the difference in O_2 affinity between the two is that hemoglobin F binds 2,3-DPG less effectively than hemoglobin A. The decrease in O_2 affinity due to the binding of 2,3-DPG is discussed in detail in Chapter 35. The quantitative aspects of gas exchange across the placenta, based on experiments on the cow, are shown in Table 32–6.

In humans, hemoglobin A first appears in the fetal circulation at about the 20th week when the bone marrow first begins to function (Fig 27–12). At birth, only 20% of the circulating hemoglobin is of the adult type. However, no more hemoglobin F is normally formed after birth, and by the age of 4 months 90% of the circulating hemoglobin is hemoglobin A.

Changes in Fetal Circulation & Respiration at Birth

Because of the patent ductus arteriosus and foramen ovale (Fig 32–20), the left and right heart pump in parallel in the fetus rather than in series as they do in the adult. At birth, the placental circulation is cut off and the peripheral resistance suddenly rises. The pressure in the aorta rises until it exceeds that in the pulmonary artery. Meanwhile, because the placental circulation has been cut off, the infant becomes increasingly asphyxic, Finally, the infant gasps several times, and the lungs expand. The markedly negative intrapleural pressure (-30 to -50 mm Hg) during the gasps contributes to the expansion of the lungs, but other poorly understood factors are also involved. The sucking action of the first breath plus constriction of the umbilical veins squeezes as much as 100 mL of blood from the placenta (the "placental transfusion").

Once the lungs are expanded, the pulmonary vascular resistance falls to less than 20% of the in utero value, and pulmonary blood flow increases markedly. Blood returning from the lungs raises the pressure in the left atrium, closing the foramen ovale by pushing the valve that guards it against the interatrial septum. The ductus arteriosus constricts within a few minutes after birth but, at least in sheep, does not close completely for 24–48 hours. Eventually, the foramen ovale and the ductus arteriosus both fuse shut in normal infants, and by the end of the first few days of life the adult circulatory pattern is established. The mechanism responsible for obliteration of the ductus arteriosus, like that responsible for expansion of the lungs, is incompletely understood, although there is evidence that a rise in arterial P_{O_2} and asphyxia are both capable of making the ductus constrict. Bradykinin has been shown to constrict the umbilical vessels and the ductus arteriosus while dilating the pulmonary vascular bed. Prostacyclin appears to have a role in maintaining the patency of the ductus arteriosus before birth, and rectal administration of one or 2 small doses of indomethacin, a drug that inhibits prostacyclin and prostaglandin synthesis (see Chapter 17), closes the ductus in many infants who would otherwise require surgical closure.

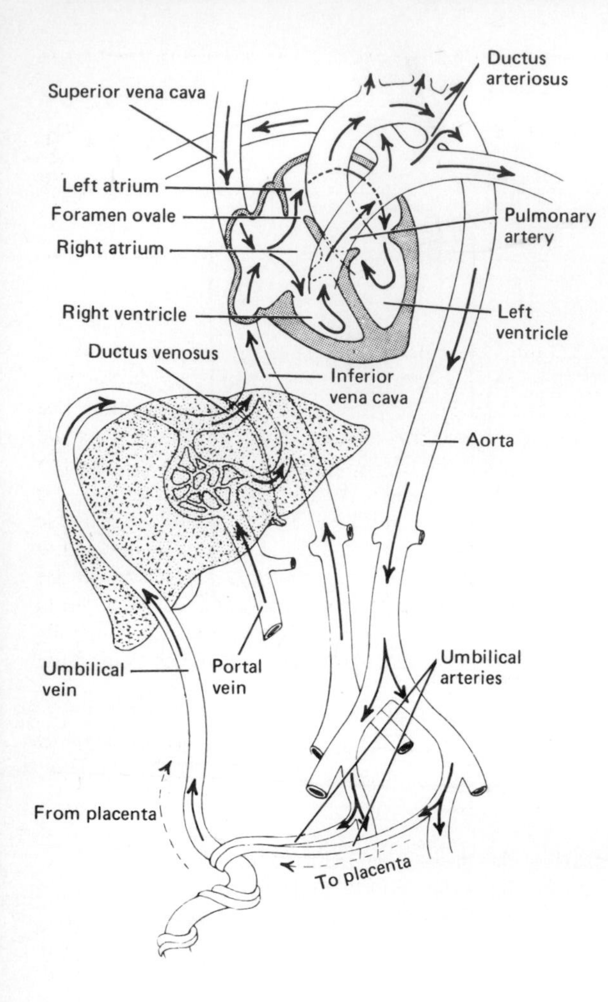

Figure 32–20. Circulation in the fetus. Most of the oxygenated blood reaching the heart via the umbilical vein and inferior vena cava is diverted through the foramen ovale and pumped out the aorta to the head, while the deoxygenated blood returned via the superior vena cava is mostly pumped through the pulmonary artery and ductus arteriosus to the feet and the umbilical arteries.

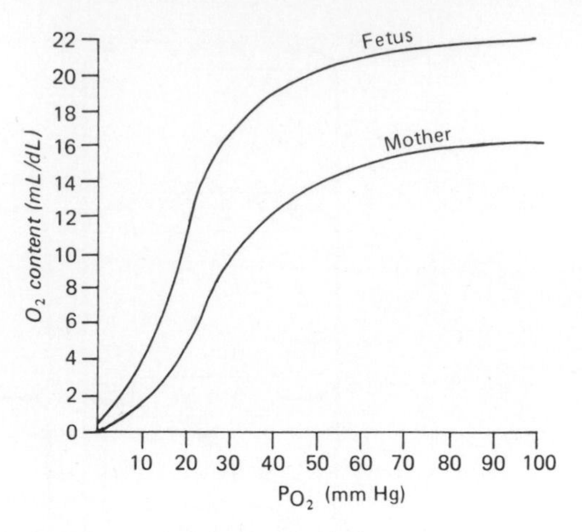

Figure 32–21. Dissociation curves of hemoglobin in human material and fetal blood. (Modified from Longo LD: Carbon monoxide: Effects on oxygenation of the fetus in utero. *Science* 1976;**194**;523.)

On the other hand, closure of the ductus before birth causes pulmonary hypertension, and in experimental animals, administration of prostaglandin D_2 during the neonatal period increases pulmonary blood flow and decreases pulmonary artery pressure. There is some evidence for an increased incidence of pulmonary hypertension in the children of women given prostaglandin inhibitors to delay the onset of labor.

Table 32–6. Summary of gaseous interchange across the placenta in the cow.*

	Hemoglobin % Saturation	Partial Pressure (mm Hg)	
		O_2	CO_2
Maternal artery	90	70	41
Uterine vein	70	41.5	46.5
Umbilical vein		11.5	48
Umbilical artery		5.5	50

*Data of Roos J, Romijn C: *J Physiol* (*Lond*) 1938;**92:**261.

Cardiovascular Homeostasis in Health & Disease

33

INTRODUCTION

The compensatory adjustments of the cardiovascular system to the challenges the circulation faces normally in everyday life and abnormally in disease illustrate the integrated operation of the cardiovascular regulatory mechanisms described in the preceding chapters. The adjustments to gravity, exercise, shock, fainting, hypertension, and heart failure are considered in this chapter.

COMPENSATIONS FOR GRAVITATIONAL EFFECTS

In the standing position, as a result of the effect of gravity on the blood (see Chapter 30), the mean arterial blood pressure in the feet of a normal adult is 180–200 mm Hg and venous pressure is 85–90 mm Hg. The arterial pressure at head level is 60–75 mm Hg, and the venous pressure is zero. If the individual does not move, 300–500 mL of blood pools in the venous capacitance vessels of the lower extremities, fluid begins to accumulate in the interstitial spaces because of increased hydrostatic pressure in the capillaries, and stroke volume is decreased up to 40%. Symptoms of cerebral ischemia develop when the cerebral blood flow decreases to less than about 60% of the flow in the recumbent position. If there were no compensatory cardiovascular changes, the reduction in cardiac output due to pooling on standing would lead to a reduction of cerebral flow of this magnitude, and consciousness would be lost.

The major compensations on assuming the upright position are triggered by the drop in blood pressure in the carotid sinus and aortic arch. The heart rate increases, helping to maintain cardiac output. There is relatively little venoconstriction in the periphery, but there is a prompt increase in the circulating levels of renin and aldosterone. The arterioles constrict, helping to maintain blood pressure. The actual blood pressure change at heart level is variable, depending upon the balance between the degree of arteriolar constriction and the drop in cardiac output (Fig 33–1).

In the cerebral circulation there are additional compensatory changes. The arterial pressure at head level drops 20–30 mm Hg, but jugular venous pressure falls 5–8 mm Hg, reducing the drop in perfusion pressure (arterial pressure minus venous pressure). Cerebral vascular resistance is reduced because intracranial pressure falls as venous pressure falls, decreasing the pressure on the cerebral vessels. The decline in cerebral blood flow increases the partial pressure of CO_2 ($P_{CO}2$) and decreases the P_{O2} and the pH in brain tissue, further actively dilating the cerebral vessels. Because of the operation of these autoregulatory mechanisms, cerebral blood flow declines only 20% on standing. In addition, the amount of O_2 extracted from each unit of blood increases, and the net effect is that cerebral O_2 consumption is about the same in the supine and the upright position.

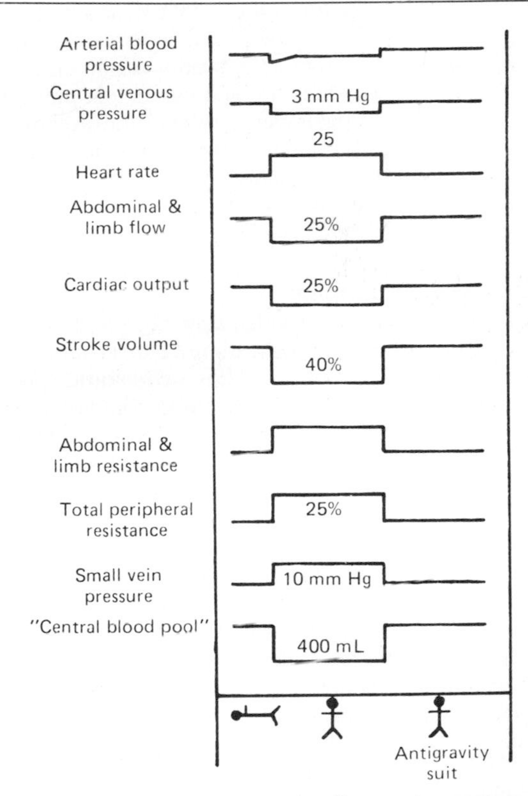

Figure 33–1. Effect on the cardiovascular system of rising from the supine to the upright position. Figures shown are average changes. Changes in abdominal and limb resistance and in blood pressure are variable from individual to individual. (Redrawn and reproduced, with permission, from Brobeck JR [editor]: *Best and Taylor's Physiological Basis of Medical Practice*, 9th ed. Williams & Wilkins, 1973.)

Prolonged standing presents an additional problem because of increasing interstitial fluid volume in the lower extremities. As long as the individual moves about, the operation of the muscle pump (see Chapter 30) keeps the venous pressure below 30 mm Hg in the feet, and venous return is adequate. However, with prolonged quiet standing (eg, in military personnel standing at attention for long periods), fainting may result. In a sense, the fainting is a "homeostatic mechanism," because falling to the horizontal position promptly restores venous return, cardiac output, and cerebral blood flow to adequate levels.

The effects of gravity on the circulation in humans depend in part upon the blood volume. When the blood volume is low, these effects are marked; when it is high, they are minimal.

The compensatory mechanisms that operate on assumption of the erect posture are better developed in humans than in quadrupeds even though these animals have sensitive carotid sinus mechanisms. Quadrupeds tolerate tilting to the upright position poorly. Of course, giraffes are an exception. These long-legged animals do not develop ankle edema despite the very large increment in vascular pressure in their legs due to gravity because they have tight skin and fascia in the lower legs—in a sense a built-in antigravity suit (see below)—and a very effective muscle pump. Perfusion in the head is maintained by a high mean arterial pressure and valves that prevent backflow in the jugular vein. When giraffes lower their heads to drink, blood is pumped up the jugular vein to the chest, presumably by rhythmic contractions of the muscles of the jaws.

Postural Hypotension

In some individuals, sudden standing causes a fall in blood pressure, dizziness, dimness of vision, and even fainting. The causes of this **orthostatic (postural) hypotension** are multiple. It is common after surgical sympathectomy and in patients receiving sympatholytic drugs. It also occurs in diseases, such as diabetes and syphilis, in which there is damage to the sympathetic nervous system, underscoring the importance of the sympathetic vasoconstrictor fibers in compensating for the effects of gravity on the circulation. Another cause of postural hypotension is **autonomic insufficiency.** In some cases of this disease, the defect is in the central nervous system, and resting plasma norepinephrine values are normal but fail to rise with standing. In other cases, the defect is peripheral, and resting norepinephrine values are low with little or no response to baroreceptor stimulation. Baroreceptor reflexes are also abnormal in patients with primary hyperaldosteronism. However, these patients generally do not have a postural hypotension, because their blood volumes are expanded sufficiently to maintain cardiac output in spite of changes in position. Indeed, mineralocorticoids are used to treat patients with postural hypotension.

Effects of Acceleration

The effects of gravity on the circulation are multiplied during acceleration or deceleration in vehicles that in modern civilization range from elevators to rockets. Force acting on the body as a result of acceleration is commonly expressed in *g* units, 1 *g* being the force of gravity on the earth's surface. "Positive *g*" is force due to acceleration acting in the long axis of the body, from head to foot; "negative *g*" is force due to acceleration acting in the opposite direction. During exposure to positive *g*, blood is "thrown" into the lower part of the body. The cerebral circulation is protected by the fall in venous pressure and intracranial pressure. Cardiac output is maintained for a time because blood is drawn from the pulmonary venous reservoir and because the force of cardiac contraction is increased. At accelerations producing more than 5 *g*, however, vision fails ("blackout") in about 5 seconds and unconsciousness follows almost immediately thereafter. The effects of positive *g* are effectively cushioned by the use of antigravity "*g* suits," double-walled pressure suits containing water or compressed air and regulated in such a way that they compress the abdomen and legs with a force proportionate to the positive *g*. This decreases venous pooling and helps maintain venous return (Fig 33–1).

Negative *g* causes increased cardiac output, a rise in cerebral arterial pressure, intense congestion of the head and neck vessels, ecchymoses around the eyes, severe throbbing head pain, and, eventually, mental confusion ("red-out"). In spite of the great rise in cerebral arterial pressure, the vessels in the brain do not rupture, because there is a corresponding increase in intracranial pressure and their walls are supported in the same way as during straining (see Chapter 32). The tolerance for *g* forces exerted across the body is much greater than it is for axial *g*. Humans tolerate 11 *g* acting in a back-to-chest direction for 3 minutes and 17 *g* acting in a chest-to-back direction for 4 minutes. Astronauts are therefore positioned to take the *g* forces of rocket flight in the chest-to-back direction. The tolerances in this position are sufficiently large to permit acceleration to orbital or escape velocity and deceleration back into the earth's atmosphere without ill effects.

Effects of Zero Gravity on the Cardiovascular System

From the data available to date, weightlessness for up to 326 days appears to have only transient adverse effects on the circulation. One would expect that the circulation in zero gravity would be essentially similar to that in the recumbent position at sea level, with possible minor variations due to changes in the position of viscera. Transient postural hypotension has been present after return to earth from orbital flights, and full readaptation to normal gravity after return from space flights to date has required 4–7 weeks. In the absence of the increases in cardiac output normally occasioned by the efforts of everyday living, there is some atrophy of the myocardium. There has been speculation that prolonged weightlessness during future trips to the planets may lead to more severe "disuse atrophy" of the heart and the cardiovascular and somatic reflex mechanisms (see Chapter 12) responsible for postural adjustments.

Other Effects of Zero Gravity

Muscular effort is much reduced when the objects to be moved are weightless, and the decrease in the extensive normal proprioceptive input due to the action of gravity on the body leads to flaccidity and atrophy of skeletal muscles. A program of regular exercises against resistance, eg, pushing against a wall or stretching a heavy rubber band appears to decrease the loss of muscle. However, the compensation is incomplete.

Other changes produced by exposure to space flight include motion sickness, a problem that has proved to be of greater magnitude than initially expected; loss of plasma volume due to headward shift of body fluids, with subsequent diuresis; loss of muscle mass; steady loss of bone mineral, with increased Ca^{2+} excretion; loss of red cell mass; and a decrease in plasma lymphocytes. The loss of body Ca^{2+} is equivalent to 0.4% of the total body Ca^{2+} per month, and although there is some evidence that the loss tapers off during prolonged space flight, loss at this rate might create problems of appreciable magnitude if continued for more than 9–12 months. A high calcium diet helps overcome this problem, but no totally effective treatment has as yet been developed. The psychologic problems associated with the isolation and monotony of prolonged space flight are also a matter of concern.

EXERCISE

Exercise is associated with very extensive alterations in the circulatory and respiratory systems. For convenience, the circulatory adjustments are considered in this chapter and the respiratory adjustments in Chapter 37. However, it should be emphasized that they occur together in an integrated fashion as part of the homeostatic responses that make moderate to severe exercise possible.

Muscle Blood Flow

The blood flow of resting skeletal muscle is low (2–4 mL/100 g/min). When a muscle contracts, it compresses the vessels in it if it develops more than 10% of its maximal tension (Fig 33–2); when it develops more than 70% of its maximal tension, blood flow is completely stopped. Between contractions, however, flow is so greatly increased that blood flow per unit of time in a rhythmically contracting muscle is increased as much as 30-fold. Blood flow sometimes increases at or even before the start of exercise, so the initial rise is probably a neurally mediated response. Impulses in the sympa-

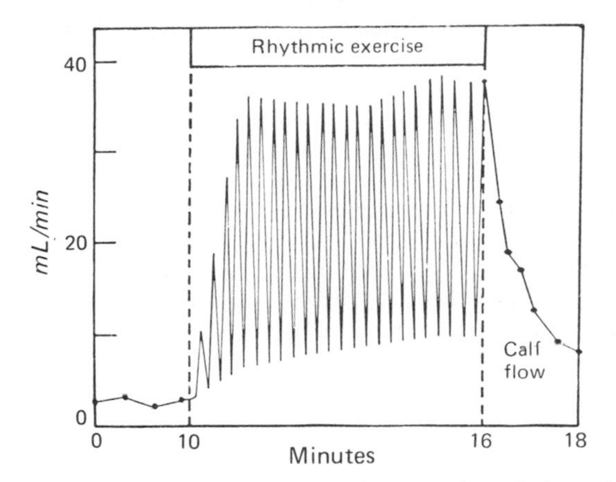

Figure 33–2. Blood flow through a portion of the calf muscles during rhythmic contraction. (Reproduced, with permission, from Barcroft H, Swan HJC: *Sympathetic Control of Human Blood Vessels.* Arnold, 1953.)

thetic vasodilator system (see Chapter 31) may be involved. The blood flow in resting muscle doubles after sympathectomy, so some decrease in tonic vasoconstrictor discharge may also be involved. However, once exercise has started, local mechanisms maintain the high blood flow, and there is no difference in flow in normal and sympathectomized animals.

Local mechanisms maintaining a high blood flow in exercising muscle include a fall in tissue P_{O2}, a rise in tissue P_{CO2}, and accumulation of K^+ and "vasodilator metabolites" (see Chapter 31). The temperature rises in active muscle, and this further dilates the vessels. Dilation of the arterioles and precapillary sphincters causes a 10- to 100-fold increase in the number of open capillaries. The average distance between the blood and the active cells—and the distance O_2 and metabolic products must diffuse—is thus greatly decreased. The dilation increases the cross-sectional area of the vascular bed, and the velocity of flow therefore decreases. The capillary pressure increases until it exceeds the oncotic pressure throughout the length of the capillaries. In addition, the accumulation of osmotically active metabolites more rapidly than they can be carried away decreases the osmotic gradient across the capillary walls. Therefore, fluid transudation into the interstitial spaces is tremendously increased. Lymph flow is also greatly increased, limiting the accumulation of interstitial fluid and in effect greatly increasing its turnover. The decreased pH and increased temperature shift the dissociation curve for hemoglobin to the right, so that more O_2 is given up by the blood. Increased concentration of 2,3-DPG in the red blood cells has been reported, and this would further decrease the O_2 affinity of hemoglobin (see Chapters 27 and 35). The net result is an up to 3-fold increase in the arteriovenous O_2 difference, and the transport of CO_2 out of the tissue is also facilitated. All of these changes combine to make it possible for the O_2

consumption of skeletal muscle to increase 100-fold during exercise. An even greater increase in energy output is possible for short periods during which the energy stores are replenished by anaerobic metabolism of glucose and the muscle incurs an O_2 debt (see Chapter 3). The overall changes in intermediary metabolism during exercise are discussed in Chapter 17.

K^+ dilates arterioles in exercising muscle, particularly during the early part of exercise. Muscle blood flow increases to a lesser degree during exercise in K^+-depleted individuals, and there is a greater tendency for severe disintegration of muscle **(exertional rhabdomyolysis)** to occur.

Systemic Circulatory Changes

The systemic cardiovascular response to exercise depends on whether the muscle contractions are primarily isometric or primarily isotonic with the performance of external work. With the start of an isometric muscle contraction, the heart rate rises. This increase still occurs if the muscle contraction is prevented by local infusion of a neuromuscular blocking drug. It also occurs with just the thought of performing a muscle contraction, so it is probably the result of psychic stimuli acting on the medulla oblongata. The increase is largely due to decreased vagal tone, although increased discharge of the cardiac sympathetic nerves plays some role. Within a few seconds of the onset of an isometric muscle contraction, systolic and diastolic blood pressures rise sharply. Stroke volume changes relatively little, and blood flow to the steadily contracting muscles is reduced as a result of compression of their blood vessels.

The response to exercise involving isotonic muscle contraction is similar in that there is a prompt increase in heart rate but different in that there is a marked increase in stroke volume. In addition, there is a fall in total peripheral resistance (Fig 33–3) due to vasodilation in exercising muscles (Table 33–1). Consequently, systolic blood pressure rises only moderately, whereas diastolic pressure may remain unchanged or even fall.

Cardiac output is increased during isotonic exercise to values that may exceed 35 L/min, the amount being proportionate to the increase in O_2 consumption. The mechanisms responsible for this increase are discussed in Chapter 29. The increase is due to an increase in heart rate and stroke volume, the heart muscle contracting more forcibly and discharging more of the end-systolic volume of blood in the ventricles. These chronotropic and inotropic effects on the heart are both due to increased activity in the noradrenergic sympathetic nerves to the heart. The activity is apparently initiated by psychic stimuli. Decreased vagal tone also contributes to the increase in heart rate. The increase in heart rate is sustained by the autonomic changes plus the stimulatory effect of the increase in P_{CO_2} on the medulla. The maximal heart rate achieved during exercise decreases with

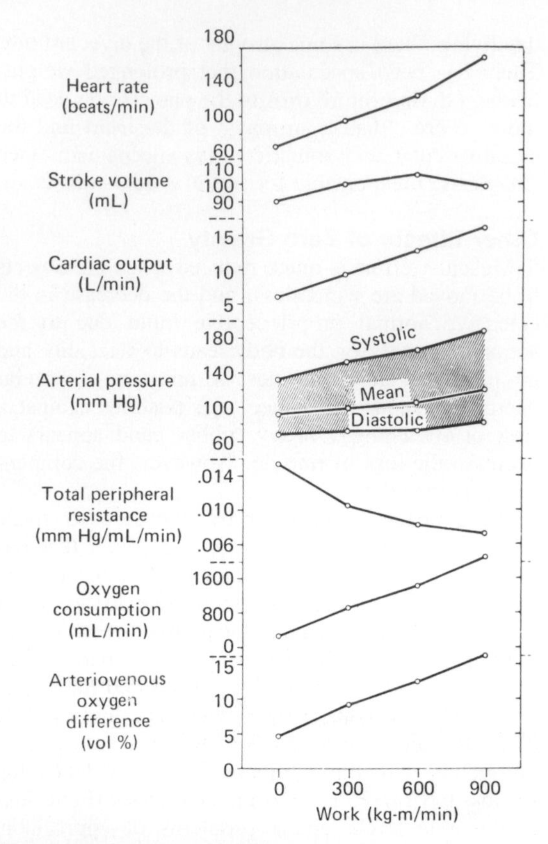

Figure 33–3. Effects of different levels of isotonic exercise on cardiovascular function. (Reproduced, with permission, from Berne RM, Levy MN: *Cardiovascular Physiology,* 5th ed. Mosby, 1986.)

age. In children, it rises to 200 or more beats per minute; in adults it rarely exceeds 195 beats per minute, and in elderly individuals the rise is even less.

There is a great increase in venous return, although the increase in venous return is not the primary cause of the increase in cardiac output. Venous return is

Table 33–1. Cardiac output and regional blood flow in a sedentary man.* Values are at rest and during isotonic exercise at maximal oxygen uptake.

	mL/min	
	Quiet Standing	Exercise
Cardiac output	5900	24,000
Blood flow to:		
Heart	250	1000
Brain	750	750
Active skeletal muscle	650	20,850
Inactive skeletal muscle	650	300
Skin	500	500
Kidney, liver, gastrointestinal tract, etc	3100	600

* Data from Mitchell JH, Blomqvist G: Maximal oxygen uptake. *N Engl J Med* 1971;**284:**1018.

increased by the great increase in the activity of the muscle and thoracic pumps; by motilization of blood from the viscera; by increased pressure transmitted through the dilated arterioles to the veins; and by noradrenergically mediated venoconstriction, which decreases the volume of blood in the veins. The amount of blood mobilized from the splanchnic area and other reservoirs may increase the amount of blood in the arterial portion of the circulation by as much as 30% during strenuous exercise.

After exercise, the blood pressure may transiently drop to subnormal levels, presumably because accumulated metabolites keep the muscle vessels dilated for a short period. However, the blood pressure soon returns to the preexercise level. The heart rate returns to normal more slowly.

Temperature Regulation

The quantitative aspects of heat dissipation during exercise are summarized in Fig 33–4. In many locations, the skin is supplied by branches of muscle arteries, so that some of the blood warmed in the muscles is transported directly to the skin, where some of the heat is radiated to the environment. There is a marked increase in ventilation (see Chapter 37), and some heat is lost in the expired air. The body temperature rises and the hypothalamic centers that control heat-dissipating mechanisms are activated. The temperature increase is due in part to inability of the heat-dissipating mechanism to handle the great increase in heat production, but there is evidence that during exercise there is in addition a "resetting of the thermostat"—ie, an increase in the body temperature at which the heat-dissipating mechanisms are activated. Sweat secretion is greatly increased, and vaporization of this sweat is the major path for heat loss. The cutaneous vessels also dilate. This dilation is primarily due to inhibition of vasoconstrictor tone, although local release of vasodilator polypeptides may also contribute (see Chapter 31).

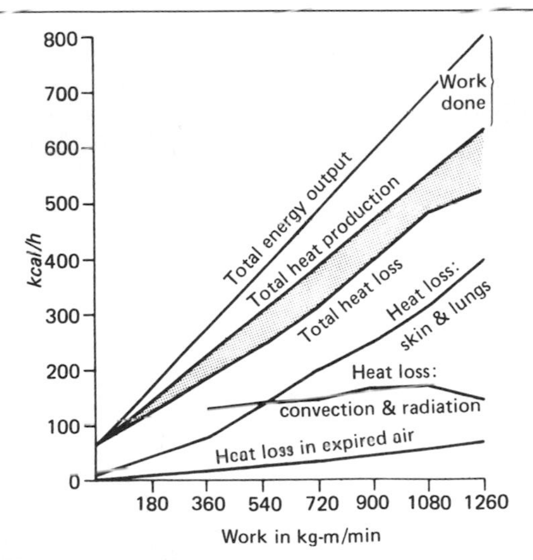

Figure 33–4. Energy exchange in muscular exercise. The shaded area represents the excess of heat production over heat loss. The total energy output equals the heat production plus the work done. (Redrawn and reproduced, with permission, from Nielsen M: Die Regulation der Korpertemperatur bei Muskelarbeit. *Skand Arch Physiol* 1938:**79**:193.)

Training

Both at rest and at any given level of exercise, trained athletes have a larger stroke volume and lower heart rate than untrained individuals (see Chapter 29), and they tend to have larger hearts. Training increases the maximal oxygen consumption (VO_2 max) that can be produced by exercise in an individual. VO_2 max averages about 38 mL/kg/min in active healthy men and about 29 mL/kg/min in active healthy women. It is lower in sedentary individuals. VO_2 max is the product of maximal cardiac output and maximal O_2 extraction by the tissues, and both increase with training.

The changes that occur in skeletal muscles with training include increases in the number of mitochondria and the enzymes involved in oxidative metabolism. There is an increase in the number of capillaries, with better distribution of blood to the muscle fibers. The net effect is more complete extraction of O_2 and consequently, for a given work load, less increase in lactate production. There is less increase in blood flow to muscles as well, and because of this, less increase in heart rate and cardiac output than in an untrained individual. This is one of the reasons that exercise is of benefit to patients with heart disease.

In view of the current vogue for jogging and other forms of exercise, it is worth noting that one of the clear-cut benefits of exercise regimens is psychologic; patients who exercise regularly "feel better." Regular physical exertion also increases the probability that an individual can remain active past the standard age of retirement. In addition, there is evidence that regular exercise decreases the incidence and severity of myocardial infarctions.

HEMORRHAGE & HEMORRHAGIC SHOCK

Effects of Hemorrhage

The decline in blood volume produced by bleeding decreases venous return, and cardiac output falls. Numerous compensatory mechanisms are activated (Table 33–2). The heart rate is increased, and with severe hemorrhage, there is always a fall in blood pressure. With moderate hemorrhage (5–15 mL/kg body weight), pulse pressure is reduced, but mean arterial pressure may be normal. The blood pressure changes vary from individual to individual, even when exactly the same amount of blood is lost. The skin is cool and pale and may have a grayish tinge because of stasis in the capillaries and a small amount of cyanosis. Respiration is rapid, and intense thirst is

Table 33–2. Compensatory reactions activated by hemorrhage.

Vasoconstriction
Tachycardia
Venoconstriction
Increased thoracic pumping
Increased skeletal muscle pumping (in some cases)
Increased movement of interstitial fluid into capillaries
Increased secretion of norepinephrine and epinephrine
Increased secretion of vasopressin
Increased secretion of glucocorticoids
Increased secretion of renin and aldosterone
Increased secretion of erythropoietin
Increased plasma protein synthesis

a prominent symptom. This combination of findings constitutes the clinical syndrome known as **hypovolemic shock. Hemorrhagic shock** is the form of hypovolemic shock due to hemorrhage. Other causes of this type of shock are discussed below.

In hypovolemic and other forms of shock, the inadequate perfusion of the tissue leads to increased anaerobic glycolysis with the production of large amounts of lactic acid. In severe cases, the blood lactate level rises from the normal value of about 1 mmol/L to 9 mmol/L or more. The resulting **lactic acidosis** depresses the myocardium, decreases peripheral vascular responsiveness to catecholamines, and may be severe enough to cause coma.

Rapid Compensatory Reactions

When blood volume is reduced and venous return is decreased, the arterial baroreceptors are stretched to a lesser degree, and sympathetic output is increased. Even if there is no drop in mean arterial pressure, the decrease in pulse pressure decreases the rate of discharge in the arterial baroreceptors, and reflex tachycardia and vasoconstriction result.

The vasoconstriction is generalized, sparing only the vessels of the brain and heart. The vasoconstrictor innervation of the cerebral arterioles is probably insignificant from a functional point of view, and the coronary vessels are dilated because of the increased myocardial metabolism secondary to the increase in heart rate (see Chapter 32). Vasoconstriction is most marked in the skin, where it accounts for the coolness and pallor, and in the kidneys and viscera.

Hemorrhage also evokes a widespread reflex venoconstriction that helps maintain the filling pressure of the heart, although the receptors that initiate the venoconstriction are unsettled. The intense vasoconstriction in the splanchnic area shifts blood from the visceral reservoir into the systemic circulation. Blood is also shifted out of the subcutaneous and pulmonary veins. Contraction of the spleen discharges more ''stored'' blood into the circulation, although the volume mobilized in humans is small.

In the kidneys, both afferent and efferent arterioles are constricted, but the efferent vessels are constricted to a greater degree. The glomerular filtration rate is depressed, but renal plasma flow is decreased to a greater extent, so that the filtration fraction (glomerular filtration rate divided by renal plasma flow) increases. There may be shunting of the blood through the medullary portions of the kidneys, bypassing the cortical glomeruli. Very little urine is formed. Sodium retention is marked, and there is retention of the nitrogenous products of metabolism in the blood **(azotemia** or **uremia).** Especially when the hypotension is prolonged, there may be severe renal tubular damage **(acute renal failure).**

Hemorrhage is a potent stimulus to adrenal medullary secretion (see Chapter 20). Circulating norepinephrine is also increased because of the increased discharge of sympathetic noradrenergic neurons. The increase in circulating catecholamines probably contributes relatively little to the generalized vasoconstriction, but it may lead to stimulation of the reticular formation (see Chapter 11). Possibly because of such reticular stimulation, some patients in hemorrhagic shock are restless and apprehensive. Others are quiet and apathetic, and their sensorium is dulled, probably because of cerebral ischemia and acidosis. When restlessness is present, increased motor activity and increased respiratory movements increase the muscular and thoracic pumping of venous blood.

The loss of red cells decreases the O_2-carrying power of the blood, and the blood flow in the carotid and aortic bodies is reduced. The resultant anemia and stagnant hypoxia (see Chapter 37) as well as the acidosis stimulate the chemoreceptors. Increased activity in chemoreceptor afferents is probaby the main cause of respiratory stimulation in shock. Chemoreceptor activity also excites the vasomotor center, increasng vasoconstrictor discharge. In fact, in hemorrhaged dogs with arterial pressures of less than 70 mm Hg, cutting the nerves to the carotid baroreceptors and chemoreceptors may cause a further fall in blood pressure rather than a rise. This paradoxic result is due to the fact that there is no baroreceptor discharge at pressures below 70 mm Hg and activity in fibers from the carotid chemoreceptors is driving the vasomotor center beyond the maximal rate produced by release of baroreceptor inhibition.

The increase in the level of circulating angiotensin II produced by the increase in plasma renin activity during hemorrhage helps to maintain blood pressure. The blood pressure fall produced by removal of a given volume of blood is greater in animals infused with drugs that block angiotensin II receptors than it is in controls. Vasopressin also raises blood pressure when administered in large doses in normal animals, but infusion of doses that produce the same plasma vasopressin levels produced by hemorrhage causes only a small increase in blood pressure, probably because of buffering by baroreceptor mechanisms. However, blood pressure falls when peptides that antagonize the effects of vasopressin are injected following hemorrhage. Thus, it appears that vasopressin also plays a significant role in maintaining blood pressure. The increases in circulating angiotensin and ACTH levels increase aldosterone secretion, and the increased circulating levels of aldosterone and vasopressin cause retention of Na^+ and water, which helps reexpand the blood volume. However, aldosterone takes about 30 minutes to exert its effect, and the initial decline in urine volume and Na^+ excretion is certainly due for the most part to the hemodynamic alterations in the kidney.

When the arterioles constrict and the venous pressure falls because of the decrease in blood volume, there is a drop in capillary pressure. Fluid moves into the capillaries along most of their course, helping to maintain the circulating blood volume. This decreases interstitial fluid volume, and fluid moves out of the cells. Decreased ECF volume has been shown to cause thirst, and thirst is intense in patients who are in shock. Thirst can occur without any change in plasma osmolality after hemorrhage and appears to be caused, at least in part, by increased concentrations of circulating angiotensin II acting on the subfornical organ (see Chapter 14).

Long-Term Compensatory Reactions

After a moderate hemorrhage, the circulating plasma volume is restored in 12–72 hours (Fig 33–5). There is also a rapid entry of preformed albumin from extravascular stores, but most of the tissue fluids that are mobilized are protein-free. They dilute the plasma proteins and cells, but when whole blood is lost, the hematocrit may not fall for several hours after the onset of bleeding. After the initial influx of preformed albumin, the rest of the plasma protein losses are replaced, presumably by hepatic synthesis, over a period of 3–4 days. Erythropoietin appears in the circulation, and the reticulocyte count increases, reaching a peak in 10 days. The red cell mass is restored to normal in 4–8 weeks. However, a low hematocrit is remarkably well tolerated because of various compensatory mechanisms. One of these is an increase in the concentration of 2,3-DPG in the red blood cells, which causes hemoglobin to give more O_2 to the tissues (see Chapter 27). In long-standing anemia in otherwise healthy individuals, exertional dyspnea is not observed until the hemoglobin concentration is about 7.5 g/dL. Weakness becomes appreciable at about 6 g/dL; dyspnea at rest appears at about 3 g/dL; and the heart fails when the hemoglobin level falls to 2 g/dL.

Irreversible Shock

Depending largely upon the amount of blood lost, some patients die soon after hemorrhage and others recover as the compensatory mechanisms, aided by appropriate treatment, gradually restore the circulation to normal. In an intermediate group of patients, shock persists for hours and gradually progresses to a state in which there is no longer any response to vasopressor drugs and in which, even if the blood volume is returned to normal, cardiac output remains depressed. This is known as **irreversible shock.** The peripheral resistance falls, the heart slows, and the patient eventually dies.

There has been much discussion about what makes shock become irreversible. It seems clear that a number of deleterious positive feedback mechanisms become operative in this stage. For example, severe cerebral ischemia leads eventually to depression of the vasomotor and cardiac areas of the brain, with slowing of the heart rate and vasodilation. These both make the blood pressure drop further, with a further reduction in cerebral blood flow and further depression of the vasomotor and cardiac areas.

Another important example of this type of positive feedback is myocardial depression. In severe shock, the coronary blood flow is reduced because of the hypotension and tachycardia (see Chapter 32), even though the coronary vessels are dilated. The myocardial failure makes the shock and the acidosis worse, and this in turn leads to further depression of myocardial function. If the reduction is marked and prolonged, the myocardium may be damaged to the point where cardiac output cannot be restored to normal in spite of reexpansion of the blood volume.

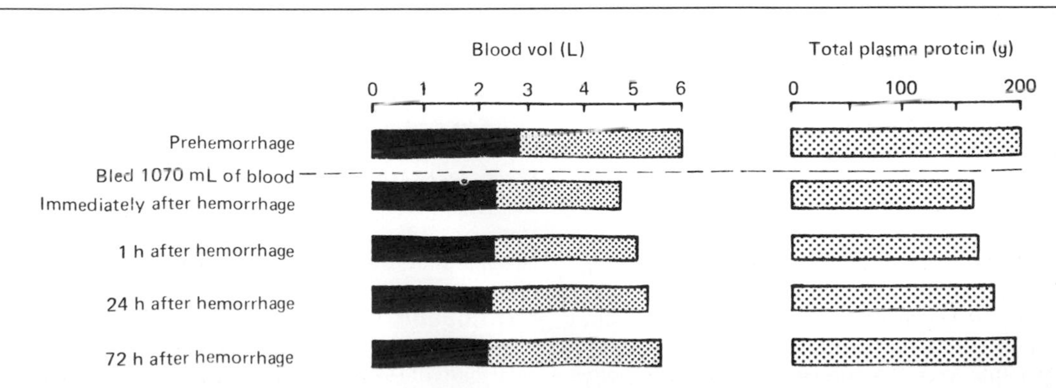

Figure 33–5. Changes in plasma volume (dotted bars), red cell volume (black bars), and total plasma protein following hemorrhage in a normal human subject. (From data in Keele CA, Neil E: *Samson Wright's Applied Physiology,* 11th ed. Oxford Univ Press, 1965.)

Spasm of the precapillary sphincters and venules, especially in the splanchnic region, is apparently a prominent feature of reversible shock. The reduced capillary perfusion due to the constriction of the precapillary sphincters leads to hypoxic tissue damage. Irreversible shock begins to develop when, after 3–5 hours, the precapillary sphincters dilate while the venules remain constricted. Blood now enters the capillaries but stagnates in these vessels, so that the tissue hypoxia continues. Capillary hydrostatic pressure rises, and fluid leaves the vascular system in increasing amounts. Circulating bacterial toxins may contribute to the paralysis of the precapillary sphincters. There is evidence that, presumably as a result of stagnant hypoxia in the gastrointestinal tract, the barriers to the entry of bacteria into the circulation from the intestinal tract break down.

OTHER FORMS OF SHOCK

General Considerations

Hemorrhage has been considered in detail as an example of a condition that produces the syndrome of shock. There has been a greal deal of confusion and controversy about this syndrome. Part of the difficulty lies in the loose use of the term by physiologists and physicians as well as laymen. **Electric shock** and **spinal shock,** for example, bear no resemblance to the condition, produced by hemorrhage. Shock, in the restricted sense of "circulatory shock," is still a collection of different entities that share certain common features. The feature that appears to be common to all cases is inadequate tissue perfusion. The cardiac output may be inadequate because the heart is damaged, or the volume of circulating blood may be less than the capacity of the circulation, either because the blood volume is reduced or because the capacity is increased. On this basis, 3 general types of shock are often delineated: (1) **hypovolemic shock,** in which blood or plasma has been lost from the circulation to the exterior or into the tissues; (2) **cardiogenic shock,** in which the pumping action of the heart is inadequate; and (3) **low-resistance shock,** in which there is vasodilation in the face of a normal cardiac output and blood volume.

Hypovolemic Shock

Hypovolemic shock is also called "cold shock." It is characterized in the typical case by hypotension; a rapid thready pulse; a cold, pale, clammy skin; intense thirst; rapid respiration; and restlessness or, alternatively, torpor. None of these findings, however, are invariably present. The hypotension may be relative. A hypertensive patient whose blood pressure is regularly 240/140, for example, may be in severe shock when the blood pressure is 120/90.

Hypovolemic shock is commonly subdivided into categories on the basis of cause. The use of terms such as hemorrhagic shock, traumatic shock, surgical shock, and burn shock is of some benefit because, although there are similarities between these various forms of shock, there are important features that are unique to each. Hemorrhagic shock is discussed above.

Traumatic shock develops when there is severe damage to muscle and bone. This is the type of shock seen in battle casualties and automobile accident victims. Frank bleeding into the injured areas is the principle cause of the shock, although some plasma also enters the tissues. The amount of blood which can be lost into an injury that appears relatively minor is remarkable; the thigh muscles can accommodate 1 L of extravasated blood, for example, with an increase in the diameter of the thigh of only 1 cm.

When there is extensive soft tissue and muscle crushing, myoglobin leaks into the circulation. It is said to precipitate in the renal tubules, causing renal damage, and the combination of traumatic shock and renal damage due to myoglobinuria has been called the **crush syndrome.** However, myoglobinuria is only one of the factors causing renal damage, and acute renal failure is a potential complication in all forms of shock.

Surgical shock is due to the combination in various proportions of external hemorrhage, bleeding into injured tissues, and dehydration.

In **burn shock,** the most apparent abnormality is loss of plasma as exudate from the burned surfaces. Since the loss in this situation is plasma rather than whole blood, the hematocrit rises and **hemoconcentration** is a prominent finding. Burns also cause complex, poorly understood metabolic changes in addition to fluid loss. For example, there is a 50% rise in metabolic rate of nonthyroidal origin, and some burned patients develop hemolytic anemia. Because of these complications, plus the severity of the shock and the problems of sepsis and kidney damage, the mortality rate when third-degree burns cover more than 75% of the body is still close to 100%.

Hypovolemic shock is a complication of various metabolic and infectious diseases. For example, although the mechanism is different in each case, adrenal insufficiency, diabetic ketoacidosis, and severe diarrhea are all characterized by loss of Na^+ from the circulation. The resultant decline in plasma volume may be severe enough to precipitate cardiovascular collapse. In severe infections, shock may occur as a result of diffuse vasculitis with leakage of plasma into the tissues (Rocky Mountain spotted fever, Korean hemorrhagic fever), circulating "toxins" that paralyze vascular smooth muscle, dehydration, or, very rarely, bilateral hemorrhage into the adrenal glands and acute adrenal insufficiency.

Cardiogenic Shock

When there is a gross decline in cardiac output due to disease of the heart rather than to inadequate blood volume, the condition that results is called cardiogenic shock. The symptoms are those of shock plus congestion of the lungs and viscera due to failure of the heart to put out all the venous blood returned to it. Consequently, the condition is sometimes called "congested shock." The incidence of this shock in patients with myocardial infarction is about 10%, and it usually occurs when there is extensive damage to the left ventricular myocardium. It has been suggested that release of chemical agents such as serotonin from the infarcted area activates ventricular receptors which trigger a reflex inhibition of vasomotor tone (Bezold-Jarisch reflex; see Chapter 31). This limits the compensatory vasoconstrictor responses and makes the shock worse.

Low-Resistance Shock

Low-resistance shock includes a number of entities in which the blood volume is normal while the capacity of the circulation is increased by massive vasodilation. For this reason, it is called **"warm shock"** or **neurogenic shock.** Examples include fainting in response to strong emotion and the reaction produced by overwhelming fear and grief. Another form is the poorly understood, immediate, stunned reaction in injured individuals. The most common form, however, is that produced by endotoxin from gram-negative bacteria **(septic shock,** or **endotoxin shock).** Endotoxin is the cell-wall lipopolysaccharide of the organisms. It causes macrophages to produce increased amounts of **cachectin (tumor necrosis factor),** a polypeptide that inhibits lipoprotein lipase and has other effects which make shock worse. Mice that congenitally produce only small amounts of cachectin are resistant to endotoxin, and the glucocorticoid dexamethasone, which helps protect animals from endotoxic shock, inhibits the translation of cachectin mRNA in macrophages.

Anaphylactic shock is a rapidly developing, severe allergic reaction that sometimes occurs when an individual who has previously been sensitized to an antigen is subsequently reexposed to it. The resultant antigen-antibody reaction releases large quantities of histamine, causing increased capillary permeability and widespread dilation of arterioles and capillaries.

In febrile patients, shock is apt to be more severe because the cutaneous blood vessels are often dilated (see Chapter 32), increasing the disparity between the capacity of the vascular system and the available circulating blood volume.

Treatment of Shock

The treatment of shock should be aimed at correcting the cause and helping the physiologic compensatory mechanisms to restore an adequate level of tissue perfusion. In hemorrhagic, traumatic, and surgical shock, for example, the primary cause of the shock is blood loss, and the treatment should include early and rapid transfusion of adequate amounts of compatible whole blood. Saline is of limited temporary value. The immediate goal is restoration of an adequate circulating blood volume, and since saline is distributed in ECF, only 25% of the amount administered stays in the vascular system. In burn shock and other conditions in which there is hemoconcentration, plasma is the treatment of choice to restore the fundamental defect, the loss of plasma. "Plasma expanders," solutions of sugars of high molecular weight and related substances that do not cross capillary walls, have some merit. Concentrated human serum albumin and other hypertonic solutions expand the blood volume by drawing fluid out of the interstitial spaces. They are valuable in emergency treatment but have the disadvantage of further dehydrating the tissues of an already dehydrated patient.

In anaphylactic shock, epinephrine has a highly beneficial and almost specific effect that must represent more than just constriction of the dilated vessels. In all types of shock, restoration of an adequate arterial pressure is important to maintain coronary blood flow. Vasopressor agents such as norepinephrine are of value for this purpose, but their use should be discontinued as soon as possible. Because it produces renal vasodilation and has a positively inotropic effect while producing vasoconstriction elsewhere (see Chapter 20), dopamine is of value in the treatment of traumatic and cardiogenic shock.

A number of measures sometimes used in the treatment of shock inhibit the operation of the physiologic compensatory mechanisms. Sedatives and other drugs that depress the central nervous system should be used as sparingly as possible, because they depress the discharge of the vasomotor center. Alcohol is particulary unphysiologic because it depresses the central nervous system and, through a central action, dilates cutaneous vessels. Care should be taken to avoid overheating and resultant cutaneous vasodilation. Further compromising the circulation by permitting the patient to sit or stand is, of course, detrimental, and gravity should be put to work to help rather than hinder the compensatory mechanisms. Raising the foot of the bed 6–12 inches (15–30 cm) with "shock blocks" is a simple but valuable therapeutic maneuver that aids venous return from the lower half of the body and improves cerebral blood flow. However, the head-down position also causes the abdominal viscera to press on the diaphragm, making adequate ventilation more difficult to maintain and favoring the development of pulmonary complications. Consequently, it should not be used for prolonged periods.

FAINTING

Fainting, or **syncope,** is sudden, transient loss of consciousness. It can be due to metabolic or neurologic abnormalities, but more commonly, it is due to peripheral vascular or cardiac abnormalities that cause inadequate cerebral blood flow. It is often benign and is most commonly due to abrupt vasodilation in association with strong emotions. This produces hypotension, generally in association with bradycardia. The term **vasovagal syncope** has been coined to denote this entity. **Postural syncope** is fainting due to pooling of blood in the dependent parts of the body on standing. **Micturition syncope,** fainting during urination, occurs in patients with orthostatic hypotension. It is due to the combination of the orthostasis and reflex bradycardia induced by voiding in these patients. Pressure on the carotid sinus, the produced, for example, by a tight collar, can cause such marked bradycardia and vasodilation that fainting results **(carotid sinus syncope).** Rarely, vasodilation and bradycardia may be precipitated by swallowing **(deglutition syncope). Cough syncope** occurs when the increase in intrathoracic pressure during straining or coughing is sufficient to block venous return. **Effort syncope** is fainting on exertion due to inability to increase cardiac output to meet the increased demands of the tissues and is particularly common in patients with aortic or pulmonary stenosis.

Syncope can also be due to more serious abnormalities. About 25% of the attacks are of cardiac origin, due to either transient obstruction of blood flow through the heart or sudden decreases in cardiac output owing to various cardiac arrhythmias. Fainting is the presenting symptom in 7% of patients with myocardial infarctions. Thus, all cases of syncope should be investigated to determine the cause.

HYPERTENSION

Hypertension is a sustained elevation of the systemic arterial pressure. **Pulmonary hypertension** can also occur, but the pressure in the pulmonary artery (see Chapter 34) is relatively independent of that in the systemic arteries.

Experimental Hypertension

The arterial pressure is determined by the cardiac output and the peripheral resistance (pressure = flow × resistance; see Chapter 30). The peripheral resistance is determined by the viscosity of the blood and, more importantly, by the caliber of the resistance vessels. Hypertension can be produced by elevating the cardiac output, but sustained hypertension is usually due to increased peripheral resistance. Some of the procedures that have been reported to produce sustained hypertension in experimental animals are listed in Table 33–3. For the most part, the procedures involve manipulation of the kidneys, the nervous system, or the adrenals. There are in addition a number of strains of rats that develop hypertension either spontaneously (SHR rats) or when fed a high-sodium diet (Dahl strain rats).

Table 33–3. Procedures that produce sustained hypertension in experimental animals.

1. Interference with renal blood flow (renal hypertension)
   a. Constriction of one renal artery; other kidney removed (one-clip, one-kidney Goldblatt hypertension)
   b. Constriction of one renal artery; other kidney intact (one-clip, 2-kidney Goldblatt hypertension)
   c. Constriction of aorta or both renal arteries (2-clip, 2-kidney Goldblatt hypertension)
   d. Compression of kidney by rubber capsules, production of perinephritis, etc
2. Interruptions of afferent input from arterial baroreceptors (neurogenic hypertension)
   a. Denervation of carotid sinuses and aortic arch
   b. Bilateral lesions of nucleus of tractus solitarius
3. Treatment with corticosteroids
   a. Deoxycorticosterone and salt
   b. Other mineralocorticoids
4. Partial adrenalectomy (adrenal regeneration hypertension)
5. Genetic
   a. Spontaneous hypertension in various strains of rats
   b. Salt-induced hypertension in genetically sensitive rats

The hypertension that follows constriction of the renal arterial blood supply or compression of the kidney is called **renal hypertension.** As noted in Chapter 24, some animals with renal hypertension have elevated plasma renin activity, whereas others do not. In general, one-clip 2-kidney Goldblatt hypertension (see Table 33–3) is renin-dependent, whereas one-clip one-kidney Goldblatt hypertension is not. An additional factor that probably contributes to renal hypertension is decreased ability of the constricted kidney to excrete Na^+.

Neurogenic hypertension is discussed in Chapter 31. Provided that salt intake is normal or high, deoxycorticosterone causes hypertension which may persist after treatment is stopped. The hypertension is more severe in unilaterally nephrectomized animals.

Hypertension in Humans

Hypertension is a very common abnormality in humans. It can be produced by many diseases (Table 33–4). It is also a prominent symptom of **toxemia of pregnancy,** a condition that may be caused by a pressor polypeptide secreted by the placenta.

Hypertension causes a number of serious disorders. When the resistance against which the left ventricle must pump (afterload) is elevated for a long period, the cardiac muscle hypertrophies. The total O_2 consumption of the heart, already increased by the work of expelling blood against a raised pressure (see Chapter 29), is increased further because there is more muscle. Therefore, any decrease in coronary blood flow has more serious consequences in hypertensive patients than it does in normal individuals, and degrees of coronary narrowing that do not produce symptoms when the size of the heart is normal may produce myocardial infarction when the heart is enlarged. There is an increased incidence of arteriosclerosis in hypertension, and myocardial infarcts are common even when the heart is not enlarged. The "Starling mechanism" (see Chapter 29) operates in hypertension, dilation of the heart stretching the muscle fibers and increasing their strength of contraction. However, the ability to compensate for the high peripheral resistance is eventually exceeded, and the heart fails. Hypertensive individuals are also predisposed to thromboses of cerebral vessels and cerebral hemorrhage. An additional complication is renal failure. However, the incidence of heart failure, strokes, and renal failure can be markedly reduced by active treatment of hypertension, even when the hypertension is relatively mild.

Malignant Hypertension

Chronic hypertension can enter an accelerated phase in which necrotic arteriolar lesions develop and there is a rapid downhill course with papilledema, cerebral symptoms, and progressive renal failure. This syndrome is known as **malignant hypertension,** and without treatment it is fatal in less than 2 years. It can be triggered by hypertension due to any cause. However, its progression can be stopped and it can be reversed by appropriate antihypertensive therapy.

Essential Hypertension

In 90% of patients with elevated blood pressure, the cause of the hypertension is unknown, and they are said to have **essential hypertension.** The remainder have blood pressure elevations due to a variety of different diseases (Table 33–4). Early in the course of essential hypertension, the blood pressure elevations are intermittent and there is an exaggerated pressor response to stimuli such as cold and excitement that produce only a moderate blood pressure elevation in normal individuals. This suggests that overactive autonomic reactions are responsible for the arteriolar spasm, and treatment with drugs that block sympathetic outflow markedly slows the progress of the disease. Later the blood pressure elevation becomes sustained. The baroreceptor mechanism is "reset" so that the blood pressure is maintained at the elevated level (see Chapter 31). Spasm of the arterioles leads to hypertrophy of their musculature, and there is some organic narrowing of the vessels. At this stage, even a normal rate of autonomic discharge is associated with an elevated pressure.

Table 33–4. Principal causes of sustained diastolic hypertension in humans. The conditions marked by asterisks are quite often curable. In all of them except Cushing's syndrome, sustained hypertension may be the only physical finding.

1. Unknown (essential hypertension)
2. Adrenocortical diseases
   a. Hypersecretion of aldosterone (Conn's syndrome)*
   b. Hypersecretion of other mineralocorticoids (hypertensive form of congenital virilizing adrenal hyperplasia; 17α-hydroxylase deficiency)
   c. Hypersecretion of glucocorticoids (Cushing's syndrome)*
3. Catecholamine-secreting tumors of adrenal medullary or paraganglionic origin (pheochromocytoma)*
4. Tumors of juxtaglomerular cells*
5. Narrowing of one or both renal arteries (renal hypertension)*
6. Renal disease
   a. Glomerulonephritis
   b. Pyelonephritis
   c. Polycystic disease
7. Narrowing (coarctation) of the aorta*
8. Severe polycythemia
9. Oral contraceptives*

The progression of untreated essential hypertension is variable. Particularly in women, it is often benign, with elevated pressure being the only finding for many years. However, it may also progress rapidly into the malignant phase.

At present, essential hypertension is a treatable but not a curable disease. On the other hand, several of the other disease processes that cause hypertension in humans are not only treatable but curable. It is therefore important to recognize these latter diseases and institute appropriate definitive treatment.

Evidence is accumulating that some humans, like Dahl salt-sensitive rats, have a relatively marked increase in blood pressure when fed a high-sodium diet, whereas others, like Dahl salt-resistant rats, have small increases in blood pressure. Since there is no easy test to distinguish salt-responsive from salt-resistant humans, the current emphasis on decreasing sodium intake in all humans makes some sense. However, the degree to which hypertension can be ameliorated or prevented will obviously vary with the salt sensitivity of the individual.

Other Causes of Hypertension

In humans, deoxycorticosterone and aldosterone both elevate the blood pressure, and hypertension is a prominent feature of primary hyperaldosteronism. It is also seen in patients who secrete excess deoxycorticosterone (see Chapter 20). The hypokalemia that these hormones produce damages the kidneys **(hypokalemic nephropathy),** and the hypertension may be partly renal in origin. Expansion of the blood volume due to Na^+ retention also plays a role.

Plasma renin activity is low in the forms of hypertension that are due to excess secretion of aldosterone and deoxycorticosterone. It has also been found to be low in the presence of normal or low aldosterone and deoxycorticosterone secretion rates in 10–15% of patients with what otherwise appears to be essential hypertension **("low-renin hypertension").** This raises the interesting possibility that an as yet unidentified mineralocorticoid is being secreted in excess in these patients.

Hypertension is also seen in Cushing's syndrome, in which aldosterone secretion is usually normal. The cause of the hypertension in this syndrome is uncertain; it may be due to the increased secretion of deoxycorticosterone produced by increased circulating ACTH, increased secretion of angiotensinogen due to the increase in circulating glucocorticoids, a direct action of glucocorticoids on the arterioles, or a combination of these factors.

Pheochromocytomas—tumors of the adrenal medulla or of catecholamine-secreting tissue elsewhere in the body—also produce hypertension. The hypertension is often episodic, and particularly when the tumors secrete predominantly epinephrine, the blood glucose level and the metabolic rate are intermittently elevated. However, in some patients with pheochromocytomas, sustained hypertension is the only finding.

Renal hypertension due to narrowing of the renal arteries is discussed above and in Chapter 24. Various other types of renal disease are associated with hypertension, but it is not known whether the association is due to renin secretion or to nonhumoral mechanisms. **Coarctation of the aorta,** a congenital narrowing of a segment of the thoracic aorta, increases the resistance to flow, producing severe hypertension in the upper part of the body. The blood pressure in the lower part of the body is usually normal but may be elevated as a result of increased renin secretion.

Increased cardiac output is sometimes a cause of diastolic as well as systolic hypertension. The elevated blood pressure seen in thyrotoxicosis and in anxious, tense patients is explained on this basis. In severe polycythemia, the increase in blood viscosity may raise the peripheral resistance enough to cause significant hypertension.

Chronic treatment with oral contraceptives containing progestins and estrogens produces significant hypertension in some women. The hypertension is due, at least in part, to an increase in circulating levels of angiotensinogen, the production of which is stimulated by estrogens (see Chapter 24). There is some evidence that women developing ''pill hypertension'' are those predisposed to hypertension in any case. Certainly the occurrence of hypertension is not a reason for normotensive women to avoid oral contraceptives, but it would seem wise for them to have their blood pressures checked at 6-month intervals.

Table 33–5. Simplified summary of pathogenesis of major findings in congestive heart failure.

Abnormality	Cause
Ankle, sacral edema	"Backward failure" of right ventricle → increased venous pressure → fluid transudation.
Hepatomegaly	Increased venous pressure → increased resistance to portal flow.
Pulmonary congestion	"Backward failure" of left ventricle → increased pulmonary venous pressure → pulmonary venous distention and transudation of fluid into air spaces.
Dyspnea on exertion	Failure of left ventricular output to rise during exercise → increased pulmonary venous pressure.
Paroxysmal dyspnea, pulmonary edema	Probably sudden failure of left heart output to keep up with right heart output → acute rise in pulmonary venous and capillary pressure → transudation of fluid into air spaces.
Orthopnea	Normal pooling of blood in lungs in supine position added to already congested pulmonary vascular system; increased venous return not put out by left ventricle. (Relieved by sitting up, raising head of bed, lying on extra pillows.)
Weakness, exercise intolerance	"Forward failure"; cardiac output inadequate to perfuse muscles; especially, failure of output to rise with exercise.
Cardiac dilation	Greater ventricular end-diastolic volume.

HEART FAILURE

Manifestations

The manifestations of heart failure range from sudden death (eg, in ventricular fibrillation or air embolism), through the shocklike state called cardiogenic shock, to chronic **congestive heart failure,** depending upon the degree of circulatory inadequacy and the rapidity with which it develops. The principal symptoms and signs of congestive failure are cardiac enlargement, weakness, edema (particulary of the dependent portions of the body), a prolonged circulation time, hepatic enlargement **(hepatomegaly),** a sensation of shortness of breath and suffocation **(dyspnea),** and distention of the neck veins. Dyspnea on exertion is a prominent symptom. In advanced cases, a common finding is dyspnea that is precipitated by lying flat and relieved by sitting up **(orthopnea).** The dyspnea may be paroxysmal and sometimes progresses to frank **pulmonary edema.** Some of these abnormalities are due to ''backward failure''—failure of the left and right heart pumps to put out all the blood returned to them in the veins, with a resultant rise in venous pressure and congestion of the lungs and viscera. Others are due to ''forward failure''—failure of the heart to maintain a cardiac output adequate for normal perfusion of all the tissues (Table 33–5). The decline in cardiac output may be relative rather than absolute. In thyrotoxicosis and thiamine deficiency, for example, output may be elevated in absolute terms, but when it is inadequate relative to the needs of the tissues, heart failure is present (''high-output failure'').

Pathogenesis

The initial pathologic event in heart failure is reduced myocardial contractility. At first, this is manifest only during exercise, and exertion is limited by dyspnea and fatigue. Later, resting cardiac output is also reduced. The ejection fraction falls from its normal value of about 0.6 to as low as 0.2.

The cause of the decreased contractility is unsettled and may vary from one disease to another. The most common diseases producing congestive heart failure are hypertension and coronary artery disease. The failing myocardium has decreased myosin ATPase activity, and decreased Ca^{2+} transport by the sarcoplasmic reticulum has been reported, but it is unknown which if either of these abnormalities is primary. There is in addition a reduced density of β-adrenergic receptors in the myocardium, but this is probably due to down-regulation of the receptors produced by elevated circulating catecholamine levels (see below).

The decrease in cardiac output triggers increased sympathetic discharge via the baroreceptors (Fig 33–6). Venous pressure also starts to rise, and the viscera become congested. Circulating catecholamine levels are increased, and there is widespread vasoconstriction. This in turn increases afterload, causing a further decrease in cardiac output.

Constriction of the vascular bed in the kidneys causes retention of Na^+ and water. In addition, the decrease in effective arterial blood volume and the increase in sympathetic discharge increase renin secretion in many patients, and the resultant increase in aldosterone secretion adds to the salt and water retention. Escape fails to occur (see Chapter 20), and edema is produced. It is interesting in this regard that the plasma ANP level is generally increased rather than decreased in the patients who have congestive heart failure, but the failure to escape may be due to end-organ unresponsiveness to the atrial peptide.

Treatment of congestive heart failure is aimed at improving cardiac contractility, treating the symptoms, and decreasing the load on the heart. Classically, myocardial contractility has been improved by administration of digitalis and related positively inotropic cardiac glycosides (see Chapter 3). These can now be supplemented with drugs such as milrinone, which increases myocardial contractility by inhibiting phosphodiesterase. This increases intracellular cyclic AMP and thus mimics the action of catecholamines. Diuretics are used to overcome sodium retention. Afterload is reduced by use of arterial vasodilators, and venodilators diminish preload. Reduction of the plasma angiotensin II level by administration of ACE inhibitors is useful, and the decline in the angiotensin II level also decreases the plasma aldosterone level.

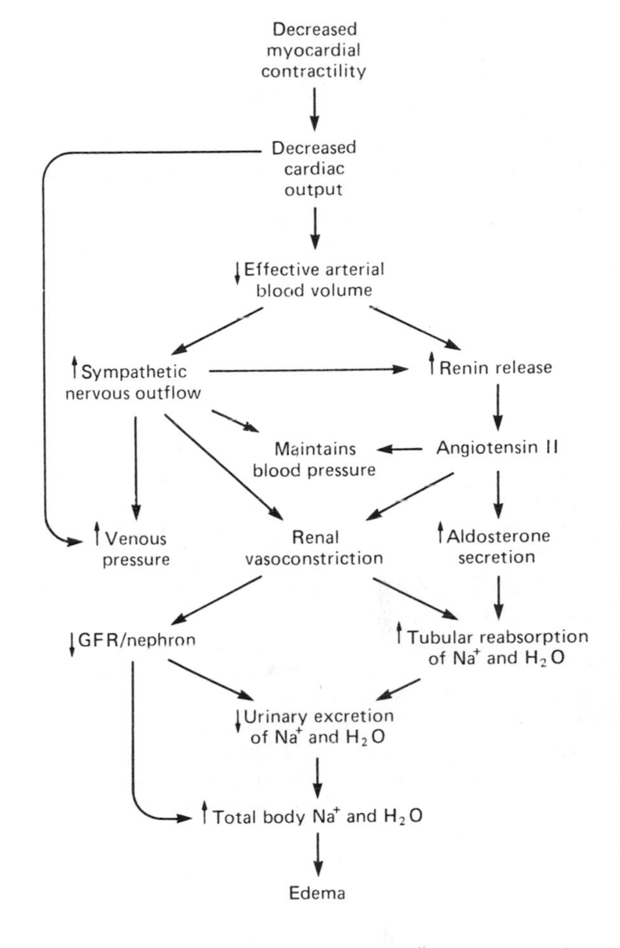

Figure 33–6. Sequence of events leading to congestive heart failure.

REFERENCES
Section VI: Circulation

Benacerraf B: Role of MHC gene products in immune regulation. *Science* 1981;**212:**1229.

Beutler B, Cerami A: Cachectin (tumor necrosis factor): a macrophage hormone governing cellular metabolism and inflammatory response. *Endo Rev* 1988;**9:**57.

Birrer RB (editor): *Sports Medicine for the Primary Care Physician.* Appleton-Century-Crofts, 1984.

Blomqvist CG. Stone HL: Cardiovascular adjustments to gravitational stress. Page 1025 in: *Handbook of Physiology: The Cardiovascular System.* Vol 3. Shepherd JT, Abboud FM (editors). American Physiological Society, 1983.

Brace RA: Progress toward resolving the controversy of positive versus negative interstitial fluid pressure. *Circ Res* 1981;**49:**281.

Braunwald E: Mechanism of action of calcium-channel-blocking agents. *N Engl J Med* 1982;**307:**1618.

Braunwald E (editor): *Heart Disease: A Textbook of Cardiovascular Medicine.* Saunders, 1980.

Brown HF: Electrophysiology of the sinoatrial node. *Physiol Rev* 1982;**62:**505.

Clark SC, Kamen R: The human colony-stimulating factors. *Science* 1987;**236:**1229.

Clouse LH, Comp PC: The regulation of hemostasis: The protein C system. *N Engl J Med* 1986;**314:**129.

Colucci WS, Wright RF, Braunwald E: New positive inotropic agents in the treatment of congestive heart failure. *N Engl J Med* 1986;**314;**290.

Connor WE, Bristow JD (editors): *Coronary Heart Disease: Prevention, Complications, and Treatment.* Lippincott, 1985.

Cooper ASL, Plum F: Biochemistry and Physiology of Brain Ammonia. *Physiol Rev* 1987;**67:**440.

Cooper MD: B lymphocytes: normal development and function. *New Eng J Med* 1987;**317:**1452.

Dinarello CA, Mier JW: Interleukins. *Annu Rev Med* 1986:**37:**173.

Dinarello CA, Mier JW: Lymphokines. *New Eng J Med* 1987;**317:**940.

Donald DH, Shepherd JT: Autonomic regulation of the peripheral circulation. *Annu Rev Physiol* 1980;**42:**429.

Doyle JT: Mechanisms and prevention of sudden death. *Mod Concepts Cardiovasc Dis* 1976;**45:**111.

Elgsaeter A et al: The molecular basis of erythrocyte shape. *Science* 1986;**234:**1217.

Esmon CT: The regulation of natural anticoagulant pathways. *Science* 1987;**235:**1348.

Farah AE: Glucagon and the circulation. *Pharmacol Rev* 1983;**35:**181.

Fauci AS: The human immuno deficiency virus: infectivity and mechanisms of pathogenesis. *Science* 1988;**299:**617.

Feigenbaum, H: *Echocardiography,* 4th ed. Lea & Febiger, 1986.

Fenstermacher JD, Rapoport SF: Blood-brain barrier. Page 969 in: *Handbook of Physiology.* Vol 4, Part 2. Renkin EM, Michel LC (editors). American Physiological Society, 1984.

Fishman RA: *Cerebrospinal Fluids in Diseases of the Nervous System.* Saunders, 1982.

Folkman J, Klagsbrun M: Angiogenic factors. *Science* 1987;**235:**442.

Fowler MB, Schroeder JS: Current status of cardiac transplantation. *Mod Concepts Cardiovasc Dis* 1986;**55:37.**

Fozzard HA et al (editors): *The Heart and Cardiovascular System: Scientific Foundations.* Raven, 1986.

Freda VJ, Pollack W, Gorman JG: Rh disease: How near the end? *Hosp Pract* (June) 1978;**13:**61.

Fridovich I, Hagen P-O, Murray JJ: Endothelium-derived relaxing factor: in search of the endogenous nitroglycerin. *News Physiol Sci* 1987;**2:**61.

Galbo H: *Hormonal and Metabolic Adaptation to Exercise.* Thieme-Stratton, 1983.

Gibbs CL: Cardiac energies. *Physiol Rev* 1978;**58:**174.

Goldman MJ: *Principles of Clinical Electrocardiography,* 12th ed. Lange, 1986.

Harpel PC: Protease inhibitors: A precarious balance. *N Eng J Med* 1983;**309:**725.

Harrison MH: Effects of thermal stress and exercise on blood volume in humans, *Physiol Rev* 1985;**65:**149.

Honig GR, Adams JG III: *Human Hemoglobin Genetics.* Springer-Verlag, 1986.

Hudlicka O, Tyler KR: *Angiogenesis.* Academic, 1987.

Kirkendall WM et al: Recommendations for human blood pressure determination by sphygmomanometers. *Hypertension* 1981;**3:**510A.

Malech HL, Gallin JI: Neutrophils in human diseases. *N Eng J Med* 1987;**317:**687.

Marcus ML: *The Coronary Circulation in Health and Disease.* McGraw-Hill, 1983.

Metcalf J et al: Gas exchange in the pregnant uterus. *Physiol Rev* 1967;**47:**782.

Milstein C: From antibody structure to immunological diversification of the immune response. *Science* 1986; **231:**1261.

Morady F, Scheinman MM: Paroxysmal supraventricular tachycardia. *Mod Concepts Cardiovasc Dis* 1982;**51:**107.

Movat HZ: *The Inflammatory Reaction.* Elsevier, 1985.

Nossal GJV: The basic components of the immune system. *N Eng J Med* 1987;**16:**1320.

Old LJ: Tumor necrosis factor *Sci Am* (May) 1988;**258:**59.

Payan DG, McGillis JP, Goelzl EJ: Neuroimmunology. *Adv Immunol* 1986;**39:**299.

Phelps ME, Mazziota JC: Positron emission tomography: Human brain function and biochemistry. *Science* 1985; **228:**799.

Putnam FW (editor): *The Plasma Proteins: Structure, Function and Genetic Control,* 2nd ed. Academic Press, 1975.

Reis DJ, Doba N: The central nervous system and neurogenic hypertension. *Cardiovasc Dis* 1974;**17:**51.

Ritman EL et al: Three-dimensional imaging of heart, lungs, and circulation. *Science* 1980;**210:**273.

Roland PE: Changes in brain blood flow and oxidative metabolism during mental activity. *News Physiol Sci* 1987;**2:**120.

Rosen FS, Cooper MD, Wedgewood RJP: The primary immunodeficiencies. *N Engl J Med* 1984;**311:**235.

Royer HD, Reinherz FC: T lymphocytes: ontogegeny, function, and relevance to clinical disorders. *N Engl J Med* 1987;**317:**1136.

Sell S: *Basic Immunology: Immune Mechanisms in Health and Disease.* Elsevier, 1987.

Shepherd JT, Vanhoutte PM: *Veins and Their Control.* Saunders, 1977.

Shoenfeld Y, Schwartz RS: Immunologic and genetic factors in autoimmune disease. *N Engl J Med* 1984; **311:**1019.

Simionescu N: Cellular aspects of transcapillary exchange. *Physiol Rev* 1983;**63:**1536.

Snyderman R, Goetzl EJ: Molecular and cellular mechanisms of leukocyte chemotaxis. *Science* 1981;**213:**830.

Sporn MB, Roberts AB: Peptide growth factors are multifunctional. *Nature* 1988;**332:**217.

Stamatoyannopoulos G et al (editors): *The Molecular Basis of Blood Diseases.* Saunders, 1987.

Staub NC, Taylor AE (editors): *Edema.* Raven, 1984.

Surgenor DM (editor): *The Red Blood Cell.* 2 vols. Academic Press, 1975.

Tonegawa S: Somatic generation of antibody diversity. *Nature* 1983;**302:**575.

Vane J, Botting R: Inflammation and the mechanism of action of anti-inflammatory drugs. *FASEB J* 1987;**1:**89.

Vassalle M: Cardiac automaticity and its control. *Am J Physiol* 1977;**233:**H625.

Zucker MB: The functioning of blood platelets. *Sci Am* (June) 1980;**242:**86.

Symposium: Autonomic control of coronary tone: Facts, interpretations, and consequences. *Fed Proc* 1984; **43:** 2855.

Symposium: Contribution of splanchnic circulation to overall cardiovascular and metabolic homeostatis. *Fed Proc* 1983;**42:**1656.

Section VII.
Respiration

Pulmonary Function

34

INTRODUCTION

Respiration, as the term is generally used, includes 2 processes: **external respiration,** the absorption of O_2 and removal of CO_2 from the body as a whole; and **internal respiration,** the utilization of O_2 and production of CO_2 by cells and the gaseous exchanges between the cells and their fluid medium. Details of the utilization of O_2 and the production of CO_2 by cells are considered in Chapter 17. This chapter is concerned with the functions of the respiratory system in external respiration, ie, the processes responsible for the uptake of O_2 and excretion of CO_2 in the lungs. Chapter 35 is concerned with the transport of O_2 and CO_2 to and from the tissues.

The respiratory system is made up of a gas-exchanging organ (the lungs) and a pump that ventilates the lungs. The pump consists of the chest wall; the respiratory muscles, which increase and decrease the size of the thoracic cavity; the centers in the brain that control the muscles; and the tracts and nerves that connect the brain to the muscles. At rest, a normal human breathes 12–15 times a minute. Five hundred milliliters of air per breath, or 6–8 L/min, is inspired and expired. This air mixes with the gas in the alveoli, and, by simple diffusion, O_2 enters the blood in the pulmonary capillaries while CO_2 enters the alveoli. In this manner, 250 mL of O_2 enters the body per minute and 200 mL of CO_2 is excreted.

Traces of other gases such as methane from the intestines are also found in expired air. Alcohol and acetone are expired when present in appreciable quantities in the body. Indeed, over 250 different volatile substances have been identified in human breath.

PROPERTIES OF GASES

Partial Pressures

Unlike liquids, gases expand to fill the volume available to them, and the volume occupied by a given number of gas molecules at a given temperature and pressure is (ideally) the same regardless of the composition of the gas.

$$P = \frac{nRT}{V}$$ **(from equation of state of ideal gas)**

where

P = Pressure
n = Number of moles
R = Gas constant
T = Absolute temperature
V = Volume

Therefore, the pressure exerted by any one gas in a mixture of gases (its **partial pressure**) is equal to the total pressure times the fraction of the total amount of gas it represents.

The composition of dry air is 20.98% O_2, 0.04% CO_2, 78.06% N_2, and 0.92% other inert constituents such as argon and helium. The barometric pressure (P_B) at sea level is 760 mm Hg (one atmosphere). The partial pressure (indicated by the symbol P) of O_2 in dry air is therefore 0.21 × 760, or 160 mm Hg at sea level. The partial pressure of N_2 and the other inert gases is 0.79 × 760, or 600 mm Hg; and the P_{CO_2} is 0.0004 × 760, or 0.3 mm Hg. The water vapor in the air in most climates reduces these percentages, and therefore the partial pressures, to a slight degree. Air equilibrated with water is saturated with water vapor, and inspired air is saturated by the time it reaches the lungs. The P_{H_2O} at body temperature (37 °C) is 47 mm Hg. Therefore, the partial pressures at sea level of the other gases in the air reaching the lungs are P_{O_2}, 149 mm Hg; P_{CO_2}, 0.3 mm Hg; and P_{N_2} (including the other inert gases), 564 mm Hg.

Gas diffuses from areas of high pressure to areas of low pressure, the rate of diffusion depending upon the concentration gradient and the nature of the barrier between the 2 areas. When a mixture of gases is in contact with and permitted to equilibrate with a liquid, each gas in the mixture dissolves in the liquid to an extent determined by its partial pressure and its solubility in the fluid. The partial pressure of a gas in a liquid is that pressure which in the gaseous phase in equilibrium with the liquid would produce the concentration of gas molecules found in the liquid.

Methods of Quantitating Respiratory Phenomena

Respiratory excursions can be recorded by using devices that measure chest expansion or recording spirometers (Fig 17–1), which also permit measurement of gas intake and output. Since gas volumes vary with temperature and pressure and since the amount of water vapor in them varies, it is important to correct respiratory measurements involving volume to a stated set of standard conditions. The 3 most commonly used standards and their abbreviations are shown in Table 34–1. Modern techniques for gas analysis make possible rapid, reliable measurements of the composition of gas mixtures and the gas content of body fluids. N_2, for example, emits light in an electrical field in vacuo, and by using a meter that continuously measures and records the light emitted, it is possible to obtain a graphic record of the fluctuations in P_{N_2} of the expired air while various gas mixtures are breathed. O_2 and CO_2 electrodes, small probes sensitive to O_2 or CO_2, can be inserted into the airway or into blood vessels or tissues and the P_{O_2} and P_{CO_2} recorded continuously. CO_2, carbon monoxide (CO), and many anesthetic gases can be measured rapidly by infrared absorption spectroscopy. Gases can also be measured by gas chromatography or mass spectrometry.

Table 34–1. Standard conditions to which measurements involving gas volumes are corrected.

STPD	0 °C, 760 mm Hg, dry (standard temperature and pressure, dry)
BTPS	Body temperature and pressure, saturated with water vapor
ATPS	Ambient temperature and pressure, saturated with water vapor

MECHANICS OF RESPIRATION

Inspiration & Expiration

The lungs and the chest wall are elastic structures. Normally, there is no more than a thin layer of fluid between the lungs and the chest wall. The lungs slide easily on the chest wall but resist being pulled away from it in the same way that 2 moist pieces of glass slide on each other but resist separation. The pressure in the "space" between the lungs and chest wall (intrapleural pressure) is subatmospheric (Fig 34–1). The lungs are stretched when they are expanded at birth, and at the end of quiet expiration their tendency to recoil from the chest wall is just balanced by the tendency of the chest wall to recoil in the opposite direction. If the chest wall is opened, the lungs collapse; and if the lungs lose their elasticity, the chest expands and becomes barrel-shaped.

Inspiration is an active process. The contraction of the inspiratory muscles increases intrathoracic volume. During quiet breathing, the intrapleural pres-

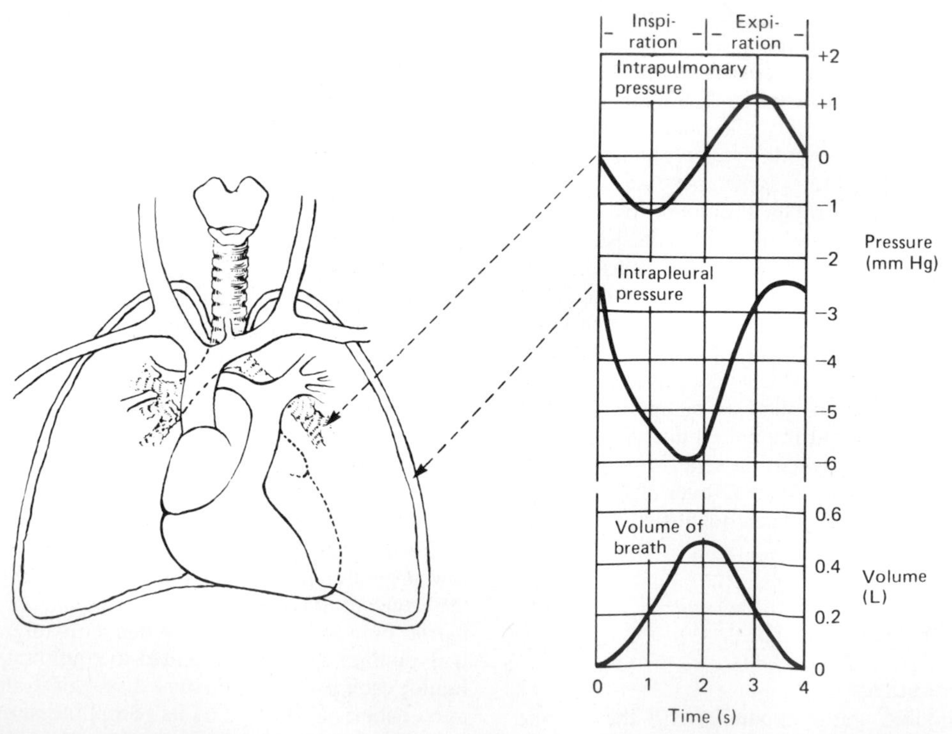

Figure 34–1. Changes in intrapleural (intrathoracic) and intrapulmonary pressure relative to atmospheric pressure during inspiration and expiration.

sure, which is about −2.5 mm Hg (relative to atmospheric) at the start of inspiration, decreases to about −6 mm Hg, and the lungs are pulled into a more expanded position. The pressure in the airway becomes slightly negative, and air flows into the lungs (Fig 34–1). At the end of inspiration, the lung recoil pulls the chest back to the expiratory position, where the recoil pressures of the lungs and chest wall balance. The pressure in the airway becomes slightly positive, and air flows out of the lungs. Expiration during quiet breathing is passive in the sense that no muscles which decrease intrathoracic volume contract. However, there is some contraction of the inspiratory muscles in the early part of expiration. This contraction exerts a braking action on the recoil forces and slows expiration.

Strong inspiratory efforts reduce intrapleural pressure to values as low as −30 mm Hg, producing correspondingly greater degrees of lung inflation. When ventilation is increased, the extent of lung deflation is also increased by active contraction of expiratory muscles that decrease intrathoracic volume. The effects of gravity on intrapleural pressure are discussed below.

Air Passages

After passing through the nasal passages and pharynx, where it is warmed and takes up water vapor, the inspired air passes down the trachea and through the bronchioles, respiratory bronchioles, and alveolar ducts to the alveoli (Fig 34–2).

Between the trachea and the alveolar sacs, the airways divide 23 times. The first 16 generations of passages form the conducting zone of the airways that transports gas from and to the exterior. They are made up of bronchi, bronchioles, and terminal bronchioles. The remaining 7 generations form the transitional and respiratory zones where gas exchange occurs and are made up of respiratory bronchioles, alveolar ducts, and alveoli. These multiple divisions greatly increase the total cross-sectional area of the airways. Consequently, the velocity of air flow in the small airways declines to very low values. It has been calculated that the aggregate circumference of the 16th generation of air passages (terminal bronchioles) is 2000 times the circumference of the trachea.

The alveoli are surrounded by pulmonary capillaries, and in most areas the structures between the air and the capillary blood across which O_2 and CO_2

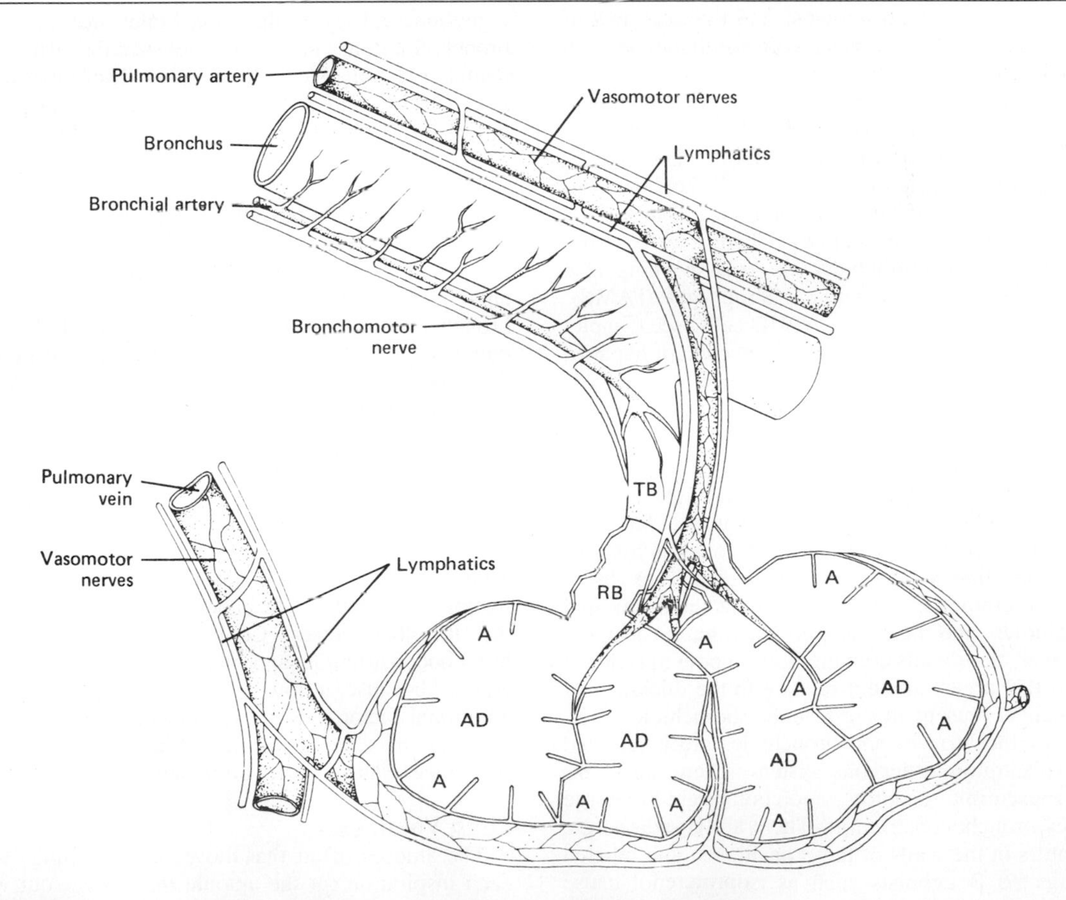

Figure 34–2. Structure of the lung. A, alveolus; AD, alveolar duct; RB, respiratory bronchiole; TB, terminal bronchiole. (Reproduced, with permission, from Staub NC: The pathophysiology of pulmonary edema. *Hum Pathol* 1970;1:419.)

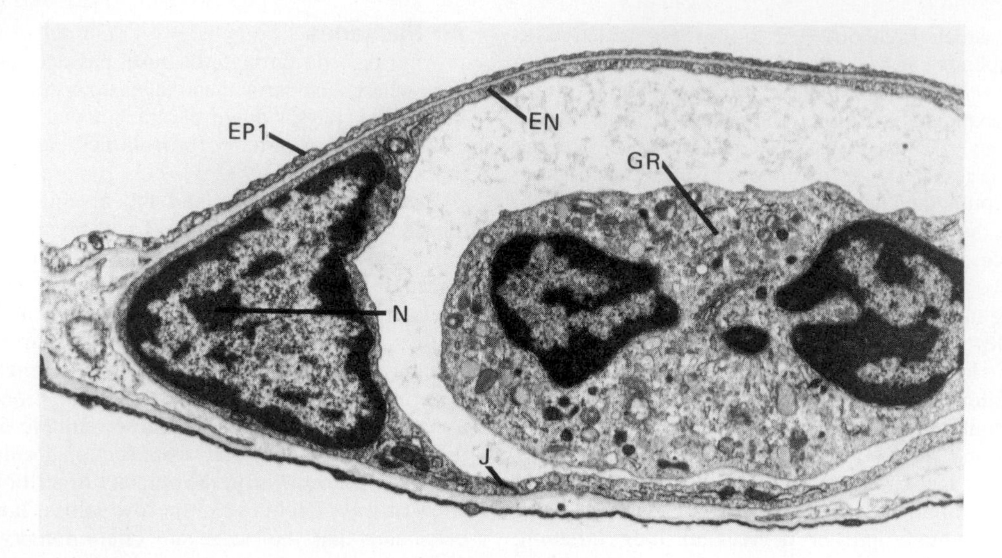

Figure 34–3. Electron photomicrograph of human lung showing a pulmonary capillary adjacent to an alveolar sac. The capillary contains a neutrophil (GR). EP1, type I pulmonary epithelial cell; EN, endothelial cell, with its nucleus (N); J, junction between 2 endothelial cells. (Reproduced, with permission, from Weibel ER: Lung cell biology. Pages 47–91 in Vol I of: *Handbook of Physiology,* Section 3: *The Respiratory System.* Fishman AP [editor]. American Physiological Society, 1985.)

diffuse are exceedingly thin (Fig 34–3). There are 300 million alveoli in humans, and the total area of the alveolar walls in contact with capillaries in both lungs is about 70 m^2.

The alveoli are lined by 2 types of epithelial cells. **Type I cells** are flat cells with large cytoplasmic extensions and are the primary lining cells. **Type II cells (granular pneumocytes)** are thicker and contain numerous lamellar inclusion bodies. These cells secrete surfactant (see below). There may be other special types of epithelial cells, and the lung also contains pulmonary alveolar macrophages (PAMS), lymphocytes, plasma cells, APUD cells (see Chapter 26), and mast cells. The mast cells contain heparin, various lipids, histamine, and polypeptides that participate in allergic reactions.

Control of Bronchial Tone

The trachea and bronchi have cartilage in their walls but relatively little smooth muscle. They are lined by a ciliated epithelium that contains mucous and serous glands. The cilia and the glands are absent from the epithelium of the bronchioles and terminal bronchioles, and their walls do not contain cartilage. However, their walls contain more smooth muscle, of which the largest amount relative to the thickness of the wall is present in the terminal bronchioles. The walls of the bronchi and bronchioles are innervated by the autonomic nervous system. There are abundant muscarinic receptors, and cholinergic discharge causes bronchoconstriction. There are β_2-adrenergic receptors in the walls of the bronchioles, and inhaled or injected β agonists such as isoproterenol cause bronchodilation. However, it does not appear that the β_2-adrenergic receptors are innervated.

There is in addition a noncholinergic, nonadrenergic innervation of the bronchioles that produces bronchodilation, and there is evidence that VIP is the mediator responsible for the dilation. The leukotrienes LTC_4, LTD_4, and LTE_4 (see Chapter 17) are potent bronchoconstrictors, particularly when administered by inhalation.

The function of the bronchial muscles is still a matter of debate, but in general, they probably help to maintain an even distribution of ventilation. Stimulation of sensory receptors in the airways by irritants and chemicals such as sulfur dioxide produces reflex bronchoconstriction that is mediated via cholinergic pathways. In addition, the bronchial muscles protect the bronchi during coughing. There is a circadian rhythm in bronchial tone, with maximal constriction at about 6:00 AM and maximal dilation at about 6:00 PM. This is why **asthma** attacks are more severe in the late night and early morning hours. β_2 stimulants such as isoproterenol are of benefit in asthma, and the muscarinic antagonist atropine produces relief in asthmatics with bronchospasm due to cholinergic hyperactivity but not in patients with bronchospasm due to other causes. Cooling the airways causes bronchoconstriction, and exercise triggers asthmatic attacks because it lowers airway temperature. An additional factor in exercise-induced asthma may be lysis of mast cells in the lung with release of substances that constrict bronchial smooth muscle.

Lung Volumes

The amount of air that moves into the lungs with each inspiration (or the amount that moves out with each expiration) is called the **tidal volume.** The air inspired with a maximal inspiratory effort in excess of the tidal volume is the **inspiratory reserve volume.** The volume expelled by an active expiratory effort after passive expiration is the **expiratory reserve volume,** and the air left in the lungs after a maximal expiratory effort is the **residual volume.** Normal values for these lung volumes, and names applied to combinations of them, are shown in Fig 34–4. The space in the conducting zone of the airways occupied by gas that does not exchange with blood in the pulmonary vessels is the **respiratory dead space.** The **vital capacity,** the largest amount of air that can be expired after a maximal inspiratory effort, is frequently measured clinically as an index of pulmonary function. It gives useful information about the strength of the respiratory muscles and other aspects of pulmonary function. The fraction of the vital capacity expired in 1 second **(timed vital capacity;** also called forced expired volume in 1 second, or FEV 1″) gives additional information; the vital capacity may be normal but the timed vital capacity greatly reduced in diseases such as asthma, in which the resistance of the airways is increased owing to bronchial constriction. The amount of air inspired per minute **(pulmonary ventilation, respiratory minute volume)** is normally about 6 L (500 mL/breath × 12 breaths/min). The **maximal voluntary ventilation (MVV),** or, as it was formerly called, the **maximal breathing capacity,** is the largest volume of gas that can be moved into and out of the lungs in 1 minute by voluntary effort. The normal MVV is 125–170 L/min.

Respiratory Muscles

Movement of the **diaphragm** accounts for 75% of the change in intrathoracic volume during quiet inspiration. Attached around the bottom of the thoracic cage, this muscle arches over the liver and moves downward like a piston when it contracts. The distance it moves ranges from 1.5 cm to as much as 7 cm with deep inspiration (Fig 34–5). The other important **inspiratory muscles** are the **external intercostal muscles,** which run obliquely downward and forward from rib to rib. The ribs pivot as if hinged at the back, so that when the external intercostals contract they elevate the lower ribs. This pushes the sternum outward and increases the anteroposterior diameter of the chest. The transverse diameter is actually changed little if at all. Either the diaphragm or the external intercostal muscles alone can maintain adequate ventilation at rest. Transection of the spinal cord above the third cervical segment is fatal without artificial respiration, but transection below the fifth cervical segment is not, because it leaves the phrenic nerves that innervate the diaphragm intact; the phrenic nerves arise from cervical segments 3–5. Conversely, in patients with bilateral phrenic nerve palsy but intact innervation of their intercostal muscles, respiration is somewhat labored but adequate to maintain

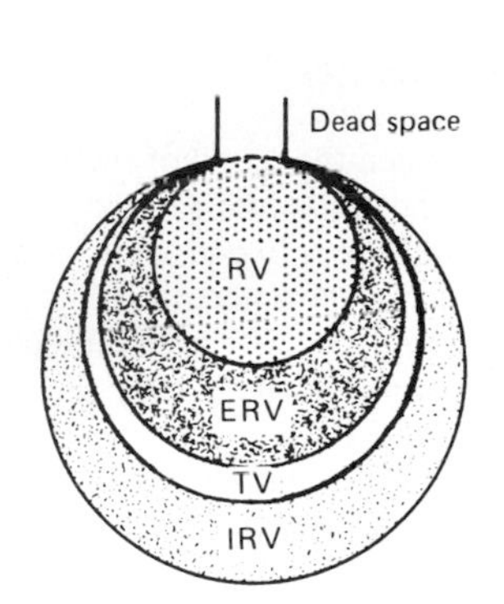

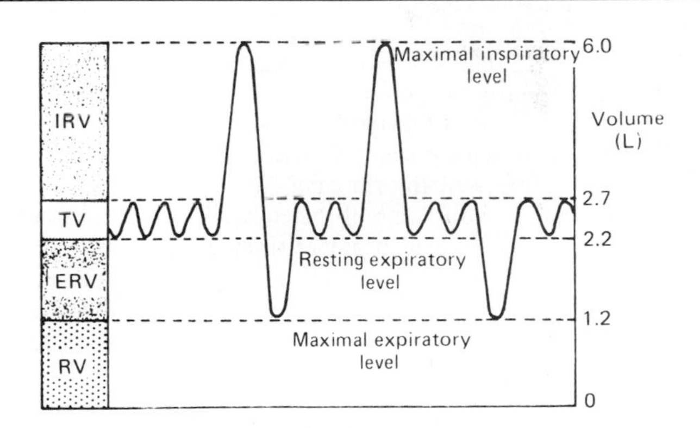

IRV = Inspiratory reserve volume
ERV = Expiratory reserve volume
TV = Tidal volume
RV = Residual volume

		Volume (L)		
		Men	Women	
Vital capacity	IRV	3.3	1.9	Inspiratory capacity
	TV	0.5	0.5	
	ERV	1.0	0.7	Functional residual capacity
	RV	1.2	1.1	
Total lung capacity		6.0	4.2	

Respiratory minute volume (rest): 6 L/min
Alveolar ventilation (rest): 4.2 L/min
Maximal voluntary ventilation (BTPS): 125–170 L/min
Timed vital capacity: 83% of total in 1 s; 97% in 3 s
Work of quiet breathing: 0.5 kg-m/min
Maximal work of breathing: 10 kg-m/breath

Figure 34–4. Lung volumes and some measurements related to the mechanics of breathing. The diagram at the upper right represents the excursions of a spirometer plotted against time. (Modified from Comroe JH Jr et al: *The Lung: Clinical Physiology and Pulmonary Function Tests,* 2nd ed. Year Book, 1962.)

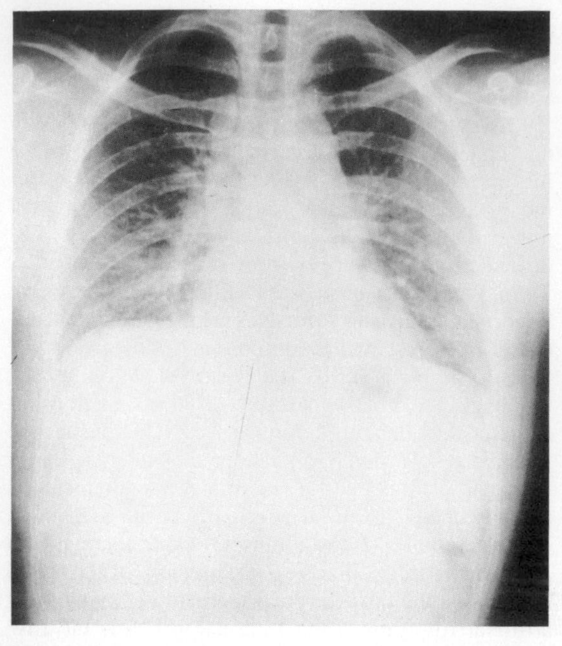

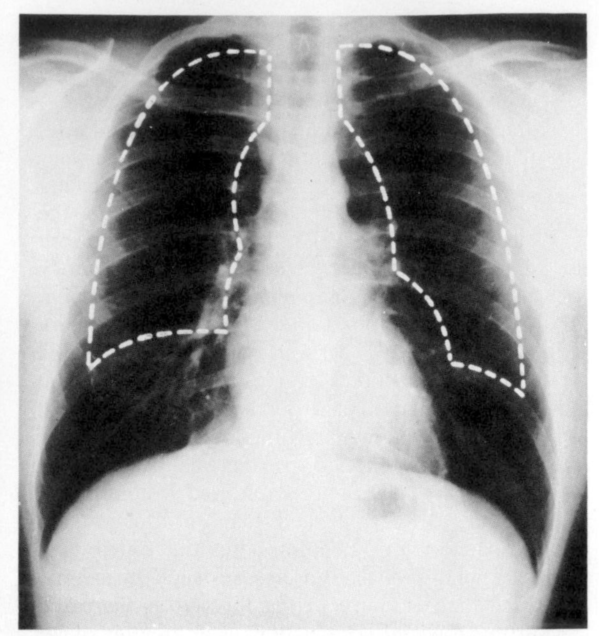

Figure 34–5. X-ray of chest in full expiration (*left*) and full inspiration (*right*). Dashed white line on right is outline of lungs in full expiration. (From Comroe JH Jr: *Physiology of Respiration,* 2nd ed. Copyright © 1974 by Year Book Medical Publishers, Inc., Chicago. Used by permission.)

life. The scalene and sternocleidomastoid muscles in the neck are accessory inspiratory muscles that help to elevate the thoracic cage during deep labored respiration.

A decrease in intrathoracic volume and forced expiration result when the **expiratory muscles** contract. The internal intercostals have this action because they pass obliquely downward and posteriorly from rib to rib and therefore pull the rib cage downward when they contract. Contractions of the muscles of the anterior abdominal wall also aid expiration by pulling the rib cage downward and inward and by increasing the intra-abdominal pressure, which pushes the diaphragm upward.

Glottis

The abductor muscles in the larynx contract early in inspiration, pulling the vocal cords apart and opening the glottis. During swallowing or gagging, there is reflex contraction of the adductor muscles that closes the glottis and prevents aspiration of food, fluid, or vomitus into the lungs. In unconscious or anesthetized patients, glottic closure may be incomplete and vomitus may enter the trachea, causing an inflammatory reaction in the lung **(aspiration pneumonia).**

The laryngeal muscles are supplied by the vagus nerves. When the abductors are paralyzed, there is inspiratory stridor. When the adductors are paralyzed, food and fluid enter the trachea, causing aspiration pneumonia and edema. Bilateral cervical vagotomy in animals causes the slow development of fatal pulmonary congestion and edema. The edema is due at least in part to aspiration, although some edema develops even if a tracheostomy is performed before the vagotomy.

Compliance of the Lungs & Chest Wall

The interaction between the recoil of the lungs and recoil of the chest can be demonstrated in living subjects. The nostrils are clipped shut, and the subject breathes through a spirometer that has a valve just beyond the mouthpiece. The mouthpiece contains a pressure-measuring device. After the subject inhales a given amount, the valve is shut, closing off the airway. The respiratory muscles are then relaxed while the pressure in the airway is recorded. The procedure is repeated after inhaling or actively exhaling various volumes. The curve of airway pressure obtained in this way, plotted against volume, is the **relaxation pressure curve** of the total respiratory system (Fig 34–6). The pressure is zero at a lung volume that corresponds to the volume of gas in the lungs at the end of quiet expiration **(relaxation volume,** which equals the functional residual capacity). It is positive at greater volumes and negative at smaller volumes. The change in lung volume per unit change in airway pressure ($\Delta V/\Delta P$) is the stretchability **(compliance)** of the lungs and chest wall. It is normally measured in the pressure range where the relaxation pressure curve is steepest, and the normal value is approximately 0.2 L/cm H_2O. However, compliance depends on lung volume; an individual with only one lung has approximately half the ΔV for a given ΔP. Compliance is also slightly greater when

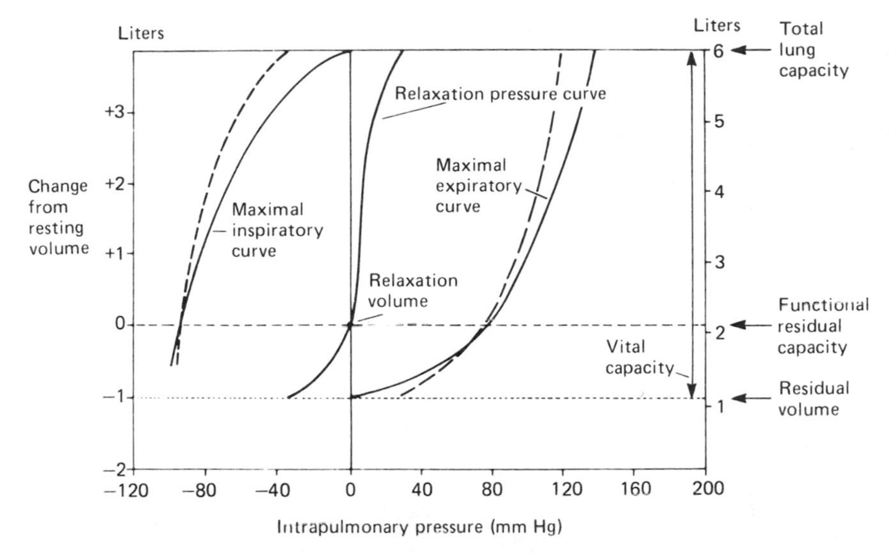

Figure 34–6. Relation between intrapulmonary pressure and volume. The middle curve is the static pressure curve of values obtained when the lungs are inflated or deflated by various amounts and the intrapulmonary pressure (elastic recoil pressure) is measured with the airway closed. The relaxation volume is the point where the recoil of the chest and the recoil of the lungs balance. The slope of the curve is the compliance of the lungs and chest wall. The maximal inspiratory and expiratory curves are the airway pressures that can be developed during maximal inspiratory and expiratory efforts. The dashed lines, the horizontal distances between the inspiratory and expiratory curves and the relaxation pressure curve, represent the net pressure developed by the inspiratory and expiratory muscles.

measured during deflation than when measured during inflation. Consequently, it is more informative to examine the whole pressure-volume curve. The curve is shifted downward and to the right (compliance is decreased) by pulmonary congestion and interstitial pulmonary fibrosis (Fig 34–7). It is shifted upward and to the left (compliance is increased) in emphysema (see Chapter 37). It should be noted that compliance is a static measure of lung and chest recoil. The **resistance** of the lung and chest is the pressure difference required for a unit of air flow, and this measurement, which is dynamic rather than static, also takes into account the resistance to air flow in the airways.

Alveolar Surface Tension

An important factor affecting the compliance of the lungs is the surface tension of the film of fluid that lines the alveoli. The magnitude of this component at various lung volumes can be measured by removing the lungs from the body and distending them alternately with saline and with air while measuring the intrapulmonary pressure. Because saline reduces the surface tension to nearly zero, the pressure-volume curve obtained with saline measures only the tissue elasticity (Fig 34–8), while the curve obtained with air measures both tissue elasticity and surface tension. The difference between the 2 curves, the elasticity due to surface tension, is much smaller at small than at large lung volumes. The surface tension is also much lower than the expected surface tension at a water-air interface of the same dimensions.

The low surface tension when the alveoli are small is due to the presence in the fluid lining the alveoli of **surfactant,** a lipid surface-tension-lowering agent. Surfactant is a mixture of dipalmitoyl-phosphatidylcholine (see Chapter 17), other lipids, and proteins (Table 34–2). If the surface tension is not kept low when the alveoli become smaller during expiration, they collapse in accordance with the law of

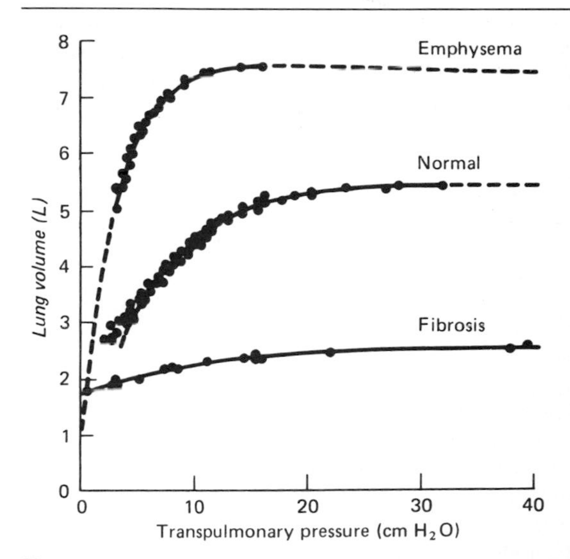

Figure 34–7. Static expiratory pressure-volume curves of lungs in normal subject and subjects with severe emphysema and pulmonary fibrosis. (Modified and reproduced, with permission, from Pride NB, Mackem PT: Lung mechanics in disease. Pages 659–692 of Vol III, Part 2, of: *Handbook of Physiology.* Section 3: *The Respiratory System.* Fishman AP [editor]. American Physiological Society, 1986.)

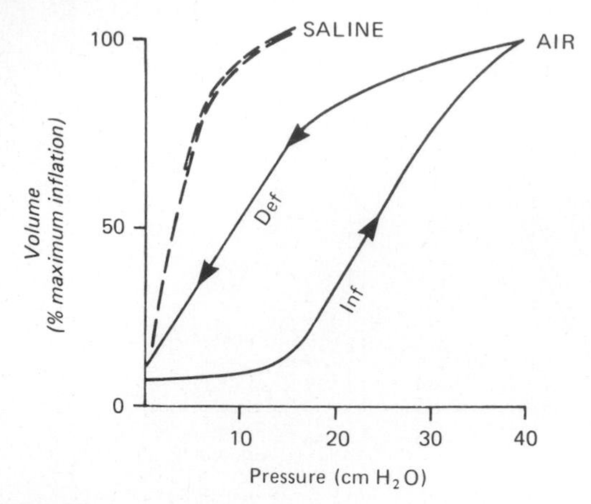

Figure 34–8. Pressure-volume relations in the lungs of a cat after removal from the body. *Air:* lungs inflated (Inf) and deflated (Def) with air. *Saline:* lungs inflated and deflated with saline. (Reproduced, with permission, from Morgan TE: Pulmonary surfactant. *N Engl J Med* 1971;**284**:1185.)

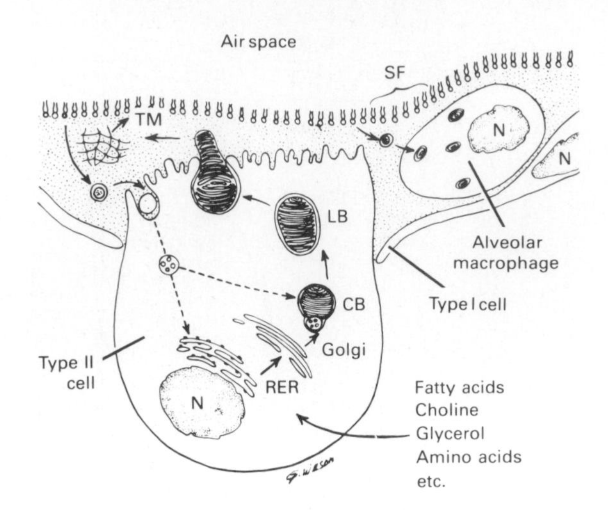

Figure 34–9. Formation and metabolism of surfactant. Lamellar bodies (LB) are formed in type II alveolar epithelial cells and secreted by exocytosis. The released lamellar body material is converted to tubular myelin (TM), and the TM is probably the source of the phospholipid surface film (SF), Some surfactant is taken up by alveolar macrophages, but more is taken up by endocytosis in type II epithelial cells. (Reproduced, with permission, from Wright JR: Metabolism and turnover of lung surfactant. *Am Rev Respir Dis* 1987;**135**:426.)

Laplace (see Chapter 30). In spherical structures like the alveoli, the distending pressure equals 2 times the tension divided by the radius (P = 2T/R); if T is not reduced as R is reduced, the tension overcomes the distending pressure. Surfactant also helps to prevent pulmonary edema. It has been calculated that if it were not present, the unopposed surface tension in the alveoli would produce a 20-mm-Hg force favoring transudation of fluid from the blood into the alveoli.

Phospholipids have a hydrophilic "head" and 2 parallel hydrophobic fatty acids "tails" like phospholipids in cell membranes (see Chapter 1). The molecules are believed to line up with their tails facing the alveolar lumen (Fig 34–9), and surface tension is inversely proportionate to their concentration per unit area. They move farther apart as the alveoli enlarge during inspiration, and surface tension increases, whereas it decreases when they move closer together during expiration. The formation of phospholipid film is greatly facilitated by the proteins in surfactant. This material contains 3 major proteins. One of these is large and collagenlike, whereas the other 2 are smaller. It is these last 2 proteins that facilitate formation of the monomolecular film of phospholipid.

Surfactant is produced by type II alveolar epithelial cells (Fig 34–9). Typical **lamellar bodies,** membrane-bound organelles containing whorls of phospholipid, are formed in these cells and secreted into the alveolar lumen by exocytosis. Tubes of lipid called **tubular myelin** form from the extruded bodies, and the tubular myelin in turn forms the phospholipid film. There is evidence that some of the protein-lipid complexes in surfactant are taken up by endocytosis in type II alveolar cells and recycled.

Table 34–2. Approximate composition of surfactant.

Component	Percent Composition
Phosphatidylcholine	62
Phosphatidylglycine	5
Other phospholipids	10
Neutral lipids	13
Proteins	8
Carbohydrate	2

Surfactant is important at birth. The fetus makes respiratory movements in utero, but the lungs remain collapsed until birth. After birth, the infant makes several strong inspiratory movements and the lungs expand. Surfactant keeps them from collapsing again. Surfactant deficiency is the cause of **hyaline membrane disease, (respiratory distress syndrome, RDS),** the serious pulmonary disease that develops in infants born before their surfactant system is functional. Surface tension in the lungs of these infants is high, and there are many areas in which the alveoli are collapsed (atelectasis). Administration of phospholipid alone by inhalation has little value in the treatment of RDS. However, encouraging results have been obtained with administration of bovine surfactant, which contains both phospholipids and proteins.

The size and number of inclusions in type II cells are increased by thyroid hormones, and RDS is more common and more severe in infants with low plasma levels of thyroid hormones than in those with normal plasma levels. Maturation of surfactant in the lungs is also accelerated by glucocorticoid hormones. There is an increase in fetal and maternal cortisol near term, and the lung is rich in glucocorticoid receptors.

Patchy atelectasis is also associated with surfactant deficiency in patients who have undergone cardiac surgery during which a pump oxygenator was used and the pulmonary circulation was interrupted. In addition, surfactant deficiency may also play a role in some of the abnormalities that develop following occlusion of a main bronchus, occlusion of one pulmonary artery, and long-term inhalation of 100% O_2. There is a decrease in surfactant in the lungs of cigarette smokers.

Work of Breathing

Work is performed by the respiratory muscles in stretching the elastic tissues of the chest wall and lungs, moving inelastic tissues (viscous resistance), and moving air through the respiratory passages (Table 34–3). Since pressure times volume ($g/cm^2 \times cm^3 = g \times cm$) has the same dimensions as work (force × distance), the work of breathing can be calculated from the relaxation pressure curve (Figs 34–6 and 34–10). In Fig 34–10, the total elastic work required for inspiration is area ABCA. Note that the relaxation pressure curve of the total respiratory system differs from that of the lungs alone. The actual elastic work required to increase the volume of the lungs alone is area ABDEA. The amount of elastic work required to inflate the whole respiratory system is less than the amount required to inflate the lungs alone because part of the work comes from elastic energy stored in the thorax. The elastic energy lost from the thorax (area AFGBA) is equal to that gained by the lungs (area AEDCA).

The frictional resistance to air movement is relatively small during quiet breathing, but it does cause the intrapleural pressure changes to lead the lung volume changes during inspiration and expiration (Fig 34–1), producing a **hysteresis loop** rather than a straight line when pressure is plotted against volume (Fig 34–11). In this diagram, area AXBYA represents the work done to overcome airway resistance and lung viscosity. If the air flow becomes turbulent during rapid respiration, the energy required to move the air is greater than when the flow is laminar.

Estimates of the total work of quiet breathing range from 0.3 up to 0.8 kg-m/min. The value rises markedly during exercise, but the energy cost of breathing in normal individuals represents less than 3% of the total energy expenditure during exercise. The work of breathing is greatly increased in diseases such as emphysema, asthma, and congestive heart failure with dyspnea and orthopnea. The respiratory muscles have length-tension relations like those of other skeletal and cardiac muscles, and when they are severely stretched, they contract with less strength.

Table 34–3. Components that make up the work of breathing during quiet inspiration, and the relative contribution of each.

Component
Nonelastic work
Viscous resistance (7%)
Airway resistance (28%)
Elastic work (65%)

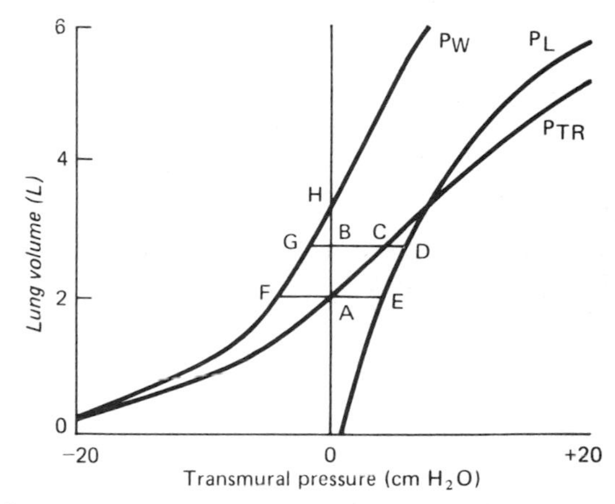

Figure 34–10. Relaxation pressure curve of total respiratory system (PTR) and its components, the relaxation pressure curves of the lungs (PL) and the chest (PW). The transmural pressure is intrapulmonary pressure (P_A) minus intrapleural pressure (P_{Pl}) in the case of the lungs, P_{Pl} minus the outside (barometric) pressure (P_B) in the case of the chest wall, and P_A minus P_B in the case of the total respiratory system. (Modified from Mines AH: *Respiratory Physiology,* 2nd ed. Raven, 1986.)

They can also become fatigued and fail (pump failure), leading to inadequate ventilation (see Chapter 37). For unknown reasons, aminophylline increases the force of contraction of the human diaphragm and is useful in the treatment of pump failure.

Effect of Gravity on Ventilation

In the upright position, ventilation per unit lung volume is greater in the lower than in the upper portions of the lungs. Ventilation is preferentially distributed to the more dependent portions of the lungs because, as a result of the weight of the lungs, the intrapleural pressure is lower (ie, less negative). The curve relating intrapleural pressure to lung vol-

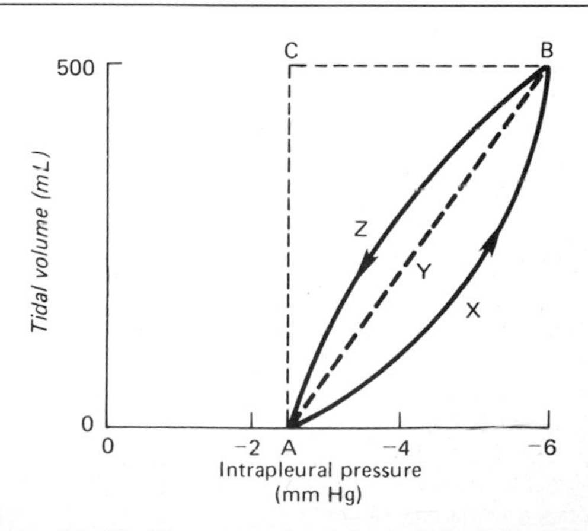

Figure 34–11. Diagrammatic representation of pressure and volume changes during quiet inspiration (line AXB) and expiration (line BZA). Line AYB is the compliance line.

ume (Figure 34–12) levels off because at higher volumes, the degree to which the lungs can be stretched is decreased—ie, the stiffness of the lungs is increased. Consequently, lung volume is smaller at the end of expiration in the lower portions of the lungs but expansion is greater because the intrapleural pressure-volume curve is steeper.

It should be noted that at very low lung volumes such as those after forced expiration, intrapleural pressure at the bases of the lungs can actually exceed the atmospheric pressure in the airways, and the small airways such as respiratory bronchioles collapse **(airway closure).** In older people and in those with chronic lung disease, some of the elastic recoil is lost, with a resulting decrease in intrapleural pressure. Consequently, airway closure may occur in the bases of the lungs in the upright position without forced expiration, at volumes as high as the functional residual capacity.

A clinical correlate of the effect of gravity on ventilation is that arterial oxygenation is improved in unilateral lung diseases when patients lie on their sides so that the good lung is in the dependent position. For reasons that are not entirely clear, the situation is opposite in infants, who do better with the diseased lung in the dependent position.

Dead Space & Uneven Ventilation

Since gaseous exchange in the respiratory system occurs only in the terminal portions of the airways, the gas that occupies the rest of the respiratory system is not available for gas exchange with pulmonary capillary blood. Normally, the volume of this dead space is approximately equal to the body weight in pounds. Thus, in a man who weighs 150 lb (68 kg), only the first 350 mL of the 500 mL inspired with each breath at rest mixes with the air in the alveoli. Conversely, with each expiration, the first 150 mL expired is gas that occupied the dead space, and only the last 350 mL is gas from the alveoli.

It is wise to distinguish between the **anatomic dead space** (respiratory system volume exclusive of alveoli) and the **total (physiologic) dead space** (volume of gas not equilibrating with blood, ie, wasted ventilation). In healthy individuals, the 2 dead spaces are identical; but in disease states, there may be no exchange between the gas in some of the alveoli and the blood, and some of the alveoli may be overventilated. The volume of gas in nonperfused alveoli and any volume of air in the alveoli in excess of that necessary to arterialize the blood in the alveolar capillaries is part of the dead space (nonequilibrating) gas volume. The anatomic dead space can be measured by analysis of the single-breath N_2 curves (Fig 34–13). From midinspiration, the subject takes as deep a breath as possible of pure O_2, then exhales steadily while the N_2 content of the expired gas is continuously measured. The initial gas exhaled (phase I) is the gas that filled the dead space and that consequently contains no N_2. This is followed by a mixture of dead space and alveolar gas (phase II) and then by alveolar gas (phase III). The volume of the dead space is the volume of the gas expired from peak inspiration to the mid portion of phase II (Fig 34–13).

Phase III of the single-breath N_2 curve terminates at the **closing volume (CV)** and is followed by phase IV, during which the N_2 content of the expired gas is increased. The CV is the lung volume above residual volume at which airways in the lower, dependent parts of the lungs begin to close off because of the lesser transmural pressure in these areas (see above).

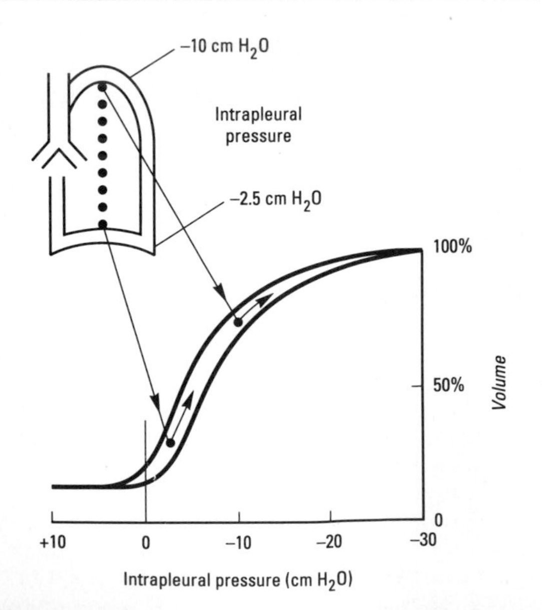

Figure 34–12. Intrapleural pressures in the upright position and their effect on ventilation. Note that during inspiration, the lower part of the lung has a greater increase in volume per unit change in intrapleural pressure. (Reproduced, with permission, from West JB: *Ventilation/Blood Flow and Gas Exchange,* 3rd ed. Blackwell, 1977.)

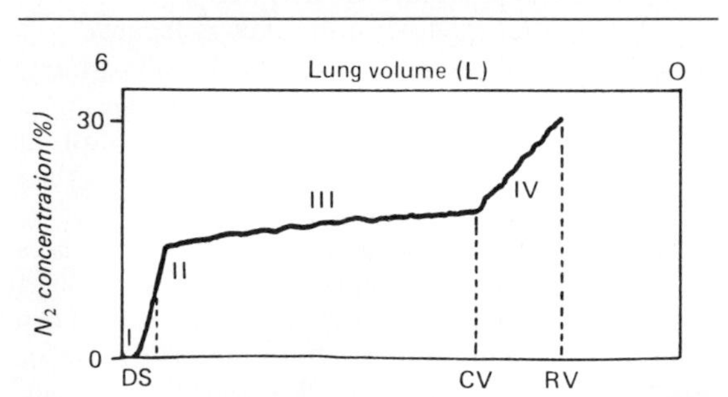

Figure 34–13. Single-breath N_2 curve. From midinspiration, the subject takes a deep breath of pure O_2, then exhales steadily. The changes in the N_2 concentration of expired gas during expiration are shown, with the various phases of the curve indicated by roman numerals. DS, dead space; CV, closing volume; RV, residual volume. (Modified from Buist AS: New tests to assess lung function: The single-breath nitrogen test. *N Engl J Med* 1975;**293**:438.)

The gas in the upper portions of the lungs is richer in N_2 than the gas in the lower, dependent portions because the alveoli in the upper portions are more distended at the start of the inspiration of O_2 (see above) and, consequently, the N_2 in them is less diluted with O_2. It is also worth noting that in most normal individuals, phase III has a slight positive slope even before phase IV is reached. This indicates that even during phase III there is a gradual increase in the proportion of the expired gas coming from the relatively N_2-rich upper portions of the lungs.

The pattern of ventilation in the lungs can be assessed by having the subject inhale a radioactive isotope of the inert gas xenon (^{133}Xe) while the chest is monitored with a battery of radiation detectors. Areas that show little radioactivity are poorly ventilated.

The total dead space can be calculated from the P_{CO_2} of expired air, the P_{CO_2} of alveolar gas, and the tidal volume. The tidal volume (V_T) times the P_{CO_2} of the expired gas ($P_{E_{CO2}}$) equals the alveolar P_{CO_2} ($P_{A_{CO2}}$) times the difference between the tidal volume and the dead space (V_D) plus the P_{CO_2} of inspired air ($P_{I_{CO2}}$) times V_D:

$$P_{E_{CO2}} \times V_T = P_{A_{CO2}} \times (V_T - V_D) + P_{I_{CO2}} \times V_D \text{ (Bohr's equation).}$$

However, the term $P_{I_{CO2}} \times V_D$ is so small that it can be ignored and the equation solved for V_D. If, for example,

$$P_{E_{CO2}} = 28 \text{ mm Hg}$$

$$P_{A_{CO2}} = 40 \text{ mm Hg}$$

$$V_T = 500 \text{ mL}$$

$$V_D = 150 \text{ mL.}$$

then,

Although it is possible to stand under water and breathe through a tube that projects above the surface, it may be worth noting that such a tube is in effect an extension of the respiratory dead space. For each milliliter of tube volume, the depth of inspiration would have to be increased 1 mL to supply the same volume of air to the alveoli. Thus, if the volume of the tube were at all large, breathing would become very laborious. Additional effort is also required to expand the chest against the pressure of the surrounding water.

Alveolar Ventilation

Because of the dead space, the amount of air reaching the alveoli **(alveolar ventilation)** at a respiratory minute volume of 6 L/min is 500 minus 150 mL times 12 breaths/min, or 4.2 L/min. Because of the dead space, rapid, shallow respiration produces much less alveolar ventilation than slow, deep respiration at the same respiratory minute volume (Table 34–4).

Table 34–4. Effect of variations in respiratory rate and depth on alveolar ventilation.

Respiratory rate	30/min	10/min
Tidal volume	200 mL	600 mL
Minute volume	6 L	6 L
Alveolar ventilation	(200 − 150) × 30 = 1500 mL	(600 − 150) × 10 = 4500 mL

GAS EXCHANGE IN THE LUNGS

Composition of Alveolar Air

Oxygen continuously diffuses out of the gas in the alveoli **(alveolar gas)** into the bloodstream, and CO_2 continuously diffuses into the alveoli from the blood. In the steady state, inspired air mixes with the alveolar gas, replacing the O_2 that has entered the blood and diluting the CO_2 that has entered the alveoli. Part of this mixture is expired. The O_2 content of the alveolar gas then falls and its CO_2 content rises until the next inspiration. Since the volume of gas in the alveoli is about 2 L at the end of expiration (functional residual capacity; Fig 34–4), each 350-mL increment of inspired and expired air changes the P_{O_2} and P_{CO_2} very little. Indeed, the composition of alveolar gas remains remarkably constant, not only at rest but in a variety of other conditions as well (see Chapter 36).

Sampling Alveolar Air

Theoretically, all but the first 150 mL expired with each expiration is alveolar air, but there is always some mixing at the interface between the dead space gas and the alveolar air (Fig 34–13). A later portion of expired air is therefore the portion taken for analysis. Using modern apparatus with a suitable automatic valve, it is possible to collect the last 10 mL expired during quiet breathing. The composition of alveolar gas is compared with that of inspired and expired air in Fig 34–14.

Diffusion Across the Alveolocapillary Membrane

Gases diffuse from the alveoli to the blood in the pulmonary capillaries or vice versa across the thin alveolocapillary membrane made up of the pulmonary epithelium, the capillary endothelium, and their fused basement membranes (Fig 34–3). Whether or not substances passing from the alveoli to the capillary blood reach equilibrium in the 0.75 second that blood takes to traverse the pulmonary capillaries at rest depends on their reaction with substances in the blood. Thus, for example, the anesthetic gas nitrous oxide does not react, and N_2O reaches equilibrium in about 0.1 second (Fig 34–15). In this situation, the amount of N_2O entering the body is not limited by diffusion but by the amount of blood flowing through the pulmonary capillaries; ie, it is **perfusion-limited.** On the other hand, carbon monoxide is taken up by

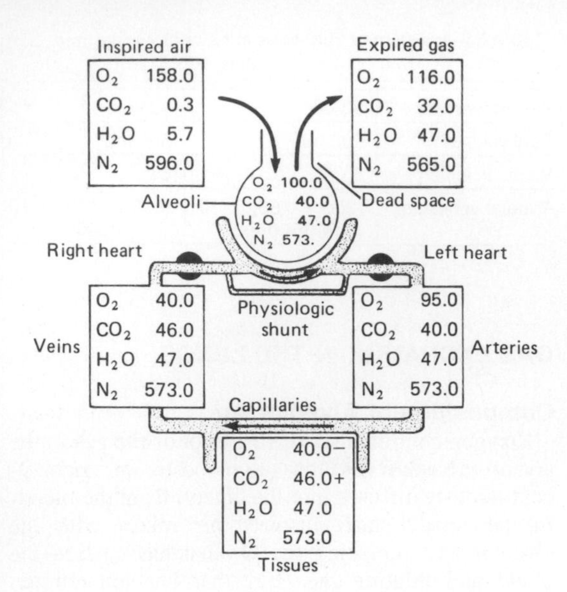

Figure 34–14. Partial pressures of gases (mm Hg) in various parts of the respiratory system and in the circulatory system.

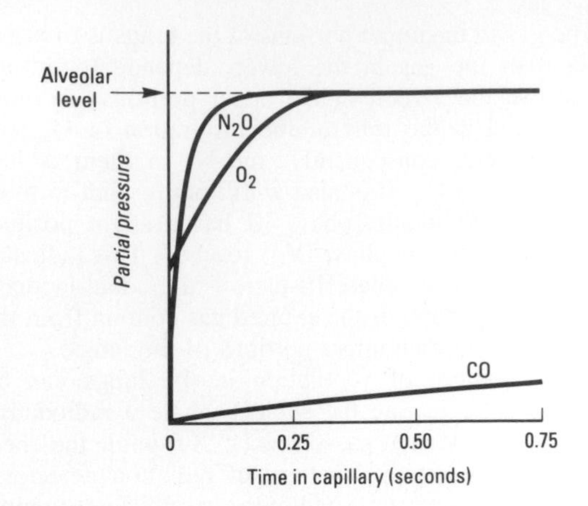

Figure 34–15. Uptake of N_2O, O_2, and CO in blood during transit through a pulmonary capillary at rest.

the hemoglobin in the red blood cells at such a high rate that the partial pressure of CO in the capillaries stays very low and equilibrium is not reached in the 0.75 second the blood is in the pulmonary capillaries. Therefore, the transfer of CO is not limited by perfusion at rest, and instead is **diffusion-limited.** O_2 is intermediate between N_2O and CO; it is taken up by hemoglobin, but much less avidly than CO, and it reaches equilibrium with capillary blood in about 0.3 second. Thus, its uptake is also perfusion-limited.

The **diffusion capacity** of the lung for a given gas is directly proportionate to the size of the alveolocapillary membrane and inversely proportionate to its thickness. The diffusion capacity for CO (D_LCO) is measured as an index of diffusion capacity because its uptake is diffusion-limited. D_LCO is proportionate to the amount of CO entering the blood ($\dot{V}_{CO}$) divided by the partial pressure of CO in the alveoli minus the partial pressure of CO in the blood entering the pulmonary capillaries. Except in habitual cigarette smokers, this latter term is close to zero, so it can be ignored and the equation becomes

$$D_{L_{CO}} = \frac{\dot{V}_{CO}}{P_{A_{CO}}}$$

The normal value of D_{LCO} at rest is about 25 ml/min/mm Hg. It increases up to 3-fold during exercise because of capillary dilation and an increase in the number of active capillaries.

The P_{O2} of alveolar air is normally 100 mm Hg (Fig 34–14), and the P_{O2} of the blood entering the pulmonary capillaries is 40 mm Hg. The diffusion capacity for O_2, like that for CO at rest, is about 25 mL/min/mm Hg, and the P_{O2} of blood is raised to 97 mm Hg, a value just under the alveolar P_{O2}. This falls to 95 mm Hg in the aorta because of the physiologic shunt (see below). D_LO_2 increases to 65 ml/min/mm Hg or more during exercise, and is reduced in diseases such as sarcoidosis and beryllium poisoning (berylliosis) that cause fibrosis of the alveolar walls. Another cause of pulmonary fibrosis is excess secretion of PDGF (see Chapter 27) by alveolar macrophages, with resulting stimulation of neighboring mesenchymal cells.

The P_{CO_2} of venous blood is 46 mm Hg, whereas that of alveolar air is 40 mm Hg, and CO_2 diffuses from the blood into the alveoli along this gradient. The P_{CO_2} of blood leaving the lungs is 40 mm Hg. CO_2 passes through all biologic membranes with ease, and the pulmonary diffusion capacity for CO_2 is much greater than the capacity for O_2. It is for this reason that CO_2 retention is rarely a problem in patients with alveolar fibrosis even when the reduction in diffusion capacity for O_2 is severe.

PULMONARY CIRCULATION

Anatomic Considerations

The pulmonary vascular bed resembles the systemic (see Chapter 30) except that the walls of the pulmonary artery and its large branches are about 30% as thick as the wall of the aorta, and the small arterial vessels, unlike the systemic arterioles, are endothelial tubes with relatively little muscle in their walls. There is also some smooth muscle in the walls of the postcapillary vessels. The pulmonary capillaries are large, and there are multiple anastomoses, so that each alveolus sits in a capillary basket. Lymphatic channels are more abundant in the lungs than in any other organ (Fig 34–2).

Pressure, Volume, & Flow

The output per minute of the right ventricle is, of course, equal to that of the left ventricle and, like that of the left ventricle, averages 5.5 L/min at rest. Thus, the pulmonary vasculature is unique in that it accommodates a blood flow equal to that of all the other organs in the body.

The entire pulmonary vascular system is a distensible low-pressure system. The pulmonary arterial pressure is about 24/9 mm Hg and the mean pressure is about 15 mm Hg. The pressure in the left atrium is about 8 mm Hg during diastole, so that the pressure gradient in the pulmonary system is about 7 mm Hg, compared with a gradient of about 90 mm Hg in the systemic circulation (Fig 34–16). It is interesting that the pressure fall from the pulmonary artery to the capillaries is relatively small and that there is an appreciable pressure drop in the veins.

The volume of blood in the pulmonary vessels at any one time is about 1 L, of which less than 100 mL is in the capillaries. The mean velocity of the blood in the root of the pulmonary artery is the same as that in the aorta (about 40 cm/s). It falls off rapidly, then rises slightly again in the larger pulmonary veins. It takes a red cell about 0.75 second to traverse the pulmonary capillaries at rest, and 0.3 second or less during exercise.

"Physiologic Shunt"

About 2% of the blood in the systemic arteries is blood that has bypassed the pulmonary capillaries. The bronchial arteries, branches of the thoracic aorta, provide blood that nourishes parts of the lung parenchyma, and some of this blood returns to the heart via the pulmonary veins. There is further dilution of the oxygenated blood in the heart by blood that flows from the coronary arteries directly into the chambers of the left side of the heart (see Chapter 32). It is because of this small "physiologic shunt" (Fig 34–14) that the blood in the systemic arteries has a P_{O_2} about 2 mm Hg less than that of the blood which has equilibrated with alveolar air.

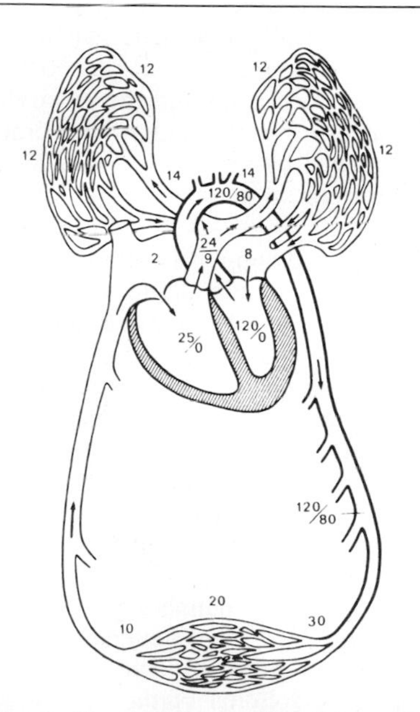

Figure 34–16. Blood pressures (mm Hg) in the pulmonary and systemic circulation. (Modified from Comroe JH Jr: *Physiology of Respiration,* 2nd ed. Copyright © 1974 by Year Book Medical Publishers, Inc., Chicago. Used by permission.)

Capillary Pressure

Pulmonary capillary pressure is about 10 mm Hg, whereas the oncotic pressure is 25 mm Hg, so that there is an inward-directed pressure gradient of about 15 mm Hg which keeps the alveoli free of fluid. When the pulmonary capillary pressure is more than 25 mm Hg—as it may be, for example, when there is "backward failure" of the left ventricle—pulmonary congestion and edema result. Patients with mitral stenosis also have a chronic, progressive rise in pulmonary capillary pressure and extensive fibrotic changes in the pulmonary vessels. Pulmonary edema is not as prominent a symptom in mitral stenosis as in frank congestive heart failure, probably because the fibrosis and constriction of the pulmonary arterial vessels "protect" the capillaries.

Effect of Gravity

Gravity has a relatively marked effect on the pulmonary circulation. In the upright position, the upper portions of the lungs are well above the level of the heart, and the bases are at or below it. Consequently, there is a relatively marked pressure gradient in the pulmonary artery (see Fig 30–16) and a resulting linear decrease in pulmonary blood flow from the bases to the apices of the lungs (Fig 34–17). At the top of the lungs, the pressure at the start of the pulmonary capillaries is close to the atmospheric pressure in the alveoli. Pulmonary arterial pressure is normally just sufficient to maintain perfusion, but if it is reduced or if alveolar pressure is increased, some

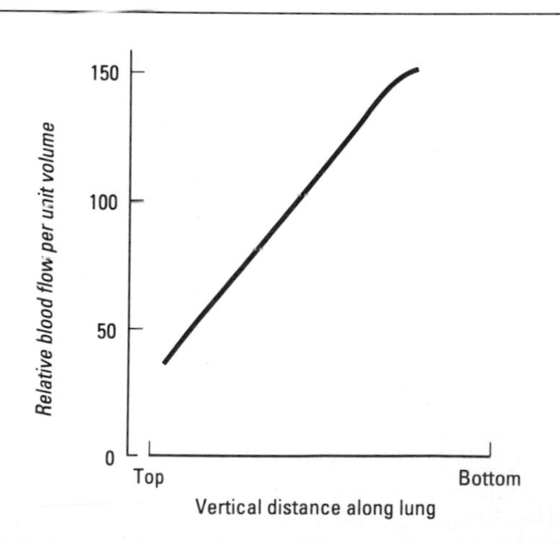

Figure 34–17. Relative blood flow from the top to the bottom of the lung. The values for lung blood flow are scaled so that if flow were uniform, the value would be 100 throughout.

of the capillaries collapse. Under these circumstances, there is no gas exchange in the affected alveoli and they become part of the physiologic dead space.

In the middle portions of the lungs, the pulmonary arterial and capillary pressure exceeds alveolar pressure, but the pressure in the pulmonary venules is lower than alveolar pressure, so they are constricted. Under these circumstances, blood flow is determined by the pulmonary artery-alveolar pressure difference rather than the pulmonary artery-pulmonary vein difference. Beyond the constriction, blood "falls" into the pulmonary veins, which are compliant and take whatever amount of blood the constriction lets flow into them. This has been called the **waterfall effect,** but the term is inappropriate because there is no restriction to flow at the top of a real waterfall.

In the lower portions of the lungs, alveolar pressure is lower than the pressure in all parts of the pulmonary circulation and blood flow is determined by the arterial-venous pressure difference.

Ventilation/Perfusion Ratios

The ratio of pulmonary ventilation to pulmonary blood flow for the whole lung at rest is about 0.8 (4.2 L/min ventilation divided by 5.5 L/min blood flow). However, there are relatively marked differences in this **ventilation/perfusion ratio** in various parts of the normal lung as a result of the effect of gravity, and local changes in the ventilation/perfusion ratio are common in disease. If the ventilation to an alveolus is reduced relative to its perfusion, the P_{O_2} in the alveolus falls because less O_2 is delivered to it and the P_{CO_2} rises because less CO_2 is expired. Conversely, if perfusion is reduced relative to ventilation, the P_{CO_2} falls because less CO_2 is delivered and the P_{O_2} rises because less O_2 enters the blood. These effects are summarized in Fig 34–18.

As noted above, ventilation, as well as perfusion, in the upright position declines in a linear fashion

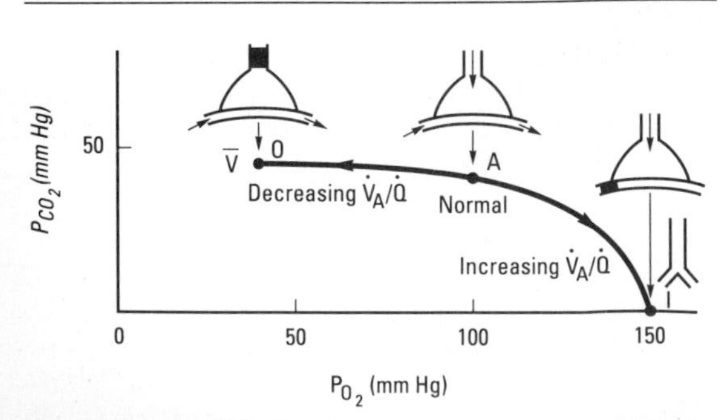

Figure 34–18. Effects of decreasing or increasing the ventilation/perfusion ratio ($\dot{V}_A/\dot{Q}$) on the P_{CO_2} and P_{O_2} in an alveolus. With complete obstruction of the airway to the alveolus. P_{CO_2} and P_{O_2} approximate the values in mixed venous blood (V), whereas with complete block of perfusion, P_{CO_2} and P_{O_2} approximate the values in inspired air. (Reproduced with permission, from West JB: *Ventilation/Blood Flow and Gas Exchange,* 3rd ed. Blackwell, 1977.)

from the bases to the apices of the lungs. However, the decline in ventilation, and consequently, the ventilation/perfusion ratios are high in the upper portions of the lungs. It is said that the high ventilation/perfusion ratios at the apices account for the predilection of tuberculosis for this area because the relatively high alveolar P_{O_2} that results provides a favorable environment for the growth of the tuberculosis bacteria.

When widespread, nonuniformity of ventilation and perfusion in the lungs can cause declines in systemic arterial P_{O_2} and P_{CO_2} retention. The consequences of nonuniformity in disease states are discussed in Chapter 37.

Pulmonary Reservoir

Because of their distensibility, the pulmonary veins are an important blood reservoir. When a normal individual lies down, the pulmonary blood volume increases by up to 400 mL, and when the person stands up this blood is discharged into the general circulation. This shift is the cause of the decrease in vital capacity in the supine position and is responsible for the occurrence of orthopnea in heart failure (see Chapter 33).

Regulation of Pulmonary Blood Flow

Pulmonary arterioles are constricted by norepinephrine, epinephrine, angiotensin II, thromboxanes, and $PGF_{2\alpha}$; they are dilated by isoproterenol, acetylcholine, and PGI_2. Pulmonary venules are constricted by serotonin, histamine, and *Escherichia coli* endotoxin. Pulmonary vessels are plentifully supplied with sympathetic vasoconstrictor nerve fibers, and stimulation of the cervical sympathetic ganglia decreases pulmonary blood flow by as much as 30%.

Despite this innervation and the reactivity of the vessels, it appears that regulation of overall pulmonary blood flow is largely passive, and local adjustments of perfusion to ventilation are determined by local effects of O_2 or its lack. With exercise, cardiac output increases and pulmonary arterial pressure rises proportionately with little or no vasodilation. Red cells move through the lungs more rapidly, without any reduction in the O_2 saturation of the hemoglobin in them, and consequently, the amount of O_2 delivered to the systemic circulation is increased. Capillaries dilate, and previously underperfused capillaries are "recruited" to carry blood. The net effect is a marked increase in pulmonary blood flow with few if any alterations in autonomic outflow to the pulmonary vessels.

Local alterations in pulmonary blood flow are produced in large part by local alterations in tissue O_2 content. ^{133}Xe can be used to survey blood flow by injecting a saline solution of the gas intravenously while monitoring the chest. The gas rapidly enters the alveoli that are perfused normally but fails to enter those that are not perfused. Another technique for locating poorly perfused areas is injection of macroaggregates of albumin labeled with radioactive iodine. These aggregates are large enough to block capillaries and small arterioles, and they lodge only in vessels in which blood was flowing when they reached the lungs. Although it seems paradoxic to study patients with defective pulmonary blood flow by producing vascular obstruction, the technique is safe because relatively few particles are injected. The particles block only a small number of pulmonary vessels and are rapidly removed by the body.

When a bronchus or a bronchiole is obstructed, hypoxia develops in the underventilated alveoli beyond the obstruction. The O_2 deficiency apparently acts directly on vascular smooth muscle in the area to produce constriction, shunting blood away from the hypoxic area. Accumulation of CO_2 leads to a drop in pH in the area, and a decline in pH also produces vasoconstriction in the lungs, as opposed to the vasodilation it produces in other tissues. Conversely, reduction of the blood flow to a portion of the lung lowers the alveolar P_{CO_2} in that area, and this leads to constriction of the bronchi supplying it, shifting ventilation away from the poorly perfused area.

Systemic hypoxia also causes the pulmonary arterioles to constrict, with a resultant increase in pulmonary arterial pressure.

Pulmonary Embolization

One of the normal functions of the lungs is to filter out small blood clots, and this occurs without any symptoms. When emboli block larger branches of the pulmonary artery, they provoke a rise in pulmonary arterial pressure and rapid, shallow respiration **(tachypnea).** The rise in pulmonary arterial pressure is apparently due to reflex vasoconstriction via the sympathetic nerve fibers, although there is controversy on this point and reflex vasoconstriction appears to be absent when large branches of the pulmonary artery are blocked. The tachypnea is a reflex response to activation of vagally innervated pulmonary deflation receptors close to the vessel walls. There is some evidence that serotonin released from platelets at the site of embolization stimulates or ''sensitizes'' these receptors.

OTHER FUNCTIONS OF THE RESPIRATORY SYSTEM

Lung Defense Mechanisms

The respiratory passages that lead from the exterior to the alveoli do more than serve as gas conduits. They humidify and cool or warm the inspired air so that even very hot or very cold air is at or near body temperature by the time it reaches the alveoli. Bronchial secretions contain secretory immunoglobulins (IgA; see Chapter 27) and other substances that help resist infections and maintain the integrity of the mucosa.

The pulmonary alveolar macrophages (PAMS, ''dust cells'') are important components of the pulmonary defense mechanisms. Like other macrophages (see Chapter 27), these cells come originally from the bone marrow. They are actively phagocytic and ingest inhaled bacteria and small particles. They also help process inhaled antigens for immunologic attack, and they secrete substances that attract polymorphonuclear leukocytes to the lungs as well as substances that stimulate granulocyte and monocyte formation in the bone marrow. Their role in the pathogenesis of emphysema is discussed in Chapter 37. When the macrophages ingest large amounts of the substances in cigarette smoke, they may also release lysosomal products into the extracellular space. This causes inflammation. Silica and asbestos particles also cause extracellular release of lysosomal enzymes.

Various mechanisms operate to prevent foreign matter from reaching the alveoli. The hairs in the nostrils strain out many particles larger than 10 μm in diameter. Most of the remaining particles of this size settle on mucous membranes in the nose and pharynx; because of their momentum, they do not follow the airstream as it curves downward into the lungs, and they impact on or near the **tonsils** and **adenoids,** large collections of immunologically active lymphoid tissue in the back of the pharynx. Particles 2–10 μm in diameter generally fall on the walls of the bronchi as the air flow slows in the smaller passages. There they initiate reflex bronchial constriction and coughing (see Chapter 14). They are also moved away from the lungs by the ''ciliary escalator.'' The epithelium of the respiratory passages from the anterior third of the nose to the beginning of the respiratory bronchioles is ciliated, and the cilia, which are covered with mucus, beat in a coordinated fashion at a frequency of 1000–1500 cycles per minute. The ciliary mechanism is capable of moving particles at a rate of at least 16 mm/min. Particles less than 2 μm in diameter generally reach the alveoli, where they are ingested by the macrophages. The importance of these defense mechanisms is evident when one remembers that in modern cities, each liter of air may contain several million particles of dust and irritants.

In **Kartagener's syndrome,** ciliary motility is defective and mucus transport virtually absent. Patients with this syndrome have chronic sinusitis and bronchiectasis.

Metabolic & Endocrine Functions of the Lungs

In addition to their functions in gas exchange, the lungs have a number of metabolic functions. They manufacture surfactant for local use as noted above. They also contain a fibrinolytic system that lyses clots in the pulmonary vessels. They release a variety of substances that enter the systemic arterial blood (Table 34–5), and they remove other substances from the systemic venous blood that reaches them via the pulmonary artery. Prostaglandins are removed from the circulation, but they are also synthesized in the lungs and released into the blood when lung tissue is stretched.

Table 34–5. Biologically active substances metabolized by the lungs.

Synthesized and used in the lungs
Surfactant
Synthesized or stored and released into the blood
Prostaglandins
Histamine
Kallikrein
Partially removed from the blood
Prostaglandins
Bradykinin
Adenine nucleotides
Serotonin
Norepinephrine
Acetylcholine
Activated in the lungs
Angiotensin I → angiotensin II

The lungs also activate one hormone; the physiologically inactive decapeptide angiotensin I is converted to the pressor, aldosterone-stimulating octapeptide angiotensin II in the pulmonary circulation (see Chapter 24). The angiotensin converting enzyme responsible for this activation is located on the surface of the endothelial cells of the pulmonary capillaries and particularly in small pits **(caveolae)** on the vascular surface of these cells. The converting enzyme also inactivates bradykinin. Circulation time through the pulmonary capillaries is less than 1 second, yet 70% of the angiotensin I reaching the lungs is converted to angiotensin II in a single trip through the capillaries. Removal of serotonin and norepinephrine reduces the amounts of these vasoactive substances reaching the systemic circulation. However, many other vasoactive hormones pass through the lungs without being metabolized. These include epinephrine, dopamine, oxytocin, vasopressin, and angiotensin II.

APUD cells and some nerve fibers (see Chapter 26) in the lungs contain biologically active peptides. The peptides that have been characterized include VIP, substance P, opioid peptides, CCK, and somatostatin. The function of most of the peptides in the lungs is unknown, but as mentioned above, VIP may be the mediator released by the nerves of the nonadrenergic noncholinergic bronchodilator system.

Gas Transport Between the Lungs & the Tissues

35

INTRODUCTION

The partial pressure gradients for O_2 and CO_2 plotted in graphic form in Fig 35–1 emphasize that they arc the key to gas movement and that O_2 "flows downhill" from the air through the alveoli and blood into the tissues whereas CO_2 "flows downhill" from the tissues to the alveoli. However, the amount of both of these gases transported to and from the tissues would be grossly inadequate if it were not that about 99% of the O_2 which dissolves in the blood combines with the O_2-carrying protein hemoglobin and that about 94.5% of the CO_2 which dissolves enters into a series of reversible chemical reactions that convert it into other compounds. Thus, the presence of hemoglobin increases the O_2-carrying capacity of the blood 70-fold, and the reactions of CO_2 increase the blood CO_2 content 17-fold (Table 35-1). This chapter reviews the mechanisms involved in O_2 and CO_2 transport.

Table 35–1. Gas content of blood.

Gas	mL/dL of Blood Containing 15 g of Hemoglobin			
	Arterial Blood (P_{O_2} 95 mm Hg; P_{CO_2} 40 mm Hg; Hb 97% Saturated)		Venous Blood (P_{O_2} 40 mm Hg; P_{CO_2} 46 mm Hg; Hb 75% Saturated)	
	Dissolved	Combined	Dissolved	Combined
O_2	0.29	19.5	0.12	15.1
CO_2	2.62	46.4	2.98	49.7
N_2	0.98	0	0.98	0

OXYGEN TRANSPORT

Oxygen Delivery to the Tissues

The O_2 delivery system in the body consists of the lungs and the cardiovascular system. O_2 delivery to a particular tissue depends on the amount of O_2 entering the lungs, the adequacy of pulmonary gas exchange, the blood flow to the tissue, and the capacity of the blood to carry O_2. The blood flow depends on the degree of constriction of the vascular bed in the tissue and the cardiac output. The amount of O_2 in the blood is determined by the amount of dissolved O_2, the amount of hemoglobin in the blood, and the affinity of the hemoglobin for O_2.

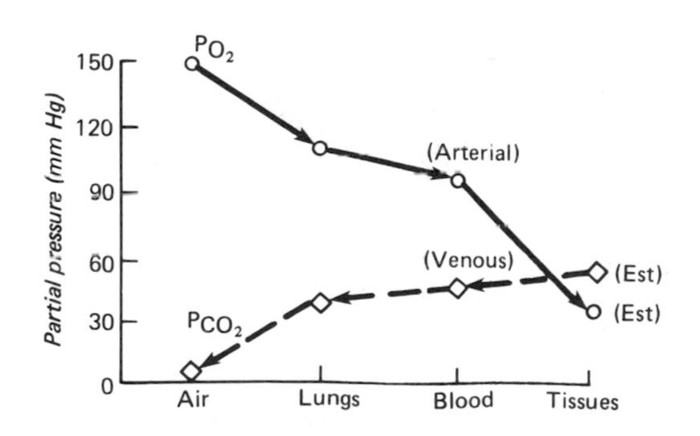

Figure 35–1. Summary of P_{O_2} and P_{CO_2} values in air, lungs, blood, and tissues, graphed to emphasize the fact that both O_2 and CO_2 diffuse "downhill" along gradients of decreasing partial pressure. (Redrawn and reproduced, with permission, from Kinney JM: Transport of carbon dioxide in blood. *Anesthesiology* 1960;21:615.)

Reaction of Hemoglobin & Oxygen

The dynamics of the reaction of hemoglobin with O_2 make it a particularly suitable O_2 carrier. Hemoglobin (see Chapter 27) is a protein made up of 4 subunits, each of which contains a **heme** moiety attached to a polypeptide chain. Heme (Fig 27–11) is a complex made up of a porphyrin and 1 atom of ferrous iron. Each of the 4 iron atoms can bind reversibly one O_2 molecule. The iron stays in the ferrous state, so that the reaction is an **oxygenation,** not an oxidation. It has been customary to write the reaction of hemoglobin with O_2 as $Hb + O_2 \quad HbO_2$. Since it contains 4 Hb units, the hemoglobin molecule can also be represented as Hb_4, and it actually reacts with 4 molecules of O_2 to form Hb_4O_8.

$$Hb_4 + O_2 \rightleftharpoons Hb_4O_2$$
$$Hb_4O_2 + O_2 \rightleftharpoons Hb_4O_4$$
$$Hb_4O_4 + O_2 \rightleftharpoons Hb_4O_6$$
$$Hb_4O_6 + O_2 \rightleftharpoons Hb_4O_8$$

The reaction is rapid, requiring less than 0.01 second. The deoxygenation (reduction) of Hb_4O_8 is also very rapid.

The quaternary structure of hemoglobin determines its affinity for O_2; by shifting the relationship of its 4 component polypeptide chains, the molecule fosters either O_2 uptake or O_2 delivery. The movement of the chains is associated with a change in the position of the heme moieties, which assume a relaxed or **R state** that favors O_2 binding or a tense or **T state** that decreases O_2 binding. The transition from one state to another involves breaking or forming bridges between the polypeptide chains, and it has been calculated that these shifts occur about 10^8 times in the life of a red blood cell.

When hemoglobin takes up a small amount of O_2, the R state is favored and additional uptake of O_2 is facilitated. This is why the **oxygen hemoglobin dissociation curve,** the curve relating percentage saturation of the O_2-carrying power of hemoglobin to the P_{O_2} (Fig 35–2), has a characteristic sigmoid shape. Combination of the first heme in the Hb molecule with O_2 increases the affinity of the second heme for O_2, and oxygenation of the second increases the affinity of the third, etc, so that the affinity of Hb for the fourth O_2 molecule is many times that for the first. When hemoglobin takes up O_2, the two β chains move closer together; when O_2 is given up, they move farther apart. This shift is essential for the shift in affinity for O_2 to occur.

When blood is equilibrated with 100% O_2 (P_{O_2}= 760 mm Hg), the hemoglobin becomes 100% saturated. When fully saturated, each gram of hemoglobin contains 1.34 mL of O_2. The hemoglobin concentration in normal blood is about 15g/dL (14 g/dL in women and 16 g/dL in men; see Chapter 27). Therefore, 1 dL of blood contains 20.1 mL (1.34 mL × 15) of O_2 bound to hemoglobin when the hemoglobin is 100% saturated. The amount of dissolved O_2 is a linear function of the P_{O_2} (0.003 mL/dL blood/mm Hg P_{O_2}).

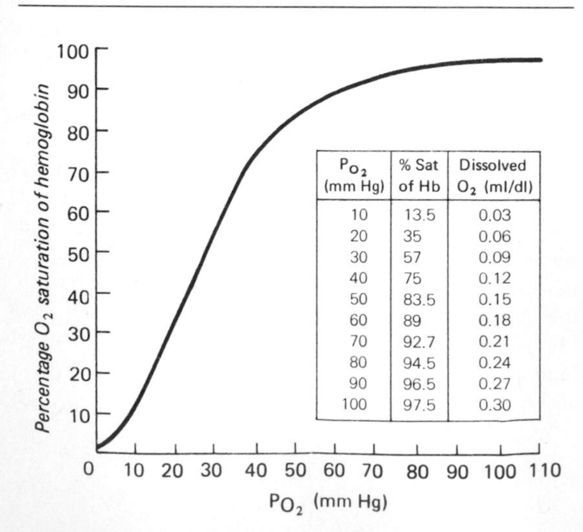

P_{O_2} (mm Hg)	% Sat of Hb	Dissolved O_2 (ml/dl)
10	13.5	0.03
20	35	0.06
30	57	0.09
40	75	0.12
50	83.5	0.15
60	89	0.18
70	92.7	0.21
80	94.5	0.24
90	96.5	0.27
100	97.5	0.30

Figure 35–2. Oxygen hemoglobin dissociation curve. pH 7.40, temperature 38° C. (Redrawn and reproduced, with permission, from Comroe JH Jr et al: *The Lung: Clinical Physiology and Pulmonary Function Tests,* 2nd ed. Year Book, 1962.)

In vivo, the hemoglobin in the blood at the ends of the pulmonary capillaries is about 97.5% saturated with O_2 (P_{O_2} = 97 mm Hg). Because of a slight admixture with venous blood that bypasses the lungs ("physiologic shunt"), the hemoglobin in systemic arterial blood is only 97% saturated. The arterial blood therefore contains a total of about 19.8 mL of O_2 per deciliter: 0.29 mL in solution and 19.5 mL bound to hemoglobin. In venous blood at rest, the hemoglobin is 75% saturated, and the total O_2 content is about 15.2 mL/dL. Thus, at rest the tissues remove about 4.6 mL of O_2 from each deciliter of blood passing through them (Table 35-1); 0.17 mL of this total represents O_2 that was liberated from hemoglobin. In this way, 250 mL of O_2 per minute is transported from the blood to the tissues at rest.

Factors Affecting the Affinity of Hemoglobin for Oxygen

Three important conditions affect the oxygen hemoglobin dissociation curve: the pH, the temperature, and the concentration of 2,3-diphosphoglycerate **(DPG; 2,3-DPG).** A rise in temperature or a fall in pH shifts the curve to the right (Fig 35–3). When the curve is shifted in this direction, a higher P_{O_2} is required for hemoglobin to bind a given amount of O_2. Conversely, a fall in temperature or a rise in pH shifts the curve to the left, and a lower P_{O_2} is required to bind a given amount of O_2. A convenient index of such shifts is the P_{50}, the higher the P_{50}, the lower the affinity of hemoglobin for O_2.

The decrease in O_2 affinity of hemoglobin when the pH of blood falls is called the **Bohr effect** and is closely related to the fact that deoxyhemoglobin binds H^+ more actively than does oxyhemoglobin. The pH of blood falls as its CO_2 content increases (see below), so that when the P_{CO_2} rises, the curve shifts to the right and the P_{50} rises. Most of the unsaturation of hemoglobin that occurs in the tissues is secondary to the decline in the P_{O_2}, but an extra 1–2% unsaturation is due to the rise in P_{CO_2} and consequent shift of the dissociation curve to the right.

2,3-DPG is very plentiful in red cells. It is formed (Fig 35–4) from 3-phosphoglyceraldehyde, which is a product of glycolysis via the Embden-Meyerhof pathway (see Chapter 17). It is a highly charged anion that binds to the β chains of deoxygenated hemoglobin. One mole of deoxygenated hemoglobin binds 1 mol of 2,3-DPG. In effect,

$$HbO_2 + 2,3\text{-DPG} \rightleftharpoons Hb\text{-}2,3\text{-DPG} + O_2$$

In this equilibrium, an increase in the concentration of 2,3-DPG shifts the reaction to the right, causing more O_2 to be liberated. ATP binds to deoxygenated hemoglobin to a lesser extent, and some other organic phosphates bind to a minor degree.

Factors affecting the concentration of 2,3-DPG in the red cells include pH. Because acidosis inhibits red

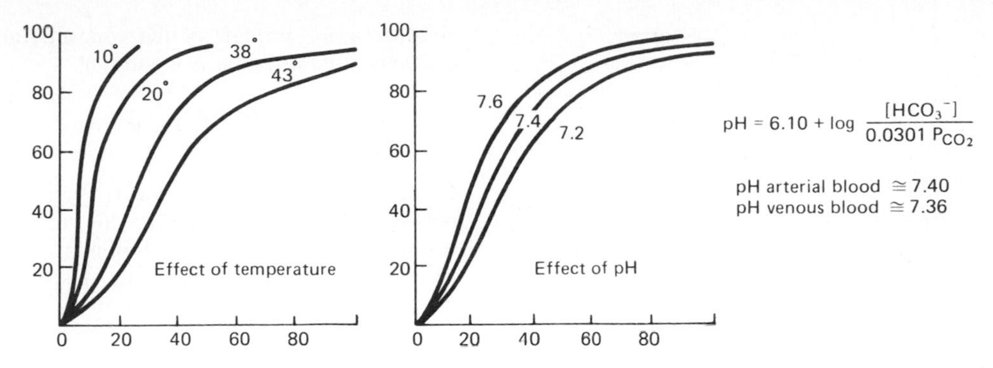

Figure 35–3. Effect of temperature and pH on hemoglobin dissociation curve. Ordinates and abscissas as in Fig 35–2. (Redrawn and reproduced, with permission, from Comroe JH Jr et al: *The Lung: Clinical Physiology and Pulmonary Function Tests,* 2nd ed. Year Book 1962.)

cell glycolysis, the 2,3-DPG concentration falls when the pH is low. Thyroid hormones, growth hormone, and androgens increase the concentration of 2,3-DPG and the P_{50}.

Exercise has been reported to produce an increase in 2,3-DPG within 60 minutes, although the rise may not occur in trained athletes. The P_{50} is also increased during exercise, because the temperature rises in active tissues and CO_2 and metabolites accumulate, lowering the pH. In addition, much more O_2 is removed from each unit of blood flowing through active tissues because the tissue P_{O_2} declines. Finally, at low P_{O_2} values, the oxygen hemoglobin dissociation curve is steep, and large amounts of O_2 are liberated per unit drop in P_{O_2}.

Ascent to high altitude triggers a substantial rise in red cell, 2,3-DPG concentration, with a consequent increase in P_{50} and increase in the availability of O_2 to tissues. The rise in 2,3-DPG, which has a half-life of 6 hours, is secondary to the rise in blood pH (see Chapter 37). 2,3-DPG levels drop to normal upon return to sea level.

Figure 35–4. Formation and catabolism of 2,3-DPG.

The greater affinity of fetal hemoglobin (hemoglobin F) than adult hemoglobin (hemoglobin A) for O_2 facilitates the movement of O_2 from the mother to the fetus (see Chapters 27 and 32). The cause of this greater affinity is the poor binding of 2,3-DPG by the γ polypeptide chains that replace β chains in fetal hemoglobin. Some abnormal hemoglobins in adults have low P_{50} values, and the resulting high O_2 affinity of the hemoglobin causes enough tissue hypoxia to stimulate increased red cell formation, with resulting polycythemia (see Chapter 24). It is interesting to speculate that these hemoglobins may not bind 2,3-DPG.

Red cell 2,3-DPG concentration is increased in anemia and in a variety of diseases in which there is chronic hypoxia. This facilitates the delivery of O_2 to the tissues by raising the P_{O_2} at which O_2 is released in peripheral capillaries. In bank blood that is stored, the 2,3-DPG level falls, and the ability of this blood to release O_2 to the tissues is reduced. This decrease, which obviously limits the benefit of the blood if it is transfused into a hypoxic patient, is less if the blood is stored in citrate-phosphate-dextrose solution rather than the usual acid citrate-dextrose solution.

Other aspects of the chemistry of hemoglobin are discussed in Chapter 27. Fetal hemoglobin and transplacental O_2 exchange are discussed in Chapter 32.

Myoglobin

Myoglobin is an iron-containing pigment found in skeletal muscle. It resembles hemoglobin but binds one rather than 4 mol of O_2 per mole. Its dissociation curve is a rectangular hyperbola rather than a sigmoid curve. Because its curve is to the left of the hemoglobin curve (Fig 35–5), it takes up O_2 from hemoglobin in the blood. It releases O_2 only at low P_{O_2} values, but the P_{O_2} in exercising muscle is close to

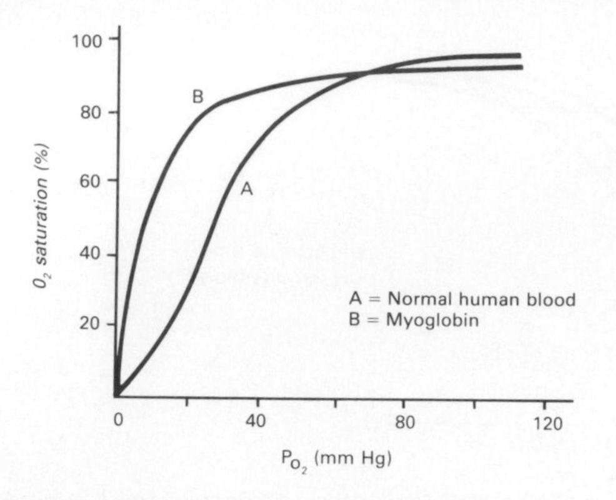

Figure 35–5. Dissociation curve of hemoglobin and myoglobin.

zero. The myoglobin content is greatest in muscles specialized for sustained contraction. The muscle blood supply is compressed during such contractions, and myoglobin may provide O_2 when blood flow is cut off. There is also evidence that myoglobin facilitates the diffusion of O_2 from the blood to the mitochondria, where the oxidative reactions occur.

Blood Substitutes

The solubility of O_2 in plasma is limited. Certain perfluoro compounds dissolve much more O_2 and have been used to totally replace blood for short periods in experimental animals. Some of these compounds have been tested in patients, and although much more testing is needed, they may be of value as substitutes for blood in emergencies until blood can be obtained and cross-matched.

BUFFERS IN BLOOD

Since CO_2 forms carbonic acid in the blood, a discussion of the buffer systems in the blood is a necessary preliminary to a consideration of CO_2 transport.

The Henderson-Hasselbalch Equation

The general equation for a buffer system is

$$HA \rightleftharpoons H^+ + A^-$$

A^- represents any anion and HA the undissociated acid. If an acid stronger than HA is added to a solution containing this system, the equilibrium is shifted to the left. Hydrogen ions are "tied up" in the formation of more undissociated HA, so the increase in H^+ concentration is much less than it would otherwise be. Conversely, if a base is added to the solution, H^+ and OH^- react to form H_2O; but more HA dissociates, limiting the decrease in H^+ concentration. By the law of mass action, the product of the concentrations of the products in a chemical reaction divided by the product of the concentration of the reactants at equilibrium is a constant:

$$\frac{[H^+][A^-]}{[HA]} = K$$

If this equation is solved for H^+ and put in pH notation (pH is the negative log of $[H^+]$), the resulting equation is that originally derived by Henderson and Hasselbalch to describe the pH changes resulting from addition of H^+ or OH^- to any buffer system **(Henderson-Hasselbalch equation):**

$$pH = pK + \log \frac{[A^-]}{[HA]}$$

It is apparent from these equations that the buffering capacity of a system is greatest when the amount of free anion is equal to the amount of undissociated HA, ie, when $[A^-]/[HA] = 1$, so that $\log [A^-/[HA] = 0$ and $pH = pK$. This is why the most effective buffers in the body would be expected to be those with pKs close to the pH in which they operate. The pH of the blood is normally 7.4; that of the cells is probably about 7.2; and that of the urine varies from 4.5 to 8.0.

It should be noted that the equilibrium constant, K, applies only to infinitely dilute solutions in which interionic forces are negligible. In body fluids, it is more appropriate to use the apparent ionization constant, K′.

Buffers in Blood

In the blood, proteins—particularly the **plasma proteins**—are effective buffers because both their free carboxyl and their free amino groups dissociate:

$$RCOOH \rightleftharpoons RCOO^- + H^+$$

$$pH = pK'_{RCOOH} + \log \frac{[RCOO^-]}{[RCOOH]}$$

$$RNH_3^+ \rightleftharpoons RNH_2 + H^+$$

$$pH = pK'_{RNH_3^+} + \log \frac{[RNH_2]}{[RNH_3^+]}$$

Another important buffer system is provided by the dissociation of the imidazole groups of the histidine residues in **hemoglobin:**

$$\text{imidazole (HN–CH=NH}^+\text{–C(R)=CH)} \rightleftharpoons \text{imidazole (HN–CH=N–C(R)=CH)} + H^+$$

In the pH 7.0–7.7 range, the free carboxyl and amino groups of hemoglobin contribute relatively little to its buffering capacity. However, the hemoglobin molecule contains 38 histidine residues, and on this basis—plus the fact that it is present in large amounts—the hemoglobin in blood has 6 times the buffering capacity of the plasma proteins. In addition, the action of hemoglobin is unique because the imidazole groups of deoxygenated hemoglobin dissociate less than those of oxyhemoglobin, making Hb a weaker acid and therefore a better buffer than HbO_2. Titration curves for Hb and HbO_2 are shown in Fig 35–6.

The third major buffer system in blood is the **carbonic acid-bicarbonate** system:

$$H_2CO_3 \rightleftharpoons H^+ + HCO_3^-$$

The Henderson-Hasselbalch equation for this system is as follows:

$$pH = pK + \log \frac{[HCO_3^-]}{[H_2CO_3]}$$

The pK for this system in an ideal solution is low (about 3), and the amount of H_2CO_3 is small and hard to measure accurately. However, in the body, H_2CO_3 is in equilibrium with CO_2.

$$H_2CO_3 \rightleftharpoons CO_2 + H_2O$$

If the pK is changed to pK′ (see above) and $[CO_2]$ is substituted for $[H_2CO_3]$, the pK′ is 6.1.

$$pH = 6.1 + \log \frac{[HCO_3^-]}{[CO_2]}$$

The clinically relevant form of this equation is as follows:

$$pH = 6.1 + \log \frac{[HCO_3^-]}{0.0301\ P_{CO_2}}$$

since the amount of dissolved CO_2, is proportionate to the partial pressure of CO_2, and the solubility coefficient of CO_2 in mmol/L/mm Hg is 0.0301. $[HCO_3^-]$ cannot be measured directly, but pH and P_{CO_2} can be measured with suitable accuracy using pH and P_{CO_2} glass electrodes, and $[HCO_3^-]$ can then be calculated.

The pK′ of this system is still low relative to the pH of the blood, but the system is one of the most effective buffer systems in the body because the amount of dissolved CO_2 is controlled by respiration. In addition, the plasma concentration of HCO_3^- is regulated by the kidneys. When H^+ is added to the blood, HCO_3^- declines as more H_2CO_3 is formed. If the extra H_2CO_3 were not converted to CO_2 and H_2O and the CO_2 excreted in the lungs, the H_2CO_3 concentration would rise. When enough H^+ had been added to halve the plasma HCO_3^-, the pH would have dropped from 7.4 to 6.0. However, not only is all the extra H_2CO_3 that is formed removed, but the H^+ rise stimulates respiration and therefore produces a drop in P_{CO_2}, so that some additional H_2CO_3 is removed. The pH thus falls only to 7.2 or 7.3 (Fig 39–5).

The reaction $CO_2 + H_2O \rightleftharpoons H_2CO_3$ proceeds slowly in either direction unless the enzyme **carbonic anhydrase** is present. There is no carbonic anhydrase in plasma, but there is an abundant supply in red blood cells. It is also found in high concentration in gastric acid-secreting cells (see Chapter 26) and in renal tubular cells (see Chapter 38). Carbonic anhydrase is a protein with a molecular weight of 30,000 that contains an atom of zinc in each molecule. It is inhibited by cyanide, azide, and sulfide. The sulfonamides also inhibit this enzyme, and sulfonamide derivatives have been used clinically as diuretics because of their inhibitory effects on carbonic anhydrase in the kidney (see Chapter 38).

The system $H_2PO_4^- \rightleftharpoons H^+ + HPO_4^{2-}$ has a pK of 6.80. In the plasma, the phosphate concentration is too low for this system to be a quantitatively important buffer, but it is important intracellularly, and it frequently plays a significant role in the urine (see Chapter 38).

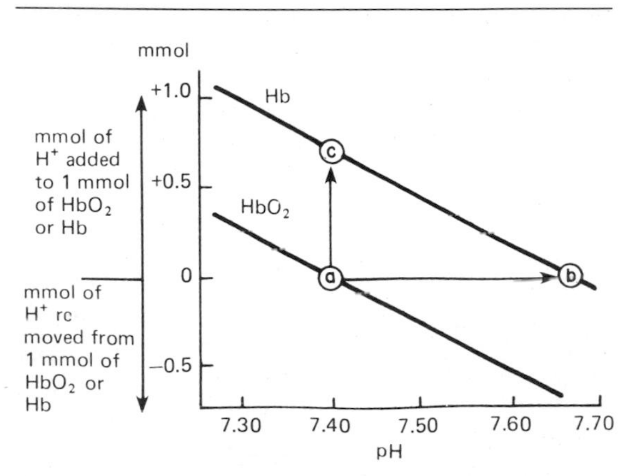

Figure 35–6. Titration curves of Hb and HbO_2. The arrow from ***a*** to ***c*** indicates the number of millimoles of H^+ that can be added without pH shift. The arrow from ***a*** to ***b*** indicates the pH shift on deoxygenation. (Modified from Davenport HW: *The ABC of Acid-Base Chemistry,* 6th ed. Univ of Chicago Press, 1974.)

Buffering in Vivo

Buffering in vivo is of course not limited to the blood. The principal buffers in the blood, the interstitial fluid, and the intracellular fluid are listed in Table 35–2. The principal buffers in cerebrospinal fluid and urine are the bicarbonate and phosphate systems. In metabolic acidosis, only 15–20% of the acid load is buffered by the H_2CO_3-HCO_3 system in the extracellular fluid, and most of the remainder is buffered in cells. In metabolic alkalosis, about 30–35% of the OH^- load is buffered in cells, whereas in respiratory acidosis and alkalosis (see Chapter 39), almost all the buffering is intracellular.

Table 35–2. Principal buffers in body fluids.

Blood	$H_2CO_3 \rightleftharpoons H^+ + HCO_3^-$ $HProt \rightleftharpoons H^+ + Prot^-$ $HHb \rightleftharpoons H^+ + Hb^-$
Interstitial fluid	$H_2CO_3 \rightleftharpoons H^+ + HCO_3^-$
Intracellular fluid	$HProt \rightleftharpoons H^+ + Prot^-$ $H_2PO_4^- \rightleftharpoons H^+ + HPO_4^{2-}$

Summary

When a strong acid is added to the blood, the major buffer reactions are driven to the left. The blood levels of the 3 "buffer anions"—Hb^- (hemoglobin), $Prot^-$ (protein), and HCO_3^-—consequently drop (Fig 35–7). The anions of the added acid are filtered into the renal tubules. They are accompanied ("covered") by cations, particularly Na^+, because electrochemical neutrality is maintained. By processes that are discussed in Chapter 38, the tubules replace the Na^+ with H^+ and in so doing reabsorb equimolar amounts of Na^+ and HCO_3^-, thus conserving the cations, eliminating the acid, and restoring the supply of buffer anions to normal. When CO_2 is added to the blood, similar reactions occur, except that since it is H_2CO_3 that is formed, the plasma HCO_3^- rises rather than falls.

CARBON DIOXIDE TRANSPORT

Fate of Carbon Dioxide in Blood

The solubility of CO_2 in blood is about 20 times that of O_2, so that there is considerably more CO_2 than O_2 in simple solution at equal partial pressures. The CO_2 that diffuses into red blood cells is rapidly hydrated to H_2CO_3 because of the presence of carbonic anhydrase. The H_2CO_3 dissociates to H^+ and HCO_3^-, and the H^+ is buffered, primarily by hemoglobin, while the HCO_3^- enters the plasma. The decline in the O_2 saturation of the hemoglobin as the blood passes through the tissue capillaries improves its buffering capacity because deoxygenated hemoglobin binds more H^+ than oxyhemoglobin (see above). Some of the CO_2 in the red cells reacts with the amino groups of proteins, principally hemoglobin, to form **carbamino compounds:**

$$CO_2 + R{-}N\begin{matrix}H\\H\end{matrix} \rightleftharpoons R{-}N\begin{matrix}H\\COOH\end{matrix}$$

Since deoxygenated hemoglobin forms carbamino compounds much more readily than HbO_2, transport of CO_2 is facilitated in venous blood. About 11% of the CO_2 added to the blood in the systemic capillaries is carried to the lungs as carbamino-CO_2.

In the plasma, CO_2 reacts with plasma proteins to form small amounts of carbamino compounds, and small amounts of CO_2 are hydrated; but the hydration reaction is slow in the absence of carbonic anhydrase.

Chloride Shift

Since the rise in the HCO_3^- content of red cells is much greater than that in plasma as the blood passes through the capillaries, about 70% of the HCO_3^- formed in the red cells enters the plasma. Normally, the protein anions cannot cross the cell membrane, and the movement of Na^+ and K^+ is regulated by the sodium-potassium pump. Electrochemical neutrality is maintained by Cl^- entering the red cells in exchange for HCO_3^-(the **chloride shift).** The Cl^- content of the red cells in venous blood is therefore significantly greater than in arterial blood. The chloride shift occurs rapidly and is essentially complete in 1 second.

Note that for each CO_2 molecule added to a red cell, there is an increase of one osmotically active particle—either an HCO_3^- or a Cl^-—in the red cell (Fig 35–8). Consequently, the red cells take up water and increase in size. For this reason, plus the fact that a small amount of fluid in the arterial blood returns via the lymphatics rather than the veins, the hematocrit of venous blood is normally 3% greater than that of

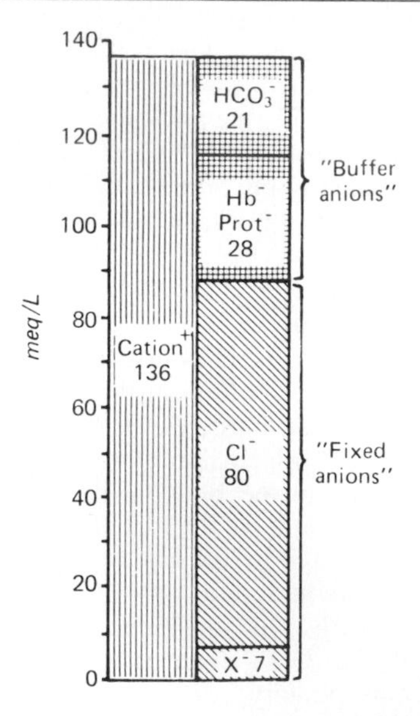

Figure 35–7. Distribution of cations and anions in whole blood, pH 7.39, P_{CO_2} 41 mm Hg. Note that there are appreciable differences in the distribution of these ions in the plasma and red blood cells that make up the blood; for example, the HCO_3^- level is 24 meq/L in plasma and 18 meq/L in red cells, and hemoglobin is present almost exclusively in the red cells. Hb^-, hemoglobin; $Prot^-$, plasma protein; X^-, remaining anions. (Reproduced, with permission, from Singer RB, in Altman PL et al: *Handbook of Respiration*. Saunders, 1958.)

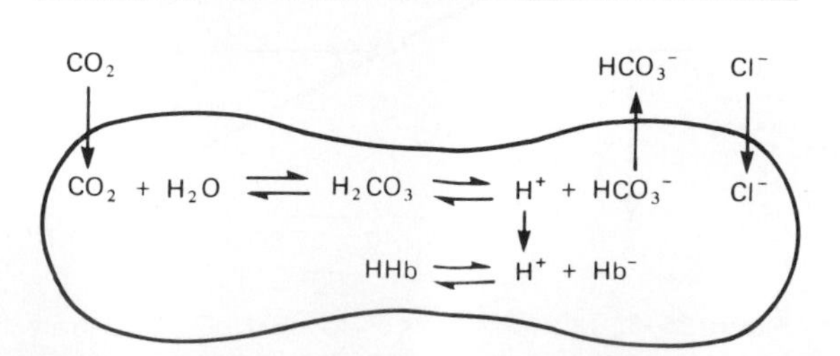

Figure 35–8. Summary of changes that occur in a red cell upon addition of CO_2 to blood. Note that for each CO_2 molecule that enters the red cell, there is an additional HCO_3^- or Cl^- ion in the cell.

Table 35–3. Fate of CO_2 in blood.

In plasma
1. Dissolved
2. Formation of carbamino compounds with plasma protein
3. Hydration, H^+ buffered, HCO_3^- in plasma
In red blood cells
1. Dissolved
2. Formation of carbamino-Hb
3. Hydration, H^+ buffered, 70% of HCO_3^- enters the plasma
4. Cl^- shifts into cells; mosm/L in cells increases

the arterial blood. In the lungs, the Cl^- moves out of the cells and they shrink.

Summary of Carbon Dioxide Transport

For convenience, the various fates of CO_2 in the plasma and red cells are summarized in Table 35–3. The extent to which they increase the capacity of the blood to carry CO_2 is indicated by the difference between the dissolved CO_2 line and the total CO_2 content lines in the dissociation curves for CO_2 shown in Fig 35–9.

Of the approximately 49 mL of CO_2 in each deciliter of arterial blood (Table 35–1), 2.6 mL is dissolved, 2.6 mL is in carbamino compounds, and 43.8 mL is in HCO_3^-. In the tissues, 3.7 mL of CO_2 per deciliter of blood is added; 0.4 mL stays in solution, 0.8 mL forms carbamino compounds, and 2.5 mL forms HCO_3^-. The pH of the blood drops from 7.40 to 7.36. In the lungs, the processes are reversed, and the 3.7 mL of CO_2 is discharged into the alveoli. In this fashion, 200 mL of CO_2 per minute at rest and much larger amounts during exercise are transported from the tissues to the lungs and excreted. It is worth noting that this amount of CO_2 is equivalent in 24 hours to over 12,500 meq of H^+.

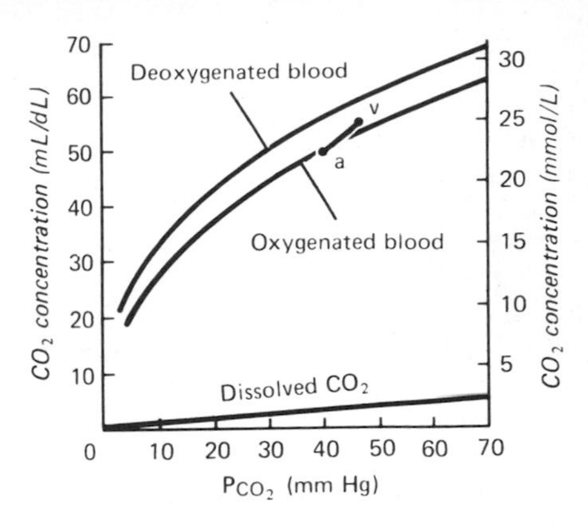

Figure 35–9. CO_2 dissociation curves. The arterial point (**a**) and the venous point (**v**) indicate the total CO_2 content found in arterial blood and venous blood of normal resting humans. (Modified and reproduced, with permission, from Schmidt RF, Thews G [editors]: *Human Physiology.* Springer-Verlag, 1983.)

36 Regulation of Respiration

INTRODUCTION

Spontaneous respiration is produced by rhythmic discharge of motor neurons that innervate the respiratory muscles. This discharge is totally dependent on nerve impulses from the brain; breathing stops if the spinal cord is transected above the origin of the phrenic nerves.

The rhythmic discharges from the brain that produce spontaneous respiration are regulated by alterations in arterial P_{O_2}, P_{CO_2}, and H^+ concentration, and this chemical control of breathing is supplemented by a number of nonchemical influences.

NEURAL CONTROL OF BREATHING

Control Systems

Two separate neural mechanisms regulate respiration. One is responsible for voluntary control and the other for automatic control. The voluntary system is located in the cerebral cortex and sends impulses to the respiratory motor neurons via the corticospinal tracts. The automatic system is located in the pons and medulla, and the efferent output from this system to the respiratory motor neurons is located in the lateral and ventral portions of the spinal cord.

The motor neurons to the expiratory muscles are inhibited when those supplying the inspiratory muscles are active, and vice versa. This **reciprocal innervation** is not due to spinal reflexes and in this regard differs from the reciprocal innervation of the limb flexors and extensors (see Chapter 6). Instead, impulses in descending pathways that excite agonists also produce inhibition of antagonists, probably by exciting inhibitory interneurons.

Medullary Centers

Rhythmic discharge of neurons in the medulla oblongata produces automatic respiration. Respiratory neurons are of 2 types: those that discharge during inspiration (I neurons) and those that discharge during expiration (E neurons). Many of these discharge at increasing frequencies during inspiration, in the case of I neurons, or during expiration, in the case of E neurons. Some discharge at decreasing frequencies, and some discharge at the same high rate during inspiration or expiration. However, expiration is passive during quiet breathing, and E neurons are quiet; they become active only when ventilation is increased.

The area in the medulla that is concerned with respiration has classically been called the **respiratory center,** but there are actually 2 groups of respiratory neurons (Figs 36–1 and 36–2). The **dorsal group** of neurons in and near the nucleus of the tractus solitarius is the source of rhythmic drive to the contralateral phrenic motor neurons. These neurons also project to and drive the **ventral group.** This group has 2 divisions. The cranial division is made up of neurons in the nucleus ambiguus that innervate the ipsilateral accessory muscles of respiration, principally via the vagus nerves. The caudal division is made up of neurons in the nucleus retroambigualis that provide the inspiratory and expiratory drive to the motor neurons supplying the intercostal muscles. The paths from these neurons to expiratory motor neurons are crossed, but those to inspiratory motor neurons are both crossed and uncrossed (Fig 36–2).

Pontine & Vagal Influences

The rhythmic discharge of the I neurons in the respiratory center is spontaneous, but it is modified by centers in the pons and by afferents in the vagus nerves from receptors in the lungs. The interactions of these components can be analyzed by evaluating the results of the experiments summarized diagrammatically in Fig 36–1. Complete transection of the brain stem below the medulla (section D in Fig 36–1) stops all respiration. When all of the cranial nerves (including the vagi) are cut and the brain stem is transected above the pons (section A in Fig 36–1), regular breathing continues. However, when an additional transection is made in the inferior portion of the pons (section B in Fig 36–1), the inspiratory neurons discharge continuously and there is a sustained contraction of the inspiratory muscles. This arrest of respiration in inspiration is called **apneusis.** The area in the pons that prevents apneusis is called the **pneumotaxic center** and is located in the nucleus parabrachialis and the Kölliker Fuse nucleus. The area in the caudal pons responsible for apneusis is called the **apneustic center.**

When the brain stem is transected in the inferior portion of the pons and the vagus nerves are left intact, regular respiration continues. In an apneustic animal, stimulation of the proximal stump of one of the cut vagi produces, after a moderate latent period,

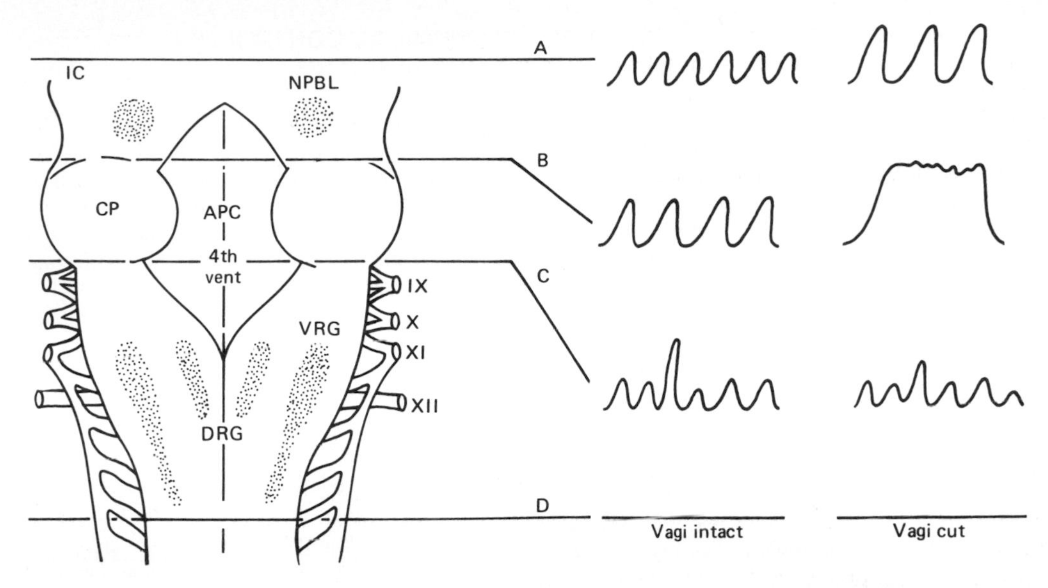

Figure 36–1. Respiratory neurons in the brain stem. Dorsal view of brain stem; cerebellum removed. The effects of transecting the brain stem at various levels are also shown. The spirometer tracings at the right indicate the depth and rate of breathing, and the letters identify the level of transection. DRG, dorsal group of respiratory neurons; VRG, ventral group of respiratory neurons; NPBL, nucleus parabrachialis (pneumotaxic center); APC, apneustic center; 4th vent, fourth ventricle; IC, inferior colliculus; CP, middle cerebellar peduncle. (Modified and reproduced, with permission, from Mitchell RA, Berger A: State of the art: Review of neural regulation of respiration. *Am Rev Respir Dis* 1975;**111**:206.

a relatively prolonged inhibition of inspiratory neuron discharge (Fig 36–3). There are stretch receptors in the lung parenchyma that relay to the medulla via afferents in the vagi, and inflation of the lung inhibits inspiratory discharge (Hering-Breuer inflation reflex; see below). Thus, stretching of the lungs during inspiration reflexly inhibits inspiratory drive, reinforcing the action of the pneumotaxic center in producing intermittency of inspiratory neuron discharge. This is why the depth of inspiration is increased after vagotomy in otherwise intact experimental animals (Fig 36–3). However, it should be emphasized that rhythmic discharge of I neurons continues in the absence of any phasic input from the lungs or any other "moving parts."

When all pontine tissue is separated from the medulla (section C in Fig 36–1), respiration continues whether or not the vagi are intact. This respiration

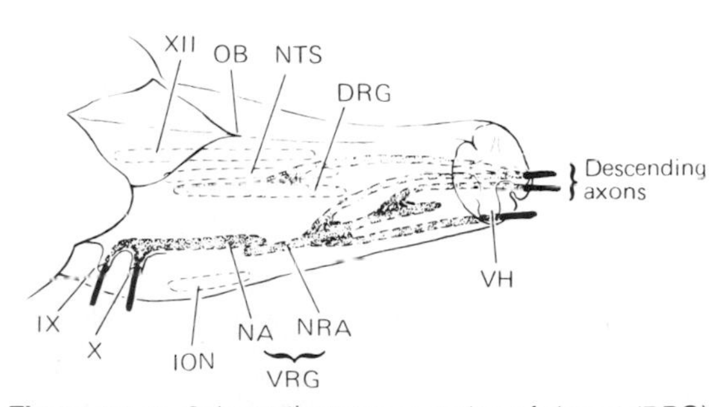

Figure 36–2. Schematic representation of dorsal (DRG) and ventral (VRG) groups of respiratory neurons and their efferent paths in the cat. ION, inferior olivary nucleus; NA, nucleus ambiguus; NRA, nucleus retroambigualis; NTS, nucleus of tractus solitarius; OB, obex; VH, ventral horn; IX, X, glossopharyngeal and vagus nerves; XII, hypoglossal nucleus. (Reproduced, with permission, from Mitchell RA, Berger A: State of the art: Review of neural regulation of respiration, *Am Rev Respir Dis* 1975;**111**:206.)

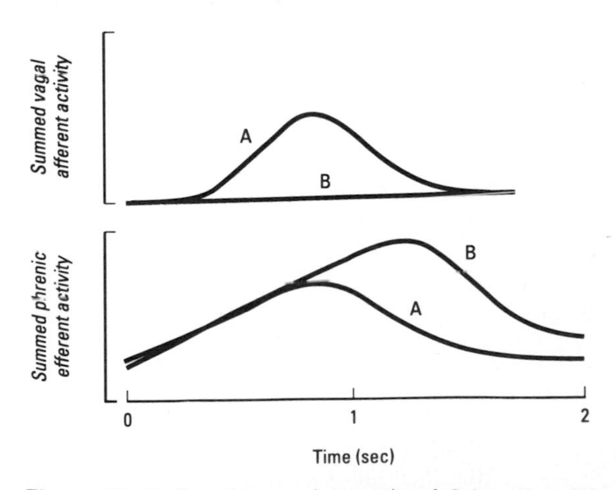

Figure 36–3. Superimposed records of 2 breaths: (A) with, and (B) without feedback vagal afferent activity from stretch receptors in the lungs. Note that the rate of rise in phrenic nerve activity to the diaphragm is unaffected but the discharge is prolonged in the absence of vagal input. (Modified from von Euler C. Central pattern generation during breathing. *Trends Neurosci* 1980;**3**:275).

is somewhat irregular and gasping, but it is rhythmic. Its occurrence demonstrates that the respiratory center neurons are capable of spontaneous rhythmic discharge.

The precise physiologic role of the pontine respiratory areas is uncertain, but they apparently make the rhythmic discharge of the medullary neurons smooth and regular. It appears that there are tonically discharging neurons in the apneustic center which drive inspiratory neurons in the medulla, and these neurons are intermittently inhibited by impulses in afferents from the pneumotaxic center and vagal afferents.

Genesis & Regulation of Rhythmicity

The exact mechanism responsible for the spontaneous discharge of the I neurons that produces automatic respiration is unsettled. The neurons of the ventral respiratory group are driven by neurons in the dorsal respiratory group, so respiratory rhythmicity does not originate in the ventral group. The I neurons appear to send an inhibitory projection to the E neurons. The I neurons are in turn inhibited during expiration, but their inhibition does not come from the E neurons, which are quiet during quiet respiration; it comes instead from an "off switch" somewhere in the medulla, and the operation of this off switch is the key to respiratory rhythmicity. Its activity is in turn affected by the chemical and nonchemical stimuli that regulate respiration.

When the activity of the inspiratory neurons is increased in intact animals, the rate and the depth of breathing are increased. The depth of respiration is increased because the lungs are stretched to a greater degree before the amount of vagal and pneumotaxic center inhibitory activity is sufficient to overcome the more intense inspiratory neuron discharge. The respiratory rate is increased because the after-discharge in the vagal and pneumotaxic afferents is rapidly overcome.

REGULATION OF RESPIRATORY CENTER ACTIVITY

A rise in the P_{CO_2} or H^+ concentration of arterial blood or a drop in its P_{O_2} increases the level of respiratory center activity, and changes in the opposite direction have a slight inhibitory effect. The effects of variations in blood chemistry on ventilation are mediated via respiratory **chemoreceptors**—the carotid and aortic bodies and collections of cells in the medulla that are sensitive to changes in the chemistry of the blood. They initiate impulses that stimulate the respiratory center. Superimposed on this basic **chemical control of respiration,** other afferents provide nonchemical controls for the "fine adjustments" that affect breathing in particular situations (Table 36–1).

CHEMICAL CONTROL OF BREATHING

The chemical regulatory mechanisms adjust ventilation in such a way that the alveolar P_{CO_2} is normally held constant, the effects of excess H^+ in the blood are combatted, and the P_{O_2} is raised when it falls to a potentially dangerous level. The respiratory minute volume is proportionate to the metabolic rate, but the link between metabolism and ventilation is CO_2, not O_2. The receptors in the carotid and aortic bodies are stimulated by a rise in the P_{CO_2} or H^+ concentration of arterial blood or a decline in its P_{O_2}. After denervation of the carotid chemoreceptors, the response to a drop in P_{O_2} is abolished; the predominant effect of hypoxia after denervation of the carotid bodies is a direct depression of the respiratory center. The response to changes in arterial blood H^+ concentration in the pH 7.3–7.5 range is also abolished, although larger changes exert some effect. The response to changes in arterial P_{CO_2}, on the other hand, is affected only slightly; it is reduced no more than 30–35%.

Carotid & Aortic Bodies

There is a carotid body near the carotid bifurcation on each side, and there are usually 2 or more aortic bodies near the arch of the aorta (Fig 36–4). Each carotid and aortic body **(glomus)** contains islands of 2 types of cells, type I and type II cells, surrounded by fenestrated sinusoidal capillaries. The type II cells, which are probably glial cells, surround the type I, or glomus, cells. Unmyelinated endings of glossopharyngeal nerve fibers are found at intervals between the type I and type II cells. There is some evidence that the chemoreceptors which sense O_2 tension are these nerve endings. The type I cells contain a catecholamine, probably dopamine, and may make reciprocal synaptic connections with the nerve endings (Fig 36–5). Dopamine inhibits discharge in the carotid body nerves, but despite this, there is evidence that the glomus cells in some way condition the nerve endings to make them sensitive to O_2.

Outside the capsule of each body, the nerve fibers acquire a myelin sheath; however, they are only 2–5

Table 36–1. Stimuli affecting the respiratory center.

Chemical control
CO_2 (via CSF and brain interstitial fluid H^+ concentration)
O_2, H^+ (via carotid and aortic bodies)
Nonchemical control
Vagal afferents from inflation and deflation receptors in the lungs
Afferents from pons, hypothalamus, and limbic system
Afferents from proprioceptors
Afferents from pharynx, trachea, and bronchi for sneezing, coughing, and swallowing
Afferents from baroreceptors: arterial, atrial, ventricular, pulmonary

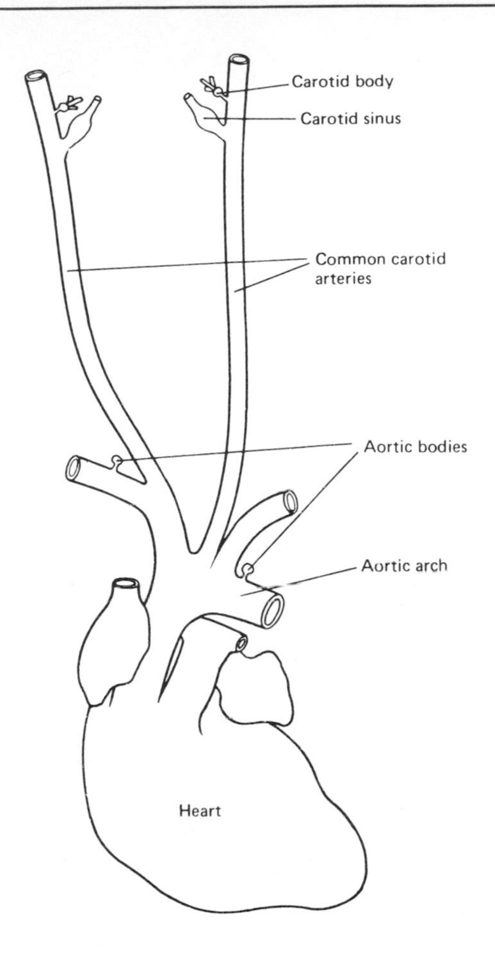

Figure 36–4. Location of carotid and aortic bodies.

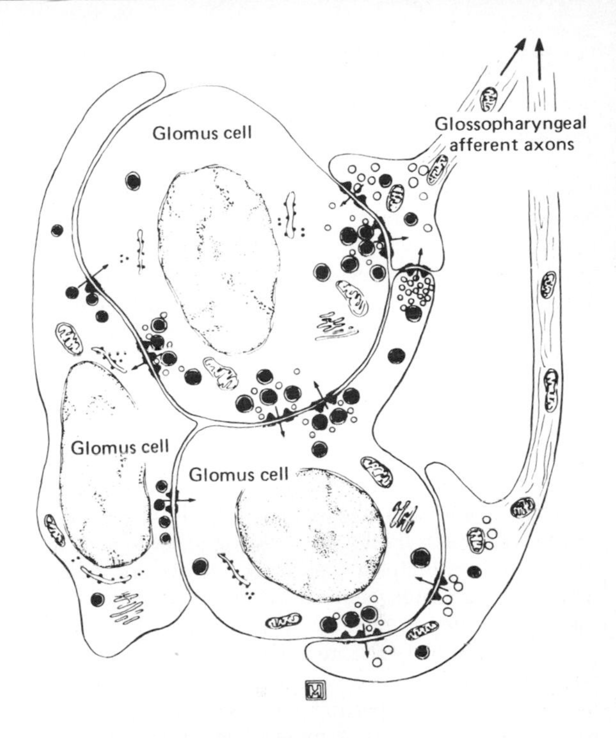

Figure 36–5. Organization of the carotid body. Type II cells are not shown. According to the view illustrated here, which is based on studies in the rat, there are reciprocal synapses (arrows) between the afferent neurons and the glomus cells. (Modified from McDonald DM, Mitchell RA: The innervation of glomus cells, ganglion cells, and blood vessels in the rat carotid body: A quantitative ultrastructural study. *J Neurocytol* 1975;4:177. By permission from Chapman & Hall Ltd.)

μm in diameter and conduct at the relatively low rate of 7–12m/s. Afferents from the carotid bodies ascend to the medulla via the carotid sinus and glossopharyngeal nerves, and fibers from the aortic bodies ascend in the vagi. Studies in which one carotid body has been isolated and perfused while recordings are being taken from its afferent nerve fibers show that there is a graded increase in impulse traffic in these afferent fibers as the P_{O_2} of the perfusing blood is lowered (Fig 36–6) or the P_{CO_2} raised.

The blood flow in each 2-mg carotid body is about 0.04 mL/min, or 2000 mL per 100 g of tissue per minute, compared with a blood flow per 100 g per minute of 54 mL in the brain and 420 mL in the kidney (Table 32–1). Because the blood flow per unit of tissue is so enormous, the O_2 needs of the cells can be met largely by dissolved O_2 alone. Therefore, the receptors are not stimulated in conditions such as anemia and carbon monoxide poisoning, in which the amount of dissolved O_2 in the blood reaching the receptors is generally normal even though the combined O_2 in the blood is markedly decreased. The receptors are stimulated when the arterial P_{O_2} is low or when, because of vascular stasis, the amount of O_2 delivered to the receptors per unit of time is decreased. Powerful stimulation is also produced by drugs such as cyanide, which prevent O_2 utilization at the tissue level. In sufficient doses, nicotine and lobeline activate the chemoreceptors.

Because of their anatomic location, the aortic bodies have not been studied in as great detail as the carotid bodies. Their responses are probably similar

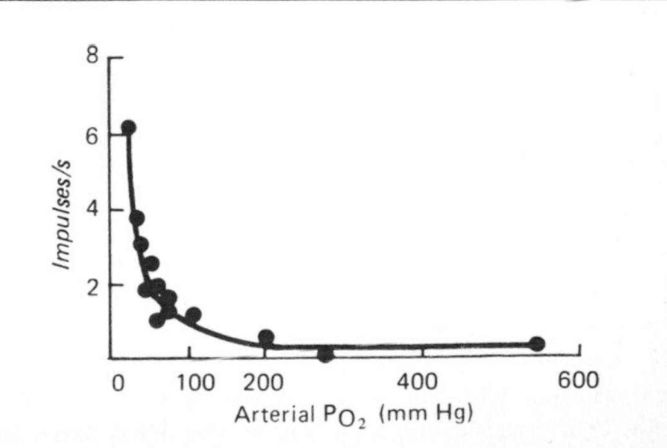

Figure 36–6. Change in the rate of discharge of a single afferent fiber from the carotid body when Pa_{O_2} is reduced. (Courtesy of S Sampson.)

but of lesser magnitude. In humans in whom both carotid bodies have been removed but the aortic bodies have been left intact, the responses are essentially the same as those following denervation of both carotid and aortic bodies in animals; there is little change in ventilation at rest, but the ventilatory response to hypoxia is lost and there is a 30% reduction in the ventilatory responses to CO_2. The humans who were studied had their carotid bodies removed in an effort to relieve severe asthma, but it now appears that bilateral carotid body resection is of little benefit in this disease.

Chemoreceptors in the Brain Stem

The chemoreceptors that mediate the hyperventilation produced by increases in arterial P_{CO_2} after the carotid and aortic bodies are denervated are located in the medulla oblongata and consequently are called **medullary chemoreceptors.** They are near the respiratory center but separate from it. The response to CO_2, for example, is depressed during anesthesia and natural sleep, but the response to hypoxia is unchanged. This experimental observation indicates that CO_2 does not act directly on the inspiratory neurons.

The location of the medullary chemoreceptors on the ventral surface of the brain stem is shown in Fig 36–7. They monitor the H^+ concentration of the cerebrospinal fluid, including the brain interstitial fluid. CO_2 readily penetrates membranes, including the blood-brain and blood-CSF barriers, whereas H^+ and HCO_3^- penetrate slowly. The CO_2 that enters the brain and CSF is promptly hydrated. The H_2CO_3 dissociates, so that the local H^+ concentration rises. The H^+ concentration in brain interstitial fluid parallels the arterial P_{CO_2}. Experimentally produced changes in the P_{CO_2} of spinal fluid have minor, variable effects on respiration as long as the H^+ concentration is held constant, but any increase in spinal fluid H^+ concentration stimulates respiration. The magnitude of the stimulation is proportionate to the rise in H^+ concentration. Thus, the effects of CO_2 on respiration are mainly due to its movement into the spinal fluid and brain interstitial fluid, where it increases the H^+ concentration and stimulates receptors sensitive to H^+.

Pulmonary & Myocardial "Chemoreceptors"

Bradycardia and hypotension produced by injections of veratridine and nicotine into the coronary circulation (Bezold-Jarisch reflex) and the pulmonary circulation are due to stimulation of "chemoreceptors" of some type in the coronary and pulmonary vessels (see Chapter 31). Such injections also produce brief periods of respiratory arrest **(apnea),** but the effects on respiration of stimuli from these receptors in physiologic situations are probably insignificant.

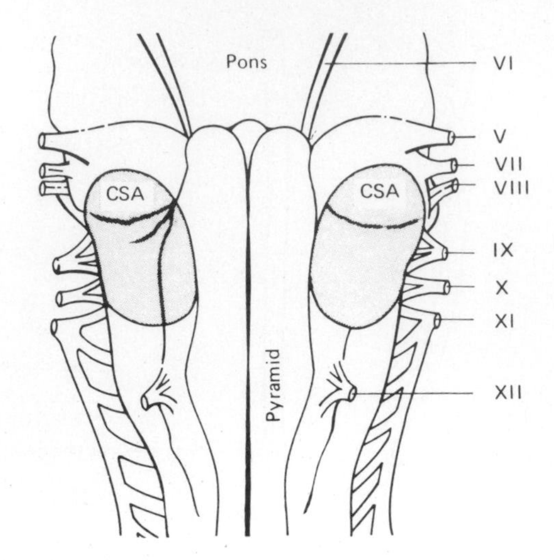

Figure 36–7. Chemosensitive areas (CSA) on the ventral surface of the medulla.

Ventilatory Responses to Changes in Acid-Base Balance

In metabolic acidosis due, for example, to the accumulation of the acid ketone bodies in the circulation in diabetes mellitus, there is pronounced respiratory stimulation (Kussmaul breathing; see Chapter 19). The hyperventilation decreases alveolar P_{CO_2} ("blows off CO_2") and thus produces a compensatory fall in blood H^+ concentration (see Chapter 39). Conversely, in metabolic alkalosis due, for example, to protracted vomiting with loss of HCl from the body, ventilation is depressed and the arterial P_{CO2} rises, raising the H^+ concentration toward normal (see Chapter 39). If there is an increase in ventilation that is not secondary to a rise in arterial H^+ concentration, the drop in P_{CO_2} lowers the H^+ concentration below normal **(respiratory alkalosis);** conversely, hypoventilation that is not secondary to a fall in plasma H^+ concentration causes **respiratory acidosis.**

Ventilatory Responses to CO_2

The arterial P_{CO_2} is normally maintained at 40 mm Hg. When there is a rise in arterial P_{CO_2} as a result of increased tissue metabolism, ventilation is stimulated and the rate of pulmonary excretion of CO_2 is increased until the arterial P_{CO_2} falls to normal, shutting off the stimulus. The operation of this feedback mechanism keeps CO_2 excretion and production in balance.

When a gas mixture containing CO_2 is inhaled, the alveolar P_{CO_2} rises, elevating the arterial P_{CO_2} and stimulating ventilation as soon as the blood that contains more CO_2 reaches the medulla. CO_2 elimination is increased, and the alveolar P_{CO_2} drops toward normal. This is why relatively large increments in the P_{CO_2} of inspired air (eg, 15 mm Hg) produce relatively slight increments in alveolar P_{CO_2} (eg, 3 mm Hg). However, the P_{CO_2} does not drop to normal, and a new equilibrium is reached at which the alveolar P_{CO_2} is slightly elevated and the hyperventilation persists as long as CO_2 is inhaled. The essentially linear relationship between respiratory minute volume and the alveolar P_{CO_2} is shown in Fig 36–8.

There is, of course, an upper limit to this linearity. When the P_{CO_2} of the inspired gas is close to the alveolar P_{CO_2}, elimination of CO_2 becomes difficult. When the CO_2 content of the inspired gas is more than 7%, the alveolar and arterial P_{CO_2} begin to rise abruptly in spite of hyperventilation. The resultant accumulation of CO_2 in the body **(hypercapnia)** depresses the central nervous system, including the respiratory center, and produces headache, confusion, and eventually, coma **(CO_2 narcosis).**

Ventilatory Response to Oxygen Lack

When the O_2 content of the inspired air is decreased, there is an increase in respiratory minute volume. The stimulation is slight when the P_{O_2} of the inspired air is more than 60 mm Hg, and marked stimulation of respiration occurs only at lower P_{O_2} values (Fig 36–9). However, any decline in arterial P_{O_2} below 100 mm Hg produces increased discharge in the nerves from the carotid and aortic chemoreceptors. There are 2 reasons why in normal individuals this increase in impulse traffic does not increase ventilation to any extent until the P_{O_2} is less than 60 mm Hg. Because Hb is a weaker acid than HbO_2 (see Chapter 35), there is a slight decrease in the H^+ concentration of arterial blood when the arterial P_{O_2} falls and hemoglobin becomes less saturated with O_2. The fall in H^+ concentration tends to inhibit respiration. In addition, any increase in ventilation that does occur lowers the alveolar P_{CO_2}, and this also tends to inhibit respiration. Therefore, the stimulatory effects of hypoxia on ventilation are not clearly manifest until they become strong enough to override the counterbalancing inhibitory effects of a decline in arterial H^+ concentration and P_{CO_2}.

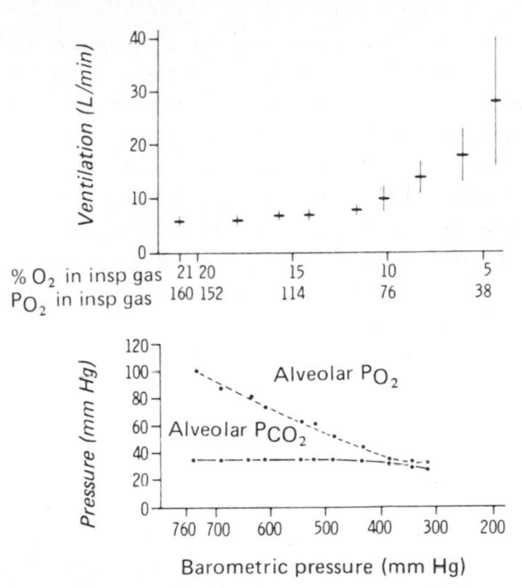

Figure 36–9. *Top:* Average respiratory minute volume during the first half hour of exposure to gases containing various amounts of O_2. The horizontal line in each case indicates the mean; the vertical bar indicates one standard deviation. *Bottom:* Alveolar P_{O_2} and P_{CO_2} values when breathing air at various barometric pressures. The 2 graphs are aligned so that the P_{O_2} of the inspired gas mixtures in the upper graph correspond to the P_{O_2} at the various barometric pressures in the lower graph. (Data from various authors, compiled by RH Kellogg.)

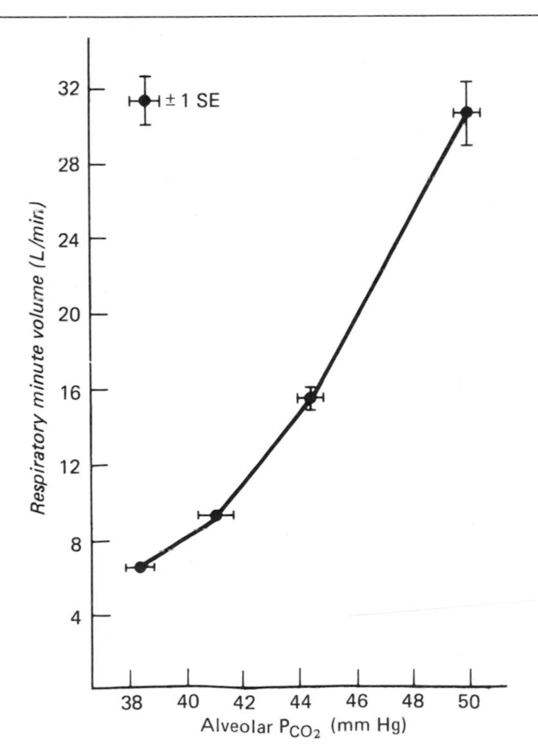

Figure 36–8. Responses of normal subjects to inhaling O and approximately 2,4, and 6% CO_2. The increase in respiratory minute volume is due to an increase in both the depth and rate of respiration. (Reproduced, with permission, from Lambertsen CJ in: *Medical Physiology,* 13th ed. Mountcastle VB [editor]. Mosby, 1974.)

The effects on ventilation of decreasing the alveolar P_{O_2} while holding the alveolar P_{CO_2} constant are shown in Fig 36–10. When the alveolar P_{CO_2} is stabilized at a level 2–3 mm Hg above normal, there is an inverse relationship between ventilation and the alveolar P_{O_2} even in the 90–110 mm Hg range; but when the alveolar P_{CO_2} is fixed at lower than normal values, there is no stimulation of ventilation by hypoxia until the alveolar P_{O_2} falls below 60 mm Hg.

Effects of Hypoxia on the CO_2 Response Curve

When the converse experiment is performed, ie, when the alveolar P_{O_2} is held constant while the response to varying amounts of inspired CO_2 is tested, a linear response is obtained (Fig 36–11). When the CO_2 response is then repeated at different fixed PO_2 values, the slope of the response curve changes, with the slope increased when alveolar PO_2 is decreased. In other words, hypoxia makes the individual more sensitive to increases in arterial P_{CO_2}. However, the alveolar P_{CO_2} level at which the curves in Fig 36–11 intersect is unaffected. In the normal individual, this threshold value is just below the normal alveolar P_{CO_2}, indicating that normally there is a very slight but definite "CO_2 drive" of the respiratory center.

Effect of H^+ on the CO_2 Response

The stimulatory effects of H^+ and CO_2 on respiration appear to be additive and not, like those of CO_2 and O_2, complexly interrelated. In metabolic acidosis, the CO_2 response curves are similar to those in Fig 36–11, except that they are shifted to the left. In other words, the same amount of respiratory stimulation is produced by lower arterial P_{CO_2} levels. It has been calculated that the CO_2 response curve shifts 0.8 mm Hg to the left for each nanomole rise in arterial H^+. About 40% of the ventilatory response to CO_2 is removed if the increase in arterial H^+ produced by CO_2 is prevented. As noted above, the remaining 60% is probably due to the effect of CO_2 on spinal fluid or brain interstitial fluid H^+ concentration.

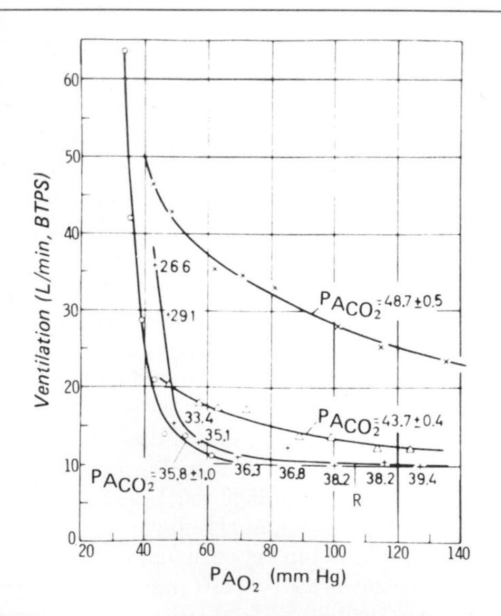

Figure 36–10. Ventilation at various alveolar P_{O2} values in a subject in whom the alveolar P_{CO2} was held constant at 48.7, 43.7, and 35.8 mm Hg and, in one experiment, varied as shown by the individual values along this curve *(R)*. (Slightly modified and reproduced, with permission, from Loeschke HH, Gertz KH: Einfluss des O_2-Druckes in der Einatmungsluft auf die Atemtätigkeit des Menschen, geprüft unter Konstanthaltung des alveolaren CO_2-Druckes. *Arch Ges Physiol Pflügers* 1958;**267**:460.)

Breath Holding

Respiration can be voluntarily inhibited for some time, but eventually the voluntary control is overridden. The point at which breathing can no longer be voluntarily inhibited is called the **breaking point.** Breaking is due to the rise in arterial P_{CO_2} and the fall in P_{O_2}. Individuals can hold their breath longer after removal of the carotid bodies. Breathing 100% oxygen before breath holding raises alveolar P_{O_2} initially, so that the breaking point is delayed. The same is true of hyperventilating room air, because CO_2 is blown off and arterial P_{CO_2} is lower at the start. Reflex or mechanical factors appear to influence the breaking point, since subjects who hold their breath as long as possible and then breathe a gas mixture low in O_2 and high in CO_2 can hold their breath for an additional 20 seconds or more. Psychic factors also play a role, and subjects can hold their breath longer when they are told their performance is very good than when they are not.

NONCHEMICAL INFLUENCES ON RESPIRATION

Pulmonary Receptors

The vagally mediated inhibition of inspiration produced by inflation of the lungs, which has been discussed above, is illustrated in Fig 36–3. The response is due to stimulation of **stretch receptors** located in the smooth muscle of the airways. Pulmonary deflation receptors that trigger inflation have also been described, and the expiratory and inspiratory

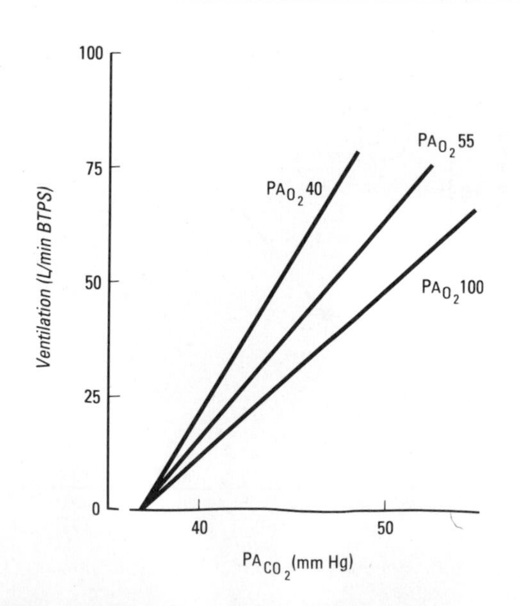

Figure 36–11. Fan of lines showing CO_2 response curves at various fixed values for alveolar P_{O2} (Modified from Keele CA, Neil E, Joels N: *Samson Wright's Applied Physiology*, 13th ed. Oxford, 1982.)

reflex responses to pulmonary inflation and deflation, respectively, have been known as the **Hering-Breuer inflation** and **deflation reflexes.** However, the deflation receptors respond better to pulmonary congestion and embolization, producing shallow, rapid breathing, and they have come to be called **J receptors** instead, because of their juxtacapillary location.

Responses to Irritation of the Air Passages

Throughout the airways, from the trachea to the respiratory bronchioles, there are endings of myelinated vagal afferents that ramify under and between the cells of the epithelium. These endings function as **irritant receptors.** In the trachea and extrapulmonary bronchi, stimulation of these receptors by chemical or mechanical irritants causes coughing. Within the lungs, stimulation of the irritant receptors does not cause coughing, but instead causes rapid, shallow breathing and bronchoconstriction.

Coughing begins with a deep inspiration followed by forced expiration against a closed glottis. This increases the intrapleural pressure to 100 mm Hg or more. The glottis is then suddenly opened, producing an explosive outflow of air at velocities up to 965 km (600 miles) per hour. Sneezing is a similar expiratory effort with a continuously open glottis (see Chapter 14). These reflexes help expel irritants and keep airways clear.

Stimulation of the irritant receptors in the lungs augments the activity of inspiratory motor neurons. The rapid, shallow breathing and bronchoconstriction that is initiated may limit the amount of noxious gases and irritants that reach the gas-exchanging surfaces. Stimulation of the receptors may also initiate the periodic sighs that occur during normal breathing, keeping the lungs expanded. The receptors are stimulated by the histamine and other chemical substances released in the lungs during allergic reactions.

Afferents From "Higher Centers"

Pain and emotional stimuli affect respiration, so there must also be afferents from the limbic system and hypothalamus to the respiratory neurons in the brain stem. In addition, even though breathing is not usually a conscious event, both inspiration and expiration are under voluntary control. The pathways for voluntary control pass from the neocortex to the motor neurons innervating the respiratory muscles, bypassing the medullary neurons.

Since voluntary and automatic control of respiration are separate, automatic control is sometimes disrupted without loss of voluntary control. The clinical condition that results has been called **Ondine's curse.** The German legend, Ondine was a water nymph who had an unfaithful mortal lover. The king of the water nymphs punished the lover by casting a curse upon him that took away all his automatic functions. In this state, he could stay alive only by staying awake and remembering to breathe. He eventually fell asleep from sheer exhaustion and his respiration stopped. Patients with this intriguing condition generally have disease processes that compress the medulla or bulbar poliomyelitis. The condition has also been inadvertently produced in patients who have been subjected to bilateral anterolateral cervical cordotomy for pain (see Chapter 7). This cuts the pathways that bring about automatic respiration while leaving the voluntary efferent pathways in the corticospinal and rubrospinal tracts intact.

Afferents From Proprioceptors

Carefully controlled experiments have shown that active and passive movements of joint stimulate respiration, presumably because impulses in afferent pathways from proprioceptors in muscles, tendons, and joints stimulate the respiratory center. This effect probably helps increase ventilation during exercise.

Respiratory Components of Visceral Reflexes

The respiratory adjustments during vomiting, swallowing, and gagging are discussed in Chapters 14 and 26. Inhibition of respiration and closure of the glottis during these activities not only prevent the aspiration of food or vomitus into the trachea but, in the case of vomiting, fix the chest so that contraction of the abdominal muscles increases the intra-abdominal pressure. Similar glottic closure and inhibition of respiration occur during voluntary and involuntary straining.

Hiccup is a spasmodic contraction of the diaphragm that produces an inspiration during which the glottis suddenly closes. The glottic closure is responsible for the characteristic sensation and sound. Yawning is a peculiar "infectious" respiratory act the physiologic basis and significance of which are uncertain. However, underventilated alveoli have a tendency to collapse, and it has been suggested that the deep inspiration and stretching open up alveoli and prevent the development of atelectasis. Yawning also increases venous return to the heart.

Respiratory Effects of Baroreceptor Stimulation

Afferent fibers from the baroreceptors in the carotid sinuses, aortic arch, atria, and ventricles relay to the respiratory center as well as the vasomotor and cardioinhibitory areas in the medulla. Impulses in them inhibit respiration, but the inhibitory effect is slight and of little physiologic importance. The hyperventilation in shock is due to chemoreceptor stimulation caused by acidosis and hypoxia secondary to local stagnation of blood flow and is not baroreceptor-mediated. The activity of inspiratory neurons affects blood pressure and heart rate (see Chapters 28 and 31), and activity in the vasomotor and cardiac areas in the medulla may have minor effects on respiration.

Effects of Sleep

Respiration is less rigorously controlled during sleep than in the waking state, and brief periods of apnea occur in normal sleeping adults. The causes of the changes include a decreased sensitivity to CO_2 during slow-wave sleep (Fig 36–12). There are variable changes in the ventilatory response to hypoxia. If the P_{CO_2} falls during the waking state, various stimuli from proprioceptors and the environment maintain respiration, but during sleep, these stimuli are decreased and a decrease in P_{CO_2} can cause apnea. During REM sleep, breathing is irregular.

In a small percentage of individuals, apnea during sleep can constitute a clinical problem of appreciable magnitude. This **sleep-apnea syndrome** can occur at any age. The symptoms include morning headaches, fatigue, and in more advanced cases, the clinical picture of respiratory failure with normal lungs. These patients are polycythemic, hypoxemic, and hypercapnic. The causes of sleep apnea are not yet clearly defined, but one cause appears to be failure during sleep of the genioglossus muscles to contract during inspiration. These muscles pull the tongue forward, and when they do not contract, the tongue falls back and obstructs the airway.

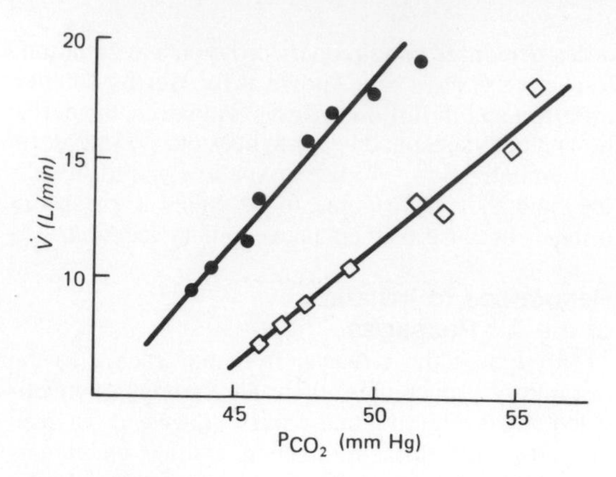

Figure 36–12. Effect of graded increases in P_{CO_2} on ventilation (V) during sleep in a healthy young adult. The solid circles are the responses while the subject was awake, and the open squares are the responses during slow-wave sleep. (Reproduced, with permission, from Cherniak NS: Respiratory dysrhythmias during sleep. *N Engl J Med* 1981;**305**:325.)

Sudden Infant Death Syndrome

The sudden infant death syndrome (SIDS) may be a form of sleep apnea. This disorder, in which apparently healthy infants are found dead, often in their cribs, has attracted a great deal of attention. Apneic spells are common in premature infants. However, periods of prolonged apnea do not correlate with the subsequent occurrence of death, and none of the known tests of chemoresponsiveness reliably predict which infants will subsequently have difficulty.

Respiratory Adjustments in Health & Disease

37

INTRODUCTION

This chapter is concerned with the respiratory adjustments to exercise, high altitude, and hypoxia. In the same way that the compensatory adjustments of the cardiovascular system to environmental changes and disease illustrate the integrated operation of the cardiovascular regulatory mechanisms (see Chapter 33), these respiratory adjustments highlight the operation of the respiratory regulatory mechanisms discussed in Chapters 34–36.

EFFECTS OF EXERCISE

Many cardiovascular and respiratory mechanisms must operate in an integrated fashion if the O_2 needs of the active tissue are to be met and the extra CO_2 and heat removed from the body during exercise. Circulatory changes increase muscle blood flow while maintaining adequate circulation in the rest of the body (see Chapter 33). There is in addition an increase in the extraction of O_2 from the blood in exercising muscles and an increase in ventilation that provides extra O_2, eliminates some of the heat, and excretes extra CO_2.

Changes in Ventilation

During exercise, the amount of O_2 entering the blood in the lungs is increased because the amount of O_2 added to each unit of blood and the pulmonary blood flow per minute are increased. The P_{O_2} of blood flowing into the pulmonary capillaries falls from 40 mm Hg to 25 mm Hg or less, so that the alveolar-capillary P_{O_2} gradient is increased and more O_2 enters the blood. Blood flow per minute is increased from 5.5L/min to as much as 20–35 L/min. The total amount of O_2 entering the blood therefore increases from 250 mL/min at rest to values as high as 4000 mL/min. The amount of CO_2 removed from each unit of blood is increased, and CO_2 excretion increases from 200 mL/min to as much as 8000 mL/min. The increase in O_2 uptake is proportionate to work load up to a maximum. Above this maximum, O_2 consumption levels off and the blood lactate level continues to rise (Fig 37–1). The lactate comes from muscles in which aerobic resynthesis of energy stores cannot keep pace with their utilization and an **oxygen debt** is being incurred (see Chapter 3).

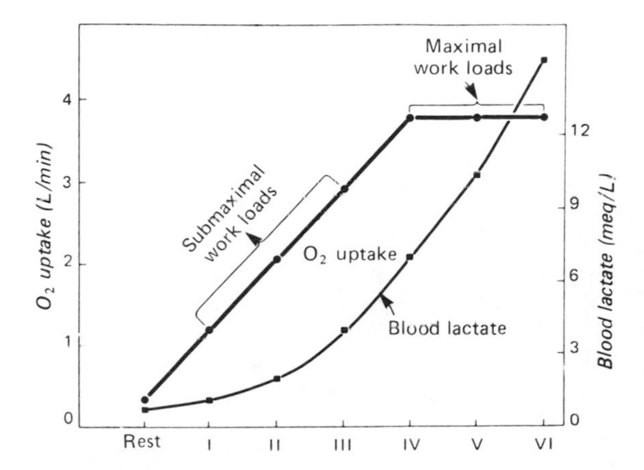

Figure 37–1. Relation between work load, blood lactate level, and O_2 uptake. I–VI, increasing work loads produced by increasing the speed and grade of a treadmill on which the subjects worked. (Reproduced, with permission, from Mitchell JH, Blomqvist G: Maximal oxygen uptake. *N Engl J Med* 1971;**284:**1018.)

There is an abrupt increase in ventilation with the onset of exercise, followed after a brief pause by a further, more gradual increase (Fig 37–2). With moderate exercise, the increase is due mostly to an increase in the depth of respiration; this is accompanied by an increase in the respiratory rate when the exercise is more strenuous. There is an abrupt decrease in ventilation when exercise ceases, followed after a brief pause by a more gradual decline to preexercise

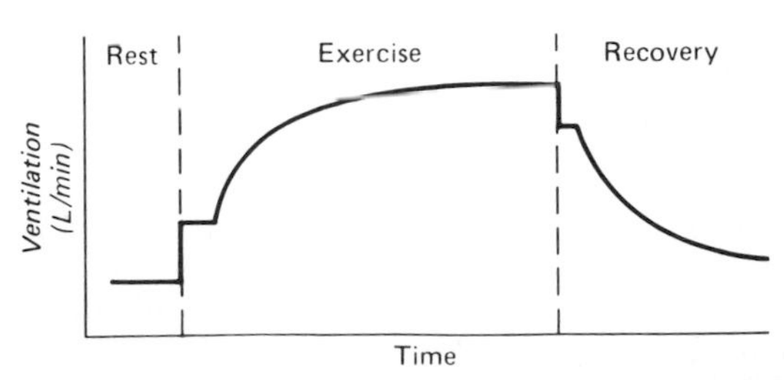

Figure 37–2. Changes in ventilation during exercise. (Modified from Wasserman K, Whipp BJ, Casaburi R: Respiratory control during exercise. In: *Handbook of Physiology* Vol II, part 2. Section 3: *The Respiratory System.* Fishman AP [editor]. American Physiological Society, 1986.)

values. The abrupt increase at the start of exercise is presumably due to psychic stimuli and afferent impulses from proprioceptors in muscles, tendons, and joints. The more gradual increase is presumably humoral even though arterial pH, P_{CO_2}, and P_{O_2} remain constant during moderate exercise. The increase in ventilation is proportionate to the increase in O_2 consumption, but the mechanisms responsible for the stimulation of respiration are still the subject of much debate. The increase in body temperature may play a role. In addition, it may be that the sensitivity of the respiratory center to CO_2 is increased or that the respiratory fluctuations in arterial P_{CO_2} increase so that, even though the mean arterial P_{CO_2} does not rise, it is CO_2 that is responsible for the increase in ventilation. O_2 also seems to play some role despite the lack of a decrease in arterial P_{O_2}, since, during the performance of a given amount of work, the increase in ventilation while breathing 100% O_2 is 10–20% less than the increase while breathing air (Fig 37–3). Thus, it currently appears that a number of different factors combine to produce the increase in ventilation seen during moderate exercise.

When exercise becomes more severe, buffering of the increased amounts of lactic acid that are produced liberates more CO_2, and this further increases ventilation. The response to graded exercise is shown in Fig 37–4. With increased production of acid, the increases in ventilation and CO_2 production remain proportionate, so alveolar and arterial CO_2 change relatively little **(isocapnic buffering).** Because of the hyperventilation, alveolar P_{O_2} increases. With further accumulation of lactic acid, the increase in ventilation outstrips CO_2 production and alveolar P_{CO_2} falls, as does arterial P_{CO_2}. The decline in arterial P_{CO_2} provides respiratory compensation (see Chapter 39) for the metabolic acidosis produced by the additional lactic acid. The additional increase in ventilation produced by the acidosis is dependent on the carotid bodies and does not occur if they are removed.

The respiratory rate after exercise does not reach basal levels until the O_2 debt is repaid. This may take as long as 90 minutes. The stimulus to ventilation is not the arterial P_{CO_2}, which is normal or low, or the arterial P_{O_2}, which is normal or high, but the elevated arterial H^+ concentration due to the lactic acidemia. The magnitude of the O_2 debt is the amount by which O_2 consumption exceeds basal consumption from the end of exertion until the O_2 consumption has returned to preexercise basal levels. During repayment of the O_2 debt, there is a small rise in the O_2 in muscle myoglobin. ATP and phosphorylcreatine are resynthesized, and lactic acid is removed. Eighty percent of the lactic acid is converted to glycogen and 20% is metabolized to CO_2 and H_2O.

Because of the extra CO_2 produced by the buffering of lactic acid during strenuous exercise, the RQ rises, values reaching 1.5–2.0. After exertion, while the O_2 debt is being repaid, the RQ falls to 0.5 or less.

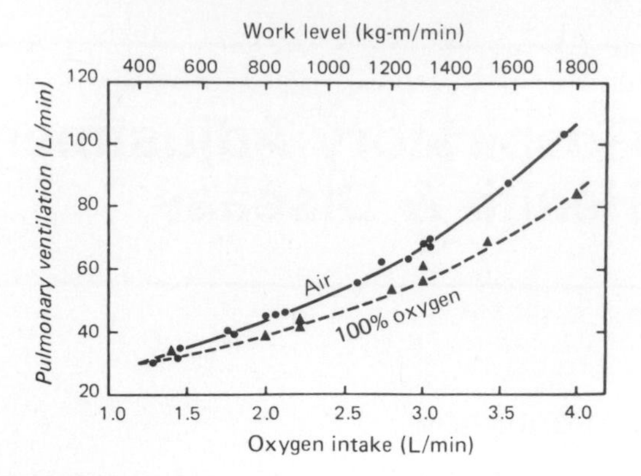

Figure 37–3. Effect of 100% O_2 on ventilation during work at various levels on the bicycle ergometer. (Reproduced, with permission, from Astrand P-O: The respiratory activity in man exposed to prolonged hypoxia. *Acta Physiol Scand* 1954;**30**:343.)

Changes in the Tissues

Maximum O_2 uptake during exercise is limited by the maximum rate at which O_2 is transported to the

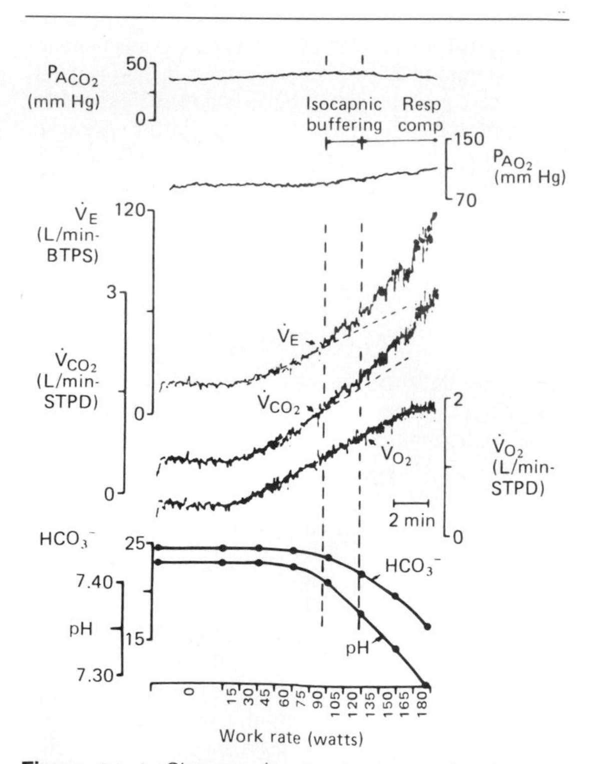

Figure 37–4. Changes in alveolar P_{CO_2}, alveolar P_{O_2}, ventilation (V_E), CO_2 production (V_{CO2}), O_2 consumption (V_{O2}), arterial HCO_3^-, and arterial pH with graded increases in work on a bicycle ergometer. The subject was a normal adult male. (Reproduced, with permission, from Wasserman K, Whipp BJ, Casaburi R: Respiratory control during exercise. In: *Handbook of Physiology.* Vol II, part 2. Section 3: *The Respiratory System.* Fishman AP [editor]. American Physiological Society, 1986.)

mitochondria in the exercising muscle. However, this limitation is not normally due to deficient O_2 uptake in the lungs, and hemoglobin in arterial blood is saturated even during the most severe exercise.

During exercise, the contracting muscles use more O_2, and the tissue P_{O_2} as well as the P_{O_2} in venous blood from exercising muscle fall nearly to zero. More O_2 diffuses from the blood, the blood P_{O_2} drops, and more O_2 is removed from hemoglobin. Because the capillary bed is dilated and many previously closed capillaries are open, the mean distance from the blood to the tissue cells is greatly decreased; this facilitates the movement of O_2 from blood to cells. The O_2 hemoglobin dissociation curve is steep in the P_{O_2} range below 60 mm Hg, and a relatively large amount of O_2 is supplied for each drop of 1 mm Hg in P_{O_2} (Fig 35–2). Additional O_2 is supplied because, as a result of the accumulation of CO_2 and the rise in temperature in active tissues—and perhaps because of a rise in red blood cell 2,3-DPG—the dissociation curve shifts to the right (Fig 35–3). The net effect is a 3-fold increase in O_2 extraction from each unit of blood. Since this increase is accompanied by a 30-fold or greater increase in blood flow, it permits the metabolic rate of muscle to rise as much as 100-fold during exercise (see Chapter 3).

Fatigue

Fatigue is a poorly understood phenomenon to which acidosis and other factors contribute. The subjective hardness or "heaviness" of exercise correlates with the rate of O_2 consumption, not with the actual work performed in kg-m/min. Barrages of impulses in afferents from proprioceptors in muscles are said to make one feel "tired." The effects of the acidosis on the brain may contribute to the sensation of fatigue. Prolonged exercise produces hypoglycemia in normal individuals, but prevention of the hypoglycemia does not affect endurance or delay the onset of exhaustion. On the other hand, there is a correlation in humans between exhaustion and the degree of depletion of muscle glycogen, as determined by muscle biopsy. Sustained muscle contractions are painful, because the muscle becomes ischemic and a substance that stimulates pain endings accumulates ("P factor"; see Chapter 7), but intermittent contractions are not painful, because the P factor is washed away. Muscle stiffness may be due in part to the accumulation of interstitial fluid in the muscles during exertion.

Heat dissipation during exercise is discussed in Chapter 33 and summarized in Fig 33–4. The changes in acid-base balance associated with respiration are reviewed in Chapter 39.

HYPOXIA

Hypoxia is O_2 deficiency at the tissue level. It is a more correct term than **anoxia,** there rarely being no O_2 at all left in the tissues.

Traditionally, hypoxia has been divided into 4 types. Numerous other classifications have been used, but the 4-type system still has considerable utility if the definitions of the terms are kept clearly in mind. The 4 categories are as follows: (1) **hypoxic hypoxia (anoxic anoxia),** in which the P_{O_2} of the arterial blood is reduced; (2) **anemic hypoxia,** in which the arterial P_{O_2} is normal but the amount of hemoglobin available to carry O_2 is reduced; (3) **stagnant** or **ischemic hypoxia,** in which the blood flow to a tissue is so low that adequate O_2 is not delivered to it despite a normal P_{O_2} and hemoglobin concentration; and (4) **histotoxic hypoxia,** in which the amount of O_2 delivered to a tissue is adequate but, because of the action of a toxic agent, the tissue cells cannot make use of the O_2 supplied to them.

Effects of Hypoxia

The effects of stagnant hypoxia depend upon the tissue affected. In hypoxic hypoxia and the other generalized forms of hypoxia, the brain is affected first. A sudden drop in the inspired P_{O_2} to less than 20 mm Hg, which occurs, for example, when cabin pressure is suddenly lost in a plane flying above 16,000 m, causes loss of consciousness in about 20 seconds (Fig 37–5) and death in 4–5 minutes. Less

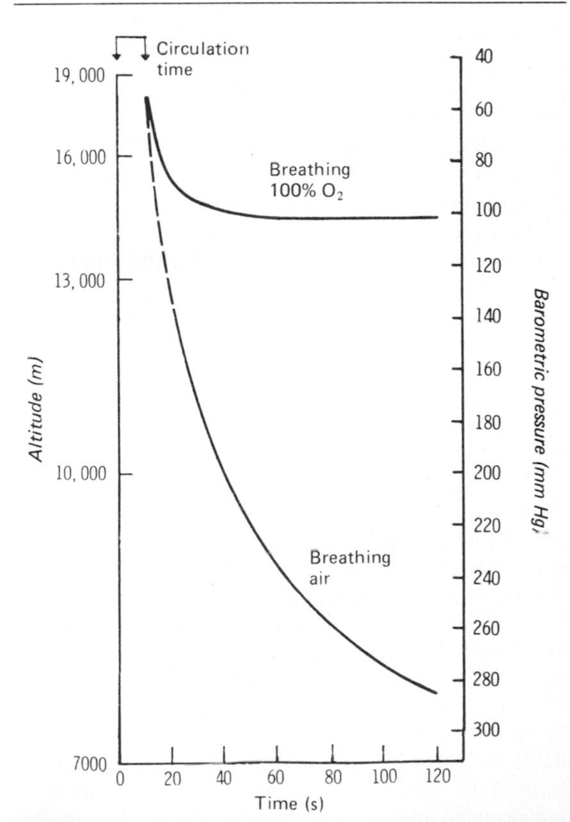

Figure 37–5. Duration of useful consciousness upon sudden exposure to the ambient pressure at various altitudes. Ten seconds is the approximate lung-to-brain circulation time.

severe hypoxia causes a variety of mental aberrations not unlike those produced by alcohol: impaired judgment, drowsiness, dulled pain sensibility, excitement, disorientation, loss of time sense, and headache. Other symptoms include anorexia, nausea, vomiting, tachycardia, and, when the hypoxia is severe, hypertension. The rate of ventilation is increased in proportion to the severity of the hypoxia of the carotid chemoreceptor cells.

Respiratory Stimulation

Dyspnea is by definition breathing in which the subject is conscious of shortness of breath; **hyperpnea** is the general term for an increase in the rate or depth of breathing regardless of the patient's subjective sensations. **Tachypnea** is rapid, shallow breathing. In general, a normal individual is not conscious of respiration until ventilation is doubled, and breathing is not uncomfortable (ie, the **dyspnea point** is not reached) until ventilation is tripled or quadrupled. Whether or not a given level of ventilation is uncomfortable also depends upon the respiratory reserve. This factor is taken into account in the **dyspneic index,** the percentage of the respiratory capacity not being used at a given respiratory minute volume.

When the dyspneic index is less than about 70%, dyspnea is usually present. An additional factor is the effort involved in moving the air in and out of the lungs (the work of breathing). In asthma, for example, the bronchi are constricted, especially during expiration and breathing against this increased airway resistance is uncomfortable work. Another factor producing dyspnea may be increased discharge in vagal afferents from lung irritant receptors.

Cyanosis

Reduced hemoglobin has a dark color, and a dusky bluish discoloration of the tissues called **cyanosis** appears when the reduced hemoglobin concentration of the blood in the capillaries is more than 5 g/dL. Cyanosis is most easily seen in the nail beds and mucous membranes and in the earlobes, lips, and fingers, where the skin is thin. Its occurrence depends upon the total amount of hemoglobin in the blood, the degree of hemoglobin unsaturation, and the state of the capillary circulation.

One might think that cyanosis would be more marked when the cutaneous vessels were dilated. However, when there is cutaneous arteriolar and venous constriction, blood flow through the capillaries is very slow and more O_2 is removed from the hemoglobin. This is why moderate cold causes cyanosis in exposed areas even in normal individuals. In very cold weather cyanosis does not develop, because the drop in skin temperature inhibits the dissociation of oxyhemoglobin and the O_2 consumption of the cold tissues is decreased. Cyanosis does not occur in anemic hypoxia when the total hemoglobin content is low; in carbon monoxide poisoning, because the color of reduced hemoglobin is obscured by the cherry-red color of carbonmonoxyhemoglobin (see below); or in histotoxic hypoxia, because the blood gas content is normal. A discoloration of the skin and mucous membranes similar to cyanosis is produced by high circulating levels of methemoglobin (see Chapter 27).

HYPOXIC HYPOXIA

Hypoxic hypoxia is a problem in normal individuals at high altitudes and is a complication of pneumonia and a variety of other diseases of the respiratory system.

Effects of Decreased Barometric Pressure

The composition of air stays the same, but the total barometric pressure falls with increasing altitude (Fig 37–6). Therefore, the P_{O_2} also falls. At 3000 m (approximately 10,000 ft) above sea level, the alveolar P_{O_2} is about 60 mm Hg and there is enough hypoxic stimulation of the chemoreceptors to definitely increase ventilation. As one ascends higher, the alveolar P_{O_2} falls less rapidly and the alveolar P_{CO_2} declines somewhat because of the hyperventilation. The resulting fall in arterial P_{CO_2} produces respiratory alkalosis.

Hypoxic Symptoms Breathing Air

There are a number of compensatory mechanisms that operate over a period of time to increase altitude tolerance **(acclimatization),** but in unacclimatized subjects, mental symptoms such as irritability appear at about 3700 m. At 5500 m, the hypoxic symptoms are severe; and at altitudes above 6100 m (20,000 ft), consciousness is usually lost (Fig 37–7).

Hypoxic Symptoms Breathing Oxygen

The total atmospheric pressure becomes the limiting factor in altitude tolerance when breathing 100% O_2. The partial pressure of water vapor in the alveolar air is constant at 47 mm Hg, and that of CO_2 is normally 40 mm Hg, so that the lowest barometric pressure at which a normal alveolar P_{O_2} of 100 mm Hg is possible is 187 mm Hg, the pressure at about 10,400 m (34,000 ft). At greater altitudes, the increased ventilation due to the decline in alveolar P_{O_2} lowers the alveolar P_{CO_2} somewhat, but the maximum alveolar P_{O_2} that can be attained breathing 100% O_2 at the ambient barometric pressure of 100 mm Hg at 13,700 m is about 40 mm Hg. At about 14,000 m, consciousness is lost in spite of the administration of 100% O_2 (Fig 37–5). However, an artificial atmosphere can be created around an individual; in a pressurized suit or cabin supplied with O_2 and a system to remove CO_2, it is possible to ascend to any altitude and to live in the vacuum of interplanetary space.

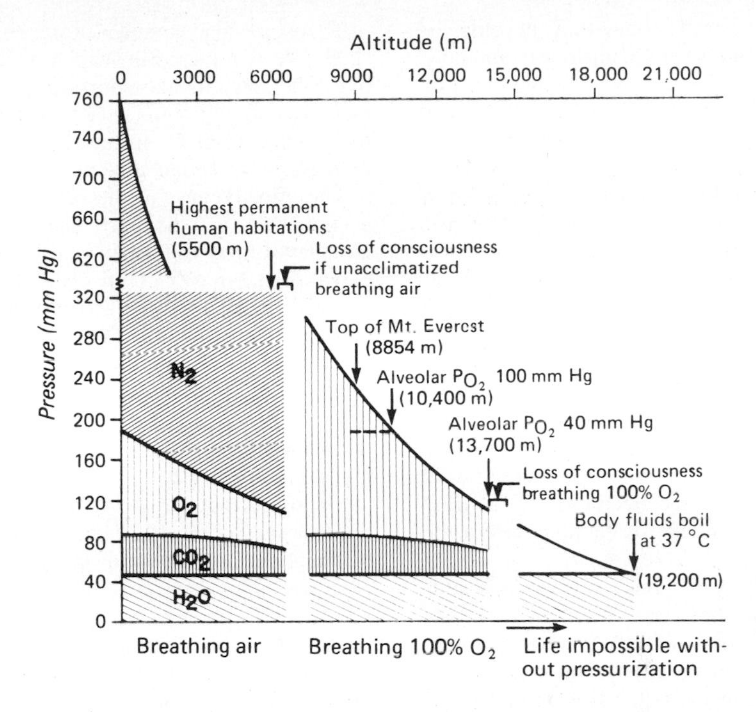

Figure 37–6. Composition of alveolar air in individuals breathing air (0–6100 m) and 100% O_2 (6100–13,700 m). The minimal alveolar P_{O_2} that an unacclimatized subject can tolerate without loss of consciousness is about 35–40 mm Hg. Note that with increasing altitude, the alveolar P_{CO_2} drops because of the hyperventilation due to hypoxic stimulation of the carotid and aortic chemoreceptors. The fall in barometric pressure with increasing altitude is not linear, because air is compressible.

At 19,200 m, the barometric pressure is 47 mm Hg, and at or below this pressure the body fluids boil at body temperature. The point is largely academic, however, because any individual exposed to such a low pressure would be dead of hypoxia before the bubbles of steam could cause death.

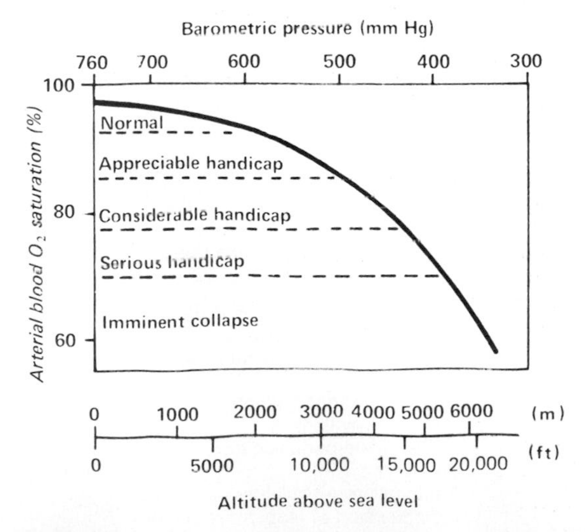

Figure 37–7. Acute effects of hypoxia in individuals breathing air at various altitudes.

Delayed Effects of High Altitude

Many individuals, when they first arrive at a high altitude, develop transient "mountain sickness." This syndrome develops 8–24 hours after arrival at altitude and lasts 4–8 days. It is characterized by headache, irritability, insomnia, breathlessness, and nausea and vomiting. Its cause is unsettled, but it appears to be associated with cerebral edema. The low P_{O_2} at high altitude causes arteriolar dilation, and if cerebral autoregulation does not compensate, there is an increase in capillary pressure that favors increased transudation of fluid into brain tissue. The symptoms are reduced if alkalosis is reduced by treatment with acetazolamide or if the cerebral edema is reduced by administration of large doses of glucocorticoids.

High altitude pulmonary edema and cerebral edema are serious forms of mountain sickness. Pulmonary edema is prone to occur in individuals who ascend quickly to altitudes above 2500 m and engage in heavy physical activity during the first 3 days after arrival. It is also seen in individuals acclimatized to high altitudes who spend 2 weeks or more at sea level and then reascend. It occurs in the absence of cardiovascular or pulmonary disease. It is associated with marked pulmonary hypertension, but left atrial pressures are normal. The protein content of the edema fluid is high, but the mechanism responsible for its production is unknown. It responds to rest and

O_2 treatment and generally does not develop in individuals who ascend to high altitudes gradually and avoid physical exertion for the first few days of high altitude exposure.

Acclimatization

Acclimatization to altitude is due to the operation of a variety of compensatory mechanisms. The respiratory alkalosis produced by the hyperventilation shifts the oxygen hemoglobin dissociation curve to the left, but there is a concomitant increase in red blood cell 2,3-DPG, which tends to decrease the O_2 affinity of hemoglobin. The net effect is an increase in P_{50} (see Chapter 35). The decrease in O_2 affinity makes more O_2 available to the tissues. However, it should be noted that the value of the increase in P_{50} is lost at very high altitudes, because when the arterial P_{O_2} is markedly reduced, the decreased O_2 affinity also interferes with O_2 uptake by hemoglobin in the lungs.

The initial ventilatory response to increased altitude is relatively small, because the alkalosis tends to counteract the stimulating effect of hypoxia. However, there is a steady increase in ventilation over the next 4 days (Fig 37–8) because the active transport of H^+ into CSF, or possibly a developing lactic acidosis in the brain, causes a fall in CSF pH that increases the response to hypoxia. After 4 days, the ventilatory response begins to decline slowly, but it takes years of residence at higher altitudes for it to decline to the initial level. Associated with this decline is a gradual desensitization to the stimulatory effects of hypoxia.

Erythropoietin secretion increases promptly on ascent to high altitude (see Chapter 24) and then falls somewhat over the following 4 days as the ventilatory response increases and the arterial P_{O_2} rises. The increase in circulating red blood cells triggered by the erythropoietin begins in 2–3 days and is sustained as long as the individual remains at high altitude.

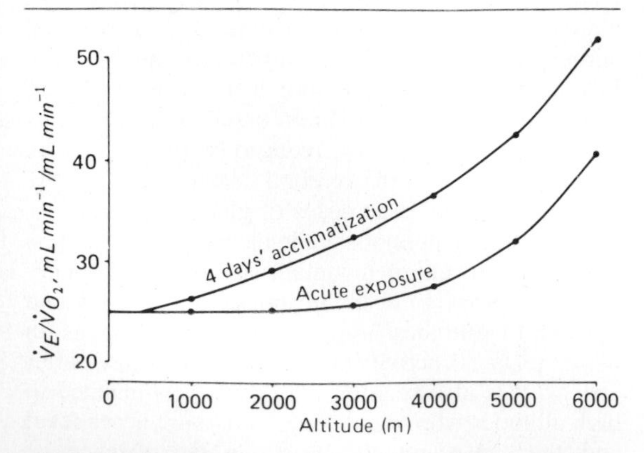

Figure 37–8. Effect of acclimatization on the ventilatory response at various altitudes. V_E/V_{O2} is the ventilatory equivalent, the ratio of expired minute volume (V_E) to the O_2 consumption (V_{O2}). (Reproduced, with permission, from Lenfant C, Sullivan K: Adaptation to high altitude. *N Engl J Med* 1971;**284**:1298.)

There are also compensatory changes in the tissues. The mitochondria, which are the site of oxidative reactions, increase in number, and there is an increase in myoglobin (see Chapter 35) that facilitates the movement of O_2 in the tissues. There is also an increase in the tissue content of cytochrome oxidase.

The effectiveness of the acclimatization processes is indicated by the fact that in the Andes and Himalayas there are permanent human habitations at elevations above 5500 m (18,000 ft). The natives who live in these villages are barrel-chested and markedly polycythemic. They have low alveolar P_{O_2} values, but in most other ways they are remarkably normal.

Diseases Causing Hypoxic Hypoxia

Hypoxic hypoxia is the most common form of hypoxia seen clinically. The diseases that cause it can be roughly divided into those in which the gas exchange apparatus fails, those such as congenital heart disease in which large amounts of blood are shunted from the venous to the arterial side of the circulation, and those in which the respiratory pump fails (Table 37–1). Lung failure occurs when conditions such as pulmonary fibrosis produce alveolar-capillary block or there is ventilation-perfusion imbalance. Pump failure can be due to depression of the respiratory center (as a result of, for example, drugs such as morphine), to fatigue of the respiratory muscles in conditions in which the work of breathing is increased, or to a variety of mechanical defects such as pneumothorax or bronchial obstruction that limit ventilation.

Ventilation-Perfusion Imbalance

Patchy ventilation-perfusion imbalance is by far the most common cause of hypoxic hypoxia in clinical situations. The physiologic effects of ventilation-perfusion imbalance and their role in producing the alterations in alveolar gas due to gravity are discussed in Chapter 34.

In disease processes that prevent ventilation of some of the alveoli, the ventilations-blood flow ratios in different parts of the lung determine the extent to which systemic arterial P_{O_2} declines. If nonventilated alveoli are perfused, the nonventilated but perfused portion of the lung is in effect a right-to-left shunt, dumping unoxygenated blood into the left side of the

Table 37–1. Disorders causing hypoxic hypoxia.

Lung failure (gas exchange failure)
Pulmonary fibrosis
Ventilation-perfusion imbalance
Shunt
Pump failure (ventilatory failure)
Depression of respiratory center
Mechanical defects
Fatigue

heart. Lesser degrees of ventilation-perfusion imbalance are more common. In the example illustrated in Fig 37–9, the underventilated alveoli (B) have a low alveolar P_{O_2}, whereas the overventilated alveoli (A) have a high alveolar P_{O_2}. However, the unsaturation of the hemoglobin of the blood coming from B is not completely compensated by the greater saturation of the blood coming from A, because hemoglobin is normally nearly saturated in the lungs and the higher alveolar P_{O_2} adds only a little more O_2 to the hemoglobin than it normally carries. Consequently, the arterial blood is unsaturated. On the other hand, the CO_2 content of the arterial blood is generally normal in such situations, since extra loss of CO_2 in overventilated regions can balance diminished loss in underventilated areas.

Venous to Arterial Shunts

When a cardiovascular abnormality such as an interatrial septal defect permits large amounts of unoxygenated venous blood to bypass the pulmonary capillaries and dilute the oxygenated blood in the systemic arteries ("right-to-left shunt"), chronic hypoxic hypoxia and cyanosis **(cyanotic congenital heart disease)** result. Administration of 100% O_2 raises the O_2 content of alveolar air and improves the hypoxia due to hypoventilation, impaired diffusion, or ventilation-perfusion imbalance (short of perfusion of totally unventilated segments) by increasing the amount of O_2 in the blood leaving the lungs. However, in patients with venous to arterial shunts and normal lungs, any beneficial effect of 100% O_2 is slight and due solely to an increase in the amount of dissolved O_2 in the blood.

Collapse of the Lung

When a bronchus or bronchiole is obstructed, the gas in the alveoli beyond the obstruction is absorbed and the lung segment collapses. Collapse of alveoli is called **atelectasis.** The atelectatic area may range in size from a small patch to a whole lung. Some blood is diverted from the collapsed area to better-ventilated portions of the lung, and this reduces the magnitude of the decline in arterial P_{O_2}.

When a large part of the lung is collapsed, there is an appreciable decrease in lung volume. The intrapleural pressure therefore becomes more negative and pulls the mediastinum, which in humans is a fairly flexible structure, to the affected side.

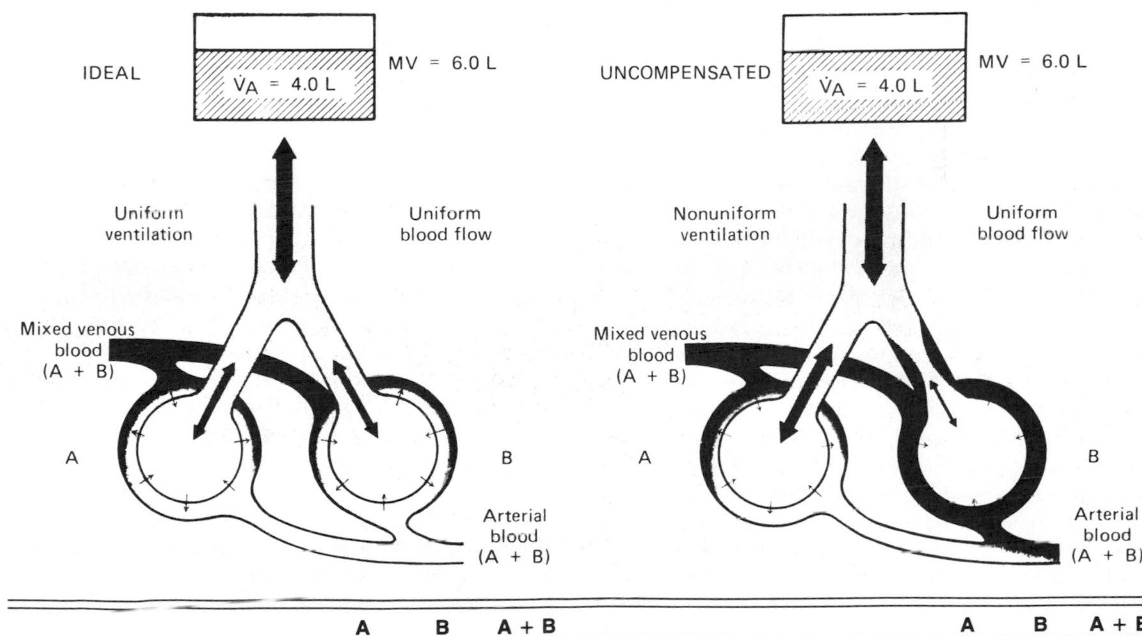

	A	B	A + B
Alveolar ventilation (L/min)	2.0	2.0	4.0
Pulmonary blood flow (L/min)	2.5	2.5	5.0
Ventilation/blood flow ratio	0.8	0.8	0.8
Mixed venous O_2 saturation (%)	75.0	75.0	75.0
Arterial O_2 saturation (%)	97.4	97.4	97.4
Mixed venous O_2 tension (mm Hg)	40.0	40.0	40.0
Alveolar O_2 tension (mm Hg)	104.0	104.0	104.0
Arterial O_2 tension (mm Hg)	104.0	104.0	104.0

	A	B	A + B
Alveolar ventilation (L/min)	3.2	0.8	4.0
Pulmonary blood flow (L/min)	2.5	2.5	5.0
Ventilation/blood flow ratio	1.3	0.3	0.8
Mixed venous O_2 saturation (%)	75.0	75.0	75.0
Arterial O_2 saturation (%)	98.2	91.7	95.0
Mixed venous O_2 tension (mm Hg)	40.0	40.0	40.0
Alveolar O_2 tension (mm Hg)	116.0	66.0	106.0
Arterial O_2 tension (mm Hg)	116.0	66.0	84.0

Figure 37–9. *Left:* "Ideal" ventilation/blood flow relationship. *Right:* Nonuniform ventilation and uniform blood flow, uncompensated. V_A, alveolar ventilation; MV, respiratory minute volume. (Reproduced, with permission, from Comroe JH Jr et al: *The Lung: Clinical Physiology and Pulmonary Function Tests,* 2nd ed. Year Book, 1962.)

Another cause of atelectasis is absence or inactivation of surfactant, the surface-tension-depressing agent normally found in the thin fluid lining the alveoli (see Chapter 34). This abnormality is a major cause of failure of the lungs to expand normally at birth. Collapse of the lung may also be due to the presence of the pleural space of air **(pneumothorax)**, tissue fluids (**hydrothorax, chylothorax**), or blood **(hemothorax).**

Pneumothorax

When air is admitted to the pleural space, through either a rupture in the lung or a hole in the chest wall, the lung on the affected side collapses because of its elastic recoil. Since the intrapleural pressure on the affected side is atmospheric, the mediastinum shifts toward the normal side. If the communication between the pleural space and the exterior remains open **(open** or **sucking pneumothorax),** more air moves in and out of the pleural space each time the patient breathes. If the hole is large, the resistance to air flow into the pleural cavity is less than the resistance to air flow into the intact lung, and little air enters the lung during inspiration. The mediastinum shifts farther to the intact side, kinking the great vessels until it flaps back during expiration. There is marked stimulation of respiration due to hypoxia, hypercapnia, and activation of pulmonary deflation receptors. Respiratory distress is severe.

If there is a flap of tissue over the hole in the lung or chest wall that acts as a flutter valve, permitting air to enter during inspiration but preventing its exit during expiration, the pressure in the pleural space rises above atmospheric pressure **(tension pneumothorax).** The hypoxic stimulus to respiration causes deeper inspiratory efforts, which futher increase the pressure in the pleural cavity, kinking the great veins and causing further hypoxia. Intrapleural pressure in such cases may rise to 20–30 mm Hg. The peripheral veins become distended, there is intense cyanosis, and the condition is potentially fatal if the pneumothorax is not decompressed by removing the air.

On the other hand, if the hole through which air enters the pleural space seals off **(closed pneumothorax),** respiratory distress is not great because, with each inspiration, air flows into the lung on the unaffected side rather than into the pleural space. Because the vascular resistance is increased in the collapsed lung, blood is diverted to the other lung. Consequently, unless the pneumothorax is very large, it does not cause much hypoxia.

The air in a closed pneumothorax is absorbed. Since it is at atmospheric pressure, its total pressure, P_{O_2} and P_{N_2} are greater than the corresponding values in venous blood (compare values for air and venous blood in Fig 34–14). Gas diffuses down these gradients into the blood, and after 1–2 weeks all of the gas disappears.

Spontaneous pneumothorax, the formation of a small closed pneumothorax due to rupture of congenital blebs on the surface of the visceral pleura, is a common benign condition. Indeed, production of an artificial pneumothorax by injecting air through the chest wall was at one time used extensively in the treatment of pulmonary tuberculosis, the aim being to collapse and thus "rest" the diseased lung.

Emphysema

In the degenerative and potentially fatal pulmonary disease called **emphysema,** the lungs lose their elasticity owing to disruption of elastic tissue and the walls between the alveoli break down so that the alveoli are replaced by large air sacs. The physiologic dead space is greatly increased, and because of inadequate and uneven alveolar ventilation and perfusion of underventilated alveoli, severe hypoxia develops. Late in the disease, hypercapnia also develops. Inspiration and expiration are labored, and the work of breathing is greatly increased. The changes in the pressure-volume curve of the lungs are shown in Fig 34–7. The chest becomes enlarged and barrel-shaped because the chest wall expands as the opposing elastic recoil of the lungs declines. The hypoxia leads to polycythemia. Pulmonary hypertension develops, and the right side of the heart enlarges **(cor pulmonale)** and then fails.

The most common cause of emphysema is heavy cigarette smoking. The smoke causes an increase in the number of pulmonary alveolar macrophages, and these macrophages release a chemical substance that attracts leukocytes to the lungs. The leukocytes in turn release proteases including elastase, which attacks the elastic tissue in the lungs. At the same time, α_1-antitrypsin, a plasma protein that normally inactivates elastase and other proteases, is itself inhibited. The α_1-antitrypsin is inactivated by oxygen radicals, and these are released by the leukocytes. In addition, there is some direct oxidation of α_1-antitrypsin by the smoke. The final result is a protease-antiprotease imbalance with increased destruction of lung tissue.

In about 2% of cases of emphysema, there is also a congenital α_1-antitrypsin deficiency. If individuals who are homozygous for this deficiency smoke, they develop crippling emphysema early in life and have a 20-year reduction in their life span. If they do not smoke, they may still develop emphysema, but they do much better and their life expectancy is much improved. Thus, α_1-antitrypsin deficiency provides an interesting example of the interaction between genetic factors and environmental factors in the production of disease.

OTHER FORMS OF HYPOXIA

Anemic Hypoxia

Hypoxia due to anemia is not severe at rest unless the hemoglobin deficiency is marked, because red blood cell 2,3-DPG increases. However, anemic patients may have considerable difficulty during exercise because of limited ability to increase O_2 delivery to the active tissues.

Carbon Monoxide Poisoning

Carbon monoxide (CO), a gas formed by incomplete combustion of carbon, was used by the Greeks and Romans to execute criminals. Today it causes more deaths than any other gas. CO poisoning is less common in the USA since natural gas, which does not contain CO, replaced artificial gases such as coal gas, which contains large amounts. However, the exhaust of gasoline engines is 6% or more CO.

CO is toxic because it reacts with hemoglobin to form **carbonmonoxyhemoglobin (carboxyhemoglobin, COHb),** and COHb cannot take up O_2. Carbon monoxide poisoning is often listed as a form of anemic hypoxia because there is a deficiency of hemoglobin that can carry O_2, but the total hemoglobin content of the blood is unaffected by CO. The affinity of hemoglobin for CO is 210 times its affinity for O_2, and COHb liberates CO very slowly. An additional difficulty is that when COHb is present the dissociation curve of the remaining HbO_2 shifts to the left, decreasing the amount of O_2 released. This is why an anemic individual who has 50% of the normal amount of HbO_2 may be able to perform moderate work, whereas an individual whose HbO_2 is reduced to the same level because of the formation of COHb is seriously incapacitated.

Because of the affinity of CO for hemoglobin, there is progressive COHb formation when the alveolar P_{CO} is greater than 0.4 mm Hg. However, the amount of COHb formed depends upon the duration of exposure to CO as well as the concentration of CO in the inspired air and the alveolar ventilation.

CO is also toxic to the cytochromes in the tissues, but the amount of CO required to poison the cytochromes is 1000 times the lethal dose; tissue toxicity thus plays no role in clinical CO poisoning.

The symptoms of CO poisoning are those of any type of hypoxia, especially headache and nausea, but there is little stimulation of respiration, since the arterial P_{O_2} remains normal and the carotid and aortic chemoreceptors are not stimulated (see Chapter 36). The cherry-red color of COHb is visible in the skin, nail beds, and mucous membranes. Death results when about 70–80% of the circulating hemoglobin is converted to COHb. The symptoms produced by chronic exposure to sublethal concentrations of CO are those of progressive brain damage, including mental changes and, sometimes, a parkinsonismlike state (see Chapter 32).

Treatment of CO poisoning consists of immediate termination of the exposure and adequate ventilation, by artificial respiration if necessary. Ventilation with O_2 is preferable to ventilation with fresh air, since O_2 hastens the dissociation of COHb. Hyperbaric oxygenation (see below) is useful in this condition.

Stagnant Hypoxia

Hypoxia due to slow circulation is a problem in organs such as the kidneys and heart during shock (see Chapter 33). The liver and possibly the brain are damaged by stagnant hypoxia in congestive heart failure. The blood flow to the lung is normally very large, and it takes prolonged hypertension to produce significant damage. However, **shock lung** can develop in prolonged circulatory collapse, particularly in those areas of the lung that are higher than the heart. In this condition, production of surfactant is decreased in the underperfused areas.

Histotoxic Hypoxia

Hypoxia due to inhibition of tissue oxidative processes is most commonly the result of cyanide poisoning. Cyanide inhibits cytochrome oxidase and possibly other enzymes. Methylene blue or nitrites are used to treat cyanide poisoning. They act by forming **methemoglobin,** which then reacts with cyanide to form **cyanmethemoglobin,** a nontoxic compound. The extent of treatment with these compounds is, of course, limited by the amount of methemoglobin that can be safely formed. Hyperbaric oxygenation may also be useful.

OXYGEN TREATMENT

Administration of oxygen-rich gas mixtures is of very limited value in stagnant, anemic, and histotoxic hypoxia because all that can be accomplished in this way is an increase in the amount of dissolved O_2 in the arterial blood. This is also true in hypoxic hypoxia when it is due to shunting of unoxygenated venous blood past the lungs. In other forms of hypoxic hypoxia, O_2 is of great benefit. However, it must be remembered that in hypercapnic patients in severe pulmonary failure the CO_2 level may be so high that it depresses rather than stimulates respiration. Some of these patients keep breathing only because the carotid and aortic chemoreceptors drive the respiratory center. If the hypoxic drive is withdrawn by administering O_2, breathing may stop. During the resultant apnea, the arterial P_{O_2} drops, but breathing may not start again, because the increase in P_{CO_2} further depresses the respiratory center. Therefore, O_2 therapy in this situation may be fatal.

When 100% O_2 is first inhaled, there may be a slight decrease in respiration in normal individuals, suggesting that there is normally some hypoxic chemoreceptor drive. However, the effect is minor and can be demonstrated only by special techniques. In addition, it is offset by a slight accumulation of H^+ ions, since the concentration of deoxygenated hemoglobin in the blood is reduced and Hb is a better buffer than HbO_2 (see Chapter 35).

Oxygen Toxicity

It is interesting that while O_2 is necessary for life in aerobic organisms, it is also toxic. Indeed, 100% O_2 has been demonstrated to exert toxic effects not only in animals but also in bacteria, fungi, cultured animal cells, and plants. The toxicity seems to be due to the production of the superoxide anion (O_2^-), which is a free radical, and H_2O_2. When 80–100% O_2 is administered to humans for periods of 8 hours or more, the respiratory passages become irritated, causing substernal distress, nasal congestion, sore throat, and coughing. Exposure for 24–48 hours causes lung damage as well. The reason O_2 produces the irritation is not completely understood, although O_2 treatment appears to inhibit the ability of lung macrophages to kill bacteria, and surfactant production is reduced. In animals, more prolonged administration without irritation is possible if treatment is briefly interrupted from time to time, but it is not certain that periodic interruptions are of benefit to humans.

Some infants treated with O_2 for respiratory distress syndrome develop a chronic condition characterized by lung cysts and densities **(bronchopulmonary dysplasia).** There is evidence that this syndrome is a manifestation of O_2 toxicity. Another complication in these infants is **retrolental fibroplasia,** the formation of opaque vascular tissue in the eyes, which can lead to serious visual defects. Current evidence indicates that the retinal receptors mature from the center to the periphery of the retina, that they use considerable O_2, and that this causes the retina to become vascularized in an orderly fashion. Oxygen treatment before maturation is complete provides the needed O_2 to the photoreceptors, and consequently, the normal vascular pattern fails to develop. There is evidence that this condition can be prevented or ameliorated by treatment with vitamin E, which exerts an antioxidant effect.

Administration of 100% O_2 at increased pressure accelerates the onset of O_2 toxicity, with the production not only of tracheobronchial irritation but also of muscle twitching, ringing in the ears, dizziness, convulsions, and coma. The speed with which these symptoms develop is proportionate to the pressure at which the O_2 is administered; eg, at 4 atmospheres, symptoms develop in half the subjects in 30 minutes, whereas at 6 atmospheres, convulsions develop in a few minutes. Administration of other gases at increased pressure also causes central nervous system symptoms (see below). Administration of O_2 at elevated pressures to rats decreases their brain GABA content (see Chapter 4) and their brain, liver, and kidney ATP content.

Exposure to O_2 at increased pressures **(hyperbaric oxygenation)** can produce marked increases in the dissolved O_2 in blood. It is for this reason that surgery for certain forms of congenital heart disease is sometimes carried out in high-pressure tanks. Hyperbaric O_2 is also of value in the treatment of gas gangrene, carbon monoxide poisoning, and probably cyanide poisoning. However, O_2 toxicity limits exposures to less than 5 hours and pressures to 3 atmospheres or less.

HYPERCAPNIA & HYPOCAPNIA

Hypercapnia

Retention of CO_2 in the body (hypercapnia) initially stimulates respiration. Retention of larger amounts produces symptoms due to depression of the central nervous system: confusion, diminished sensory acuity, and, eventually, coma with respiratory depression and death. In patients with these symptoms, the P_{CO_2} is markedly elevated, there is severe respiratory acidosis, and the plasma HCO_3^- may exceed 40 meq/L. Large amounts of HCO_3^- are excreted, but more HCO_3^- is reabsorbed, raising the plasma HCO_3^- and partially compensating for the acidosis (see Chapter 39).

CO_2 is so much more soluble than O_2 that hypercapnia is rarely a problem in patients with pulmonary fibrosis. However, it does occur in ventilation-perfusion inequality and when for any reason alveolar ventilation is inadequate in the various forms of pump failure.

Hypocapnia

Hypocapnia is the result of hyperventilation. During voluntary hyperventilation, the arterial P_{CO_2} falls from 40 to as low as 15 mm Hg while the alveolar P_{O2} rises to 120–140 mm Hg.

The more chronic effects of hypocapnia are seen in neurotic patients who chronically hyperventilate. Cerebral blood flow may be reduced 30% or more because of the direct constrictor effect of hypocapnia on the cerebral vessels (see Chapter 32). The cerebral ischemia causes light-headedness, dizziness, and paresthesias. Hypocapnia also increases cardiac output. It has a direct constrictor effect on many vessels, but it depresses the vasomotor center, so that the blood pressure is usually unchanged or only slightly elevated.

Other consequences of hypocapnia are due to the associated respiratory alkalosis, the blood pH being increased to 7.5 or 7.6. The plasma HCO_3^- level is low, but bicarbonate reabsorption is decreased because of the inhibition of renal acid secretion by the low P_{CO_2}. The plasma total calcium level does not change, but the plasma Ca^{2+} level falls and hypocapnic individuals develop carpopedal spasm, a positive Chvostek sign, and other signs of tetany (see Chapter 21).

OTHER RESPIRATORY ABNORMALITIES

Asphyxia

In asphyxia produced by occlusion of the airway, acute hypercapnia and hypoxia develop together. There is pronounced stimulation of respiration, with violent respiratory efforts. Blood pressure and heart rate rise sharply, and blood pH drops. Catecholamine secretion is stimulated, the stimulation of norepinephrine secretion being proportionately greater than the stimulation of epinephrine secretion. Eventually the respiratory efforts cease, the blood pressure falls, and the heart slows. Asphyxiated animals can still be revived at this point by artificial respiration, although they are prone to ventricular fibrillation, probably owing to the combination of hypoxic myocardial damage and high circulating catecholamine levels. If artificial respiration is not started, cardiac arrest occurs in 4–5 minutes.

Drowning

In about 10% of drownings, the first gasp of water after the losing struggle not to breathe triggers laryngospasm, and death results from asphyxia without any water in the lungs. In the remaining cases, the glottic muscles eventually relax and the lungs are flooded. Fresh water is rapidly absorbed, diluting the plasma and causing intravascular hemolysis. Ocean water is markedly hypertonic and draws fluid from the vascular system into the lungs, decreasing plasma volume. The immediate goal in the treatment of drowning is, of course, resuscitation, but long-term treatment must also take into account the circulatory effects of the water in the lungs.

Periodic Breathing

The acute effects of voluntary hyperventilation demonstrate the interaction of the chemical respiratory regulating mechanisms. When an individual hyperventilates for 2–3 minutes, then stops and permits respiration to continue without exerting any voluntary control over it, there is a period of apnea. This is followed by a few shallow breaths and then by another period of apnea, followed again by a few breaths (**periodic breathing**). The cycles may last for some time before normal breathing is resumed (Fig 37–10). The apnea apparently is due to CO_2 lack because it does not occur following hyperventilation with gas mixtures containing 5% CO_2. During the apnea, the alveolar P_{O_2} falls and the P_{CO_2} rises. Breathing resumes because of hypoxic stimulation of the carotid and aortic chemoreceptors before the CO_2 level has returned to normal. A few breaths eliminate the hypoxic stimulus, and breathing stops until the alveolar P_{O_2} falls again. Gradually, however, the P_{CO_2} returns to normal, and normal breathing resumes.

Cheyne-Stokes Respiration

Periodic breathing occurs in various disease states and is often called **Cheyne-Stokes respiration.** It is seen most commonly in patients with congestive heart failure and uremia, but it occurs also in patients with brain disease and during sleep in some normal individuals (Fig 37–11). Some of the patients with Cheyne-Stokes respiration have been shown to have increased sensitivity to CO_2. The increased response is apparently due to disruption of neural pathways that normally inhibit respiration. In these individuals, CO_2 causes relative hyperventilation, lowering the arterial P_{CO_2}. During the resultant apnea, the arterial P_{CO_2} again rises to normal, but the respiratory mechanism again overresponds to CO_2. Breathing ceases, and the cycle repeats.

Another cause of periodic breathing in patients with cardiac disease is prolongation of the lung-to-brain circulation time, so that it takes longer for changes in arterial gas tensions to affect the respiratory center. When individuals with a slower circulation hyperventilate, they lower the P_{CO_2} of the blood in their lungs, but it takes longer than normal for the blood with a low P_{CO_2} to reach the brain. During this time, the P_{CO_2} in the pulmonary capillary blood continues to be lowered, and when this blood reaches the brain, the low P_{CO_2} inhibits the respiratory center, producing apnea. In other words, the respiratory control system oscillates because the negative feedback loop from lungs to brain is abnormally long.

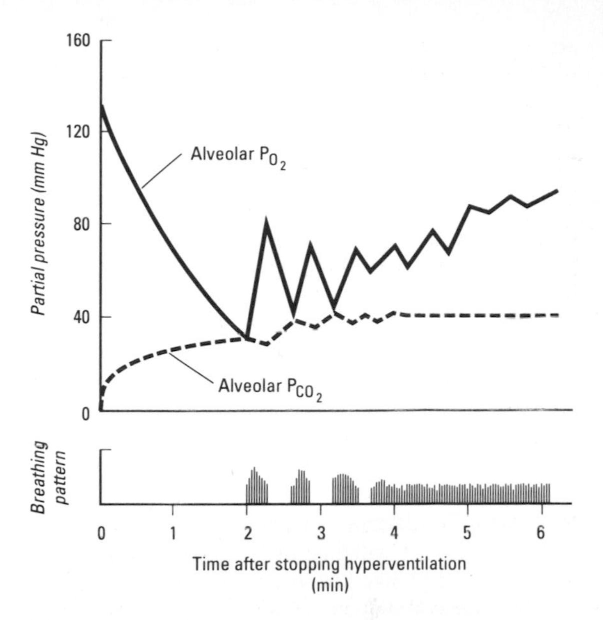

Figure 37–10. Changes in breathing and composition of alveolar air after forced hyperventilation for 2 minutes.

EFFECTS OF INCREASED BAROMETRIC PRESSURE

The ambient pressure increases by 1 atmosphere for every 10 m of depth in seawater and every 10.4 m of depth in fresh water. Therefore, at a depth of 31 m (100 ft) in the ocean, a diver is exposed to a pressure of 4 atmospheres. Those who dig underwater tunnels are also exposed to the same hazards because the

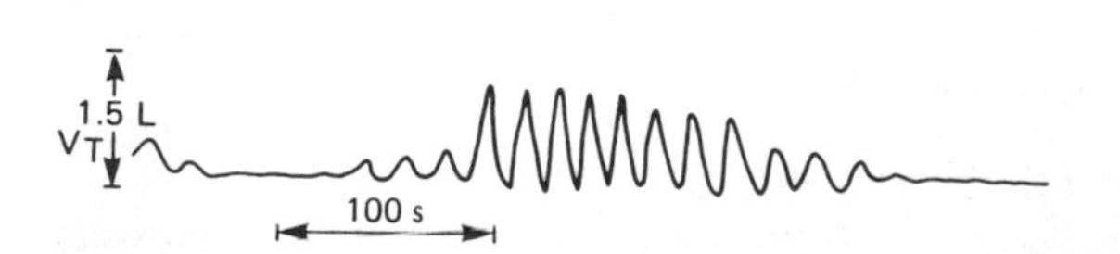

Figure 37–11. Cheyne-Stokes breathing during sleep. Two periods of apnea are separated by an increase and then a smooth decrease in tidal volume (V_T). (Reproduced, with permission, from Cherniack NS: Respiratory dysrhythmias during sleep. *N Engl J Med* 1981;**305**:325.)

pressure in the chambers (caissons) in which they work is increased to keep out the water.

The hazards of exposure to increased barometric pressure used to be the concern largely of the specialists who cared for deep-sea divers and tunnel workers. However, the invention of SCUBA gear (self-contained underwater breathing apparatus, a tank and valve system carried by the diver) has transformed diving from a business into a sport. The popularity of skin diving is so great that all physicians should be aware of its potential dangers. The interest in performing certain surgical procedures in high-pressure tanks (see above) also focuses attention on the effects of increased barometric pressure.

Table 37–2. Potential problems associated with exposure to increased barometric pressure.

Oxygen toxicity
Lung damage
Convulsions
Nitrogen narcosis
Euphoria
Impaired performance
High-pressure nervous syndrome
Tremors
Somnolence
Decompression sickness
Pain
Paralyses
Air embolism
Sudden death

Nitrogen Narcosis & the High-Pressure Nervous Syndrome

A diver must breathe air or other gases at increased pressure to equalize the increased pressure on the chest wall and abdomen. CO_2 is routinely removed to prevent its accumulation. At increased pressure, 100% O_2 causes central nervous system symptoms of oxygen toxicity (Table 37–2). Since the harmful effects of breathing O_2 (see above) are proportionate to the P_{O_2}, they can be prevented by decreasing the concentration of O_2 in the gas mixture to 20% or less.

If a diver breathes compressed air, the increased P_{N_2} can cause **nitrogen narcosis,** a condition also known as "rapture of the deep" (Table 37–2). At pressures of 4–5 atmospheres (ie, at depths of 30–40 m in the ocean), 80% N_2 produces definite euphoria. At greater pressures, the symptoms resemble alcohol intoxication. Manual dexterity is maintained, but intellectual functions are impaired.

The problem of nitrogen narcosis can be avoided by breathing mixtures of O_2 and helium, and deeper dives can be made. However, the **high-pressure nervous syndrome** (HPNS) develops during deep dives with such mixtures. This condition is characterized by tremors, drowsiness, and a depression of the α activity in the EEG. Unlike nitrogen narcosis, intellectual functions are not severely affected but manual dexterity is impaired. The cause of HPNS is not settled, but it is worth noting that a variety of gases that are physiologically inert at atmospheric pressure are anesthetics at increased pressure. This is true of N_2 and also of xenon, krypton, argon, neon, and helium. Their anesthetic activity parallels their lipid solubility, and the anesthesia may be due to an action on nerve cell membranes.

Decompression Sickness

As a diver breathing 80% N_2 ascends from a dive, the elevated alveolar P_{N_2} falls. N_2 diffuses from the tissues into the lungs along the partial pressure gradient. If the return to atmospheric pressure (decompression) is gradual, no harmful effects are observed; but if the ascent is rapid, N_2 escapes from solution. Bubbles form in the tissues and blood, causing the symptoms of **decompression sickness (the bends, caisson disease).** Bubbles in the tissues cause severe pains, particularly around joints, and neurologic symptoms that include paresthesias and itching. Bubbles in the bloodstream, which occur in more severe cases, obstruct the arteries to the brain, causing major paralyses and respiratory failure. Bubbles in the pulmonary capillaries ar apparently responsible for the dyspnea that divers call "the chokes," and bubbles in the coronary arteries may cause myocardial damage.

Treatment of this disease is prompt recompression in a pressure chamber, followed by slow decompression. Recompression is frequently lifesaving. Recovery is often complete, but there may be residual neurologic sequelae due to irreversible damage to the nervous system.

It should be noted that ascent in an airplane is equivalent to ascent from a dive. The decompression during a climb from sea level to 8550 m in the unpressurized cabin of an airplane (pressure drop from 1 down to one-third atmosphere) is the same as that during ascent to the surface after spending a period of time in 20 m of seawater (pressure drop from 3 down to 1 atmosphere). A rapid ascent in either situation can cause decompression sickness.

Air Embolism

If a diver breathing from a tank at increased pressure during a dive holds his or her breath and suddenly heads for the surface—as may occur if the diver gets into trouble and panics—the gas in the lungs may expand rapidly enough to rupture the pulmonary veins. This drives air into the vessels, causing air embolism. Fatal air embolism has occurred during rapid ascent from as shallow a depth as 5 m. The consequences of air in the circulatory system are discussed in Chapter 30. This cannot happen, of course, to an individual who takes a breath on the surface, dives and returns to the surface still holding that breath, no matter how deep the dive.

Air embolism also occurs as a result of rapid expansion of the gas in the lungs when the external pressure is suddenly reduced from atmospheric to subatmospheric, as it is when the wall of the pressurized cabin of an airplane or rocket at high altitude is breached **(explosive decompression).**

ARTIFICIAL RESPIRATION

In acute asphyxia due to drowning, CO or other forms of gas poisoning, electrocution, anesthetic accidents, and other similar causes, artificial respiration after breathing has ceased may be lifesaving. It should always be attempted, because the respiratory center fails before the vasomotor center and heart. There are numerous methods of emergency artificial respiration, but the method presently recommended to produce adequate ventilation in all cases is mouth-to-mouth breathing.

Mouth-to-Mouth Breathing

In this form of resuscitation, the operator first places the victim in the supine position and opens the airway by placing a hand under the neck and lifting, while keeping pressure with the other hand on the victim's forehead. This extends the neck and lifts the tongue away from the back of the throat. The victim's mouth is covered by the operator's mouth while the fingers of the hand already on the forehead occlude the nostrils (Fig 37–12). About 12 times a minute, the operator blows into the victim's mouth a volume about twice the tidal volume, then permits the elastic recoil of the victim's lungs to produce passive expiration. The victim's neck is kept extended. Any gas blown into the stomach can be expelled by applying upward pressure on the abdomen from time to time. In apneic individuals in whom there is no detectable heartbeat, mouth-to-mouth breathing should be alternated with cardiac massage (see Chapter 28).

The advantages of mouth-to-mouth resuscitation lie not only in its simplicity but also in the fact that it works by expanding the lungs. In the prone or supine pressure methods of artificial respiration, the lungs are compressed by pressure on the thorax, and passive recoil of the chest wall pulls air into them (Fig 37–13). Some expansion of the lungs can be achieved by raising the arms, but even when arm raising and pressure are alternated, the effective ventilation is not so great as that produced by the mouth-to-mouth method.

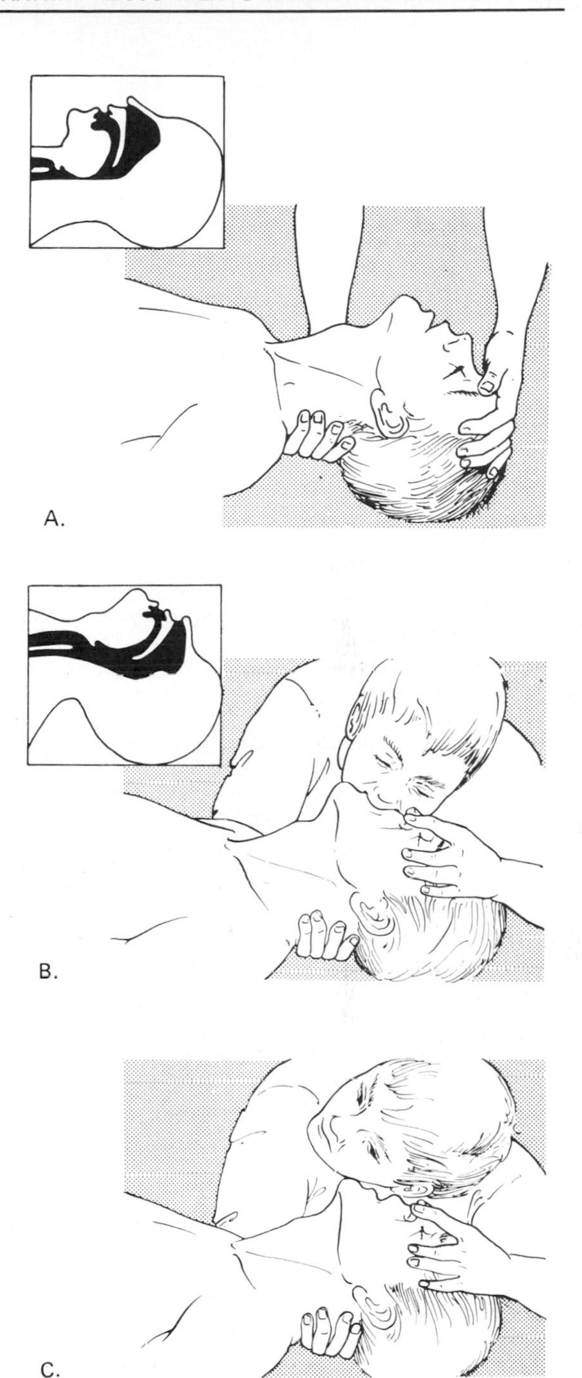

Figure 37–12. Proper performance of mouth-to-mouth resuscitation. *A:* Open airway by positioning neck anteriorly in extension. Inserts show airway obstructed when the neck is in resting flexed position and open when neck is extended. *B:* Close victim's nose with fingers, seal mouth around victim's mouth, and deliver breath by vigorous expiration. *C:* Allow victim to exhale passively by unsealing mouth and nose. Rescuer should listen and feel for expiratory air flow. (Reproduced, with permission, from Schroeder SA, Krupp MA, Tierney LM Jr. [editors]: *Current Medical Diagnosis & Treatment 1988.* Appleton & Lange, 1988.)

Mechanical Respirators

For treatment of chronic respiratory insufficiency due to inadequate ventilation, mechanical respirators are available. These are airtight metal or plastic containers that used to enclose the body except for the head but are now portable and cover only the chest. By means of a motor, negative pressure is applied to the chest at regular intervals, moving the chest wall in a way that resembles normal breathing. Intermittent positive pressure breathing machines are also used. These machines produce intermittent increases in

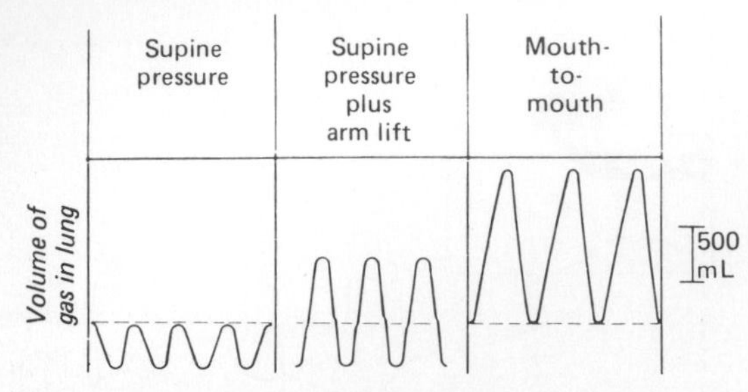

Figure 37–13. Techniques of artificial respiration. The prone pressure plus arm lift method moves about the same amount of air as the supine pressure plus arm lift method. The dotted line indicates the volume of gas in the lung in the subject whose respiration has ceased.

intrapulmonary pressure by means of "pulses" of air delivered via a face mask.

Breathing against a positive end-expiratory pressure (PEEP) is of value in pulmonary edema, probably because it recruits previously unventilated alveoli. It also improves oxygenation in **adult respiratory distress syndrome.** This condition, which has a high mortality rate, is characterized by noncardiac pulmonary edema and inflammation produced by aspiration or other direct injuries to the lung or, in some instances, by sepsis elsewhere in the body.

REFERENCES
Section VII: Respiration

Berger AJ: Properties of medullary respiratory neurons. *Fed Proc* 1981;**40:**2378.

Bourbon JR, Rieutort M: Pulmonary surfactant: biochemistry, physiology and pathology. *News Physiol Sci* 1987;**2:**129.

Burrows B et al: *Respiratory Disorders: A Pathophysiologic Approach,* 2nd ed. Year Book, 1983.

Carrell RW et al: Structure and function of human α_1-antitrypsin. *Nature* 1982;**298:**329.

Deneke SM, Fonburg BL: Normobaric oxygen toxicity of the lung. *N Engl J Med* 1980;**303:**76.

Fels AOS, Cohn ZA: The alveolar macrophage. *J Appl Physiol* 1986;**60:**353.

Finch CA, Lenfant C: Oxygen transport in man. *N Engl J Med* 1972;**286:**407.

Fishman AP (editor): *Circulation and Nonrespiratory Functions.* Vol I of: Handbook of Physiology, Section 3: *The Respiratory System.* American Physiological Society, 1985.

Fishman AP, Renkin EM (editors): *Pulmonary Edema.* American Physiological Society, 1979.

Golding J, Limerick S, Macfarlane A: *Sudden Infant Death: Patterns, Puzzles and Problems.* Univ of Washington Press, 1985.

Johnson TS, Rock PB: Acute mountain sickness. *N Engl J Med* 1988;**319:**841.

Kilmartin JN, Rossi-Bernardini L: Interaction of hemoglobin with hydrogen ions, carbon dioxide, and organic phosphates. *Physiol Rev* 1973;**53:**836.

Lambertsen CJ (editor): *Underwater Physiology.* Academic Press, 1971.

Macklem PT: Respiratory mechanics. *Annu Rev Physiol* 1978;**40:**157.

McCord JM. Oxygen-derived free radicals in postischemic tissue injury. *N Engl J Med* 1985;**312:**159.

Meechan RT, Zavala DC: The pathophysiology of high-altitude illness. *Am J Med* 1982;**73:**395.

Millhorn DE, Eldridge FL: Role of ventrolateral medulla in regulation of respiratory and cardiovascular systems. *J Appl Physiol* 1986;**61:**1249.

Moffat K, Deatherage JF, Seybert DW: A structural model for the kinetic behavior of hemoglobin. *Science* 1979;**206:**1035.

Murray, JF: *The Normal Lung: The Basis for Diagnosis and Treatment of Pulmonary Disease,* 2nd ed. Saunders, 1985.

Murray JF, Nadel J (editors): *Respiratory Medicine.* Saunders, 1986.

Newhouse M, Sanchis J, Bienenstock J: Lung defense mechanisms. *N Engl J Med* 1976;**295:**990.

Rannels DE et al: Use of radioisotopes in quantitative studies of lung metabolism. *Fed Proc* 1982;**41:**2833.

Richter DW, Ballantine D, Remmers JE: How is the respiratory rhythm generated? A model. *News in Physiological Sciences* 1986;**1:**109.

Rigatto H: Control of ventilation in the newborn. *Annu Rev Physiol* 1984;**46:**661.

Rounds S, Brody JS: Putting PEEP in perspective. *N Engl J Med* 1984;**311:**323.

Roussos C, Macklem PT (editors): *Lung Biology in Health and Disease: The Thorax.* Marcel Dekker, 1985.

Said SI (editor): *The Pulmonary Circulation and Acute Lung Injury.* Futura, 1985.

Sandberg LB, Soskel NT, Leslie JG: Elastin structure, biosynthesis, and relation to disease states. *N Engl J Med* 1981;**304:**566.

Sears TA, Berger AJ, Phillipson EA: Reciprocal tonic activation of inspiratory and expiratory motoneurones by chemical drives. *Nature* 1982;**299:**728.

Sugar O: In search of Ondine's curse. *JAMA* 1978;**240:**236.

Wagner PD: Diffusion and chemical reaction in pulmonary gas exchange. *Physiol Rev* 1977;**57:**257.

Weibel ER: *The Pathway for Oxygen: Structure and Function in the Mammalian Respiratory System.* Harvard Univ Press, 1984.

Welsh MJ: Electrolyte transport by airway epithelial. *Physiol Rev* 1987;**67:**1143.

West JB: Human physiology at extreme altitudes on Mount Everest. *Science* 1984;**223:**784.

West JB: *Respiratory Physiology: The Essentials,* 2nd ed. Williams & Wilkins, 1979.

Whittenberger JL: Myoglobin-facilitated oxygen diffusion: Role of myoglobin in oxygen entry into muscle. *Physiol Rev* 1970;**50:**559.

Wright JR, Clements JA: Metabolism and turnover of lung surfactant. *Am Rev Respir Dis* 1987;**135:**426.

Symposium: Mucus secretion and ion transport in airways. *Fed Proc* 1980;**39:**3061.

Symposium: Recent advances in carotid body physiology. *Fed Proc* 1980;**39:**2626.

Section VIII. Formation & Excretion of Urine

Renal Function & Micturition — 38

INTRODUCTION

In the kidneys, a fluid that resembles plasma is filtered through the glomerular capillaries into the renal tubules **(glomerular filtration).** As this glomerular filtrate passes down the tubules, its volume is reduced and its composition altered by the processes of **tubular reabsorption** (removal of water and solutes from the tubular fluid) and **tubular secretion** (secretion of solutes into the tubular fluid) to form the urine that enters the renal pelvis. A comparison of the composition of the plasma and an average urine specimen illustrates the magnitude of some of these changes (Table 38–1) and emphasizes the manner in which wastes are eliminated while water and important electrolytes and metabolites are conserved. Furthermore, the composition of the urine can be varied, and many homeostatic regulatory mechanisms minimize or prevent changes in the composition of the ECF by changing the amount of water and various specific solutes in the urine. From the renal pelvis, the urine passes to the bladder and is expelled to the exterior by the process of urination, or **micturition.** The kidneys make prostaglandins and kinins, and they are also endocrine organs, secreting renin (see Chapter 24) and renal erythropoietic factor (see Chapters 24 and 27) and forming 1,25-dihydroxycholecalciferol (see Chapter 21).

FUNCTIONAL ANATOMY

The Nephron

Each individual renal tubule and its glomerulus is a unit **(nephron).** The size of the kidneys in various species is largely determined by the number of nephrons they contain. There are approximately 1.3 million nephrons in each human kidney. The parts of the nephron are shown in diagrammatic fashion in Fig 38–1. A more detailed drawing is shown in Fig 38–2,

Table 38–1. Urinary and plasma concentrations of some physiologically important substances.

Substance	Concentration in		U/P Ratio
	Urine (U)	Plasma (P)	
Glucose (mg/dL)	0	100	0
Na^+ (meq/L)	90	150	0.6
Urea (mg/dL)	900	15	60
Creatinine (mg/dL)	150	1	150

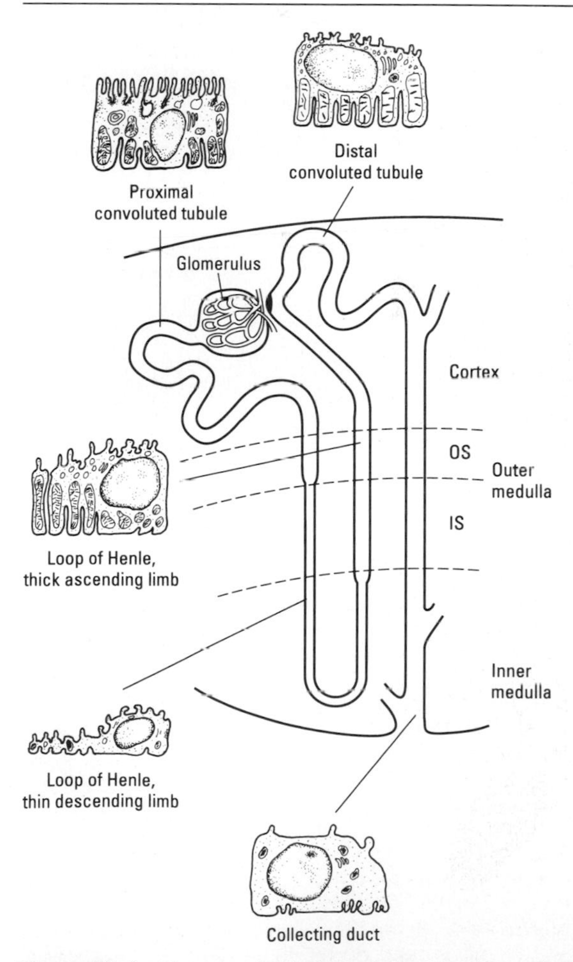

Figure 38–1. Diagram of a juxtamedullary nephron. The main histologic features of the cells that make up each portion of the tubule are also shown. IS, inner stripe, OS, outer stripe.

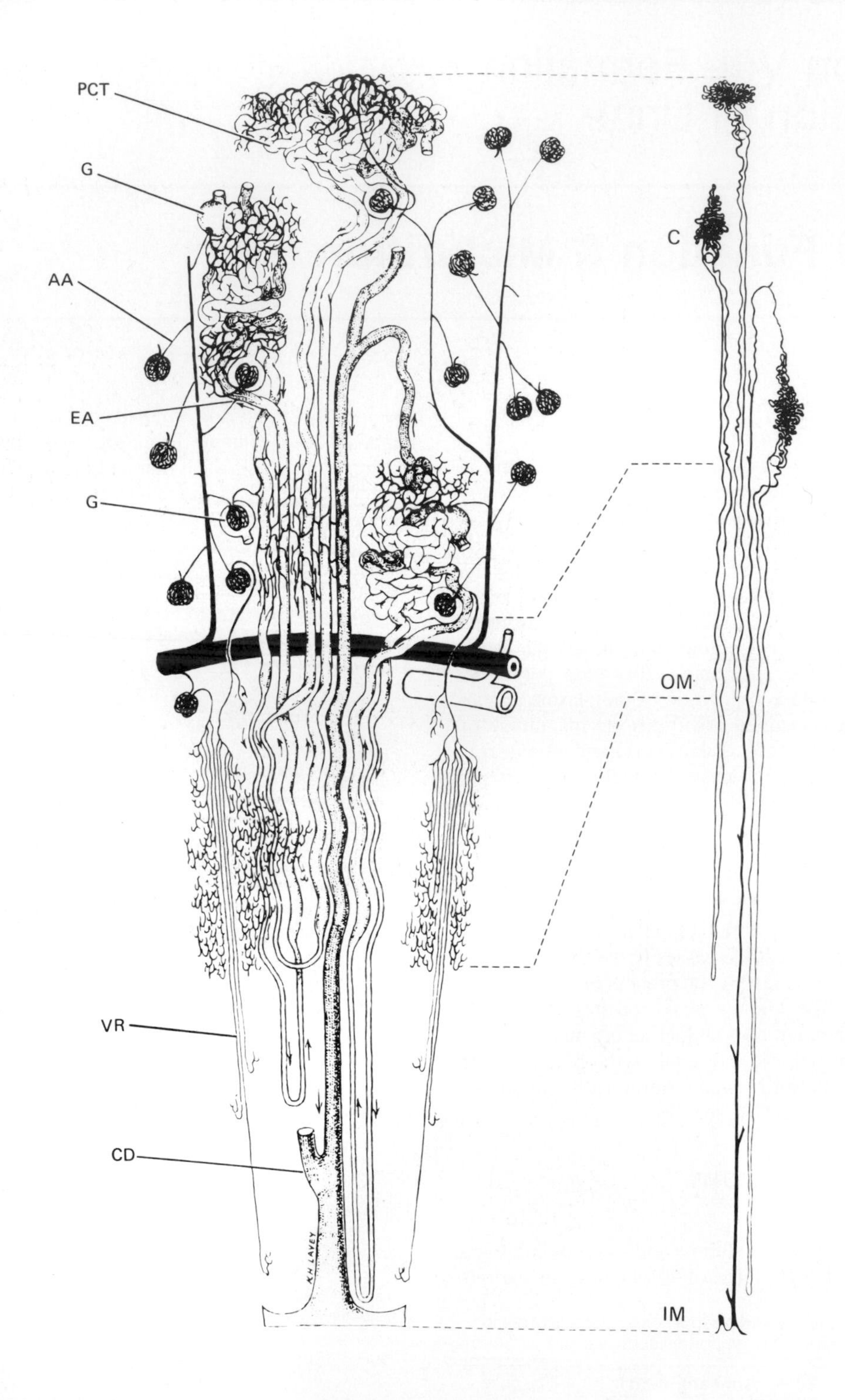

Figure 38–2. Renal circulation in the dog. The pattern in human kidneys is essentially identical. The vascular structures are simplified, and on the left, the horizontal scale is expanded to show details. The same nephrons are shown undistorted on the right. AA, afferent arteriole; C, cortex; CD, collecting duct; EA, efferent arteriole; G, glomerulus; IM, inner medulla; OM, outer medulla; PCT, proximal convoluted tubule; VR, vasa recta. The distal convoluted tubules are shaded. (Reproduced, with permission, from Beeuwkes R III: The vascular organization of the kidney. *Annu Rev Physiol* 1980;**42**:531. © 1980 by Annual Reviews, Inc.)

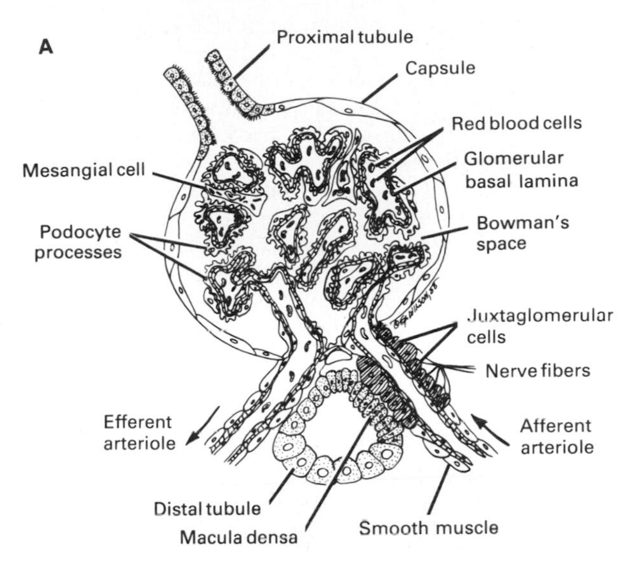

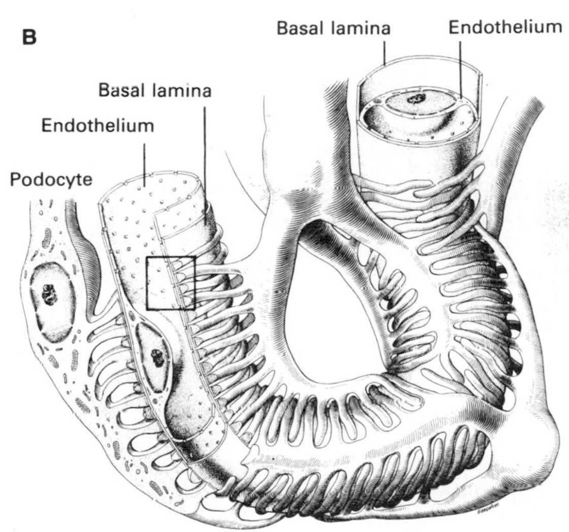

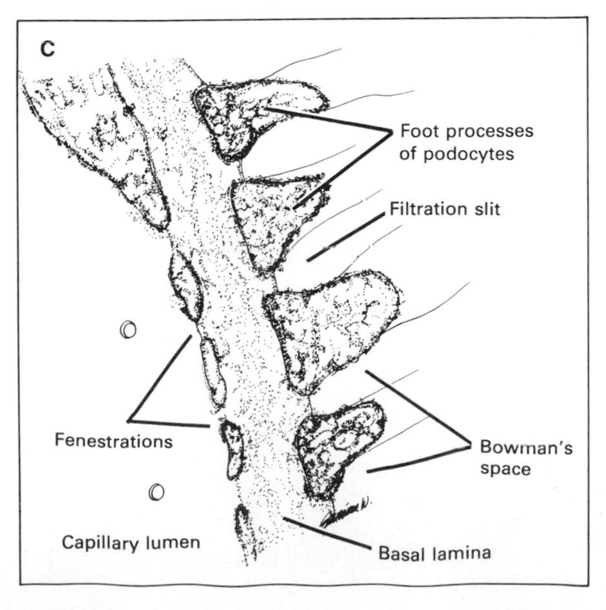

which also illustrates the long, thin nature of the nephron.

The glomerulus, which is about 200 μm in diameter, is formed by the invagination of a tuft of capillaries into the dilated, blind end of the nephron **(Bowman's capsule).** The capillaries are supplied by an **afferent arteriole** and drained by a slightly smaller **efferent arteriole** (Figs 38–1 and 38–2). There are 2 cellular layers separating the blood from the glomerular filtrate in Bowman's capsule: the capillary endothelium and the specialized epithelium that lies on top of the glomerular capillaries. These layers are separated by a basal lamina. In addition, stellate cells called **mesangial cells** send processes between the endothelium and the basal lamina. Similar cells called pericytes are found in the walls of capillaries elsewhere in the body. The mesangial cells in the kidneys are often found between capillary loops (Fig 38–3). They are contractile and play a role in the regulation of glomerular filtration (see below). They also secrete various substances, take up immune complexes, and are involved in the production of glomerular disease.

The endothelium of the glomerular capillaries is fenestrated, with pores that are approximately 100 nm in diameter. The cells of the epithelium **(podocytes)** have numerous pseudopodia that interdigitate (Fig 38–3) to form **filtration slits** along the capillary wall. The slits are approximately 25 nm wide, and each is closed by a thin membrane. The basal lamina does not contain visible gaps or pores.

Functionally, the glomerular membrane permits the free passage of neutral substances up to 4 nm in diameter and almost totally excludes those with diameters greater than 8 nm. However, the charges on molecules as well as their diameters affect their passage into Bowman's capsule (see below). The average capillary area in each glomerulus is about 0.4 mm^2, and the total area of glomerular capillary endothelium across which filtration occurs in humans is about 0.8 m^2.

The general features of the cells that make up the walls of the tubules are shown in Fig 38–1. However, there are cell subtypes in all segments, and the anatomic differences between them correlate with differences in function (see below).

The human **proximal convoluted tubule** is about 15 mm long and 55 μm in diameter. Its wall is made up of a single layer of cells that interdigitate with one

Figure 38–3. Structure of glomerulus. *A:* Section through vascular pole, showing capillary loops and mesangial cells. *B:* Detail of the way processes of the podocytes form filtration slits on the basal lamina, and the relation of the lamina to the capillary endothelium. *C:* The rectangle in B is enlarged to show the processes of the podocytes coated with glomerular polyanion. (Modified from Ham AW: *Histology,* 7th ed, Lippincott, 1974; and Wyngaarden JB, Smith LN Jr [editors]: *Cecil Textbook of Medicine,* 18th ed, Saunders, 1988).

another and are united by apical tight junctions. Between the bases of the cells, there are extensions of the extracellular space called the **lateral intercellular spaces.** The luminal edges of the cells have a striate **brush border** due to the presence of innumerable 1 × 0.7 μm microvili.

The convoluted portion of the proximal tubule (pars convoluta) drains into the straight portion (pars recta), which forms the first part of the **loop of Henle** (Figs 38–1 and 38–2). The proximal tubule terminates in the thin segment of the descending limb of the loop of Henle, which has an epithelium made up of attenuated, flat cells. The nephrons with glomeruli in the outer portions of the renal cortex have short loops of Henle **(cortical nephrons),** whereas those with glomeruli in the juxtamedullary region of the cortex **(juxtamedullary nephrons)** have long loops extending down into the medullary pyramids. In humans, only 15% of the nephrons have long loops. The total length of the thin segment of the loop varies from 2 to 14 mm in length. It ends in the thick segment of the ascending limb, which is about 12 mm in length. The cells of the thick ascending limb are cuboid. They have numerous mitochondria, and the basilar portions of their cell membranes are extensively invaginated.

The thick ascending limb of the loop of Henle reaches the glomerulus of the nephron from which the tubule arose and passes close to its afferent arteriole and efferent arteriole. The walls of the afferent arterioles contain the renin-secreting juxtaglomerular cells. At this point, the tubular epithelium is modified histologically to form the **macula densa.** The juxtaglomerular cells, the macula densa, and the granulated lacis cells near them are known collectively as the **juxtaglomerular apparatus** (Fig 24–3).

The **distal convoluted tubule** is about 5 mm long. Its epithelium is lower than that of the proximal tubule, and although there are a few microvilli, there is no distinct brush border. The distal tubules coalesce to form **collecting ducts** that are about 20 mm long and pass through the renal cortex and medulla to empty into the pelvis of the kidney at the apexes of the medullary pyramids. The total length of the nephrons, including the collecting ducts, ranges from 45 to 65 mm.

Cells in the kidneys that appear to have a secretory function include not only the juxtaglomerular cells but also some of the cells in the interstitial tissue of the medulla. These cells are called **type I medullary interstitial cells.** They contain lipid droplets and probably secrete prostaglandins, predominantly prostaglandin PGE_2 (see Chapter 17). PGE_2 is also secreted by the cells in the collecting ducts, and prostacyclin (PGI_2) as well as other prostaglandins is secreted by the arterioles and glomeruli.

Blood Vessels

The renal circulation is diagrammed in Fig 38–2. The afferent arterioles are short, straight branches of the interlobular arteries. Each divides into multiple capillary branches to form the tuft of vessels in the glomerulus. The capillaries coalesce to form the efferent arterioles, which in turn break up into capillaries that supply the tubules **(peritubular capillaries)** before draining into the interlobular veins. The arterial segments between glomeruli and tubules are thus technically a portal system, and the glomerular capillaries are the only capillaries in the body that drain into arterioles. However, there is relatively little smooth muscle in the efferent arterioles.

The capillaries draining the tubules of the cortical nephrons form a peritubular network, whereas the efferent arterioles from the juxtamedullary glomeruli drain not only into a peritubular network but into vessels that form hairpin loops (the **vasa recta).** These loops dip into the medullary pyramids alongside the loops of Henle (Fig 38–2). The efferent arteriole from each glomerulus breaks up into capillaries that supply a number of different nephrons. Thus, the tubule of each nephron does not necessarily receive blood solely from the efferent arteriole of that nephron. In humans, the total surface of the renal capillaries is approximately equal to the total surface area of the tubules, both being about 12 m^2. The volume of blood in the renal capillaries at any given time is 30–40 mL.

Lymphatics

The kidneys have an abundant lymphatic supply that drains via the thoracic duct into the venous circulation in the thorax.

Capsule

The renal capsule is thin but tough. If the kidney becomes edematous, the capsule limits the swelling, and the tissue pressure **(renal interstitial pressure)** rises. This decreases the glomerular filtration rate and is claimed to enhance and prolong the anuria in acute renal failure (see Chapter 33).

Innervation of the Renal Vessels

The renal nerves travel along the renal blood vessels as they enter the kidney. They contain many sympathetic efferent fibers and a few afferent fibers of unknown function. There also appears to be a cholinergic innervation via the vagus nerve, but its function is uncertain. The sympathetic preganglionic innervation comes primarily from the lower thoracic and upper lumbar segments of the spinal cord, and the cell bodies of the postganglionic neurons are in the sympathetic ganglion chain, in the superior mesenteric ganglion, and along the renal artery. The sympathetic fibers are distributed primarily to the afferent and efferent arterioles. However, noradrenergic nerve fibers also end in close proximity to the renal tubular cells and the juxtaglomerular cells.

RENAL CIRCULATION

Blood Flow

In a resting adult, the kidneys receive 1.2–1.3 L of blood per minute, or just under 25% of the cardiac output (Table 32–1). Renal blood flow can be measured with electromagnetic or other types of flow meters, or it can be determined by applying the Fick principle (see Chapter 29) to the kidney—ie, by measuring the amount of a given substance taken up per unit of time and dividing this value by the arteriovenous difference for the substance across the kidney. Since the kidney filters plasma, the **renal plasma flow** equals the amount of a substance excreted per unit of time divided by the renal arteriovenous difference as long as the amount in the red cells is unaltered during passage through the kidney. Any excreted substance can be used if its concentration in arterial and renal venous plasma can be measured and if it is neither metabolized, stored, nor produced by the kidney and does not itself affect blood flow.

Renal plasma flow can be measured by infusing para-aminohippuric acid (PAH) and determining its urine and plasma concentrations. PAH is filtered by the glomeruli and secreted by the tubular cells, so that its **extraction ratio** (arterial concentration minus renal venous concentration/arterial concentration) is high. For example, when PAH is infused at low doses, 90% of the PAH in arterial blood is removed in a single circulation through the kidney. It has therefore become commonplace to calculate the "renal plasma flow" by dividing the amount of PAH in the urine by the plasma PAH level, ignoring the level in renal venous blood. Peripheral venous plasma can be used because its PAH concentration is essentially identical to that in the arterial plasma reaching the kidney. The value obtained should be called the **effective renal plasma flow (ERPF)** to indicate that the level in renal venous plasma was not measured. In humans, ERPF averages about 625 mL/min.

Effective renal plasma flow (ERPF) =

$$\frac{U_{PAH}\dot{V}}{P_{PAH}} = \text{Clearance of PAH } (C_{PAH})$$

Example:

Concentration of PAH in urine (U_{PAH}): 14 mg/mL

Urine flow ($\dot{V}$): 0.9 mL/min

Concentration of PAH in plasma (P_{PAH}): 0.02 mg/mL

$$\text{ERPF} = \frac{14 \times 0.9}{0.02} = 630 \text{ mL/min}$$

It should be noted that the ERPF determined in this way is the **clearance** of PAH. The concept of clearance is discussed in detail below.

ERPF can be converted to actual renal plasma flow (RPF):

Average PAH extraction ratio: 0.9

$$\frac{\text{ERPF}}{\text{Extraction ratio}} = \frac{630}{0.9} = \text{Actual RPF} = 700 \text{ mL/min}$$

From the renal plasma flow, the renal blood flow can be calculated by dividing by one minus the hematocrit:

Hematocrit (Hct): 45%

$$\text{Renal blood flow} = \text{RPF} \times \frac{1}{1 - \text{Hct}} =$$

$$700 \times \frac{1}{0.55} = 1273 \text{ mL/min}$$

Regional Blood Flow

The blood flow in the renal cortex is much greater than that in the medulla. Values obtained in dogs are 4–5 mL/g of kidney tissue per minute in the cortex, 0.2 mL/g/min in the outer medulla, and 0.03 mL/g/min in the inner medulla. However, even the medullary flow is relatively large on a tissue weight basis. For comparison, the normal blood flow of the brain is 0.5 mL/g/min (see Chapter 32).

Pressure in Renal Vessels

The pressure in the glomerular capillaries has been measured directly in rats and has been found to be considerably lower than predicted on the basis of indirect measurements. When the mean systemic arterial pressure is 100 mm Hg, the glomerular capillary pressure is about 45 mm Hg. The pressure drop across the glomerulus is only 1–3 mm Hg, but there is a further drop in the efferent arteriole so that the pressure in the peritubular capillaries is about 8 mm Hg. The pressure in the renal vein is about 4 mm Hg. Pressure gradients are similar in the squirrel monkey and presumably in humans, with a glomerular capillary pressure that is about 40% of systemic arterial pressure.

Renal Vasoconstriction

Catecholamines constrict the renal vessels, with the greatest effect of injected norepinephrine being exerted on the interlobular arteries and the afferent arterioles. Angiotensin II exerts a selective constrictor effect on the efferent arterioles. Prostaglandins increase blood flow in the renal cortex and decrease blood flow in the renal medulla. In conscious dogs, moderate hemorrhage actually causes renal vasodilation, and this response is prevented by indomethacin, a drug that blocks prostaglandin synthesis. Acetyl choline also produces renal vasodilation. For unknown reasons, a high-protein diet increases renal blood flow.

Stimulation of the renal nerves causes a marked decrease in renal blood flow. This effect is mediated by α_1-adrenergic receptors and to a lesser extent by postsynaptic α_2-adrenergic receptors. Similar renal vasoconstriction can be produced by stimulating the vasomotor area in the medulla oblongata, parts of the brain stem, and parts of the cerebral cortex. There is some tonic discharge in the renal nerves at rest in animals and humans. When systemic blood pressure falls, the vasoconstrictor response produced by decreased discharge in the baroreceptor nerves includes renal vasoconstriction. Renal blood flow is decreased during exercise and, to a lesser extent, on rising from the supine to the standing position.

Other Functions of the Renal Nerves

In addition to producing renal vasoconstriction, stimulation of the renal nerves increases renin secretion from the juxtaglomerular cells, primarily via β-adrenergic receptors located on the membranes of these cells (see Chapter 24). In addition, stimulation of the renal nerves increases the reabsorption of Na^+ and water in the tubules. Since these actions are independent of the renin-angiotensin system and prostaglandins and can, at least in part, be produced in vitro, they are probably due to a direct action of catecholamines on the renal tubules. It is still unsettled whether the effects on Na^+ and water reabsorption are mediated via α- or β-adrenergic receptors, and they may be mediated by both. The physiologic role of the renal nerves is also unsettled, in part because most renal functions appear to be normal in patients with transplanted kidneys, and it takes some time for transplanted kidneys to acquire a functional innervation.

Autoregulation of Renal Blood Flow

When the kidney is perfused at moderate pressures (90–220 mm Hg in the dog), the renal vascular resistance varies with the pressure so that renal blood flow is relatively constant (Fig 38–4). Autoregulation of this type occurs in other organs, and several factors contribute to it (see Chapter 31). Renal autoregulation is present in denervated and in isolated, perfused kidneys but is prevented by the administration of drugs that paralyze vascular smooth muscle. It is probably produced in part by a direct contractile response of the smooth muscle of the afferent arteriole to stretch. At low perfusion pressures, angiotensin II also appears to play a role by constricting the efferent arterioles, thus maintaining the glomerular filtration rate. This is believed to be the explanation of the renal failure that sometimes develops in patients with poor renal perfusion who are treated with drugs which inhibit angiotensin converting enzyme.

Renal Oxygen Consumption

The O_2 consumption of the human kidney is about 18 mL/min. The renal blood flow per gram of tissue is very large, and it is therefore not surprising that the arteriovenous O_2 difference is only 14 mL/L of blood, compared with 62 mL/L for the brain and 114 mL/L for the heart (Table 32–1). The renal function with which the O_2 consumption correlates best is the rate of active transport of Na^+. The O_2 consumption of the cortex is about 9 mL/100 g/min, whereas that of the inner medulla is only 0.4 mL/100 g/min. The tubular fluid passes through the medulla on its way to the renal pelvis, and the P_{O_2} of the urine is also low.

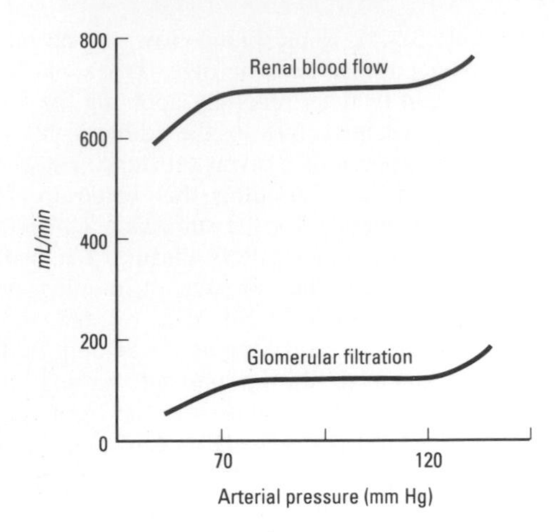

Figure 38–4. Autoregulation in the kidneys.

GLOMERULAR FILTRATION

Measuring GFR

The **glomerular filtration rate (GFR)** can be measured in intact animals and humans by measuring the excretion and plasma level of a substance that is freely filtered through the glomeruli and neither secreted nor reabsorbed by the tubules. The amount of such a substance in the urine per unit of time must have been provided by filtering exactly the number of milliliters of plasma that contained this amount. Therefore, if the substance is designated by the letter x, the GFR is equal to the **concentration** of x in urine (U_x) times the **urine flow** per unit of time (V) divided by the **arterial plasma level** of x (P_x), or $U_x V/P_x$. This value is called the **clearance** of x (C_x). P_x is, of course, the same in all parts of the arterial circulation, and if x is not metabolized to any extent in the tissues, the level of x in peripheral venous plasma can be substituted for the arterial plasma level.

Substances Used to Measure GFR

In addition to the requirement that it be freely filtered and neither reabsorbed nor secreted in the tubules, a substance suitable for measuring the GFR should meet other criteria (Table 38–2). Inulin, a polymer of fructose with a molecular weight of 5200 that is found in dahlia tubers, meets these criteria in

Table 38–2. Characteristics of a substance suitable for measuring the GFR by determining its clearance.

Freely filtered (ie, not bound to protein in plasma or sieved in the process of ultrafiltration)
Not reabsorbed or secreted by tubules
Not metabolized
Not stored in kidney
Not toxic
Has no effect on filtration rate
Preferably easy to measure in plasma and urine

humans and most animals and is extensively used to measure GFR. In practice, a loading dose of inulin is administered intravenously, followed by a sustaining infusion to keep the arterial plasma level constant. After the inulin has equilibrated with body fluids, an accurately timed urine specimen is collected and a plasma sample obtained halfway through the collection. Plasma and urinary inulin concentrations are determined and the clearance calculated.

Example:

$$U_{in} = 35 \text{ mg/mL}$$

$$\dot{V} = 0.9 \text{ mL/min}$$

$$P_{in} = 0.25 \text{ mg/mL}$$

$$C_{in} = \frac{U_{in}\dot{V}}{P_{in}} = \frac{35 \times 0.9}{0.25}$$

$$C_{in} = 126 \text{ mL/min}$$

In dogs, cats, rabbits, and a number of other mammalian species, clearance of creatinine (C_{Cr}) can also be used to determine the GFR, but in primates, including humans, some creatinine is secreted by the tubules, and some may be reabsorbed. In addition, plasma creatinine determinations are inaccurate at low creatinine levels because the method for determining creatinine measures small amounts of other plasma constituents. In spite of this, the clearance of endogenous creatinine is frequently measured in patients. The values agree quite well with the GFR values measured with inulin because, although the value for U_{Cr} $\dot{V}$ is high as a result of tubular secretion, the value for P_{Cr} is also high as a result of nonspecific chromogens, and the errors thus tend to cancel. Endogenous creatinine clearance is easy to measure and is a worthwhile index of renal function, but when precise measurements of GFR are needed it seems unwise to rely on a method that owes what accuracy it has to compensating errors.

Normal GFR

The GFR in an average-sized normal man is approximately 125 mL/min. Its magnitude correlates fairly well with surface area, but values in women are 10% lower than those in men even after correction for surface area. It should be noted that 125 mL/min is 7.5 L/h, or 180 L/d, whereas the normal urine volume is about 1 L/d. Thus, 99% or more of the filtrate is normally reabsorbed. At the rate of 125 mL/min, the kidneys filter in 1 day an amount of fluid equal to 4 times the total body water, 15 times the ECF volume, and 60 times the plasma volume.

Control of GFR

The factors governing filtration across the glomerular capillaries are the same as those governing filtration across all other capillaries (see Chapter 30), ie, the size of the capillary bed, the permeability of the capillaries, and the hydrostatic and osmotic pressure gradients across the capillary wall. For each nephron:

$$GFR = K_f[(P_{GC} - P_T) - (\Pi_{GC} - \Pi_T)]$$

K_f, the glomerular ultrafiltration coefficient, is the product of the glomerular capillary wall hydraulic conductivity (ie, its permeability) and the effective filtration surface area. P_{GC} is the mean hydrostatic pressure in the glomerular capillaries, P_T the mean hydrostatic pressure in the tubule, Π_{GC} the osmotic pressure of the plasma in the glomerular capillaries, and Π_T the osmotic pressure of the filtrate in the tubule.

Permeability

The permeability of the glomerular capillaries is about 50 times that of the capillaries in skeletal muscle. Neutral substances with effective molecular diameters of less than 4 nm are freely filtered, and the filtration of neutral substances with diameters of more than 8 nm approaches zero (Fig 38–5). Between these values, filtration is inversely proportionate to diameter. However, sialoproteins in the glomerular capillary wall are negatively charged, and studies with anionically charged and cationically charged dextrans indicate that the negative charges repel negatively charged substances in blood, with the result that filtration of anionic substances 4 nm in diameter is less than half that of neutral substances of the same size. This probably explains why albumin, with an effective molecular diameter of approximately 7 nm, normally has a glomerular concentration only 0.2% of its plasma concentration rather than the higher concentration that would be expected on

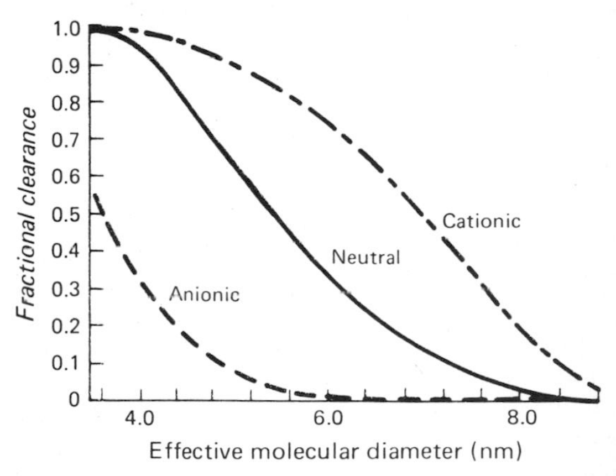

Figure 38–5. Effect of electrical charge on the fractional clearance of dextran molecules of various sizes in rats. The negative charges in the glomerular membrane retard the passage of negatively charged molecules (anionic dextran) and facilitate the passage of positively charged molecules (cationic dextran). (Reproduced, with permission, from Brenner BM, Beeuwkes R: The renal circulations. *Hosp Pract* [July] 1978;**13**:35.)

the basis of diameter alone; circulating albumin is negatively charged. Filtration of cationic substances is slightly greater than that of neutral substances.

The presence of albumin in the urine is called **albuminuria.** In nephritis, the negative charges in the glomerular wall are dissipated, and albuminuria can occur for this reason without an increase in the size of the ''pores'' in the membrane.

Because of the nondiffusibility of plasma proteins, there is a Donnan effect (see Chapter 1) on the monovalent diffusible ion distribution, the concentration of anions being about 5% greater in the glomerular filtrate than in plasma and the concentration of monovalent cations 5% less. However, for most purposes this effect can be ignored, and in other respects the composition of the filtrate is similar to that of plasma.

Size of the Capillary Bed

It is now clear that K_f can be altered by the mesangial cells, contraction of these cells producing a decrease in K_f that is largely due to a reduction in the area available for filtration. Contraction of points where the capillary loops bifurcate probably shifts flow away from some of the loops, and elsewhere, contracted mesangial cells distort and encroach on the capillary lumen. Agents that have been shown to affect the mesangial cells are listed in Table 38–3. Angiotensin II is an important regulator of mesangial contraction, and there are angiotensin II receptors in the glomeruli. In addition, there is some evidence that mesangial cells make renin.

Table 38–3. Agents causing contraction or relaxation of mesangial cells.*

Contraction	Relaxation
Angiotensin II	ANP
Vasopressin	Dopamine
Norepinephrine	PGE_2
Platelet-activating factor	Cyclic AMP
Platelet-derived growth factor	
Thromboxane A_2	
PGF_2	
Leukotrienes C and D	
Histamine	

*Modified from Schlondorff D: The glomerular mesangial cell: an expanding role for a specialized pericyte. *FASEB J* 1987; **1**:272.

Hydrostatic & Osmotic Pressure

The pressure in the glomerular capillaries is higher than that in other capillary beds because the afferent arterioles are short, straight branches of the interlobular arteries. Furthermore, the vessels ''downstream'' from the glomeruli, the efferent arterioles, have a relatively high resistance. The capillary hydrostatic pressure is opposed by the hydrostatic pressure in Bowman's capsule. It is also opposed by the osmotic pressure gradient across the glomerular capillaries (Π_{GC}–Π_T). Π_T is normally negligible, and the gradient is equal to the oncotic pressure of the plasma proteins.

The actual pressures in rats are shown in Fig 38–6. The corresponding pressures in humans are probably comparable. The net filtration pressure in rats (P_{UF}) is 15 mm Hg at the afferent end of the glomerular capillaries, but it falls to zero—ie, **filtration equilibrium** is reached—proximal to the efferent end of the glomerular capillaries. This is because fluid leaves the plasma and the oncotic pressure rises as blood passes through the glomerular capillaries. The calculated change in $\Delta\Pi$ along an idealized glomerular capillary is shown in Fig 38–6. It is apparent that portions of the glomerular capillaries do not normally contribute to the formation of the glomerular ultrafiltrate; ie, exchange across the glomerular capillaries is flow-limited rather than diffusion-limited (see Chapter 30). It is also apparent that a decrease in the rate of rise of the $\Delta\Pi$ curve produced by an increase in renal plasma flow would increase filtration without any change in ΔP because it would increase the distance along the capillary in which filtration was taking place.

Changes in GFR

Variations in the factors listed in the preceding paragraphs have predictable effects on the GFR (Table 38–4). Changes in renal vascular resistance due to autoregulation tend to stabilize filtration pressure, but when the mean systemic arterial pressure drops below 90 mm Hg, there is a sharp drop in GFR. The GFR tends to be maintained when efferent arteriolar constriction is greater than afferent constriction, but either type of constriction decreases the tubular blood flow.

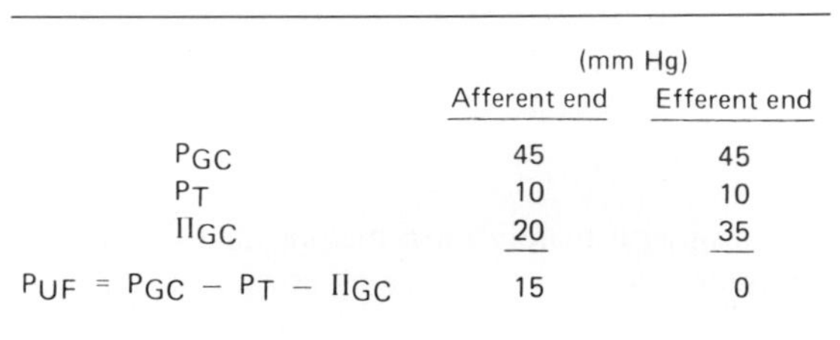

	Afferent end (mm Hg)	Efferent end (mm Hg)
P_{GC}	45	45
P_T	10	10
Π_{GC}	20	35
$P_{UF} = P_{GC} - P_T - \Pi_{GC}$	15	0

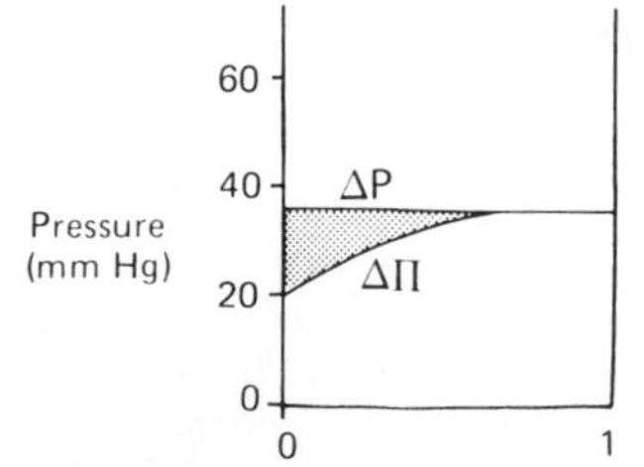

Figure 38–6. Hydrostatic pressure (P_{GC}) and osmotic pressure (Π_{GC}) in a glomerular capillary in the rat. P_T, pressure in Bowman's capsule; P_{UF}, net filtration pressure. Π_T is normally negligible, so $\Delta\Pi = \Pi_{GC}$. $\Delta P = P_{GC} - P_T$. (Reproduced, with permission, from Mercer PF, Maddox, DA, Brenner BM: Current concepts of sodium chloride and water transport by the mammalian nephron. *West J Med* 1974;**120**:33.)

Table 38–4. Factors affecting the glomerular filtration rate.

1. Changes in renal blood flow
2. Changes in glomerular capillary hydrostatic pressure
   a. Changes in systemic blood pressure
   b. Afferent or efferent arteriolar constriction
3. Changes in hydrostatic pressure in Bowman's capsule
   a. Ureteral obstruction
   b. Edema of kidney inside tight renal capsule
4. Changes in concentration of plasma proteins: dehydration, hypoproteinemia, etc (minor factors)
5. Changes in K_f
   a. Changes in glomerular capillary permeability
   b. Changes in effective filtration surface area

Filtration Fraction

The ratio of the GFR to the renal plasma flow (RPF), the **filtration fraction,** is normally 0.16–0.20. The GFR varies less than the RPF. When there is a fall in systemic blood pressure, the GFR falls less than the RPF because of efferent arteriolar constriction, and consequently the filtration fraction rises.

TUBULAR FUNCTION

General Considerations

The amount of any substance that is filtered is the product of the GFR and the plasma level of the substance ($C_{In}P_X$). The tubular cells may add more of the substance to the filtrate (tubular secretion), may remove some or all of the substance from the filtrate (tubular reabsorption), or may do both. The amount of the substance excreted (U_XV) equals the amount filtered plus the **net amount transferred** by the tubules. This latter quantity is conveniently indicated by the symbol T_X. When there is net tubular secretion, T_X is positive; when there is net tubular absorption, T_X is negative (Fig 38–7). Since the clearance of any substance is UV/P, the clearance equals the GFR if there is no net tubular secretion or reabsorption; exceeds the GFR if there is net tubular secretion; and is less than the GFR if there is net tubular reabsorption. It should be noted that in the latter 2 cases the clearance is a "virtual volume," an index of renal function, rather than an actual volume.

Much of our knowledge about glomerular filtration and tubular function has been obtained by the use of micropuncture techniques. Micropipettes can be inserted into the tubules of the living kidney and the composition of aspirated tubular fluid determined by the use of microchemical techniques. Techniques such as insertion of 2 pipettes in a single nephron have made it possible to study tubular function in great detail. It has also been possible to study isolated perfused segments of renal tubules.

Membrane potentials of tubular cells have been measured with microelectrodes. The potential difference of −70 mV between the tubular lumen and the interior of proximal tubular cells is the same, or nearly the same, as that between the cells and the ECF. The potential between the tubular lumen and the ECF is −2 mV (tubular lumen negative) in the first part of the proximal tubule and +2 mV (tubular lumen positive) in the remainder of the proximal tubule. The lumen of the loop of Henle is negative to the ECF except for the thick ascending limb, where the potential difference is +7 mV. In the distal tubule, the potential difference again becomes lumen negative, and it increases to a peak of −45 mV in the late portion. The potential difference in the collecting duct is −35 mV (lumen negative).

The effects of tubular reabsorption and secretion on substances of major physiologic interest are summarized in Table 38–5.

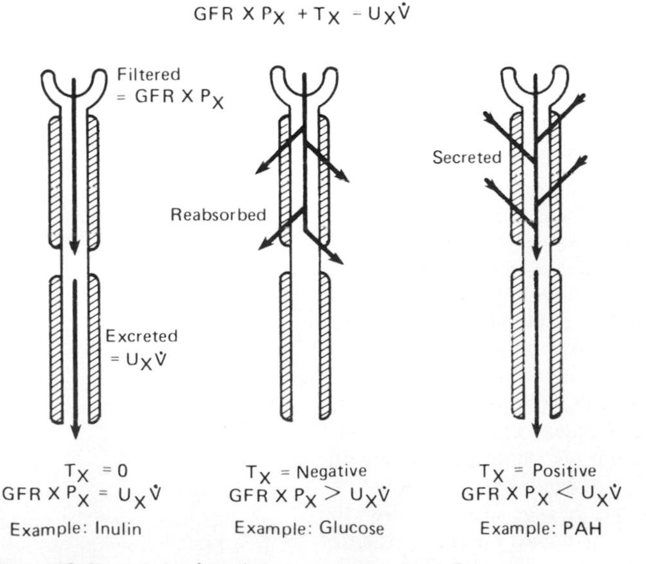

Figure 38–7. Tubular function. For explanation of symbols, see text.

Table 38–5. Renal handling of various plasma constituents in a normal adult human on an average diet. P, proximal tubules; L, loops of Henle; D. distal tubules; C, collecting ducts.

Substance	Per 24 Hours				Percentage Reabsorbed	Location
	Filtered	Reabsorbed	Secreted	Excreted		
Na^+ (meq)	26,000	25,850		150	99.4	P, L, D, C
K^+ (meq)	600	560*	50*	90	93.3	P, L, D, C
Cl^- (meq)	18,000	17,850		150	99.2	P, L, D, C
HCO_3^- (meq)	4,900	4,900		0	100	P, D
Urea (mmol)	870	460†		410	53	P, L, D, C
Creatinine (mmol)	12	1‡	1‡	12	. . .	. . .
Uric acid (mmol)	50	49	4	5	98	P
Glucose (mmol)	800	800		0	100	P
Total solute (mosm)	54,000	53,400	100	700	87	P, L, D, C
Water (mL)	180,000	179,000		1000	99.4	P, L, D, C

*K^+ is both reabsorbed and secreted.
†Urea diffuses into as well as out of some portions of the nephron.
‡Variable secretion and probable reabsorption of creatinine in humans.

Mechanisms of Tubular Reabsorption & Secretion

Small proteins and some peptide hormones are reabsorbed in the proximal tubules by endocytosis. Other substances are secreted or reabsorbed in the tubules by passive or facilitated diffusion down chemical or electrical gradients or actively transported against such gradients (see Chapter 1). Like transport systems elsewhere, renal active transport systems have a maximal rate, or **transport maximum (Tm),** at which they can transport a particular solute. Thus, the amount of a particular solute transported is proportionate to the amount present up to the Tm for the solute, but at higher concentrations, the transport mechanism is **saturated** and there is no appreciable increment in the amount transported. However, the Tm's for some systems are high, and it is difficult to saturate them.

Not only does the reabsorption of Na^+ and Cl^- play a major role in body electrolyte and water metabolism, but Na^+ transport is coupled to the movement of H^+, other electrolytes, glucose, amino acids, organic acids, phosphate, and other substances across the tubule walls.

Na^+ Reabsorption

In the proximal and distal tubules and the collecting ducts, Na^+ diffuses passively from the tubular lumen into the tubular epithelial cells down its concentration and electrical gradients and is actively pumped from these cells into the interstitial space. Na^+ is pumped into the interstitium by Na^+-K^+ ATPase. The operation of this ubiquitous Na^+ pump is considered in detail in Chapter 1. It extrudes 3 Na^+ in exchange for 2 K^+ that are pumped into the cell.

The tubular cells are connected by tight junctions at their luminal edges, but there is space between the cells along the rest of their lateral borders. Much of the Na^+ is actively transported into these extensions of the interstitial space, the **lateral intercellular spaces** (Fig 38–8). Proximal tubular reabsorbate is slightly hypertonic, and water moves passively along the osmotic gradient created by its absorption. The rate at which solutes and water move into the capil-

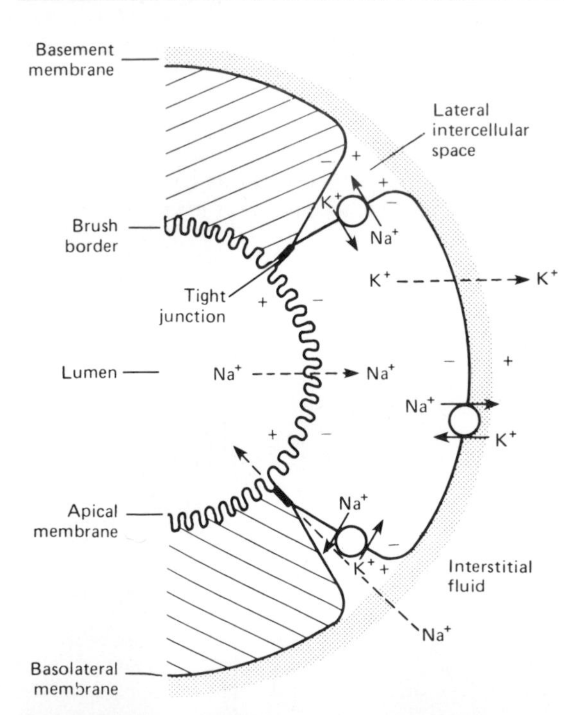

Figure 38–8. Mechanism for Na^+ reabsorption in the proximal tubule. Solid lines indicate active transport, and dashed lines indicate diffusion. Note that some Na^+ and H_2O leak back into the renal tubule through the tight junctions between the epithelial cells. (Reproduced, with permission, from Sullivan LP, Grantham JJ: *Physiology of the Kidney,* 2nd ed. Lea & Febiger, 1982.)

laries from the lateral intercellular spaces and the rest of the interstitium is determined by the Starling forces determining movement across the walls of all capillaries, ie, the hydrostatic and osmotic pressures in the interstitium and the capillaries (see Chapter 30). When the capillary hydrostatic pressure is increased or the plasma protein pressure is decreased, the movement of solute and water into the capillaries slows, and the lateral intercellular spaces expand. Water can leak across the tight junctions even in the hydropenic state, and when the lateral intercellular spaces are expanded, the "leak" back into the tubules via this paracellular pathway is large. In this way, net Na^+ reabsorption is decreased. In the thick portion of the ascending limb of the loop of Henle, Na^+ is cotransported from the lumen into the tubular epithelial cells with K^+ and Cl^- (see below) and is then actively transported into the lateral intercellular spaces. Thus, Na^+ is actively transported out of all parts of the renal tubule except the thin portions of the loop of Henle.

Glucose Reabsorption

Glucose, amino acids, and bicarbonate are reabsorbed along with Na^+ in the early portion of the proximal tubule (Fig 38–9). Farther along the tubule,

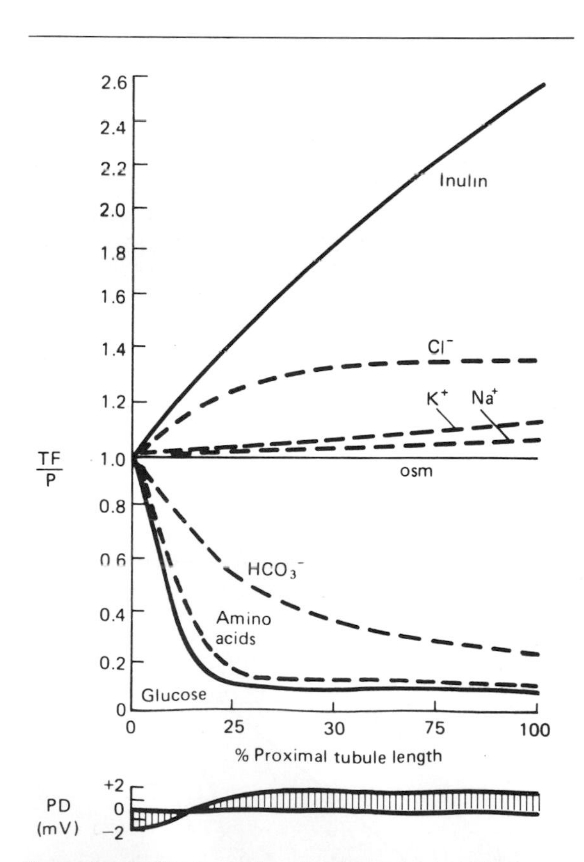

Figure 38–9. Reabsorption of various solutes in the proximal tubule in relation to the potential difference (PD) along the tubule. TF/P, tubular fluid to plasma concentration ratio. (Courtesy of FC Rector Jr.)

Na^+ is reabsorbed with Cl^-. Glucose is typical of substances removed from the urine by secondary active transport. It is filtered at a rate of approximately 100 mg/min (80 mg/dL of plasma × 125 mL/min). Essentially all of the glucose is reabsorbed, and no more than a few milligrams appear in the urine per 24 hours. The amount reabsorbed is proportionate to the amount filtered and hence to the plasma glucose level (P_G) times the GFR up to the transport maximum (Tm_G); but when the Tm_G is exceeded, the amount of glucose in the urine rises (Fig 38–10). The Tm_G is about 375 mg/min in men and 300 mg/min in women.

The **renal threshold** for glucose is the plasma level at which the glucose first appears in the urine in more than the normal minute amounts. One would predict that the renal threshold would be about 300 mg/dL—ie, 375 mg/min (Tm_G) divided by 125 mL/min (GFR). However, the actual renal threshold is about 200 mg/dL of arterial plasma, which corresponds to a venous level of about 180 mg/dL.

Fig 38–10 shows why the actual renal threshold is less than the predicted threshold. The "ideal" curve shown in this diagram would be obtained if the Tm_Gs in all the tubules were identical and if all the glucose were removed from each tubule when the amount filtered was below the Tm_G. The curve of data actually obtained in humans is not sharply angulated and deviates considerably from the "ideal" curve. This deviation is called **splay.** It is present for 2 reasons. In the first place, not all of the millions of nephrons in the kidneys have exactly the same Tm_G or filtration rate; in some, Tm_G is exceeded at low

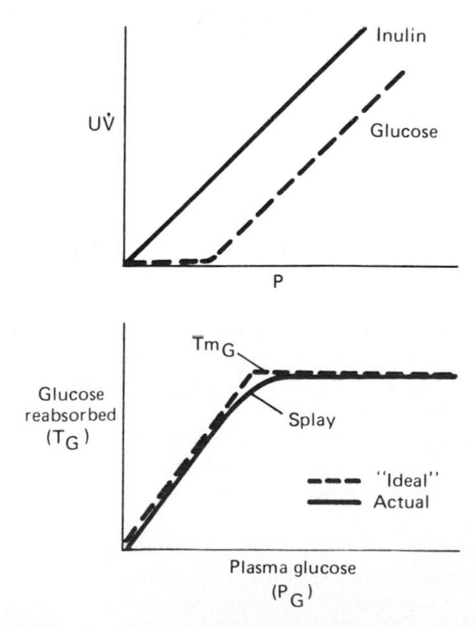

Figure 38–10. *Top:* Relation between plasma level (P) and excretion (UV) of glucose and inulin. ***Bottom:*** Relation between plasma glucose level (P_G) and amount of glucose reabsorbed (T_G).

levels of P_G. In the second place, some glucose escapes reabsorption when the amount filtered is below the Tm_G, because the reactions involved in glucose transport are not completely irreversible. Indeed, the dynamics of all the active transport systems that remove substances from the tubules are probably such that, for each, the curve relating the amount transported to the plasma level of the substance is rounded rather than sharply angulated. The degree of rounding is inversely proportionate to the avidity with which the transport mechanism binds the substance it transports.

Glucose Transport Mechanism

Glucose reabsorption in the kidneys is similar to glucose reabsorption in the intestine (see Chapter 25). Glucose and Na^+ bind to a common carrier (symport) in the luminal membrane (Fig 25–3), and glucose is carried into the cell as Na^+ moves down its electrical and chemical gradient. The Na^+ is then pumped out of the cell into the lateral intercellular spaces, and the glucose moves into the interstitial fluid by simple diffusion. Thus, glucose transport in the kidney as well as in the intestine is an example of secondary active transport; the energy for the active transport is provided by the Na^+-K^+ ATPase that pumps the Na^+ out of the cell.

The symport specifically binds the D isomer of glucose, and the rate of transport of D-glucose is many times greater than that of L-glucose. Glucose transport in the kidney is inhibited, as it is in the intestine, by the plant glucoside **phlorhizin,** which competes with D-glucose for binding to the symport.

Additional Examples of Secondary Active Transport

A variety of other substances are transported by secondary active transport via symports, with the energy provided by active transport of Na^+ out of the renal tubular cells. These substances include amino acid, lactate, citrate, phosphate, H^+, and Cl^-.

Like glucose reabsorption, amino acid reabsorption is most marked in the early portion of the proximal convoluted tubule. There appear to be separate carriers for (1) the neutral amino acids, (2) the dibasic amino acids, (3) the dicarboxylic amino acids, and (4) the imino acids and glycine; however, all appear to be symports and to transport amino acids with Na^+. The Na^+ is pumped out of the cells by Na^+-K^+ ATPase, and the amino acids leave by passive or facilitated diffusion to the interstitial fluid.

It appears that Cl^- is transported in 3 ways: by passive diffusion when the Cl^- concentration in the lumen is elevated or when the tubular lumen is negative relative to the interstitial fluid; by cotransport with Na^+ and K^+; and by a Cl^--OH^- antiport which operates in concert with a Na^+-H^+ antiport, putting H^+ and OH^- into the tubular lumen.

Reabsorption of Other Substances

Other substances that are actively reabsorbed include creatine, sulfate, uric acid, ascorbic acid, and the ketone bodies acetoacetate and β-hydroxybutyrate. The Tm's of these substances range from very low to very high. Some of the transport mechanisms appear to share a common step. Some transport in either direction, depending on circumstances, and some may exchange one substance for another. Most of the active transport mechanisms responsible for reabsorbing particular solutes are located in the proximal tubules.

Many renal active transport mechanisms, like active transport systems elsewhere, can be inhibited. For example, the mechanism responsible for the reabsorption of uric acid can be inhibited by probenecid (Benemid) and phenylbutazone (Butazolidin), a fact of practical importance in the treatment of gout (see Chapter 17).

PAH Transport

The dynamics of PAH transport illustrate the operation of the active transport mechanisms that secrete substances into the tubular fluid. The filtered load of PAH is a linear function of the plasma level, but PAH secretion increases as P_{PAH} rises only until a maximal secretion rate (Tm_{PAH}) is reached (Fig 38–11). When P_{PAH} is low, C_{PAH} is high; but as P_{PAH} rises above Tm_{PAH}, C_{PAH} falls progressively. It eventually approaches the clearance of inulin (C_{In}) (Fig 38–12), because the amount of PAH secreted becomes a smaller and smaller fraction of the total amount excreted. Conversely, the clearance of glucose is essentially zero at P_G levels below the renal threshold; but above the threshold, C_G rises to approach C_{In} as P_G is raised.

The use of C_{PAH} to measure effective renal plasma flow (ERPF) is discussed above. It should be emphasized again that the excretion of any substance that is not stored or metabolized in the kidney can be used to measure renal plasma flow if the level in arterial and renal venous plasma can be measured, and the only advantage of PAH is that its extraction ratio is so high

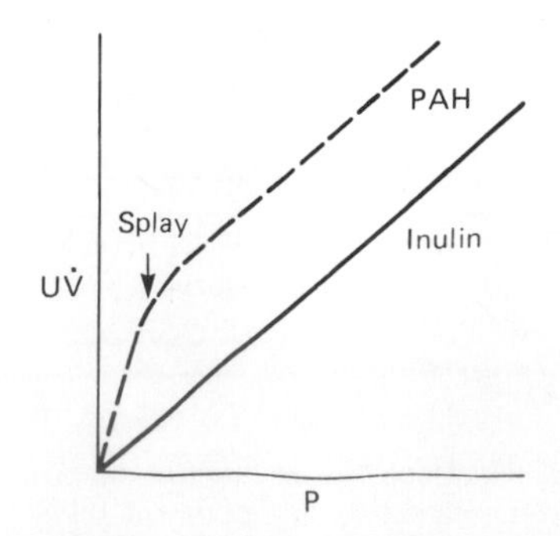

Figure 38–11. Relation between plasma levels (P) and excretion (UV) of PAH and inulin.

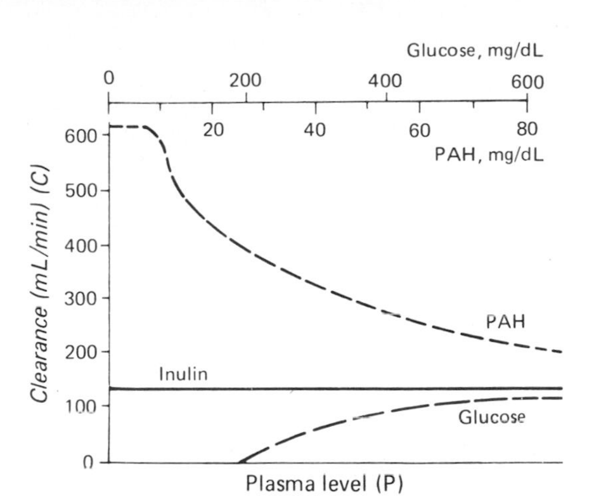

Figure 38–12. Clearance of inulin, glucose, and PAH at various plasma levels of each substance in humans.

that renal plasma flow can be approximated without measuring the PAH concentration in renal venous blood.

Other Substances Secreted by the Tubules

Derivatives of hippuric acid in addition to PAH, phenol red and other sulfonphthalein dyes, penicillin, and a variety of iodinated dyes such as iodopyracet (Diodrast) are actively secreted into the tubular fluid. Substances that are normally produced in the body and secreted by the tubules include various ethereal sulfates, steroid and other glucuronides, and 5-hydroxyindoleacetic acid, the principal metabolite of serotonin (see Chapter 4). All of these secreted substances are weak anions and compete with each other for secretion; for example, PAH reduces the excretion of both the 18-glucuronide and the 3-glucuronide derivatives of aldosterone (see Chapter 20). Thus, a single transport system appears to be involved. This transport system is limited to the proximal tubule. PAH transport and, presumably, the transport of the other weak anions are facilitated by acetate and lactate and inhibited by various Krebs cycle intermediates, probenecid, phenylbutazone, dinitrophenol, and mercurial diuretics. A separate proximal tubular transport mechanism secretes certain drugs that are organic bases. A third, which secretes ethylenediaminetetraacetic acid (EDTA), has also been described.

Tubuloglomerular Feedback

Signals from the renal tubules feed back to affect glomerular filtration. As the rate of flow through the ascending limb of the loop of Henle and first part of the distal tubule increases, glomerular filtration in the same nephron decreases, and, conversely, a decrease in flow increases GFR (Fig 38–13). This tends to maintain the constancy of the load delivered to the distal tubule. The sensor for the response appears to be the macula densa, and GFR is probably adjusted by constriction or dilation of the afferent arteriole. Constriction may be mediated by the renin-angiotensin system, prostaglandins, or cyclic AMP. The urinary component responsible for the feedback may be Cl^-, and the degree of constriction is probably proportionate to the rate of Cl^- reabsorption across the macula densa. The sensitivity of the feedback is increased when ECF volume is decreased and decreased when ECF volume is expanded.

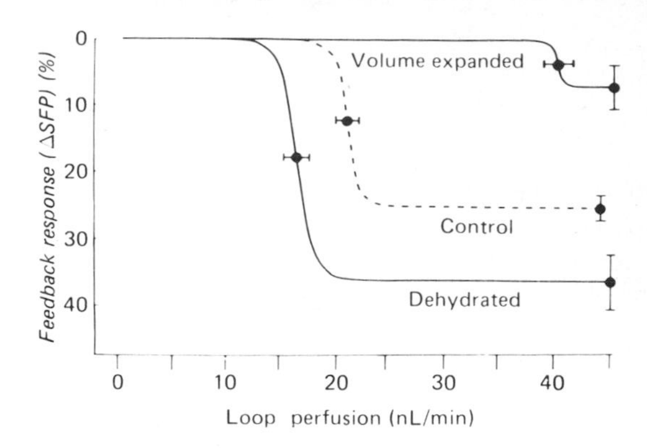

Figure 38–13. Percent decrease in filtration (measured in this case by change in proximal stop flow pressure, ΔSFP) produced by increased perfusion of the loop of Henle in rats. The sensitivity of the feedback effect on GFR is increased in dehydrated animals and decreased when ECF volume is expanded. (Reproduced, with permission, from Persson AEG et al: *Kidney Int* 1982;**22:[Suppl 12]**:S122.)

WATER EXCRETION

Normally, 180 L of fluid is filtered through the glomeruli each day, while the average daily urine volume is about 1 L. The same load of solute can be excreted per 24 hours in a urine volume of 500 mL with a concentration of 1400 mosm/L or in a volume of 23.3 L with a concentration of 30 mosm/L (Table 38–6). These figures demonstrate 2 important facts: first, that at least 87% of the filtered water is reabsorbed, even when the urine volume is 23 L; and second, that the reabsorption of the remainder of the filtered water can be varied without affecting total solute excretion. Therefore, when the urine is concentrated, water is retained in excess of solute; and when it is dilute, water is lost from the body in excess of solute. Both facts have great importance in the body economy and the regulation of the osmolality of the body fluids.

Proximal Tubule

Many substances are actively transported out of the fluid in the proximal tubule, but fluid obtained by micropuncture remains essentially isoosmotic to the end of the proximal tubule (Fig 38–9). Therefore, in

Table 38–6. Alterations in water metabolism produced by vasopressin in humans. In each case, the osmotic load excreted is 700 mosm/d.

	GFR (mL/min)	Percentage of Filtered Water Reabsorbed	Urine Volume (L/d)	Urine Concentration (mosm/L)	Gain or Loss of Water in Excess of Solute (L/d)
Urine isotonic to plasma	125	98.7	2.4	290	. . .
Vasopressin (maximal antidiuresis)	125	99.7	0.5	1400	1.9 gain
No vasopressin ("complete" diabetes insipidus)	125	87.1	23.3	30	20.9 loss

the proximal tubule, water moves passively out of the tubule along the osmotic gradients set up by active transport of solutes, and isotonicity is maintained. The osmotic gradient between renal interstitial fluid and proximal tubular fluid is very small, but it is effective in moving water out of the tubular lumen. Since the ratio of the concentration in tubular fluid to the concentration in plasma (TF/P) of the nonreabsorbably substance inulin is 2.5–3.3 at the end of the proximal tubule, it follows that 60–70% of the filtered solute and 60–70% of the filtered water have been removed by the time the filtrate reaches this point (Fig 38–14).

Loop of Henle

As noted above, the loops of Henle of the juxtamedullary nephrons dip deeply into the medullary pyramids before draining into the distal convoluted tubules in the cortex, and all of the collecting ducts descend back through the medullary pyramids to drain at the tips of the pyramids into the renal pelvis. There is a graded increase in the osmolality of the interstitium of the pyramids, the osmolality at the tips of the papillae normally being about 1200 mosm/L, approximately 4 times that of plasma. The descending limb of the loop of Henle is permeable to water, but the ascending limb is impermeable (Table 38–7). Na^+, K^+, and Cl^- are cotransported out of the thick segment of the ascending limb (see below). Therefore, the fluid in the descending limb of the loop of Henle becomes hypertonic as water moves into the hypertonic interstitium. In the ascending limb, it becomes more dilute, and when it reaches the top, it is hypotonic to plasma because of the movement of Na^+ and Cl^- out of the tubular lumen (Fig 38–15). In passing through the loop of Henle, another 15% of the filtered water is removed, so approximately 20% of the filtered water enters the distal tubule, and the TF/P of inulin at this point is about 5.

Distal Tubule

The distal tubule, particularly its first part, is in effect an extension of the thick segment of the ascending limb. It is relatively impermeable to water,

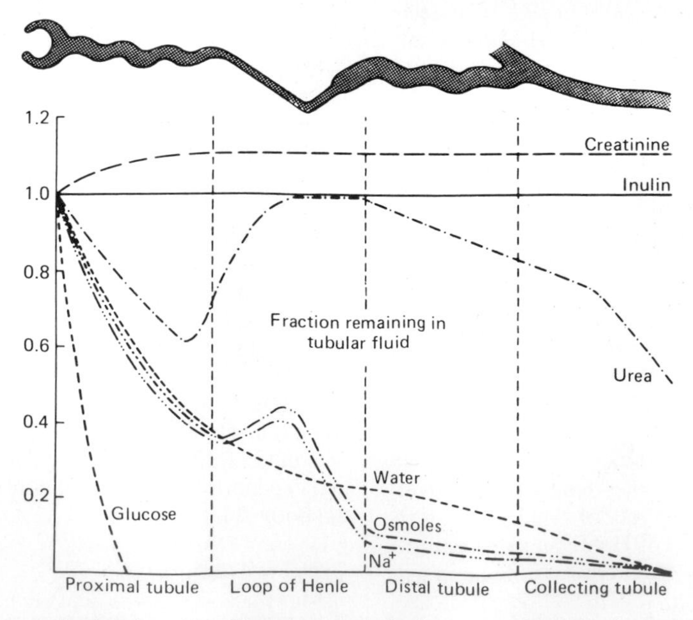

Figure 38–14. Changes in the fraction of the filtered amount of substances remaining in the tubular fluid along the length of the nephron. (Reproduced, with permission, from Sullivan LP, Grantham JJ: *Physiology of the Kidney,* 2nd ed. Lea & Febiger, 1982.)

Table 38–7. Permeability and transport in various segments of the nephron. Data based on studies of rabbit and human kidneys. Values indicated by asterisks are in the presence of vasopressin. These values are 1+ in absence of vasopressin.†

	Permeability			Active Transport of Na^+
	H_2O	Urea	NaCl	
Loop of Henle				
Thin descending limb	4+	+	±	0
Thin ascending limb	0	3+	4+	0
Thick ascending limb	0	±	±	4+
Distal convoluted tubule	±	±	±	3+
Collecting tubule				
Cortical portion	3+*	0	±	2+
Outer medullary portion	3+*	0	±	1+
Inner medullary portion	3+*	3+	±	1+

†Modified and reproduced, with permission, from Kokko JP: Renal concentrating and diluting mechanisms. *Hosp Pract* (Feb) 1979;**110:**14.

and continued removal of the solute in excess of solvent further dilutes the tubular fluid. However, the reabsorption of Na^+ in this segment is variable, being regulated by aldosterone. About 5% of the filtered water is removed in this segment.

Collecting Ducts

The collecting ducts have 2 portions: a cortical portion and a medullary portion through which the filtrate flows from the cortex to the renal pelvis. The changes in osmolality and volume in the collecting ducts depend on the amount of vasopressin acting on

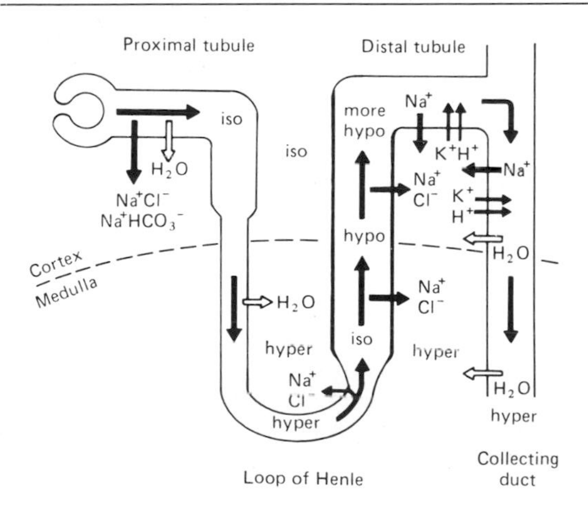

Figure 38–15. Summary of changes in the osmolality of tubular fluid in various parts of the nephron. The thickened wall of the ascending limb of the loop of Henle indicates relative impermeability of the tubular epithelium to water. In the presence of vasopressin, the fluid in the collecting ducts becomes hypertonic, whereas in the absence of this hormone, the fluid remains hypotonic throughout the collecting duct. Aldosterone promotes reabsorption of Na^+ and secretion of H^+ and K^+ in the distal convoluted tubule. (Reproduced, with permission, from Cannon PJ: The kidney in heart failure. *N Engl J Med* 1977;**296:**26.)

the ducts. This antidiuretic hormone from the posterior pituitary gland (see Chapter 14) increases the permeability of the collecting ducts to water. In the presence of enough vasopressin to produce maximal antidiuresis, water moves out of the hypotonic fluid entering the cortical collecting ducts into the interstitium of the cortex, and the tubular fluid becomes isotonic. In this fashion, as much as 10% of the filtered water is removed. The isotonic fluid then enters the medullary collecting ducts with a TF/P inulin of about 20. An additional 4.7% or more of the filtrate is reabsorbed into the hypertonic interstitium of the medulla, producing a concentrated urine with a TF/P inulin of over 300. In humans, the osmolality of urine may reach 1400 mosm/L, almost 5 times the osmolality of plasma, with a total of 99.7% of the filtered water being reabsorbed and only 0.3% appearing in the urine (Table 38–6). In other species, the ability to concentrate urine is even greater. Maximal urine osmolality in dogs is about 2500 mosm/L; in laboratory rats, about 3200 mosm/L; and in certain desert rodents, as high as 5000 mosm/L.

When vasopressin is absent, the collecting duct epithelium is relatively impermeable to water. The fluid therefore remains hypotonic, and large amounts flow into the renal pelvis. In humans, the urine osmolality may be as low as 30 mosm/L. The impermeability of the distal portions of the nephron is not absolute; along with the salt that is pumped out of the collecting duct fluid, about 2% of the filtered water is reabsorbed in the absence of vasopressin. However, as much as 13% of the filtered water may be excreted, and urine flow may reach 15 mL/min or more. The effects on daily water metabolism of the absence and the presence of vasopressin are summarized in Table 38–6.

It is much easier to drop a hydrometer into a urine specimen than to determine its osmolality, and the **specific gravity** is still occasionally measured as an index of urine concentration. The specific gravity of an ultrafiltrate of plasma is 1.010, while that of a maximally concentrated urine specimen is about 1.035. However, it should be remembered that the specific gravity of a solution depends upon the nature as well as the number of solute particles in it. Thus, for example, a subject excreting radiographic contrast medium may have a urine specific gravity of 1.040–1.050 with relatively little increase in osmolality. Consequently, it is more accurate to measure osmolality.

The Countercurrent Mechanism

The concentrating mechanism depends upon the maintenance of a gradient of increasing osmolality along the medullary pyramids. This gradient exists because of the operation of the loops of Henle as **countercurrent multipliers** and the vasa recta as **countercurrent exchangers.** A countercurrent system is a system in which the inflow runs parallel to, counter to, and in close proximity to the outflow for some distance. The operation of such a system in increasing the heating at the apex of a loop of pipe is shown in Fig 38–16. The heat in the outgoing fluid heats the incoming fluid so that by the time it reaches the heater its temperature is 90 degrees instead of 30 degrees, and the heater therefore raises the fluid temperature from 90 degrees to 100 degrees rather than from 30 degrees to 40 degrees.

The loop of Henle operates in an analogous fashion. The descending limb of the loop of Henle is relatively impermeable to solute but highly permeable to water. Consequently, water moves into the interstitium, and the concentration of Na^+ in the tubular fluid rises markedly (Fig 38–15). In addition, it appears that some urea enters the descending limb, adding to the increase in osmolality. The thin ascending limb of the loop is relatively impermeable to water and relatively permeable to Na^+ and urea, but more permeable to Na^+ than to urea. Consequently, Na^+ moves passively into the interstitium along a concentration gradient. The thick ascending limb is relatively impermeable to both water and solute, but in this segment, there is a carrier that transports 1 Na^+, 1 K^+, and 2 Cl^- from the tubular lumen into the cells (Fig 38–17). The Na^+ is actively transported out of the cells into the interstitium by Na^+-K^+ ATPase, and this portion of the renal tubule has a higher Na^+-K^+ ATPase content than any other part. The K^+ diffuses passively back into the tubular lumen, maintaining the transtubular potential difference of +7 mV (see above). One Cl^- diffuses passively into the interstitium, and one is cotransported with K^+.

The outer portions of the collecting duct are relatively impermeable to urea but permeable to water in the presence of vasopressin. Consequently, water leaves the lumen, and the urea concentration of the fluid increases markedly. Finally, the inner medullary portion of the collecting ducts is permeable to urea and, in the presence of vasopressin, to water. Urea moves passively into the interstitium, maintaining the high osmolality of the medullary pyramid. Additional water is also removed, and the fluid in the tubule becomes highly concentrated.

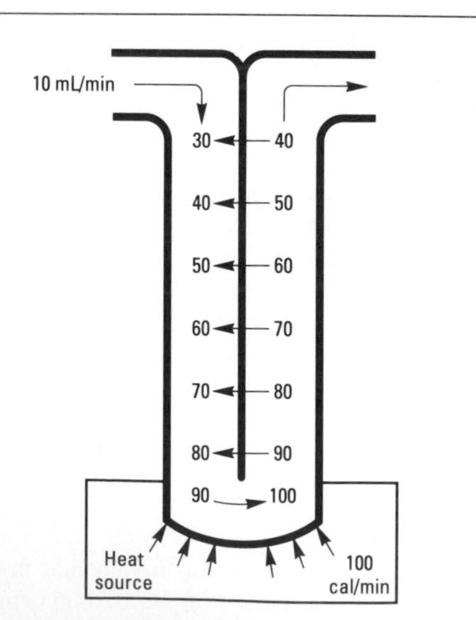

Figure 38–16. Operation of a countercurrent system. The heater raises the temperature of the water 10 degrees, but the heated water flowing away from the heater warms the inflow. A gradient of temperature is thus set up along the pipe, so that at the bend the temperature is raised not from 30 to 40 degrees but from 90 to 100 degrees.

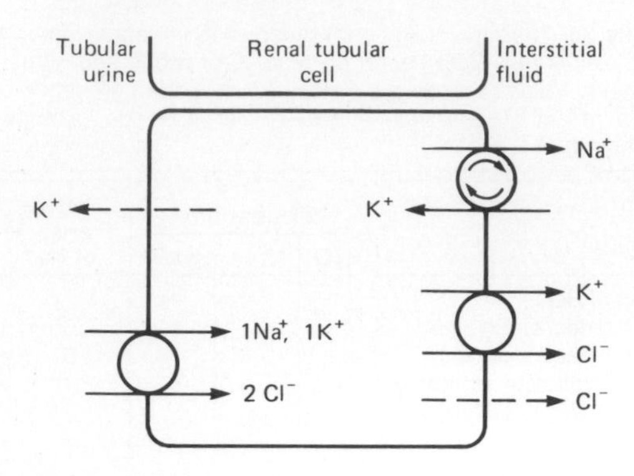

Figure 38–17. Cotransport of Na^+, K^+, and Cl^- by secondary active transport in cells in the thick ascending limb of Henle. Solid lines across membrane indicate active transport or secondary active transport; dashed lines indicate diffusion.

The osmotic gradient in the medullary pyramids would not last long if the Na^+ and urea in the interstitial spaces were removed by the circulation. These solutes remain in the pyramids primarily because the vasa recta (Fig 38–2) operate as countercurrent exchangers. The solutes diffuse out of the vessels conducting blood toward the cortex and into the vessels descending into the pyramid. Conversely, water diffuses out of the descending vessels and into the ascending vessels. Therefore, the solutes tend to recirculate in the medulla and water tends to bypass it, so that hypertonicity is maintained. The water removed from the collecting ducts in the pyramids is also removed by the vasa recta and enters the general circulation. It should be noted that countercurrent exchange is a passive process; it depends upon diffusion of water and solutes in both directions across the permeable walls of the vasa recta and could not maintain the osmotic gradient along the pyramids if the process of countercurrent multiplication in the loops of Henle were to cease.

It is also worth noting that there is a very large osmotic gradient in this portion of the renal tubules, and it is the countercurrent system that makes this gradient possible by spreading it along a system of tubules 1 cm or more in length rather than across a single layer of cells that is only a few micrometers thick. There are other examples of the operation of countercurrent systems in animals. One is the heat exchange between the arteries and venae comitantes of the limbs. To a minor degree in humans but to a major degree in mammals living in cold water, heat is transferred from the arterial blood flowing into the limbs to the adjacent veins draining blood back into the body, making the tips of the limbs cold while conserving body heat.

Role of Urea

The urea in the glomerular filtrate moves out of the tubules as its concentration is increased by the progressive reduction of filtrate volume. It may cross renal membranes by simple diffusion or possibly by facilitated diffusion. When urine flows are low, there is more opportunity for urea to leave the tubules, and only 10–20% of the filtered urea is excreted; at high urine flows, 50–70% is excreted. Urea accumulates in the interstitium of the medullary pyramids, where it tends to remain trapped by the countercurrent exchange in the vasa recta. The urea concentration of the tubular fluid is high when the fluid enters the inner medullary portions of the collecting ducts, and urea diffuses into the interstitium along with water. The amount of urea in the medullary interstitium and, consequently, in the urine varies with the amount of urea filtered, and this in turn varies with the dietary intake of protein. Therefore, a high-protein diet increases the ability of the kidneys to concentrate the urine.

Water Diuresis

The feedback mechanism controlling vasopressin secretion and the way vasopressin secretion is stimulated by a rise and inhibited by a drop in the effective osmotic pressure of the plasma are discussed in Chapter 14. The **water diuresis** produced by drinking large amounts of hypotonic fluid begins about 15 minutes after ingestion of a water load and reaches its maximum in about 40 minutes. The act of drinking produces a small decrease in vasopressin secretion before the water is absorbed, but most of the inhibition is produced by the decrease in plasma osmolality after the water is absorbed.

Water Intoxication

While excreting an average osmotic load, the maximal urine flow that can be produced during a water diuresis is about 16 mL/min. If water is ingested at a higher rate than this for any length of time, swelling of the cells because of the uptake of water from the hypotonic ECF becomes severe, and rarely, the symptoms of **water intoxication** may develop. Swelling of the cells in the brain causes convulsions and coma and leads eventually to death. Water intoxication can also occur when water intake is not reduced after administration of exogenous vasopressin or secretion of endogenous vasopressin in response to nonosmotic stimuli such as surgical trauma.

Osmotic Diuresis

The pressure of large quantities of unreabsorbed solutes in the renal tubules causes an increase in urine volume called **osmotic diuresis.** Solutes that are not reabsorbed in the proximal tubules exert an appreciable osmotic effect as the volume of tubular fluid decreases and their concentration rises. Therefore, they "hold water in the tubules." In addition, there is a limit to the concentration gradient against which Na^+ can be pumped out of the proximal tubules. Normally, the movement of water out of the proximal tubule prevents any appreciable gradient from developing, but Na^+ concentration in the fluid falls when water reabsorption is decreased because of the presence in the tubular fluid of increased amounts of unreabsorbable solutes. The limiting concentration gradient is reached, and further proximal reabsorption of Na^+ is prevented; more Na^+ remains in the tubule, and water stays with it. The result is that the loop of Henle is presented with a greatly increased volume of isotonic fluid. This fluid has a decreased Na^+ concentration, but the total amount of Na^+ reaching the loop per unit time is increased. In the loop, reabsorption of water and Na^+ is decreased because the medullary hypertonicity is decreased. The decrease is due primarily to decreased reabsorption of Na^+, K^+, and Cl^- in the ascending limb of the loop because the limiting concentration gradient for Na^+ reabsorption is reached. More fluid passes through the distal tubule, and because of the decrease in the osmotic gradient along the medullary pyramids, less water is reabsorbed in the collecting ducts. The result is a marked increase in urine volume and Na^+ excretion and in excretion of other electrolytes.

Osmotic diuresis is produced by the administration of compounds such as mannitol and related polysaccharides that are filtered but not reabsorbed. It is also produced by naturally occurring substances when they are present in amounts exceeding the capacity of the tubules to reabsorb them. In diabetes, for example, the glucose that remains in the tubules when the filtered load exceeds the Tm_G causes polyuria. Osmotic diuresis can also be produced by the infusion of large amounts of sodium chloride or urea.

It is important to recognize the difference between osmotic diuresis and water diuresis. In water diuresis, the amount of water reabsorbed in the proximal portions of the nephron is normal, and the maximal urine flow that can be produced is about 16 mL/min. In osmotic diuresis, increased urine flow is due to decreased water reabsorption in the proximal tubules and loops and very large urine flows can be produced. As the load of excreted solute is increased, the concentration of the urine approaches that of plasma (Fig 38–18) in spite of maximal vasopressin secretion, because an increasingly large fraction of the excreted urine is isotonic proximal tubular fluid. If osmotic diuresis is produced in an animal with diabetes insipidus, the urine concentration rises for the same reason.

Relation of Urine Concentration to GFR

The magnitude of the osmotic gradient along the medullary pyramids is increased when the rate of

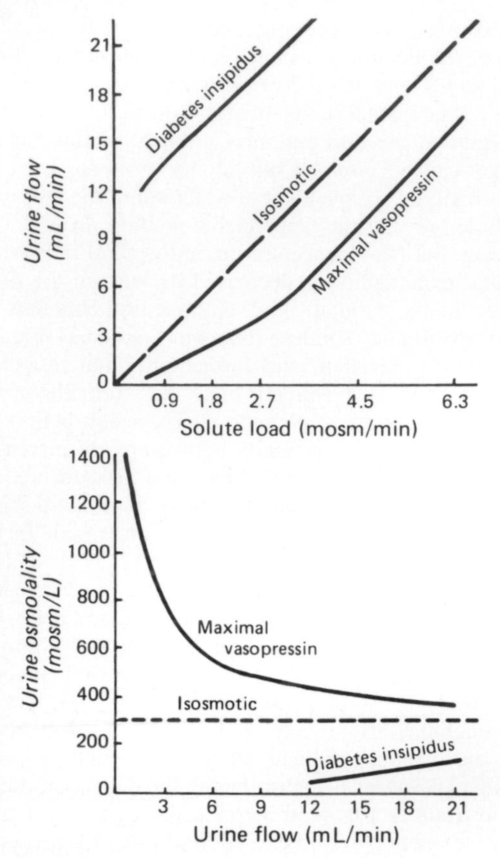

Figure 38–18. Approximate relationship between urine concentration and urine flow in osmotic diuresis in humans. The dashed line in the lower diagram indicates the concentration at which the urine is isoosmotic with plasma. (Reproduced, with permission, from Berliner RW, Giebisch G in: *Best and Taylor's Physiological Basis of Medical Practice,* 9th ed. Brobeck JR [editor]. Williams & Wilkins, 1979.)

flow of fluid through the loops of Henle is decreased. A reduction in GFR such as that caused by dehydration produces a decrease in the volume of fluid presented to the countercurrent mechanism, so that the rate of flow in the loops declines and the urine becomes more concentrated. When the GFR is low, the urine can become quite concentrated in the absence of vasopressin; if one renal artery is constricted in an animal with diabetes insipidus, the urine excreted on the side of the constriction becomes hypertonic because of the reduction in GFR, whereas that excreted on the opposite side remains hypotonic.

"Free Water Clearance"

In order to quantitate the gain or loss of water by excretion of a concentrated or dilute urine, the "free water clearance" (C_{H_2O}) is sometimes calculated. This is the difference between the urine volume and the clearance of osmoles (C_{osm}):

$$C_{H_2O} = \dot{V} - \frac{U_{osm}\dot{V}}{P_{osm}}$$

where V is the urine flow rate and U_{osm} and P_{osm} the urine and plasma osmolality. C_{osm} is the amount of water necessary to excrete the osmotic load in a urine that is isotonic with plasma. Therefore, C_{H_2O} is negative when the urine is hypertonic and positive when the urine is hypotonic. In Table 38–6, for example, the values for C_{H_2O} are -1.3 mL/min (-1.9 L/d) during maximal antidiuresis and 14.5 mL/min (20.9 L/d) in the absence of vasopressin.

ACIDIFICATION OF THE URINE & BICARBONATE EXCRETION

H^+ Secretion

The cells of the proximal and distal tubules, like the cells of the gastric glands, secrete hydrogen ions (see Chapter 26). Acidification also occurs in the collecting ducts. The reaction that is primarily responsible for H^+ secretion in the proximal tubules is Na^+-H^+ exchange (Fig 38–19). This is an example of secondary active transport; extrusion of Na^+ from the cells into the interstitium by Na^+-K^+ ATPase lowers intracellular Na^+, and this causes Na^+ to enter the cell from the tubular lumen, with coupled extrusion of H^+. The H^+ comes from intracellular dissociation of H_2CO_3, and the HCO_3^- that is formed diffuses into the interstitial fluid. Thus, for each H^+ secreted, one Na^+ and one HCO_3^- enter the interstitial fluid.

Carbonic anhydrase catalyzes the formation of H_2CO_3, and drugs that inhibit carbonic anhydrase depress both secretion of acid by the proximal tubules and the reactions which depend on it.

There is some evidence that H^+ is secreted in the proximal tubules by other types of pumps, but the evidence for these additional pumps is controversial, and in any case, their contribution is small relative to that of the Na^+-H^+ exhange mechanism. This is in

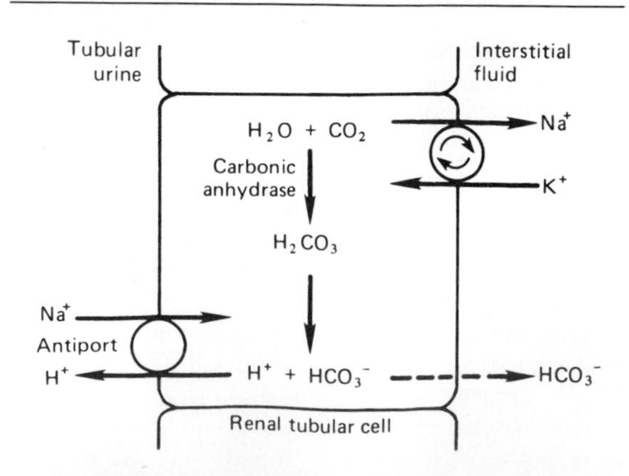

Figure 38–19. Chemical reactions involved in secretion of H^+ by proximal tubular cells in the kidney. Solid arrows crossing cell membranes indicate primary active transport in the case of Na^+ and K^+ and secondary active transport in the case of Na^+ and H^+. Dashed arrows indicate diffusion. Compare with Fig 26–8.

contrast to what occurs in the distal tubules and collecting ducts, where H^+ secretion is not dependent on Na^+ in the tubular lumen. In this part of the tubule, H^+ is secreted by an ATP-driven proton pump. Aldosterone acts on this pump to increase distal H^+ secretion. The epithelium of the collecting ducts is made up of **principal cells (P cells)** and **intercalated cells (I cells),** and there are I cells in the distal tubules. The I cells secrete acid, and like the parietal cells in the stomach, the I cells contain abundant carbonic anhydrase and numerous tubulovesicle structures. There is evidence that the H^+-translocating ATPase which produces H^+ secretion is located on these vesicles as well as on the luminal cell membrane and that, in acidosis, the number of H^+ pumps is increased by insertion of these tubulovesicles into the luminal cell membrane. These cells also contain **band 3,** an anion exchange protein, in their basolateral cell membranes, and this protein may function as a Cl^--HCO_3^- exchanger for the transport of HCO_3^- to the interstitial fluid.

Fate of H^+ in the Urine

The amount of acid secreted depends upon the subsequent events in the tubular urine. The maximal H^+ gradient against which the transport mechanisms can secrete in humans corresponds to a urine pH of about 4.5, ie, an H^+ concentration in the urine that is 1000 times the concentration of plasma. pH 4.5 is thus the **limiting pH.** If there were no buffers that "tied up" H^+ in the urine, this pH would be reached rapidly, and H^+ secretion would stop. However, 3 important reactions in the tubular fluid remove free H^+, permitting more acid to be secreted (Fig 38–20). These are the reactions with HCO_3^- to form CO_2 and H_2O, with HPO_4^{2-} to form $H_2PO_4^-$, and with NH_3 to form NH_4^+.

Reaction With Buffers

The dynamics of buffering are discussed in Chapter 35. The pK′ of the bicarbonate system is 6.1 (see Chapter 35), that of the dibasic phosphate system is 6.8, and that of the ammonia system is 9.0. The concentration of HCO_3^- in the plasma, and consequently in the glomerular filtrate, is normally about 24 meq/L, whereas that of phosphate is only 1.5 meq/L. Therefore, in the proximal tubule, most of the secreted H^+ reacts with HCO_3^- to form H_2CO_3 (Fig 38–20). The H_2CO_3 breaks down to form CO_2 and H_2O. In the proximal (but not in the distal) tubule, there is carbonic anhydrase in the brush border of the cells; this facilitates the formation of CO_2 and H_2O in the tubular fluid. The CO_2, which diffuses readily across all biologic membranes, enters the tubular cells, where it adds to the pool of CO_2 available to form H_2CO_3. Since most of the H^+ is removed from the tubule, the pH of the fluid is changed very little. This is the mechanism by which HCO_3^- is reabsorbed; for each mole of HCO_3^- removed from the tubular fluid, 1 mole of HCO_3 diffuses from the tubular cells into the blood, even though it is not the same mole that disappeared from the tubular fluid.

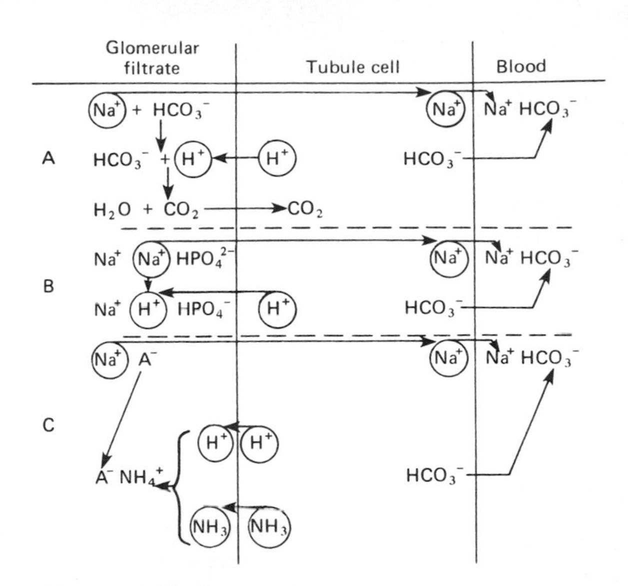

Figure 38–20. Fate of H^+ secreted into tubule in exchange for Na^+. *A:* Reabsorption of filtered bicarbonate. *B:* Formation of monobasic phosphate. *C:* Ammonium formation. Note that in each instance one Na^+ and one HCO_3^- enter the bloodstream for each H^+ secreted. A^-, anion.

Secreted H^+ also reacts with dibasic phosphate (HPO_4^{2-}) to form monobasic phosphate ($H_2PO_4^-$). This happens to the greatest extent in the distal tubules and collecting ducts, because it is here that the phosphate which escapes proximal reabsorption is greatly concentrated by the reabsorption of water. The reaction with NH_3 occurs in the proximal and distal tubules. H^+ also combines to a minor degree with other buffer anions.

Each H^+ that reacts with the buffers contributes to the urinary **titratable acidity,** which is measured by determining the amount of alkali that must be added to the urine to return its pH to 7.4, the pH of the glomerular filtrate. However, the titratable acidity obviously measures only a fraction of the acid secreted, since it does not account for the H_2CO_3 that has been converted to H_2O and CO_2. In addition, the pK′ of the ammonia system is 9.0, and the ammonia system is only titrated from the acidity of the urine to pH 7.4, so it contributes very little to the titratable acidity.

Ammonia Secretion

Several reactions in the renal tubular cells produce NH_4^+. NH_4^+ is in equilibrium with $NH_3 + H^+$ in the cells. Since the pK′ of this reaction is 9.0, the ratio of NH_3 to NH_4^+ at pH 7.0 is 1:100 (Fig 38–21). However, NH_3 is lipid-soluble and diffuses across the cell membranes down its concentration gradient into the interstitial fluid and tubular urine, whereas NH_4^+ does not cross cell membranes but stays in the cells.

The principal reaction producing NH_4^+ in cells is conversion of glutamine to glutamate. This reaction is catalyzed by the enzyme **glutaminase,** which is

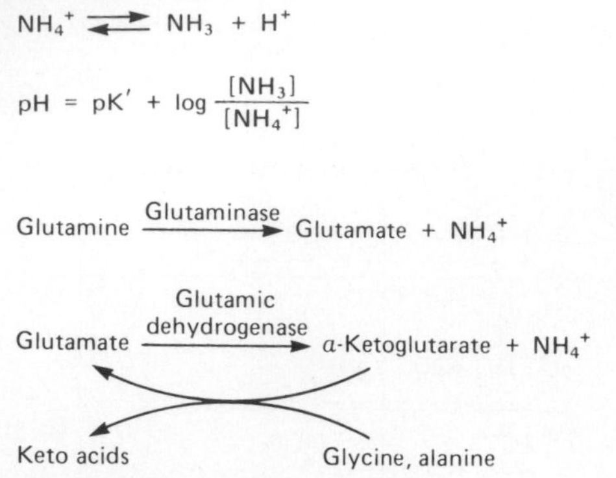

$$NH_4^+ \rightleftharpoons NH_3 + H^+$$

$$pH = pK' + \log \frac{[NH_3]}{[NH_4^+]}$$

$$\text{Glutamine} \xrightarrow{\text{Glutaminase}} \text{Glutamate} + NH_4^+$$

$$\text{Glutamate} \xrightarrow{\text{Glutamic dehydrogenase}} \alpha\text{-Ketoglutarate} + NH_4^+$$

Keto acids — Glycine, alanine

Figure 38–21. Major reactions involved in ammonia production in the kidney. See also Chapter 17.

abundant in renal tubular cells (Fig 38–21). **Glutamic dehydrogenase** catalyzes the conversion of glutamate to α-ketoglutarate, with the production of more NH_4^+. Some NH_4^+ is also produced by transamination of glycine and alanine to keto acids, with transfer of the amino group to α-ketoglutarate; this forms glutamate, which in turn is deaminated to form more NH_4^+.

In the interstitial fluid and tubular urine, NH_3 combines with H^+ to form NH_4^+, removing NH_3 and maintaining the concentration gradient that facilitates diffusion of NH_3 out of the cell. The NH_4^+ stays in the urine. When the urine pH is 7.0, as it is in the proximal tubule, the ratio of NH_3 to NH_4^+ is 1:100; but the ratio decreases 10-fold for each unit decrease in pH. Consequently, in the collecting ducts, where the urine is more acidic, the equilibrium changes even further to NH_4^+.

In chronic acidosis, the amount of NH_4^+ excreted at any given urine pH also increases, because more NH_3 enters the tubular urine. The effect of this **adaptation** of NH_3 secretion, the cause of which is unsettled, is a further removal of H^+ from the tubular fluid and consequently a further enhancement of H^+ secretion.

The process by which NH_3 is secreted is called **nonionic diffusion** (see Chapter 1). Salicylates and a number of other drugs that are weak bases or weak acids are also secreted by nonionic diffusion. They diffuse into the tubular fluid at a rate that is dependent upon the urine pH, and the amount of each drug excreted therefore varies with the pH of the urine.

pH Changes Along the Nephrons

Micropuncture studies indicate that an appreciable drop in pH occurs in the proximal tubule in mammalian kidneys, although there is further acidification in the distal tubule. Since most of the HCO_3^- is absorbed in the proximal tubule and the H^+ secretion responsible for its reabsorption has only a small effect on the pH of the fluid, it is apparent that there is considerably more acid secretion in the proximal than in the distal tubule. The acid-secreting mechanism in the distal tubules, and particularly in the collecting ducts, has a lower secretory capacity but is capable of generating a large pH difference between the tubular lumen and the cells.

Factors Affecting Acid Secretion

Renal acid secretion is altered by changes in the intracellular P_{CO_2}, K^+ concentration, carbonic anhydrase level, and adrenocortical hormone concentration. When the P_{CO_2} is high (respiratory acidosis), more intracellular H_2CO_3 is available to buffer the hydroxyl ions and acid secretion is enhanced, whereas the reverse is true when the P_{CO_2} falls. K^+ depletion enhances acid secretion, apparently because the loss of K^+ causes intracellular acidosis even though the plasma pH may be elevated. Conversely, K^+ excess in the cells inhibits acid secretion. When carbonic anhydrase is inhibited, acid secretion is inhibited, because the formation of H_2CO_3 is decreased. Aldosterone and the other adrenocortical steroids that enhance tubular reabsorption of Na^+ also increase the secretion of H^+ and K^+.

Bicarbonate Excretion

Although the process of HCO_3^- reabsorption does not actually involve transport of this ion into the tubular cells. HCO_3^- reabsorption is proportionate to the amount filtered over a relatively wide range. There is no Tm, but by an unknown mechanism, HCO_3^- reabsorption is decreased when the ECF volume is expanded (Fig 38–22). When the plasma HCO_3^- concentration is low, all the filtered HCO_3^- is reabsorbed; but when the plasma HCO_3^- concen-

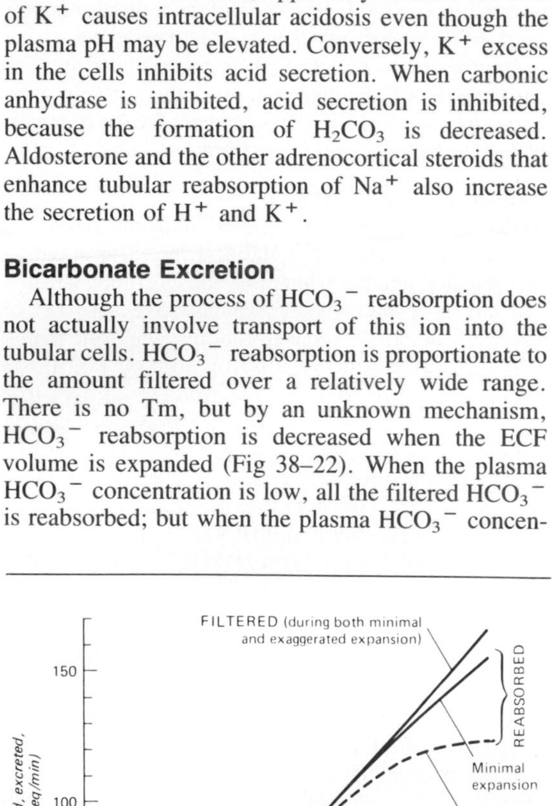

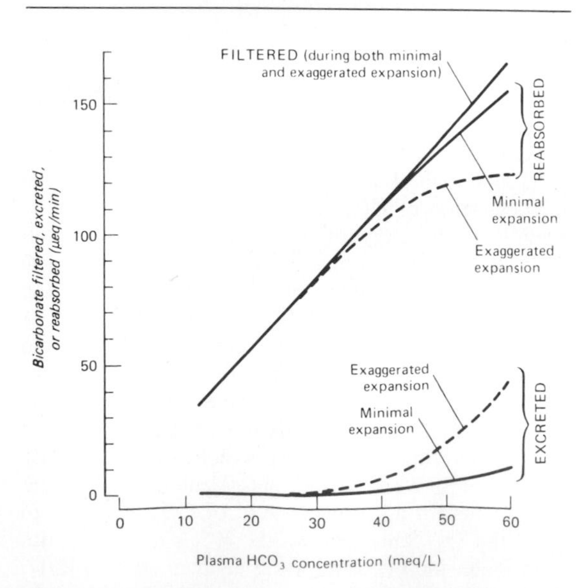

Figure 38–22. Effect of ECF volume on HCO_3^- filtration, reabsorption, and excretion. The plasma HCO_3^- concentration is normally about 24 meq/L. (Reproduced, with permission, from Valtin H: *Renal Function,* 2nd ed. Little, Brown, 1983. Copyright Little, Brown and Co., 1983.)

tration is high, HCO_3^- appears in the urine and the urine becomes alkaline.

When the plasma HCO_3^- level is 28 meq/L, H^+ is being secreted at its maximal rate and all of it is being used to reabsorb HCO_3^-; but as the plasma HCO_3^- level drops, more H^+ becomes available to combine with other buffer anions. Therefore, the lower the plasma HCO_3^- concentration drops, the more acidic the urine becomes and the greater its NH_4^+ content.

Implications of Urinary pH Changes

Depending upon the rates of the interrelated processes of acid secretion, NH_4^+ production, and HCO_3^- excretion, the pH of the urine in humans varies from 4.5 to 8.0. Excretion of a urine that is at a pH different from that of the body fluids has important implications for the body's electrolyte and acid-base economy which are discussed in detail in Chapter 39. Acids are buffered in the plasma and cells, the overall reaction being $HA + NaHCO_3$ $NaA + H_2CO_3$. The H_2CO_3 forms CO_2 and H_2O, and the CO_2 is expired, while the NaA appears in the glomerular filtrate. To the extent that the Na^+ is replaced by H^+ in the urine, Na^+ is conserved in the body. Furthermore, for each H^+ excreted with phosphate or as NH_4^+, there is a net gain of one HCO_3^- in the blood, replenishing the supply of this important buffer anion. Conversely, when base is added to the body fluids, the OH^- ions are buffered, raising the plasma HCO_3^-. When the plasma level exceeds 28 meq/L, the urine becomes alkaline and the extra HCO_3^- is excreted in the urine. Because the rate of maximal H^+ secretion by the tubules varies directly with the arterial P_{CO_2}, HCO_3^- reabsorption also is affected by the P_{CO_2}. This relationship is discussed in detail in Chapter 39.

REGULATION OF Na^+ & Cl^- EXCRETION

Na^+ is filtered in large amounts, but it moves passively out of some portions of the nephron and is actively transported out of the proximal tubule, the thick ascending limb, the distal tubule, and the collecting duct. Normally, 96% to well over 99% of the filtered Na^+ is reabsorbed. Most of the Na^+ is reabsorbed with Cl^- (Table 38–8), but some is reabsorbed in the processes by which one Na^+ enters the bloodstream for each H^+ secreted by the tubules, and in the distal tubules, a small amount is actively reabsorbed in association with the secretion of K^+.

Regulation of Na^+ Excretion

Because Na^+ is the most abundant cation in ECF and because sodium salts account for over 90% of the osmotically active solute in the plasma and interstitial fluid, the amount of Na^+ in the body is a prime determinant of the ECF volume. Therefore, it is not

Table 38–8. Quantitative aspects of Na^+ reabsorption in a normal man.

GFR = 125 mL/min
Plasma HCO_3^- = 27 meq/L
Plasma Na^+ = 145 meq/L

Na^+ filtered per minute	18,125 μeq
Reabsorbed with Cl^-	14,585 μeq
Reabsorbed while reabsorbing 3375 μeq HCO_3^-	3,375 μeq
Reabsorbed in association with formation of titratable acidity and ammonia	50 μeq
Reabsorbed in association with secretion of K^+	50 μeq
Total Na^+ reabsorbed per minute	18,060 μeq

surprising that multiple regulatory mechanisms have evolved in terrestrial animals to control the excretion of this ion. Through the operation of these regulatory mechanisms, the amount of Na^+ excreted is adjusted to equal the amount ingested over a wide range of dietary intakes, and the individual stays in Na^+ balance. Thus, urinary Na^+ output ranges from less than 1 meq/d on a low-salt diet to 400 meq/d or more when the dietary Na^+ intake is high. In addition, there is a natriuresis when saline is infused intravenously and a decrease in Na^+ excretion when ECF volume is reduced. Variations in Na^+ excretion are effected by changes in the amount filtered and the amount reabsorbed in the tubules. The factors affecting the GFR are discussed above. The factors affecting Na^+ reabsorption include the oncotic and hydrostatic pressures in the peritubular capillaries, the circulating level of aldosterone and other adrenocortical hormones, the circulating level of ANP and possibly other natriuretic hormones, and the rate of tubular secretion of H^+ and K^+.

Glomerulotubular Balance

The amount of Na^+ filtered is so large (over 26,000 meq/d) that if the total amount of Na^+ reabsorbed stayed constant, a rise in GFR of only 2 mL/min would more than double the amount excreted (Table 38–9). Conversely, a small fall in GFR would reduce Na^+ excretion to zero. However, the total amount reabsorbed rises when the GFR rises and falls when it falls. Most of the change in reabsorption occurs in the proximal tubule. The compensation is usually not complete, but it is appreciable. The proximal tubular reabsorption of a number of other substances is also proportionate to the load delivered to the tubule by filtration; ie, the tubule tends to reabsorb a constant fraction of the amount filtered rather than a constant amount. The proportionality, which is especially apparent in the case of Na^+ reabsorption, has come to be referred to as **glomerulotubular balance.** This term should not be confused with tubuloglomerular feedback, which refers to tubular regulation of GFR (see above). The mechanism by which the change in proximal reabsorp-

Table 38–9. Changes in Na^+ excretion that would occur as a result of changes in GFR if there were no concomitant changes in Na^+ reabsorption.

GFR (mL/min)	Plasma Na^+ (μeq/mL)	Amount Filtered (μeq/min)	Amount Reabsorbed (μeq/min)	Amount Excreted (μeq/min)
125	145	18,125	18,000	125
127	145	18,415	18,000	415
124.1	145	18,000	18,000	0

tion is brought about is currently the subject of intensive research. The change in Na^+ reabsorption occurs within seconds after a change in filtration, so it seems unlikely that an extrarenal humoral factor is involved. One factor is the oncotic pressure in the peritubular capillaries. When the GFR is high, there is a relatively large increase in the oncotic pressure of the plasma by the time it reaches the efferent arterioles and their capillary branches. This increases the reabsorption of Na^+ from the tubule. However, all the intrarenal mechanisms that bring about the change have not as yet been identified.

Hemodynamic Effects on Tubular Reabsorption

Changes in renal capillary hydrostatic and oncotic pressure appear to be important in the response to salt loading. Intravenous saline produces a natriuresis with decreased Na^+ reabsorption in the proximal tubule that may be due in part to increased circulating ANP (see Chapter 24). In addition, saline dilutes the blood, lowering the oncotic pressure, and volume expansion also appears to decrease renal vascular resistance, with a consequent increase in hydrostatic pressure. Conversely, conditions that reduce ECF volume cause renal vasoconstriction, with a consequent reduction in hydrostatic pressure in the capillaries, and proximal tubular Na^+ reabsorption is increased. Changes in hydrostatic and osmotic pressures may also play a role in glomerulotubular balance and in the ''escape phenomenon'' (see Chapter 20), although ANP is also involved. Angiotensin II formed in the kidneys by renin secreted by the juxtaglomerular cells reduces Na^+ and H_2O excretion. As noted in Chapter 24, ANP appears to increase Na^+ excretion by increasing the GFR.

Effects of Adrenocortical Steroids

Adrenal mineralocorticoids such as aldosterone increase tubular reabsorption of Na^+ in association with secretion of K^+ and H^+ (Fig 38–15) and also Na^+ reabsorption with Cl^- (see Chapter 20). When these hormones are injected into adrenalectomized animals, there is a latent period of 10–30 minutes before their effects on Na^+ reabsorption become manifest. They act on the distal convoluted tubules and collecting ducts, and there is some evidence that they cause the reabsorption of a small amount of Na^+ from the bladder. The glucocorticoids such as cortisol also increase Na^+ reabsorption; but, unlike the mineralocorticoids, they increase the GFR. They may therefore increase rather than decrease Na^+ excretion, because the increase they produce in the filtered load of Na^+ may exceed the increase in reabsorption.

Reduction of the dietary intake of salt increases aldosterone secretion (Fig 20–30). Changes in aldosterone secretion can explain changes in Na^+ excretion that occur over days or even hours, but, because of the latent period before its effects are manifest, rapid changes in Na^+ excretion cannot be attributed to this hormone.

Relation to Acid & K^+ Secretion

Na^+ excretion is increased by drugs that decrease renal acid secretion by inhibiting carbonic anhydrase. After CO_2 or acid is buffered in the blood, Na^+ filtered with acid anions is lost in the urine if the amount in the filtrate exceeds the capacity of the tubules to exchange the Na^+ for H^+. Changes in Na^+ excretion due to changes in the rate of K^+ secretion are small (Table 38–8).

Chloride Excretion

Chloride reabsorption is increased when HCO_3^- reabsorption is decreased, and vice versa, so that Cl^- concentration in plasma varies inversely with the HCO_3^- concentration, keeping the total anion concentration constant. Passive diffusion can explain the movements of Cl^- in some situations. However, Cl^- also leaves the tubular lumen by secondary active transport, as described above.

REGULATION OF K^+ EXCRETION

Much of the filtered K^+ is removed from the tubular fluid by active reabsorption in the proximal tubules (Table 38–5), and K^+ is then secreted into the fluid by the distal tubular cells. The rate of K^+ secretion is proportionate to the rate of flow of the tubular fluid through the distal portions of the nephron, because with rapid flow there is less opportunity for the tubular K^+ concentration to rise to a value that stops further secretion. In the absence of complicating factors, the amount secreted is approximately equal to the K^+ intake, and K^+ balance is maintained. In the distal tubules, Na^+ is generally reabsorbed and K^+ is secreted. There is no rigid one-for-one exchange, and much of the movement of K^+ is passive. However, there is electrical coupling in the sense that intracellular migration of Na^+ tends to lower the potential difference across the tubular cell, and this favors movement of K^+ into the tubular lumen. Since Na^+ is also reabsorbed in association with H^+ secretion, there is competition for the Na^+ in the tubular fluid. K^+ excretion is decreased when the amount of Na^+ reaching the distal tubule is low, and K^+ excretion is also decreased when H^+ secretion is increased. When total body K^+ is high, H^+ secretion is inhibited, apparently because of intracellular alkalosis; and K^+ secretion and excretion are therefore facilitated. Conversely, the cells are acid in K^+ depletion, and K^+ secretion declines. Apparently the K^+ secretory mechanism is capable of "adaptation," because the amount of K^+ excreted gradually increases when a constant large dose of a potassium salt is administered for a prolonged period.

DIURETICS

Although a detailed discussion of diuretic agents is outside the scope of this book, consideration of their mechanisms of action constitutes an informative review of the factors affecting urine volume and electrolyte excretion. These mechanisms are summarized in Table 38–10. Water, alcohol, osmotic diuretics, xanthines, and acidifying salts have limited clinical usefulness, and the vasopressin antagonists are used primarily for research. However, many of the other agents on the list are used extensively in medical practice.

Table 38–10. Mechanism of action of various diuretics.

Agent	Mechanism of Action
Water	Inhibits vasopressin secretion.
Ethanol	Inhibits vasopressin secretion.
Large quantities of osmotically active substances such as mannitol and glucose	Produce osmotic diuresis.
Xanthines such as caffeine and theophylline	Probably decrease tubular reabsorption of Na^+ and increase GFR.
Acidifying salts such as $CaCl_2$ and NH_4Cl	Supply acid load; H^+ is buffered, and anion is excreted with Na^+ when the ability of the kidney to replace Na^+ with H^+ is exceeded.
Carbonic anhydrase inhibitors such as acetazolamide (Diamox)	Decrease H^+ secretion, with resultant increase in Na^+ and K^+ excretion.
Metolazone (Zaroxolyn), thiazides such as chlorothiazide (Diuril)	Inhibit Na^+-K^+-$2Cl^-$ cotransport in the early portion of the distal tubule.
Furosemide (Lasix), ethacrynic acid (Edecrin), and bumetanide	Inhibit Na^+ and Cl^- reabsorption in the medullary thick ascending limb of the loop of Henle.
K^+-retaining natriuretics such as spironolactone (Aldactone), triamterene (Dyrenium), and amiloride (Colectril)	Inhibit Na^+-K^+ "exchange" in the distal portion of the distal tubule and the collecting duct by inhibiting the action of aldosterone (spironolactone) or by inhibiting Na^+ reabsorption (triamterene, amiloride).
Antagonists of V_2 vasopressin receptors	Inhibit action of vasopressin on collecting duct.

Water diuresis and osmotic diuresis have been discussed above. Antagonists to V_2 vasopressin receptors mimic the effects of water loading. Ethanol acts directly on the hypothalamus. The diuretic action of the xanthines is weak. When NH_4Cl is ingested, the NH_4^+ is converted to urea, and this reaction generates protons:

$$2NH^+_4 + CO_2 \rightarrow 2H^+ + \text{Urea} + H_2O$$

When the capacity of the hepatic cells to buffer these protons is exceeded, they enter the ECF, where they are also buffered. The Cl^- is filtered along with Na^+, thus maintaining electrical neutrality. To the extent that any of the Na^+ is not replaced by H^+ from the renal tubules, Na^+ and water are lost in the urine. Other acidifying salts produce diuresis in a similar way.

The carbonic anhydrase-inhibiting drugs are only moderately effective as diuretic agents, but because they inhibit acid secretion by decreasing the supply of carbonic acid, they have far-reaching effects. Not only is Na^+ excretion increased because H^+ secretion is decreased, but HCO_3^- reabsorption is also depressed; and because H^+ and K^+ compete with each other and with Na^+, the decrease in H^+ secretion facilitates the secretion and excretion of K^+.

Another determinant of the rate of K^+ secretion is the amount of Na^+ delivered to the Na^+-K^+ "exhange" site in the distal tubule. Thiazides, furosemide, ethacrynic acid, and bumetanide act proximal to this site, and the resultant increase in Na^+ delivery increases K^+ secretion. The K^+ loss is appreciable, and K^+ depletion is one of the common complications of treatment with these agents. The thiazides inhibit Na^+ and Cl^- transport in the proximal portion of the distal tubule (Fig 38–23). Furosemide, ethacrynic acid, and bumetanide inhibit Na^+ and Cl^- reabsorption in the medullary thick ascending limb of the loop of Henle by inhibiting Na^+-K^+-$2Cl^-$ transport (Fig 38–17). Spironolactone, triamterene, and amiloride act on the Na^+-K^+ exchange mechanism itself, causing K^+ retention and, in some cases, mild hyperkalemia.

EFFECTS OF DISORDERED RENAL FUNCTION

A number of abnormalities are common to many different types of renal disease. The secretion of renin by the kidneys and the relation of the kidneys to hypertension are discussed in Chapters 24 and 33. A frequent finding in various forms of renal disease is the presence in the urine of protein, leukocytes, red

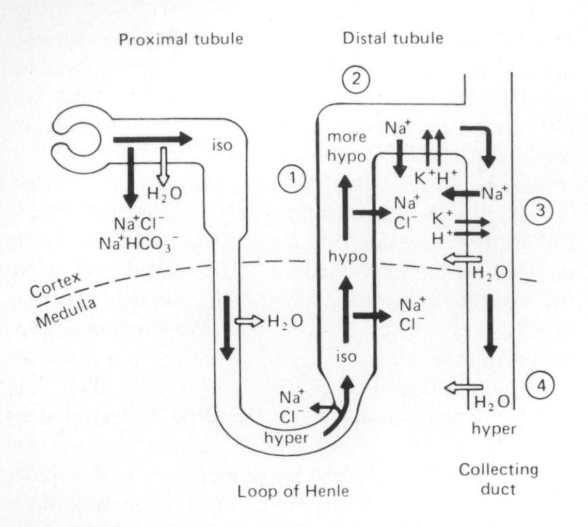

Figure 38–23. Sites of action of various diuretics. ①: Furosemide and other loop diuretics act on the thick ascending limb of the loop of Henle. ②: Thiazides act on the early portion of the distal convoluted tubule. ③: Aldosterone antagonists, triamterene, and amiloride act primarily on the collecting ducts. ④: Antagonists to V_2 vasopressin receptors act on the collecting ducts.

cells, and **casts,** which are bits of proteinaceous material precipitated in the tubules and washed into the bladder. Other important consequences of renal disease are loss of the ability to concentrate or dilute the urine, uremia, acidosis, and abnormal retention of Na^+.

Proteinuria

In many renal diseases and in one benign condition, the permeability of the glomerular capillaries is increased, and protein is found in the urine in more than the usual trace amounts **(proteinuria).** Most of this protein is albumin, and the defect is commonly called **albuminuria.** The relation of charges on the glomerular membrane to albuminuria is discussed above. The amount of protein in the urine may be very large, and, especially in nephrosis, the urinary protein loss may exceed the rate at which the liver can synthesize plasma proteins. The resulting hypoproteinemia reduces the oncotic pressure, and the plasma volume declines, sometimes to dangerously low levels, while edema fluid accumulates in the tissues.

A benign condition that causes proteinuria is a poorly understood change in renal hemodynamics which in some otherwise normal individuals causes protein to appear in urine formed when they are in the standing position **(orthostatic albuminuria).** Urine formed when these individuals are lying down is protein-free.

Loss of Concentrating & Diluting Ability

In renal disease, the urine becomes less concentrated and urine volume is often increased, producing the symptoms of **polyuria** and **nocturia** (waking up at night to void). The ability to form a dilute urine is often retained, but in advanced renal disease, the osmolality of the urine becomes fixed at about that of plasma, indicating that the diluting and concentrating functions of the kidney have both been lost. The loss is due in part to disruption of the countercurrent mechanism, but a more important cause is a loss of functioning nephrons. When one kidney is removed surgically, the number of functioning nephrons is halved. The number of osmoles excreted is not reduced to this extent, and so the remaining nephrons must each be filtering and excreting more osmotically active substances, producing what is in effect an osmotic diuresis. In osmotic diuresis, the osmolality of the urine approaches that of plasma (see above); the same thing happens when the number of functioning nephrons is reduced by disease. Of course, when most of the nephrons are destroyed, urine volume falls and **oliguria** or even **anuria** is present.

Uremia

When the breakdown products of protein metabolism accumulate in the blood, the syndrome known as **uremia** develops. The symptoms of uremia include lethargy, anorexia, nausea and vomiting, mental deterioration and confusion, muscle twitching, convulsions, and coma. Because erythropoiesis is depressed, anemia is a prominent feature of chronic uremia. The blood urea nitrogen (BUN) and creatinine levels are high, and the blood levels of these substances are used as an index of the severity of the uremia. It is often stated that it is not the accumulation of urea and creatinine per se but rather the accumulation of other toxic substances—possibly organic acids or phenols—that produces the symptoms of uremia. Because urea crosses the blood-brain barrier slowly, it is administered intravenously to neurosurgical patients to make the ECF outside the brain hypertonic and shrink the brain during surgery. However, urea infusions cause changes in the electrical activity of the brain in experimental animals and may not be as innocuous as they appear to be. In dogs, prolonged infusions of urea cause anorexia, weakness, vomiting, and diarrhea.

The toxic substances that cause the symptoms of uremia can be removed by dialyzing the blood of uremic patients against a bath of suitable composition in an **artificial kidney.** By repeated dialysis, patients can be kept alive and in reasonable health for many months, even when they are completely anuric or have had both kidneys removed.

Acidosis

Acidosis is common in chronic renal disease because of failure to excrete the acid products of digestion and metabolism (see Chapter 39). In the rare syndrome of **renal tubular acidosis,** there is specific impairment of the ability to make the urine acidic, and other renal functions are usually normal. However, in most cases of chronic renal disease, the urine is maximally acidified, and acidosis develops because the total amount of H^+ that can be secreted is reduced because of impaired renal tubular production of NH_3

Abnormal Na^+ Metabolism

Many patients with renal disease retain excessive amounts of Na^+ and become edematous. There are at least 3 causes of Na^+ retention in renal disease. In acute glomerulonephritis, a disease that primarily affects the glomeruli, there is a marked decrease in the amount of Na^+ filtered. In the nephrotic syndrome, an increase in aldosterone secretion contributes to the salt retention. The plasma protein is low in this condition, and so fluid moves from the plasma into the interstitial spaces and the plasma volume falls. The decline in plasma volume triggers the increase in aldosterone secretion via the renin-angiotensin system. A third cause of Na^+ retention and edema in renal disease is **heart failure** (see Chapter 33). Renal disease predisposes to heart failure, partly because of the hypertension it frequently produces.

Variations in the Response to Aldosterone

An interesting problem is why aldosterone-treated normal individuals and patients with primary hyperaldosteronism become severely K^+ depleted, whereas edematous patients with the nephrotic syndrome, cirrhosis, or heart failure who have secondary hyperaldosteronism do not. One factor seems to be the amount of Na^+ reaching the distal portions of the nephrons; Na^+ in the tubular fluid helps maintain the potential difference between the tubular lumen and the cells, and this favors K^+ secretion. In patients with primary hyperaldosteronism, "escape" has taken place and proximal tubular Na^+ reabsorption is decreased, so that large quantities of Na^+ reach the distal tubules. In patients with edema, little Na^+ reaches the distal tubules. The filtered Na^+ load is small because the plasma Na^+ is low as a result of "dilutional hyponatremia," or the GFR is low, or both. In addition, there is increased Na^+ reabsorption in the more proximal portions of the nephron. Another factor in patients with edema is the flow of fluid in the distal portions of the nephrons; K^+ excretion is proportionate to flow, and in these patients, the amount of fluid reaching the distal tubules is often reduced.

FILLING OF THE BLADDER

The walls of the ureters contain smooth muscle arranged in spiral, longitudinal, and circular bundles, but distinct layers of muscle are not seen. Regular peristaltic contractions occurring 1–5 times per minute move the urine from the renal pelvis to the bladder, where it enters in spurts synchronous with each peristaltic wave. The ureters pass obliquely through the bladder wall and, although there are no ureteral sphincters as such, the oblique passage tends to keep the ureters closed except during peristaltic waves, preventing reflux of urine from the bladder.

EMPTYING OF THE BLADDER

Anatomic Considerations

The smooth muscle of the bladder, like that of the ureters, is arranged in spiral, longitudinal, and circular bundles. Contraction of this muscle, which is called the **detrusor muscle,** is mainly responsible for emptying the bladder during urination (micturition). Muscle bundles pass on either side of the urethra, and these fibers are sometimes called the **internal urethral sphincter,** although they do not encircle the urethra. Farther along the urethra is a sphincter of skeletal muscle, the sphincter of the membranous urethra **(external urethral sphincter).** The bladder epithelium is made up of a superficial layer of flat cells and a deep layer of cuboidal cells. The innervation of the bladder is summarized in Fig 38–24.

Micturition

The physiology of micturition and its disorders are subjects about which there is much confusion. Micturition is fundamentally a spinal reflex facilitated and inhibited by higher brain centers and, like defecation, subject to voluntary facilitation and inhibition. Urine enters the bladder without producing much increase in intravesical pressure until the viscus is well filled. In addition, like other types of smooth muscle, the bladder muscle has the property of plasticity; when it is stretched, the tension initially produced is not maintained. The relation between

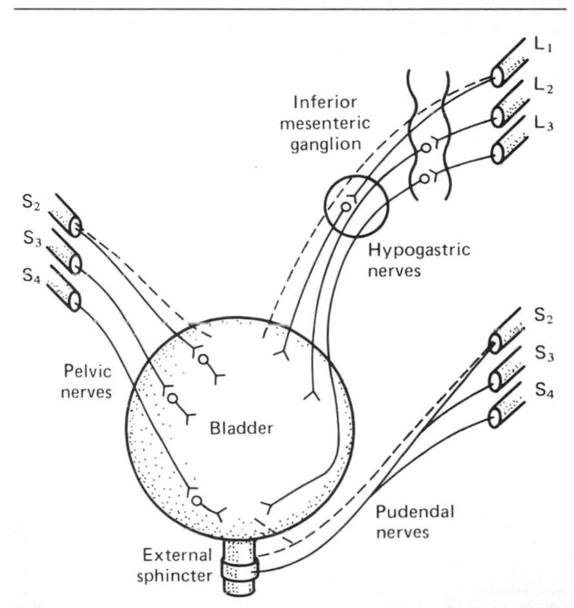

Figure 38–24. Innervation of the bladder. Dashed lines indicate sensory nerves. Parasympathetic innervation is shown at left, sympathetic at upper right, and somatic at lower right.

intravesical pressure and volume can be studied by inserting a catheter and emptying the bladder, then recording the pressure while the bladder is filled with 50-mL increments of water or air **(cystometry).** A plot of intravesical pressure against the volume of fluid in the bladder is called a **cystometrogram** (Fig 38–25). The curve shows an initial slight rise in pressure when the first increments in volume are produced; a long, nearly flat segment as further increments are produced; and a sudden, sharp rise in pressure as the micturition reflex is triggered. These 3 components are sometimes called segments Ia, Ib, and II (Fig 38–25). The first urge to void is felt at a bladder volume of about 150 mL, and a marked sense of fullness at about 400 mL. The flatness of segment Ib is a manifestation of the law of Laplace (see Chapter 30). This law states that the pressure in a spherical viscus is equal to twice the wall tension divided by the radius. In the case of the bladder, the tension increases as the organ fills, but so does the radius. Therefore, the pressure increase is slight until the organ is relatively full.

During micturition, the perineal muscles and external urethral sphincter are relaxed; the detrusor muscle contracts; and urine passes out through the urethra. The bands of smooth muscle on either side of the urethra apparently play no role in micturition, and their main function is believed to be the prevention of reflux of semen into the bladder during ejaculation.

The mechanism by which voluntary urination is initiated remains unsettled. One of the initial events is relaxation of the muscles of the pelvic floor, and this may cause a sufficient downward tug on the detrusor muscle to initiate its contraction. The perineal muscles and external sphincter can be contracted voluntarily, preventing urine from passing down the urethra or interrupting the flow once urination has begun. It is through the learned ability to maintain the external sphincter in a contracted state that adults are able to delay urination until the opportunity to void presents itself. After urination, the female urethra empties by gravity. Urine remaining in the urethra of the male is expelled by several contractions of the bulbocavernosus muscle.

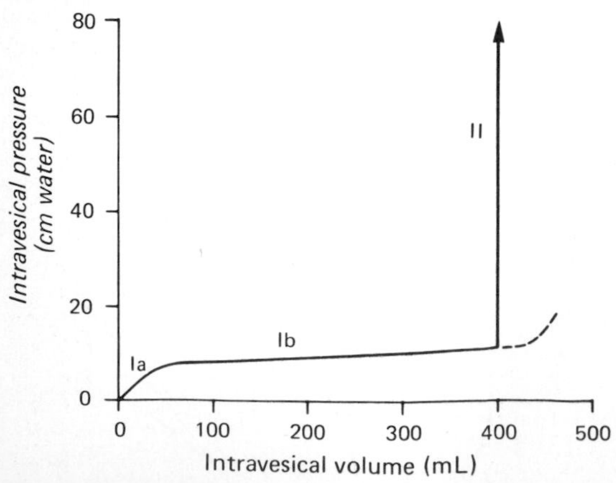

Figure 38–25. Cystometrogram in normal human. The numerals identify the 3 components of the curve described in the text. The dotted line indicates the pressure-volume relations that would have been found had micturition not occurred and produced component II. (Modified and reproduced, with permission, from Tanagho EA, McAninch JW: Smith's General Urology, 12th ed., Appleton & Lange, 1988.)

Reflex Control

The bladder smooth muscle has some inherent contractile activity; however, when its nerve supply is intact, stretch receptors in the bladder wall initiate a reflex contraction that has a lower threshold than the inherent contractile response of the muscle. Fibers in the pelvic nerves are the afferent limb of the voiding reflex, and the parasympathetic fibers to the bladder that constitute the efferent limb also travel in these nerves. The reflex is integrated in the sacral portion of the spinal cord. In the adult, the volume of urine in the bladder that normally initiates a reflex contraction is about 300–400 mL. The sympathetic nerves to the bladder play no part in micturition, but they do mediate the contraction of the bladder muscle that prevents semen from entering the bladder during ejaculation (see Chapter 23).

There is no small motor nerve system to the stretch receptors in the bladder wall; but the threshold for the voiding reflex, like the stretch reflexes, is adjusted by the activity of facilitatory and inhibitory centers in the brain stem. There is a facilitatory area in the pontine region and an inhibitory area in the midbrain. After transection of the brain stem just above the pons, the threshold is lowered, and less bladder filling is required to trigger it; whereas after transection at the top of the midbrain, the threshold for the reflex is essentially normal. There is another facilitatory area in the posterior hypothalamus. In humans with lesions in the superior frontal gyrus, the desire to urinate is reduced, and there is also difficulty in stopping micturition once it has commenced. However, stimulation experiments in animals indicate that other cortical areas also affect the process. The bladder can be made to contract by voluntary facilitation of the spinal voiding reflex when it contains only a few milliliters of urine. Voluntary contraction of the abdominal muscles aids the expulsion of urine by increasing the intra-abdominal pressure, but voiding can be initiated without straining even when the bladder is nearly empty.

ABNORMALITIES OF MICTURITION

There are 3 major types of bladder dysfunction due to neural lesions: (1) the type due to interruption of the afferent nerves from the bladder; (2) the type due to interruption of both afferent and efferent nerves; and (3) the type due to interruption of facilitatory and inhibitory pathways descending from the brain. In all

3 types the bladder contracts, but the contractions are generally not sufficient to empty the viscus completely, and residual urine is left in the bladder.

Effects of Deafferentation

When the sacral dorsal roots are cut in experimental animals or interrupted by diseases of the dorsal roots such as **tabes dorsalis** in humans, all reflex contractions of the bladder are abolished. The bladder becomes distended, thin-walled, and hypotonic, but there are some contractions because of the intrinsic response of the smooth muscle to stretch.

Effects of Denervation

When the afferent and efferent nerves are both destroyed, as they may be by tumors of the cauda equina or filum terminale, the bladder is flaccid and distended for a while. Gradually, however, the muscle of the "decentralized bladder" becomes active, with many contraction waves that expel dribbles of urine out of the urethra. The bladder becomes shrunken and the bladder wall hypertrophied. The reason for the difference between the small, hypertrophic bladder seen in this condition and the distended, hypotonic bladder seen when only the afferent nerves are interrupted is not known. The hyperactive state in the former condition suggests the development of denervation hypersensitization even though the neurons interrupted are preganglionic rather than postganglionic.

Effects of Spinal Cord Transection

During spinal shock, the bladder is flaccid and unresponsive. It becomes overfilled, and urine dribbles through the sphincters **(overflow incontinence).** After spinal shock has passed, the voiding reflex returns, although there is, of course, no voluntary control and no inhibition or facilitation from higher centers when the spinal cord is transected. Some paraplegic patients train themselves to initiate voiding by pinching or stroking their thighs, provoking a mild mass reflex (see Chapter 12). In some instances, the voiding reflex becomes hyperactive. Bladder capacity is reduced, and the wall becomes hypertrophied. This type of bladder is sometimes called the **spastic neurogenic bladder.** The reflex hyperactivity is made worse by, and may possibly be caused by, infection in the bladder wall.

39 Regulation of Extracellular Fluid Composition & Volume

INTRODUCTION

This chapter is a review of the major homeostatic mechanisms that operate, primarily through the kidneys and the lungs, to maintain the **tonicity,** the **volume,** and the **specific ionic composition,** particularly the $\mathbf{H^+}$ **concentration,** of the ECF. The interstitial portion of this fluid is the fluid environment of the cells, and life depends upon the constancy of this ''internal sea'' (see Chapter 1).

DEFENSE OF TONICITY

The defense of the tonicity of the ECF is primarily the function of the vasopressin-secreting and thirst mechanisms. The total body osmolality is directly proportionate to the total body sodium plus the total body potassium divided by the total body water, so that changes in the osmolality of the body fluids occur when there is a disproportion between the amount of these electrolytes and the amount of water ingested or lost from the body (see Chapter 1). When the effective osmotic pressure of the plasma rises, vasopressin secretion is increased and the thirst mechanism is stimulated. Water is retained in the body, diluting the hypertonic plasma, and water intake is increased (Fig 39–1). Conversely, when the plasma becomes hypotonic, vasopressin secretion is decreased and ''solute-free water'' (water in excess of solute) is excreted. In this way, the tonicity of the body fluids is maintained within a narrow normal range. In health, plasma osmolality ranges from 280 to 295 mosm/L, with vasopressin secretion maximally inhibited at 285 mosm/L and stimulated at higher values (Fig 14–13). The details of the way the regulatory mechanisms operate and the disorders that result when their function is disrupted are considered in Chapters 14 and 38.

DEFENSE OF VOLUME

The volume of the ECF is determined primarily by the total amount of osmotically active solute in the ECF. The composition of the ECF is discussed in Chapter 1. Since Na^+ and Cl^- are by far the most abundant osmotically active solutes in ECF, and since changes in Cl^- are to a great extent secondary to changes in Na^+, the amount of Na^+ in the ECF is the most important determinant of ECF volume, and the mechanisms that control Na^+ balance are the major mechanisms defending ECF volume. There is, however, a volume control of water excretion as well; a rise in ECF volume inhibits vasopressin secretion, and a decline in ECF volume produces an increase in the secretion of this hormone. Volume stimuli override the osmotic regulation of vasopressin secretion. Angiotensin II stimulates aldosterone and vasopressin secretion. It also causes thirst and constricts blood vessels, which help to maintain blood pressure. Thus, angiotensin II plays a key role in the body's response to hypovolemia (Fig 39–2). In addition, expansion of the ECF volume increases the secretion of ANP by the heart, and this causes natriuresis and diuresis (see Chapter 24).

In disease states, loss of water from the body **(dehydration)** causes a moderate decrease in ECF volume, because water is lost from both the intracellular and extracellular fluid compartments; but loss of Na^+ in the stools (diarrhea), urine (severe acidosis, adrenal insufficiency), or sweat (heat prostration) decreases ECF volume markedly and eventually leads to shock. The immediate compensations in shock operate principally to maintain intravascular volume (see Chapter 33), but they also affect Na^+ balance. In adrenal insufficiency, the decline in ECF volume is due not only to loss of Na^+ in the urine but also to its movement into cells (see Chapter 20).

Because of the key position of Na^+ in volume homeostasis, it is not surprising that more than one mechanism has evolved to control the excretion of this ion. The filtration and reabsorption of Na^+ in the

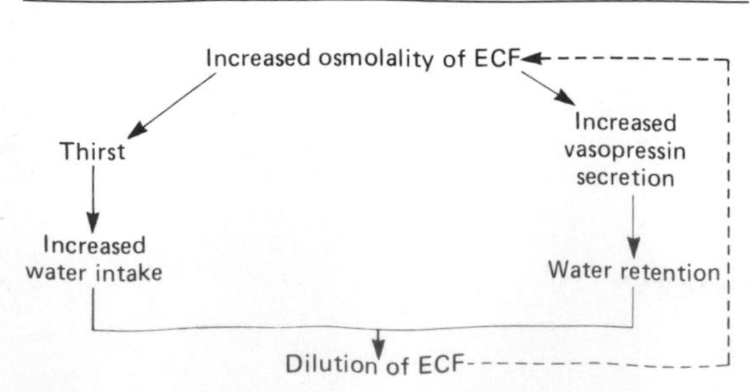

Figure 39–1. Mechanisms defending ECF tonicity. The dashed arrow indicates inhibition. (Courtesy of J Fitzsimons.)

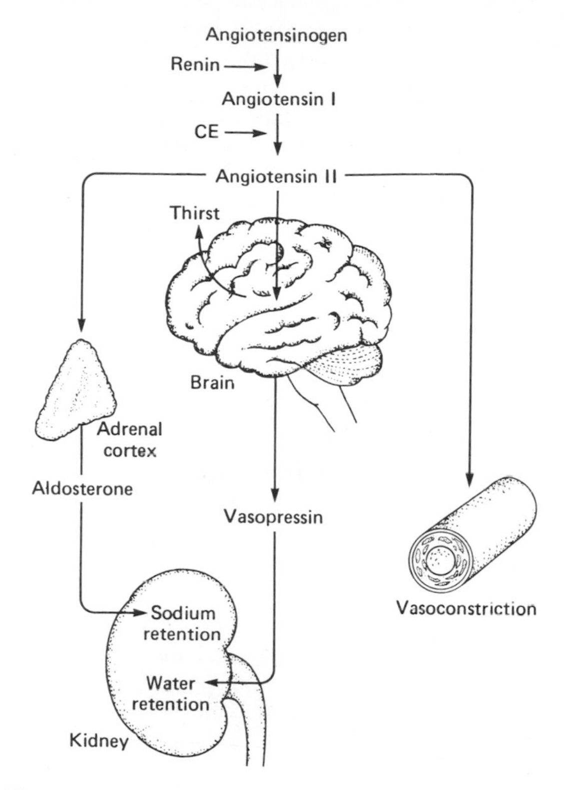

Figure 39–2. Defense of ECF volume by angiotensin II. CE, angiotensin-converting enzyme. (Modified from Ramsay DJ, Ganong WF: CNS regulation of salt and water intake. In: *Neuroendocrinology.* Krieger DT, Hughes JC [editors]. Sinauer Associates, 1980.)

kidneys and the effects of these processes on Na^+ excretion are discussed in Chapter 38. When ECF volume is decreased, blood pressure falls. Glomerular capillary pressure declines, and the GFR therefore falls, reducing the amount of Na^+ filtered. Tubular reabsorption of Na^+ is increased, in part because the secretion of aldosterone is increased. Aldosterone secretion is controlled in part by a feedback system in which the change that initiates increased secretion is a decline in mean intravascular pressure (see Chapters 20 and 24). Other changes in Na^+ excretion occur too rapidly to be due solely to changes in aldosterone secretion. For example, rising from the supine to the standing position increases aldosterone secretion. However, Na^+ excretion is decreased within a few minutes, and this rapid change in Na^+ excretion occurs in adrenalectomized subjects. It is probably due to hemodynamic changes and possibly to increased ANP secretion.

DEFENSE OF SPECIFIC IONIC COMPOSITION

Special regulatory mechanisms maintain the levels of certain specific ions in the ECF as well as the levels of glucose and other nonionized substances important in metabolism (see Chapters 17 and 19). The feedback of Ca^{2+} on the parathyroids and the calcitonin-secreting cells to adjust their secretion maintains the ionized calcium level of the ECF (see Chapter 21). Magnesium concentration is subject to close regulation, but the mechanisms controlling Mg^{2+} metabolism are incompletely understood.

The mechanisms controlling Na^+ and K^+ content are part of those determining the volume and tonicity of ECF and are discussed above. The levels of these ions are also dependent upon the H^+ concentration, and pH is one of the major factors affecting the anion composition of ECF.

DEFENSE OF H^+ CONCENTRATION

The mystique that has grown up around the subject of acid-base balance makes it necessary to point out that the core of the problem is not "buffer base" or "fixed cation" or the like but simply the maintenance of the H^+ concentration of the ECF. The mechanisms regulating the composition of the ECF are particularly important as far as this specific ion is concerned, because the machinery of the cells is very sensitive to changes in H^+ concentration. Intracellular H^+ concentration, which can be measured by using microelectrodes, pH-sensitive fluorescent dyes, and phosphorus nuclear magnetic resonance, is different from extracellular pH and appears to regulate a variety of intracellular processes. However, it is sensitive to changes in ECF H^+ concentration.

The pH notation is a useful means of expressing H^+ concentrations in the body, because the H^+ concentrations happen to be low relative to those of other cations. Thus, the normal Na^+ concentration of arterial plasma that has been equilibrated with red blood cells is about 140 meq/L, whereas the H^+ concentration is 0.00004 meq/L (Table 39–1). The pH, the negative logarithm of 0.00004, is therefore 7.4. Of course, a **decrease** in pH of 1 unit, eg, from 7.0 to 6.0, represents a 10-fold increase in H^+ concentration. It is also important to remember that the pH of blood is the pH of **true plasma**—plasma that has been in equilibrium with red cells—because the red cells contain hemoglobin, which is quantitatively one of the most important blood buffers (see Chapter 35).

Table 39–1. H^+ concentration and pH of body fluids.

		H^+ Concentration		
		meq/L	mol/L	pH
Gastric HCl		150	0.15	0.8
Maximal urine acidity		0.03	3×10^{-5}	4.5
Plasma	Extreme acidosis	0.0001	1×10^{-7}	7.0
	Normal	0.00004	4×10^{-8}	7.4
	Extreme alkalosis	0.00002	2×10^{-8}	7.7
Pancreatic juice		0.00001	1×10^{-8}	8.0

H^+ Balance

The pH of the arterial plasma is normally 7.40 and that of venous plasma slightly lower. Technically, **acidosis** is present whenever the arterial pH is below 7.40, and **alkalosis** is present whenever it is above 7.40, although variations of up to 0.05 pH unit occur without untoward effects. The H^+ concentrations in the ECF that are compatible with life cover an approximately 5-fold range, from 0.00002 meq/L (pH 7.70) to 0.0001 meq/L (pH 7.00).

Amino acids are utilized in the liver for gluconeogenesis, leaving as products NH_4^+ and HCO_3^- from their amino and carboxyl groups (Fig 39–3). The NH_4^+ is incorporated into urea and the protons that are formed are buffered intracellularly by HCO_3^-, so little NH_4^+ and HCO_3^- escape into the circulation. However, metabolism of sulfur-containing amino acids produces H_2SO_4, and metabolism of phosphorylated amino acids such as phosphoserine produces H_3PO_4. These strong acids enter the circulation and present a major H^+ load to the buffers in the ECF. The H^+ load from amino acid metabolism is normally about 50 meq/d. The CO_2 formed by metabolism in the tissues is in large part hydrated to H_2CO_3 (see Chapter 35), and the total H^+ load from this source is over 12,500 meq/d. However, most of the CO_2 is excreted in the lungs, and only small quantities of the H^+ remain to be excreted by the kidneys. Common sources of extra acid loads are strenuous exercise (lactic acid), diabetic ketosis (acetoacetic acid and β-hydroxybutyric acid), and ingestion of acidifying salts such as NH_4Cl and $CaCl_2$, which in effect add HCl to the body. Failure of diseased kidneys to excrete normal amounts of acid is also a cause of acidosis. Fruits are the main dietary source of alkali. They contain Na^+ and K^+ salts of weak organic acids, and the anions of these salts are metabolized to CO_2, leaving $NaHCO_3$ and $KHCO_3$ in the body. $NaHCO_3$ and other alkalinizing salts are sometimes ingested in large amounts, but a more common cause of alkalosis is loss of acid from the body due to vomiting of gastric juice rich in HCl. This is, of course, equivalent to adding alkali to the body.

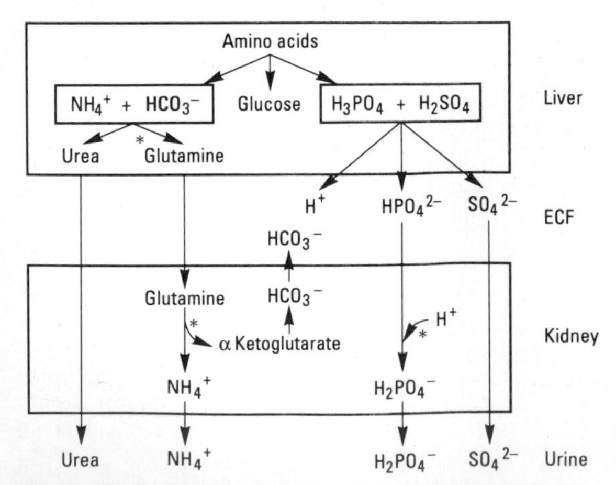

Figure 39–3. Role of the liver and the kidneys in the handling of metabolically produced acid loads. Sites where regulation occurs are indicated by asterisks. (Modified and produced, with permission, from Knepper MA et al: Ammonium, urea, and systemic pH regulation. *Am J Physiol* 1987;**235:**F199.

Buffering

Buffering and the buffer systems in the body are discussed in Chapters 1 and 35. The principal buffers (Table 35–2) are hemoglobin (Hb), protein (Prot), phosphate ($H_2PO_4^-$), and H_2CO_3:

$$HHb \rightleftharpoons H^+ + Hb^-$$

$$HProt \rightleftharpoons H^+ + Prot^-$$

$$H_2PO_4^- \rightleftharpoons H^+ + HPO_4^{2-}$$

$$H_2CO_3 \rightleftharpoons H^+ + HCO_3^-$$

The position of H_2CO_3 is unique, because it is converted to H_2O and CO_2 and the CO_2 is then excreted in the lungs. As noted in Chapter 35, the clinically relevant form of the Henderson-Hasselbalch equation for the equilibrium in this system is

$$pH = 6.10 + \log \frac{[HCO_3^-]}{0.0301\ P_{CO_2}}$$

Respiratory Acidosis & Alkalosis

It is apparent from the Henderson-Hasselbalch equation for the bicarbonate system that the primary changes in arterial P_{CO_2} which occur with disorders of respiration change the $[HCO_3^-]/[CO_2]$ ratio and, therefore, the pH. A rise in arterial P_{CO_2} due to decreased ventilation causes **respiratory acidosis.** The CO_2 that is retained is in equilibrium with H_2CO_3, which in turn is in equilibrium with HCO_3^-, so that the plasma HCO_3^- rises and a new equilibrium is reached at a lower pH. This can be indicated graphically on a plot of plasma HCO_3^- concentration versus pH (Fig 39–3). Conversely, a decline in P_{CO_2} causes **respiratory alkalosis.**

The initial changes shown in Fig 39–4 are those which occur independently of any compensatory mechanism; ie, they are those of **uncompensated** respiratory acidosis or alkalosis. In either situation, changes are produced in the kidneys, which then tend to **compensate** for the acidosis or alkalosis, adjusting the pH toward normal.

Renal Compensation

HCO_3^- reabsorption in the renal tubules depends not only on the filtered load of HCO_3^-, which is the product of the GFR and the plasma HCO_3^- level, but also on the rate of H^+ secretion by the renal tubular cells, since HCO_3^- is reabsorbed by exchange for H^+. The rate of H^+ secretion—and hence the rate of HCO_3^- reabsorption—is proportionate to the arterial

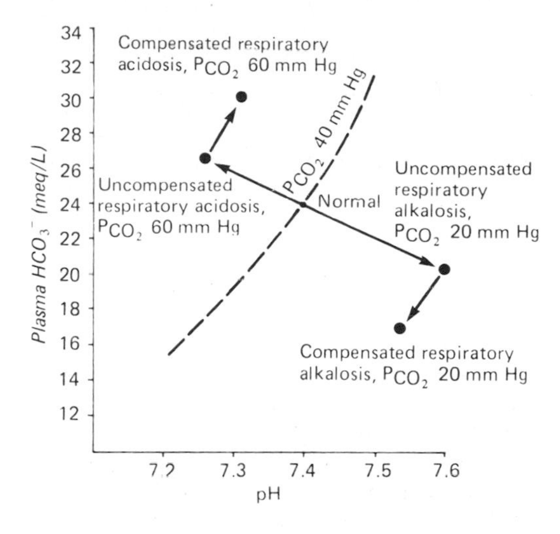

Figure 39–4. Changes in true plasma pH, HCO_3^-, and P_{CO_2} in respiratory acidosis and alkalosis. Note that with this plot the addition of CO_2 moves the point up and to the left along the line indicated by the arrow, and removal of CO_2 moves it down and to the right. On the other hand, addition of stronger acid moves the point down the dashed "iso-CO_2" line to the left, whereas removal of stronger acid (or addition of alkali) moves the point up the iso-CO_2 line to the right. This diagram and Fig 39–6 are called Davenport diagrams and are based on Davenport HW: *The ABC of Acid-Base Chemistry,* 6th ed. Univ of Chicago Press, 1974.

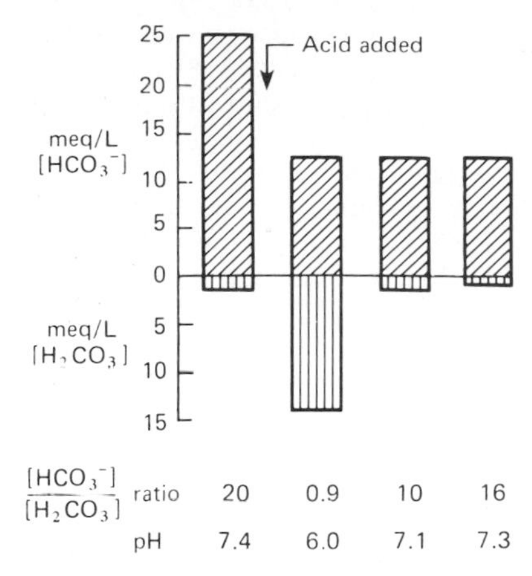

Figure 39–5. Buffering by the H_2CO_3-HCO_3^- system in blood. The bars are drawn as if buffering occurred in separate steps in order to show the effect of the initial reaction, the reduction of H_2CO_3 to its previous value, and its further reduction by the increase in ventilation. In this case, $[H_2CO_3]$ is actually the concentration of dissolved CO_2, so that the meq/L values for it are arbitrary.

P_{CO_2} probably because the greater the amount of CO_2 that is available to form H_2CO_3 in the cells, the greater the amount of H^+ that can be secreted (see Chapter 38). Furthermore, when the P_{CO_2} is high, the interior of most cells becomes more acidic (see Chapter 35). In respiratory acidosis, renal tubular H^+ secretion is therefore increased, removing H^+ from the body; and even though the plasma HCO_3^- is elevated, HCO_3^- reabsorption is increased, further raising the plasma HCO_3^-. This renal compensation for respiratory acidosis is shown graphically in Fig 39–3. Cl^- excretion is increased, and plasma Cl^- falls as plasma HCO_3^- is increased. Conversely, in respiratory alkalosis, the low P_{CO_2} hinders renal H^+ secretion, HCO_3^- reabsorption is depressed, and HCO_3^- is excreted, further reducing the already low plasma HCO_3^- and lowering the pH toward normal (Fig 39–3).

Metabolic Acidosis

When acids stronger than HHb and the other buffer acids are added to blood, **metabolic acidosis** is produced; and when the free H^+ level falls as a result of addition of alkali or removal of acid, **metabolic alkalosis** results. If, for example, H_2SO_4 is added, the H^+ is buffered and the Hb^-, $Prot^-$, and HCO_3^- levels in plasma drop. The H_2CO_3 formed is converted to H_2O and CO_2, and the CO_2 is rapidly excreted via the lungs. The importance of this fact is demonstrated in Fig 39–5. If enough acid were added to halve the HCO_3^- in plasma and CO_2 were not formed and excreted, the pH would fall to approximately 6.0 and death would result. If the H_2CO_3 level were so regulated that it remained constant, the pH would fall only to about 7.1. This is the situation in **uncompensated** metabolic acidosis (Fig 39–6). Actually, the rise in plasma H^+ stimulates respiration, so that the P_{CO_2}, instead of rising or remaining constant, is reduced. This **respiratory compensation** raises the pH even further. The **renal** compensatory mechanisms then bring about the excretion of the extra H^+ and return the buffer systems to normal.

Renal Compensation

The anions that replace HCO_3^- in the plasma in metabolic acidosis are filtered, each with a cation (principally Na^+), thus maintaining electrical neutrality. The renal tubular cells secrete H^+ into the tubular fluid in exchange for Na^+; and for each H^+ secreted, 1 Na^+ and 1 HCO_3^- are added to the blood (see Chapter 38). The limiting urinary pH of 4.5 would be reached rapidly and the total amount of H^+ secreted would be small if there were no buffers in the urine that "tied up" H^+. However, secreted H^+ reacts with HCO_3^- to form CO_2 and H_2O (bicarbonate reabsorption); with HPO_4^{2-} to form $H_2PO_4^-$; and with NH_3 to form NH_4^+. In this way large amounts of H^+ can be secreted, permitting correspondingly large amounts of HCO_3^- to be returned to (in the case of bicarbonate reabsorption) or added to the depleted body stores and large numbers of the cations to be reabsorbed. It is only when the acid load is very large that cations are lost with the anions, producing diuresis and depletion of body cation stores. In chronic acidosis, glutamine synthesis in the liver is increased, using some of the NH_4^+ that usually is converted to urea (Fig 39–3), and the glutamine provides the kidneys with an additional source of NH_4^+ (see Chapter 38). NH_3 secretion-increases over a period of days (adaptation of NH_3 secretion; see Chapter 38), further improving the renal compensation for acidosis. In addition, the metabolism of glutamine in the kidneys produces α-ketoglutarate, and this in turn is decarboxylated, producing HCO_3^-, which enters the bloodstream and helps buffer the acid load (Fig 39–3).

The overall reaction in blood when a strong acid such as H_2SO_4 is added is

$$2NaHCO_3 + H_2SO_4 \rightarrow Na_2SO_4 + 2H_2CO_3$$

For each mole of H^+ added, 1mol of $NaHCO_3$ is lost. The kidney in effect reverses the reaction

$$Na_2SO_4 + 2H_2CO_3 \rightarrow 2NaHCO_3 + 2H^+ + SO_4^{2-}$$

and the H^+ and SO_4^{2-} are excreted. Of course, H_2SO_4 is not excreted as such, the H^+ appearing in the urine as titratable acidity and NH_4^+.

In metabolic acidosis, the respiratory compensation tends to inhibit the renal response in the sense that the induced drop in P_{CO_2} hinders acid secretion, but it also decreases the filtered load of HCO_3^- and so its net inhibitory effect is not great.

Metabolic Alkalosis

In metabolic alkalosis, the plasma HCO_3^- level and pH rise (Fig 39–6). The respiratory compensation is a decrease in ventilation produced by the decline in H^+ concentration, and this elevates the P_{CO_2} This brings the pH back toward normal while elevating the plasma HCO_3^- level still further. The magnitude of this compensation is limited by the carotid and aortic chemoreceptor mechanisms, which drive the respiratory center if there is any appreciable fall in the arterial P_{O_2}. In metabolic alkalosis, more renal H^+ secretion is expended in reabsorbing the increased filtered load of HCO_3^-; and if the HCO_3^- level in plasma exceeds 28 meq/L, HCO_3^- appears in the urine. The rise in P_{CO_2} inhibits the renal compensation by facilitating acid secretion, but its effect is relatively slight.

Clinical Evaluation of Acid-Base Status

In evaluating disturbances of acid-base balance, it is important to know the pH and HCO_3^- content of arterial plasma. Reliable pH determinations can be made with a pH meter and a glass pH electrode. The HCO_3^- content of plasma cannot be measured directly, but the P_{CO_2} can be measured with a CO_2 electrode and the HCO_3^- concentration calculated, as noted above. The P_{CO_2} is 7–8 mm Hg higher and the pH 0.03-0.04 unit lower in venous than arterial plasma because venous blood contains the CO_2 being carried from the tissues to the lungs. Therefore, the calculated HCO_3^- concentration is about 2 mmol/L higher. However, if this is kept in mind, free-flowing

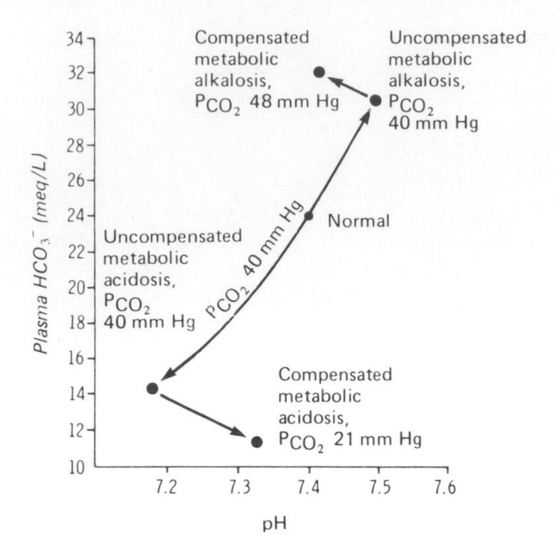

Figure 39–6. Changes in true plasma pH, HCO_3^-, and P_{CO_2} in metabolic acidosis and alkalosis.

venous blood can be substituted for arterial blood in most clinical situations.

A measurement that is of some value in the differential diagnosis of metabolic acidosis is the **anion gap.** This gap, which is somewhat of a misnomer, refers to the difference between the concentration of cations other than Na^+ and the concentration of anions other than Cl^- and HCO_3^- in the plasma. It consists for the most part of proteins in the anionic form, HPO_4^{2-}, SO_4^{2-}, and organic acids, and a normal value is about 12 meq/L. It is increased when the plasma concentration of K^+, Ca^{2+}, or Mg^{2+} is decreased; when the concentration or the charge on plasma proteins is increased; or when organic anions such as lactate or foreign anions accumulate in blood. It is decreased when cations are increased or when plasma albumin is decreased. The anion gap is increased in metabolic acidosis due to ketoacidosis, lactic acidosis, and other forms of acidosis in which organic anions are increased. It is not increased in hyperchloremic acidosis due to ingestion of NH_4Cl or carbonic anhydrase inhibitors.

The Siggaard-Andersen Curve Nomogram

Use of the Siggaard-Andersen curve nomogram (Fig 39–7) to plot the acid-base characteristics of arterial blood is helpful in clinical situations. This nomogram has log P_{CO_2} on the vertical scale and pH on the horizontal scale. Thus, any point to the left of a vertical line through pH 7.40 indicated acidosis, and any point to the right indicated alkalosis. The position of the point above or below the horizontal line through a P_{CO_2} of 40 mm Hg defines the effective degree of hypoventilation or hyperventilation.

If a solution containing $NaHCO_3$ and no buffers were equilibrated with gas mixtures containing various amounts of CO_2, the pH and P_{CO_2} values at equilibrium would fall along the dashed line on the left in Fig 39–7 or a line parallel to it. If buffers were present, the slope of the line would be greater; and the greater the buffering capacity of the solution, the steeper the line. For normal blood containing 15 g of hemoglobin per deciliter, the **CO_2 titration line** passes through the 15 g/dL mark on the hemoglobin scale (on the underside of the upper curved scale) and the point where the P_{CO_2} = 40 mm Hg and pH = 7.40 lines intersect, as shown in Fig 39–7. When the hemoglobin content of the blood is low, there is significant loss of buffering capacity, and the slope of the CO_2 titration line diminishes. However, blood of course contains buffers in addition to hemoglobin, so that even the line drawn from the zero point on the hemoglobin scale through the normal P_{CO_2}-pH intercept is steeper than the curve for a solution containing no buffers.

The nomogram is very useful clinically. Arterial blood or arterialized capillary blood is drawn anaerobically and its pH measured. The pHs of the same blood after equilibration with each of 2 gas mixtures containing different known amounts of CO_2 are also determined. The pH values at the known P_{CO_2} levels are plotted and connected to provide the CO_2 titration line for the blood sample. The pH of the blood sample before equilibration is plotted on this line, and the P_{CO_2} of the sample is read off the vertical scale. The **standard bicarbonate** content of the sample is indicated by the point at which the CO_2 titration line intersects the bicarbonate scale on the P_{CO_2} = 40 mm Hg line. The standard bicarbonate is not the actual bicarbonate concentration of the sample but rather what the bicarbonate concentration would be after elimination of any respiratory component. It is a measure of the alkali reserve of the blood, except that it is measured by determining the pH rather than the total CO_2 content of the sample after equilibration. Like the alkali reserve, it is an index of the degree of metabolic acidosis or alkalosis present.

Additional graduations on the upper curved scale of the nomogram (Fig 39–7) are provided for measuring **buffer base** content; the point where the CO_2 calibration line of the arterial blood sample intersects this scale shows the meq/L of buffer base in the sample. The buffer base is equal to the total number of buffer anions (principally $Prot^-$, HCO_3^-, and

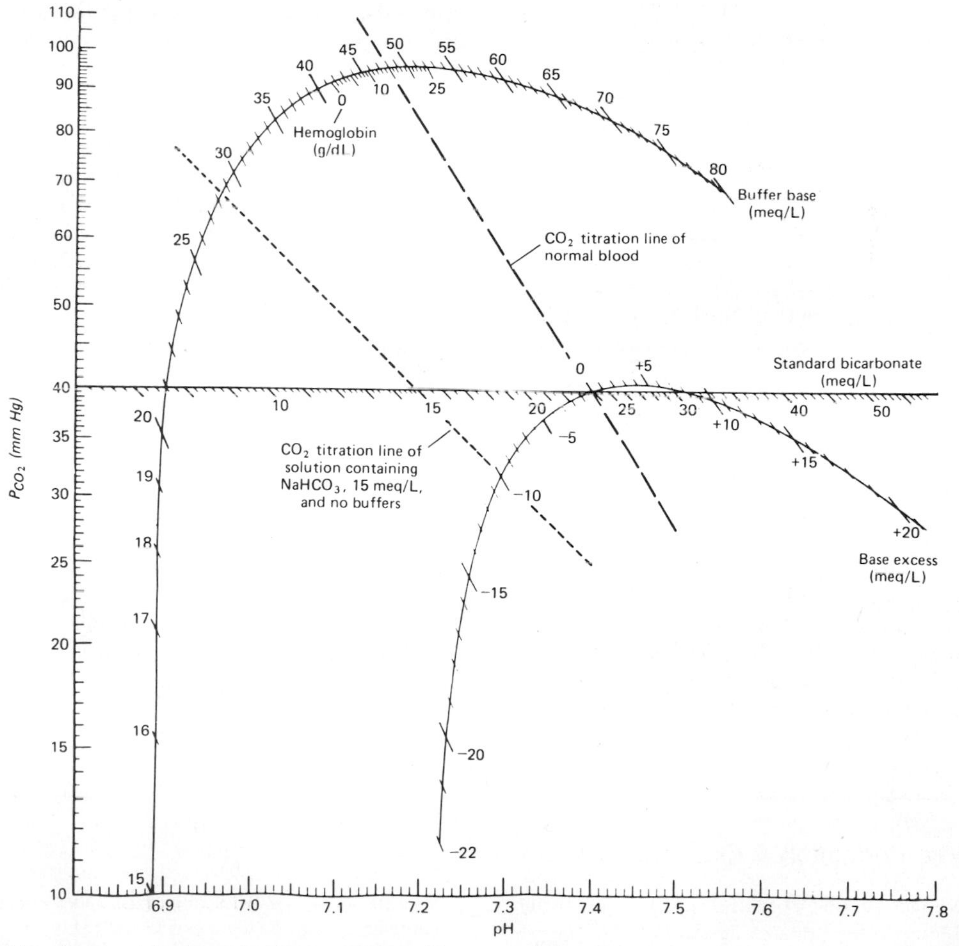

Figure 39–7. Siggaard-Andersen curve nomogram. (Courtesy of O Siggaard-Andersen and Radiometer, Copenhagen, Denmark.)

Table 39–2. Plasma pH, HCO_3^-, and P_{CO_2} values in various typical disturbances of acid-base balance. In the diabetic acidosis and prolonged vomiting examples, respiratory compensation for primary metabolic acidosis and alkalosis has occurred, and the P_{CO_2} has shifted from 40 mm Hg. In the emphysema and high altitude examples, renal compensation for primary respiratory acidosis and alkalosis has occurred and has made the deviations from normal of the plasma HCO_3^- larger than they would otherwise be. Data from various authors.

	Arterial Plasma			
Condition	pH	HCO_3^- (meq/L)	P_{CO_2} (mm Hg)	Cause
NORMAL	7.40	24.1	40	
Metabolic	7.28	18.1	40	NH_4Cl ingestion
acidosis	6.96	5.0	23	Diabetic acidosis
Metabolic	7.50	30.1	40	$NaHCO_3$ ingestion
alkalosis	7.56	49.8	58	Prolonged vomiting
Respiratory	7.34	25.0	48	Breathing 7% CO_2
acidosis	7.34	33.5	64	Emphysema
Respiratory	7.53	22.0	27	Voluntary hyperventilation
alkalosis	7.48	18.7	26	Three-week residence at 4000 m altitude

Hb^-; see Chapter 35) that can accept hydrogen ions in the blood. The normal value in an individual with 15 g of hemoglobin per deciliter of blood is 48 meq/L.

The point at which the CO_2 calibration line intersects the lower curved scale on the nomogram indicates the **base excess.** This value, which is positive in alkalosis and negative in acidosis, is the amount of acid or base that would restore 1 L of blood to normal acid-base composition at a P_{CO_2} of 40 mm Hg. It should be noted that a base deficiency cannot be completely corrected simply by calculating the difference between the normal standard bicarbonate (24 meq/L) and the actual standard bicarbonate and administering this amount of $NaHCO_3$ per liter of blood; some of the added HCO_3^- is converted to CO_2 and H_2O, and the CO_2 is lost in the lungs. The actual amount that must be added is roughly 1.2 times the standard bicarbonate deficit, but the lower curved scale on the nomogram, which has been developed empirically by analyzing many blood samples, is more accurate.

Some examples of clinical acid-base disturbances are shown in Table 39–2.

In treating acid-base disturbances, one must, of course, consider not only the blood but all the body fluid compartments. The other fluid compartments have markedly different concentrations of buffers. It has been determined empirically that administration of an amount of acid (in alkalosis) or base (in acidosis) equal to 50% of the body weight in kilograms times the blood base excess per liter will correct the acid-base disturbance in the whole body. At least when the abnormality is severe, however, it is unwise to attempt such a large correction in a single step; instead, about half the indicated amount should be given and the arterial blood acid-base values determined again. The amount required for final correction can then be calculated and administered. It is also worth noting that, at least in lactic acidosis, $NaHCO_3$ decreases cardiac output and lowers blood pressure, so it should be used with caution.

Relation Between Potassium Metabolism & Acid-Base Balance

The K^+ and H^+ concentrations of the ECF parallel each other, in part because of the effects of K^+ on renal H^+ secretion (see Chapter 38). K^+ depletion apparently produces an intracellular acidosis, promoting H^+ secretion into the urine. H^+ is thus removed from the body and HCO_3^- reabsorption increased, producing an extracellular alkalosis. Conversely, K^+ excess increases K^+ secretion by the renal tubular cells. Since both H^+ and K^+ are secreted in exchange for Na^+ and compete for the available Na^+ in the tubular fluid, H^+ secretion is inhibited, promoting extracellular acidosis.

REFERENCES
Section VIII: Formation & Excretion of Urine

Aronson PS: Mechanisms of active H^+ secretion in the proximal tubule. *Am J Physiol* 1983;**245:**F647.

Bayliss C, Blantz RC: Glomerular hemodynamics. *News Physiol Sci* 1986;**1:**86.

Boyle J III, Robertson G: Acid-base balance: an educational computer game. *Bioscience* 1987;**37:**511.

Brenner BM: Nephron adaptation to renal injury or ablation. *Am J Physiol* 1985;**249:**F324.

Brenner BM, Meyer TW, Hostetter TH: Dietary protein intake and the progressive nature of kidney disease. *N Engl J Med* 1982;**307:**652.

Brenner BM, Rector FC Jr (editors): *The Kidney,* 3rd ed. 2 vols. Saunders, 1986.

Brenner BM, Stein JH [editors]: *Body Fluid Homeostasis.* Churchill Livingstone, 1987.

Burg MB: The nephron in transport of sodium, amino acids, and glucose. *Hosp Pract* (Oct) 1978;**13:**99.

Davenport HW: *The ABC of Acid-Base Chemistry,* 6th ed. Univ of Chicago Press, 1974.

DiBona GF: The functions of the renal nerves. *Rev Physiol Biochem Pharmacol* 1982;**94:**76.

Dunn MJ (editor): *Renal Endocrinology.* Williams & Wilkins, 1983.

Forte JG, Reenstra WW: Plasma membrane proton pumps in animal cells. *Bioscience* 1985;**35:**38.

Frömter E: Viewing the kidney through microelectrodes. *Am J Physiol* 1984;**247:**F695.

Gebow PA: Disorders associated with an altered anion gap. *Kidney Int* 1985;**27:**472.

Hayslett JP: Functional adaptation to reduction in renal mass. *Physiol Rev* 1979;**59:**137.

Hood I, Campbell EJM: Is pK OK? *N Engl J Med* 1982;**306:**864.

Hutch JA: *Anatomy and Physiology of the Bladder, Trigone and Urethra.* Appleton-Century-Crofts, 1972.

Jamison RL, Kriz W: *Urinary Concentrating Mechanism: Structure and Function.* Oxford Univ Press, 1982.

Loeb JN: The hyperosmolar state. *N Engl J Med* 1974;**290:**1184.

Madsen KM, Tisher CC: Structural-functional relationships along the distal nephron. *Am J Physiol* 1986;**250:**F1.

Mandel LJ: Metabolic substrates, cellular energy production, and the regulation of proximal tubular transport. *Annu Rev Physiol* 1985;**47:**85.

Morel F: Sites of hormone action in the mammalian nephron. *Am J Physiol* 1981;**240:**F159.

Orci L et al: Membrane ultrastructure in urinary tubules. *Int Rev Cytol* 1981;**73:**183.

Rector FC Jr: Sodium bicarbonate and chloride absorption by the proximal tubule. *Am J Physiol* 1983;**244:**F461.

Roos A, Boron WF: Intracellular pH. *Physiol Rev* 1981;**61:**296.

Schafer JA: Mechanisms coupling the absorption of solute and water in proximal nephron. *Kidney Int* 1984;**25:**708.

Schlondorff D: The glomerular mesangial cell: an expanding role for a specialized pericyte. *FASEB J* 1987;**1:**272.

Schmidt-Nielson K: Countercurrent systems in animals. *Sci Am* (May) 1981;**244:**118.

Schuster VC, Stokes JB: Chloride transport by the cortical and outer medullary collecting duct. *Am J Physiol* 1987;**253:**F203.

Segal S: Disorders of renal amino acid transport. *N Engl J Med* 1976;**294:**1044.

Seldin DW, Giebisch G (editors): *The Kidney: Physiology and Pathophysiology,* 2 vols. Raven, 1985.

Siggaard-Andersen O: *The New Acid-Base Status of Blood.* Williams & Wilkins, 1965.

Stanton B, Giebisch G: Mechanism of urinary potassium excretion. *Miner Electrolyte Metab* 1981;**5:**100.

Tannen RL: Ammonia metabolism. *Am J Physiol* 1978;**235:**F265.

Tisher CC: Structural-functional relationships along the distal nephron. *Am J Physiol* 1986;**250:**F1.

Valtin H: *Renal Function,* 2nd ed. Little, Brown, 1983.

Valtin H, Gennari FJ: *Acid-Base Disorders: Basic Concepts and Clinical Management.* Little, Brown, 1986.

Wright FS: Flow-dependent transport processes: Filtration, absorption, secretion. *Am J Physiol* 1982;**243:**F1.

Wright FS: Intrarenal regulation of glomerular filtration rate. *J Hypertension* 1984;**2(Suppl 1):**105.

Symposium: Renal concentrating mechanism. *Fed Proc* 1983;**42:**2377.

Appendix

GENERAL REFERENCES

Many large, comprehensive textbooks of physiology are available. The following are among the best of the recently published or revised volumes:

Berne RM, Levy MN (editors): *Physiology,* 2nd ed. Mosby, 1988.
Guyton AC: *Textbook of Medical Physiology,* 7th ed. Saunders, 1986.
Keil CA, Neil E, Joels N: *Samson Wright's Applied Physiology,* 13th ed. Oxford Univ Press, 1982.
West JB (editor): *Best and Taylor's Physiological Basis of Medical Practice,* 11th ed. Williams & Wilkins, 1985.

Textbooks of pathophysiology include the following:

Frohlich ED (editor): *Pathophysiology,* 3rd ed. Lippincott, 1984.
Sodeman WA Jr, Sodeman WA: *Pathologic Physiology,* 6th ed. Saunders, 1979.

Excellent summaries of current research on selected aspects of physiology can be found in the *FASEB Journal,* and in the *Annals of the New York Academy of Sciences.* Summary articles appear in *Hospital Practice,* and various types of valuable reviews appear in the *New England Journal of Medicine.* These include articles that review current topics in physiology and biochemistry with the aim of providing up-to-date information for practicing physicians. The most pertinent serial review publications are *Physiological Reviews, Pharmacological Reviews, Recent Progress in Hormone Research* (the proceedings of each annual Laurentian Hormone Conference), *Annual Review of Physiology,* and other volumes of the Annual Review series. The *Handbook of Physiology,* which is now being published by Oxford University Press, New York, has separate volumes that cover all aspects of physiology. The volumes cover the nervous system, the cardiovascular system, respiration, adaptation to the environment, adipose tissue, the alimentary canal, endocrinology, and renal physiology. The articles in the handbook are valuable but extremely detailed reviews. The *Biology Data Book,* published in 3 volumes by the Federation of American Societies for Experimental Biology, contains tables and lists that are useful for looking up specific items of information.

Powerful new techniques of imaging have added to the advance of physiology as well as clinical medicine. These include **nuclear magnetic resonance imaging (NMRI),** a method that permits not only the production of pictures of the interior of the body but also noninvasive analysis of metabolic and other phenomena in living animals and humans. Various forms of **emission tomography** are also very valuable for localizing compounds such as dopamine and various sugars in living individuals. References that describe these important techniques include the following:

Eli PJ, Holman BL (editors): *Computed Emission Tomography,* Oxford Univ Press, 1983.
Young SW: *Nuclear Magnetic Resonance Imaging: Basic Principles.* Raven, 1984.

NORMAL VALUES & THE STATISTICAL EVALUATION OF DATA

The approximate ranges of values in normal humans for some commonly measured plasma constituents are summarized in the table on the inside back cover. A worldwide attempt has been under way to convert to a single standard nomenclature by using SI (Système International) units. The system is based on the 7 dimensionally independent physical quantities summarized in Table 1. Units derived from the basic units are summarized in Table 2, and the prefixes used to refer to decimal fractions and multiples of these and other units are listed in Table 3. There are a number of complexities involved in the use of these units—for example, the problem of expressing enzyme units—and they are only slowly making their way into the medical literature. In this book, the values in the text are in traditional units, but they are followed in key instances by values in SI units. In

Table 1. Basic units.

Quantity	Name	Symbol
Length	meter	m
Mass	kilogram	kg
Time	second	s
Electric current	ampere	A
Thermodynamic temperature	kelvin	K
Luminous intensity	candela	cd
Amount of substance	mole	mol

Table 2. Some derived SI units.

Quantity	Unit Name	Unit Symbol
Area	square meter	m^2
Clearance	liter/second	L/s
Concentration		
Mass	kilogram/liter	kg/L
Substance	mole/liter	mol/L
Density	kilogram/liter	kg/L
Electric potential	volt	$V = kg\,m^2/s^3\,A$
Energy	joule	$J = kg\,m^2/s^2$
Force	newton	$N = kg\,m/s^2$
Frequency	hertz	Hz = 1 cycle/s
Pressure	pascal	$Pa = kg/m\,s^2$
Temperature	degree Celsius	°C = °K − 273.15
Volume	cubic meter	m^3
	liter	$L = dm^3$

addition, values in SI units are listed beside values in more traditional units in the table on the inside back cover.

The accuracy of the methods used for laboratory measurements varies. It is important in evaluating any single measurement to know the possible errors in making the measurement. In the case of chemical determinations on body fluids, these include errors in obtaining the sample and the inherent error of the chemical method. However, the values obtained using even the most accurate methods vary from one normal individual to the next as a result of what is usually called **biologic variation.** This variation is due to the fact that in any system as complex as a living organism or tissue there are many variables which affect the particular measurement. Variables such as age, sex, time of day, time since last meal, etc, can be controlled. Numerous other variables cannot, and for this reason the values obtained differ from individual to individual.

Table 3. Standard prefixes.

Prefix	Abbreviation	Magnitude
exa-	E	10^{18}
peta-	P	10^{15}
tera-	T	10^{12}
giga-	G	10^{9}
mega-	M	10^{6}
kilo-	k	10^{3}
hecto-	h	10^{2}
deca-	da	10^{1}
deci-	d	10^{-1}
centi-	c	10^{-2}
milli-	m	10^{-3}
micro-	μ	10^{-6}
nano-	n, mμ	10^{-9}
pico-	p, $\mu\mu$	10^{-12}
femto-	f	10^{-15}
atto-	a	10^{-18}

These prefixes are applied to SI and other units. For example, a micrometer (μm) is 10^{-6} meter (also called a micron); a picoliter (pL) is 10^{-12} liter, and a kilogram (kg) is 10^3 grams. Also applied to seconds, units, moles, hertz, volts, farads, ohms, curies, equivalents, osmoles, etc.

The magnitude of the normal range for any given physiologic or clinical measurement can be calculated by standard statistical techniques if the measurement has been made on a suitable sample of the normal population (preferably more than 20 individuals). It is important to know not only the average value in this sample but also the extent of the deviation of the individual values from the average.

The average (**arithmetic mean, M**) of the series of values is readily calculated:

$$M = \frac{\Sigma X}{n}$$

where

Σ = **Sum of**
X = **The individual values**
n = **Number of individual values in the series**

The average deviation is the mean of the deviations of each of the values from the mean. From a mathematical point of view, a better measure of the deviation is the **geometric mean** of the deviations from the mean. This is called the **standard deviation** (σ):

$$\sigma = \sqrt{\frac{\Sigma (M - X)^2}{n - 1}}$$

The term n–1 rather than n is used in the denominator of this equation because the σ of a sample of the population rather than the σ of the whole population is being calculated. The following form of the equation for σ can be derived algebraically and is convenient for those computing σ.

$$\sigma = \sqrt{\frac{\Sigma X^2 - \frac{(\Sigma X)^2}{n}}{n - 1}}$$

Here, ΣX^2 is the sum of the squares of the individual values, and $(\Sigma X)^2$ is the square of the sum of the items, ie, the items added together and then squared.

Another commonly used index of the variation is the **standard error of the mean** (SE, SEM):

$$SE = \frac{\sigma}{\sqrt{n}}$$

Strictly speaking, the SE indicates the reliability of the sample mean as representative of the true mean of the general population from which the sample was drawn.

A **frequency distribution** curve can be constructed from the individual values in the sample by plotting the frequency with which any particular value occurs in the series against the values. If the group of individuals tested was homogeneous, the frequency distribution curve is usually symmetric (Fig 1), with the highest frequency corresponding to the mean and the width of the curve varying with σ (curve of **normal distribution**). Within an ideal curve of normal distribution, the percentage of observations

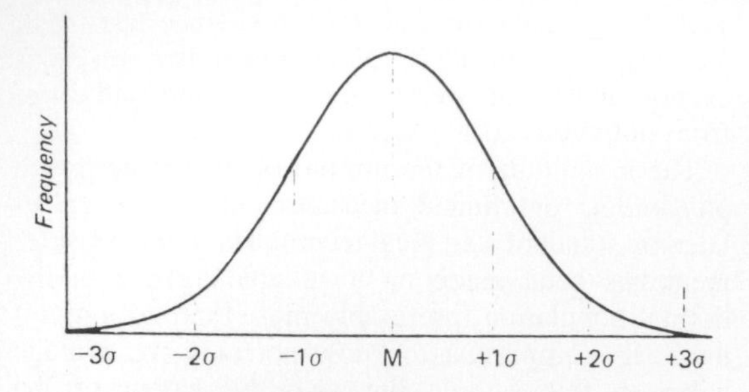

Figure 1. Curve of normal distribution (frequency distribution curve of values from a homogeneous sample of a population).

that fall within various ranges is shown in Table 4. The mean and σ of a representative sample are approximately those of the whole population. It is therefore possible to predict from the mean and σ of the sample the probability that any particular value in the general population is normal. For example, if the difference between such a value and the mean is equal to 1.96 σ, the chances are 1 out of 20 (5 out of 100) that it is normal. Conversely, of course, the chances are 19 out of 20 that it is abnormal. It is unfortunate that data on normal means and σs are not more generally available for the important plasma and urinary constituents in humans.

Statistical analysis is also useful in evaluating the significance of the difference between 2 means. In physiologic and clinical research, measurements are often made on a group of animals or patients given a particular treatment. These measurements are compared with similar measurements made on a control group that ideally has been exposed to exactly the same conditions except that the treatment has not been given. If a particular mean value in the treated group is different from the corresponding mean for the control group, the question arises whether the difference is due to the treatment or to chance variation. The probability that the difference represents chance variation can be estimated in many instances by using **"Student's" t test.** The value t is the ratio of the difference in the means of 2 series (M_a and M_b) to the uncertainty in these means. The formula used to calculate t is

$$t = \frac{M_a - M_b}{\sqrt{\dfrac{(n_a + n_b)[(n_a - 1)\sigma_a^2 + (n_b - 1)\sigma_b^2]}{n_a n_b (n_a + n_b - 2)}}}$$

where n_a and n_b are the number of individual values in series a and b, respectively. When $n_a = n_b$, the equation for t becomes simplified to

$$t = \frac{M_a - M_b}{\sqrt{(SE_a)^2 + (SE_b)^2}}$$

The higher the value of t, the less the probability that the difference represents chance variation. This probability also decreases as the number of individuals (n) in each group rises, because the greater the number of measurements, the smaller the error in the measurements. A mathematical expression of the probability (P) for any value of t at different values of n can be found in tables in most texts on statistics. P is a fraction which expresses the probability that the difference between 2 means was due to chance variation. Thus, for example, if the P value is 0.10, the probability that the difference was due to chance is 10% (1 chance in 10). A P value of <0.001 means that the chances that the difference was due to random variation are less than 1 in 1000. When the P value is <0.05, most investigators call the difference "statistically significant," ie, it is concluded that the difference is due to the operation of some factor other than chance—presumably the treatment. The conditions under which it is appropriate to use Student's t test and the conditions under which other tests should be used to calculate P are discussed in statistics texts.

These elementary methods and many others available for statistical analysis in the research laboratory and the clinic provide a valuable objective means of evaluation. Statistical significance does not arbitrarily mean physiologic significance, and the reverse may sometimes be true; but replacement of evaluation by subjective impression with analysis by statistical methods is an important goal in the medical sciences.

Table 4. Percentage of values in a population which will fall within various ranges within an ideal curve of normal distribution.

Range	Percentage
Mean ± 1 σ	68.27%
Mean ± 1.96 σ	95.00%
Mean ± 2 σ	95.45%
Mean ± 3 σ	99.73%

REFERENCES: APPENDIX

Rosner B: *Fundamentals of Biostatistics.* Duxbury, 1982.

The SI for the Health Professions: World Health Organization, 1977.

Winer GJ: *Statistical Principles in Experimental Design,* 2nd ed. McGraw-Hill, 1971.

Zar JH: *Biostatistical Analysis.* Prentice-Hall, 1974.

ABBREVIATIONS & SYMBOLS COMMONLY USED IN PHYSIOLOGY

[ ]: Concentration of
Δ: Change in. (*Example:* ΔV = change in volume.) In steroid nomenclature, Δ followed by a number (eg, Δ^4-) indicates the position of a double bond
σ: Standard deviation
Ia, Ib, II, III, IV nerve fibers: Types of fibers in sensory nerves (see Chapter 2)
μ: Micro, 10^{-6}; see Table 3, above
a: atto-, 10^{-18}; see Table 3, above
A(A): Angstrom unit(s) (10^{-10} m, 0.1 nm); also alanine
A^-: General symbol for anion
A_1, A_2, A_1B, A_2B, B, O: Major blood groups
AASH: Adrenal androgen-stimulating hormone
A, B, and C nerve fibers: Types of fibers in peripheral nerves (see Chapter 2)
ABP: Androgen-binding protein
ACE: Angiotensin converting enzyme
Acetyl-CoA: Acetyl-coenzyme A
ACH, Ach: Acetylcholine
ACTH; Adrenocorticotropic hormone
Acyl-CoA: General symbol for an organic compound-coenzyme A ester
ADH: Antidiuretic hormone (vasopressin)
ADP: Adenosine diphosphate
AHG: Antihemophilic globulin
Ala: Alanine
AMP: Adenosine-5-monophosphate
ANP: Atrial natriuretic peptide
APUD cells: Amine precursor uptake and decarboxylation cells that secrete hormones
Arg: Arginine
NH_2
|
Asn, Asp: Asparagine
Asp: Aspartic acid
atm: Atmosphere: 1 atm = 760 torr = mean atmospheric pressure at sea level
ATP: Adenosine triphosphate
A-V difference: Arteriovenous concentration difference of any given substance
AV node: Atrioventricular node
aVR, aVF, aVL: Augmented unipolar electrocardiographic leads
AV valves: Atrioventricular valves of heart
BER: Basic electrical rhythm
BMR: Basal metabolic rate
BSP: Sulfobromophthalein
BUN: Blood urea nitrogen
c: Centi-, 10^{-2}; see Table 3, above
C: Celsius; also cysteine
C followed by subscript: Clearance; eg, C_{In} = clearance of inulin
C_{19} steroids: Steroids containing 19 carbon atoms
C_{21} steroids: Steroids Containing 21 carbon atoms
cal: The calorie (gram calorie)
Cal: 1000 calories; kilocalorie
cAMP: Cyclic adenosine-3′, 5′-monophosphate
CAT: computer-assisted tomography
CBF: Cerebral blood flow
CBG: Corticosteroid-binding globulin, transcortin
cc: Cubic centimeters
CCK, CCK-PZ: Cholecystokinin-pancreozymin
CFF: Critical fusion frequency
CGP: Chorionic growth hormone-prolactin (same as hCS)
CGRP: Calcitonin-gene-related peptide
C_{H_2O}: "Free water clearance"
Ci: Curie
CLIP: Corticotropinlike intermediate lobe peptide
$CMRO_2$: Cerebral metabolic rate for oxygen
CNS: Central nervous system
CoA: Coenzyme A
COHb: Carbonmonoxyhemoglobin
Compound A: 11-Dehydrocorticosterone
Compound B: Corticosterone
Compound E: Cortisone
Compound F: Cortisol
Compound S: 11-Deoxycortisol
COMT: Catechol-O-methyltransferase
cps: Cycles per second, hertz
CR: Conditioned reflex
Cr: Creatinine
CRH, CRF: Corticotropin-releasing hormone
CRO: Cathode-ray oscilloscope
CrP: Creatinine phosphate (phosphocreatine, PC)
CS: Conditioned stimulus
CSF: Cerebrospinal fluid
C terminal: End of peptide or protein having a free −COOH group
CTP: Cytidine triphosphate
CV: Closing volume
CVR: Cerebral vascular resistance
cyclic AMP: Cyclic adenosine-3′, 5′-monophosphate
CyS: Half-cystine
CZI: Crystalline zinc insulin
d: Day
D: Geometric isomer of L form of chemical compound
D: Aspartic acid
Da: dalton, a unit of mass, equal to one-twelfth the mass of the carbon-12 atom, or about 1.66 x 10^{-24} g
dB: Decibel
DDAVP: 1-Deamino-8-D-arginine vasopressin
DDD: Derivative of DDT that inhibits adrenocortica function
DEA, DHEA: Dehydroepiandrosterone
DEAS, DHEAS: Dehydroepiandrosterone sulfate
DFP: diisopropyl fluorophosphate
DHT: Dihydrotestosterone
DIT: Diiodotyrosine
DNA: Deoxyribonucleic acid
D/N ratio: Ratio of dextrose (glucose) to nitrogen in the urine
D_2O: Deuterium oxide (heavy water)
DOCA: Desoxycorticosterone acetate
DOMA: 3,4-Dihydroxymandelic acid
Dopa: Dihydroxyphenylalanine, L-dopa
DOPAC: 3,4-Dihydroxyphenylacetic acid
DOPEG: 3,4-Dihydroxyphenylglycol
DOPET: 3,4-Dihydroxyphenylethanol
DPG, 2,3-DPG: 2,3-Diphosphoglycerate

DPL: Dipalmitoyl lecithin
DPN: Diphosphopyridine nucleotide
DPNH: Reduced diphosphopyridine nucleotide
DPPC: Dipalmitoyl-phosphatidyl-choline
e: Base for natural logarithms = 2.7182818 . . .
E: Glutamic acid
E followed by subscript number: Esophageal electrocardiographic lead, followed by number of cm it is inserted in esophagus
E cells: Expiratory neurons
E_1: Estrone
E_2: Estradiol
EACA: Epsilon-aminocaproic acid
ECF: Extracellular fluid
eCG: Equine chorionic gonadotropin
ECG: Electrocardiogram
ECoG: Electrocorticogram
EDRF: Endothelium-derived relaxing factor
EDTA; Ethylenediamine-tetraacetic acid
EEG: Electroencephalogram
EGF: Epidermal growth factor
EJP: Excitatory junction potential
EKG: Electrocardiogram
EMG: Electromyogram
EP: Endogenous pyrogen
EPSP: Excitatory postsynaptic potential
eq: Equivalent(s)
ERG: Electroretinogram
ERPF: Effective renal plasma flow
ETP: Electron transport particle
f: Femto-, 10^{15}; see Table 3, above
F: Fahrenheit; also phenylalanine
FDP: Fibrinogen degradation products
FeV 1″: Forced expiratory volume in 1 second
FFA: Unesterified free fatty acid (also called NEFA, UFA)
FGF: Fibroblast growth factor
FRH, FSH-RH, FRF: FSH-releasing hormone
FSH: Follicle-stimulating hormone
ft: Foot or feet
g, gm: Gram(s)
g: Unit of force, 1 *g* equals the force of gravity on the earth's surface
G: Glucose; also giga-, 10^9; see Table 3, above; also glycine
GABA: Gamma-aminobutyate
GAD: Glutamate decarboxylase
GBG: Gonadal steroid-binding globulin
G-CSF: Macrophage colony-stimulating factor
GFR: Glomerular filtration rate
GH: Growth hormone
GIH, GIF: Growth hormone-inhibiting hormone
GIP: Gastric inhibitory peptide
Gla: γ-Carboxyglutamic acid
GLI: Glucagonlike immuno-reactive factor from a gastrointestinal mucosa

NH_2
|
Gln, Glu: Glutamine
Glu: Glutamic acid
Gly: Clycine
GM-CSF: Granulocyte-macrophage colony-stimulating factor
GnRH: Gonadotropin-releasing hormone; same as LHRH
G6PD: Glucose 6-phosphate dehydrogenase
GRH, GRF: Growth hormone-releasing hormone
GTP: Guanosine triphosphate
h: Hour(s)
H: Histidine
HA: General symbol for an acid
Hb: Deoxygenated hemoglobin
HBE: His bundle electrogram
HbO_2: Oxyhemoglobin
HCC, 25-HCC: 25-Hydroxycholecalciferol, an active metabolite of vitamin D_3
hCG: Human chorionic gonadotropin
hCS: Human chorionic somatomammotropin
Hct: Hematocrit
HDL: High-density lipoprotein
hGH: Human growth hormone
HIOMT: Hydroxyindole-O-methyltransferase
His: Histidine
HIV: Human immunodeficiency virus.
HLA: Human leukocyte antigens
hPL: Human placental lactogen (same as hCS)
HS-CoA: Reduced coenzyme A
H substance: Histaminelike capillary vasodilator
5-HT: Serotonin
HVA: Homovanillic acid
Hyl: Hydroxylysine
Hyp: 4-Hydroxyproline
Hz: Hertz, unit of frequency. 1 cycle per second = 1 hertz
I: Isoleucine
I cells: Inspiratory neurons; also intercalated cells in the renal tubules
IDL: Intermediate-density lipoprotein
IGF-I, IGF-II: Insulinlike growth factors I and II
^{123}I-IMP: ^{123}I-labeled iodoamphetamine
IJP: Inhibitory junction potential
IL: Interleukin
Ile, Ileu: Isoleucine
In: Inulin
IPSP: Inhibitory postsynaptic potential
ITP: Inosine triphosphate
IU: International unit(s)
IUD: Intrauterine device
J: Joule (SI unit of energy)
k: Kilo-, 10^3; see Table 3, above
K: Lysine; also kilodalton
kcal (Cal): Kilocalorie (1000 calories)
K_E: Exchangeable body potassium
K_f: Glomerular ultrafiltration coefficient
L: Geometric isomer of D form of chemical compound
L: Leucine
LATS: Long-acting thyroid stimulator
LDL: Low-density lipoprotein
LES: Lower esophageal sphincter
Leu: Leucine
LH: Luteinizing hormone
ln: Natural logarithm
log: Logarithm to base 10
LPH: Lipotropin
LRH, LHRH, LRF: Luteinizing hormone-releasing hormone
LSD: Lysergic acid diethylamide
LTH: Luteotropic hormone (prolactin)
LVET: Left ventricular ejection time
Lys: Lysine
m: Meter(s); also milli-, 10^{-3}; see Table 3, above

M: Molarity (mol/L); also mega-, 10^6; see Table 3, above; also methionine
MAO: Monoamine oxidase
MBC: Maximal breathing capacity (same as MVV)
Met: Methionine
MHC: Major histocompatibility complex
mho: Unit of conductance; the reciprocal of the ohm
min: Minute(s)
MIS: Müllerian inhibiting substance; same as MRF, müllerian regression factor
MIT: Monoiodotyrosine
mol: Mole, gram-molecular weight
MOPEG: 3-Methoxy-4-hydroxyphenylglycol
MPR: Mannose-6-phosphate receptor
mRNA: Messenger RNA
MSH: Melanocyte-stimulating hormone
MT: 3-Methoxytyramine
multi-CSF: Multipotential colony-stimulating factor
MVV: Maximal voluntary ventilation
n: nano-, 10^{-9}; see Table 3, above
N: Normality (of a solution): also newton (SI unit of force); also asparagine
NAD$^+$: Nicotinamide adenine dinucleotide; same as DPN
NADH: Dihydronicotinamide adenine dinucleotide; same as DPNH
NADP$^+$: Nicotinamide adenine dinucleotide phosphate; same as TPN
NADPH: Dihydronicotinamide adenine dinucleotide phosphate; same as TPNH
Na$_E$: Exchangeable body sodium
NEFA: Unesterified (nonesterified) free fatty acid (same as FFA)
NGF: Nerve growth factor
NMRI: Nuclear magnetic resonance imaging
NPH insulin: Neutral protamine Hagedorn insulin
NPN: Nonprotein nitrogen
NREM sleep: Nonrapid eye movement (spindle) sleep
NSILA: Nonsuppressible insulinlike activity
NSILP: Nonsuppressible insulinlike protein
N terminal: End of peptide or protein having a free $-NH_2$ group
NTS: Nucleus of the tractus solitarius
O: Indicates absence of a sex chromosome, eg, XO as opposed to XX or XY
OGF: Ovarian growth factor
osm: Osmole(s)
OVLT: Organum vasculosum of the lamina terminalis
p: Pico-, 10^{-12}; see Table 3, above
P: Proline
P followed by subscript: Plasma concentration, eg, P_{Cr} = plasma creatinine concentration; also permeability coefficient, eg, P_{Na^+} = permeability coefficient for Na^+; also pressure (see respiratory symbols, below)
P$_{50}$: Partial pressure of O_2 at which hemoglobin is half-saturated with O_2
Pa: Pascal (SI unit of pressure)
PAF: Platelet activating factor
PAH: Para-aminohippuric acid
PAM: Pulmonary alveolar macrophage
PBI: Protein-bound iodine
P cells: Principal cells in the renal tubules
PDGF: Platelet-derived growth factor
PEEP: Positive end-expiratory pressure breathing
PEP: Preejection period
PET: Positron emission tomography
P factor: Hypothetical painproducing substance produced in ischemic muscle
PGO spikes: Ponto-geniculo-occipital spikes in REM sleep
pH: Negative logarithm of the H^+ concentration of a solution
Phe: Phenylalanine
PHM-27: Peptide histidyl methionine, produced along with VIP from prepro VIP in humans
Pi: Inorganic phosphate
PIH, PIF: Prolactin-inhibiting hormone
pK: Negative logarithm of the equilibrium constant for a chemical reaction
PMN: Polymorphonuclear neutrophilic leukocyte
POMC: Pro-opiomelanocortin
PRA: Plasma renin activity
PRC: Plasma renin concentration
PRH, PRF: Prolactin-releasing hormone
Pro: Proline
Prot$^-$: Protein anion
5-PRPP: 5-Phosphoribosyl pyrophosphate
PTA: Plasma thromboplastin antecedent (clotting factor XI)
PTC: Plasma thromboplastin component (clotting factor IX); also phenylthiocarbamide
PTH: Parathyroid hormone
(pyro)Glu: Pyroglutamic acid
PZI: Protamine zinc insulin
Q: Glutamine
QS$_2$: Total electrochemical systole
R: General symbol for remainder of a chemical formula, eg, an alcohol is R–OH; also gas constant; also respiratory exchange ratio; also Reynolds' number; also arginine
RAS: Reticular activating system
rbc: Red blood cell(s)
RDS: Respiratory distress syndrome
REF: Renal erythropoietic factor
REM sleep: Rapid eye movement (paradoxical) sleep
Reverse T$_3$: 3,3′, 5′-Triiodothyronine; isomer of triiodothyronine
Rh factor: Rhesus group of red cell agglutinogens
RMV: Respiratory minute volume
RNA: Ribonucleic acid
RPF: Renal plasma flow
RQ: Respiratory quotient
R state: State of heme in hemoglobin that increases O_2 binding
R unit: Unit of resistance in cardiovascular system; mm Hg divided by mL/s
s: Second(s)
S: Serine
SA: Specific activity
SA node: Sinoatrial node
SCUBA: Self-contained underwater breathing apparatus
SDA: Specific dynamic action
SE (SEM): Standard error of the mean
Ser: Serine

S_f units: Svedberg units of flotation
SGOT: Serum glutamic-oxaloacetic transaminase
SH: Sulfhydryl
SI units: Units of the Système International d'Unités
SIADH: Syndrome of inappropriate antidiuretic hormone (vasopressin) secretion
SIDS: Sudden infant death syndrome
SIF cells: Small, intensely fluorescent cells in sympathetic ganglia
SPCA: Proconvertin (clotting factor VII)
SPECT: Single proton emission computed tomography
sq cm: Square centimeter(s), cm^2
SRIF: Somatotropin release-inhibiting factor; same as GIH
sRNA: Soluble or transfer RNA
SS 14: Somatostatin 14
SS 28: Somatostatin 28
SS 28 (1-12): Polypeptide related to somatostatin that is found in tissues
STH: Somatotropin, growth hormone
Substance P: Polypeptide found in brain
T: Absolute temperature; also threonine
T_3: 3,5,3′-Triiodothyronine
T_4: Thyroxine
TBG: Thyroxine-binding globulin
TBPA: Thyroxine-binding prealbumin
TBW: Total body water
^{99m}Tc-PYP: Technetium 99m stannous pyrophosphate
TDF: Testis-determining factor
TEA: Tetraethylammonium
TETRAC: Tetraiodothyroacetic acid
TF/P: Concentration of a substance in renal tubular fluid divided by its concentration in plasma
TGF-α: Transforming growth factor-α
Thr: Threonine
Tm: Renal tubular maximum
torr: 1/760 atm = 1.00000014 mm Hg; unit for various pressures in the body
t-PA: Tissue plasminogen activator
TPN: Triphosphopyridine nucleotide
TPNH: Reduced triphosphopyridine nucleotide
TRH, TRF; Thyrotropin-releasing hormone
tRNA: Transfer RNA; same as sRNA
Trp, Try, Tryp: Tryptophan
TSF: Thrombopoietic stimulating factor, thrombopoietin
TSH: Thyroid-stimulating hormone
TSI: Thyroid-stimulating immunoglobulins
T/S ratio: Thyroid/serum iodide ratio
T state: State of heme in hemoglobin that decreases O_2 binding
TTX: Tetrodotoxin
Tyr: Tyrosine
U: Unit(s)
U followed by subscript: Urine concentration, eg, U_{Cr} = urine creatinine concentration
UDPG: Uridine diphosphoglucose
UFA: Unesterified free fatty acid (same as FFA)
URF: Uterine-relaxing factor; relaxin
US: Unconditioned stimulus
UTP: Uridine triphosphate
UWL: Unstirred water layer
V: Urine volume/unit time; volt; also valine
V_1, V_2, etc: Unipolar chest electrocardiographic leads
Val: Valine
VF: Unipolar left leg electrocardiographic lead
VIP: Vasoactive intestinal polypeptide
VL: Left arm unipolar electrocardiographic lead
VLDL: Very low density lipoprotein
VMA: Vanillylmandelic acid (3-methoxy-4-hydroxymandelic acid)
VOR: Vestibulo-ocular reflex
VR: Unipolar right arm electrocardiographic lead
W: Tryptophan
wbc: White blood cell(s)
X chromosome: One of the sex chromosomes in humans
X zone: Inner zone of adrenal cortex in some young mammals
Y: Tyrosine
Y chromosome: One of the sex chromosomes in humans

STANDARD RESPIRATORY SYMBOLS

(See *Fed Proc* 1950;**9:**602.)

General Variables

V Gas volume

$\dot{V}$ Gas volume/unit of time. (Dot over a symbol indicates rate.)

P Gas pressure

$\bar{P}$ Mean gas pressure

f Respiratory frequency (breaths/unit of time)

D Diffusing capacity

F Fractional concentration in dry gas phase

C Concentration in blood phase

R Respiratory exchange ratio = V_{CO_2}/V_{O_2}

Q Volume of blood

$\dot{Q}$ Volume flow of blood/unit of time

Localization (Subscript letters)

I Inspired gas

E Expired gas

A Alveolar gas

T Tidal gas

D Dead space gas

B Barometric

a Arterial blood

c Capillary blood

v Venous blood

Special Symbols

STPD Standard temperature and pressure, dry (0 °C, 760 mm Hg)

BTPS Body temperature and pressure, saturated with water vapor

ATPD Ambient temperature and pressure, dry

ATPS Ambient temperature and pressure, saturated with water vapor

Molecular Species

Indicated by chemical formula printed as subscript

Examples

P_{IO_2} = Pressure of oxygen in inspired air

V_D = Dead space gas volume

Equivalents of Metric, United States, and English Measures

(Values rounded off to 2 decimal places.)

Length

1 kilometer = 0.62 mile
1 mile = 5280 feet = 1.61 kilometers
1 meter = 39.37 inches
1 inch = 1/12 foot = 2.54 centimeters

Volume

1 liter = 1.06 US liquid quart
1 US liquid quart = 32 fluid ounces = 1/4 US gallon = 0.95 liter
1 milliliter = 0.03 fluid ounce
1 fluid ounce = 29.57 milliliters
1 US gallon = 0.83 English (Imperial) gallon

Weight

1 kilogram = 2.20 pounds (avoirdupois) = 2.68 pounds (apothecaries')
1 pound (avoirdupois) = 16 ounces = 453.60 grams
1 grain = 65 milligrams

Energy

1 kilogram-meter = 7.25 foot-pounds
1 foot-pound = 0.14 kilogram-meters

Temperature

To convert Celsius degrees into Fahrenheit, multiply by 9/5 and add 32
To convert Fahrenheit degrees into Celsius, subtract 32 and multiply by 5/9

Greek Alphabet

Symbol		Name	Symbol		Name
Α	α	alpha	Ν	ν	nu
Β	β	beta	Ξ	ξ	xi
Γ	γ	gamma	Ο	o	omicron
Δ	δ	delta	Π	π	pi
Ε	ϵ	epsilon	Ρ	ρ	rho
Ζ	ζ	zeta	Σ	σ, ς	sigma
Η	η	eta	Τ	τ	tau
Θ	θ	theta	Υ	υ	upsilon
Ι	ι	iota	Φ	ϕ	phi
Κ	κ	kappa	Χ	χ	chi
Λ	λ	lambda	Ψ	ψ	psi
Μ	μ	mu	Ω	ω	omega

Index